Thermal Performance of the Exterior Envelopes of Whole Buildings XII International Conference

Clearwater, Florida, USA
1 - 5 December 2013

ISBN: 978-1-5108-2783-7

Printed from e-media with permission by:

Curran Associates, Inc.
57 Morehouse Lane
Red Hook, NY 12571

Some format issues inherent in the e-media version may also appear in this print version.

Printed by Curran Associates, Inc. (2016)

For permission requests, please contact Oak Ridge National Laboratory at the address below.

Oak Ridge National Laboratory
P.O. Box 2008, Bldg. 3147
Oak Ridge, TN 37831-6070

Phone: (865) 576-7658
Fax: (865) 576-5812

Additional copies of this publication are available from:

Curran Associates, Inc.
57 Morehouse Lane
Red Hook, NY 12571 USA
Phone: 845-758-0400
Fax: 845-758-2633
Email: curran@proceedings.com
Web: www.proceedings.com

TABLE OF CONTENTS

The Hygrothermal Performance of Ventilated Stucco Walls Using Spray Polyurethane Foam Insulations and a Smart Vapor Retarder in the Pacific Northwest

Sam Yuan, PhD
Member ASHRAE

ABSTRACT

Spray polyurethane foam insulations used in residential wood frame buildings have superior thermal performance compared with traditional fiberglass materials. Close-cell foam especially also has low water-vapor transmission property, and at the same time offers good airtightness to building envelope systems. Contrary to common building practice, the application of spray foam to the interior wall cavities requires careful moisture analysis and design to prevent potential moisture related damages in humid climates such as the US Pacific Northwest. This unique climate area has moderate temperatures and high moisture levels caused by precipitation and high relative humidity throughout the year. Recent research on wood framed wall systems with exterior stucco and cavity spray polyurethane foams is being conducted in this climate zone. Two ventilated stucco clad wall systems have been installed side by side in a natural exposure testing facility in the Seattle, Washington area. One wall is insulated with a hybrid insulation system comprised of two-inch spray polyurethane closed cell foam and unfaced fiberglass batts. The other wall is insulated with five and half inches of open cell foam and has a variable vapor resistance smart vapor retarder attached to the interior surfaces of wood studs. The previous study on these walls without ventilation between the stucco and the building paper showed high relative humidity and high moisture content in the exterior sheathing for a long period. The field test data from this study are discussed with respect to the previous results. This paper will describe the hygrothermal performance of these two walls and emphasize the need for clear design guidance for the application of spray polyurethane foam insulation in the wood frame wall systems in the Pacific Northwest.

INTRODUCTION

As spray polyurethane foam (SPF) insulation materials are gaining use in the building industry, the evaluation of the hygrothermal performance of building envelopes adopting SPF as a system becomes critical in many climates. When used in residential wood frame buildings, SPF insulations have superior thermal resistance compared with conventional fiberglass materials. Close cell SPF (CCF) insulation, in particular, has high thermal resistance and low water-vapor transmission rates. When installed in the wall cavities in the Pacific Northwest region, CCF may cause potential moisture related damages. This region is identified as a "Marine Climate" with a unique combination of high precipitation and high relative humidity that creates a challenge to the moisture performance of the building envelope systems. The design of the wall systems with CCF and Open Cell Foam (OCF) requires careful consideration and moisture analysis to avoid long term structural failures.

To accomplish safe insulating with spray foam, system design engineers and architects incorporate hygrothermal modeling into the design process to optimize the building envelope systems under consideration. In this design process, hygrothermal modeling tools such as WUFI (WUFI® 2012) have been extensively used in the building industry for the purpose of building envelop design and performance comparison. As they have been improved substantially over the last decade, these modeling tools can be more easily incorporated into the decision-making process. Also the improved accuracy of the tools has reduced the discrepancy between the modeling results.

***Sam Yuan** is a senior principal research engineer in the Laveque Technical Center for Insulation Group of Saint-Gobain/Certain Teed Corporation, Blue Bell, PA.*

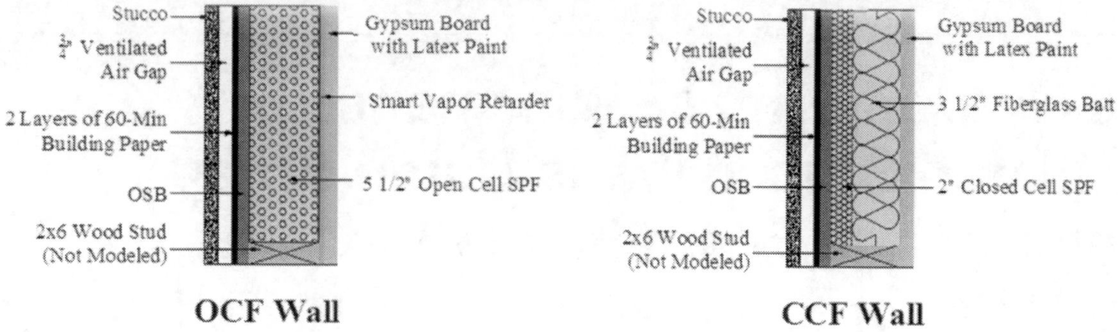

Figure 1 *Cross sections of modeled wall systems (wood studs not modeled).*

One important feature that has been added was the modeling of the ventilated air space behind the cladding material (WUFI® 2012). This feature has been used in the building science community for a few years. Straube and Finch (2009) have shown that the satisfactory agreement between the modeling results and field test data can be achieved with the appropriate use of the air change rate in the ventilated air space behind claddings. Their study also concluded that the determination of the air change rate in the ventilated air space may become complicated for the hourly rate; an estimated annual average rate is sufficient for most modeling purposes. This approach of using an annual average rate is adopted in this study of the SPF insulated walls.

A previous article on this project (Yuan and Karagiozis 2010) presented the hygrothermal modeling and field test results on the SPF insulated walls. The two 2 × 6 wood frame walls had stucco installed on oriented strand board (OSB) sheathing covered with building paper. Two insulation systems were incorporated in the walls: the open-cell SPF and closed-cell SPF with R-13 fiberglass batts. The open-cell SPF wall incorporated a vapor retarding layer on the interior while the closed-cell SPF walls did not have any vapor retarder. The vapor retarding layer used was a smart vapor retarder which has a relative humidity-dependent water vapor permeance (Gatland et al. 2009). Interior 0.5 in. (12.7 mm) thick gypsum wall boards were finished with one coat of primer and one coat of latex paint. Both modeling and field test results showed potential moisture problems with the SPF walls installed in the Seattle, WA climate. The article recommended adding ventilation in the exterior stucco cladding of the walls (Yuan and Karagiozis 2010).

The purpose of this paper is to investigate the hygrothermal performance of the same two residential SPF walls with ventilated stucco in the Pacific Northwest US climate zone. The walls installed in Puyallup, WA were modified with openings at the top and bottom to create ventilated air spaces behind the stucco claddings. These walls were monitored for over a year and the field test data were compared with the modeling results.

HYGROTHERMAL MODELS

Two walls using ventilated stucco cladding, OCF and CCF insulations were modeled. These walls, as shown in Figure 1, consisted of the following layers (from exterior to interior):

- OCF Wall: Stucco, 0.75 in. (19.1 mm) ventilated air gap, 2 layers of 60-minute building paper, 0.5 in. (12.7 mm) OSB sheathing, 2 × 6 wood framing, 5.5 in. (139.7 mm) open-cell SPF, one layer of smart vapor retarder and 0.5 in. (12.7 mm) gypsum plaster board with one layer of primer and one layer of latex paint,
- CCF Wall: Stucco, 0.75 in. (19.1 mm) ventilated air gap, 2 layers of 60-minute building paper, 0.5 in. (12.7 mm) OSB sheathing, 2 × 6 wood framing, 2 in. (50.8 mm) closed cell SPF, 3.5 in. (88.9 mm) fiberglass batt (R-13), and 0.5 in. (12.7 mm) gypsum plaster board with one layer of primer and one layer of latex paint.

One-dimensional hygrothermal modeling was conducted on these four walls for three-year duration beginning January 1, 2011.

Materials Properties

Most material properties used in the models are from the Generic North America Materials Database in the hygrothermal modeling software (WUFI® 2012). The SPF insulation materials are supplied by a major US building insulation manufacturer and their relevant properties are summarized in Table 1 (Yuan and Karagiozis 2010). These materials properties were obtained from the product specification sheets.

A constant annual air exchange rate was imposed in the models for the ventilated air gap behind the stucco cladding according to the recommendations in Straube and Finch (2009). A conservative value of 10 air change per hour (ACH) was applied to a heat and moisture source in the air gap of the models.

The initial and boundary conditions for the models are the default values in the software tool and also listed in (Yuan and Karagiozis 2010). The weather data file in the software for Seattle, WA was imposed on the walls as exterior boundary conditions.

Table 1. OCF and CCF Materials Properties Used in the Modeling

Material Property	OCF (Open-Cell SPF)		CCF (Closed-Cell SPF)	
Density	0.5 pcf	8.0 kg/m^3	2.0 pcf	32.0 kg/m^3
Thermal Conductivity (Aged)	0.278 Btu·in./(h·ft^2·°F)	0.0401 W/(m·K)	0.172 Btu·in./(h·ft^2·°F)	0.0248 W/(m·K)
Thermal Resistivity (Aged)	3.6 h·ft^2·°F/(Btu·in.)	25.0 m·K/W	5.8 h·ft^2·°F/(Btu·in.)	40.2 m·K/W
Water Vapor Permeability (RH_{mean}=25 %)	23.8 perm·in.	34.5 ng/(Pa·s·m^2)	1.51 perm·in.	2.19 ng/(Pa·s·m^2)

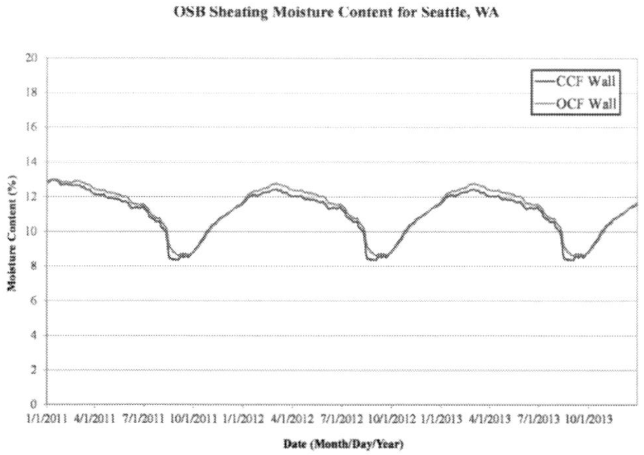

Figure 2 *Modeled OSB Moisture Content For SPF walls in Seattle, WA.*

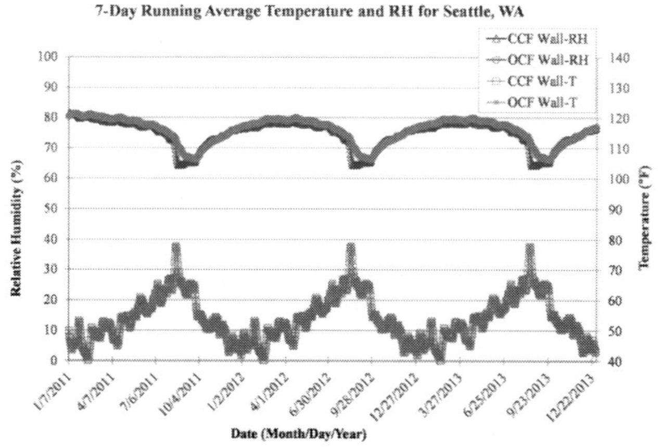

Figure 3 *OSB 7-day running average temperature and RH for Seattle, WA.*

MODELING RESULTS

In the wall systems, wood components are most likely subject to mold growth. Since the wood studs were not modeled, only the OSB sheathing water content is evaluated. Also the surface temperature and relative humidity results for the OSB sheathing are important factors that impact condensation in the wall structure. In this paper, the wall moisture performance evaluation criteria for wood components for avoiding mold growth are as follows (ASHRAE 2009):

1. OSB or wood-stud cavity-moisture content is less than 20% (not referenced in ASHRAE Standard 160 [ASHRAE 2009]);
2. 7-day running average surface RH is less than 98% when the 7-day running average surface temperature is between 40°F (4°C) and 100 °F (38°C);
3. 30-day running average surface RH is less than 80% when the 30-day running average surface temperature is between 40°F (4°C) and 100°F (38°C).

The modeling results show that the OSB sheathing has less than 13% moisture content for both walls for the entire three-year period. Figure 2 shows the OSB moisture content graphically. In the figure, the OSB moisture contents are very close to each other for both walls, due to the addition of the ventilated stucco cladding. Both walls have high resistance to the moisture flow from indoors reaching the OSB sheathing.

Both SPF walls modeled have highest moisture content and relative humidity levels at the OSB sheathing layer compared with the other components in the same systems. The 7-day and 30-day moving average values of OSB (oriented stranded board) surface temperature and relative humidity (RH) are calculated and plotted in Figures 3 and 4. In Figure 3, the peak RH values for both walls are much less than 90%, while in Figure 4, the RH 30-day running average peak values are reaching 80% but not over 80% when the OSB surface temperatures are within 40°F (4°C) to 100°F (38°C). These results predict no danger of moisture related mold growth problems for the OSB sheathing. Therefore, both walls met the proposed moisture performance criterion for the 7-day and 30-day moving average values of relative humidity.

FIELD TEST

Originally these two walls were constructed in the summer of 2008 in the Washington State University Natural Exposure Testing (NET) facility constructed by Tichy and Murray (2003) through a collaborative research (Dr. Carolyn Roos, WSU investigator) contract with the Oak Ridge National Laboratory (Dr. Karagiozis, Principal investigator). The NET facility is located in Puyallup, WA, approximately 30 miles (48 km) south of Seattle, WA. They were installed on the southern side of the facility. Following the recommendation proposed in Yuan and Karagiozis (2010), the outside claddings were rebuilt to have a 0.75 in. (19.1 mm) ventilated air gap behind the stucco in 2010. Figure 5shows the finished stucco cladding systems with top and bottom air vents. The

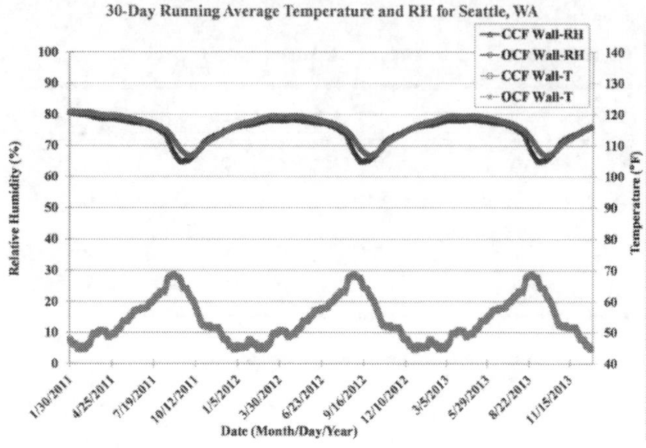

Figure 4 *OSB 30-day running average temperature and RH for Seattle, WA.*

Figure 5 *Photo 1 field test SPF walls in the NET facility (south facing).*

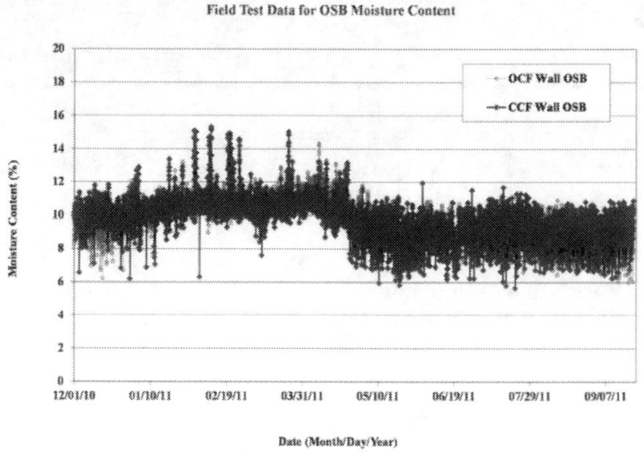

Figure 6 *OSB moisture content field test data for OCF and CCF walls.*

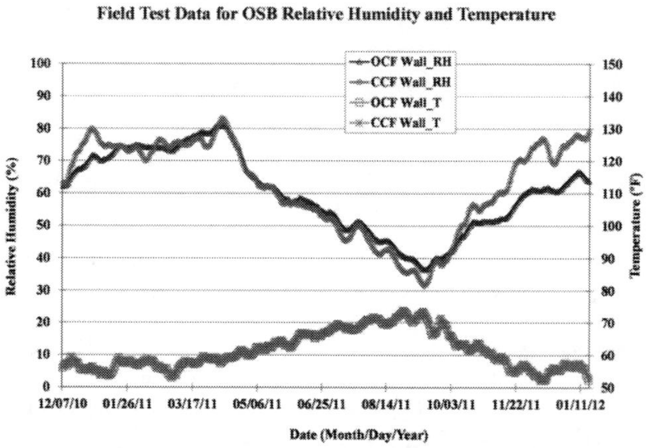

Figure 7 *Field test data of seven-day moving average for OSB relative humidity and temperature for OCF and CCF walls.*

rest of the walls including instrumentations was kept the same as the original walls. The relative humidity in the Net facility was kept constant at 45%, and the temperature was maintained at 68°F (20°C).

FIELD TEST RESULTS

After the modification of the stucco claddings, both walls were monitored for a period of approximately 14 months. Field test data including OSB moisture content, surface temperature and relative humidity (RH) were analyzed for performance determination. The moisture content sensor is physically located approximately 0.125 in. (3 mm) from the exterior of the OSB surface. The hourly moisture content measurements for the OSB are plotted in Figure 8. For both walls, the OSB moisture content never exceeded 16% for the entire 14-month period, indicating no potential mold growth in the OSB. When compared with the OSB moisture contents shown in Figure 2 with the modeling results, the modeling and field test data match very well except, for some sporadic high field moisture content readings (15%) in Figure 8.

The 7-day and 30-day moving average test data for the RH and temperature of the OSB are plotted in Figures 6 to 7 for both walls. The data in these figures are obtained from the RH and temperature sensors on the interior side of the OSB sheathing (Yuan and Karagiozis 2010). For the whole field test period recorded, the interior OSB surface for both walls had temperatures between 50°F (10°C) to 75°F (24°C), matching the modeling results as shown in Figures 3 and 4.

The 7-day moving average OSB RH values for both OCF and CCF walls (with the maximum RH being approximately 82%) did not exceed 85% as shown in Figure 8, meeting ASHRAE Standard 160 (ASHRAE 2009). The 30-day moving average of OSB RH values are shown in Figure 8. It shows that the OCF OSB RH reached 80% in April 2011 and dried out to 40% in the middle of September. The CCF OSB RH never reached 80%. Thus both OCF and CCF walls met the

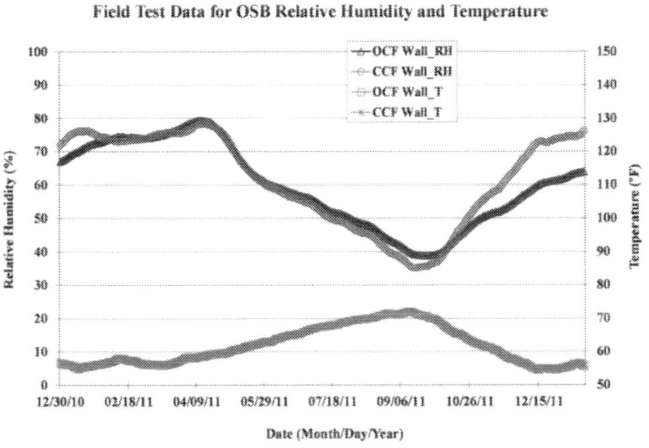

Figure 8 *Field test data of 30-day moving average for OSB relative humidity and temperature for OCF and CCF walls.*

moisture performance evaluation criteria used in this paper following ASHRAE Standard 160 (ASHRAE 2009).

Compared with modeling results, Figures 3 and 4, the field test data showed much lower RH values in Summer 2011. This difference is caused by the application of the conservative 10 ach annual rate in the ventilated air gap behind the stucco claddings in the hygrothermal models. Since low RH values on the OSB interior surface decrease the potential OSB mold growth, this difference between the modeling and the actual field performance does not bring the negative effect on the system design criteria. Thus the models with ventilated stucco cladding predict well the hygrothermal performance of the SPF wall systems.

CONCLUSIONS

Wood frame wall systems insulated with SPF materials may have serious mold growth problems when stucco is applied to the wood sheathing of residential buildings in the Pacific Northwest climates (Yuan and Karagiozis 2010). The design change of ventilating the stucco cladding improves the moisture performance of the wall systems significantly. The two SPF walls described in this paper met all three satisfactory moisture performance criteria for over a 14-month monitoring period after the creation of the ventilated stucco claddings. The field test data showed substantial drying effect on the OSB sheathing interior surface. The ventilated air gap between the stucco and building paper attached to the OSB layer allows moisture flow with air out of the wall instead of penetrating into and accumulating in the OSB and wood studs. This reduces the impact of water penetration through the cladding and also reduces solar driven moisture.

Hygrothermal modeling for building envelopes can quickly predict the actual field performance in specific climate zones. The added feature of the modeling tool that incorporates the sources and sinks for both heat and moisture into the building envelope systems improves the accuracy of such predictions. Using a conservative air change rate in the ventilated air gap behind the stucco cladding for the SPF wall systems tested in the in Seattle, WA area, the modeling results achieved highly reasonable agreement with the field test data. The modeled OSB moisture content results match well the actual field data, while the RH modeling results had agreement on the peak values with the field data. The modeling data did not show the quick drying during summer months as seen in the field data due to the modeled conservative air change rate applied in the ventilated air gap behind the stucco.

REFERENCES

ASHRAE. 2009. ANSI/ASHRAE Standard 160, *Criteria for Moisture Design Analysis in Buildings*. Atlanta: ASHRAE.

Gatland, S.D. II, A.N. Karagiozis, C. Murray, and K. Ueno, 2007. The hygrothermal performance of wood-framed wall systems using a relative humidity-dependent vapor retarder in the Pacific Northwest. *ASHRAE Thermal X Conference*, Clearwater Beach, FL.

Straube, J.F. and G. Finch. 2002. Ventilated wall claddings: Review, field performance, and hygrothermal modeling. Research Report 0906, 2009 Building Science Press, buildingscience.com.

Tichy, R. and Murray, C. 2003. Hygrothermal performance research program development innovative wall systems that improve hygrothermal performance of residential buildings. WSU/ORNL/DOE/Industry Cooperative Research Project Report.

WUFI®. 2012. WUFI Pro 5.0, A PC program for analyzing the one-dimensional heat and moisture transport in building components; Fraunhofer Institut Bauphysik and Oak Ridge National Laboratory (ORNL).

Yuan, S. and A.N. Karagiozis. 2010. The Hygrothermal performance of wood-framed wall systems using spray polyurethane foam insulations and a smart vapor retarder in the Pacific Northwest. *ASHRAE Thermal XI Conference,* Clearwater Beach, FL.

ACKNOWLEDGMENTS

The authors wish to thank Mr. Phillip Childs at Oak Ridge National Laboratory for his efforts in troubleshooting instrumentation in Puyallup, WA, and providing the field test data as well. The authors would like to acknowledge The Washington State University Extended Energy Program for providing the NET facility and monitoring the site for this research, especially Mr. Chris Fuess and Mr. Doug Koenen.

Practical Methods for Ensuring Energy-Efficient Window-to-Wall Connections

Cezary Misiopecki

Larisa Marinova Decheva

Bjørn Petter Jelle, PhD

Arild Gustavsen, PhD
Member ASHRAE

ABSTRACT

Nowadays, saving energy and reducing carbon emissions is one of the top priorities in each sector of human activity. The building sector accounts for a significant part of the world's total energy use and greenhouse gas emissions. Windows are inseparable components of the building envelope. In modern designs the trend of increasing the area of fenestration products for residential and commercial constructions is seen worldwide. Windows contribute to a better standard of living with daylight and useful heat gains, but, on the other hand, cause higher heat losses or non-desirable heat gains. Currently, the thermal transmittance of fenestration components is still much higher than for walls. Recent improvements of the walls' and windows' thermal properties contribute to a better building performance and hence cause the thermal bridging effects occurring in the building envelope to be more important. Among the other thermal bridges the one of the window-to-wall connection appears to be especially important in energy context and should not be underestimated.

Window placement in the window opening can be accomplished in several ways. Well insulated walls are relatively thick, so the window can be placed at several locations in the window opening. Moreover, different strategies for sealing and insulating the connection between window and wall can be used. This study focuses on investigating different solutions for window-to-wall connections and their potential of ensuring a better thermal performance (reducing the thermal bridge effect). Several heat transfer simulations have been done for various solutions of the window-to-wall connection, using the THERM 6.3 computing software. Based on those calculations, a parametric study is carried out to show the importance of various factors on the linear thermal transmittance.

Results show that different window placements influence the linear thermal transmittance values significantly. The optimal window position, considering the reduction of heat loss, was found to be about 35 mm from the surface of the external wind barrier to the edge of the window frame (for timber wall case, regardless of the wall thickness). Furthermore, the lowest surface temperatures of inside window sill were compared for different window locations. The lowest temperature for optimal solution was found to be only 1°C lower than that for other placements. In addition, alternative designs of inside window framing, which can substantially lower the linear thermal transmittance, are presented.

INTRODUCTION

Nowadays, saving energy and reducing carbon emissions is one of the top priorities in each sector of human activity. Buildings are an important part of human life, since people use them for the majority of the day. Energy use in buildings worldwide accounts for over 40% of primary energy use and for around 24% of greenhouse gas emissions. Energy and emissions include both direct usage of fossil fuels (on-site) and indirect use of energy in the form of electricity, district heating, district cooling, and embodied energy in construction materials (Voss and Musall 2011).

Windows are inseparable components of the building envelope. In modern designs a trend of increasing the area of fenestration products for residential and commercial construc-

***Cezary Misiopecki** is a PhD candidate, **Larisa Marinova Decheva** is a trainee at Veidekke Entreprenør AS, Trondheim, Norway, and **Bjørn Petter Jelle** and **Arild Gustavsen** are professors at the Norwegian University of Science and Technology, Trondheim, Norway.*

tions is seen worldwide. Windows contribute to a better standard of living and useful heat gains but on the other hand can cause higher heat losses or non-desirable heat gains. Currently, the thermal transmittance of fenestration components is still much higher than those for walls. Studies show that up to 60% of the total energy loss through the building envelope can be due to windows (Gustavsen et al. 2008a), but this is of course very dependent on the building geometry/configuration, assemblies' and windows' thermal performance, and amount of windows.

Recent improvements of walls' and windows' insulating properties contribute to better building performance, and addressing thermal bridging effects occurring in the building envelope can thus be more important. Among other thermal bridges, the one in the window-to-wall connection appears to be especially important. The study by Gustavsen et al. (2008b) shows that for a typical 160 m^2 Norwegian dwelling, the window-to-wall interface can be responsible for about 40% of the total heat loss caused by thermal bridging effects. For the same building but with the focus on reducing thermal bridges, the window-to-wall interfaces were responsible for 17% of the total heat loss through thermal bridges. Similar results were reported in the ISO 14683:2007 standard (ISO 2007d) regarding thermal bridges calculation methods, where it was concluded that for relatively low performing generic buildings, thermal bridges caused 36% of the total energy loss through the building envelope, while the window-to-wall connection was responsible for 38% of this value. This demonstrated that the heat loss through the window-to-wall connection is very important in energy context and should hence not be underestimated.

Window openings are created in order to give support for window placement in a wall and should have the ability to protect from water intrusion and air leakage into the building envelope. The study of Maref et al. (2011) concluded that this interface has the lowest temperature indices throughout building envelope, which makes it the most likely location for condensation to occur. Window placement in the window opening can be accomplished in several ways. Well insulated walls are relatively thick, therefore a window can be placed at several locations in the window opening. Moreover, different strategies for sealing and insulating the connections can be used. However, there are usually conflicting recommendations and requirements for the diverse solutions driven by thermal performance, moisture resistance, structure durability, solar heating, daylight distribution, and architectural conditions.

This study focuses on investigating different solutions for window-to-wall connections and their potential of ensuring better thermal performance (reducing thermal bridge effect). Thermal bridge effect for different connections depends on the wall type, window position in the opening, thermal resistance between frame and window opening, and internal window framing construction. The work focuses on these factors and aims to find the optimal solution for a window-to-wall connection. In order to accomplish this goal, several heat flux calculations have been carried out for various solutions of window-to-wall connections, using the THERM 6.3 software. Based on calculations, a parametric study is carried out to show the importance of various factors on the linear thermal transmittance. The linear thermal transmittance defines the additional heat loss of thermal bridge in comparison to the reference one-dimensional heat loss of the undisturbed element.

LITERATURE REVIEW

The topic of window-to-wall connections has been studied before in literature, however the issue of thermal bridging effect on window-to-wall connections has not been widely explored yet. Experimental studies by Maref et al. (2012, 2011) investigated the influence of air leakage through window-to-wall connections with respect to condensation risk. These studies concluded that filling the gap between the window frame and window opening with spray-in-foam insulation lowered the possibility of condensation—since less cold air is transported to the warm side of the window. Another study by Lacasse et al. (2009) concluded that the window-to-wall connection is important from a water intrusion point of view. Water intrusion can lead to premature failure of the building envelope. Many solutions have been designed over the years in order to guard the envelope from water intrusion. The study assessed effectiveness of water intrusion in the window-to-wall connection by laboratory testing. The investigation was performed for typical Canadian window-to-wall connection solutions.

SINTEF, which is the largest non-commercial and independent research organization in Scandinavia, has been a major contributor to the research on thermal bridging effects. In various reports and guidelines, the window position in the window opening is referred to as an important parameter for minimizing thermal bridges. SINTEF building research design guideline 471.015 (Gustavsen and Roald 2008) gave an example of the relationship between the window position and the linear thermal transmittance for a timber wall with 250 mm insulation. The simulated wall was constructed with exterior vented cladding, wind barrier, gypsum board, and interior lining. The geometry is illustrated in Figure 1 and values of linear thermal transmittance for the various window positions are shown in Table 1.

The study shows certain window location (35 mm from the outside wall edge) is the most favorable in terms of reducing the thermal bridge effect (among the positions studied). For cases in which a window was placed at an outside position (hanging out –42 mm from the wind barrier to line up with the exterior cladding), the linear thermal transmittance is five times higher. Choosing the solution for window placement is important not only from the thermal point of view. Other factors must be taken into account, such as moisture conditions, solar heating, daylight distribution, window durability, maintenance needs, and architectural aspects. The optimum window location according to some other SINTEF recommen-

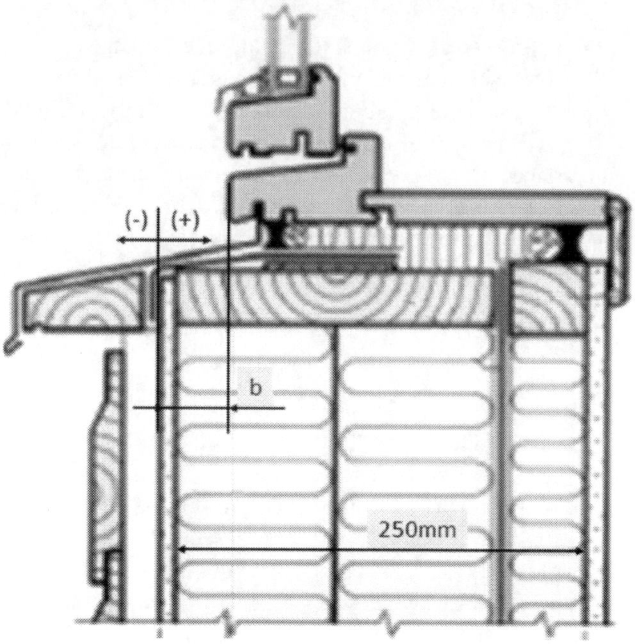

Figure 1 *Examples of window position regarding linear thermal transmittance (Gustavsen and Roald 2008).*

Table 1. Linear Thermal Transmittance for Different Window Positions (Gustavsen and Roald 2008)

Distance from Wind Barrier to Frame b, mm	Ψ, W/(m·K)
–42	0.05
0	0.02
35	0.01
85	0.02
140	0.03

dations differs from presented solution. SINTEF building research design guideline 523.701 (SINTEF 2003b) (concerning window installation details in the timber walls) and 523.702 (SINTEF 2003c) (concerning window installation details in masonry and concrete walls) recommend two ways of installing windows depending on the climate. Outer window position (equivalent to –42 mm in Table 1) is recommended for coastal Norwegian regions—according to the guidelines such a position makes the sealing process against rain water easier. Inner position (equivalent to 85–140 mm in Table 1) is suggested for midland climates—with limited amounts of wind driven-rain. The justification for this position is the fact that placing a window inside the opening gives higher inner surface temperatures on the window—which lowers the risk of condensation. This paper aims to determine if the inside surface temperatures for highly insulating windows depend strongly on window position in the opening.

Another study by Cappelletti et al. (2011) investigated the influence of window installation details on the thermal bridging effect on window-to-wall connections for clay-block walls. Wooden windows have been numerically tested with two different wall constructions (brick wall insulated from outside and brick wall with insulated cavity) at three different positions (outside, intermediate, inside) for each wall design. For each case the linear thermal transmittance based on external dimensions was calculated (in accordance with EN ISO 10211:2007 [ISO 2007b]). It was found that window positioning and connection construction details of the window opening have an impact on linear thermal transmittance, which differs by 70%—75% between cases. In addition, the actual method used to include the thermal bridging effect of window-to-wall connections with the window U-factors has been proposed.

SCOPE AND LIMITATION OF WORK

The scope of this study is to do the following:

- determine the most efficient position of a window in window openings in terms of minimizing the linear thermal transmittance value for typical wooden wall constructions with insulation of three different thicknesses (198, 246, 298 mm),
- determine how an internal window framing thickness and design influence the linear thermal transmittance on window-to-wall interfaces for a typical wooden wall,
- show the influence of the thermal bridging effect on the U-factor for different window sizes, and
- determine temperature changes of internal window surfaces with regard to window position in the opening.

It should be noted that this study does not investigate the air leakage or water drainage capabilities of the modeled solutions. Moreover, flashing is not included in the simulating models. Flashing made of high conductivity materials may have influence on the linear thermal transmittance value. However, the study compares the linear thermal transmittance between different cases rather than determining its actual value. Air flow along the surfaces was not simulated in detail; simplified convection coefficients were used. Presented optimal solutions for each case must be further investigated in terms of the other aspects of window-to-wall connections. The connection of window sills and window openings was the only connection modeled.

NUMERICAL SIMULATIONS

The objective of the heat flow calculations was to evaluate thermal bridges at the window-to-wall connection. Heat flow calculations during the study were carried out using the computer program THERM, version 6.3. The software is developed by Lawrence Berkeley National Laboratory

(LBNL) in California. THERM is a two-dimensional finite element program that calculates conduction, radiation heat exchange, and also estimates the influence of convection phenomena in the air cavities (LBNL 2011). The software was used to prepare geometry and conduct two-dimensional heat transfer simulations. Window frame geometry was prepared in accordance with ISO 10077-2 (ISO 2012) standard and overall geometry of window-to-wall connection in accordance with ISO 10211 (ISO 2007b).

This study investigates a typical wooden wall structure with ventilated cladding (a sketch of the wall intersection is presented in Figure 1). Wall geometry was simplified and external cladding was not modeled. This influenced outside boundary conditions described below. Wall geometry is presented and described in Table 2. Influence of non-continuous (in horizontal direction) wall construction was neglected in the study. The simulations for all cases were performed using a well thermally insulated window produced by a Norwegian manufacturer (an aluminum-clad wood window with polyurethane thermal braked, total window U-factor of about 0.7 W/[m^2·K]). For simulation purposes, the glazing unit was replaced with an insulating panel (thermal conductivity of 0.035 W/[m·K])—value typical for 3P insulated glazing unit employing two low-e coatings and argon gas fill) and glazing spacers were neglected in the model. Moreover, no window fixings elements were modeled in the study. The edge U-factor of windows equals 0.82 W/m^2·K (with insulating panel as glazing). The window sketches and THERM geometry are presented in Figure 2 and Figure 3. Wall and window material properties are shown in Table 3. Only the connections of window openings and window sills were considered in this study.

Boundary conditions (BC) used for simulations depended on the simulation scope. For simulations used to determine linear thermal transmittance values, simplified ISO 10077-2:2012 (ISO 2012) boundary conditions were applied to the window unit. Since external cladding and ventilated air cavities were not included in the wall model, the outside surface of the gypsum board was treated as a surface facing a ventilated air cavity. BCs on the surface were set according to the procedure described in the standard ISO 6946:2007 (ISO 2007a). In order to determine the lowest temperatures on the window inside surfaces, the boundary conditions for surfaces facing the outside environment were set to values recommended by National Fenestration Rating Council (NFRC 2012a) for both wall and window geometry. The reason for using different BCs was to determine the actual temperatures on the inside surfaces at severe winter conditions in order to assess risk of condensation (although condensation usually should not occur for such a good window). Boundary conditions are described in Table 4.

Wall structures were drawn with a length of 1.2 meters and a window was inserted in the desired position. The sketch showing the geometry of the model is presented in Figure 4

The linear thermal transmittance was calculated according to the formula (1).

$$\Psi = L^{2D} - \sum_i U_i \cdot l_i \ \mathrm{W/(m \cdot K)} \qquad (1)$$

where

L^{2D} = the thermal coupling coefficient obtained from a two-dimensional calculation of the component separating the two considered environments.

U_i = the thermal transmittance of the one-dimensional component separating the two considered environments (see Figure 4).

l_i = the length within the two-dimensional geometrical model over which the value of U_i applies (see Figure 4).

RESULTS AND DISCUSSION

For all simulations, typical wooden wall structures were used, which consisted of thermal insulation enclosed between two gypsum boards and a wooden opening window frame. Any divisions in the materials are not taken into account. Vapor barrier has not been considered in this study due to its low impact on thermal resistance..

First, simulations were conducted for five window locations in the wall with 246 mm insulation (see Case 1 in

Table 2. Wall Construction Used in the Study

Geometry	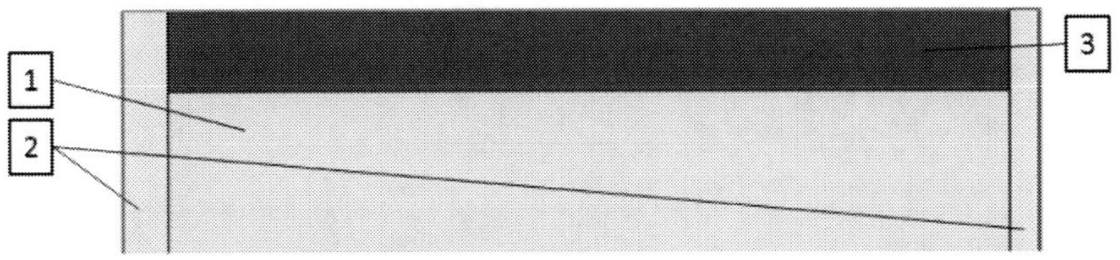	
	Material (from inside to outside)	**Thickness [mm]**
Construction	Gypsum board Insulation Gypsum board	13 198/246/296: differs for cases 9

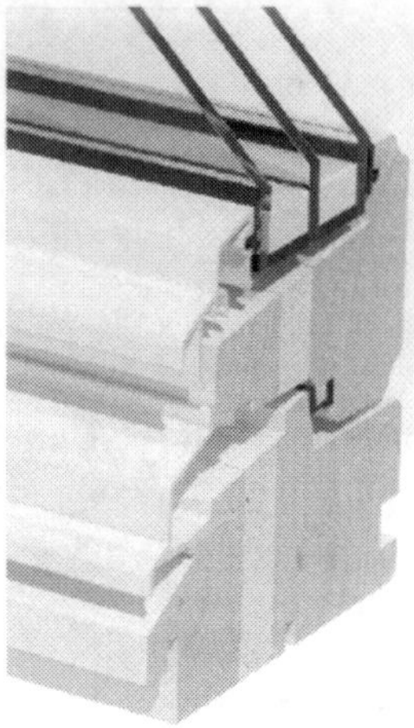

Figure 2 *Sketch of window intersection used in the study. (NorDan 2010).*

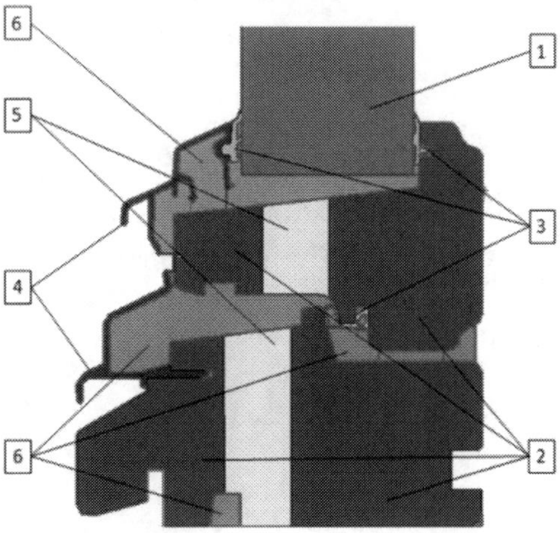

Figure 3 *THERM model of the simulated window. Number 6 indicates frame air cavities, other labels are described in Table 3.*

Table 5). For this case the calculated wall U-factor equals 0.19 W/(m^2·K). Placement of the window was based on the distance from the outside surface of the external gypsum board to the outside edge of the window sill (this distance is represented by value b [mm]). Additional 14 mm thick wooden elements/shims were added to the window elevation for the following still positions: b = +85 mm and b = +140 mm. This was necessary to fulfill the requirements of minimum slope of the outside flashing, according to the SINTEF building research design guideline 523.701 (SINTEF 2003b).

The next two variations of the same wall construction with different insulation thickness were considered. In the first case, the insulated wall with 198 mm thick insulation (U-factor 0.22 W/[m^2·K]) was analyzed. Calculations were also performed for a wall with a 296 mm thick insulation layer (U-factor 0.15 W/[m^2·K]). Linear thermal transmittance values were calculated for all cases (refer to Table 5). In order to investigate the influence of insulation thermal conductivity (between window frame and window opening) on linear thermal transmittance, the insulation thermal conductivity was set to 0.02 W/(m·K) (Case 4). Case 5 included standard wall construction (246 mm insulation thickness) but with thicker inside window framing (the thickness was changed from 19 mm to 32 mm). In Case 6 the alternative internal window framing design was presented. Details of the connection are shown in the graph for Cases 6–8 (Table 5). The proposed design allows more insulation than the standard design. The two last cases (Cases 7 and 8) involve the same geometry as used in Case 6, but insulation thermal conductivity between the window frame and the window opening is set to 0.02 and 0.01 W/(m·K), respectively (the 0.01 W/[m·K] material is included to investigate possible improvements if new insulation materials are found). Moreover, different exterior boundary conditions for Cases 1, 3, 6, and 8 were applied to determine the lowest temperature of the inside frame surfaces.

Table 5 describes and presents results for different cases of the timber wall. The results show that different window placements are crucial parameters for the linear thermal transmittance values. Theoretically for this wall type, window-to-

Table 3. Material Properties Used in the Study*

Material	Figure or Table Number/ Number of Element	Conductivity, W/(m·K)	Emissivity
Aluminum alloys	Fig 3 / 4	160	0.8
Ethylene propylene diene monomer (EPDM)	Fig 3 / 3	0.25	0.9
Foam panel	Fig 3 / 1	0.035	0.9
Mineral wool	Table 2 / 1	0.037	0.9
Gypsum board	Table 2 / 2	0.25	0.9
Polyurethane (PUR)	Fig 3 / 5	0.03	0.9
Wood (pine)	Fig 3 / 2, Table 2 / 3	0.13	0.9

* Material data used in the calculations are derived from ISO 10456 (ISO 2007c), SINTEF building research design guideline 471.010 (SINTEF 2003a) or NFRC 101 (NFRC 2012b).

Table 4. Boundary Conditions Applied in the Simulations

Description →	Simulations Determining Linear Thermal Transmittance —Window Geometry	Simulations Determining Linear Thermal Transmittance— Wall Geometry	Simulations Determining the Lowest Inside Surface Temperatures	Units
Boundary Condition ↓	**Values**			
Indoor temperature	20	20	20	°C
Outdoor temperature	0	0	–18	°C
Combined convection and radiation surface coefficient of heat transfer for the indoor frame and vision sections	7.692	7.692	7.692	$W/(m^2 \cdot K)$
Combined convection and radiation surface coefficient of heat transfer for the outdoor frame and vision section	25.0	7.692	26.0 (radiation model: BLACKBODY)	$W/(m^2 \cdot K)$
References	ISO 10077-2:2012	ISO 6946:2007	NFRC 100	—

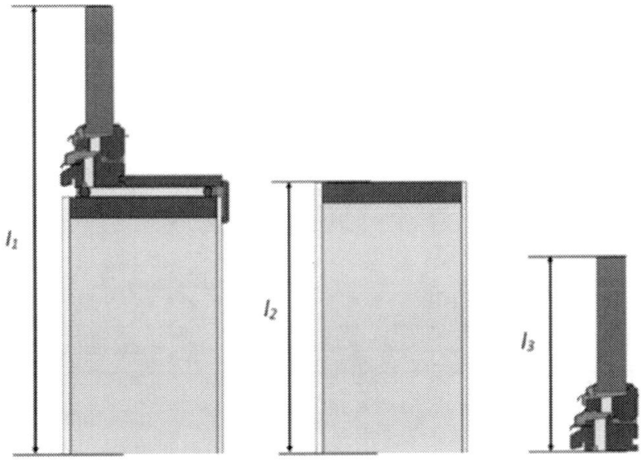

Figure 4 *Sketch of geometry modeled in the study (the model is not to scale).*

wall connection causes the lowest thermal bridging if the window is installed at the middle of the insulation as is suggested in Table A.2/Case W5 in ISO 14683 Standard (ISO 2007d). This was also confirmed by a simple simulation prior to study. However, if wooden shims providing the required flashing slope are added at the middle position, it is not anymore optimal in terms of thermal performance. The optimal window placement was found to be around 35 mm from the outside surface of external gypsum board to the edge of the window frame (for all timber wall cases). Windows mounted more to the outside or inside of the wooden wall provided larger values of linear thermal transmittance. The lowest inside surface temperatures of the window frame were compared for different window positions in order to address recommendations from SINTEF building research design guideline 523.701 and 523.702. For Case 1 the difference between positions b = +140 mm and optimal position (b = +35 mm). equals 1°C, while actual temperatures are equal to 13.6 and 14.6°C, respectively. For an indoor temperature equal to 20°C, condensation might occur for window position b = +35 mm at the level of indoor air relative humidity (RH) 66%, while for position b = +140 mm, at the level of rh 70%. It was found that for the thicker wall (296 mm of insulation) the difference between the same cases is lower (0.4°C). The proposed new window framing design (Case 6) not significantly lowers the temperature to 13.1°C for optimal position (b = +35 mm)—what in this particular case may be caused by frame design. Filling the gap with improved insulation (Case 8) gives the same temperature values as for the traditional solution (Case 1) while lowering thermal bridging value around 87% (actual change of 0.013 W/[m·K]).

It was also found that the thermal bridging effect increases with increasing thickness of the wall construction—the exceptions are window positions in the window opening very close to the interior. In such a case the heat paths to the internal wall are shorter, thus causing a larger heat loss compared to the thicker walls with the same window distance from the external wall (dimension b). The thickness of the insulating layer in the wall is crucial for the wall U-factor and the wall thickness. Higher thermal resistance of the wall leads to a relatively larger heat transfer through the window-to-wall connection. Because the wall is thicker, heat flow paths are relatively shorter for the window-to-wall connection compared with the rest of the structure. It causes relatively higher thermal bridging effect, even if the total heat transfer through wall and window will be less. This effect is greatest for a window placed outermost in the window opening.

It was found that the linear thermal transmittance is reduced when the space between the window and the wall is insulated with a material of lower thermal conductivity. Changing conductivity from 0.032 to 0.020 W/(m·K) resulted in decreased linear thermal transmittance by around

Table 5. Simulation Results for Wooden Wall Cases*

Geometry

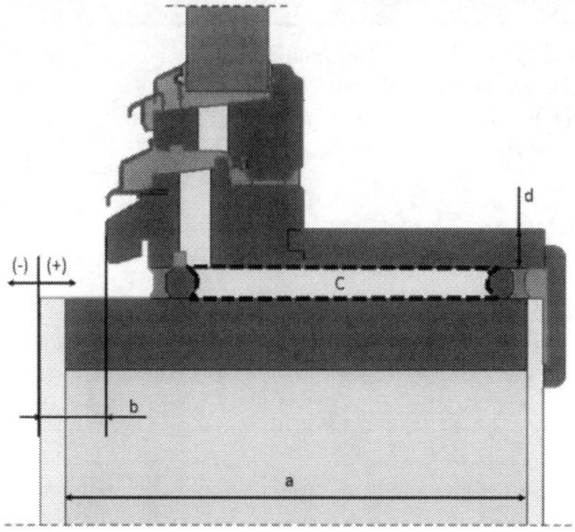

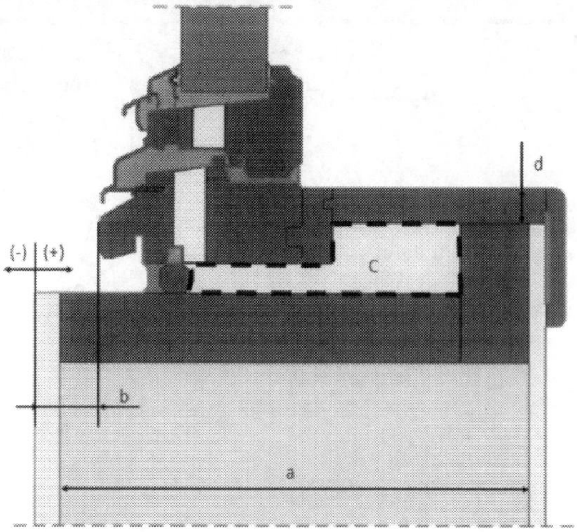

Case Number		1	2	3	4	5	6	7	8
Insulation Thickness a, mm		246	198	296	246	246	246	246	246
Internal Sill/Flashing Thickness d, mm		19	19	19	19	32	18	18	18
Insulation Conductivity C, W/(m·K) (dotted area)		0.032	0.032	0.032	0.02	0.032	0.032	0.02	0.01
Wall U-factor, W/(m²·K)		0.19	0.22	0.15	0.19	0.19	0.19	0.19	0.19
		Linear Thermal Transmittance Ψ, W/(m·K)					Linear Thermal Transmittance Ψ, W/(m·K)		
Position in Window Opening b, mm	–42	0.047	0.040	0.052	0.041	0.032	0.025*	0.019*	0.014*
	00	0.019	0.014	0.024	0.015	0.014	0.009*	0.005*	0.002*
	+ 35	0.015	0.012	0.019	0.012	0.011	0.007*	0.005*	0.002*
	+ 85	0.025	0.031	0.026	0.023	0.022	—	—	—
	+ 140	0.041	—	0.032	0.040	0.039	—	—	—
		Lowest Inside Surface Temperature, °C					Lowest Inside Surface Temperature, °C		
Position in Window Opening b, mm	–42	12.5	—	11.3	—	—	11.8	—	12.3
	00	13.4	—	13.3	—	—	13.0	—	13.4
	+ 35	13.6	—	13.5	—	—	13.1	—	13.6
	+ 85	13.8	—	13.5	—	—	—	—	—
	+ 140	14.6	—	13.9	—	—	—	—	—

* These results are calculated with the assumption that the window frame U- factor is not changed due to changed projected area of window frame.

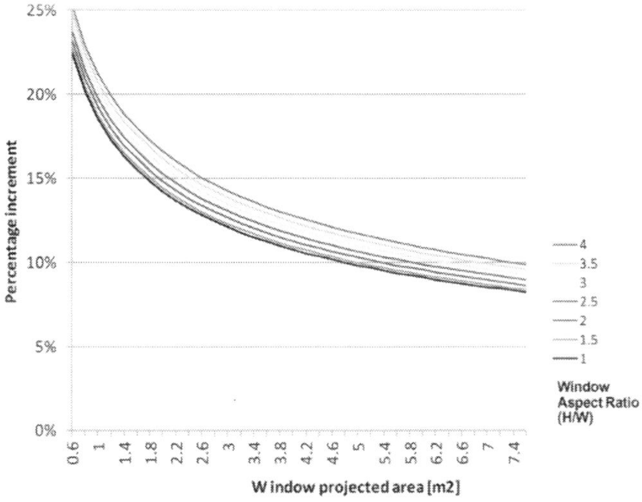

Figure 5 *Window U-factor increment for Case 1 (outer window position b = –42 mm).*

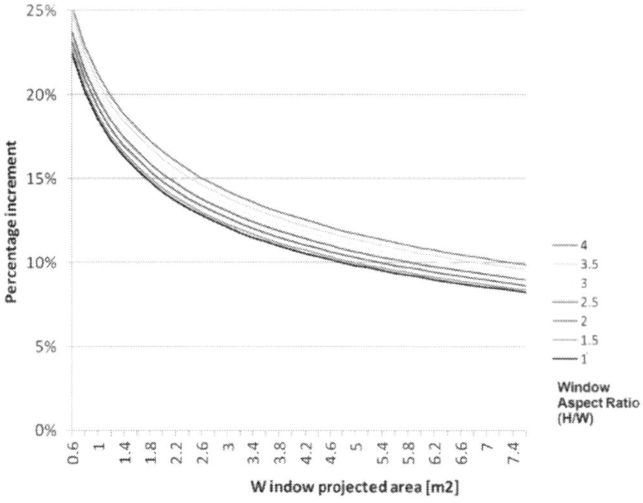

Figure 6 *Window U-factor increment for Case 1 (optimal window position b = +35 mm).*

20% (actual change of 0.003 W/[m·K]) (for an optimal window position, b = +35 mm). Placing a thicker inside window framing (32 mm instead of 19 mm) improved the thermal performance of the connection by around 26% (actual change of 0.004 W/[m·K]) for the optimal window position (b = +35 mm). A different design of inside window framing allows for accommodating additional insulation and has a great effect on thermal linear transmittance, due to increased volume of insulation and material in the region of window-to-wall connection and sill of the window frame. Linear thermal transmittance may be lower by 53% (actual change of 0.008 W/[m·K]) (using insulation with conductivity of 0.037 W/[m·K]) and 87% (actual change of 0.013 W/[m·K]) (using insulation with conductivity 0.010 W/[m·K]) for an optimal window positioning (b = +35 mm). The results were calculated with the assumption that the window edge U-factor is not changed due to changed window projected area for Cases 6–8. This way of calculation was applied as in Norway window manufacturers are not providing window inside framing as an integral part of the window. Usually, contractors take care of window wall connections and inside framing. The purpose of this assumption was to show how the wall connection can be improved by different ways of arranging inside framing using the same window in the same wall construction.

In order to show how the window position in the window opening influences the window U-factor by including the thermal bridging effect into window performance, percentage increment was calculated. The calculations intend to show how the window-to-wall issue is important, hence they should not be confused with U-values defined by ISO (ISO 2006) or NFRC (NFRC 2012b) standards. The methodology for this calculation was published in the study by Cappelletti et al. (2011). Window U-factor percentage increment was calculated according to the following formula:

$$\Delta U_w = \frac{U_{2D} - U_w}{U_w} \times 100\% \tag{2}$$

where

$$U_{2D} = \frac{U_w A_w + \Psi_w (2H + 2W)}{A_w}$$

U_w = windows U-value, W/(m^2·K)

A_w = window projected area, m^2

Ψ_w = linear thermal transmitance on window – wall connection, W/(m^2·K)

H, W = respectively projected height and width of the window, m

In order to accomplish the calculation with available simulation results the following assumptions were made:

- Linear thermal transmittance for jambs and head has the same value as for sill for different window positions.
- Jambs and head of window frame have the same U-factors as a window sill. Foam insulation panel properties were used as a glazing (U-factor calculated for foam panel equals to 0.70 W/m^2·K).
- Window U-factors were calculated with accordance to ISO standard 10077-1:2006 (ISO 2006). Linear thermal transmittance describing thermal bridge caused by spacer was also derived from ISO 10077-1:2006 and set to the value of 0.06 W/(m·K).

Figure 5 and Figure 6 show increment of window U-factor with included thermal bridging effect, for different window positions in Case 1 (b = –42 mm and b = +35 mm). It can be seen that higher increments are observed for smaller windows, since edge region performance has a higher influence on the overall fenestration rating value. Moreover it appears that windows with aspect ratio equal to 1 (window height to window width) are the least influence by thermal bridging effect on the window-to-wall connection.

Table 6 presents window U-factor increments calculated for three different window sizes for installing scenarios. Table 6 shows that U-factor increment calculated according to the assumptions for standard window size is relatively high – 15% for the window positioned –42 mm from the wind barrier which is recommended for coastal climates in SINTEF building research design guideline 523.701 and 523.702. Positioning windows in the thermally optimal location (b = +35 mm) reduces the increment to 5%. Arranging inside window sills in the proposed way (Cases 6–8 in Table 5) gives an increment value of 2% (standard size window). For the proposed solution if the insulation between window frame and window opening is replaced with insulation with conductivity equal to 0.1 W/(m·K) then the window U-factor increment is lower than 1% (Case 8).

UNCERTAINTIES

A number of geometrical simplifications in the developed model and material properties may influence the accuracy of the results. For the purpose of modeling, several intermittent layers were assumed continuous and simplified. Some other details were not included such as window flashing and vapor barrier. Moreover, simplified boundary conditions were used. Simulation accuracy was assessed using so-called energy error norm defined by THERM. The computational software estimates error in heat flux calculations using the fact that the superconvergent points coincide with Gauss integration points. Using this fact, the algorithm makes a "guess" at all the nodal points. Least squares method is used to fit a smooth function to the values of the gradient at these points. Values obtained from the smooth function are used to calculate difference between actual value of heat flux and the value calculated on the fitted smooth function. The ratio between this difference and predicted heat flux value (based on the smooth function) is used to estimate the accuracy of the numerical solution. The 10% criterion is usually used for mesh refinement. If the estimated error is higher than 10%, the numerical mesh is refined to obtain higher accuracy (i.e. decrease discretization error). According to numerical tests performed in THERM, a 10% limit for energy error norm ensures maximum 1% error in U-factor calculation—for more details refer to THERM technical documentation (LBNL 2011). For all simulations the energy error norm was kept around 6% which yields a U-factor uncertainty of less than 1%.

CONCLUSIONS

A well-insulated building envelope is essential for thermal comfort and energy savings. The individual building components and the connections should be carefully designed. Window-to-wall connections appear to be important in building envelope energy balance.

The study shows that the position of the window in the window opening is a complex subject. There are many factors and conditions related to the insertion of windows that need to be taken into account when selecting a connection solution. The current study is focused mainly on thermal aspects of the window-to-wall interface.

Results show that placement plays a crucial role in linear thermal transmittance. It was also found that thermal bridge increases with increasing thickness of wall construction for all window positions in the opening, except positions in which window is very close to the interior. The optimal window placement position, considering the reduction of heat loss, was found to be around 35 mm from the outside surface of external wind barrier to the edge of the window frame (for timber wall case, regardless of wall thickness). Moreover, the lowest interior surface temperatures were compared for different window positions. The lowest temperature for the optimal solution was found to be only 1°C lower than that for other placements.

It was also found that linear thermal transmittance is reduced when the space between the window and the wall is insulated with a material of lower thermal conductivity or when the window framing thickness is increased. Alternative designs of inside window framing were presented which can lower the linear thermal transmittance value by 53% (using

Table 6. Percentage U-factor Increments for Different Window Sizes

Window Size	Small Window	Standard Window	Large Window
Dimension (W × H), m	0.7 × 0.9	1.24 × 1.48	3 × 2.5
Base window U-factor, W/(m²·K)	1.09	0.95	0.83
Case	**Increment (%) / Window U-factor (W/[m²·K])**		
Case 1 (position –42 mm)	22% / 1.33	15% / 0.95	12% / 0.90
Case 1 (position +35 mm)	7% / 1.16	5% / 0.99	3% / 0.85
Case 6 (position +35 mm)	3% / 1.12	2% / 0.97	1% / 0.84
Case 8 (position +35 mm)	1% / 1.10	0.6% / 0.95	0.4% / 0.83

insulation with conductivity of 0.037 W/[m·K]) and 87% (using insulation with conductivity of 0.010 W/[m·K]).

It was shown that for a standard size window (1.23 m × 1.46 m) the outermost window position caused a window U-factor increment of 15%, while for the optimal location (b = +35 mm) the increment was reduced to 5%. Considering inside window sill in the proposed alternative way gives increment values of 3% and 1% for the case with improved insulation.

FUTURE WORK

The work presented in this paper is a good starting point for further investigation of window-to-wall connection properties. Further work might focus on the following:

- investigating window-to-wall connection including window jambs and head,
- investigating window-to-wall connection for walls with different materials/constructions, and
- investigating window-to-wall connection in three dimensions to capture influence of geometric thermal bridges (corners of the window).

ACKNOWLEDGMENTS

This work has been funded by the Research Council of Norway, Lian Trevarefabrikk and Lawrence Berkeley National Laboratory (LBNL) through the NTNU and SINTEF research project *Improved Window Technologies for Energy Efficient Buildings* (EffWin).

REFERENCES

Cappelletti, F., Gasparella, A., Romagnoni, P., and Baggio, P. (2011) Analysis of the influence of installation thermal bridges on windows performance: The case of clay block walls. *Energy and Buildings,* 43 (6), 1435–1442.

Gustavsen, A., Arasteh, D., Jelle, B. P., Curcija, C., and Kohler, C. (2008a) Developing Low-conductance Window Frames: Capabilities and Limitations of Current Window Heat Transfer Design Tools, State-of-the-Art Review. *Journal of Building Physics,* 32 (2), 131–153.

Gustavsen, A. and Roald, B. K. (2008) Byggforskserien, Byggdetaljer 471.015 / Kuldebroer. Konsekvenser og dokumentasjon av energibruk. (Thermal bridges. Impact and documentation on energy usage.). Oslo, SINTEF Byggforsk.

Gustavsen, A., Thue, J. V., Blom, P., Dalehaug, A., Aurlien, T., Grynning, S., and Uvsløkk, S. (2008b) Kuldebroer: beregning, kuldebroverdier og innvirkning på energibruk. (Thermal bridges: Calculation, thermal bridge values and impact on energy use.).

ISO, I. O. f. S. (2006) ISO 10077-1:2006. Thermal performance of windows, doors and shutters, Calculation of thermal transmittance - Part 1: General. Wacetob.

ISO, I. O. f. S. (2007a) ISO 6946:2007. Building components and building elements, Thermal resistance and thermal transmittance. Calculation method. Wacetob.

ISO, I. O. f. S. (2007b) ISO 10211:2007. Thermal bridges in building construction, Heat flows and surface temperatures. Detailed calculations.

ISO, I. O. f. S. (2007c) ISO 10456:2007 - Building materials and products. Hygrothermal properties, Tabulated design values and procedures for determining declared and design thermal values.

ISO, I. O. f. S. (2007d) ISO 14683:2007. Thermal bridges in building construction, Linear thermal transmittance. Simplified methods and default values.

ISO, I. O. f. S. (2012) ISO 10077-2:2012. Thermal performance of windows, doors and shutters, Calculation of thermal transmittance, Part 2: Numerical method for frames. Wacetob.

Lacasse, M. A., Armstrong, M., Ganapathy, G., Rousseau, M., Cornick, S. M., Bibee, D., Shuler, D., and Hoffee, A. (2009) Results on Assessing the Effectiveness of Wall-Window Interface Details to Manage Rainwater. *Journal of ASTM International,* 6 (9).

LBNL, L. B. N. L. (2011) THERM 6.3/WINDOW 6.3 NFRC Simulation Manual. Windows and Daylighting Group, US.

Maref, W., Van de Bossche, N., Armstrong, M., Lacasse, M. A., Elmahdy, H., and Glazer, R. (2011) Laboratory Tests of Window-Wall Interface Details to Evaluate the Risk of Condensation on Windows. *Journal of Testing and Evaluation,* 39 (4), 562–575.

Maref, W., Van Den Bossche, N., Armstrong, M., Lacasse, M. A., Elmahdy, H., and Glazer, R. (2012). Condensation risk assessment on box windows: the effect of the window-wall interface. *Journal of Building Physics,* 36 (1), 35–56.

NFRC, N. F. R. C. (2012a) Procedure for Determining Fenestration Product U-factors, 100.

NFRC, N. F. R. C. (2012b) Procedure for Determining Thermophysical Properties of Materials For Use in NFRC-Approved Software.

SINTEF (2003a) 471.010 - Varmekonduktivitet og varmemotstand for bygningsmaterialer. (Thermal conductivity and thermal resistance of building materials.).

SINTEF (2003b) 523.701 - Innsetting av vindu i vegger av bindingsverk. (Installation of windows in wooden frame walls.).

SINTEF (2003c) 523.702 - Innsetting av vindu i mur- og betongvegger. (Installation of windows in brick and concrete walls.).

Voss, K. and Musall, E. (2011) Towards Net Zero Energy Solar Buildings. International Energy Agency.

Insulating and Air Sealing Low-Pitch Residential Attic Spaces: Cost-Effectiveness Evaluation

Gary Parsons
Member ASHRAE

Mae Drzyzga

ABSTRACT

Rising energy costs and attractive incentive programs established by the government and/or energy companies continue to fuel the need and desire to increase the energy efficiency of residential homes. Air sealing and insulating attic spaces remains one of the best opportunities to attain these energy-efficiency improvements.

Low-pitch attics, though, present unique challenges to effectively insulate and air seal them, especially in retrofit situations where the rafter/truss design is fixed and space constraints exist at the perimeter of the structure. The perimeter space constraints limit accessibility to these areas as well as restrict the amount of insulation that can be installed.

This paper quantitatively and qualitatively summarizes the cost-effectiveness of various solutions for sealing and insulating attics under low-pitch roofs, with an emphasis on evaluating solutions in areas with space constraints.

Insulation and air sealing materials as well as methods for installing those materials were evaluated in a laboratory setting to determine the cost-effectiveness of each method. In addition to the cost versus benefit of each insulating/air sealing solution, the practicality of each approach is compared.

INTRODUCTION

The Importance of Adequately Insulating and Air Sealing Attics

While adequately insulating and air sealing attics conserves energy and reduces energy costs to homeowners, these actions also result in several other positive outcomes.

First, air sealing slows air movement through the structure due to stack effects. In the cooler months, stack effects invite air containing moisture from the outside in through leaks at the bottom of the house. The air travels up into the attic. In a ventilated attic cool roof design, the moisture quickly condenses on cold objects in the attic and/or the underside of the roof and the potential for moisture damage is heightened. Similarly, stack effects may drag radon gases from the ground into a home. Adequately air sealing the attic space can help limit the stack effect, thereby limiting moisture in the attic and slowing radon intrusion into a home. The overall comfort of the home is also improved since drafts are reduced as well.

Second, in colder climates, if the attic is not sufficiently insulated or has too much warm air leaking in, the roof deck may become warm enough to melt snow. The water from the melted snow runs downward and freezes on the eaves, creating ice dams. As this phenomenon is repeated, the ice dams grow. The water is unable to flow downward to the eaves and instead moves up under the shingles and roof covering. Ultimately, water leaks into the attic and along the exterior walls.

Walls present one of the largest opportunities to improve the air sealing of a building. Between 18% to 50% of a building's air leakage is through the walls, which includes the leaks into the attic at the top plate (ASHRAE 2005).

Gary Parsons is a research and development fellow and Mae Drzyzga is a research and development manager with Dow Building Solutions, Midland MI.

Building Code and Government Program Requirements

The International Codes, ENERGY STAR® Qualified Homes Program, and United States Department of Energy (DOE) have all acknowledged the importance of insulating and air sealing attic spaces and have strived to address the issue by mandating minimum requirements and/or providing recommendations to home owners and builders. The requirements and recommendations for insulating and air sealing attics have increased over the last decade.

R-Value Minimums for Attic Spaces. The International Codes and the ENERGY STAR Qualified Homes Program outline minimum R-value targets for insulating attics in new homes. The DOE has published recommended R-value minimums for insulating attics in new and existing homes, indicating a desire and need to address insulation levels in existing residential structures as well as new structures. Table 1 summarizes these minimum R-value requirements.

The International Codes and ENERGY STAR program allow mechanisms other than those identified in Table 1 to satisfy the attic insulation requirements; Table 1 lists the recommendations for a common compliance mechanism providing a sense of the R-value expectations for an energy-efficient home. Note the trend towards higher R-value thresholds in the International Codes. Also, note that the International Codes and ENERGY STAR program requirements openly recognize the space constraints along the perimeter of attics and provide some reduction in the minimum R-value required for this area.

Air Sealing Requirements for Attic Spaces. As with insulating attic spaces, the International Codes, ENERGY STAR program, and DOE have placed similar and growing emphasis on air sealing the building envelop of a home.

For example, the 2006 *International Residential Code*'s (*IRC*'s) early call for action requiring that "the building thermal envelope shall be sealed to limit infiltration" and listing ten locations to be "caulked, gasketed, weatherstripped or otherwise sealed with an air barrier material, suitable film or solid material" (ICC 2006) was quickly and vastly improved upon. The 2009 and 2012 editions of *International Energy Conservation Code* (*IECC*) identify two additional locations to be sealed and outline compliance verification requirements (ICC 2009, 2012).

The 2009 *IECC* requires a home to have ≤ 7 air changes per hour (ACH) at 50 Pa or be visually inspected by the code official or approved party independent of the installer (ICC 2009). The 2012 *IECC* mandates even tighter homes and more rigorous compliance verification. Under the 2012 *IECC*, homes must have ≤ 5 ACH and ≤ 3 ACH at 50 Pa for climate zones 1–2 and climate zones 3–8, respectively. Further, the option to only visually inspect for air sealing is removed in the 2012 *IECC*, and all homes must be tested (ICC 2012).

The ENERGY STAR program devotes an entire section, Section 5, of its Thermal Enclosure System Rater Checklist to air sealing. Section 5.2.3 specifically requires that the top plate at all unconditioned attics/wall interfaced be sealed to the drywall. Further, the ENERGY STAR program requires ACH at 50 Pa to be less than or equal to 6, 5, 4, and 3 in climate zones 1–2, 3–4, 5–7, and 8, respectively (EPA 2012).

Lastly, the DOE has sponsored and published many guides and fact sheets to educate and encourage homeowners and home builders to air seal homes (DOE 2000a, 2000b, 2000c, 2008, 2010).

Table 1. Minimum Attic R-Value Requirements and Recommendations in *IRC* (ICC 2006), *IECC* (ICC 2009, 2012), ENERGY STAR Qualified Homes Program (EPA 2012), and DOE Fact Sheet (2008)

Climate Zone	2006 IRC	2009 IECC	2012 IECC	ENERGY STAR, Version 3 (Rev. 06)	DOE Insulation Fact Sheet		
					New Homes	Uninsulated Existing Attic	3-4 in. of Insulation in Existing Attic
1	30	30	30	30	30 to 49	30 to 49	25 to 30
2	30	30	38	30	30 to 60	30 to 60	25 to 38
3	30	30	38	30	30 to 60	30 to 60	25 to 38
4	38	38	49	38	38 to 60	38 to 60	38
5	38	38	49	38	38 to 60	49 to 60	38 to 49
6	49	49	49	49	49 to 60	49 to 60	38 to 49
7	49	49	49	49	49 to 60	49 to 60	38 to 49
8	49	49	49	49	49 to 60	49 to 60	38 to 49

NOTE: Under the 2006 *IRC*, 2009 and 2012 *IECC*, and ENERGY STAR Qualified Homes Program, R-30 is deemed to satisfy requirements for R-38 whenever the full height of uncompressed R-30 insulation extends over the wall top plate at eaves. Similarly, R-38 is deemed to satisfy requirements for R-49 whenever the full height of uncompressed R-38 insulation extends over the wall top plate at eaves.

Low-Slope Attics: Where Space Constraints Limit Insulating and Air Sealing

While strong driving forces exist to adequately insulate and air seal attics, low-slope attics pose space constraints and make insulating and air sealing a challenge. While many roof designs exist, most designs closely resemble either a hip or a gable roof system. In a gable roof system, assuming a rectangular footprint, the perimeter space constraints presented by the truss/rafter design are largely present on only two sides of the home. By contrast, in a low-slope hip roof system, the attic has truss/rafter design constraints on all four sides and the corners.

Sources conflict on the exact definition of a low-slope roof, with some defining a rise of 3 units or less for every 12 units of horizontal distance (i.e., 3/12 slope) as a low slope and others citing a 2/12 slope (ASTM 2006; Feirer and Feirer 2004; CoolFlatRoof.com 2012). Regardless of the exact definition, roofs with even as much as a 5/12 slope are still significantly space constrained and pose accessibility challenges for installing insulation and air sealing.

Figure 1 illustrates the space constraints. It provides a visual of how the space for installing insulation and physically maneuvering to do installation work in an attic shrinks as the roof slope decreases from 8/12 to 3/12. It also shows the approximate distance from the exterior of the top plate to the interior of the attic required before enough vertical space exists to insulate to an R-30 and R-49 level using a generic R-3.5/inch insulation.

The 2006 *IRC*, 2009 *IECC*, 2012 *IECC*, and ENERGY STAR program all extend some latitude in their insulation requirements for space-constrained areas around the perimeter of the attic (see the note in Table 1). But, even a densely packed insulation can require 8 in. of vertical space before reaching an R-30 level, which is not available along the perimeter of many low-slope attics.

In new construction, several approaches can be used to build a home with a low-slope roof and provide sufficient space to physically access for air sealing and adequately insulating the attic perimeter. Three popular approaches include raised heel truss systems, extended bottom chords, and dropped chord trusses (see Figure 2).

Of the three approaches, the raised heel truss system has been the most popular in recent years. It creates a slightly higher wall as viewed from the exterior of the structure. By contrast, the extended bottom chord approach maintains the traditional wall height appearance; however, it does result in eaves that project further beyond the home's walls than "normal." Finally, the dropped chord truss, like the raised heel

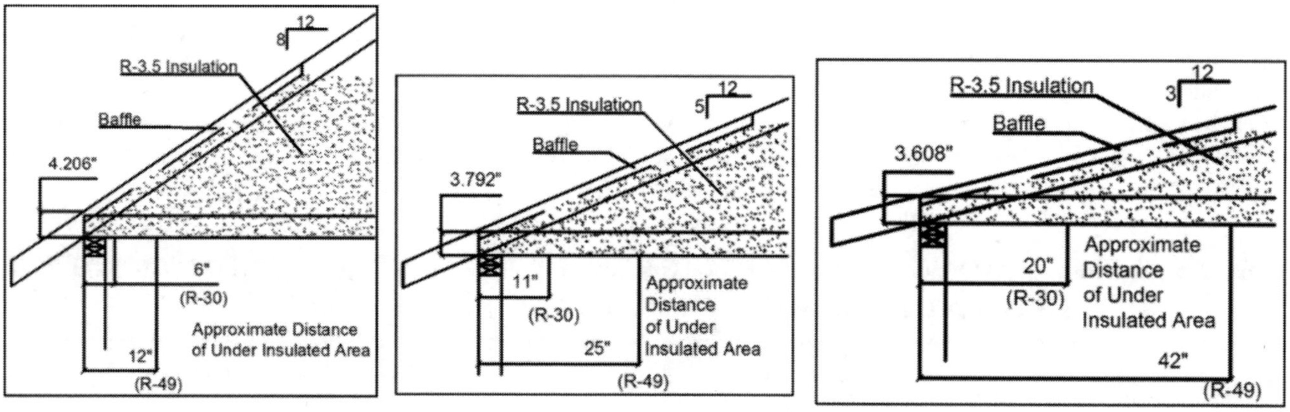

Figure 1 *Attic space constraints increasing with decreasing roof slope, negatively impacting the attic insulation levels.*

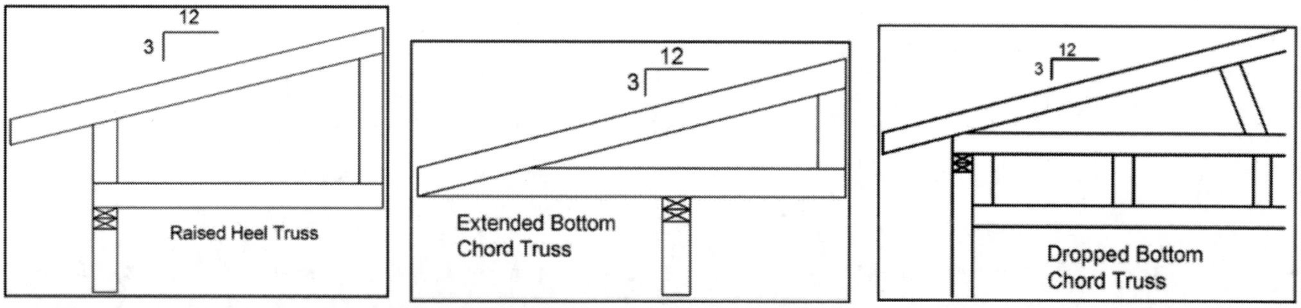

Figure 2 *New construction methods for increasing vertical space along the perimeter of a low slope attic: raised heel truss, extended bottom chord, and dropped bottom chord.*

truss, results in a slightly taller wall appearance from the exterior of the home.

In new construction, low-slope roofs and their sealing and insulating challenges can be avoided through the design of steeper roof slopes. Similarly, a hip roof system with its space-constrained corners is an option that can be ruled out in the design phase of new construction. Unlike new construction, though, retrofitting an existing home with a low-slope roof to ensure adequate insulation and air sealing is a challenge that cannot be designed out.

Accordingly, the challenge of air sealing and insulating existing low-slope roofs is the subject of the research presented here. The objectives of this research were to

- understand current methods employed to address this challenge,
- conduct a series of simulated installations to evaluate the methods known today, and
- begin a data-based discussion on the most effective methods to address space-constrained attic areas.

The authors acknowledge that a potential option to address air sealing and insulating an existing, unconditioned, vented low-slope attic is to convert the attic into a nonvented, conditioned space. While this is a viable option that should be considered in a broad discussion on air sealing and insulating homes, for this research nonvented roofs were left out of the scope.

Current Air Sealing and Insulating Methods Recommended/Employed for Constrained Attic Spaces

Literature Review and Internet Search. While the challenge is broadly recognized, a literature review and internet search found limited information on recommendations specific to addressing the challenges of air sealing and insulating existing low-slope attics. The review did reveal generic attic air sealing and insulation instructions specific to the top plate area, which are relevant to the research topic since, as discussed previously, the attic perimeter typically represents the most space-constrained area of the attic.

After first instructing to pull back any existing insulation in the attic and install baffles along the roof rafters when missing, the instructions offered three options, discussed in detail in the following subsections. Only one source offered the second option listed, which, interestingly, offers some potential to increase the R-value around the perimeter of the attic using foam boards, which typically have a higher R/in. than traditional blown-in or loose-fill attic insulation materials (e.g., R-5 to 6.5 verses approximately R-3.5 to 3.8).

OPTION 1: Air Seal without Improving Insulation (DOE 2010; Lstiburek 2012)

i. Apply a continuous bead of either one-component foam or caulk sealant to the gaps between the drywall, oriented strand board (OSB), and top plate, with most references only calling for the sealing of the gap between the drywall and top plate.
ii. Place a fiberglass roll at the top plate to provide additional support to the baffle.

**OPTION 2:
Air Seal and Improve Insulation (North 2012)**

i. Apply a continuous bead of either one-component foam or caulk sealant to the gaps between the drywall and top plate.
ii. Fit a foam sheet/block extending from over the exterior sheathing to at least the interior drywall.
iii. Install a second foam sheet/block vertically, flush against the outside edge of the attic baffle. Seal around the attic baffle and the vertical foam block seams.

**OPTION 3:
Abandon Ventilation in Some Areas to Improve Air Seal (DOE 2010; Lstiburek 2012; Straube 2012)**

i. Place fiberglass in large opening between roof deck, rafters, ceiling joists, and top plate, where baffles cannot be installed.
ii. Apply two-component air sealing spray foam to the top plate area.

All three options concluded with instructions to reinstall previously existing insulation and/or add new loose-fill or blown-in insulation.

Phone Interviews with and Completed Questionnaires from Experienced Home Performance Retrofitters. Experienced home performance retrofitters' phone interviews and completed questionnaires revealed several consistent themes:

- Experience matters; solutions often need to be customized to the situation
- One- and two-component spray polyurethane foams are popular, often with customized dispensing guns
- Air sealing is a priority over insulation

Other methods employed by home performance retrofitters to address the top plate and corners of attics included:

i. Reconstructing the roof
ii. Conducting air sealing via the exterior eave area as an access point
iii. Removing/replacing interior drywall
iv. Sealing/insulating the wall cavities with urea formaldehyde urethane foam or dense pack cellulose
v. Doing nothing.

Some retrofitters indicated that in many homes the cost and effort to air seal and insulate attics, especially those with access challenges or located in southern climates, did not return cost savings as high as those from making improvements to other areas of the home.

Like the option to convert an unconditioned, vented attic into a conditioned, nonvented attic, the methods identified in this numbered list to air seal and insulate the attic are viable approaches that should be considered in a broad discussion on air sealing and insulating a home. However, to adequately consider all of these approaches is too large a task with the resources available and under the scope of one research project.

Accordingly, the scope of this research project was limited to employable techniques for air sealing and insulating constrained, vented attic spaces that require no renovations (e.g., replacing the roof deck) and are accomplished from the interior of the attic.

EXPERIMENTAL DESIGN

Test Equipment

The laboratory work conducted for this research employed a duct blaster with a digital pressure and flow gauge and two mock attic structures designed to conduct qualitative and quantitative research.

The duct blaster had three flow-ring configurations available, each corresponding to a different airflow range and minimum fan pressure. Based on the differential pressures targeted and the airflows generated during this research, a ring with a flow range of 10–125 cfm and a minimum fan pressure of 3 Pa was employed for all duct blaster testing (EC 2012).

The digital pressure and flow gauge used in conjunction with the duct blaster is capable of pressure measurements that are accurate to ±1% of the reading from –1250 to 1250 Pa. The duct blaster, when used with this specific digital pressure and flow gauge, has a flow accuracy of ±3% (EC 2011).

The first of the two mock attic structures was designed to replicate a section of low-slope, hip roof attic space so that qualitative research could be done on the relative ease of various attic air sealing methodologies as well as some very basic quantitative research on the effectiveness of the air sealing methodologies. Air leakage was not measured using the first mock attic structure. The structure's footprint was 12 × 20 ft with a 3/12 roof pitch and included a simulated hip roof section (see Figure 3).

Construction of the structure in a laboratory setting created differences from in-field practices (e.g., there is an extra 2 × 4 framing layer below the bottom chord). The subtle differences are not believed to materially impact the results of the research.

The second of the two mock attic structures was also designed to replicate a section of a low-slope attic; however, this second structure was designed to perform quantitative air sealing measurements for the top plate area of an attic. Figure 4 shows this second structure.

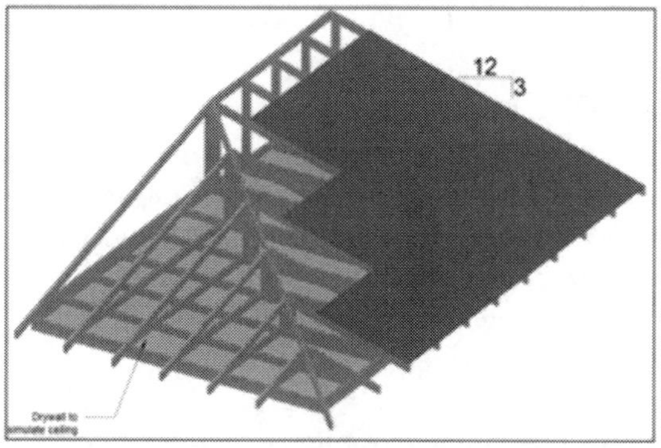

Figure 3 *Mock attic structure with 3/12 hip roof and 12 × 20 ft footprint.*

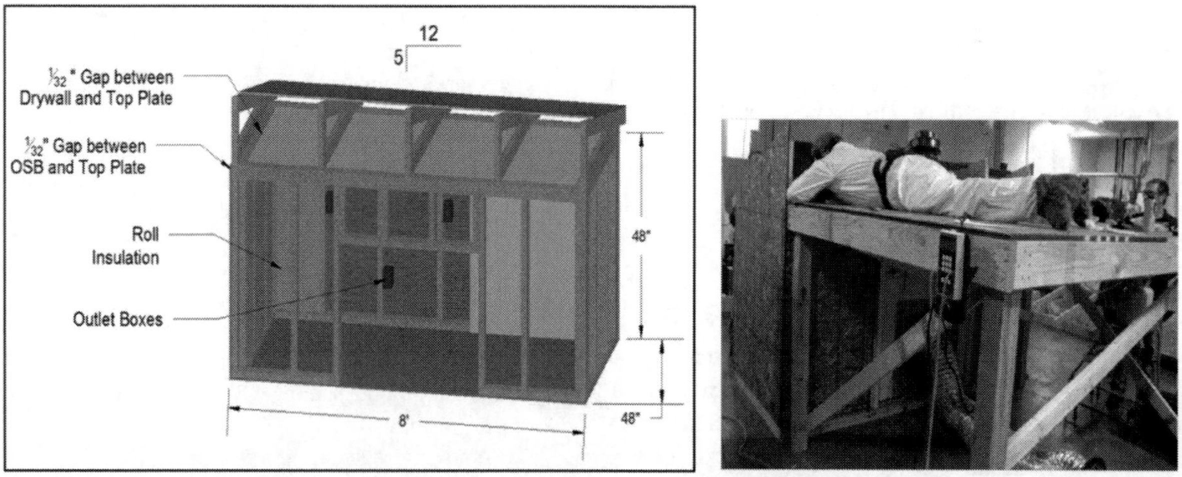

Figure 4 *Mock attic structure with 5/12 gable roof, 4 × 8 × 4 ft chamber, door equipped with duct blaster, and working deck.*

The deck on the right-hand side of the structure was strictly to simulate accessing the attic top plate as an installer would do in the field (i.e., it enabled the installer to stay low on his belly to do the installation work). Air infiltration studies were done on the hut-like structure to the left of the installer deck. The hut consisted of a small chamber below an attic. The small chamber provided an area to allow suction by the duct blaster.

The chamber had a 4 × 8 ft footprint. It was built on a 2 × 4 framed floor, with one single 4 × 8 ft sheet of OSB making up the solid floor surface.

The walls were 4 ft in height and framed 16 in. on center with 2 × 4 studs. Kraft-faced R-13 fiberglass batts were installed in the wall cavities with the kraft facer closest to the interior of the wall. Three rectangular electrical outlets were installed in one interior wall. All four walls were sheathed with solid sheets of drywall on the interior and solid sheets of OSB on the exterior. Similar to common construction practices, the exterior OSB sheathing was not flush with the top face of the top plate; rather, the top of the OSB sat approximately 1/2 in. below the top face of the top plate. Washers were placed between the sheathing and the top and bottom plates to introduce a ~1/32 in. air gap to simulate air gaps found in field construction that occur as a result of lumber drying (Onysko and Jones 1989). All drywall seams were taped and mudded. These features were all included to simulate actual construction.

The roof was approximately a 5/12 slope and comprised of ceiling joists and roof rafters spaced 24 in. o.c. The two short sides were sheathed with OSB to the top of the roof deck and constructed to simulate gable ends. One long side had the rafters protruding over the wall edge to create an eave area. The other long side was left open to access the attic from the deck.

The mini-door with the port for attaching the duct blaster was installed using a standard door frame. The floor sill was sealed with silicone and the other edges were sealed with one-component foam. All penetrations used for tubing inserts as part of the pressure measurements were sealed with silicone.

Air Sealing and Thermal Insulating Techniques Evaluated. Based on the results of the literature search and home performance retrofit contractor information gathered, three strategies were selected to evaluate:

i. Air sealing without improving insulation
ii. Air sealing and improving insulation
iii. Air sealing and improving insulation at the expense of ventilating portions of the attic

A matching, abbreviated description of the technique found in the literature to undertake each strategy is described in Figure 5 and identified as either1, 2, or 3.

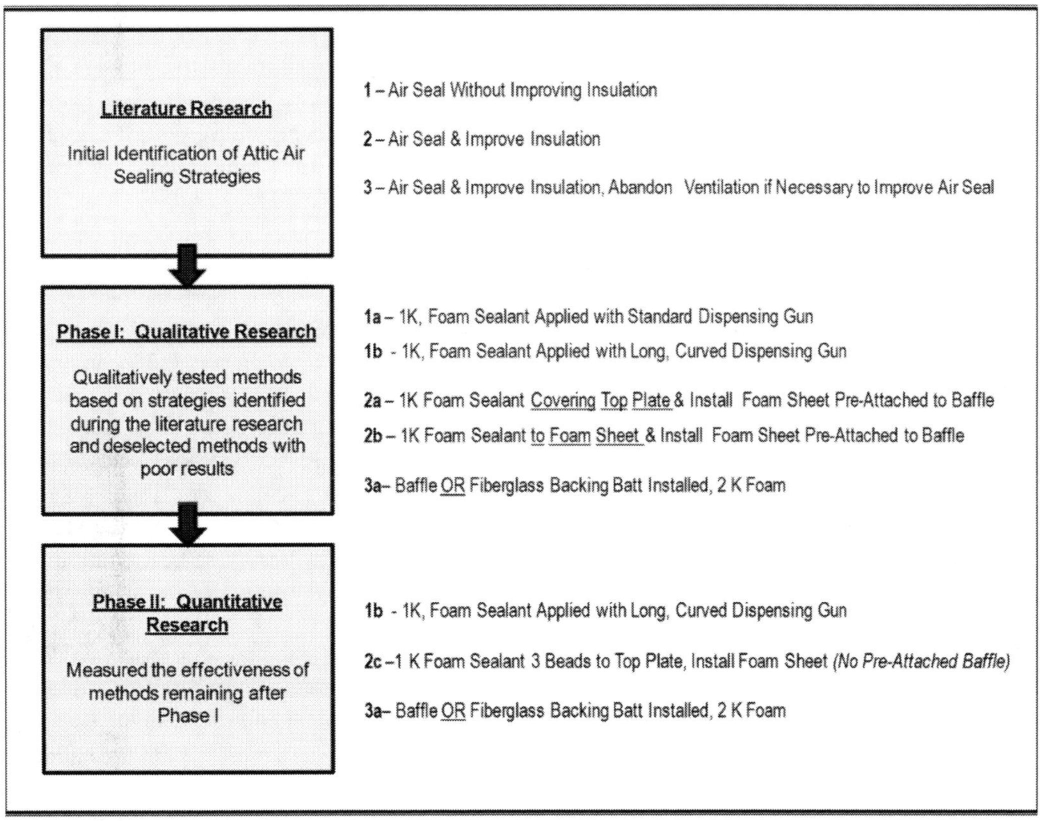

Figure 5 *Experimental strategy for investigating air-sealing and thermal-insulating techniques.*

As shown in Figure 5, as the research progressed beyond the literature search into the qualitative and quantitative phases, each technique originally found in the literature was refined. New information was learned in each phase, helping to optimize techniques. Optimized techniques are identified with their original technique number followed by a chronological alpha-character, representing the order the technique was integrated into the research project.

Note that for the first two strategies the literature was fairly broad in recommending a "sealant." To keep the scope of this research manageable, a one-component polyurethane foam sealant was selected to conduct the qualitative and quantitative research under techniques 1 and 2.

Figure 6 shows the difference between the standard length and long, curved dispensing gun used with the one component (1 K) polyurethane foam during phase I.

As explained previously, the foam board used in techniques 2a, 2b and 2c can enhance the R-value along the attic perimeter over more traditional attic insulation materials. In addition, it was hypothesized that the foam board could be pressed down upon by the installer to help compress and distribute the one-component foam, thereby further aiding in air sealing. A 1 in. thick foil-faced polyisocyanurate foam was used in all testing.

The dispensing system for the two-component foam sealant used in phases I and II consisted of a set of hoses, a mixing nozzle, and a cone tip.

TEST PROCEDURES

As mentioned previously, this research involved the construction of two mock attic structures, which were each used to conduct two separate phases of experiments.

Figure 6 *Standard-length dispensing gun compared to long, curved dispensing gun.*

Phase I Test Procedures

Phase I utilized the 12 × 20 ft mock attic structure, which had a tarp installed on the open 12 ft end to help simulate the dark, limited visibility conditions of an attic. The long 20 ft end was left open, as a flood light would most likely be placed in the center of an attic during real-life retrofit work. The research installer's access was constrained to simulate an actual attic access situation. Again, this ground rule was established to help simulate real-life retrofit work. Most of the data collected in phase I was qualitative.

The first task in phase I involved having the installer enter the attic area and mark the point of his furthest physical reach along the perimeter of the entire attic structure. The roof deck was then removed and the distance between the top plate and ceiling drywall joint and the point of reach was measured. Photos were taken to visually document the reach points.

Next, the techniques outlined in Figure 5 for phase I were executed. Note that the same experienced installer was used for all installation tasks in this research program. In addition, all necessary materials for each technique were pre-staged in the attic structure.

Techniques 1a and 1b required the installer to apply one-component foam to the top plate. Two beads were applied. One was targeted between the exterior sheathing and top plate. The second bead was targeted between the top plate and interior drywall. The installer used the standard dispensing gun in two cavities and the long, curved gun in six cavities.

For techniques 2a and 2b, the foam sheet was pre-cut to ~24 in. wide by 16–20 in. long. Foam baffles were pre-attached to the foam sheet by pushing four or five drywall screws through the foam baffle and foam insulation board (see Figure 7). The hypothesis was that having the baffle pre-attached to the foam would ease baffle installation by acting as an extension arm to get the baffle into place. Foam sheets were limited to the 16–20 in. length based on the assumption that greater lengths would be difficult to maneuver in a confined attic space.

For technique 2a, where the one-component foam was applied to the top plate using the long, curved dispensing gun, four cavities were employed. For technique 2b, three cavities were employed and the standard dispensing gun was used to apply the sealant to the foam sheet (see Figure 8).

For technique 3a, foam baffles were installed in all rafter spaces that the installer was able to access. The installer attempted to stuff ~36 in. long by 3.5 in. thick fiberglass into the ends of the cavities where he was not able to install a baffle. A broom stick was used to ball up the fiberglass and push it out over the soffit area. Once the baffles and fiberglass were in place, the installer left the attic space, suited up in a protective suit and full respirator, re-entered the attic space, and installed the two-component polyurethane foam sealant.

For each installation step and technique, installer feedback was immediately captured after the work was completed. Foam was given adequate time to cure, and then the roof deck

Figure 7 Foam attic baffle pre-attached to 1 in. polyisocyanurate foam insulation board using drywall screws.

was removed. A qualitative analysis of air sealing effectiveness and photographic documentation was then conducted.

Phase II Test Procedures

Phase II employed the 4 × 8 ft by 4 ft tall chamber with the ~5/12 slope roof and duct blaster to quantitatively measure air infiltration improvements after three different air sealing techniques were applied to gaps on either side of the top plate (refer to Figure 5). The installer had the same ground rules as in phase I, with no body part allowed above the highest point of the attic structure and all materials required to be pre-staged either in the attic or on the work deck.

For technique 1b, two beads of one-component foam were applied using the long, curved dispensing gun. The first bead targeted the gap between the interior ceiling drywall and the top plate. The second bead targeted the gap between the top plate and exterior OSB sheathing

For technique 2c, the first bead of the one-component foam was applied onto the interior ceiling drywall, slightly forward of the gap between the drywall and top plate. Phase I had shown that as the foam sheet was pushed out over the top plate area it also pushed the foam sealant. Accordingly, by placing the bead slightly forward of the gap intended to be air sealed, it was hypothesized that bead would move to the ultimately intended location. A second bead was still applied along the gap between the interior ceiling drywall and top plate, though. The third bead applied targeted the outer edge of the top plate. This area was identified as the target after technique 1b was employed during phase II. During that step, it was not possible to maneuver the dispensing gun to apply the foam sealant directly to the gap between the top plate and the OSB sheathing with the sheathing sitting 1/2 in. below the top face of the top plate. Thus, the hypothesis was that it might be possible for a bead of foam to be applied to the outer edge of the top plate to "fall into place" as the foam sheet was installed, sealing this outer gap.

Figure 8 Application of 8–10 in. of one-component foam sealant to polyisocyanurate foam board with attic baffle pre-attached.

No air sealing difference had been seen in phase I between using baffles or fiberglass batts as a backer material for the two-component foam. Foam attic baffles were selected for phase II, technique 3a, since they would typically be used in most attic cavities.

The measurement procedure for phase II first involved taking an air infiltration measurement with both the interior and exterior sheathing to top plate gaps sealed with construction tape to establish what a very good air infiltration rate would be for the chamber if both gaps around the top plate were sealed. This measurement was noted as the targeted performance level. Note that all other gapped areas were also sealed with construction tape to establish the targeted performance level as well as to determine the performance level of each of the three techniques tested (i.e., researchers taped gaps around the bottom plate, gaps between the OSB sheathing, gaps between rafter ends and OSB sheathing, etc.).

Once the targeted performance level measurements were complete, a baseline measurement with no tape on either side of the top plate was conducted. This was important to do before each new air sealing method was installed since the same structure was to be used for each air sealing method. While the previous air sealing material would be removed prior to installing the next method, no guarantee could be made that all the air sealing material would be removed or that the gap size might not be increased in some areas as the old material was chipped away.

Each technique outlined in Figure 5, as well as the targeted performance level, was tested at 25 and 50 Pa pressure differentials. At each pressure differential, five airflow measurements were collected at 20 s apart. The average of the five measurements was rounded to the nearest whole number and reported.

RESULTS

Phase I Results

Phase I results are summarized in Figure 9 and Table 2. Figure 9 illustrates how far the installer was able to physically reach into various areas of the low-slope roof assembly. It provides a visual guide to understanding the constraints an installer faces when air sealing such structures. Table 2 captures both the air sealing observations and the installer feedback for each phase I air sealing and thermal insulating technique. Figures 10–12 provide visual examples of the air seal achieved for the techniques employed in phase I.

Figure 10 exemplifies what a good-quality sealant installation around the top plate looked like using technique 1a or 1b relative to poor installation quality around the top plate. Similarly, Figure 11 shows what a quality sealant installation looked like using technique 2a and a poor-quality installation using technique 2b. In the case of technique 3a, if an installer was able to reach into a given cavity to apply the two-compo-

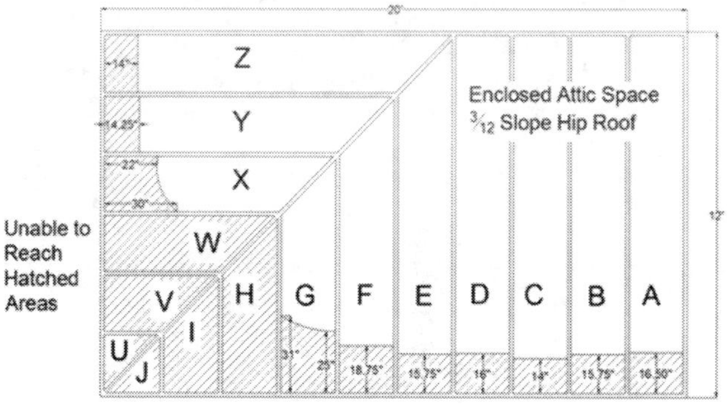

Figure 9 *Bird's-eye view of attic mock up: distance between the furthest physical reach point to the interior top plate edge.*

Table 2. Phase I Testing Summary, Techniques 1a, 1b, 2a, 2b, and 3a

Phase I Technique	Sealant Installation Quality Results Summary	Key Learnings
1a	• Sealant installation quality was poor.	• Extended-length, curved dispensing gun works better than standard-length dispensing gun. • Design improvement opportunities exist for the extended gun to enable more flexible positioning of the gun in a confined area. • Installation could be optimized if the one-component foam can product could be interchangeably positioned vertically and horizontally.
1b	• Reach was to cavity I. • Sealant installation quality was good to cavity D.	
2a	• Sealant installation quality was good.	• Pre-attached foam baffles to a foam sheet are cumbersome to maneuver, become easily damaged, and limit the reach of the technique.
2b	• Sealant installation quality was poor.	
3a	• Reach was to cavity E installing foam baffle. • Reach was to cavity I installing rolled fiberglass batt. • Reach was to cavity G installing two-component foam. • Sealant installation quality was good in *all* cavities reached, installing the two-component foam (e.g., to cavity G).	• An extension wand at the end of the dispensing hose would enable accessing more areas.
Overall	• **Easiest installation, physically:** One-component with the extended, curved dispensing gun. • **Longest reach:** One-component with the extended, curved dispensing gun. • **Best sealant installation quality:** i) Pre-attached baffle to foam sheet with one-component applied to top plate OR ii) two-component spray foam. • **Best sealant installation quality/reach balance:** Two-component spray foam.	

Figure 10 Examples of good- and poor-quality foam sealant installations using techniques 1b and 1a, respectively.

Figure 11 Examples of good- and poor-quality foam sealant installations using techniques 2a and 2b, respectively.

nent foam, the top plate area appeared completely air sealed during post-installation inspection (see Figure 12).

Figure 12 Example of a good-quality foam sealant installation using technique 3a.

Phase II Results

Phase II results are summarized in Figure 13. Recall that only the good-quality installation approaches from phase I were tested in phase II. The reduction in air infiltration was generally the same for all three air sealing techniques employed at pressure differentials of 25 and 50 Pa.

All three techniques had a large performance gap relative to the targeted performance level, having air leakage rates 20% to 35% greater than the target. Recall that both the interior and exterior top plate gaps were sealed with construction tape and air infiltration was measured at the outset of phase II, representing the target performance level for each technique employed. A large portion of the delta from the target can most likely be attributed to the inability of techniques 1b, 2c, and 3a to reach and seal the gap between the exterior sheathing and the top plate.

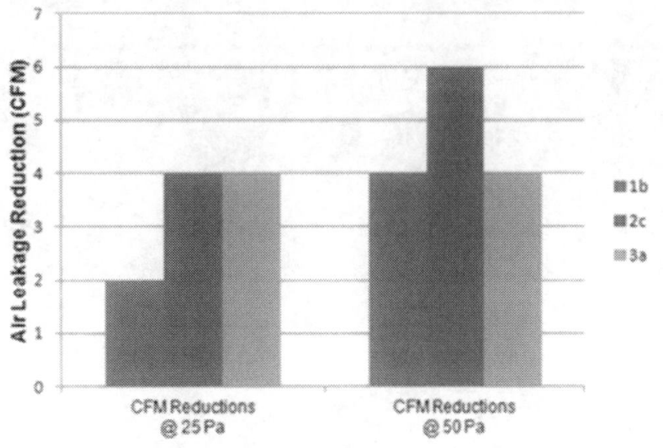

Figure 13 *Reduction in air infiltration in cubic feet per minute (cfm) obtained by sealing the top plate using techniques 1b, 2c, and 3a.*

DISCUSSION

Practicality and Effectiveness of Air Sealing Methods Explored in Phase II

Generally, the installer could not physically reach closer than within about 14 in. of the interior edge of the top plate; however, it was possible to apply air sealing materials to most of the perimeter of a 3/12 slope roof using various reach extension tools and techniques. With the 3/12 hip roof, the installer was generally able to *apply* air sealing materials within 6 ft of the corner.

Various techniques were explored during the early phases of this research program; however, they were eventually refined and narrowed to three techniques that were quantitatively tested in phase II, yielding similar performance results. Since these three techniques were deemed most suitable for use in low-slope attics relative to the other techniques explored, only phase II air sealing techniques are discussed at greater length in this paper.

With three significantly different techniques for air sealing the top plate yielding similar quantitative air infiltration reductions in a laboratory setting during phase II, it is easy to understand why home performance retrofitters consistently stated that a "one size fits all" approach does not exist. Each installation technique employed in phase II was equal in terms of air infiltration reduction; however, the practicality and feasibility study in phase I revealed pluses and minuses of the techniques that would need to be considered for an in-field attic retrofit.

First, while the one-component foam applied using the long, curved dispensing gun was physically the easiest technique to execute and gave the installer the longest reach, the quality of the air seal was not as good at extended distances as it was using other techniques. The dispensing gun was designed for pinpointed application of the foam sealant. The installer, though, could not see if he was applying the foam along the gaps. The installer had to rely on intuition or, in the case of a larger, well-defined gap, feeling the dispensing gun setting into the gap and then moving along the gap.

Further, the one-component foam sealant technique was also plagued by limited maneuverability of the can and long, curved dispensing gun. The can must be in the vertical position to apply the product. The long, curved dispensing gun could not be adequately rotated in the confined space to enable applying one long, continuous bead of foam sealant in a given cavity. Rather, the installer had to stop short of the full cavity width, reposition the dispensing gun, and finish applying the foam sealant the remainder of the cavity width.

The second technique of applying one-component foam to the top plate area and then installing a pre-sized foam sheet extending over the top plate area and slightly into the main attic floor space also faced maneuverability issues. The foam sheet was cumbersome to maneuver around rafters and objects penetrating through the attic. While not completely resolved, some of these issues were reduced when the pre-attached baffle used in phase I was removed in phase II. Logistics aside, one significant disadvantage of this method was the need to pre-cut the foam sheets to size. For in-field applications, depending on the attic layout, this could be tedious.

Despite its in-field limitations, this technique was still considered after phase I largely since it presented a viable method to improve the thermal insulation levels around the perimeter of an attic. Assuming a 3 in. thick R-6.5 foam insulation and a generic R-3.5 insulation installed over the foam, Figure 14 illustrates the shrinking distance of the underinsulated area for R-30 and R-49 targets using this technique with a 3/12 roof. Overall, for a 32 × 40 ft attic, the technique enables ~10% more of the attic area to be insulated to the target value.

The third and final technique explored, using a two-component foam to seal the top plate area, seemed to offer the best balance between reach and air seal quality. While the reach was not as far as the long, curved dispensing gun used with the one-component foam, the two-component foam technique was able to effectively air seal cavities further out than the one-component technique. The reach of the two-component foam might be improved if an extension wand was available to attach to the end of the dispensing hose. The disadvantage of using a two-component polyurethane foam sealant in field is that it requires a specially trained installer who must wear a respirator during the installation.

For all three techniques, the dispensing equipment in-field cleaning factor was a disadvantage. The one-component gun had to be cleaned after application to approximately every fifth cavity, otherwise the nozzle would have a large glob of foam build-up on its end. The tip on the two-component foam dispensing equipment had to be replaced every time the installer had to reposition and relocate himself on the OSB boards laid over the ceiling joists. Otherwise, the foam would harden in the tip and block dispensing. This worked out to be approximately every third cavity. Overall, in-field cleaning

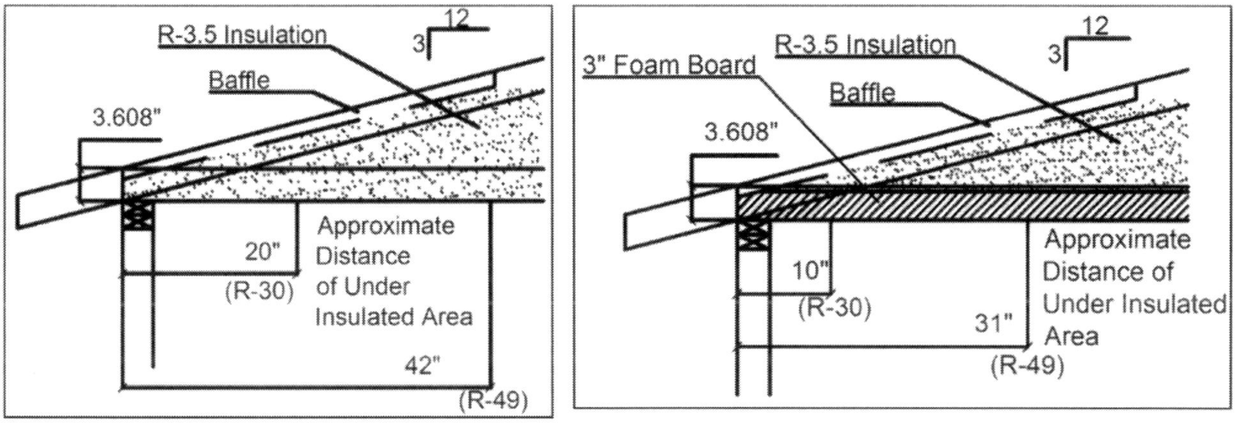

Figure 14 *Shrinking distance of underinsulated area using technique 2c in a 3/12 roof.*

requirements were cumbersome to execute in the confined attic space and slowed the installation work.

Unfortunately, all three air sealing techniques shared a quantitative performance gap. None were able to address sealing between the top plate and exterior sheathing when the exterior sheathing was not flush with the top plate. While few sources specifically recommended sealing this gap and several home performance contractors voiced consciously not working to seal this gap, this gap does offer significant potential for air leakage in a home. Assuming a 1/32 in. gap on each side of the top plate, an unsealed gap between the top plate and the exterior sheathing represents slightly more than 50% of the 107 in.^{2} of open space around the top plate in a 32 × 40 ft attic space.

Air Sealing and Insulating Limitations for 3/12 Slope Hip Roof

Most home performance retrofit contractors commented that they deem air sealing to be more important than insulating. Some of their emphasis on air sealing may be due to the inability to insulate low-slope attics due to a lack of materials with R-values high enough to insulate height-constrained attic spaces of attics to the targeted values. Blown-in cellulose (R-3.8/in.) generally offers a higher R-value than blown-in fiberglass (typically ~R-3/in.) and is most commonly used (CIMA 2012; CertainTeed 2011; GBP 2008a, 2008b; Greenfiber 2012; JM 2012; Owens Corning 2012). If a 3/12 slope roof attic has a targeted R-49 insulation level, though, blown-in cellulose can only insulate approximately 66% of a 32 × 40 ft attic area to this target level.

Foam insulation board products can offer a higher R/in. (e.g., up to R-6.5/in.). However, they are cumbersome to install in an attic and their impact on achieving targeted R-values in attics is minimal unless relatively thick boards are used. For a 32 × 40 ft home with a 3/12 roof, adding a 1 in. thick R-6.5 foam sheet under blown-in cellulose only enables ~68% of the attic area to be insulated to a R-49 target. Adding a 3 in. thick R-6.5 foam sheet under blown-in cellulose enables ~73% of the attic area to be insulated to an R-49 target.

Two-component spray polyurethane high pressurized drum foams can offer R-values comparable to foam boards (e.g., R-6.5). In a 32 × 40 ft attic with a 3/12 slope, it can be installed at heights that would result in the R-49 target being realized in ~86% of the attic area.

Cost Considerations

Recognizing every attic space is different and costs fluctuate (i.e., over time, by geographic regions, between manufacturers, etc.), some basic considerations of cost are appropriate.

First, a 24 oz can of one-component polyurethane foam sealant can typically be purchased for less than $20 per can. By contrast, a ~100 board foot two-component spray foam sealant costs between $200 to 300, and a 55 gal drum set of high-pressure drum foam runs around $2000 (Diamond Tool 2012; FarmTeck 2012a, 2012b; Home Depot 2012; Industrial Products 2012a, 2012b; Lowe's 2012; Menards 2012a; SD 2012; SoyThane 2013). Accordingly, the initial sticker price of both the two-component foams can be an immediate deterrent to an individual inexperienced with the relative effectiveness of attic air sealing and insulating of various materials. Of course, the 24 oz can will not cover as much area as the 100 board foot kit or 55 gal drum set, but its functionality is limited to air sealing whereas many two-component foams offer insulation.

On a board foot (BDF) basis, the material cost of the two-component low-pressure foam can be five times the cost of the two-component high-pressure drum foam. At the low end of the material cost for two-component low-pressure foam, the cost is typically 40% higher than the cost of the two-component high-pressure drum foam on a BDF basis.

Next, moving beyond the sticker shock, the labor costs for each technique might vary. It has been reported that labor costs often outweigh material cost considerations (Keefe 1995). While this research indicated that it was physically easier to install the one-component foam in the laboratory attic than the low-pressure two-component foam, the two-component foam was less tedious, which may equate to less time on the job. On the other hand, a trained installer is required for both the low-pressure and high-pressure two-component foam as well as a supplied-air or air-purifying respirator. These measures are aimed at protecting workers from potential health risks; however, it should be noted that health risks are under continuous evaluation (Yost 2011). The need for trained installers and additional personal protective equipment may increase labor costs for two-component foams.

Although each of these products can be used to achieve the desired air sealing result, a more important consideration than cost is the pairing of the product to work-space access, equipment availability, and installer expertise. All can cost-effectively air seal in the situation appropriate to the product.

CONCLUSIONS

The following conclusions pertaining to air sealing and insulating an existing low-slope attic were drawn from the work documented in this report.

- **The tools are as important as the product.** Phase II results showed that both one-component and two-component foam sealants can effectively seal the gaps surrounding the top plate; however, the tools to reach and apply the products ultimately impact the quality of the air seal. When using the one-component foam sealant, more confined attic space areas were accessible and effectively air sealed using the long, curved dispensing gun than the standard dispensing gun. Yet, the long, curved gun still had some maneuvering challenges and could not seal a gap in a continuous bead. Further, it often did not achieve an air seal in many areas, as it was difficult for the installer to tell where the product was being applied with the gun. Similarly, the two-component foam could have reached even more areas if it had an extension wand.
- **Keep it simple in the attic.** Using foam boards with baffles pre-attached to foam sheets was cumbersome and resulted in damage to the baffles during installation. It was easier and more successful installing the two items independently. That said, maneuvering 24 × 20 in. foam boards around rafters and other obstructions was not trivial either. Accordingly, assembling smaller pieces as they are installed in the attic is preferred over pre-assembly and large pieces.
- **High-pressure two-component spray polyurethane foam offers the highest thermal insulating potential.** Existing thermal materials, independent of cost, do not have the R-value/inch to meet the targets along the perimeter of the attic. When strictly insulation value is considered, high-pressurized two-component spray polyurethane foams offer the best potential to meet target R-values in low-slope attics. Yet, they still cannot meet higher (e.g., R-49, R-60) R-value targets at the perimeter of the lower-sloped attics. While not tested for air sealing during research, they are assumed to have capabilities similar to their low-pressure, small-kit versions.
- **Insulating the exterior gap is very difficult from the interior.** Air sealing the joint between the exterior sheathing and top plate when the exterior sheathing was not flush with the top of the top plate was problematic with the approaches investigated. It is highly probable that this gap is best addressed from the exterior of a low-slope attic home.
- **Two-component foam sealants have the best air sealing capacity.** While it is possible to achieve a good air seal over the top plate area with all three of the air sealing methods explored in this paper, the air sealing achieved with the two-component foam is superior as it is able to adequately air seal more confined-spaced areas than either of the one-component foam techniques investigated.

RECOMMENDATIONS

Overall, the work presented in this paper lays a foundation to begin work on developing better methods and materials to address air sealing and insulating space-constrained areas of low-slope attics. The following recommendations are drawn from the work documented in this report.

- Manufacturers of one-component foam sealants should continue to refine the design of their extended dispensing guns to maximize the ability to efficiently and effectively apply these products in space-constrained, non-visible areas.
- Manufacturers of two-component foam sealants should explore developing extended-reach dispensing wands to increase the accessibility to more confined attic areas.
- Manufacturers of both one- and two-component foam sealants should consider extension dispensing guns to address accessing the joint between the exterior sheathing and top plate when the exterior sheathing is not flush with the top of the plate.
- Manufacturers of one-component foam sealants should consider enabling product dispensing from cans oriented in more directions to improve the ability to effectively apply these products.
- Further research should be conducted to better understand the cost and practicality of using high-pressure two-component spray polyurethane foams in low-slope attics, since they currently offer the greatest insulating potential for such attics.
- Research should be conducted to explore the cost-effectiveness of attic air sealing approaches from the exterior of the home.
- Research should be conducted to understand the ability of various blind wall cavity fill insulations to simultane-

ously insulate the wall cavities and air seal the top plate area.

ACKNOWLEDGMENTS

The authors would like to extend their sincere gratitude and thanks to Brandon Bills, Dan Darling, and Brian Walther of The Dow Chemical Company's Building Solutions Systems Development Laboratory for performing all the construction and installation work encompassed in this research program.

REFERENCES AND BIBLIOGRAPHY

Abrey, D. 2012. Phone interview with Dave Abrey, Director of Quality Assurance at Green Homes America, December 10.

ASHRAE. 2005. *ASHRAE Handbook—Fundamentals*, p. 7.13.

ASTM. 2006. ASTM E631-06, *Standard Terminology of Building Construction*." West Conshohocken, PA: ASTM International.

CertainTeed. 2011. Specification Sheet: TrueComfort® Fiber Glass Blowing Insulation. November. Valley Forge, PA: CertainTeed Saint-Gobain. www.certainteed.com/resources/30-49-189%20TrueComfort.pdf

CIMA. 2012. Insulation Performance Value, Key Performance Features Table. Dayton, OH: Cellulose Insulation Manufacturers Association. www.cellulose.org/HomeOwners/KeyPerformanceFeaturesTable.php

CoolFlatRoof.com. 2012. Ice Dams Prevention & Ventilation of Low-Slope Roofs. www.coolflatroof.com/flat-roofing-blog/ice-dams-prevention-ventilation-of-low-slope-roofs

Diamond Tool. 2012. 11/12/12 Online Pricing for Handi-Foam Quick Cure, Fomo Sealant 105 BDF Yield.

DOE. 2000a. Technology Fact Sheet: Air Sealing, Seal Air Leaks and Save Energy, February. Energy Efficiency & Renewable Energy, U.S. Department of Energy, Washington, DC.

DOE. 2000b. Technology Fact Sheet: Attic Access, Providing Adequate Insulation Coverage and Air Sealing for the Access Between Living Space and Unconditioned Attic, February. Energy Efficiency & Renewable Energy, U.S. Department of Energy, Washington, DC.

DOE. 2000c. Technology Fact Sheet: Ceilings and Attics, Install Insulation and Provide Ventilation, February. Energy Efficiency & Renewable Energy, U.S. Department of Energy, Washington, DC.

DOE. 2008. Department of Energy's Insulation Fact Sheet, Version 2008. Prepared by Oak Ridge National Laboratory for U.S Department of Energy, Washington, DC.

DOE. 2010. Build America Best Practices: Retrofit Techniques and Technologies: Air Sealing A Guide for Contractors to Share with Homeowners, April 12. Prepared by Pacific Northwest National Laboratories and Oak Ridge National Laboratory for the office of Energy Efficiency & Renewable Energy, U.S. Department of Energy, Washington, DC.

EC. 2011. DG-700 Pressure and Flow Gauge Specification Sheet. Minneapolis, MN: The Energy Conservatory. www.energyconservatory.com/system/files/documents/dg700brochure.pdf

Drum, H. 2012. E-mail correspondence with Hayden Drum, Garland Insulating, Ltd, December 3, 4, and 5.

EC. 2012. Minneapolis Duct Blaster® Operation Manual (Series B Systems). Minneapolis, MN: The Energy Conservatory. www.energyconservatory.com/sites/default/files/documents/duct_blaster_manual_series_b_-_dg700_revised.pdf

EPA. 2012. ENERGY STAR® Qualified Homes: Version 3 (Rev. 06). Inspection Checklists for National Program Requirements. Washington, DC: U.S. Environmental Protection Agency.

FarmTeck. 2012a. 11/1/12 Online Pricing for Handi-Foam Fireblock Gun by Fomo.

FarmTeck. 2012b. 11/12/12 Online Pricing for Handi-Foam Quick Cure, Fomo Sealant 105 BDF Yield.

Feirer, M.D., and J.L. Feirer. 2004. *Glencoe Carpentry & Building Construction*. New York: McGraw Hill.

GBP. 2008a. E-Z Attic Loose-fill Fiberglass Insulation. September. Greer, SC: Guardian Building Products.

GBP. 2008b. Guardian Attic Insulation. May. Greer, SC: Guardian Building Products.

GreenBuildingAdvisor.com. 2011. Spray Foam Insulation: Open and Closed Cell, Spray Foam Is The King of Insulation—Stellar Performance on a Price to Match, updated June 22.

GreenFiber. 2012. Attic Card: This is GREENFIBER® Loose Fill Cellulose Insulation, INS510 Loose Fill 30 lb Bag. Charlotte, NC: GreenFiber.

Home Depot. 2012. 11/1/12 Online Pricing for GREAT STUFF™ Gaps & Cracks Pro by Dow Chemical.

ICC. 2006. *International Residential Code*. Washington, DC: International Code Council.

ICC. 2009. *International Energy Conservation Code*. Washington, DC: International Code Council.

ICC. 2012. *International Energy Conservation Code*. Washington, DC: International Code Council.

ICC. 2012. *International Residential Code*. Washington, DC: International Code Council.

Industrial Products. 2012a. 11/2/12 Online Pricing for Touch 'n Seal Gun Foam II Fire Blocking Foam by Convenience.

Industrial Products. 2012b. 11/12/12 Online Pricing for Touch 'n Seal Standard U2-110 Sealant by Convenience.

JM. 2012. Attic Protector® Formaldehyde-free™ Blow-in Loose-fill Fiber Glass Insulation Product Data Sheet. July. Denver, CO: Johns Manville. www.jmhomeowner.com/products/files/AtticProtector.pdf

Keefe, D. 1995. Air Sealing in Occupied Homes. *Home Energy Magazine Online*, November/December.

Knauf. 2012. Jet Stream® Ultra Blowing Insulation Data Sheet. Knauf Insulation, Shelbyville, IN.

Lowe's. 2012. 11/1/12 Online Pricing for GREAT STUFF™ Gaps & Cracks Pro by Dow Chemical.

Lstiburek, J. 2012. Guide to Attic Air Sealing: Identifying and Blocking Air Leakage Pathways Providing Airtight Closure. www.buildingscience.com/documents/guides-and-manuals/gm-attic-air-sealing-guide

Menards. 2012a. 11/1/12 Online Pricing for GREAT STUFF™ Gaps & Cracks Pro by Dow Chemical.

Menards. 2012b. 11/12/12 Online Pricing for FROTHPAK™ 120 Sealant by Dow Chemical.

North, E. 2012. How to Install and Air Seal Attic Baffles. www.energyauditingblog.com, post dated March 30.

Onysko, D.M., and S.K. Jones. 1989. Airtightness of wall sheathing as a function of lumber drying. *Proceedings of the ASHRAE/DOE/BTECC/ CIBSE Conference, Thermal Performance of the Exterior Envelopes of Buildings IV.*

Owens Corning. 2012. AttiCat® Expanding Blown-In PINK® FIBERGLAS™ Insulation Manufacturers Fact Sheet. December. Toledo, OH: Owens Corning. www2.owenscorning.com/literature/pdfs/10011287AttiCatManufacturersFactSheet.pdf

Resnick, D. 2012. E-mail correspondence with David Resnick, Operations Manager at Dr. Energy Saver, December 13.

SFD. 2012. Product Improvement & ICC Approvals, Online Discussion Forum, Spray Foam Distributors, Austin, TX. Viewed in January. http://sprayfoamdistributors.com/?page_id=12

SoyThane. 2013. 1/14/13 Online Pricing for 55 gallon drum set of SoyThane 2000.

Smith, H. 2012. Phone interview with Hal Smith, Halco Plumbing & Heating, December 4.

Straube, J. 2012. Fixing the Lid. Presented at the 2012 Affordable Comfort Conference, March 29.

Yost, P. 2011. EPA Takes Action on Spray-Foam Health Risks. www.buildinggreen.com/auth/article.cfm/2011/5/3/EPA-Takes-Action-on-Spray-Foam-Health-Risks

The Experimental Hygrothermal Performance of a Robust Cold-Climate Retrofit Interior Foundation Wall Insulation System

Louise F. Goldberg, PhD **Mark L. Stender**

ABSTRACT

A durable interior foundation wall retrofit insulation system, including a water management system in compliance with the foundation wall performance option in the 2009 Minnesota Energy Code, *has been developed, patented, and tested. The system was tested in an interior foundation wall insulation retrofit application on a masonry block foundation wall using two different types of interior insulation. An interior basement insulation system without a water management system was used as a control. The system was tested under a severe internal boundary condition regime in which the relative humidity reached a maximum value of 72% during the heating season. During the cooling season, water was injected into the system via a spray bar to mimic bulk water leakage through the foundation wall. The results confirm that the water management system was effective in draining bulk water and condensate from both sides of the water separation plane in compliance with the requirements of the performance option under the imposed severe boundary conditions.*

INTRODUCTION

A core difficulty in designing durable interior basement wall insulation systems in cold climates (Department of Energy Climate Zones 6 and 7 [ICC 2006]) is the management of bulk water generated on the condensing planes of the insulation system. The condensate is generated from interior-, soil-, and ambient-sourced vapor, although usually not simultaneously.

The peer-reviewed literature on the experimental hygrothermal performance of foundation insulation systems in cold climates is sparse (other than some of the references cited herein). There are studies based on simulations (for example, Chuangchild et al. [2004]), and there is a fairly extensive discussion in the *Building Foundation Design Handbook* (Labs et al. 1988) that also is not based strongly on empirical data. Specific guidance on the durability requirements for foundation wall insulation assemblies in cold climates is not included in the 2006 *International Energy Conservation Code* (ICC 2006). However, specific guidance is provided in, for example, the 2009 *Minnesota Energy Code*. In this code, novel basement wall insulation system designs that are not compliant with the "standard" prescriptive rules are required to comply with the performance option as stipulated in Section N1102.2.6.12. In particular, Section N1102.2.6.12.1 states that "The foundation shall be designed and built to have a continuous water separation plane between the interior and the exterior" (State of Minnesota 2009).

Unfortunately, the statute does not explicitly define a water separation plane (WSP).[1] However, as the statute was based on a research report prepared for the Minnesota Department of Labor and Industry (Goldberg and Huelman 2005), the definition given in the report on which the language of the performance option rules was based is:

> **Bulk water separation plane:** A single component or system of components creating a plane that prevents capillary water flow and water flow caused by hydrostatic pressure.

[1] A definition of the WSP is expected to be included in the 2013 revision of the statute.

Louise F. Goldberg *is the principal of Lofrango Engineering, Minneapolis, MN.* ***Mark L. Stender*** *is president of Moisture Management, LLC, Chaska, MN.*

In a later development (discussed in Goldberg et al. [2010]), this definition was modified to the following:

> A *water separation plane* (WSP) is a single component or a system of components creating a plane that effectively resists capillary water flow and water flow caused by hydrostatic pressure and provides a water vapor permeance of 0.1 perms (5.75 ng/s.m^2.Pa) or less to retard water vapor flow by diffusion.

From a design perspective, therefore, the WSP separates the interior and exterior sides of the foundation insulation system from a water transport perspective (in all phases). Hence, condensate from water vapor originating on the exterior side (soil, ambient environment) of the WSP must be disposed of to the exterior and condensate from interior water vapor sources on the interior side. This creates a difficulty for managing interior bulk water using traditional methodologies such as an interior footing drain tile, since it is not possible to route interior condensate runoff across the WSP without perforating it in some way, nominally violating the performance rule requirement that the WSP be continuous.

The problem is exacerbated for interior foundation wall retrofit insulation systems, particularly those in applications where there is no existing interior footing drain tile and the wall experiences significant bulk water leakage. In order to reduce installation costs and, in particular, avoid the high cost of retrofitting the foundation with an interior footing drain tile, a means of routing the exterior-sourced bulk water to an interior disposal within the WSP is required. This drainage must not result in leakage of significant quantities of water vapor from vapor-saturated wall drainage cavities on either side of the WSP to the foundation interior. The drainage system should also not permit vapor transport across the WSP.

A foundation wall insulation system embodying an integrated water management system (WMS) that addresses these issues and that is compliant with the performance criteria in the *Minnesota Energy Code* has been developed, patented (Goldberg and Stender 2011a, 20llb, 2012), and tested under severe boundary conditions in a Minneapolis, Minnesota, climate. The results of the experimental research are reported herein.

THE WATER MANAGEMENT SYSTEM

The WMS as tested is shown in Figure 1. In the experiment, the base plate of the system was installed on the upper surface of an isolating stud frame base plate and sealed to this plate, the frame studs, and the masonry block wall surface.

With reference to Figure 1, the system comprises a WSP, a bulk water drainage cavity, fixtures to retain the insulation and WSP, and an interior finishing system comprising, in this case, of 2 × 2 in. (51 × 51 mm) nominal wood studs 24 in. (609 mm) on center covered with unpainted 0.5 in. (13 mm) thick gypsum board. A clear gap between the studs and the insulation surface is provided to accommodate utilities such as electrical wiring and plumbing.

Two different insulation systems were tested with the same WMS; see Figure 2, Panels A and B. In the experimental configuration, the WSP was simply a dimpled high-density polyethylene foundation waterproofing sheet with a manufacturer's reported permeance of 0.05 U.S. perms (2.86 ng/s·m^2·Pa) per ASTM Test Method E96 (2005). In the case of Panel A, with semi-rigid fiberglass board insulation, a single sheet was used with the dimples facing toward the wall. In the case of Panel B, with spray foam insulation, two sheets were laminated together so that the dimples projected outward to the interior as well as to the exterior. This was done for convenience because single-foundation waterproofing sheets with dimples protruding on both sides are quite readily available.

The water vapor sealing and bulk water transport system has two components: wicking material adhered to both sides of the bottom edge of the WSP and a pair of seals on either side of the WSP that contact the wicking layers. The WSP assembly is inserted into the receiver cavity and held in place with a retention assembly. When installed, the retention assembly causes the inboard and outboard seals to be compressed against the wicking material to the extent necessary to provide a robust water vapor seal while not impeding the flow of bulk water by capillary action through the wicking sheets.The bulk water from both sides of the WSP is collected in the receiver cavity, from which it passes into the bulk water drainage cavity via weep holes drilled at intervals through the base of the receiver cavity. In the experiment, the drainage cavities in Panels A and B were drained to interior receivers via clear plastic tubing.

When operating as intended, the WMS seals both the interior and exterior cavities on either side of the wall WSP from each other as well as from the bulk water drainage cavity. As the drainage cavity can be connected to a vapor-sealed sump, the system isolates the bulk water collection system as well as the cavities on either side of the WSP from the interior. As long as the WSP below the drainage cavity (typically the sub-slab vapor retarder) is physically adhered to the drainage cavity, the overall system meets the requirements for a continuous WSP as required by the performance option in the *Minnesota Energy Code*. Further, the WSP also meets the requirements for an air barrier (Chapter 1322, Section N1102.2.6.12.4; State of Minnesota 2009) without necessitating an additional layer.

The top plate of the WMS was designed so that the insulation, studs, and gypsum board spanning the wall surface between the top and bottom plates are aligned vertically. This configuration permits easy assembly of the insulation and interior finish components without the need for any additional components.

The top plate assembly includes a 0.75 in. (19 mm) deep channel into which a material that qualifies as a fire break (such as gypsum board) is inserted. This provides the code-required fire break above any combustible insulation (such as polystyrene or polyurethane). The WSP is inserted into the top receiver cavity that is pre-filled with a suitable sealant (such as butyl caulk) sealing the WSP on both sides as well its top edge.

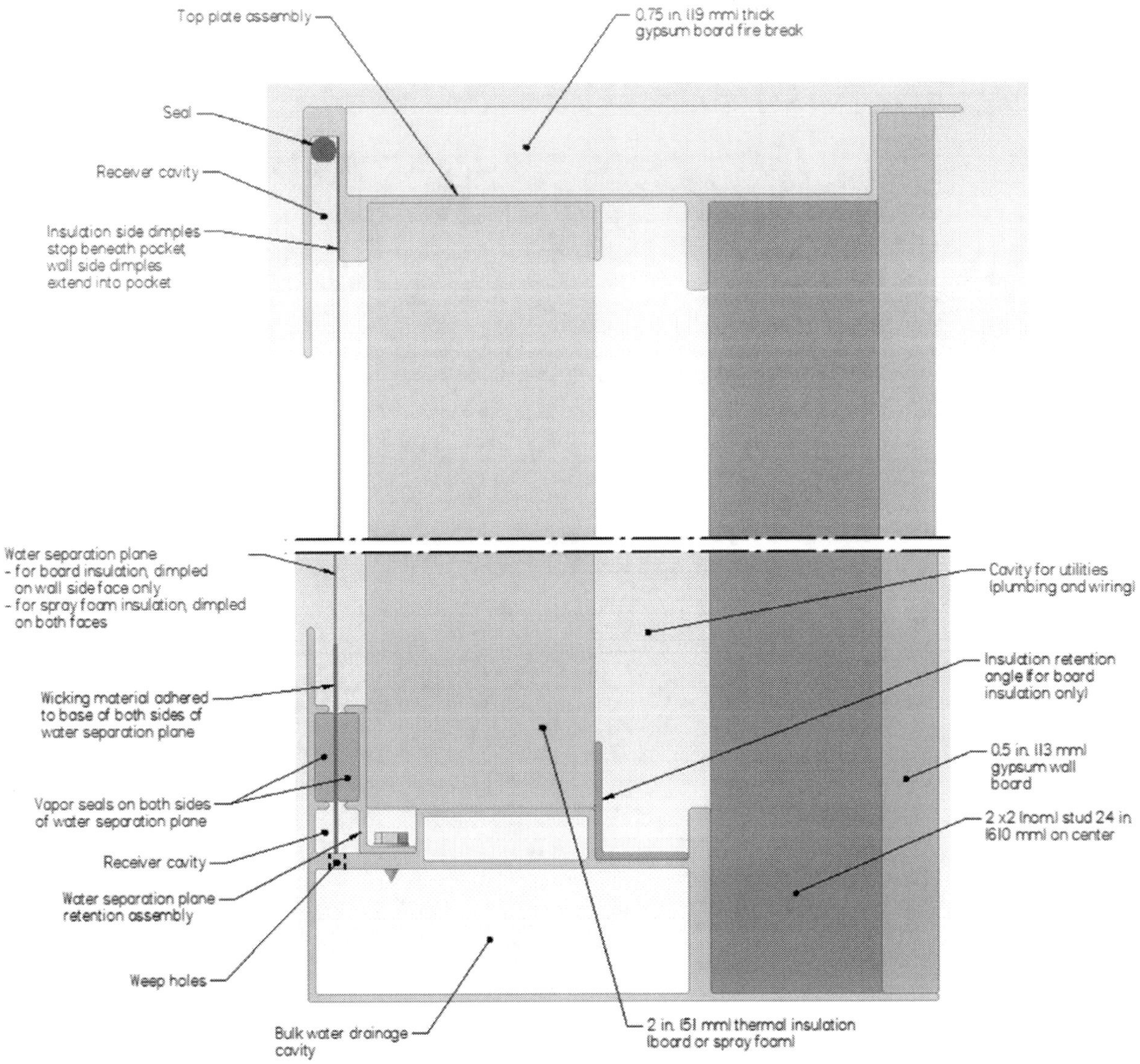

Figure 1 *Water management system (U.S. Patent nos. 8,001,736, 8,074,409 and 8,316,597).*

The only thermal bridging that occurs in the WMS assembly is through the three polyvinylchloride horizontal components of the top and bottom assemblies (each with a thickness of ~1/16 in. [1.6 mm]) and the gypsum fire break in the top plate (0.75 in. [19 mm]). With a bridging depth of about 2.5 in. (63.5 mm), the four bridging elements together yield a nominal steady-state thermal resistance (R-value) of 2.14 $ft^2 \cdot °F \cdot h/Btu$ (0.377 $m^2 \cdot K/W$). As the net thickness of the bridging elements is less than 1 in. (25.4 mm), the overall thermal bridge in the WMS is significantly less than the typical 24 in. (610 mm) on center 2 × 4 interior stud frame used to support fiberglass batt foundation wall insulation. In cases where a continuous plastic material insulation system is used (such as extruded polystyrene board), a fire break is required in any case above the insulation and is typically a 2 × 4 softwood plate. The bridging of this latter system has an R-value of 4.24 $ft^2 \cdot °F \cdot h/Btu$ (0.747 $m^2 \cdot K/W$). Thus, the net heat flow *per unit width* of the WMS thermal bridge is 0.014 Btu/ft·°F·h (0.024 W/m·K) and that of the fire break in a continuous rigid-board system is 0.029 Btu/ft·°F·h (0.051 W/m·K). The thermal bridging of the WMS is therefore less than that of a typical continuous rigid-board based insulation system.

It is important to note that the WMS tested in the experiment was not the final production version that will be fabricated from extrusions that intrinsically will yield a watertight bulk water drainage system (Figure 1).[2] Hence, of necessity, the system tested was built up from individual polyvinylchloride parts (angles, channels, and sheets) fastened together with adhesive. Thus, though every effort was made to make the bottom plate, in particular, watertight, leaks were expected to and eventually did occur over the duration of the experiment.

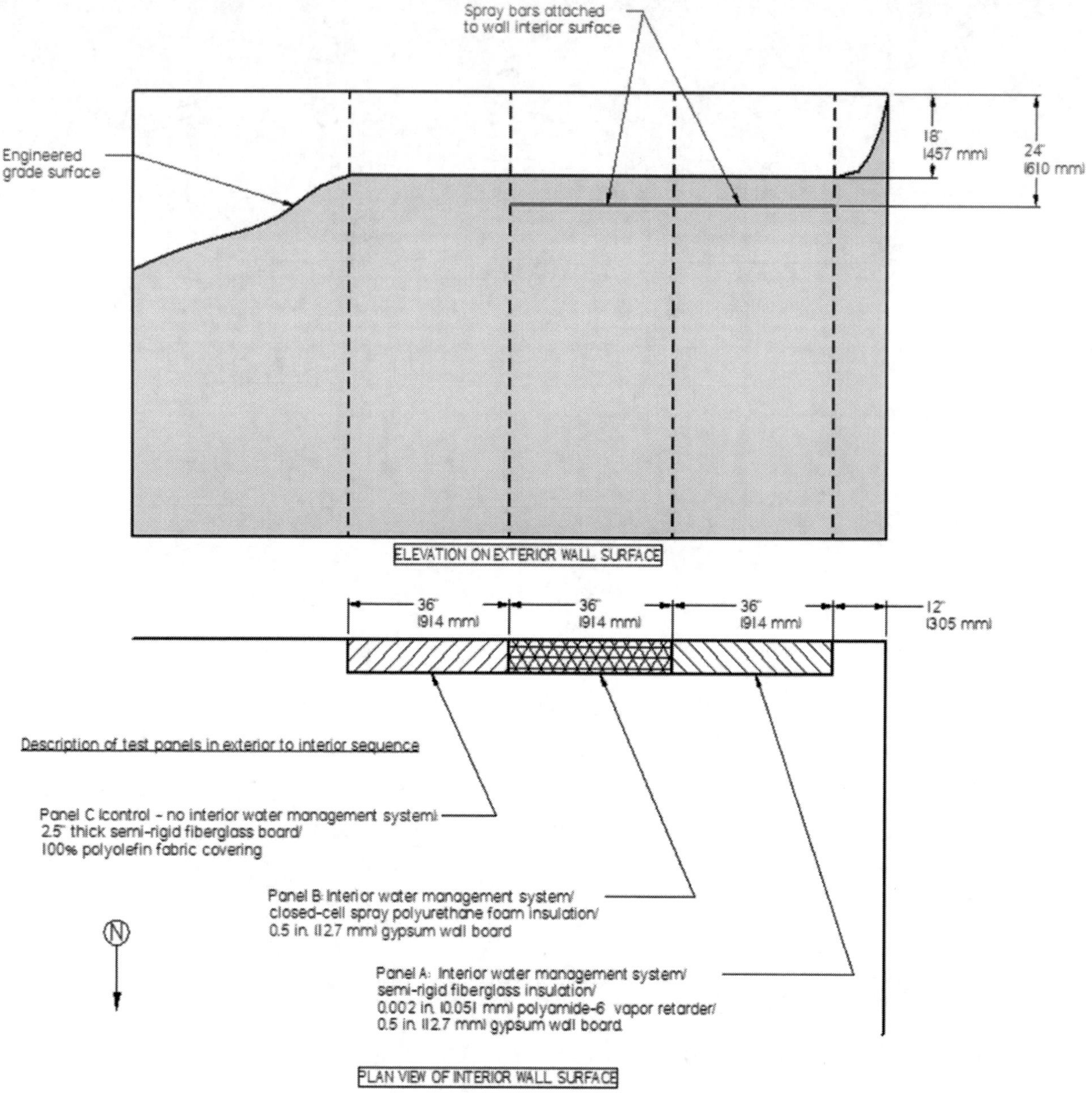

Figure 2 *Experimental schematic design.*

The observed leaks cannot occur in the production WMS fabricated from extruded components.

DESCRIPTION OF THE EXPERIMENT

The experiment was conducted in a foundations laboratory approximately 14 × 8 ft (4.3 × 2.4 m) in size located in Bloomington, Minnesota. The south and west walls were of masonry block construction fabricated from standard two-core, unfilled masonry blocks nominally 8 × 16 × 10 in. (203 × 406 × 254 mm) in size. The north and east walls were interior stud frame partition walls fitted with exterior extruded polystyrene insulation. The laboratory was in the lower level of a two-story structure and the floor joist cavities were filled with encapsulated fiberglass batt insulation. The laboratory was equipped with an electrical resistance convection heater, an air conditioner, and a steam humidifier, all of which were controlled digitally by the laboratory data acquisition and control system. A freestanding dehumidifier was introduced into the laboratory midway through the experiment with an independent digital control (that is, it was not managed by the data acquisition and control system as was the case for the other plant).

[2] Extruded components were not available at the time of the experiment.

Schematic Design

The schematic design of the experiment is shown in Figure 2. The experiment comprised three test panels each 36 in. (914 mm) wide by approximately 92 in. (2.34 m) tall. Each test panel was mounted into a cavity of a stud frame located on the south wall of the laboratory. The stud frame was sealed to the wall and the floor. The rim joist cavities above the south wall were filled with closed-cell spray polyurethane foam to air-seal the top of the foundation wall and rim board. The east masonry block wall and south wall adjacent to the stud frame containing the test panels were left bare for the beginning of the experiment and then covered with 2 in. (51 mm) of extruded polystyrene insulation.

Three panels were tested as shown in Figure 3:

- Panel A consisted of the WMS mounted on the interior surface of the wall with 2 in. (51 mm) of semi-rigid fiberglass board insulation covered with a 0.002 in. (0.051 mm) thick polyamide-6 warm-side vapor retarder beneath 0.5 in. (12.7 mm) unpainted gypsum board.
- Panel B also included the interior wall surface WMS with 2 in. (51 mm) of closed-cell spray polyurethane foam insulation (with no additional warm-side vapor retarder), again beneath 0.5 in. (12.7 mm) unpainted gypsum board.
- Panel C acted as the control without a WMS, with 2.5 in. (64 mm) semi-rigid fiberglass board insulation

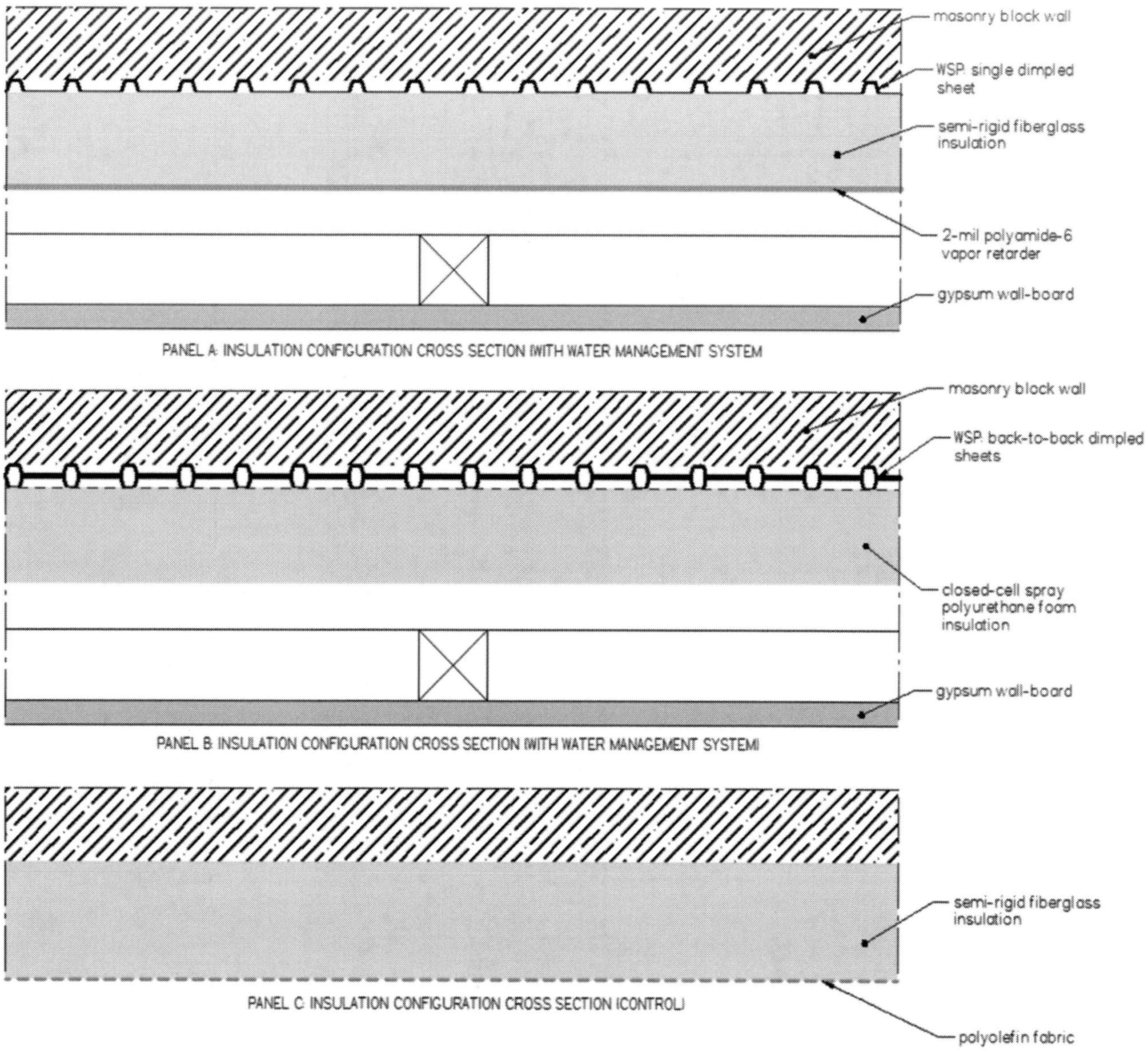

Figure 3 *Test panel insulation cross sections.*

covered with 100% polyolefin fabric on its interior surface as the finish. No warm-side vapor retarders or gypsum were installed. This system approximated a basement insulation system tested previously (Goldberg and Aloi 2001).

The primary purpose of the experiment was to test the ability of the WMS to drain bulk water from both sides of the WSP. In order to approximate bulk water leakage from the exterior, identical spray bars were mounted on the interior wall surface of Panels A and B about 24 in. (610 mm) below the top of the wall (6 in. [152 mm] below grade). These spray bars were operated in parallel in a pulsatile mode with both the water ejection period and the period between ejections being controlled independently by the laboratory data acquisition and control system so that Panels A and B received the same volume of water. As control Panel C essentially had no resistance to vapor flow from the interior, all of the condensate from the diffusion of interior humidity was absorbed into the masonry block wall or ran down to the floor.

Instrumentation

The transducer layout is described in terms of Figure 4, which is an image captured from the real-time data display system for Panel A that had the greatest number of sensors. The transducers are color-coded according to the data readout background as follows:

- Gray: relative humidity
- Blue: temperature
- Pink: masonry block moisture content (qualitative)
- Red: drainage cavity liquid level (qualitative: both indicator lights off implies dry, yellow indicator light only implies damp, yellow and red indicator lights on implies wet)

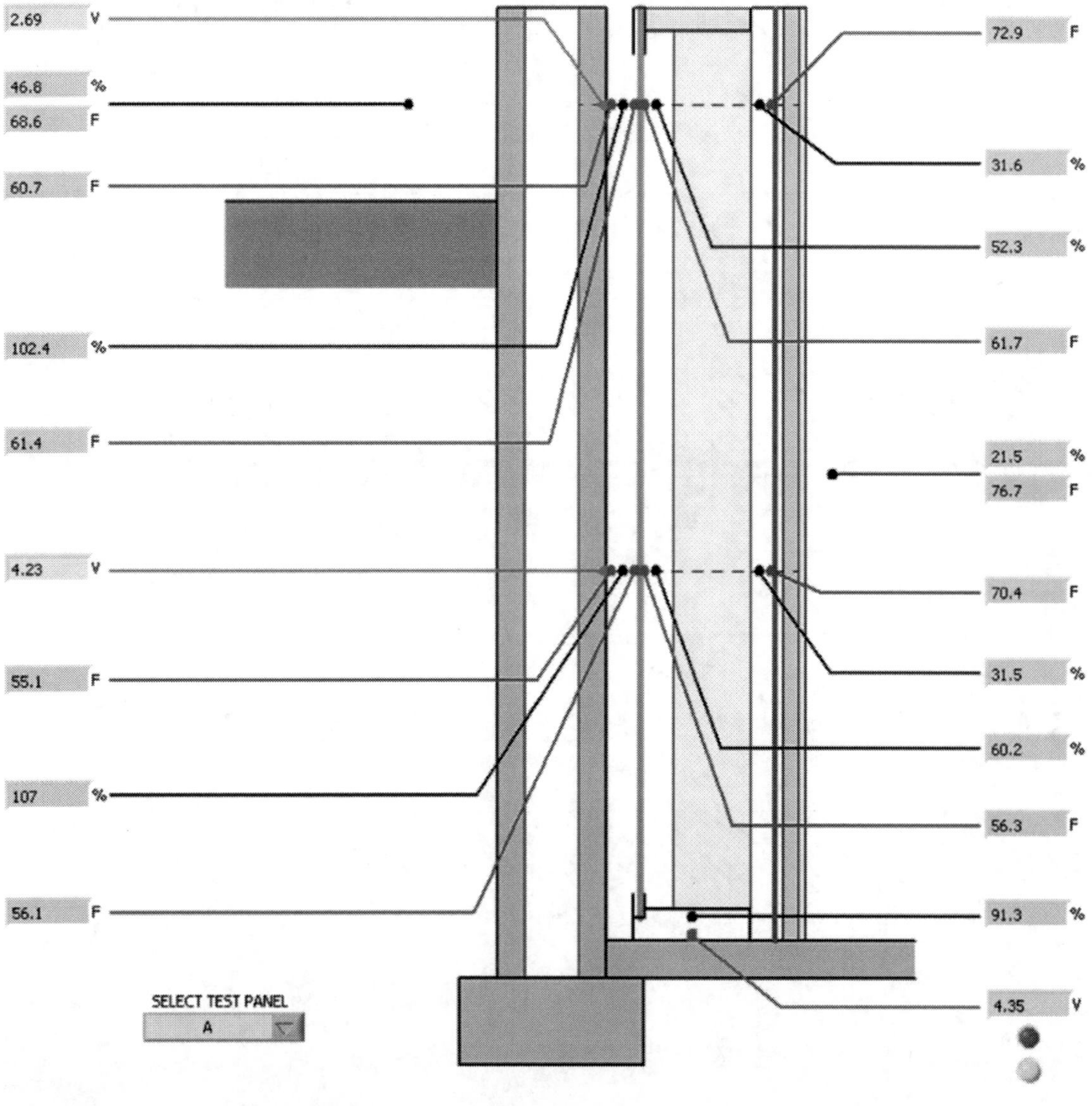

***Figure 4** Real-time data display for Panel A.*

The transducers were located at the mid-heights of the above- and below-grade wall sections, respectively, and located at the horizontal midpoint of the test cavity. The temperature and relative humidity (RH) were measured on either side of the WSP (shown schematically in Figure 4 as the lighter green vertical line between the wall and the insulation) as well as on the exterior face of the warm-side vapor retarder (shown in dark green). The drainage cavity contained a RH sensor as well as a water detector.

In addition, an experimental transducer was deployed to qualitatively assess the moisture content of the interior masonry block wall surface. The transducer sensor consisted very simply of two sheet metal screws (forming electrodes) inserted into the wall surface at a spacing of about 14 in. (356 mm) (see Figure 13). The electrodes were connected to the poles of a standard wood moisture content transmitter via a relay multiplexer. The moisture content was inversely proportional to the potential difference across the electrodes with a completely dry wall registering about 10 V. These sensors operated successfully for about the first 80% of the experimental duration but eventually failed owing to corrosion of the electrodes. This produced a galvanic corrosion cell that generated sufficient voltage to burn out the input stage of the moisture transmitter.

The exterior temperature and humidity were also measured about 8 in. (303 mm) away from the above-grade wall exterior surface as well as about 8 in. (203 mm) from the interior panel surface at the mid-wall position. The barometric pressure also was recorded.

The sensor layout for Panel B was identical to that of Panel A except, since there was no interior vapor retarder in this panel, the temperature and RH transducers between the insulation and the gypsum board were deleted. In Panel C, without a WMS, all of the transducers to the right of the light green vertical line in Figure 4 were deleted with the exception of those for interior temperature and RH.

Relative humidity was measured with individually calibrated thin-film capacitance transducers with an accuracy of ±3.5% up to about 92% RH. Above this humidity, this type of transducer becomes unreliable (note the readings above 100% adjacent to the wall in Figure 11), particularly next to materials (such as masonry blocks) exhibiting a "hockey-stick" sorption isotherm above 90% RH (a small change in RH is produced by a large change in surface moisture content). This was the motivation for deploying the experimental masonry block surface moisture content sensors even though these sensors were only capable of yielding qualitative data. The temperatures were measured exclusively with T-type (copper-constantan) thermocouples with an overall accuracy of 1.8°F (±1°C).

Data were collected continuously with a scan interval of about 3 s and were time-integrated so that a temporal average was recorded every 6 min. Data capture commenced on December 23, 2011, and continued through June 8, 2012, spanning a heating season and part of a cooling season. The raw data were post-processed to yield relevant psychrometric variables such as condensing surface dew-point temperatures and cavity vapor pressures.

Experimental Methodology

The primary purpose of the experiment was to test the bulk-water-handling capability of the WMS under very severe boundary conditions in a cold climate. The intention was to test the system to the point of failure to determine which aspects of the design required improvement or modification prior to manufacturing on a mass-production basis. Thus, the interior RH boundary condition in particular was driven to a maximum of 72% at 71°F (21.6°C) when the exterior ambient temperature was at freezing. Clearly, such severe conditions usually are not encountered in cold-climate residential basements during the heating season.[3]

Further, it was desired to test the WMS with the WSP being wetted from both sides simultaneously during the heating season. This could not be accomplished safely using the spray bars owing to the potential of water freezing in the pipes. Thus, the approach adopted was to leave the walls on either side of the test panels bare for about the first two months of the test period, allowing the high levels of interior humidity to be transported through the masonry blocks to the rear of Panels A and B. Thereafter, the bare walls were covered with 2 in. (51 mm) of extruded polystyrene insulation yielding a Class II vapor retarder with a permeance of about 0.6 perms (0.32 ng/Pa·s·m^2). This strategy was effective in producing saturated vapor conditions during the heating season on both sides of the WSP simultaneously.

Water injection using the spray bars commenced on May 14, 2012, after the laboratory interior had dried out (RH less than 24%) and proceeded until the termination of the experiment. The chosen injection rate was a 1 s pulse every two hours, corresponding roughly to a high-volume seep through the walls (test Panels A and B were wetted across their full 36 in. [914 mm] width).

The insulation systems tested were chosen to represent two of the more common approaches to foundation insulation in a cold climate with different vapor permeances and diffusion phenomenologies. Semi-rigid fiberglass board was chosen for Panel A (rather than the more conventional R-13 fiberglass batts) largely for convenience since it was easily accommodated by the experimental WMS components and has about the same vapor permeance as a batt; that is, they are both essentially vapor open. Previous research (Goldberg 2006) showed that an R-13 fiberglass batt/2-mil (0.05 mm) thick polyamide-6 warm-side vapor retarder combination yielded significant condensation on the interior foundation wall surface, thus providing a good test of the interior drainage capability of the WMS (although with asymmetric wetting/

3. The laboratory interior resembled a very wet cave with condensate dripping prodigiously from the floor joists and every metallic surface (such as the door knob).

drying performance—lower rates of wetting and higher rates of drying were produced by the relative-humidity-dependent permeance of polyamide-6). Closed-cell polyurethane foam was chosen as the alternate since it affords nominally symmetric wetting/drying performance. The 2 in. (51 mm) thickness installed yielded a permeance of about 0.76 perms (per ASTM Test Method E96 [2005][4]) according to the manufacturer's specification (a Class II vapor retarder).

Boundary Conditions

The experimental boundary conditions are shown in Figure 5. The changes in the boundary conditions may be annotated as:

[4] Test method A or B is not stated.

1. **From day 0 (12/23/11) through day 35 (1/27/12):** No humidification with 68°F (20°C) interior temperature heating setpoint.
2. **From day 35 (1/27/12) through day 60 (2/21/12):** Humidification with the RH setpoint increased in steps to 72% and the temperature heating setpoint to 70°F (21°C).
3. **From day 60 (2/21/12) through day 98 (3/30/12):** No humidification with a 70°F (21°C) interior temperature heating setpoint through day 70 (3/2/12) and thereafter 68°F (20°C) (no further humidification).
4. **From day 98 (3/30/12) through day 136 (5/7/12):** Dehumidification with a 68°F (20°C) interior temperature heating setpoint through day 98 (3/30/12) and thereafter 72°F (22°C). Note the steady increase in interior temperature during this period produced by the heat

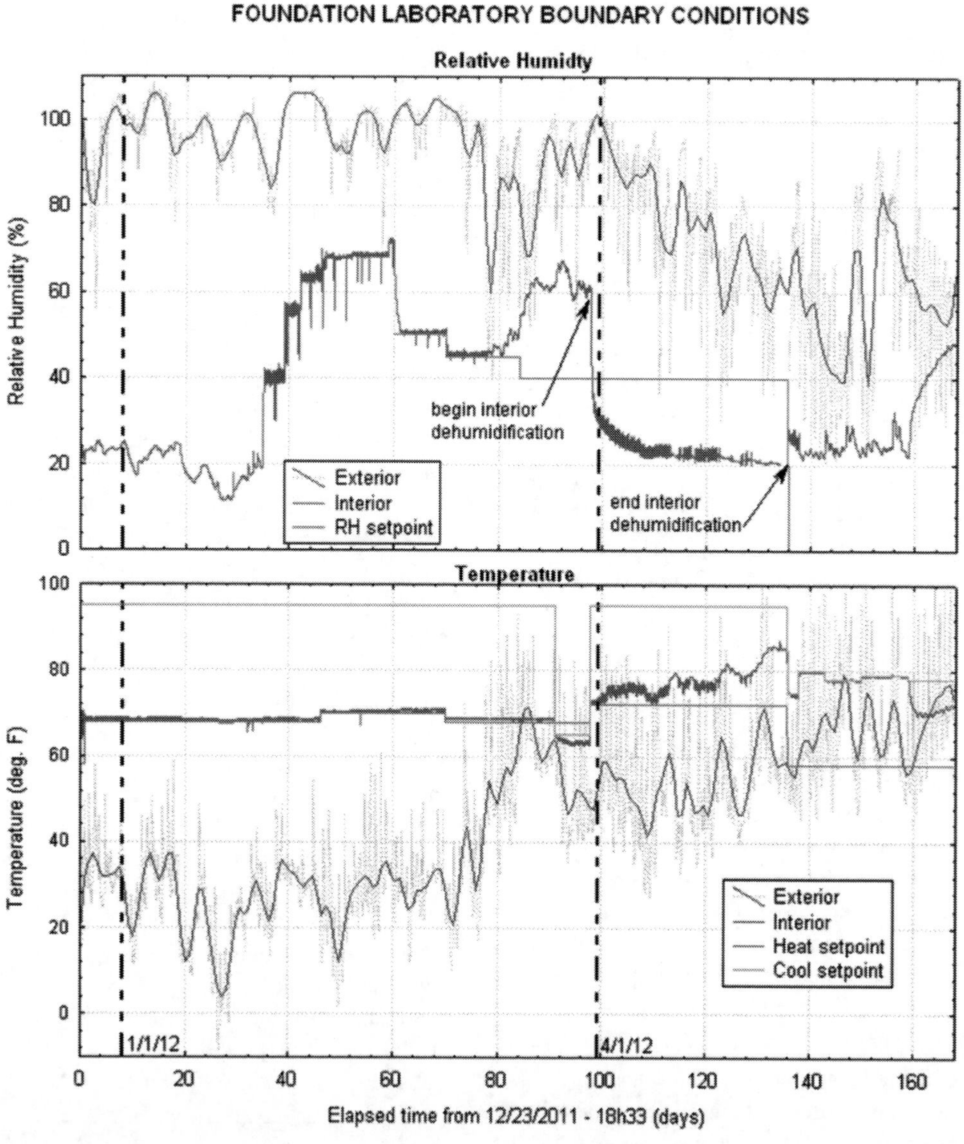

Figure 5 *Experimental boundary conditions.*

rejected to the laboratory interior by the stand-alone dehumidifier.

5. **From day 136 (5/7/12) through day 143 (5/14/12):** Air conditioning with a temperature cooling setpoint of 75°F (24°C) increasing to 80°F (27°C) on day 138 (5/9/12).
6. **From day 143 (5/14/12) through day 168 (6/8/12) (end of experiment):** Water injection at wall surface of Panels A and B (1 s pulse every 2 h) and air conditioning with a temperature cooling setpoint of 78°F (26°C).

In summary, the experiment sequence was a thermal equilibration period to provide a "normal basement" operating baseline, a period of interior humidification to wet the WSPs in Panels A and B from both sides, a drying-out period, and lastly, a cooling season operating period with interior air conditioning and bulk water injection at the wall surface in Panels A and B.

DISCUSSION OF THE EXPERIMENTAL RESULTS

Relative Humidity Phenomenology

The RH profiles for the duration of the experiment are shown in Figures 6, 7, and 8 for Panels A, B, and C, respectively. The bottom panel in each figure duplicates the RH boundary conditions for referential convenience. All the profiles are smoothed using a negative exponential curve fitting procedure, and the surface relative humidities are the actual humidities on the surface based on the adjacent cavity vapor pressure and surface temperature (and not the measured cavity relative humidities).

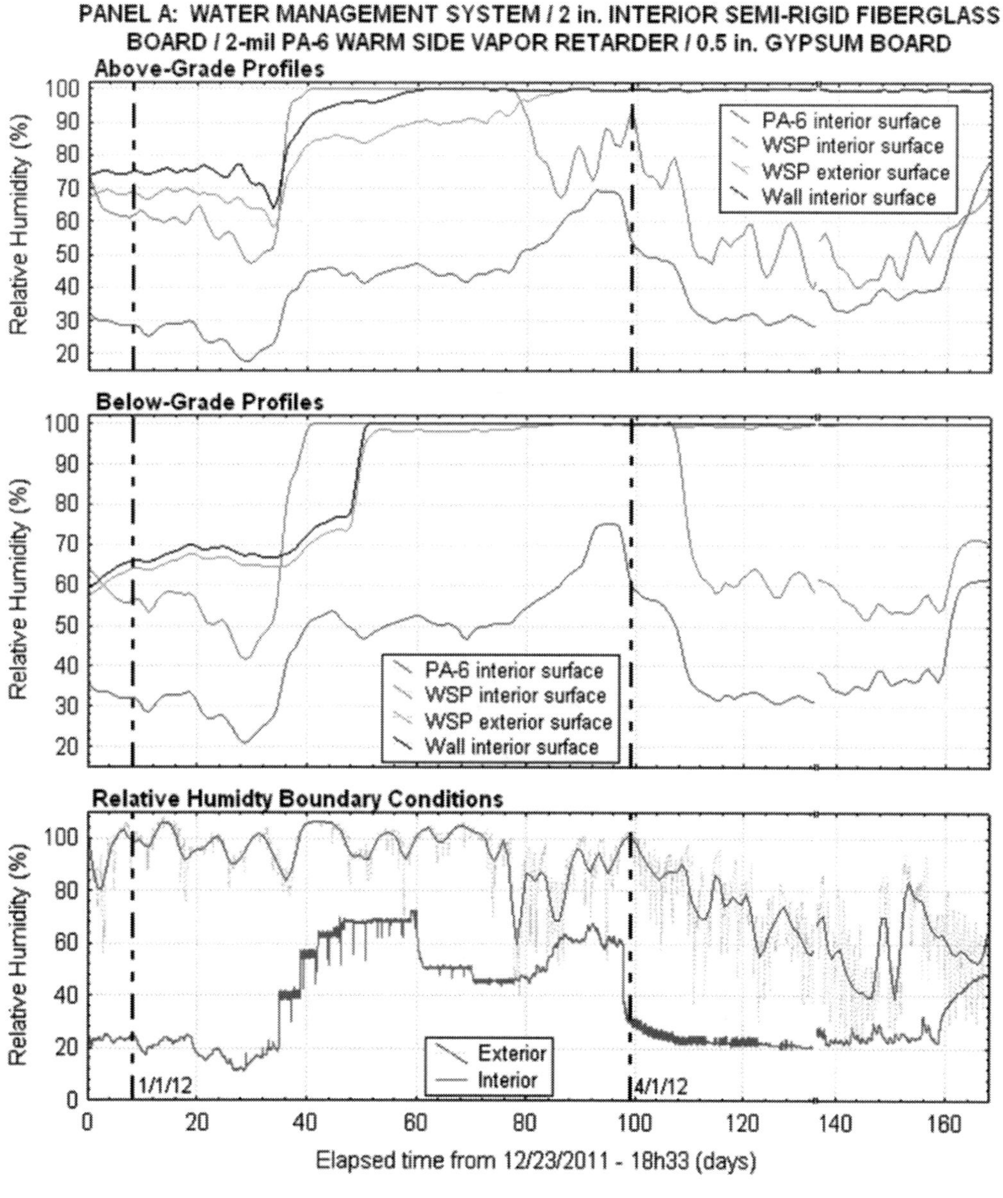

Figure 6 *Panel A relative humidity profiles.*

Figure 6 shows the sharp rise of the interior WSP RH to condensing conditions upon the start of interior humidification on day 35. Below grade, the wall and WSP exterior surface RHs reached saturation about 11 days after the WSP interior surface RHs reached saturation. Note the sharp rise in the below-grade WSP exterior's RH at about day 44, when the high-humidity incident on the bare wall adjoining Panel A reached the sensor plane by diffusion through the masonry block wall. Prior to this, the RH rise was gradual in response to soil-sourced vapor diffusion.

Above grade, the interior WSP RH increase was in phase with that below grade. But, in this case, the spike in the WSP exterior's RH occurred simultaneously with that of the interior surface RH, although after the initial spike the response was far more gradual, with the wall surface reaching saturation at day 60 and the exterior WSP surface reaching it much later at day 84. The simultaneity of the initial interior and exterior spikes is an indication of vapor leakage across the WSP at the vertical seams abutting the enclosing frame studs. The damped response of the exterior RH profiles is a consequence of a combination of freezing (diminished source strength) and drying to the exterior.

Above grade, the WSP interior surface commenced drying on day 76, while below grade this occurred much later, on day 106. The delay likely was caused by the time taken for the condensate to drain down (open surface or droplet trickle flow) to the level of the below-grade transducer as well as warmer above-grade wall surface temperatures. However, it is

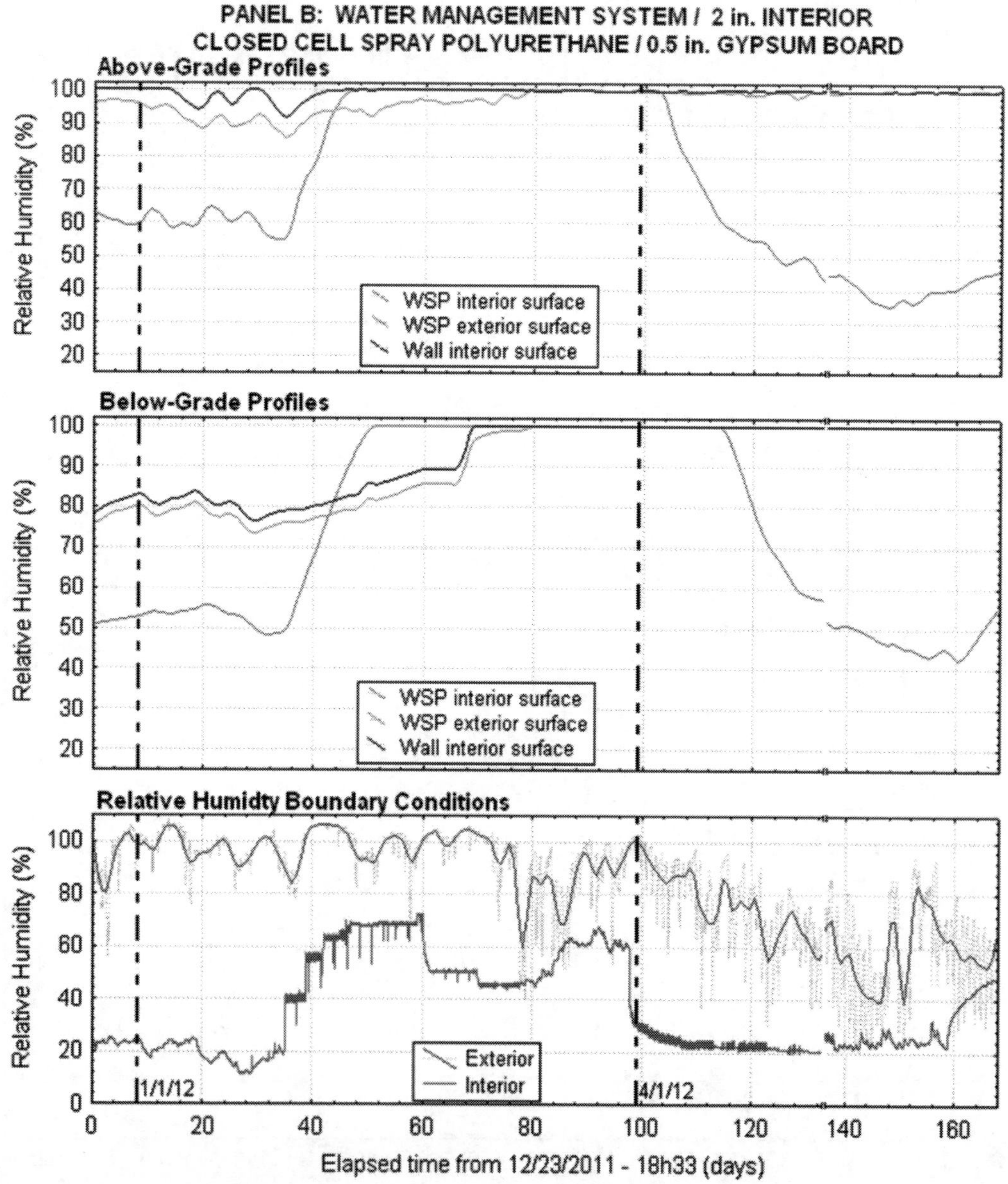

Figure 7 *Panel B relative humidity profiles.*

clear the wall surface and exterior WSP did not dry out completely. Below grade, this might be expected since bulk water injection started at day 143, but is noteworthy above grade where there was no water injection. Clearly, a risk in using non-attached interior WSPs is the creation of a perpetually wet exterior drainage cavity under the appropriate exterior boundary conditions.

Above and below grade, Figure 6 shows no condensation occurring on the exterior face of the warm-side vapor retarder, with the surface RH not exceeding 80% over the duration of the experiment. This is markedly different from the behavior observed in a previous experiment (Goldberg 2006) in which there was significant condensation on the wall side of the vapor retarder during the wall-drying phase.

Similar overall patterns are shown in Figure 7 for Panel B; however, there are some interesting divergences. First, the WSP interior RH reached saturation on days 44 and 45 at the above- and below-grade locations, respectively, compared with day 40 for Panel A. This is a consequence of the smaller effective permeance of the spray polyurethane foam in Panel B compared with that of the polyamide-6 vapor retarder in Panel A. Secondly, below grade, the RH spike on the wall and WSP exterior surfaces occurred on day 65, 21 days after it occurred in Panel A. This was doubtless a result of the vertical measurement plane in Panel B being effectively farther from the bare side walls than that of Panel A. Finally, the data show that at the beginning of the experiment the wall behind Panel B was wetter than those of either Panels A or C. Below

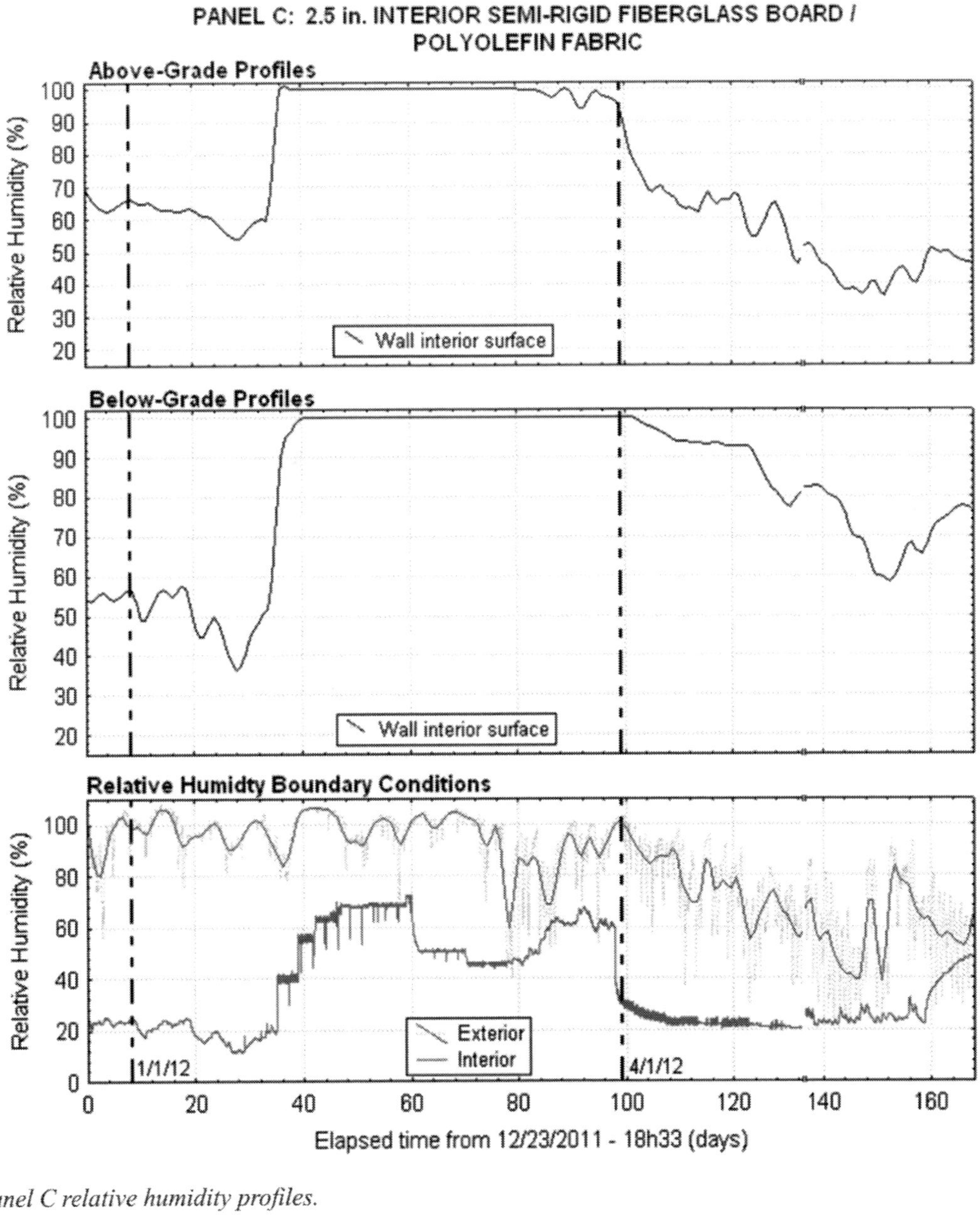

Figure 8   *Panel C relative humidity profiles.*

grade, the surface RH difference was about 20% and 25% relative to Panels A and C, respectively. Above grade, the differences were about 30%, with the above-grade wall surface already at saturation at the start of the experiment. It should be noted that data acquisition only commenced about a week after the walls were closed in, so this difference can be ascribed to edge drying effects (as there was no evidence that there were any soil moisture horizontal gradients). In other words, the drying potential to the interior from Panels A and C was larger than that of Panel B because the walls adjacent to Panels A and C were bare initially by intention. In the standard residential case where there are no bare walls around the foundation perimeter, the Panel B data would be the norm.

The above- and below-grade wall surface RH profiles in the control Panel C (Figure 8) were in conformance with previous results (Goldberg and Aloi 2001), with the wall reaching saturation on days 36 and 40 above and below grade, respectively. Comparing these results with those of Panel A in Figure 6, where the WSP interior surface reached saturation on day 40 at both locations, indicates that the polyamide-6 warm-side vapor retarder was largely ineffective (compared with no warm-side vapor retarder, as the polyolefin fabric covering Panel C was vapor open) for the applied interior boundary conditions. Thus, Panel C confirms that the condensate drainage on the WSP interior surface was as effective a condensate management strategy as the more conventional approach of using the masonry block interior surface as a heating season condensate store (Goldberg and Aloi 2001; Goldberg 2004, 2006).

Water Transport Performance

The water drainage performance of Panels A and B is shown in Figures 9 and 10. The upper two panels show the wetting/drying profiles at the below-grade location since this also captures the condensate drainage from the upper portion of the wall. The bottom panel shows the RH and wetness conditions in the bulk water drainage cavity.

For Panel A, Figure 9 shows that wetting (sensible temperature less than or equal to the dew-point temperature) began at day 46 (per Figure 6) on the exterior WSP surface, which remained wet for the duration of the experiment. On the interior surface, wetting began on day 40 and ended on day 120. In response, the drainage cavity became wet on day 52. In other words, the interior WSP surface condensate required at least six days to trickle down to the drainage cavity, indicating open surface (droplet) flow. However, the RH in the cavity stayed below 96% until after day 143, when wall-side bulk water injection began, superficially (owing to the limitations of the RH transducer discussed previously) indicating a rather small volume of condensate rundown. Of interest is that the RH only reached 100% on day 160, 17 days after the commencement of water injection. This is a strong indication that the bulk water drainage cavity was leaking, preventing any accumulation of water. This was confirmed when the experiment was dismantled (discussed in a subsequent section).

The phenomenology for Panel B as shown in Figure 10 is somewhat different. Unfortunately, the wetness sensor in the Panel B drainage trough failed at the outset of the experiment, so its data have been omitted. Wetting commenced on the exterior WSP surface on day 72 and on the interior surface at day 45. However, it is clear that as the RH in the drainage cavity remained below 76% for the entire period of the experiment prior to commencement of water injection at day 143, there was no bulk water present in the drainage cavity during this period. In comparison with Figure 9, where there was an RH of 90% or greater from day 80 onwards, this shows the following:

- Even though the exterior WSP surface reached saturation, the amount of condensate rundown on the exterior of the WSP was insufficient to wet the drainage cavity, despite the attempt to create artificially wet wall surfaces during the heating season.
- The bulk of the condensate rundown during the heating season originated from the interior side of the WSP as the volume of diffused vapor reaching the WSP from the interior in Panel A was significantly greater than that of Panel B, where it was negligible (RH peak of about 76% on day 96). This arises directly from the much lower effective permeance of the Panel B insulation system (closed-cell spray polyurethane) than that of Panel A (semi-rigid fiberglass plus polyamide-6 vapor retarder). This highlights the point that existence of saturated water vapor conditions on a vertical condensing surface does not allow inference about the quantity of condensate.

After commencement of water injection on day 143, the Panel B drainage cavity reached saturation at day 160, exactly the same behavior as in Panel A. Thus, the WMS was effective in draining the injected bulk water from the exterior drainage cavity in both Panels A and B.

Wall Surface Sorption Phenomenology

Of great interest in deploying the WMS is the wall surface moisture content. As discussed previously, RH measurements alone generally are inadequate to make inferences about the actual moisture content of the wall surface given the characteristics of masonry sorption isotherms (the "hockey stick" profile). Figure 11 shows the sorption phenomenology at the above- and below-grade locations for Panel A (the RH boundary conditions again are included for reference). Plotted are the WSP exterior *cavity* calibrated raw RH as read off the transducer and the uncalibrated voltage produced by the experimental moisture content sensor. In the case of the moisture content sensor, the moisture content is inversely proportional to the voltage with a dry condition indicated by a value of 10 V (as noted previously).

Of immediate note is that the RH transducer recorded a physically impossible RH greater than 100% at both the above- and below-grade locations, although the extent of the over-reading (about 8% at a maximum) was much larger below grade. This simply confirms the unreliability of this type of transducer, as previously discussed, at RHs above about 92%. At the same time, the wall moisture content exhibited a gradual and stable change in response to the steeply increasing cavity RH. For example, at the above-grade location, while the cavity RH was at 100% from day 94 onward, the moisture content increased monotonically and gradually from about 3.5 V to about 1.75 V prior to the transducer failing on day 133. Similarly, below grade, after day 52 when nominal cavity saturation was reached, the moisture content sensor showed the wall surface moisture content increasing from about 6.1 V to about 3.7 V at day 88 and staying in the range from 3.7 to 4.3 V until the transducer failed on day 133.

The wall surface moisture content data show that the wall was wetter above grade than below grade, in agreement with previous experiments (Goldberg and Aloi 2001; Goldberg 2004). However, as these results are uncalibrated, any absolute inferences about the physical moisture content of the wall surface cannot be made.

Thus, in principle, these results confirm the effectiveness of the wall surface experimental moisture content measurement technique deployed to obtain a deeper understanding of the hygrothermal performance of closed drainage cavities on the interior surfaces of foundation walls.

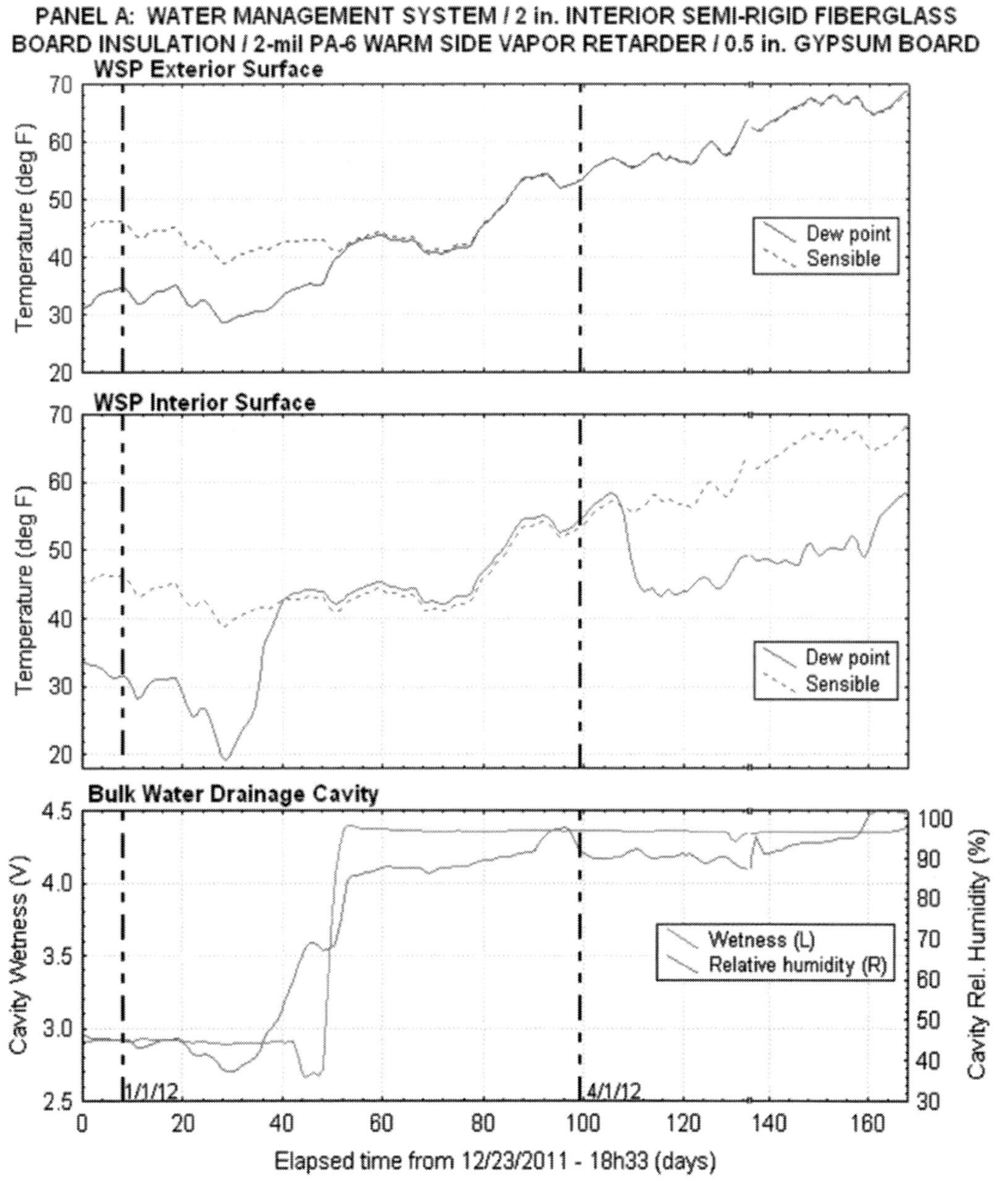

Figure 9 *Panel A below-grade water transport phenomenology.*

DISMANTLING OF THE EXPERIMENT

The experiment was dismantled on September 2, 2012, 86 days after the experiment was terminated on June 8, 2012. The long delay allowed the impact of any natural drying of the exterior wall cavity to be assessed.

The wetness of the Panel A WSP exterior wall cavity may be assessed in terms of Figure 12. This shows a partially wet wall both above and below the spray bar.

Above the level of the spray bar, there were still condensate droplets on the WSP exterior surface, but the surface was dry below the level of the spray bar, indicating that all the injected bulk water had drained away. There was no evidence of mold on either the wall or WSP surfaces.

Again, this is entirely consistent with previous experimental work (Goldberg and Aloi [2001], for example), in which insolation on the above-grade portion of a south-facing foundation wall drives moisture from the wall to the interior, where it condenses on the first condensing plane encountered.

Figure 13 shows the rear face of the WSP removed from the upper section of the wall above the spray bar in Panel B. In this case as well, the wall surface was partially wet and the rear face of the WSP was covered with condensate droplets (except for the upper part that was inserted into the top plate receiver cavity.

Thus, Figures 12 and 13 confirm that prolonged natural drying of sealed wall surface drainage cavities during the cooling season was insufficient to completely dry out these cavi-

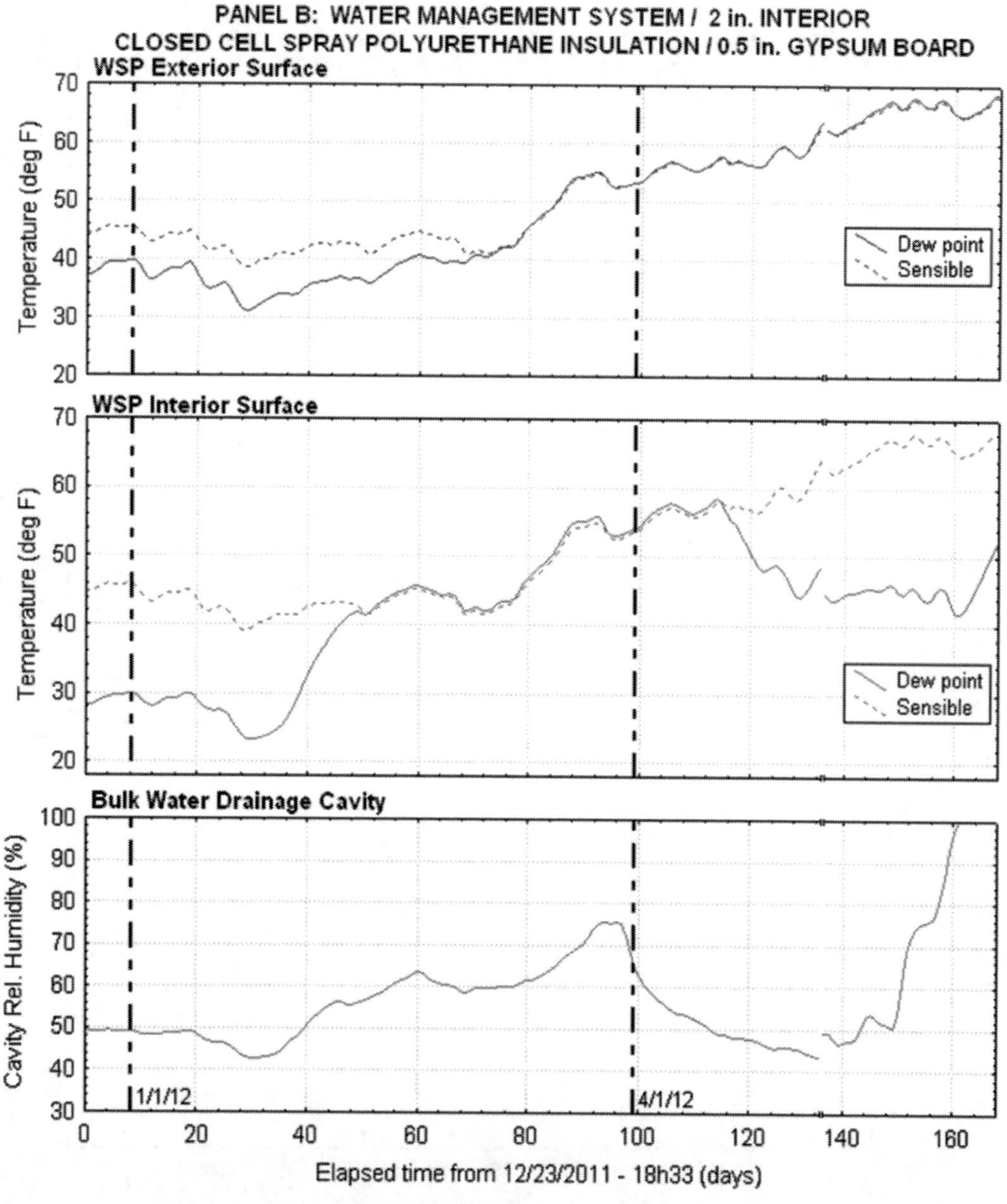

Figure 10 *Panel B below-grade water transport phenomenology.*

ties, as some residual level of moisture remains. Whether this is enough to cause mold/rot or structural problems (such as freeze/thaw-cycle-derived wall spalling) under typical basement continuous operating conditions is a question for future research. Absent such data, the possibility of a perpetually wet sealed interior wall drainage cavity must be considered in the design of foundation wall insulation systems. including the WMS tested.

Further findings during the dismantling (not shown) include the following:

- Both bulk water drainage cavities leaked where the assembly adhesive had failed (as expected).
- The caulk used to seal the WSP to the vertical studs was wet and soggy and had lost its water-sealing capacity. This accounted for the vapor leakage effects across the WSP noted in the experimental results.

CONCLUSION

The experimental results confirm that the water management system tested was effective in draining condensate from both sides of the water separation plane and bulk water from the exterior side. Attention needs to be paid to sealing the vertical edges of the water separation plane to avoid vapor bypass around these edges. Further, the results confirm that the system was in compliance with the water separation plane

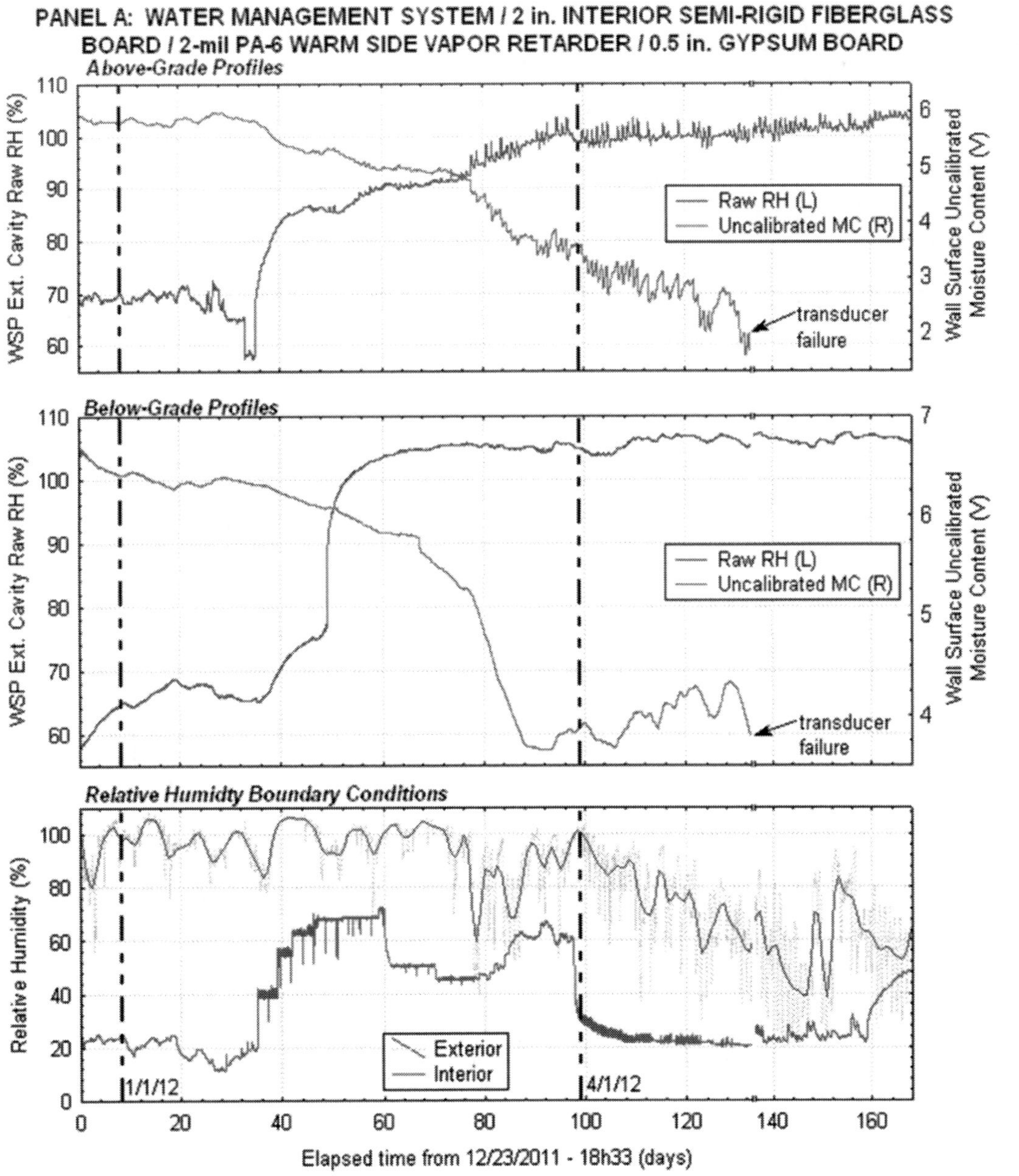

Figure 11 *Panel A wall surface sorption phenomenology.*

Figure 12 *Panel A WSP exterior cavity (dismantled on September 2, 2012).*

Figure 13 *Panel B WSP exterior cavity (dismantled on September 2, 2012).*

requirements of the performance option in the 2009 *Minnesota Energy Code* (State of Minnesota 2009).

Future work will include installing the production water management system (fabricated from extruded components) around the basement perimeter of a house that experiences significant bulk water leakage through the walls and monitoring the system for several years. In addition, the system will be provided to retrofit contractors for trial installations to isolate and correct any issues arising from improper installation.

REFERENCES

ASTM. 2005. ASTM Test Method E96/E 96M, *Standard Test Methods for Water Vapor Transmission of Materials*. West Conshohocken, PA: ASTM International.

Chuangchild, P., P. Ihm, and M. Krarti. 2004. Analysis of heat and moisture transfer beneath freezer foundations—Part I. *ASME Transactions, J. Solar Energy Society* 126(2):716–25.

Goldberg, L.F., and T. Aloi. 2001. Space humidity/interior basement wall insulation moisture content relationships with and without vapor retarders. *ASHRAE IAQ 2001 Conference Proceedings*.

Goldberg, L.F. 2004. Icynene foundation insulation project final report, Energy Systems Design Program Research Report, University of Minnesota, Minneapolis. www.buildingfoundation.umn.edu/IcynenFinalReport/default.htm

Goldberg, L.F., and P.H. Huelman. 2005. Minnesota Energy Code Building Foundation Rule: Amendment Proposal Development Project Final Report, Energy Systems Design Program Research Report, University of Minnesota, Minneapolis. www.buildingfoundation.umn.edu/FinalReportWWW/default.htm

Goldberg, L.F. 2006. Polyamide-6 based interior foundation insulation system: Experimental evaluation. Energy Systems Design Program Research Report, University of Minnesota, Minneapolis. www.buildingfoundation.umn.edu/CT-FTF/default.htm

Goldberg, L.F., P.H. Huelman, and S.D. Gatland II. 2010. A prototype universal building envelope hygrothermal performance standard for successful net-zero energy building design. *Proceedings of Thermal Performance of the Envelopes of Whole Buildings XI International Conference* [CD].

Goldberg, L.F,. and M.L. Stender. 2011a. Exterior Wall Assembly Including Moisture Transportation Feature. United States Patent, no. 8,001,736.

Goldberg, L.F,. and M.L. Stender. 2011b. Exterior Wall Assembly Including Moisture Removal Feature. United States Patent, no. 8,074,409.

Goldberg, L.F,. and M.L. Stender. 2012. Method of Removing Moisture from a Wall Assembly. United States Patent, no. 8,316,597.

ICC. 2006. *International Energy Conservation Code.* Section 301.1. Washington, DC: International Code Council, Inc.

Labs, K., L. Shen, J. Carmody, Y.J. Huang, R.L. Sterling, and D. Parker. 1988. *Building Foundation Design Handbook*. Minneapolis: University of Minnesota.

State of Minnesota. 2009. *Minnesota Energy Code*. MN Statutes. Residential Energy Code, Chapter 1322. Saint Paul: Department of Labor and Industry, State of Minnesota.

Building Envelope and Duct Airtightness of New US Dwellings

Wanyu R. Chan, PhD

Max H. Sherman, PhD
Fellow ASHRAE

ABSTRACT

We analyzed the building envelope and duct system airtightness of US single-family detached homes built since 2000, where the data was part of the Lawrence Berkeley National Laboratory Residential Diagnostics Database (ResDB). The airtightness of homes is compared with the IECC guidelines, which are the basis of many state building codes. A large number of homes are considered in this analysis, representing many states: 26,000 building envelope and 11,000 duct system air leakage data. Our analysis shows that the majority of US homes built in the past ten years from ResDB are meeting IECC 2009. About 80% of all homes built since 2000 met the IECC building envelope airtightness guideline of 7 ACH50. Over 90% of the homes met the IECC 2009 duct system airtightness guideline of 12 cfm25. However, fewer homes are reaching the 2012 levels, especially in terms of building envelope airtightness. Only 30% of the homes met IECC 2012 building envelope airtightness of 5 ACH50 in climate zones 1 and 2, and 10% meeting 3 ACH50 in climate zones 3–8. Overall, only 12% of the homes in ResDB met the IECC 2012 building envelope airtightness guidelines. On the other hand, slightly over half of the US homes in ResDB are meeting the IECC 2012 duct system airtightness guideline of 4 cfm25. These comparisons of airtightness with respect to guidelines show the current status of homes built recently in the US. The data presented here are also important for estimating the energy consumption on residential heating and cooling and the extra cost of energy spent because of air leakage.

INTRODUCTION

Building envelope and duct system airtightness is important to residential energy use and estimating ventilation needs. In 2005, the total consumption by all end uses in US households was estimated to be 11.13 EJ (RECS 2005). Space heating (4.54 EJ) and air conditioning (0.93 EJ) made up half of the energy use. To minimize the energy loss due to air infiltration, the IECC 2012 now requires pressure testing to demonstrate whole-house air leakage to <5 ACH50 (air changes per hour at 50 Pa) for homes in climate zones 1 and 2, and <3 ACH50 in climate zones 3–8. In 2009, the IECC requirement on whole-building envelope was less stringent at <7 ACH50, and homes also had the option of meeting the requirement by visual inspection only. Many states in the US have adopted the IECC 2009 equivalent or more energy efficient building codes; see DOE (2012) for a US map showing the adoption status of state energy code. Moreover, some states are projected to adopt IECC 2012 or equivalent by the end of 2015. So even though not all states follow IECC (e.g., California and Washington have developed their own codes), it is useful for assessing new homes as a common set of guidelines that are widely used. But, the implementation status of state energy codes is far more complicated in reality because local jurisdictions often have the flexibility to adopt codes as they see fit (Cort and Butner 2012).

Duct system pressure testing has been mandatory in the IECC since 2009. About three-quarter of US homes use forced-air or other heating and cooling systems that are installed with ducts (AHS 2011). IECC 2012 limits the duct leakage rate, both total and to-outside, to <4 cfm25 (ft^3 per minute at 25 Pa) per 100 ft^2 of conditioned floor. In 2009, the requirement was <12 cfm25 if testing is performed post-

Wanyu R. Chan *is a research scientist in the Environmental Energy Technologies Division and* ***Max H. Sherman*** *is a group leader at Lawrence Berkeley National Laboratory, Berkeley, CA.*

Published by ASHRAE 2013

construction, or <8 cfm25 if the duct-leakage-to-outside is measured instead of the total duct leakage through the system.

A general trend that new US homes are being built more airtight has been observed in recent studies (Nelson 2012; Offermann 2009; Proctor et al. 2011). However, there is considerable variability in the airtightness of US homes, as observed in an analysis of the Residential Diagnostics Database (ResDB) that considered 134,000 measurements collected by many contributors across the US (Chan et al. 2012). Using the data from ResDB, we focus here specifically on the airtightness of single-family homes that were built since 2000. Of interest is the question if these US homes are approaching the stringent airtightness requirements of IECC 2012. As homes are built more airtight, it is increasingly important to provide adequate ventilation to comply with ASHRAE 62.2 (2010). While it is beyond the scope of our discussion here, airtightening of the building envelope and duct system will impact many aspects of building performance, including indoor air quality, comfort, as well as energy consumptions.

RESDB AIR LEAKAGE DATABASE

ResDB contains air leakage and other diagnostic measurements of US homes that are contributed voluntarily by various energy auditors, building contractors, energy efficiency program managers, and researchers. The data are gathered and compiled into a database by Lawrence Berkeley National Laboratory. Over the years, more data have been gathered and re-analyzed to support calculations of air infiltration and its implications to residential energy use (Sherman and Dickerhoff 1998; Sherman and Matson 2001; Chan et al. 2005; McWilliams and Jung 2006). In 2011, a large number of whole-building envelope air leakage data from more than 100,000 homes were added to ResDB. Chan et al. (2012) described the air leakage data of single-family homes and presented a regression model that relates normalized leakage (*NL*) to house characteristics, such as climate zone, year built, and floor area. This model will be utilized in this analysis to evaluate if the building envelope airtightness of US homes continues to improve in the past ten years since 2000.

In addition, our analysis will also consider duct system air leakage data that were collected in 2011. Air leakage in the duct system can significantly impact home energy performance (Walker et al. 1999; Modera 2005). This type of data is a new addition to ResDB (previously, ResDB only contained building envelope air leakage data). To date, ResDB contains duct leakage data on approximately 20,000 homes. Our analysis will focus on homes built since 2000, which make up the majority of the data.

Whole-Building Envelope Air Leakage

E779 (ASTM 2010) is the standard used in the US to measure building envelope air leakage. Air leakage is measured by the airflow rate, $Q_{\Delta P}$ (m^3/s) through the building envelope as a function of the pressure difference, ΔP (Pa), across the building envelope. The most common pressure difference used is 50 Pa, which is low enough for standard blower doors to achieve in most houses, and at the same time high enough to be reasonably independent of weather influences. Many metrics are used to describe whole-building envelope air leakage normalized to building volume or surface area, such as ACH50, NL (normalized leakage), ELA (effective leakage area), and SLA (specific leakage area). There is no consensus from buildings codes or other energy efficiency guidelines on which one metric is preferable to the others. We selected *NL* because it has been the choice in previous data analyses of ResDB. *NL* is also used in ASHRAE 62.2 (2012) to calculate the mechanical ventilation needs of homes. Measurements of air leakage by blower door, typically available as Q_{50} (m^3/s), are converted to *NL* as follows:

$$NL = 1000\left(\frac{ELA_4}{\text{Area}}\right)\left(\frac{H}{2.5\text{ m}}\right)^{0.3}$$
$$ELA_4 = \sqrt{\frac{\rho}{2(4\text{ Pa})}}Q_{50}\left(\frac{4\text{ Pa}}{50\text{ Pa}}\right)^n \qquad (1)$$

where ELA_4 (m^2) is the area of an orifice that would result in the same airflow through the building envelope at a pressure difference of 4 Pa, and $\rho = 1.2$ kg/m^3. n is the pressure coefficient, assumed to equal 0.65 if it is not measured directly; see Chan et al. (2012) for a distribution of n from ResDB. *NL* is normalized to floor area, Area (m^2), and house height, H (m), according to Equation 1. Assumptions were made to compute *NL* if Area or H is unavailable. For instance, H is assumed to equal 3 m for one-story, and 5.5 m for two-story; see Chan et al. (2012) for more details. Of the 134,000 single-family detached homes with estimates of *NL*, 26,000 were new homes built since 2000. Note that the definition of *NL* has been modified slightly in ASHRAE Standard 62.2 (2012) since the analysis presented here: the exponent is changed from 0.3 to 0.4. This change has a negligible effect on the results presented in this paper.

Duct System Air Leakage

Duct leakage is commonly measured following E-1554 (ASTM 2007). A calibrated fan delivers air into the duct system to pressurize it to either 25 Pa or 50 Pa, with all registers closed. For measuring duct leakage, 25 Pa is commonly used to represent a pressure difference that more closely resembles the condition during system operation. Some building codes and energy efficiency guidelines scaled the duct system leakage with respect to floor area (e.g., $Q_{d,25}$ = cfm25 per 100 ft^2 of conditioned floor area). Most of the duct system air leakage measurements in ResDB are provided in units of cfm25, and so $Q_{d,25}$ is the metric used in the analysis of approximately 11,000 homes. In an additional 8,000 homes, duct leakage was measured by pressurizing the house with a blower door to the same pressure as the duct system during the test. The duct leakage to-outside, $Q_{d,25,\text{to-outside}}$, is the flow required to equalize the house and duct pressures. In this analysis, $Q_{d,25,\text{to-outside}}$ is scaled to 100 ft^2 of conditioned floor area, in the same way as $Q_{d,25}$.

California, Florida, Nevada, Texas, and Washington are states with the most number of duct system air leakage measurements in ResDB. In Washington, some of the duct leakage measurements were collected at 50 Pa pressure difference. Those data are converted to 25 Pa assuming a pressure exponent n of 0.6 (Walker 1998) as follows:

$$Q_d \propto (\Delta P)^n$$

$$\frac{Q_{d,25}}{Q_{d,50}} = \left(\frac{25}{50}\right)^{0.6} \tag{2}$$

For estimating the energy consumption and delivery efficiency of thermal distribution systems, as in ASHRAE Standard 152 (2004), the duct system leakage as a percentage of the system fan airflow is needed as an input parameter. Unfortunately, duct leakage as a percentage of fan airflow can only be computed in about 800 homes because relatively few homes in ResDB provided the fan airflow rates. The percentages of fan airflow data are collected mostly by weatherization and other retrofit programs from existing homes. Because those data are typically from older homes built before 2000, they will not be discussed here.

ANALYSIS OF NORMALIZED LEAKAGE DATA

Measurements of building envelope air leakage in California (Offermann 2009; Proctor et al. 2011) and in Minnesota (Nelson 2012) suggest that homes built since 2000 met the IECC 2009 airtightness threshold of 7 ACH50, especially in colder climates where the state energy code tends to be more stringent (e.g., Minnesota energy code is <1.5 ACH50 in 2000). Figure 1 shows the *NL* of homes built since 2000 available in ResDB from 14 states. The median *NL* is 0.3, which corresponds to approximately 5 ACH50. Even though this data set is not a statistically representative sample of the US housing stock built since 2000, it includes homes that are located in many climate zones, ranging from cold areas like Alaska and Wisconsin, to warmer areas like Florida and Texas.

It is important to note that despite the large number of data depicted in Figure 1, the data are likely skewed towards being more airtight than the typical homes. This is because roughly one-third of the homes shown in Figure 1 are rated for energy efficiency. Many of these homes are rated following ENERGY STAR® guidelines for new homes. The current ENERGY STAR guidelines, version 3, specify ACH50 to be less than 3 to 6 (EPA 2012), depending on the climate zone. Consequently, there is a selection bias in the data shown in Figure 1 towards better-than-average airtightness. The energy-efficiency rated homes shown in Figure 1 likely followed earlier versions of such guidelines (versions 1 and 2 were adopted in 1995 and 2007, respectively), or their equivalent, which had less stringent airtightness guidelines. Nonetheless, the regression analysis by Chan et al. (2012) found that energy-efficiency rated homes are 30% lower in *NL* compared to conventional homes in ResDB, regardless of which ENERGY STAR guidelines these homes most likely followed. Note that the 30% describes an overall difference between energy-efficiency rated homes and conventional homes from all available data in ResDB, and not just the subset of the data that is shown in Figure 1.

Comparison with IECC Airtightness Guidelines

Many states have adopted the airtightness guideline of 7 ACH50 following IECC 2009 or have building codes that are more stringent in the range of 3 to 5 ACH50, similar to the levels specified in IECC 2012. Equation 3 is used to convert the air leakage data shown in Figure 1 from units of *NL* to ACH50 for comparison with these airtightness guidelines, so that we can evaluate if homes built since 2000 in ResDB are meeting these IECC levels.

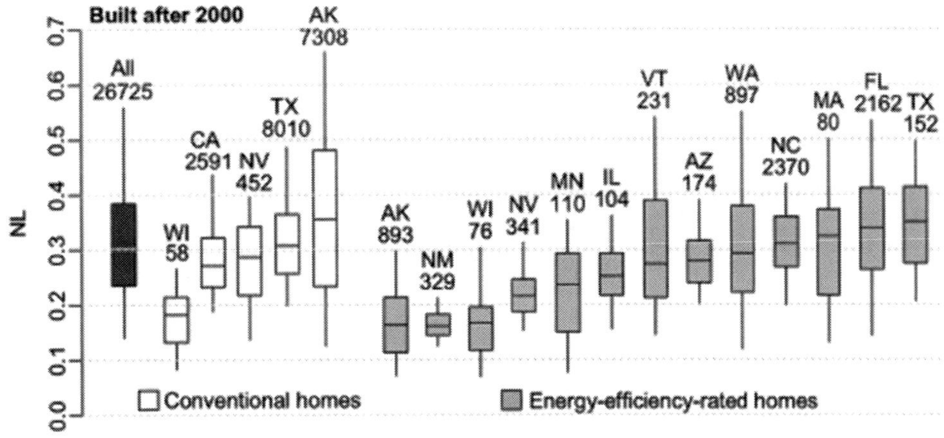

Figure 1 *Whole-building envelope air leakage of homes in ResDB built after 2000. Each box plot shows the median and interquartile range, and the whiskers show 5th and 95th percentiles, of the* NL *of homes from different states. The number shown beneath each state is the house count.*

$$\text{ACH50} = \frac{NL}{\frac{1000}{3600}\left(\frac{\sqrt{\frac{\rho}{2(4\text{ Pa})}}\left(\frac{4\text{ Pa}}{50\text{ Pa}}\right)^{n}}{2.5\text{ m}}\right)\left(\frac{H}{2.5\text{ m}}\right)^{0.3}} \quad (3)$$

Figure 2 shows the cumulative percentiles of ACH50 computed using Equation 3. Overall, 81% of all the homes, regardless of their climate zones, met the IECC 2009 airtightness guideline of 7 ACH50. The IECC 2012 airtightness guideline of <5 ACH50 for climate zones 1 and 2 (i.e., some parts of Florida and Texas) are met in 29% of the homes. But just 10% of the homes in the other climate zones 3–8 meeting the most stringent guideline of 3 ACH50. Altogether, only 12% of the homes met IECC 2012 guidelines. These statistics suggest that an airtightness level of <7 ACH50 is often achieved across the US. Figure 2 also shows that about 40% of the homes built in the past ten years are already sufficiently airtight to reach <5 ACH50. Reaching <3 ACH50 is not an unattainable goal, but it is a challenge. Consequently, only a small fraction of the homes (12%) built since 2000 reached that level of airtightness specified in IECC 2012.

Improvement in Airtightness of Homes Built Since 2000

There are other attributes of the homes also known to be associated with *NL* (Chan et al. 2012), such as climate zone, floor area, and so on. By first removing the influences of these other attributes, we can evaluate if the data from ResDB show continuing improvement in the airtightness of US houses in the past ten years since 2000. In Equation 4, the expected values of *NL* are computed using regression estimates from Chan et al. (2012), where the coefficient estimate βs are reproduced in Table 1. After adjusting for all the relevant attributes, the model residuals $\ln(NL_{obs}) - \ln(NL_{pred})$ are regressed with respect to year built of the homes, as shown in Equation 5.

$$\ln(NL_{\text{pred}}) = \beta_{CZ} \cdot I_{CZ} + \beta_{\text{post-2000}} + \beta_{\text{area}}\text{Area} + \beta_H H + \beta_e \cdot I_e \quad (4)$$

$$\ln(NL_{obs}) - \ln(NL_{pred}) = \beta_{\text{year}} \cdot (2005 - \text{Year}) \quad (5)$$

In Equation 4, *NL* is adjusted by climate zone (*cz*), floor area, house height, and if the home is rated for energy efficiency (*e*) or not. The *I*s in Equation 4 are indicator variables:

Table 1. Regression Coefficients Used in Equation 4 to Compute NL_{pred}

Variable	Coefficient
Climate Zone Humid (A) Zone 1, 2	$\beta_{A\text{–}1,2} = 0.473$
(A) Zone 3	$\beta_{A\text{–}3} = 0.253$
(A) Zone 4	$\beta_{A\text{–}4} = 0.326$
(A) Zone 5	$\beta_{A\text{–}5} = 0.112$
(A) Zone 6, 7	$\beta_{A\text{–}6,7} = 0$
Dry (B) Zone 2, 3	$\beta_{B\text{–}2,3} = -0.038$
(B) Zone 4, 5	$\beta_{B\text{–}4,5} = -0.009$
(B) Zone 6	$\beta_{B\text{–}6} = 0.019$
Marine (C) Zone 3	$\beta_{C\text{–}3} = 0.048$
(C) Zone 4	$\beta_{C\text{–}4} = 0.258$
Alaska (AK) Zone 7	$\beta_{AK\text{–}7} = 0.026$
(AK) Zone 8	$\beta_{AK\text{–}8} = -0.512$
Year Built (homes built after 2000)	$\beta_{\text{post-2000}} = -1.058$
Floor Area	$\beta_{\text{area}} = -0.00208\ (\text{m}^{-2})$
House Height	$\beta_H = 0.064\ (\text{m}^{-1})$
Energy-Efficiency Rated Homes	$\beta_e = -0.384$

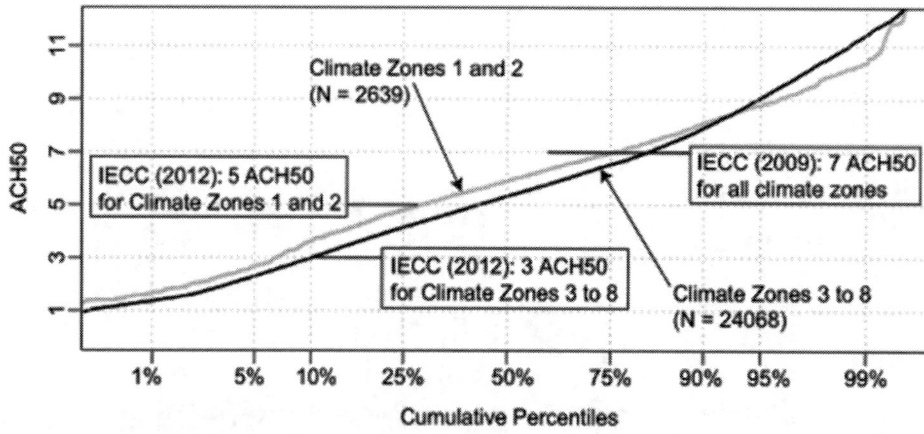

Figure 2 *Cumulative percentiles of building envelope air leakage of homes in ResDB built after 2000, shown separately for climate zones 1 and 2 and climate zones 3 to 8 (N = house count).*

$I = 1$ if the condition is true, and $I = 0$ if not. The coefficient $\beta_{post\text{-}2000}$ is a constant used to describe all homes built since 2000 (as opposed to homes built in 1990–99, 1980–89, etc. that are part of the regression analysis by Chan et al. [201]), but are not considered here in this paper). Other coefficients estimated by Chan et al. (2012) are excluded because they do not apply to the subset of data considered here (i.e. none of the homes considered here participated in a weatherization assistance program, and too few data are available on foundation type and duct system location to account for their influences on *NL* in this analysis).

The regression shown in Equation 5 gives $\beta_{year} = -0.016$ (95% C.I. –0.014 to –0.017), suggesting that there is an overall reduction in *NL* all through the years since 2000 at approximately 1.6% per year (Figure 3). But R^2 of the regression is only 0.02, meaning that besides the year built, there may be other factors that contribute to the variability in *NL*. Nonetheless, this regression found a relationship between *NL* and the year built, and it estimates that homes built in 2011 are 20% more airtight than those built in 2000, all else held equal. Our finding suggests that recent attention paid to airtightness, likely in response to tightening of building codes, has led to persisting reduction in building envelope air leakage. This agrees with the recent findings in California homes. Proctor et al. (2011) found a median ACH50 = 4.7 (range min–max: 2.0–8.2) in 40 single-family homes that were first occupied in 2007. Offermann (2009) found a median ACH50 = 4.8 (3.6–8.4) in 108 homes built between 2002 and 2004. Hoeschele et al. (2002) found an average ACH50 = 5.5 (2.6–8.7) in 30 new homes tested in 1999 and 2001. These levels are more airtight than an earlier study cited in Offermann (2009): a median of 8.6 (6.2–13.2) in 13 homes built before 1987.

DUCT SYSTEM AIR LEAKAGE

There are fewer duct system air leakage data in ResDB and as a result, the available data are not as representative of the US housing stock. The vast majority of duct leakage data are from homes built after 2000, which are shown here. The western states (e.g., California and Washington) are well represented, but less so for the US overall, as shown in Figure 4. Because there are significant differences in the heating and cooling systems used in the different parts of the US, the discussions here are limited to areas that are represented by the duct leakage data in ResDB.

Figure 5 shows the distribution of the total duct leakage and compared the data with IECC duct systems airtightness guidelines. The median total duct leakage $Q_{d,25}$ of homes built since 2000 in ResDB is 3.5 cfm25 per 100 ft^2 of conditioned floor area, meaning that over half of the homes would meet the IECC 2012 total duct leakage guideline of <4 cfm25. Nearly all the homes (greater than 90%) met IECC 2009 total duct leakage guideline of <12 cfm25. These data show significant improvements in duct systems airtightness compared to the analysis by Neme et al. (1999) that found an average of 270 cfm25, or about 20 cfm25 per 100 ft^2 of conditioned floor area for a typical 1500 ft^2 house, from 19 studies of duct leakage and its effects on electric HVAC efficiency.

Duct Leakage To-Outside

Duct leakage to-outside $Q_{d,25,\text{to-outside}}$ was measured in a small number of states. Figure 6 shows the ratio of the duct leakage to-outside and total duct leakage, $Q_{d,25,\text{to-outside}}/Q_{d,25}$, for the Florida data shown in Figure 4 and for a few other states. The Florida data has a median ratio of 0.36, with 90% of the data range between 0.16 and 0.64 (Figure 6(a)). This

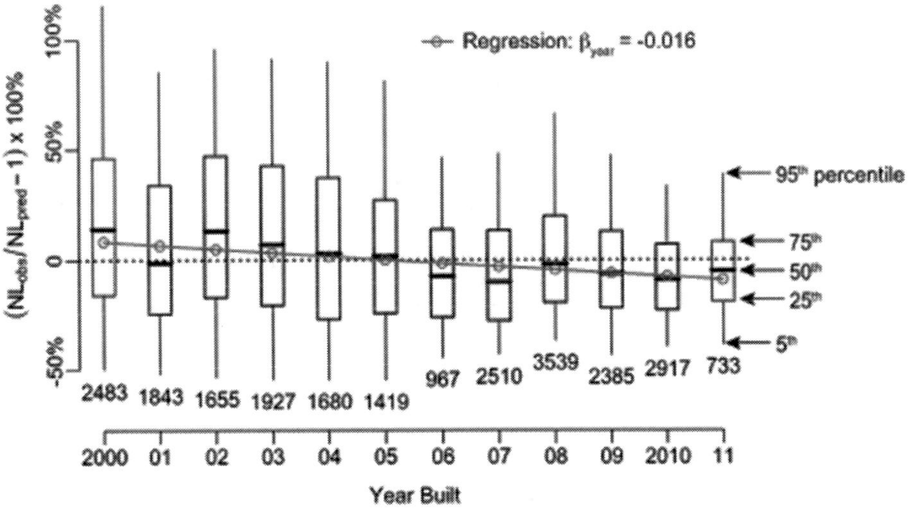

Figure 3 *Relationship between* NL *and year built, where* NL *has been adjusted by first accounting for other attributes that influence air leakage, as described in Equation 4. Each box plot shows the median and interquartile range, and the whiskers show 5th and 95th percentiles. The regression line determined via Equation 5, as shown here as circles with lines through them, describes this relationship.*

ratio is computed from paired data only, where $Q_{d,25,\text{to-outside}}$ and $Q_{d,25}$ were measured in the same house. Figure 6(b) shows this ratio from two other sources of data in ResDB where the ratio of $Q_{d,25,\text{to-outside}}/Q_{d,25}$ can be calculated. The study teams that were part of the Department of Energy Building America Program collected most of these measurements. The other source of data, also included in Figure 6(b), is from Proctor et al. (2011) who measured the duct systems air leakage and conducted other diagnostic tests in 40 California homes.

Figure 6(b) shows that the ratio of $Q_{d,25,\text{to-outside}}/Q_{d,25}$ varies significantly by region: median = 0.072 in Connecticut, Illinois, and Minnesota, 0.44 in Florida and Texas, and 0.74 in Arizona, California, and Nevada. Hoeschele et al. (2002) measured this ratio in 30 California homes and found an average value of 0.81, which agrees with the range shown in Figure 6(b). The differences between states are due to the locations where ducts are commonly placed in the house. Placing the duct system inside the conditioned space means a lower ratio, which is more common in states with colder climate; whereas placing the duct system in vented attic or crawlspace would tend to result in a higher ratio, as found in states with more temperate climate.

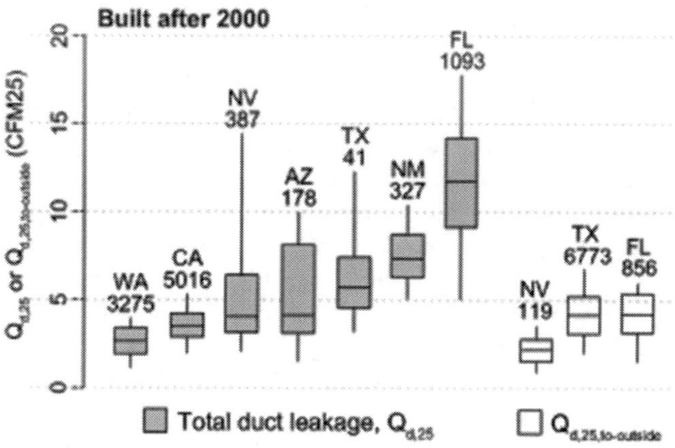

Figure 4 *Duct system leakage of homes in ResDB built after 2000, where the total duct leakage* $Q_{d,25}$, *and the total duct leakage to-outside* $Q_{d,25,\text{to-outside}}$, *are in units of cfm25 per 100 ft*^{2} *of conditioned floor area. Each box plot shows the median and interquartile range, and the whiskers show 5*th *and 95*th *percentiles. The number shown beneath each state is the house count.*

Regression of Duct Leakage and Year Built

Similar to building envelope air leakage, continuing improvement in airtightness is also observed from the duct leakage data collected over the past decade. Approximately 3,000 total duct leakage measurements shown in Figure 4 also provided the exact year built of the house (other year built data in ResDB are categorical, i.e., built between certain years). These include data mostly from California (60%), about 10% each from New Mexico, Nevada, and Florida, and the remaining 10% from 16 states. Equation 6 is the linear regression used to see if there is a relationship between duct leakage and year built. For this regression, the total duct leakage data $Q_{d,25}$ is log-transformed to more closely resemble a normal distribution. Further, 2005 is subtracted from the year built to give meaning to the intercept term such that exp(k) equals $Q_{d,25}$ of homes built in 2005 as estimated by the regression model.

$$\ln(Q_{d,25}) = k + \beta_{d,\text{year}} \cdot (\text{Year} - 2005) \qquad (6)$$

The regression analysis found a negative relationship between year built and duct system air leakage, where $\beta_{d,\text{year}} = -0.068$ (95% C.I. = −0.063 to −0.073), and exp(k) = 5.21 cfm25 per 100 ft^2 of conditioned floor area (95% C.I. 5.10 to 5.31 cfm25). The regression has a R^2 of 0.20, meaning that the year built explains some of the variability in duct leakage, but there are other factors that matter as well. Together, these coefficients estimated a 50% improvement in

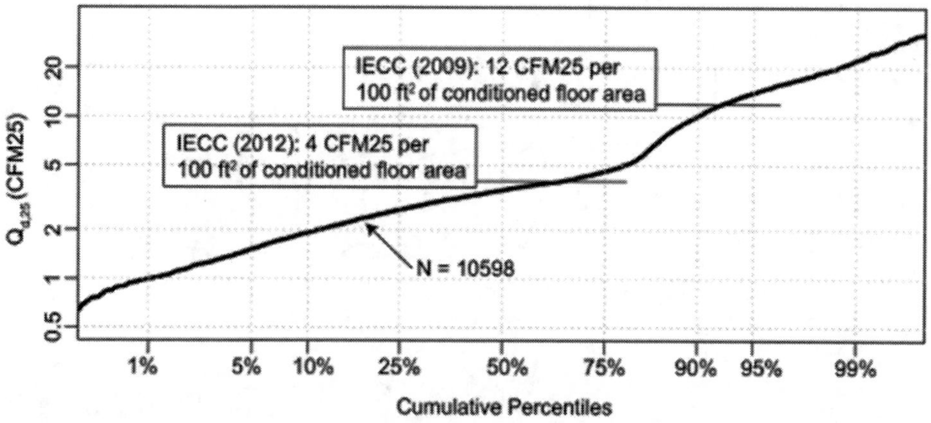

Figure 5 *Cumulative percentiles of total duct leakage* $Q_{d,25}$ *in units of cfm25 per 100 ft*^{2} *of conditioned floor area, of homes in ResDB built after 2000 (N = house count).*

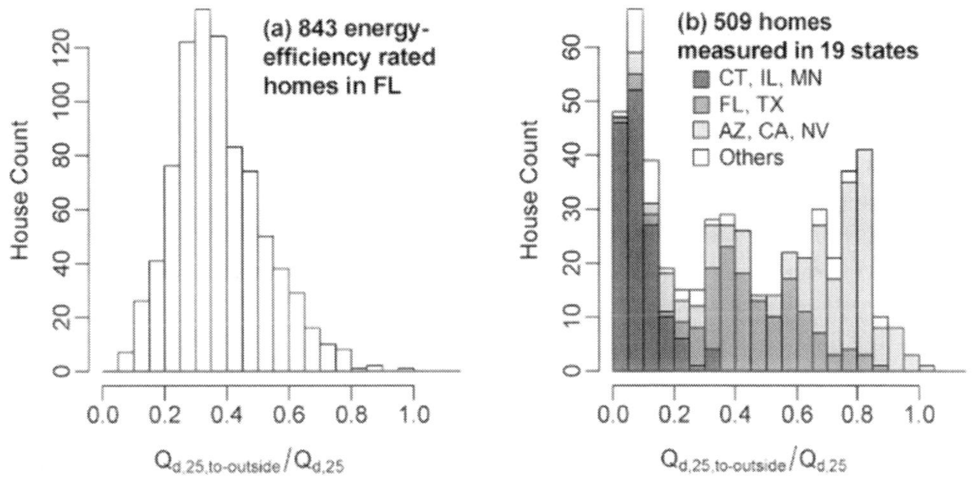

Figure 6 *Ratio of duct leakage to-outside and total duct leakage of homes built after 2000 in ResDB. Data are plotted separately by sources: (a) all Florida homes contributed by an energy rating company, (b) data from many states collected by Building America project teams.*

duct airtightness overall in the past ten years, from $Q_{d,25}$ = 7.3 cfm25 in 2000, to $Q_{d,25}$ = 3.7 cfm25 in 2010. Based on this analysis, the average new homes that are built nowadays would meet the IECC 2012 guideline of 4 cfm25. Again, this analysis also concludes that new homes built in the US since 2000 had more success in meeting IECC duct system airtightness guidelines than the whole-building envelope airtightness, where only 10% of the homes met the 2012 IECC guideline.

CONCLUSION

A large fraction of the homes built since 2000 from ResDB have airtightness that met the IECC 2009 guidelines, both for building envelope and duct systems. This is perhaps not surprising because parts of the US already have energy codes that are equivalent or more stringent than the requirements of IECC 2009. Moreover, many of the homes considered in this analysis from ResDB are built to meet certain energy efficiency targets. Even so, compliance with state building code is not guaranteed, especially because air leakage testing was not mandatory for example, in IECC until 2009 for duct systems and 2012 for building envelope. The adoption rate of state energy code by local jurisdictions also varies across the US (Cort and Butner 2012). Thus, it is an important finding that the 26,000 measurements of building envelope air leakage, and 11,000 measurements of duct system leakage, which together represent many US states, collectively show airtightness that is at the IECC 2009 level.

Over half of the homes built since 2000 from ResDB would meet the IECC 2012 total duct leakage guideline. On the other hand, only a small fraction of the homes met the IECC 2012 building envelope airtightness guidelines. This is important because mandating building envelope airtightness is part of many states' energy efficiency strategy. Moving forward, in order to reach these energy efficiency goals, building envelope airtightness must continue to improve such that more homes would meet tighter guidelines. Logue et al. (2012) estimated that there are substantial energy savings from air sealing of the building envelope to IECC 2012 levels, which shows the importance of airtightness as part of the energy efficiency strategy in the US residential sector.

By pooling together existing data that were collected by various sources, ResDB provides the quantity of data needed to evaluate the state of airtightness of single-family detached homes in the US. Trends such as the gradual changes over time, for example, are often difficult to observe in small data sets because of this large variability in a housing stock. Analysis of residential consumption data over the years has found energy savings as state-level building codes are being adopted (Aroonruengsawat et al. 2012). Potentially, energy savings would increase by setting even more stringent airtightness limits on new homes. Yet another opportunity to realize these energy savings is by implementing policies that can enhance conformance to airtightness guidelines and, by doing so, reduce the variability in airtightness in the new homes being built. Still, what remains to be proven is how persistent is airtightness as the homes are being occupied. If building envelope and duct systems soon lose their airtightness, this may adversely impact the home energy performance over time.

ACKNOWLEDGMENTS

This work was supported by the Assistant Secretary for Energy Efficiency and Renewable Energy, Building Technologies Program, of the US Department of Energy under contract No. DE-AC02-05CH11231, and the California Energy Commission Public Interest Energy Research

Program award number CEC-500-07-006. We greatly appreciate the organizations and individuals who shared their blower door and duct blaster measurements, and other diagnostic data with us.

NOMENCLATURE

ACH50	=	air changes (h^{-1}) at 50 Pa pressure difference
Area	=	house floor area (m^2)
β	=	regression coefficient estimates
CFM50	=	blower door airflow (ft^3/min) at 50 Pa pressure difference
ELA_4	=	effective leakage area (m^2) at 4 Pa pressure difference
H	=	house height (m)
I	=	indicator variables (–)
k	=	linear regression intercept
n	=	pressure coefficient (–)
NL	=	normalized leakage (–)
ΔP	=	pressure difference (Pa) across building envelope or duct system
Q_{50}	=	building envelope leakage airflow (ft^3/min) at 50 Pa pressure difference
$Q_{d,25}$	=	duct system leakage airflow (ft^3/min) at 25 Pa pressure difference per 100 ft^2 of conditioned floor area
$Q_{d,25,\text{to-outside}}$	=	duct system leakage airflow (ft^3/min) to outside at 25 Pa pressure difference per 100 ft^2 of conditioned floor area
ρ	=	air density 1.2 kg/m^3
Year	=	house year built

REFERENCES

Aroonruengsawat, A., M. Auffhammer, and A.H. Sanstad. 2012. The impact of state level building codes on residential electricity consumption. *The Energy Journal* 33(1): 31–52.

AHS. 2011. American Housing Survey for the United States, Series H-150, National Summary Data on Heating, Air Conditioning, and Appliances. Washington DC: US Census Bureau.

ASHRAE. 2004. *Method of Test for Determining the Design and Seasonal Efficiencies of Residential Thermal Distribution Systems*. ANSI/ASHRAE 152-2004. Atlanta: ASHRAE.

ASHRAE. 2012. *Ventilation and Acceptable Indoor Air Quality in Low-Rise Residential Buildings*. Addendum n to ANSI/ASHRAE Standard 62.2-2010. Atlanta: ASHRAE.

ASTM. 2007. *Standard Test Methods for Determining Air Leakage of Air Distribution Systems by Fan Pressurization E1554*. Washington DC: American Society for Testing and Materials.

ASTM. 2010. E779 *Standard Test Method for Determining Air Leakage Rate by Fan Pressurization*. Washington DC: American Society for Testing and Materials.

Chan, W.R., W.W. Nazaroff, P.N. Price, M.D. Sohn, and A.J. Gadgil. 2005. Analyzing a database of residential air leakage in the United States. *Atmospheric Environment* 39(19): 3445–3455.

Cort, K.A., and R.S. Butner. 2012. An Analysis of Statewide Adoption Rates of Building Energy Code by Local Jurisdictions. Richland: Pacific Northwest National Laboratory. PNNL Report No. 21963.

DOE. 2012. Status of State Energy Code Adoption. Washington DC: Energy Efficiency & Renewable Energy, US Department of Energy. http://www.energycodes.gov/adoption/states. Accessed 2012/12/05.

EPA. 2012. ENERGY STAR Qualified Homes, Version 3 (Rev. 06) National Program Requirements. Washington DC: US Environmental Protection Agency. http://www.energystar.gov/ia/partners/bldrs_lenders_raters/ES_Combined_Path_v_65_clean_508.pdf?12c7-46ca. Accessed 2012/12/07.

Hoeschele, M., R. Chitwood, and B. Pennington. 2002. Diagnostic performance assessment of 30 new California homes. Proceedings of ACEEE Summer Study on Energy Efficiency in Buildings.

IECC. 2009. *International Energy Conservation Code*. Washington DC: International Code Council.

IECC. 2012. *International Energy Conservation Code*. Washington DC: International Code Council.

Logue, J.M., M.H. Sherman, I.S. Walker, and B.C. Singer. 2012. Energy Impacts of Envelope Tightening and Mechanical Ventilation for the U.S. Residential Sector. Manuscript submitted to *Energy and Buildings* in October 2012.

McWilliams, J., and M. Jung. 2006. Development of a Mathematical Air-Leakage Model from Measured Data. Berkeley: Lawrence Berkeley National Laboratory. LBNL Report No. 59041.

Modera, M. 2005. ASHRAE Standard 152 and duct leaks in houses. *ASHRAE Journal* 47(3): 28–33.

Nelson, B.D. 2012. Successful implementation of air tightness requirements for residential buildings. Proceedings of Best2 Conference—A New Design Paradigm for Energy Efficient Buildings.

Neme, C., J. Proctor, and S. Nadel. 1999. Energy Savings Potential from Addressing Residential Air Conditioner and Heat Pump Installation Problems. Washington: American Council for Energy-Efficiency Economy. ACEEE Research Report No. A992.

Offermann, F.J. 2009. Ventilation and Indoor Air Quality in New Homes. Collaborative Report. Prepared for California Air Resources Board and California Energy Commission. Publication No. CEC-500-2009-085.

Proctor, J., R. Chitwood, and B.A. Wilcox. 2011. Efficiency Characteristics and Opportunities for New California Homes. Final Project Report. Prepared for California Energy Commission. Publication No. CEC-500-2012-062.

RECS. 2005. Residential Energy Consumption Survey 2005, Table US12: Total consumption by energy end uses. Washington DC: Energy Information Administration.

Sherman, M.H., and D.J. Dickerhoff. 1998. Airtightness of US dwellings. *ASHRAE Transactions* 104(2): 1359–1367.

Sherman, M.H., and N.E. Matson. 2001. Air tightness of new houses in the U.S. Proceedings of 22nd Air Infiltration and Ventilation Center.

Walker, I.S. 1998. Technical Background for Default Values Used for Forced Air Systems in Proposed ASHRAE Standard 152P. Berkeley: Lawrence Berkeley National Laboratory. LBNL Report No. 40588.

Walker, I., M. Sherman, J. Siegel, and M. Modera. 1999. Effects of Duct Improvement and EnergyStar Equipment on Comfort and Energy Efficiency. Berkeley: Lawrence Berkeley National Laboratory. LBNL Report No. 43723.

Field Monitoring and Simulation of a Historic Mass Masonry Building Retrofitted with Interior Insulation

Kohta Ueno
Associate Member ASHRAE

John Straube, PhD, PEng
Associate Member ASHRAE

Randy Van Straaten
Student Member ASHRAE

ABSTRACT

Load-bearing masonry buildings are a significant portion of the existing building stock, and there is a great deal of interest in adding thermal insulation to the walls of these structures. Exterior insulation provides the ideal conditions for building durability; however, many buildings cannot be retrofitted with insulation on the exterior for reasons such as historic preservation, aesthetics, zoning, or space restrictions. Adding insulation to the interior side of walls of such masonry buildings in cold, and particularly cold and wet, climates may increase risks of performance and durability problems.

A circa 1917 construction mass masonry building located on a Boston-area university campus was retrofitted with interior polyurethane spray foam insulation. Sensors were installed in the retrofitted walls to measure temperature and moisture conditions within the assembly; interior and exterior boundary conditions were also monitored. Experimental variables included orientation, spatial location of monitoring, and assembly type (insulated experimental versus uninsulated control). Hygrothermal simulations were run on the original and retrofitted assemblies using measured site environmental data, both to assess durability risks, and for comparison with the measured data.

Measured data indicated that the insulated masonry walls were colder and had higher moisture contents than uninsulated assemblies, as would be expected. Hygrothermal simulations had good correlation to temperature measurements, but moisture measurements were less consistent. These differences may be due to sensor response, driving rain exposure, or anomalies within the mass masonry wall assembly. The simulations indicated a low risk of freeze-thaw damage.

The effect of thermal bridging through structural elements (uninsulated floor slabs) was examined with cold weather infrared thermography.

INTRODUCTION

Load-bearing mass masonry buildings are a significant portion of the existing building stock in the East Coast and Midwest regions of the United States. Continued use of these buildings (as opposed to demolition) retains the historic character of the building and its associated neighborhood, and utilizes the embodied energy built into these structures. However, in cold climates the need for improved energy performance and greater requirements for occupant comfort are motivations to retrofit insulation in these buildings. The thermal resistance of even thick masonry walls is far below modern code requirements: common moderate density brickwork (80 to 110 lb/ft^3 or 1280–1760 kg/m^3) can be assumed to provide an R-value from R-0.25 to R-0.33 per inch (k ≈ 0.43 to 0.58 W/[m·K]). Even at a thickness of 12–16 in. (0.3–0.4 m), this only provides insulation levels of R-3 to R-5 (RSI 0.5–0.9). The use of hollow clay block may add an additional R-1 to R-2 (RSI 0.1 to 0.4).

However, adding insulation to the interior side of walls of such masonry buildings in cold—and particularly cold and wet—climates has the potential to cause or accelerate performance and durability problems. These issues include freeze-thaw damage of the masonry wall, corrosion of embedded metal elements, and potential moisture damage to wood struc-

***Kohta Ueno** is a senior associate at Building Science Corporation, Somerville, MA and Waterloo, ON. **John Straube** is a principal of Building Science Corporation and a professor of building science in the Civil Engineering Department and School of Architecture at the University of Waterloo, Canada. **Randy Van Straaten** is an NSERC Industrial Partnership Doctoral Candidate with Building Science Corporation. at the University of Western Ontario, London, ON.*

tural members embedded in the masonry. Exterior insulation provides the ideal conditions for building durability; however, many buildings cannot be retrofitted with insulation on the exterior for reasons such as historic preservation, cost, zoning or space restrictions, or aesthetics.

The authors worked with a Boston-area academic institution that owns a large inventory of existing mass masonry buildings. This institution has been moving into capital renovation projects which include the retrofit of interior insulation; to gain a greater understanding of the potential risks, the institution implemented wall monitoring on an initial project. The intent was that the data collected from this site would be used to inform upcoming projects, including the magnitude of potential risks. This work covers the instrumentation, collected data, and analysis of this masonry insulation retrofit project.

BACKGROUND AND PREVIOUS WORK

An overview of the building physics and durability risks associated with interior insulation of mass masonry buildings is covered by Straube and Schumacher (2007). Interior insulation reduces heat flow through the assembly, thus changing the existing moisture balance of wetting and drying. Given this reduced drying, the retrofit design should reduce wetting in a commensurate manner. The winter temperatures of parts of the inner layers of masonry are also significantly lowered, thereby raising the risk of freeze-thaw damage. The document covers the risks associated with interior-sourced air leakage condensation; this problem is likely when fibrous air permeable (e.g., fiberglass batt) insulation is used with an inadequate interior air barrier. This is a particular problem if the insulation is not applied tightly to the masonry wall, resulting in an air space which allows convective looping. A preferred approach, instead, is to use air impermeable, vapor semi-permeable spray foam, which can achieve airtightness levels required to avoid condensation issues.

Wilkinson et al. (2009) installed wall instrumentation in a circa 1950 retrofitted mass masonry three-story building, located in Toronto, Ontario, Canada; their data compared the performance of insulated and non-insulated wall sections. Wilkinson also provided a review of the current relevant literature, including case studies showing good durability of buildings with interior insulation retrofits, guidance on methods to minimize risks in retrofit design, and specific durability concerns raised during retrofits (freeze-thaw damage of masonry and corrosion of embedded metal components). One conclusion was the fact that although the brick would not meet modern freeze-thaw performance standards, the retrofitted wall had low freeze-thaw deterioration risk (based on both monitored data and simulations). Monitored data suggested that there might be a minor increase in embedded metal corrosion risk, but simulations indicated no added risk; this was ascribed to an artifact of the sensor measurements. Another conclusion was that the embedded wall moisture sensors used were not suitable for detecting moisture contents in the range causing freeze-thaw risk; slow sensor time response also reduced the utility of these sensors.

Mensinga et al. (2010) and Lstiburek (2010) describe masonry material property testing that provides better prediction of freeze-thaw resistance than currently used methods, such as the cold soak/boil ratio or saturation coefficient (ASTM 2012) and rapid freeze thaw (ASTM 2008). The metric used in the recent work is the critical degree of saturation (Fagerlund 1977), or S_{crit}; masonry can experience unlimited freeze-thaw cycling below this critical moisture content without damage, while above this moisture level, damage will occur quickly. Unlike previous methods, which are simple pass/fail criteria (suitable versus unsuitable for exposure), this metric results in a limit state design process for assessing the retrofit risk of the assembly. The measured S_{crit} is used as a risk threshold in hygrothermal simulations; local climate conditions, building exposures, and enclosure design are used as inputs to determine the durability risk associated with the retrofit of interior insulation.

Straube et al. (2012) provides an overview of the topic of interior insulation retrofits of mass masonry buildings. The document includes background on the building physics and potential durability risks, and presents a variety of recommended (and non-recommended) interior retrofit assemblies. This is followed by guidance on problematic details, typically linked with water concentrations on the face of the building, and methods to reduce the moisture loading. The final section provides steps to mitigate risk in these retrofits, including site assessment of the building, various material property tests and computer simulations, site assessment of the driving rain loads, monitoring of prototype insulated assemblies, and recommendations for ongoing post-retrofit maintenance and repair of the building.

SITE DESCRIPTION AND MONITORING PACKAGE

A research project was initiated by a Boston-area academic institution to monitor the in-situ hygrothermal performance of a mass masonry building retrofitted with interior insulation. The test site for this research is the Arthur D. Little Building, which is a three-story (with basement) building (overviews in Figure 1; close up of north and south façades in Figure 2) located in Cambridge, MA (DOE Zone 5A). The building was constructed in 1917, and is on the National Historic Register. The structural elements are site-cast reinforced concrete; the existing walls included (from exterior to interior) face brick, fill brick, hollow clay block, an asphalt-based dampproofing coating, and a direct-applied cementitious interior plaster rendering (see Figure 3 for representative wall sections). A major renovation project in 2010–2011 included the installation of insulation on the interior side of the mass masonry assembly. The interior plaster was removed, and open cell polyurethane spray foam (ocSPF) was applied to the interior face of the masonry wall. Insulation was approximately 3 to 3 ½ in. (75 to 90 mm) thick, partially encasing the 2 ½ in. (63.5 mm) steel studs, which were spaced approximately 1 ¾ in. (44 mm) off the masonry wall. This interior wall retrofit assembly addresses air leakage condensation risks; spacing the framing members off the wall reduces the impact of thermal bridging associated with the steel studs. The overall R-value

Figure 1 *Exterior views of the Arthur D. Little Building (left to right): front/south elevation, partial rear/north elevation, and overhead rendering from the south, showing surroundings.*

Figure 2 *South (front) and north (rear) monitoring locations and wall type nomenclature.*

of the retrofit assembly is in the R-10.6 (RSI 1.9) range, using algorithms from Kosny and Christian (1995). The building was occupied in late 2011; the building's primary function is office space.

Wall Designations and Nomenclature

The wall instrumentation has been clustered into the following test walls; they are shown on the building elevation in Figure 2. Instrumentation was installed on north and south orientations; project north and south are the rear and front elevations of the building, respectively; actual orientations are 335° and 155°. Greater emphasis was placed on the north-facing walls, given the colder conditions, which results in greater risks of freeze-thaw damage.

- South 1 ("Thin" Wall): The walls that surround the window openings are roughly 12 in. (305 mm) thick and are composed of three layers (interior hollow core clay block, fill brick, and face brick; see Figure 3).
- South 2 ("Thick" Wall): At the main body of the wall or pilasters between windows, the masonry wall is thicker, with two fill brick layers instead of one, resulting in a wall thickness of approximately 17 in. (430 mm).
- North 1 ("Thin" Wall): As per south side, at analogous location.
- North 2 ("Thick" Wall): As per south side, at analogous location.
- North 3 ("False Parapet"): Given the higher rain exposure at the top of the building, additional instrumentation was installed high on the walls, above the third floor's finish ceiling. Note that this building does not have an actual parapet (roof slab is cast directly on top of the wall); however, this "false parapet" space has high exposure to rain.
- North 4 (Uninsulated): The rear service stairwell has been left uninsulated; therefore, it provides an opportunity to compare the behavior of walls with and without insulation, but with similar exposure. The other remaining walls are insulated. One problem with the installation was that the instrumentation was installed in a location that is sheltered from rain by the roof cornice detail, due to interior details (stairwell landing). In addition, stairwell temperature conditions were not as tightly controlled as interior main space setpoints, including no summertime cooling.

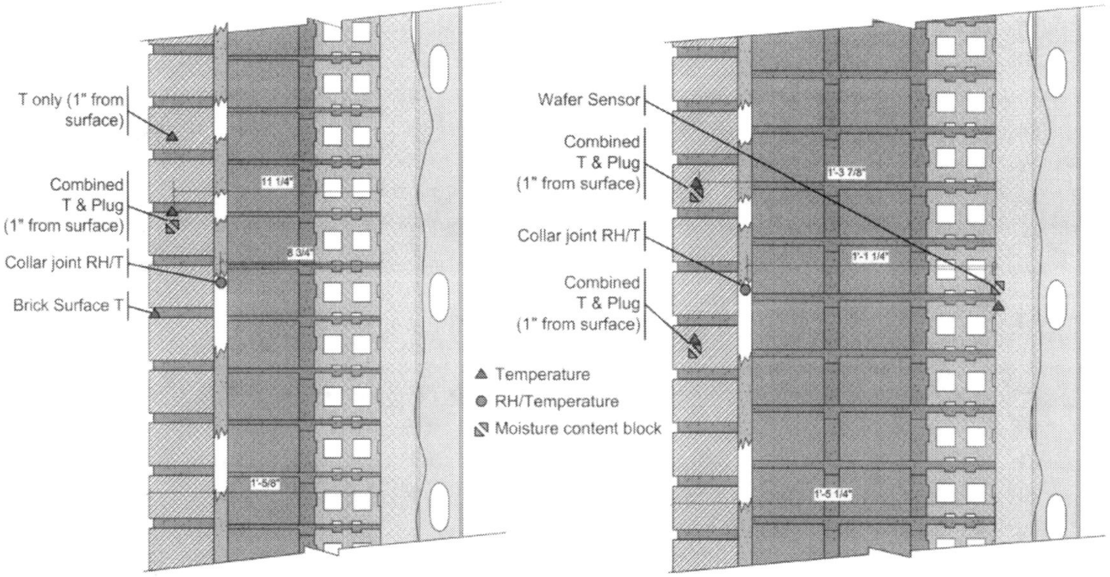

Figure 3 *Sensor package for "thin" wall (N1/S1) left; "thick" wall (N2/S2) right.*

Figure 4 *(Left) Installation of combined temperature and "plug" (wood cylinder) sensor; (middle) sensor encapsulated in clay for capillary contact; (right) boroscope image of installation hole.*

Sensor Arrangement and Data Acquisition System

A variety of sensors were embedded through the thickness of the test walls; the layout varies depending on wall configuration (Figure 3). The sensors included temperature sensors (±0.2°F [0.1°C] NTC thermistor), relative humidity sensors (thermoset polymer capacitive based sensors; ±3% between 10% and 90% rh), and "moisture content blocks." The latter sensors are wood-based relative humidity surrogate sensors, described in detail by Ueno and Straube (2008); the moisture content of the wood (eastern white pine) is measured via electrical resistance.

There are two configurations of moisture content blocks: the "plug" sensor (a cylindrical sensor, embedded in a hole drilled into the masonry) and the "wafer" sensor (a flat, thin rectangle of wood which measures conditions at an interface surface, such as between the insulation and the masonry wall). These surrogate sensors were chosen in lieu of relative humidity sensors due to (a) greater longevity in condensing or liquid water wetted environments, (b) greater resolution of bulk water wetting events, and (c) some indication of the accumulation of moisture (or storage) over time. Further information on these monitoring methods is provided in Straube et al. (2002).

Some sensors are duplicates, given the variation that can occur in conditions within a wall assembly. The uninsulated wall was wired with the same scheme as the "thick" wall.

Embedded sensors were installed from the interior, at holes drilled through the masonry to the appropriate layer (Figure 4, left). The holes were then sealed from the interior (and other layers/air spaces) using low-expansion foam. The plug sensors were encapsulated with clay before installation, to improve capillary contact between the sensor and the surrounding material (Figure 4, middle). An alternate installation method (not used here) is to inject a clay slurry into the hole;

this ensures consistent capillary contact, but adds an initial moisture signal to disaggregate from the wall behavior. The conditions of the installation hole were typically examined with a boroscope before sensor installation, to identify anomalies (Figure 4, right). The installation locations included in masonry joints (i.e., higher rainwater absorptivity), within the solid brick body, and intersecting with the air-filled brick cores (which could allow some drying). The lateral location of the sensors within the masonry wall could not be tightly controlled due to this "blind" (interior side) installation.

One caution when interpreting the results of the surrogate sensors is that their time response is very slow, based on Ueno and Straube (2008) and Wilkinson et al. (2009). Therefore, they would have difficulty resolving short-term (daily) rain events. Specifically, the sensors have an asymmetric response: slow response to increasing RH (adsorption/wetting), and fast response to decreasing RH (desorption/drying). However, the sensors are still useful to track seasonal trends when comparing wall assemblies.

A weather station was installed on the building's roof measuring temperature, relative humidity, wind speed and direction, horizontal rain, and horizontal solar radiation. Solar radiation sensors were mounted on the front and rear walls to directly measure radiation on the vertical surfaces.

Data are being collected by a measurement and control system installed in the service stairwell; data are measured at five minute intervals, and average values are recorded hourly. No battery backup for the data logger is provided; however, the unit has non-volatile memory, and resumes data collection after a power failure.

MONITORING RESULTS

Data have been collected from early October 2011 through early June 2013, providing over one and a half years of results. The intent is to collect data for a minimum of two years.

Boundary Conditions

Exterior and interior boundary conditions were recorded; exterior and interior temperatures are shown in Figure 5. Interior conditions were measured in the north-facing room, south-facing room, and the north-facing stairwell with uninsulated walls. North and south interior temperatures were typically controlled to the 70°F to 77°F (21°C to 25°C) range, but with some excursions to the high side in the south-facing room (likely due to solar gain and/or thermostat setup control issues). The stairwell was semi-conditioned (heating-only), with temperatures between interior setpoints and exterior temperatures.

Winter 2011–2012 conditions were exceptionally mild (4400 HDD Base 65°F [2440 HDD18°C] versus 5600 HDD65°F [3110 HDD18°C] climate normal). Winter 2012–2013 was closer to normal conditions (5400 HDD Base 65°F [3000 HDD18°C]).

Wintertime relative humidities have been exceptionally low, often in the 5%–15% range. Summertime measurements show control of interior relative humidity levels, typically in the 30%–50% range. When dewpoint temperatures are calculated (Figure 6), wintertime interior dewpoints are almost identical to exterior dewpoints, which might indicate an excessively high ventilation rate (or air leakage, less likely based on test results), low interior moisture generation, or a combination of both. Some dewpoint data are missing, due to failures of relative humidity sensors.

Figure 7 (left) shows driving rain, as calculated from Boston airport (KBOS) data, using methods described by Straube (2005). Site-measured weather station rainfall data appeared to be missing significant rainfall events, so it was not used in this analysis. The plot shows the cumulative driving rainfall for the monitored period. The test wall orientations are superimposed on the driving rain rosette for reference. The plot clearly shows that the test walls were not at the highest exposures to driving rain, and that the rear/project north side had much lower rainfall than the front/project south side.

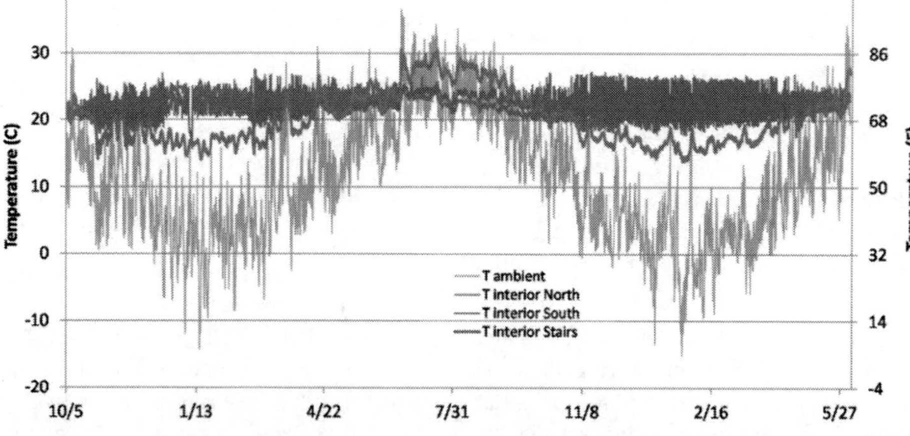

Figure 5 *Exterior and interior temperatures.*

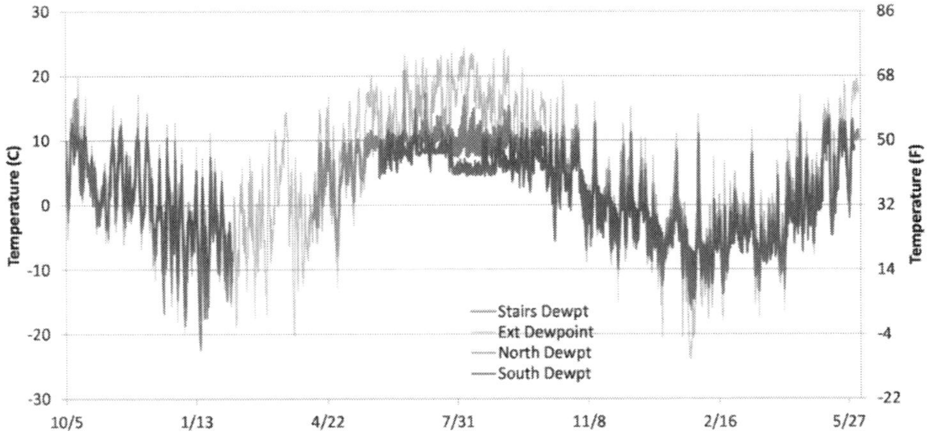

Figure 6 *Exterior and interior dewpoint temperatures.*

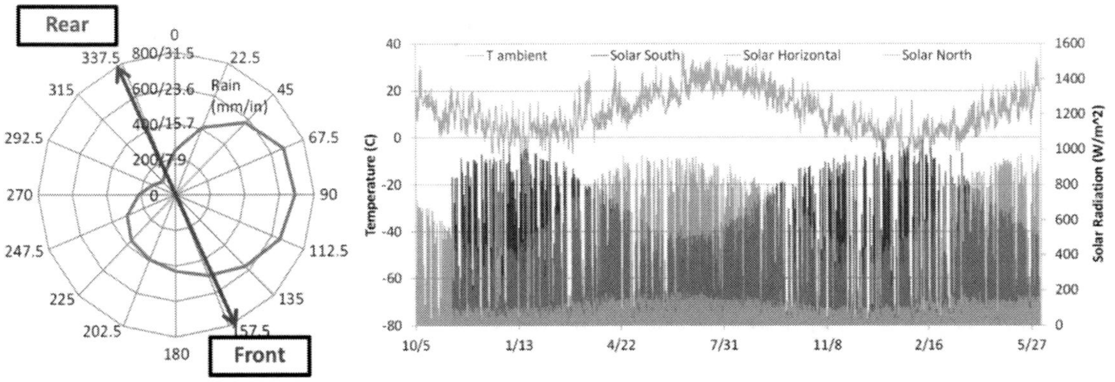

Figure 7 *(Left) Driving rain rosette in mm (in.) (calculated from Boston airport data), with building faces shown; (right) solar radiation at building for horizontal, north wall, and south wall.*

Solar radiation was plotted with exterior temperature for reference including horizontal, north wall, and south wall data (Figure 7 right). It shows the expected pattern of a seasonal rise and fall with higher solar gain on the south wall in wintertime (due to low sun angle). The north wall experiences consistently low solar gain (diffuse radiation only).

Temperature Measurements

Temperatures were measured through the thickness of the wall, including at the exterior surface, in the exterior brick wythe, in the collar joint (inboard of the exterior wythe), and at the masonry-insulation interface (see Figure 3). Selected temperatures for north-facing insulated walls are shown in Figure 8; it shows that the wall temperatures basically follow exterior conditions (within several degrees) regardless of position within the wall (albeit with some damping due to thermal mass, especially at the innermost wafer temperature).

The south-facing walls have much higher temperature peaks than the north-facing walls (due to solar gain), with peak temperatures over 104°F (40°C). However, coldest temperatures are roughly as low as on the north side, during night periods.

The north-facing uninsulated wall is markedly different from the north-facing insulated walls. Instead of tracking outdoor conditions, the temperatures fall in a range between indoors and outdoors, depending on position through the assembly thickness (Figure 9). The sensor at the interior surface of the masonry (but hidden by drywall) is roughly halfway between indoor and outdoor temperatures. The wintertime temperatures within the uninsulated masonry wall are warmer than the insulated case.

The temperature behavior of these walls is consistent with a steady-state calculation of the temperature gradient through the wall. In the uninsulated case, the masonry provides the majority of the insulation value of the wall, so the temperature drops gradually through the thickness of the masonry. In contrast, post-insulation, the interior insulation (spray foam) layer provides the majority of the insulation value in the wall. Therefore, the entire thickness of the masonry wall operates close to exterior temperatures.

In addition, the number of freeze-thaw cycles at several locations in the assembly was summed for the recorded data (Table 1). A freezing temperature of 23°F (–5°C) was used (as

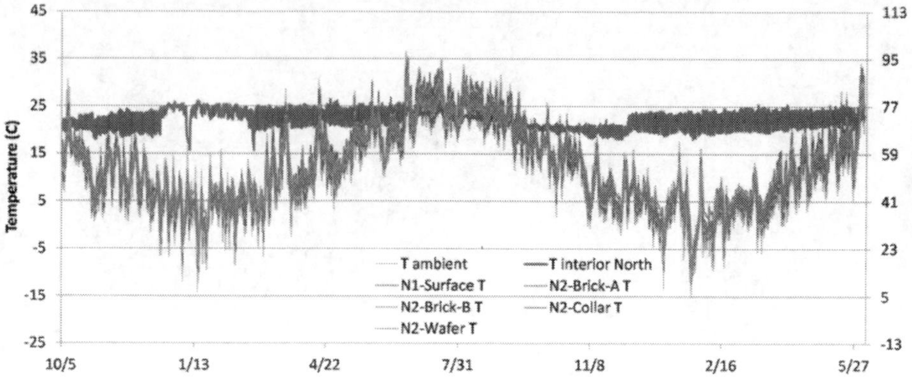

Figure 8 North insulated wall representative temperatures, with interior and exterior temperatures.

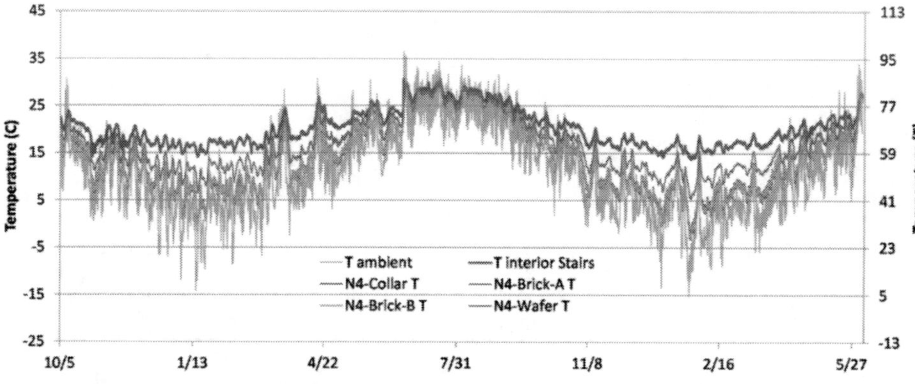

Figure 9 North uninsulated wall representative temperatures, with interior and exterior temperatures.

opposed to 32°F (0°C), which reflects the depressed freezing point within porous media (Straube and Burnett 2005), although even lower temperatures may be more appropriate. The number of ambient air cycles below the critical temperature is shown in Table 1 for reference.

Although ambient air temperature shows multiple sub-23°F (–5°C) cycles, only a fraction of them occur within the wall, due to thermal mass effects. The surface of the masonry wall experiences more cycling than inner portions ("collar joints"). Insulated north facing wall collar joints (N1, N2, and N3) have more cycles than uninsulated (N4) or south (S1) walls.

In order to provide greater differentiation between N4 and S1 (both zero counts in Table 1), the summed freezing cycles using a warmer temperature of 32°F (0°C) are shown in Table 2. The results are consistent with Table 1, but show that the collar joint of the uninsulated wall (N4) experiences the fewest cycles through 32°F (0°C). In addition, the count of freeze-thaw cycles in the north and south insulated walls (N1, S1) can be compared. At the surface, the south wall has a larger number of cycles, which is consistent with wintertime solar gain temperature cycling. But deeper in the wall (at the collar joint), the north wall has more cycles than the south wall.

Overall, these results should not be taken in isolation to indicate greater risk due to insulation retrofits. As described by Mensinga et al. (2010), freeze-thaw damage only occurs when freezing temperature cycles occurs while masonry moisture content is above S_{crit}; moisture behavior is discussed further below.

Moisture Plug (Brick Moisture) Measurements

The results from the plug surrogate relative humidity sensors placed in the exterior brick wythe were plotted; the graphs include calculated driving rain on the wall (in mm/h and in./h) and a dotted line showing the approximate equivalence of 100% rh (27%–32% wood moisture content, as calculated by previous calibration; see Ueno and Straube 2008).

The north-facing insulated wall responses (Figure 10) vary widely, and are not necessarily correlated with location. For instance, most sensors start within a similar band of moisture contents (MCs), but over the summer, the "thin" wall and one "false parapet" sensor dry down to 10%–15% MC (55%–80% rh equivalent). But other walls stay much wetter, and/or dry only at the end of the summer ("thick" wall sensors, N2). There was no indication that the "false parapet" sensors (N3) consistently experienced more wetting than other portions of the wall.

Most plug sensors show a response to some driving rain events; for instance, Hurricane Sandy (October 29, 2012) resulted in a major increase in moisture content across all sensors. However, some significant driving rain events did not result in a noticeable wetting response. This might be expected, given the complication of calculating and predicting driving rain on the face of a building in an urban environment. The general pattern for most of the north-facing sensors was in the 22%–35% MC range (93%–100% rh equivalent).

The south-facing sensors (Figure 11) had similar patterns; one "thick" wall sensor (S2-Plug B) showed "spiky" behavior closely matching driving rain events, while other sensors were less responsive. The general range for the south-facing sensors was dryer than the north-facing sensors, at 15%–28% MC (78%–100% rh equivalent). The moisture contents appear to rise higher in winter/spring 2013, which is not directly correlated to greater amounts of calculated driving rain.

The sensors in the north-facing uninsulated wall (Figure 12) have a markedly drier response than the insulated walls, typically in the range of 9%–13% MC (50%–70% rh equivalent). There is a slight seasonal rise and fall (greater moisture in winter), but the response is small compared to the insulated walls.

One puzzling aspect, though, is that there is no discernible response to any driving rain wetting events, unlike the insulated north- and south-facing walls. Looking at the installation (Figure 2), it seems plausible that the sensors have

Table 1. Number of Freeze-Thaw Cycles (Through 23°F [–5°C]) within Wall Assemblies

Location	Number of Occurrences	Location	Number of Occurrences
Exterior (Ambient Air)	34	N1 (Thin)-Collar Joint	8
		N2 (Thick)-Collar Joint	5
N1 (Thin)-Surface	11	N3 (Parapet)-Collar Joint	2
S1 (Thin)-Surface	12	N4 (Uninsulated)-Collar Joint	0
		S1 (Thin)-Collar Joint	0

Table 2. Number of Freeze-Thaw Cycles (Through 32°F [0°C]) within Wall Assemblies

Location	Number of Occurrences	Location	Number of Occurrences
Exterior (Ambient Air)	88	N1 (Thin)-Collar Joint	19
		N2 (Thick)-Collar Joint	21
N1 (Thin)-Surface	44	N3 (Parapet)-Collar Joint	6
S1 (Thin)-Surface	65	N4 (Uninsulated)-Collar Joint	4
		S1 (Thin)-Collar Joint	10

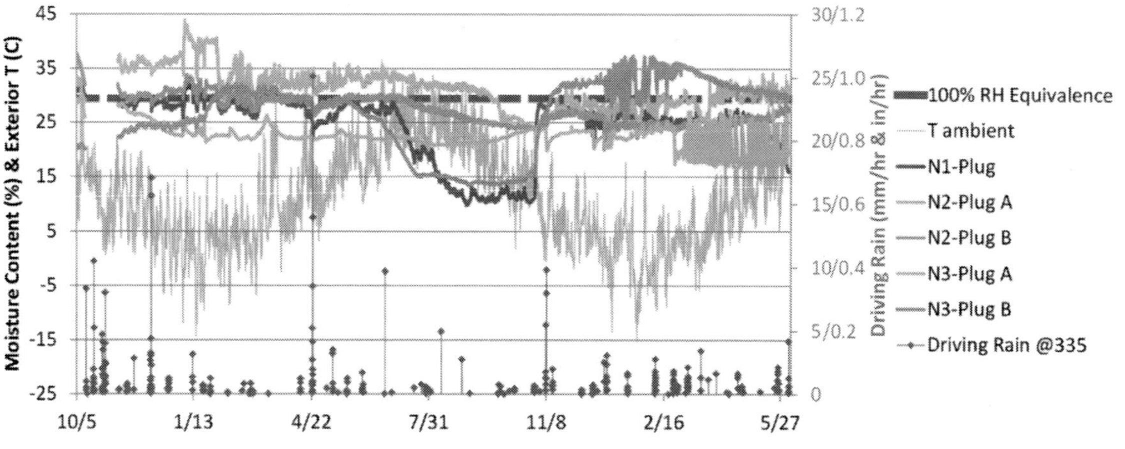

Figure 10 *North insulated wall plug sensor moisture contents with exterior temperatures and driving rain.*

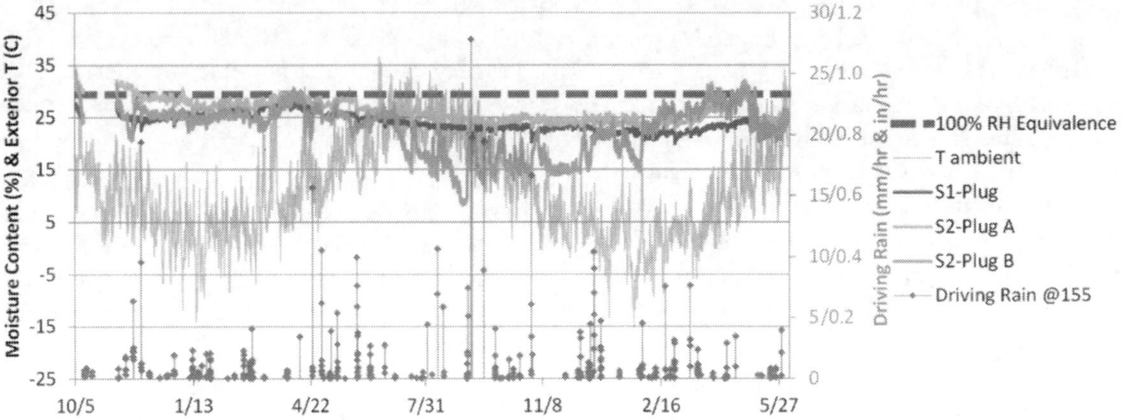

Figure 11 South insulated wall plug sensor moisture contents with exterior temperatures and driving rain.

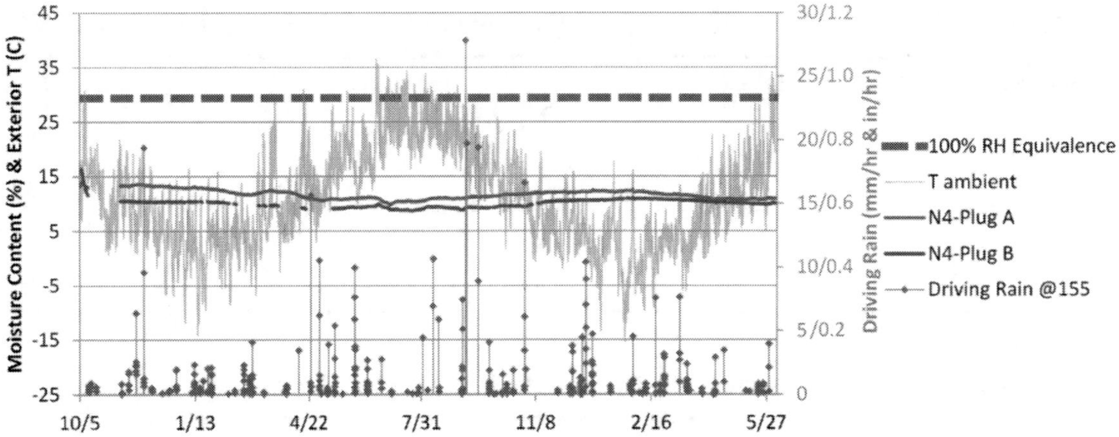

Figure 12 North uninsulated wall plug sensor moisture contents with exterior temperatures and driving rain.

been installed close enough to the exterior cornice detail that they are largely shielded from driving rain. This would explain both the lack of response as well as the exceptional dryness of the masonry wall. Overall, it is unclear how much of the sensor patterns can be attributed to the lack of thermal insulation, compared to the lack of rain deposition on the wall.

Although it should be a concern that the insulated walls are measurably wetter than the uninsulated wall, it is important to understand the mechanism of potential failure modes. The primary worry is freeze-thaw damage of the outer face of the masonry, which occurs when brick moisture contents exceed the critical degree of saturation (S_{crit}; see Mensinga et al. 2010). This topic is covered later in the synthesis of measured and simulation results.

Relative Humidity Measurements

Temperature and relative humidity sensors were placed in the "collar joint" region (between the outer brick wythe and the backup wall) of all of the test walls. However, there was a systemic failure of the relative humidity sensors after the first winter, so the results are not presented here. The results were generally consistent with the surrogate humidity plug sensors: for instance, the north insulated walls remained at 100% rh for most of the monitored period. The south insulated wall was drier than the north insulated wall (70%–80% rh), and the uninsulated north had substantially lower RH levels than either insulated wall (45%–60%). When interpreting this data, it should be noted that relative humidity measurements do not resolve moisture contents in the high range needed to determine whether masonry is below or above critical degree of saturation (S_{crit}).

Moisture Wafer Measurements (Masonry-Insulation Interface Conditions)

Temperature and surrogate relative humidity sensors (wafer sensors) were installed at the interior surface of the masonry to measure conditions at the insulation-masonry interface (Figure 13, left). The intent of these measurements is to capture the occurrence of interior-sourced moisture that may be transported by diffusion and condense on the cold ma-

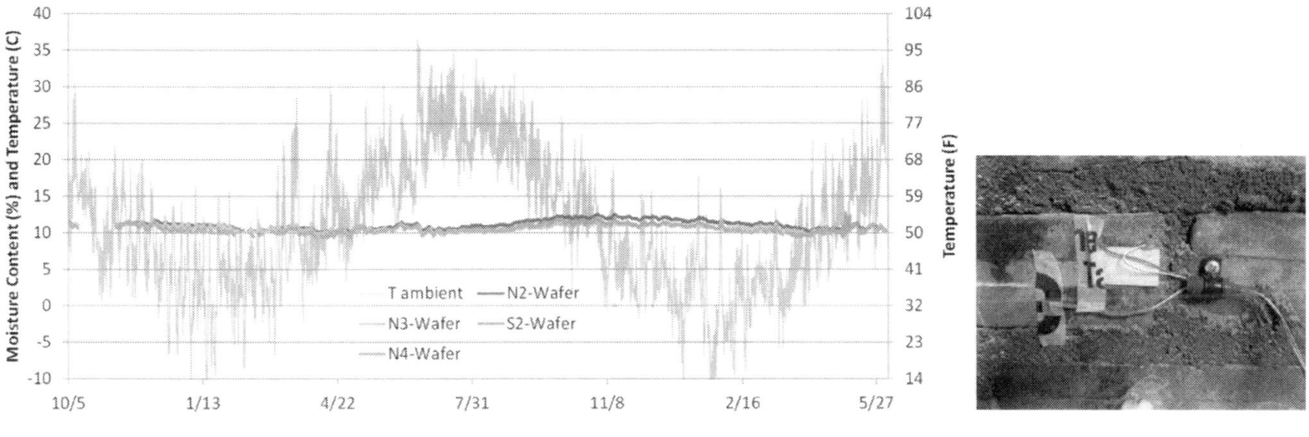

Figure 13 (L) Moisture wafer results, with exterior temperature for reference; (R) example of wafer sensor.

sonry surface during wintertime. This risk was considered greater due to the use of open-cell foam, which has higher vapor permeability than closed cell foam.

The measurements (Figure 13, left) remained in the 10%–13% MC range (50%–70% rh equivalent), which is a safe range for avoiding moisture-induced degradation of building materials. Before concluding that this retrofit wall assembly is safe from interstitial condensation, it must be noted that interior relative humidity levels were exceptionally low (5%–20%) during both winters, which reduces the risk of wintertime condensation.

The risk could also be assessed by comparing the temperature of the masonry-insulation interface to the interior dewpoint: during the winter, interior dewpoints were almost always below the interface temperature, which essentially eliminates condensation risks. These low dewpoints could be due to high ventilation rates, air leakage, low interior moisture generation, or a combination thereof. Overventilation due to the active beam heating and cooling system is suspected: a common problem is overspecifying primary airflow (beyond ventilation and latent load requirements) to meet space conditioning requirements (Livchak and Lowell 2012).

HYGROTHERMAL SIMULATIONS

Simulation Setup

One-dimensional hygrothermal simulations were run using WUFI 5.2 (IBP 2012); a variety of wall assemblies and the two orientations were simulated; the site-measured interior and airport weather data provided interior and exterior boundary conditions. Simulations were run for a period matching the monitored data (October 2011–May 2013).

Material properties for the masonry materials were taken from an outside masonry testing consultant's report on samples collected from the building (22 samples of four material types). Properties included dry density, capillary uptake, calculated porosity, and an estimate of the critical degree of saturation (S_{crit}). These values were used to modify existing materials in the WUFI database. The masonry consultant reported mortar consistent with Type N composition; the WUFI database Type N mortar was used. Open cell polyurethane foam from the WUFI database (based on ASHRAE 1018-RP data) was used (11.8 perms [670 ng/(Pa·s·m^2)]) for the 4.5 in. (116 mm layer).

A typical wall section is shown in Figure 14, which models the wall layers pictured in Figure 3; the image also shows the location of temperature/relative humidity monitors, which mostly correspond to site measurements.

The collar joint and the joint between the face and fill bricks were modeled as solid mortar; in reality, these joints are incompletely filled with mortar, with a variety of cracks and voids in the fill. In addition, the hollow clay block is shown as solid (as per the block's web condition), as opposed to having an air space (hollow core).

A sensitivity analysis was done in WUFI to determine whether replacing these mortar joints and clay block cores with air spaces would result in a significant difference in performance. A comparison was run between these two cases (no air spaces versus all air spaces): in the uninsulated wall, temperatures through the wall were identical. In the insulated wall, there was a small (at most 0.9°F [0.5°C]) difference; adding air spaces resulted in a colder outer wythe, due to thermally decoupling the face brick from the backup wall. In terms of moisture performance, replacing mortar with an air space eliminates capillary contact between the two layers, increasing risk for the wetted layer (e.g., exterior), and reducing risk for the non-wetted layer (e.g., backup wall). However, modeling showed a relatively small difference between these cases.

Simulation Results and Comparison to Measured Data

The modeled masonry temperatures generally followed the measured values: during the winter, the simulated brick and collar temperatures tracked several degrees colder than measured. One possible reason is that the "true" thermal conductance of the brick outboard of these sensors is lower than

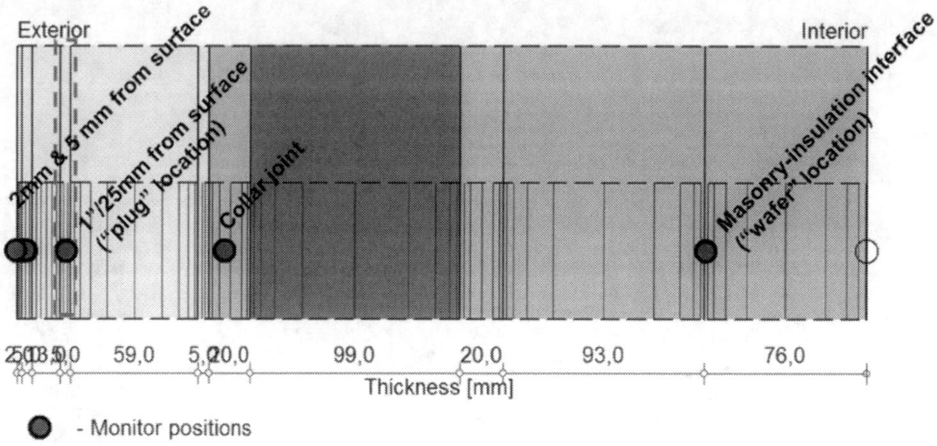

Figure 14 WUFI model of Wall S1/N1 (insulated "thin" wall) with materials and monitor positions.

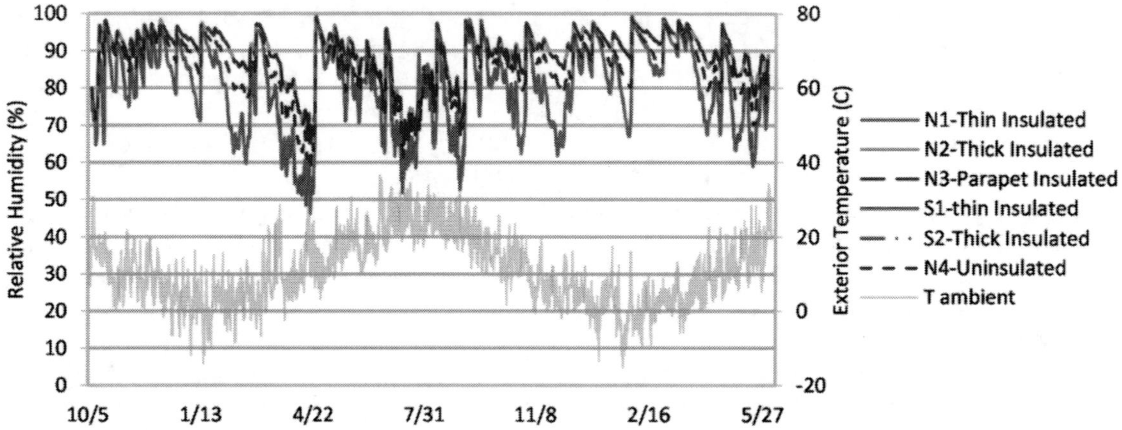

Figure 15 Simulation brick layer 1 in. (25 mm) from exterior face relative humidity; exterior temperature for reference.

in the model (i.e., the outboard brick assembly is more insulating than simulated). The brick may be less dense or thicker, the sensor may be located further away from the outside surface than modeled, or the presence of air voids around the sensors might reduce apparent thermal conductance. Similarly, the wafer temperatures tracked several degrees colder as well. This could be in part due to the brick conductivity and in part due to a conservative value for the open cell foam conductivity used in the model.

The modeled response of the outer layer of brick, 1 in. (25 mm) from the surface (layer highlighted in Figure 14) was examined closely as its position corresponds to the plug sensor in the monitored wall. The response is plotted in Figure 15 in terms of relative humidity (as opposed to moisture content) to allow a simpler comparison to the plug sensor response.

The response of the model is a common pattern seen in hygrothermal simulations of mass masonry assemblies. Moisture conditions rise sharply (to the 90%–100% range) in response to driving rain events and then dry slowly over time. Further driving rain events cause additional jumps in RH. During the winter, south-facing walls dry at a faster rate than the north-facing walls, and the uninsulated north-facing wall dries faster than the insulated walls.

However, the simulated behavior contrasts sharply with the measured data (see Figures 10, 11, and 12). For instance, very few driving rain events caused discrete responses in measured data but were evident in the simulation. The general trend in the measured data was consistently high humidity levels in the wintertime, followed by drying (in some cases) in the summertime.

In addition, the measured moisture levels mostly fell into overlapping bands that correlated with the wall category. The insulated north walls were consistently the wettest, the south-facing insulated walls the driest, and the north-facing uninsulated wall somewhere in between.

Another comparison is to the relative humidity levels at the collar joint between the face brick and the backup wall. The simulated response at this layer is shown in Figure 16. The order of the responses is consistent with the order seen in the measured plug data. In addition, the RH levels are roughly consistent with the limited measured collar joint RH data, although measured RHs in the uninsulated wall (N4) are drier (45%–67% rh) than in the simulation (80%–85% rh). This difference may be ascribed to the rain sheltering caused by the overhanging cornice detail at the N4 monitoring location.

Simulation results for the relative humidity at the masonry-insulation interface (equivalent to the wafer placement) are shown in Figure 17. The simulation results show RH levels in the 65%–85% range, with variations correlating with outdoor temperature. In comparison, the measured data showed a more damped response, as well as drier conditions; the measurements centered at 50%–70% rh equivalent (10%–13% MC).

Part of the difference between simulation and measured responses may be due to the slow response of the wood-based surrogate humidity sensors. However, the cycles seen in these simulations are not simple diurnal variations; they are variations over the course of multiple days. It is plausible that the simulation does not correctly capture the vapor permeance characteristics of the wall assembly installed inboard of the masonry wall (open-cell foam and painted gypsum board).

The simulations also examined the moisture content of outer brick layers, to determine freeze-thaw risks. Assuming typical rain exposure levels, and S_{crit} values as presented in the masonry consultant's report, simulation peak moisture contents (under 3% by mass) never approached levels which would cause freeze-thaw damage (5%–7% by mass). This held true for the thin outermost layers, centered at 0.1 in. (2 mm) and 0.2 in. (5 mm) from the surface. Furthermore, the comparison to measured data showed that the simulation was conservative, assuming colder temperatures and more moist conditions.

ANALYSIS AND SYNTHESIS

Both the measured data and the modeling indicate that insulation of the masonry wall results in colder temperatures (through the thickness of the wall) and higher moisture levels. These changes are entirely expected given the fundamental physics. Although the north-facing insulated walls have higher moisture contents than the uninsu-

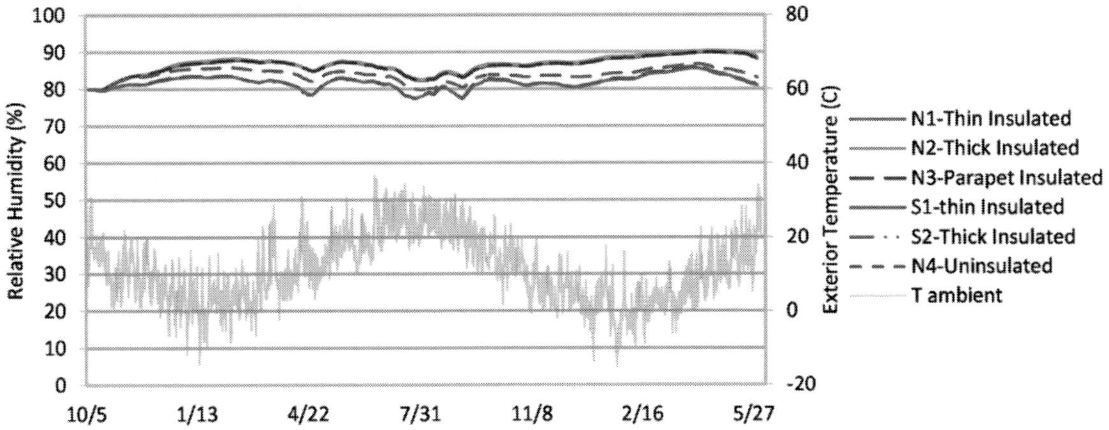

Figure 16 *Simulation collar joint (mortar layer) relative humidity; exterior temperature for reference.*

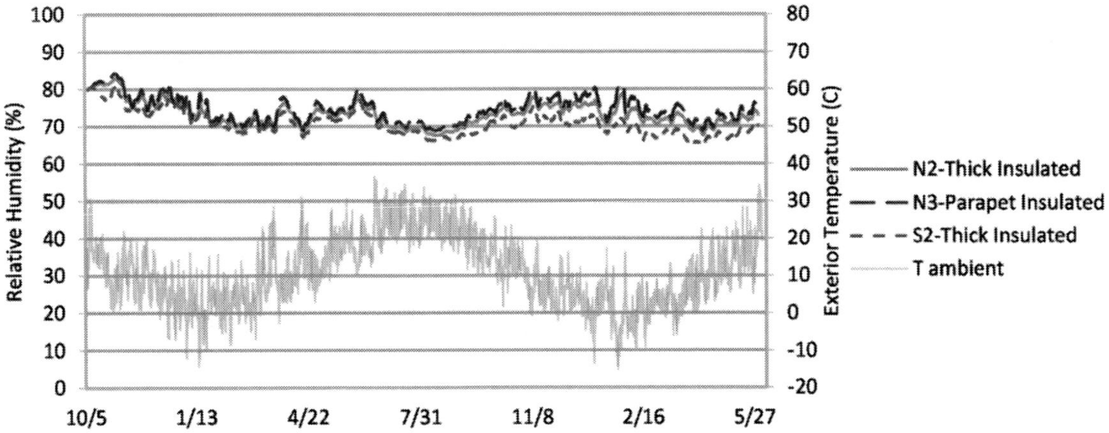

Figure 17 *Simulation wafer (masonry-insulation interface) relative humidity; exterior temperature for reference.*

lated north-facing wall, this difference might reflect both insulation levels and rain exposure, due to shielding by the overhanging cornice.

Although temperature agreement between the model and measured data was good, there were significant differences in moisture responses. Part of this can be ascribed to the properties of the wood-based humidity surrogate sensor; they are unlikely to indicate short-term humidity fluctuations. The difference might also be due to rain exposure: driving rain on the walls was calculated based on measured horizontal rain and wind data. Localized effects—in particular, surrounding buildings—can have an effect on rain deposition. Driving rain gauges on the exterior walls, near the monitored locations, would reduce the uncertainty of this boundary condition. A potential confounding factor is that the capillary contact between the plug sensors and the surrounding masonry may not be consistent from installation to installation.

Another potential cause of anomalies is the construction of built-up mass masonry wall assemblies. In a light wood frame wall, thermal and moisture anomalies can typically be characterized (e.g., framing members, air leakage at sheathing joints or roof-floor connection), and monitoring locations can be chosen to avoid them (e.g., mid-height/center of a stud bay). In contrast, in a mass masonry assembly built of multiple layers and materials (such as these walls), the assembly has a network of voids (incompletely filled mortar joints, brick cores, clay block cores) and capillary active cracks which can result in redistribution of moisture by bulk water drainage, capillarity, air movement, or vapor diffusion. These types of anomalies would be difficult to capture in a one-dimensional hygrothermal simulation. For instance, these networks could result in drainage of bulk water away from one portion of the wall, resulting in concentration in another portion (Laska 1997). This is consistent with the varying responses of plug sensors that are nominally in the same type of wall (i.e., duplicated sensors). Some sensors show strong wetting and drying responses, while others show minimal response to outdoor conditions.

As discussed by Wilkinson et al. (2009), the plug sensors are not suitable for determining whether masonry layers are reaching moisture contents above S_{crit}. One reason is that S_{crit} is a high moisture content (high fractions of saturation), which is above the resolution range of this sensor. An example is given below in Table 3, showing the relative humidity, equivalent wood moisture content (based on the sorption isotherm and correlated by Ueno 2008), and equivalent brick moisture content (for the face brick material properties measured at this site).

The table shows that although the plug sensors are useful for general trending data of moisture levels, it cannot resolve the high moisture contents that are critical for freeze-thaw performance. However, the sensor can provide an accurate indication that the wall is below the critical moisture levels associated with freeze-thaw damage.

Table 3. Correlation of Relative Humidity, Wood Moisture Content, and Brick Moisture Content

RH, %	Wood MC (Weight %)	Face Brick MC (Weight %)	Notes
50	~9	0.02	Lower limit of resolution for wood surrogate sensors
80	16	0.09	Reference water content (W_{ref} or W_{80})
90	20	0.19	
95	24	0.38	
100	29+	4.3	Free water saturation (W_f)
100	n/a	5–7	Critical degree of saturation (S_{crit}) for face brick samples

Another sensor issue is that freeze-thaw failures typically occur when S_{crit} is exceeded in a thin layer, roughly ~1/16 to 1/8 in. (1.6 to 3 mm) from the surface. This is the layer that is (a) sufficiently cold to experience freeze-thaw cycling (near the surface), (b) experiences outdoor precipitation events, but (c) has limited drying to the exterior (unlike the surface layer), resulting in the highest moisture contents. These sensors do not have spatial resolution to distinguish this type of thin layer.

This raises the question, though, of whether there are any effective in-situ measurements that can measure the moisture content of masonry materials in the S_{crit} range. Physical removal of samples by cutting (for gravimetric testing) will result in localized heating, and thus a change in moisture content. An easily removable surface sample will have capillary contact dissimilar to the body of the wall. Künzel and Holm (2009) describe the use of in-situ nuclear magnetic resonance (NMR) scanning to obtain one-dimensional moisture profiles in mass masonry walls. This method is promising, but limitations include high equipment cost, limited penetration (0.4–0.8 inches [10-20 mm]), and interference from embedded ferrous materials. In addition, it may be worthwhile to perform laboratory tests of handheld impedance-based moisture meters or similar instruments to determine whether they can differentiate between the moisture contents in the S_{crit} range.

EXTERIOR INFRARED OBSERVATIONS

In addition to the measured data and hygrothermal simulations, cold-weather infrared thermography was conducted at the building in January of 2012; the exterior temperature at the time of these measurements was roughly 25°F (–4°C). The front/south elevation is shown in Figure 18.

There is conspicuous evidence of thermal bridging through the exposed floor slab edges. Due to the construction of the building, interior insulation of the masonry walls does not address the thermal bridging through the slabs (including

Figure 18 *Front (south) elevation visual and infrared images (composite infrared image). There is a significant thermal anomaly (indicating heat loss) at the entryway; this is consistent with the heated but uninsulated (due to historic finishes) entryway, and possibly air leakage around the entry doors.*

the roof slab) cast into the structure. The effect of thermal bridging on overall thermal performance of the wall can be approximated using a parallel paths method: assuming R-10.6/RSI 1.9 (over a 10 ft [3.0 m] wall), and R-3/RSI 0.5 (for a 1 ft [0.3 m] floor slab), this results in an overall R-8.6/RSI 1.5. The penalty for thermal bridging through the uninsulated slab edge becomes greater as wall insulation levels are increased. Calculations using linear transmittance (Ψ) values for similar assemblies from Morrison Hershfield (2011) gave comparable results, of R-7.3/RSI 1.3 for a nominal R-10 wall (73% effective), and R-13.3/RSI 2.3 for a nominal R-25 wall (53% effective).

There is a significant thermal anomaly (indicating heat loss) at the entryway; this is consistent with the heated but uninsulated (due to historic finishes) entryway, and possibly air leakage around the entry doors.

There is also a significant thermal anomaly at grade, with warmer temperatures at the half-buried floor than at the wall above. The basement walls at the front of the building are shown as insulated on the plans. The thermal anomaly might reflect the thermal connection between the wall and the surrounding soil. Based on previous measurements at multiple sites (Ueno et al. 2007), soil temperatures often stay above freezing (32°F [0°C]) even during extended freezing or subfreezing weather, due to the latent heat of fusion of the bound water in the soil. The warmer soil temperatures tend to keep interior-insulated foundation walls warmer, even above grade. The surface temperatures observed at the front of the building are consistent with this phenomenon.

Some windows appear as low temperature surfaces; however, this should not be interpreted as a lack of heat loss (i.e., good insulation value). Instead, the cold surface temperatures are due to reflection of the cold night sky temperatures from the glass to the infrared camera.

The rear or north elevation clearly shows thermal bridging at the slab edges (Figure 19). In addition, a comparison can be drawn between the uninsulated wall (stairwell, with lights on; third bay from left, highlighted in yellow) and the insulated walls. Although the uninsulated wall has arguably warmer exterior surface temperature, the two walls are very close in temperature, as would be predicted by a steady-state thermal gradient analysis. However, another confounding factor is different interior setpoints of 76°F (25°C) insulated north room and 59°F (15°C) uninsulated stairwell. The difference in setpoints is also consistent with the temperature difference between the window frames in the insulated and uninsulated portions.

In addition, there is a large thermal anomaly at the rear elevation curtain wall (signature stairwell) intersection with the masonry wall; this is consistent with an air leak between the masonry and curtain wall conditions. This appears to be a localized air leak at a specific detail; the body of the building shows no evidence of significant distributed air leakage.

CONCLUSIONS

The measured data from this mass masonry building retrofitted with interior insulation indicates that the masonry wall experiences colder temperatures than uninsulated walls, as would be expected. Monitoring also indicates that the insulated wall experiences higher moisture contents; however, this might reflect both the insulation retrofit and rain exposure at the sensor location. In addition, the moisture measurements in the walls varied in nominally identical wall sections: some sensors measured seasonally steady moisture levels, while others measured wetting responses consistent with driving rain events, followed by drying in warmer/drier conditions.

Hygrothermal simulations of the wall assemblies show good correlation to temperature measurements; however, there were significant differences in the moisture responses. These differences may be due to sensor response, driving rain exposure, or anomalies within the mass masonry wall assembly (redistribution of moisture due to voids and cracks).

The hygrothermal simulations indicate a low risk of freeze-thaw damage, based on predicted brick moisture content levels and insulation levels. The installed sensors cannot resolve moisture contents in the high range (critical degree of saturation or S_{crit}) at which freeze-thaw damage occurs. However, these instruments indicate seasonal trends of wetting and drying.

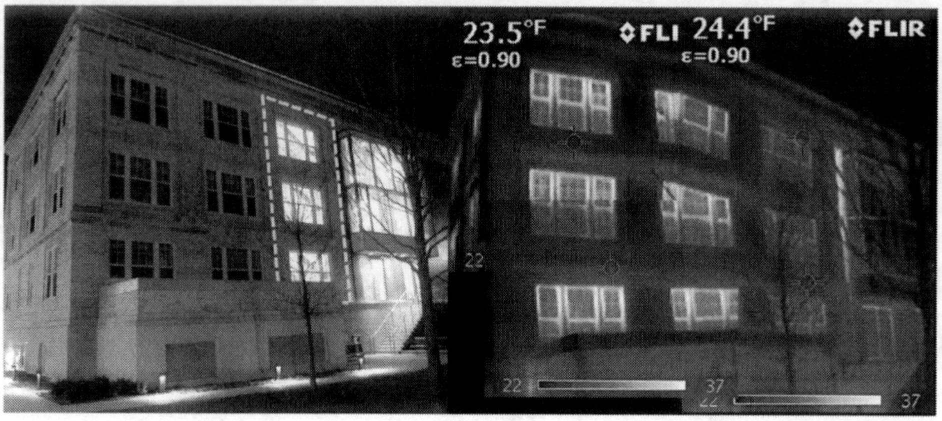

Figure 19 *Rear (north) elevation visual and infrared images (composite infrared image) with uninsulated stairwell wall highlighted.*

Although the measured moisture levels were highly variable, and did not have high correlation with modeled results, it still may be useful to install instrumentation in other mass masonry buildings retrofitted with interior insulation to gain understanding of the variables that affect the results. Direct measurement of driving rain on the instrumented wall surface may reduce the uncertainty.

In future work on insulated mass masonry buildings, the assessment of water shedding and water concentrations on the exterior face and improving the water shedding details are the key requirements before considering interior insulation. Material property testing and hygrothermal simulations are useful for assessing the risk in a more rigorous manner, based on localized climate and assembly type. Site load monitoring (driving rain, climate conditions) and building assembly monitoring are also useful tools—albeit more costly, intrusive, and time consuming—to consider in critical cases.

ACKNOWLEDGMENTS

This work was done with extensive cooperation from the Massachusetts Institute of Technology's Department of Facilities, with special thanks to Walter E. Henry, P.E., Director of Engineering, MIT Department of Facilities (retired).

The authors are also grateful to Daniel Bergey for his assistance with the monitoring installation and Christopher Schumacher for his key role in data acquisition system programming and instrument design.

REFERENCES

ASHRAE. 2009. *ASHRAE Handbook-Fundamentals*. Atlanta, GA: ASHRAE.

ASTM. 2012. *ASTM Standard C 67-12, Standard Test Methods for Sampling and Testing Brick and Structural Clay Tile.* West Conshohocken, PA: American Society for Testing and Materials.

ASTM. 2008. *ASTM Standard C 666-03, Standard Test Methods for Resistance of Concrete to Rapid Freezing and Thawing*. West Conshohocken, PA: American Society for Testing and Materials.

IBP. 2012. WUFI (Wärme und Feuchte instationär [Transient Heat and Moisture]), Version 5.2. Fraunhofer Institute for Building Physics IBP, Holzkirchen, Germany.

Fagerlund, G., 1977. The critical degree of saturation method of assessing the freeze/thaw resistance of concrete, *Journal of Structures and Materials,* 10(58):217–229.

Kosny, J., and J.E. Christian. 1995. Reducing the Uncertainties Associated with Using the ASHRAE Zone Method for R-value Calculations of Metal Frame Walls. *ASHRAE Transactions* 1995 v.101, Pt 2. Atlanta, GA: ASHRAE.

Künzel, H. and A. Holm. 2009. Moisture Control and Problem Analysis of Heritage Constructions. Congreso de Patologia y Rehabilitacao de Edificios (PATORREB) 2009, pp.85–102.

Laska, W. 1997. Understanding Collar Joints. *Magazine of Masonry Construction* October, 1997. Chicago, IL: Hanley Wood Business Media.

Livchak, A., and C. Lowell. 2012. "Don't Turn Active Beams Into Expensive Diffusers" *ASHRAE Journal* vol. 54; pp. 52–60. Atlanta, GA: ASHRAE.

Lstiburek, J.W. May 2010. Building Sciences: Thick as a Brick. *ASHRAE Journal* vol. 52; pp. 50–56. Atlanta, GA: ASHRAE.

Mensinga, P., J. Straube, and C. Schumacher. 2010. Assessing the Freeze-Thaw Resistance of Clay Brick for Interior Insulation Retrofit Projects, *Performance of the Exterior Envelopes of Whole Buildings XI.* Atlanta, GA: ASHRAE.

Morrison Hershfield. 2011. Thermal Performance of Building Envelope Details for Mid- and High-Rise Buildings (ASHRAE Research Project 1365-RP) Report No. 5085243.01. Morrison Hershfield, Vancouver, BC.

Straube, J.F., Onysko, D., and Schumacher, C.J. 2002. Methodology and Design of Field Experiments for Monitoring the Hygrothermal Performance of Wood Frame Enclosures, *Journal of Thermal Env. & Bldg. Sci.*, Vol.26, No.2—October 2002. Thousand Oaks, Calif.: SAGE Publications.

Straube, J.F., and Burnett, E.F.P., 2005. *Building Science for Building Enclosure Design*, Building Science Press: Westford, MA.

Straube, J.F., and C.J. Schumacher. 2007. Interior Insulation Retrofits of Load-Bearing Masonry Walls in Cold Climates. *Journal of Green Buildings* 2(2):42–50.

Straube, J.F., K. Ueno, and C.J. Schumacher. 2012. Measure Guideline: Internal Insulation of Masonry Walls. Straube, Washington, DC: Building America Building Technologies Program, Office of Energy Efficiency and Renewable Energy, U.S. Department of Energy. www.nrel.gov/docs/fy12osti/54163.pdf

Ueno, K., C. Schumacher, and J. Straube. 2007. Field Monitoring and Hygrothermal Modeling of Interior Basement Insulation Systems, *Performance of the Exterior Envelopes of Whole Buildings X*. Atlanta, GA: ASHRAE.

Ueno, K., and Straube J. 2008. Laboratory Calibration and Field Results of Wood Resistance Humidity Sensors, *Proceedings of BEST 1 Conference*, Minneapolis, June 10–12, 2008.

Wilkinson, J., D. DeRose, J.F. Straube, and B. Sullivan. 2009. Measuring the Impact of Interior Insulation on Solid Masonry Walls In A Cold Climate. *Proceedings of Canadian Building Science & Technology Conference*, Montreal, pp. 97–110.

Characterizing Thermal Performance of Buildings Using Dynamic Model Identification

Geert Bauwens **Staf Roels**

ABSTRACT

The energy performance of a building is, apart from installed systems and building usage, primarily determined by the thermal characteristics of the building envelope. An important distinction needs to be made between the theoretical building energy performance as calculated in the design phase and its actual as-built performance. Several studies have shown that these can differ rather significantly.

As a first step in bridging this energy performance gap, this paper investigates the possibility of characterizing the as-built thermal performance of building envelopes using dynamic data analysis of appropriate experiments. Essentially, the investigated building is modeled as a simplified thermal system. The parameters governing the workings of the thermal system are estimated by fitting it to measured thermal performance data of the building, by means of maximum likelihood estimation.

A scale model of a building serves as a case study. Identical measurement campaigns are simulated in winter and summer. Models of increasing complexity are fitted to the simulated measurement data of both campaigns. The fitted models are evaluated and compared on the basis of their log-likelihood, output residuals, and corresponding auto-correlation function and the cumulated periodogram. Using likelihood ratio tests, it is assessed whether models of larger complexity perform significantly better than smaller models. The best-performing models of different model orders are cross-validated. Hence, a model order tailored to the investigated fabric, the imposed experiment, and the weather conditions is determined.

INTRODUCTION

Background

The energy performance of a building is essentially determined by the thermal characteristics of the building fabric, the installed services, and the building usage. As the latter is not easily predicted nor controlled, the first two are decisive in achieving the envisaged building energy performance, for both newly built and renovated buildings. An important distinction needs to be made here between the theoretical building fabric thermal performance as calculated in the design phase and its actual as-built performance. Several studies have shown that these can differ rather significantly (Hens et al. 2007; Bell et al. 2010; Lowe et al. 2007). This difference, among others, can be attributed to the applied materials differing from the designated ones, poor detailing and/or workmanship issues, and physical phenomena such as thermal bridging, wind washing, and air looping. Hence, the thermal performance characterization of a building envelope, as investigated in this paper, constitutes an important step in bridging the gap between the designed and the as-built energy performance of buildings.

Buildings essentially constitute dynamic thermal systems, which can be modeled as a series of thermal nodes (e.g., an indoor air temperature node or a node in the capacitive building fabric, among others). Such nodes, often referred to as *states,* bear a certain capacity (C) to store or release heat and are interlinked through thermal resistances (R). The system dynamics are determined by its time constants (e.g., how fast the indoor air temperature responds when the outdoor air

Geert Bauwens *is a research assistant and* ***Staf Roels*** *is a professor in the Building Physics Section, Department of Civil Engineering, KU Leuven, Leuven, Belgium.*

temperature rises with a given step), which directly relate to the heat capacities associated with the thermal nodes and the thermal resistances between them. Depending on the investigated building and the way in which it is excited (e.g., by a specifically designed experiment), its dynamics will be marked by one or more time constants. The aim of this paper is to evaluate the suitability of certain simplified thermal models to grasp the building dynamics, based on the analysis of experiments performed on the building of interest. In a next step, these models can be used to characterize the physical parameters of the building (R and C, among others) or simulate the building behavior, given known boundary conditions or inputs.

The analysis methodology used in this paper is based on a data-driven identification of models with physical relevance, referred to as *grey-box modeling*. Essentially, this holds the middle between white-box modeling, where the thermal behavior is assumed to be known and deterministic, and black-box modeling, in which prior physical knowledge is discarded and the main aim is to describe the output of the thermal system (e.g., indoor air temperature) as a function of the measured input (e.g., outdoor air temperature, solar radiation, etc.).

A grey-box model mathematically describes the investigated building as a simplified thermal system, based on a physical interpretation of that building. The description consists of a system of continuous-time stochastic differential equations and discrete-time measurement equations. As such, the model comprises a certain configuration of parameters (thermal resistances, heat capacities, solar apertures, etc.) that govern the links between temperature nodes in a thermal system. The unknown parameters are then estimated on the basis of information that is embedded in the measurement data, using maximum likelihood estimation (Kristensen and Madsen 2003). The obtained parameter values serve to maximize the probability that, assuming respective models, the measured response is actually observed.

Instead of applying the methodology on a real building, a very simple, small scale model of a building, the Knauf Insulation Test Box (KI Test Box) is used as a case study. This box was designed as an object of study for the Round Robin Experiment organized in the framework of the currently ongoing IEA ECBCS Annex 58 project (Roels 2011). The KI Test Box allows for great control of the experiment conditions, facilitating the investigation of the reliability and practicality of the applied data analysis methods.

Instead of acquiring measurement data by performing real experiments, different measurement campaigns are simulated on the KI Test Box. Experiments are simulated during a winter month (January 2011) and a summer month (July 2011). Both experiments excite the indoor environment of the box using an identical predetermined heating power that is made to exhibit the characteristics of a pseudo random binary signal (PRBS). Hence, the only difference between winter and summer experiments lies in the particular weather conditions.

Grey-box models of increasing complexity are fitted to the simulated measurement data. The performance of the fitted models is compared using statistical methods. More specifically, the model log-likelihoods are compared, likelihood ratio tests (Thyregod and Madsen 2006) are used to assess whether a more complex model performs significantly better than a smaller (less complex) sub-model (Bacher and Madsen 2011), the residuals and corresponding autocorrelation functions (ACFs) are plotted to investigate how well the models are able to predict the response of the building (referred to as the *states*) based on the imposed boundary conditions (referred to as the *inputs*), and finally, the cumulated periodogram serves to test whether the residuals exhibit white noise behavior. A white noise signal represents a sequence of mutually uncorrelated identically distributed random variables with zero mean and constant variance.

Outline

The investigated building scale model and the simulated measurement campaigns are discussed in the section "Simulated Measurement Campaigns." In a following section, the candidate grey-box models are described and the statistical methods to select the most appropriate one are introduced. In the section "Identification of Simulated Box," these models are fitted to the simulated measurement data in winter and summer. A forward selection strategy reveals the most suitable model based on the performance. The best-performing models of consecutive orders are cross-validated. Conclusions are drawn in the last section.

SIMULATED MEASUREMENT CAMPAIGNS

Knauf Insulation Test Box

Instead of applying the methodology on real experiments applied on a real building, a small scale model of a building, the KI Test Box (Figure 1) of IEA ECBCS Annex 58 (Roels 2011), is investigated. The investigated test box has a cubic form, with exterior dimensions of 120 × 120 × 120 cm^3. The floor, roof, and wall components are all identical and exhibit a thickness of 12 cm, leaving an inner volume of 96 × 96 × 96 cm^3. One wall contains a window component with dimensions 60 × 60 cm^2. The fixed window frame has the same thickness as the other façade elements. A structure is provided around the box that allows it to remain free from the thermal influence of the ground. Hence, the box can be considered as floating in free air.

As such, the test box allows for rather simple, well-controlled experiments to investigate different scenarios and to determine the influence of different aspects that come to play in the thermal behavior of buildings.

Simulation of Test Box

Instead of acquiring measurement data by performing real experiments, different measurement campaigns are simulated, using Physibel software VOLTRA 7.0w (Physibel 2011).

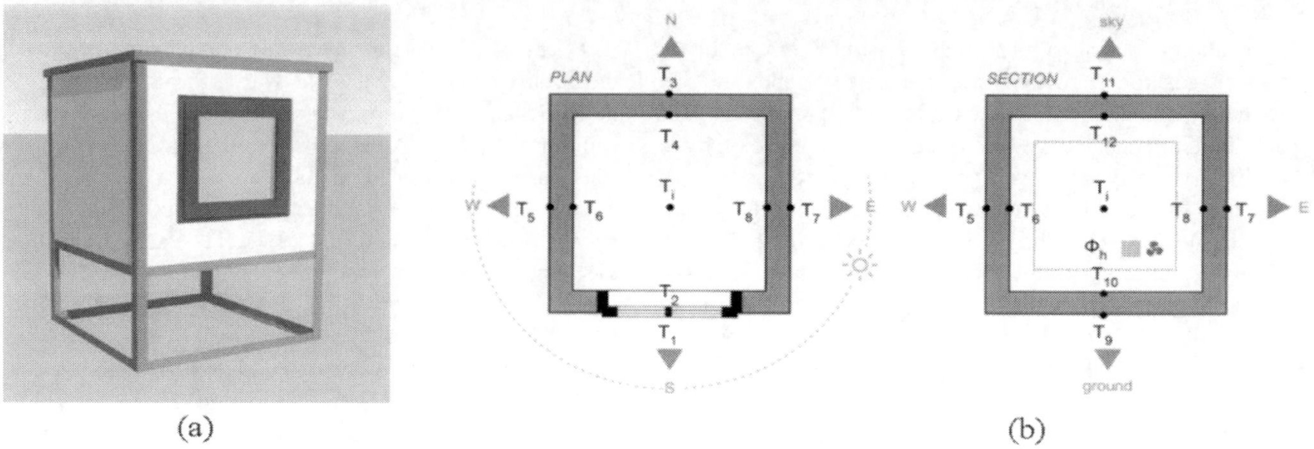

Figure 1 *Investigated test box: (a) three-dimensional picture of simulated test box and (b) plan and section of simulated test box. The numbered nodes,* T_1, T_2, ..., T_{12}, *refer to the inside (even numbers) and outside (odd numbers) surface temperatures (in the middle of the surface).* T_i *indicates the air temperature inside the box.* Φ_h *represents the heating power injected into the indoor environment of the box.*

Table 1 shows the boundary conditions assumed as a basis for the simulations. The simulation time step is 1 min. The thermal conductivity, density, and specific heat of all materials are assumed constant. Inside and outside surface heat fluxes are decoupled in a convective and radiative component. Additionally, the radiative heat exchange to the sky and environment are decoupled. The south-oriented window is modeled as a gas-filled double-glazed window. Ventilation heat losses have not been incorporated in the simulations. The box fabric is assumed perfectly airtight. Hence, this study focuses only on the characterization of the test box with respect to transmission heat losses.

Simulated Experimental Conditions

The test box is simulated to be subjected to measurement campaigns in January 2011 and July 2011. Figure 2 shows the applied heating power signal, the resulting indoor air temperatures, indoor and outdoor surface temperatures, and governing weather conditions (for Limelette, Belgium) during both campaigns. Assuming Equation 1 holds, the sky temperature T_{sky} is calculated from infrared radiation measurements on a horizontal surface $q_{ir,hor}$ (with σ as the Stefan-Boltzmann constant).

$$T_{sky} = \left(\frac{q_{ir,\,hor}}{\sigma}\right)^{1/4} \tag{1}$$

As suggested by Rabl (1988), a more sensible assessment of the solar apertures can be obtained when direct and diffuse solar radiation are separated for each of the building's window orientations. The global and diffuse solar radiation on a horizontal surface are available as measured input. In combination with accurate knowledge of the sun position (Reda and Andreas 2008) throughout the experiment, all five projections (horizontal, north, east, south, and west) of the direct solar radiation in all orientations are calculated. In what follows, $q_{sw,*}$ denotes the global solar radiation projection perpendicular to orientation * (e.g., south).

Applied Excitation Signal

During both measurement campaigns, the KI Test Box is heated with a predetermined power corresponding to a PRBS. The PRBS properties of interest are that it is uncorrelated with the weather conditions and has an autocorrelation that approximates that of white noise (Godfrey 1980).

The PRBS effectively shifts between two levels: full power and no power. The time period λ indicates the shortest time step at which the signal is maintained at a certain level. As suggested by Madsen and Schultz (1993), λ is chosen to closely relate to the smallest expected time constant of the investigated box, and the signal order n may be selected so that $N\lambda = (2^n - 1)\lambda$ closely relates to the largest time constant of the box.

The system is excited using two consecutive PRBSs. The first part is designed in such a way that the system is excited around time constants of 2 h (λ) and 30 h ($N\lambda$). This signal has periods at one level between 2 h and 8 h ($n\lambda$). The second part has a λ of 4 h and $n\lambda$ of 16 h. Hence, the system is excited around time constants of 4 h and 60 h. Both PRBSs are repeated 4 times, resulting in a total experiment duration of 15 days. The parameters governing the heating power signal generation are collected in Table 2. The resulting signal is incorporated in Figure 2a and is identical for both January and July experiments.

Perturbed Simulated Measurement Data

In order to yield the simulated measurement data as more representative, the data are perturbed by adding random noise proportional to the original data (up to ±1%). This effectively

Table 1. Assumed Indoor and Outdoor Boundary Conditions During Simulations

Outdoor Boundary Conditions		
$q_{esc} = h_{ec}(\theta_a - \theta_{es})$	q_{esc}	External surface convective heat flux
	h_{ec}	Variable convective heat transfer coefficient $= 5.85\ v^{0.52}$ (leeward side) $= 7.18\ v^{0.57}$ (windward side)
	θ_a	Ambient air temperature
	θ_{es}	External surface temperature
$q_{esr} = \xi\sigma F_e(T_{es}^4 - T_{e,rad}^4)$ $q_{esr,sky} = \xi\sigma F_{sky}(T_{es}^4 - T_{sky,rad}^4)$	q_{esr}	External surface radiative heat flux to environment
	$q_{esr,sky}$	External surface radiative heat flux to sky
	ξ	External surface emissivity = 0.9
	σ	Stefan-Boltzmann constant
	T_{es}	Absolute external surface temperature
	$T_{e,rad}$	Average radiative temperature environment
	F_e	Environment view factor
	$T_{sky,rad}$	Average radiative temperature sky
	F_{sky}	Sky view factor
$q_{solar} = \alpha E_{st}$	q_{solar}	Solar irradiation
	α	Solar absorption factor of external surface = 0.97
	E_{st}	Solar radiation on surfaces, cfr. Muneer (1989)
Indoor Boundary Conditions		
$q_{isc} = h_{ic}(\theta_i - \theta_{is})$	q_{isc}	Internal surface convective heat flux
	h_{ic}	Variable convective heat transfer coefficient $= 1.46\ (\theta_i - \theta_{is})^{1/3}$
	θ_i	Internal air temperature (assumed uniform)
	θ_{is}	Internal surface temperature
$q_{isr} = \xi\sigma(T_{is}^4 - T_{i,rad}^4)$	q_{isr}	Internal surface radiative heat flux
	ξ	Internal surface emissivity = 0.9
	σ	Stefan-Boltzmann constant
	T_{is}	Absolute averaged internal surface temperature
	$T_{i,rad}$	Average radiative temperature interior room

simulates measurement noise, frequently specified as a percentage of the measured quantity.

SIMPLIFIED MODELING OF THE BUILDING ENVELOPE THERMAL PERFORMANCE

Lumped State-Space Modeling

As mentioned previously, the investigated building is described by a series of simplified thermal systems of varying complexity, referred to as *grey-box models*. All of the candidate models considered in this paper are linear time-invariant and can be mathematically described as a system of continuous-time stochastic differential equations (Equation 2) and discrete-time measurement equations (Equation 3):

$$dT = (AT + BU)dt + d\omega(t) \tag{2}$$

$$Y_k = CT_k + DU_k + \varepsilon_k \tag{3}$$

where

T = state vector (response of the KI Test Box, e.g., indoor air temperature)

U = input vector (e.g., outdoor air temperature, solar radiation, heating power)

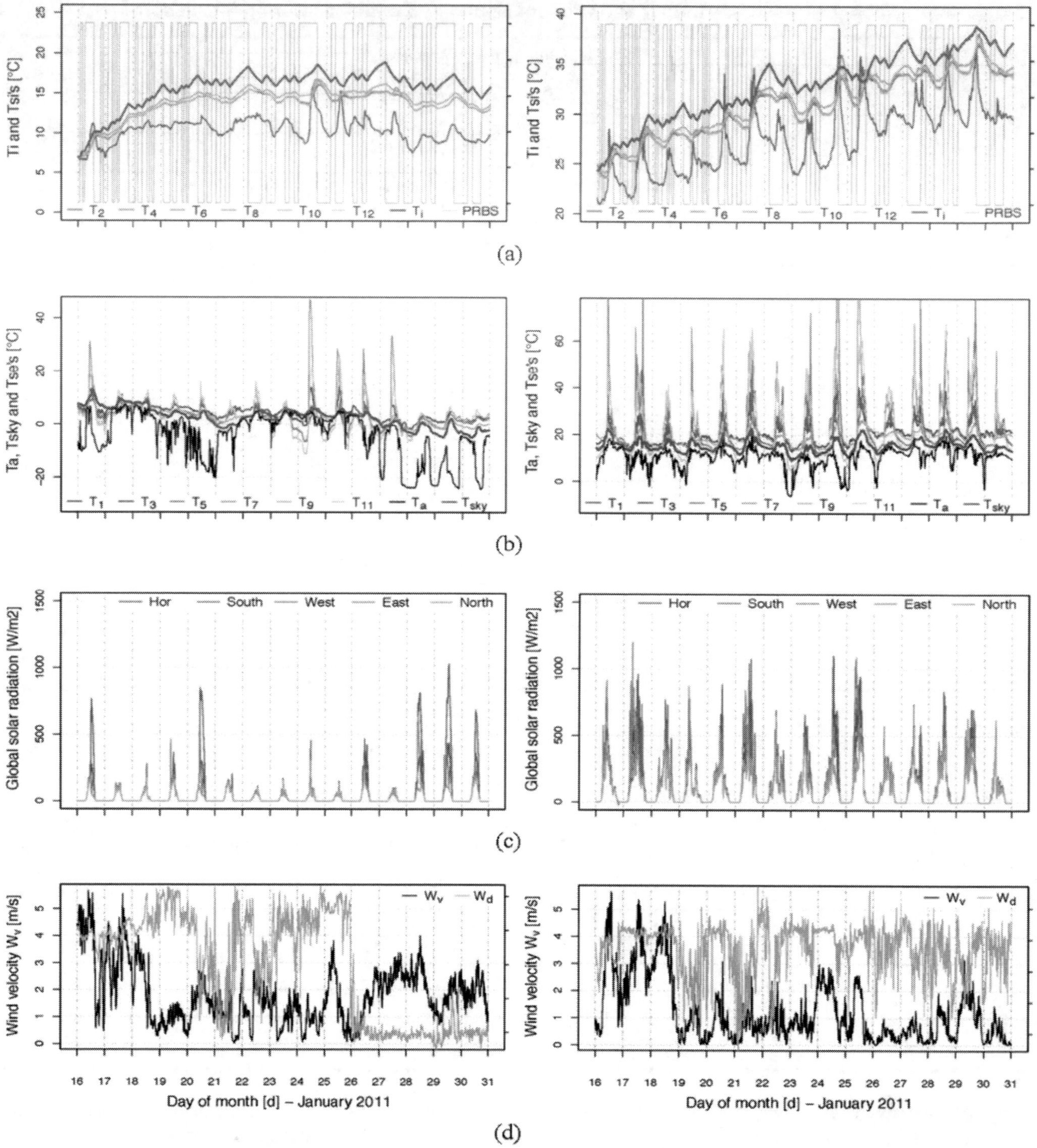

Figure 2 *Simulated measurement campaigns in January 2011 (left) and July 2011 (right): (a) simulated indoor air temperature, composite PRBS as heating power, and indoor surface temperatures; (b) measured ambient air temperature, calculated sky temperature, and simulated exterior surface temperatures; (c) measured global solar radiation on horizontal surface and calculated global solar radiation projections; and (d) measured wind speed and direction.*

A, B = parameter matrices that govern the dynamics of the states and how inputs alter the system, respectively.

Hence, A and B consist of physical parameters that serve to characterize the heat dynamics of the test box and are of particular interest here.

In this study, depending on the complexity of the considered model, the input vector comprises up to four variables: outdoor air temperature T_a, heating power Φ_h, solar radiation projection $q_{sw,*}$, and sky temperature T_{sky}.

$$U = \left[T_a \ \phi_h \ q_{sw,*} \ T_{sky}\right]^T \quad (4)$$

In all of the considered models, the indoor air temperature T_i is considered to be the sole output or response of the system. The candidate models considered in this paper, of increasing complexity, are described in more detail in the following subsections on model Ti, model TiTe, and model TiTeiTea.

Model Ti

As illustrated in Figure 3a and described mathematically in Equations 5–7, the indoor air temperature T_i [K] is considered the sole state of the system. The outdoor ambient temperature T_a (Equations 5–7), the heating power Φ_h (Equations 6–7), and the solar radiation $q_{sw,*}$ (Equation 7) are added consecutively as inputs to the system. The identified heat capacity C_i will, aside from the limited heat capacity associated with the indoor air, include the heat capacity of the building fabric that is effectively excited during the experiment. R_{ia} [K/W] represents the thermal resistance between indoor and outdoor air (Equations 5–7). The fraction of solar radiation that actually constitutes a heat input to the indoor air of the KI Test Box is denoted A_{sw} [m^2·K] (Equation 7). In this case, as illustrated in Figure 3a, it includes the fraction entering through the transparent part $A_{sw,t}$ and opaque part $A_{sw,o}$.

Model Ti.1

$$dT_i = \left(\frac{1}{C_i R_{ia}}(T_a - T_i)\right)dt + d\omega_i(t) \quad (5)$$

Model Ti.2

$$dT_i = \left(\frac{1}{C_i R_{ia}}(T_a - T_i) + \frac{\phi_h}{C_i}\right)dt + d\omega_i(t) \quad (6)$$

Model Ti.3

$$dT_i = \left(\frac{1}{C_i R_{ia}}(T_a - T_i) + \frac{\phi_h}{C_i} + \frac{A_{sw} q_{sw,*}}{C_i}\right)dt + d\omega_i(t) \quad (7)$$

Model TiTe

To more accurately take into account the dynamic response of the fabric, second-order models are suggested in Equations 8–10. Essentially, as depicted in Figure 3b, an addi-

Table 2. Parameters of PRBS Generation

#	Name	n	λ	N	$n\lambda$	$N\lambda$
4	PRBS1	4	2 h	15	8 h	30 h
4	PRBS2	4	4 h	15	12 h	60 h

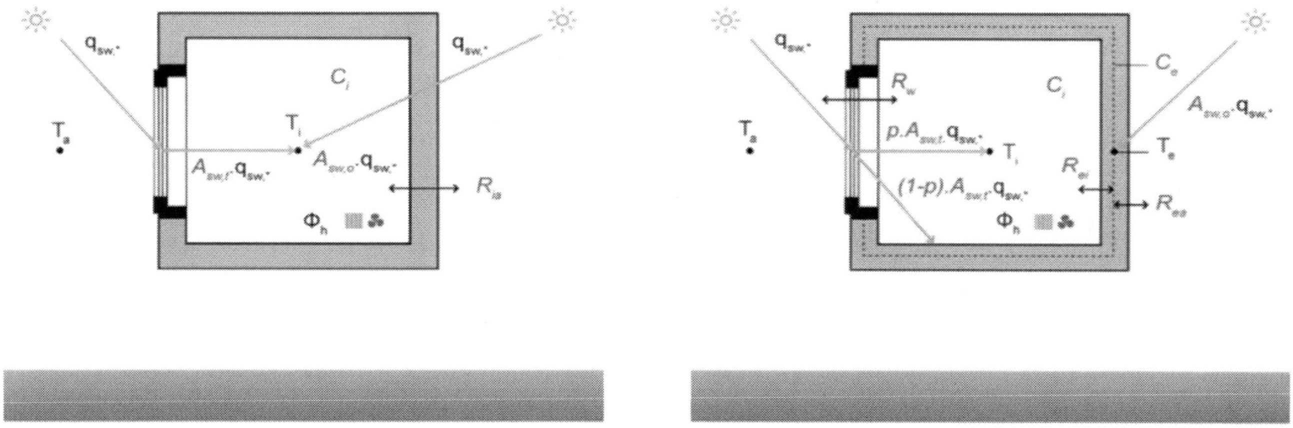

Figure 3 *Diagrams of the assumed thermal models Ti (a) and TiTe (b). System states (indicated in blue):* T_i *[K] = indoor air temperature state;* T_e *[K] = envelope temperature state. Inputs (black):* T_a *[K] = outdoor air temperature;* $q_{sw,*}$ *[W/m^2] = global solar radiation projection perpendicular to orientation *;* Φ_h *[W] = heating power. Estimated parameters (red):* C_i *and* C_e *[J/(kg·K)] = heat capacity associated with thermal nodes* T_i *and* T_e*, respectively;* R_{ia}, R_{ei}, *and* R_{ea} *[K/W] = thermal resistance between* $T_i - T_a$, $T_i - T_e$, *and* $T_e - T_a$*, respectively;* R_w *[K/W] = thermal resistance of window;* $A_{sw,t}$ *and* $A_{sw,o}$ *[m^2·K] = fraction of solar radiation entering through transparent window area and opaque parts, respectively.*

tional state T_e [K] is considered between thermal nodes T_a and T_i, comprising the larger fraction C_e of the total heat capacity $C_i + C_e$ effectively charged or discharged during the experiment. R_{ei} and R_{ea} [K/W] represent the thermal resistances between $T_i - T_e$ and $T_e - T_a$, respectively.

Model TiTe.1

$$dT_i = \left(\frac{1}{C_iR_{ei}}(T_e - T_i) + \frac{\phi_h}{C_i}\right)dt + d\omega_i(t)$$

$$dT_e = \left(\frac{1}{C_eR_{ea}}(T_a - T_e) + \frac{1}{C_eR_{ei}}(T_i - T_e)\right)dt + d\omega_e(t) \tag{8}$$

Model TiTe.2

$$dT_i = \left(\frac{1}{C_iR_{ei}}(T_e - T_i) + \frac{\phi_h}{C_i} + \frac{pA_{sw,t}q_{sw,*}}{C_i}\right)dt + d\omega_i(t)$$

$$dT_e = \left(\begin{array}{c}\frac{1}{C_eR_{ea}}(T_a - T_e) + \frac{1}{C_eR_{ei}}(T_i - T_e)\\ + \frac{((1-p)A_{sw,t})q_{sw,*} + A_{sw,o}q_{sw,*}}{C_e}\end{array}\right)dt + d\omega_e(t) \tag{9}$$

As a base case (Equation 8), model TiTe.1 does not consider solar radiation $q_{sw,*}$ as an input. Figure 3b and Equation 9 illustrate how the influence of the latter is modeled in model TiTe.2. A fraction p of the solar radiation entering the box through the south-facing window ($A_{sw,t}\, q_{sw,*}$) is considered to affect the indoor temperature state T_i directly. The other fraction $1 - p$ is expected to consequently strike the fabric components and influence the newly considered envelope state T_e. In addition, a small fraction $A_{sw,o}$ of solar radiation is expected to enter the state T_e through the opaque envelope surfaces.

To differentiate between the transparent and opaque components, the indoor air can be coupled directly with the ambient air (via R_w), without any associated heat capacity (Equation 10). This can be reasonably assumed due to the fact that the heat capacity of the window is likely to be negligible in comparison with the opaque fabric parts (Rabl 1988).

Model TiTe.3

$$dT_i = \left(\begin{array}{c}\frac{1}{C_iR_{ei}}(T_e - T_i) + \frac{1}{C_iR_w}(T_a - T_i)\\ + \frac{\phi_h}{C_i} + \frac{pA_{sw,t}q_{sw,*}}{C_i}\end{array}\right)dt + d\omega_i(t)$$

$$dT_e = \left(\begin{array}{c}\frac{1}{C_eR_{ea}}(T_a - T_e) + \frac{1}{C_eR_{ei}}(T_i - T_e)\\ + \frac{((1-p)A_{sw,t})q_{sw,*} + A_{sw,o}q_{sw,*}}{C_e}\end{array}\right)dt + d\omega_e(t) \tag{10}$$

Model TiTeiTea

In a last step, a more sophisticated third-order model is considered by assuming two states T_{ei} and T_{ea} between thermal nodes T_a and T_i, as illustrated in Figure 4. This assumption is likely to be valid in many practical cases, e.g., when considering a dwelling with insulated walls, where the heat capacity is concentrated at either side of a light insulating layer.

C_{ei} and C_{ea} [J/(kg·K)] represent the heat capacities associated with T_{ei} and T_{ea}. R_{ei}, R_e, and R_{ea} [K/W] represent the thermal resistances between the respective thermal nodes T_i, T_{ei}, T_{ea}, and T_a. The fraction $1 - p$ of the solar radiation that enters the box through the window ($A_{sw,t}\, q_{sw,*}$) is assumed to affect the indoor fabric state T_{ei}, whereas the outdoor fabric state T_{ea} is independently affected ($A_{sw,o}q_{sw,*}$) by solar radiation striking the outside surface of the fabric.

With T_{ea} representing a state at the outer layers of the fabric, it becomes sensible to additionally inject this state with radiative heat exchange to the sky.

Hence, two three-order models are considered in the identification, only differing in the state equation for T_{ea}. Model TiTeiTea.1 is described by Equations 11–12. Model TiTeiTea.2 is governed by Equation 11 and the modified T_{ea} state equation in Equation 13, including the influence of longwave radiative heat exchange to the sky, governed by the radiative heat transfer coefficient f [W/K].

Model TiTeiTea

$$dT_i = \left(\frac{1}{C_iR_{ei}}(T_{ei} - T_i) + \frac{\phi_h}{C_i} + \frac{pA_{sw,t}q_{sw,*}}{C_i}\right)dt + d\omega_i(t)$$

$$dT_{ei} = \left(\begin{array}{c}\frac{1}{C_{ei}R_e}(T_{ea} - T_{ei}) + \frac{1}{C_{ei}R_{ei}}(T_i - T_{ei})\\ + \frac{((1-p)A_{sw,t})q_{sw,*}}{C_{ei}}\end{array}\right)dt + d\omega_{ei}(t) \tag{11}$$

Model TiTeiTea.1

$$dT_{ea} = \left(\begin{array}{c}\frac{1}{C_{ea}R_{ea}}(T_a - T_{ea})\\ + \frac{1}{C_{ea}R_e}(T_{ei} - T_{ea}) + \frac{A_{sw,o}q_{sw,*}}{C_{ea}}\end{array}\right)dt + d\omega_{ea}(t) \tag{12}$$

Model TiTeiTea.2

$$dT_{ea} = \left(\begin{array}{c}\frac{1}{C_{ea}R_{ea}}(T_a - T_{ea})\\ + \frac{1}{C_{ea}R_e}(T_{ei} - T_{ea}) + \frac{A_{sw,o}q_{sw,*}}{C_{ea}}\\ + \frac{f}{C_{ea}}(T_{sky} - T_{ea})\end{array}\right)dt + d\omega_{ea}(t) \tag{13}$$

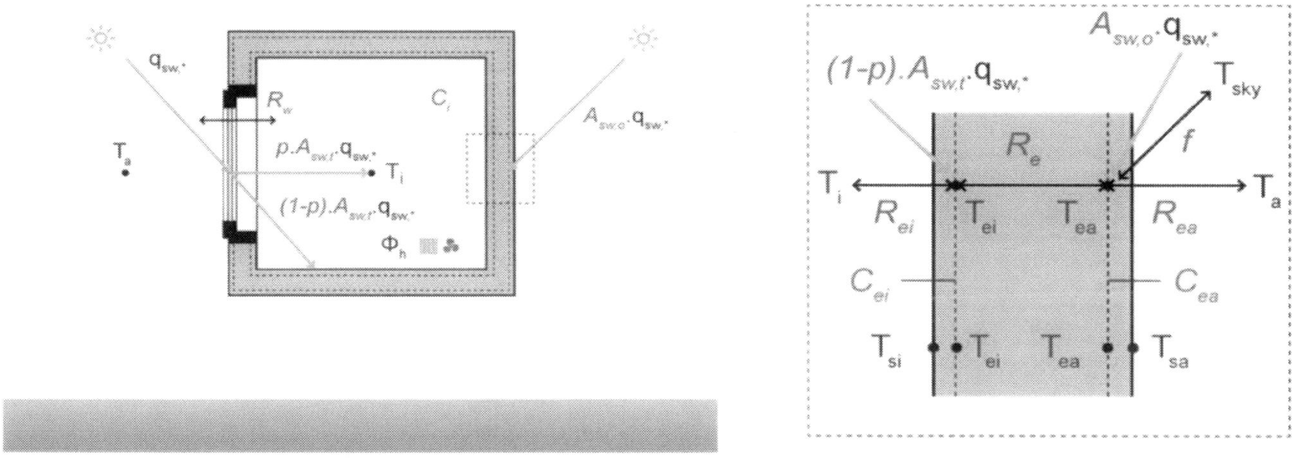

(a) (b)

Figure 4 *Diagrams of model TiTeiTea: (a) overview of the considered heat dynamics and (b) detailed view of considered temperature states and assumed heat dynamics in the opaque fabric components. System states (indicated in blue):* T_i *[K] = indoor air temperature state;* T_{ei} *[K] = inner envelope temperature state;* T_{ea} *[K] = outer envelope temperature state. Inputs (black):* T_a *[K] = outdoor air temperature;* $q_{sw,*}$ *[W/m²] = global solar radiation projection perpendicular to orientation *;* Φ_h *[W] = heating power;* T_{sky} *[K] = sky temperature. Estimated parameters (red):* C_i, C_{ei}, *and* C_{ea} *[J/(kg·K)] = heat capacity associated with thermal nodes* T_i, T_{ei}, *and* T_{ea}, *respectively;* R_{ei}, R_e, *and* R_{ea} *= thermal resistance between* $T_i - T_{ei}$, $T_{ei} - T_{ea}$, *and* $T_{ea} - T_a$, *respectively;* R_w *[K/W] = thermal resistance of window;* $A_{sw,t}$ *and* $A_{sw,o}$ *[m²·K] = fraction of solar radiation entering through transparent window area and opaque parts, respectively;* p *[dimensionless] = fraction of solar radiation affecting* T_i*;* f *[W/K] = radiative heat transfer coefficient governing radiative heat exchange with sky.*

Model Selection Using Statistical Tests

The main aim of the modeling task is to retain a comprehensible model that effectively describes the thermal behavior of the investigated building in all its frequency parts.

The log-likelihood value associated with a certain candidate model (and given a certain data set) refers to the probability that the model actually has the observed response of the real system as a result. Evidently, the observed measurement data is more likely to result from well-performing models than from models of lower performance. To evaluate the descriptiveness of a certain model (alternative model) compared to that of a sub-model (null model), the likelihood ratio test is used. In addition to considering the log-likelihood values of respective models, it takes into account the number of estimated parameters.

The residuals constitute the difference between the modeled response and the actual measured response. Hence, the residuals associated with a suitable candidate thermal model should ideally show white noise behavior. A white noise signal represents a sequence of mutually uncorrelated, identically distributed random variables with zero mean and constant variance.

In a periodogram, the signal energy is decomposed into contributions relating to different frequencies. In the case of white noise, the residual periodogram should ideally vary around some constant. A cumulated periodogram plots the accumulation of a signal's energy contributions, hence depicting in which frequency range the model possibly tends to misrepresent the true thermal system's dynamic response. As the cumulated periodogram of theoretical white noise is a given, it offers a valuable tool to identify residual whiteness.

IDENTIFICATION OF SIMULATED BOX

In this section, the full set of identified models is discussed. The system identification is performed using CTSM-R, developed at DTU Informatics (DTU 2013). As a basis for the parameter estimation, the simulated measurement data are aggregated to 15 min averages. Tables 3, 4, and 5 collect the log-likelihoods for models Ti, TiTe, and TiTeiTea, respectively. The model complexity is gradually built up, with the final aim of obtaining a model of physical relevance, which has a complexity adapted to the investigated building and the data-embedded information.

Tables 3, 4, and 5 include *p*-values resulting from likelihood ratio tests. Here, the sub-model of lower complexity (lower model order or same model order with lower number of estimated parameters) that was seen to be most suitable in a previous step is assumed to be the null model, whereas the model to be evaluated is seen as the alternative. Hence, using a forward selection strategy, the most suitable model is retained from the set of candidate models defined in the previous section of this paper.

Table 3. Log-Likelihoods, Ti Models

	Ti.1	p-value	Ti.2	p-value	Ti.3
# Parameters	4		5		6
January	2048	$< 10^{-16}$	2783	1	2783
July	1487	$< 10^{-16}$	1867	1	1866

Table 4. Log-Likelihoods, TiTe Models

	p-value	TiTe.1	p-value	TiTe.2	p-value	TiTe.3
# Parameters		9		12		13
January	$< 10^{-16}$	2965	0.475	2966	0.742	2966
July	$< 10^{-16}$	1981	$< 10^{-16}$	2049	1	2047

Table 5. Log-Likelihoods, TiTeiTea Models

	p-value	TiTeiTea.1	p-value	TiTeiTea.2
# Parameters		16		17
January	0.845	2966	7.185e-12	2999
July	1.556e-12	2080	0.024	2082

For each model order (models Ti, TiTe, TiTeiTea), the best-performing models are indicated in grey in Tables 3, 4, and 5. The associated residuals and corresponding ACFs, both for winter and summer measurement campaigns, are shown in Figure 5. Figure 6 shows the log-likelihoods, and finally, Figure 7 depicts the respective cumulated periodograms of all considered candidate models.

Model Ti

Table 3 shows the log-likelihoods of models Ti.1, Ti.2, and Ti.3 and the p-values resulting from the likelihood ratio test statistic.

It is evident that the addition of the heating power signal Φ_h as an input (models Ti.2 and Ti.3) greatly increases the log-likelihood of the model. As seen in Figure 6, this represents the most important leap forward in the case of this model performance indicator. At the same time, the cumulative periodogram (Figure 7) shows that an important step is taken toward residual whiteness.

The addition of solar radiation as an input (model Ti.3), however, does not seem to yield an important improvement to the model fit. Intuitively, this seems plausible for the winter experiment, with little associated solar radiation. It is, however, confirmed by the results relating to the summer experiment. It is important to note that the global solar radiation perpendicular to a south-oriented vertical surface is selected as the sole solar radiation projection $q_{sw,*}$, serving as an input to the current and higher-order models.

The ACF of model Ti.2 shows a certain diurnal wave pattern for both winter and, above all, summer experiments. In the latter case, the peaks in the ACF are also clearly seen to cross the 5% confidence bands. Hence, model Ti.2 inadequately describes the output of the system. The wavy pattern and the important peaks indeed reveal that the model output T_i is partly dependent on earlier values of itself, whereas the model should explain the output as a function of the inputs and other eventual system states.

Model TiTe

As evident from the log-likelihoods and the p-value test statistics collected in Table 4 and plotted in Figure 6, the model description is greatly improved when extended by a capacitive thermal node between T_a and T_i. From the ACF (Figure 5) and the cumulated periodogram (Figure 7), the residuals are shown to exhibit white noise behavior for both winter and summer measurements. Hence, the model is able to adequately describe the investigated system in all its frequency components.

From the analysis of models Ti, it was seen that solar radiation does not necessarily represent a significant heat input when modeling the indoor temperature of the investigated KI Test Box. This is confirmed here for the winter experiments (model TiTe.2 vs. model TiTe.1). On the other hand, for the July experiment, $q_{sw,*}$ does serve as an important input to accurately describe the KI Test Box response. Hence, in the latter case, the most suitable model is model TiTe.2, whereas for the winter experiment model TiTe.1 represents the best option in the class of second-order candidate models.

For both measurement campaigns, modeling the transparent window component separately (model TiTe.3, Equation 10) does not yield a significant improvement. Hence, the low-capacitive window is no longer distinguished in models TiTeiTea.

Inspecting the ACF, the diurnal wave pattern seems to persist and become slightly more apparent in the case of the January experiment, whereas in the case of the July experiment it is seen to have been completely disentangled.

Model TiTeiTea

At first sight, the log-likelihoods of models TiTe and TiTeiTea (Figure 6, Tables 4 and 5) seem to suggest that the second-order models (models TiTe) already describe the thermal system with sufficient accuracy. However, the likelihood ratio test results in Table 5 clearly show that by adding another capacitive thermal node between T_a and T_i a significant improvement of the model is attained. Especially in the case of the summer experiment (models TiTe.2 to TiTeiTea.1 and TiTeiTea.1 to TiTeiTea.2), but the improvement from models TiTe.1 to TiTeiTea.2 in the case of the winter experiment is also seen to be significant.

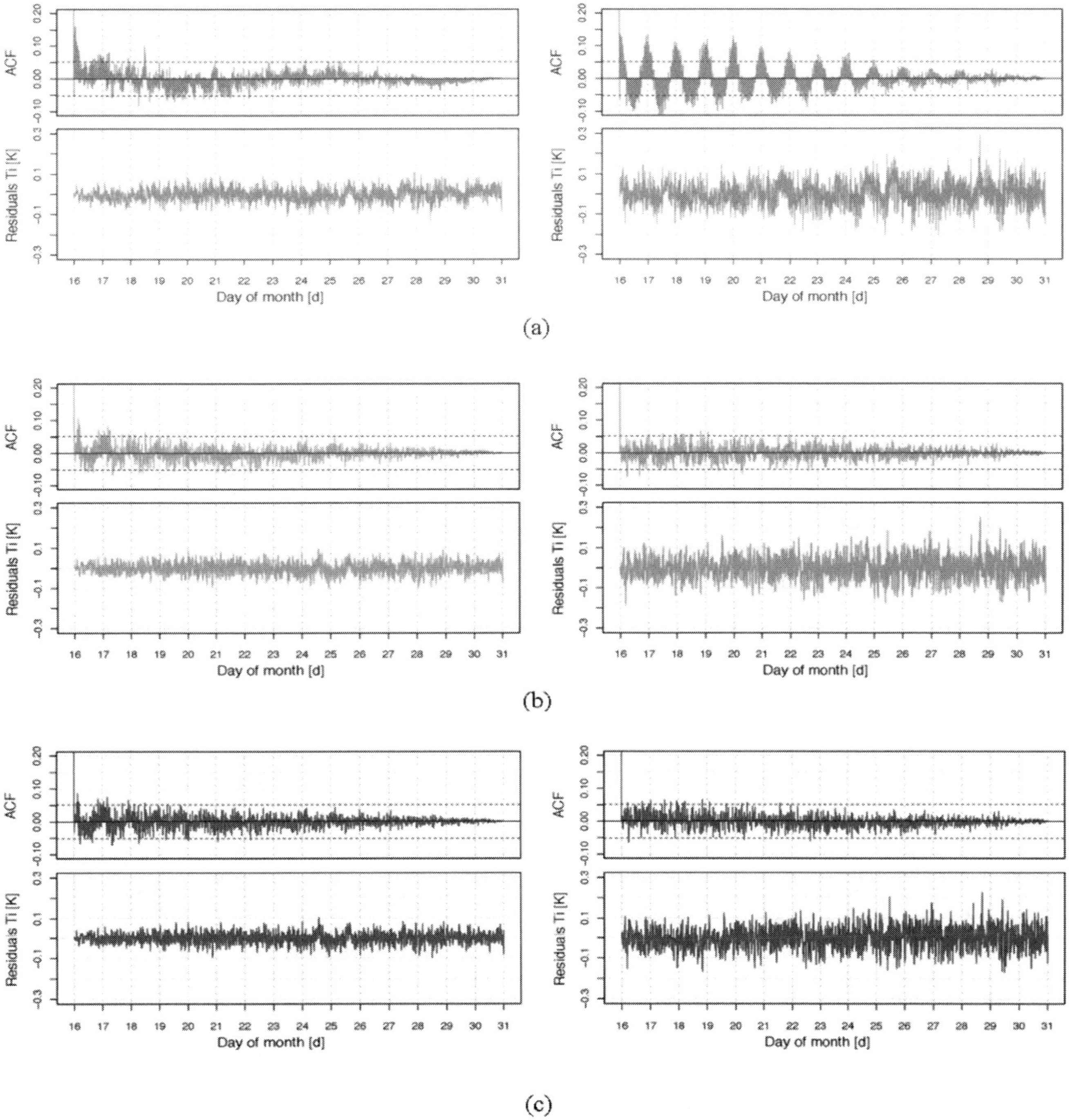

Figure 5 *Residual plots and corresponding ACFs for selected models, fitted to simulated measurement data collected in January 2011 (left) and in July 2011 (right), respectively: (a) residuals output* T_i *of model Ti.2, (b) residuals output* T_i *of model TiTe.1 for January and model TiTe.2 for July, and (c) residuals output* T_i *of model TiTeiTea.2. The stapled blue lines indicate a 5% confidence band.*

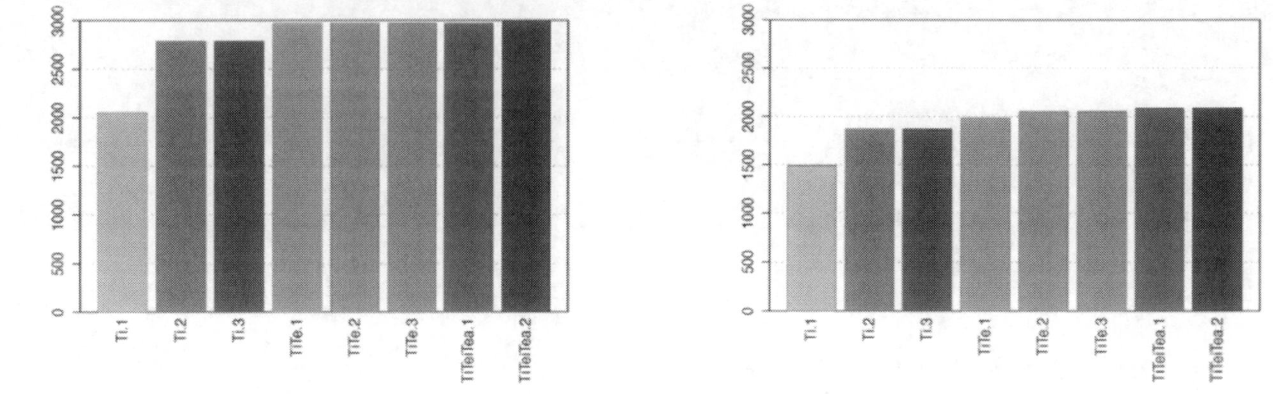

(a) (b)

Figure 6 *Log-likelihoods of the considered candidate models, fitted to measurement data: (a) January 2011 and (b) July 2011.*

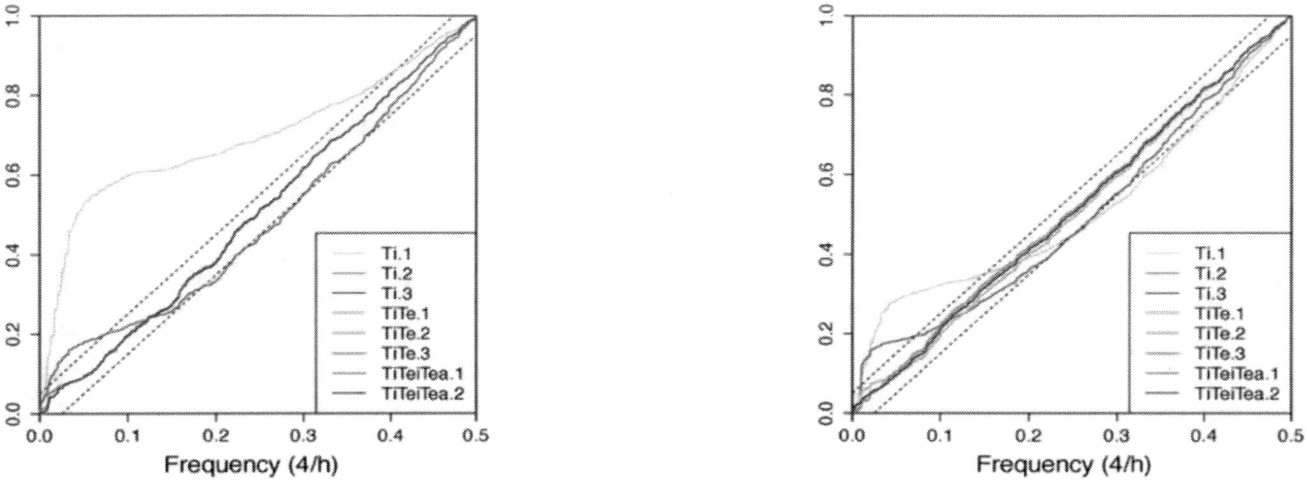

(a) (b)

Figure 7 *Cumulated periodograms of the considered candidate models, fitted to measurement data: (a) January 2011 and (b) July 2011. Residual whiteness is apparent when the plotted line falls within the blue stapled bounds.*

In the case of the experiment in January, the model improvement TiTe.1 to TiTeiTea.1 is proven not to be significant. However, skipping a step and going from TiTe.1 to TiTeiTea.2 does prove to be a step forward in the modeling task (*p*-value: 7.185e-12). As expected from the outset, looking at the sky temperature time series in Figure 2b, the long-wave radiative heat exchange with the sky (the additional term in Equation 13) is seen to have an important influence on the KI Test Box behavior. This is confirmed by the log-likelihood of model TiTeiTea.2 and the likelihood ratio tests with TiTe.1 (January) and TiTe.2 (July) as null models and TiTeiTea.2 (January and July) as the alternative model (Table 5).

Cross-Validation of Identified Models

An important distinction needs to be made between the mere characterization of the building dynamics, the characterization of the thermal performance (e.g., *R* and *C*) of buildings, and the model-based prediction of its response to certain inputs or imposed boundary conditions. In this section, the predictive performance of the identified models is tested by simulating the KI Test Box response in another period than the period over which the model was trained—i.e., the models are cross-validated.

The root mean square errors plotted in Figures 8e and 8f confirm the third-order models perform best. Additionally, it is seen that the higher-order models identified on the basis of the summer experiment outperform the ones identified on the basis of the winter experiment when it comes to prediction. This is contrary to what might be expected from the log-likelihoods discussed previously. As illustrated by Figures 8b and 8d, the models trained using the January data set, although grasping the KI Test Box dynamics, have not learned to prop-

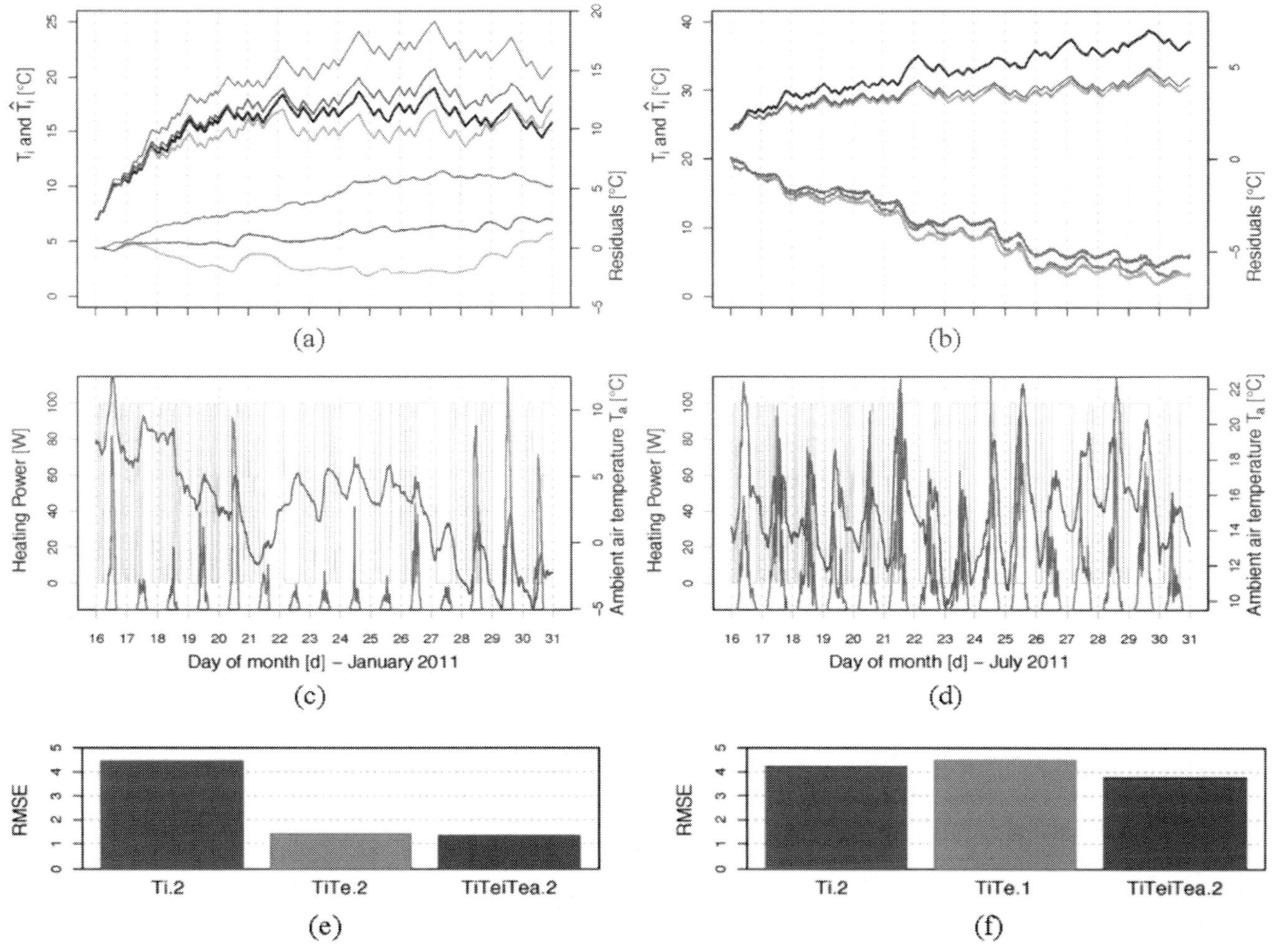

Figure 8 *(a) KI Test Box response simulated in January, using models Ti.2 (red), TiTe.2 (green), and TiTeiTea.2 (blue) identified on the basis of the July training data set; direct comparison with the actual measured response during the January experiment, complemented with residual plots. (b) KI Test Box response simulated in July, using models Ti.2 (red), TiTe.1 (green), and TiTeiTea.2 (blue) identified on the basis of the January training data set; direct comparison with the actual measured response during the July experiment, complemented with residual plots. (c) Ambient air temperature* T_a *[K], PRBS as heating power* Φ_h *[W], and global solar radiation on a vertical south-oriented wall* $q_{sw,south}$ *[W/m²] during January experiment. (d) Ambient air temperature* T_a *[K], PRBS as heating power* Φ_h *[W], and global solar radiation on a vertical south-oriented wall* $q_{sw,south}$ *[W/m²] during July experiment. (e) and (f) Root mean square errors.*

erly cope with the heat input to the indoor environment due to solar radiation. This is likely to be attributed to both the limited window surface of the KI Test Box and the fact that the solar irradiance is rather low during the winter experiment. The actual solar gain is seen to be underestimated, which is revealed by the steady overall increase of the absolute residual values with simulation time, as well as the modest reductions followed by steep rises in the absolute residual values during each day with considerable solar radiation.

CONCLUSIONS

The aim of this research was to investigate the ability to identify the thermal dynamics of buildings using grey-box modeling. In this framework, simplified thermal dynamic models were defined based on prior physical knowledge. These models were consecutively fitted to *simulated* data sets acquired from *simulated* experiments performed on a *simulated* small scale model of a building, the KI Test Box of IEA ECBCS Annex 58 (Roels 2011). The performance of the fitted models was compared using well-known statistical tests. The identified thermal dynamic models can be used to characterize the actual thermal performance of buildings and to simulate its response to boundary conditions to which it might be subdued.

In general, it is seen to be crucial to inspect both the log-likelihoods and the likelihood ratio test when judging whether an alternative (more complex) model actually performs better.

The main aim of the modeling task is to obtain a model with associated residuals exhibiting white noise characteristics. Residual whiteness becomes apparent in the auto-correlation function (ACF) plot and the cumulated periodograms.

The order of the candidate models found to be most suitable depends on the perceivable or apparent time constants of the investigated building. In other words, the time constants associated with the model reflect the accessible heat capacity that is effectively excited during the experiment. Whereas the accessibility of the thermal mass is solely determined by the composition of the building fabric, the effectively charged/discharged part additionally depends on the imposed experiment (T_i, Φ_h) and the outside weather conditions (T_a and $q_{sw,*}$, among others).

The simulated measurement campaigns in winter are seen to systematically yield better model fits. The standard deviation of the residuals is considerably smaller than its counterparts associated with models fit to measurement data in summer. Model fits for both winter and summer experiments show residual whiteness as soon as two or more states are considered (models TiTe and TiTeiTea). In the case of the second-order TiTe models, it is seen that the most suitable model for the winter and summer experiments differs, even though the imposed experiments are identical. This clearly illustrates that, aside from the investigated building and the imposed experiment (heating power signal Φ_h, e.g., PRBS), the most suitable model depends on the particular weather conditions (T_a, $q_{sw,*}$, etc.) taking place during the experiment. The latter effectively excite the building in their respective prevalent frequency components. In order to obtain reliable and separated parameter estimates, however, a heating power signal that is uncorrelated with the outside weather conditions, persistently exciting the building, is required.

Although the summer experiment yields model fits with lower log-likelihoods than the winter experiment, it is seen to deliver models that are better able to simulate the KI Test Box response in other circumstances than over which they were fitted. Cross-validation clearly shows that the solar gains are not properly modeled when models are trained on measurement data in January.

Influential weather conditions need to be accounted for in a suitable way. Taking into account solar radiation, however significant in the higher-order models, is seen to only modestly improve the quality of the model fit. The limited surface area of the KI Test Box window partly serves to explain this. Additionally, as illustrated by the results for the winter experiment, the intensity of the solar radiation input signal will partly determine its significance.

All solar radiation projections could potentially serve as input to the models. As opposed to model Ti.3, the choice of the solar radiation projection that serves as input to models TiTe.2, TiTe.3, TiTeiTea.1, and TiTeiTea.2 becomes less straightforward. Future research should investigate the possibility to sensibly relate different solar radiation projections to different thermal nodes considered in the system, or how to take into account different projections depending on the time of day.

The KI Test Box was simulated to be perfectly airtight. Hence, the wind speed and direction are not influential. Their influence on the convective heat transfer coefficient hence is negligible, given the fact that the KI Test Box is relatively well insulated.

It is important to note that all of the models considered in this paper are linear time-invariant models, whereas the thermal behaviour of the real system is essentially nonlinear, due to such factors as radiative heat exchange with sky, dependence of thermal resistance of building fabric on temperature, influence of solar radiation, and dependence of convective heat transfer coefficients on wind speed and direction. Nonetheless, it is seen that good fits are attained and reasonable characterizations are within reach. This greatly depends, however, on the thermal performance of the investigated building, the weather conditions, the imposed experiment, and its purpose.

NOMENCLATURE

λ	=	shortest time period at which the heating power (PRBS) shifts from one level to the other, h
R	=	thermal resistance, K/W
C	=	heat capacity, J/(kg·K)
A	=	solar aperture, $m^2 \cdot K$
T	=	absolute temperature, K
θ	=	temperature, °C
Φ	=	power, W
q	=	heat flux, W/m^2
p	=	fraction of solar radiation entering T_i, dimensionless
f	=	radiative heat transfer coefficient (radiative heat exchange with sky), W/K
α	=	solar absorption factor, dimensionless
ω	=	process noise, K
ε	=	measurement noise, K
ξ	=	emissivity, dimensionless
τ	=	time constant, s

Subscripts

i	=	indoor environment
a	=	ambient/outdoor environment
e	=	envelope
ei	=	inside of the envelope
ea	=	outside of the envelope
g	=	ground
h	=	heating
sw	=	shortwave
ir	=	infrared
c	=	convective

t	=	transparent
o	=	opaque
sky	=	sky
w	=	window
*	=	projection perpendicular to horizontal, north-, east-, south-, and west-oriented surfaces
k	=	discrete time step

REFERENCES

Bacher, P., and H. Madsen. 2011. Identifying suitable models for the heat dynamics of buildings. *Energy and Buildings* 43:1511–22.

Bell, M., J. Wingfield, and D. Miles-Shenton. 2010. Low carbon housing. Lessons from Elm Tree Mews. Joseph Rowntree Foundation.

DTU. 2013. CTSM-R. Software used for parameter estimation: Implementation of the Continuous Time Stochastic Modeling (CTSM) software in R. DTU Informatik, Denmark. www.ctsm.info

Godfrey, K.R. 1980. Correlation methods. *Automatica* 16(5): 527–34.

Hens, H., A. Janssens, W. Depraetere, J. Carmeliet, and L. Lecompte. 2007. Brick cavity walls: A performance analysis based on measurements and simulations. *Journal of Building Physics* 31(2):95–124.

Kristensen, N.R., and H. Madsen. 2003. *Continuous Time Stochastic Modelling. CTSM 2.3—Mathematics Guide.* Denmark: DTU Informatik.

Lowe, R.J., et al. 2007. Evidence for heat losses via party wall cavities in masonry construction. *Building Service Engineers Research and Technology* 28(2):161–81.

Madsen, H., and J.M. Schultz. 1993. Short time determination of the heat dynamics of buildings. Technical report, DTU Informatik, Denmark.

Muneer, T. 1989. Algorithms for estimating hourly solar irradiation on slopes. *Building Serv. Eng. Res. Technol.* 10(2):81–83

Rabl, A. 1988. Parameter estimation in buildings: Methods for dynamic analysis of measured energy use. *Journal of Solar Energy Engineering—Transactions of the ASME* 110(1):52–66.

Reda, I., and A. Andreas. 2008. Solar Position Algorithm for Solar Radiation Applications, NREL Report No. TP-560-34302; 2003, revised January 2008. Golden, CO: National Renewable Energy Laboratory.

Roels, S. 2011. Reliable building energy performance characterisation based on full scale dynamic measurements. Annex text, November 2011. www.kuleuven.be/bwf/projects/annex58/index.htm

Thyregod, P., and H. Madsen. 2006. *General and Generalized Linear Models*. Lyngby, Denmark: IMM, DTU Informatik.

Physibel. 2011. VOLTRA version 7.0w. Software used for simulations. Physibel, Maldegem, Belgium.

Evaluation of Experimental Parameters in the Accelerated Aging of Closed-Cell Foam Insulation: Results after Five Years of Full-Thickness Aging

Therese Stovall, PE
Member ASHRAE

Michael Vanderlan

Jerald Atchley

ABSTRACT

The thermal conductivity of many closed-cell foam insulation products changes over time as production gases diffuse out of the cell matrix and atmospheric gases diffuse into the cells. Thin slicing has been shown to be an effective means of accelerating this process in such a way as to produce meaningful results. Efforts to produce a more prescriptive version of the ASTM C1303 standard test method led to the ruggedness test (a "test the test" process) described here. This test program included the aging of full-size insulation specimens for time periods of five years for direct comparison to the predicted results. Experimental parameters under investigation include: slice thickness, slice origin (at the surface or from the core of the slab), thin slice stack composition, product facings, original product thickness, product density, and product type. This paper provides a detailed evaluation of the impact of the test parameters on the accuracy of the five-year thermal conductivity prediction.

BACKGROUND

This study took place over a six-year time span and multiple papers have been published with intermediate results. The test protocol is described by Stovall and Bogdan (2007) and Stovall et al. (2012). The early material characterization results for the polyisocyanurate (PIR) and extruded polystyrene (XPS) are summarized by Stovall (2009) and Stovall et al. (2012).

Heat transfer through closed-cell foam insulation occurs via radiation, solid conduction, and gaseous conduction (Scheutz and Glicksman 1983). The radiation and solid conduction change little over time, but the gaseous conduction is determined by the composition of the gas mixture within the foam cells. This project focuses on the changes in thermal conductance due to the changes that occur over time as atmospheric gases diffuse into the cells and the blowing agent gas(es) diffuse out into the surrounding environment. These changes are governed by the diffusion coefficient for each gas for the cell wall polymer, the foam thickness, and time (Isberg 1988). For insulation sheets where the thickness is small relative to the width and length, this diffusion process has been shown to follow Fick's Law for one dimensional diffusion (Kumaran and Bomberg 1990; Bomberg 1990; Edgecombe 1989; Ball et al. 1978; Mullenkamp and Johnson 1983; Booth 1980; McElroy et al. 1991; Hoogendoorn 1994).

Appliance manufacturers and builders have expressed interest in the long-term thermal resistance values of existing foam, as opposed to the thermal resistance values of new foam, so accelerated aging methods were developed to test long-term thermal resistance values. An ASTM task group was formed and the original version of the ASTM C1303 test method was published in 1995 (Graves et al.1995; ASTM 1995). In 2000, a prescriptive test method based on ASTM C1303, but expanded to include permeably-faced products, was published in Canada (CAN/ULC-S770) and was required for foam insulation products sold in Canada (ULC 2000). An extensive interlaboratory comparison showed that the original CAN/ULC-S770 procedure produced biased results and that the magnitude of the bias varied according to the material tested and the slice thickness (Drouin 2009; Stovall et al. 2002). During the multiyear effort to improve C1303 (ma-

***Therese Stovall**, retired, was a distinguished research scientist and **Jerald Atchley** is a senior technician at Oak Ridge National Laboratory, Oak Ridge, TN. **Michael Vanderlan** is an operations research analyst for Link Analytics, Knoxville, TN.*

Published by ASHRAE 2013

jor updates were issued in 2000, 2007, and 2012), questions were raised by members of the ASTM C1303 task group at working group meetings. Many of these questions regarded the applicability of accelerated aged performance values derived from measurements on 50 mm (2 in.) products to products of other thicknesses. Within C1303 and in this paper, this is called an *alternate thickness* or *alternate product thickness* prediction. (Cell morphology differences have been postulated to explain differences in aging behavior for products of differing thicknesses.) There were also questions regarding the influence of variations within each material class, such as density, manufacturing process, and facer material. Also, since real foam insulation materials are not perfectly homogenous, slices taken from different locations in the cross section may exhibit different thermal conductivity and may age at different rates. Another question was whether or not adhered or intrinsic facings at the outer surfaces were more or less resistant to gas diffusion.

Two previous ruggedness tests had addressed questions regarding the cutting methods used to prepare the foam thin slices and the thickness of the destroyed surface layer that results from the cutting process (Fabian et al. 1997; Stovall 2007). The ruggedness test reported here was organized to answer questions regarding product differences (material, class or density, and thickness) and stack composition that were considered to be the most important for the 2007 prescriptive version of ASTM C1303. Another variable, slice thickness (in particular, the differences for slice thicknesses of 8, 10, and 12 mm [0.31, 0.39, and 0.47 in.]), was examined in a more limited fashion and was previously considered in the evaluation of CAN/ULC-S770 (Drouin 2009). This ruggedness test explored the applicability of accelerated aged performance values derived from measurements on one product thickness to products of other thicknesses for the same material and material class, i.e., alternate thickness or alternate product thickness predictions. In addition to the evaluation of experimental factor effects, this test was designed to provide an estimate of the test method accuracy. For this purpose, full-thickness uncut insulation specimens were aged within a conditioned laboratory space for five years so that the measured thermal conductivity at that point could be compared to the prediction produced via the accelerated test method. The full-thickness specimens were cut to fit within the test apparatus at the end of the aging period.

The thin slices were arranged in stacks for thermal conductivity measurements. This is done to ensure the thermal conductivity measurements are outside the regime of the "thickness effect," that is, to ensure the test specimen is optically thick (Glicksman 1994). The stacks differ in composition according to the origin of the slices, which may be taken from the core of the product or at the surface. For the PIR specimens, which were faced with permeable materials, the facers were included in the surface slices, but

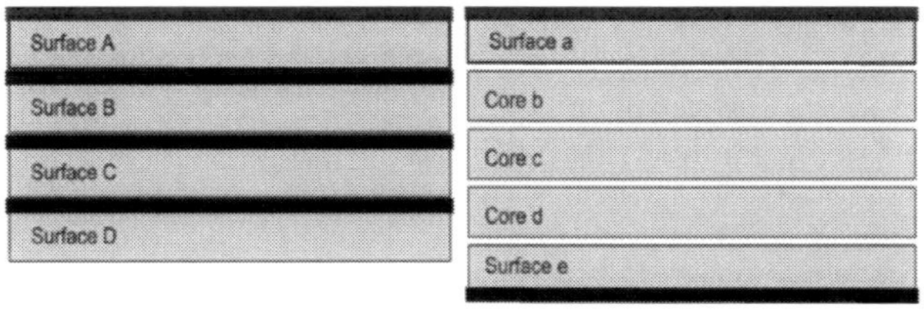

Figure 1 *Diagram of stack design: (left) Surface stack and (right) Profile Mixed stack.*

the slice thickness used for all calculations did not include the thickness of the facers. Five stacks were included in the ruggedness test:

- The "Surface" stack is a simple stack of surface slices, all with surface side up, as shown on the left in Figure 1. The Surface stack includes the effect of any diffusion resistance provided by a densified surface or a facer, although each surface slice also includes an opposite cut surface without this extra diffusion resistance.
- The "Core" stack consists of only slices taken from the core of the product, that is, excluding the surfaces. The Core stack is composed of that portion of the foam that is relatively undisturbed by any diffusion that occurs between the time the foam is manufactured and when the thin slices are prepared.
- The "Profile Mixed" stack, shown on the right in Figure 1, is a replication of an actual cross-product section with two surface sides out and enough cores to make the stack height equal to the original product thickness, minus the thickness of the material destroyed during the slicing process.
- The "Four-Slice Mixed" stack consists of two outward-facing surface slices with two core slices in the center.
- The "Math" stack is a mathematical derivation, using the standard series resistance expression to weight the measured values from the Core and Surface stacks to represent the overall product structure, as described by Stovall et al. (2012).

During the course of this project, more than 1000 ASTM C518 thermal conductivity measurements were made (ASTM 2010). For the data used in the five-year comparison, there were 214 thermal conductivity measurements for slice stacks and 100 for full-thickness specimens. The thermal conductivity of all 20 full-thickness products was measured. All the thermal conductivity measurements were made on one of four heat flux meter apparatus. For most materials, there were at least three full thickness measurements made after five years of aging. For all full thickness materials, at least one measurement was made using a 600 × 600 mm (24 × 24 in.) specimen in a Fox 605 heat flux meter apparatus with a 20 × 20 mm (8 × 8 in.) metering area. For 25 and 50 mm (1 and 2 in.) thick specimens, that test specimen was then cut into four equal pieces so that multiple measurements could also be made in a Fox 304 heat flux meter apparatus with a 10 × 10 mm (4 × 4 in.) metering area. That same Fox 304 machine was used for most of the thin slice measurements; a Fox 300 machine was used for the remaining thin slice measurements.

ANALYSIS

In order to produce a robust statistical analysis, every individual prediction was compared to every corresponding full-thickness measurement. That is, the value predicted for five years from each of the five thin-slice stacks was compared to each of the replicate full-thickness measurements individually. In the cases where replicate measurements were available for the thin-slice stacks, each predicted value was individually compared to the full-thickness five-year measurement. The comparisons include predictions made using original specimens from both the same product thickness and from other product thicknesses. The resulting data set therefore reflects unbalanced multiple effects. The final data set contains a family of multiple predicted values for each measured full-thickness value, resulting in a total of ~1300 observations. The error, or difference between the predicted value and the measured value as defined in Equation 1, was calculated for each observation.

$$\text{Error (\%)} = \frac{\left(k_{Predicted} - k_{Full\ Thickness}\right)}{k_{Full\ Thickness}} \times 100\% \quad (1)$$

where k is the thermal conductivity, W/m·K.

Using this definition:

- A positive error indicates the predicted thermal conductivity was too high—and therefore the predicted thermal resistance was too low.
- A negative error indicates the predicted thermal conductivity was too low—and therefore the predicted thermal resistance was too high.

The data set also included corresponding values for multiple possible effect classes. Some of these classes are associated with possible variations in the test methodology, such as stack type, whether or not predictions can be made using products of alternate thickness, or slice thickness. Other classes are useful in examining whether the test method accuracy varies between different applications, that is, whether it works equally well for different material varieties or products from different manufacturers. A few variables were considered as continuous numerical values rather than as classes of distinct values. A partial listing of the classes included in the data set are summarized in Table 1. Screening classes for homogeneity and alternate product thickness were examined as described by Stovall et al. (2012).

Statistical analysis was used to compare the response, or dependent variable, to the independent model variables. General linear models were used to identify which experimental factors were most important. The population marginal means (PMM; also called *LSMeans* in SAS [2007]) were used to determine the influence of each parameter in the model (SAS 2007). The PMM is a function of both the structure of the parametric model and the data (Searle et al. 1980). For example, the PMM for the effect of a Core stack will be different depending on which other effects were included in the model. If the model looked at the effects of stack type (Surface, Core, Mixed, Profile, and Math) and product variation (Class1 and Class 2), the PMM for Core stack effect would be the average of independent averages for Core stacks in Class 1 and Core stacks in Class 2. If the data set were perfectly balanced, there would be an equal number of test data points in each of those averages and the overall mean for Core stacks would equal the PMM for Core stacks. However, in an unbalanced data set, there will be more data points in one class than in the other. In that case, the PMM serves to balance an unbalanced data set. Note that the PMM can only be calculated if there is at least one observation for each possible combination of the effects in the model.

Table 1. Experimental Parameter Classes Used in Analysis

Classes	Possible Values
	Methodology Classes
Product thickness	25, 50, 100 mm (1, 2, 3 in.); note 75 mm (3 in.) included in 100 mm (4 in.) class
Original product thickness	Applies when alternates product thickness is used to make the prediction
Stack type	Surface, Core, Four-Slice Mixed, Profile Mixed, and Math
Machine comparison	Same or Different, whether the heat flux meter apparatus used to measure the full-thickness specimen was the same as the one used to measure the thin-slice stack that produced the predicted value
Thickness comparison	Same or Alternate, whether the original specimen used to produce the prediction was of the same thickness as the full-thickness specimen
Slice thickness	8, 10, or 12 mm (0.31, 0.39, or 0.47 in.)
	Application Classes
Variety	Class 1 or Class 2 for PIR, Low or High density for XPS
Manufacturer	Four included, two each for PIR and XPS

The analysis also produces the F-statistics, the explained variance divided by the unexplained variance as shown in Equation 2, which is useful in determining which experimental parameters are most important within the test protocol. The Type III ANOVA used here is unbiased, that is, the results are independent of the order in which the class variables are considered (Gill 2001; Pasta 2011). Also, the Bonferroni t-test form was used to adjust for the large number of parameters in the model (Miller 1981). This form of t-test provides a more rigorous test of significance for the influence of each parameter in the model.

$$F = \frac{\text{explained variance}}{\text{unexplained variance}} \tag{2}$$

RESULTS

The raw data are summarized in Figures 2 to 4. Figure 2 shows that the bulk of the predictions are within ±5% of the full thickness values, but there are also some outliers. In Figure 3, the same data set is broken down further to show the impact of stack type and whether predictions are based upon slices taken from the same product thickness or an alternate product thickness. Looking at both of these figures, it is apparent that some sort of screening is needed for the alternate thickness predictions. Further, it appears that Surface stacks would not be a good choice for PIR and that Core stacks would not be the best choice for XPS.

The test plan called for most of the test slices to be 10 mm (0.39 in.) thick, with some at 8 mm (0.31 in.) and others at 12 mm (0.47 in.) to evaluate the impact of that test parameter. Previous unpublished work done during the development of CAN/ULC S770 has shown that it is difficult to maintain the necessary slice flatness, that is, the uniformity of the slice thickness, for slice thicknesses smaller than 8 mm (0.31 in.) (Drouin 2012). Figure 4 shows that the actual execution represents a more continuous spectrum of slice thickness. Therefore, statistical models were explored to consider slice thickness both as a continuous numerical value and as a class variable with three levels.

The PIR and XPS were treated separately for the general linear model analyses, as were the alternate product thickness predictions and the same product thickness predictions. Before conducting the linear model analysis, the distributions of the results for these four groupings of interest were evaluated as shown in Figure 2. The distributions are sufficiently close to a normal distribution to justify the linear model approach. The final models had correlation coefficients of 0.84 and 0.86 for alternate thickness comparisons and 0.55 and 0.63 for same thickness comparisons, for PIR and XPS, respectively.

The F-statistics (see Equation 2) indicate which experimental parameters (see Table 1) are more closely correlated with the experimental error. The original product thickness, i.e., the thickness of the product used to produce the thin slices, appeared in every statistically significant correlation. The manufacturer variable showed little to no effect on its own, but for alternate product thickness comparisons the combination of manufacturer and product thickness was important. For PIR, the product variety (Class 1 or Class 2) was not correlated with the experimental error. However, the XPS product variety (standard or high density) was influential for same thickness predictions. Some variables were confounded by their close association with other variables. For example, the test duration for all 75–100 mm (3–4 in.) products was less than 40 days, so those two factors were ineffective when placed in the same model.

Two experimental factors were of special interest: stack type and slice thickness. Stack type plays a critical role in test method accuracy for both of these two materials. Slice thicknesses between 8 and 12 mm (0.31 and 0.47 in.) were statis-

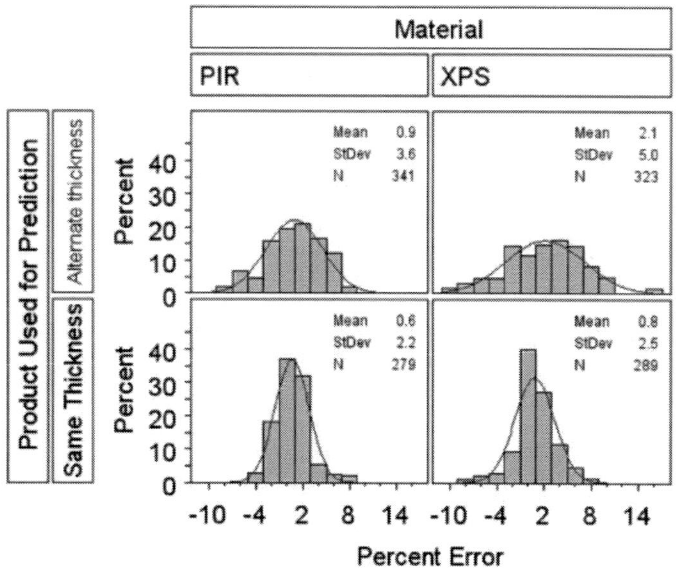

Figure 2 *Unscreened data set, range of values for PIR and XPS for alternate thickness comparisons and same thickness comparisons.*

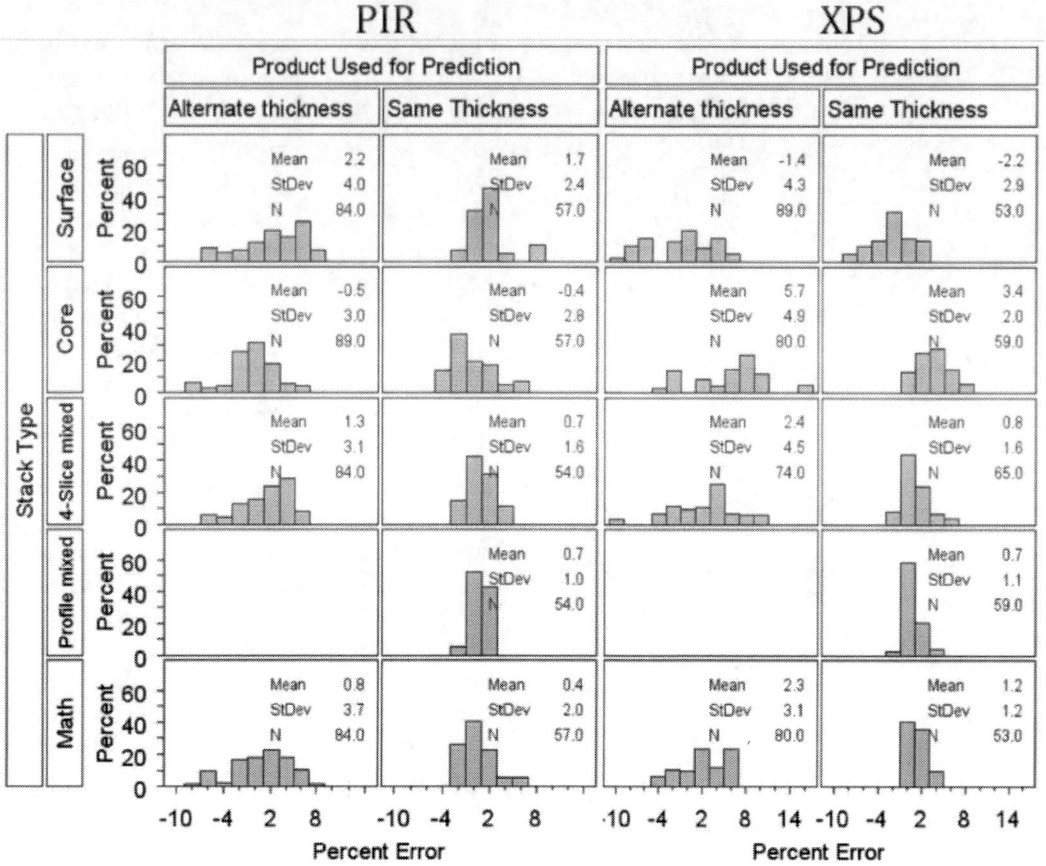

Figure 3 *Unscreened data set, range of values for PIR and XPS for both comparisons (same and alternate thickness) for five stack types.*

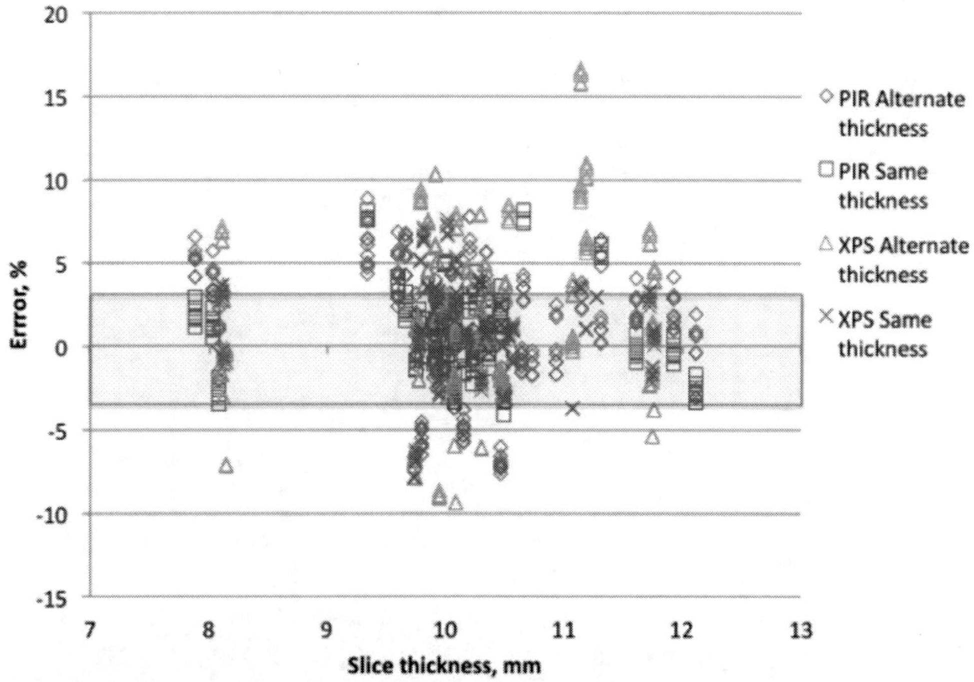

Figure 4 *Unscreened data set, range of values for both materials and both comparisons as a function of average slice thickness within the test stack.*

tically significant only for PIR predictions made from the same product thickness.

PIR, Alternate and Same Thickness Predictions

The PMM values for PIR, shown in Table 2, indicate that (1) the Core and Math stacks produce more accurate results; (2) using 100 mm (4 in.) products produces large errors, even when used to make predictions for products of that same thickness; and (3) results for slice thickness, significant only for same thickness predictions, were the same for 8 and 10 mm (0.31 and 0.39 in.) slices and much worse for 12 mm (0.47 in.) slice thickness, which is contrary to expectations based upon the relative magnitude of the errors introduced by the thickness of the destroyed surface layer (TDSL) (Stovall et al. 2012).

While the PMM values correct for the unbalanced nature of the data set, the means and standard deviations for the actual test values support these conclusions, except that the results for 12 mm (0.47 in.) slices appear to be very nearly the same as for 10 mm (0.39 in.) slices. Also, the Profile Mixed stack produced good results for the same thickness prediction.

Using these trends to guide the selection of test methodology options by deleting the Surface stack and the predictions based on the original product thickness of 100 mm (4 in.), the means for subsets of the actual test values are shown in Table 3. Considering the means and standard deviations, the most robust approach to the test methodology would appear to be to use Core stacks from 50 mm (2 in.) products to produce alternate thickness predictions and to use Profile Mixed stacks to produce same thickness predictions. The Math stack also offers reasonable results for same thickness and for alternate thickness if combined with the alternate thickness qualification criteria.

Several application factors were also statistically significant in the general linear models. For alternate thickness predictions, the combination of product thickness and manufacturer was the most important factor in the model, based on the Type III F-values as discussed above. Similarly, the combination of product thickness and variety were the third most important factor for same thickness predictions. The impact of the selected test methodology options for these application factors are shown in Table 4.

XPS, Alternate and Same Thickness Predictions

The PMM values in Table 2 for XPS indicate the following.

- The most accurate stacks for same thickness predictions are the two Mixed stacks and the Math stack.
- The most accurate stacks for alternate thickness predictions are the Surface and Math stacks.
- Using 75–100 mm (3–4 in.) products produces large errors for alternate thickness predictions.

Table 2. Multi-Factor Analysis Results: Population Marginal Means (PMM) of "Error"*

Class	Class Value	Alternate Thickness		Same Thickness	
		PIR	XPS	PIR	XPS
Stack	Surface	2.2	–1.0	1.6	–2.2
	Core	–0.8	5.5	–0.5	3.4
	Four-Slice Mixed	1.5	2.4	0.9	0.8
	Profile Mixed	NA	NA	0.9	0.7
	Math	0.9	2.1	0.3 (89% CL)	1.1
Slice origin product thickness	25 mm (1 in.)	–1.0	Not sig.		
	50 mm (2 in.)	0.6	0.8		
	75–100 mm (3–4 in.)	3.5	4.9		
Product thickness, Class 1 for PIR and standard density for XPS (other variations not significant)	25 mm (1 in.)			–2.1	2.4
	50 mm (2 in.)			2.4	0.7
	75–100 mm (3–4 in.)			5.0	2.3
Nominal slice thickness	8 mm (0.31 in.)			1.3	
	10 mm (0.39 in.)			1.2	
	12 mm (0.47 in.)			–3.9	

* Greater than 99% confidence level (CL) except where indicated otherwise.

Table 3. PIR Data Set Multiple-Effect Results for Error (%) for Test Methodology and Screening Criteria Classes

Original Product Thickness, mm (in.)	Stack Type	Alternate Thickness Criteria B*	Alternate Thickness			Same Thickness		
			Error	Std. Dev.	No. of Comparisons	Mean	Std. Dev.	No. of Comparisons
25 (1)	Core	Pass	0.1	2.6	10	–0.1	2.4	23
		Fail	–7.0	0.5	9			
	Four-Slice Mixed	Pass	1.0	2.2	13	0.4	2.0	23
		Fail	–5.0	0.6	9			
	Profile Mixed			NA**		0.6	1.1	23
	Math	Pass	–0.3	2.6	10	0.5	1.3	23
		Fail	–6.6	0.5	9			
50 (2)	Core	Pass	–0.6	1.6	41	–1.1	2.7	28
		Fail	–0.3	0.9	6			
	Four-Slice Mixed	Pass	1.6	2.5	38			
		Fail	2.6	0.9	6			
	Profile Mixed			NA**		0.6	0.9	28
	Math	Pass	1.0	2.6	41	–0.2	1.6	28
		Fail	4.0	0.9	6			

*Alternate thickness criteria B is based on taking the average of the thermal conductivity comparisons (core to core and surface to surface) at 30 days, with a 95% to 105% passing criteria. See Stovall et al. (2012).
**Profile Mixed stack is not applicable when slices from one product thickness are used to predict the aged thermal conductivity for products of another thickness.

- The same thickness predictions for 50 mm (2 in.) products were more accurate than for other product thicknesses.

While the PMM values correct for the unbalanced nature of the data set, the means and standard deviations for the actual test values support these conclusions. Looking at the actual test values, it would appear that the best stack selections would be the Math stack for alternate thickness predictions and either the Profile Mixed or Math stack for same thickness predictions. For alternate thickness predictions, it appears that only 50 mm (2 in.) products should be used.

Several application factors were also statistically significant in the general linear models. For alternate thickness predictions, product thickness was the most important factor in the model, and the combination of product thickness and manufacturer was the fourth most important, based on the Type III F-values. Similarly, variety (standard or high density) was the second most important, and the combination of product thickness and variety were the third most important factor for same thickness predictions. Table 5 shows how the recommended test method options of Math and Profile Mixed stacks perform for these application categories.

Data Distributions for PIR and XPS Boardstock for Selected Test Method Options

In addition to looking at the PMM and the data set means and standard deviations, histograms are useful in examining whether the results are sufficiently clustered near small error values. Core stacks were found useful for PIR products, while Math stacks were found useful for XPS products. For same thickness predictions, Profile Mixed stacks worked well for both products.

Figure 5 shows the relevant data distributions for these test method options. Looking at the top two quadrants of this figure, remember that the alternate product thickness method is only recommended when the slice origin product thickness is 50 mm (2 in.). The bottom right quadrant of Figure 5 shows the data distribution when the Math stack is used to produce predictions for both boardstock products using a slice origin product thickness of 50 mm (2 in.). Another accelerated aging protocol, CAN/ULC S770, also uses the Math stack approach but is based on an aging factor that is then applied to the initial full-thickness value, so the error values reported here will not be applicable to that test method (ULC 2000).

DISCUSSION

Test method recommendations for the two products included in this ruggedness test were presented based on both

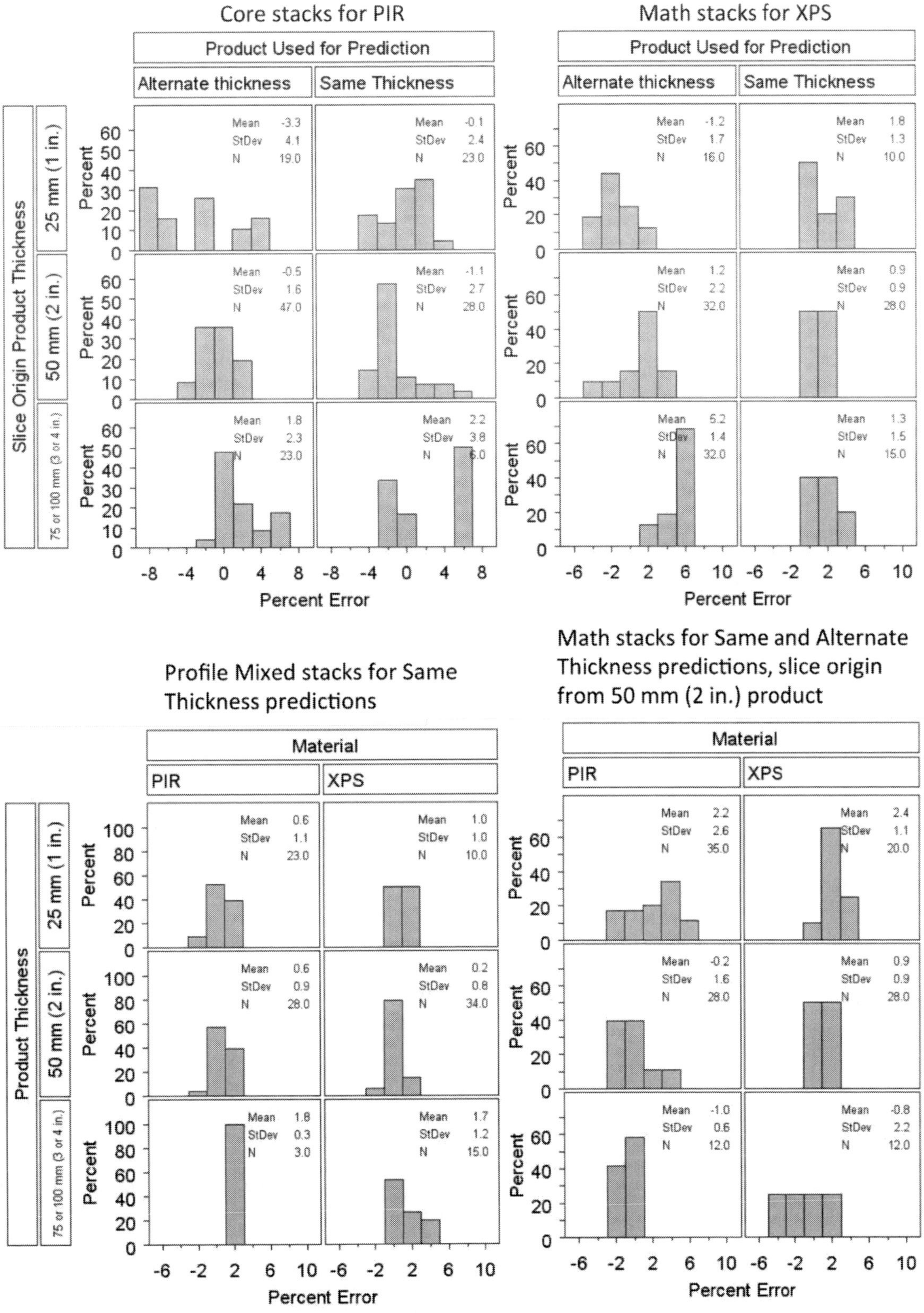

Figure 5 *Measured experimental error when selected stacks are used to produce both same and alternative thickness predictions.*

Table 4. Evaluating Applications for Selected Test Methodology Options for PIR Boardstock

	Product Thickness, mm (in.)	Mean of Error	Standard Deviation of Error	Number of Comparisons
*Alternate thickness prediction using **Core** stacks from **50 mm (2 in.)** original product thicknesses*				
Manufacturer a	25 (1)	0.7	1.1	23
	100 (4)	−2.6	0.6	9
Manufacturer b	25 (1)	−1.3	0.6	12
	100 (4)	−1.0	0.4	3
*Alternate thickness prediction using **Math** stacks from **50 mm (2 in.)** original product thicknesses, including only those that pass Alternate Thickness Criteria B**				
	25 (1)	1.8	2.7	29
	100 (4)	−1.0	0.6	12
*Same thickness comparison using **Core** stack type*				
Class 1	25 (1)	−0.6	3.0	12
	50 (2)	−2.7	0.5	18
	100 (4)	2.2	3.8	6
Class 2	25 (1)	0.6	1.2	11
	50 (2)	1.6	2.7	10
*Same thickness comparison using **Profile Mixed** stack type*				
Class 1	25 (1)	−0.2	0.9	12
	50 (2)	0.5	0.9	18
	100 (4)	1.8	0.3	3
Class 2	25 (1)	1.5	0.7	11
	50 (2)	0.8	1.0	10
*Same thickness comparison using **Math** stack type for 25–50 mm (1–2 in.) products*				
	25 (1)	0.5	1.3	23
	50 (2)	−0.2	1.6	28

*Alternate thickness criteria B is based on taking the average of the thermal conductivity comparisons (core to core and surface to surface) at 30 days, with a 95% to 105% passing criteria.

Table 5. Evaluating Applications for Selected Test Methodology Options for XPS Boardstock

	Product Thickness, mm (in.)	Mean of Error	Standard Deviation of Error	Number of Comparisons
*Alternate thickness prediction using **Math** stack type from 50 mm (2 in.) original product thicknesses*				
Manufacturer c	25 (1)	4.2	0.0	5
	75–100 (3–4)	1.2	0.8	6
Manufacturer d	25 (1)	1.8	0.5	15
	75–100 (3–4)	−2.8	0.3	6
*Same thickness comparison using **Profile Mixed** stack type*				
Standard Density	25 (1)	1.0	1.0	10
	50 (2)	0.4	0.8	22
	75–100 (3–4)	3.1	0.2	6
Higher Density	50 (2)	−0.2	0.6	12
	75–100 (3–4)	0.8	0.2	9
*Same thickness comparison using **Math** stack type*				
Standard Density	25 (1)	1.8	1.3	10
	50 (2)	1.1	0.7	16
	75–100 (3–4)	2.7	0.6	6
Higher Density	50 (2)	0.5	0.9	12
	75–100 (3–4)	0.4	1.0	9

the population marginal mean analysis (see Table 2) and the raw data analysis (see the error distributions in Figure 5). These recommendations are summarized in Table 6. Some recommendations are common to both materials:

- The Profile Mixed stack is excellent for same thickness predictions.
- The Math stack is acceptable for alternate thickness predictions.
- Don't use slices from 75–100 mm (3–4 in.) products for alternate thickness predictions—slices taken from a 50 mm (2 in.) product gave the best results.

The current edition of C1303 allows the use of the three test stack configurations as replicates. As the stack configuration is refined based on this ruggedness test, the issue of sufficient replicate measurements is obviously important and must be addressed by the task group.

ACKNOWLEDGMENTS

This work was supported by the U.S. Department of Energy through the Building Envelope Technology Program under the guidance of Marc Lafrance. Planning support was provided by a committee including Mary Bogdan of Honeywell, Inc.; Gary Chu of the Dow Chemical Company; Michel Drouin, an independent consultant; Barbara Fabian of Owens Corning Corporation; and David Yarbrough of R&D Services Inc. Four foam insulation manufacturers supported the project by supplying foam boards on an ambitious and exacting schedule. Owens Corning provided all the destroyed surface layer measurements. More than a thousand thermal conductivity measurements were made, many on precise time schedules, with help from Phil Childs and Joanna Miller. The slice

Table 6. Summary of Recommendations for Both Product Types Included in the Ruggedness Test

	PIR	XPS
Same thickness	Math or Profile mixed stack except for 100 mm (4 in.) products Core stack except for 100 mm (4 in.) products only	Profile mixed stack Math stack
Alternate thickness	Math stacks from 50 mm (2 in.) products that pass the alternate thickness criteria Core stacks from 50 mm (2 in.) products.	Math stack from 50 mm (2 in.) products. Surface stack from 50 mm (2 in.) products for 25 mm (1 in.) products

preparation is critical to this test method, and Jerry Atchley accomplished most of that work with assistance from Dr. Thomas Petrie.

REFERENCES

ASTM. 1995. ASTM C1303, *Standard Test Method for Predicting Long-Term Thermal Resistance of Closed-Cell Foam Insulation*. West Conshohocken, PA: ASTM International. (Relevant versions published in 1995, 2000, 2007, and 2012).

ASTM. 2010. ASTM C518, *Standard Test Method for Steady-State Thermal Transmission Properties by Means of the Heat Flow Meter Apparatus*. West Conshohocken, PA: ASTM International.

Ball, J.S., Healey, G.W., and Partington, J.B. 1978. Thermal conductivity of isocyanate-based rigid cellular plastics: Performance in practice. *European Journal of Cellular Plastics*, pp. 50–62.

Bomberg, M.T. 1990 Scaling factors in aging of gas-filled cellular plastics. *Journal of Thermal Insulation* 13:149.

Booth, J.R. 1980. R-value aging of rigid urethane foam products. *Proceedings, Society of Plastics Industry of Canada*.

Drouin, M. 2009. Supporting data for ASTM C1289-02 specification for faced rigid cellular polyisocyanurate thermal insulation board. ASTM Research Report C16-1035, March 31, ASTM International, West Conshohocken, PA.

Drouin, M. 2012. Private communications regarding UL study results.

Edgecombe, F.H. 1989. Progress in evaluating long-term thermal resistance of cellular plastics. *CFCS & Polyurethane Industry: Volume 2, A Compilation of Technical Publications 1988–1989*, pp. 17–24. F.W. Lichtenburg, ed. Lancaster, PA: Technomic Publishing Co.

Fabian, B.A., Graves, R.S., Hofton, M.R., and Yarbrough, D.W. 1997. A variability study on the ASTM thin slicing and scaling test method for evaluating the long-term performance of an extruded polystyrene foam blown with HCFC-142b. *Insulation Materials: Testing and Applications*, Third Volume, ASTM STP 1320, pp. 197–215. West Conshohocken, PA: ASTM International.

Gill, J. 2001. Interpreting interactions and Interaction Hierarchies in Generalied Linear Models: Issues and Applications. American Political Science Association Annual Meeting, San Francisco, CA, September 2001.

Glicksman, L.R. 1994. Heat transfer in foams. Chapter 5 in *Low Density Cellular Plastics, Physical Basis of Behaviour*. Hilyard, N.C., and Cunningham, A., Eds. London: Chapman & Hall.

Graves, R. S., McElroy, D.L., Weaver, F.J., and Yarbrough, D.W. 1995. Interlaboratory comparison on estimating the long-term thermal resistance of unfaced, rigid, closed-cell, polyisocyanurate (PIR) foam insulation—A cooperative industry/government project. Report ORNL/M-3976, Oak Ridge National Laboratory, Oak Ridge, TN. January.

Hoogendoorn, C.J. 1994. Thermal ageing. Chapter 6 in *Low Density Cellular Plastics Physical Basis of Behaviour*, pp. 153–86. N.C. Hilyard and A. Cunningham, eds. London: Chapman & Hall.

Isberg, J. 1988. Thermal insulation—Conditioning of rigid cellular plastics containing a gas with lower thermal conductivity than air prior to determination of thermal resistance and related properties. No. 698, Chalmers University of Technology, Goteborg, Sweden.

Kumaran, M.K., and Bomberg, M.T. 1990. Thermal performance of sprayed polyurethane foam insulation with alternative blowing agents. *Journal of Thermal Insulation* 14:43–58.

McElroy, D.L., Graves, R.S., Weaver, F.J., and Yarbrough, D.W. 1991. The technical viability of alternative blowing agents in polyisocyanurate roof insulation, Part 3: Acceleration of thermal resistance aging using thin boards. *Polyurethanes 90 Conference Proceedings*, Orlando, FL.

Miller, R.G.J. 1981. *Simultaneous Statistical Inference*. New York: Springer-Verlag.

Mullenkamp, S.P., and Johnson, S.E. 1983. In-place thermal aging of polyurethane foam roof insulations. *7th Conference on Roofing Technology*, National Roofing Contractors Association.

Pasta, D.J. 2011. Those confounded interactions: Building and interpreting a model with many potential confounders and interactions. *Proceedings of the SAS Global Forum 2011*, 347-2011.

SAS. 2007. *SAS User's Guide*. Cary, NC: SAS Institute, Inc.

Scheutz, M.A., and Glicksman, L.R. 1983 A basic study of heat transfer through foam insulations. *Proceedings of the Sixth International Polyurethane Conference*, San Diego, CA, pp. 341–47.

Searle, S.R., Speed, F.M., and Milliken, G.A. 1980. Population marginal means in the linear model: An alternative to least squares means. *The American Statistician* 34(4):216–21.

Stovall, T.K. 2007. Interlaboratory comparison of the thickness of the destroyed surface layer of closed-cell foam insulation specimens. *Journal of ASTM International* 4(1), Paper ID JAI100321.

Stovall, T.K. 2009. Measuring the impact of experimental parameters upon the estimated thermal conductivity of closed-cell foam insulation subjected to an accelerated aging protocol: Two-year results. *Journal of ASTM International* 6(5), Paper ID JAI102025.

Stovall, T.K., and Bogdan, M. 2007. Measuring the impact of experimental parameters upon the estimated thermal conductivity of closed-cell foam insulation subjected to an accelerated aging protocol. *Proceedings of the 29th International Thermal Conductivity Conference*, Birmingham, AL, June, 2007.

Stovall, T.K., Fabian, B.A., Nelson, G.E., and Beatty, D.R. 2002. A comparison of accelerated aging test protocols for cellular foam insulation. *Insulation Materials: Testing and Applications*, 4th Volume, ASTM STP 1426. A.O. Desjarlais and R.R. Zarr, Eds., West Conshohocken, PA: ASTM International.

Stovall, T.K., Vanderlan, M., and Atchley, J. 2012. Evaluation of experimental parameters in the accelerated aging of closed-cell foam insulation. ORNL/TM-2012/214, Oak Ridge National Laboratory, Oak Ridge, TN.

ULC. 2000. CAN/ULC S770, *Standard Test Method for Determination of Long-Term. Thermal Resistance (LTTR) of Closed-Cell Thermal Insulating Foams*. Toronto, ON, Canada: Underwriters Laboratories of Canada.

Measured Performance of an R-40 Double-Stud Wall in Climate Zone 5A

Lois B. Arena
Associate Member ASHRAE

Doug Owens

Pallavi Mantha

ABSTRACT

Moisture problems within the building shell can be caused by a number of factors including excess interior moisture, bulk water intrusion, capillary action from concrete to wood connections, and through wetted building materials such as siding wetted from rain splash back. Depending on the temperature of the surfaces and the permeability of the materials, that moisture may get trapped inside of the walls and could potentially lead to mold growth and/or decay of the buildings materials.

High-R wall assemblies are gaining popularity due to programs promoting high performance building shells for both new construction and retrofit activities. While the reduction in heat transfer is desirable from an energy standpoint, decreased heat transfer also results in decreased drying potential. With the increasing thickness of walls, moisture related failures could become much bigger problems.

To evaluate the potential for moisture problems in high R-value walls, the project team conducted a combination of modeling and field monitoring. Climate Zones 4A, 5A, 6A, and 7 were the focus of the modeling exercise. This paper presents the results of the modeling compared to data collected from a double cellulose wall in Climate Zone 5A. Several different failure criteria and their validity are discussed.

INTRODUCTION

Moisture problems within the building shell can be caused by a number of factors including excess interior moisture which is transported into the wall through air leakage and vapor drive, bulk water intrusion from leaks and wind driven rain, capillary action from concrete to wood connections, and through wetted building materials such as siding wetted from rain splash back. Depending on the temperature of the surfaces and the permeability of the materials, moisture may get trapped inside of the walls and could potentially lead to mold growth and/or decay of the buildings materials.

High-R wall assemblies and assemblies employing a combination of insulation products are gaining popularity in the market due to programs like Passive House (PH) and Net Zero Energy Home (NZEH) challenges in several states and highly incentivized retrofit programs. New insulation products, code changes, and the desire to reduce costs to achieve these new efficiency levels are also factors behind the various assemblies available today.

While several researchers have been conducting hygrothermal analysis for years, there are few published works on high R-value walls, and there has been even less field research to validate the results. Climate zones of the greatest interest are 4A, 5A, 6A, and 7 as defined in the 2009 *International Energy Conservation Code* (ICC 2009). These zones experience both cooling and heating seasons as well as considerable humidity during the summer, which will likely reduce the drying potential of the wall assemblies. The research presented here is intended to fill some of these gaps.

The first phase of this research project involved extensive modeling (Arena and Mantha 2012). Several different failure criteria were used to analyze the hygrothermal performance of the assemblies, including condensation potential,

***Lois B. Arena** is a senior building systems engineer and **Doug Owens** is a residential energy analyst for Steven Winter Associates, Inc., Norwalk, CT. **Pallavi Mantha** is a sustainability consultant for Arup, New York, NY.*

moisture content thresholds, drying capacity of the assembly, and potential for mold growth.

The goals of the second phase of this study were to monitor an R-40, double-stud cellulose assembly in Climate Zone 5A to determine the accuracy of the moisture modeling and make recommendations to ensure durable, efficient assemblies. Results of the field monitoring are presented here along with a comparison to predicted results.

MODELING

Hygrothermal modeling of the test home wall assembly was conducted using WUFI 5.2. Wall construction consisted of double-stud framing, full depth dense-packed cellulose insulation, an exterior sheathing with an integrated weather resistant barrier (WRB), and vinyl siding (Figure 1).

The hygrothermal properties for the exterior sheathing were provided by the manufacturer. The permeability of the WRB increases with increasing RH, thereby, acting like a smart vapor retarder (SVR). According to the manufacturer's data, the permeance ranges from 0.3 perms at 0% rh to 13 perms at 100% rh. SVRs are typically installed on the interior of the wall just behind the drywall in cold climates to prevent warm moist interior air from entering the wall in the winter while still allowing the wall to dry to the interior in the summer. Even though the main benefits of this type of product are lost due to its placement on the exterior of the wall assembly being studied, its location shouldn't result in poor hygrothermal performance because it will become more vapor open as RH increases.

The exterior vinyl siding is applied directly to the sheathing and is considered by code to be a vented cladding. Therefore, an air change rate of 10 ach per hour was assumed behind the siding. This value was chosen based on results from ASHRAE research report 1091 which states that the equivalent permeance of the wall assembly is drastically increased once the minimum threshold of 10 ach is reached. Since vinyl siding is considered by code to act as a vented cladding, this minimum threshold was used.

The interior of the drywall was coated with a vapor retarder primer. The manufacturer's specifications state that this product results in a film with a water vapor permeance of 0.5 perms when the coating has dried. For the purposes of the initial modeling, it was assumed that the coating was uniform and that this permeance was achieved. Therefore, this value was applied to the interior side of the sheetrock in the model.

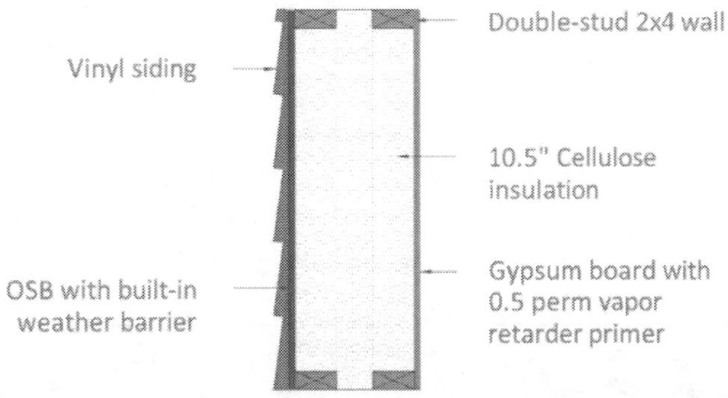

Figure 1 *Typical double stud high-R wall with cellulose insulation.*

The interior conditions were determined in accordance with the criteria for design parameters (intermediate method) as outlined in ASHRAE Standard 160 (2009) and all current, applicable addenda. Interior and exterior boundary conditions are displayed in Table 1.

Experimental Method

The test house was a new construction project located in Climate Zone 5A. Two test bays were identified: one on the north and one on the south facade. Data recorded during this study included: (1) temperature, relative humidity (RH), and moisture content (MC) of the studs at different heights and depths within the exterior walls; (2) MC, temperature, and RH of the sheathing in the center of the bay as well as temperature and RH in the center of the wall cavity and just behind the sheetrock; (3) interior temperature and RH; (4) exterior temperature and RH; and (5) insolation on the south wall. All interior sensors and wall sensors were hard wired to a data logger located in the basement, which was in turn connected to a wireless modem for data collection. The data logger was programmed to read temperature, relative humidity, and wood resistance data every 5 minutes and output the average of those readings every hour. The data was filtered for zero and negative values and the remaining outputs were averaged hourly (Karagiozis and Childs 2006). Data from the pyranometer and exterior temperature/RH sensors were downloaded manually every couple of months.

Table 1. Boundary Conditions

Parameter	Setting
Wall Orientation	North (worst-case scenario) and South
Initial Temperature and RH	68°F and 80%
Heating Setpoint	70°F
Cooling Setpoint	75°F
Floating Indoor Temperature Shift	5°F
RH	Generated in WUFI per ASHRAE Standard 160 criteria for design
Moisture Generation Rate (lb/h)	0.83 (2 bedrooms)
Air Exchange Rate (1/h)	0.2 (equivalent of standard construction)
Climate Zone	5A: Boston, MA
Simulation Start Date	10/1/2011
Initial MC of OSB (lb/ft^3)	5.62 (≈14% MC)
Rain Penetration	1% of driving rain deposited on exterior of WRB

Initial Conditions

During the sensor install in March of 2012, a handheld Delmhorst BD 2100 moisture meter was used to take multiple measurements of the studs and sheathing. According to these tests, initial MC in the sheathing ranged from 7.5%–9.0% and averaged 9.5% in the above grade studs, indicating that the materials were very dry and had been protected from getting wet during construction.

In July 2012, the data logging equipment was installed and interior conditions recorded (Table 2). The interior RH was elevated because the house had been closed up since completion of construction, and the air conditioner was not running. No additional dehumidification equipment was present to remove foundation moisture.

Sensors Types, Calibration, and Accuracy

A total of fifty-six temperature, relative humidity, and moisture content sensors were installed. A description of each of the sensors installed for this study along with its function, accuracy, and model number is located in the Appendix.

To verify temperature sensor accuracy prior to installation, all thermistors were connected to a Campbell Scientific CR10X data logger, allowed to settle to room conditions, and then tested and compared to a high accuracy Omega Engineering thermistor (model 44031) sensor which was used as the reference temperature sensor. The Omega 44031 element is rated at ±0.1°C (0.18°F) interchangeability.

Pre-installation testing of the moisture content sensors and measurement circuit was accomplished by measuring known, fixed resistance values which included both very low (shorted pins) and high resistances. The high resistance values chosen were lower than desired, but were limited by the in-house verification equipment's measurement range.

Sensor Placement

Two test bays were identified: one bay under the window facing north (hereafter referred to as the north bay), while the second bay is located next to a window in the south facing wall (hereafter referred to as the south bay). Both test bays were located on the first floor of the house.

Table 2. Measured Initial Indoor Conditions

Level	Parameter Measured	Result
1st Floor	Interior Temperature	78°F
	RH	69%
	Drywall MC	0.5% to 1%, no mold visible
Basement	Temperature	72°F
	RH	83%
	Stud MC	18%, mold visible

Because windows have been known to be a source of bulk moisture leakage in wall assemblies, a test bay under a window was chosen. Even if best practice methods were applied, damage to flashing or failure of materials could result in leaks. If there was a bulk water leak, the effects of that leak on the assembly's hygrothermal performance would be captured.

In the north bay, sensors were clustered in 11 different locations (four clusters total directly under the window) within the bay [Figure 2]). At each of these locations temperature, RH, and moisture resistance data was collected. The locations are shown in Figure 2. Configuration was similar for the south bay sensors.

The temperature sensors were held firmly against the stud or OSB with a zip tie base so that they would not be pulled away from the wood during installation of the blown-in insulation. The RH sensors were covered with moisture permeable house wrap to prevent dust and bulk moisture from damaging them while allowing moisture vapor to penetrate for accurate readings of operating conditions. The house wrap product used has a moisture vapor permeance of 58 perms (Figure 3).

Before installation, heat shrink tape was installed over the length of the moisture pins, leaving only approximately ¼ in. of the tip exposed and the very top where the wires had been soldered. They were gently tapped into the wood studs and the sheathing to a depth of approximately ½ in. in the direction of the grain and were spaced 1 in. apart. A template was used to ensure consistent spacing. To prevent short circuiting of the pins, electrical tape was secured over the portion of the pin not embedded in the wood.

The interior temperature/RH sensor was installed in the living room on the first floor to record indoor operating conditions. This sensor is located at the same height as the thermostat on a wall in the center of the room. An exterior

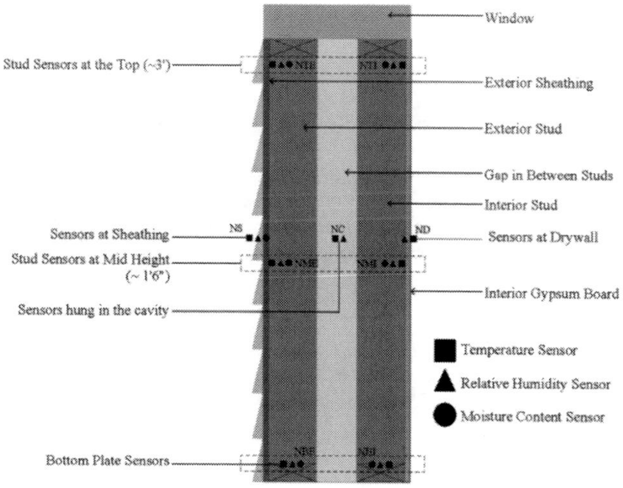

Figure 2 *Sectional view showing typical sensors locations for each bay.*

temperature and relative humidity sensor was installed on the south side of the house along with a pyranometer to collect solar radiation data (Figure 4). The data logging equipment that collects data from the pyranometer and the external HOBO was located on the underside of the front porch overhang to protect it from the elements.

MATHEMATICAL METHODS

Wood resistance based sensors were used to measure moisture content in the sheathing and wood studs. Moisture measurements using wood electrical resistance sensors have been in practice for years and have been well documented (Pfaff 1986; Garrahan 1988; Carll 1996; Ueno 2008; Straube et al. 2002).

An error analysis was performed on previously published resistance and MC data for Douglas fir provided by the Forest Products Laboratory (FPL). It was discovered that, due to the logarithmic relationship between the measured resistance and the corresponding MC, the resulting error in the calculated MC is rather small, especially in the range of 8%–20% MC, which is the range of concern for this study. A 20% error in a resistance measurement results in a 2% to 3% error in the resulting MC value. These accuracies are highly acceptable.

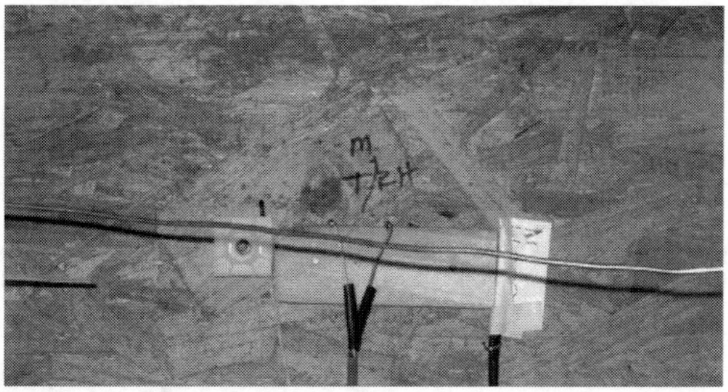

Figure 3 North bay OSB sensors: temperature, moisture content pins, and RH from left to right.

To convert the electrical resistance field measurements to MC values, a regression was performed on the data from the FPL listed in James (1988). The resulting equation from the curve fit is as follows:

$$MC = (-0.1071)\log(R)^3 + (1.3325)\log(R)^2 - (7.2625)\log(R) + 22.298 \quad (1)$$

where

MC = Douglas fir moisture content in % mass

R = electrical resistance in ohms (Ω)

Most DC resistance sensors are calibrated or designed for Douglas fir and a correction factor is applied to that measurement to correct for species as well as temperature. Because the studs in the test home were labeled "SPF" (spruce/pine/fir), it was assumed that the studs were Fir, and therefore, no correction factor was applied.

Though a wide range of correction factors are available for wood species from the FPL, this is not the case with wood products like OSB and plywood. In this study, correction factors for sheathing were obtained from Straube et al. (2002). It should be noted that these factors were developed over a very limited number of samplings from one supplier.

Sensor readings must also be corrected for temperature of the material when using resistance based sensors. The resistance of the wood is dependent on temperature, and as the temperature of the material rises, the resistance decreases. According to Straube et al. (2002), the electrical conductivity of wood approximately doubles for every 10°C temperature change. Affected moisture content is in the range of 0.10%–0.15% MC/°C. Equation 3 was used (Pfaff and Garrahan 1986) to correct for both species and temperature of the sheathing. In Equation 3, constants a and b are specific to each species.

Figure 4 Location of exterior HOBO, pyranometer, and associated data logger.

$$MC_c = \left[\frac{MC + (0.567 - 0.0260t) + 0.000051t^2}{0.881(1.0056)^t} - b\right] + a \quad (2)$$

where

MC_c = corrected moisture content

MC = moisture content in megaohms

t = temperature of material at the same depth as the MC reading in °C

a = 1.1114

b = 0.366

It is also recommended that the temperature (t) in Equation 3 be the temperature of the material at the same depth as the moisture pins. However, in this study the temperature at the surface of the sheathing was used. THERM models were run for measured conditions to verify accuracy of the installed temperature sensors and to identify temperature at the depth of the moisture pins. It was found that the temperature varied less than half a degree at 0°F outside. This difference lessened as the exterior temperature increased. Therefore, the surface temperature of sheathing was deemed sufficient for the corrections.

MONITORING RESULTS AND DATA ANALYSIS

This section presents analysis of the measured data and draws comparisons between the measured and predicted modeling results. The model was first compared to the measured performance without calibration for differences in actual vs. modeled weather conditions. This was done to draw conclusions about the accuracy of the modeling software if a user does not have access to measured data to perform such calibrations. The model was then run using actual measured interior and exterior boundary conditions as well as the measured initial sheathing MC.

Boundary Conditions

The weather file for Boston, MA was used in the modeling. Table 3 compares the average monthly outdoor temperature, RH, and corresponding humidity ratio from the measured data and WUFI's weather file for the monitoring period (mid July 2012 through mid April 2013). In general, the monitoring period was warmer and more humid than that represented by Boston's cold year weather file. The average monthly humidity ratios show that the moisture content in the exterior air was higher for most of the monitoring period than presented in the weather file.

The interior conditions for both the predicted and actual cases are shown in Figures 5 and 6. It should be noted that the home was unoccupied until the middle of September and the cooling system was not running until the beginning of August.

Solar radiation on the south façade of the test home was also monitored. To prevent damage to they vinyl siding, the sensor was attached to a piece of wood trim to the right of the entry instead of directly over the test bay. This resulted in a situation where the pyranometer would be shaded more than the actual test bay.

The shading function in SketchUp was used to determine the extent to which the pyranometer was shaded, as compared to the test bay (Figure 7). July and August show little to no additional shading. As the sun gets lower in the sky from September to December it appears that the pyranometer gets about an hour to an hour and a half less sun than the test bay during the afternoon between 2:30 and 4 pm. Based on solar intensity values for Boston generated with Solrad, a solar position and solar radiation calculator from the Washington

Table 3. Comparison of Measured Outdoor Temperature and RH to Predicted Values from WUFI and TMY2 Data.

	WUFI			Measured		
	Temperature, °F	RH, %	Humidity Ratio, lb_W/lb_{da}	Temperature, °F	RH, %	Humidity Ratio, lb_W/lb_{da}
July	70	69	0.0110	76	63	0.0122
August	69	74	0.0104	78	64	0.0132
September	62	66	0.0077	67	69	0.0097
October	53	67	0.0057	56	75	0.0072
November	41	66	0.0036	41	65	0.0042
December	29	65	0.0024	36	76	0.0034
January	28	59	0.0019	30	64	0.0022
February	27	72	0.0020	31	65	0.0026
March	39	65	0.0033	38	63	0.0032
April	46	57	0.0037	48	55	0.0038

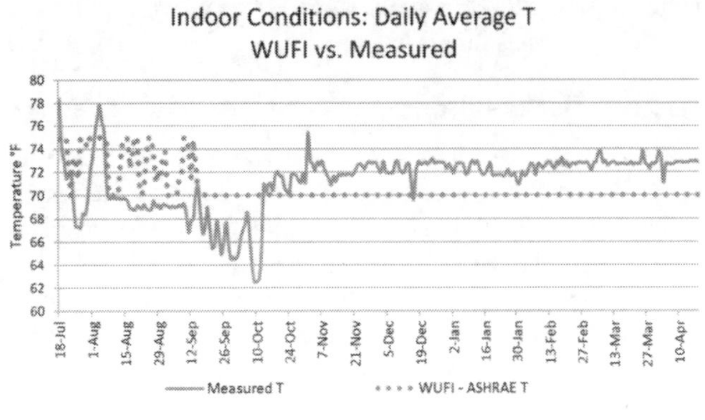

Figure 5 *Measured and predicted interior temperature in the test house.*

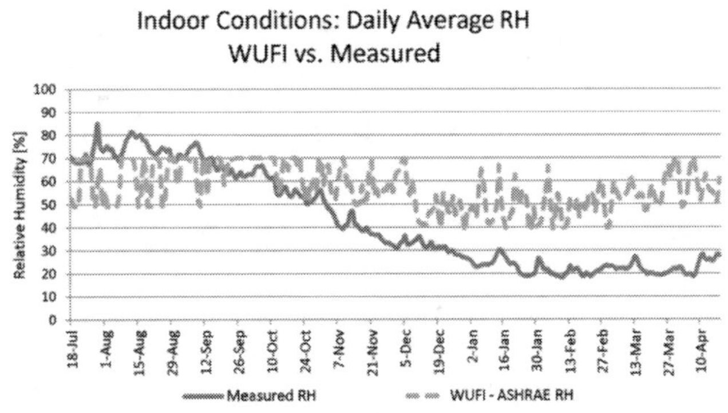

Figure 6 *Measured and predicted interior RH in the test house.*

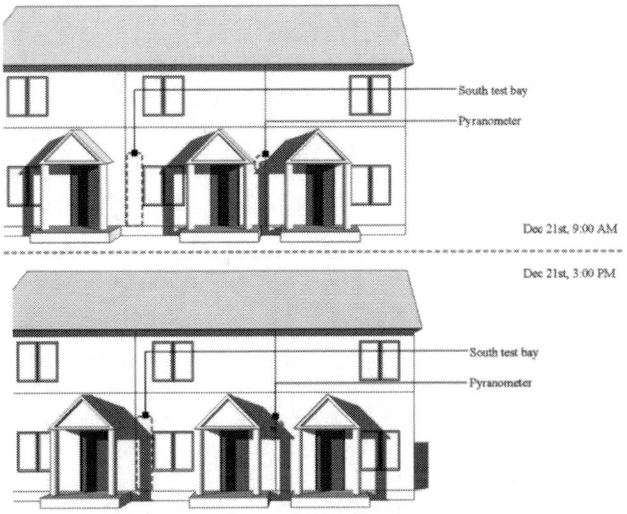

Figure 7 *Solar shadow analysis for pyranometer located on the south wall.*

State Department of Ecology (Pelletier 2013), the amount of radiation hitting the pyranometer is estimated to be 10% to 15% lower from September to December respectively than the that reaching the façade of the south test bay.

Table 4 lists the average daily solar radiation by month for Boston as assumed by WUFI and PVWatts compared to the measured data.

This comparison of measured and predicted data shows that the actual insolation was less than that in the WUFI data file and the PVWatts predictions during the monitored period. This is logical since no shading was assumed on the vertical surface for either the WUFI or PVWatts analysis, while the porches on the south façade definitely shade the sensor and the test bay in the morning and afternoon.

Hygrothermal Performance of the Assembly

Several failure criteria were identified against which wall assemblies should be evaluated. The criteria evaluated for this study include condensation potential, moisture content thresholds, drying potential of the assembly, and the potential for mold growth as determined by ASHRAE Standard 160, *Criteria for Moisture-Control Design Analysis in Buildings* (2009).

Moisture Content

It is often quoted that the minimum moisture content (MC) requirement for the growth of fungi is approximately 20% in wood, corresponding to about 80%–90% rh (Siau 1984). MC values were calculated for each pair of moisture pins installed in the studs and sheathing and evaluated against this threshold.

As of mid-April 2013, the sheathing MC in the south bay has remained below the 20% threshold peaking at approximately 16% (Figure 8). In the north bay, the MC rose slightly

Table 4. Comparison of Actual and Predicted Solar Radiation on the South Wall

Month	Solar Radiation, kWh/m^2 per day		
	WUFI	PVWatts	Actual*
July	2.75	2.76	1.99
August	2.81	3.27	2.64
September	3.33	3.59	2.49
October	2.90	3.92	1.55
November	2.82	2.96	2.50
December	2.77	3.06	1.41
January	3.08	3.27	2.22
February	3.24	4.19	2.02
March	3.54	3.60	2.17
April	3.13	3.03	2.43

*Actual solar radiation reaching the test bay on the south façade is estimated to be approximately 10% to 15% higher than the values shown in the table.

above 20% during most of February and a portion of March. The difference in these moisture levels was anticipated since the south bay receives direct solar radiation which should result in enhanced drying potential. Measured temperature and RH levels confirm that the sheathing in the south bay was warmer and had a lower RH than the sheathing in the north bay supporting the MC results. Predicted differences in MC between the two walls also supports these findings. Figure 8 shows the predicted MC using the WUFI weather file and the ASHRAE Standard 160 method for generating the boundary conditions versus the MC predictions using the measured boundary conditions. The predictions using the measured conditions map the measured MC values slightly better at the beginning of the monitoring period, but neither set of assumptions result in the same MC peaks, nor do the peaks occur at the same time of the year as the measured values in the south bay.

To investigate the reasons for these differences, several different parameters in the model were altered, including permeance of the VR paint, permeance of the WRB on the exterior sheathing, moisture penetration into the wall assembly, and various combinations of these values. The values analyzed include:

- Constant permeance of the sheathing using the upper and lower boundaries of the permeability curve: 0.3 and 13 perm.
- Increased permeance of the VR paint, assuming it was not applied per the manufacturer's recommendations: 10 perm.
- A rain leak into the OSB layer behind the WRB equal to 1% of the driving rain.

Thirteen different combinations of these parameters were modeled and analyzed. Five different curves have been presented in Figure 9 and are compared to the measured data and the predicted MC using the ASHRAE Standard 160 boundary conditions. Combinations that did not appear to change the predictions when compared to the results using the original material properties and measured boundary conditions have been eliminated. The most significant match appears to be the case where a 1% moisture leak was assumed in the north bay. Those assumptions result in a much closer matching of the curves, peak MC, rate of moisture increase, and a closer match of the timing of the peak MC. Considering these sensors are located directly under a window, a moisture leak could be very possible.

A similar study was conducted for the south bay (Figure 10). The results displayed in Figure 10 show that the measured MC peaks in January at approximately 17% MC and begins to dry out again in February. The modeling results show peaks of 14% in March and drying beginning in April.

To validate the MC results for the sheathing, the MC values for the stud sensors were also calculated and evaluated. Figures 11 and 12 show the MC readings for all the moisture

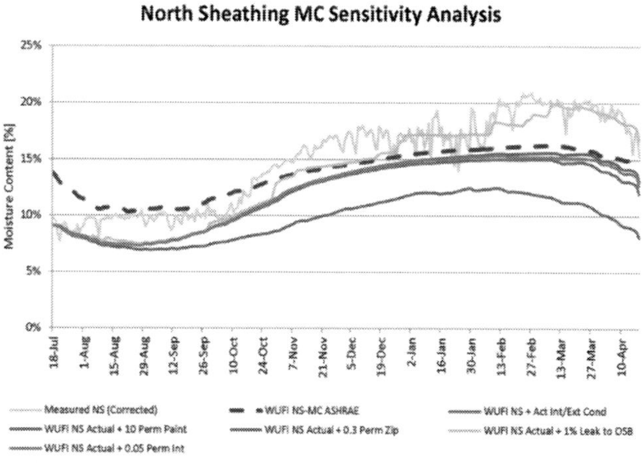

Figure 9 *Comparison of predicted MC levels in the north bay using different boundary conditions.*

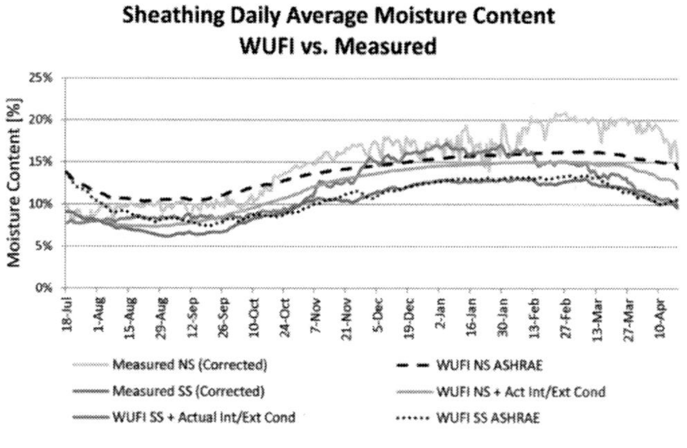

Figure 8 *Measured and predicted daily MC for the north and south sheathing.*

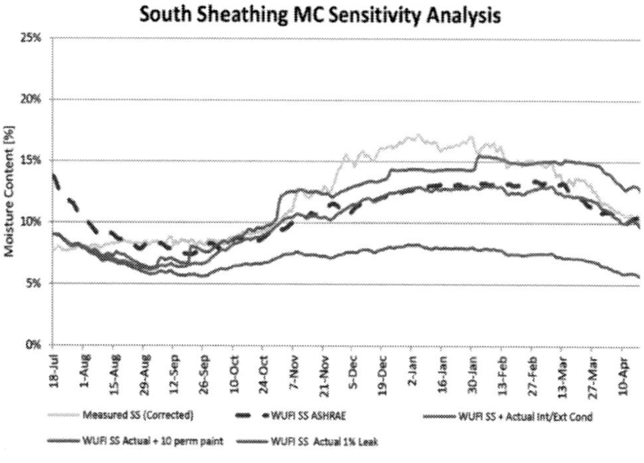

Figure 10 *Comparison of predicted MC levels in the south bay using different boundary conditions.*

pins in the south and north bays respectively. The north bay exterior stud MC readings are elevated like the sheathing MC and follow a similar pattern. In the south bay, most of the sensors show decreasing MC levels, except for the lower exterior stud position. Since it follows the same pattern as the north bay sensors, this could indicate a leak or concentrated drainage pattern at the corner of the window bordering that bay.

Measurements from both bays support the findings from the sheathing analysis and showed several consistent trends. In general, the monitored locations in the walls appear to be drying out from July until the end of September, at which time the sensors located in the sheathing and exterior studs show an increase in MC. The sensors on the interior studs show a decrease in MC in both bays during the entire monitoring period. Lastly, the bottom plate sensors (exterior and interior) were the wettest of all the locations at the beginning of the monitoring period.

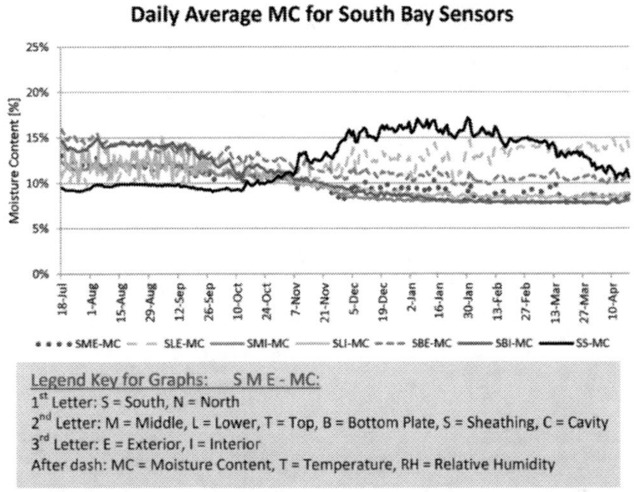

Figure 11 *Daily average MC readings for the sensors in the south test bay.*

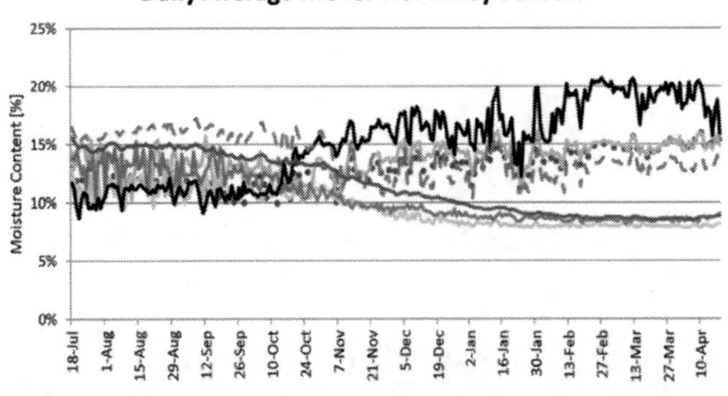

Figure 12 *Daily average MC readings for the sensors in the north test bay.*

Condensation Potential

Condensation potential within the wall was evaluated by comparing the interior air dew-point temperature to the surface temperature of the potential condensing surface. This method generally represents a worst case scenario as it assumes that moisture through air leakage enters the wall assembly from the interior. If the surface temperature of the material is lower than the dew-point temperature of the air (Straub et al. 2009), condensation is likely to occur: the longer the period during which the surface temperature falls below the air dew-point temperature, the greater the risk for damage. The critical juncture analyzed was the inside surface of the exterior sheathing in both bays.

As shown in Figure 13, the measured north and south sheathing temperatures started to drop below the indoor dew point in September. According to the predicted model (Figure 14) the sheathing surface temperature should start to drop below the indoor dew point in late October. The duration of time the surface temperatures stay below the dew point increases as the weather gets colder.

As with MC, condensation potential was also evaluated at the studs. The interior studs in both bays are consistently above the dew point, however, the exterior studs in both bays

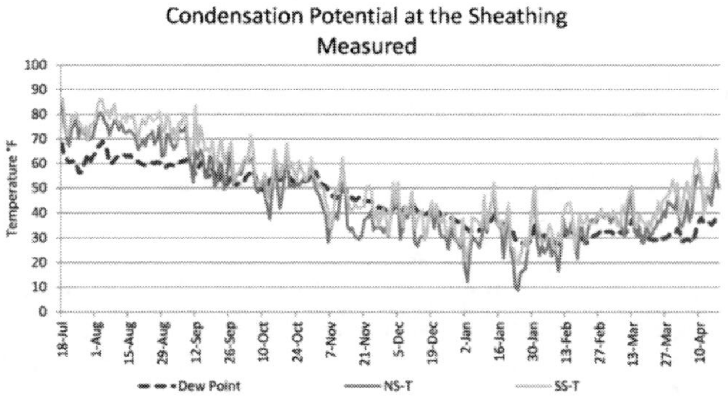

Figure 13 *Condensation potential in the north and south sheathing.*

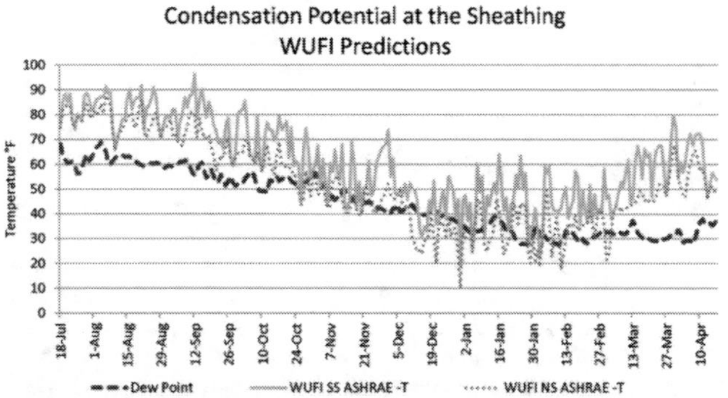

Figure 14 *Predicted condensation potential in the sheathing.*

fall below, closely mapping the sheathing and further validating the results.

Based on the indoor boundary conditions, the dew point was calculated on an hourly basis for the monitoring period for both the predicted (ASHRAE Standard 160 conditions) and measured data. The actual and predicted surface temperatures of the sheathing were then compared to the respective dew points. The percentage of time that the surface temperature fell below the dew point was calculated and tabulated for each bay. A summary of the results is shown in Table 5.

Because the predicted interior RH using ASHRAE Standard 160 is higher than the measured data, the calculated condensation potential is significantly higher than that of the measured data.

ASHRAE Standard 160 Criteria

ASHRAE recently published ASHRAE Standard 160, *Criteria for Moisture-Control Design Analysis in Buildings* which sets the performance criteria to minimize problems associated with moisture in building envelope assemblies. The standard specifies that the following conditions be met: the 30 day running average surface RH must be less than 80% when the 30-day running average surface temperature is between 41°F and 104°F. This threshold applies to all materials and surfaces in the building envelope except the exterior surface. If just one 30-day average fails to meet these conditions, the assembly is considered to have failed.

Table 5. Condensation Potential for Exterior Sheathing—Measured vs. Predicted

Orientation	Case	Hours of Potential Condensation (Jul–April)[1]	% of Collection Period
North	Measured	2866	44%
	Predicted	4776	61%
South	Measured	2255	34%
	Predicted	4777	44%

[1]Collection period consisted of 6573 hours.

Table 6. Analysis Results for ASHRAE Standard 160 Criteria for Double Stud Walls with Cellulose Insulation: Measured vs. Predicted

Orientation	Case	% of 30-Day Averages that Fail (Jul–Mar)[1]	Pass/Fail
North	Measured	23%	Fail
	Predicted	36%	Fail
South	Measured	18%	Fail
	Predicted	54%	Fail

[1]Collection period resulted in 244 30-day averages.

The interior face of the sheathing was chosen as the surface to be evaluated with this criterion. Surface temperature and RH of the sheathing was averaged over a 30-day period and compared to the above limits. A summary of the results can be found in Table 6. Even though it only takes one noncompliant 30-day period to result in failure, the magnitude or number of failures should be considered. Therefore, the percentage of 30-day averages for the monitoring period that have failed these criteria have been presented for both the measured data and the predicted results using the ASHRAE Standard 160 boundary conditions. According to the data, both the north and south wall fail to meet this criterion, but as discovered in the first phase of this research (Arena and Mantha 2012), most walls with sheathing outside the insulation will fail if the interior face of the sheathing is the surface evaluated.

Note that the initial assumptions for the sheathing RH start at 80% when using the ASHRAE Standard 160 design criteria. The actual RH of both bays was lower than this, resulting in fewer 30-day failures from the measured data compared to the predicted data. RH levels recorded for each of the sensors on the studs show agreement with the sheathing values. The south sheathing sees a higher potential for mold growth according to the predicted data because the predicted south sheathing temperatures stay above 41°F for the majority of the monitored period whereas the north sheathing temperatures are predicted to fall below 41°F for several months.

DISCUSSION

Moisture Content

The difference between the measured results of the two facades is significantly different and is supported by the modeling predictions, however predicted peaks in MC are lower than actual for both walls. The south bay sheathing shows the MC values peaking around 17%. Drying of the assembly starts around January and continues gradually through the end of the modeling period where it ends up at about 10% MC. The north bay sheathing peaks at just over 21% MC. Drying of the assembly didn't begin until late March. At the end of the monitoring period, the MC was still above 15%.

Except for the beginning of the monitoring period, predictions from WUFI using the ASHRAE Standard 160 design criteria closely mapped the predictions when using the actual measured boundary conditions. The difference at the beginning of the period is due to the materials being drier than is assumed when using the Standard 160 criteria. However, neither resulted in peak MC levels as high as was measured in the field.

Based on the parametric study conducted, it appears that the elevated MC in the north wall could be due to a moisture leak under the window. When modeled with a 1% leak to the OSB, the predicted results very closely mapped the measured performance. Little effect was seen from varying the perm rating of the WRB or from decreasing the perm rating of the interior VR paint to 0.05 perm. Assuming a 10 perm VR paint, on the other hand, resulted in even lower MC predictions and, therefore, was determined not to be likely, considering the measured results.

When evaluating this assembly over a three year period using the initial modeling assumptions and the ASHRAE Standard 160 method, it is predicted that the sheathing will dry out seasonally and peak MC levels will decrease over time (Figure 15). The monitored results from the south bay support this conclusion. The results from the north bay are not so definitive. While the sheathing does appearing to be drying out at the rate and time shown in Figure 15, the measured MC levels exceed the predictions by about 4%, rising above the 20% MC threshold for mold growth. Another concern is that several of the positions in the exterior studs show slightly increasing MC levels through the end of the monitoring period. Additionally, introducing a 1% rain leak to the OSB layer resulted in much better agreement of the predicted and measured results. These three factors raise concerns about the durability of the north wall.

Condensation Potential

When using the ASHRAE Standard 160 boundary conditions, the measured condensation potential for the monitoring period is lower than predicted. This is due to the fact that the interior RH is significantly lower than that generated by that method. The Standard 160 method of generating RH resulted in values ranging from 70% rh in the summer to 40% rh in the winter. This home's actual interior RH dropped under 30% from January through March, while the Standard 160 method produced RH levels of 50% to 60% for the same time period. This has a drastic affect on calculation of the dew point and hence, condensation potential. Even though the percentage of hours of potential condensation is lower for the actual assemblies than predicted, it is still very high at 34% for the south wall and 44% for the north.

ASHRAE Standard 160 Criteria

Both the north and south walls fail the ASHRAE Standard 160 30-day criterion using the monitored data or the predicted results. When comparing actual versus predicted results the north and south sheathing failure rate for the measured data was significantly lower than that predicted with the modeling. This is because the starting RH of the materials as recommended in Standard 160 was 80% when the actual measured RH started at approximately 70% for the north wall and 60% for the south. All the temperature and RH sensors in the bays support the readings captured for the sheathing.

Past modeling work in this area has indicated that very few walls in cold climates will pass this criterion if the inside face of the exterior sheathing is evaluated. It is even less likely if the modeling period is started in October and the ASHRAE recommendations for initial surface temperature and RH are assumed. Although the probability of mold growth should be evaluated, there should be further investigation into whether failing this criterion does actually result in mold growth. One instance of failure of all the 30-day averages through the year appears to be much too restrictive of a metric.

When the failures occur is also important. Did they take place at the very beginning of the modeling period due to the assumptions for the initial conditions and then not again? Do they occur if the modeling is started at a different time of year? Has there been any forensic proof that exceeding these conditions on the interior of the exterior sheathing does in fact result in mold growth? Do the thousands of homes constructed with the sheathing outside of the insulation in cold moist climates have mold problems?

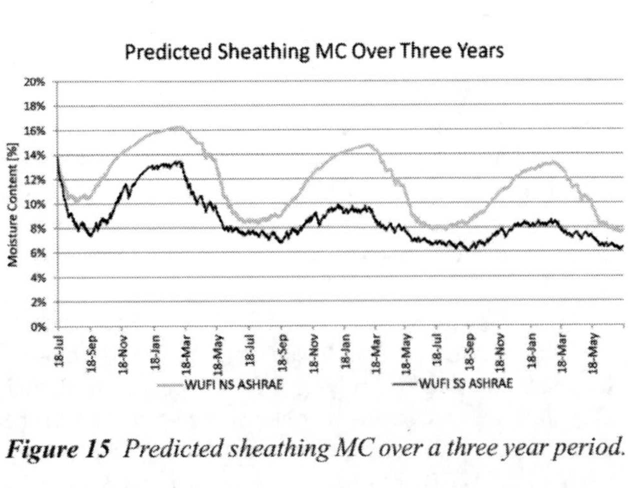

Figure 15 Predicted sheathing MC over a three year period.

CONCLUSIONS

The main focus of this study was to evaluate the predicted performance of the specified wall assemblies against the measured performance. The goals were to monitor an actual high R-value wall assembly to determine the accuracy of the moisture modeling and make recommendations to ensure durable, efficient assemblies. The wall assembly monitored and analyzed in this report was an R-40, double-stud cellulose assembly in Climate Zone 5A.

Several failure criteria were analyzed including MC, condensation potential, drying potential, and the potential for mold growth. No one failure criteria should be evaluated on its own. Several key performance metrics need to be evaluated before making a decision on the performance of a particular assembly. The most useful combination of criteria will look at long term and short term drying potential. Overall assembly moisture content should never increase over time and should ideally decrease. Each component in a given assembly should dry out to a safe level every year. The condensation potential and thresholds for mold growth should be evaluated. If hydrophobic materials exist in the assembly, condensation potential should be minimized at all costs, while materials with moisture absorption capabilities will be more forgiving. Finally, the thresholds for decay should be avoided entirely.

Data collected from July to mid-April of 2013 indicates reasonable agreement with the modeling and shows the following:

- Long term modeling results indicate that this assembly should dry out over the course of the year and experience decreasing peak moisture content levels for the following years. These are major indicators that a wall is durable and shows good hygrothermal performance.
- Measured data for the south wall shows good agreement with predicted peak MC levels using the ASHRAE Standard 160 design criteria. The increase and decrease of MC over the monitoring period was faster for the actual wall than predicted, resulting in less time spent at elevated moisture conditions than predicted.
- Measured data for the north bay indicates a similar rate of response for moisture build up and drying compared to predicted response. However, peak MC levels were higher than predicted and reached approximately 21%.
- Condensation potential, while high, is lower than that predicted, due primarily to indoor assumptions used during the modeling.
- Both walls fail the ASHRAE Standard 160 30-day criterion for mold growth. Although it is important to evaluate the potential for mold, this criterion may be too restrictive and may not be appropriate for all surfaces of the assembly. Further investigation into its validity is recommended.
- A parametric study indicates that the north bay may be experiencing moisture intrusion from driving rain.
- A more vapor open interior surface (10 perm) on this high-R assembly results in lower predicted peak MC levels and faster drying than the initial assumption of 0.5 perm for the VR paint.

Overall, the wall assembly is predicted to perform well. Measured data in the south wall supports these findings, while measured data in the north wall suggests the performance is marginal to fair.

ACKNOWLEDGMENTS

Steven Winter Associates, Inc. acknowledges the United States Department of Energy Building America program and their funding and support for the development of this technical report as well as research that informed it. SWA would like to thank William Anderson from Metric Corp for his assistance with finding and setting up the test homes for this project.

APPENDIX—SENSOR TYPE AND DESCRIPTIONS

All monitoring equipment specifications and sensor accuracies are provided in Table A1. The temperature sensor used was an Omega Engineering TH-44006-40-T fast-response, 10KΩ thermistor with an exposed element and an interchangeability of ±0.2°C (0.36°F).

The measurement circuit was a DC half-bridge circuit. The two resistors making up the half-bridge consisted of the TH-44006-40-T thermistor and a fixed-value 10KΩ, 0.1% precision resistor. An excitation voltage of 2.5VDC was used.

To verify temperature sensor accuracy prior to installation, all thermistors were connected to a Campbell Scientific CR10X data logger, allowed to settle to room conditions, then tested and compared to a high accuracy Omega Engineering thermistor (model 44031) sensor which was used as the reference temperature sensor. The Omega 44031 element is rated at ±0.1°C (0.18°F) interchangeability.

The humidity sensor used was a Honeywell HIH-4021 RH sensor. It contains a thermoset polymer capacitive sensing element with on-chip integrated signal conditioning. The sensing element is covered and is condensation resistant. As a

Table A1. Equipment Used for Measurements

Measurement	Equipment and Specification
Moisture Content: Studs and Sheathing	Custom Wood Moisture Sensors Manufactured from: GE Protimeter Moisture Meter Replacement Pins, 1 in. Long, Grainger P/N 3YYE5, Insulated with Heat Shrink, Soldered to Shielded, Stranded, 22AWG Sensor Cable
Relative Humidity: Inside Walls	Honeywell Humidity Sensor Model HIH-4021-002 Measurement Range = 0%–100% rh, Accuracy = ±3.5%, Sensor Is Covered and Condensation Resistant, Supply Voltage = 5Vdc, Device is Calibrated at 5Vdc and 25°C
Temperature: Inside Walls	Omega Engineering P/N TH-44006-40-T, Fast Response Thermistor Sensor with Exposed Element, Operating Range = –80°C 120°C, Interchangeability = ±0.2°C, 10K-Ohm Bead, Equipped with 40 in. of 26AWG PFA Insulated Cable
Interior Temperature/RH	Humirel HTM2500LF Temperature and Relative Humidity Module, 0–100% rh, ±3% from 10%–95% rh Typical
Exterior Temperature/RH	HOBO U23-002 Exterior T/RH Data Logger
Incident Solar Radiation	HOBO Micro Station Logger, H21-002, HOBO S-LIB-M003 Silicon Pyranometer
One-Time Moisture Content Reading of Building Materials	Delmhorst BD 2100 Moisture Meter

further means of protecting the sensor, the HIH-4021 was enclosed in a small bag made from house wrap prior to its installation in the wall assembly. The sensor's tight, laser-trimmed interchangeability helps to reduce or eliminate calibration requirements.

The HIH-4021 humidity sensor requires a supply voltage of between 4 and 5.8Vdc. The supply voltage used was 5Vdc (the sensor is calibrated at 5 V dc and 25°C). Sensor output is a near linear DC voltage. At higher humidity levels the output voltage exceeds the data logger's 2.5Vdc maximum input voltage. A voltage divider reduced the data logger's input voltage to an acceptable level and consisted of two 2MΩ, 0.1% precision resistors which also helped satisfy the sensor's high input impedance measurement circuit requirement.

To verify humidity sensor accuracy, prior to installation all sensors were connected to a Campbell Scientific CR10X data logger, allowed to settle to room conditions then tested and compared to both an Onset Computer Corporation U14-001 Temperature/Relative Humidity Data Logger (±2.5% rh from 10%–90% typical) as well as a Humirel HTM2500LF (calibrated within ±2% at 55% rh).

Each moisture content sensor consisted of two one-inch, stainless steel, GE moisture meter replacement pins soldered to stranded, shielded, 22AWG sound and security cable. The pins were insulated except for the tip to eliminate measuring the highest moisture content over the pin's length and were driven into clear wood at a distance of 1 in. from each other.

The measurement circuit used was a DC half-bridge circuit similar to that used for the wall temperature sensor measurements. The two resistors making up the half-bridge circuit consisted of the moisture content pin-pair and a fixed-value 2MΩ, 0.1% precision resistor. An excitation voltage of 2.5Vdc was used and a two second delay allowed the resistance reading to stabilize before measurement.

The data logger program's scan rate (execution interval) was originally set to fifteen minutes but was subsequently changed to five minutes to provide for more data points due to the need to filter random negative values from the final resistance averages.

REFERENCES

Arena, L; Mantha, P; (2012). Technical Report: Moisture Research—Optimization of Wall Assemblies. US Department of Energy Building America Technical Report, Washington, DC.

ASHRAE 1091, Development of Design Strategies for Rainscreen and Sheathing Membrane Performance in Wood Frame Walls: Synthesis Report and Guidelines; Report #12; (2004); ASHRAE, 1791 Tullie Circle, Atlanta, GA 30329

ASHRAE Standard 160-2009, *Criteria for Moisture-Control Design Analysis in Buildings*. Atlanta: ASHRAE.

ASTM MNL 18–2nd. (2009) *Moisture Control in Buildings: The Key Factor in Mold Prevention*. West Conshohocken, PA. ASTM International.

ASTM MNL 40. (2001) *Moisture Analysis and Condensation Control in Building Envelopes*. West Conshohocken, PA. ASTM.

Carll, C. and TenWolde, A., (1996) Accuracy of Wood Resistance Sensors for Measurement of Humidity, *JTEVA*, Vol. 24, No. 3, May 1996, pp. 154–160.

Garrahan P. (1988). Moisture Meter Correction Factors. Forintek Canada Corp. Proceedings of a seminar on "In-grade testing of structural lumber", held at USDA Forest Products Laboratory, Madison WI.

ICC. 2009. *International Residential Code for One- and Two-Family Dwellings*. Washington, DC: International Code Council.

James, W.L.; (1988). Electric Moisture Meters for Wood. United States Department of Agriculture, Forest Service, Forest Products Laboratory, General Technical Report FPL-GTR-6, Madison, WI.

Karagiozis, A.N.; Childs, P; (2006). They Hygrothermal Performance of the Net Exterior Insulation Walls. BTC/ORNL 05-UF-05-550, Engineering Science and Technology Division, Oak Ridge National Laboratory, Oak Ridge, TN.

Pelletier, G. Solrad ver. 1.2. Washington State Department of Ecology, Olympia, WA. Models for Total Maximum Daily Load Studies, www.ecy.wa.gov/programs/eap/models.html

Pfaff, F and P. Garrahan. (1986), New temperature correction factors for the portable resistance-type moisture meter. Technical Note. *Forest Products Journal*. 36(3):28–30.

Siau, J.F., (1984). Transport Processes in Wood. Springer-Verlag, Berlin Heidlberg New York.

Straube, J., D. Onysko, and C. Schumacher. 2002. Methodology and Design of Field Experiments for Monitoring the Hygrothermal Performance of Wood Frame Enclosures. *Journal of Thermal Env. & Bldg. Sci.*, Vol. 26, No.2—October 2002.

Straube, J.; Finch, G.; (2009). Ventilated Wall Claddings: Review, Field Performance and Hygrothermal Modeling. Research Report – 0906. Somerville, MA. Building Science Corporation.

Straube, J.; Smegal, J; (2009) Building America Special Research Project: High-R Walls Case Study Analysis. Research Report – 0903. Somerville, MA. Building Science Corporation.

Straube, J.F.; (2003) Controlling Cold-Weather Condensation Using Insulation.

Ueno, K and Straube, J. Laboratory Calibration and Field Results of Wood Resistance Humidity Sensors. BEST 1 Conference.

Comparative Energy and Wall Performance of Twelve Residential Houses Constructed in a Cold Climate

Gary Parsons
Member ASHRAE

Brian Lieburn

ABSTRACT

Many construction options exist to build homes which comply with energy code requirements. Dow Building Solutions partnered with Cobblestone Homes to build case study, single family research houses in Midland, Michigan (climate zone 5) in 2011 to compare the performance of four building energy efficiency strategies. Energy code compliance strategies ranged from 2006 International Energy Conservation Code *(IECC), 2012 IECC, and beyond 2012 IECC high performance enclosures. A total of twelve houses were built for this case study, with three house plans replicated for each of the four energy efficient strategies. All twelve homes are in very close proximity with a similar geographic orientation and will remain occupied through the multiyear study period.*

This paper analyzes the cost and benefit of each enclosure construction method, plus a comparison of overall energy consumption. HVAC equipment energy use was separated from plug loads to understand its impact on the entire energy use of the homes.

Additionally, this paper addresses durability and moisture concerns raised about some high performance envelope strategies. Hygrothermal performance comparisons of the above grade and below grade walls was measured by relative humidity, temperature, and moisture content instruments installed in each house. This data was used to evaluate relative durability risks associated with each energy performance strategy and construction.

INTRODUCTION

Dow Building Solutions and Cobblestone Homes partnered to build single family research homes in Midland, Michigan to quantify the construction cost and energy performance benefit of residential wood frame homes built according to the 2012 *International Energy Conservation Code* (IECC) compared to the less demanding 2006 IECC code (ICC 2012, 2006). Midland, Michigan is located interface between climate zones 5 and 6 and has 6645 annual heating degree days (NCDC) and can be characterized as a cold climate. The project included twelve homes and utilized three floor plans of similar size. A base case cost was established by constructing each floor plan to comply prescriptively with the 2006 IECC. Six homes were then constructed to comply with the 2012 IECC using the performance approach. Finally three additional homes were constructed with energy performance exceeding the 2012 IECC requirements, essentially capable of net zero energy performance with the addition of renewable energy sources. The homes sizes are considered typical representatives of homes built for a family of four. Home areas are shown in Table 1. Living space includes the above grade space and additional living space in the partially finished basement. Figure 1 shows the subdivision layout and photographs of the home built.

DESCRIPTION OF HOUSES AND ENERGY EFFICIENCY FEATURES

Several strategies to control the energy performance of the test homes were investigated. For foundation walls, faced fibrous insulation was compared to both rigid extruded polystyrene foam insulation and rigid polyisocyanurate insulation. Use of continuous exterior insulation on above grade walls was compared to

***Gary Parsons** is a research and development fellow at Dow Building Solutions in Midland, MI. Brian Lieburn is a senior research specialist at Dow Building Solutions in Columbus, OH.*

walls built with housewrap and insulation only in the cavity space, along with a comparison of 2 × 4 and 2 × 6 constructions as a component of the wall design. Additionally, fibrous insulation and closed cell spray polyurethane foams were compared as cavity insulation and rim joist insulation. Descriptions and pictures of the energy performance strategies can be found in Tables 2–3 and Figures 2–7.

80% annual fuel utilization efficiency (AFUE), 71,000 Btu natural gas furnaces were used in the 2006 IECC homes. 92% AFUE furnaces were used for both the 2012 IECC minimum and 2012 IECC premium homes with sizes ranging from 69,000—93,000 Btu. The high performance homes used 96% AFUE, 57,000 Btu furnaces. Although envelope insulation and air sealing improvements

Table 1. Description of Test Houses

	House Plan A	House Plan B	House Plan C
Conditioned floor area (ft^2)	2800	3100	3100
Conditioned space volume (ft^3)	23900	27600	27100
Floors above grade	2	1	1

Table 2. Foundation Wall and Basement Floor Design

	2006 IECC	2012 IECC—Minimum	2012 IECC—Premium	High Performance
Under Floor Slab	None	None	None	R-10 XPS
Rim Joist—Interior	R-19 Kraft Faced Fiberglass batt	R-19 Kraft Faced Fiberglass batt	R-16 cc SPF	R-16 cc SPF
Rim Joist—Exterior	None	None	R-5 XPS	R-10 XPS
Basement Wall—Interior Finished	R-13 Kraft Faced Fiberglass batt	R-19 Kraft Faced Fiberglass batt	R-5 XPS	R-10 XPS
Basement Wall—Interior Unfinished	R-10 Fiberglass vinyl faced	R-15 Fiberglass vinyl faced	R-5 PIR	R-10 PIR
Basement Wall—Exterior	None	None	R-10 XPS	R-10 XPS

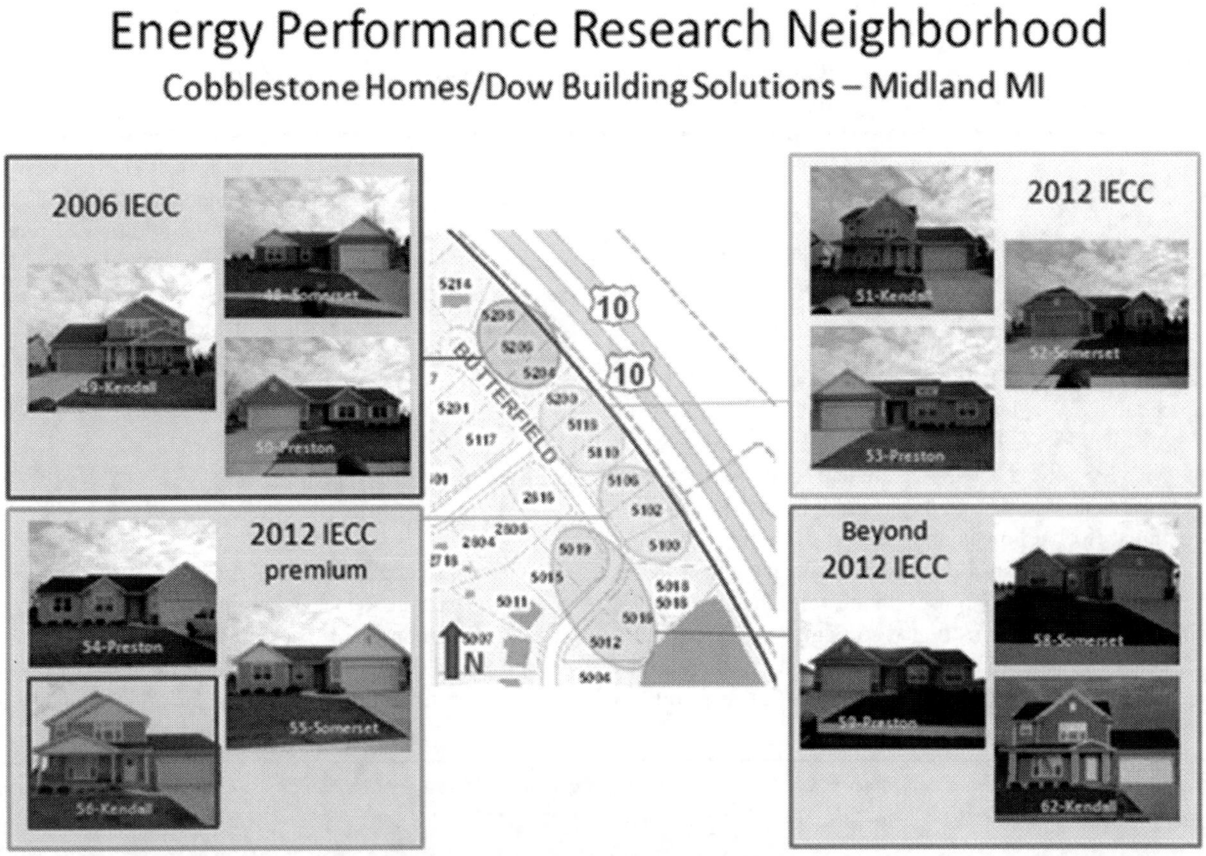

Figure 1 Energy performance research neighborhood.

in both 2012 IECC and high performance homes should have allowed reduced furnace sizes, HVAC contractors did not do so.

U-0.35 windows were used in the 2006 IECC homes. U-0.32 windows were used for both the 2012 IECC minimum and 2012 IECC premium homes. U-0.28 windows were used in the high performance homes.

Air Leakage Details

Above grade walls of 2006 IECC and 2012 IECC minimum homes were air sealed using minimal expanding spray foam sealant around windows and doors. Mechanical penetrations within and through the walls, rim joist, and ceilings were sealed using one-component expanding spray foam sealant. Wall plates and wall junctions were air sealed with latex caulk. The gypsum wall boards were installed using an adhesive at the top plate of the walls. Foundation air sealing in these homes was provided by taping the seams of the vinyl facing of the insulation for the unfinished area and drywall installation in the finished areas.

Table 3. Above Grade Wall and Ceiling Design

	2006 IECC	2012 IECC—Minimum	2012 IECC—Premium	High Performance
Stud Configuration	2 × 6 – 16 in. o.c.	2 × 6 – 16 in. o.c.	2 × 4 – 16 in. o.c.	2 × 6 – 24 in. o.c.
Interior	R-19 Fiberglass batt	R-19 Fiberglass batt	R-16 cc SPF	R-31 cc SPF
Exterior	Housewrap	Housewrap	R-5.5 structural insulated sheathing	R-5.5 structural insulated sheathing + R-5 XPS
Ceiling	R-38 Dry Blown Cellulose	R-49 Dry Blown Cellulose	R-49 Dry Blown Cellulose	R-12 2 in. cc SPF and R-49 Dry Blown Cellulose

Above grade walls of 2012 IECC premium and high performance homes were air sealed with continuous insulating sheathing with taped seams and closed cell spray foam in the wall cavities and rim joist. The penetrations in the ceiling of the 2012 premium homes were sealed with one-component expanding spray foam sealant while the high performance ceilings had 2 in. of closed cell spray foam applied. Windows and doors were sealed using minimal expanding spray foam sealant. Wires, pipes, and other penetrations in the wall framing were sealed using one-component expanding spray foam sealant. Wall plates and wall junctions were air sealed with latex caulk. Gypsum wall boards were installed with adhesive at the top plate of the walls. Foundation air sealing was provided by taping the seams of the rigid extruded polystyrene (XPS) or polyisocyanurate (PIR) insulating sheathing installed on the interior of the foundation wall. Closed cell spray foam in the rim joist was sprayed onto the foundation wall insulation to provide continuity.

INSTRUMENTATION

Each house was instrumented to collect hygrothermal and energy use data. Comfort data is collected using temperature and relative humidity data in above grade and below grade living spaces. Hygrothermal data including moisture content, humidity, and temperature is collected in three locations: one above grade wall cavity in a common living area such as a living room or eating area and two below grade locations in both the finished and unfinished areas of the basement. Additionally rim joist conditions are collected in both the finished and unfinished basement areas. Figures 8 and 9 show the instrumentation configuration for an example below grade measured cavity and a close up of cavity sheathing and stud moisture content and temperature measurements. Table 4 details the instrumentation used for data collection. Electrical circuit level monitoring is accomplished using a Powerhouse Dynamics eMonitor system wired to the electrical distribution box of

Figure 2 *Unfinished below grade area with vinyl faced fiberglass insulation.*

Figure 3 *Below grade finished basement area with extruded foam insulation.*

Figure 4 *Below grade unfinished area with polyisocyanurate insulation.*

Figure 5 *House 3: 2006 IECC, OSB walls prior to housewrap installation.*

Figure 6 *2012 IECC premium home with continuous insulation.*

Figure 7 *Above grade wall with spray polyurethane foam cavity and top plate insulation.*

Figure 8 *Below grade cavity instrumentation.*

Figure 9 *Upper cavity instrumentation including sheathing and stud temperature and moisture content instruments.*

each home. All energy consumption is electrical, including domestic hot water, with the exception of natural gas used for gas forced air space heating.

COMPARISON OF CONSTRUCTION COSTS

Invoices for all construction costs were compiled to calculate a total cost of construction for each of the homes. The total cost included cost variances due to site and elevation variations as well as other common construction variances such as price fluctuations, theft, damage, and weather related costs. This required a calibration process to equalize and isolate the energy performance costs. The following describes the calibration process used to eliminate cost variations not related to the energy performance strategies.

Price Variations

Construction of the twelve homes began in April of 2011 and completed in December 2011. Price changes occurred over the duration of the project. To establish a uniform price for each component, the unit price paid for all pertinent components was recorded for all the jobs. The average price of each component was used to calculate the final cost comparison.

Table 4. Instrumentation

Measurement	Device	Description
Temperature	Type T Thermocouple	
Moisture Content	Pin Resistance	Brass Screws, 1 in. apart
Relative Humidity	RH Probe	Campbell Scientific CS210/10162
	Data Logger	Campbell Scientific CR1000
	Multiplexer	Campbell Scientific AM 16/30

The installed cost for the 80% AFUE furnaces was higher than the 92% AFUE furnaces due to the increased cost for B-venting. Additionally 92% AFUE furnaces are the norm for the region, which resulted in premiums to purchase the 80%

AFUE furnaces. The HVAC cost for the 2012 IECC minimum and 2012 IECC premium homes should have been lower if smaller furnaces had been installed. Two different HVAC contractors were used on the project. An average price of the two contractors was used in the final cost comparison.

Quantity Variations

Consistent area values were used for each of the four different energy performance levels for all of the various components used on the enclosure. Area values were determined for each of the following locations on the thermal enclosure:

1. Basement floor
2. Floor over exterior (cantilever)
3. Basement wall—finished space
4. Basement wall—unfinished space
5. Rim joist
6. Above grade exterior wall between conditioned space and outside
7. Above grade exterior wall between conditioned space and garage
8. Above grade exterior wall between conditioned space and attic
9. Ceiling to attic

An analysis of the 2 × 4 and 2 × 6 materials invoiced for above grade walls revealed quantity variations between houses of the same plan type. These variations can be attributed to elevation differences, theft, and varying waste factors, among other factors. Adjustments were made to the quantity of these components to eliminate the variations using the following assumptions:

- The quantity of 2 × 6 studs (exterior walls) should be the same in 2006 IECC and 2012 minimum homes.
- The quantity of 2 × 6 studs in the high performance homes should be less than in 2006 IECC and 2012 minimum homes due to 24 o.c. spacing.
- The quantity of 2 × 4 studs (exterior and interior walls) in the 2012 IECC premium homes should equal the combined total of 2 × 6 studs (exterior walls) and 2 × 4 studs (interior walls) in the 2006 IECC homes.
- The quantity of 2 × 6 lineal material should be the same between 2006 IECC, 2012 minimum, and high performance homes.
- The quantity of 2 × 4 lineal material in the 2012 premium homes should equal the combined total of 2 × 6 and 2 × 4 lineal material in the 2006 IECC homes.

Basement framing material was tallied separately. The 2006 IECC baseline, 2012 IECC premium and high performance homes all used 2 × 4 16 in. on center interior wall framing on the inside of the foundation for the finished portion of the basement. Basement framing materials were equalized for these home types. The 2012 IECC minimum used 2 × 6 exterior and 2 × 4 partition walls. The combined 2 × 4 and 2 × 6 stud quantities were calibrated to match the baseline 2 × 4 total.

Table 5. Average Cost Premium Comparison

	2012 IECC—Minimum	2012 IECC—Premium	High Performance
Frame, Insulation, and Air Seal	$540	$4760	$12,080
Fenestration	$1380	$1380	$2330
HVAC	($410)	($410)	$890
Lighting	$100	$100	$350
Total Premium over Baseline	$1610	$5830	$15,650
Range of Premiums	$1250–$2040	$5420–$6590	$15,480–15,880

Elevation Variations

Elevation changes were made to provide variety to the neighborhood. Roof lines were altered by adding or clipping gables. Front porch styles also varied. Combinations of exterior cladding materials provided further variety. All labor and material costs related to these variations were considered and excluded from the cost comparison.

The thermal envelope, including fenestration, remained unchanged for each of the house types. The energy performance may be impacted due to minor changes in sunlight exposure and shading characteristics as a result of the elevation variations.

Lot Variations

Variations in soil conditions required different heights of foundations walls for the porches, garages, and basements. Excavation, foundation, and other lot related cost variables were excluded from the comparison. The height from the top of basement floor to the top of foundation was used to calculate the insulated area.

Job Cost Variances

Common job cost variances due to theft, damage, weather, or invoicing errors were excluded from the comparison.

COMPARISON OF PREDICTED AND ACTUAL ENERGY USE

A significant aspect of this research is to find out the impact for each of the energy performance strategies on the total amount of energy actually used. The scope of this paper covers the energy used to heat for the entire first heating season. The energy used includes natural gas only. The furnace is the only natural gas appliance in the homes, therefore 100% of the natural gas used was by the furnace. The furnace electric use was excluded to make a direct comparison to the HERS modeled Heating Btu amount.

Predicted Energy Use

Energy use was predicted using AEC REM/Rate™ Version 12.97. Specifications for 2012 minimum and 2012 premium homes were intentionally designed to have similar HERS scores. Table 6 shows both the planned and confirmed or "as built" HERS score averages for the three homes built to each energy performance strategy.

Cost Effectiveness

Table 7 compares the cost of improvements with the modeled savings over the baseline 2006 IECC homes. The baseline homes were built to common building practices in the area used to meet the 2006 IECC. These homes had an average confirmed HERS score of 77, which means their energy performance is considerably better than a home built to 2006 IECC code minimum.

Assuming a 30 year term mortgage at 4% interest, a positive cash flow can be achieved by building to the 2012 IECC minimum. Building to the 2012 IECC premium strategy is essentially cash flow neutral. We believe the 2012 IECC minimum and 2012 IECC premium represent the extremes of strategies most builders will use to comply with 2012 IECC. The optimum value strategy most likely would fall between the 2012 IECC minimum and premium and would be cash flow positive. Considering only economics, building to the high performance level does not produce a positive cash flow over a HERS Index of 77.

Actual Energy Use

Natural gas data was collected for the 2012–2013 winter. Four of the homes, one home from each of the performance strategies, were vacant for a significant portion of the winter. Data from these four homes was excluded in the comparison of actual versus modeled energy use. Figure 10 shows the total of two homes built to each energy performance strategy.

Table 6. Average Annual Modeled Energy Use

	2006 IECC	2012 IECC—Minimum	2012 IECC—Premium	High Performance
HERS Score (Planned)	82	57	57	45
HERS Score (Confirmed)	77	59	54	48
Heating (kBtu)	76	52	45	30
Heating Cost[1]	$733	$502	$434	$287
Heating Savings from Baseline	$0	$231	$299	$446

[1] Heating cost based on $.963/kBtu natural gas.

Observations

- 2006 IECC significantly outperformed the model.
- 2012 IECC minimum, 2012 IECC premium and high performance strategies all aligned well with the model.
- Three out of the four energy performance strategies outperformed the model.
- Occupants provided day and night temperature setpoints. Below are the average day and night setpoints for both homes in each performance level:
  - 2006 IECC 68.3
  - 2012 minimum 64.3
  - 2012 premium 69.5
  - high performance 73.0

The reason the 2006 IECC significantly outperformed the model is not understood. The exact impact of the thermostat setpoint on the energy use cannot be discerned. Most of the rental agreements are short term, therefore, we expect over the five year research period, the range of temperature setpoints should narrow. The high performance home used less energy

Table 7. Cash Flow Analysis of Modeled Energy Savings

	2012 IECC—Minimum	2012 IECC—Premium	High Performance
Average Cost of Improvements over Baseline	$1612	$5828	$15,644
Monthly Payment 30 Yr Term 4%	$7.70	$27.82	$74.69
Modeled Energy Savings/Month	$19.67	$25.40	$37.81
Cash Flow	$11.97	($2.42)	($36.88)

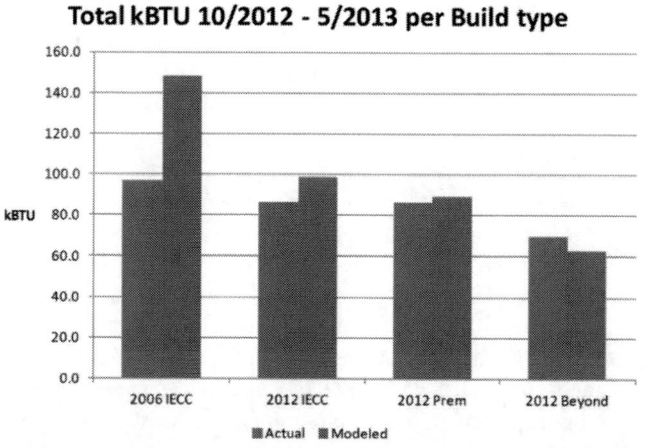

Figure 10 Winter 2012–13 actual versus modeled natural gas use.

than the than the 2012 minimum even though the high performance home maintained an 8.7°F higher interior temperature.

COMPARISONS OF HOME PERFORMANCE

Comparison of Air Leakage

Air leakage comparisons for all homes are shown in Table 8. Hygrothermal performance comparisons are shown for the two different 2012 strategies for above grade cavity wall, rim joist, and foundation wall.

Comparison of Air Leakage Details

Blower door test results showed air sealing measures in all but the 2012 IECC minimum met the 2012 IECC code requirement of <3 air changes per hour (ACH) @ 50 Pa for climate zone 5. The 2006 IECC does not have a performance requirement for air leakage. The 2012 premium and high performance homes with continuous exterior sheathing and closed cell spray foam in the stud cavity produced the lowest air leakage rate.

Table 8. Whole House Air Leakage

	2006 IECC	2012 IECC—Minimum	2012 IECC—Premium	High Performance
CFM @50 Pa Average	1230	1340	960	770
CFM @50 Pa Range	1180–1270	1270–1420	880–1040	740–800
ACH @50 Pa Average	2.8	3.1	2.2	1.8

Comparison of Wall Cavity Hygrothermal Performance of Homes with Continuous Insulation and Homes with Housewrap

We chose to make the most direct hygrothermal comparison between the two strategies designed to meet the 2012 IECC as they are most directly comparable. Temperature and moisture content of both continuously insulated walls and walls with cavity insulation alone are shown in Figure 11.The continuously insulated walls are constructed with 2 × 4 framing, a sheathing layer consisting of a structural layer laminated to a R5 insulation layer, which provides the continuous thermal break, and the cavities are filled with approximately R16 closed cell spray polyurethane foam (houses 7–9). The comparison walls are constructed with 2 × 6 framing, a layer of oriented strand board (OSB) sheathing, and spun bonded polyolefin housewrap. The cavity insulation is R19 Kraft faced fiberglass batt insulation. Both wall configurations meet the requirements of the 2012 IECC. The temperature of the cavity side, interior, of the sheathing for the six homes compared is shown in Figure 11. Although the temperature movement trends for all homes is similar, the average temperature of the continuously insulated wall cavities is higher than those with cavity insulation, and, as expected, the exterior sheathing temperature of the non-continuous insulation (CI) walls approaches the exterior conditions. The average temperature of the non-CI houses was 40°F compared to the continuously insulated house group average of 45°F. The moisture content of the walls also follows the same trend but the CI walls generally have 2%–4% lower moisture content, with non-CI wall reaching a maximum moisture content of 23% in the coldest months. In comparison, the CI walls reached a lower moisture content of 21%. During the winter season, the non-CI walls had greater than 20% moisture content for 5% of the period whereas the CI walls were above 20% moisture content for 1% of the same period. It has been long noted that

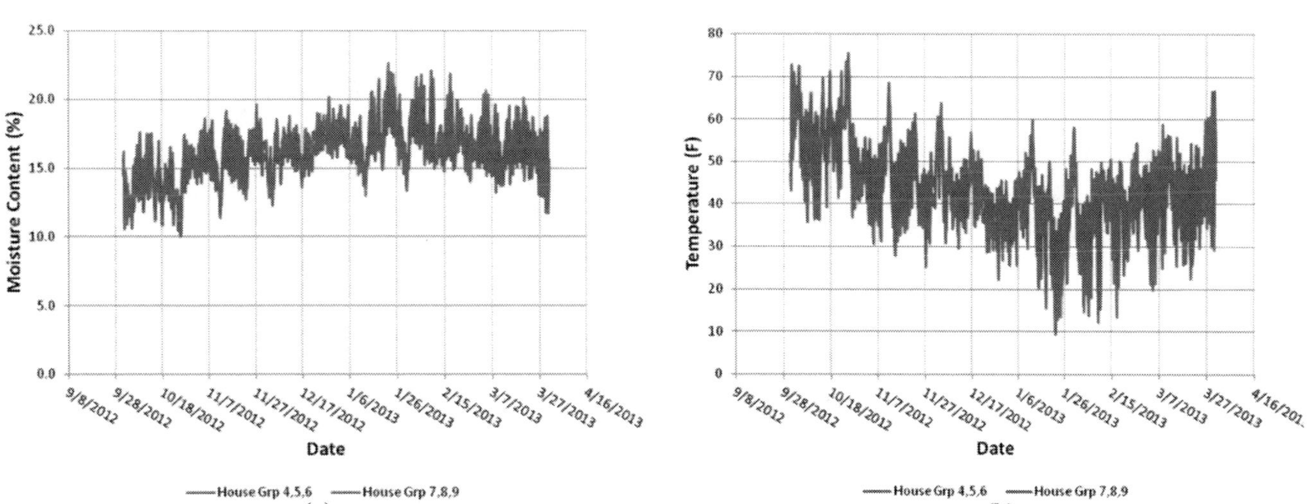

(a) (b)

Figure 11 (a) Moisture content and (b) temperature: above grade wall sheathing at mid-height of wall.

wood should be protected from fungal decay by keeping the moisture content below 20%.

Comparison of Foundation Wall Performance

All foundations were 8 in. poured concrete with brick pattern on the interior and exterior. Foundation insulation was compared in both finished and unfinished areas of the research homes. Homes 4–6 utilize vinyl faced fiber batt insulation hung from the top of the basement wall to the basement floor in the unfinished basement area. The finished basement area of homes 4–6 is insulated by installing fiber batt insulation between 2 × 6 stud frame walls. All of the foundation insulation of homes 4–6 is installed on the interior of the basement wall. The unfinished area of homes 7–9 are insulated by adhering polyisocyanurate insulation directly to the interior of the basement wall, which does not require gypsum board or other thermal barrier for fire protection. The finished part of the basements of homes 7–9 are insulated with a layer of extruded polystyrene foam insulation applied to the basement wall prior to 2 × 4 framing and drywall. No additional insulation was installed in the framed wall cavities. In homes 7–9, a layer of continuous extruded polystyrene foam insulation was installed on the exterior of the basement wall. Instrumentation of moisture content and temperature is placed at the interface between the interior basement wall and the interior insulation system as it is of interest to measure the conditions at this location which could be prone to moisture build up. The average moisture content of the interface between the basement wall and insulation is higher for the fiber insulation compared to the rigid board insulation in both finished and unfinished basement areas. The 4, 5, 6 home group average moisture content in the unfinished basement area is 15% and reaches a maximum of 18%; in comparison, the 7, 8, 9 group average moisture is 12% and reaches a maximum of 13% (Figures 12 and 13). The temperature of the interface between

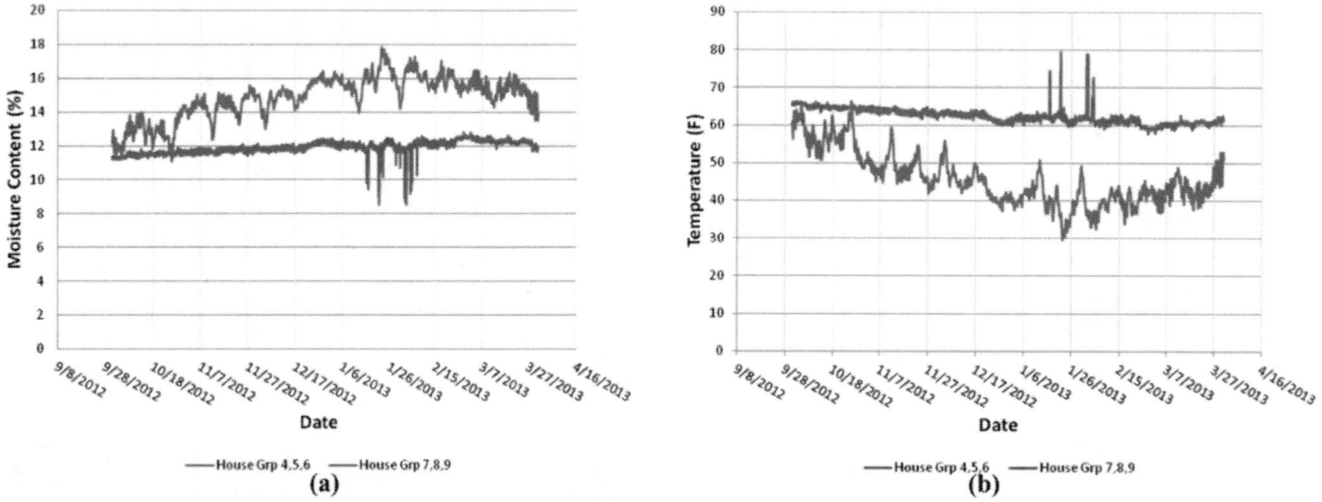

Figure 13 *(a) Moisture content and (b) temperature: basement unfinished wall 1 in. from top of wall.*

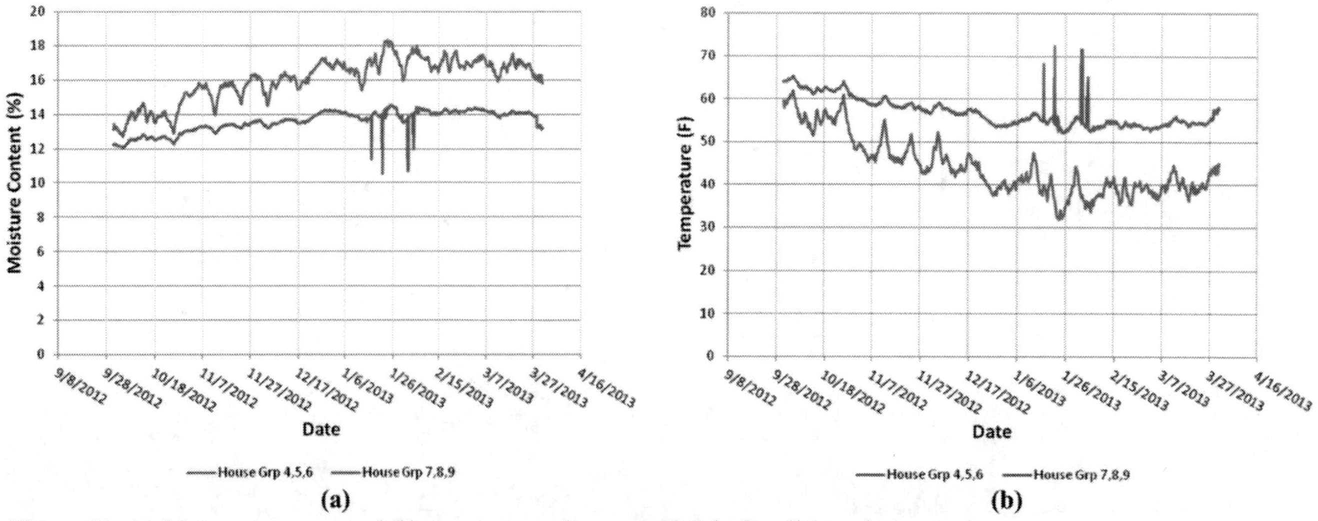

Figure 12 *(a) Moisture content and (b) temperature: basement finished wall 1 in. from top plate.*

the basement wall and insulation tends to more closely follow the interior basement conditions for the rigid board insulation cases whereas the temperature of the interface between the wall and fibrous insulation more closely follows the exterior temperature and in winter conditions tends to be much colder, clearly impacted by the lack of an exterior insulation layer. The location of the temperature shown is slightly below the basement sill plate as this is a worse case situation. The 4, 5, 6 group average temperature is 46°F and reaches a minimum of 30°F compared to the 7, 8, 9 group average of 62°F and minimum 58°F.

Comparison of Rim Joist Performance

Rim joists of homes 4–6 were insulated using R-19 Kraft faced batt insulation cut to fit between floor joists. Closed cell spray polyurethane foam (cc SPF) was installed in the interior band joists of homes 7–9 with additional insulation provided by the continuous insulation layer installed on the enclosure exterior, note that this would be considered a double vapor barrier situation around the OSB rim joist as both the cc SPF and XPS are low vapor permeance insulations. Figure 14 shows moisture content and temperature were measured at the interior rim joist surface under the interior insulation. Homes 7–9 demonstrate lower rim joist moisture content than homes 4–6. The average of the 4, 5, 6 group reaches moisture content above 20% for periods of time during January and February whereas the 7, 8, 9 group average remains less than 20% during the same period. It has been long noted for protection of wood and wood products from fungal decay moisture content should be kept below 20% moisture content. Temperature of the interior band of homes 7–9 is higher than the same measurement in homes 4–6.

CONCLUSION

Although this multi-year research project is in its early stages, some key conclusions can be made:

- Detailed cost analysis shows the premium Cobblestone Builders paid to build to 2012 IECC was considerably lower than many estimates.
- A comparison of HERS scores based on plan versus confirmed HERS score based on field verification and whole house air leakage testing showed a marked improvement of the 2102 CI homes over the 2012 non-CI homes.
- Actual energy used in three out of the four strategies outperformed the model.
- High performance strategy used the least amount of energy to heat the home even though it had an average thermostat set point 8.7°F higher than the 2012 minimum.
- Exterior foundation insulation significantly reduced moisture content at the interior concrete surface.
- Insulating both sides of the OSB rim joist with low vapor permeance materials (double vapor barrier) had lower moisture content than vapor open insulation strategies.
- The foam insulated walls in general had lower moisture content than those without foam insulation.

ACKNOWLEDGMENTS

The authors would like to acknowledge Cobblestone Homes for their partnership. We also express our appreciation to IBACOS for their assistance with instrumentation and data collection systems.

REFERENCES

ICC. 2006. *International Energy Conservation Code*. Washington, DC: International Code Council.

ICC. 2012. *International Energy Conservation Code*. Washington, DC: International Code Council.

NCDC. 2012. National Climatic Data Center, U.S. Department of Commerce, Washington, DC.

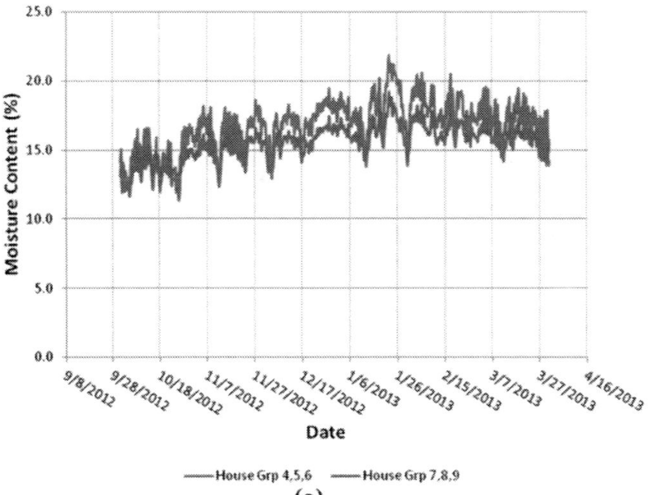

(a)

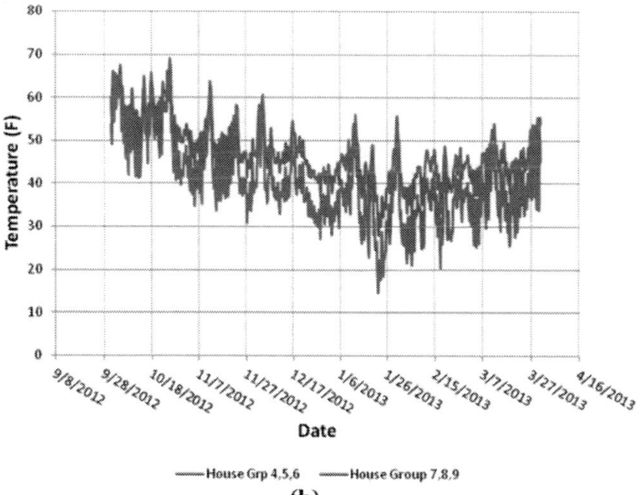

(b)

Figure 14 *(a) Moisture content and (b) temperature: rim joist at midpoint in each direction.*

Investigating Wind-Driven Rain Intrusion in Walls with the CARWASh

C.R. Boardman
Associate Member ASHRAE

Samuel V. Glass, PhD
Member ASHRAE

ABSTRACT

Wind-driven rain provides the primary external moisture load for exterior walls. Water absorption by the cladding, runoff, and penetration through the cladding or at details determine how a wall system performs. In this paper we describe a new laboratory facility that can create controlled outdoor and indoor conditions and use it to investigate the water management performance of a typical exterior wall assembly. Specifically we have quantified how much water gets past the cladding and hits the water-resistive barrier under particular conditions for a common lap siding.

The CARWASh (Chamber for Analytical Research on Wall Assemblies exposed to Simulated weather) was used to deliver wind-driven rain at two wind conditions and a range of rain intensities. The amount of wind-driven rain was measured on an exposed wall in the CARWASh. Results are presented as a function of wall location, wind speed, and rain intensity. Next, defects were introduced in the lap siding and absorbent pads were placed in the drainage cavity between the cladding and the sheathing to capture any moisture that got past the siding. Results are presented as a percentage of moisture getting past the cladding to incident wind-driven rain and are correlated with wind pressure across the cladding for various wind speeds and rain intensities.

INTRODUCTION

Wind-driven rain provides the primary external moisture load for exterior walls (Trechsel and Bomberg 2009; Straube and Burnett 2005). Hygrothermal and durability analysis of façades requires the quantification of the wind-driven rain loads (Blocken and Carmeliet 2004; Teasdale-St-Hilaire and Derome 2005). Software tools for transient hygrothermal analysis typically allow the user to specify a moisture source to represent wind-driven rain that intrudes past the cladding to simulate defects in construction and ensure a robust exterior wall system. ASHRAE Standard 160 (2009) provides criteria for inputs to transient hygrothermal models and includes a provision for rain penetration past the cladding, allowing a default value of 1% of the total wind-driven rain to get past the cladding, given limited measured data (Tenwolde 2011).

This study introduces a new laboratory facility that can create controlled outdoor and indoor conditions and thus help investigate the water management performance and moisture dynamics of exterior wall assemblies: the CARWASh (Chamber for Analytical Research on Wall Assemblies exposed to Simulated weather). One of the first uses for the CARWASh was to begin to supply measured data to better inform the modeling suggestion that 1% of wind-driven rain get past the cladding. The CARWASh was designed to provide realistic wind-driven rain under typical conditions. It is thus distinguished from other test facilities with different goals, such as the Wall of Wind, which is used to test under hurricane conditions (Bitsuamlak et al. 2009). It is further distinguished from spray tests, which rely on large pressure differences to force rapid water intrusion through wall defects (Sahal and Lacasse 2005). Spray rack and pressure systems are commonly used to test water penetration resistance for windows, with recent interest in dynamic pressure loading (Van Straaten et al. 2010; Lopez et al. 2011; ASTM 2011). While the CARWASh can apply modest pressure loads to a wall system, the primary intent was to use the force of the simulated wind to deliver

***C.R. Boardman** is a general engineer and **Samuel Glass** is a research physical scientist at the Forest Products Laboratory, U.S. Forest Service, Madison, WI.*

Published by ASHRAE 2013

Figure 1 *Outside the CARWASh showing the wind blower.*

water to the wall from falling raindrops. The kinetic energy thus comes from the wind, and the dynamic wind pressure is the primary force moving water into wall defects.

This study reports on basic CARWASh capabilities, with a focus on the distribution of wind-driven rain on the exterior wall, and presents intrusion results from one cladding, a simple horizontal lap siding, with and without intentional defects. Future work will compare modeling and experimental results for newer wall systems and continue exploring wind-driven rain intrusion in other cladding systems.

MATERIALS AND METHODS

The CARWASh is a temperature- and humidity-controlled chamber large enough to mount a 10 × 10 ft (3.05 × 3.05 m) wall assembly between the inner and outer sections of the chamber; see Figure 1 for a photograph of part of the laboratory in which the CARWASh functions and Figure 2 for a schematic of the inside.

The small inner section (Figure 3) can simulate indoor conditions, ranging in humidity from 30%–70% relative humidity (RH) and temperatures (*T*) from 60°F to 80°F (15.5°C to 26.6°C). It also houses the measurement systems that can probe the wall for *T*, RH, and wood moisture content in various locations within an assembly, along with pressure differentials across wall assembly elements. The wall assembly can float on an air-cushioned platform that allows transport of a wall from the assembly area located near the CARWASh. The platform and wall fit into an opening between the two sections of the chamber, and any remaining gap is closed with an inflatable seal. Figure 4 shows a wall in place.

The larger outer section can simulate outdoor conditions, including wind, rain, and sun, ranging in dew point from 25°F to 80°F (–4°C to 27°C) and temperature from 30°F

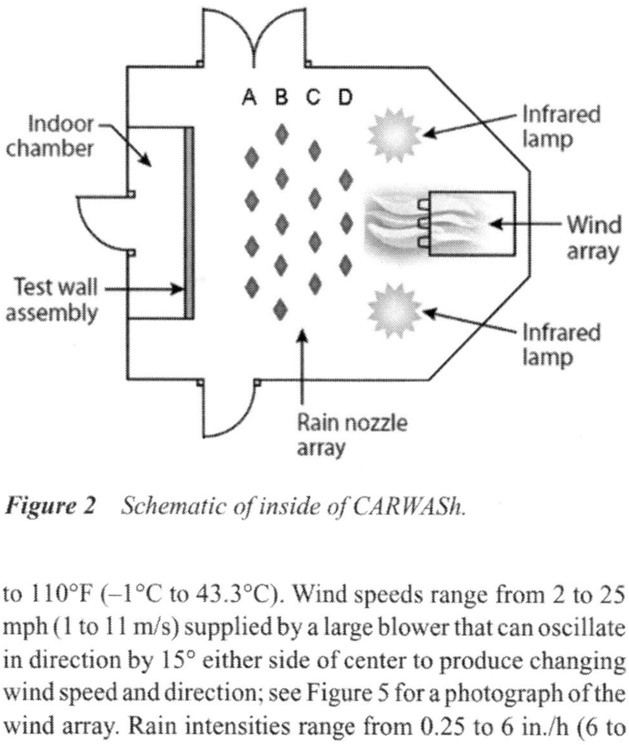

Figure 2 *Schematic of inside of CARWASh.*

to 110°F (–1°C to 43.3°C). Wind speeds range from 2 to 25 mph (1 to 11 m/s) supplied by a large blower that can oscillate in direction by 15° either side of center to produce changing wind speed and direction; see Figure 5 for a photograph of the wind array. Rain intensities range from 0.25 to 6 in./h (6 to 150 mm/h) supplied by four rows of nozzles (labeled A through D in the Figure 2 schematic) designed to produce realistic-sized raindrops. Three different nozzle sizes were used to achieve a smaller flow in the front and a larger flow in the back. The spray pattern is a wide-angle flat fan with uniform distribution and medium-sized drops, typical diameter near 0.03 in. (0.0007 m). The nozzles point straight down, intended to create a sheet of water drops that could be picked up by the wind so that the wind delivers the water to the wall. Each row can individually be turned on and off

Figure 3 Small indoor chamber in CARWASh.

Figure 4 Wall installed between indoor and outdoor chambers, with wind and rain gauges.

through computer control, and individual nozzles can be manually turned off if desired. Figure 6 shows a nozzle in action. The outer chamber also has infrared lamps (seen in Figure 5) that can simulate the drying effects of the sun, capable of heating from 0 to 100 W/ft^2 (0 to 1100 W/m^2). Return air ducts on each side of the wall assembly return air to the conditioning system for air recirculation, and grates in the floor return rainwater to a sump for water recirculation. By partially blocking the return air ducts a wind gust can be simulated. These gusts can last up to 5 s and apply an additional pressure up to 125 Pa (0.5 in. water).

Measuring Wind-Driven Rain

As a baseline for all work in the CARWASh we needed to characterize how much wind-driven rain was actually delivered to the wall under a variety of operating conditions. A number of wind-driven rain gauges were created out of polymethyl methacrylate (PMMA) with a collection area of 0.2 × 0.2 m (7.8 × 7.8 in.) following the recommendations of Blocken and Carmeliet (2006); see Figure 4 for one installed on a wall and Figure 7 for a close up of the latest design.

Captured rainwater was drained through the wall and collected in plastic containers that were weighed periodically. The CARWASh was set to operate at a variety of constant conditions so we could record the average rate of wind-driven rain deposited in each gauge. At any one time only four gauges

Figure 5 CARWASh wind array, also showing IR heaters and rain spray nozzles overhead.

Figure 6 Rain nozzle in action inside CARWASh.

were available, so the data was collected in five different runs and grouped into three sets of results. The first set of gauges (set I) was placed at mid-height of the wall, such that the bottom of the collector area was 50.25 in (1.27 m) from the top of the platform on which the wall sits; see Figure 8, Set I. Four gauges collected rain simultaneously, with the first positioned in the center of the wall such that the center of the collector area was 57.5 in. (1.46 m) from the left side as facing the wall. The remaining gauges were on the right side of the wall as shown in Figure 8, Set I. Set II was done with the four gauges in the center of the wall and extending down to allow us to study the effect of height, as shown in Figure 8, Set II. The lowest had its lowest collection level 22.25 in. (0.56 m) from the platform surface, with the others as shown in Figure 8, Set II. Set III explored wind-driven rain both left and right of center at a constant height of 30 in. (0.76 m) from the platform, as shown in Figure 8, Set III low, and at a height of 54.25 in. (1.38 m) across the full width of the wall, as shown in Figure 8, Set III high. The 4 gauges are numbered 2, 3, 6, and 7 in Figure 8, Set III low, to correspond to the same gauges when moved a bit higher in Figure 8, Set III high. Further, the final Set III high measurements were done in two runs (four gauges at a time) such that positions 4 and 5 are the same physical location in the center of the wall. Note also that Set III high included locations where defects were later introduced, namely at positions 1, 4, and 8.

The wind-driven rain data was analyzed using the theory from Straube and Burnett (2005) as outlined in their equation 12.1, but neglecting the angle between the wind and normal to the wall, since this is close to head-on in the CARWASh. Further, our Equation 1 assumes a core rain deposition factor (RDF) of 0.35 and introduces a location-specific factor (LF) that modifies the RDF to account for wind flow patterns at that location on the wall:

$$\mathrm{WDR} = \mathrm{LF} \times \mathrm{RDF} \times \mathrm{DRF} \times V \times r \qquad (1)$$

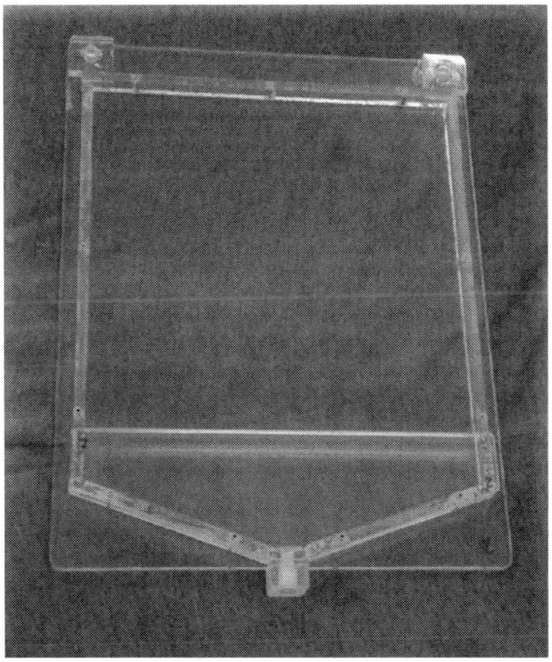

Figure 7 Wind-driven rain gauge close up.

Here WDR is the wind-driven rain at the surface location (kg/m^2·h), V is the velocity of the wind (m/s), and r is rain intensity on a horizontal surface (mm/h). The driving rain factor, DRF, accounts for the speed of falling raindrops,[1] which depends on the rain intensity and is calculated using Straube and Burnett's equation 12.2 in our Equation 2:

$$DRF = 0.22 \times r^{-0.123} \qquad (2)$$

In this analysis, the key is the LF, which should be near 1 if the RDF is actually 0.35. Of course the RDF in real buildings ranges widely depending on wind flow patterns, so we expect the LF to show divergence from our base case in the CARWASh.

Defects in the Siding and Measurement of Water Intrusion

The simple wall used for baseline wind-driven rain and intrusion measurements was an 8 in. (0.2 m) wide lap siding installed over a vertical strapping to form a rain screen. Generally there was 2 in. (0.05 m) of overlap, leaving a 6 in. (0.15 m) reveal. The strapping was attached to 7/16 in. (11 mm) oriented strand board sheathing on nominal 2 × 8 in. (38 × 184 mm) framing 16 in. (0.4 m) on center. Access panels were created, generally 11.75 × 18.75 in. (0.3 × 0.47 m), allowing absorbent pads to be placed on the outside of the sheathing (facing the siding, inside the rain screen gap). Four of these panels were used to detect any water getting past the siding by measuring

1. The units work out for Equation 1 because DRF has units of inverse velocity (m/s)$^{-1}$ and a rain intensity of 1 mm over a 1 m^2 surface yields 1 L of water, which weighs 1 kg.

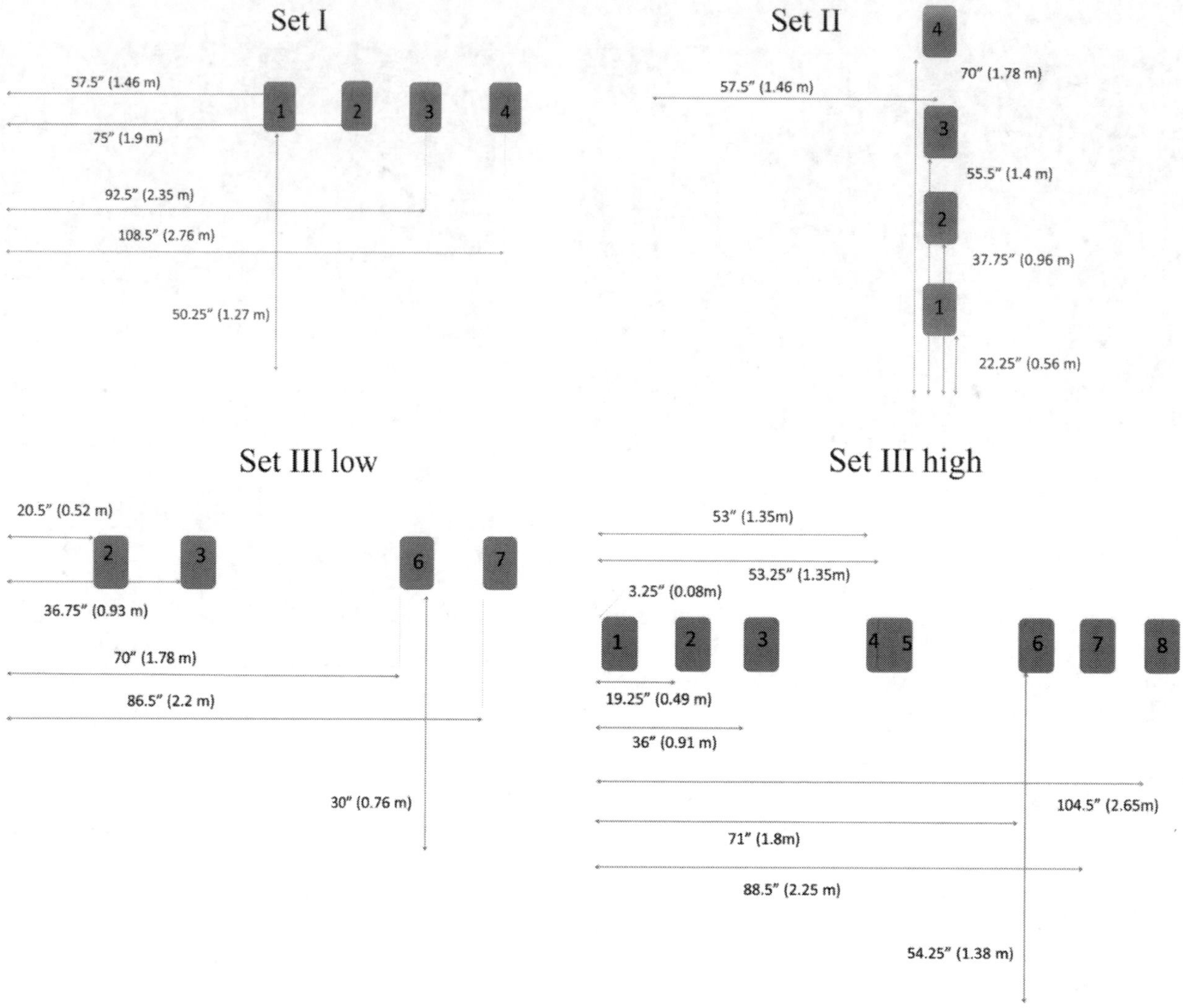

Figure 8 *Positions of wind-driven rain gauges.*

weight gain after a rain event. Two of the access panels were located at the edges of the wall, at middle height, with the bottom of panel 54 in. (1.37 m) from the platform and the outermost edge of the panel 4 in. (0.1 m) from the wall edge. The two middle panels were centered on the wall (again center of panel was 57 in. [1.44 m] from the left side), one at the same middle height as the other side panels and one just below it. The access panel locations can be visualized using Figure 8, Set III high, as capturing water entering at locations 1, 4, and 8, where the defects (described in the following paragraph) were introduced, and just below location 4, where there was no defect.

Defects in the siding were introduced by sliding a wedge up between the lap siding, creating a small, irregularly shaped gap (see Figure 9). Three of these defects were introduced at locations corresponding to the center of the access panels behind them, one at each of the two edges and one in the center, all at middle height such that the wedges were 61 in. (1.55 m) up from the platform. The gap thus created was largest in the center, from 0.177 to 0.24 in. (4.5 to 6 mm) tapering down to approximately a 0.118 in. (3 mm) gap when 7 in. (0.18 m) away from the center on either side. Under pressure from wind-driven rain, these gaps did allow water into the drainage cavity to be captured by the absorbent pads on the sheathing behind them.

Pressure Monitoring across the Siding

A number of experiments were performed to measure pressure differences across elements of the wall assembly at a variety of wind speeds. The differential pressure sensors could resolve pressure differences to 0.1 Pa (0.0004 in. water), and data acquisition was done at 20 Hz. Results are presented showing measurements directly across the cladding (see Figure 10).

RESULTS

Overall the wind-driven rain that falls on a particular location of a wall in the CARWASh depends on a large number of factors and is not simple to characterize. This reflects the

Figure 9 *Wedge to create defect in lap siding.*

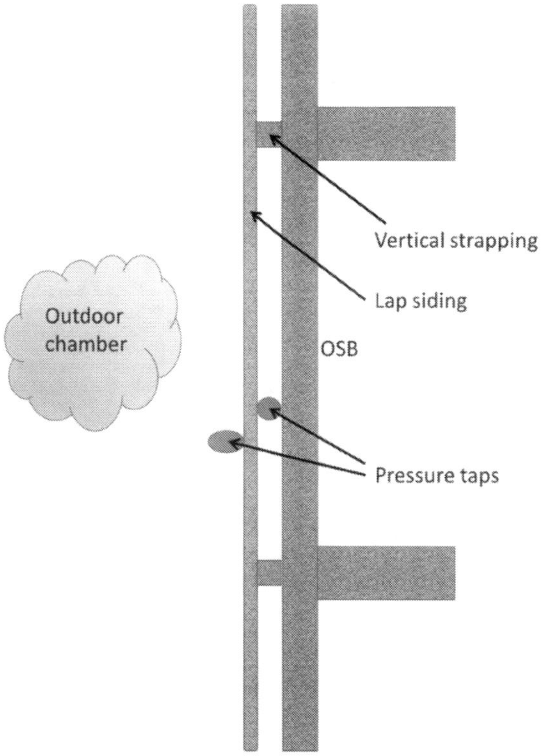

Figure 10 *Wall schematic with pressure taps across the cladding.*

complex fluid dynamics that control the wind flow and rain distribution and is approximated by Straube and Burnett (2005) in the RDF. We decided to limit the wind speeds investigated to 12 and 20 mph (5.3 and 8.9 m/s) to reduce the problem complexity. This reflects our rough observation that two wind flow patterns occur in the CARWASh. In the high-velocity case there is a more uniform pattern of wind-driven rain deposition on the wall and on other locations inside the chamber. However, the rain intensity is also a factor since it corresponds to which row of nozzles is turned on. For low-intensity rain events the front rows (A to B, front to back in Figure 2) are turned on, while for high-intensity rain events all the rows get turned on, with the back rows (C and D, furthest from the wall) providing more water. A further complicating factor is the ability of the CARWASh to shift the wind direction by sweeping the wind nozzle back and forth across the face of the wall by ±15°. To provide more realistic simulation of outdoor wind conditions we typically allowed this oscillation to vary the wind direction and arbitrarily set the oscillation period to 20 s. The wind-driven rain results shown in the following sections are average collection rates over sample periods greater than one hour so the wind direction differences are all averaged out.

WDR Set I—Locations Center to Right at Mid-Height

Table 1 presents the measured wind-driven rain (in $kg/m^2 \cdot h$) at the four locations for the first set of measurements investigating variation from center (WDR1) to the right side of the wall (WDR4) at mid-height. Table 2 helps interpret this data by providing the LF for each location when optimized over all the data to predict the WDR (first row), and when optimized only over similar wind speed and rain nozzle rows activated.

Inspection of Table 2 shows that the high-speed wind flow pattern provides more uniform distribution across the wall surface. To see this, first notice that LF4, the location to the far right, receives less wind-driven rain, as would be expected because the wind is less often blowing in that direction due to the blower oscillation. But also note that LF4 drops off more sharply for the low-speed conditions, indicating the less uniform distribution. Further, in the case with just row A where very little water is available, the high-speed case delivers less water to the wall because water is spread to other locations in the chamber. Finally, inspection of Table 2 also shows that our choice for overall RDF was reasonable, since the LF factors are close to 1.

WDR Set II—Locations in Center of Wall from Bottom to Mid-Height

Table 3 presents the measured wind-driven rain (in $kg/m^2 \cdot h$) at the four locations for the second set of measurements investigating variation from bottom (WDR1) to mid-height on the wall (WDR4) in the center. Table 4 helps interpret this data by providing the LF for each location when optimized over all the data to predict the wind-driven rain (first row) and when optimized only over similar wind speeds and rain nozzle rows activated.

Table 1. WDR: Mid-Height Center to Right—Set I

WDR1	WDR2	WDR3	WDR4	V, mph	r, in./h	Rows
9.0	8.2	5.2	2.4	12	1/2	A
8.7	8.1	5.5	2.7	12	1/2	A
9.5	9.2	6.1	3.6	20	1/2	A
10.8	10.0	6.9	4.1	12	3/4	A
13.3	11.8	8.1	5.1	12	1	A
6.2	6.3	6.1	5.5	20	1	A
11.8	9.7	6.2	2.8	12	3/2	A-B
22.6	20.7	19.8	15.8	20	3/2	A-B
25.3	27.7	25.1	18.3	20	2	A-B
25.5	27.7	25.4	18.4	25	2	A-B
28.7	24.5	18.9	9.2	15	3	A-C
11.2	9.5	8.7	6.5	5	4	A-C
25.4	27.3	23.6	11.6	10	4	A-C
41.3	42.2	37.4	18.5	15	4	A-C
54.4	56.9	53.6	38.7	20	4	A-C
54.8	57.0	53.5	37.9	20	4	A-C
59.6	61.5	57.5	39.4	20	4	A-C
56.1	59.0	50.4	35.7	25	4	A-C
67.4	68.8	66.0	40.5	20	5	A-D
85.0	90.1	81.9	55.4	25	6	A-D
76.8	80.8	74.6	49.4	25	6	A-D
28.0	31.7	26.0	12.0	12	6	A-D

Table 2. LF: Mid-Height Center to Right—Set I

LF1	LF2	LF3	LF4	Case
1.217	1.267	1.152	0.760	Overall fit
2.023	1.834	1.243	0.715	A low speed
0.743	0.745	0.618	0.490	A high speed
1.094	1.149	1.057	0.785	A-B all speed
1.314	1.288	1.102	0.553	A-C low speed
1.322	1.378	1.259	0.888	A-C high speed
1.193	1.250	1.157	0.760	A-D high speed
0.825	0.936	0.767	0.354	A-D low speed

Table 3. WDR: Center from Bottom to Mid-Height—Set II

WDR1	WDR2	WDR3	WDR4	V, mph	r, in./h	Rows
16.6	9.6	7.7	4.3	12	1/2	A
26.2	13.6	15.8	16.4	12	1	A
28.3	14.6	15.6	16.5	12	1	A
8.1	5.0	7.9	12.6	20	1	A
42.4	25.3	18.5	11.1	12	2	A-B
51.7	40.3	30.3	21.0	12	3	A-B
32.4	28.3	27.2	21.3	20	2	A-B
38.3	34.4	41.3	39.3	20	3	A-B
41.7	29.7	18.2	6.6	12	3	A-C
63.4	47.3	32.4	15.1	12	4	A-C
57.7	47.9	43.6	25.5	20	3	A-C
66.3	60.1	59.9	42.6	20	4	A-C
67.8	60.5	59.5	42.2	20	4	A-C
53.8	36.1	17.4	4.5	12	5	A-D
67.1	45.6	27.8	9.1	12	6	A-D
85.9	75.7	63.0	29.4	20	5	A-D
97.0	87.8	78.8	38.7	20	6	A-D
28.6	18.4	7.9	1.4	12	4	B-D
33.5	19.3	7.8	1.5	12	4	A-D

Table 4. LF: Center from Bottom to Mid-Height—Set II

LF1	LF2	LF3	LF4	Case
1.786	1.460	1.233	0.717	Overall fit
3.927	2.065	2.195	2.177	A low speed
0.688	0.424	0.674	1.075	A high speed
2.958	2.107	1.571	1.046	A-B low speed
1.759	1.426	1.383	1.146	A-B high speed
2.513	1.847	1.221	0.531	A-C low speed
1.734	1.530	1.486	1.014	A-C high speed
1.744	1.561	1.357	0.653	A-D high speed
1.929	1.305	0.728	0.221	A-D low speed

Table 5. Summary LF: Full Width at Two Heights —Set III

LF1	LF2	LF3	LF4	LF5	LF6	LF7	LF8	Height
1.014	1.438	1.364	1.224	1.232	1.277	1.154	0.768	54.25 in.
	1.261	1.268			1.126	1.021		30 in.

Inspection of Table 4 again confirms that high-speed flows show more uniform distribution, but there is more variation from bottom to top than from center to side, with the lower sections of wall generally receiving more wind-driven rain.

WDR Set III—Left and Right Asymmetry

Finally, a series of experiments at two different heights across the full width of the wall, measurement set III, showed that the wind-driven rain distribution was not as symmetric as expected, with the left side getting slightly more rain. This is illustrated in Table 5, which shows the overall summary fit LF (over all wind speeds and rain intensities) for each location. The top row corresponds to the height 54.25 in. (1.38 m) above the platform, with positions shown in Figure 8, Set III high. The second row of Table 5 corresponds to the height 30 in. (0.76 m) above the platform, with distribution as shown in Figure 8, Set III low.

Defects and High Wind Required for Rain Intrusion

Regardless of the wind speed or rain intensity, we never measured any rain intrusion past well-installed lap siding. The typical test was 20 mph (8.9 m/s) wind and 4 in./h (102 mm/h) rain intensity. One such test lasted 23 h with no weight gain in the absorbent pads. However, a similar near 8 h test with the gaps created by wedges at three locations did show intrusion. The amounts collected were very small, with 6.9 g (0.21 oz) at location 1 on the left side of the wall, 2.8 g (0.098 oz) at location 2 in the center of the wall, 2.8 g again at location 3 on the right, and 0 at location 4 just below location 2 but with no gap in the siding. These results are compared to a similar test done at 12 mph (5.3 m/s) wind speed and 2 in./h (51 mm/h) rain intensity in Table 6. That table shows the percent intrusion (intrusion ÷ incident of wind-driven rain × 100%) for the four locations. To area normalize the intrusion, each defect was assumed to represent a 6 × 11 in. (0.043 m^2) area corresponding to the revealed height of the lap siding and the width of the absorbent pad capturing the water behind the defect.

We are near the intrusion detection limit (around 0.05 g (0.002 oz) mass gain in the pad) with this method in the 12 mph test, with hardly any water getting past the siding, even with the defects. The wind speed appears to be the major factor allowing intrusion. Three more tests, all at rain intensity of 4 in./h but with increasing wind speeds of 12, 16, and 18 mph (5.3, 7.1, and 8 m/s), show this effect in Table 7. The only significant water intrusion (greater than 1 g [0.035 oz] of water in the pad) starts at 18 mph (8 m/s).

Table 6. Percent Intrusion—Four Locations

1	2	3	4	Case
0.046	0.015	0.026	0.000	4 in./h, 20 mph
0.001	0.000	0.001	0.000	2 in./h, 12 mph

Table 7. Percent Intrusion—Wind Speed Effects

1	2	3	4	Case
0.000	0.000	0.000	0.000	12 mph
0.001	0.000	0.001	0.000	16 mph
0.005	0.002	0.002	0.000	18 mph
0.046	0.015	0.026	0.000	20 mph

Pressure Measurements across the Cladding

Figure 11 shows a graph of pressure differential across the cladding (as in Figure 10) at two locations, near the defects, one in the center and one on the left side. These pressures are from 18 mph winds where we start to see rain intrusion, and they show peaks over 40 Pa (0.16 in. water) when the wind is blowing toward the left side of the wall.

The maximum pressure near 40 Pa at a wind speed of 18 mph is in the range we would expect from the stagnation pressure formula from Bernoulli's equation; see Equation 3 and Table 8.

$$P = \frac{\rho}{2} \times V^2 \tag{3}$$

Note that the wind speeds reported were measured less than 3 ft (0.9 m) from the wall, 11 ft (3.3 m) from the nozzle, so they should be close to the wind speed normal to the wall just before it changes direction to flow along the wall surface. Further, most of the total pressure drop occurs across the cladding. Figure 12 compares the pressure drop across the whole assembly (i.e., from the outdoor chamber to the indoor chamber), with the drop only across the sheathing (the only resistance left after the cladding), for a wind speed of 20 mph, with the pressure taps on the left side of the wall.

It is not entirely clear why most of the drop is across the cladding. The access holes in the oriented strand board sheath-

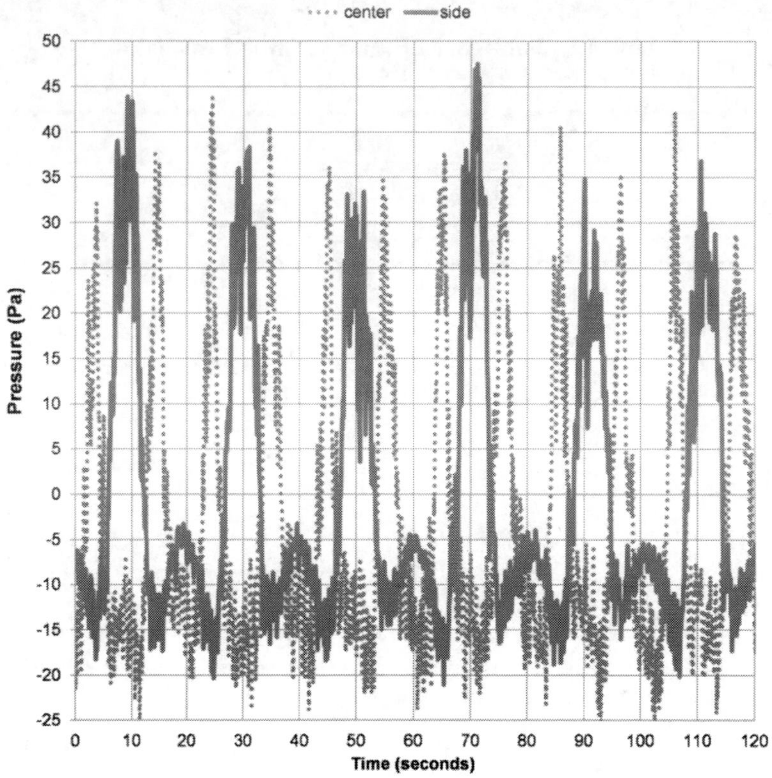

Figure 11 *Differential pressure across cladding at 18 mph.*

Table 8. Stagnation Pressure for Various Wind Speeds

P, Pa	*V*, mph
17.1	12
30.4	16
38.5	18
47.6	20

ing were gasketed, but there was no air barrier installed. Meanwhile, the cladding was fairly tight because the sides of the cladding were well sealed to keep out rain and the gaps at the top and bottom of the rain screen cavity were covered with metal flashing. Finally, in Figure 13 we can see the full range of the pressure variation across the cladding by comparing 12 and 20 mph pressure differentials in the center of the wall.

The maximum pressure values in Figure 13 again compare well with the stagnation pressures for the relevant wind speeds in Table 8.

DISCUSSION

In order to make further progress on specifying what percentage of water gets past the cladding we need to study more cladding types, and perhaps more importantly get a better handle on the kind of defects we expect. For example,

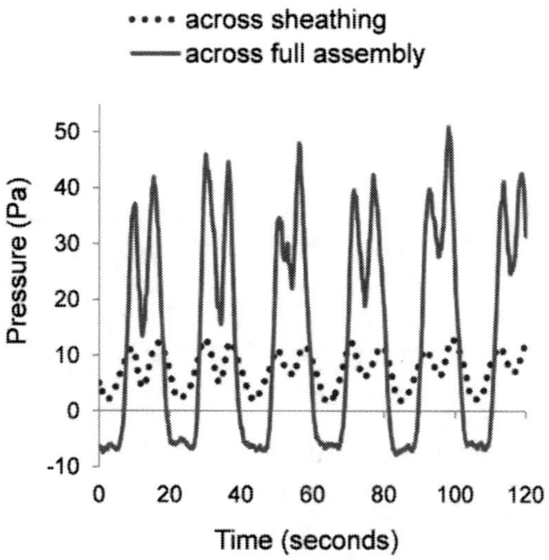

Figure 12 *Wind pressure for 20 mph across the full wall assembly was much greater than across only the sheathing.*

much larger amounts were measured by Sahal and Lacasse (2005). They worked with hardboard siding similar to our lap siding but introduced a number of defects, the most relevant of which was a penetration for a ventilation duct that was imperfectly flashed both at the cladding and at the sheathing. These defects allowed water to flow directly into the wall cavity.

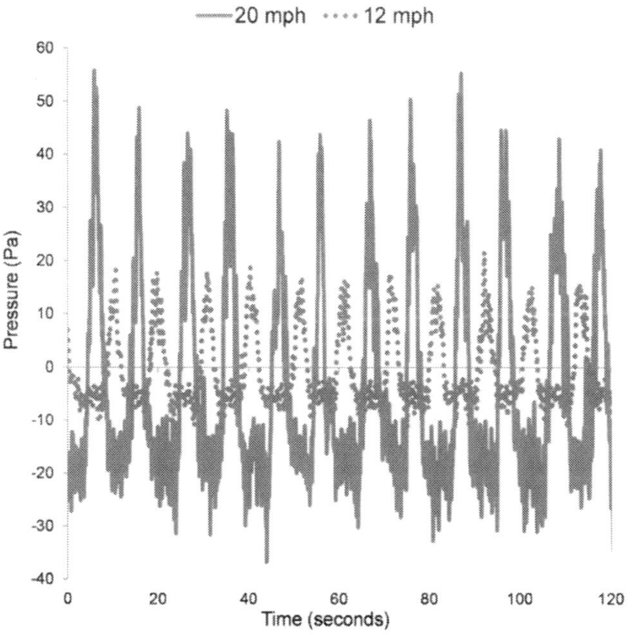

***Figure 13** Pressure across the cladding at two wind speeds.*

Sahal and Lacasse created a water entry function that relates the water intrusion at one defect, Q, with the spray rate (R_p, similar to the WDR) and a water entry potential, m_p, which depends on the pressure difference across the assembly. Their improperly flashed duct defect allowed 1000 times more water intrusion than our simple wedge that created a small gap in the lap siding. It may be helpful to follow an example calculation in which we produce an equivalent water entry potential for our defect.

Sahal and Lacasse (2005) define the water entry function:

$$Q = m_p \times R_p \tag{4}$$

where Q is the water entry rate (kg/min) and R_p is the spray rate (kg/min·m^2). For an example pressure difference of 50 Pa (0.2 in. water), their water entry potential, m_p, has a value of 1.3×10^{-2} (kg· min^{-1}/kg·min^{-1}·m^{-2}). This 50 Pa pressure roughly corresponds to our 20 mph wind speed, so we can use the 20 mph wind (8.9 m/s, 32.2 km/h) and 4 in./h (102 mm/h) rain intensity example to find in Table 6 that the average intrusion at a defect is near 0.03%. From Table 5 we see the typical LF is 1.2, so we can calculate the WDR from Equation 1 as about 48 kg/m^2·h, which we can convert to a spray rate R_p of 0.8 kg/min·m^2. Our Q, calculated as 0.03% of 0.8 kg·min^{-1}·m^{-2} times the area of 0.043 m^2, is 1×10^{-5} kg/min. Thus our mp value is 1.3×10^{-5}, or 1/1000 of the value of Sahal and Lacasse (2005).

This large difference in measured water intrusion from a defect reflects two different approaches to delivery of the water to the wall and different selections for the severity of defect. Sahal and Lacasse (2005) typically use large pressure differences, from 40 to 600 Pa, across the wall to drive water into the defect, and their spray rates (up to 300 kg/m^2·h) were typically higher than in the CARWASh, which had maximum wind-driven rain around 100 kg/m^2·h (3.9 in./h). We posit that the CARWASh conditions are closer to typical wind-driven rain patterns for most residential buildings. However, in this case the primary difference from our results is in the defect. Their defect fails to prevent water entry simply due to gravity. Even with no pressure difference across the wall, water cascading down the wall will enter their defects. Our defect requires wind-driven rain with a significant wind speed to see any intrusion.

An experimental setup and result similar to Sahal and Lacasse's (2005) can be seen in the window defect testing of Teasdale-St-Hilaire and Derome (2006). Acknowledging that high spray rate and large pressure differential tests represent extreme conditions, they used a WDR of 200 kg/m^2·h and pressure of 137 Pa to drive water into a window pane, where it cascaded down to small, 25 mm^2 area holes in the windowsill, representing defects in flashing. Each hole captured about 5% of the wind-driven rain, again about 1000 times the amount of water our defects saw. The window defects again allow water to run in by gravity.

Of course all wall systems should be designed to shed water cascading down the surface by gravity. If the façade allows water entry as if into a drain, and this water is not channeled back out through use of a secondary drainage plane, then modeling the wall system with 1% wind-driven rain getting past the cladding will not be enough. The local failures will be severe. However, if the wall defects are more like the gaps in our lap siding, then our first results hint that modeling 1% intrusion may be a bit high.

CONCLUSION

The general purposes for and capabilities of the CARWASh have been outlined. It can deliver realistic wind-driven rain to a full-size wall assembly, and we have presented details regarding the local distribution of that rain on the wall under different wind speeds and rain intensities. Details on rain intrusion into a particular kind of façade defect have been presented and contrasted with previous studies. Results suggest that we need to reflect further on the kinds of defects we intend to model and test with facilities such as the CARWASh.

ACKNOWLEDGMENTS

We acknowledge the support of the Forest Products Laboratory in carrying out this work. We thank Robert Munson for overall support and help in running the CARWASh. We benefitted from helpful reviews by Samuel Zelinka, Marshall Begel, and one anonymous reviewer.

NOMENCLATURE

LF = location factor; accounts for wind flow at location on wall, modifies RDF

RDF	=	rain deposition factor, fixed at 0.35 for our calculations but generally accounts for complex wind interactions with the wall
DRF	=	driving rain factor; accounts for speed of falling raindrops, m/s^{-1}
WDR	=	wind-driven rain, $kg/m^2 \cdot h$
V	=	wind velocity, m/s
r	=	rain intensity, mm/h
P	=	pressure, Pa
ρ	=	density, kg/m^3 (1.19 for air in Equation 3)
Q	=	water entry at a defect, kg/min
R_p	=	spray rate, similar to WDR, $kg/m^2 \cdot min$ (used in Equation 4)
m_p	=	water entry potential, m^2

REFERENCES

ASHRAE. 2009. ANSI/ASHRAE Standard 160, *Criteria for Moisture-Control Design Analysis in Buildings*. Atlanta: ASHRAE.

ASTM. 2011. ASTM Standard E2268, *Standard test method for water penetration of exterior windows, skylights, and doors by rapid pulsed air pressure difference*. West Conshohocken PA: ASTM International.

Bitsuamlak, G.T., A.G. Chowdhury, and D. Sambare. 2009. Application of a full-scale testing facility for assessing wind-driven-rain intrusion. *Building and Environment* 44:2430–2441.

Blocken, B., and J. Carmeliet. 2004. A review of wind-driven rain research in building science. *Journal of Wind Engineering and Industrial Aerodynamics* 92(13):1079–1130.

Blocken, B., and J. Carmeliet. 2006. On the accuracy of wind-driven rain measurements on buildings. *Building and Environment* 41:1798–1810.

Lopez, C., F.J. Masters, and S. Bolton. 2011. Water penetration resistance of residential window and wall systems subjected to steady and unsteady wind loading. *Building and Environment* 46:1329–42.

Sahal, N., and M. Lacasse. 2005. Water entry function of a hardboard siding-clad wood stud wall. *Building and Environment* 40:1479–91.

Straube, J.F., and E.F.P. Burnett. 2005. *Building Science for Building Enclosures*. Westford, MA: Building Science Press Inc.

Teasdale-St-Hilaire, A., and D. Derome. 2005. State-of-the-art review of simulated rain infiltration and environmental loading for large-scale building envelope testing. *ASHRAE Transactions* 111(2):389–401.

Teasdale-St-Hilaire, A., and D. Derome. 2006. Methodology and application of simulated wind-driven rain infiltration in building envelope experimental testing. *ASHRAE Transactions* 112(2):656–70.

Tenwolde, A. 2011. A review of ASHRAE Standard 160—Criteria for moisture control design analysis in buildings. *Journal of Testing and Evaluation* 39(1). DOI: 10.1520/JTE102896.

Trechsel, H.R., and M. Bomberg (eds). 2009. *Moisture control in Buildings: The Key Factor in Mold Prevention*, 2nd Ed. MNL 18. West Conshohocken, PA: ASTM International.

Van Straaten, R.A., G.A. Kopp, and J.F. Straube. 2010. Testing water penetration resistance of window systems exposed to "realistic" dynamic air pressures. *Proceedings of International Conference of Building Envelope Systems and Technology (ICBEST)*, Vancouver, June 27–30.

Smart Vapour Barriers in Compact Wood Frame Roofs

Stig Geving, PhD **Markus Stellander** **Sivert Uvsløkk**

ABSTRACT

Compact wood frame roofs typically need a vapor and air barrier at the warm side to avoid interstitial condensation due to vapor diffusion and air leakage from the interior. A vapour barrier such as a polyethylene foil does not, however, allow drying of moisture to the interior, making the construction vulnerable to moisture damages. A so-called smart vapour barrier or retarder could allow condensed moisture, built-in moisture, or moisture from minor leakages to dry to the interior. The concept of smart vapour barriers has been known for some time; such a barrier consists of a material that changes its vapour resistance according to the level of relative humidity in the surrounding air.

The application of smart vapour barriers and retarders in compact wood frame roofs was investigated in this study. The study was performed by laboratory measurements. The concept was to investigate how four different types of smart vapour barriers/ retarders performed during typical summer conditions, including an innovative product that has a much broader variation in its vapour resistance than previously known commercial products. The rate of drying was measured and internal humidity conditions within the roof structures were monitored. The various smart vapour barriers were found to function quite differently in regard to their ability to enhance the drying to the interior.

INTRODUCTION

In Nordic climates a vapour and air barrier is typically used on the warm side of the building envelope to avoid interstitial condensation during the heating season due to vapour diffusion and air leakages from the interior. Often a polyethylene (PE) foil is used, which has a very high vapour resistance. Such a vapour-tight product does not, however, allow moisture to dry to the interior. A more vapour-open product could allow condensed moisture, built-in moisture, or moisture from minor leakages to dry to the interior. This could be especially useful for constructions with reduced or no possibility to dry to the exterior, such as unventilated flat roofs. It might allow cheaper solutions, making it possible to use organic (wooden) materials in such constructions or skip the use of wooden preservatives.

The term *vapour retarder*, as opposed to *vapour barrier*, is often used for products that have a lower vapour resistance than recommended for vapour barriers. Such products may allow for some drying to the interior. Different commercial products vary quite a lot in regard to their level of vapour resistance, typically somewhere between s_d = 2–10 m (0,34–1,7 Perm) for products sold on the European market. For North American climates, Lstiburek (2002) defines vapour retarders as having the following rage: 0,34 m < s_d < 3,4 m (1–10 Perm).

However, Geving and Holme (2012) concluded that vapour retarders had little effect on the total drying of ordinary wood frame walls with relatively vapour-open exterior sides. To get an inward drying of built-in moisture of some significance (25% or more of the total drying), the vapour resistance on the warm side should be rather low, typically with an s_d value lower than 1–2 m (1,7–3,4 Perm). However, with such a low vapour resistance on the warm side the risk for interstitial condensation increases. While a wood frame wall has good drying possibilities to the exterior, and thereby needs very

***Stig Geving** is a professor and **Markus Stellander** is a master's student in the Department of Civil and Transport Engineering, Norwegian University of Science and Technology, Trondheim, Norway. **Sivert Uvsløkk** is a senior researcher at SINTEF Building and Infrastructure, Trondheim, Norway.*

vapour-open vapor retarders to have a significant drying to the interior, it seems obvious that for constructions with low or no drying to the exterior one might get a significant drying to the interior even with higher s_d values on the warm side. In a simulation study using climate data for various places in Northern Europe and Northern America, Bludau and Künzel (2009) found for instance that for an unvented flat roof a vapour retarder with a constant s_d value down to 3 m (1,1 Perm) could be acceptable for moderate heating climates (not too far north) without risk of interstitial condensation and mould growth.

While a vapour retarder has a given constant vapour resistance, some vapour barriers or retarders[1] are sold on the European and North American markets with adaptable vapour resistance. Popular terms for these products are *smart*, *intelligent*, *moisture-adaptive*, and *humidity-dependent* vapor barriers. The physical behavior of these products varies, but the main principle is that the vapour barrier should function as an ordinary vapour-tight vapour barrier most of the time, preventing vapour diffusion into the construction from the indoor air. If, on the other hand, the construction is wet, for example due to built-in moisture or leakages, so that the relative humidity (RH) on the exterior side of the vapour retarder gets high, the vapour resistance will be reduced so that there may be possibilities for drying inwards. Another possibility may be that the barrier layer has capillary properties with possibilities to transport condensed water to the interior surface.

Probably the first commercial product that can be put under the term *smart vapour barrier* (SVB) was a capillary active product that was developed in Denmark during the 1980s (Korsgaard 1985). It consisted of synthetic fibres sandwiched between stripes of polyethylene, giving it a constant very high vapour resistance similar to the polyethylene film. If, on the other hand, moisture condensed on the exterior side of the vapour barrier, the water would be transported sideways through the fibres by the wicking action and be allowed to evaporate to the interior side. Research on the practical use of this product was performed in both Denmark and the U.S. in the late 1980s and early 1990s (Korsgaard and Pedersen 1990; Rode et al. 1993; Pedersen et al. 1992), and it was found to give a significant drying effect of unvented lightweight roofs.

In the mid 1990s another type of SVB was developed in Germany (Künzel 1996). It is now sold both in Europe and in North America under several different names. It consists of a polyamide film with no capillary properties but with a pronounced difference in vapour resistance for high and low ambient RH. The s_d value is approximately 4–5 m (0,68–0,85 Perm) when the RH is below about 40% and 0,1–0,2 (17–34 Perm) when the RH is above 80% (Künzel 1996). Also this product was first introduced as beneficial for unvented roofs. Most research since then has been focused on the use of this product. There are a lot of studies on unvented roof systems with this product (Künzel 1996; Künzel 1998; Künzel 1999; Künzel and Leimer 2001; Ghazi Wakili et al. 2004; Kaufmann et al. 2006; Bludau and Künzel 2009; Bludau et al. 2010).

Since then there have been developed various similar products with varying levels of vapour resistance. Some of these new products may have a higher resulting vapour resistance for winter conditions compared to the polyamide SVB, reducing the risk for interstitial conditions. Some may have a lower resulting vapour resistance for summer conditions, resulting in higher drying rate to the interior. This study investigates and compares the performance of three newer products with the polyamide SVB.

LABORATORY TESTS

General

The purpose of the laboratory tests was to investigate and compare the drying conditions in compact wood frame roofs when applying different types of commercially available SVBs. The drying typically takes place during the summer season, when the external surface is heated by the sun and the driving potential for inward vapour flow is highest. A typical diurnal variation of external roof surface temperatures during a sunny summer day was repeated during 144 days.

The following type of compact unventilated wood frame roof was tested (from the exterior side):

- Roofing membrane
- 21 mm (0,83 in.) plywood (spruce) sheathing
- 200 mm (7,9 in.) glass wool or expanded polystyrene (EPS)
- SVB or PE foil
- 13 mm (0,5 in.) gypsum board
- Latex paint (high permeance)

The measurements took place in the laboratories of the Department of Civil and Transport Engineering at the Norwegian University of Science and Technology during spring 2012.

Experimental Set-Up/Test Rig

Six different configurations of the wood frame roof were tested where the only differences were the type of SVB or PE foil used and whether glass wool or EPS were used as insulation. Five configurations had glass wool; four types of SVBs and PE foil were used as the barrier layer among these five. One configuration with SVB had EPS insulation (see later description). For better accuracy each configuration was tested with two specimens, i.e., there was a total of 12 specimens. The roof specimens were built up in storage boxes of polypropylene, where the bottom of the box was used to imitate the roofing membrane and the rest of the materials were adjusted to fit the box (see Figure 1a). The boxes had interior dimensions $b = 0{,}28$ m (10,6 in.), $l = 0{,}40$ m (15,3 in.), and $h = 0{,}22$ m (8,7 in.) and flanges onto which the SVBs and gypsum board were taped. The corners of the boxes were curved so the resulting "light opening" of the boxes had an

[1]. *Vapour barrier* is used in the rest of this paper for simplification.

area of 0,1053 m^2 (1,13 ft^2). The boxes were placed in a lifted test rig with the bottoms up, with insulation on all sides to get isothermal conditions and a heating foil that could be controlled to give the wanted surface temperature on the roofing membrane (i.e., the bottom of the polypropylene box) (see Figure 1b). Each test box could be dismounted during the experiment to monitor the continuous weight loss. See details in Figure 2.

Materials

The four types of SVBs are anonymized: A, B, C, and D. SVB-A is a polyethylene copolymer, where the copolymer is an acrylic with hygroscopic properties. The s_d value ranges from 0,25–26 m (0,13–13,6 Perm). SVB-B is a polyamide with an s_d value ranging from 0,2–5 m (0,68–17 Perm). This product is manufactured under many different names in Europe and North America. It was originally developed in Germany in the late 1990s (Künzel 1996) and was the first commercially available humidity-dependent vapour barrier product for building applications. Most scientific studies on use of SVBs in building applications are limited to this product only, together with studies on the previously mentioned capillary active product (Korsgaard 1985). SVB-C is composed of a fabric of synthetic fibres with good capillary properties laminated with stripes of perforated PE foil and polyamide foil. It should be noted that this product is a further development of the original capillary active SVB (Korsgaard 1985) mentioned previously. The s_d value ranges from 1–20 m (0,17–3,4 Perm). SVB-D is a polyvinylalcohol film (with spunbonded polypropylene as a reinforcement and protecting layer) with an s_d value ranging from 0,02–102 m (0,03–170 Perm).

The PE foil has a thickness of 0,15 mm (6 mil) and an estimated s_d value of 70 m (0,05 Perm). The EPS has an estimated vapour resistance factor μ = 50 (5,45E–6 Perm-in). The gypsum board and latex paint has a combined s_d value that was measured with the cup method of 0,13 m (26 Perm).

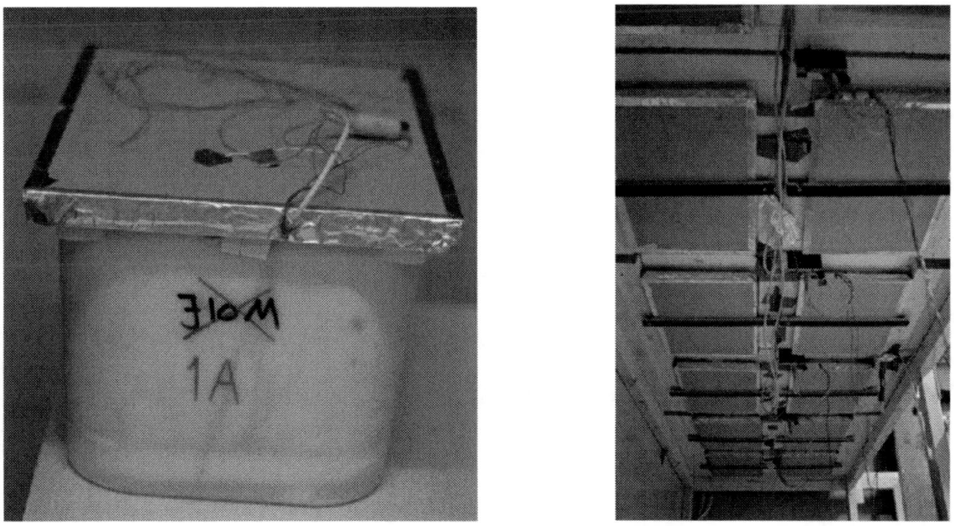

Figure 1 *Test box and test rig: a) test box with gypsum board and measurement cables on top and b) test rig seen from below.*

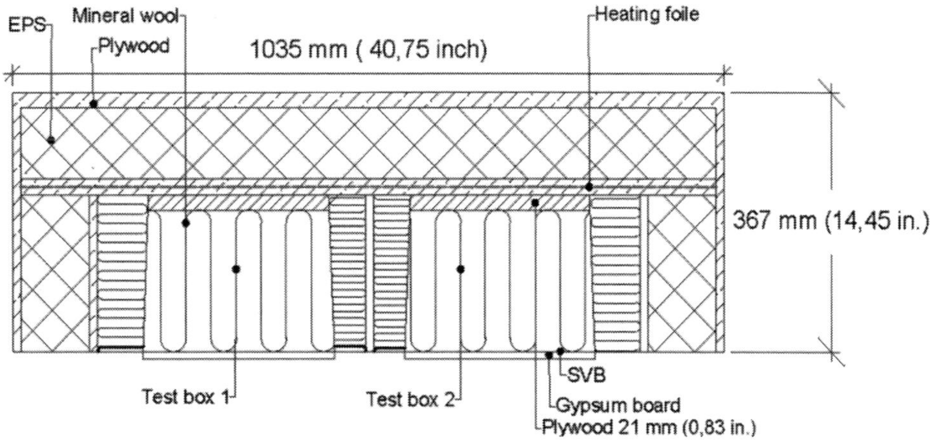

Figure 2 *Section of the test rig showing the two parallel test boxes.*

Boundary and Initial Conditions

The external boundary conditions were given by the temperature controlled by the heating foil in the test rig. During sunny summer days the temperature on flat roofs may reach 50°C–60°C (122°F–140°F). However, to use a constant value would not be realistic. To find a realistic diurnal variation for the external temperature, measured external surface temperature on an unventilated flat wood frame roof at a test house in Trondheim, Norway, was used as a basis (see Figure 3). It was decided to try to imitate the diurnal temperature cycle given by the roof with a light membrane on June 7 with a maximum surface temperature of 49°C (120°F). It was, however, difficult to use the heating foil to recreate the exact temperature cycle given in Figure 3. Figure 4 compares the temperature profile that was chosen to the test house profile from June 7. The driving potential for inward vapour flow was calculated for the two profiles in Figure 4, assuming an RH of 100% both at the external and internal parts of the insulation cavity, and the laboratory test cycle was adjusted to give it almost similar inward driving potential as that of the test house cycle. It was difficult to get a sufficiently quick temperature drop at the heating foil due to the thermal mass of the test rig, so a slightly lower maximum temperature was chosen, 45°C (113°F), so that the evening and nighttime temperatures would not be too high.

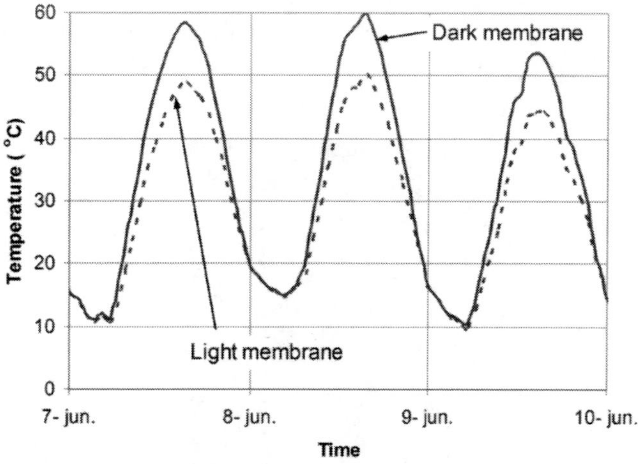

Figure 3 *Typical measured external temperature variations of a flat roof at a test house during three sunny summer days for an element with a dark roofing membrane and a light (solar absorption factor ≈ 0.65) roofing membrane (Geving and Holme 2010).*

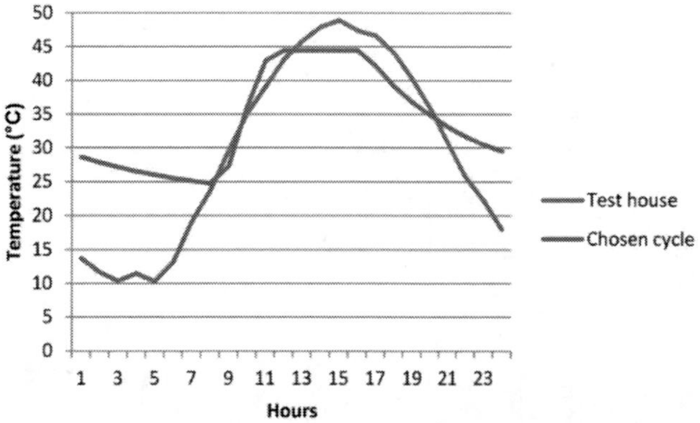

Figure 4 *Comparison of chosen temperature cycle with the measured surface temperature at a test house in Trondheim.*

The test rig was placed in a climatic chamber with a constant indoor temperature of 23°C (73°F) and 50% RH.

To simulate built-in moisture or an accidental leak, the plywood sheathing was given an initial moisture content varying between 23 and 32 weight % by immersion in water for 20 h; see the detailed initial moisture content of each plywood specimen in Table 1. The sides of the plywood were sealed to get a uniform moisture uptake in the specimens. This initial moisture content was determined by the gravimetric method by measuring both wet weight and finally the dry weight when the measurement series were finished.

Measurements

The rig and test boxes were constructed so that the boxes could be dismounted and weighed regularly, to follow the total drying of the configurations. In addition the temperatures and RHs at the interfaces between the SVB and insulation were measured (logged each sixth minute) in one of the two test boxes of each configuration. The moisture content of the plywood board was measured manually once a day by traditional resistance measurements, using screws (dimensions 3.5 × 9.5 mm) with a distance of 25 mm (1 in.) as electrodes—

Table 1. Initial Moisture Content in Each Plywood Specimen

Configuration/ Specimen	Weight %, %	Total Moisture, kg
SVB-A #1	24,7	0,199
SVB-A #2	25,0	0,197
SVB-B #1	29,7	0,238
SVB-B #2	28,9	0,226
SVB-C #1	30,6	0,245
SVB-C #2	32,5	0,255
SVB-D #1	26,9	0,214
SVB-D #2	29,7	0,238
PE foil #1	23,7	0,190
PE foil #2	27,5	0,219
SVB-D/EPS #1	22,8	0,184
SVB-D/EPS #2	29,7	0,230

one measurement point with the two screws mounted from the top of the board and another point with two screws mounted from the bottom (see Figure 5). The screws were sealed with epoxy on top to avoid extra capillary water uptake during the immersion in water. As for the RH measurements, the wood moisture content was only measured in one of the boxes for each roof configuration. The wood moisture meter was set at Scandinavian Spruce. Both the weighing of the boxes and the measurement of wood moisture content were made just before the heating foil was turned on, i.e., when the temperature of the boxes was uniform and close to laboratory temperature.

The RHs and temperatures of the indoor climate were also continuously logged.

RESULTS

The drying of the built-in moisture of the six roof configurations are illustrated in Figures 6, 7, and 8. Figures 6 and 7 show the total water content of the boxes during the test period, i.e. the total water content (kg) divided by drying opening (area) at the SVB level. The total water content is determined from the initial water content in Table 1 and the weight loss determined by weighing the boxes throughout the test period. Figure 6 shows the average of the two parallel specimens for each configuration, while Figure 7 shows each specimen to give an impression of the variance. From Figure 6 we see that the configurations with SVB-B and SVB-D have a much faster rate of drying than the rest of the configurations, followed by SVB-C and SVB-A. As expected, the PE foil has a very slow rate of drying. When comparing the two configurations with SVB-D, one with glass wool insulation and the other with EPS insulation, we see that the one with EPS has a much slower rate of drying. This is as expected since the vapour resistance of the EPS is rather high, slowing down the vapour flow from the plywood layer to the interior.

We also see that at the end of the test period all configurations are approaching the same level of total water content.

Figure 8 shows the total drying from the start at several different points of time. We see that the configuration with SVB-D has a faster drying rate after 10 days compared to the one with SVB-B. This is logical since SVB-D has a lower

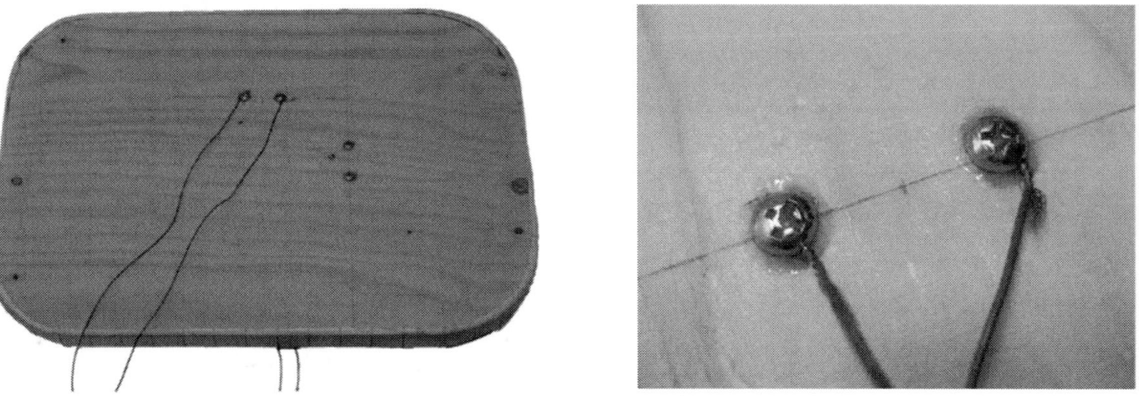

Figure 5 *Screws used as electrodes for wood resistance measurements.*

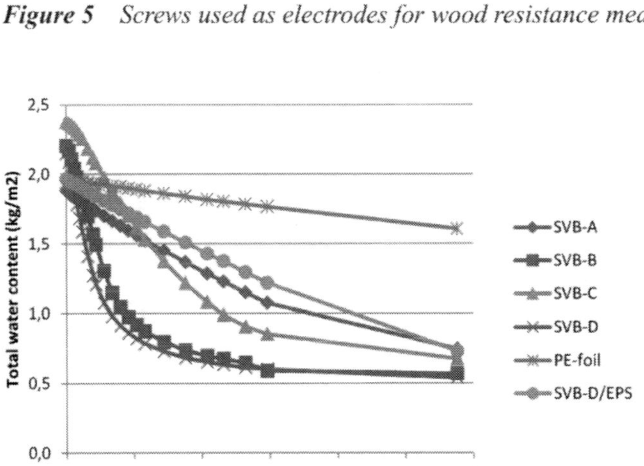

Figure 6 *Total water content in the six roof configurations during the test period. Each curve represents the average of two test boxes.*

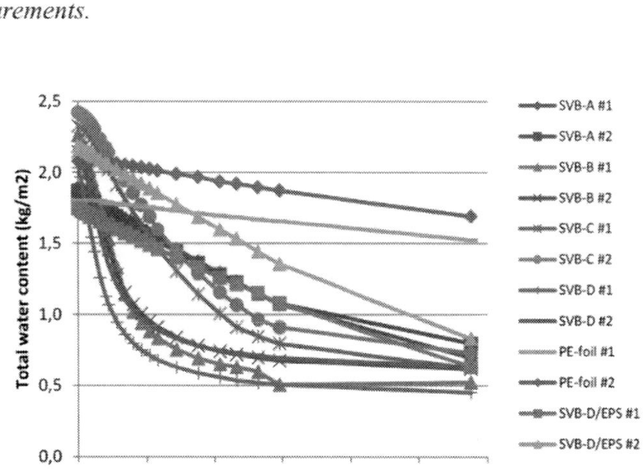

Figure 7 *Total water content for all the 12 specimens during the test period.*

s_d value than SVB-B at high RH levels. We also note that both configurations have a relative high part of the total drying during the first 10 days. That fact makes them very effective in regard to drying of built-in moisture in a Nordic climate, where the number of summer days with high external surface temperatures may be limited. Even though the other configurations finally reach the same level of total moisture content, this will take much more time, increasing the risk for mould growth and rot.

Figure 9 shows the measured RH at the interface between the SVB and insulation level for the first month of the test. We see that the configurations with SVB-D (with glass wool insulation) and SVB-B reach an RH of 100% in one to two days. The RH in the roof with SVB-D does, however, start to drop after 1 week, while for SVB-B it stays at 100% for 2,5 weeks. This high initial RH is probably due to condensation at the SVB, which dries up after the initial period. After this initial period we see that the RH is fluctuating, being high at midday

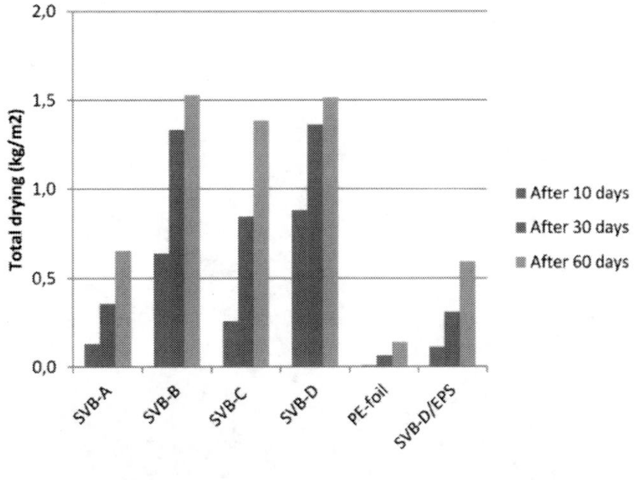

Figure 8 *Total drying from the start and after 10, 30, and 60 days for the six roof configurations; average values for two test boxes.*

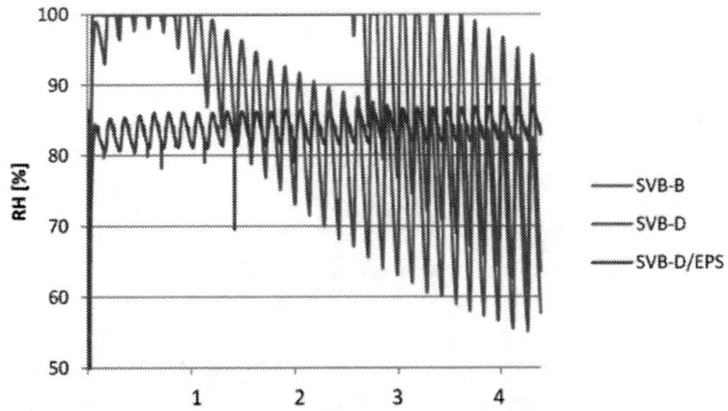

Figure 9 *RH at the interface between the insulation material and SVB.*

(when the external temperature is high) and low during nighttime. After approximately two months the RH curves for SVB-B and SVB-D are practically equal (not shown here). A more detailed illustration of the diurnal variation of the RH, compared to the external surface temperature, is given for one day after four weeks of testing in Figure 10. We can see that the RH follows the temperature curve with only one to two hours' delay.

It should be noted that the RH curves for PE foil, SVB-C, and SVB-A are not shown in Figure 9 since they reached 100% after only approximately eight hours and stayed at 100% during the whole measurement period. That is partly as expected, due to their higher levels of vapour resistance, but may also be caused by a malfunction of the RH sensors after a prolonged period at 100%. All RH sensors were, however, controlled after finishing the experiment, and they all functioned well. Especially for SVB-A you could expect some dropping in RH during the later part of the test period. But it should be mentioned that the sensor at the SVB-C started having daily fluctuations between 0 and 100% after 75 days, indicating at least that the RH level was not 100% during "nighttime."

The moisture content (measured by the electrical resistance method) in the plywood layer is shown in Figure 11. Except for the configuration with EPS insulation, which has a slow drying rate, the difference between the various SVBs is not big. We see, for instance, that the roof with the PE foil has one of the highest drying rates, indicating that internal redistribution of moisture within the insulation cavity is an important factor. Since the external surface temperature always is higher or equal to the indoor air temperature, the plywood will tend to dry to the interior side of the insulation layer—even for the PE foil configuration. What we can read out of Figure 11, in combination with Figures 6 and 9 is the following: due to the high temperatures on the external side, the vapour pressure in the plywood layer will be very high the first weeks. This will

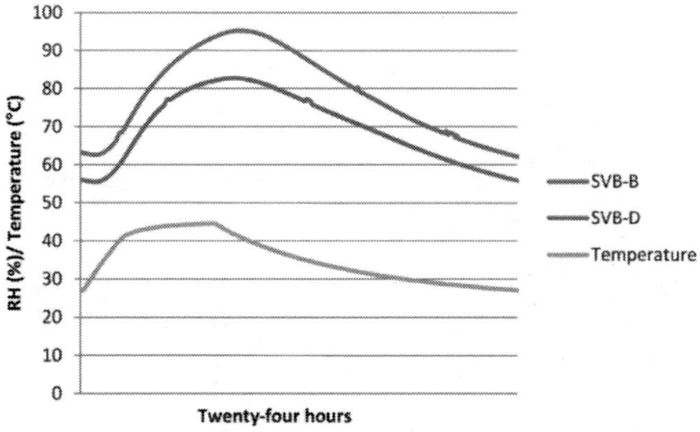

Figure 10 *RH at interface between glass wool and SVB with temperature at exterior surface; shown for one day after four weeks of testing. The starting point is when the heating foil is turned on.*

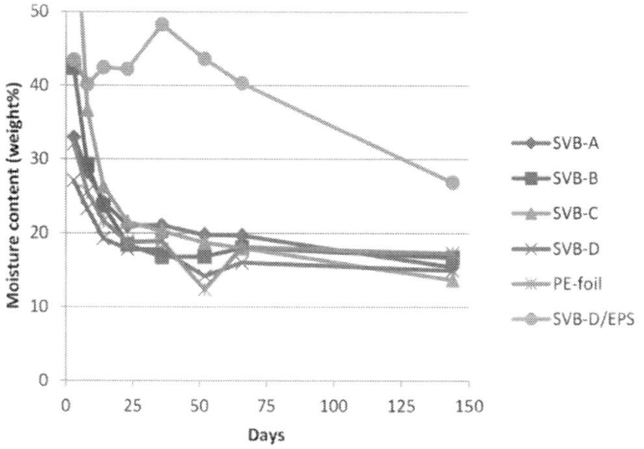

Figure 11 *Moisture content in plywood board (measured with the electrical resistance method); average of measurement points on the top and bottom of the board.*

give a high inward-directed vapour flux—even if the RH at the interface between the SVB/PE foil is somewhere close to 100% as shown in Figure 9. After a few weeks most of the moisture in the plywood layer will probably either have dried to the indoor air, as we see for the vapour-open SVB-B and SVB-D in Figure 6, or partly have condensed in the interior part of the mineral wool for the SVBs with higher vapour resistance and the PE foil. This explains why we do not see the differences between the different SVBs in Figure 11 as we see in Figure 6. In a real case this may be different, however, due to nighttime undercooling of the roof surface caused by black-body radiation to the sky. As shown in Figure 4, the exterior surface temperatures drops to 10°C (50°F) during the night on the test house measurements. This will give an outward-directed vapour flow during the night, which will probably be higher for those configurations with an RH close to 100% in the interface between the insulation and SVB, such as the one with PE foil and SVB-A. It should be mentioned that the measurement uncertainties are relatively high when measuring in plywood. Even if the wood moisture meter is set on a calibration curve for spruce, you will usually measure higher moisture contents than the real ones, this difference being higher for higher moisture contents (Geving and Holme 2010). This means that the high moisture contents measured at the start of the test period, most well above 30 weight %, in reality will be considerably lower and closer to the initial moisture contents given in Table 1.

DISCUSSION

An aspect that is not investigated in this study is the RH occurring between the SVB and the interior lining during the most intense drying period. While it may be advantageous having an SVB with a high permeance (at a high RH level) giving a fast drying, it may also be a disadvantage if this leads to an increased RH level and mould risk for the interior lining during periods with inward-directed drying. The most vapour-open SVB will obviously give a lower RH level and mould risk at the exterior side of the SVB, as illustrated in Figure 9, while at the same time it may lead to a higher RH level and mould risk between the SVB and interior lining. Whether there will be any problems also depend strongly on the vapour permeance of the interior lining itself. In general the lining should be of high-permeance materials to decrease this risk, such as gypsum board with high-permeance paint.

An interesting aspect demonstrated by this experiment is that if only looking at the moisture content of the plywood layer in Figure 11 there seems to be no difference in the rate of drying between the various SVBs and PE foil. The most reasonable explanation for this is that there will be an internal redistribution of moisture from the plywood to the internal part of the insulation (as condensation) the first few weeks, the size of this redistribution being almost independent of the vapour resistance of the vapour barrier. The drying of the roof can therefore only be observed when looking on the total water content in the roof as given in Figure 6—or the RH at the interface between the insulation and vapour barrier, where we see that the condensed moisture in the interior part of the insulation dries to the interior air at very different rates depending on the type of SVB or PE foil used.

There are several limitations or shortcomings with this laboratory experiment. One is that the effect of undercooling of the external surface during the night caused by black-body radiation to the sky is not included. This overestimates the drying to the interior. In addition, the geometry of the experiment is a simplification, since rafters are not included in the test boxes. The rafters will represent an absorptive material in the insulation layer and may influence the moisture distribution and diurnal fluctuations. It should also be mentioned that in this study we have repeated a summer day continuously, which is far from reality. On the other hand, a flat roof with a very dark roofing membrane may have higher maximum external roof temperatures than used in this study. How fast the drying will be in reality will be investigated in a separate simulation study.

The most efficient SVBs for use in flat roofs seem to be the most vapour-open (at high RH levels) products such as SVB-B and SVB-D. Which one of those two performs best cannot, however, be concluded before drying at lower external surface temperatures (shaded summer conditions) and the risk for interstitial condensation during winter conditions are investigated.

In Autumn 2012 a similar laboratory study was performed, this time using lower external surface temperature, as if the roof was shaded from direct solar radiation. Preliminary results seem to indicate slower drying, approximately a reduction of 50% of the drying rate. We are currently also investigating the risk for interstitial condensation during wintertime. This is also very interesting, since one of the benefits of the newer SVBs such as SVB-A and SVB-D, as compared to the older polyamide product SVB-B (Künzel 1996), is that they have a higher resulting vapour resistance during winter conditions.

CONCLUSIONS

Nowadays there exist several types of SVBs. They have various properties in regard to their range of vapour resistance at various RH levels. In this study a laboratory experiment was performed to investigate the drying rates for various types of SVBs during summer conditions for an unventilated compact wood frame roof. The results showed that the various SVBs performed differently in regard to drying rate, depending on the vapour resistance and its dependance on RH.

REFERENCES

Bludau, C., and H.M. Künzel. 2009. Flat roofs in cold climates—Climatic limits for building flat roofs with a permeable vapour retarder. *Proceedings of the 6th International Conference on Cold Climate HVAC,* Sisimiut, Greenland.

Bludau, C., T. Schmidt, and H.M. Künzel. 2010. Hygrothermal effects in lightweight roofs shaded by PV-elements. *Proceedings of Thermophysics 2010*, Valtice, Czech Republic, November 3–5, pp. 56–62.

Geving, S., and J. Holme. 2010. The drying potential and risk for mould growth in compact wood frame roofs with built-in moisture. *Journal of Building Physics* 33(3):249–69.

Geving, S., and J. Holme. 2012. Vapor retarders in wood frame walls—Do they have any effect on the drying capability? *Proceedings of the 5th International Building Physics Conference*, Kyoto, Japan, May 28–31.

Ghazi Wakili, K., and T. Frank. 2004. A humidity dependent vapour retarder in non-ventilated flat roofs. In situ measurements and numerical analysis. *Indoor and Built Environment* 13:433–41.

Kaufmann, A., H.M. Künzel, and J. Radoń. 2006. Preventing moisture problems in retrofitted pitched roofs. *Architectura* 5(1):69–79.

Korsgaard, V. 1985. Hygro diode membrane: A new vapor barrier. *ASHRAE/DOE/BTECC Conference: Thermal Performance of the Exterior Envelopes*, Clearwater Beach, Florida, December 2–5.

Korsgaard, V., and C.R. Pedersen. 1990. Moisture content and distribution in flat roofs with polyethylene or Hygro Diode vapour retarder. *Proceedings of the 2nd Symposium on Building Physics in the Nordic Countries*, Trondheim, Norway, August 20–22, pp. 253–58.

Künzel, H.M. 1996. Humidity controlled vapour retarders reduce risk of moisture damages. *Proceedings of the 4th Symposium on Building Physics in the Nordic Countries*, Espoo, Finland, September 9–10, pp. 447–54.

Künzel, H.M. 1998. More moisture load tolerance of construction assemblies through the application of a smart vapor retarder. *Proceedings of ASHRAE VII International Conference on Thermal Performance of the Exterior Envelopes of Whole Buildings*, Clearwater Beach, Florida, USA, 129–32.

Künzel, H.M. 1999. Flexible vapor control solves moisture problems of building assemblies—Smart retarder to replace the conventional PE-film. *Journal of Thermal Envelope and Building Science* 23(July):95–102.

Künzel, H.M., and H.-P. Leimer. 2001. Performance of innovative vapor Retarders under summer conditions. *ASHRAE Transactions* 107(1).

Lstiburek, J. 2002. Moisture control for buildings. *ASHRAE Journal*, February.

Pedersen, C.R., T.W. Petrie, G.E. Courville, A.O. Desjarlais, P.W. Childs, and K.E. Wilkes. 1992. Moisture effects in low-slope roofs: Drying rates after addition with various vapor retarders. Report, Oak Ridge National Laboratory, Oak Ridge, TN, USA.

Rode, C., G. Christensen, and M. Vesterløkke. 1993. Moisture performance of flat, non-ventilated, wooden roof cassettes. *Proceedings of the 3rd Symposium on Building Physics in the Nordic Countries*, Copenhagen, Denmark, September 13–15, pp. 623–31.

The Influence of Energy Performance Levels on the Heating Demand in Dwellings: Case-Study Analyses on Neighbourhoods

Marc Delghust **Jelle Laverge, PhD** **Arnold Janssens, PhD**
Member ASHRAE

Elisabeth Cnockaert **Tracy Davidson**

ABSTRACT

Unforeseen variations in user behaviour can cause large differences between real and predicted heating demands in dwellings. While these discrepancies have been extensively documented in the literature, taking user behaviour into account in energy calculations remains challenging and could benefit from additional knowledge of the inhabitants' interactions with their houses and systems.

In response to this need for additional knowledge, case-study analyses were carried out on two different neighbourhoods, both consisting of nearly identical houses: one old neighbourhood with uninsulated houses and one recent neighbourhood built according to contemporary energy regulations. Information on user behaviour, on the building characteristics, and on the resulting performance was gathered both by in-situ measurements and by surveys of the inhabitants.

Correlations and contradictions between presence, heating, and ventilation profiles were identified as decisive elements for a better estimation of real heating demands. However, the analysis also illustrated that it is hardly possible to fully isolate user behaviour from building characteristics and assess these as distinct entities. This is caused by the direct correlation between the household typologies and the building typologies as well as by the influence of building characteristics on the user behaviour.

INTRODUCTION

The most commonly used tool for the evaluation of the energy performance of dwellings in Belgium, the governmental EPBD calculation tool (VEA [2013]; based on EN ISO 13790 [ISO 2007]), assumes an average user behaviour as well as average comfort criteria. However, field studies have demonstrated the large dependency of both energy use and indoor comfort on varying user profiles. Furthermore, shifts in user behaviour have also been noted depending on the buildings' characteristics, as indicated by Andersen (2009) and by Hens et al. (2009). This added complexity and variation in real life, when compared to EPBD calculations, is of huge concern for many actors within the building sector, from policy makers and architects to their building clients. This is certainly the case for the Belgian residential building sector. It consists mainly of small housing projects in which there is often no time or budget available for more advanced, tailor-made studies than mere EPBD calculations based on a quasi-steady-state heat balance with a single average user profile.

The aim of this study is to gain further insight info and knowledge of the relation between user behaviour (e.g., use of windows, ventilation, and heating devices), perceived comfort, and the resulting heating energy use in Belgian single-family houses. The data for this study were collected through surveys and measurements on housing neighbourhoods of different energy performances. The houses range from old, barely insulated houses and current standard houses to low-energy and passive houses. This paper presents findings from the first two neighbourhoods of single-family houses, comparing old, barely insulated houses with recently constructed houses built according to the current Belgian energy regulation standards.

***Marc Delghust** is a doctoral fellow of architectural engineering, Department of Architecture & Urban Planning, Ghent University, Ghent, Belgium, and holds a PhD fellowship from the Research Foundation Flanders, Brussels, Belgium (FWO) and the Flemish Institute for Technological Research, Mol, Belgium (VITO). **Jelle Laverge** is a post-doctoral assistant, **Arnold Janssens** is a full professor, and **Elisabeth Cnockaert** and **Tracy Davidson** have Master's degrees in architectural engineering, Department of Architecture & Urban Planning, Ghent University.*

MATERIALS AND METHODS

Case Description

The variation in building characteristics within the Belgian building stock is as vast as the variation in household typologies, resulting in an amount of possible situations that is too large to be encompassed by a very thorough field study. The challenge, therefore, lies in pursuing this field investigation through a pragmatic though comprehensive approach on a limited number of houses. On the one hand, for the cases to be complementary and usable for cross-comparisons, they were selected to be illustrative of the heterogeneity in thermal performance within the Belgian housing sector while still being houses of comparable size and typology. On the other hand, to differentiate the variations due to the users' behaviour from those due to the characteristics of buildings, uniform neighbourhoods were selected. Each selected neighbourhood had been built by one single construction company and one architect, in one time period, according to the standards of that time and with similar materials and services across all houses. As such, each neighbourhood is a separate cluster of nearly identical houses with different households.

This paper presents results from the analysis on two neighbourhoods of single-family houses, which include 62 households or 186 people. Both samples (neighbourhoods) share some basic characteristics that are typical for most Belgian houses. They both consist of two-storey single-family houses with brick cavity walls and tiled roofs. The living area (including a living room, a kitchen, and a toilet) is located on the ground floor. The sleeping area on the first floor consists of three bedrooms and one bathroom. Despite these similarities, the theoretical energy performances of both neighbourhoods are distinctively different.

The first sample (cs1: 36 houses) is an old social housing neighbourhood, dating from the 1960s. The houses are barely insulated, have no mechanical ventilation system, and are mainly heated by a single gas furnace in the living room. Small additional electric heaters are used in the bathrooms and sometimes in the bedroom. Over their 50 year lifespan, only a few houses have been refurbished and only to a limited extent.

The second sample (cs3: 26 houses) is a five-year-old neighbourhood of privately owned houses. The houses are representative of the present Belgian building standards, complying to the new energy performance regulations. All the houses have standard insulation levels, central exhaust ventilation systems, and central hydronic heating via modern gas boilers. While 16 out of the 26 houses are based on one single design, the remaining 10 houses have slightly different layouts, causing some limited building variation within this sample.

Methodology

Information on the buildings and households was gathered by in-situ measurements, by a thorough survey of the inhabitants, and by collecting utility bills and meter readings for gas and electricity consumption. The focus of this study was on the heating energy demand and, therefore, measurements and surveys took place during the winter season. The measurements focused both on the building characteristics (through infrared thermography, heat-flux measurements, and airtightness measurements) as well as on the indoor climate (indoor temperature, humidity, carbon dioxide level). The aim was to gather data not only for cross-comparisons between houses but also for comparisons between real values and designed values (e.g., U-values of walls). The surveys of the inhabitants focused on the users themselves, their habits, and their interactions with the building as well as on their motivations, understanding of the building, and appreciation of the resulting performance (energy use and comfort). This way, measurements and surveys could be compared and their complementarity exploited.

Three considerations were made when setting up the survey strategy. First, as the neighbourhoods were relatively small (approximately 30 to 60 houses), reaching a good participation rate was crucial to end up with samples large enough for further analyses. Second, to be representative for the neighbourhoods' varying households, the final samples should not be composed merely of motivated inhabitants willing to participate actively in intrusive studies. For these first two reasons, the measurement and surveying procedures could not be too demanding and a balance had to be found between the researchers' scientific eagerness for data and the resulting intrusion into the inhabitants' daily routines and privacy. Third, care had to be taken that the surveys would not influence the inhabitants' behaviour during the measurements and yet that the gathered information would correspond as closely as possible to the measured conditions. Because of this third consideration, the surveys were submitted directly after the measurement periods, when collecting the measurement equipment. The participants were not asked to keep detailed diaries of all activities and behavioural aspects of interest at a high frequency over the whole week. Instead, they were asked to draw daily profiles for an average weekday and an average weekend day (e.g., directly concerning presence, heating, ventilation), and the survey also included specific questions to complement these profiles (e.g., Do you close the windows when it rains? Do you close the windows when the heating system is on?). When applicable, these questions allowed input per individual inhabitant as well as per different room or appliance. To avoid misinterpretations and to lower the workload for the participants, the surveys were adapted to each specific neighbourhood (e.g., the surveys omitted questions on the operation of mechanical ventilation systems for the old housing neighbourhoods without those systems).

RESULTS

The Dwellings

Due to the much higher recent building standards, the more recent houses are supposed to reach a much better energy performance than the old, barely renovated houses. Some differences were confirmed by in-situ measurements (airtightness and heat flux measurements, infrared thermography). By way of example, Figure 1a compares the airtightness for both samples with reference data for Belgian single-family houses. The reference data includes a random sample of houses built in the late 1980s and early 1990s from the Senvivv project (WTCB et al. 1998), a random sample of standard dwellings from the past five years (UGent), and recent measurement data from private-party consultants (BD), of which the explicit low-energy houses (LEH) are separated as a representation of today's "engaged" market segment, as discussed by Laverge et al. (2010). The measured airtightness in both neighbourhoods appeared to be in line with the expectations for their respective building periods and standards. The spread in airtightness within both neighbourhoods can be attributed to workmanship, occasional window replacements in the old houses, and the presence of a few semi-detached houses.

The heating energy demand derived from the meter readings, however, does not correspond to the theoretical predictions. Figure 1b compares the real heating energy demand with the theoretical demand according to the EPBD method. The real energy use is clearly less than the predictions. Finding a lower real energy demand might be considered to be a good thing. However, as the overestimation of the energy demand is higher in the old dwellings, the reduction in energy demand when shifting towards the new dwellings appeared to be much lower in practice than assumed: on average, approximately 75 $kWh/(m^2{\cdot}a)$ (24 000 $Btu/[ft^2{\cdot}yr]$) instead of 225 $kWh/(m^2{\cdot}a)$ (71 000 $Btu/[ft^2{\cdot}yr]$). Furthermore, within each neighbourhood, the real spread in energy use was much larger than in theory. While the accumulation of small aforementioned variations between the individual buildings does create some spread within the theoretical energy use according to the EPBD method, the much larger spread in real energy use can only be fully explained by taking the behaviour of the inhabitants into account.

The Inhabitants

Demographics. The first neighbourhood consists of rented social houses, built and owned by a social housing company. The second neighbourhood was also built by a single construction firm, but many houses are inhabited by private owners. This neighbourhood mainly houses young families, while the first set of households is more heterogeneous, with, as shown in Figures 2a and 2b, more older people, lower education levels, and more retired and unemployed people. This correlation between the ages and performances of the buildings and their respective population sets is quite symptomatic of underlying sociodemographic factors such as the ages and incomes of the inhabitants. As will be shown here, these differences both in households as well as in building typologies coincide with differences in user behaviour.

Presence. Each inhabitant reported his or her presence in each room for an average weekday and for an average weekend day. This data allowed us to build daily profiles representing the probability of the presence of people in each room. Presence is directly related to the daily occupation of the inhabitants and thus on the type of household. As a reference, statistical data for Belgium is depicted in Figure 3a (Laverge et al. 2011), next to the results from both neighbourhoods. The stacked coloured areas represent the probability for one specific person to be present in a certain room of the house over the course of the day. A probability of 1 means that

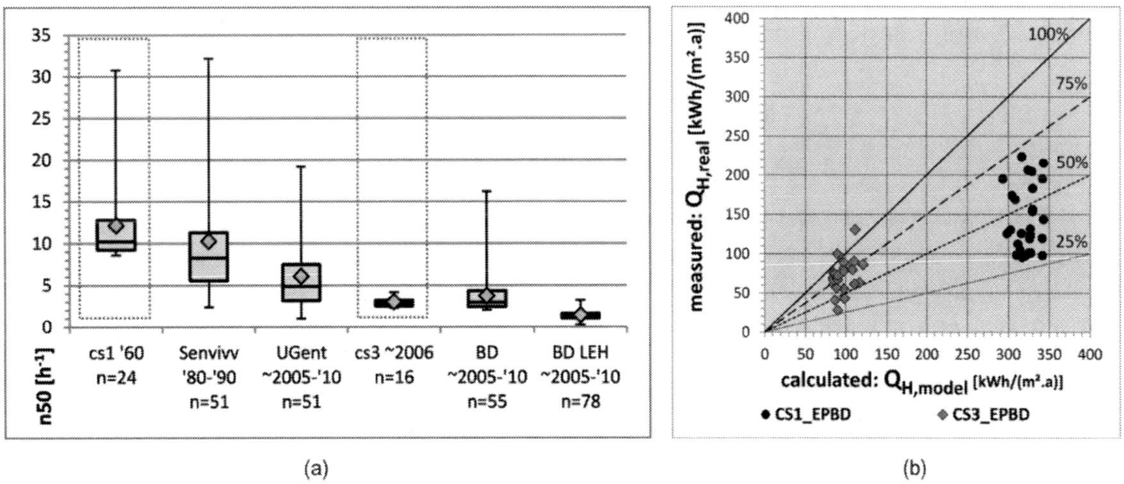

Figure 1 *(a) Measured airtightness (air change rate per hour at 50 Pa [0,0073 psi] pressure difference) and (b) measured vs. theoretical yearly heating energy demand (normalized per unit of floor area).*

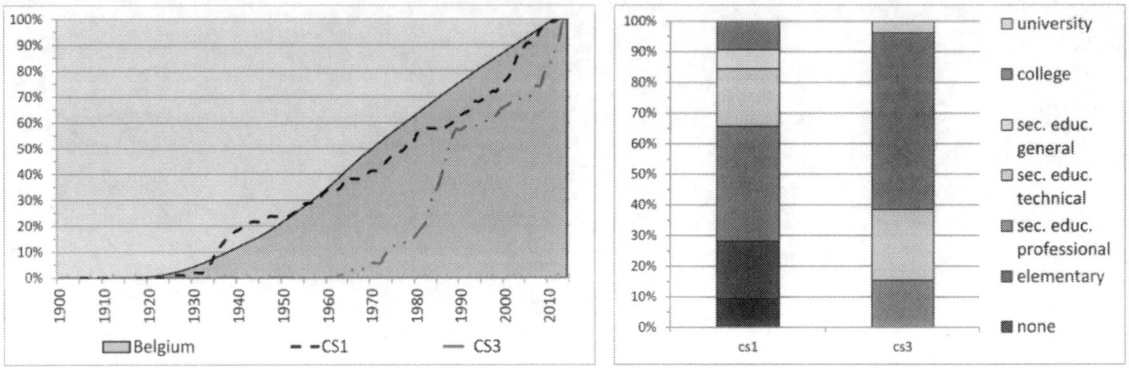

(a) (b)

Figure 2 *(a) Date of birth: cs1 and cs3, compared to the Belgian population (cumulative distribution), and (b) degree of education: highest diploma within each household.*

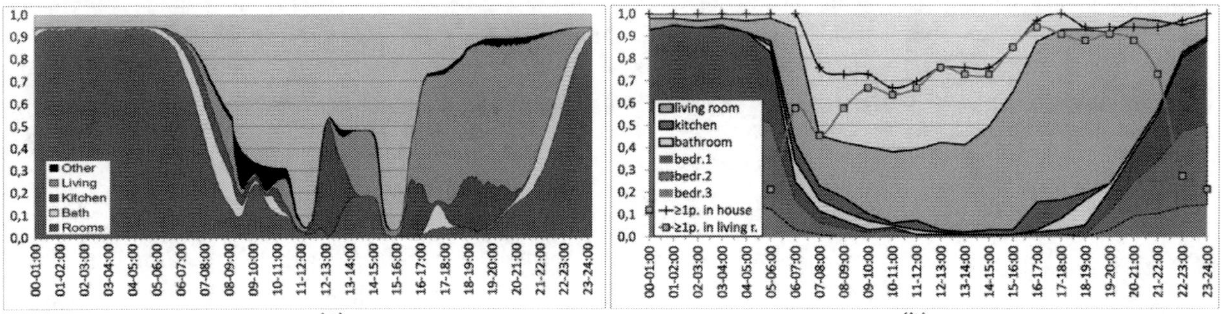

(a) (b)

Figure 3 *Probability of presence: (a) Belgian reference data (Laverge et al. 2011) and (b) cs1.*
(Figure 3a reproduced from Laverge et al. [2011] with permission of ELSEVIER SCIENCE.)

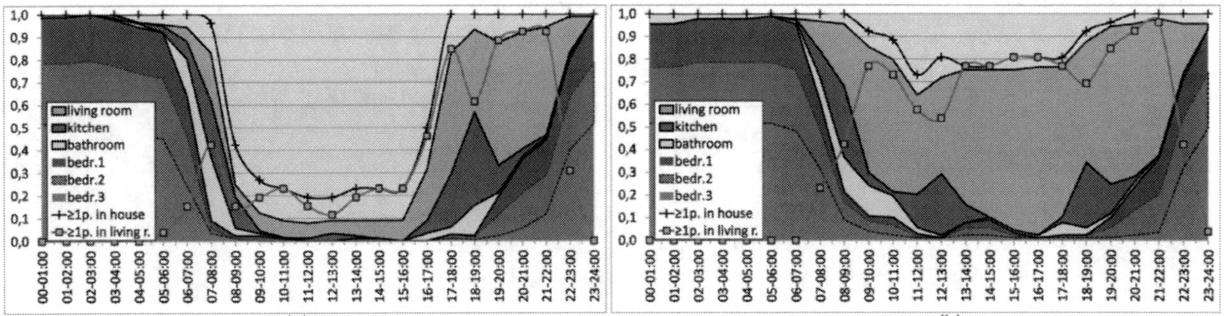

(a) (b)

Figure 4 *Probability of presence in cs3 (a) weekday (b) weekend.*

everyone is stated to be present in the house at that time of the day, while reaching 0 means that no one is stated to be at home at that time. Additionally, the lines show the probability of at least one person being present in the room and are thus also dependent on the size of the household. In cs1 (Figure 3b), the probability of presence in the houses (and more specifically in the living room) during weekdays was much higher than the Belgian reference level. The opposite was true for cs3 (Figure 4a). This could be explained by the higher number of unemployed and retired people in cs1, as indicated by Guerra-Santin (2010). Also, the increased presence in the kitchen and the living room during lunchtime in the reference data was not found in either of the two samples. While no direct cause was reported in the surveys, one explanation of the limited use of the kitchens in cs1 could come from their very poor state and their small size, making the directly heated and thus more comfortable living room the preferred eating place. During the weekend, the occupancy profiles in cs3 changed dramatically, as shown in Figure 4b. Small time-shifts in the morning and in the evening, increased presence in the living room, and a wider spread of use of the kitchen over lunchtime and dinnertime were the most obvious and noticeable differences.

Heating. Daily probability profiles for the heating of each room were derived from the questionnaires, in a similar way as the presence probability profiles. Both profiles proved to be strongly correlated, especially for the living rooms. This is obvious when comparing the heating profile of the living room, as shown in Figure 5a for cs1 and Figure 6 for cs3, with the probability of having at least one person present in the living room (the green lines in Figures 3 and 4). For cs3, the heating profile of the kitchens was much higher than that of cs1 and closely followed the profile of the living rooms. The higher functional and qualitative value of the kitchens in cs3 was not the only reason. The close fit in heating profiles between living rooms and kitchens in cs3 is also caused by the frequent occurrence of open kitchens and the use of one centralized heating system, in most cases controlled by one single clock-thermostat. In the other rooms, the direct link between presence and heating was not always as obvious as in the living room. The electric heaters in the bathrooms of cs1 were reported to be turned on only when the bathroom was in use. However, in most bathrooms of cs3, the heating appeared to remain on from the moment someone was present in the house during the daytime, synchronously to the heating in the living area. In general, in the bedrooms, the heating devices were barely used, merely occasionally to heat up the room before going to sleep. Although the higher thermal performance of the new dwellings should improve the thermal comfort in the bedrooms of cs3, even without heating them directly, the bedroom radiators were used more frequently in cs3 than in cs1. The higher expense of electric heating and the idea of a lowered heating demand due the higher insulation levels in cs3, further emphasized by differences in income between the two groups of households, might explain these apparent rebound-effects with regard to the heating in the bathrooms and the bedrooms, as illustrated by Hens et al. (2009).

The large spread in daily heating profiles is illustrated even better in Figure 5b. This figure shows the average number of heating hours per day as well as the heating setpoints for each living room in both neighbourhoods. These were deduced by cross-analysis of the heating period data from the surveys together with the continuous indoor temperature measurements. Heating times ranged from 4 hours to 24 hours per day. The overall gap between cs1 and cs3 is shown to be

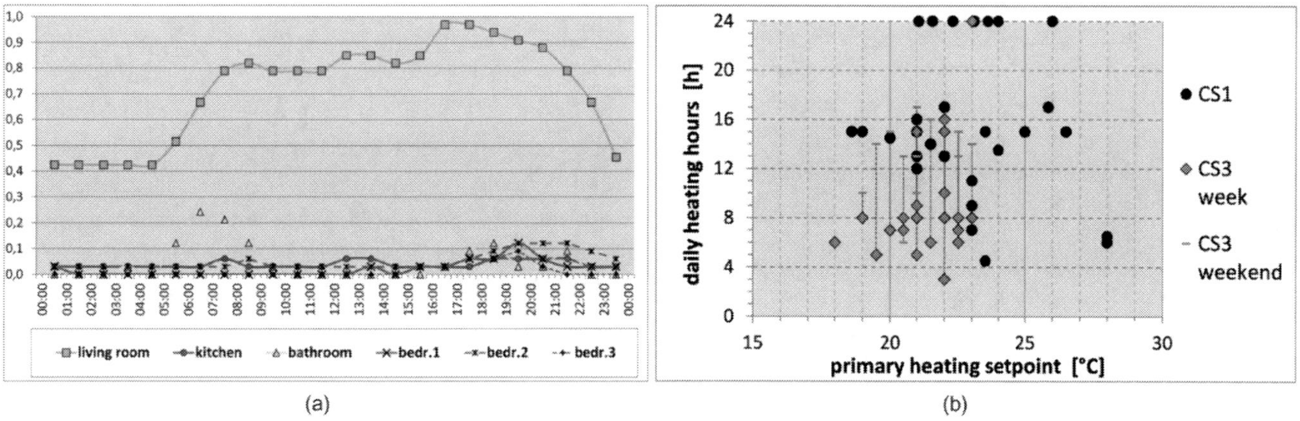

Figure 5 *(a) Probability of heating in cs1 and (b) daily heating hours and primary setpoints for the living rooms (cs1 and cs3).*

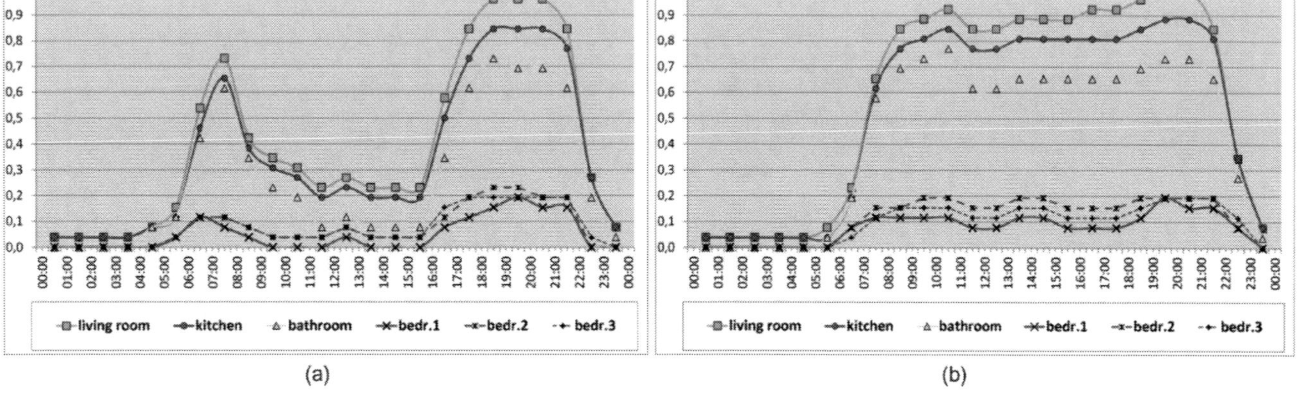

Figure 6 *Probability of heating in cs3: (a) weekday and (b) weekend.*

largely bridged during the weekend (error bars). The variation in heating behaviour was even more striking when taking into account the heating setpoints, which ranged from 18°C (64°F) to 28°C (82°F). It is also important to note that the highest values were found in the old neighbourhood. Most likely, the reasons for this are to be found both in the different users (e.g., older people requiring higher indoor temperatures, as concluded by Mishra and Ramgopal [2013]) as well as in the different building characteristics. The poor thermal performance of the building envelope of the old houses (as pointed out by the airtightness and heat flux measurements and the presence of large single-glazing areas), in combination with the centrally located gas furnace, may have caused draught, radiation asymmetry, and cold floor temperatures. Increasing the heating setpoint can partially compensate these forms of local discomfort. It is therefore impossible to give one unique and indisputable cause for these higher setpoints.There seems to be a contradiction between, on the one hand, the higher setpoints and longer heating durations in the living rooms of the old houses and, on the other, the larger overestimation of energy use in these same houses. The differences in heating profiles, between different houses and between different rooms, are not the only missing factors in the standardized boundary conditions of the EPBD calculation. Other related factors include the real, user-influenced ventilation heat losses in the different zones.

Ventilation. For both the old and the recent dwellings, the surveys showed a recurrent asynchrony, in time and space, between the opening of windows on the one hand and both the presence in and the heating of rooms on the other hand. As shown in Figure 7, windows in the most heated rooms, the living rooms, were reported to be opened only very rarely, while the windows in the rarely heated bedrooms were often opened during the day, especially in the morning when leaving the bedroom. Except for a small shift in time, no significant difference was noted between the two neighbourhoods with regard to the use of the bedroom windows. However, the use of the bathroom and kitchen windows did differ significantly between the old and recent neighbourhoods. The bathroom and kitchen windows were opened much more frequently in the old houses than in the new houses. Only the new houses, however, have an exhaust ventilation system and an improved insulation level. Therefore, in the bathrooms of the older houses, high humidity levels were measured and condensation spots were reported on the uninsulated walls. In the absence of any mechanical ventilation system, opening the windows became the inhabitants' logical approach to tackling the problem.

The mechanical exhaust ventilation in cs3 could be thought of as a more effective ventilation solution, technically more robust than window airing and still freely controllable by the users. However, the measurements demonstrated large divergences in installed flow rates between the different houses, while the surveys showed a very uniform use of the ventilation systems across all households. According to the local ventilation standard, for these houses the highest total exhaust flow must reach 150 m^3/h (88,3 cfm) (75 m^3/h [44,1 cfm] for the kitchen, 50 m^3/h [29,4 cfm] for the bathroom, and 25 m^3/h [14,7 cfm] for the toilet) and the user must be able to select one of three preset ventilation flow rates. However, almost every household stated that they left the ventilation system at its lowest flow rate all the time, while the total installed exhaust flows ranged from 81 m^3/h (47,7 cfm) to 206 m^3/h (121 cfm) at full power (Figure 8a). The discrepancy between installed and prescribed flow rates was even more striking when looking at the rooms separately. The prescribed 75 m^3/h (44,1 cfm) could not be reached in any of the kitchens (Figure 8b), while in most toilets the 25 m^3/h (14,7 cfm) target was reached at the second or even the lowest flow rate (Figure 9b). These findings about installation and use of ventilation systems appeared to be in line with findings from previous field studies in Belgium and the Netherlands by Rosseel (2008) and by van Dijken and Boerstra (2010).

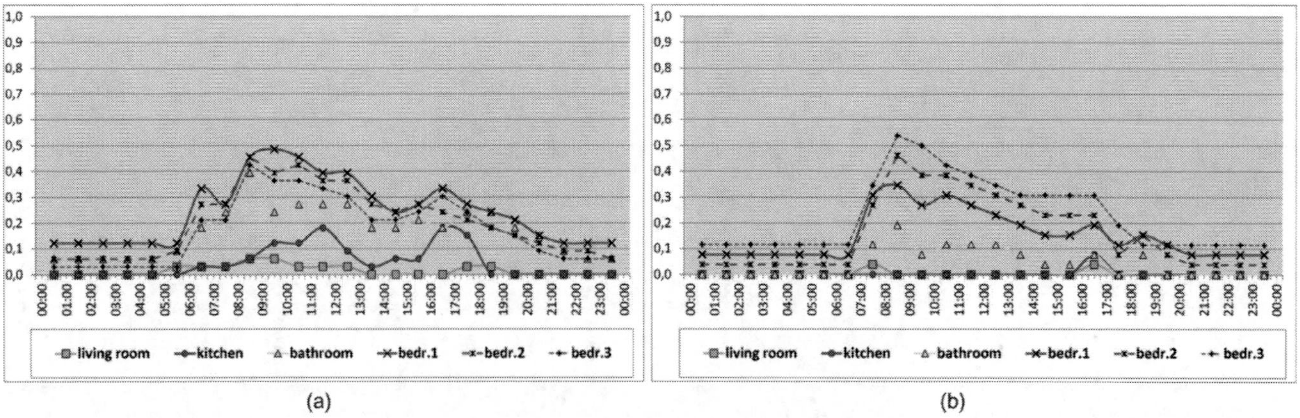

Figure 7 Probability of airing the windows in (a) cs1 and (b) cs3.

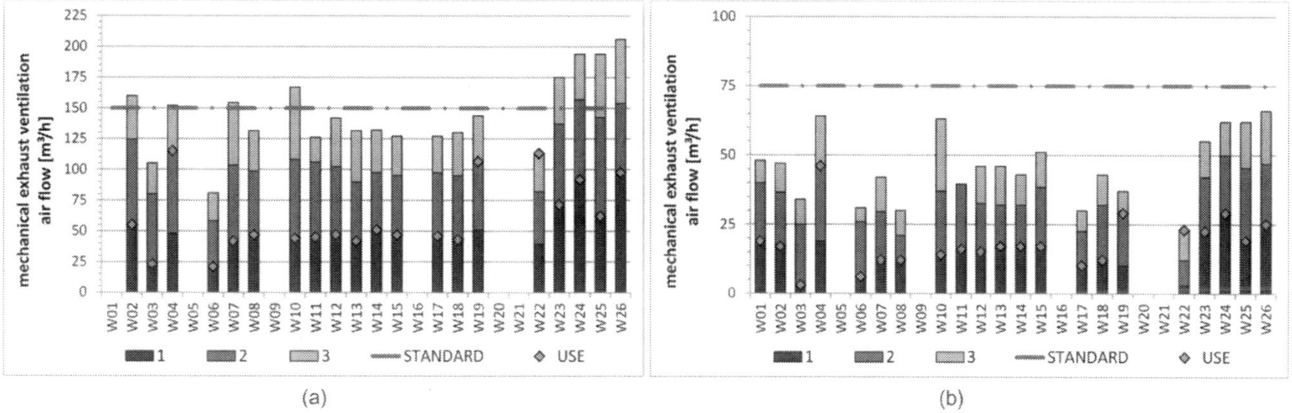

Figure 8 *Exhaust ventilation airflows in cs3: (a) total and (b) kitchen.*

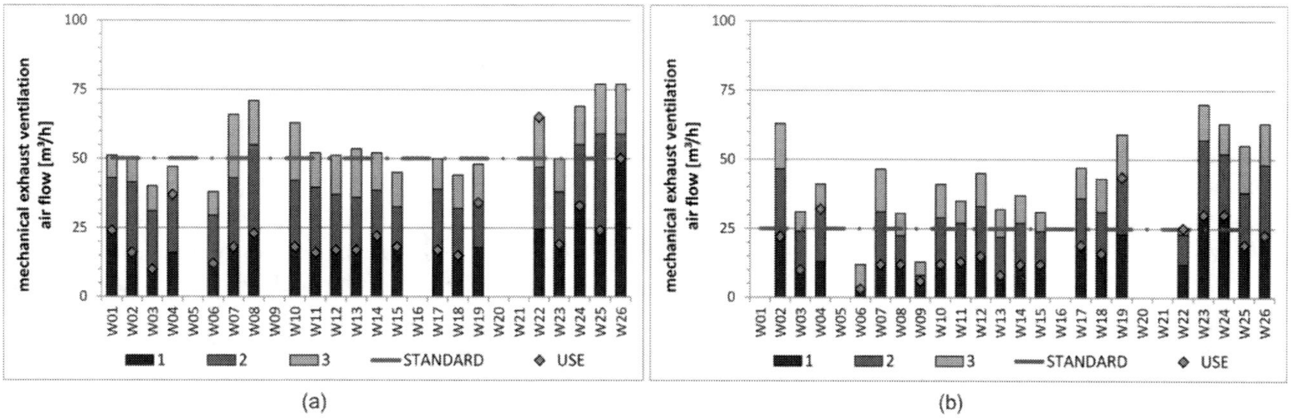

Figure 9 *Exhaust ventilation airflows in cs3: (a) bathroom and (b) toilet.*

DISCUSSION: PREDICTING AND ASSESSING HEATING ENERGY USE

Ultimately, the goal of increasing knowledge on user behaviour and real energy use is to improve the accuracy of calculated predictions of heating demands (before building the houses) and to enable a correct assessment of the real building performances (after completion). Big challenges arise from the huge variations in user behaviour both within a house (between different rooms) as well as between different houses (depending on the properties of the building envelope, the HVAC systems, and their controls).

Prediction

To come to a fairly accurate prediction of the energy use, beyond the accuracy of simple calculation methods using standardized boundary conditions (such as the EPBD software), one has to take into account both the discrepancies between theoretical and real technical characteristics (e.g., the airtightness and the installed ventilation flow rates) as well as the specific user profiles (e.g., heating duration and setpoints, use of windows, and mechanical ventilation systems). Advanced, dynamic simulation tools, by their complex physical algorithms and input flexibility, allow thorough modelling of many of these parameters, as illustrated by de Meester et al. (2013). However, more pragmatic, simplified models can also be used to reach vastly improved predictions. To illustrate this, Figure 10 adds to Figure 1b the calculated values for cs1 from an improved yet simple model. Similarly to the EPBD software, this fast calculation tool is based on a monthly quasi-steady-state method. The improved predictions arise mainly from the extension to a multi-zone model in which the differences in heating setpoints and daily heating times are taken into account as well as the lower, more realistic ventilation airflows, as discussed by Delghust et al. (2012). The remaining errors result both from the limitation of the model as well as from the remaining uncertainties about the building and the user behaviour. It is worth considering whether the benefit of a more complex advanced model will be valuable in daily building practice when making predictions on yearly energy use in small single-family houses. The use of more advanced modelling tools would increase the calculation time while the improved model's precision would be made partly

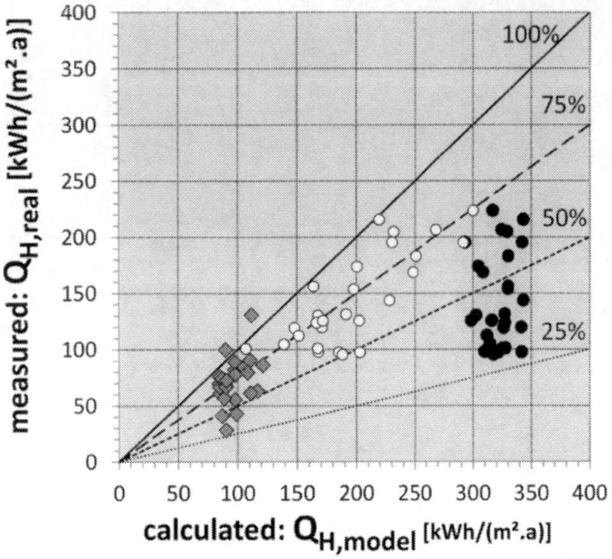

Figure 10 Yearly heating energy demand, normalized per unit of floor area: real vs. calculated values.

superfluous due to limited available data on the building and the user behaviour before construction starts and thus before the users move in. Balancing these considerations might result in different modelling choices dependent on the type of building, the available time and expertise, the required calculation accuracy, and the goal of the analyses (e.g., estimating peak loads instead of total yearly energy use, focusing on individual buildings or building stocks).

Assessment

When carrying out an assessment of existing buildings and their real energy use, through measurements and/or surveys, we often want to differentiate the *performance of the buildings* from the effect of the specific user's behaviour on the resulting energy bill. This seems a justifiable goal when aiming at assessing the workmanship and the reliability of the design performance, when taking into account a certain predefined use, agreed upon between all parties at the start of the project. However, the findings from in-situ analyses question whether this is possible, even in theory (assuming perfect measurement and simulation tools), from the moment the real use differs from the predefined one. As the properties of the buildings influence users' behaviours, while user behaviour is also influenced by other parameters (e.g., socio-economical parameters), it is doubtful whether this maze of causal connections could be fully disentangled. Can we fully attribute divergent heating setpoints and use of windows to the user on the one hand, or to the building on the other, when local discomfort, health conditions, etc., can also play a role? Applying statistical methods to large-scale samples might give a solution when looking, for example, at larger building stock levels. Findings from these analyses might, in turn, be used for improved assessment models on smaller samples or individual buildings. The robustness of such approaches for the assessment of individual projects, e.g., for energy labelling, might become an interesting challenge for future research.

CONCLUSIONS

While the collected data and the findings from these case studies are much more extensive than presented here, this paper illustrates the complexity of the problem. First, the variation in user behaviours has a large impact on the real energy use in old and new houses, adding an important uncertainty to the predicted energy use. Second, both the tuning and the use of the ventilation systems differ significantly from their design targets. The installed flow rates do not correspond with the design values, and the inhabitants almost never use the control buttons. Third, the technical differences between old and new houses also influence user behavior. The most striking example was noted in the bathroom, where different heating and window airing profiles coincided with the differences in heating and ventilation systems.

The paper also demonstrates the valuable complementarity between the different data sources (measurements and surveys) in order to investigate these interactions between the buildings and their users. It appeared very challenging to differentiate the influences of building choices from those of a specific household's behaviour, e.g., with regard to heating setpoints and the use of ventilation systems and windows in the different rooms. This is a major challenge both for making realistic prediction models on future energy use as well as for assessments of energy performances of buildings in use. Statistical analyses on large though detailed datasets would be of considerable help in understanding and quantifying the underlying phenomena. However, the steps towards detailed assessment of the performance of individual buildings will probably remain very challenging.

ACKNOWLEDGMENTS

The authors would like to thank the Flemish Association for Social Housing (VMSW), the local social housing company (MGW Mechelen), and all the inhabitants who participated willingly in this project.

REFERENCES

Andersen, R.V. 2009. Occupant behaviour with regard to control of the indoor environment. PhD thesis, Department of Civil Engineering, Technical University of Denmark.

Delghust, M., et al. 2012. The influence of user behaviour on energy use in old dwellings: Case-study analysis of a social housing neighbourhood. *Proceedings of the 5th International Building Physics Conference (IBPC 2012), Kyoto, Japan*, pp. 809–16

de Meester, T., et al. 2013. Impacts of occupant behaviours on residential heating consumption for detached houses in a temperate climate in the northern part of Europe. *Energy and Buildings* 57:313–23.

Guerra Santin, O., 2010. Actual energy consumption in dwellings. The effect of energy performance regulations and occupant behavior. PhD thesis, IOS Press BV, Amsterdam, The Netherlands.

Hens, H., et al. 2009. Energy consumption for heating and rebound effects. *Energy and Buildings* 42(1):105–10.

ISO. 2007. EN ISO 13790, *Energy performance of buildings – Calculation of energy use for space heating and cooling*. Geneva: International Organization for Standardization.

Laverge, J., et al. 2010. Airtightness assessment of newly built single family houses in Belgium, Building and ductwork air-tightness. *Proceedings of the 5th International Symposium on Building and Ductwork Air-tightness (Buildair 2010)*, Copenhagen/Lyngby, Denmark.

Laverge, J., et al. 2011. Energy saving potential and repercussions on indoor air quality of demand controlled residential ventilation strategies. *Building and Environment* 46: 1497–1503.

Mishra, A.K,. and Ramgopal, M., 2013. Field studies on human thermal comfort—An overview. *Building and Environment* 64:94–106.

Rosseel, S. 2008. Optimalisatie van concepten voor vraaggestuurde residentiële ventilatie. (In Dutch. Optimalisation of concepts for demand-controlled residential ventilation.) Master's thesis, Ghent University, Ghent, Belgium.

van Dijken, F., and A.C. Boerstra. 2010. Onderzoek naar de kwaliteit van ventilatiesystemen in nieuwbouw eengezinswoningen. (In Dutch. Research on the quality of ventilation systems in new single family houses.) BBA Binnenmilieu, Netherlands.

VEA. 2013. EPB-software Vlaanderen, Version 1.7.2. Vlaams Energieagentschap (Flemish Energy Agency), Brussels, Belgium.

WTCB, WENK, VLIET. 1998. *Senvviv: Study on the energy related aspects of new built houses in Flanders: Insulation, ventilation, heating*. (In Dutch. Senvviv: Studie van de energieaspecten van nieuwbouwwoningen in Vlaanderen: Isolatie, ventilatie, verwarming.)

Demonstration of the Performance of Highly Insulating (R-5) Windows in a Matched Pair of Homes

Sarah Widder
Associate Member ASHRAE

Graham Parker

ABSTRACT

Improving the insulation and solar heat gain characteristics of a home's windows has the potential to significantly improve the home's overall thermal performance by reducing heat loss (in the winter) and cooling loss and solar heat gain (in the summer) through the windows. A high-quality installation will also minimize or reduce air leakage through the building envelope, decreasing infiltration and thus reducing heat transmission through building envelope. These improvements all decrease overall annual home energy use. In addition to improvements in energy efficiency, highly insulating windows can have important impacts on occupant comfort by minimizing or eliminating the cold draft many homeowners experience near window surfaces that are noticeably cooler than the room temperature. Energy efficiency measures, such as highly insulating windows, also have the potential to decrease peak energy use in a home, which can lead to measurable peak load decreases for a utility service territory if implemented on a large scale.

High-performance windows now feature triple-pane glass, double low-e coatings, and vinyl insulated frames to achieve U-factors around 0.2 (the U-factor is the inverse of R-value, with units of Btu/h·ft^2·°F, it represents the rate of heat transfer through a material), as compared to double-pane clear glass windows with a U-factor of 0.67, which are common in existing homes across the United States. The highly insulating windows (as they will be referred to in this document) are now available from several manufacturers and have demonstrated the ability to yield considerable energy savings and thermal comfort improvements in homes.

INTRODUCTION

To examine the energy, air leakage, and thermal comfort performance of highly insulating (R-5) windows, a field evaluation was undertaken in a matched pair of "Lab Homes" located on the Pacific Northwest National Laboratory (PNNL) campus in Richland, Washington, during the winter heating and summer cooling seasons in 2011–2012.[1] In this field test, the energy savings from highly insulating windows in the experimental home (Lab Home B) were compared to those of the standard double-pane clear glass windows in the baseline home (Lab Home A).

In addition to the improved energy and thermal comfort performance, highly insulating windows must prove to be cost-effective compared to baseline, clear glass windows to enable significant market penetration. Based on measured and modeled energy savings, as well as installed cost data from window manufacturers, the cost-effectiveness of windows in new construction and retrofit scenarios was examined.

The PNNL Lab Homes

The Lab Homes are factory-built,[2] all-electric, 1493 ft^2 (139 m^2) homes with 3 bedrooms/2 bathrooms and have

1. The final report for the heating and cooling season experiments can be downloaded at http://labhomes.pnnl.gov/resources.stm.

2. The homes were built at the Marlette Industries factory in Hermiston, Oregon.

Sarah Widder *is a research engineer and* ***Graham Parker*** *is a senior research engineer at Pacific Northwest National Laboratory, Richland, WA.*

Published by ASHRAE 2013

approximately 196 ft^2 (18.2 m^2) of window area. The matched pair of Lab Homes were constructed at a Marlette® Homes manufacturing plant in Hermiston, Oregon. The floor plan of the homes as constructed is shown in Figure 1.

Features of the PNNL Lab Homes that are relevant to the highly insulating windows evaluation include the following:

- all-electric
- window area: 196 ft^2 (18.2 m^2) or ~13% of the floor area
- frame: 2 in. × 6 in. (0.05 m × 0.15 m) walls; 2 in. × 10 in. (0.05 m × 0.25 m) floor and ceiling
- insulation: R-22 floors (with belly wrap), R-11 walls, R-22 vaulted ceiling, R-4.2 exterior ducts
- heating/cooling: non-setback central thermostat-controlled ducted 2.5 ton heat pump; 13 Seasonal Energy Efficiency Ratio (SEER)/8.0 Heating Seasonal Performance Factor (HSPF) with single-speed air handler; three 5 kilo-watt (kW) electrical elements, plus alternative heating provided by Cadet Model #RMC151W 120V fan wall heaters with individual thermostats[3]

3. Note that the locations of the Cadet wall heaters are not shown in Figure 1. These heaters were not used/not activated during any part of the highly insulating windows winter evaluation.

- underfloor ducting (in belly of manufactured home)
- lighting: 100% incandescent
- 6 in. (0.15 m) overhang vented eaves all around
- bath fans, kitchen range hood, and whole-house exhaust fan
- wood panel siding
- carpet and vinyl flooring
- vented crawl space with Rapid Wall 2 in. (0.05 m) thick expanded polystyrene foam-backed aluminum (R-9 equivalent insulation value) with access panels
- foam-core exterior doors including an exterior access door to water heater

The homes were sited 90 ft (27 m) apart on the PNNL campus in a flat, open area with no buildings or trees near them that would create shade on the homes during any time of the year. The homes were oriented in an east/west direction with the main entry doors to the living room facing north. This orientation resulted in one sliding glass patio door facing west and the other sliding glass patio door facing south. The homes were sited such that neither house shaded the other or was affected by the other home's prevailing wind shadow.

The homes were fully instrumented to collect energy end-use and environmental data. One of the homes (Lab Home A—the baseline home) was retrofitted with standard double-pane

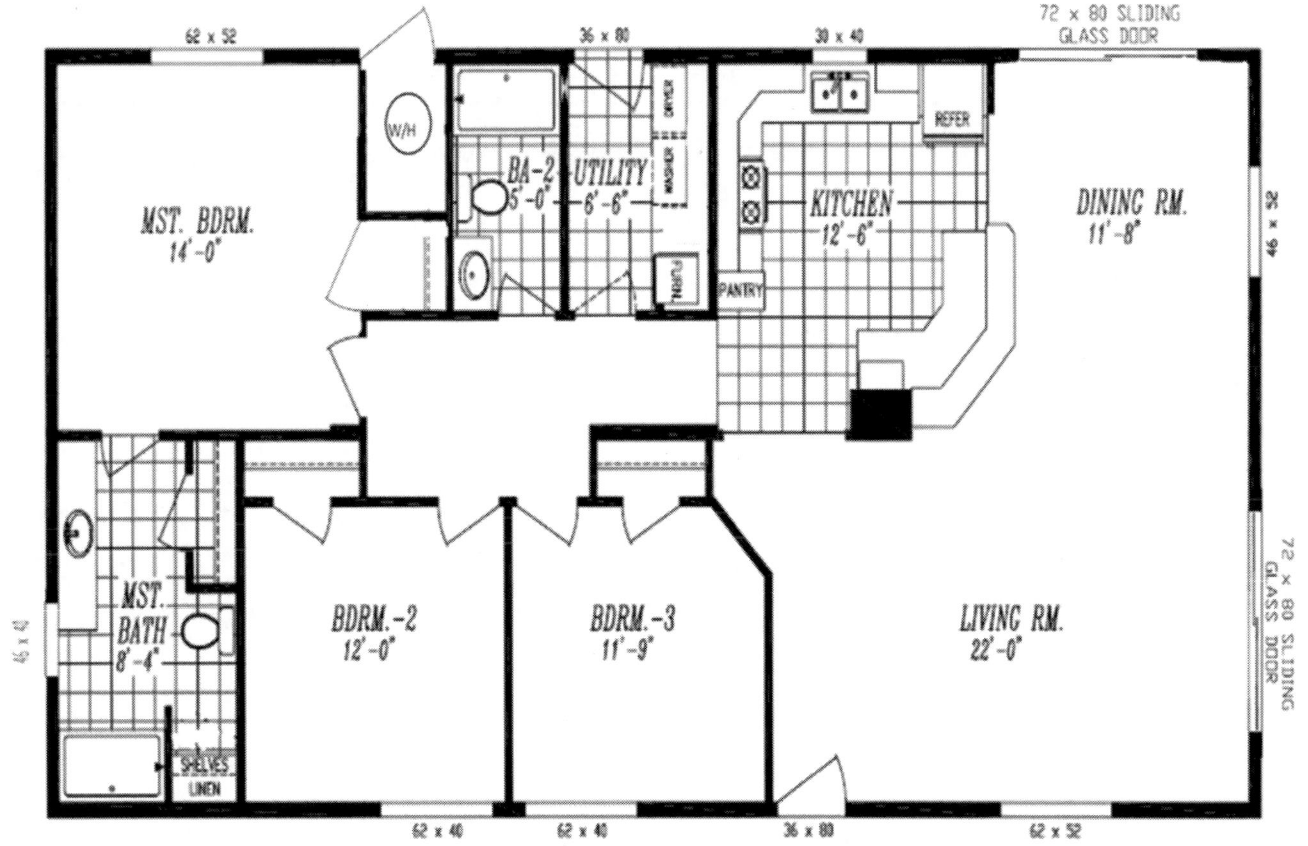

Figure 1 *Floor plan of the Lab Homes as constructed.*

clear aluminum-frame slider windows and patio doors typically found in many existing homes built in the decades of the 1960s through 1980s throughout the Pacific Northwest (PNW) region and across the country.[4] The other home (Lab Home B—the experimental home) was retrofit with Jeld-Wen® triple-pane vinyl-frame slider windows and patio doors. The retrofit windows' characteristics are given in Table 1. The homes are identical in every way except for the windows, including simulated occupancy and weather variations. Thus, all variations in energy use and temperature distributions are directly a result of the difference in window performance. This type of side-by-side experiment provides a degree of accuracy and precision in the results that pre-/post-assessments do not afford.

[4] These double-pane clear windows are typical of many windows installed in homes across the country in the late 60s through the late 90s. They were also the 2006–2012 Federal Housing and Urban Development (HUD)-mandated nationwide minimum code for manufactured housing (see http://www.gpo.gov/fdsys/pkg/CFR-2003-title24-vol1/content-detail.html).

Metering and Monitoring

The approach to the metering included metering and system-control activities taking place at both the electrical panel and at the end-use or point of use. Monitoring was broken into electrical (Table 2) and temperature/other (Table 3). Tables 2 and 3 highlight the performance metric (the equipment/system being monitored), the monitoring method and/or point, the monitored variables, and the data application

Table 1. Window Performance Characteristics of the Windows in Lab Home A (Standard Retrofit Windows and Patio Doors) and Lab Home B (Highly Insulating Windows and Patio Doors) Retrofit in the Lab Homes

Value	Lab Home A Standard Retrofit Windows		Lab Home B Highly Insulating Retrofit Windows	
	Windows	Patio Doors	Windows	Patio Doors
U-factor	0.68	0.66	0.20	0.20
SHGC[(a)]	0.70	0.66	0.19	0.19
VT[(b)]	0.73	0.71	0.36	0.37

(a) Solar heat gain coefficient
(b) Visible transmittance

Table 2. Electrical Points Monitored

Performance Metric	Monitoring Method/Points	Monitored Variables	Data Application
Whole Building Energy Use	Electrical panel mains	kW, amps, volts	Comparison and difference calculations between homes of: • power profiles • time-series energy use • differences and savings
Heating/ Cooling Energy Use (Heat Pump)	Panel metering compressor Panel metering air handling unit End-use metering condensing unit (CU) fan/controls	kW, amps, volts kW, amps, volts kW, amps, volts	Comparison and difference calculations between systems of: • power profiles • time-series energy use • differences and savings
Ventilation Energy Use (Ventilation)	Panel metering of 3 ventilation breakers (2 bathroom and whole-house fans)	kW, amps, volts	Comparison and difference calculations between systems of: • power profiles • time-series energy use • differences and savings
Water Heating	Panel metering of water heater breakers	kW, amps, volts	Comparison and difference calculations between systems of: • power profiles • time-series energy use • differences and savings
Appliances and Lighting	Panel metering of each individual appliance and lighting breakers	kW, amps, volts	Comparison and difference calculations.

Table 3. Temperature and Environmental Points Monitored

Performance Metric	Monitoring Method/Points	Monitored Variables	Data Application
Space Temperatures	13 Ceiling-hung thermocouples/1–2 sensors per room/area, and 1 duct supply temperature per home	Temperature, °F (°C)	Comparison and difference calculations between homes of: • temperature profiles • time-series temperature changes
	2 mean radiant sensors per home (main living area, master bedroom)	Temperature, °F (°C)	
Space Relative Humidity (RH)	2 percent-relative-humidity sensors per home (main living area, hall outside of bathroom)	% rh	Comparison and difference calculations between homes of: • RH profiles • time-series RH changes
Glass Surface Temperatures	22 thermocouples (2 sensors per window interior/exterior center of glass); west window with 6 sensors	Temperature, °F (°C)	Comparison and difference calculations between homes of: • temperature profiles • time-series temperature changes
Mean Radiant Temperature	2 mean radiant temperature sensors (black body) per home (master bed-room and living room)	Temperature, °F (°C)	Comparison and difference calculations between homes of: • temperature profiles • time-series temperature changes
Through-Glass Solar Radiation	1 pyranometer sensor per home trained on west-facing window	Watts/m^2	Comparison and difference calculations between homes of profiles by window and location
Meteorological Station	Package station mounted on Lab Home B	Temperature, °F (°C) Humidity, % Wind speed, m/s (ft/s) Wind direction Barometric pressure, mm Rainfall, in. (m)	Analytical application to quantify setting and develop routines for application to other climate zones

for the electrical monitoring points and the temperature and environmental monitoring points, respectively.

All metering was done using Campbell® Scientific data loggers and matching sensors. Two Campbell data loggers were installed in each home, one allocated to electrical measurements and one to temperature and other data collection. Data from all sensors were collected via cellular modems that were individually connected to each of the loggers. The polling computer located in the metering lab on the PNNL campus connected to each logger using Campbell Scientific software.

In a parallel effort, occupancy in the homes was simulated via programmed access to a custom-designed 42-circuit breaker panel (one per home) using motorized breakers. These breakers were programmed to activate connected loads on schedules to simulate human occupancy.

EXPERIMENT AND RESULTS

Building shell air leakage was measured prior to and after the windows retrofit to determine whether there was any difference in the air leakage of the windows or installation techniques. Prior to the windows retrofit and after the metering equipment and sensors were installed in the homes, the blower door test results show the air leakage of the two homes to be statistically the same, with 95% confidence. The homes are both fairly tight, which is typical of the manufactured housing industry in the PNW. Lab Home A has leakage of 657.6 ± 27.8 cubic feet/minute (CFM) at 50 Pascals depressurization (CFM50) and Lab Home B has an air leakage of 701.4 ± 26.7 CFM50, normalized for exterior weather conditions. These values correspond to 0.15 ± 0.01 natural air changes per hour (ACH_n) in Lab Home A and 0.16 ± 0.01 ACH_n in Lab Home B.

The windows in Lab Home B were retrofitted by the general contractor (GC) with technical assistance and materials (foam sealant and window drain mat material) provided by Jeld-Wen, Inc. The Jeld-Wen, Inc. staff ensured installation was in accordance with the documented recommendations for these windows. The windows in Lab Home A were retrofitted by the same GC staff using standard caulking technique and no special flashing or additional materials.

In Lab Home B, air leakage, as characterized by the CFM50 depressurization with respect to the outside, decreased 46.4 ± 34.9 cfm50 or 6.9%. Conversely, air leakage in Lab Home A increased 50.3 ± 34.1 cfm50. While the total error is large in comparison to the magnitude of the change, the overall impact of these changes was statistically significant with 95% confidence.[5] This decrease in air leakage in Lab Home B is attributed primarily to the quality of the installation given that the air leakage (AL) rating of the highly insulating windows and patio doors (AL = 0.3 cfm/ft^2) is greater than the factory-supplied windows and patio doors (AL=0.1 cfm/ft^2).[6] This is, therefore, compelling evidence of the positive impact of the installation of the retrofit windows on home air leakage.

Prior to initiating the summer cooling season experiment, the air leakage in both homes was retested to determine the persistence of improved air leakage resulting from the windows installation. The measured air leakage was 660.1 ± 21.2 CFM50 in Lab Home A and 622.9 ± 21.9 CFM50 in Lab Home B after normalizing for outdoor weather conditions. These numbers show that the statistically significant difference in air leakage between Lab Home A and Lab Home B was maintained, with the leakage in Lab Home B being less than Lab Home A. However, the values have decreased slightly in both homes, as shown in Table 4.

Winter Heating Season Results

During the heating season study period, the Lab Homes were operated in a number of configurations including with and without occupancy simulation, in heat pump (HP) and electric resistance (ER) heating modes, and at two different interior thermostat setpoints. Comparing the periods without occupancy simulation to comparable periods after occupancy simulation was introduced, the savings decrease approximately 2.5% when occupancy is introduced, as shown in Table 5. This decrease is logical, as one would expect increased energy use of equivalent magnitudes from the identical occupancy simulations in both homes and the resulting decrease in relative whole house energy savings because the magnitude of savings is diluted by increased internal electrical loads. However, this difference is not statistically significant due to the variation in savings based on other factors, such as outdoor air temperature and solar insolation.

When comparing savings from periods when the HP mode was providing heating versus ER mode, no statistically significant difference was observed. In HP mode, the ER elements were still enabled and cycled to provide heating during extremely cold periods.

Combined Heating and Cooling Energy Savings

The measured energy savings in Lab Home B averaged 5821 watt-hours per day (Wh/day) during the heating season and 6518 Wh/day during the cooling season. The overall whole-house energy savings of Lab Home B compared to Lab Home A are 11.6% ± 1.53% for the heating season and 18.4% ± 2.06% for the cooling season. Extrapolating these energy savings numbers based on typical average annual heating degree days (HDD) and cooling degree days (CDD) for Pasco, Washington yields an estimated annual energy savings of 12.2%, or 1784 kWh/yr (see Table 6).

In addition to increasing the thermal performance of the windows, low-e coatings[7] also affect the solar heat gain through the windows. The highly insulating windows

5. Assuming a normal distribution of the data, the null hypothesis is rejected.

6. The AL rating of the retrofit windows in Lab Home A is unknown because this attribute was not included in the windows certification data the manufacturer provided to PNNL.

Table 4. Building Shell Leakage in the Baseline and Experimental Homes Before Window Retrofits, After Window Retrofits, and Prior to Initiation of the Summer Experiment

	Baseline Home	± Error	Experimental Home	± Error
		Null Data		
CFM25	477.4	30.4	478.5	30.5
CFM50	638.5	27.8	681.1	26.7
ACH50	3.07	0.13	3.28	0.13
ACH_n	0.14	0.01	0.15	0.01
		Post-Windows Install		
CFM25	446.9	19.0	372.8	15.7
CFM50	690.8	24.6	639.3	22.5
ACH50	3.32	0.12	3.08	0.11
ACH_n	0.15	0.01	0.14	0.01
		Prior to Summer Experiment		
CFM25	432.8	17.7	401.8	26.3
CFM50	660.1	21.2	622.9	21.9
ACH50	3.18	0.10	3.00	0.11
ACH_n	0.15	0.005	0.14	0.005

Table 5. Average Heating Season Energy Savings and 95% Confidence Interval from Highly Insulating Windows in Different Operating Scenarios: With and Without Occupancy Simulation and in HP Versus ER Heating Modes

Operating Scenario	Average Savings, Wh*	95% Confidence Interval, Wh	Average Savings, %	95% Confidence Interval, %
Without Occupancy Simulation	9392	4611	14.3	5.23
With Occupancy Simulation	4420	1025	11.7	2.08
ER Mode (with occupancy simulation)	7137	2348	10.4	2.96
HP Mode (with occupancy simulation)	4420	1023	11.7	2.08

* Watt-hours (Wh)

Table 6. Average Daily Energy Use and Energy Savings in the Heating Season and Cooling Season

		Average Daily Energy Use, (Wh)	Average Daily Energy Savings, (Wh)	Average Daily Energy Savings, (%)
Heating Season	Lab Home A (Baseline)	47,599	5821 ± 1054	11.6 ± 1.53
	Lab Home B (Experimental)	41,896		
Cooling Season	Lab Home A (Baseline)	35,572	6518 ± 842	18.4 ± 2.06
	Lab Home B (Experimental)	29,055		

installed in Lab Home B have a very low SHGC compared to the clear glass windows in Lab Home A, as shown in Table 1, which was found to decrease the solar insolation measured on the inside of window by 83.2% in Lab Home B compared to Lab Home A. An analysis of the whole-house energy use in Wh versus solar insolation in watts per square meter (W/m^2) for Lab Home A and Lab Home B revealed that increased solar insolation decreased whole-house energy use in the winter and increased whole-house energy use in the summer. The difference in solar heat gain also affected the whole-house energy savings.

Data analysis of whole-house energy savings on overcast days during the heating season showed energy savings for Lab Home B compared to Lab Home A of 14.6% ± 1.86%, while the energy savings on clear days are 8.9% ± 1.42%. In the summer, days are much more consistently sunny in Richland, Washington, so the impact of solar insolation cannot be as clearly observed. However, the energy savings appear to be more correlated to outdoor air temperatures, with higher temperatures yielding higher savings. Also, the high savings observed in the cooling season (18.4% ± 2.06%) indicate that both the low U-factor and low SHGC are contributing to savings during the hot summer days.

Energy Modeling

A representative EnergyPlus model was created for the Lab Homes to compare modeled savings from the highly insulating windows to measured results. The EnergyPlus analysis was run with typical weather data for the nearby Pasco, Washington, airport station to show average whole-building energy savings in Lab Home B from the highly insulating windows to be 13.9% during the heating and cooling season experimental periods (February 2012 to April 2012 and July to August 2012). These results agree fairly well with measured data, which indicated weighted average[8] savings of 14.4% ± 2.57% in Lab Home B versus Lab Home A for the combined heating and cooling test periods. The EnergyPlus model was also used to predict annual savings from the highly insulating windows using typical occupancy patterns and thermostat setpoints. The EnergyPlus model predicts 13.2% annual savings, or 1370 kWh/yr.

7. Low-emittance (low-e) coatings are microscopically thin, virtually invisible, metal or metallic oxide layers deposited on a window surface primarily to reduce the heat loss through the glass by suppressing radiative heat flow. The principal mechanism of heat transfer in multilayer glazing is thermal radiation from a warm pane of glass to a cooler pane. Coating a glass surface with a low-e material and facing that coating into the gap between the glass layers of multi-pane windows blocks a significant amount of this radiant heat transfer, thus lowering the total heat flow through the window.

8. Due to the fact that the heating season and cooling season test periods were different length, the average of the heating and cooling season is weighted based on the number of days in each test period.

Peak Load Reduction

Another impact of reduced energy use that is important to mention from a utility and resource planning perspective is the ability to reduce peak load (kW). Figure 2 depicts peak load impacts due to installation of the highly insulating windows in the experimental home over a representative one-week time period in the summer. The impact on peak load is significantly more than the average daily energy savings, reducing the peak power consumed on an hourly basis in the home with highly insulating windows (Lab Home B) by 24.7% ± 0.1% over the entire cooling experiment study period. This is higher than the average observed daily savings because the peak load reduction represents a maximum real-time savings, rather than an average savings. Peak power savings occur, typically, between 10:00 a.m. and 6:00 p.m. when the overall grid demand is highest, as shown in Figure 2.

Peak load savings were also measured in the heating season. Average hourly peak savings for the heating season were 33.9% ± 0.6%. However, these savings occurred in the nighttime hours when outdoor air temperatures were coldest, which does not typically correspond with utility peak demand periods. The grey line shows the load curve, in megawatts (MW), for the Bonneville Power Administration service territory, for July 6 through July 13, 2011, since 2012 data was not available at the time of publication. While the load shape changes in the winter months, peaks still occur during the day, which juxtaposes the periods of peak demand in the Lab Homes during the winter heating season.

Thermal Comfort

In addition to yielding energy savings, the highly insulating windows in Lab Home B exhibited much more consistent indoor temperatures than Lab Home A. For example, on a sunny day in February, the average indoor temperatures in the kitchen of Lab Home A reached a balmy 84°F (28.9°C), while Lab Home B reached only 78°F (25.6°C), within 3°F (1.67°C) of the 75°F (23.9°C) thermostat setpoint. The average outside temperature on this day was 39°F (3.9°C). This could cause a comfort problem for any occupants at home during this time, even during these cold winter months.

In the summer, comfort problems also were observed. Measured indoor temperatures reached 73°F (22.8°C), 3°F (1.67°C) above the 70°F (21.1°C) thermostat setpoint, on several days in Lab Home A, indicating that the heat pump is undersized for the load on this home with clear glass windows and no window coverings, while the heat pump in Lab Home B (also with no window coverings) was able to maintain an average interior temperature of 70°F (21.1°C) on most days, which was consistent with the 70°F (21.1°C) thermostat setpoint. Also, the temperature rise in Lab Home A caused increased overcooling in some rooms with the shortest duct runs, where temperatures as cool as 63°F (17.2°C) were observed. The outdoor temperature typically reached 90°F to 100°F (32.2°C to 37.8°C) during the cooling season experimental period.

The window surface temperatures also affect comfort in the home felt by occupants, particularly in the winter. A window with a colder surface temperature is noticeable to an

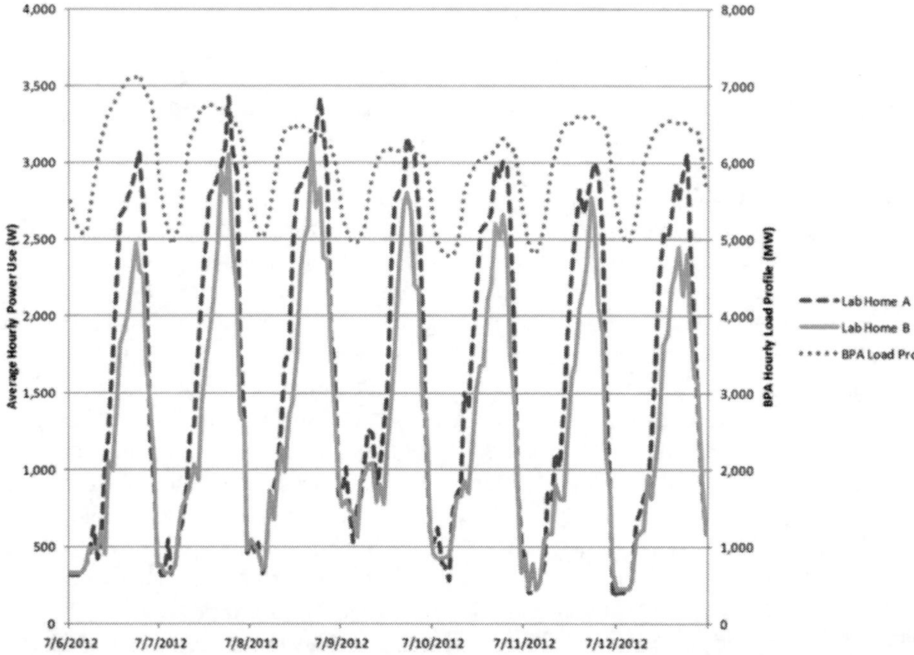

Figure 2 *Average hourly power use for Lab Home A (dashed), Lab Home B (solid), and the Bonneville Power Administration Load Curve (dotted) during a 1-week Period from July 6 to July 13, 2012 (Note: the Bonneville Power Administration data is for the same days in 2011 since 2012 data was not available at the time of publication).*

occupant (near the window area) even though the room dry bulb temperature may be at a comfortable level, due to convective and radiative heat transfer from the occupant to the cooler air near the window or cooler window surface temperatures. This effect is apparent in the Lab Homes, as illustrated by the dramatically cooler window surface temperatures observed in Lab Home A (baseline home); temperatures as low as 50°F (10°C) recorded on the west-facing living room window. In the Lab Home B (experimental home), the highly insulating window's surface temperature never dropped below 60°F (15.6°C)—the lowest temperature measured on the same window. When considering the average interior glass surface temperature measurement of all the windows over the heating season, a difference of 7°F (3.89°C) is observed, with an average interior glass surface temperature of 68.7°F ± 0.05°F (20.4°C ± 0.03°C) in the baseline home compared to 75.7°F ± 0.04°F (24.3°C ± 0.02°C)—almost exactly the interior thermostat setpoint—in the experimental home.

Mean radiant temperature (MRT) is also a measured proxy for thermal comfort/discomfort resulting from the radiant heat exchange between an occupant (a body) and surrounding surface temperatures such as the surface temperature of a window or a wall. Each Lab Home has two MRT sensors: one located in the master bedroom and one in the northwest corner of the living room. The average indoor MRT in Lab Home A and Lab Home B was determined after the windows retrofit over a 1-week time period when the radiant temperature loss was most extreme. During this week the outside temperature averaged 49°F (9.4°C), with a maximum of 65°F (18.3°C) and a minimum of 35°F (1.67°C), and the interior thermostat setpoint was 80°F (26.7°C).[9]

The average MRT in Lab Home A during the nighttime period during the week when radiant heat loss was most extreme was 79.0°F ± 0.02°F (26.1°C ± 0.01°C) in both the master bedroom and the living room compared to an average room temperature of 80.2°F ± 0.11°F (26.8°C ± 0.06°C). The average MRT of both the living room and master bedroom was 1.63°F ± 0.01°F (0.91°C ± 0.01°C) lower than the average interior room temperature in Lab Home A. The maximum MRT recorded in Lab Home A was 86.3°F (30.2°C) and the minimum MRT was 77.0°F (25.0°C).

In Lab Home B the average MRT in the living room and master bedroom was 1.64°F ± 0.01°F (0.91°C ± 0.01°C) lower than the average room temperature of 80.4°F ± 0.01°F (26.9°C ± 0.01°C) in the living room and master bedroom. The maximum MRT recorded in Lab Home B was 81.7°F (27.6°C) and the minimum MRT was 77.7°F (25.4°C). Overall, the average nighttime MRT in Lab Home B was slightly warmer than the average MRT in Lab Home A during the nighttime periods when radiant heat loss is most extreme, but the difference was small. However, the combined impact of decreased mean radiant temperature and surface glass temperatures can have a profound impact on occupant comfort. Also, the maximum MRT recorded in Lab Home B was much lower than Lab Home A during sunny afternoons, even when the outdoor temperatures where quite cool.

Condensation/Moisture on Windows

Condensation on windows is of concern to homeowners and can be a health issue because moisture can lead to mold growth. A direct measurement was not made of condensation or moisture on windows in either Lab Home A or Lab Home B.[10] However, using the windows' inside surface temperature measurements allowed a calculation of the potential for condensation to form on the windows. An interior window temperature as low as 50°F (10°C) was recorded in Lab Home A, while the lowest interior window temperature was 60°F (16°C) in Lab Home B. With these temperatures and an average interior room temperature of 75°F (24°C), the relative humidity of the air in Lab Home B would have to exceed 70% to cause condensation on the highly insulating windows, while a relative humidity of 40% or more would cause condensation on the double-pane windows in Lab Home A. However, condensation was not a concern due to the dry climate in Richland. The average measured relative humidity in the heating season was 20.7%, with a maximum of 31.7%, in Lab Home A, and it was 21.7%, with a maximum of 28.1% in Lab Home B. During the cooling season, similarly low indoor relative humidities were observed. In addition, the warm interior glass temperatures do not cause concerns related to the potential for condensation in the cooling season.

Windows Cost

While the highly insulating windows show significant energy savings compared to the baseline windows, the capital cost of windows must also be considered to determine the cost-effectiveness of highly insulating windows as an energy efficiency measure. The capital cost of highly insulating windows, as delivered, was used to determine the cost effectiveness of windows in a retrofit scenario. An incremental cost, compared to code minimum windows,[11] was also considered for new construction or if windows are being replaced for another reason (e.g., safety, functionality, aesthetics). In the incremental cost scenario, installation cost is not included and only the material cost difference between code minimum and highly insulating (triple-pane) windows is used.

The costs of purchasing the highly insulating windows was obtained from the window distributor as \$6,243 or \$32/ft^2.[12] However, window costs are highly variable and, thus, these

9. The thermostat setpoint was increased for some periods of the study to maintain a consistent temperature differential between the indoors and outdoors (approximately 70°F) throughout the heating season data collection period.

10. Humidity to represent occupants and occupant activity was not generated in the Lab Homes for this experiment.

11. Minimum code for the State of Washington Climate Zone 1 is 0.32 U-factor and 0.40 SHGC.

costs were compared with other manufacturers and other windows types, to ensure the costs were representative of the range of window costs observed in the retrofit market. A delivered (to the PNW) cost range for highly insulating slider windows and patio glass doors in small quantities for retrofit was obtained from the vendors participating in the US Department of Energy (DOE) Windows Volume Purchase Program who sell a product line of sliders and patio doors.[13] Observed window costs were highly variable and ranged from approximately \$10/ft^2 to \$100/ft^2 (\$108/m^2 to \$1,076/m^2). Although manufacturers were asked for data representing delivered costs (not including installation), it is assumed that higher costs represent window costs which include installation costs. Removing extreme values, average windows cost data for highly insulating windows and sliding glass doors in small quantities for retrofit from the Windows Volume Purchase Program ranged from \$25/ft^2 to \$46/ft^2, with an average cost of \$34/ft^2.[14] Installation and materials costs were estimated at \$3/ft^2 (\$32/m^2) based on the demonstrated retrofit costs of installing the windows at the Lab Homes. However installation costs, similar to all contractor bids, can also be considerably variable. Using an installation cost of \$3/ft^2 (\$32/m^2), the total installed cost for the highly insulating windows and sliding glass doors in Lab Home B was estimated to be \$4,900 to \$9,000. This range is consistent with the cost for retrofitting (purchasing and installing) double-pane, clear glass, aluminum-frame windows with triple-pane, low-e, vinyl-framed windows listed in the National Residential Efficiency Measures Database, maintained by the National Renewable Energy Laboratory of \$21/ft^2 to \$55/ft^2 (\$226/m^2 to \$592/m^2), with an average of \$33/ft^2 (\$355/m^2).

Based on these costs and the modeled annual savings of 1370 kWh/yr, the payback period (PBP) for highly insulating windows ranges from 30 to 55 years, for the low to high cost estimates, respectively, as shown in Figure 3. Using the extrapolated measured data to estimate annual savings, a savings of 1784 kWh/yr, the PBP ranges from 23 to 42 years.

If the highly insulating windows were installed in a new home or windows are already being replaced for another reason (i.e., safety, operability, aesthetics), the incremental cost of the windows over a code minimum window should be considered and installation costs can be ignored. The incremental cost of highly insulating windows is estimated to be approximately \$4/ft^2 (\$42/m^2) and ranges from \$2/ft^2 to \$6/ft^2 (\$17/m^2 to \$63/m^2). Using the average value, the total incremental cost for highly insulating windows is estimated to be \$1,372 for a home with the same window area as the Lab Homes (196 ft^2; 18.2 m^2). Based on the modeled incremental annual energy savings of highly insulating windows over code minimum windows of 363 kWh, the PBP is 32 years.

12. This price was obtained from the invoice from the local window supplier for the specified windows and sliding glass doors.

13. The window vendors selected for estimating costs from the qualified vendors in the Windows Volume Purchase Program were vendors who offered horizontal sliders and patio doors and who sold or delivered products in the Pacific Northwest. An average cost per square foot across the multiple sizes of windows in Lab Home B was determined.

14. To provide an average range for window costs, extreme outliers were removed from the cost data in the windows volume purchase program, and lower and higher averages (averages of lower 50% of data and of higher 50% of data, respectively) were taken to bound the average value.

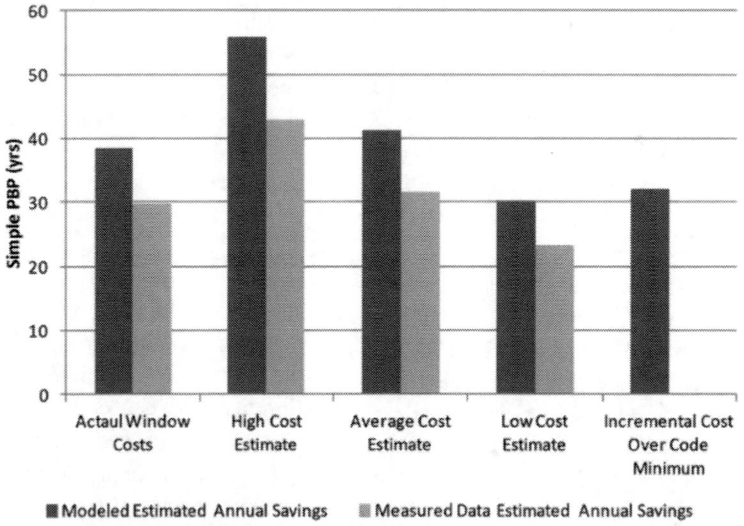

Figure 3 *Simple PBP for R-5 windows for a retrofit or in an incremental cost scenario. Costs considered are the actual windows costs as retrofit in the Lab Homes; a high, average, and low cost estimate based on data in the DOE Windows Volume Purchase Program; and incremental cost over code minimum windows. Modeled and an estimate of annual energy savings are calculated.*

CONCLUSIONS

The side-by-side assessment in the PNNL Lab Homes demonstrates that highly insulating windows show considerable energy savings when compared to double-pane clear glass windows. These energy savings may also contribute to reduced peak energy load, especially in the cooling season, if implemented on a large scale. Highly insulating windows show more consistent interior temperature distributions and improved thermal comfort because interior glass surface temperatures are much closer to interior dry bulb temperatures. Increased glass surface temperatures in the winter could also decrease the risk of condensation and mold issues in regions where high humidity exists. Based on windows cost data available from manufacturers via the Windows Volume Purchase Program and a local cost of electricity, highly insulating windows have a simple PBP of 23 to 55 years. The range is primarily due to the large variability in primary window costs.

These data suggest that highly insulating windows are an effective energy-saving measure that should be considered for high-performance new homes and in existing retrofits. However, the cost effectiveness of the measure, as determined by the simple PBP, suggests that highly insulating window costs continue to make windows difficult to justify on a cost effectiveness basis alone. Additional reductions in costs via improvements in manufacturing and/or market penetration that continue to drive down the manufacturing cost of highly insulating windows will make this technology much more viable as a cost-effective energy efficiency measure.

This study also illustrates that highly insulating windows have important impacts on peak load, occupant comfort, and condensation potential, which are not captured in the energy savings calculation. In addition, more consistent and uniform interior temperature distributions suggest that highly insulated windows, as part of a high performance building envelope, may enable more centralized duct design and downsized heating, ventilation, and air-conditioning systems. Shorter, more centralized duct systems and smaller heating and cooling equipment could yield additional cost savings, making highly insulating windows more cost effective as part of a package of new construction or retrofit measures that achieve significant reductions in home energy use.

ACKNOWLEDGMENTS

This project was funded by the US Department of Energy, Energy Efficiency and Renewable Energy Building America Program (Eric Werling/David Lee, Program Managers) and Emerging Technology Windows R&D Program (Marc LaFrance, Program Manager); the Bonneville Power Administration (Kacie Bedney, Project Manager); and the US Department of Energy, Office of Electricity Delivery and Energy Reliability (Dan Ton, Program Manager). Additional support was provided by Battelle Memorial Institute; Northwest Energy Works (Tom Hews/Brady Peeks); Washington State University Extension Energy Program (Ken Eklund/Emily Salzberg); GE Appliances (Dave Najewicz); Tri Cities Research District (Diahann Howard); the City of Richland, Washington (Bob Hammond); and Jeld Wen, Inc. (Ray Garris/Mike Westfall).

Other stakeholders on this project include the US Department of Energy Pacific Northwest Site Office, Northwest Energy Efficiency Alliance, the Energy Trust of Oregon, and Marlette Industries.

PNNL staff significantly contributing to this project include facilities, operations, contracting, and financial staff: Raul Carreno, Garret Hyatt, Sheena Kanyid, David Koontz, Bob Turner, Ray Sadesky, T.R. Hensyel, Jim Bixler, Vicki Stephens, Sam Martinez, and Jamie Spangle; media relations staff: Annie Haas, Franny White, Dawn Zimmerman, Megan Neer, Tim Ledbetter, and Kevin Kautzky; editor: Susan Ennor; management staff: Evan Jones, Sean McDonald, Mark Morgan, Todd Samuel, Sriram Somasundaram, Terry Brog, Rob Prat,t and Gary Spanner; and technical staff: Vrushali Mendon, Susan Sande, Jeremy Blanchard, Eric Richman, Susan Loper, Michael Kintner-Meyer, Michael Baechler, and Laura Van Kolck (student intern).

The authors also acknowledge the technical support provided by Greg Sullivan, Principal, Efficiency Solutions, LLC, Richland, Washington, for specification development, occupancy simulation development, metering, and data collection.

REFERENCES

Federal Energy Regulatory Commission (FERC). 2012. Bonneville Power Administration Peak Load Data. Form 714 for respondent 122, available at http://www.ferc.gov/docs-filing/forms/form-714/data.asp.

National Renewable Energy Laboratory (NREL). 2012. *National Residential Efficiency Measures Database*. Available at: http://www.nrel.gov/ap/retrofits/measures.cfm?gId=1&ctId=47&scId=558.

FOR FURTHER INFORMATION

Widder, SH; GB Parker; MC Beachler; and NN Bauman. 2012. Side-by-Side Evaluation of Highly Insulating Windows in the PNNL Lab Homes. PNNL-21678. Pacific Northwest National Laboratory, Richland, WA. Available at: http://labhomes.pnnl.gov/documents/Field_Evaluation_Highly_Insulating_Windows_Lab_Homes.pdf.

Net Zero Building Enclosure Retrofits for Houses: An Analysis of Retrofit Strategies

Graham Finch, PEng
Associate Member ASHRAE

Brittany Hanam, PEng, BEMP
Associate Member ASHRAE, Building Energy Modeling Professional

ABSTRACT

Efforts are underway across North America to improve the energy performance of new building stock whether through building code or energy efficient building programs; however, it is also important to find ways to address energy performance in existing buildings. A research study was completed to identify and evaluate building enclosure retrofit strategies to use in near net zero retrofits for houses across all Canadian climate zones (ASHRAE/DOE Zones 4 through 8). This paper will present the key findings of this research study.

General thermal performance targets for net zero house enclosures in heating dominated climates were developed through energy simulations of four existing archetype houses constructed between the 1950s and 1980s. Retrofit strategies to achieve these targets were then identified and analyzed.

Retrofit measures are grouped into five categories: the attic or roof, above grade walls, windows and doors, below grade walls, and slab. Each retrofit was evaluated in terms of its thermal and hygrothermal performance. As improving the thermal performance of an assembly affects its hygrothermal performance, findings and retrofit design strategies for the control of air, vapour, and moisture in highly insulated enclosure assemblies were developed and are presented.

A cost analysis was completed for each retrofit strategy to determine the financial feasibility of near net zero enclosure retrofits. It was found that such retrofits are difficult to justify based on energy savings alone in most areas of North America; however, if retrofits are being completed for other reasons, such as to upgrade aesthetics or to repair water penetration issues, the incremental cost to complete a near net zero retrofit can have good financial return.

INTRODUCTION

In Canada, approximately 30% of all energy is consumed in buildings, with about 16% used in residential buildings and 14% in other buildings (NRCan 2005). Efforts are underway to improve the energy performance of new building stock, however, it is also important to find ways to address energy performance in existing buildings.

A large amount of information is available to the public on simple ways to save energy at home, for example: turning off the lights, basic air sealing, topping up attic insulation, purchasing EnergyStar® or other high efficiency appliances. However, less information is available on deep or very high energy savings retrofits or methods to approach net zero energy consumption in existing houses. A research study was performed for Canada Mortgage and Housing Corporation (CMHC) focusing on building enclosure energy efficiency retrofit measures that will enable near net zero energy consumption for existing houses and this paper provides a summary of the key findings (RDH 2012). The study evaluates various building enclosure retrofit alternatives using criteria that include thermal performance, hygrothermal performance and durability, cost, constructability, and environmental considerations for near net zero housing retrofits in the order of 90% space conditioning savings.

Graham Finch *is a principal and building science research engineer and* ***Brittany Hanam*** *is a building science engineer with RDH Building Engineering Ltd., Vancouver, BC, Canada.*

Background

A net zero energy house is a house that produces as much energy as it consumes on an annual basis. Net zero buildings are often grid-tied, meaning they draw energy from the grid as needed and supply energy back to the grid as it is produced, with the balance over the course of the year being zero or net supply. Net zero is typically achieved by reducing the heating, cooling, and other electrical loads as much as possible through high performance enclosures, efficient mechanical and electrical systems, and generating the balance of energy required to be considered net zero through on-site renewable means such as solar photovoltaics.

Other terms that are often associated with net zero are near net zero and net zero ready. Near net zero simply means a house that produces almost as much energy as it consumes annually, though there is no quantified measure for how close to net zero the house should be. Net zero ready means a house that has minimized heating, cooling, and other loads, the balance of which could be generated by on-site renewable energy. Net zero carbon would be a scenario where carbon offsets would be purchased to balance consumption, but is not discussed here.

The building enclosure is one component of a near net zero house. A highly insulated and airtight enclosure with appropriately selected glazing is needed to minimize heating and cooling loads and maximize solar heat gain during the winter. Without a high performance building enclosure, reaching near net zero energy consumption is very difficult. The location and orientation of the house also affects near net zero design by affecting the amount of renewable energy that can be generated at the site (for example, available solar energy) and also the solar heat gain benefits, shading, ventilation, etc. The building enclosure defines the heating and cooling loads on a building. The building enclosure design should reduce the loads to a level that is as low as practically possible and should make sense from a cost and functionality standpoint.

The research presented in this paper is limited to retrofits to the building enclosure to achieve near net zero energy use. Analysis of mechanical, electrical, and renewable energy generation systems were beyond the scope of this project.

Methodology

A literature review was completed to identify a comprehensive list of potential Energy Efficiency Measures (EEMs) for roofs, above grade walls, windows, below grade walls and slab-on-grade floors. Many, but not all, of the EEMs have been implemented by builders and renovators across North America. These EEMs were then evaluated for each of four existing Canadian housing archetypes (one and a half storey post-war era house, two storey raised bungalow with attached garage, two storey house with basement and garage, two storey row house/duplex end unit with basement and garage). Each house was evaluated in 14 major Canadian cities (shown in Figure 1) relative to construction feasibility, energy savings potential, cost to implement, and cost effectiveness. EEM strategies that were considered feasible and included in more detailed evaluation are summarized in Table 1.

Energy usage characteristics were evaluated using the energy modeling program HOT2000 Version 10.51 (NRCan 2008). Building enclosure EEM packages (groups of EEMs suitable for a particular housing archetype) that provided an EnerGuide for Houses (EGH) rating of 83 were used as a benchmark of a high-performance and a near net zero building enclosure. The Canadian EnerGuide

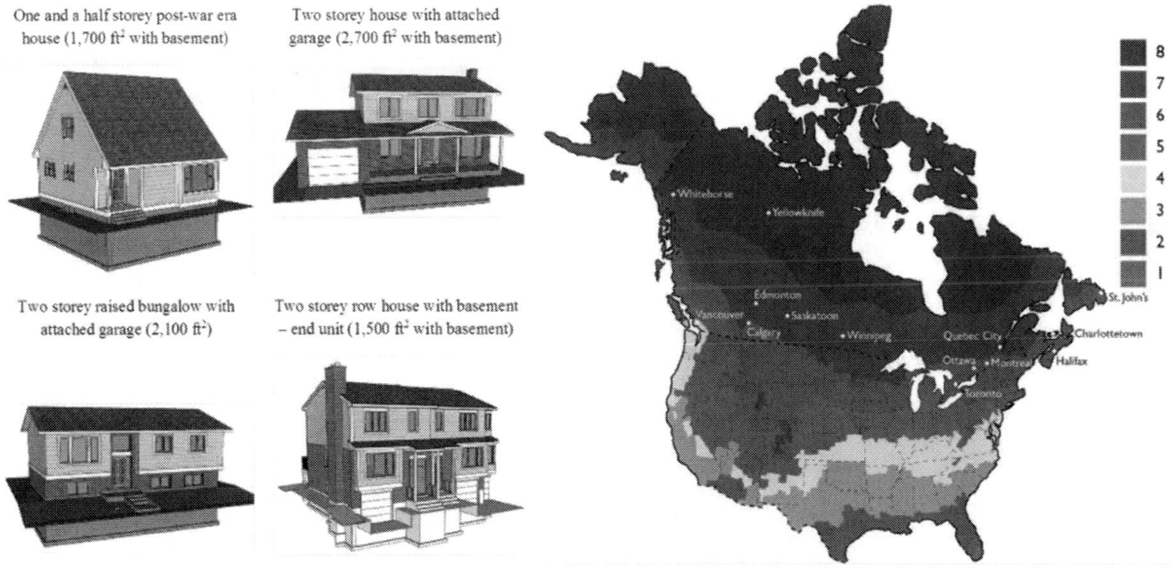

Figure 1 *a) Schematic of existing Canadian Archetype houses b) ASHRAE/DOE Climate Zone Map of Canada/US showing location of 14 Canadian cities and climate zones which were analyzed within this study.*

Table 1. Summary of Building Enclosure Retrofit Energy Efficiency Measures (EEMs) Evaluated

Attic and Roof Assemblies	Above Grade Wall Assemblies	Below Grade Wall and Slab Assemblies	Windows and Doors
• Top up existing attic insulation • Spray foam flash and fill attic insulation • Roof with exterior insulation • Improve airtightness	• Double stud (interior retrofit) • Larsen truss (exterior retrofit) • Exterior rigid or semi-rigid mineral fiber insulation (exterior retrofit) • Exterior rigid foam insulation (exterior retrofit) • Interior vacuum insulation panels (interior retrofit) • Exterior vacuum insulation panels (exterior retrofit) • Improve airtightness	• Interior slab XPS insulation • Interior slab vacuum insulation panels • Interior wall XPS or SPF insulation or in combination with fiberglass batt • Interior wall vacuum insulation panels • Exterior wall insulation (rigid XPS, SPF or mineral fiber) • Improve airtightness	• Thermally efficient frames and glazing • Insulated doors • Improve airtightness

rating system was developed by NRCan as a standard measure to measure a home's energy performance with a scale from 0 to 100 with a rating of 100 being a net zero house (NRCan 2011). A rating of EGH-83 was used since it approximates the highest energy performance level that can be achieved with a building enclosure retrofit using mid-efficiency mechanical and electrical systems (i.e., without major mechanical or electrical retrofits or renewable energy) (CMHC 2011). All simulations therefore used mid-level efficiency mechanical equipment (80% efficient furnace) and standard EnerGuide assumptions so that the only variables were the building enclosure retrofit measures. Note that while the building enclosure is important, there are many other factors that contribute to achieving net zero or near net zero energy in housing, and it takes a whole-house approach to achieve this goal. The retrofit strategies identified in this paper will therefore improve the energy efficiency of a house but will not achieve net zero energy performance on their own.

In addition to the energy analysis and assessment of cost effectiveness, the evaluation of EEMs considered other building science performance issues including hygrothermal behaviour and relative risk as it is critical that the implementation of EEMs does not compromise other aspects of the performance of the building enclosure.

Conceptual retrofit details for EEM packages were developed for each of the four housing archetypes and are contained within the full report. An overview of the retrofit strategies and details are provided within this paper.

FINDINGS

Retrofit Strategies

As previously discussed, retrofit strategies are grouped into five areas: roof or attics, above grade walls, windows and doors, below grade walls, and slab. For each area there are a number of alternative retrofit strategies that are feasible for a near net zero enclosure retrofit, provided that appropriate control of air, water, and vapour are addressed. The most appropriate retrofit strategies will vary based on specific project variables; however, general comparisons are made based on the parameters and assumptions of this study.

One of the biggest considerations for a building enclosure retrofit, primarily for the exterior walls, is whether to perform the work from the interior side or the exterior side of the house. This decision is something that needs to be considered carefully by the homeowner and has significant impacts on the feasibility, cost, house size constraints, aesthetics, lifestyle disruption, and long term durability. Regardless of the approach, the research finds that the best opportunity in terms of cost and disruption is to perform an enclosure retrofit when other major work is already necessary or planned. Exterior building enclosure retrofits are very common when replacing cladding, installing new windows, or repairing moisture damaged buildings and can be expanded easily to address other building enclosure issues and improve thermal performance. Occupants can continue to live within buildings undergoing an exterior retrofit which is a significant advantage in terms of convenience and cost. When performing interior retrofits, significant disruption of space is required, access to exterior walls is necessary, the work is generally dusty (and may be hazardous if exposing lead paint or asbestos), electrical work is required, and the decision may result in a loss of interior floor space.

From a building enclosure durability standpoint, exterior retrofits which replace the cladding are preferred so that water shedding details and interface details can be improved along with overall airtightness. Interior retrofits may, however, be necessary or preferred due to a heritage cladding, house aesthetic, or a building lot set-back restraint.

Figure 2 shows the typical interior and exterior building enclosure retrofit measures which were studied and are applicable to most of the housing archetypes. The two storey raised bungalow is shown in this paper and encompasses many of the common details for each archetype. The discussion in the following paragraphs covers each of the five major components from the roof to the below grade floor.

Attic or roof retrofit measures improve the attic or roof insulation and wall interface airtightness. The top-up attic insulation retrofit approach provides a high level of thermal performance at a low cost. The airtightness must be improved by sealing all ceiling penetrations to the existing polyethylene (where present) including partition walls, lights, bathroom fans, ducts, etc. to create a continuous ceiling air barrier. Alternative roof and attic retrofit strategies include a flash and fill insulation application in which a few inches of spray foam are directly applied to the existing ceiling drywall/plaster (important if there is no existing ceiling air barrier) topped with lower cost cellulose or fiberglass fill to meet the R-value needed or an exterior insulation approach (may be cost effective if the roof requires replacement and is well suited for the one and a half storey post world war two "victory house"). Figure 3 shows conceptual sketches for interior and exterior insulated roof and attic retrofits.

Above grade walls may be retrofitted from the exterior or the interior. If an exterior retrofit is chosen, exterior rigid or semi-rigid mineral wool insulation provides good thermal and hygrothermal performance with the lowest cost of the exterior insulation retrofit options. This strategy involves the removal of the existing cladding down to the sheathing. A continuous air barrier and water resistant barrier (WRB) membrane is installed over the sheathing with proper detailing at all joints/interfaces to improve the airtightness and water penetration control. Mineral wool insulation is then installed and new cladding is attached with screws through vertical wood strapping.

Where an interior retrofit is chosen, the double stud wall retrofit is the most cost effective. With this strategy the existing drywall is removed, a second row of non-load bearing 2×4 wood framing is installed and filled with dense pack fiberglass/cellulose or spray foam insulation, and new drywall is installed to finish. A continuous air barrier system is detailed utilizing either an airtight drywall, sealed polyethylene or spray foam approach with proper detailing at all joints and interfaces. This also involves removal of some ceiling drywall to insulate and air seal the floor joist cavities. Alternative above grade wall retrofit strategies include exterior foam insulation, an exterior Larsen truss filled with fiberglass or spray foam insulation, or vacuum insulated panels at the interior (Mattock 2012). Figure 4 shows conceptual level sketches for four exterior and interior insulated retrofit wall assemblies.

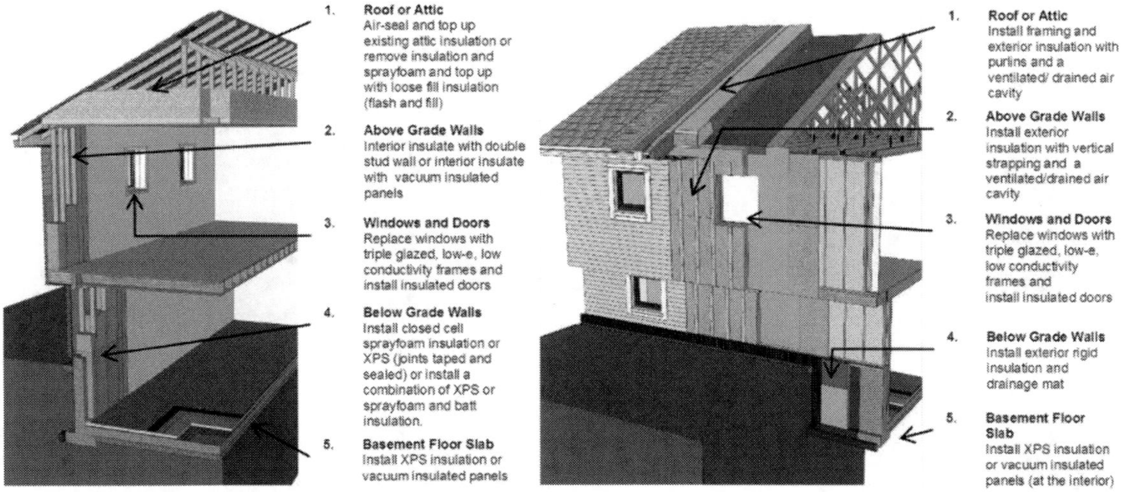

Figure 2 *(Left) Section through a house showing a full building enclosure interior retrofit and (right) exterior retrofit for the two storey raised bungalow archetype.*

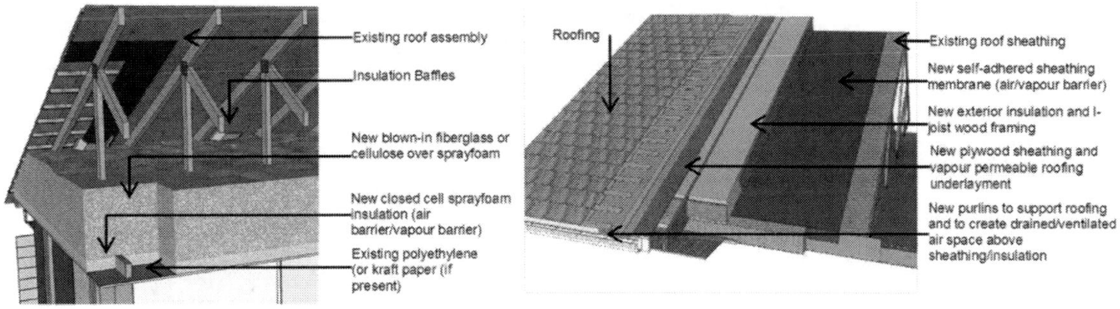

Figure 3 *(Left) Example of interior flash and fill ceiling insulation and (right) exterior roof insulation retrofit strategies.*

Window retrofits will improve the existing window thermal performance and the interface and component airtightness. Triple glazed windows with two low-e coatings, argon gas fill, low conductivity edge spacers, and low conductivity frames (vinyl, wood, or insulated fiberglass) with overall performance in the R-6 (U-0.17) range provide significant energy savings. Proper detailing is important at the window to wall interface to provide a continuous air barrier and effective water penetration control. The solar heat gain coefficient (SHGC) of the glazing should be selected based on project specific conditions (generally in the moderate to high range, SHGC 0.4 to 0.5), and operable or fixed exterior shading devices should be used to optimize seasonal solar heat gain. If triple glazed windows are cost prohibitive, installing new double glazed windows can still have significant energy savings. Entry doors with an insulated core can provide additional energy savings.

The near net zero retrofit will result in a thicker wall assembly. This means the window can be installed in one of three positions: in line with the exterior surface, in line with the interior surface, or in the middle of the rough opening. Any of these positions can work; however, aligned with the insulation gives a higher R-value and installing the window aligned with the interior surface will help to minimize the potential for condensation. Conceptual details for a window installed into an exterior insulated wall and interior double stud insulated wall assembly are shown in Figure 5.

Below grade wall retrofits should improve the wall insulation as well as the airtightness at the below grade to above grade wall interface. Like above grade walls, below grade walls can be retrofitted to add insulation either from the interior or the exterior. Interior retrofits are typically less expensive since exterior retrofits require excavating the foundation; however, interior retrofits typically result in a loss of floor space. With below grade wall retrofits it is important to follow

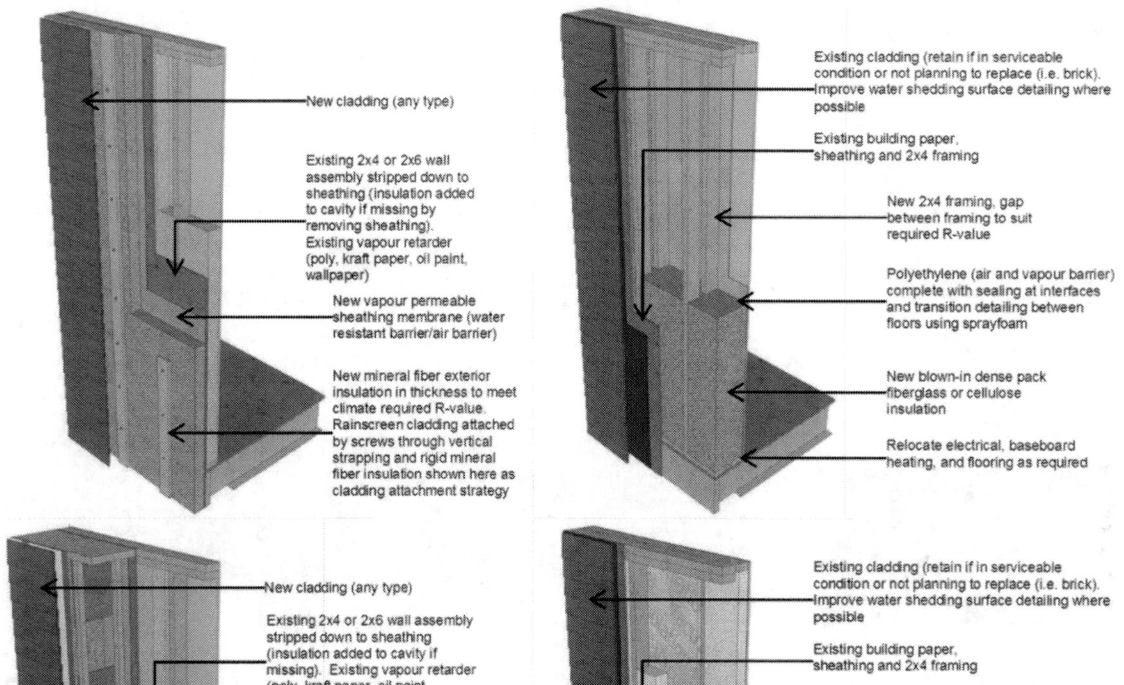

Figure 4 *Conceptual wall retrofit assemblies: (top left) exterior mineral wool insulation, (top right) interior double stud wall, (bottom left) exterior Larsen truss, and (bottom right) interior vacuum insulation panel.*

best practice guidelines for the control of air and water vapour in order to prevent moisture problems. The interior closed cell spray foam insulation retrofit provides the best hygrothermal performance of the interior retrofits based on the study results (exterior insulation is the best but is often cost prohibitive). Other retrofits have comparable cost but require more attention to installation and details to provide good hygrothermal performance. Alternative below grade wall retrofit strategies include installing XPS insulation with sealed and taped joints, installing only 1 to 2 in.of closed cell spray foam insulation followed by 2×4 framing with batt insulation to reduce the cost, installing vacuum insulation panels, or excavating the foundation and installing XPS insulation at the exterior. Figure 6 presents conceptual sketches for an exterior and interior insulated below grade retrofit.

Floor slab retrofits should improve the insulation and be airtight to prevent moisture problems. Insulating the slab at the exterior is generally not practical in retrofits unless the basement is being deepened for other purposes, and therefore the slab must be insulated at the interior. This limits the thickness of insulation that can be used since added interior slab insulation results in loss of floor to ceiling height in the interior space. The slab can be insulated by installing XPS insulation with taped joints below the flooring. An alternative but more expensive slab retrofit is to use vacuum insulated panels and may be desirable in basements with lower ceiling clearances.

Insulating the slab at the interior can create colder conditions at the bottom plate of partition walls. To protect the wood at these locations, consider the following precautions:

- Structural partition walls: replace the bottom plate with pressure treated wood.
- Non-structural partition walls: raise the bottom plate to sit on top of the floor sheathing. If VIPs are used, strips of XPS must be installed beneath the bottom plate to allow for fastening to the concrete slab.

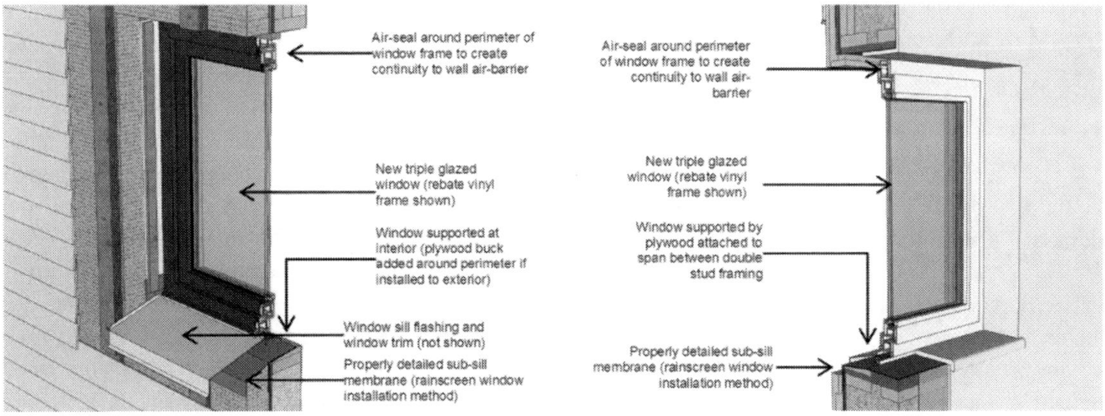

Figure 5 *Conceptual detailing for (left) windows installed into exterior insulated wall and (right) interior double stud insulated wall assembly.*

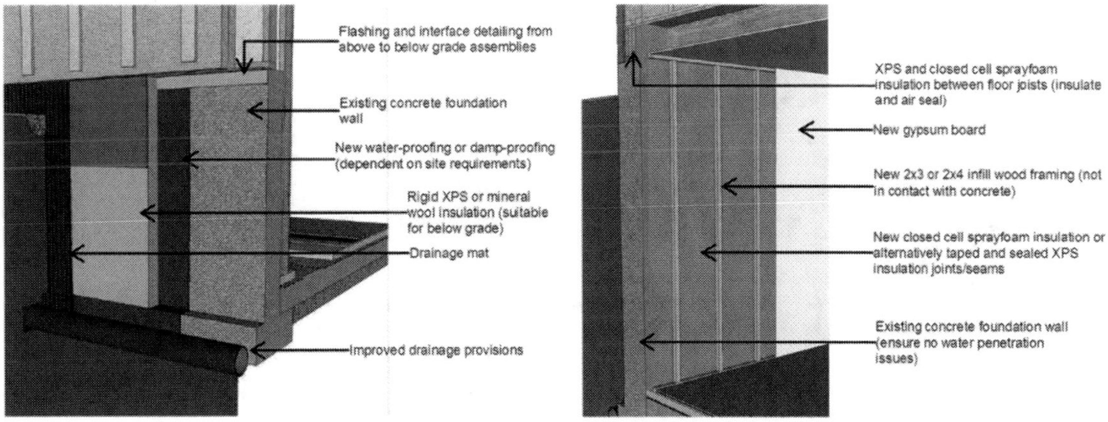

Figure 6 *Conceptual below grade wall assemblies: (left) exterior insulated for half height basement in split level bungalow and (right) interior insulated for other archetypes.*

Figure 7 presents a conceptual sketch for a basement slab and partition wall details. Adding insulation in this configuration over a drainage/ventilation matt also decouples slab moisture issues, particularly in slabs that do not have an existing vapour retarder or capillary break.

Thermal Performance of Near Net Zero Enclosures

To achieve near net zero energy consumption, regardless of geographic location, more insulation always reduces the heating and cooling load. However, increasing insulation has diminishing returns on energy savings. Each retrofit project should be evaluated for how much insulation is required and makes sense on a case-by-case basis considering the location, mechanical and electrical systems, economics of the owners, renewable energy, and overall energy goals for the project.

The following general guidelines for effective R-values (i.e., including the impact of thermal bridging) and whole house airtightness for near net zero houses were identified through this research, and this finding is fairly consistent with other research findings in the literature review (i.e., Passivhaus Standard and CMHC EQuilibrium™ Project Profiles):

- Ceiling or roof: R-60 h·ft^2·°F/Btu (R_{SI}-10.6 m^2·K/W)
- Above grade walls: R-40 h·ft^2·°F/Btu (R_{SI}-7.0 m^2·K/W)
- Windows: R-6 h·ft^2·°F/Btu (R_{SI}-0.9 m^2·K/W)
- Doors: R-6.5 h·ft^2·°F/Btu (R_{SI}-1.1 m^2·K/W)
- Below grade walls: R-20 h·ft^2·°F/Btu (R_{SI}-3.5 m^2·K/W)
- Foundation slab: R-10 h·ft^2·°F/Btu (R_{SI}-1.8 m^2·K/W)
- Airtightness: 1 ach @ 50 Pa

Specific building enclosure R-value requirements for each of the different housing archetypes in each city and climate zone are provided within the full report (RDH 2012). For example in Vancouver, BC (Climate Zone 4) it is possible to avoid insulating the basement slab, and lower wall R-values may be feasible for some archetypes which can allow for flexibility in the retrofit.

Hygrothermal Considerations

Near net zero energy building enclosure retrofits for houses require significantly higher enclosure R-values than is typical for new or retrofit construction. The net result on construction is thicker insulated walls and roofs, split insulated walls, and the use of alternate insulation materials which all reduce heat flow through the building enclosure. The reduction in heat flow through these highly insulated enclosure assemblies is beneficial to whole building energy consumption but does introduce additional considerations with respect to heat, air, and moisture control.

Analysis was undertaken to evaluate the heat, air, and moisture control functions for each of the proposed retrofit strategies to assess how the retrofit may change the moisture performance of the wall assembly. Thermal performance was assessed by determining the overall effective R-value of the assemblies using the three-dimensional heat transfer simulation program HEAT3. Hygrothermal analysis was performed qualitatively for each assembly using industry accepted best practices, and for more complex assemblies, hourly hygrothermal simulations were performed using the WUFI 5.1 Pro (WUFI) computer model. WUFI was used to simulate the performance of the wall assemblies under climatic loads for representative Canadian cities including Vancouver, Calgary, Edmonton, Winnipeg, Toronto, Ottawa, Montreal, Quebec, and St. John's. Exterior boundary conditions were assessed to model different orientations and a range of exposure and driving rainfall conditions. The interior climate was modeled in WUFI based on the exterior climatic conditions with typical interior moisture generation rates and house air-exchange rates. A range of interior moisture generation rates were simulated to assess the impact of varying low to high indoor relative humidity levels on the performance of the enclosure assemblies here. As expected, higher indoor RH levels put some of the assemblies at greater risk for condensation and wetting, and decreased the drying potential.

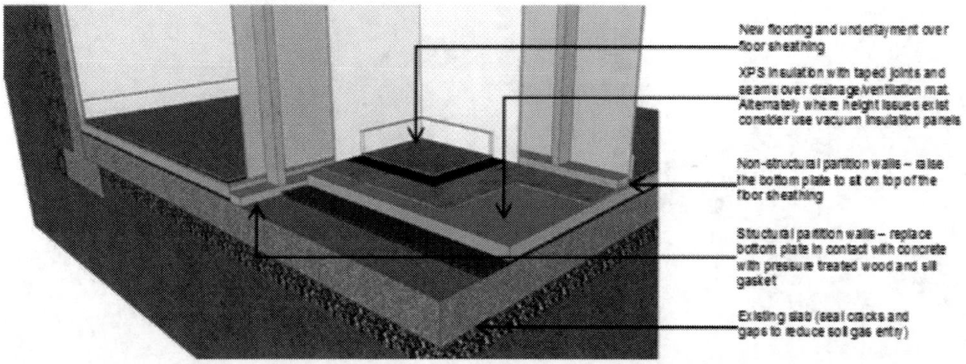

Figure 7 *Conceptual level floor slab insulation retrofit and partition wall considerations.*

The hygrothermal performance of each of the proposed highly insulated enclosure assemblies was compared to a calibrated baseline wall or roof assembly that is known to provide satisfactory performance. Assemblies are calibrated to field testing and experience in the assessment of performance of standard 2×4 or 2×6 insulated wood-frame assemblies. The performance of these baseline assemblies under normal Canadian climate conditions is well understood and modeling using WUFI has found good agreement with field measurements and historical performance data for these baseline assemblies. Supplemental data from field monitoring studies of highly insulated assemblies was also used where available to calibrate the results.

Three hygrothermal modeling scenarios were simulated to assess the relative performance and risk of each proposed enclosure assembly. For each assembly, at least a two year simulation is performed to assess the year over year moisture content, relative humidity levels, and the potential for long-term moisture accumulation. The three scenarios include:

- Normal performance
- Assembly exposed to air-leakage
  - An exfiltration air leakage rate that results in condensation and wetting of the interior surface of the wall or roof sheathing is selected, which brings the moisture content up to 30% MC for a few weeks in the winter but dries out in the spring. This air leak, which is then known to be seasonably tolerable within the baseline wall or roof, is introduced into the models for the alternate assemblies to compare relative performance. The quantity of air leakage is not important since the analysis is intended to be a relative comparison to the baseline assembly.
- Assembly exposed to a driving rain leak
  - A driving rain leak that results in wetting of the sheathing is selected which brings the moisture content in the baseline wall or roof sheathing up to 30% MC for a few weeks in the winter but dries out in the spring. The leakage rate is selected as a percentage of the driving rain and is introduced to wet the exterior surface of the sheathing. This rain leak, which is known to be seasonably tolerable within the baseline wall or roof, is introduced into the models for the alternate assemblies to assess relative performance. The quantity of rain leakage is not important since the analysis is intended to be a relative comparison to the baseline assembly.

While all parameters and results from the hygrothermal simulation were reviewed to assess the hygrothermal performance, the moisture content of the sheathing is used as a primary indicator of performance within the above grade wall and roof assemblies. The sheathing is moisture sensitive and is typically the component most likely to be become damaged by moisture ingress from vapour diffusion, air leakage, or water leakage.

Results demonstrating the hygrothermal performance and the relative sensitivity of the interior double stud retrofit and exterior insulation retrofit (with foam or rigid mineral wool) to both air leakage and rain water penetration are shown in Figure 8 and 9 for Vancouver, BC. In addition, the total number of hours per year that the sheathing is above 20% MC in six Canadian cities is shown in Table 2 for both of the wall assemblies.

The results demonstrate that the interior double stud retrofit increases the risk of air leakage or rainwater leakage wetting and damage to the exterior sheathing as compared to the baseline wall, largely because of the colder sheathing temperatures from the high levels of interior insulation. This means that detailing of both the air barrier and water shedding surface/water resistant barrier is crucial to the long-term durability of this wall. This highlights the importance of water penetration control at the interface details and can be an issue in an interior retrofit if these details are not being improved.

The results also show that the type of insulation in an exterior retrofit is very important. Rigid foam insulation is vapour impermeable and coupled with the existing vapour impermeable interior finishes (oil paint, tiles, kraft paper, or polyethylene) in an exterior retrofit can result in a situation where moisture that gets into the assembly through a defect is unable to dry out and may lead to damage. The use of vapour permeable exterior insulation such as rigid mineral fiber allows for drying outwards (beneficial in all Canadian climate zones) regardless of the interior finish. Inward vapour drive through the mineral fiber insulation is not an issue if the cladding is ventilated (i.e., rainscreen), though it could potentially be problematic in more humid and southern climate zones (i.e., Zones 1–3) with unventilated claddings due to inward vapour drive. In these climate zones, the usual lack of interior vapour control allows for the more safe use of exterior foam insulation with direct applied claddings, which reduces the inward vapour drive.

The hygrothermal analysis indicates that most of the assemblies perform well under normal conditions when care is taken to follow best practices for the control of heat, air, and moisture. Some assemblies perform better than the baseline wall (i.e., less moisture accumulation occurs), while other assemblies see greater moisture accumulation. All of the assemblies, including the baseline assemblies, will perform adequately provided that moisture risks related to rain or air leakage control can be minimized through the use of good water managed assemblies (i.e., rainscreen walls) with appropriate detailing, interfaces, flashings, and continuous air barriers. Care must be taken to avoid a dual vapour barrier situation including the use of exterior foam insulation with interior polyethylene, kraft paper, wallpaper, oil paint, or other impermeable finishes.

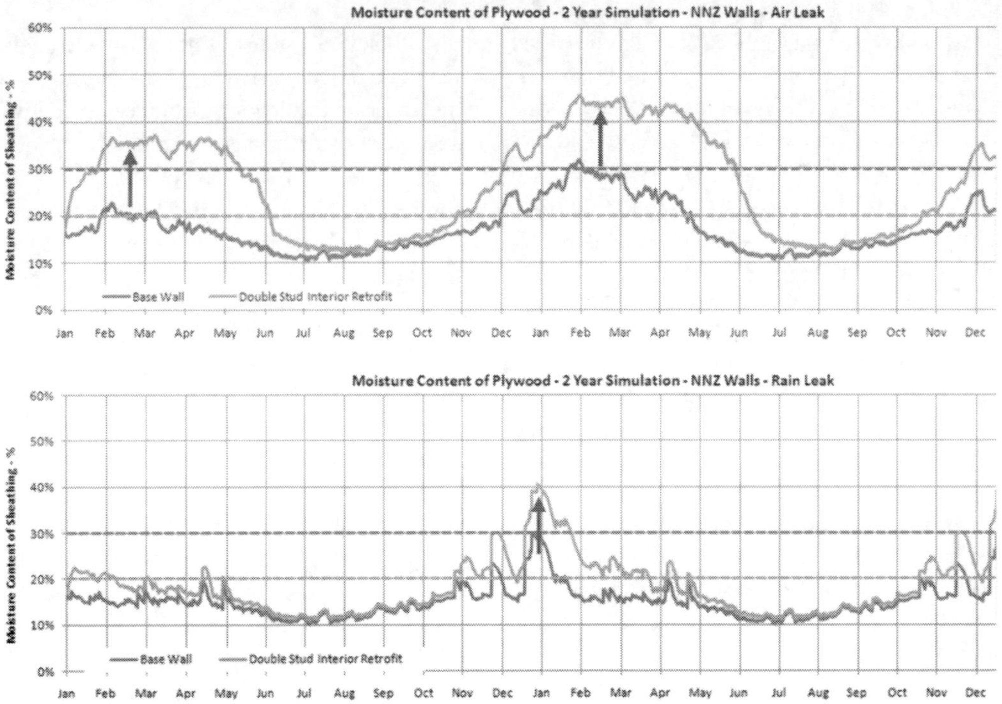

Figure 8 *Moisture content of exterior plywood sheathing (existing) in a double stud retrofit wall and baseline 2×4 insulated wall exposed to (upper) an air leak and (lower) rain water leak for Vancouver, BC.*

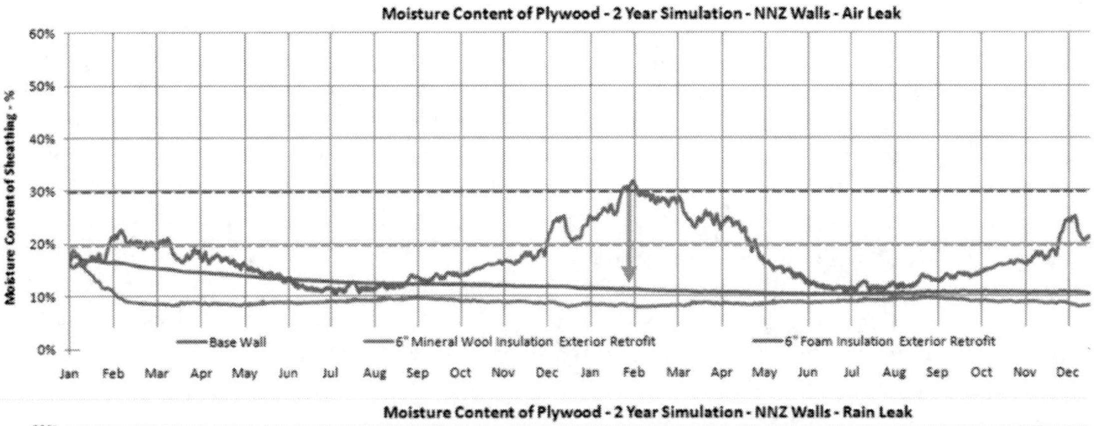

Figure 9 *Moisture content of exterior plywood sheathing (original) in a exterior insulation retrofit with either mineral fiber or foam insulation and baseline 2×4 insulated wall exposed to (upper) an air leak and (lower) rain water leak for Vancouver, BC.*

Energy Savings

A near net zero building enclosure retrofit can significantly reduce the heating energy consumption when high insulation values are used, thermal bridging is minimized, and airtightness is improved. Specific retrofit costs, energy savings, and payback periods depend on many factors, including the location, the design and construction of the existing house, and the level of retrofit performed. Every retrofit project will be different, so it is important to look at costs and savings on a case-by-case basis.

Figure 10 shows annual heating energy consumption for a typical raised bungalow house in various Canadian climates with various levels of enclosure thermal performance (all with the same mid-efficiency mechanical equipment). If a near net zero building enclosure retrofit is completed with highly insulated enclosure components, heating energy consumption could be reduced by up to 90%, depending on the existing construction and the level of enclosure retrofit performed.

Financial Analysis

A financial analysis was performed to estimate the initial construction costs for a near net zero enclosure retrofit and to estimate the annual savings in energy costs. Many home retrofit projects are undertaken for reasons other than energy consumption, for example to repair damaged assemblies, to replace assemblies at the end of their service life (such as roofs), or to improve aesthetics. If retrofits are done for reasons other than energy savings, the incremental cost of incorporating energy efficiency measures (EEMs) will always have a better financial payback than the cost of renovating for energy savings alone. Base costs were estimated for an exterior wall assembly retrofit and roof retrofit to determine the incremental payback on the additional cost of incorporating EEMs.

The four baseline archetypes were simulated using HOT2000 to determine the baseline or typical energy consumption of each of the four archetype houses. The energy consumption of the house with each individual EEM was compared to the baseline energy consumption to determine

Table 2. Summary of the Number of Hours per Year that the Exterior Sheathing Moisture Content is Above 20% when Retrofit and Baseline Walls are Exposed to Calibrated Air and Rain Water Leaks, Six Canadian Cities

City	Vancouver, BC	Edmonton, AB	Winnipeg, MB	Ottawa, ON	Quebec, QC	Saint John's, NL
Air Leaks						
Baseline 2×4 wall	3510	2650	3463	2910	2842	3820
Interior double stud	5453	4420	4864	4205	6176	5889
Exterior mineral fiber	0	0	0	0	0	0
Exterior foam	0	0	0	0	0	0
Rainwater Leaks						
Baseline 2×4 wall	853	346	342	276	1024	436
Interior double stud	3832	425	445	1169	2058	772
Exterior mineral fiber	0	27	125	45	0	19
Exterior foam	8760	6667	8760	8760	8760	544

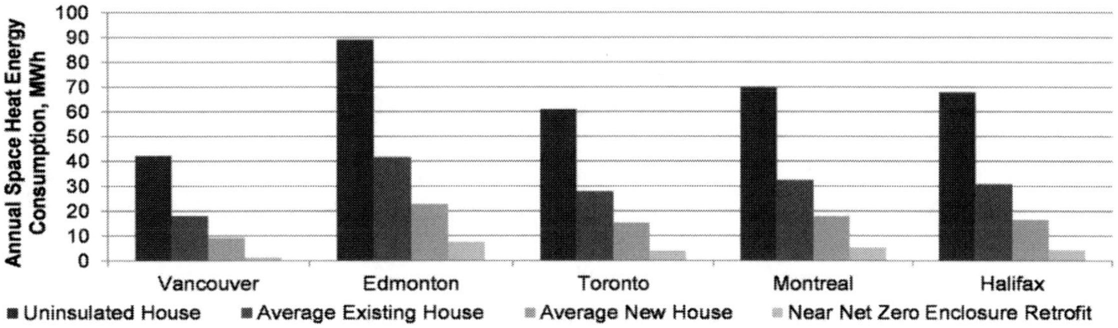

Figure 10 Annual space heat energy consumption of houses with various insulation and airtightness levels in five different cities across Canada.

the energy savings resulting from each EEM. For EEMs where a base retrofit was identified, the financial metrics were calculated for both the total cost of the full retrofit and the incremental cost of the EEM. To determine the energy savings of a whole house enclosure retrofit, a group of EEMs for one house were identified and simulated. The payback period and other financial metrics were calculated for the whole house near net zero retrofit. This analysis was done for locations across Canada. The construction cost of each EEM was estimated using published cost data from various sources, tempered by knowledge and experience of the retrofit market in 2011. This information is contained within the full report (RDH 2012). The energy cost savings from each EEM was based on 2011 utility rates for those cities analyzed (Hydro Quebec 2011).

Previous research shows that many deep energy retrofit measures have long payback periods (though this could change as energy costs rise). Despite high initial costs it may still be desirable to perform a near net zero energy retrofit for other non-financial reasons. An important consideration when designing a near net zero energy retrofit is to consider the cost of retrofits compared to the cost of renewable energy; at some point it may be cheaper to use renewable energy than to lower the energy consumption through enclosure EEMs (e.g., photovoltaics or solar domestic hot water). A metric to compare the cost of EEMs to the cost of renewable energy is the Value Index (Proskiw and Parekh 2010), which is the initial cost (or incremental cost) of the EEM divided by the annual energy savings attributed to it.

The Value Index can be compared to the cost of renewable energy that may also be used to achieve energy efficiency targets. In urban settings a common renewable energy source used for near net zero houses is photovoltaics. Solar PV prices have significantly dropped in the past few years from about $8/kWh per year in 2008 (Proskiw and Parekh 2010) to less than $3.75/kWh per year in 2011 (Mattock 2011). PV costs will continue to decrease as the technology becomes more common, and this value index can be calculated on a project basis based on local PV prices. The Value Index comparison can be thought of as comparing the cost per kWh saved (i.e., building enclosure EEMs) to the cost per kWh generated (i.e., renewable energy). This metric can help determine whether it is more cost effective to achieve near net zero energy by reducing the loads with enclosure retrofits or by generating renewable energy.

The individual EEM analysis showed that many individual or packaged house EEMs have payback periods greater than 20 years. The whole house analysis showed that where a retrofit is being completed for reasons other than energy savings (e.g., to repair water penetration issues or to upgrade aesthetics), the incremental cost to complete a near net zero retrofit can have a reduced payback period, particularly in those cities with high utility rates or rural houses still utilizing fuel oil for heating. Further, if a near net zero retrofit is desired, many of the enclosure EEMs are more cost effective than the installation of a photovoltaics system in order to achieve near net zero energy consumption. The financial analysis by city showed that energy prices have a significant impact on payback periods. High energy costs can make near net zero retrofits more financially feasible. Table 3 provides a brief summary of the total and incremental retrofit costs (as well as other financial metrics) to retrofit the building enclosure of a two storey raised bungalow archetype in six Canadian cities to near net zero.

Environmental Considerations

The primary focus of the near net zero retrofits research is to reduce the energy use in single family residential buildings. Other important considerations are occupant health and safety, resource efficiency, and environmental effect. With regards to health and safety, important issues to address as part of a near net zero enclosure retrofit include hazardous materials (asbestos, vermiculite, and lead paint), carbon monoxide, ventilation, and radon gas. The need to consider the impact of improved airtightness on the potential for back-drafting of naturally aspirated gas/oil appliances is critical, and may necessitate dedicated appliance makeup air or upgraded direct-vent appliances.

A near net zero building enclosure retrofit will provide a more airtight house, which is important for reducing energy consumption; however, this will also result in less incidental ventilation that was otherwise provided by a leaky enclosure. Therefore, the retrofit should also include energy efficient mechanical ventilation. This can be accomplished by adding a heat recovery ventilator (HRV) or an energy recovery ventilator (ERV). This ventilation should prevent condensation in the house and improve occupant health and comfort.

CONCLUSIONS

The results of this study indicate that near net zero retrofits are technically feasible but cannot be justified in the short term on the basis of a simple payback analysis alone. However, most retrofit projects are initiated to address a variety of other parameters such as moisture related performance issues, updating of the appearance of the building, or thermal comfort issues. The cost associated with minimum building enclosure retrofits to address a variety of drivers is more reasonably considered the baseline for near net zero cost payback analysis. In addition, it may be cheaper to use a combination of renewable energy (photovoltaics or solar domestic hot water) together with some of the more cost effective building enclosure retrofits to meet overall energy efficiency goals. However, as energy becomes more expensive, near net zero enclosure retrofits will become increasingly financially viable. This study has several implications for the housing industry:

- Building enclosure retrofits can lead to significant reductions in energy consumption in existing homes.
- High performance building enclosure retrofits can significantly decrease the energy loads for houses in Canada to the point where, in combination with efficient mechanical systems and on-site renewable energy generation, net zero energy can be achieved.

Table 3. **Financial Analysis of Incremental Cost for Full Near Net Zero Retrofit in 6 Canadian Cities, Bungalow Archetype, 2011 Construction Cost and Utility Rates**

City	Total Versus Incremental Retrofit	Cost	Annual Savings	Value Index	Simple Payback (Years)	Discounted Payback (Years)	Return on Investment
Vancouver	Total	$73,900	$1,400	$1.82	55	33	1.8%
	Incremental	$37,400	$1,400	$0.92	28	20	3.6%
Edmonton	Total	$75,800	$1,300	$0.94	56	33	1.8%
	Incremental	$38,500	$1,300	$0.48	29	21	3.5%
Winnipeg	Total	$68,700	$3,700	$0.82	18	14	5.4%
	Incremental	$34,800	$3,700	$0.42	9	8	10.7%
Toronto	Total	$74,800	$1,300	$1.32	56	33	1.8%
	Incremental	$37,900	$1,300	$0.67	29	21	3.5%
Montreal	Total	$72,100	$2,100	$1.13	34	23	2.9%
	Incremental	$36,600	$2,100	$0.57	17	14	5.8%
Halifax	Total	$67,200	$6,300	$1.07	11	9	9.4%
	Incremental	$34,000	$6,300	$0.54	5	5	18.5%

- Short-term financial benefits should not be overstated. Low energy prices mean that it will take longer for energy savings to offset retrofit costs than in places with higher energy costs.
- Consider the opportunity to make incremental energy efficiency improvements if performing such renewal or rehabilitation work as re-cladding, window replacement, or repairs to address moisture damage, as these will have the greatest payback.
- Building enclosure energy efficiency retrofits can be undertaken with little risk of moisture damage, provided care is taken to control water, air, and vapour.

REFERENCES

Blocon. 2012 *HEAT 3 Version 5.2*. www.blocon.se

CMHC. 2011. Equilibrium Sustainable Housing Demonstration Initiative. www.cmhc.ca

Fraunhofer Institut Bauphysik. 2012. WUFI Pro v 5.1. www.wufi.de.

Hydro Quebec. 2011. *Comparison of Electricity Prices in Major North American Cities*. [Online] [Cited: April 12, 2011] http://www.hydroquebec.com/publications/en/comparison_prices/pdf/comp_2010_en.pdf

Mattock, C. 2011. *Harmony House*.

Mattock, C. 2012. Final Report: Review and Evaluation of Vacuum Panel Insulation for use in Net Zero or Near Net Zero Energy Low-Rise Residential Construction. CMHC.

Natural Resources Canada (NRCan). 2008. *HOT2000 v10.51*.

Natural Resources Canada (NRCan). EnerGuide Rating System (Existing Homes). [Online] [Cited: October 11, 2011] http://oee.nrcan.gc.ca/residential/personal/home-improvement/service/rating.cfm?attr=4

Natural Resources Canada (NRCan). 2005. National Energy Use Database – Energy Use Handbook. Natural Resources Canada. Available online: http://oee.nrcan.gc.ca/corporate/statistics/neud/dpa/data_e/databases.cfm?attr=0

Proskiw, G., and Parekh, A. 2010. *Optimization of Net Zero Energy Houses*. Portland: BEST2.

Straube, J. 2011. Towards Net Zero Housing. [Online] [Cited: April 12, 2011] http://www.buildingscienceconsulting.com/presentations/documents/2011-01-13%20ENERhouse%202011%20Towards%20Net%20Zero%20Housing.pdf

RDH Building Engineering. 2012. Research Report: Near Net Zero Energy Retrofits for Houses. CMHC.

Energy Ratings for Windows: Balancing Energy Consumption and Thermal Comfort

Brittany Hanam, PEng
Associate Member ASHRAE

Graham Finch, PEng
Associate Member ASHRAE

Susan Hayes, PEng
Associate Member ASHRAE

ABSTRACT

Energy rating (ER) is a method used to compare the energy performance of windows. Its purpose is to establish the ranking of energy performance so that consumers can rate the relative energy efficiency of a window with a single number. The ER formula is based on the balance between heat loss due to conduction and air leakage and solar heat gain through the window. Since these heat gain and heat loss parameters have opposite effects on heating and cooling energy, each window can have a separate ER for heating and cooling conditions.

A number of countries use some variation of an ER to rate windows. The ISO provides an international standard method for calculating the ER for windows in ISO 18292: Energy Performance of Fenestration Systems for Residential Buildings, *and in Canada, the ER formula is specified in the CSA A440 Standard.*

Over the years, some potential issues have been brought forward by industry that could affect the validity of the ER as a tool for properly ranking window energy performance in a predominantly heating climate. These concerns mainly deal with the impact on thermal comfort, the selection of a reference house, and advancements in glazing technology.

This paper will present an analysis of whole house energy consumption and thermal comfort for different fenestration configurations. Computer simulations of thermal comfort parameters will be used to show considerations for balancing energy consumption with thermal comfort when comparing and selecting windows. In addition, the limitations of using a single ER to rank windows based on energy consumption will be identified.

INTRODUCTION

Windows have a significant impact on both the energy consumption and thermal comfort of a building. Choosing the right windows for a building can save energy and improve the indoor environmental quality of a space. The energy performance of windows is typically specified based on the product's U-factor and solar heat gain coefficient (SHGC). However, another metric that can be used to rate energy performance of windows is the energy rating (ER). The ER formula is based on the balance between heat loss due to thermal transmittance and air leakage and solar heat gain through the window. Since these heat gain and heat loss parameters can have opposite effects on heating and cooling energy, each window can have a separate ER for heating and cooling conditions. The ER used in the rating of windows in Canada, and discussed in this paper, refers to the heating ER only. The purpose of the ER is to help consumers rate the relative energy efficiency of windows (and glazed sliding doors) with a single number.

There are some limitations to the applicability of the ER. As explained in Clause I.6 of A440.3-09 (*User Guide to CSA A440.2-09*), the ER is intended for low-rise residential buildings under heating conditions. The ER is based on a reference house, with equal fenestration areas in each of the four cardinal directions, a standard fenestration system size, and climate conditions averaged for several cities across Canada and the Northern United States. Separate equations are available for a cooling season rating (ERC) and a specific energy rating (ERS), though these are rarely used in practice.

***Brittany Hanam** is a building science engineer, **Graham Finch** is a principal and building science research engineer, and **Susan Hayes** is a senior project engineer with RDH Building Engineering Ltd., Vancouver, BC, Canada.*

A number of countries use some variation of an ER to rate windows. The ISO provides an international standard method for calculating the ER for windows in ISO 18292, *Energy Performance of Fenestration Systems for Residential Buildings*. In Canada, the ER formula is specified in CSA A440.2-09 *Fenestration Energy Performance* standard. The Canadian ER can only be used for fenestration in low-rise residential buildings (three stories or less), installed in a vertical orientation. The ER can be used to certify fenestration products for ENERGY STAR® in Canada (see Table 1), and is also being incorporated into building codes and standards in certain jurisdictions, including the National Building Code. Note that ENERGY STAR does not have an ER compliance path in the United States.

Over the years industry members in Canada have questioned the reliability of using the ER alone to select energy efficient windows and doors. It was observed that in certain regions, products with good ER ratings and high solar heat gain characteristics resulted in overheating discomfort and customer complaints. In these regions, the market preferred products that achieved equally good ER ratings with lower solar heat gain characteristics. There is also concern that the ER does not give a complete picture of energy use and/or thermal comfort due to its lack of a cooling component. Because of the changes in house archetypes and advances in glass coating and window framing technology in the decades since 1989 (when the ER was first developed [Enermodal 1989]), it was determined that additional research was necessary to confirm the validity of the ER for use in all regions of Canada.

This study was conducted in order to provide a thorough review of the Canadian ER and to determine if the ER in its current form is still appropriate for selecting energy efficient windows and doors for all areas of Canada. Both energy and thermal comfort implications are considered.

LITERATURE REVIEW

Energy Rating Standards

The calculation for the Canadian ER is presented in CSA A440.2-09/A440.3-09 *Fenestration Energy Performance/User Guide to CSA A440.2-09* Fenestration energy performance (Canadian Standards Association 2009). This standard also provides a calculation procedure for a specific energy rating (ERS) and a cooling energy rating (ERC). The ERS is an ER value calculated for a specific house, location, orientation, and window size. The ERC can be used to compare fenestration systems to avoid overheating or to reduce cooling energy consumption, and was developed in a similar manner to the ER (by averaging weather dependent values for Canadian locations). However, the ERS and ERC are rarely used in Canada.

The International Organization for Standardization (ISO) has a methodology for determining window energy performance ratings as outlined in ISO18292, *Energy Performance of Fenestration Systems for Residential Buildings—Calculation Procedure*. The standard suggests a similar rating procedure to the Canadian ER, where solar gain, air infiltration, and transmission (conduction) losses are added to determine an overall rating. The rating can be calculated for both heating and cooling conditions.

The primary difference between the ISO standard and the CSA ER calculations is in how the solar heat gain term is calculated. With the ISO standard there are two different approaches to the solar heat gain component: a gain utilization factor approach and a degree-day approach. The degree day approach does not directly incorporate solar and internal heat gains, but factors them in by calculating heat losses at hours below a certain temperature. The utilization factor approach determines energy needs for heating based on the difference between gains and losses.

The gain utilization factor is based on a ratio of total heat gains over transmission and infiltration losses. If the internal and solar heat gains are smaller than the losses, the gain utilization factor is 1 since all solar and internal heat gains are used to help offset the heating demand. However, if the internal and solar heat gains are greater than the losses, the gain utilization factor is less than 1.

While the ISO standard uses the gain utilization factor approach for its energy rating, the Canadian ER uses a combination of the degree-day approach and the gain utilization factor. The degree day approach was initially calculated by averaging the hours where the outdoor temperature was less

Table 1. Canadian ENERGY STAR Qualification Criteria for Windows (Natural Resources Canada 2013)

Zone	Heating Degree-Day Range	Compliance Paths			
		Energy Rating (ER)		U-Factor	
		Max. U-Factor of 2.0 W/m²·K (0.35 Btu/h·ft²·°F)		Maximum U-Factor, W/m²·K (Btu/h·ft²·°F)	Minimum ER
A	≤3500	21	or	1.80 (0.32)	13
B	>3500 to ≤5500	25	or	1.60 (0.28)	17
C	>5500 to ≤8000	29	or	1.40 (0.25)	21
D	>8000	34	or	1.20 (0.21)	25

than 12°C (i.e., assuming that above this temperature heating would not be required due to internal and solar gains). Adding a solar gain reduction factor (R = 0.8) makes this approach similar to a simplified form of the gain utilization approach.

International Energy Ratings

In addition to Canada, several countries have fenestration ER ratings. The United Kingdom, for example, has a rating that is similar to the Canadian ER and was established by the British Fenestration Rating Council (BFRC). The BFRC rating is translated to a letter rating, A to G, and is displayed on a reader-friendly label along with the rating number, window U-factor, solar factor, and effective air leakage rate.

Other European countries have window energy rating programs as well. In Denmark and Sweden, windows are solely rated on their U-factor (Avasoo 2003), and in Germany the German Fenestration Institute has adopted an energy rating system that follows the ISO standard.

Australia and New Zealand have fenestration energy rating systems called the Window Energy Rating Scheme (WERS) and NZ-WERS, respectively (WERS 2012; Burgess and Skates 2011). The WERS rating system is different from the Canadian ER system. To determine the WERS rating, windows are simulated using a software program called CHENATH, using the window U-factor, SHGC, and air leakage rate. The energy consumption of a particular window is compared to that of a reference window with single glazing, clear glass, and a non-thermally broken aluminum frame. The reference single glazed window is given one star, with higher performing windows scaled accordingly between 1 and 10. Three different ratings are given, for heating, cooling, and interior fading.

In the United States, windows are rated for the ENERGY STAR program based on their U-factor and SHGC (Table 2). There is no ER approach, however, the latest criteria for certifying windows includes an option for windows in Northern climates that allows for a higher U-factor and SHGC (ENERGY STAR 2009). A 2004 study by Lawrence Berkeley National Laboratory (LBNL) looked at the impact of relaxing U-factor requirements and balancing the increase in energy consumption with changes to the SHGC (Huang, et al. 2004). The study found that in northern zones, increasing the U-factor would result in values above code-minimum. However, the study determined that if in the future ENERGY STAR U-factor requirements are lowered (i.e., made more stringent), it could be feasible to permit higher U-factors with higher SHGCs.

ENERGY ANALYSIS

The ER is intended to assist consumers in choosing energy efficient windows or to specify a minimum level of energy performance in codes and standards such as ENERGY STAR. An energy analysis was completed to determine how the ER rates energy consumption in houses.

Several parameters were established for use throughout the analysis work for this study. Archetypical houses, a range of geographic locations, and a selection of window configurations (defined by U-factor and solar heat gain coefficient, SHGC) were selected through a review of codes and standards, climate data, and the ENERGY STAR Canada database of windows. Hourly energy simulations were completed using the whole building energy simulation program DesignBuilder (an interface for EnergyPlus) to compare the annual energy consumption of different combinations of variables. The goal of the analysis was to simulate a wide range of archetype house variables to assess how the ER ranks energy consumption for different types of windows in different Canadian climates. The windows investigated included typical double glazed (U_{SI}-2.0 W/m^2·K or U-0.35 Btu/h·ft^2·°F, U_{SI}-1.8 W/m^2·K or U- 0.31 Btu/h·ft^2·°F, and U_{SI}-1.5 W/m^2·K or U-0.26 Btu/h·ft^2·°F) and triple glazed (U_{SI}-1.2 W/m^2·K or U-0.21 Btu/h·ft^2·°F, and U_{SI}-0.9 W/m^2·K or U-0.16 Btu/h·ft^2·°F) windows with low and high solar heat gain (SHGC values of 0.50, 0.35, 0.20). Several additional combinations were also included for comparison. A complete list of energy simulation inputs, as well as verification of simulation results, can be found in the research report (RDH 2013).

Table 2. United States ENERGY STAR Qualification Criteria (ENERGY STAR 2009)

Zone	Windows and Doors		Skylights	
	U-factor, W/m^2·K (Btu/h·ft^2·°F)	SHGC	U-factor, W/m^2·K (Btu/h·ft^2·°F)	SHGC
Northern			≤3.1 (0.55)	Any
Prescriptive	≤1.70 (0.30)	Any		
Equivalent Performance	1.8 (0.31)	≤0.35		
Equivalent Performance	1.8 (0.32)	≤0.40		
North/Central	≤1.8 (0.32)	≤0.55	≤3.1 (0.55)	≤0.40
South/Central	≤2.0 (0.35)	≤0.40	≤3.2 (0.57)	≤0.40
Southern	≤3.4 (0.60)	≤0.40	≤4.0 (0.70)	≤0.40

The window energy analysis shows that in general, windows with a higher ER use less heating energy. On the other hand, there were a number of simulated windows that showed that having a slightly higher ER resulted in higher heating energy consumption. Figure 1 shows a plot of the annual heating energy consumption versus ER value for the fenestration configurations that were simulated in a typical house, in five Canadian climates. The house shown here is a single story house with a basement, gas furnace, and a window to wall ratio of 15%. The results show a good linear correlation between heating energy consumption and ER; in other words, as the ER number increases, annual energy consumption decreases for most cases simulated. By comparison, Figure 2 shows the annual heating energy consumption plotted versus U-factor. The correlation between energy consumption and U-factor is not as good as the ER plot (which is apparent in the lower R^2 values in Figure 2) because of the impact of solar heat gain. Several groupings are visible where products with the same U-factor but different SHGCs have different annual energy consumption. This highlights the impact of the SHGC on fenestration energy performance where higher SHGC products have lower heating energy consumption in the heating-dominated Canadian climates. This effect is not captured when windows are rated by U-factor only.

Energy simulation results indicated that the ER does not rank cooling energy appropriately, which was expected since the ER is designed to reduce heating energy (i.e., favors high solar heat gain). However, in houses that have mechanical cooling, cooling energy consumption is very low compared to heating energy in all Canadian locations under worst-case cooling conditions (e.g., no natural ventilation). In general, cooling energy was found to be roughly 10% of overall space conditioning energy in houses that have mechanical cooling (Figure 3). Therefore, when looking at space conditioning energy consumption alone, heating energy considerations far outweigh cooling. Note that this is for low-rise residential buildings with typical window to wall ratios in the range of 15% to 20%. Higher window to wall ratios, though uncommon in houses, would likely generate different results as cooling energy could become significant.

Figure 4 shows a plot of the total annual energy consumption (including lighting, plug loads, and domestic hot water) versus ER. This plot generally shows a good correlation between total energy and the ER, though with higher error in Yellowknife.

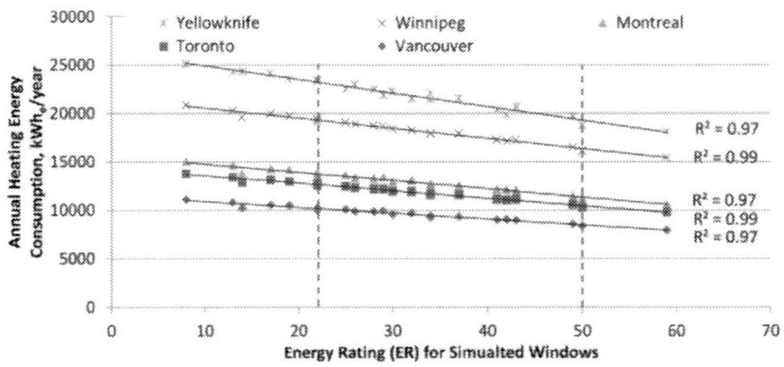

Figure 1 *Plot of annual heating energy consumption versus ER for several fenestration configurations in a typical house in five Canadian climates. Dashed lines indicate typical range of ENERGY STAR windows.*

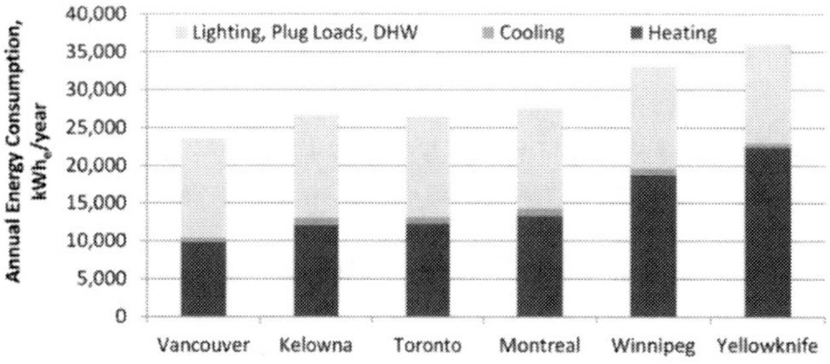

Figure 3 *Plot of annual energy consumption for a typical house with air conditioning simulated in six Canadian climates, obtained from simulations. Even climates with warmer summers (Kelowna and Toronto) cooling energy is low compared to annual heating energy.*

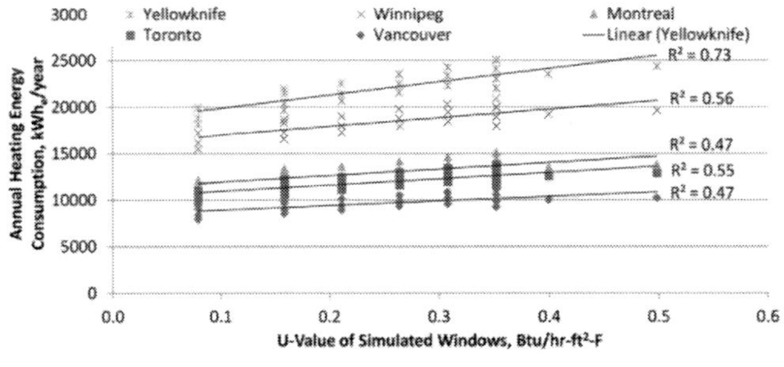

Figure 2 *Plot of annual heating energy consumption versus U-factor for several fenestration configurations in a typical house in five Canadian climates.*

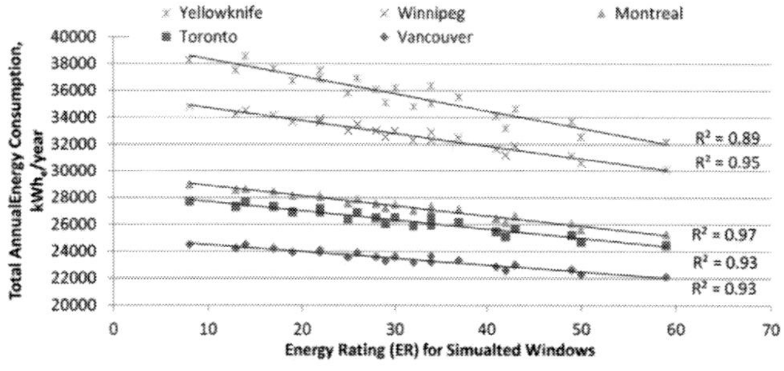

Figure 4 *Plot of total annual energy consumption versus ER for several fenestration configurations in a typical house in five Canadian climates.*

As noted previously, having a higher ER window generally resulted in lower heating energy use. This was observed in the majority of the geographic locations simulated with the exception of Yellowknife (i.e., the far north). Here, there was a significant variance in heating energy trends compared to other locations due to the low amount of solar radiation that Yellowknife receives in the winter. As a result, there is less of a benefit to using high SHGC windows.

The same trends, where higher ER correlates directly with lower heating energy consumption, were also observed in all of the archetype house parameters that were simulated. This included several variables not directly related to the windows, such as house size, presence of a basement, enclosure thermal performance, and mechanical system type. These observations suggest that a representative archetype house can be used to develop an ER system to rank windows (which was done for the Canadian ER development in Enermodal [1989]). However, when looking at different window orientations, window to wall ratios, and shading strategies, the simulations show varying trends in the ability of the ER number to rank energy consumption. This analysis suggests that rating windows with a single ER number may not necessarily indicate lower energy consumption for houses with non-typical window orientations (i.e., primarily facing one direction), high window to wall ratios, and exterior winter shading (see Figure 5 and Figure 6).

THERMAL COMFORT ANALYSIS

A thermal comfort analysis was completed to evaluate whether the energy objectives of the ER system are consistent with improving thermal comfort, and, if not, what impact this can have on the thermal comfort of a space.

A quantitative analysis of thermal comfort is challenging since it is impacted by many occupant-dependent variables. The ASHRAE Standard 55-2010, *Thermal Environmental Conditions for Human Occupancy* was used as a guideline for this analysis. Two parameters were used to assess comfort: the operative temperature, the mean of the internal air and radiant temperatures; and the window surface temperature, a proxy for radiant asymmetry. A range of comfortable temperatures was established for both operative temperature and surface temperature based on ASHRAE Standard 55, and the number of hours outside of this range were counted from hourly energy simulations. This allowed the comparison of various window types based on how many hours were outside of the comfort range for each window.

The comfort range for operative temperatures was set between 19°C to 25°C (66°F to 77°F), and the comfort range for surface temperatures was set between 15°C to 30°C (59°F to 86°F). Five windows with different combinations of U-factors and SHGCs were simulated; window parameters are shown in Table 3. Windows A, B, and C are double glazed windows, and windows D and E are triple glazed windows.

When assessing the operative temperature of a space, simulation results show that the number of warm hours, or overheating hours recorded, is much greater than the number of cold hours. Figure 7 shows a plot of the percent of time in a year that the operative temperature is outside of the defined comfort range, illustrating time that is both too warm and too cold. The percent of time overheating (dashed lines) is far greater than the percent of time underheated (solid lines) in all climates except the far north, Yellowknife. Further, windows with a high SHGC had a greater number of overheating hours than windows with a low SHGC. The window with a SHGC of 0.64 had the greatest percentage of time overheating, followed by the windows with a SHGC of 0.5. The windows with a SHGC of 0.2 had the lowest amount of overheating. The number of cold hours is relatively low for most locations except the

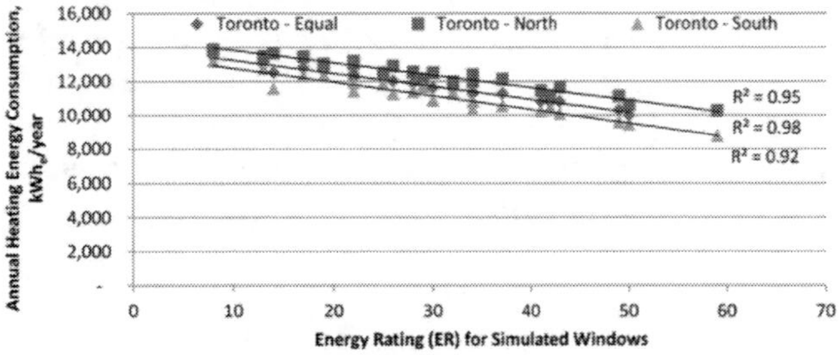

Figure 5 *Plot of annual heating energy consumption versus ER for several fenestration configurations in a typical house in Toronto, with different window orientation distributions. Results are shown with windows oriented equally in all four cardinal directions, windows primarily facing north, and windows primarily facing south. The correlation for north- and south-facing windows is not as good as the correlation with windows distributed equally.*

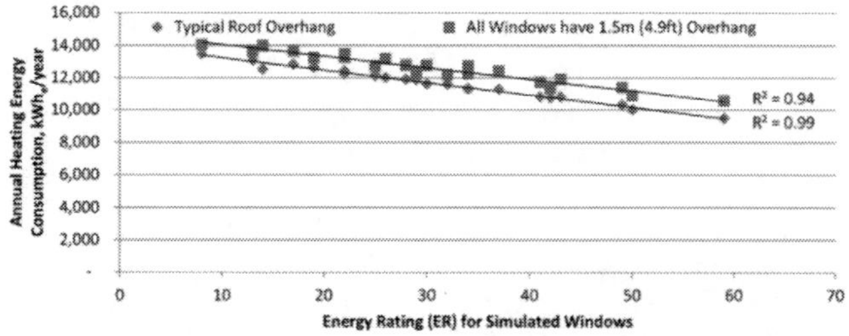

Figure 6 *Plot of annual heating energy consumption versus ER for several fenestration configurations in a typical house in Toronto, with different window shading. Results are shown with a typical roof overhang (minimal shading of windows) and a 1.5 m (4.9 ft) overhang above all windows (some winter shading). The correlation is better with minimal shading than with a 1.5 m (4.9 ft) overhang.*

far north. U-factor has a greater impact on the number of cold hours, with low U-factor windows performing best (i.e., triple glazing).

Broadly speaking, the correlation between ER and comfort was not as clear as the correlation between SHGC and operative temperature discomfort or the correlation between U-factor and surface temperature discomfort (Figure 8). The thermal comfort analysis generally showed that a better U-factor resulted in less cold surface and operative temperatures, but did not significantly impact overheating operative temperatures. On the other hand, the analysis did show that a higher SHGC, though it did not significantly impact high surface temperatures, had an impact on creating a higher portion of warm operative temperature hours. This is in contrast to the energy analysis, where a high SHGC resulted in lower energy consumption.

One particular concern with the ER system has been whether higher ER windows increase the occurrence of overheating in a space. Figure 9 shows a plot of the percent of time operative temperature is too high versus the ER value for the five windows presented above. The results show no correlation between ER and overheating. A higher ER does not necessarily result in overheating, though the window with the second highest ER has the second greatest percentage of time overheating. This illustrates that even though there is no direct correlation between overheating and the ER, having a higher ER will not necessarily lead to better thermal comfort, particularly overheating. Therefore, if the ER is used to select windows for a typical single family dwelling, additional measures may need to be taken to prevent overheating, depending on the SHGC of the window.

Table 3. U-Factor, SHGC, and ER of Windows Shown in Thermal Comfort Analysis

Window	A	B	C	D	E
U-Factor, W/m^2·K (Btu/h·ft^2·°F)	2.8 (0.50)	2.0 (0.35)	2.0 (0.35)	0.9 (0.16)	0.9 (0.16)
SHGC	0.64	0.2	0.5	0.2	0.5
ER	14	8	26	32	49

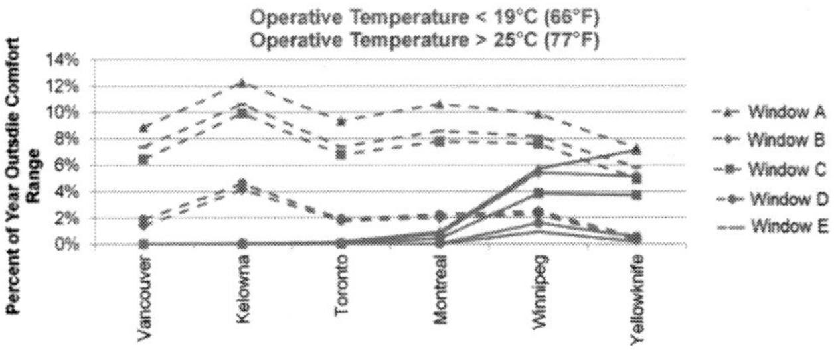

Figure 7 Plot showing the percentage of hours in a year where the operative temperatures are outside of the thermal comfort range for five windows simulated. Dashed lines show percentage of the time when the operative temperature is too warm (greater than 25°C [77°F]) and solid lines show percentage of the time when the operative temperature is too cold (less than 19°C [66°F]).

CONCLUSIONS AND RECOMMENDATIONS

Energy

The ER is an appropriate tool for comparing windows on the basis of energy consumption for typical Canadian houses. The ER provides a better ranking of window energy consumption than U-factor alone. Where anomalies occur (i.e., a window with a higher ER uses more energy), the differences in both the ER value and energy consumption are typically small. The ER, however, should not be the only metric that one considers when choosing windows for

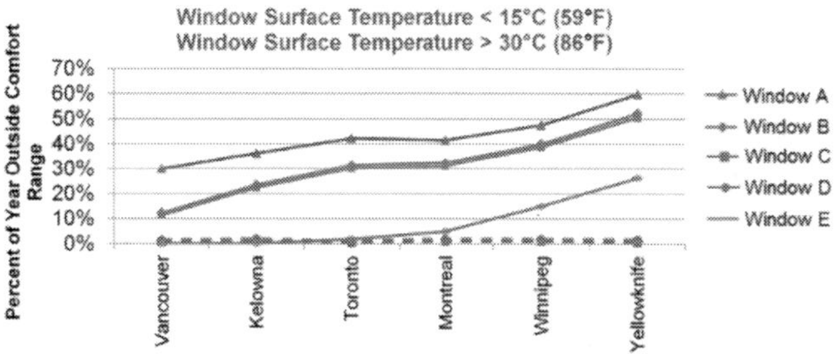

Figure 8 Plot showing the percentage of hours in a year where the window surface temperatures are outside of the thermal comfort range for five windows simulated. Dashed lines show percentage of the time when the surface temperature is too warm (greater than 30°C [86°F]) and solid lines show percentage of the time when the surface temperature is too cold (less than 15°C [59°F]).

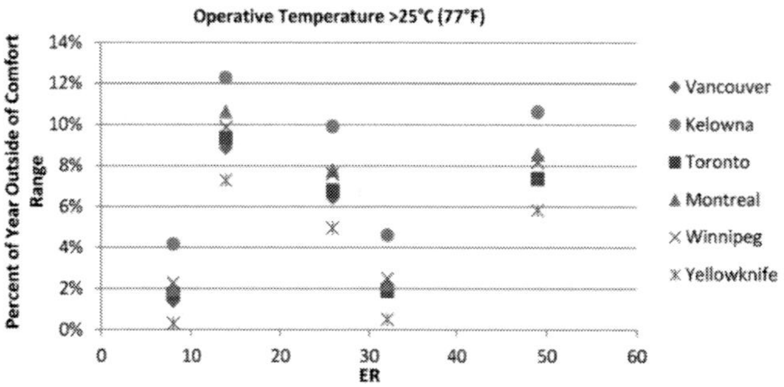

Figure 9 Plot showing the percentage of hours in a year where the operative temperatures are higher than the thermal comfort range of 25°C (77°F) versus ER for five windows simulated.

a non-typical home. In particular, the ER on its own is not an appropriate measure to compare windows under the following non-typical conditions for houses and low-rise residential buildings:

1. Far north locations, including the Canadian Territories
2. Windows with significant winter exterior shading
3. A house with windows oriented primarily in one direction

Despite these minor exceptions, the current ER formula works in most common house situations, in most locations in Canada, and is therefore an appropriate metric for rating the relative energy performance of windows. However, for houses that are non-typical, that have more site-specific design or energy efficient design, it would be best to select windows based on U-factor and SHGC rather than ER alone. If the ER is incorporated into standards then it should be accompanied by explanatory text regarding when it is appropriate and when it is not appropriate. Likewise, if the U-factor alone is used to select energy efficient windows, explanatory text regarding the potential energy savings of a high or a moderate SHGC should also be provided. While the ER should be maintained, provisions to keep the alternate U-factor compliance path are necessary because of these non-typical conditions.

Thermal Comfort

Although the ER is appropriate for selection based on lower energy consumption, a high SHGC window can result in overheating, which may be a concern even in typical houses with typical (15% to 20%) window to wall ratios. The ER cannot be used to assess thermal comfort performance, particularly overheating. Overheating should be evaluated on a case-by-case basis, depending on project-specific factors such as location, orientation, shading, and window to wall ratio. Additional guidance is needed to assist consumers in selecting appropriate windows (and SHGC factors) for energy consumption and thermal comfort.

Recommendations for Future Work

Although the ER metric performs relatively well for rating heating energy consumption of windows in Canada, several opportunities for improvements and further work were identified in this study.

The far north was consistently identified as an anomaly where the ER may not appropriately rate windows. However, the ER still resulted in a better ranking of heating energy consumption than U-factor alone. A different ER calculation, one with solar heat gain weighted differently, would likely provide better results for the far north region. One option would be to have a separate ER calculation (i.e., different location-dependent parameters) for different climate zones. However, this would require the maintenance of multiple ER values for each window configuration.

The possibility of adding either a standardized cooling energy rating (ERC) or some other rating to indicate thermal comfort could be investigated, particularly for warmer climate zones. This would be an additional rating to the current ER. There would still be some trade-off in balancing this ERC value with the current ER to help select a window for both summertime and wintertime conditions.

The ER serves as a good indication of energy performance for the average consumer; however, the best energy performance will be obtained with building-specific design. Published education and guideline documentation would assist consumers in understanding the issues and would help them make informed decisions regarding the best windows for their situation. One or two brief bulletins could be developed to assist consumers, as well as manufacturers, installers, sales professionals, and consultants in selecting the best windows for their situation beyond just the ER. Other things can be considered such as unique building features and thermal comfort.

This work could also be used to inform studies in other countries or jurisdictions that are considering the use of an ER system to rate the energy efficiency of windows, or considering incorporating a minimum SHGC to offset a better U-factor in codes or standards.

ACKNOWLEDGMENTS

This project was completed with the support of the following funding partners: Natural Resources Canada (NRCan), Homeowner Protection Office (HPO) branch of BC Housing, Fenestration Canada, Window and Door Manufacturers Association of BC (WDMA-BC), Glazing Contractors Association of BC (GCABC), Canadian Glass Association (CGA), Association des industries de produits de vitrerie et de fenestration du Québec (AIPVFQ), BC Hydro, Manitoba Hydro, and Hydro-Québec.

REFERENCES

Avasoo, D. 2007. The European window energy labelling challenge. European Council for an Energy Efficient Economy (ECEEE) Summer Study, (pp. 1347-1352). La Colle sur Loup, France.

ASHRAE. 2010. ASHRAE 55-2010, *Thermal Environmental Conditions for Human Occupancy*. Atlanta: ASHRAE.

Burgess, J., and Skates, H. 2001. A New Zealand Window Efficiency Rating System. International Conference on Building Envelope Systems and Technology. Ottawa.

Canadian Standards Association. 2009. CSA A440.2-09/ A440.3-09 *Fenestration Energy Performance/User Guide to CSA A440.2-09 Fenestration Energy Performance*. Mississauga.

ENERGY STAR. 2009. ENERGY STAR Program Requirements for Residential Windows, Doors, and Skylights. Washington DC.

Enermodal Engineering Ltd. 1989. *Calculating the Window Energy Rating*. Waterloo.

Huang, J., Mitchell, R., Selkowitz, S., and Arasteh, D. 2004. Analysis Results for Performance-based Ratings for the ENERGY STAR Windows Program. Windows and Daylighting Group, Lawrence Berkeley National Laboratory.

International Organization for Standardization (ISO). 2011. ISO18292 *Energy Performance of Fenestration Systems for Residential Buildings–Calculation Procedure*. Geneva.

Natural Resources Canada. 2013. http://oee.nrcan.gc.ca/equipment/windows-doors/1066. Ottawa.

RDH Building Engineering Ltd. 2013. Review of Window Energy Rating Procedure in Canada: Final Research Report. Vancouver.

Window Energy Rating Scheme (WERS). 2012. www.wers.net. Gordon, Australia.

The Impact of Occupants' Exterior Shading Use and Window Opening on Summer Comfort in a Lowest-Energy Apartment Building

Naomi Morishita **Thomas Bednar, Dr techn**

ABSTRACT

A Viennese apartment building, KAM, was built in 2007 to minimum passive house requirements primarily designed to perform for heat conservation in winter. Less emphasis was placed on cooling in the warm months when planning the building due to the temperate continental climate of Vienna, Austria. Due to apartment owner complaints of overheating, exterior blinds were installed on apartment windows and doors on the topmost floor. Indoor temperature and humidity in four apartments have been monitored over three summer periods. The monitoring results are presented where mechanical ventilation with heat recovery is used in combination with exterior sun shading. The users in the case studies prefer using natural window ventilation to changing ventilation system settings. Indoor daily temperatures remain constant when windows are left closed and shading is used. Results from monitoring and interviews indicate the importance of considering summer cooling strategies in combination with user behavior in highly insulated buildings to control overheating in summer.

INTRODUCTION AND BACKGROUND

This paper analyses the effect of user behavior on the summer indoor comfort conditions in a passive house apartment building (KAM) in Vienna, Austria and continues monitoring of the indoor climate as described in (Bednar, Korjenic, et al. 2010).

A passive house has a super-insulated exterior envelope (including windows and doors) and is extremely airtight. To ensure good indoor environmental quality (IEQ), the buildings are equipped with highly efficient compact ventilation systems with heat recovery and air supply preconditioning using ground-source heat exchangers. Since 1995, the number of residential buildings using the passive house calculation method (PHPP) has increased exponentially in Austria. From 1995 to 2010, 11,800 buildings including both new construction and thermal renovations of existing buildings have been constructed to the passive house construction standard.

The majority of residences in Austria are in apartment buildings or row houses. One fifth of the population resides in Vienna, of which the majority lives in low-rise apartment buildings (Statistik Austria 2012). Thus, the studied building is selected as a representative case study for observing the effects of inhabitants' interactions with external shading devices on summer comfort.

The highest recorded summer air temperature during the three summer seasons was 36.1°C (97°F). However, the operative temperature is higher, taking the mean radiant temperature into account, and can reach over 40°C (104°F). During the hottest heat waves, the outdoor temperatures sink by 10 K (18°F) in the night. Theoretically, the residents can maintain cooler indoor temperatures by taking advantage of the night cooling by opening all their windows during the night.

BUILDING DESCRIPTION

KAM is a seven story condominium building with 88 flats built to the passive house standard as seen in Figure 1.

The building envelope is highly insulated with high airtightness (0.28 h^{-1}). The thermal transmittance values are summarized in Table 1. The building is a combination of a reinforced concrete shell with a prefabricated wood façade. In-

Naomi Morishita *is a project assistant and* ***Thomas Bednar*** *is a professor in the Research Center for Building Physics and Sound Protection, Institute for Building Construction and Technology, Vienna University of Technology, Vienna, Austria.*

terior partitions, insulated walls with 0.25 m (9 ¾ in.) mineral wool to an unheated interior atrium, and floor slabs are prefabricated reinforced concrete. The exterior envelope is highly insulated with two layers of mineral wool insulation, 30 cm + 5 cm (11.8 in. + 2 in.).

The ventilation system has a central pre-conditioned air supply through a ground source heat exchanger. Fresh air intakes and exhaust air outlets are located on the roof of the building. The air supply is distributed to each apartment containing a compact ventilation unit with a plate heat exchanger. Air supply volume can be regulated in the individual apartments (low, medium, and high). The building has been designed to take advantage of hybrid ventilation during warm periods, combining both mechanical and natural ventilation/shading strategies to cool the apartment interiors. The natural ventilation strategy is especially dependent on night cooling, taking advantage of the cooler evening temperatures and prevailing breezes from the southwest.

The monitored apartments are on the west and east facades, are stacked atop each other, and are located on the fifth and sixth (top) floors. The flats on the west façade have the same floor plan, as do the flats on the east side. Each flat has a balcony, and the glazed balcony doors to the bedrooms on the west side (W2) have exterior blinds. All windows and glazed doors are insulated triple-glazed with tilt and turn opening. The glazed area of the west apartments is 11.7 m² (125.7 ft²); the glazed area of the east apartments is 14.9 m² (160.8 ft²). The west flats are larger with 3 bedrooms and a floor area of 97 m² (1044 ft²); the east flats have two bedrooms and a floor area of 90 m² (969 ft²); the average flat size is 94 m² (1012 ft²).

Figure 1 *South (street view) and east facades of KAM.*

Table 1. Building Envelope Thermal Resistance Values (teamgmi 2004)

Building Assembly	Thermal Resistance SI, $m^2 \cdot K/W$	Thermal Resistance US, $ft^2 \cdot {}^\circ F \cdot h/Btu$
Roof	14	81
Exterior Wall	8	47
Floor Slab to Unheated Underground Garage	7	38
Windows	1	7

HOUSEHOLDS

Table 2 summarizes the household sizes of the participating households in 2012. A post occupancy evaluation was conducted by Keul in May 2008, and it was determined that the average household size in the building is 2.1 people with a mean age of 37.6 years (Keul 2009). It has been observed that some residents moved into KAM to establish families. Two of the participating households increased in size from 2010 to 2012.

In Austria, there is no tradition of air conditioning or mechanical ventilation in residences. Cooling in hot months is usually provided by a combination of window ventilation, fans, and using interior and exterior window shading.

In 2011, W1 installed a set of large outdoor curtains hung at the edge of their loggia. All windows, except one living room window are shaded by the loggia above. Similarly, E1 does not have any exterior shading as all windows except one to the living room are also shaded by the loggia above. The occupants of E2 installed new exterior blinds in May 2011 stating that the shading from the originally installed blinds were insufficient for maintaining comfortable summer temperatures. In this paper, the flats are coded according to orientation, west (W) and east (E); and floor, 5th Floor (1) and 6th/Top Floor (2).

APPROACH

Self-Reported Cooling Behavior

In September 2012, personal interviews were conducted with the inhabitants to determine the occupants' daily self-reported natural ventilation cooling habits: window opening and shading use. It is interesting to note that all households reported owning indoor/outdoor thermo hygrometers, and that window opening decisions in summer are partially dependent upon outdoor temperature readings and time of day when direct sunlight enters the individual

Table 2. KAM Household Sizes in 2012

Apartment	Household Size
W1	4
E1	4
W2	4
E2	2

apartments. Generally, when the outdoor temperature is cooler than the indoor temperature and when the sun shone directly in the apartment, the inhabitants opened their windows.

The window opening and shading use for a typical day as determined from the interviews are summarized for each apartment in Figures 2, 3, 4, and 5.

E1 is the only household that reported short periodic ventilation after meals, where the window is fully open for five to 10 minutes, Figure 3.

W2 originally had outdoor blinds on only the bedroom windows. The homeowners retrofitted outdoor blinds on the balcony door from the living room in 2010. The occupants have additional air movement during the day with the tilted windows while blocking solar irradiation, Figure 4.

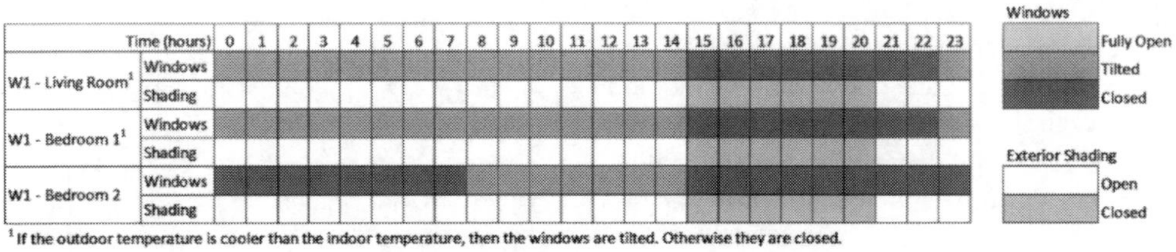

Figure 2 *Apartment W1 window opening and shading schedule for a typical day.*

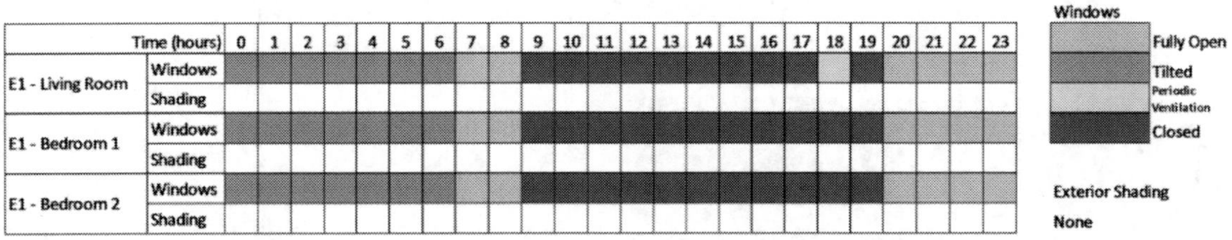

Figure 3 *Apartment E1 window opening and shading schedule for a typical day.*

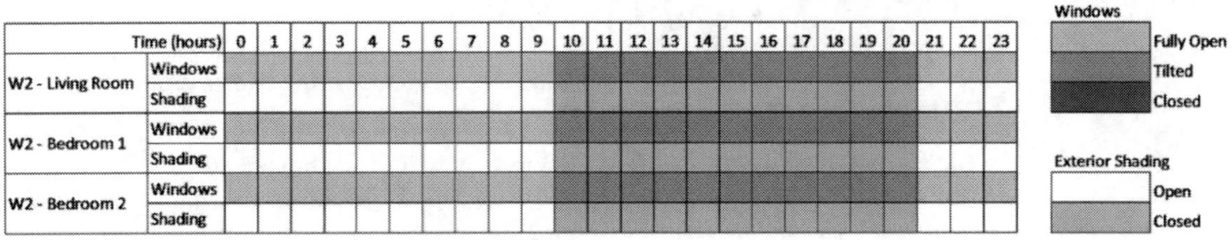

Figure 4 *Apartment W2 window opening and shading schedule for a typical day.*

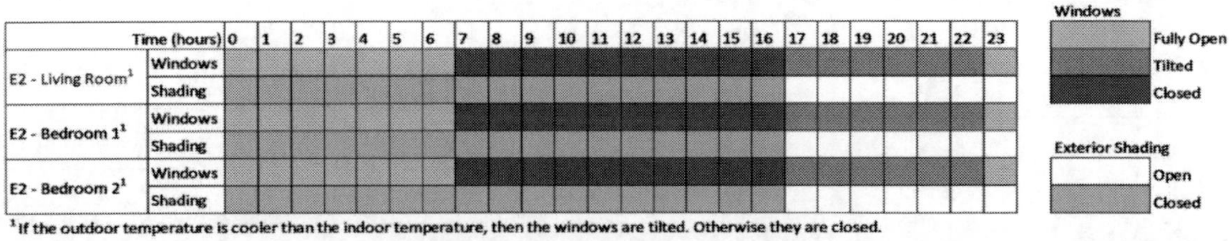

Figure 5 *Apartment E2 window opening and shading schedule for a typical day.*

Both the windows and exterior blinds are closed for the majority of the day. The apartment did not have any shading devices initially. In May 2011, exterior security blinds were installed. Not only heat is blocked but also daylight when the shading devices are used, Figure 5. Due to the exposed position on the topmost floor, the residents reported that it was necessary to keep the blinds down for the duration of the day to maintain cooler temperatures.

The general preference in all apartments is to leave windows open for as long as the occupants believe is possible as a fresh air supply is important, and fresh air from window ventilation is preferred to changing the ventilation system settings.

Longitudinal Monitoring in Four Apartments

Twelve T&D RTR-5 data loggers were placed in the four apartments measuring the indoor temperatures and relative humidities in three rooms: living room, master bedroom, and second bedroom/study. In this paper, temperature data from the summer seasons from June 1st to August 31st for 2010, 2011, and 2012 are compared, representing the third to fifth years of residency. Relative humidity values have been excluded as the indoor values range between 30% and 60%. Figure 6 shows the plans of the apartments where the data loggers were placed.

The results from 2012 are discussed in detail in Figures 8 to10. Data is shown in two groups: living rooms, and bedrooms/study. Outdoor conditions were obtained from a weather station in Vienna located 9 km (5.6 miles) from the apartment building. The daily ventilation and shading habits were compared to the measurement results.

Monitoring Results Analysis

The temperatures for the monitored apartments were visually depicted in two sets of graphs comparing outdoor temperatures to living room temperatures, and outdoor temperatures to bedrooms/secondary rooms for the hottest week of a chosen period. The data is then analyzed in more detail for the hottest day of the selected period. The temperature data was then compared in the two groups with indoor temperatures versus outdoor temperatures for each year.

RESULTS

The summer 2010 and 2012 outdoor temperatures are compared to the indoor temperatures in 1 K (1.8°F) temperature bins in Figures 7 and 8. In 2010, the living room temperatures in E2 are constantly in a range of 1 K (1.8°F) to 1.5 K (2.7°F) higher than all other apartments. The temperatures reflect a combination of factors: lack of exterior shading, predominantly closed windows, and exposed location at the highest level in the building. Bedroom temperatures are more closely grouped together for all apartments with the highest temperatures in study E2, Figure 7.

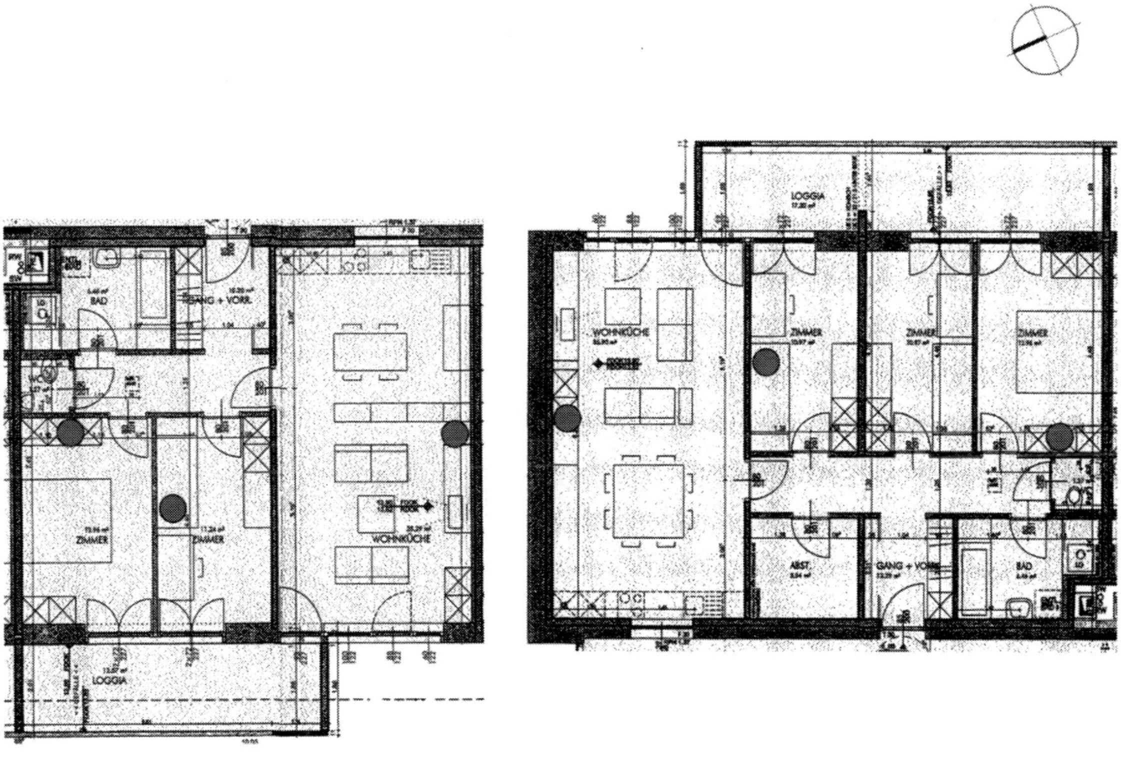

Figure 6 *Plans of apartments, W1 and W2 (left) and E1 and E2 (right), 5th floor shown. Data logger locations shown in red (Kaufmann 2007a, 2007b).*

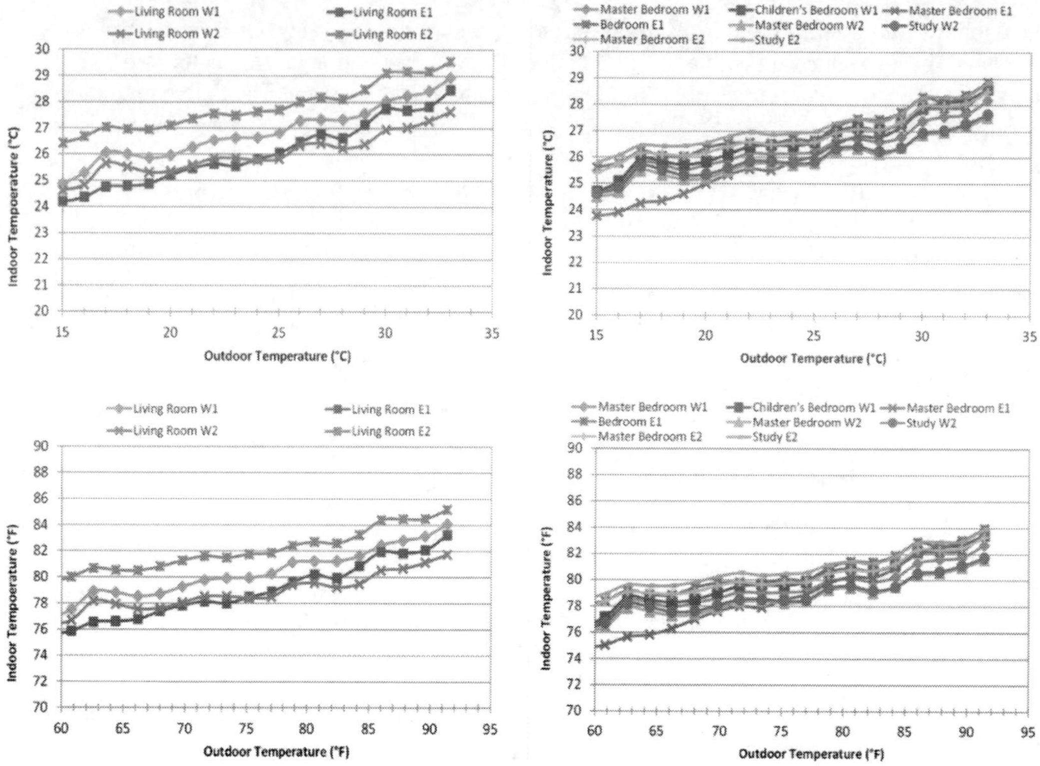

Figure 7 *Average indoor living room (left) and average bedrooms/study temperatures (right) from hourly measurements for 1 K intervals of outdoor temperature, 2010.*

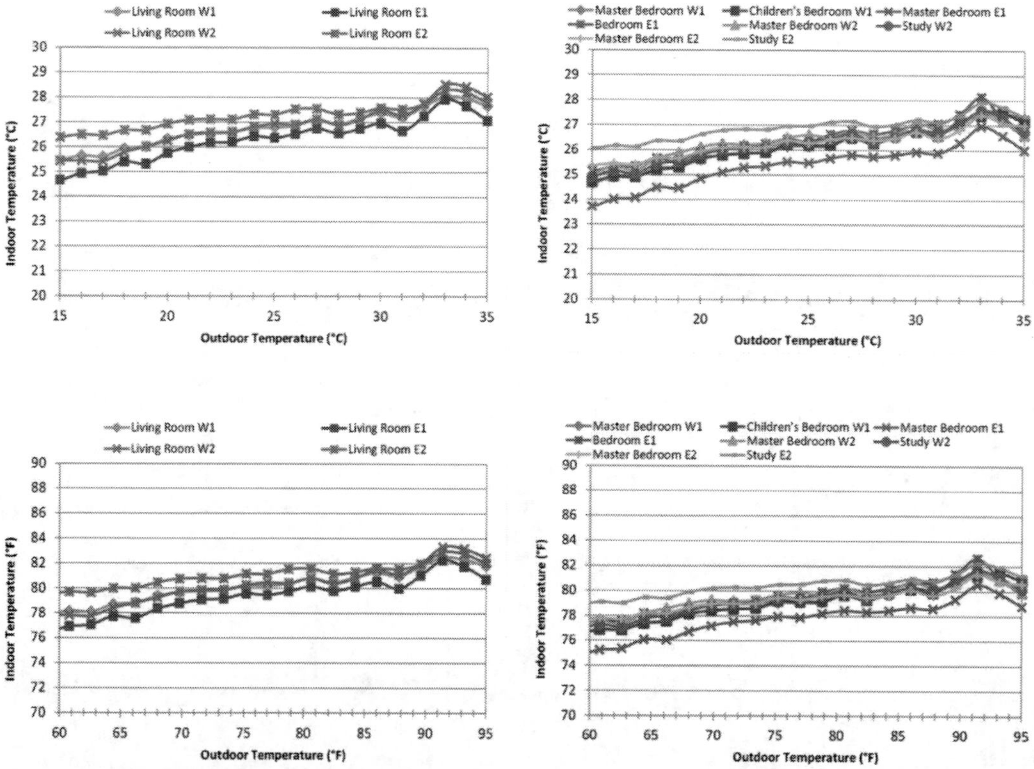

Figure 8 *Average indoor living room (left) and (right) average bedrooms/study temperatures from hourly measurements for 1 K (1.8°F) intervals of outdoor temperature, 2012.*

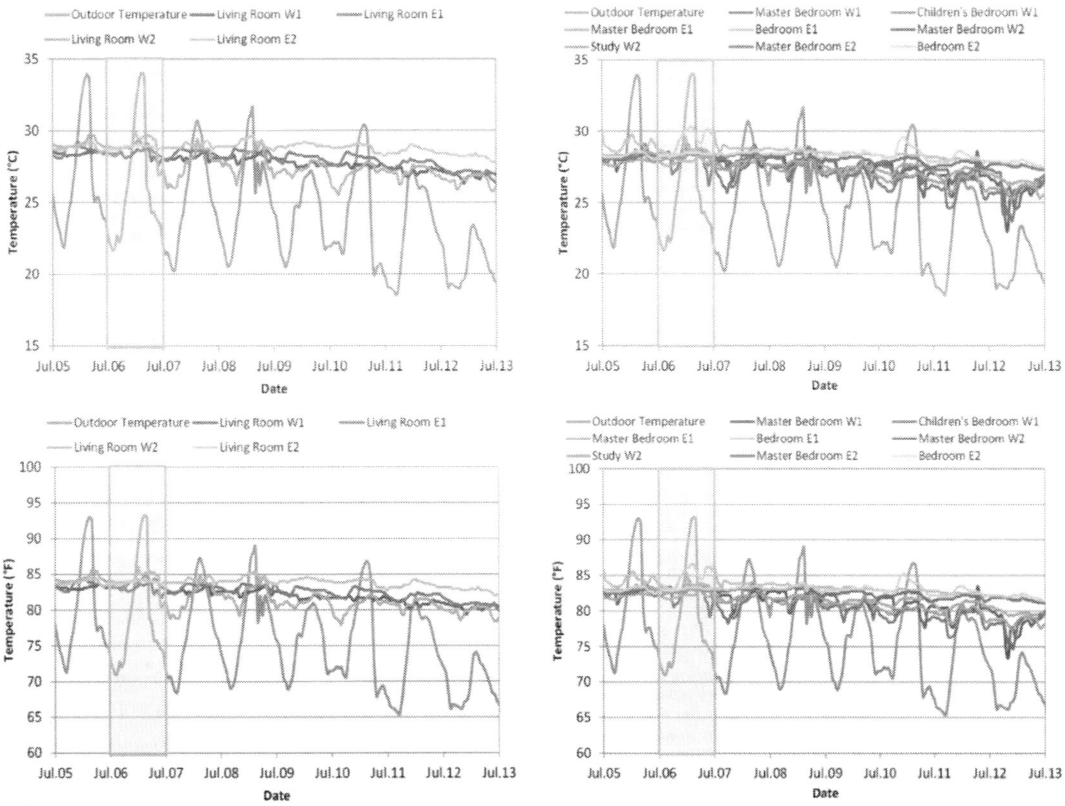

Figure 9 *Living room (left) and bedroom/study temperatures (right) in all apartments, July 5th to 13th, 2012.*

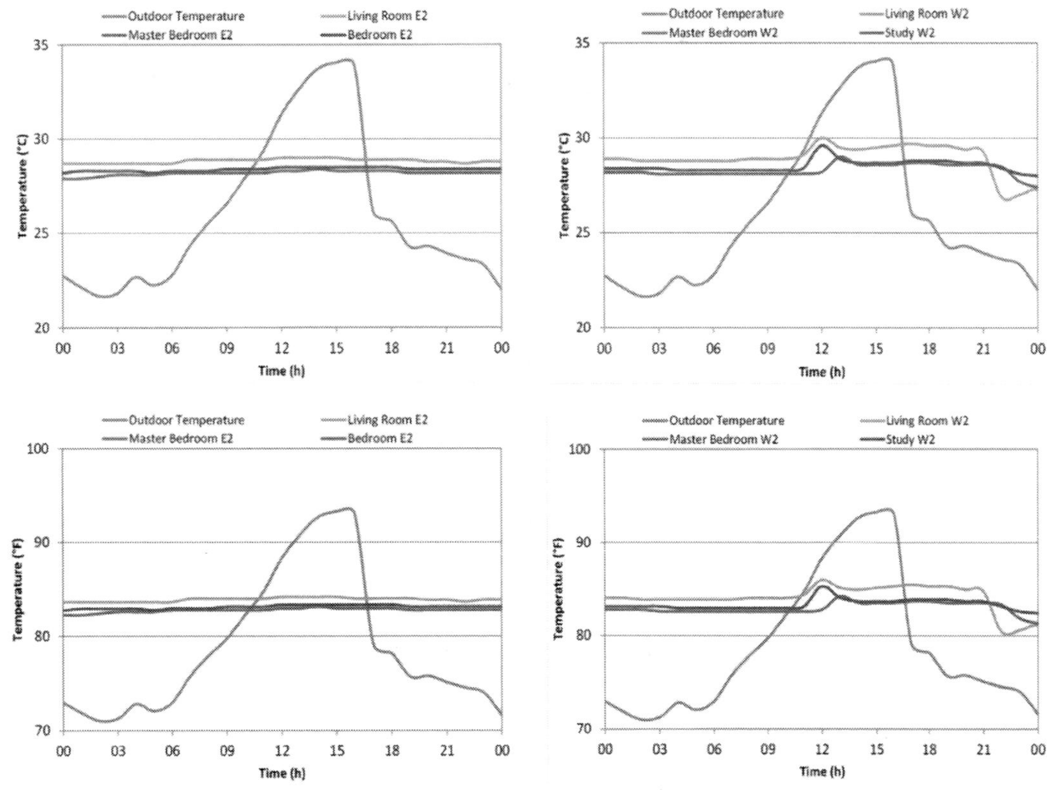

Figure 10 *Hourly values for all rooms in apartment E2 (left) and W2 (right) for July 6th, 2012.*

Figure 8 compares the indoor to outdoor temperatures in 2012. Similar to the living room temperatures, the living room temperatures in E2 are higher than the other apartments. The flatter slopes of both living room and bedroom/study graphs indicate that the rate of increase in indoor temperature is less than in 2010. Especially for temperatures between 26°C (79°F) and 30°C (86°F), the average indoor temperatures stabilize despite the rising outdoor temperatures. Living room temperatures are generally higher than in bedrooms.

2012 was the hottest of the three summers. The apartments show the same living room temperature patterns in Figure 8 (left) relative to the previous years. The hottest week is indicated in gray.

Figure 9 examines the heat wave from July 5th to 13th, 2012 in greater detail. W2 shows the greatest decrease in nightly temperature, however rises to the same daily temperature range as the other apartments which may be partially explained by open windows during the day with closed shading as described in Figure 4. E2 again shows the warmest temperatures, and especially shows continual constant indoor temperatures despite cooler outdoor temperatures on July 12th and 13th, reflecting that the occupants were on holiday.

The hourly indoor temperatures for July 6th, 2012 are compared in the two 6th floor apartments, E2 and W2, in Figure 10. None of the rooms in E2 reflect coupling to outdoor conditions indicating that the highly insulated and airtight envelope effectively decouples indoor from outdoor conditions, and that windows were not opened in this period.

The warmest to coolest rooms are the same in both apartments. However, increases in temperature at midday, and falling temperatures in the evening indicate that W2 has opened the windows. It is possible that the windows in the living room were fully open.

DISCUSSION

The measurement results show differences from the self-reported behavior. Following the adaptive comfort theory (Nicol and Humphres 2002), the occupants should be actively adjusting their environment to establish comfortable conditions. The homeowners received a "summer box", a unit to bypass the heat exchanger in the ventilation unit. Through experience, the homeowners have decided not to use the summer box, meaning that heat recovery always takes place within the apartment. The effect of the ground-source heat exchanger seems to have a larger effect than the heat recovery in the individual apartments in summer.

Measurements show that the homeowners leave windows closed at night during heat waves, causing indoor temperatures to rise up to 30°C (86°F). All homeowners were informed of the importance of opening windows at night, keeping windows closed during the day, and keeping exterior blinds closed during the day for maintaining cooler indoor summer temperatures. However, when interviewed, the homeowners gave reasons to keep windows tilted or closed at night such as outdoor noise (e.g., noise from rooftop fans from the ventilation system, public transport, traffic, and people's voices), perceived security, and insects. It is also possible that windows were kept fully open with closed blinds to maintain darkness in bedrooms while cooling. The combination of the above factors potentially limited the effectiveness of night cooling. On the other hand, the temperatures in E1 are constantly lower than all other apartments in 2012, indicating that the combined actions of the occupants, shading from the balcony above, and ventilation system use maintained cooler temperatures, Figure 9.

The external blinds that were installed in E2 blocks most of the sunlight. The homeowners were dismayed by the need to close the blinds in summer for the majority of the day in order to maintain comfortable indoor temperatures as the interior becomes very dark during the day and no view to the outdoors is available.

CONCLUSIONS

The combined influence of the airtight and highly insulated envelope, thermal mass of the reinforced concrete structure, and the preconditioned air supply maintained stable indoor temperatures that were decoupled from outdoor conditions. User behavior to open and close windows and blinds was not necessarily influenced by outdoor climate conditions as a primary driving force, but rather other aspects such as access to daylight, night noise, and individual psychosocial factors (e.g., perceived safety), influenced behavior. Steady patterns of individual households were visible from year to year, even though behavior between households differed paralleling findings about window opening behavior by (Fabi, Andersen et al. 2012). More detailed investigations are necessary to determine detailed information about window opening behavior and shading use for longer periods, individual window opening monitoring, and ventilation system settings.

The measurement data reflects the actions taken by each household. However, the driving forces behind the reasons users decided to interact with windows and blinds cannot be obtained from the data; but rather requires further study into the factors about the context at physical, sociological, and psychological levels. IEA-ECBCS Annex 53, "Total Energy Use in Buildings—Analysis and Evaluation Methods" explores the contextual interaction of people and buildings. The possibility to categorize energy uses according to lifestyle and values for understanding users' actions and motivations are being explored in a dissertation which will also consider the influence of situational factors such as outdoor night noise, access to daylight, and temperature tolerance on summer cooling behavior.

ACKNOWLEDGMENTS

The authors gratefully acknowledge the participation by the families in this study.

REFERENCES

Bednar, T., A. Korjenic, et al. (2010). "Performance and Experiences with Austrian demonstration projects for lowest energy houses (passive houses) in social housing." *ASHRAE Transactions* 116, Part 2.

Fabi, V., R. Andersen, et al. (2012). Main physical environmental variables driving occupant behaviour with regard to natural ventilation. 5th International Building Physics Conference—The Role of Building Physics in Resolving Carbon Reduction Challenge and Promoting Human Health in Buildings. t.I.O. Committee. Kyoto, Japan: 921–928.

Kaufmann, H. (2007a). Residential Building Kammelweg. As Built Drawings—6th Floor. J. Kaufmann. Vienna, Kaufmann, Johannes.

Kaufmann, H. (2007b). Residential Building Kammelweg. As Built Drawings—5th Floor. J. Kaufmann. Vienna, Kaufmann, Johannes.

Keul, A. (2009). Umweltpsychologische Evaluation (POE) von sechs Wiener Passivhaussiedlungen (225 Wohneinheiten) im Vergleich zu konventionellen Bauten (156 Wohneinheiten). Wien:1–19.

Nicol, J.F., and M.A. Humphreys (2002). "Adaptive thermal comfort and sustainable thermal standards for buildings." *Energy and Buildings* 34(6): 563–572.

Statistik Austria (2012). Bevölkerungsstand 1.1.2012. Statistik Austria. Vienna, Statistik Austria:307.

team gmi (2004). Passivhausgeschosswohnbau Kammelweg Wien, Bauplatz E. Vienna:1–8.

Development of Industry Guidelines for Standard Practice of Fenestration Product Installation

James D. Katsaros, PhD

ABSTRACT

The requirements set out in building codes in the U.S. oblige window and door manufacturers to provide guidance on the installation of fenestration products, which includes proper integration with the building envelope. Historically, installation standards in the U.S. such as the ASTM E2112 Installation Practice for Window, Doors and Skylights and AAMA Installation Masters™ did not adequately address the wide variety of window and wall construction styles or the impact of exposure level on the installation methodology (2007). In addition, these standards are not written in a format that is user friendly for field-use, as might be required of a general contractor or builder when providing instructions to a window or door installer. To address this need, the three major fenestration industry organizations in the U.S., the Fenestration Manufacturers Association (FMA), the American Architectural Manufacturers Association (AAMA), and the Window and Door Manufacturing Association (WDMA), formed a joint committee to develop a series of illustrated, step-by-step installation guidelines for windows and doors in specific wall configurations, initially focusing on extreme wind and rain exposure conditions particular to the southeastern U.S. Representative installations from each standard practice were installed by committee members and tested to extreme exposure conditions using ASTM fenestration performance standards. Key discoveries were made as to the effectiveness and practicality of the methods that influenced the content of the guidelines. This paper provides details of the work of the FMA, AAMA, and WDMA installation committee in the development of guidelines for standard practices in the U.S. that have been published and also offers a brief summary of the future plans of this committee.

INTRODUCTION

The building envelope is defined by the American Society for Materials and Testing (ASTM) as the boundary or barrier separating the interior volume of a building from the outside environment (ASTM 2006). As such, the building envelope is the primary air and water protection to the building interior as well as the moisture sensitive components of the wall system. Thus, great attention must be given to ensure that the building envelope features continuous, redundant air and water protection consisting of durable materials applied with robust water-management principles and attention to workmanship detail. Water penetration through the building envelope, particularly with a building envelope discontinuity, can result in catastrophic damage to the building, especially if the water is not properly managed through drainage to the exterior. There is no more prevalent or complex building envelope discontinuity than creating a large hole in the building envelope containing an item made up of highly complex and variable materials, such as a window or door (fenestration product). The interface between the fenestration product and the wall is recognized as the one of the most problematic areas for water intrusion into the building envelope (RDH 2002). Adding to this concern is the fact that the installation of fenestration products into walls features a vast set of considerations key to the proper integration with the building envelope. These considerations include the following:

- The fenestration product attachment system (integral flange frontal location, integral flange nonfrontal location,

***James D. Katsaros** is R&D Team Leader, Residential Building Envelope at DuPont Building Innovations, Richmond, VA.*

field applied flange, box-frame with jamb fasteners, brick mold casing, etc.);

- The fenestration product material of construction (vinyl, wood, aluminum, fiberglass, composite), which greatly impacts the structural and thermal performance of the building envelope integration;
- The flashing system used for integration (self-adhered, mechanically attached, liquid applied, among other types of flashing);
- The type of wall system (wood framed, concrete block, metal frame/commercial, open stud, insulated concrete forms, among others), as well as the wall dimensions/depth;
- The exterior cladding system (vinyl or wood composite siding, brick, various stucco finishes including direct applied, 3-coat, and exterior insulated finishing system, among others);
- The use of exterior insulating materials, such as foam plastic insulated sheathing, to comply with increasing energy code requirements;
- The anticipated environmental exposure of the installed product, including thermal cycles, wind driven rain and dust;
- Regional and local building practices;
- Continuity of the drainage plane at wall to window/door interface;
- Differential movement, based on the material variations noted above.

Each of the considerations listed above have the potential to significantly impact the air, structural, and moisture performance of the installed fenestration product and the building envelope. This does not consider the case of replacement of existing fenestration products, which adds an additional level of complexity of how to manage around the existing fenestration product frame, cladding when not removed, interior finish, and building envelope integration.

Consequently, the task for the fenestration product installer to achieve consistent performance with this multitude of installation variations is extremely challenging. In order for the fenestration industry to provide goodness to installers, one could propose to develop a comprehensive list of prescriptive installation details based on a matrix of the above considerations, but this is not, from a logistical point of view, possible. It was subsequently decided by the industry representatives to take two alternative approaches to resolve such concerns. Specifically:

1. Develop guiding principles for installation that can be applied generically to any combination of the above installation considerations;
2. Develop guidelines for specific combinations of fenestration/wall systems in given environmental exposure conditions, aiming to address specific common practices of most concern to the installers.

The first approach was initiated through a task group of the ASTM E06.51.11 committee responsible for the ASTM E2112 (2007) installation standard. This study will focus on approach number two through the work of the Fenestration Manufacturers Association (FMA), the American Architectural Manufacturers Association (AAMA), and the Window and Door Manufacturing Association (WDMA) installation committee to develop guidelines specific to windows and doors installed in wood frame and surface barrier concrete masonry unit (CMU) construction with extreme exposure to wind and water loads, which is typical of southeastern U.S. residential buildings.

STANDARD PRACTICES FOR EXTREME EXPOSURE IN THE FLORIDA WALL

After the extreme hurricane season in Florida in 2004, widespread water damage in residential buildings was reported, particularly through the rough openings of fenestration products (Lstiburek 2005). To address this concern, the Florida Building Commission called upon the fenestration industry to develop robust standard practices to enhance the water-management capabilities for the installation of fenestration products into residential construction common to Florida and the Gulf Coast. The FMA took on this challenge and formed the FMA installation committee. Soon after, as a means of enhancing consensus in the fenestration industry, the AAMA and then the WDMA joined the effort, and the FMA/AAMA/WDMA Installation Committee was officially formed.

In order to initiate the development of installation standard practices for the southeastern U.S. region, the considerations previously noted above needed to be carefully evaluated. The original FMA installation committee recognized the unique characteristics of residential construction common to the southeastern U.S. and the challenges associated with these types of wall systems. Specifically, whereas some parts of the region build with standard wood frame construction, there is also a tendency to build residential homes from CMU. To further complicate matters, as shown in Figure 1, it is quite common for two-story residential structures to feature both types of walls, with wood frame construction installed on the second floor on top of a CMU wall. While the façade system for these buildings is predominantly stucco, the application of the stucco is dramatically different in the two-wall systems. In the case of wood frame construction, the stucco is applied on a lath with a WRB (water-resistent barrier plane behind the stucco, creating a standard-membrane drainage wall system. However, for the CMU walls, the stucco is directly applied onto the CMU with no drainage plane, thus creating a surface-barrier mass wall system. The combination of a drainage wall installed on top of a surface-barrier mass wall creates a challenging environment in order to maintain building-envelope protection. Given that the scope of this effort is to focus on the fenestration product integration with the building envelope and not the interaction between the two-wall systems, the FMA Installation committee decided the best approach was to consider the wood frame construction separate from the surface-barrier CMU and develop independent practices for each of these wall systems.

However, of importance is the fact that the FMA/AAMA/WDMA installation committee set forth common features for all wall types. First, in all standard practices, representative installations are to be water tested according to either the standard test method given in ASTM E331 (2009) or that of E547 (2009) up to 575 Pa (12 lb/ft^2) pressure differential in order to simulate extreme exposure conditions, as installed by those developing the guideline. This practice has proven to be highly valuable in not only testing the effectiveness of the water management of the installations, but also in evaluating the practicality of the installation method—i.e., is this installation one that a typical installer is likely to undertake? Second, all standard practices contain a set of related issues that provide valuable guidance to any installer, such as principles for continuity between the fenestration product and the building envelope/drainage plane, sealant selection and application, and compatibility concerns. And third, all standard practices include an Appendix that provides guidance as to trade responsibilities for the various steps covered in the documents. These common features provide reality-based, valuable information to the installer for any fenestration product installation system.

Wood Frame Construction Exposed to Extreme Wind and Water Exposure

The first wall system addressed by the original FMA installation committee was wood frame construction exposed to extreme wind and water loads. There are two distinct standard practices published for this type of wall system, one focused exclusively for windows (FMA/AAMA-100, *Standard Practice for the Installation of Windows with Flanges or Mounting Fins in Wood Frame Construction for Extreme Wind/Water Exposure*) (2012) and the other focused exclusively on doors (FMA/AAMA/WDMA-300, *Standard Practice for the Installation of Exterior Doors in Wood Frame Construction*) (2012). The attributes of both these practices are as follows.

Figure 1 *Typical two-story residential construction in southeastern U.S./Florida.*

FMA/AAMA 100 (2012). Whereas the well known ASTM E2112, *Standard Practice for the Installation of Exterior Windows, Doors and Skylights* provides a wealth of essential information to guide a design professional or general contractor for the proper installation of fenestration products, it is not presented in a user friendly format that can be given to an installer to follow step-by-step illustrated instructions (2007). This, in addition to the need to address extreme exposure considerations, as noted above, became a primary objective for the development of these FMA standard practices. As such, the installation committee set forth to make the FMA/AAMA 100 document a brief, yet detailed guide for specific window systems (flanged or mounting fins) subjected to specific exposure conditions (e.g., extreme wind and water) in wood framed construction having exterior wood sheathing panels (e.g., not open stud) (2012). While the original emphasis was on the Gulf Coast/Florida region, it was considered that other climates with similar attributes, such as the coastal Pacific Northwest, may benefit from these installation principles. Because the scope of this guidelines is to address extreme wind and water exposure, preventative measures were taken that are above normal installation practices, such as those referenced in ASTM E2112 (2007).

What were the preventative measures that were taken that are above normal installation practices? First and foremost, the committee focused on implementing the principle of robust water-management protection through the practical application of a drainable installation. This, as was learned through the water testing of representative installations for both the FMA/AAMA 100 (2012) and FMA/AAMA/WDMA 300 (2012) standard practices, is not as simple as installing sill pan flashing, as recommended for all installations in ASTM E2112, Section 5.16.3 (2007). As will be described below in more detail, two key observations will be highlighted that were obtained from results of wall testing and that aid in the practical application of drainable installations for protection to extreme exposure conditions:

1. Taking steps to prevent incidental water intrusion under extreme exposure that could migrate behind the WRB due to penetrations, fasteners, or cladding attachment and may migrate into the window rough opening.
2. It is critical that the perimeter interior air and water seal between the fenestration product and the sill-pan flashing is able to withstand the anticipated pressure differential without air and water leakage.

Preventative Measure Number 1: WRB Wrap Methods for Resisting Exposure to Extreme Wind and Water Loads. Results from wall testing for water intrusion of representative installations for FMA/AAMA/WDMA 300 (2012) differed than that which was reported for FMA/AAMA 100 (2012) testing as summarized in an earlier publication

(Katsaros 2007), in that vinyl siding cladding was applied to the wall, creating numerous unsealed penetrations into the WRB. It was found that at water-test pressure differentials roughly above the typical threshold for standard windows and doors in residential applications (e.g., 300 Pa [6.24 lb/ft^2]), which corresponds to a design pressure 45 rated window, water intrusion through the cladding fasteners could penetrate into the rough opening behind the WRB (note: air barrier performance considerations of the WRB were not taken into account). Although this was considered out of scope of the fenestration product installation itself, which focused on the integration of the fenestration product with the WRB and drainage plane, this observation was considered of consequence to address in the FMA/AAMA 100 (2012) and FMA/AAMA/WDMA 300 (2012) documents.

As such, given the extreme exposure scope of these documents, it was recommended that the underside of the WRB be sealed at the rough opening according to three prescribed methods that provide protection beyond that described in standard installation practices such as ASTM E2112 (2007) or the AAMA Installation Masters™ course (2004–2010). These three methods are illustrated as follows:

WRB Method A. This method is similar to what is often done in commercial construction in the U.S., where the WRB is cut around the perimeter of the rough opening and sealed with self-adhered flashing. The detailed steps with illustrations taken from FMA/AAMA 100 (2012) include: cut the WRB flush at the rough-opening perimeter (Figure 2), and seal the WRB to the rough opening with the application of 100 or 150 mm (4 or 6 in.) self-adhered flashing between the jamb corner at the head, such that self-adhered flashing covers 50 mm (2 in.) on the WRB as well as into the rough opening return at least to the depth of the window (Figure 3). Apply sealant at jamb/head interface (Figure 3). See Schematic 1 for jamb detail.

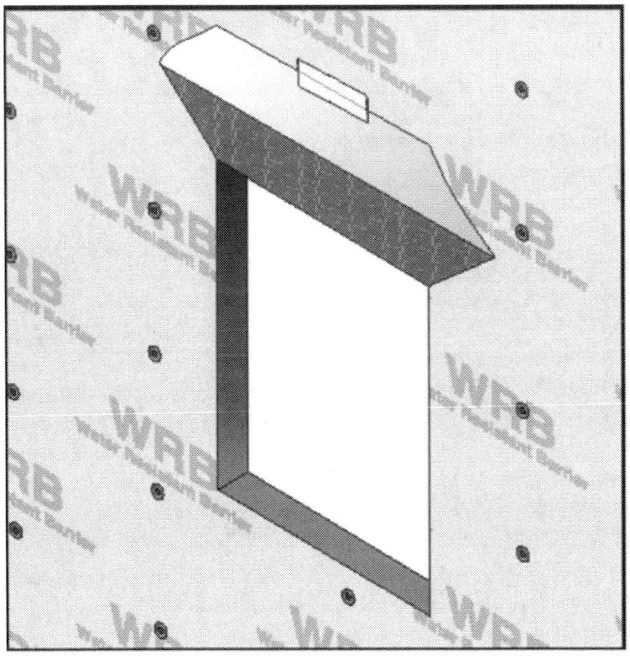

Figure 2 *WRB wrap method A, step 1.*

WRB Method B. This method (for use with mounting flanged windows only) treats the side jambs of the window in the same fashion as the head, where the WRB is cut away and then integrated after the window is installed and the flashing is applied. This is also a modified wrap after the WRB method, as described in ASTM E2112 (2007). The detailed steps from FMA/AAMA 100 (2012) include: cut the WRB flush at the rough-opening perimeter, and make a cut onto the face of the wall at each jamb corner and fold back the jamb as with head flap, ensuring that the jamb cuts at the sill are angled upwards (Figure 4). After the window is installed, apply sealant along the jamb and bring over the previously-folded WRB jamb flap

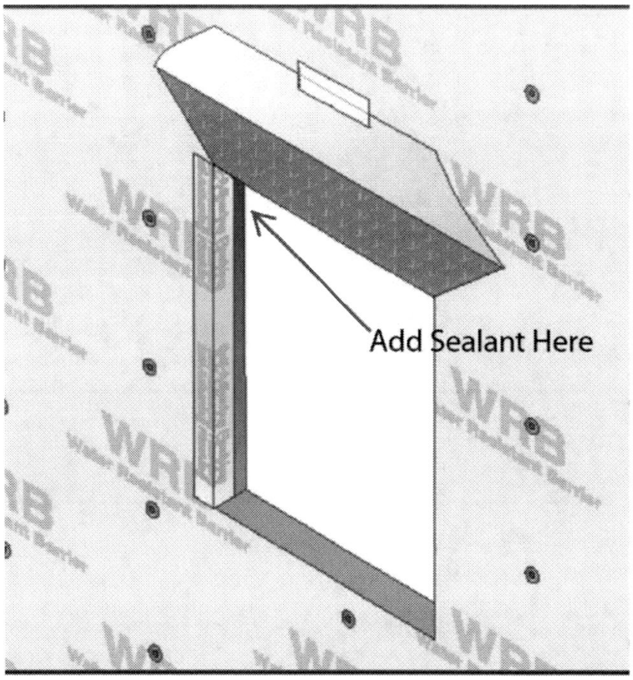

Figure 3 *WRB wrap method A, step 2*

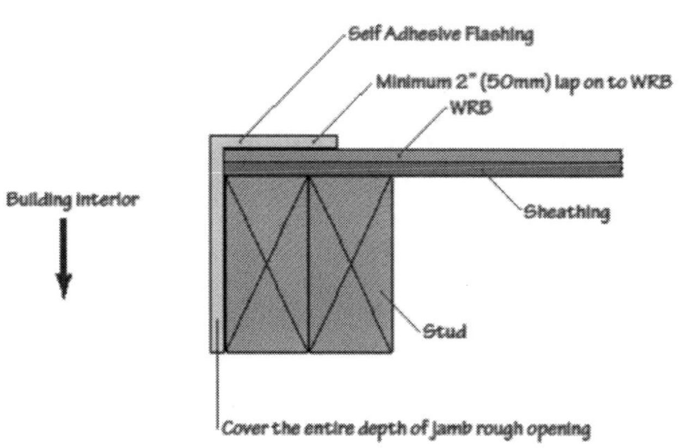

Schematic 1 *Top view: WRB wrap method A.*

allowing it to integrate with the sheathing window frame (Figures 5a and 5b). Press down on sealant bead under the WRB. Integrate the WRB to the window with sheathing tape or self-adhering flashing. See Schematic 2 for jamb detail.

WRB Method C. This method is very similar to the standard WRB wrap methods described in ASTM E2112, except that sealant is applied under the WRB at the exterior side jambs (2007). The detailed steps from FMA/AAMA 100 (2012) include: cut the WRB in an I-cut (Figure 6a) or modified I-cut (Figure 6b), apply sealant onto sheathing under WRB at jambs (Figure 7), and wrap into cavity and secure (Figure 8). Attach the WRB into position on the inside of the rough opening and trim any excess as required (Figure 9). Press down on sealant bead below WRB. See Schematic 3 for jamb detail.

Extreme Exposure WRB Wrap Methods—Summary. Note that in the three methods described above, only the side jambs of the rough opening differ from the standard WRB wrap methods described in ASTM E2112 (the head and sill details are the same in each case) (2007). Thus, these methods describe the situation where the exterior wind pressure is high enough to drive incidental water through fastener penetrations in the WRB and horizontally into the rough opening. This pressure thus

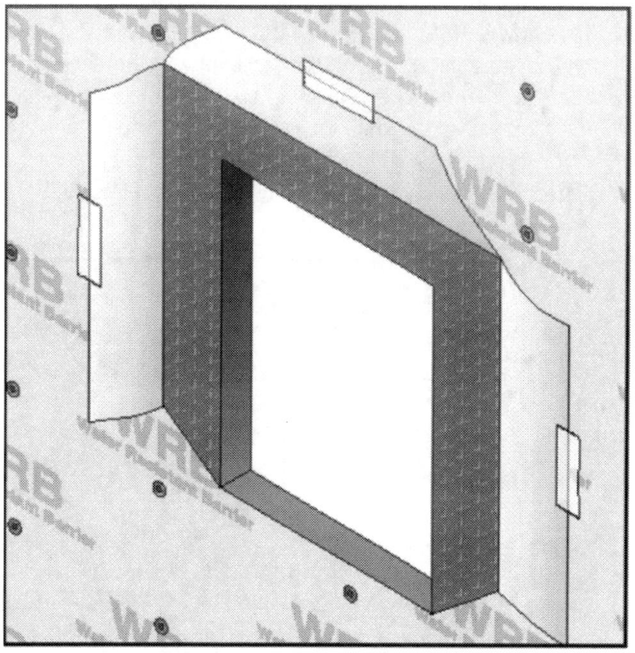

Figure 4 *Wrap method B, step 1.*

Figure 5a *Wrap method B, step 2a.*

Figure 5b *Wrap method B, step 2b.*

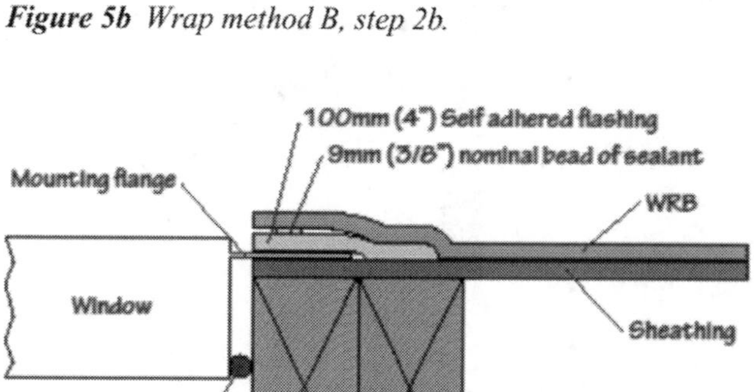

Schematic 2 *WRB wrap method B.*

helps define what is extreme and justifies the extra preventative measures above normal practices. Representative installations for all three extreme exposure WRB wrap methods were successfully tested (which will be explained in more detail for the FMA/AAMA/WDMA 300 testing for doors) up to 575 Pa (12 lb/ft^2) of pressure differential, which correspond to level 1 hurricane force winds, as installed by members of the FMA/AAMA/WDMA Installation Committee (2012).

Preventative Measure Number 2: Interior Air/Water Seal. A drainable installation of a fenestration product is one that enables an exit path for any water intrusion that could occur either though the fenestration product itself (such as joinery, welds, and seams), or by entering the rough-opening cavity from the interior wall or building exterior. If not properly managed or drained, as with a nondrainable or barrier installation, such water intrusion will tend to collect and be

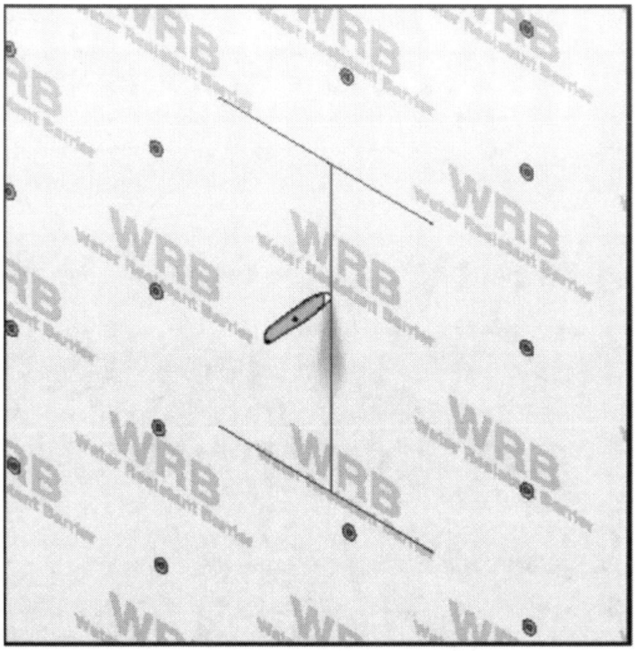

Figure 6a *WRB wrap method C, step 1.*

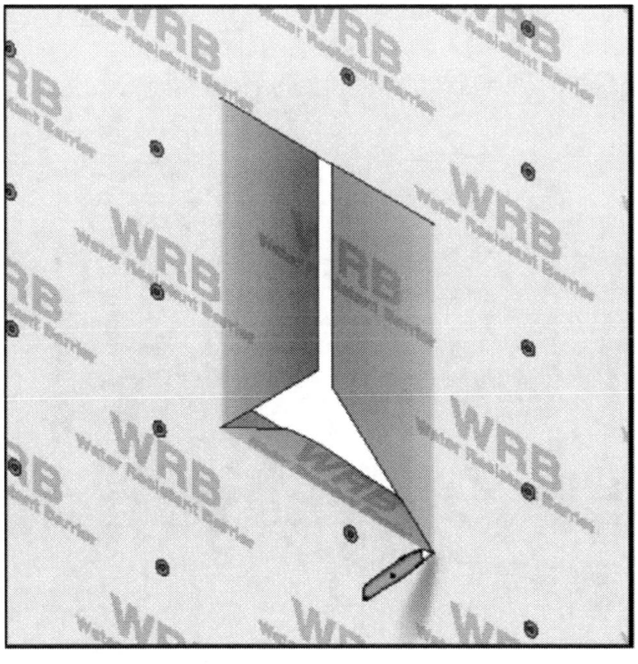

Figure 6b *WRB wrap method C, step 1 alternative.*

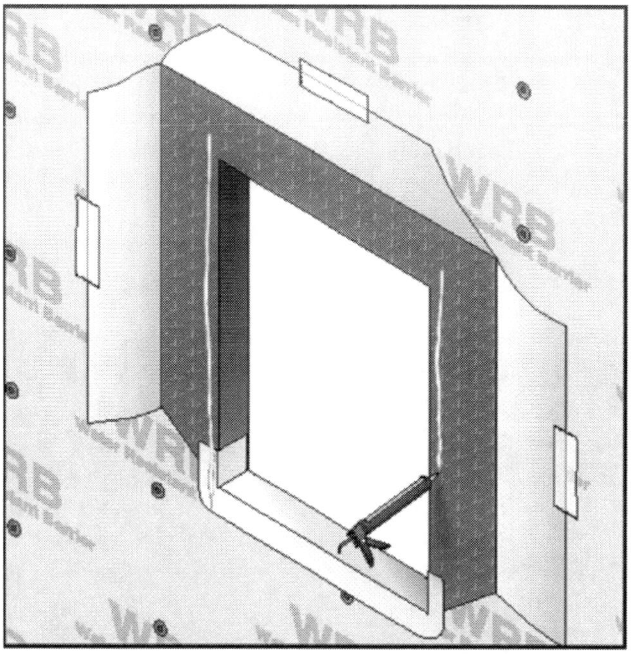

Figure 7 *WRB wrap method C, step 2.*

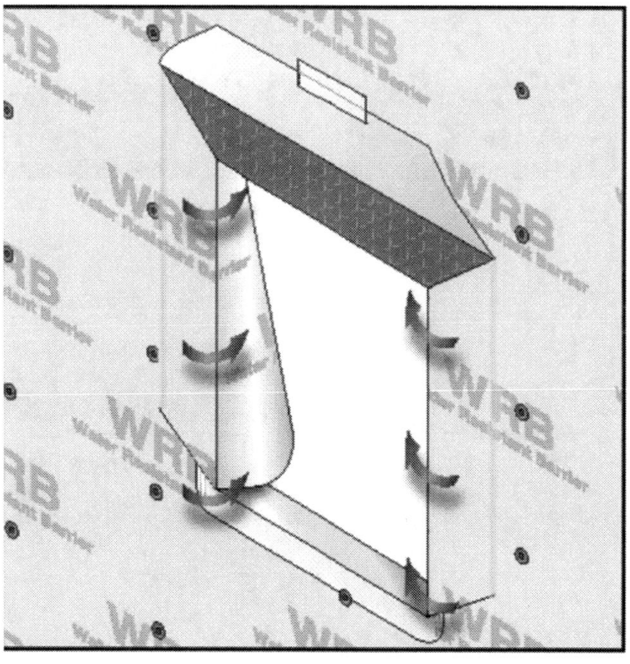

Figure 8 *WRB wrap method C, step 3.*

trapped at the sill corner of the rough opening, which is a leading cause of water damage in buildings (Figures 10 and 11). Note that while water damage at the sill of the rough opening is often perceived to be caused by a window leak, the source of the intrusion is often unrelated to the fenestration product. However, water intrusion may find its way to the sill of the rough opening due to a disruption in the continuity of the building envelope and thus be associated as a fenestration product water intrusion concern.

In the case of extreme exposure conditions, where the risk of water intrusion is greater than normal, the drainable installation option is mandated according to that given in FMA/AAMA 100 (2012). An essential feature that defines a drainable installation for a fenestration product is to maintain an unhindered exit path for any incidental water intrusion into the cavity of the rough opening. As illustrated in the schematic in Figure 12, taken from ASTM E2112, the drainable installation has three critical features (2007):

1. The application of durable, waterproof sill-pan flashing (either rigid or self-adhered flashing is acceptable) below the sill of the fenestration product,
2. Leaving drainage openings in the sealant bead below the sill flange of the window (FMA/AAMA 100 [2012] requires a minimum of two 50 mm (2 in.) openings within 100mm (4 in.) of each side jamb),
3. Application of a perimeter air and water seal around the interior perimeter of the gap between the fenestration product and WRB or rough opening.

A drainable installation adds complexity to the window installation when compared to a traditional full barrier installation (which features a continuous bead of sealant around entire exterior perimeter of fenestration product attachment), but the system is far more forgiving to water intrusion into the rough opening. As such, a second key element that was understood

Figure 9 *WRB wrap method C, step 4.*

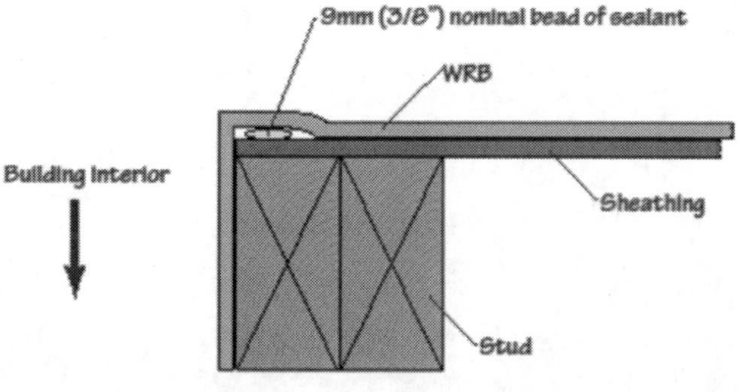

Schematic 3 *WRB wrap method C.*

Figure 10 *Moisture damage at window sill.*

Figure 11 *Moisture damage near window/wall interface.*

from the results of representative wall testing undertaken for FMA/AAMA 100 (2012) and FMA/AAMA/WDMA 300 (2012) is that the interior seal (critical feature number 3 noted above) can often be the most complex and problematic preventative measure for extreme wind and water-exposure conditions. Achieving this continuous air and water seal around this perimeter can be challenging due to obstructive features of the installation, such as shims and anchors (Figure 13). In addition, even small dimensional variability can lead to irregularities in the gap distance between the fenestration product and the interior perimeter of the rough opening. As such, it is difficult to use standard air and water-seal materials such as foam backer rod or low expansion foam to continuously seal the rough-opening perimeter. Any discontinuity in the interior perimeter gap that results from these obstructions will increase the potential for air and water intrusion. Furthermore, even if there are no obstructions or irregularities in the gap, achieving an adequate sealant bond between the fenestration product and the sill flashing, WRB, and rough-opening cavity is critical, as well as careful workmanship in the application and tooling of the sealant. As was discovered in the FMA/AAMA 100 (2012) and FMA/AAMA/WDMA 300 (2012) wall testing, even tiny pin hole gaps under extreme exposure conditions could lead to water intrusion for drainable installations. As such, both documents make the following statement:

> To ensure adequate protection against extreme wind-driven water, it is critical that the perimeter interior air and water seal between the window and the sill-pan flashing is able to withstand this pressure load without air and water leakage. Special caution needs to be used when applying perimeter air and water seals to the interior corners (FMA/AAMA 100 [2012] Section 7.7.3).

Results from testing representative wall installations for FMA/AAMA 100 (2012) were provided in an earlier publication by Katsaros (2007).

FMA/AAMA/WDMA 300 (2012). A related document to FMA/AAMA 100 (2012) is FMA/AAMA/WDMA 300 (2012), for which the installation of doors subjected to extreme exposure conditions is addressed (note that WDMA joined the effort after the original window document was completed, for which there was a corresponding change in the document notation). There are two key differences between the FMA/AAMA 100 (2012) and FMA/AAMA/WDMA 300 (2012) documents. One is that three different attachment systems are addressed: mounting flanges (Section 7.6 per Figure 14), exterior casing/brick molding (Section 7.7 per Figure 15), and box frame/nonflanged (Section 7.8 per Figure 16). While these attachment systems can be found in window units, particularly with box frame in heavy commercial buildings, it is more common in residential buildings to find brick mold casing and box frame system with doors. However, the guidance provided in FMA/AAMA/WDMA 300 for these types of door systems can be applicable for any fenestration products (2012).

A second important difference between information provided in the FMA/AAMA/WDMA 300 (2012) and FMA/AAMA 100 (2012) guidelines is the sill detail. In the FMA/AAMA/WDMA 300 (2012) guideline, sill details for three types of flooring are described: wood floor (Figure 17), concrete flat/slab on grade (Figure 18), and concrete recessed

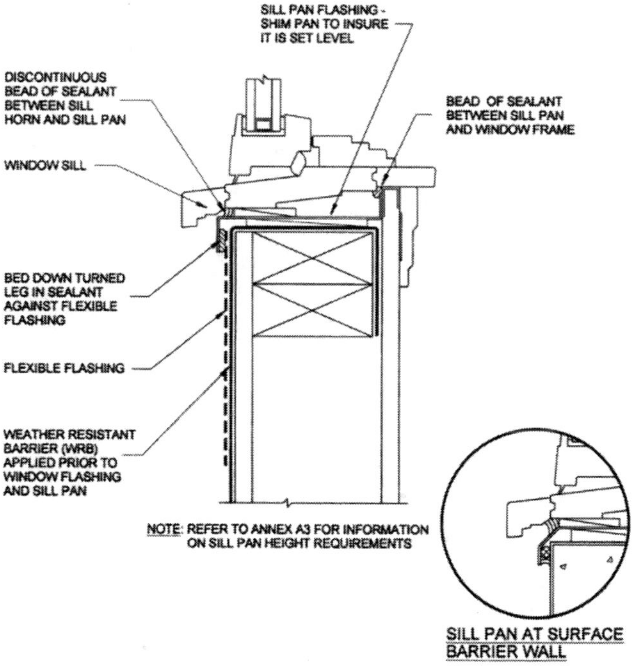

Figure 12 *Taken from ASTM E2112,* Standard Practice for Installation of Exterior Windows, Doors and Skylights. *Figure 15.*

Figure 13 *Removal of potential obstruction to interior air/water seal.*

(Figure 19). For doors installed into a wood floor, the detail differs from the FMA/AAMA 100 guideline for windows in that a rigid sill pan with raised-leg back dam is utilized (2012). For concrete floors, both flat and recessed, a liquid-applied flashing is illustrated to prevent absorption of liquid water in the sill of the rough opening; although, a rigid sill pan can be used in these floor systems as well.

Each type of door mounting-flange system and sill detail were successfully tested for water intrusion up to 575 Pa (12 lb/ft^2) pressure differential as installed by members of the FMA/AAMA/WDMA Installation Committee. Unlike the FMA/AAMA 100 (2012) testing reported in Katsaros (2007), these installations included the application of vinyl siding. The key findings from this testing led to the preventative measures described in the previous section.

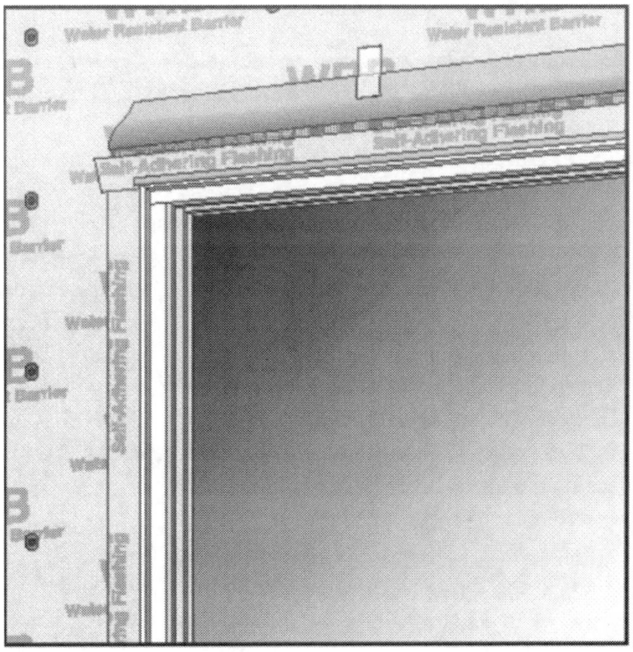

Figure 14 *Mounting-flange door installation step.*

Figure 15 *Brick mold/exterior casing door-installation step.*

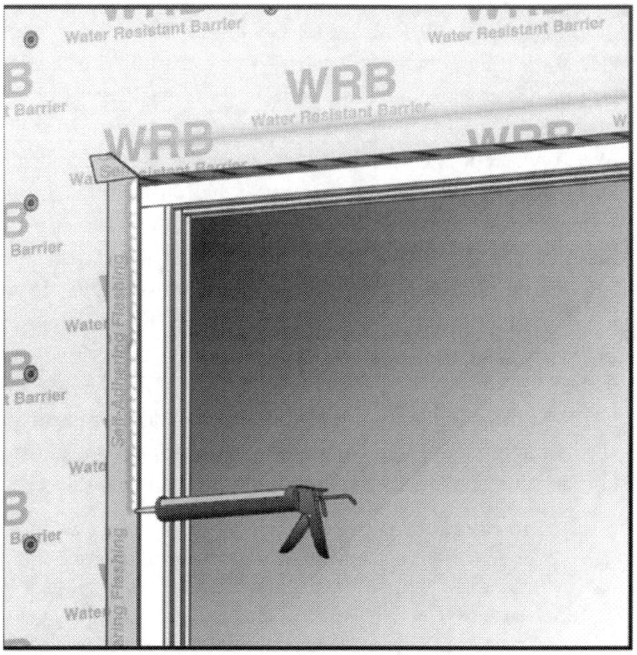

Figure 16 *Box-frame door-installation step.*

Figure 17 *Sill detail for wood-floor door installation.*

SURFACE-BARRIER CMU WALLS IN EXTREME EXPOSURE

The original FMA installation committee took on the challenge of developing a window installation guideline specific for surface barrier wall systems having direct-applied stucco, wall systems common to the southeastern U.S. In this system, a concrete masonry block wall typically incorporating a precast concrete sill at the window openings (Figure 20) is finished with a layer of stucco that is directly applied to the CMU (Figure 21) and thereafter covered with a water resistant acrylic coating (Figure 22).

The water-management strategy for this wall system is completely different than for the membrane drainage wood

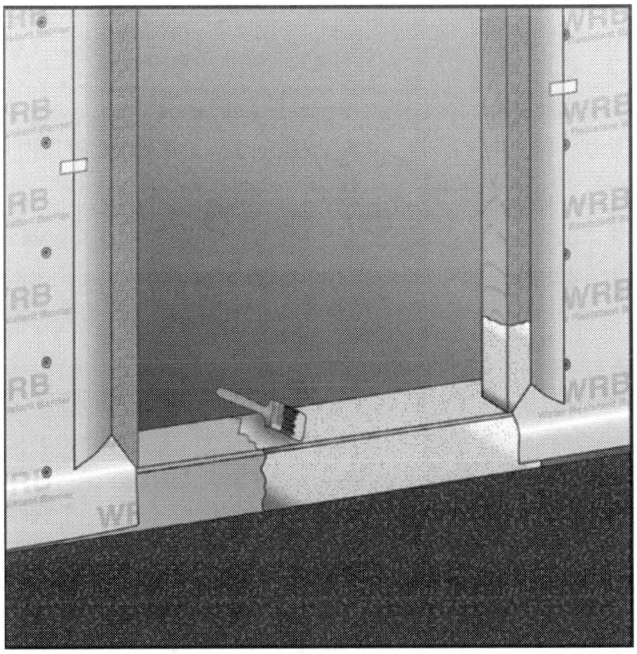

Figure 18 *Sill detail for level-concrete sill-door installation.*

Figure 19 *Sill detail for recessed-concrete sill-door installation*

Figure 20 *CMU wall with precast sill for residential construction.*

Figure 21 *Direct-applied stucco on CMU wall.*

frame wall system described in FMA/AAMA 100 (2012) and FMA/AAMA/WDMA 300 (2012), as well as other installation guidelines such as ASTM E2112 (2007). In this case, the first line of defense against water intrusion is the acrylic coating, which typically has to be applied in several layers to effectively seal through the inevitable cracks that form in the stucco finish. As noted in the report by Lstiburek, "Highly textured surfaces are almost impossible to coat in a manner to seal microcracks" (2005). This report goes on to note:

> The nature of stucco application, materials, and substrates make it impossible to construct crack free monolithic stucco renderings—in other words, shrinkage cracks and settlement cracks are to be expected. Based on the field observations, it is our belief that it is not possible to construct stucco assemblies without cracks. It is also our belief that paint coating systems—even the high build coatings—are unable to span the typical stucco cracks encountered, both from initial drying and from subsequent settlement (Lstiburek 2005).

Based on the above findings, it is expected that this surface-barrier CMU wall construction will encounter water intrusion through the exterior surface. In addition, since this CMU wall will often have a wood frame wall incorporating a drainage plane and sheathing membrane installed above it, drainage from the wall above is another possible source of water intrusion into this mass wall.

Figure 22 *Acrylic finish on direct-applied stucco CMU wall.*

So what happens to the water that gets into the CMU mass wall? For the most part, liquid water will be absorbed through cracks in the stucco into the masonry and will be driven into the building interior from heat derived from external exposure to this hot/humid climate. This moisture driven to the interior will be removed by the building HVAC system, adding to the HVAC loads, assuming there is no vapor barrier installed on the inside of the wall, such as foil-faced insulation in which case it will condense causing moisture along the interior of the building wall. When the moisture content exceeds the storage capacity of the CMU, it will flow down through the wall by gravity to the concrete floor where it is expected to be directed to the exterior by flashing located at the intersection between the wall and floor. To summarize, the water-management strategy for the direct-applied stucco surface barrier building envelope contains the following key features:

1. Use of the storage capacity of the masonry wall to contain moisture ingress;
2. Rely on HVAC to manage inward driven moisture given conditions of a hot and humid climate; and
3. Drain excess moisture out the bottom of the wall.

The common practice to prevent this moisture intrusion in this region is to fill cracks in the stucco and joints with paint or sealant (need to do this at least twice); although, Lstiburek found this to be an ineffective means of controlling surface moisture (2005).

Given that the scope of this study is extreme exposure of fenestration installations to wind and water loads, item number three above, "drain excess moisture to the bottom of the wall," is the focus of further evaluation. This exit path for excess moisture is only suitable when there are no obstructions in the wall, or if there is a possible breach of the building envelope. As illustrated in Figure 23, excess liquid water in a CMU wall that drains through a continuous set of CMU blocks can escape at the bottom of the wall, but what happens when it finds a large opening in the wall, such as at a window or door? Water that reaches the head of the rough opening will either enter the building interior at the window head or migrate around the rough-opening cavity and perhaps enter the interior at the sill. Note that it is recognized that in such type of construction, the head of the rough opening typically consists of a dense concrete header that will be more water resistant than the standard CMU block. However, this header is not considered a waterproof element of construction.

Nonetheless, this latter scenario for water intrusion is especially likely if the path to the exterior is blocked by a fenestration product in the wall that acts as a disruption to the normal drainage path in the wall. In this case, the water intrusion into the interior at this disruption may be considered a leak of the fenestration product; thus, making the fenestration product manufacturer possibly responsible for the repair and potential damage to the interior. The realization of this scenario was a primary concern for the original FMA Installation Committee and the ensuing standard practices, FMA/AAMA 200 (2012) and FMA/WDMA 250 (2010).

FMA/AAMA 200 (2012) and FMA/WDMA 250 (2010). As noted above, the original FMA Installation committee recognized the inherent challenges of developing guidelines of standard practices for fenestration product installations in barrier CMU walls systems having direct-applied stucco subject to extreme wind and water exposure.

As such, the committee resolved that addressing water-management concerns for the entire wall system was out of scope of the effort. Alternatively, the scope was set to focus only on the rough-opening area, where gravity-driven drainage within the wall system is disrupted by the discontinuity in the building envelope due to the window or door opening, and with this discontinuity, a potential exists to cause water intrusion into the building. As such, the committee selected the approach to protect the rough opening with a liquid-applied flashing, as specified in AAMA 714 (2011). In this approach it is proposed that when the head of the rough opening is exposed to water intrusion, the liquid-applied flashing will act as a water protective barrier that will divert water to the side jambs and subsequently drain to the bottom of the wall (Figure 24). The side jambs are also protected by liquid-applied flashing, thus not allowing water penetration along the jambs. In addition, any water intrusion into the rough-opening cavity from the exterior, either through the window joinery or attachment, will be managed by the liquid-applied flashing treatment of the rough opening. This includes water intrusion to the sill area where water tends to collect. As long as water at the sill is directed to the exterior through the use of weeps at the exterior sill and there exists an interior air and water seal located at the back of the sill as installed in accordance with the normal installation procedures that promote drainage, then the installation should perform adequately.

As a best practice, it is recommended that the liquid-applied flashing coating also be applied on the interior surface of the head, acting as an effective back dam for water intrusion that enters from above. In addition, treating the interior of the sill area is recommended.

The FMA/AAMA 200 (2012) and FMA/WDMA 250 (2010) documents are divided by type of fenestration attachment system for which: FMA/AAMA 200 (2012) addresses windows with frontal flanges (e.g. aluminum windows typically used in the southeastern U.S.); whereas, FMA/WDMA 250 (2010) addresses nonfrontal-flanged windows with mounting flanges (e.g. conventional wood windows). The key difference in the installation methods is with the design of the sill, in which the frontal flange window system (FMA/AAMA 200) typically utilizes a precast sill (both protruding and flush with the exterior wall), whereas the nonfrontal flange windows utilize an insert buck for attachment (2012). For the purpose of illustration the figures and steps shown below are taken from the FMA/AAMA 200 standard practice (2012).

The primary steps for installation of frontal-flanged windows into direct-applied surface-barrier CMU walls according to FMA/AAMA 200 are as follows (2012):

1. After checking tolerances, prepare the rough opening by applying a liquid-applied flashing to the rough-opening interior (Figure 25). Note: it is recommended not to coat the exterior surface of the CMU unless a durable bond between the liquid-applied flashing and the direct-applied stucco has been demonstrated.

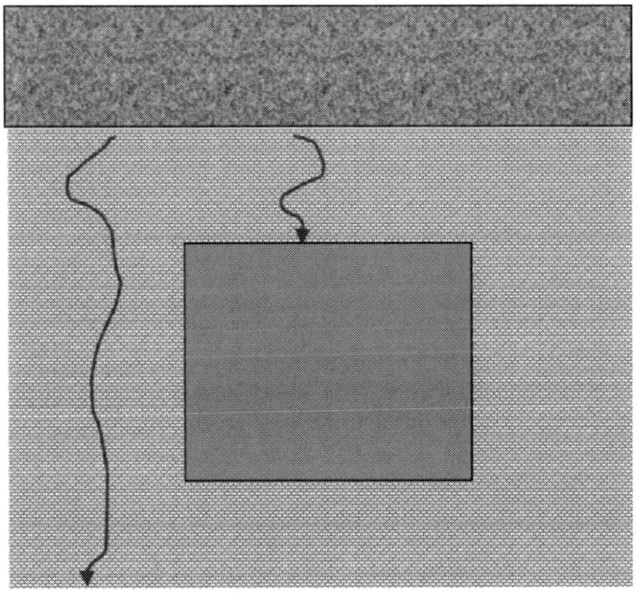

Figure 23 *Potential excess water intrusion path in CMU wall due to excess water migration from above the wall assembly.*

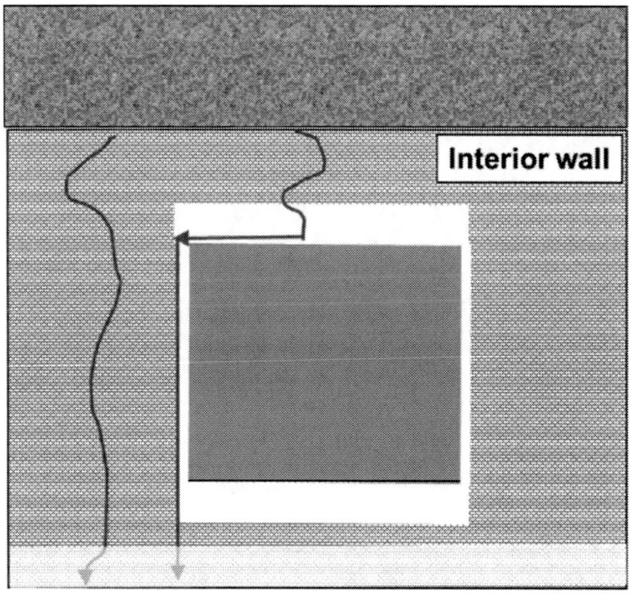

Figure 24 *Application of liquid-applied flashing as moisture protection for CMU wall.*

2. Seal all corners and cracks around the rough opening with liquid-applied flashing or compatible sealant (Figure 26).
3. Apply sealant to the interior side of a buck to act as a structural element for the attachment of the window (Figure 27).
4. Attached the buck per structural recommendations (Figure 28).
5. Treat the remaining, exposed portion of the buck with a liquid-applied flashing (Figure 29).
6. Apply sealant to the perimeter of the frontal-flanged window, leaving two 50 mm (2 in.) gaps at the sill for

Figure 25 *Treatment of rough opening with liquid-applied flashing.*

Figure 26 *Corner and crack treatment with liquid-applied flashing or sealant.*

Figure 27 *Sealant application on structural buck for rough opening.*

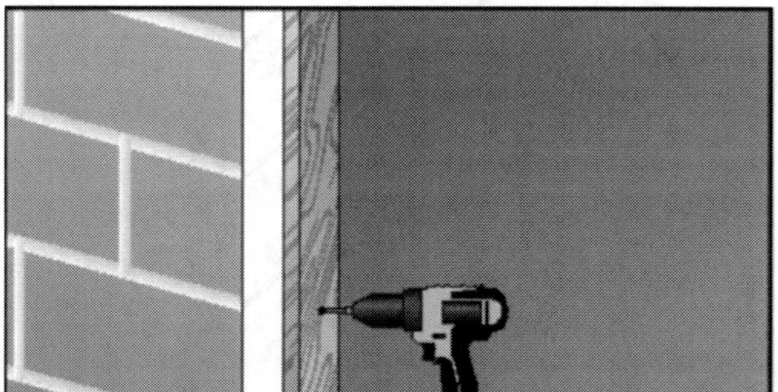

Figure 28 *Application of structural buck in rough opening.*

drainable installations; barrier installations are also acceptable in this standard practice (Figure 30).

7. Install the window into the structural buck elements according to the manufacturer's recommendations (Figure 31).
8. Apply an interior air and water seal that is designed to resist the anticipated level of pressure differential from extreme exposure conditions (Figure 32).

As given in detail in Katsaros (2007), representative installations from both FMA/AAMA 200 (2012) and FMA/WDMA 250 (2010) as installed by committee members were successfully tested for water intrusion according to the methods given in ASTM E457 (2009) up to 575 Pa (12 lb/ft^2) water pressure differential to simulate exposure to extreme conditions (2009).

Figure 29 *Water-resistant treatment of structural buck in rough opening.*

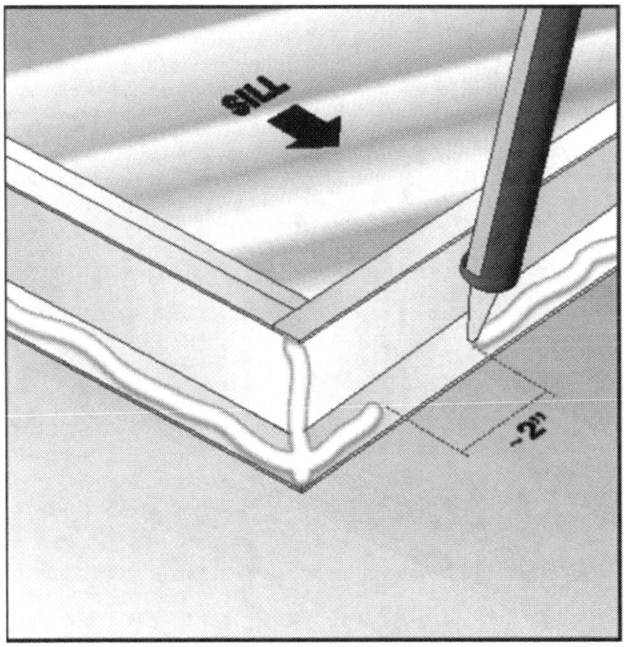

Figure 30 *Application of sealant on window flange.*

Figure 31 *Installation of window into CMU wall system.*

Figure 32 *Application of interior air and water seal.*

A version of the FMA/AAMA 200 (2012) document for door installations having the same mounting variations as given in the FMA/AAMA/WDMA 300 (2012) guideline, is now under development as FMA/AAMA/WDMA 400, and is expected to be published in 2013. Six installations, representing two flanged, two brick mold, and two nonflanged doors in a concrete flat and recessed sill (one for each attachment type), from FMA/AAMA/WDMA 400 were tested for water intrusion by members of the installation committee in May 2011. These walls, however, contained an additional important distinguishing feature from the FMA/AAMA 200 test walls for windows, in that a stucco finish was direct applied onto the exterior of the CMU wall (2012). After the doors were installed and the liquid-applied flashing and sealants were allowed to cure (roughly two weeks), the stucco with an acrylic paint finish was applied to the CMU wall. Water testing began after a week of cure time for the stucco finish. Figures 33–38 provide an illustration of representative walls from this testing. Figure 33 illustrates the application of liquid-applied flashing around a recessed-concrete sill area; Figure 34 shows the rough-opening application, including the interior. Figure 35 illustrates a completed wall assembly from the exterior with stucco applied and Figure 36 shows the interior side of that wall. The water testing process is illustrated in Figure 37. All walls passed the ASTM E547 test protocol at 575 Pa (12 lb/ft^2), run in three test pressure steps at 15 min. each up to 575 Pa (2009).

Figure 33 Application of liquid-applied flashing on recessed-concrete sill.

Figure 34 Application of liquid-applied flashing in the rough opening and around the interior perimeter.

CONCLUSION

This report provides a summary of the efforts of members of the FMA/AAMA/WDMA installation committee, over the period from 2006–2012, to publish five installation standard practice documents representing specific window and door systems in both wood frame and CMU wall construction in extreme wind and water-exposure conditions. A key guiding principle applied to the development of each of these practices was the hands-on installation and water-intrusion testing of representative installations by committee members. This provided key insights in respect to the effectiveness and practical

Figure 35 Exterior side of FMA/AAMA/WDMA 400 wall assembly with direct-applied stucco.

application of installation practices being developed, and in many cases influenced content changes in the documents.

FUTURE WORK

A new effort by the FMA/AAMA/WDMA installation committee has been initiated for the integration of windows and doors into walls that include foam plastic-insulated sheathing (FPIS) in wood frame construction. This effort was initiated in order to address performance requirements set out in the new energy code starting in 2012. The committee has addressed three primary concerns: structural attachment to support the windows, maintaining proper water-management principles through integration with the building envelope, and finishing details with the façade or siding. In addition, a new FMA/AAMA/WDMA task group has been formed to address installation details for replacement window systems. As such, this committee will continue to develop valuable tools for the fenestration industry to enhance the effectiveness and practical application of the installation of fenestration products into a variety of wall and exposure conditions.

Figure 36 *Interior side of FMA/AAMA/WDMA 400 wall assembly.*

Figure 37 *Water-testing process for FMA/AAMA/WDMA 400 wall assembly.*

ACKNOWLEDGMENTS

The author would like to acknowledge that this study represents the work of many contributors to the FMA/AAMA/WDMA Installation Committee over the past several years. In particular, the FMA/AAMA/WDMA Advisory and Oversight Committee, which is made up of six representatives (two from each organization) and is co-led with the author by Kelly Broker from Dow Corning as AAMA Chair. In addition, Lori Marino from Schnee Moorehead also represents AAMA, Steve Strawn from Jeld Wen Windows and Doors and Joe Hayden from Pella Windows represent WDMA, and Jim Krahn from Marvin Windows and the author, Jim Katsaros from DuPont, represent FMA. Also, document and ballot administration by Andrea Rhodes from AAMA and Becky Buice representing FMA is greatly appreciated. Third-party testing of the installations reviewed in this report was performed at and partially sponsored by the Architectural Testing Institute (ATI). The initial vision for this effort was set by Mark Daniels of Sika and was made a reality through the leadership of Freddie Cole of Schnee Moorehead (then FMA President) and the endless tenacity of Dick Wilhelm, FMA Executive Director. Lastly, this work is dedicated to the memory of Barry G. Hardman of National Building Sciences, past chairman of ASTM E2112 and key inspiration for the FMA/AAMA/WDMA effort (2007).

REFERENCES

AAMA. 2011. AAMA Standard 714, *Voluntary Specification for Liquid Applied Flashing Used to Create a Water-Resistive Seal around Exterior Wall Openings in Buildings*. Schaumberg, IL: AAMA.

AAMA. 2004–2010. InstallationMasters® Training and Certification Program. InstallationMastersUSA.com. Schaumberg, IL: AAMA.

ASTM. 2007. ASTM Standard E2112, *Standard Practice for Installation of Exterior Windows, Doors, and Skylights*. Philadelphia: ASTM.

ASTM. 2006. ASTM Standard E631, *Standard Terminology of Building Constructions*. Philadelphia: ASTM.

ASTM. 2009. ASTM Standard E331, *Standard Test Method for Water Penetration of Exterior Windows, Skylights, Doors, and Curtain Walls by Uniform Static Air Pressure Difference*. Philadelphia: ASTM.

ASTM. 2009 ASTM Standard E547, *Standard Test Method for Water Penetration of Exterior Windows, Skylights, Doors, and Curtain Walls by Cyclic Static Air Pressure Difference*. Philadelphia: ASTM.

FMA/AAMA. 2012. FMA/AAMA Standard 100, *Standard Practice for the Installation of Windows with Flanges or Mounting Fins in Wood Frame Construction for Extreme Wind/Water Conditions*. Tallahassee, FL: FMA and Schaumberg, IL: AAMA.

FMA/AAMA. 2012. FMA/AAMA Standard 200, *Standard Practice for the Installation of Windows with Frontal Flanges for Surface Barrier Masonry Construction for Extreme Wind/Water Conditions*. Tallahassee, FL: FMA and Schaumberg, IL: AAMA.

FMA/AAMA/WDMA. 2012. FMA/AAMA Standard 300, *Standard Practice for the Installation of Doors in Wood Frame Construction with Extreme Wind/Water Conditions*. Tallahassee, FL: FMA, Schaumberg, IL: AAMA, and Chicago, IL: WDMA.

FMA/WDMA. 2010. FMA/WDMA Standard 250, *Standard Practice for the Installation of Non-Frontal Flange Windows with Mounting Flanges for Surface Barrier Masonry Construction for Extreme Wind/Water Contains*. Tallahassee, FL: FMA and Chicago, IL:WDMA.

Katsaros, J.D. 2007. Extreme exposure fenestration installations—the Florida challenge. ASTM E06 Symposium, "Up Against the Wall: An Examination of the Building envelope Interface Techniques and Systems, November 1, 2007, Tampa, Florida.

Lstiburek, J.W. 2005 Rainwater management performance of new constructed residential building enclosures during August and September 2004. Home Builders Association of Metro Orlando and the Florida Home Builders Association Report, Building Science Corporation, www.buildingscience.com.

RDH. 2002. *Building Engineering Limited, Vancouver, BC, Water Penetration Resistance of Windows – Study of Manufacturing, Building Design, Installation and Maintenance Factors*. Ottawa, Ontario: submitted to Canada Mortgage and Housing Corporation.

Wooden Stud Walls with Aerogel Thermal Insulation— Material Properties and HAM Simulations

Axel Berge **Carl-Eric Hagentoft, PhD** **Paula Wahlgren, PhD**

ABSTRACT

This study evaluates the heat and moisture performance of wooden stud walls with aerogel blankets as the main insulation. There is always a risk that a new material will create problems if it is applied in the same way as conventional materials. To predict and prevent such problems, it is of essential interest to analyze aerogel applications from a heat and moisture perspective. In this study, thermal conductivity and the vapor diffusivity of aerogel blankets were measured. The work found that the aerogel blankets can be compressed and that compression actually lowers the thermal conductivity. The measured results were used as input in numerical simulations. Four different wall layouts were modeled and compared to a wall with conventional insulation. The simulations showed that the thickness of a wooden stud wall could be decreased to 60% with aerogel blankets instead of conventional insulation, with the same thermal transmittance. Also, the position of the aerogel layers had a large influence on the moisture condition in the wall. The aerogel wall in the study with the best moisture performance has a homogeneous aerogel layer outside of the wooden studs and the studs are in line, forming a continuous thermal bridge through the wall. The moisture performance of the best aerogel wall in the study was as good as the studied wall with conventional insulation.

INTRODUCTION

Today there is a lot of concern regarding how humans affect the environment. One large source of impact on the environment is energy production; therefore, a large effort is made to lower the energy consumption through all of society. The Swedish Government (2013) aims to decrease energy consumption by 20% until the year 2020; similar goals can be seen throughout Europe.

The environmental concern has led to new concepts to lower the energy consumption of new buildings. One of the main factors for energy loss in buildings is the heat transmission through the building walls. On one hand, the thermal transmittance (U-factor) of a wall can be decreased by adding new layers of insulating materials, thus decreasing the heat losses. On the other hand, the wall thickness will increase, which reduces the available building area, which may lead to architectural problems such as decreased natural lightning. A change in wall layout will also lead to a change in the moisture conditions in the wall. Therefore, the moisture performance of a new wall should be evaluated to reveal any risks of moisture damage such as mold or rot.

During the last century, some new insulation materials have been developed with thermal properties superior to those of conventional insulation materials (Berge and Johansson 2012). One of these materials is aerogel, invented by Kistler (1931) during the thirties. An aerogel is a former gel which has been dried at supercritical conditions so that the original gel structure is kept intact. This can lead to a material with high porosity and a pore diameter in the range of 20–40 nm ($1 \cdot 10^{-6}$–$2 \cdot 10^{-6}$ in.) (Soleimani Dorcheh et al. 2008). The pore diameter is of the same magnitude as the mean free path of air, where the mean free path is the average length a gas molecule travels before colliding with another gas molecule. This leads to a decreased probability of gas-gas collisions compared to gas-solid collisions. Gas-solid collisions transfer less energy than gas-gas collisions, which leads to a situation where an air-filled

__Axel Berge__ is a doctoral student, __Carl-Eric Hagentoft__ is a full professor, and __Paula Wahlgren__ is a senior lecturer in the Division of Building Technology, Department of Civil and Environmental Engineering, Chalmers University of Technology, Göteborg, Sweden.

porous material can have a lower conductivity than still air with unlimited space. This has been shown by Fricke et al. (1992), who found a thermal conductivity for aerogel as low as 13 m·W/m/K (0.083 Btu·in./h/ft^2/°F), which is substantially lower than the conductivity of example mineral wool or expanded polystyrene insulation with a conductivity above 30 m·W/m/K (0.21 Btu in./h/ft^2/°F).

A problem with monolithic aerogels is that they are very fragile, which creates issues with handling the material. This has led to the creation of aerogel composites, where some kind of reinforcement fibers are mixed with the gel before drying (Stepanian et al. 2006). One type of composite is the so-called aerogel blanket, shown in Figure 1, which can be stored in rolls and can be handled like a thick fabric.

Today aerogel blankets are very expensive compared to other insulation materials, but the price is decreasing and has been forecast to decrease even more in the future. Therefore, it is of interest to find the benefits that a low thermal conductivity could yield and to investigate possible problems that could occur.

Aim

This work aims to investigate aerogel blankets as a possible insulation material for walls. The work has been divided into two parts: the first part is an analysis of the material properties to understand the material and the second part is numerical simulations of heat and moisture conditions in a load-bearing wooden stud wall element. This work generates the first insight into possible benefits from using aerogel blankets as the primary insulation in wooden stud walls.

Figure 1 *A roll of aerogel blanket with a nominal thickness of 10 mm (0.39 in.).*

MATERIAL PROPERTIES OF AEROGEL BLANKETS

To simulate the use of aerogel blankets in a wall structure, various material properties have to be known to use as input data in the model. For this study, the most important properties are those related to heat and moisture transport. One problem with determining the properties is the variation through the specimens. Since aerogel is a fragile material, the aerogel in the outer layers of the blanket will break and fall off as dust, leaving only reinforcement fibers. Deeper into the material, the fiber reinforcement keeps the aerogel matrix together. For simulation input you have to choose between apparent properties, representing the whole material piece, or you have to try to measure the properties of the different layers of the specimen.

The measurements were focused on one type of aerogel blanket with a low thermal conductivity. The results were compared to another aerogel blanket covered with a dust protection coating. The coating consisted of some kind of polymer applied to the surface of the blanket to stop aerogel particles from falling off. The properties of the blankets differ to some extent, and while no clear conclusions about the effect of adding a dust coating can be drawn, it can give indications. Some basic properties of the aerogel blankets are shown in Table 1. The specimen thickness was measured according to EN 823, *Thermal insulation products for building applications – determination of thickness* (CEN 1994).

Thermal Conductivity Measurements of Aerogel Blankets

The thermal conductivity was measured in a guarded heat flowmeter apparatus. A 300 × 300 mm^2 (11.8 × 11.8 in.^{2}) large sample was put in between two plates with different temperatures. To decrease the measurement error from the sample thickness, three layers of aerogel blanket, stacked on top of each other, were measured corresponding to a total nominal thickness of 30 mm (1.18 in.). A heat flowmeter measured the heat flow from the hot plate into the material, through the middle 100 × 100 mm^2 (3.94 in.^{2}) part of the sample. Spacers were used to lock the sample thickness during the measurement. With the area, A, temperature difference, ΔT, heat

Table 1. Density and Thickness of Two Types of Aerogel Blanket Without (A) and With (B) a Dust Coating

Sample	Density, ρ		Measured Thickness, t		Nominal Thickness, t_{nom}	
	kg/m^3	lb/ft^3	mm	in.	mm	in.
A	157	9.81	10.6	0.413	10	0.394
B	147	9.19	10.2	0.401	10	0.394

flow, Q, and thickness, d, the thermal conductivity, λ, can be calculated from Equation 1.

$$\lambda = \frac{Q \cdot d}{A \cdot \Delta T} \quad (1)$$

Temperature Dependence of the Thermal Conductivity. Thermal conductivity measurements were performed at different mean temperatures, from 10°C to 40°C (50°F to 104°F), to examine the variation in thermal conductivity due to temperature. The results for blanket type A are shown in Figure 2, which gives an average conductivity of 15.4 m·W/m/K (0.107 Btu·in./h/ft²/°F) at 10°C (50°F) and a temperature variation for each degree of 0.02 (m·W/m/K)/K ($8 \cdot 10^{-5}$ [Btu·in./h/ft²/°F]/°F). The temperature influence on the thermal conductivity is quite small and can probably be neglected in a wall context. The result can be compared to the results by Pietruszka et al. (2012), who measured a thermal conductivity of aerogel blankets between 15.0 and 16.6 m·W/m/K (0.104 and 0.115 Btu·in./h/ft²/°F) at 10°C (50°F).

The thermal conductivity of the dust-coated aerogel blanket (B) was also measured; it gave a resulting thermal conductivity of 19 m·W/m/K (0.13 Btu·in./h/ft²/°F) at 10°C (50°F). This is much higher than the aerogel blanket without the coating, but it is not the same product so it cannot be compared directly.

Compression Dependence of the Thermal Conductivity. Aerogel is a fragile material, held together by the reinforcement fibers. On the surface of the material, the fibers cannot entangle the aerogel as much as in the core, therefore aerogel falls off as dust, leaving a surface layer with less aerogel and more air. Since the aerogel itself has a lower conductivity than still air, a compression of the specimen might decrease the measured conductivity by reducing the air-filled outer layers.

From the product technical guide for the aerogel blankets, the measured conductivity is 13.1 m·W/m/K for a sample compressed to 83% of the nominal thickness. To test the influence of thickness, three specimens of aerogel blanket were tested at a mean temperature of 10°C (50°F). The maximum thickness was approximately 32 mm (1.26 in.) and the minimum thickness was limited to approximately 29 mm (1.14 in.) by the possible compression in the metering equipment. The thermal conductivity of each specimen was measured at several different thicknesses, from the minimum thickness up to the measured sample thickness.

The results are shown in Figures 3 and 4 (in SI and I-P units, respectively), where it is shown that 10% compression gives 1 m·W/m/K (0.06 Btu·in./h/ft²/°F) lower thermal conductivity. Figures 3 and 4 also show the conductance through the whole thickness of the specimen. The conductance is defined as the heat flow through the material layer, normalized over area and temperature. Even though the thermal conductivity decreases with a lower thickness, the total flow still increases,

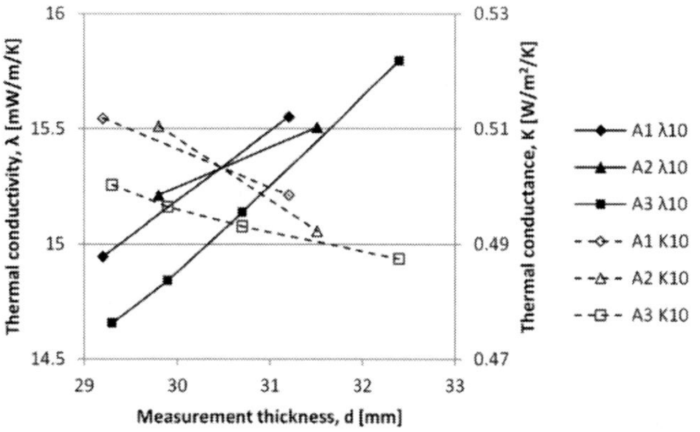

Figure 3 *Thermal conductivity and thermal conductance of samples A1–A3 as a function of the sample compression represented by the measured thickness (SI units).*

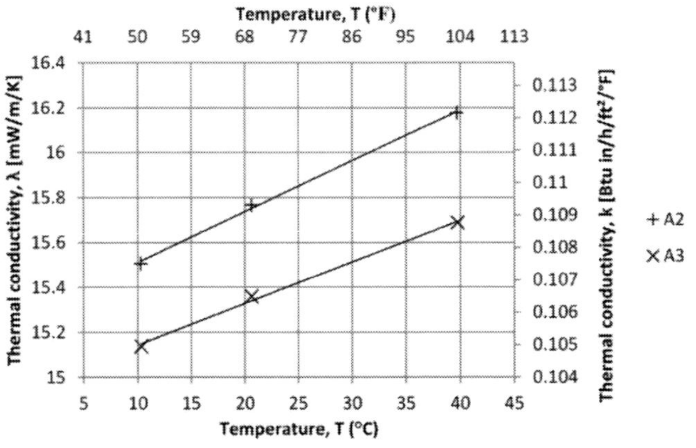

Figure 2 *Thermal conductivity of aerogel blankets as a function of temperature for two different specimens of the same type.*

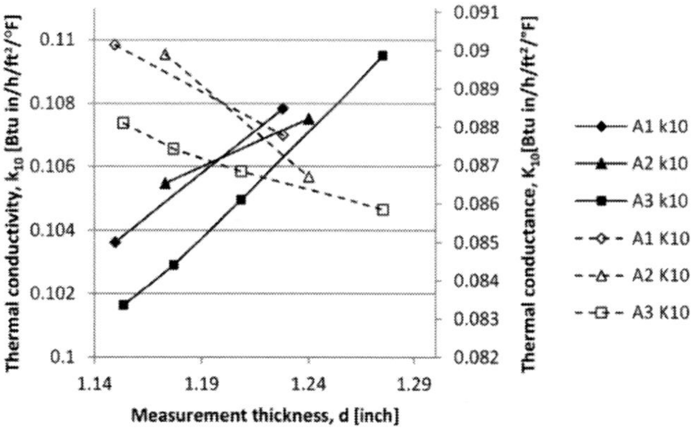

Figure 4 *Thermal conductivity and thermal conductance of samples A1–A3 as a function of the sample compression represented by the measured thickness (I-P units).*

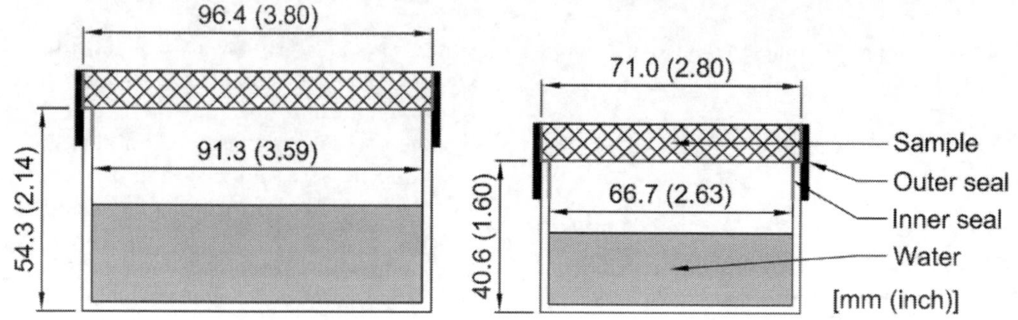

Figure 5 *Set-up for wet cup measurements. The cups are cylindrical.*

which means that an uncompressed aerogel blanket still gives a higher thermal resistance per invested dollar.

Vapor Diffusion Coefficient for Aerogel Blankets

The vapor diffusion properties of the aerogel blankets were measured by the cup method. Samples of aerogel blanket were mounted on a glass cup half filled with water. The samples were placed in a climate chamber at 20°C (68°F) and 50% relative humidity (RH). The samples were continuously weighed, and with the weights together with the measured time, the mass flow of vapor leaving through the sample could be calculated.

To analyze the possible increased flow through the cut circumference of the samples, two different sample sizes where tested. The two sample sizes had different relations between the area and the circumference so that a quotient of flow per edge length could be calculated.

The setups for the measurements are shown in Figure 5. The vapor diffusivity can then be calculated from Equation 2. From the diffusivity, the permeability can be calculated using Equation 3.

$$G = \frac{\Delta\varphi \cdot v_s \cdot A}{\dfrac{d_a}{D_{v,a}} + \dfrac{d_{aeb}}{D_{v,aeb}}} \tag{2}$$

$$\delta_v = \frac{\Delta\varphi \cdot v_s}{\Delta P_v} \cdot D_v \tag{3}$$

where

G	=	mass flow, kg/s (lb/h)
$\Delta\varphi$	=	difference in relative humidity, dimensionless
v_s	=	saturation vapor content, kg/m^3 (lb/ft^3)
A	=	sample area, m^2 (ft^2)
d_a , d_{aeb}	=	air gap thickness and aerogel sample thickness, respectively, m (in.)
$D_{v,a}$, $D_{v,aeb}$	=	vapor diffusivity of air and aerogel, respectively, m^2/s
δ_v	=	vapor permeability, kg/Pa/s/m (perm in.)
ΔP	=	difference in partial pressure, Pa (mm Hg)

Table 2. Measurement of Permeability on Small and Large Samples of Aerogel Blanket, Without (A) and With (B) a Dust Coating

Sample	Diffusivity, D_v, $\cdot 10^{-6}$ m^2/s	Permeability, δ_v	
		$\cdot 10^{-12}$ kg/Pa/s/m	Perm in.
A1 small	5.93	45	30.7
A2 small	5.54	42	28.6
A3 small	5.12	38	26.5
A_{avg} large	5.53	42	28.6
A4 large	5.32	40	27.5
A5 large	5.15	39	26.6
A6 large	5.67	43	29.3
A_{avg} large	5.38	40	27.8
$A_{tot.avg}$	5.46	41	28.2
B1 large	6.32	48	32.7
B2 large	6.61	50	34.2
B3 large	5.78	43	29.9
B_{avg}	6.24	47	32.2

The results from the measurements are shown in Table 2. The variation between the different sample sizes is of the same magnitude as the variation between samples of the same size, which indicates that the extra vapor flow through the edge is small compared to the total vapor flow through the sample. This gives an average permeability of $5.46 \cdot 10^{-6}$ m^2/s (28.2 perm in.).

The influence from a dust coating was also tested. The coated blanket (B) was of another type, with a higher thermal conductivity, but the results could indicate influence of a coating. The measurement showed that the coated samples had a higher permeability than the samples without the coating. This indicates that the type of coating used does not interrupt the vapor flow through the material. It can be assumed that it would be possible to coat the other aerogel blanket (A) with the same coating without any large influence on the moisture transport through the wall.

WALL ASSEMBLY SIMULATIONS

A common Swedish concept of a wall for low-energy buildings with conventional insulation was found from an insulation material producer and is shown as walls 1a and 1b in Figure 6. From the interior, the layers of the structure are:

- 13 mm (1/2 in.) gypsum board
- 45 mm (1.77 in.) layer for service penetrations with mineral wool and wooden studs
- vaportight film
- homogenous layer of mineral wool
- load-bearing layer with mineral wool and wooden studs
- homogenous mineral wool board
- 25 mm (0.99 in.) ventilated air gap
- wooden façade

The load-bearing studs are 120 × 45 mm^2 (4.73 × 1.77 in.^{2}) with a center distance of 600 mm (1.97 ft). This is based on the design load from snow on a 15 m (49 ft) wide roof placed in Göteborg, Sweden (Johannesson 1992).

When a wall gets thinner, the influence of the thermal bridge through structural elements increases. Therefore, two different placements of the wooden studs were analyzed: alternating as in the "a" cases or in line as in the "b" cases, shown in Figure 6. Walls 1a and 1b are the reference walls with conventional insulation materials. These are compared to two versions of aerogel walls with the same U-factor; wall type 2, where a homogenous aerogel layer is placed in between the layer for service penetrations and the load-bearing layer, and wall type 3, where a homogenous aerogel layer is placed on the outside of the construction, before the façade.

Thermal Analysis of the Wall Assemblies. The heat flows through the wall assemblies were simulated with the Heat Transfer Module of COMSOL Multiphysics® 4.2b (COMSOL 2013), a finite element method software, with constant bound-

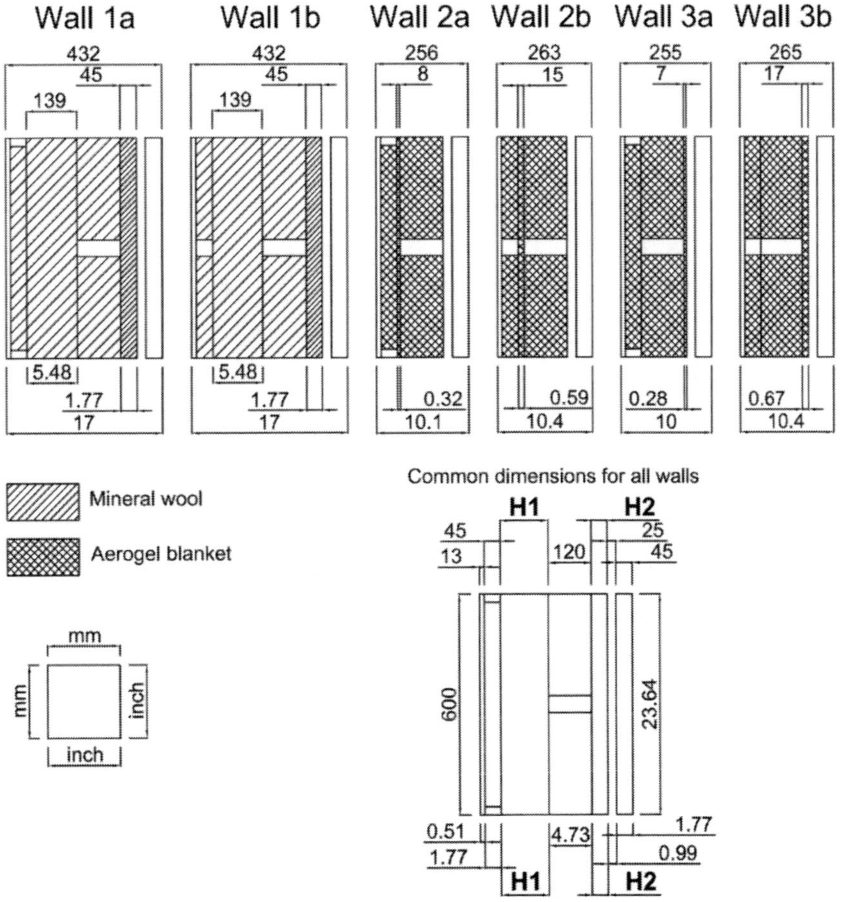

Figure 6 *Horizontal incisions of the insulated wall constructions. Walls 1a and 1b are insulated with conventional insulation and walls 2a, 2b, 3a, and 3b with aerogel blanket insulation. The wall to the bottom right shows the dimensions common for all walls, such as the load-bearing layer, and also the position of the two layers H1 and H2, for which the thickness was varied in the thermal analysis. The measures up and to the left are in millimeters and the measures down and to the right are in inches.*

ary temperatures. The U-factor was calculated and the thickness was altered until each wall had a U-factor of 0.100 W/m^2/K (0.0176 Btu/h/ft^2/°F) to represent a modern low-energy building. The thickness was varied for layer H1 for walls 1a, 1b, 2a, and 2b and for layer H2 for walls 3a and 3b. The positions of layers H1 and H2 are shown in the bottom right of Figure 6.

For the choice of walls, similarity in U-factor was prioritized rather than actual available material dimensions. Therefore, the adjusted homogenous layers have unconventional thicknesses.

The input data for the thermal analysis is shown in Table 3 and were taken from the material database in the simulation software WUFI 2D (IBP 2013), which was used for the moisture analysis. The air gap and the façade were treated as an extra resistance of 0.20 m^2·K/W (1.14 ft^2·h·°F/Btu) added to the exterior surface resistance. This gave internal and external heat transfer coefficients of 25 and 3 W/m^2/K (4.4 and 0.53 Btu/h/ft^2/°F), respectively (Petersson 2009).

The resulting wall thicknesses are shown in Table 4 together with the relative thickness compared to wall 1. With aerogel blanket insulation, the total wall thickness could be reduced to 60% of the original thickness for the same U-factor. For both the conventional walls and the aerogel blanket walls, the influence of the relative position of studs is small. For the conventional wall the position did not affect the thickness at all, and for the aerogel walls the thickness changed a maximum of 10 mm (0.39 in.), which correspond to a difference of around 4%.

Moisture Performance Analysis. The wall constructions were analyzed in a combined heat and moisture simulation in WUFI 2D (IBP 2013). The outdoor climate was varied over a repeated year of climate data for Göteborg, Sweden. The indoor climate was based on Standard EN 13788 (CEN 2012), one of the possible climate inputs in WUFI 2D (Zirkelbach et al. 2013). This gave an indoor temperature of 22°C (72°F) and an indoor moisture load of 4 g/m^3 (0.23 gr/gal) for outdoor temperatures below 0°C (32°F), decreasing linearly with temperature to zero at 20°C (68°F). The vaportight film, placed after the service gap in the walls, was chosen to have an equivalent air diffusion layer thickness of 1500 m (4900 ft).

The daily average temperature and relative humidity in the wall were monitored at eight different points, shown in Figure 7. From the temperature, the critical relative humidity for the wood elements was calculated according to Huuka and Viitanen (2009) and the mold growth potential was calculated as the quotient between the relative humidity and the critical relative humidity. Thus, a higher value indicates worse moisture conditions and a value above 1 indicates a risk of mold growth.

The results from the simulations are shown in Table 5. Walls 2a and 2b are the only walls for which the mold growth

Table 3. Thermal Conductivities Used for the Thermal Analysis of the Wall Assemblies

Material	Thermal Conductivity	
	m·W/m/K	Btu·in./h/ft^2/°F
Gypsum board	200	1.39
Mineral wool	36	0.25
Wooden studs	90	0.62
Mineral wool board	33	0.23
Aerogel blankets	15.4	0.107

Table 4. Simulated Wall Thickness and Relative Thickness Compared to Wall 1

Wall	Total Wall Thickness, *t*		Thickness Relative to Wall 1 (%)
	mm	in.	
1a	432	17.0	100.0
1b	432	17.0	100.0
2a	256	10.1	59.3
2b	263	10.4	60.9
3a	255	10.0	59.0
3b	265	10.4	61.3

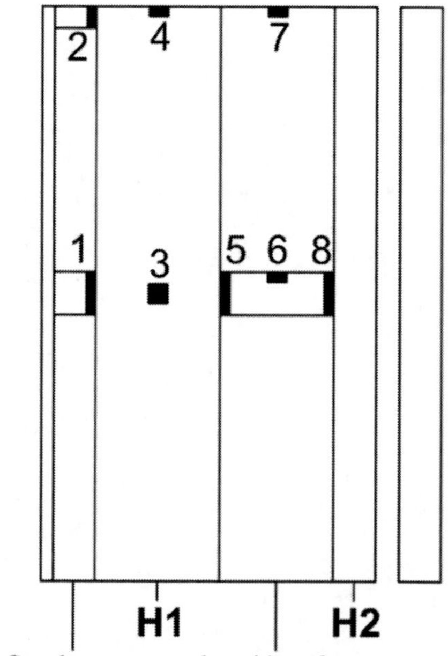

Figure 7 *The eight measurement positions in the heat and moisture simulations for the analyzed walls. This is a schematic version of the walls in Figure 6. The measurement points on the wooden studs were measured on the surface of the studs. Walls 3a and 3b do not have the layer H1, which is why there are no results for positions 3 and 4.*

Table 5. Resulting Minimum and Maximum Values for the Mold Growth Potential and the Relative Humidity for the Different Positions in the Analyzed Walls

	Wall Position							
	1	2	3	4	5	6	7	8
Wall 1a								
MP_{max}, dimensionless	0.74	0.71	0.74	0.75	0.79	0.82	0.85	0.92
MP_{min}, dimensionless	0.56	0.56	0.31	0.28	0.45	0.51	0.43	0.47
RH_{max}, %	60	57	60	60	63	65	68	75
RH_{min}, %	45	45	25	22	36	41	35	39
Wall 1b								
MP_{max}, dimensionless	0.71	0.74	0.74	0.75	0.78	0.82	0.85	0.92
MP_{min}, dimensionless	0.56	0.56	0.30	0.28	0.45	0.51	0.44	0.47
RH_{max}, %	57	59	59	60	63	65	68	75
RH_{min}, %	45	44	24	23	36	41	35	39
Wall 2a								
MP_{max}, dimensionless	0.93	0.72	0.72	0.75	0.74	0.82	0.84	1.03
MP_{min}, dimensionless	0.68	0.59	0.46	0.28	0.49	0.59	0.48	0.41
$MP > 1$, d	—	—	—	—	—	—	—	7 (3)[1]
RH_{max}, %	75	58	58	60	59	65	67	85
RH_{min}, %	54	47	37	22	40	48	39	36
Wall 2b								
MP_{max}, dimensionless	0.74	0.77	0.71	0.76	0.73	0.81	0.85	1.02
MP_{min}, dimensionless	0.62	0.62	0.40	0.34	0.47	0.58	0.50	0.41
$MP > 1$, d	—	—	—	—	—	—	—	6 (3)[1]
RH_{max}, %	59	62	57	61	59	65	68	84
RH_{min}, %	50	49	32	27	38	47	41	36
Wall 3a								
MP_{max}, dimensionless	0.96	0.72	—	—	0.69	0.78	0.82	0.95
MP_{min}, dimensionless	0.69	0.59	—	—	0.47	0.55	0.47	0.42
RH_{max}, %	77	58	—	—	55	62	66	77
RH_{min}, %	55	47	—	—	38	44	38	37
Wall 3b								
MP_{max}, dimensionless	0.74	0.77	—	—	0.65	0.74	0.82	0.88
MP_{min}, dimensionless	0.63	0.61	—	—	0.38	0.47	0.48	0.40
RH_{max}, %	59	62	—	—	52	59	65	70
RH_{min}, %	50	49	—	—	30	38	39	35

[1]Described as *x (y)* where *x* is the total number of days and *y* is the longest sequence of consecutive days.

potential exceeds 1—it occurs for 7 days and 6 days, respectively, with a maximum of 3 consecutive days for both walls. This occurs at position 8, which for these walls is next to the ventilated cavity behind the façade and thus close to outdoor conditions. For the outdoor climate the mold growth potential is more than 1 during 49 days of the year, with a maximum of 7 consecutive days. Therefore, outdoor wood is usually treated against moisture damage and requires more maintenance than wood inside the walls.

The mold growth potential maximums are visualized in Figure 8. Both the various walls and the different data collection positions are shown.

Walls 2a and 3a characterize themselves with a high mold growth potential maximum in position 1, close to the indoor condition. In these walls, the wooden studs are positioned alternately and, therefore, heat is let out from the point through the load-bearing wooden stud while the indoor heat is stopped by an aerogel layer. The temperature at the point decreases and consequently the relative humidity goes up. A higher indoor moisture load would increase the mold growth potential, possibly above 1.

Position 8 seems to be the worst position for all walls except wall 3a, where positions 1 and 8 have almost the same maximum. It is in position 8 that walls 2a and 2b exceed the mold growth potential limit of 1.

Wall 3b is the aerogel wall with the best moisture performance in the simulations. Compared to the conventional walls, walls 1a and 1b, the mold growth potential maximum in the outer parts of the wall is lower for wall 3b but higher in the inner parts of the wall. Though, as for walls 1a and 1b, the mold growth potential is far below 1 in all positions of wall 3b.

CONCLUSION

For aerogel blankets, the apparent thermal conductivity increases with thickness. This means that a compressed blanket has a better thermal conductivity. On the other hand, the total resistance of a blanket also decreases with compression and, consequently, an uncompressed blanket has a higher thermal resistance relative to the material cost. To model a wall assembly with aerogel you would have to describe how the blankets are simulated, compressed or noncompressed.

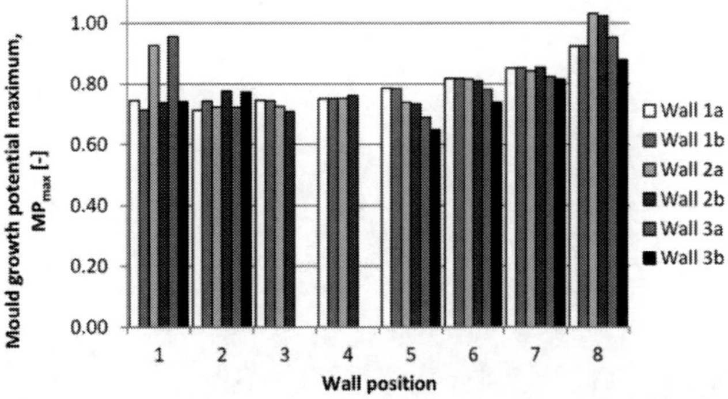

Figure 8 *Mold potential maximums for the different positions, shown in Figure 7, of the analyzed walls.*

From a pure thermal analysis of the aerogel blankets, the wall thickness of the studied load-bearing wooden stud walls could be decreased by around 60% by using aerogel blanket insulation instead of conventional insulation. The comparisons were made for a wall with a U-factor of 0.100 W/m^2/K (0.0176 Btu/h/ft^2/°F) with 120 × 45 mm^2 (4.7 × 1.8 in.^{2}) load-bearing wooden studs.

The heat and moisture simulations show that the position of the aerogel blankets in relation to the thermal bridges is relevant. The different aerogel blanket walls had a large variation in the moisture performance; for some positions the moisture condition was much worse than for the corresponding position in the walls with conventional insulation.

Wall 3b is the aerogel blanket wall that has the best moisture performance in this study (shown in Figure 6). The wooden studs in the service gap and in the load-bearing layer are positioned in line and a homogeneous layer of aerogel is put on the outside of the studs, just behind the façade.

The worst wall layout was found to be wall 2a (shown in Figure 6), for which the wooden studs in the service gap and the load-bearing layer were alternated, with a homogeneous aerogel layer placed in between. This construction had a high mold growth potential, both in the outmost parts of the wall and on the inside of the vaportight barrier. The performance was much worse than the performance for corresponding positions in the walls with conventional insulation.

ACKNOWLEDGMENTS

This work was a part of the project Homes for Tomorrow, mainly funded by FORMAS.

NOMENCLATURE

A	=	area, m^2 (ft^2)
D	=	vapor diffusivity, m^2/s
d	=	thickness, m (in. or ft)
G	=	mass flow, kg/s (lb/h)
K	=	thermal conductance, W/m^2/K (Btu/h/ft^2/°F)
MP	=	mold potential, dimensionless
P	=	pressure, Pa (mm Hg)
Q	=	heat flow, W (Btu/h)
RH	=	relative humidity, %
T	=	temperature, K or °C (°F)
U	=	thermal transmittance, W/m^2/K (Btu/h/ft^2/°F)
v	=	vapor content, kg/m^3 (lb/ft^3)
δ	=	permeability, kg/Pa/s/m (perm in.)
λ	=	thermal conductivity, W/m/K (Btu·in./h/ft^2/°F)
φ	=	relative humidity, dimensionless

Subscripts

a	=	air
aeb	=	aerogel blanket
$crit$	=	critical value
s	=	saturation
v	=	vapor

REFERENCES

Berge, A., and P. Johansson. 2012. Literature Review of High Performance Thermal Insulation (Report 2012:2), Department of Civil and Environmental Engineering, Chalmers University of Technology, Göteborg, Sweden.

COMSOL. 2013. COMSOL Multiphysics® 4.2b, Heat Transfer Module. COMSOL Inc.

CEN. 1994. Standard EN 823:1994, *Thermal insulation products for building applications – determination of thickness*. Brussels: European Committee for Standardization.

CEN. 2012. CEN Standard EN ISO 13788, *Hygrothermal performance of building components and building elements - Internal surface temperature to avoid critical surface humidity and interstitial condensation - Calculation methods*. Brussels: European Committee for Standardization.

Fricke, J., X. Lu, P. Wang, D. Büttner, and U. Heinemann. 1992. Optimization of monolithic silica aerogel insulants. *International Journal of Heat and Mass Transfer* 35(9):2305–2309.

Hukka E., and H.A. Viitanen. 1999. A mathematical model of mould growth on wooden material. *Wood Science and Technology* 33.

Johannesson, C. 1992. *Träbyggnadshandbok 3: Väggar* (*Wood Building Handbook 3: Walls*). Tryckeriteknik i Malmö AB, Malmö. (ISBN 91-85576-22-0)

Kistler, S.S. 1931. Coherent expanded aerogels. *Journal of Physical Chemistry* 36(1):52–64.

Petersson, B.-Å. 2009. *Tillämpad byggnadsfysik*, 4th ed. Lund: Studentlitteratur. (ISBN: 978-91-44-05817-7)

Pietruszka, B., J. Babinska, and R. Gerylo. 2012. Aerogel-based thermal insulation materials - structure and properties. *Proceedings of the 5th IBPC. International Building Physics Conference, Kyoto*, pp. 117–23.

Soleimani Dorcheh, A., and M. Abbasi. 2008. Silica aerogel: Synthesis, properties and characterization. *Journal Of Materials Processing Technology* 199(1–3):10–26.

Stepanian, C., G. Gould, and R. Begag. 2006. Aerogel composite with fibrous batting. Pat. No. US 7,504,346.

Swedish Government. 2013. Mål och åtgärder (Objectives and measures). www.regeringen.se Ansvarsområden > Miljö, energi och klimat > Klimat > Nationellt klimatarbete (available: 2013-01-17).

IBP. 2013. WUFI® 2D. Stuttgart: Fraunhofer IBP. www.wufi.com

Zirkelbach, D., T. Schmidt, H. Künzel, M. Kehrer, and C. Bludau. 2013. WUFI 2D Calculation example step by step. www.wufi.com > Downloads > WUFI 2D example [available: 2013-01-31].

Performance of a Retrofitted 1950s Multi-Unit Residential Building—Measurements and Calculated Transient Hygrothermal Behaviour

Carl-Magnus Capener, PhD **Kenneth Sandin, PhD**

ABSTRACT

In this study, a Swedish multi-unit residential building built in the 1950s is investigated. The house, situated in the Gothenburg metropolitan area, with a lightweight concrete structure and brick cladding, was severely affected by weathering and had accumulated moisture in its structure, leading to high moisture content throughout the walls. In addition, the energy performance of the building was poor. To improve its hygrothermal performance, an exterior insulation and finishing (EIF) system with mineral wool and a thick layer of mineral scratch rendering (stucco) was chosen to not alter the time typical aesthetics of the 1950s house and cultural historic neighbourhood.

The walls most affected by moisture were instrumented with wireless hygrothermal gauges and monitored before, during, and for more than one year after the retrofitting. This paper presents the final results from the in-situ measurements and compares the findings to calculated transient hygrothermal simulations of the façade as well as the whole building. The thermal properties of the building components and their impact on heat losses are evaluated as well as the energy performance for the whole building by the energetic building simulation.

The results show that the chosen EIF system facilitated a quick drying of the accumulated moisture in the autoclaved aerated concrete (AAC) and that simulations are in accordance with monitored hygrothermal behaviour, both for the façade and the whole building simulations. It is also shown that the effect of the moisture taken up by the façades on energy performance can be estimated using a holistic whole building simulation model.

INTRODUCTION

This paper deals with the refurbishment of a 1950s multi-dwelling residential building situated in the Gothenburg metropolitan area. The energy performance of multi-dwelling blocks from the 1950s in Sweden is quite poor and external thermal insulation composite systems are often proposed to improve the thermal performance of the walls of these buildings, but the hygrothermal effects caused by the retrofitting have not been fully investigated. The studied building, see Figure 1, has a brick cladding façade with an autoclaved aerated concrete (AAC) structure behind. The façade was in a poor state and severely affected by weathering with deteriorated mortar joints and moisture accumulating in its structure. The building supervisor reported high moisture content throughout the walls in some apartments with possible hygienic problems and health risks for the tenants due to mould growth at the interior.

Apart from moisture related damages, high moisture content in the walls also negatively affect the thermal properties of the AAC, as thermal and hygric behavior of porous building materials are closely interconnected (Sandin 1984). In particular, the south facing walls suffered from moisture damages and several of the buildings apartments had been affected. The property owners had over the previous years unsuccessfully attempted to treat the exterior surface with different hydrophobing agents to reduce the water uptake and also tried to raise the indoor temperature of the apartments to

***Carl-Magnus Capener** is a researcher at the Department of Energy Technology at SP Technical Research Institute of Sweden, Boras, Sweden. **Kenneth Sandin** is an associate professor at the Department for Building and Environmental Technology at Lund University in Sweden, Lund, Sweden.*

a) b)

Figure 1 *The investigated building before retrofitting (photos taken early March 2010): (a) west façade with gable and (b) details of the south façade's bay windows.*

drive out the moisture. This resulted in a high energy demand and proved not to be a sustainable solution.

With the aim of stopping the water uptake by the AAC, rapidly improving the façades hygrothermal performance, and facilitating drying of the lightweight concrete, an exterior insulation and finishing (EIF) system (EIFS; in Europe referred to as ETICS, or external thermal insulation composite system) with 50 mm mineral wool and a thick layer of mineral scratch rendering was chosen after a first hygrothermal investigation. This investigation revealed a much faster drying process of the selected EIFS compared to traditional EIFS using expanded polystyrene (EPS) insulation. Another reason for the choice of cladding with the mineral scratch rendering was not to alter the time-typical aesthetics of the 1950s house and its place in the cultural historic neighborhood. The original windows were also replaced during the renovation as they were in a poor state and had low thermal performance. The building after the completed retrofit is shown in Figure 2.

This study presents the final results of a three-year study where the hygrothermal properties of the retrofitted walls depending on choice of EIF system have been investigated. The final results from the field measurements in the monitored building are shown and compared with modeled hygrothermal simulations. The effects of moisture on the transient thermal transmission through the walls are also presented, elaborating on the difference found between different EIF systems and choice of materials. The building's energy demand for space heating supplied by district heating, pre- and post-retrofit, is presented along with the results of the whole building simulation.

FIELD MEASUREMENTS

The original building's walls consisted of 20 cm of AAC and had a half stone (65 mm) brick cladding façade more or less directly adjacent without any proper air gap in between. With deteriorating bricks and brick mortar, this contributed to the strong water uptake by the lightweight concrete and subsequent indoor problems.

The renovation started August 2010 when the building was weather protected, see Figure 3, and the refurbishment of the walls began. Using hydrophobic and vapor permeable insulation materials for exterior insulation and in EIFS is a very effective way of preventing further rainwater from reaching the inner walls and will enable the walls to dry out. The installed EIF system consisted of a 50 mm dense mineral wool board (85 kg/m^3), and two layers of render (total 20 mm) on top, a base coat with a stainless steel mesh embedded, and a scratch coat.

The original dual glazed windows with an approximate U-factor of $U = 3$ W/(m^2·K) were replaced by more energy efficient triple glazed windows, $U = 1.0$ W/(m^2·K).

The field measurements included an inventory performed end of March 2010 with reference core samples taken from the inside of the building through the aerated concrete of the south and west facing walls at the second and third floor. The moisture content was then established for different depths of the walls by the laboratory at the Department of Building Mate-

a) b)

Figure 2 *The investigated building after completion of the retrofitting (photos taken September 2011): (a) complete with new windows and EIF system and (b) details of the south facades bay windows.*

a) b)

Figure 3 *(a) Weather protection during renovations and (b) selected EIF system used for the retrofit of the building. Two layers of stucco on a mineral wool board attached by mechanical fasteners and metal mesh.*

rials at Lund Technical University. To minimize the disturbance of the tenants, a wireless system for continuous monitoring was installed at the same time allowing instant access to sensor readings.

Field measurements and hygrothermal monitoring of autoclaved aerated concrete have in the past been carefully investigated (Sandin 1987) and the experience from these results guided the placement of the monitoring equipment. The south and west facing walls most affected by weathering were instrumented in two apartments: a south facing wall in the first apartment on the third floor and two walls in the second apartment on the second floor, facing south and west respectively, see Figure 4. Temperature and relative humidity at different depths of the lightweight concrete was measured by wireless hygrothermal gauges as well as moisture content (wood moisture equivalent, WME) at 5 cm depth in the AAC.

The equipment used for the wireless monitoring was the GE HygroTrac v2 sensors. They allow for three gauges at

Figure 4 *Original drawings of the building's (a) south and (b) west facing facades. The apartments' locations and approximate monitoring points are marked with red.*

each transmitter: two RH/temperature sensors and a moisture (WME) 2-pin probe. The sensors were calibrated by laboratory staff at Lund University. For very high relative humidities in the AAC, >95%, the sensors are not very precise, as is shown later in the results. The data from the wireless sensors were received by a gateway placed in the basement of the studied building and forwarded by a 3G router to a server for storage.

In order to monitor the conditions pre- as well as post-retrofit, the in-situ measurements started in March 2010 and were finalized early 2013. Energy for space heating and domestic hot water by district heating was also monitored and data was delivered by the property owner.

HYGROTHERMAL SIMULATIONS

The demonstration building has also been modeled by the hygrothermal software WUFI Pro and WUFI Plus. This software has been experimentally verified for many types of building component assemblies (Künzel 1995; Karagiozis 2001) and similar setups (Tariku 2006) as the studied building and is well established. Before the renovations began and in order to select a suitable EIF system, the existing building façade walls and two possible EIF system solutions were analyzed, including the installed one. The alternative EIF system (ETICS) is a more common EPS insulation and acrylic render system. Material data and initial moisture conditions were supplied by material producers as well as from material databases such as MASEA Datenbank and the IBP Fraunhofer Material Database and from measurements.

Precise historic material data is always lacking and, as the studied building originates from the early 1950s, appropriate assumptions had to be made. As many materials age, their properties will change with time. For instance, there occur microstructure changes in AAC during carbonation as reported by Matsushita et al. (2004) and these will most likely affect hygrothermal properties such as sorption isotherms and thermal conductivity.

For the comparative hygrothermal analysis of the walls and subsequent retrofitting, normalized climatic data supplied by Lund University was used for the Gothenburg outdoor climate. For initial moisture conditions of the AAC, the results from the core sample measurements were used as shown in Table 1. To simulate the leaky mortar joints of the brick façade, a moisture source of 2% of the driving rain was added to the outer layer of the AAC to redistribute to the wall and raise the moisture levels to values presented in Table 1, at least once a year. For the interior, a vapor resistance was added to simulate a few layers of paint or vapor-tight wallpaper, as was found in the actual apartments.

Energy calculations were also performed with WUFI Plus. This software can simulate whole buildings and introduce the effect of built-in moisture on heat flows and heating demand. The studied building in Gothenburg was modeled from its original state (Figure 5), and with different renovation measures: window change, additional insulation, and combined window change with EIFS. The different measures were analyzed and the results studied along with the effect of moisture in the walls. Again, the measured moisture contents were used for the calculations.

RESULTS AND DISCUSSION

From the field monitoring, core sample analysis, and the simulations, interesting results were revealed. First, the autoclaved aerated concrete core sample from the south wall that had been segmented, sealed in glass containers, and brought to laboratory for testing showed elevated moisture contents throughout the wall, see Table 1. Moisture contents reaching almost 30% by weight was found in the outer parts of the wall, indicating leakage and water uptake by the AAC. This was well anticipated as some apartments with south facing walls had reported much more severe moisture problems earlier with damp surfaces on the inside of the apartments' walls. This was not seen during the inventory which took place in early spring, and some drying had probably occurred during the winter and heating season.

Table 1. Results from AAC Core Sample—Moisture Content by Weight of the AAC Measured at Different Depths from the Interior of the South Facing Wall

Depth from Interior, cm	Average Moisture Content, M-%
0–3	19,1
4–8	24,1
8–12	25,0
13–16	26,7
17–19	29,0

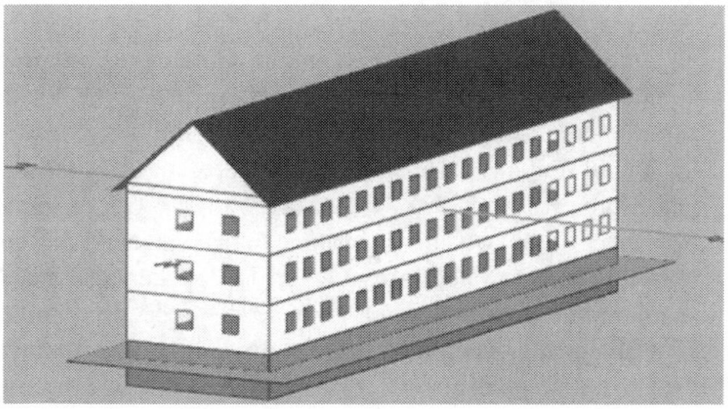

Figure 5 *Model in WUFI Plus of the studied building pre-retrofit, seen from southwest.*

From the hygrothermal simulations study, it is clear that the choice of EIFS assembly has a great effect on the drying of moisture from the AAC, see Figure 6, but also on the transient U-factors. The drying rate was much slower and moisture content higher for the EPS and acrylic stucco because both materials are less permeable.

The field monitoring revealed varying moisture conditions in the walls with height of building and direction of exposed outer surface. Figure 7 shows the measured relative humidity and temperature as well as the simulated relative humidity of the south facing façade at the outer most part of the AAC, which was also the most humid point of all the measurement points. It is clear that the EIF system installed allows for a rapid drying of the AAC. The rate of drying is visibly affected by seasonal changes with different outdoor moisture content in the air. A high rate of drying was seen in the first winter period and this was repeated the next year. As the relative humidity in the AAC decreases, the corresponding moisture content goes down as well. The reason for the apparently slower drying rate in the first winter is partly due to the limitations of the RH sensors for very high relative humidities, but also due to the high specific moisture capacity of AAC at high relative humidities compared to a much lower at relative humidities below ~90%.

To illustrate the amounts of moisture this correlates to, Figure 8 display a desorption isotherm for a typical AAC based on data from Kumaran et al. (2002). The corresponding moisture contents (M-%) of the AAC from the RH measurements in Figure 7 are highlighted from the beginning of the first drying to the end of the measurements (i.e., from approximately 30% to less than 5%).

Heating of domestic hot water (DHW) is seldom measured separately in older buildings. Based on previous research on tenants' energy behaviour it is possible to estimate approximately the use of DHW of the tenants as suggested by Warfvinge (2008) who studied Swedish residents' use of DHW and give approximate values on the energy for producing DHW. In Figure 9 the

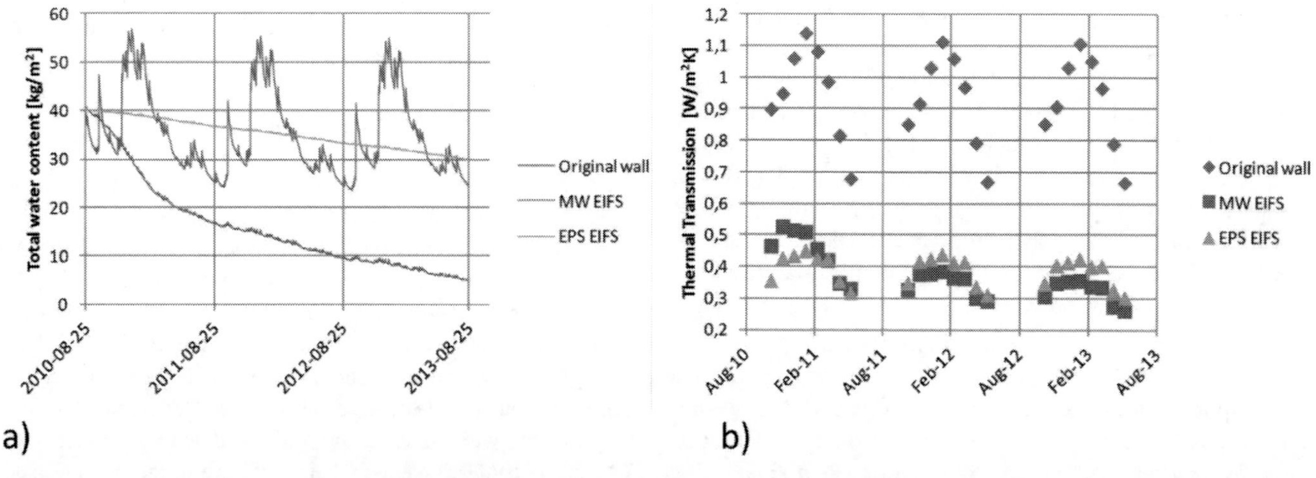

Figure 6 *(a) Total water content of the walls as a function of time for all three cases simulated and (b) corresponding monthly transient U-factors.*

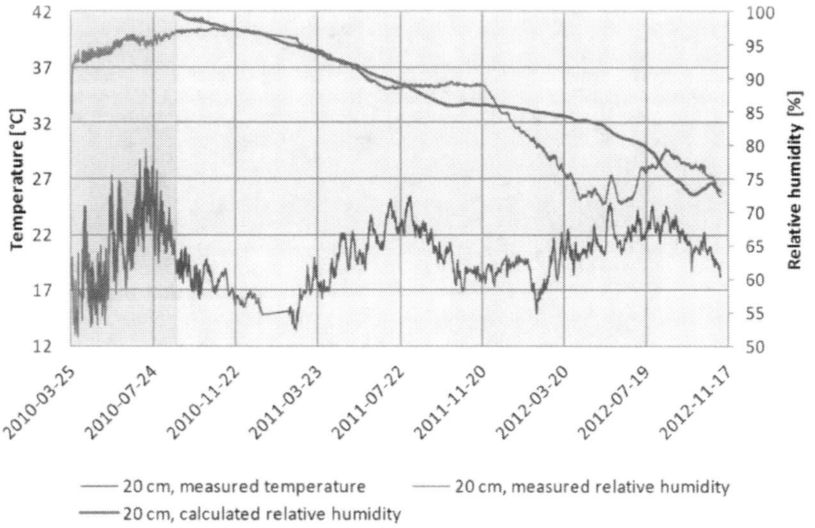

Figure 7 *Measured relative humidity and temperature as well as calculated relative humidity for the outermost point (20 cm) in the AAC of the south wall, third floor.*

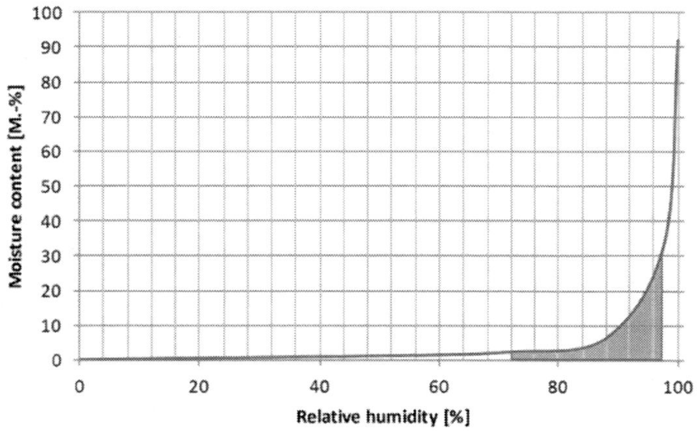

Figure 8 *Example of an AAC desorption isotherm. The corresponding moisture content of the AAC from the RH measurements in Figure 7 is highlighted from the beginning of the first drying to the end of the measurement.*

degree day corrected heating demand (without DHW) is presented for a five year period, including both pre- and post-retrofit. It is clear that approximations of DHW are not precise and vary over the building stock and individual houses depending on the demographics of tenants. With the above approximation even the summer month of July shows a need for space heating which is, of course, not so. The July value is most likely a part of the DHW and should be added to this and deducted from the heating demand.

The degree day corrected values indicate a reduction in supplied energy by 25% from 2010 to 2011. The measured average 2009/2010 to 2011/2012 shows a reduction by 22%. The cold month of January 2011 (post-retrofit) compared to the same month 2010 (pre-retrofit) showed a drop in energy by 41%, showing a significant reduction in peak power.

In Table 2 the heating demand (without DHW) for the whole building is shown; the measured demand supplied by the energy company (in red), pre- and post-retrofit, but also the calculated, with and without the effect of moisture on the heat flows, as simulated by WUFI Plus. Individual measures are also shown. To study the effect of the high moisture contents of the south facing wall, an average moisture content equivalent measured in the AAC was applied for some cases.

CONCLUSION

AAC can hold large amounts of water and it is crucial that excess moisture is allowed to dry out. The material's thermal conductivity will be much higher as long as the material contains excess water and proper drying must be facilitated. The chosen EIF-system facilitated a quick drying of the accu-

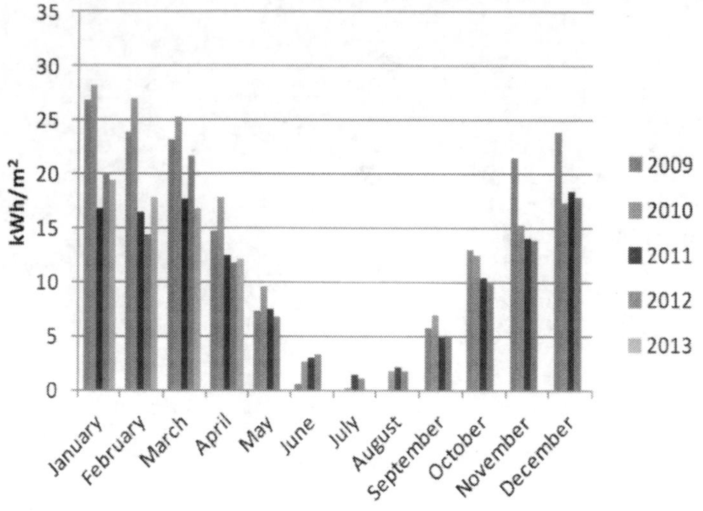

Figure 9 *Measured values of supplied energy by district heating for space heating (district heating net of DHW). Values 2009–2010 are mainly pre-retrofit and are marked in green. However, for the months of November and December 2010 the walls were already insulated and can be considered as post-retrofit. Values 2011–2013 are post-retrofit and marked in blue.*

Table 2. Measured and Calculated Heating Demand with Different Renovation Measures and Considering the Effect of Moisture on Heat Flows

Case	kWh/m²·year
Pre-retrofit, measured average 2009/2010	161,5
Pre-retrofit, calculated, with moisture	162,3
Pre-retrofit, calculated, without moisture	157,1
Window change, calculated, with moisture	147,9
Window change, calculated, without moisture	142,7
With 50 mm EIFS, calculated, without moisture	137,7
Window change and 50 mm EIFS, calculated, without moisture	123,3
Post-retrofit, with window change and 50 mm EIFS, measured average 2011/2012	126,0

mulated moisture in the AAC and simulations are well in accordance with monitored hygrothermal behaviour with some seasonal differences. For very high relative humidities, >95%, the RH sensors only give an indication of the moisture content and the actual amount can be much higher. Core samples can supplement this data for reference and initial values.

There was a clear reduction in heating demand after the retrofit and a significant reduction in peak power. Calculated and measured energy figures are well in accordance, but the effect of moisture on energy demand is more difficult to estimate precisely for older buildings as a more comprehensive picture of the walls moisture content needs to be established. The results do give a good indication of the effect of elevated moisture contents in the walls and it is clear that excess moisture in AAC significantly increases the energy demand.

ACKNOWLEDGMENTS

The authors would like to acknowledge those who funded this study, including the support from the research and innovation programme CERBOF initiated by the Swedish Energy Agency, as well as funding from Weber Saint-Gobain Construction Products Sweden. SBUF, the Swedish construction industry's organisation for research and development has also contributed with funding and the property owner, Göteborgs stads bostadsaktiebolag, Bostadsbolaget AB, has been very helpful facilitating the field measurements before, during and after the renovation of their building.

REFERENCES

Karagiozis, A., Künzel, H.M., Holm, A. 2001. WUFI ORNL/IBP – A North American Hygrothermal Model: Proceedings of the Thermal Performance of the Exterior Envelopes of Whole Buildings VIII (Buildings VIII), Clearwater Beach, Florida, December 2-7, 2001.

Kumaran, M.K., Lackey, J.C., Normandin, N., Tariku, F., v.Reenen, D. 2002. A Thermal and Moisture Transport Property Database for Common Building and Insulating Materials – Final Report from ASHRAE Research Project 1018-RP.

Künzel, H.M. 1995. Simultaneous Heat and Moisture Transport in Building Components – One- and two-dimensional calculation using simple parameters, IRB Verlag, Germany.

Künzel, H.M. 1998. Effect of interior and exterior insulation on the hygrothermal behaviour of exposed walls, Materials and Structures/Materiaux et Constructions, Vol. 31, March 1998, pp 99-103.

Sandin, K. 1984. Fuktens inverkan på energiflödet genom ytterväggar, Avdelningen för Byggnadsmateriallära, Lunds Tekniska Högskola, Rapport TVBM-3019, Lund, Sweden.

Sandin, K. 1987. Fukttillstånd i autoklaverade lättbetongväggar – Fältmätning av slagregnets och ytskiktets inverkan, Avdelningen för Byggnadsmateriallära, Lunds Tekniska Högskola, Rapport TVBM-3026, Lund, Sweden.

Tariku, F., Kumaran, M.K. 2006. Hygrothermal modelling of aerated concrete wall and comparison with field experiment, Proceedings of the 3rd International Building Physics Conference, Montreal, Canada, August 27, 2006, pp 321-328.

Warfvinge, C. 2008. Mycket energi att spara i miljonprogrammet, VVS-Forum, April 2008, pp 7-12.

Monitored Performance of Ultra-High Performance Residential Envelopes in the Pacific Northwest Marine Climate

David G. Hales
Member ASHRAE

Luke Howard

ABSTRACT

Case studies of three ultra-high performance residential envelopes monitored as part of the Building America program are examined. Three occupied homes in the Pacific Northwest Marine climate were monitored for at least a year. Two homes met the European Passive House standard. Monitoring included total energy use, heating, cooling, DHW, room temperatures, and relative humidity. The challenge of economically meeting very low design loads while meeting comfort expectations is discussed. The homes were all approximately 2000 ft^2 (186 m^2) with three bedrooms and two baths. They all used slab-on-grade construction with either SIPS panels or double wall construction. The studies present a cost benefit analysis of the high performance envelope measures including high R-value building assemblies, ultra-tight air leakage control (< 0.6 ACH50), and high performance fenestration. Total energy use ranged to less than 20.5 kBtu (6,000 kWh) annually. Space conditioning represented less than 25% of total energy use while loads other than space conditioning and DHW were over 60% of total annual energy use.

INTRODUCTION

The Washington State University Extension Energy Program (WSU-EEP) has worked with builders in the marine and cold climates of the Pacific Northwest to develop exceptionally efficient residential construction practices providing energy efficiency for over 30 years. With support from the BIRA and BA-PIRC teams funded by Building America, the performance of three high performance homes built in 2009 to 2012 was monitored. The monitoring documented actual energy performance, environmental conditions affecting occupant comfort, and, when possible, incremental costs of measures enhancing performance. Design assistance was not provided by the Building America teams. Monitoring was initiated to evaluate the output of the Passive House design process as supported by the Passive House Institute of the United States (PHIUS) in the Olympia and Hood River projects and to evaluate alternative building practices in a beyond code home in the Portland case.

The homes were constructed in Portland and Hood River, Oregon and Olympia, Washington by three different custom home builders with very different backgrounds and levels of experience. The Portland project was driven by the homeowner whose contractor's experience had been primarily as a remodeler. This was the contractor's first new home. The Hood River project was built by a contractor for himself whose experience started over 25 years ago in Europe. The Olympia project was built by an established design build firm as their first effort to meet the European Passive House standard after training from the Passive House Institute of the United States. All the designs were developed by licensed architects and the Hood River and Olympia projects were designed using the 2007 American version of the Passive House Planning Package (PHPP) with the intent of achieving PHIUS certification.

All three projects shared a similar approach in attempting to reduce overall space conditioning loads by enhancements to the building enclosure: increased insulation levels for all building assemblies, improved fenestration, reduced thermal bridging through assemblies, and extreme reductions in air leakage. The projects shared the further assump-

***David G. Hales** is a building systems and energy specialist at the Washington State University Extension Energy Program in Spokane, WA. **Luke Howard** is a building science specialist at the Washington State University Extension Energy Program in Olympia, WA.*

Published by ASHRAE 2013

tion that with a sufficient reduction in the design loads, offsetting savings might be obtained from the use of alternative, extremely downsized space conditioning systems.

Pictures of the three completed projects are seen in Figure 1.

EXPERIMENTAL DESIGN

Monitoring at each site was designed to determine total site energy use, space conditioning energy, domestic hot water energy, and other loads. The Portland and Hood River sites both use ductless mini-split heat pumps for space conditioning with zonal electrical resistance backup. Monitoring was designed to determine the amount of energy used by each heat source. The Olympia site provided heating with a water to air heat exchanger coil integrated with the heat recovery ventilation system. The heat source is a gas fired condensing tankless water heater that also provides the domestic hot water.

The Portland and Hood River sites both utilize solar thermal domestic hot water systems with conventional electric tank type water heaters providing backup. Monitoring tracked total energy use for hot water and the fraction provided by the solar systems.

The Olympia project also has on-site generation from a photo electric system. The contribution of this system was also tracked.

To assess comfort levels provided by these non-conventional space conditioning systems, temperature and relative humidity levels were monitored at multiple points within all three homes. Primary comfort questions concern the possibility of overheating during summer months and the ability of the point source space conditioning systems in the Portland and Hood River projects to maintain comfort throughout the entire structure.

Data collection was to run for at least 1 year.

The builders were asked to provide construction costs broken out by measures that were employed to enhance the overall energy performance of the projects. Cost of the building enclosure was broken down to fenestration, wall assemblies, foundations, ceilings/roofs, and air sealing. Costs of components of the mechanical systems including space conditioning, ventilation, and water heating affecting energy performance were also collected.

CLIMATE

The average climatic conditions at the three sites are quite similar as seen in Table 1. The values displayed are a current five year average at the nearest airport weather site. The areas are all characterized by wet, relatively mild winters and dry summers.

BUILDING CHARACTERISTICS

All three projects share a number of similarities. They are all two story structures, have approximately 2000 ft^2 (186 m^2) of conditioned space with three bedrooms, are built with slab-on-grade foundations and use heat recovery ventilators. The building enclosure in Portland uses structurally insulated panels (SIP) for the walls, floor and ceiling. The Hood River project also uses SIP for walls and ceiling but adds an additional 4 in. (102 mm) layer of expanded polystyrene sheeting for increased insulation. The Olympia project employs double wall construction, rigid foam sub-slab insulation with a trussed roof, and blown attic insulation. Table 2 compares the overall characteristics of the three projects. Figure 2 shows a cross sectional detail of the insulation placement for each project.

Portland

Hood River

Olympia

Figure 1 *Completed homes.*

Table 1. Climate Conditions

Site	$HDD_{65(18)}$	$CDD_{70(21)}$	Annual Precipitation, in. (mm)
Hood River Passive House	5902 (3279)	309 (172)	30.0 (762)
Olympia Passive House	5944 (3302)	161 (89)	50.6 (1285)
Portland SIPs House	4785 (2658)	292 (162)	43.1 (1095)

Table 2. Building Characteristics

Component	Hood River	Portland	Olympia
Slab-on-grade, U-factor	0.023 (0.131)*	0.033 (0.189)	0.020 (0.114)
Walls, U-factor	0.0198 (0.112)*	0.033 (0.189)	0.0211 (0.119)
Ceiling, U-factor	0.0130 (0.189)*	0.022 (0.126)	0.0125 (0.0710)
Windows, U-factor/SHGC	0.20 (1.14)/.63 south 0.14 (0.795)/.36 north	0.31 (1.76)/.30	0.20 (1.14)/0.55
Doors, U-factor	0.13 (0.738)*	.20 (1.14)	0.14 (0.795)
Air Sealing, ACH50	0.35	2.4	0.46
Heating	DHP single head HSPF = 11 + ER	(2) DHP HSPF = 10 + ER	96% Gas Tankless w/circulation through HRV
Cooling	SEER 22	SEER 18	None
Domestic Hot Water	Solar Thermal	Solar Thermal	96% Gas Tankless
Ventilation	HRV	HRV	HRV
Lighting	100% CFL + LED	100% CFL	80% CFL
Conditioned Area, ft^2 (m^2)	2004(186)	2089 (194)	1935 (180)
Glazing Area, ft^2 (m^2)	600(55.7)	371 (34.5)	295 (27.4)
Energy Source	Electric + Solar	Electric + Solar	Electric + Gas + Solar
Occupants	2	2	4

* Fenestration U-values for the Hood River site were taken from PHPP inputs and are not National Fenestration Rating Council values.

DATA COLLECTION

The Portland project was completed in the fall of 2009 and the reported data were collected during the 2010 calendar year. The Olympia house was completed in May of 2011 and the reported data were collected from September 2011 through July 2012. Started in 2009, the Hood River project was not completed until August of 2012 because of adverse market conditions. Monitoring was initiated in December 2012 and is expected to run through the end of 2013 with only partial data available here.

Energy Use

Table 3 shows actual monitored energy use for the three projects and estimated energy use of a similarly sized home built to the 2009 International Energy Code for Climate Zone 4 as a rough comparison as modeled in BEopt.

Both the Portland and Hood River projects use ductless mini-split heat pumps for primary space conditioning with electric resistance zonal heat in the peripheral rooms as back up. Table 3 breaks out space conditioning by type of appliance for these sites. The Olympia site uses a tankless gas water heater with a water to air heat exchanger in the supply side of the heat recovery ventilator (HRV) for supplemental heating.

Environmental Data

Heating distribution in all three projects was considered unconventional compared to standard practice in the region. A design goal for both the mini-split heat pump homes was to minimize the use of the zonal electric resistance backup and rely on convective mixing enhanced by continuous HRV operation. The Olympia site used the HRV for all distribution.

Temperature and relative humidity data were collected at least hourly in the habitable areas in each home to assess comfort levels. Figures 3a, 3b, and 3c show typical results for peripheral zones during extreme cold weather conditions at each site. Table 4 shows a comparison of comfort conditions at the three sites under similar typical extreme cold conditions on a single day.

Figure 4 shows monthly average interior relative humidity conditions for the three sites. Wintertime outside conditions often exceed 80% rh at all the sites. Summertime outside conditions are typically 50% rh or less.

Cost Data

Cost data reported here were provided by the builders. Project costs include all the costs of construction but not the cost of land. Where incremental costs are indicated, they were either based on estimates by the builder based on their standard practices or by comparison of the builder's reported costs to the library of standard measure costs maintained as part of the BEopt program for Building America. Total construction costs and the incremental costs for all measures included to enhance energy performance beyond 2009 IECC levels are shown in Table 5.

The costs associated with the upgrades to the building enclosures are shown in Table 5.

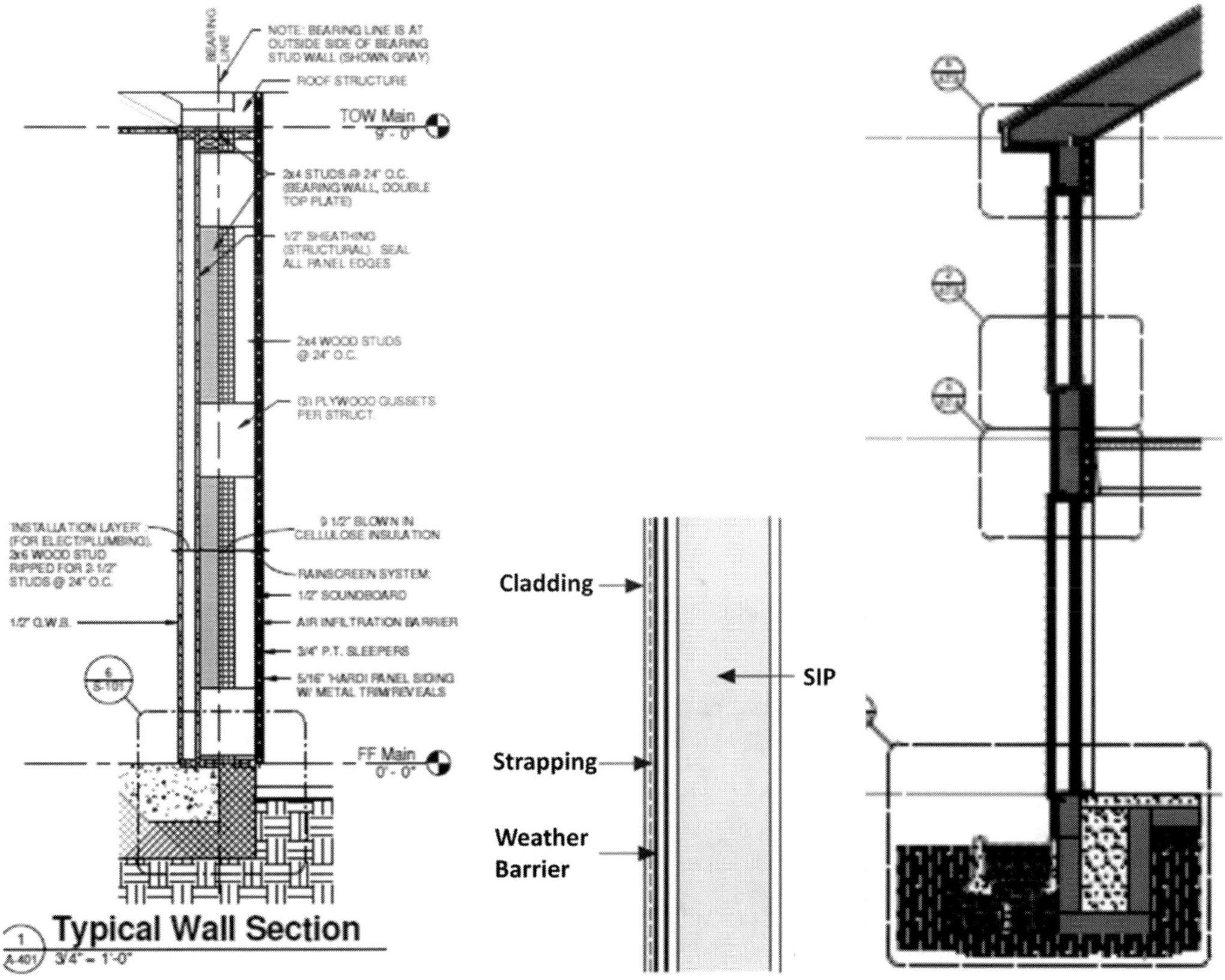

Olympia *Portland* *Hood River*

Figure 2 *Cross section detail for each project.*

Table 3. Annual Energy Use

Item	2009 IECC	Hood River	Portland	Olympia
Total Site Energy, kWh	16569	5458[1]	6643	13003 (8467)[2]
Space Conditioning, kWh	5474	1343 total 1092 heat pump 251 electric resistance	1760 total 997 heat pump 763 electric resistance	4389 (1625)[2]
Utility Domestic Hot Water Energy, kWh	3509	0	982	2813 (1041)[2]
All Other Loads, kWh	7586	4115	3901	5801
% Total site Energy Reduction from 2009 IECC	0%	65.6%	59.9%	48.9%

[1] Estimated based on 8 months of data (September to April).
[2] Adjusted for comparison for source energy conversion for gas and electricity.

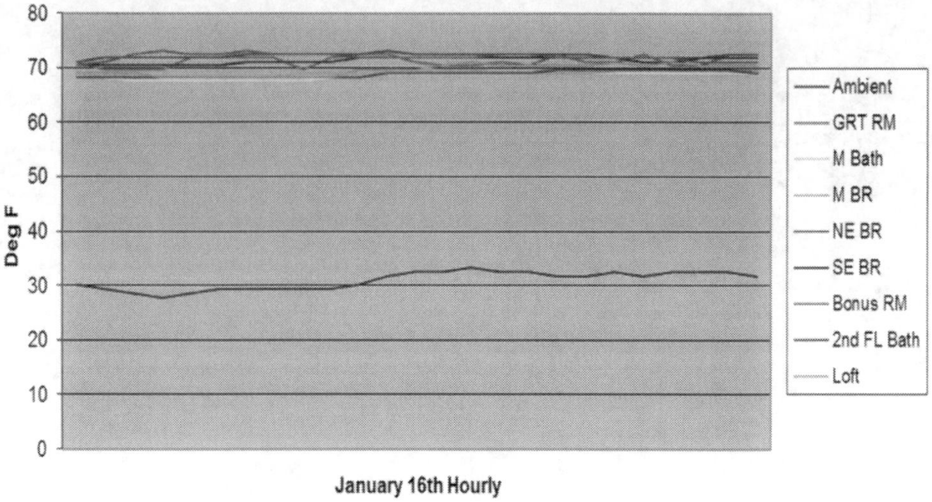

(a)

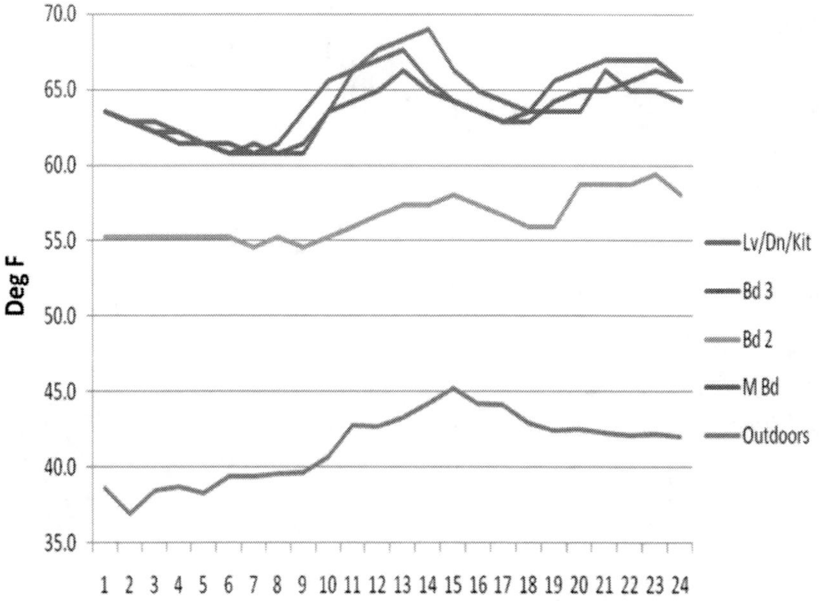

(b) February 1, 2010

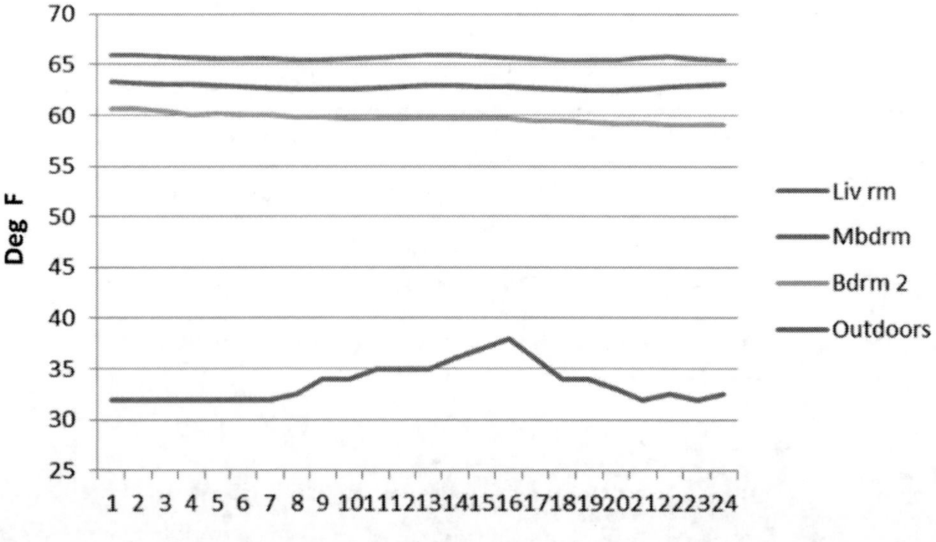

(c) January 1, 2013

Figure 3 *Typical interior cold weather temperature conditions: (a) Olympia site, (b) Portland site, and (c) Hood River site.*

Table 4. Comparison of Comfort Conditions

Site	Outdoor Temperature Range, °F (°C)	Maximum Indoor Temperature °F (°C)	Location	Minimum Indoor Temperature °F (°C)	Location	Max Indoor ΔT °F (°C)
Olympia	27 (–2.8) to 33 (0.6)	73 (22.8)	NE Bedroom	68 (20)	Master Bedroom	4 (2.2)
Portland	37 (2.8) to 45 (7.5)	69 (20.6)	Living room	55 (12.8)	Bedroom #2	10 (5.)
Hood River	32 (0.0) to 38 (3.3)	66 (18.9)	Living room	59 (15)	Bedroom #2	6.5 (3.6)

DISCUSSION

Energy Use

The Olympia and Hood River projects were both designed to meet the European Passive House standard. The differences in total energy use can be attributed to differences in occupancy (two versus four) and because the Hood River project gains in apparent site energy efficiency with the use of a heat pump instead of natural gas. Hood River also has a significant reduction for energy used for hot water because of the use of a solar thermal system again compared to natural gas. Adjusting for the differences for source energy versus site energy for electricity and natural gas as seen in the values in parentheses for the Olympia house shows that the performance of the Olympia house is quite comparable to the Hood River house for space conditioning.

The Portland project was designed to a significantly lower level of enclosure performance but had energy use levels below modeled predictions driven by a motivated homeowner who used very low setpoints for heating and allowed little used bedrooms to coast below generally accepted comfort levels. As seen here the impact of occupant behavior can have a large impact on overall performance and may be considered a very low cost to no cost measure.

Comfort Levels

The heating system in the Olympia project that integrated heating distribution with the HRV provided the most uniform temperature conditions throughout the entire living area and was able to meet the load at design conditions.

The goal of using a point source heating system (ductless mini-split system) worked well in both the Portland and Hood River projects but in both cases required the use of zonal electric resistance heating in peripheral zones during colder weather and especially when doors to the zones were closed for extended periods. Even though the heat pumps were sized to carry the entire load of the house, convective mixing from a point source and air movement induced by the HRV could not maintain comfort levels under all conditions in the peripheral zones. Because the loads are so small, it's hard to find cost effective alternatives to the zonal electric resistance supplemental heat.

Costs

Establishing a baseline for costs for standard construction practices is challenging and will vary by region as well as from builder to builder. As indicated here, the total incremental costs reported for all energy performance enhancements ranged from 6% to 20% of the total construction costs. Building enclosure enhancements ranged from 2.5% to 16% of the total construction costs.

Advanced air leakage control was by far the least-cost building enclosure enhancement employed at all three sites. Enhanced insulation levels resulted in the highest incremental costs, especially at the Olympia and Hood River sites where the extreme levels needed for Passive House qualification required alternative construction practices to accommodate insulation thickness. Insulation and air sealing practices should arguably be considered a lost opportunity in new construction and considered in the context of the expected life of the structure.

With reduced costs in the marketplace for on-site renewable generation, such as PV, the balance point between enhanced enclosure measures and on-site generation should come into play.

CONCLUSIONS

The long term monitoring of the energy performance of these homes shows that ultra-high performance building enclosures can produce excellent performance and significant reductions in space conditioning energy use. Coupled with high performance space conditioning appliances, heating and cooling site energy use was reduced by as much as 75% below the estimated use for a similar sized home meeting the 2009 IECC in Climate Zone 4. These enclosures with additional measures to assure domestic hot water, lighting, refrigeration, and other appliance efficiencies had total site energy consumption as low as 5500 kWh a year.

Appropriate space conditioning equipment can be a challenge. Loads are so low that conventional approaches and equipment are not ideal. Cost effective alternative systems don't always deliver the level of comfort expected from traditional systems. The combination of mini-split ductless heat pumps with zonal electric resistance heating in peripheral zones can meet comfort needs but the supplemental electric resistance heating can require up to 50% of the total space

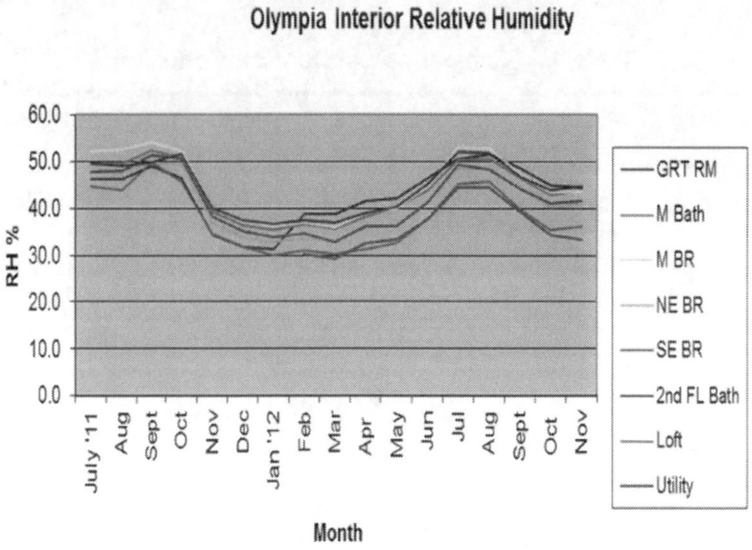

(a)

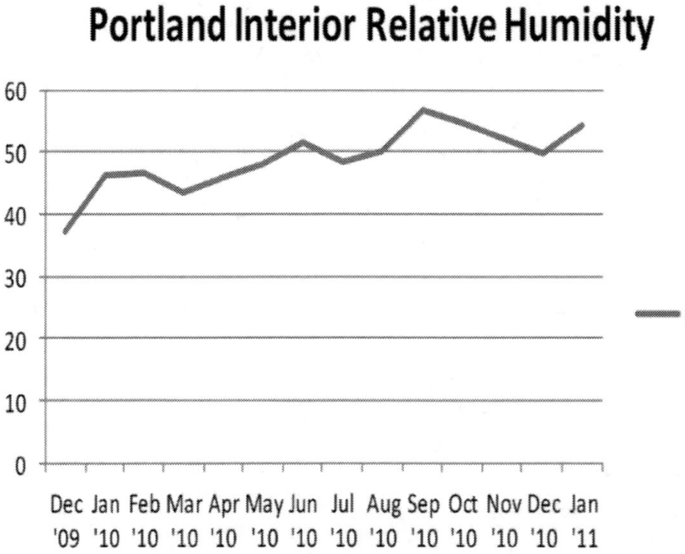

(b)

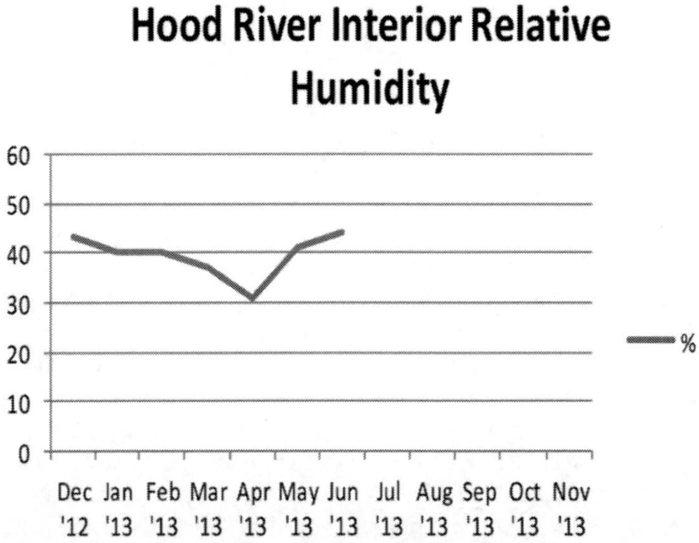

(c)

Figure 4 *Average monthly interior relative humidity: (a) Olympia site, (b) Portland site, and (c) Hood River site.*

Table 5. Incremental Costs of Building Enclosure Enhancements

Project	Air Sealing	Insulation	Fenestration	Total
Portland	$705	$9,570	$0.00	$10,275
Olympia	$6,442	$18,274	$12,913	$37,629
Hood River	$4,800	$34,237	$11,650	$50,687

conditioning energy. Because the loads are so low, energy use remains very low compared to more conventional houses. This approach has shown the best cost to benefit in the homes monitored.

These enclosures can be very sensitive to overheating. Proper design analysis, glazing choices, and shading strategies are essential. The PHPP design process helps assess overheating potential and provides an integrated systematic approach for approaching very low load design.

Reported costs to achieve ultra-high level enclosure performance vary widely. The incremental costs on these projects ranged from 10% to 30% of the construction budgets. The enclosures work, integration with mechanical systems will improve with experience, but value engineering to reduce costs is needed to make the approach more widely accepted beyond the niche market for exceptionally efficient custom homes to the broader market served by production builders.

New construction offers the least cost opportunity for enhancements to the performance of the building enclosure. Maximization of insulation levels, fenestration performance, and air sealing present a potential lost opportunity affecting performance for the life of the structure. Cost benefit analysis, however, would indicate that at some price point alternatives such as on-site renewable energy generation can achieve similar goals at the same or lower cost.

ACKNOWLEDGMENTS

This work was made possible by the cooperation of the homeowners, the project management teams (Root Design, the Artisan Group and David Cohan), and funding from Building America through the BIRA and BA-PIRC teams.

REFERENCES

2009 *International Energy Conservation Code*. International Code Council, Country Club Hills, IL.

BEopt E+, Version 1.2. National Renewable Energy Laboratory, Golden, CO.

Passive House Institute U. S. (2007). Passive House Planning Package 2007, Urbana, Il.

Thermal Performance of Exterior Insulation and Finish Systems Containing Vacuum Insulation Panels

Kenneth Childs

Therese Stovall
Member ASHRAE

Kaushik Biswas
Associate Member ASHRAE

Lawrence D. Carbary

ABSTRACT

A high-performance wall system is under development to improve wall thermal performance to a level of U-factor of 0.19 $W/(m^2{\cdot}K)$ (R-30 $[h{\cdot}ft^2{\cdot}F]/Btu$) in a standard wall thickness by incorporating vacuum insulation panels (VIPs) into an exterior insulation finish system (EIFS). Such a system would be applicable to new construction and will offer a solution to more challenging retrofit situations as well. Multiple design options were considered to balance the need to protect the VIPs during construction and building operation, while minimizing heat transfer through the wall system. The results reported here encompass an in-depth assessment of potential system performances including thermal modeling, detailed laboratory measurements under controlled conditions on the component, and system levels according to ASTM C518 (ASTM 2010). The results demonstrate the importance of maximizing the VIP coverage over the wall face. The results also reveal the impact of both the design and execution of system details, such as the joints between adjacent VIPs. The test results include an explicit modeled evaluation of the system performance in a clear wall.

INTRODUCTION

According to the US Department of Energy, buildings in the commercial and residential environment use 40% of the energy consumed in the US (DOE 2008). The global awareness of excessive energy use and the resulting emissions of carbon dioxide have created an environment where businesses and individuals alike are striving for reduced energy consumption. Building codes around the world continue to require reduced energy use of commercial and residential structures.

This paper focuses on increasing the insulating value of walls without increasing the thickness. Structures that reach a passive-house wall-construction standard of a U-factor of 0.15 $W/(m^2{\cdot}K)$ (R-38 $[h{\cdot}ft^2{\cdot}°F]/Btu$) have extremely thick walls when using conventional insulation (Wikipedia 2010). Thick walls result in additional construction detailing of the fenestration components along with eaves and overhangs. The design of an increased insulating wall system without increased thickness is the goal of this paper. Incorporation of superinsulating fumed-silica vacuum insulation panels (VIPs) into expanded polystyrene (EPS) units will allow an increased efficiency system to be installed in the typical exterior insulation-finishing system (EIFS) construction wall system.

DESCRIPTION OF VIP

The VIP is a superinsulating product consisting of a core material vacuum-packaged in a barrier envelope, Figure 1. The performance and life expectancy of the VIP depends on the conductivity and stability of the core material as well as the level of vacuum maintained in service. These products have been used commercially in appliances and shipping containers for many years. While VIP technology is not new, the products have had limited application in the construction industry due in large part to concerns for the limited service life and fragility of the vacuum-packaging system. The combination of paying attention to VIP panel protection by incorporation into a workable and user-friendly system has shown promise for greatly extended service life, suitable for consideration in construction

Lawrence D. Carbary is an Industry Scientist, Dow Corning Corporation, Midland MI, Kaushik Biswas is an R&D Associate, Kenneth Childs is a Senior Scientist, and Therese Stovall is a Distinguished Research Engineer, Retired, Oak Ridge National Laboratory, Oak Ridge, TN.

applications. Although new product and manufacturing process innovations will be required to incorporate this form of insulation into EIFS, the technology can be customized using new adhesive and barrier technologies.

This project documents the steps to evaluate and develop the feasibility of incorporating more highly-insulated materials (VIP) into a standard overall insulation thickness of 75 mm (3 in.) of insulating materials, ultimately removing marketing barriers to the use of EIFS on current buildings that desire high-insulation walls but are limited with current technology/materials.

EXTERIOR INSULATING AND FINISH SYSTEMS (EIFS)

Current insulation materials in the marketplace are limited to material R-0.6–0.9 $m^2 \cdot K/W$ (3.5–5 $h \cdot ft^2 \cdot °F/Btu$) and would require thicknesses of 200–275 mm (8–11 in.) to achieve a U-factor of 0.14 $W/m^2 \cdot K$ (R-40 $h \cdot ft^2 \cdot °F/Btu$). This creates significant implementation issues in both new and retrofit construction. The thickness required using existing materials poses a major barrier to retrofit applications because windows, doors, roofs, and other building details are not designed to accommodate such a drastic change in envelope thickness. Existing forms of EIFSs siding using an average of 50 mm (2 in.) thickness of EPS insulation are already sold today; however, these systems only deliver an average U-factor of 0.3 $W/m^2 \cdot K$ (R-12.9 $h \cdot ft^2 \cdot °F/Btu$).

A new form of insulation made using VIP technology encapsulated into EPS was studied and is expected to achieve 0.14 $W/m^2 \cdot K$ (R-40 $h \cdot ft^2 \cdot °F/Btu$) values with a much smaller thickness.

Higher values of insulation in building envelopes can be achieved with currently available building practices; however, the techniques are not widely used because they either significantly increase the cost of construction or they result in a reduction of the living space available. Current solutions to increase the energy efficiency of building envelopes do exist for retrofit applications; however, these solutions require the use of multiple components such as EPS insulation and siding. In addition to increased labor costs to install the multiple components associated with these solutions, they are also limited in their introduction because EPS insulation can only achieve a limited improvement in thermal resistance.

To enable 0.14 $W/m^2 \cdot K$ (R-40 $h \cdot ft^2 \cdot °F/Btu$) envelope construction, a minimum thickness of 250 mm (10 in.) of EPS foam would be required and this is technically and economically unfeasible in retrofit construction due to the high labor costs required, the poor aesthetics, and the technical capability of the solution. A solution to this is the incorporation of vacuum-insulated panel insulation into a significantly improved form of EIFS to increase the U-factor of a typical wall from 0.3 $W/m^2 \cdot K$ to 0.14 $W/m^2 \cdot K$ (R-12.9 $h \cdot ft^2 \cdot °F/Btu$ to R-40 $h \cdot ft^2 \cdot °F/Btu$). While EIFSs outperform most cladding systems, this new system creates significant performance improvements in terms of increased thermal resistance per dimension of thickness of insulation delivered versus current envelope systems.

A cross section of typical EIFSs using EPS as the insulation is shown in Figure 2. EIFSs offers four distinct advantages: any desired thickness of the insulation can be used to achieve higher insulation walls, incorporates an integral air barrier component, completion of the exterior with minimal interior disruption, and minimal thermal bridging. The technology is so versatile that it is possible to prefabricate the insulated panels in a fabrication plant and assemble them at the construction site. This not only speeds up the construction process, but minimizes the impact of climatic conditions on the construction. As it stands today, EIFS technology is both versatile and cost-effective. By integrating VIP panels into EIFSs, a new capability is brought to the construction market that helps new or existing buildings achieve close to zero

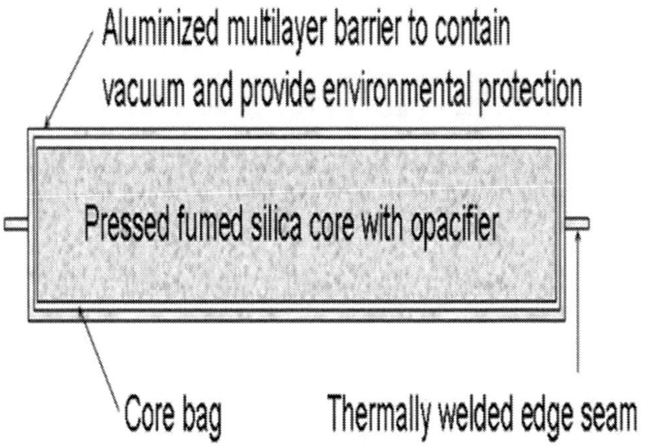

Figure 1 *Cross section of VIP.*

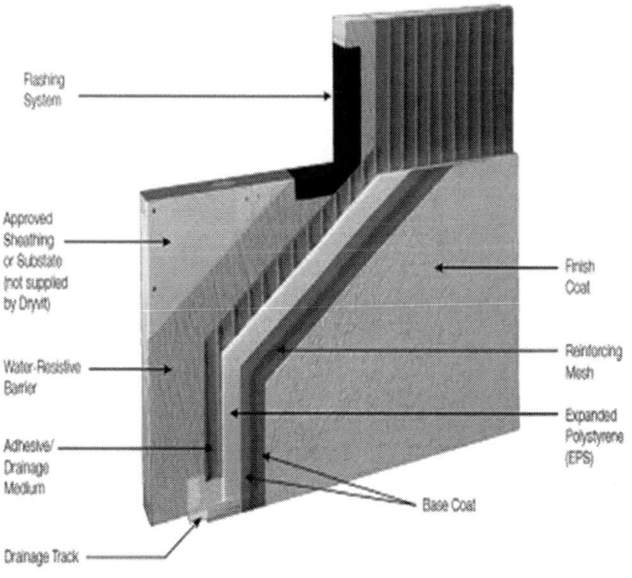

Figure 2 *Example of EIFS.*

energy levels through a superinsulating exterior system, ultimately helping to minimize the energy needs of these buildings.

There are three general classes of core materials that can be considered for use in VIP: polymeric foams; fiberglass; and nanoporous silica. The first two are characterized by cell/pore sizes in excess of 5 μm, whereas the pore size of nanoporous silica has a pore size on the 50 nm (0.05 μm) range or 100 times smaller. The larger pore size of foams and fiberglass means that the vacuum panel pressure must be maintained at ~100 times lower absolute pressure level than a nanoporous silica panel for the same thermal performance. The change in thermal conductivity for the three types of cores is illustrated in Figure 3.

For construction applications, the normal range of ambient water vapor partial pressure (even at 100% rh) is limited and on the order of tens of millibar. Water vapor permeates through barrier films much faster than other atmospheric gases due to the molecular size. For a silica VIP, this pressure of 10–30 mbar only causes a minor loss in thermal performance; whereas, for foam and fiberglass VIP applications have decreased their insulating value significantly.

Older generations of VIP insulation failed to offer long-term performance due to the fact that organic chemical materials were used in the VIP products. These organic materials tended to lose mass due to the loss of vapor over time and this resulted in reduced performance. Vapor loss within the vacuum system will elevate the internal pressure resulting in reduced thermal performance. The form of VIP insulation described here is made from an inorganic silica (silicon dioxide) chemistry which is inert and will not change over time. Long-term

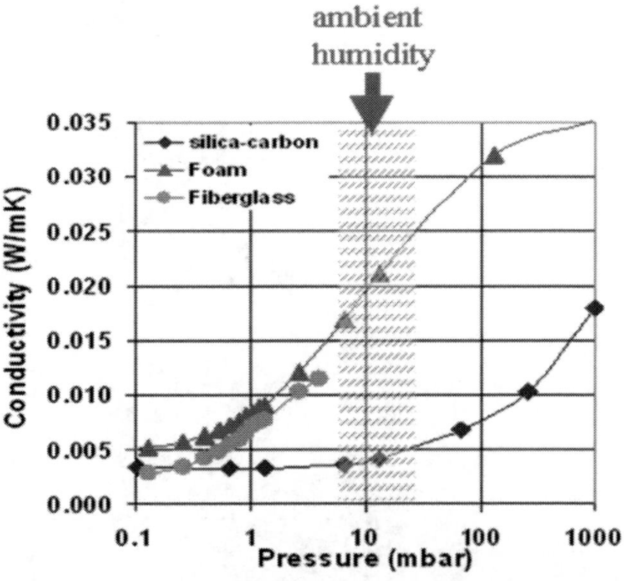

Figure 3 *Change in thermal performance of VIP during loss of pressure when using different core materials.*

aging studies of this material suggest that it will offer effective performance over a 25+ year period (Wegger et al. 2011).

The current drawback associated with all forms of VIP insulation is the fact that the insulation value is significantly decreased if the insulation panel loses its vacuum seal due to a puncture or rupture. While this can be a significant drawback in construction applications that involve nailing and fastening, it is possible to overcome with innovative design techniques such as encapsulating the VIPs into rigid boards and adhesively fastening them to a building structure. It is critical to have nonhazardous, long-lasting safe adhesives for the attachment of VIP panels. Adhesives with very low shrinkage, such as neutral curing silicone materials, are needed so that the curing process will not compromise the foil bag. The adhesive and VIP systems have to be able to resist the windloads and gravity loads associated with the construction application and should be tested to industry standards for such applications.

MEASUREMENT METHODOLOGY

The heat flux meter (HFM) apparatus is a secondary measurement method, often used to evaluate the thermal conductivity of homogenous materials. The typical construction and use are described in ASTM C518 (ASTM 2010). The two plates, shown in Figure 4, are carefully controlled at fixed temperatures. As the heat flows from the hot plate and through the test specimen into the cold plate, a small temperature difference across the transducers generates an electrical signal that is proportional to the heat flux through the transducer. For homogenous specimens with uniform plate temperatures, the heat flux is unidirectional and can therefore be used with Fourier's Law to calculate the thermal conductivity of the material.

Most HFM apparatus use a single transducer on each of the temperature-controlled plates. A special-purpose apparatus has been designed to facilitate the characterization of nonhomogenous specimens. This apparatus has an array of transducers on both the top and bottom plates, so that nonuniform heat flux patterns can be measured. The design of this array is shown in Figure 5. The test produces a map of heat transfer over this collection of thirty (15 top and 15 bottom)

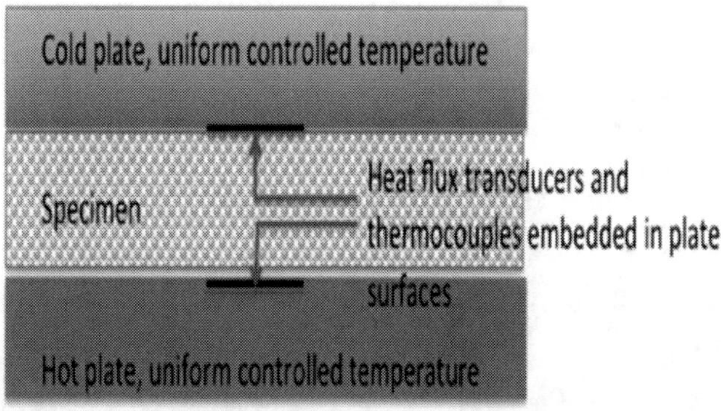

Figure 4 *Conceptual cross section of a HFM apparatus.*

7.5 × 7.5 cm (3 × 3 in.) transducers. Even with this extensive measurement scheme, it is typically necessary to merge the test measurements with a mathematical model to fully characterize complex specimens. This apparatus has previously been used to examine VIPs as a part of a refrigerator design project (Wilkes et al. 1997; Wilkes et al. 1999; Stovall and Brzezinski 2002).

A nonstandard calibration method, similar to that described in ASTM C1667, was used for this project to more accurately measure very small heat flux values (ASTM 2009). The calibration process, along with other test practice details, are not described here. Using this calibration process, the heat flux values reported here have an extended combined uncertainty between 5% and 10% at a 95% confidence level, with the greater uncertainty corresponding to the lower heat flux values found over the center of a VIP (Taylor and Kuyatt 1994).

DESCRIPTION OF SPECIMENS TESTED

In order to use VIPs in a building, it is necessary to protect them from puncture and abrasion during the construction process. In the proposed designs, that is accomplished by encasing the vacuum panels within closed-cell foam insulation. Within this general design framework, there are a large number of possible permutations with regard to: the arrangement of panel seams relative to joints where panels are adjacent to each other, the type of foam insulation, the thickness of the foam separation between VIPs, and other possible protective materials.

Ten special purpose test specimens were prepared to place these variations within the monitored area of the HFM apparatus. Figure 6 demonstrates how these test specimens are representative of panel intersections within a full wall. The joints between VIPs in the test specimens were located along the lines of central transducers. All of the assemblies cover an area slightly less than 610 × 610 mm (24 × 24 in.) and consist of a layer containing 2.5 cm (1 in.) thick VIPs sandwiched between layers of rigid foam insulation, either EPS or extruded polystyrene (XPS). The assemblies differ in (1) the number and size of VIPs they contain (one large panel, two half panels, or one half panel with two quarter panels), (2) the type and thickness of the rigid foam insulation layers, and (3) the material adjacent to the edges of the VIPs. An example of an assembly construction is illustrated in the exploded view of design option 1 shown in Figure 7. A summary of the key features of the VIPs tested is given in Table 1. Some internal details of design options 11 and 12 could not be determined by visual examination of the assemblies.

In addition to the above VIP assemblies, tests were also performed on two bare VIPs that were approximately 59.6 cm^2 (23.5 in^2). The thermal conductivity of EPS and XPS foam boards similar to those used in the assemblies was measured.

TEST DATA

Each of the assemblies contains EPS insulation board, XPS insulation board, or a combination of both. Since the conductivity of these materials influences the performance of the assemblies, a series of HFM apparatus tests were performed to determine the thermal conductivity of both EPS and XPS as a function of temperature. The temperature-dependent thermal conductivity of EPS and XPS from these measurements is given by Equations 1 and 2, respectively, and shown in Figure 8.

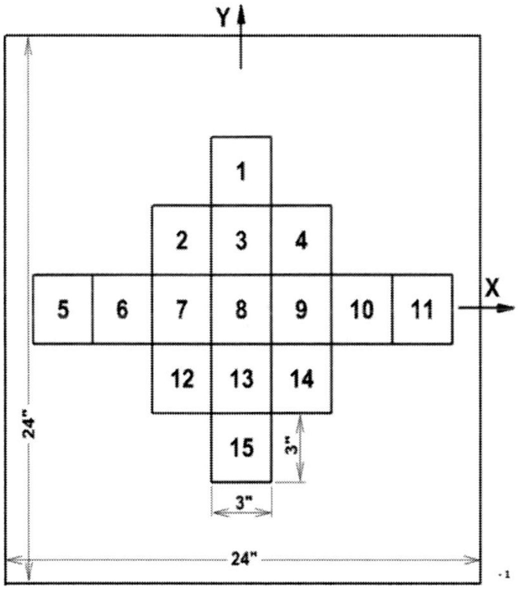

Figure 5 *HFT numbering scheme: 1 (back of HFM), 5 (left edge), 11 (right edge), 15 (front).*

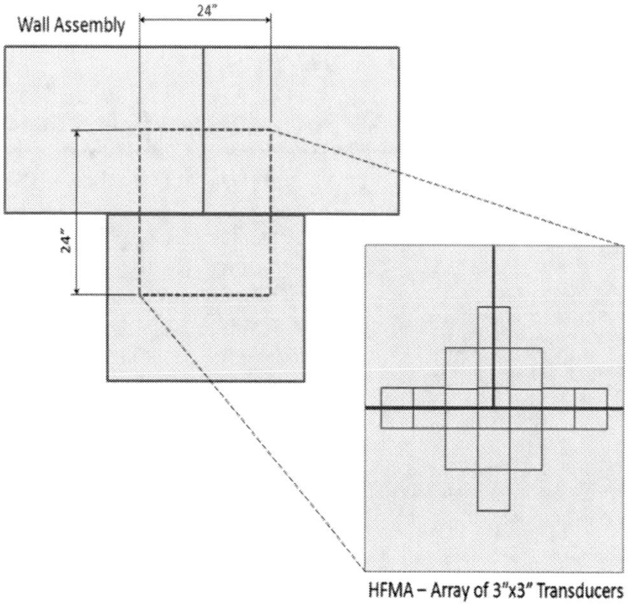

Figure 6 *Test specimen represents joints between panels on a full wall.*

$$k = 3.4020 \times 10^{-2} + 1.4591 \times 10^{-4} T \quad (1)$$

$$k = 2.5638 \times 10^{-2} + 1.1688 \times 10^{-4} T \quad (2)$$

where

T = temperature, °C

K = thermal conductivity, W/m·K.

In the series of the tests performed on the ten VIP assemblies and the two large individual VIPs, the bottom-plate temperature is set to 35°C and the top-plate temperature is set to 12.78°C. Some of the specimens were measured multiple times. The following pages give the measured fluxes from a representative test for each of the specimens. Figures 9, 11, 13, 15, 17, 19, 21, 23, 25, and 27 give the measured flux (W/m^2) for each heat flux transducer (HFT) location. The top number for each location is the flux on the upper surface, and the bottom number is the flux on the lower surface. Note that HFT 12 on the bottom plate is inoperable, and HFT 11 on the upper plate operates only intermittently. To aid in interpreting the fluxes, the location of each VIP in the assembly is also outlined in the figure. The white space represents the material surrounding the VIPs. At the bottom of each page are two plots of the heat fluxes along the two centerlines of the transducer array. These are Figures 10, 12, 14, 16, 18, 20, 22, 24, 26, and 28. Uncertainty bounds of 7.5% are shown on these plots.

Look at these results for design option 1 in Figure 9 and Figure 10. As you travel across the x-axis from left to right, you see the flux is slightly greater near the edges, but the greatest

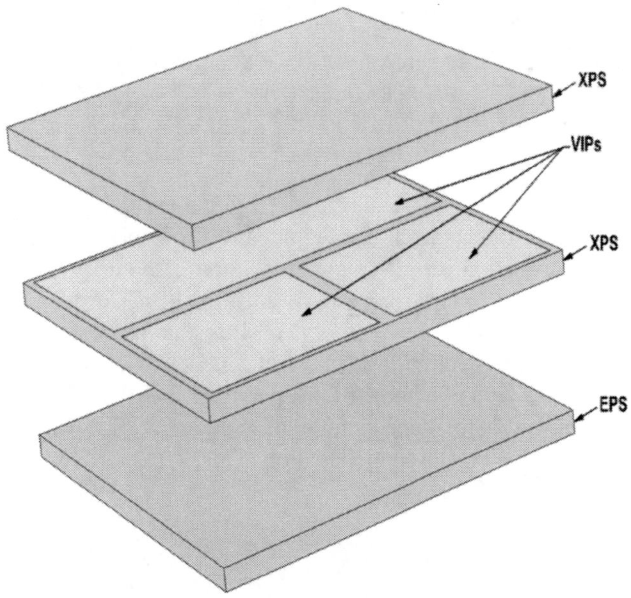

Figure 7 *Exploded view of design option 1.*

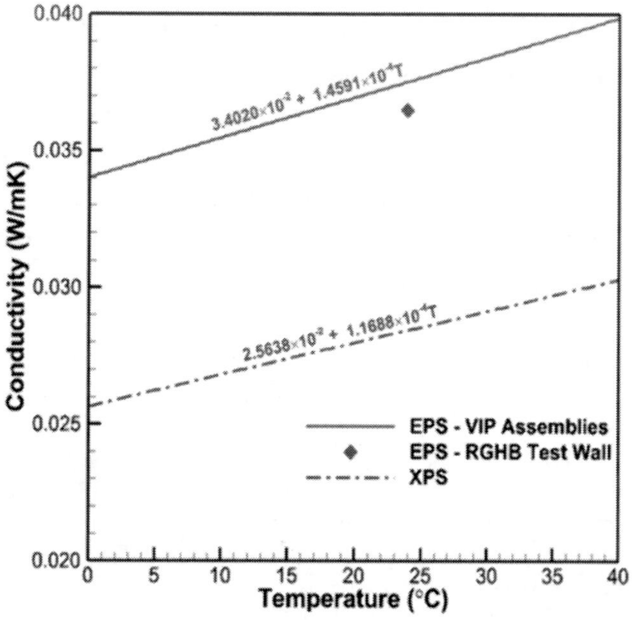

Figure 8 *Foam board thermal conductivity.*

Table 1. Test Specimen Construction Details

Design Option	Number VIPs	Material Surrounding VIPs		Layers (from Top to Bottom)			
		Assembly Edges	Between VIPS	1	2	3	4
1	3	0.5 in. XPS	1 in. XPS	1 in. XPS	1 in. VIP	1 in.EPS	—
2	3	0.5 in. XPS	1 in. XPS	1 in. EPS	0.5 in. XPS	1 in. VIP	0.5 in. EPS
3	3	0.5 in. XPS	1 in. XPS	1 in. EPS	0.5 in. XPS	1 in. VIP	0.5 in. XPS
4	3	0.04 in. PVC*	0.08 in. PVC*	1 in. EPS	1 in. VIP	1 in. EPS	—
5	1	0.5 in. XPS	1 in. XPS	1 in. EPS	1 in. VIP	1 in. XPS	—
6	3	0.5 in. XPS	1 in. XPS	1 in. XPS	1 in. VIP	1 in. XPS	—
8	3	0.5 in. EPS	1 in. XPS	1 in. EPS	1 in. VIP	1 in. EPS	—
10	2	0.5 in. XPS	Butt joint, edge to edge (not flap to flap)	1 in. XPS	1 in. VIP	1 in. XPS	—

* PVC was approximately 50 mm (2 in.) high, covering the side of the VIP and the side of the bottom foam so that it touched the bottom plate of the apparatus but not the top plate.

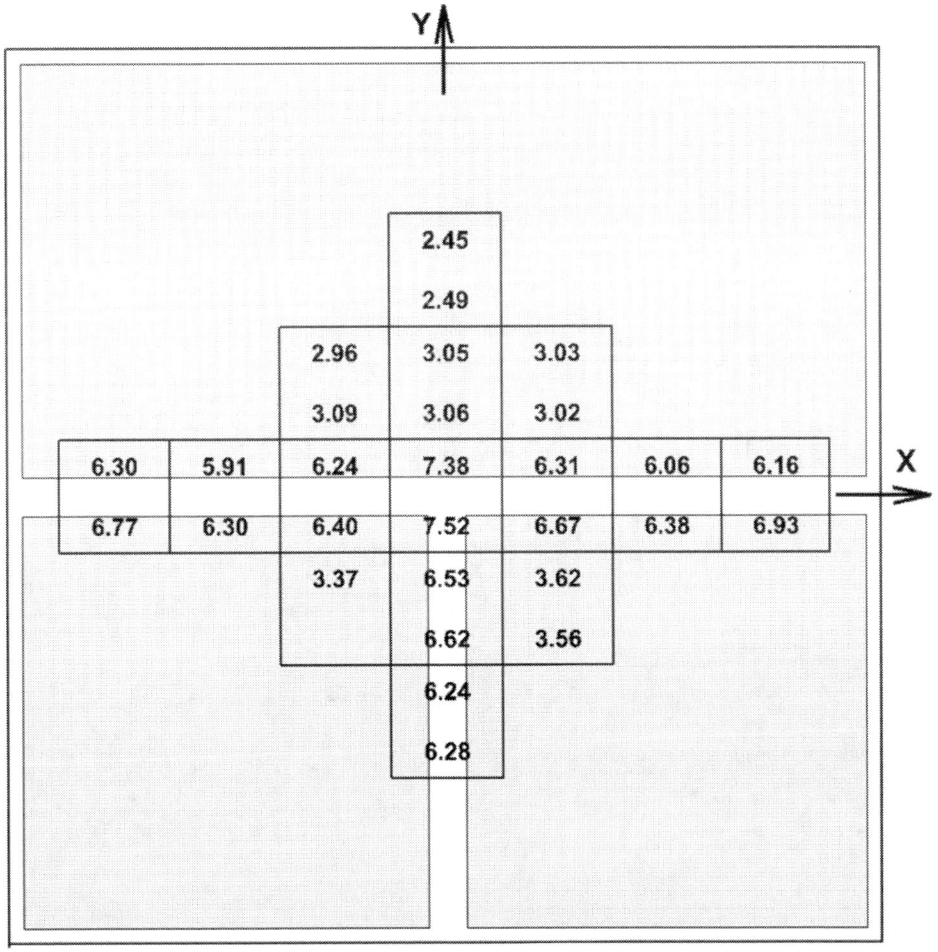

Figure 9 *Fluxes: Design option 1, EPS on bottom, XPS on top.*

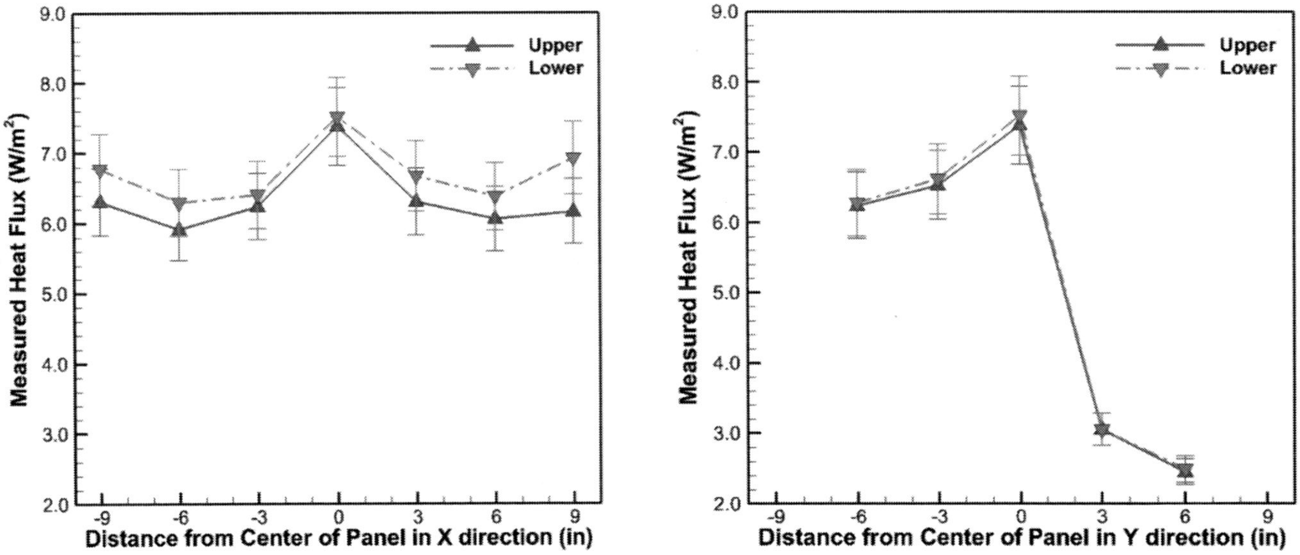

Figure 10 *Upper and lower fluxes along center axes: design option 1.*

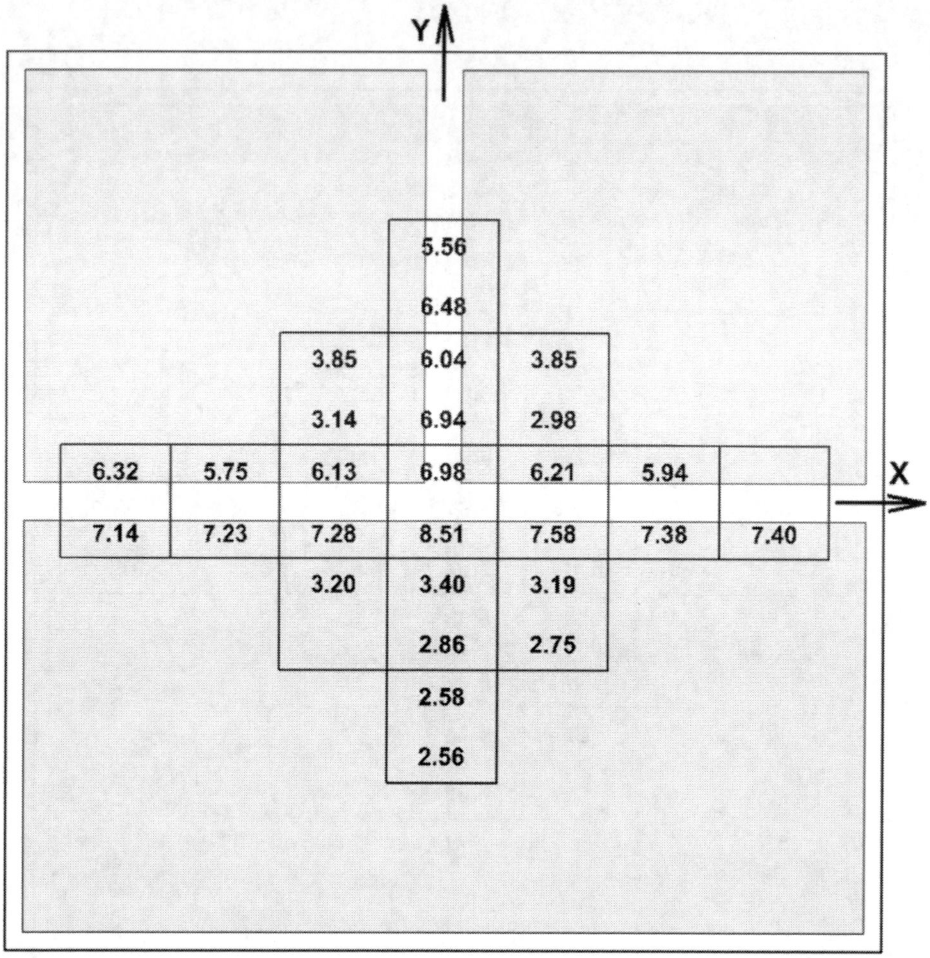

Figure 11 *Design option 2, thin EPS on bottom, EPS and XPS on top.*

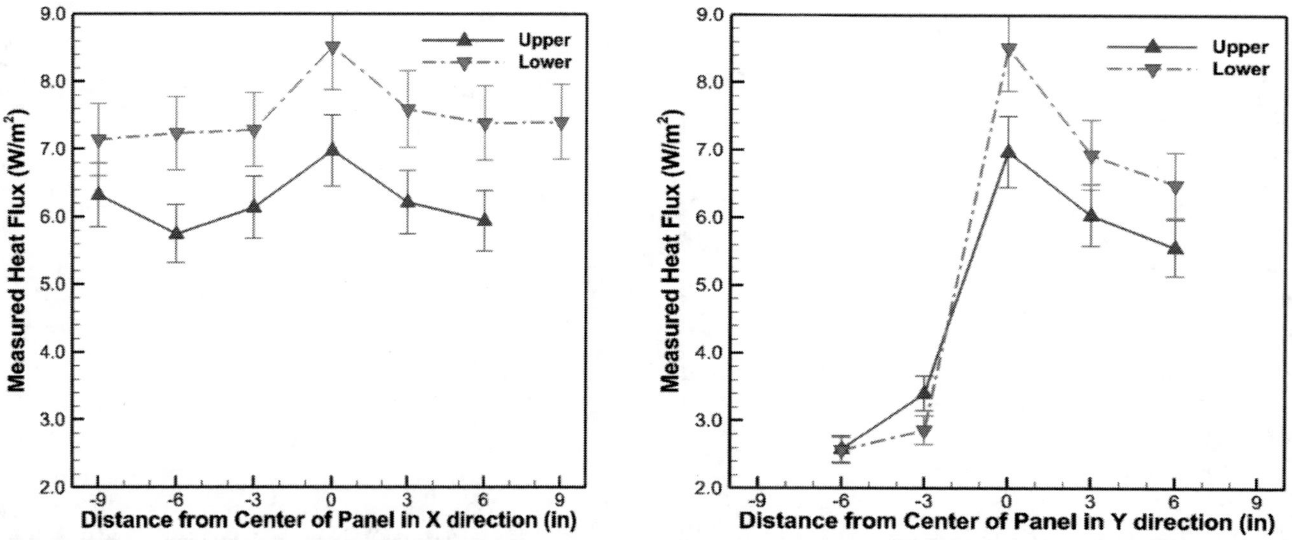

Figure 12 *Upper and lower fluxes along center axes: design option 2.*

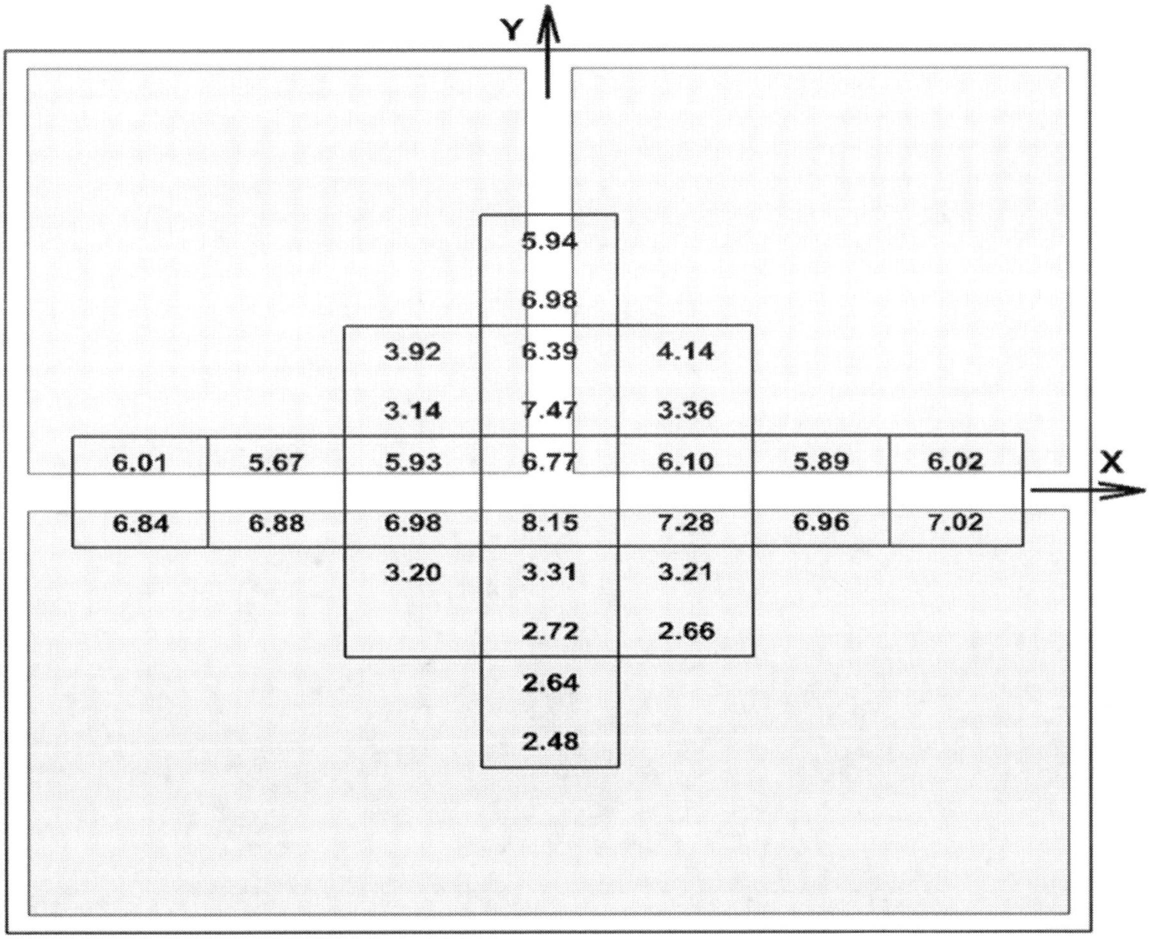

Figure 13 *Fluxes: Design option 3, thin XPS on bottom, EPS and XPS on top.*

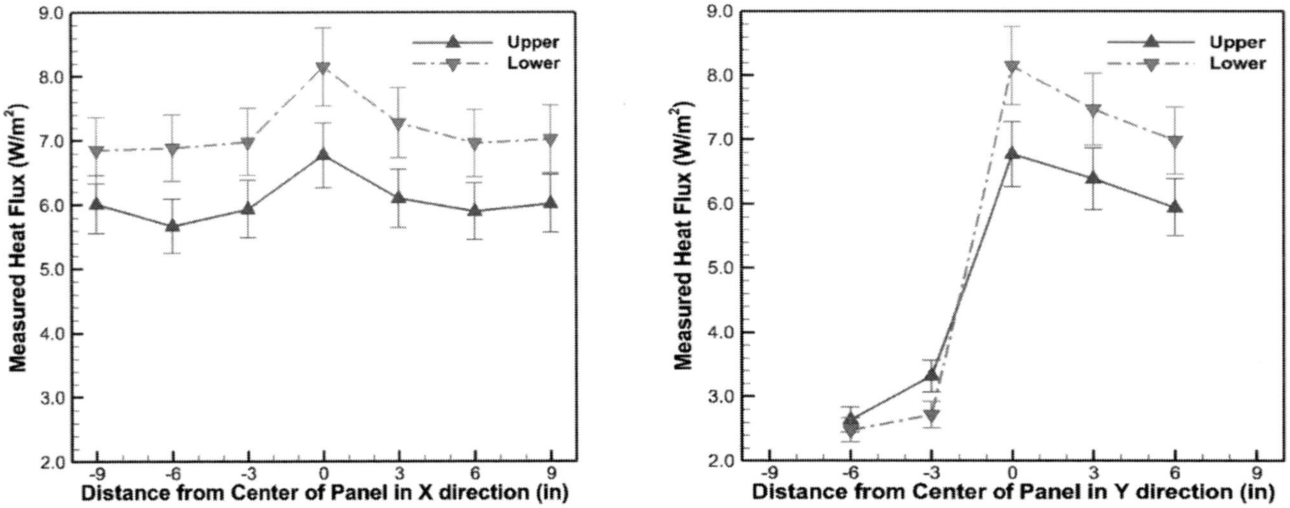

Figure 14 *Upper and lower fluxes along center axes: design option 3.*

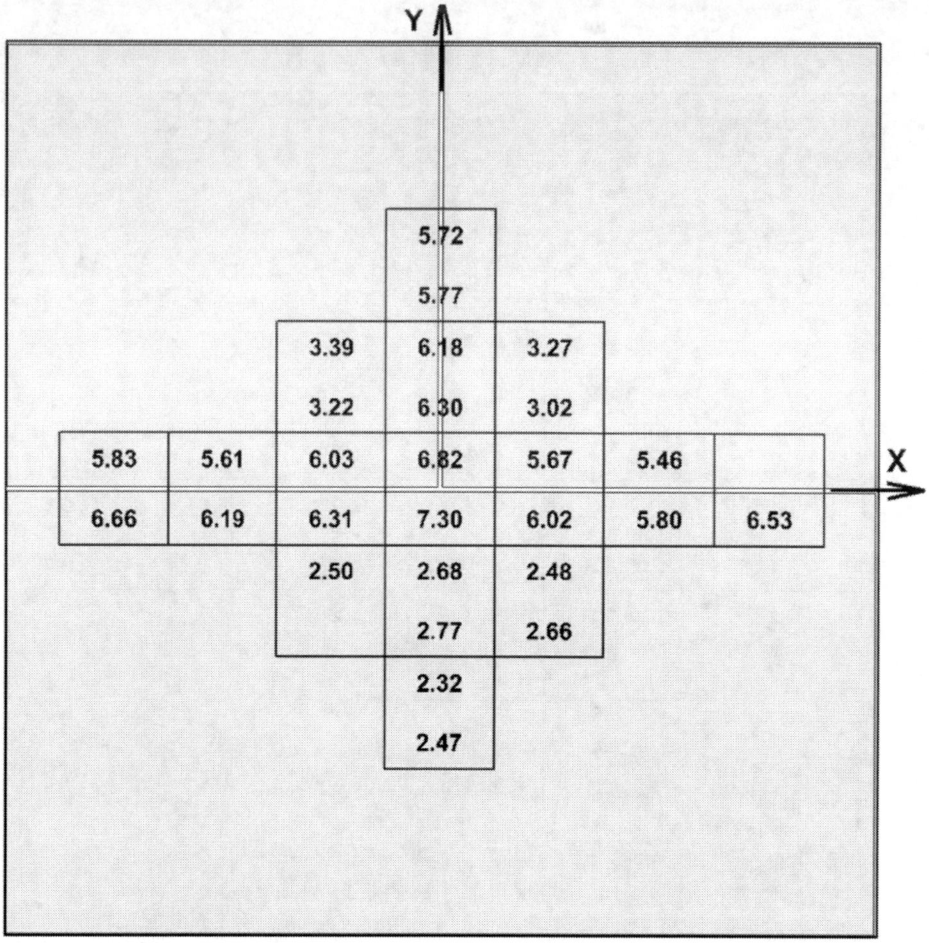

Figure 15 Fluxes: Design option 4, EPS top and bottom, thin PVC between panels and at edges.

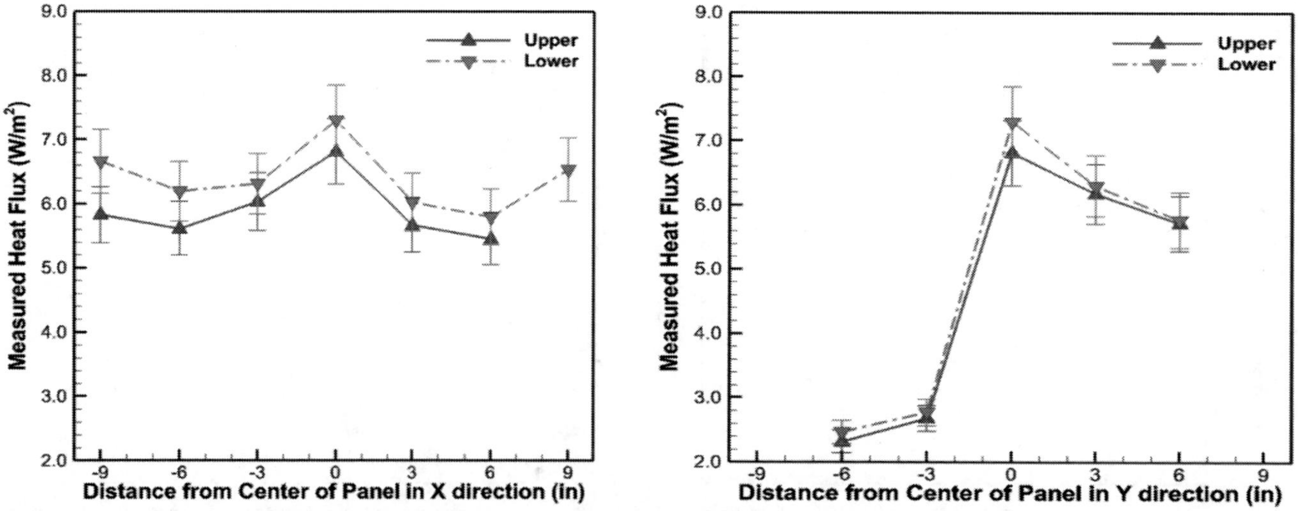

Figure 16 Upper and lower fluxes along center axes: design option 4.

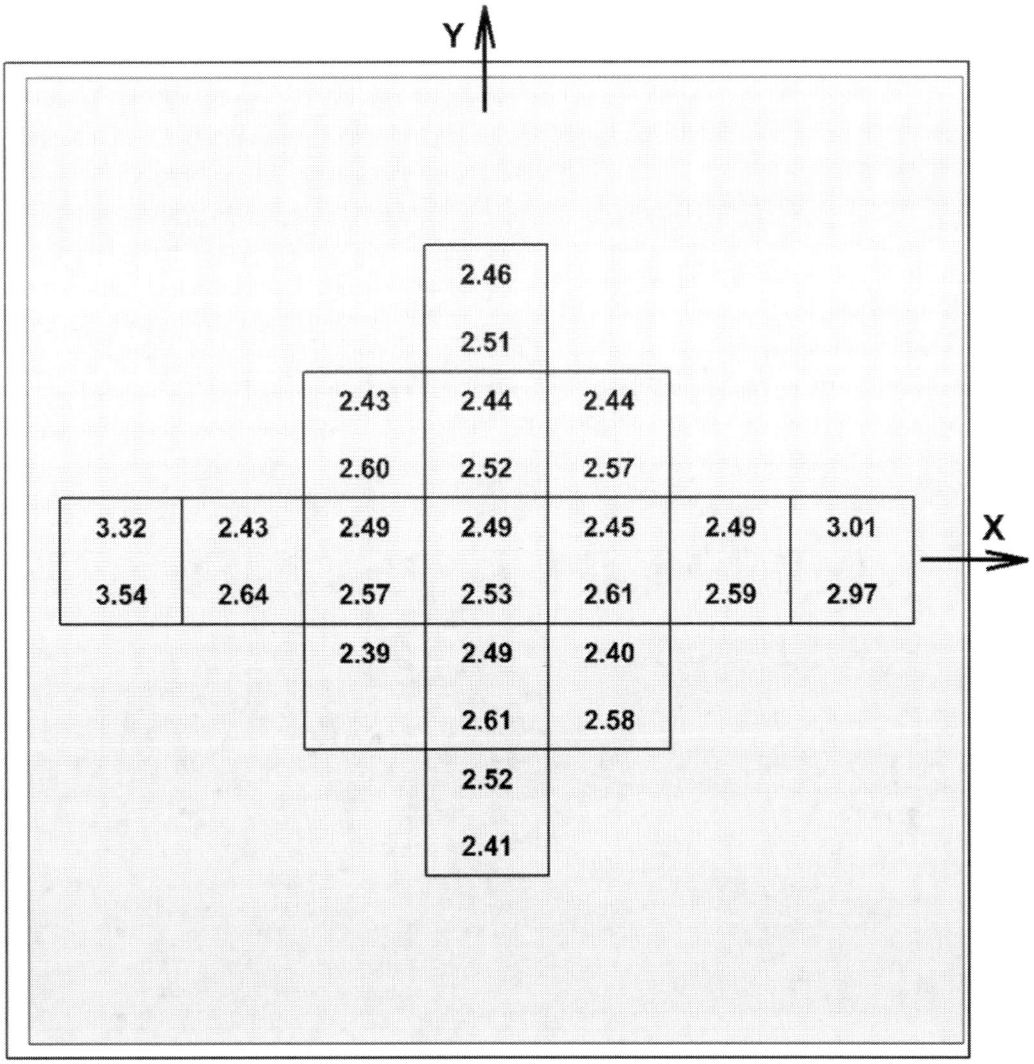

Figure 17 *Design option 5, XPS bottom and EPS top, single large VIP.*

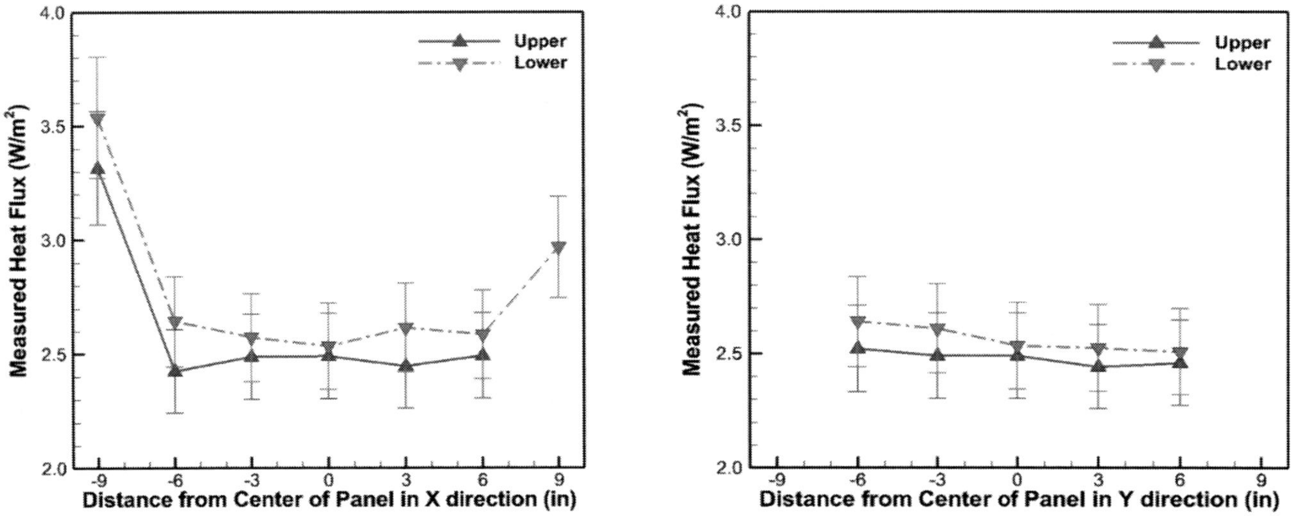

Figure 18 *Upper and lower fluxes along center axes: design option 5.*

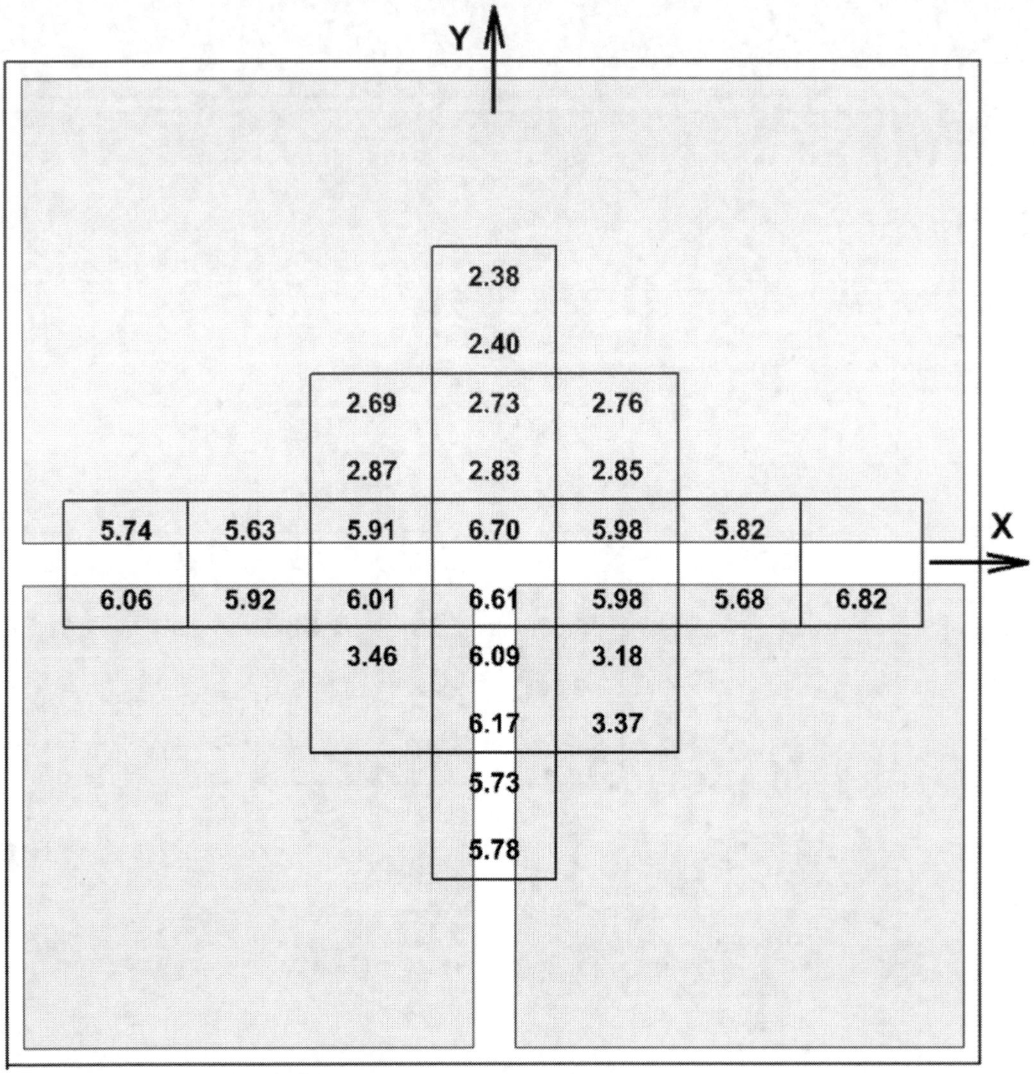

Figure 19 *Fluxes: Design option 6, XPS top and bottom.*

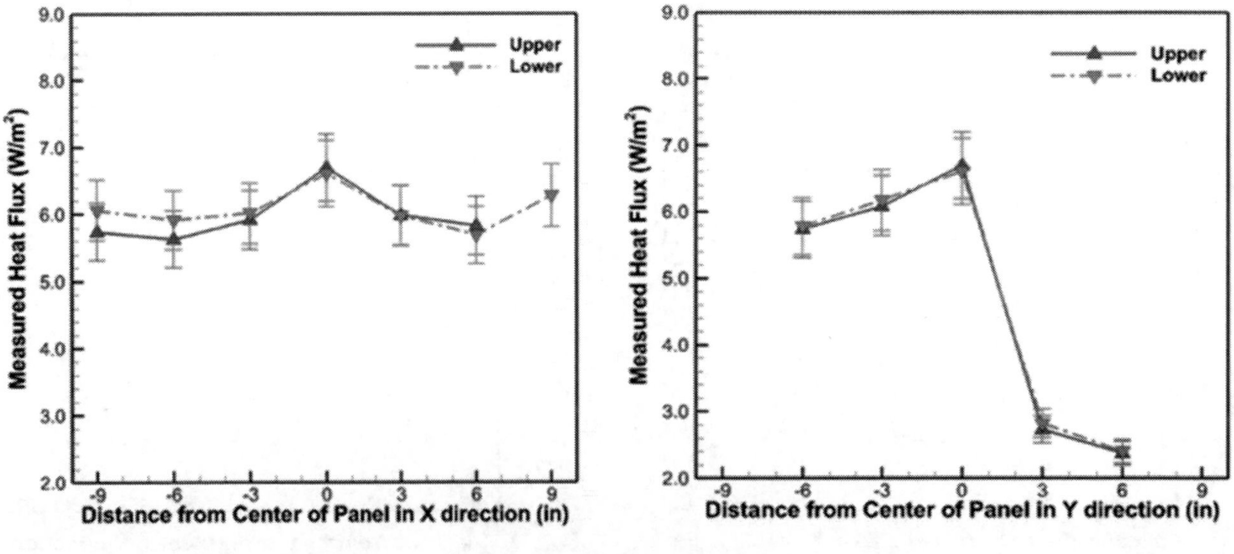

Figure 20 *Upper and lower fluxes along center axes: design option 6.*

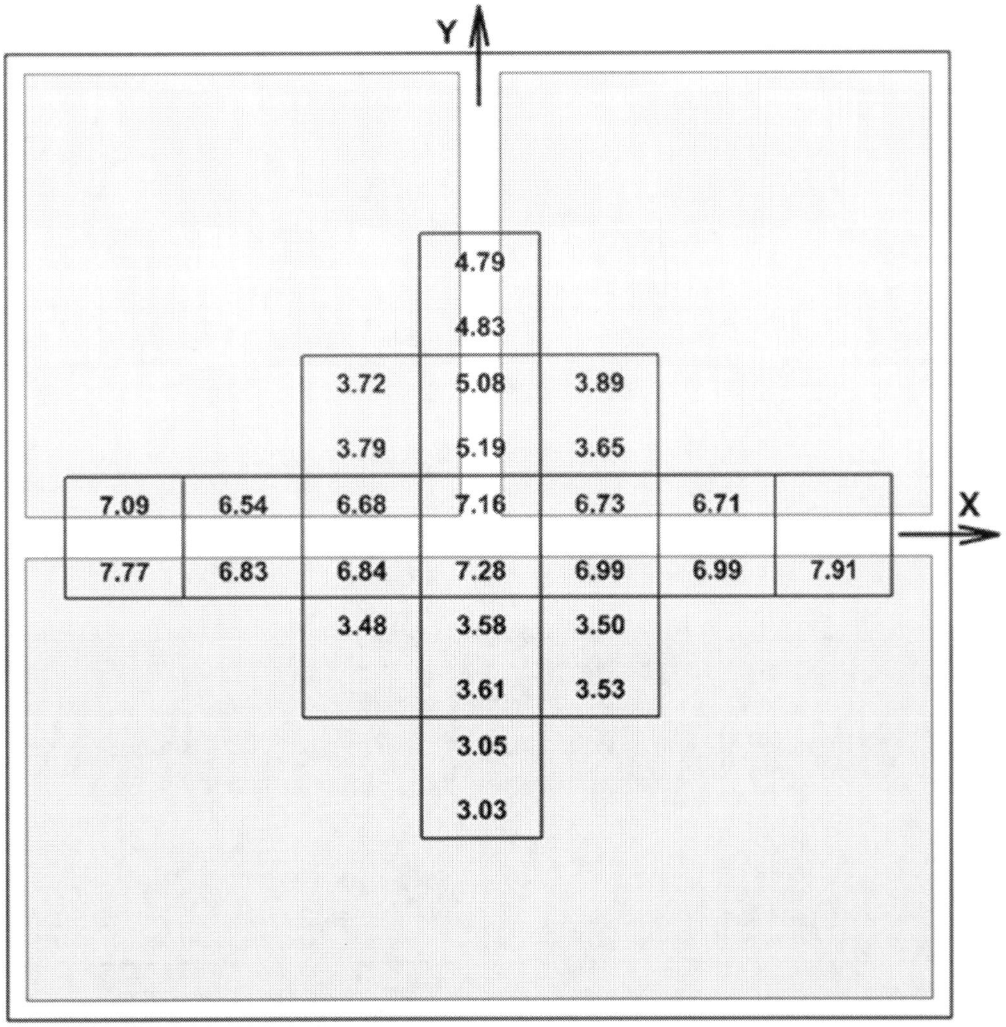

Figure 21 *Fluxes: design option 8, EPS top and bottom.*

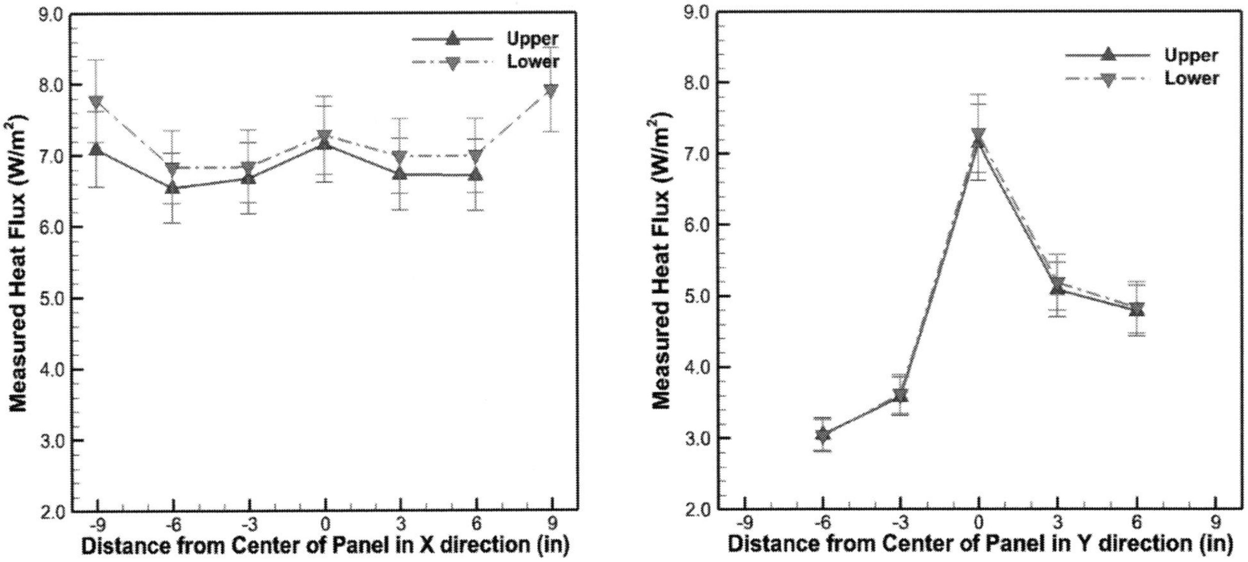

Figure 22 *Upper and lower fluxes along center axes: design option 8.*

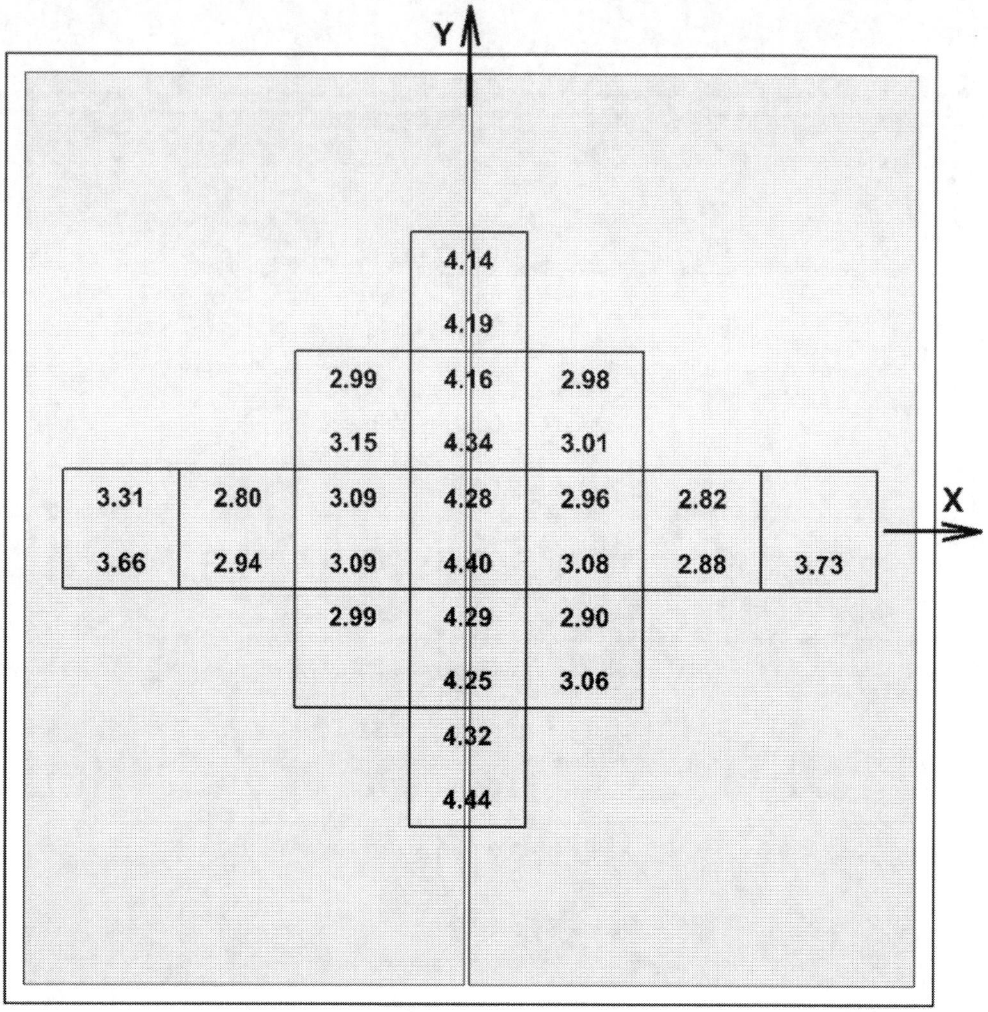

Figure 23 *Design option 10, XPS top and bottom, two half-size VIPs butted together edge to edge in the center.*

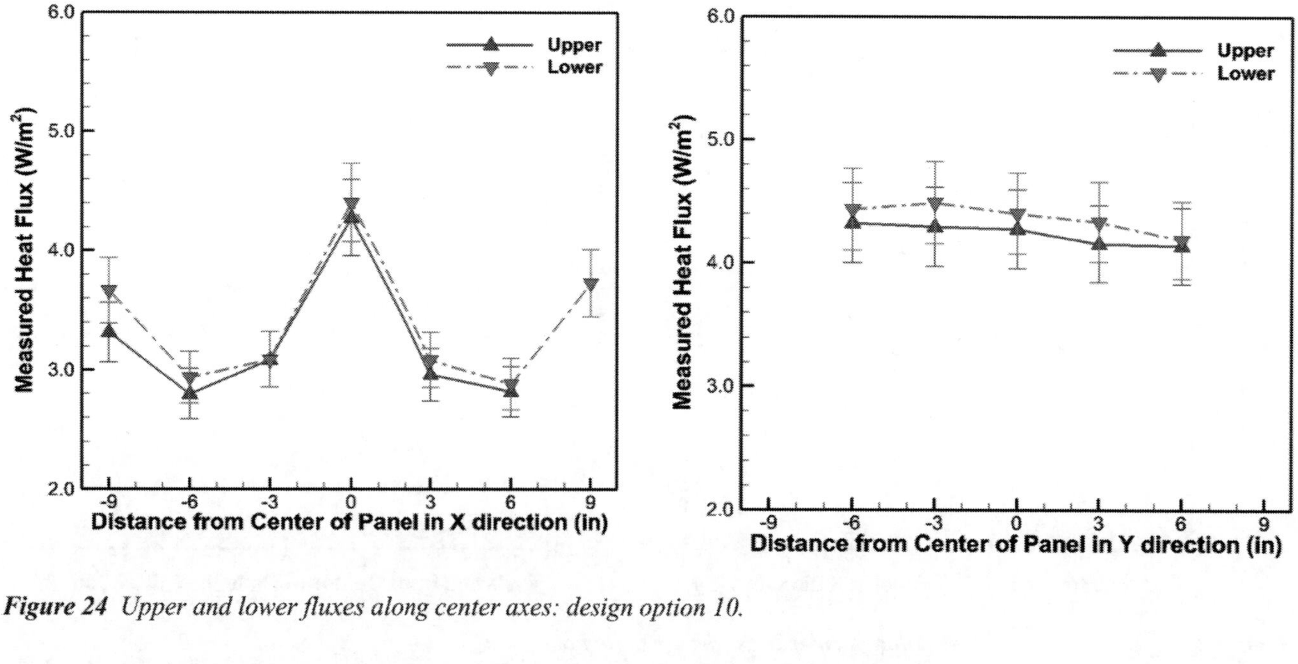

Figure 24 *Upper and lower fluxes along center axes: design option 10.*

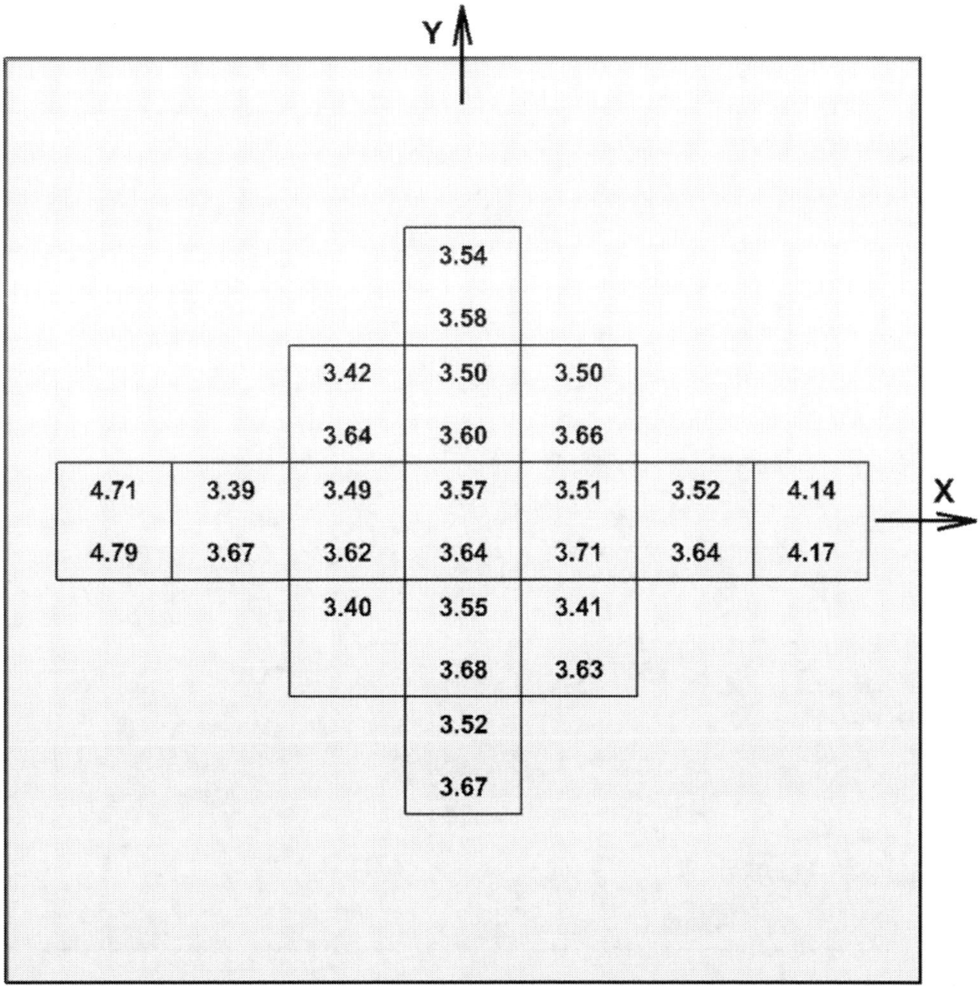

Figure 25 *Fluxes: 23.5 × 23.5 in. VIP #1.*

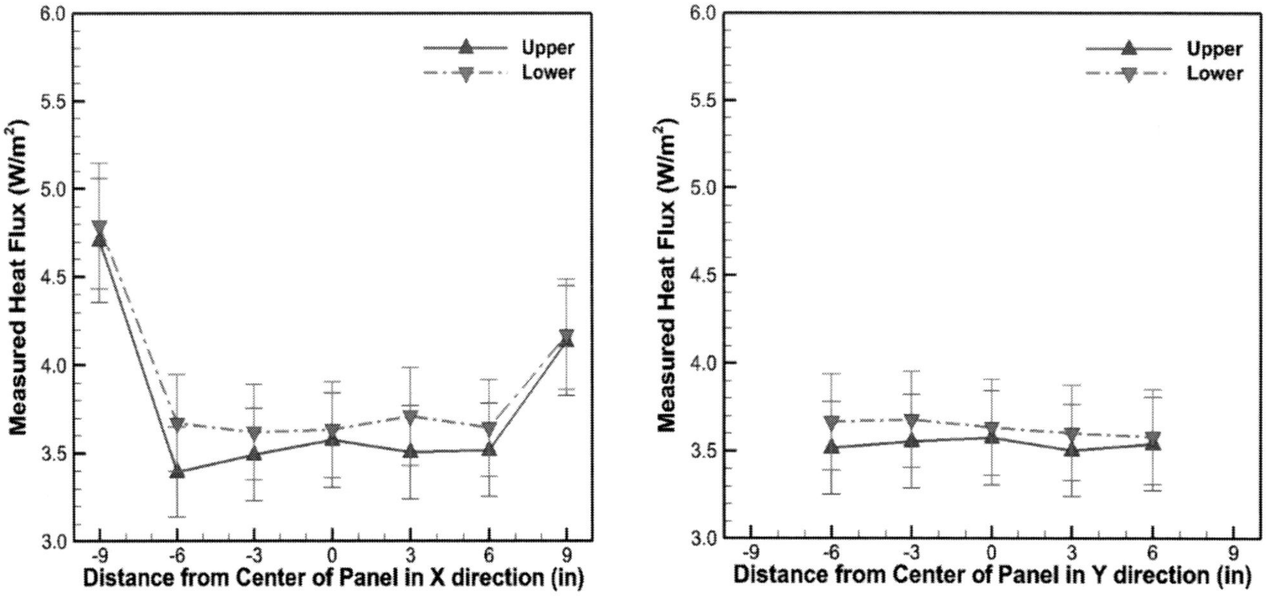

Figure 26 *Upper and lower fluxes along center axes: 23.5 × 23.5 in. VIP #1.*

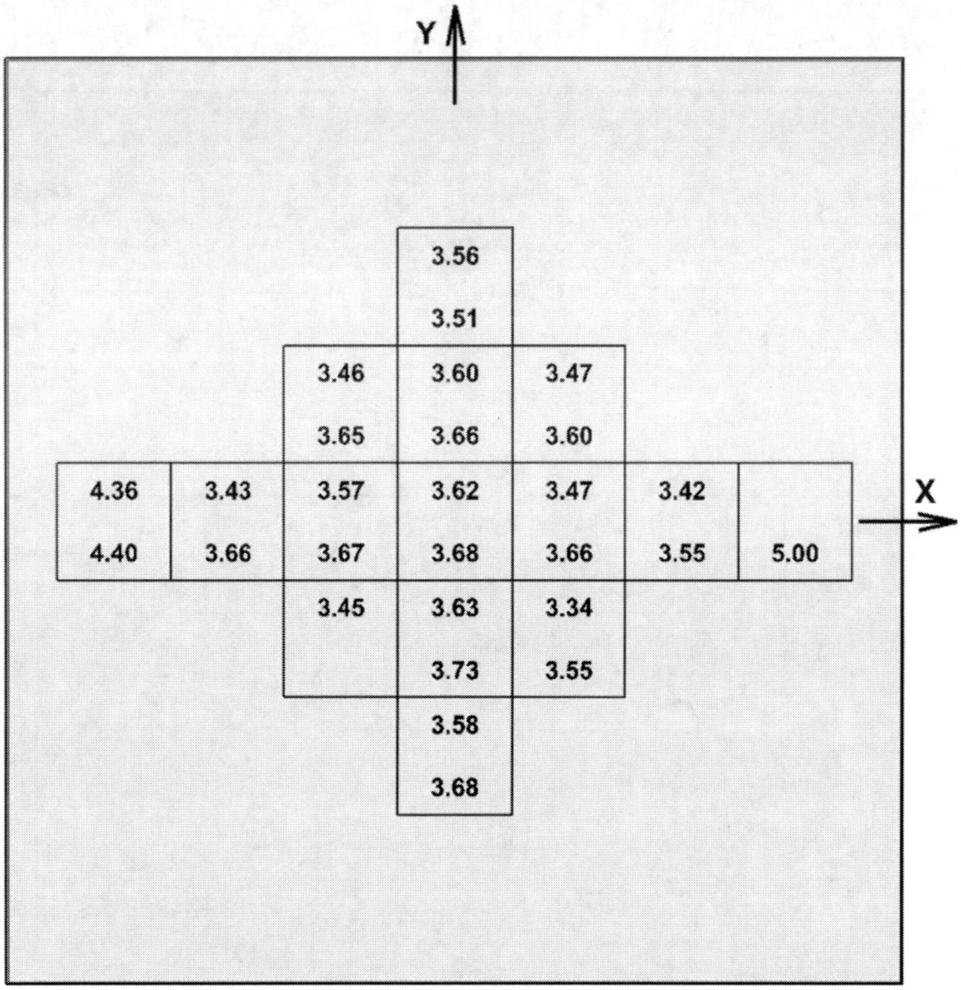

Figure 27 *Fluxes: 23.5 × 23.5 in. VIP #2.*

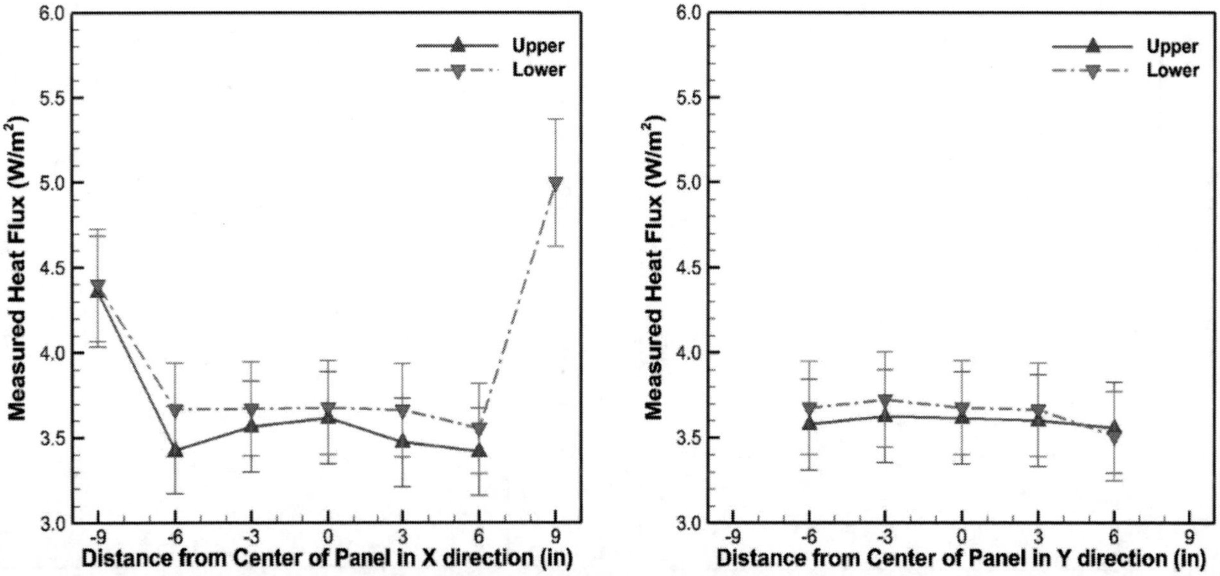

Figure 28 *Upper and lower fluxes along center axes: 23.5 × 23.5 in. VIP #2.*

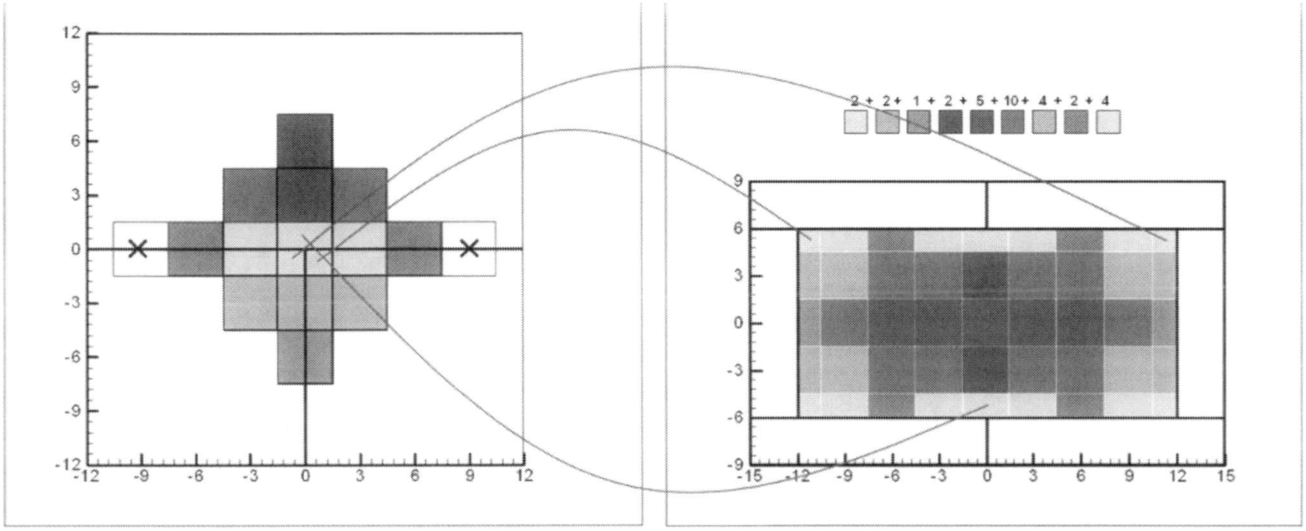

Figure 29 *Mapping HFM data to a staggered array of 12 × 24 in. VIP assemblies.*

heat flux occurs at the center of the specimen, where *x* and *y* are both equal to zero. This is the transducer location with the maximum area of foam insulation and the minimum amount of VIP coverage. As you travel along the *y*-axis in the right half of Figure 10, notice the heat flux drops significantly as the *y*-value increases, which corresponds to the positive *y*-values, or the top half of Figure 9, where the transducer is located over the center of a VIP and farther away from the foam insulation sections.

Design option 1 is symmetric in the *z*-direction (i.e., the thickness of the insulation on the top and the bottom of the VIP are the same), and the top and bottom heat fluxes are very close. Design option 2 on the other hand has only 12.7 mm (0.5 in.) of insulation on the bottom of the VIP and 38.1 mm (1.5 in.) of insulation on the top. Comparing Figure 12 to Figure 10, the difference between the top and bottom plate heat fluxes is much greater for option 2. In option 2, the heat flux along the foam insulation seams is greater on the lower plate, but the heat flux near the center of the panels is less. This can be seen on the *y*-axis profile in Figure 12 and in the numbers displayed for the VIP areas (in yellow) in Figure 11. Notice that the difference between the minimum and maximum flux is less on the upper side where the insulation thickness is greater.

This more uniform heat flux would produce a more uniform temperature distribution on the outside of the thicker foam surface. This may prove useful if there is any concern with regard to ghosting phenomena.[1] However, panels constructed in this manner would have to be clearly marked to ensure proper installation during the construction process.

1. Ghosting may occur when some portions of a wall are cooler than others, leading to uneven patterns of condensation, which can then lead to uneven patterns of dust accumulation. This occurred on the inside of metal-framed walls in some buildings before exterior insulation became more common for this framing system.

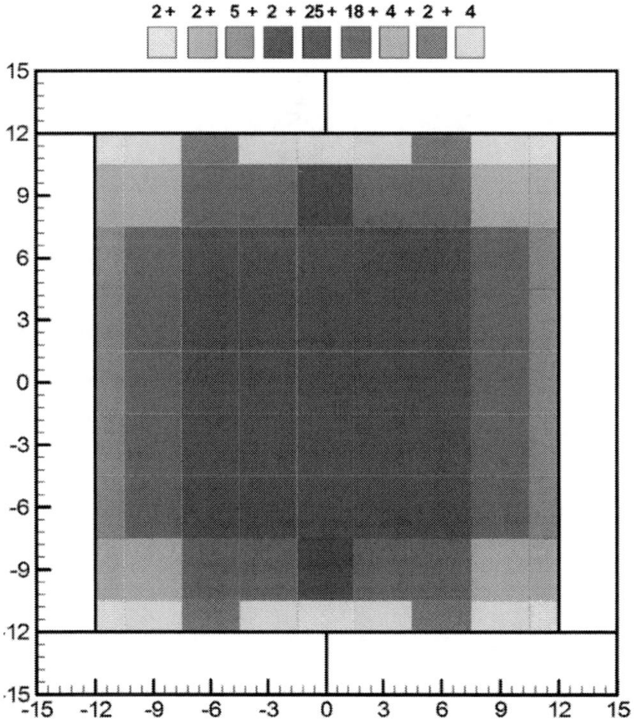

Figure 30 *Mapping HFM data to staggered array of 24 × 24 in. VIP assemblies.*

MAPPING HFT READINGS TO FULL VIP ASSEMBLIES

The test specimens were designed to place the junctions between VIPs directly within the metered areas, and specifically, to center the junctions along a centerline of the transducers. Each transducer reports the average heat flux over a 75 × 75 mm (3 × 3 in.) area. The collection of test specimens has therefore produced information typical of heat transfer through the center of a panel (purple in the left half of Figure 33), near the corner of a panel (tan), along a straight junction of panels (blues and greens), and at a junction where two corners come together beside the continuous edge of a third panel (yellow). There is no information for a four-corner junction, but such an arrangement is not anticipated in a wall where the units would be staggered to reduce edge losses.

The data from these tests is therefore sufficient to estimate the effective R-value of a VIP-foam panel applied in a staggered array of panels to form a wall. Figures 33 and 34 demonstrate how the HFM readings can be mapped to a 30 × 60 cm

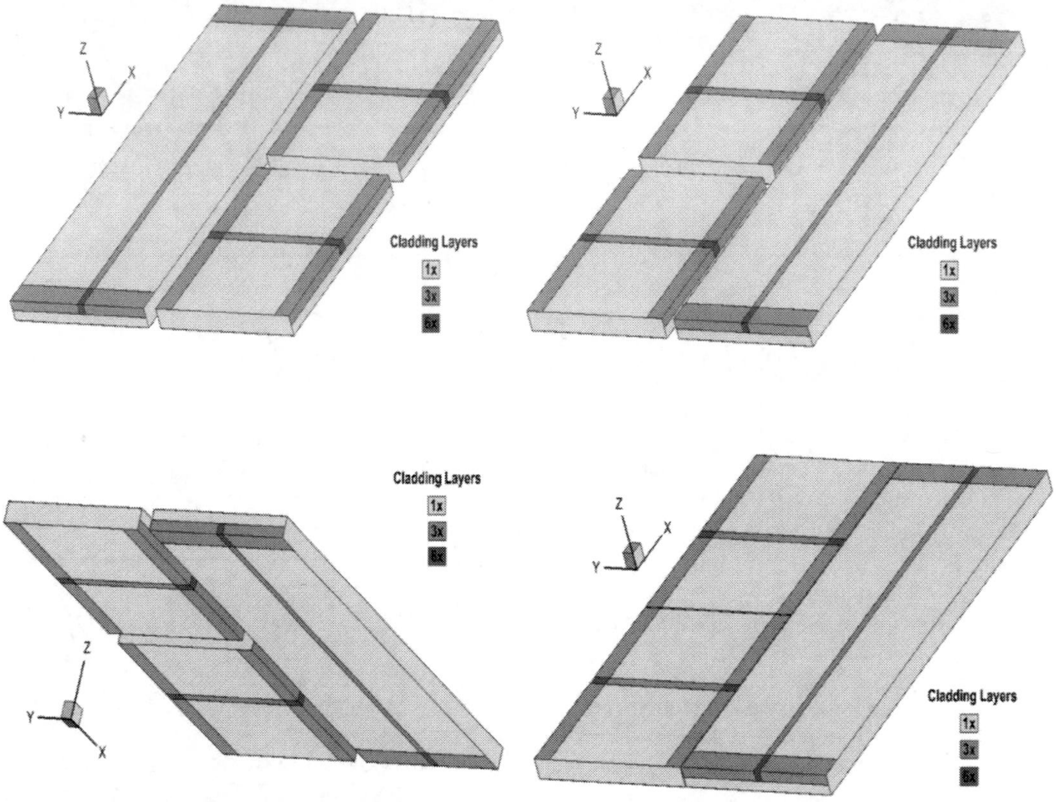

Figure 31 *Number of layers of cladding due to overlap where edges are sealed and wrapping of barrier material around panel: design option 1 (top left), 2 (top right), 3 (bottom left), 4 (bottom right).*

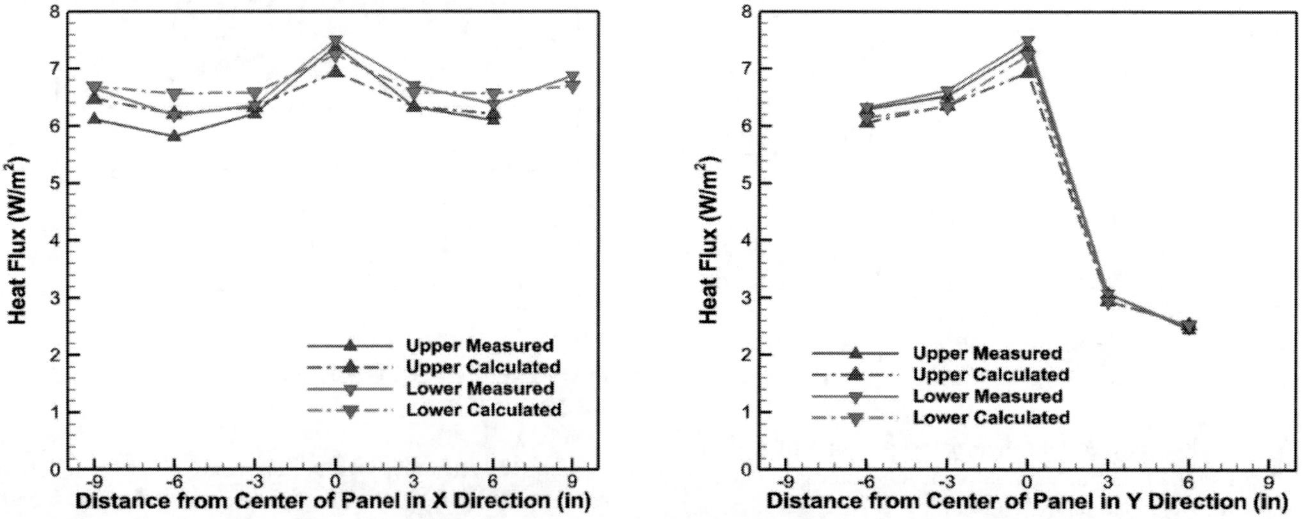

Figure 32 *Measured and calculated fluxes: design option 1.*

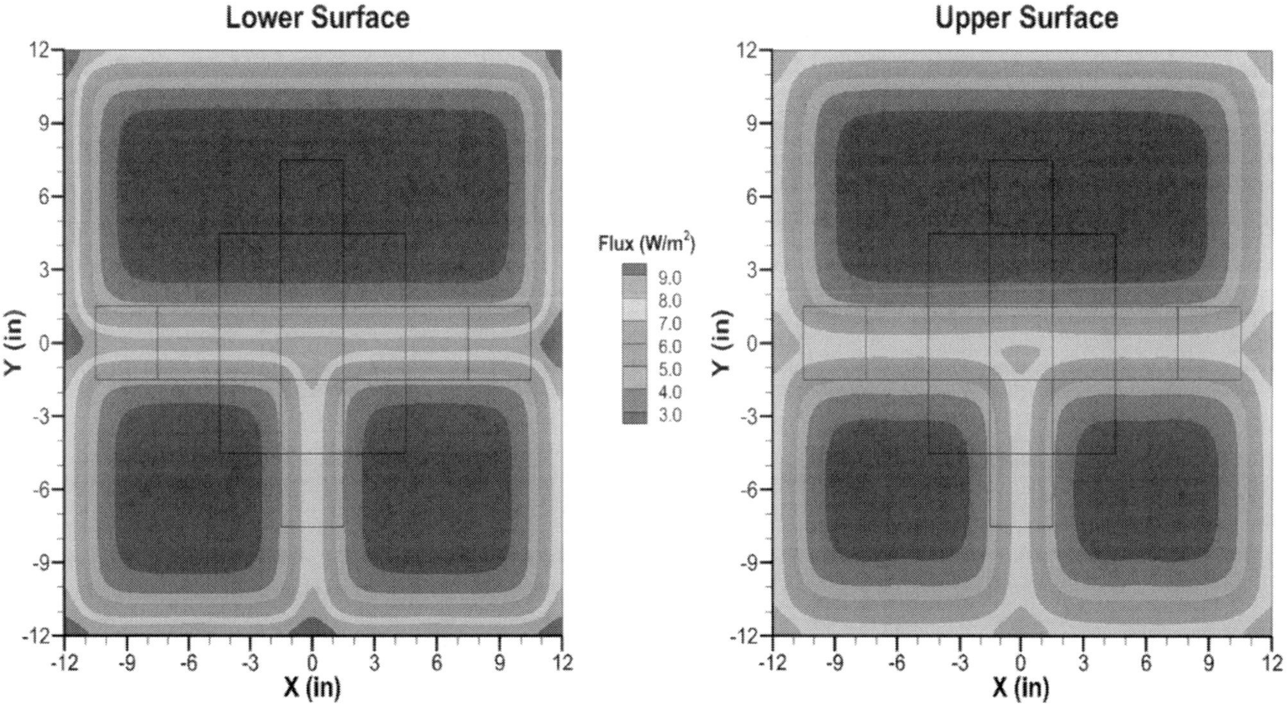

Figure 33 Calculated flux distributions on lower and upper surfaces: design option 1.

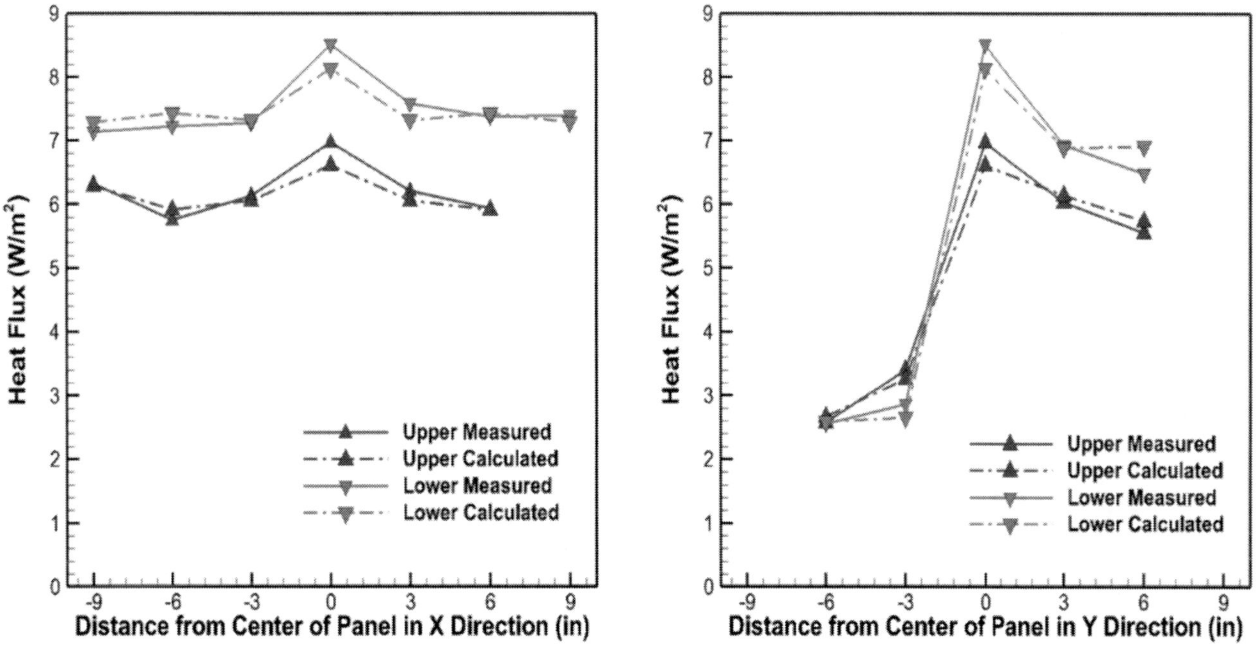

Figure 34 Measured and calculated fluxes: design option 2.

Table 2. Mapped Thermal Resistance for Design Option 1 (Test 9716*)

Transducer Color Code	Transducer(s)	Average Flux	12 × 24 in. Panel Assembly		24 × 24 in. Panel Assembly	
		(W/m²)	Area Factor	Product	Area Factor	Product
Yellow	8	7.25	2	14.50	2	14.50
Light green	13	6.37	2	12.74	2	12.74
Medium blue	15	6.11	1	6.11	5	30.56
Dark blue	3	3.06	2	6.11	2	6.11
Purple	1	2.41	5	12.05	25	60.26
Magenta	2, 4	3.00	10	30.04	18	54.07
Orange	12, 14	3.38	4	13.52	4	13.52
Medium green	6, 10	5.99	2	11.97	2	11.97
Light blue	7, 9	6.26	4	25.04	4	25.04
	Area-Weighted Average Flux (W/m^2)			4.13		3.57
	Assembly R-value ($m^2 \cdot K/W$)			5.38		6.22
	Assembly R-value ($h \cdot ft^2 \cdot °F/Btu$)			30.6		35.3

*Test 9716 was determined to have run for an insufficient time; it is only used here as an example of the influence of experimental differences upon the mapping results.

Table 3. Mapped Thermal Resistance for Design Option 1 (Test 9938)

Transducer Color Code	Transducer(s)	Average Flux	12 × 24 in. Panel Assembly		24 × 24 in. Panel Assembly	
		(W/m²)	Area Factor	Product	Area Factor	Product
Yellow	8	7.35	2	14.70	2	14.70
Light green	13	6.74	2	13.47	2	13.47
Medium blue	15	6.40	1	6.40	5	31.98
Dark blue	3	2.92	2	5.83	2	5.83
Purple	1	2.52	5	12.60	25	62.99
Magenta	2, 4	2.81	10	28.12	18	50.62
Orange	12, 14	3.61	4	14.45	4	14.45
Medium green	6, 10	6.29	2	12.59	2	12.59
Light blue	7, 9	6.53	4	26.11	4	26.11
	Area-Weighted Average Flux (W/m^2)			4.20		3.64
	Assembly R-value ($m^2 \cdot K/W$)			5.30		6.11
	Assembly R-value ($h \cdot ft^2 \cdot °F/Btu$)			30.1		34.7

Table 4. Mapped Thermal Resistance for Design Option 2 (Test 9953)

Transducer Color Code	Transducer(s)	Average Flux (W/m²)	12 × 24 in. Panel Assembly		24 × 24 in. Panel Assembly	
			Area Factor	Product	Area Factor	Product
Yellow	8	7.75	2	15.49	2	15.49
Light green	3	6.49	2	12.97	2	12.97
Medium blue	1	6.02	1	6.02	5	30.10
Dark blue	13	3.13	2	6.26	2	6.26
Purple	15	2.57	5	12.85	25	64.26
Magenta	12, 14	2.97	10	29.74	18	53.53
Orange	2, 4	3.46	4	13.83	4	13.83
Medium green	6, 10	6.57	2	13.15	2	13.15
Light blue	7, 9	6.80	4	27.21	4	27.21
	Area-Weighted Average Flux (W/m²)			4.30		3.70
	Assembly R-value (m²·K/W)			5.17		6.01
	Assembly R-value (h·ft²·°F/Btu)			29.4		34.1

Table 5. Mapped Thermal Resistance for Design Option 3 (Test 9954)

Transducer Color Code	Transducer(s)	Average Flux (W/m²)	12 × 24 in. Panel Assembly		24 × 24 in. Panel Assembly	
			Area Factor	Product	Area Factor	Product
Yellow	8	7.46	2	14.92	2	14.92
Light green	3	6.93	2	13.86	2	13.86
Medium blue	1	6.46	1	6.46	5	32.31
Dark blue	13	3.01	2	6.03	2	6.03
Purple	15	2.56	5	12.79	25	63.95
Magenta	12, 14	2.93	10	29.30	18	52.74
Orange	2, 4	3.64	4	14.57	4	14.57
Medium green	6, 10	6.35	2	12.70	2	12.70
Light blue	7, 9	6.57	4	26.29	4	26.29
	Area-Weighted Average Flux (W/m²)			4.28		3.71
	Assembly R-value (m²·K/W)			5.19		5.99
	Assembly R-value (h·ft²·°F/Btu)			29.5		34.0

Table 6. Mapped Thermal Resistance for Design Option 4 (Test 9763)

Transducer Color Code	Transducer(s)	Average Flux (W/m²)	12 × 24 in. Panel Assembly		24 × 24 in. Panel Assembly	
			Area Factor	Product	Area Factor	Product
Yellow	8	7.06	2	14.12	2	14.12
Light green	3	6.24	2	12.47	2	12.47
Medium blue	1	5.74	1	5.74	5	28.72
Dark blue	13	2.72	2	5.45	2	5.45
Purple	15	2.39	5	11.95	25	59.77
Magenta	12, 14	2.57	10	25.73	18	46.32
Orange	2, 4	3.23	4	12.91	4	12.91
Medium green	6, 10	5.76	2	11.53	2	11.53
Light blue	7, 9	6.01	4	24.03	4	24.03
	Area-Weighted Average Flux (W/m²)			3.87		3.36
	Assembly R-value (m²·K/W)			5.74		6.60
	Assembly R-value (h·ft²·°F/Btu)			32.6		37.5

Table 7. Mapped Thermal Resistance for Design Option 6 (Test 9955)

Transducer Color Code	Transducer(s)	Average Flux (W/m²)	12 × 24 in. Panel Assembly		24 × 24 in. Panel Assembly	
			Area Factor	Product	Area Factor	Product
Yellow	8	6.65	2	13.30	2	13.30
Light green	13	6.13	2	12.26	2	12.26
Medium blue	15	5.75	1	5.75	5	28.77
Dark blue	3	2.78	2	5.56	2	5.56
Purple	1	2.39	5	11.94	25	59.69
Magenta	2, 4	2.80	10	27.95	18	50.32
Orange	12, 14	3.35	4	13.39	4	13.39
Medium green	6, 10	5.76	2	11.52	2	11.52
Light blue	7, 9	5.97	4	23.88	4	23.88
	Area-Weighted Average Flux (W/m²)			3.92		3.42
	Assembly R-value (m²·K/W)			5.66		6.50
	Assembly R-value (h·ft²·°F/Btu)			32.2		36.9

Table 8. Mapped Thermal Resistance for Design Option 8 (Test 9956)

Transducer Color Code	Transducer(s)	Average Flux (W/m^2)	12 × 24 in. Panel Assembly Area Factor	12 × 24 in. Panel Assembly Product	24 × 24 in. Panel Assembly Area Factor	24 × 24 in. Panel Assembly Product
Yellow	8	7.22	2	14.43	2	14.43
Light green	3	5.13	2	10.27	2	10.27
Medium blue	1	4.81	1	4.81	5	24.04
Dark blue	13	3.60	2	7.20	2	7.20
Purple	15	3.04	5	15.21	25	76.07
Magenta	12, 14	3.51	10	35.07	18	63.12
Orange	2, 4	3.76	4	15.05	4	15.05
Medium green	6, 10	6.77	2	13.54	2	13.54
Light blue	7, 9	6.81	4	27.24	4	27.24
		Area-Weighted Average Flux (W/m^2)		4.46		3.92
		Assembly R-value (m^2·K/W)		4.98		5.67
		Assembly R-value (h·ft^2·°F/Btu)		28.3		32.2

Table 9. Summary and Ranking of Mapped Thermal Resistance

Ranking (Best to Worst)	Design Option	Assembly R-value 12 × 24 in. Panel	Assembly R-value 24 × 24 in. Panel
1	4	32.6	37.5
2	6	32.2	36.9
3	1	29.9	34.5
4/5	2	29.4	34.1
	3	29.5	34.0
6	8	28.3	32.2

(12 × 24 in.) or a 60 × 60 cm (24 × 24 in.) panel. For this technique to work, the gaps between VIPs have to be centered over the HFTs, as was done here. Also there is an inherent assumption that the edge effects are not significant for a transducer that is centered 15 cm (6 in.) from the edge of a VIP.

The following tables give estimates of the effective R-value for design options 1, 2, 3, 4, 6, 8, 11, and 12. The heat flux listed is the average of the top and bottom fluxes for the transducer locations listed. Option 1 was tested multiple times. Two sets of these results are given for design option 1 to get an indication of how much the variability in measurements affects the estimated R-value. The two cases that are compared are the ones with the lowest measured fluxes (9716) and the highest measured fluxes (9938). There is a difference in estimated R-value from the two tests of about 2.1% for a 30 × 60 cm (12 × 24 in.) panel and about 1.8% for a 60 × 60 cm (24 × 24 in.) panel.

The design options are ranked from best to worst based on effective R-value in Table 11.

EXTRACTION OF VIP THERMALPROPERTIES FROM TEST DATA

While mapping HFT values, as presented in the previous section, can be useful in determining the performance of specific design options that have been tested, it does not allow the exploration of alternate design options that have not been tested. Numerical modeling is an option for the exploration of alternate designs that have not been tested, but this is only possible if thermal properties of the VIPs are known. The key VIP properties needed for the numerical model are: (1) the effective core thermal conductivity, which influences the amount of heat conducted through the VIP and (2) the product of the barrier material's thickness and thermal conductivity, which influences the amount of heat conducted around the edges of the VIP. This section presents a methodology for extracting these properties utilizing the data collected from HFM experiments.

The main purpose of running a test with a single large VIP is to obtain a center-of-panel effective core thermal conductivity. Design option 5 contains a single large VIP. The measured heat flux and temperature difference across the assembly can be used to obtain a center-of-panel R-value for the assembly. Since the thickness and temperature-dependent thermal conductivities of the rigid foam layers are known, their contribution to the total R-value can be removed to give the center-of-panel R-value for the VIP. The VIP center-of-panel R-value is then used to calculate an effective core thermal conductivity for the VIP. The effective core thermal conductivity was determined using the data from four separate tests of design option 5. The conductivity was first calculated using the measured heat flux at the center of the panel (transducer #8). The conductivity was calculated three ways: using the flux measured on the upper plate, using the flux measured on the lower plate, and using the average of the flux on the

Table 10. Effective Core Thermal Conductivity (W/m·K) Derived from Tests of Design Option 5

Test	Center Transducer Only (Transducer 8)			Average of Central Nine Transducers (Transducers 2, 3, 4, 7, 8, 9, 12, 13 ,14)		
	Upper	Lower	Average U and L	Upper	Lower	Average U and L
9767	0.00344	0.00351	0.00348	0.00337	0.00358	0.00348
9907	0.00349	0.00355	0.00352	0.00341	0.00360	0.00350
9909	0.00350	0.00354	0.00352	0.00343	0.00359	0.00351
9913	0.00350	0.00353	0.00351	0.00343	0.00358	0.00350
Average	0.00348	0.00353	0.00351	0.00341	0.00359	0.00350

Table 11. Effective Core Thermal Conductivity (W/m·K) Derived from Tests of 600 × 600 mm (23.5 × 23.5 in.) VIPs

Test	Center Transducer Only (Transducer 8)			Average of Central Nine Transducers (Transducers 2–4, 7–9, 12–14)		
	Upper	Lower	Average U and L	Upper	Lower	Average U and L
9822	0.00427	0.00438	0.00433	0.00417	0.00391	0.00403
9825	0.00432	0.00443	0.00438	0.00420	0.00391	0.00405
Average	0.00430	0.00441	0.00436	0.00419	0.00391	0.00404

upper and lower plate. Using this center transducer, the difference between the thermal conductivities calculated from the upper plate flux and lower plate flux is less than 2% for all tests. When using the average of the central nine transducers, the upper to lower difference is as much as 6%; but this difference is still within the experimental uncertainty the measured heat fluxes. The average of the central nine transducers is probably a better indication of the true flux than a single value at the center, so the best estimate of the core conductivity from these tests is 0.0035 W/m·K.

Two bare VIPs were also tested in the same HFM apparatus. These VIPs are not part of an assembly and are nominally 600 mm (23.5 in.) square. The effective core thermal conductivity calculated from these tests is given in Table 11, and is about 15% greater than that of the VIPs within the composite test specimens. These VIPs were manufactured in May 2011 by the same manufacturer as those used in the test assemblies. The difference between the two sets of VIPs is unknown. Design option 5 was disassembled, and the VIP was tested by itself to see if there was something about the assembly that may have skewed the calculated core conductivity for the panel. The resulting core conductivity from the test of the VIP alone was within 2% of the value obtained from testing of the assembly.

In addition to direct measurements of VIP properties, numerical models can be used to extract this information from the array of heat flux data generated by the HFM apparatus test. Therefore, detailed three-dimensional models of several of the VIP test assemblies were developed using the Oak Ridge National Library-developed (ORNL) heat transfer program HEATING (Childs 1993). All of the assemblies consist of one or more VIPs sandwiched between layers of rigid foam insulation.

These models include details such as the multiple layers of barrier material present on some portions of the VIPs due to the way edges are sealed and wrapped around the panel. It is not known a priori if these details significantly impact the overall performance of the assembly, but they may significantly impact the reading of a transducer placed directly over the edge. Therefore it is important to include them in the model used to extract properties from transducer readings. Unfortunately there are other features of the VIP assemblies that are not included in the model. For example, the amount, distribution, and properties of the silicone adhesive used to adhere the VIP to the foam within the test samples are not known. The size and location of small gaps between components and misalignment of the VIPs are also not known.

The models calculate the average heat flux that would be seen by each of the 30 transducers in the HFM apparatus. The HEATING model is coupled with an optimization procedure that varies the VIP properties (effective core thermal conductivity and the conductivity-thickness product of the barrier) to find the combination that gives the best match between the measured and calculated heat fluxes. The best match is defined as the combination that minimizes the function shown in Equation 3.

$$F = \frac{\sum_{i=1}^{n} W_i (q_{i-\text{meas}} - q_{i-\text{calc}})^2}{\sum_{i=1}^{n} W_i} \quad (3)$$

where W_i is a weighting factor. The weighting factors are assigned based on the approximate proportion of the overall area that the area covered by a transducer represents.

Detailed models were developed for the first four design options as shown in Figure 35.

Table 12. **Effective Thermal Conductance of Vacuum Insulation Panel Components**

Design Option (Test ID)	Use All Available Transducers		Exclude Edge Transducers (5 & 11)	
	Barrier $k \times t$ (W/K)	Effective Core k (W/m·K)	Barrier $k \times t$ (W/K)	Effective Core k (W/m·K)
1 (9936)	2.28×10^{-4}	3.49×10^{-3}	2.31×10^{-4}	3.48×10^{-3}
2 (9953)	1.89×10^{-4}	3.59×10^{-3}	1.89×10^{-4}	3.59×10^{-3}
3 (9954)	2.54×10^{-4}	3.53×10^{-3}	2.74×10^{-4}	3.51×10^{-3}
4 (9763)	1.25×10^{-4}	3.25×10^{-3}	1.18×10^{-4}	3.25×10^{-3}
Average of 1–3	2.24×10^{-4}	3.54×10^{-3}	2.31×10^{-4}	3.53×10^{-3}

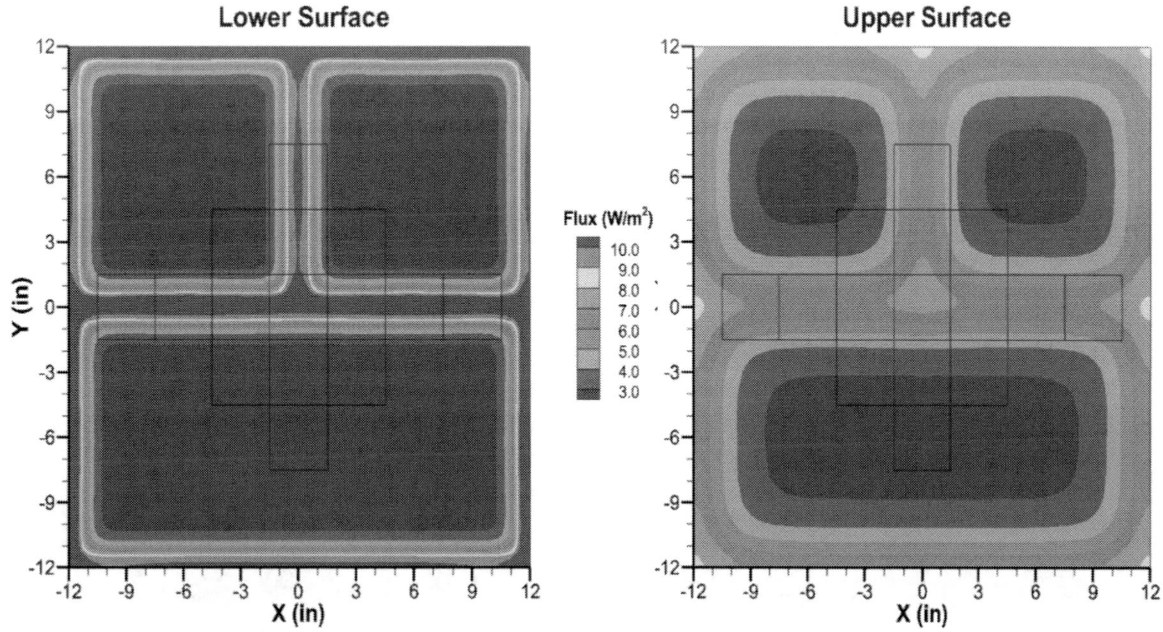

Figure 35 Calculated flux distributions on lower and upper surfaces: design option 2.

Since the two transducers nearest the edges of the device (5 and 11) may be influenced by the surroundings (i.e., the edges are not perfectly adiabatic) the optimization was performed two ways: once with all available transducers used and once with edge transducers removed. Removal of transducers 5 and 11 is accomplished by setting their weighting factors to zero. The results for these two optimizations were nearly identical. The calculated and measured heat fluxes are compared for both the upper and lower surfaces of the test specimens, along with contour plots of calculated flux distribution for design options 1 through 4 in Figure 32 to Figure 39.

The resulting properties are given in Table 12. Since some features of the assemblies were not included in the model, the calculated properties may not represent the actual physical properties of the materials, but rather are effective properties that compensate for the impact of the omitted features. Estimated properties from the analyses of design options 1, 2, and 3 are fairly similar to each other: the minimum and maximum estimates of effective core conductivity differ by about 3%. Also, the average calculated value of effective core thermal conductivity matches very well with that found when measuring the center of design option 5, 0.0035 W/m·K, as was shown in Table 12. There is a greater variation in the estimated properties of the barrier material. However, design option 4 produces much lower estimates of the thermal conductivity for both the core material and the barrier. Design option 4 has a band of PVC covering the edges of each of the VIPs and covering the bottom layer of XPS, but not covering the top layer of XPS, as shown in Figure 44. The thermal conductivity of the PVC was obtained from literature values, not measured. It is possible that the value for the PVC conductivity used in the model is too high, resulting in the model predicting lower core and barrier conductivities to compensate. The model assumes that the PVC is in perfect thermal contact with the VIP and the foam insulation—examination of panel after the tests showed that this was not the case. The model also assumes a VIP thickness of 25.4 mm, whereas

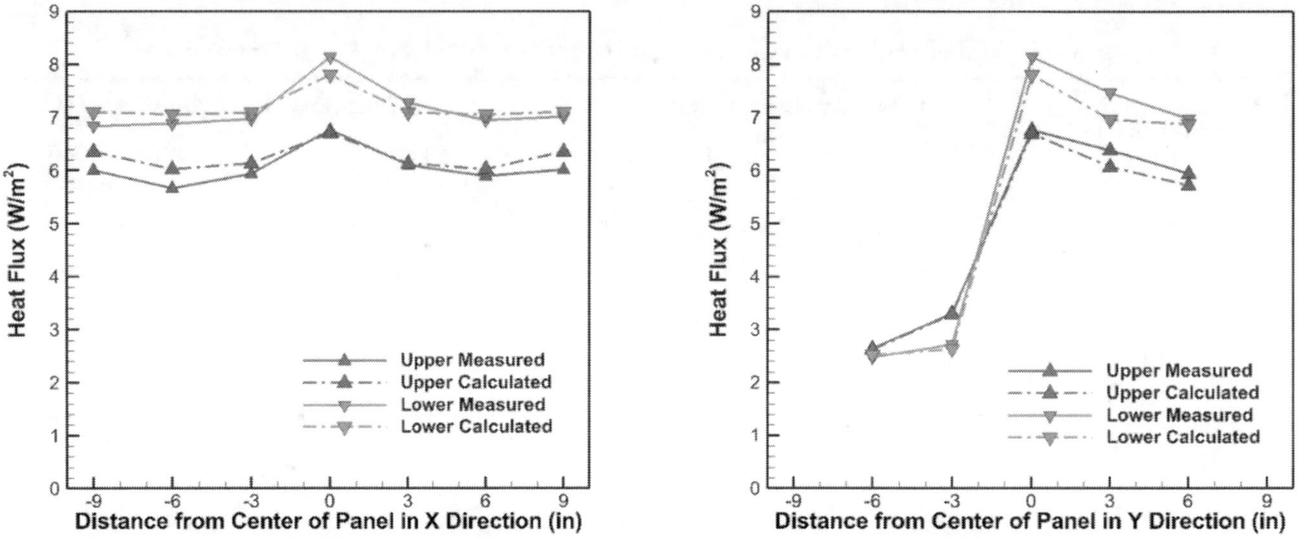

Figure 36 *Measured and calculated fluxes:design option 3.*

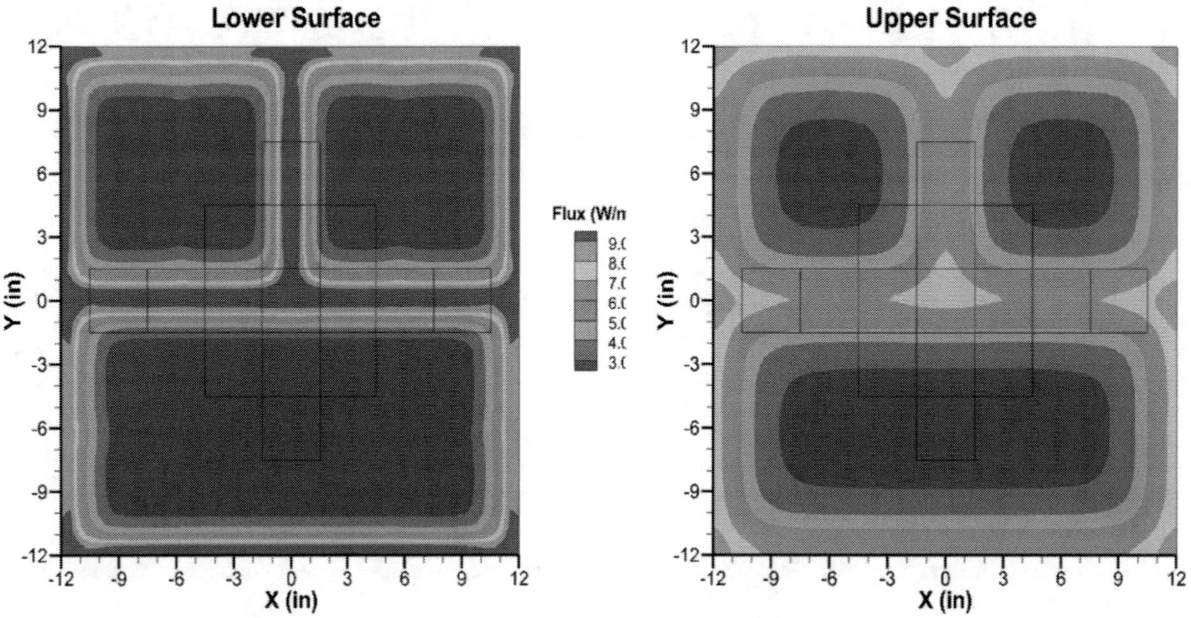

Figure 37 *Calculated flux distributions on lower and upper surfaces: design option 3.*

measurements of two bare VIPs gave values of 26.23 mm and 25.14 mm. The variation in the VIP thickness is on the same order as the variability in the calculated core conductivities for tests on design options 1–3. These factors illustrate the importance of obtaining accurate data as input to a model being used to estimate other parameters. There is greater confidence in the property estimates from design options 1–3.

Earlier studies at ORNL (Wilkes et al. 1997; Wilkes et al. 1999; Stovall and Brzezinski 2002) have indicated that it is difficult to determine the properties of both the core and barrier from testing of the assembly only. The heat flux at any transducer location results from a combination of heat conducted directly through the core and heat conducted around the core in the barrier. An increase/decrease in the core conductivity can be offset by a decrease/increase in barrier conductivity to produce about the same average heat flux at a transducer location. When trying to minimize the difference between the measured and calculated heat fluxes there is not a single, narrowly-defined combination but rather a range of combinations of conductivities that give nearly the same result. The optimizer will find the combination that gives the best fit, but there may be other combinations near this one that are almost as good.

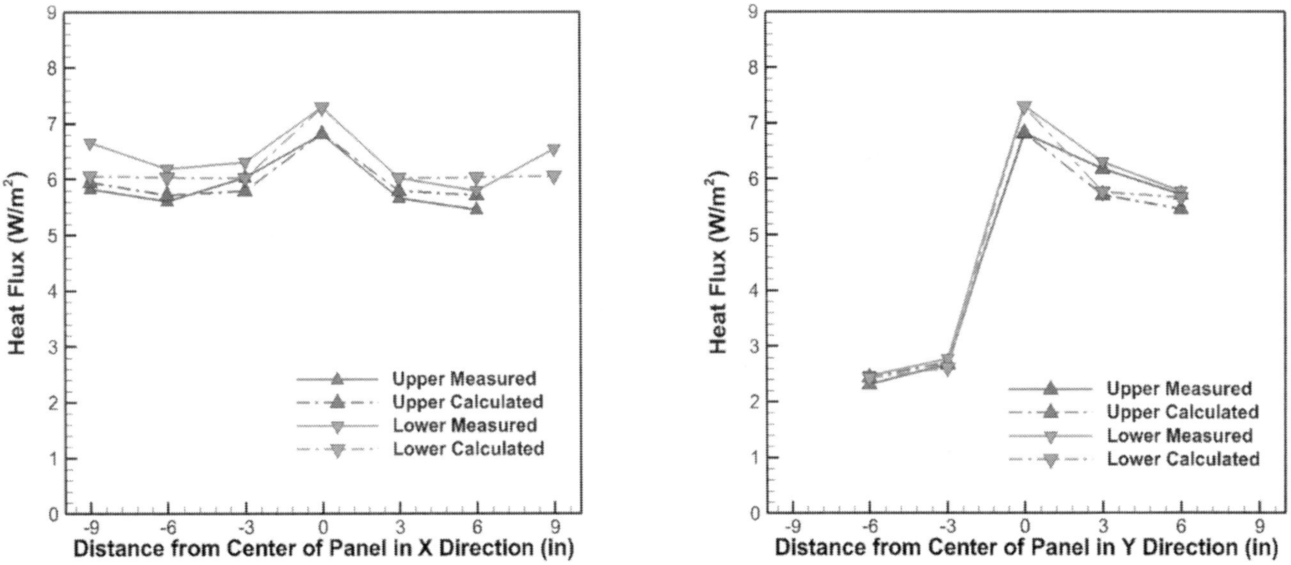

Figure 38 *Measured and calculated fluxes: design option 4.*

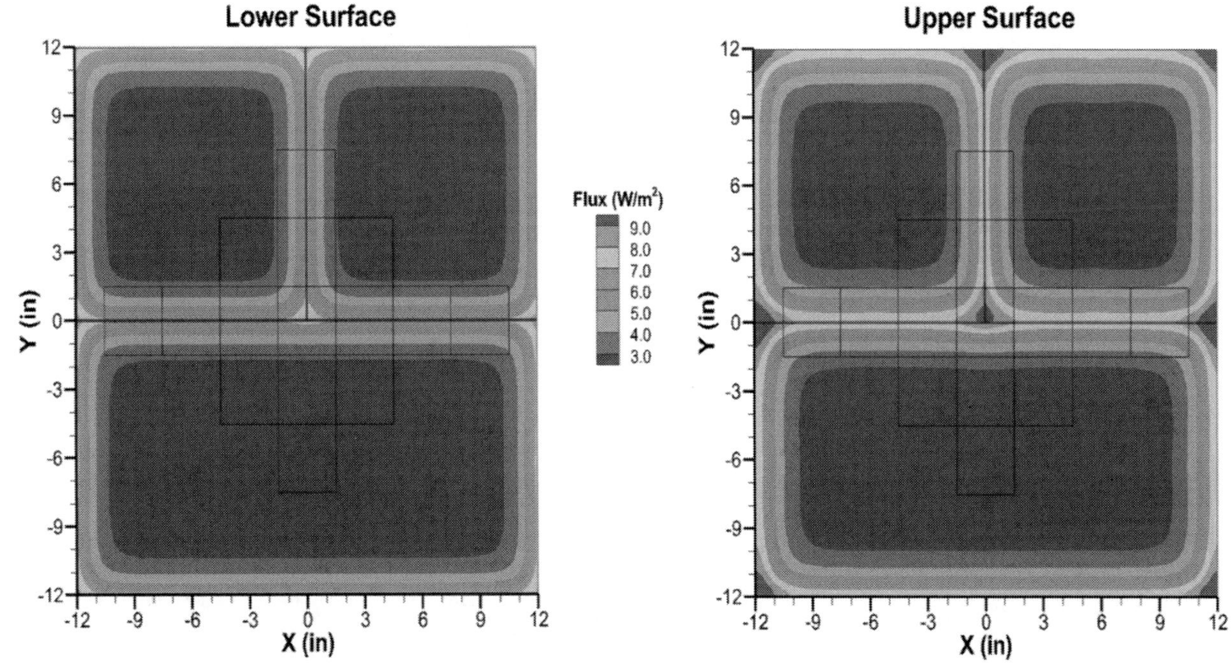

Figure 39 *Calculated flux distributions on lower and upper surfaces: design option 4 (PVC extends from bottom plate to top of VIP, but does not extend upwards to the top plate).*

Since there is an independent measurement of the core thermal conductivity, there might be more confidence in the ability to calculate the barrier properties if the core conductivity is assumed to be fixed. However there would still be a problem with the impact of the features that are not included in the model. With the core conductivity fixed at the value obtained from testing design option 5 (3.50×10^{-3} W/m·K) the barrier conductivity/thickness product was varied to determine the value that gives the best fit to the measured heat flux data. The results of this exercise are given in Figure 41 for design options 1, 2, and 3. This graph shows the optimization function defined in Equation 3 and the results are very similar to those obtained when simultaneously optimizing on both the barrier and core properties, shown in Table 12. Given the uncertainties in heat flux measurement and geometry details, the scatter shown in Table 12 may be indicative of the uncertainty in the calculated properties.

Figure 40 Corner view of design option 4 test specimen.

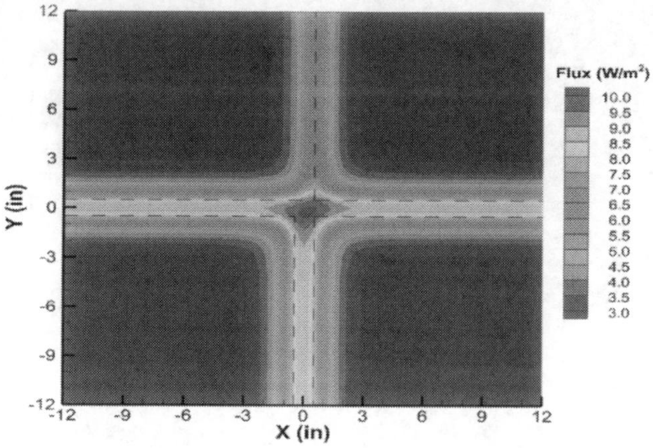

Figure 42 Surface flux around VIP joints for assembly tested in hot box.

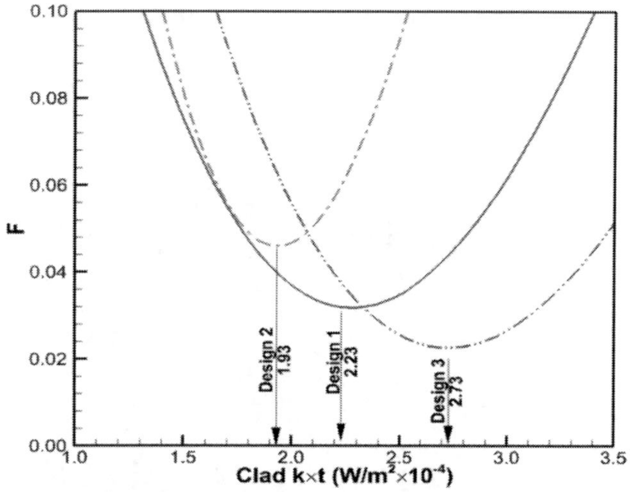

Figure 41 Optimal clad properties with core conductivity fixed at 0.0035 W/m·K.

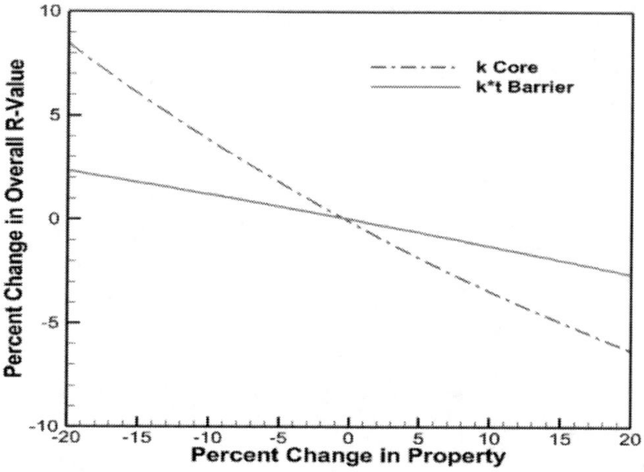

Figure 43 Sensitivity of R-value to changes in VIP's constituent properties.

NUMERICAL MODELING OF VIPS

A model was developed of a 24 in.^{2} portion of the VIP assemblies used in the 10 ×10 ft wall tested in the rotatable guarded hot box. This portion is representative of the whole wall because the pattern repeats to cover the entire wall surface. The VIP properties used in the model were those derived from tests of design options 1, 2, and 3, that is 0.00353 W/m^2 for the effective core region thermal conductivity and 0.00023 W/K for the barrier conductance. The same temperature boundary conditions used in the HFM tests were applied to the surfaces. The resulting heat flux is given in the Figure 42. In this figure, the top two vacuum panels are butted against each other, while the bottom two panels are separated by a piece of EPS foam insulation. To achieve this design, the top left vacuum panel is actually slightly larger than the other vacuum panels, as can be seen in the offset between the upper joint parallel to the *y*-axis and the lower joint. This is also reflected in the heat flux map, with a heat flux of about 7 W/m^2 at the butt joint and about 9 W/m^2 at the EPS joint.

The model gives an R-value of 5.79 m^2·K/W (32.9 h·ft^2·°F/Btu) for 24 in.^{2} model. However, if the effective core conductivity is increased to 0.0041 W/m·K (as was measured in tests 9822 and 9825 for the 600 mm^2 [23.5 in.^{2}] VIPs) is used then the R-value is 5.41 m^2·K/W (30.7 h·ft^2·°F/Btu). Since the model only considers the VIP assembly and not the sheathing, stud wall, gypsum board face, and surface film coefficients, the R-value for the wall tested in the hot box would be higher.

Once developed, a model can be used to determine the impact of various factors on the wall's performance. For instance, Figure 43 shows the impact changes in the VIP properties have on the assembly's R-value. The R-value is much more sensitive to changes in the effective core conductivity than it is to changes in the barrier conductivity-thickness product. The model was also used to examine the impact of the barrier seals that wrap around the ends of a VIP (as previously shown in Figure 32). When the multiple layered areas in the model were changed to a single layer of barrier material the R-value improved by 2.7%.

SUMMARY AND CONCLUSIONS

A proposed wall system incorporates VIPs enclosed within closed-cell insulating foam. In addition to adding some thermal resistance the foam serves to protect the vacuum panels during construction and to provide a surface appropriate for an adhesive joint on both sides of the foam VIP unit.

Multiple configurations of a composite VIP foam insulation structure were evaluated using small sub-sections that could be placed within a HFM apparatus. Through careful test specimen construction, all component joints were located within the range of an array of HFTs installed within the upper and lower plates of that apparatus. Special calibration procedures were used to ensure accurate heat flux measurements for these very-low thermal conductivity specimens. The resulting measurements were combined with modeling efforts to investigate the impact of vacuum-panel size, the type of foam used to encase the vacuum panels, the thickness and shape of the foam sections between panels, and possible adhesive effects.

Some of the proposed configurations placed a greater amount of foam insulation on one side of the VIP panels than on the other. For a given total amount of foam insulation, this arrangement was found to produce a more uniform heat flux, and therefore more uniform temperature distribution, on the outside of the thicker foam surface. This may prove useful if there is any concern with regard to uneven surface soiling phenomena. However, panels constructed in this manner would have the inner and outer faces clearly marked to ensure proper installation during the construction process.

A new procedure was developed to map the HFM measurement results onto full-scale wall designs to predict the system thermal performance. This new method for mapping transducer measurements onto an array of larger panels shows promise, and was used to quantify the expected performance of several candidate construction arrangements. Several potential arrangements of the foam VIP unit were also evaluated using a finite-difference methodology.

The major conclusions of the project can be summarized as:

- It is possible to develop a wall with an overall performance of U-factor of 0.19 W/(m^2·K) (R-30 [h·ft^2·F]/Btu) or greater with multiple VIP arrangements.
- The best arrangements are those that maximize VIP coverage.
- The wall thermal performance is much more sensitive to the effective core thermal conductivity than to the barrier thermal conductance.
- Distributing a greater portion of the foam insulation on the side of the panel facing the exterior environment may be advantageous.

ACKNOWLEDGMENTS

This report was prepared as an account of work sponsored by an agency of the United States Government and Dow Corning. Neither the United States Government, Dow Corning, nor any agency thereof, nor any of their employees, makes any warranty, express or implied, or assumes any legal liability or responsibility for the accuracy, completeness, or usefulness of any information, apparatus, product, or process disclosed, or represents that its use would not infringe privately owned rights. Reference herein to any specific commercial product, process, or service by trade name, trademark, manufacturer, or otherwise, does not necessarily constitute or imply its endorsement, recommendation, or favoring by the United States Government or any agency thereof. The views and opinions of authors expressed herein do not necessarily state or reflect those of the United States Government or any agency thereof.

REFERENCES

ASTM. 2010. ASTM Standard C177, *Standard Test Method for Steady-State Heat Flux Measurements and Thermal Transmission Properties by Means of the Guarded-Hot-Plate Apparatus*. West Conshohocken, PA: ASTM.

ASTM. 2010. ASTM Standand C518, *Standard Test Method for Steady-State Thermal Transmission Properties by Means of the Heat Flow Meter Apparatus*. West Conshohocken, PA: ASTM.

ASTM. 2010. ASTM Standard C1667, *Standard Test Method for Using Heat Flow Meter Apparatus to Measure the Center-of-Panel Thermal Resistivity of Vacuum Panels*. West Conshohocken, PA: ASTM.

Childs, K.W. 1993. *HEATING 7.2 User's Manual*, ORNL/TM-12262. Oak Ridge, Tennessee: Oak Ridge National Laboratory.

Stovall, T.K., Brzezinski, A. 2002. *Vacuum Insulation Round-Robin to Compare Different Methods of Determining Effective Vacuum Insulation Panel Thermal Resistance, Insulation Materials: Testing and Applications,* Fourth Volume STP 1426. A. Desjarlais, ed. West Conshohocken, PA: ASTM.

Taylor, B.N., and Kuyatt, C.E. 1994. Guidelines for evaluating and expressing the uncertainty of NIST measurement results, NIST Technical Note 1297.

US Department of Energy. 2008. *Energy Efficiency Trends in Residential and Commercial Buildings*. http://apps1.eere.energy.gov/buildings/publications/pdfs/corporate/bt_stateindustry.pdf.

Wegger, E. P.J. Bjørn, E. Sveipe, S. Grynning, A. Gustavsen, R. Baetens and J.V. Thue. 2011. Aging effects on thermal properties and service life of vacuum insulation panels. *Journal of Building Physics* 35:128.

Wilkes, K.E., et al. 1997 *Development of Metal-Clad Filled Evacuated Panel Superinsulation*, ORNL/M-5871.

Wilkes, K. E., et al. 1999. *Development of Cladding Materials for Evacuated Panel Superinsulation*, C/ORNL 92-0123.

A Study of the Energy-Saving Potential of Metal Roofs Incorporating Dynamic Insulation Systems

Kaushik Biswas, PhD
Associate Member ASHRAE

William Miller, PhD
Member ASHRAE

Scott Kriner

Gary Manlove

ABSTRACT

This article presents various metal roof configurations that were tested at Oak Ridge National Laboratory in Tennessee, U.S. between 2009 and 2013, and describes their potential for reducing the attic-generated space-conditioning loads. These roofs contained different combinations of phase-change material, rigid insulation, low emittance surface, and above-sheathing ventilation with standing-seam metal panels on top. These roofs were designed to be installed on existing roofs' decks, or on top of asphalt shingles for retrofit construction.

All the tested roofs showed the potential for substantial energy savings compared to an asphalt shingle roof, which was used as a control for comparison. The roofs were constructed on a series of adjacent attics separated at the gables using thick foam insulation. The attics were built on top of a conditioned room. All attics were vented at the soffit and ridge. The test roofs and attics were instrumented with an array of thermocouples. Heat flux transducers were installed in the roof deck and attic floor (ceiling) to measure the heat flows through the roof and between the attic and conditioned space below. Temperature and heat flux data were collected during the heating, cooling and swing seasons over a three-year period. Data from previous years of testing have been published. Here, data from the latest roof configurations being tested in year three of the project are presented. All test roofs were highly effective in reducing the heat flows through the roof and ceiling, and in reducing the diurnal attic-temperature fluctuations.

INTRODUCTION

This article describes phase three of a study that began in 2009 to evaluate the energy benefits of a sustainable reroofing technology utilizing standing-seam metal roofing panels combined with energy efficient features like above-sheathing ventilation (ASV), phase-change material (PCM) and rigid insulation board. The data from phases one and two have been previously published and reported (Kosny et al. 2011; Biswas et al. 2011; Biswas and Childs 2012; Kosny et al. 2012). Based on previous data analysis and discussions within the research group, additional test roofs were installed in May 2012 to test new configurations and further investigate different components of the dynamic insulation systems.

It has been well-documented that roofs and attics experience higher temperature fluctuations than other building envelope components. A study by Huang et al. found roofs contributed 12%–14% of the heating and cooling loads in residential buildings (1999). A great deal of research has been devoted to reducing the roof and attic-generated space-conditioning loads. Parker et al. compared the heat transfer through direct-nailed and counter-batten tile roofs to an asphalt shingle roof and found 50% reductions in heat transfer (1995). Akbari et al. found that a roof surface with high solar reflectance and high thermal emittance resulted in cooling energy savings in moderate and hot climates (2004). Miller and Kosny studied prototype roof designs that combined strategies like infrared reflective roofs, radiant barriers, ASV, low emittance (low-e) surfaces, insulation, and thermal mass to regulate the attic temperature and reduce the ceiling heat transfer (2007). Attic radiant barriers have shown the pote5ntial for reducing radiation heat exchange across roof cavities and attic spaces (Medina 2010). In addition to the aforementioned technologies, PCMs

Kaushik Biswas *is an R&D associate (building scientist) and* ***William Miller*** *is an R&D staff member at Oak Ridge National Laboratory, Oak Ridge, TN.* ***Scott Kriner*** *is techhnical director at Metal Construction Association, Glenview, IL.* ***Gary Manlove*** *is president of Metanna, Monument, CO.*

are increasingly being investigated for use in building envelopes, including roofs for thermal storage, and a consequent reduction in space-conditioning loads. Zalba et al. (2003), Sharma et al. (2009) and Xin et al. (2009) have provided detailed reviews of the thermal storage systems incorporating PCMs, including those for building envelope applications.

The current test roofs combined PCM, fiberglass insulation, a low-e surface, and ASV in different configurations to evaluate their potential in reducing the heating and cooling loads in buildings. In addition to the steady thermal resistance of the fiberglass insulation, these roofs utilize the dynamic effects of the PCM, ASV airflow, and the low-e surface to reduce the heat flow through the roof. The roofs were designed and built by a collaboration between Metal Construction Association (MCA), CertainTeed Corporation, Phase Change Energy Solutions, and Oak Ridge National Laboratory (ORNL). MCA is a North American trade association of metal building manufacturers, builders, and material suppliers; CertainTeed is a manufacturer of thermal insulation and building envelope materials; and Phase Change Energy Solutions manufactured the PCM.

The current dynamic roofs were designed such that they can be installed in new construction or on top of existing roofs. According to a 2002 report, asphalt shingles covered more than 85% of the residential roofing area in the U.S (Dodge 2002). Reroofing at the end of their service lives generates an estimated 6.8 million tons of waste asphalt shingles per year in the U.S., requiring large disposal areas (Sengoz and Topal 2005). Using the present dynamic roofing systems for reroofing and retrofitting applications precludes the asphalt shingle waste generation, in addition to improving the energy efficiency of the building.

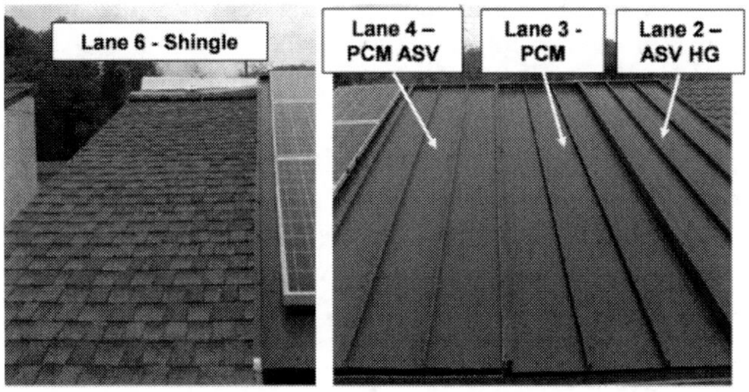

Figure 1 *Control asphalt shingle roof (left) and the test roofs (right).*

DESCRIPTION OF TEST ROOFS

These test roofs were built on side by side attics on the Envelope Research Systems Apparatus (ESRA) facility in Oak Ridge, Tennessee. The attics are thermally isolated from each other by using foam insulation at the gable ends and are vented at the soffit and ridge. A conventional asphalt shingle roof was used as a control to evaluate the energy benefits of the test roofs. Further details of the attic construction are provided by Miller (2006).

Figure 1 shows the attic roofs located on the ESRA facility. Three test roofs are described in this article and will be referred to as Lane 2 ASV HG (heat gain), Lane 3 PCM and Lane 4 PCM ASV. Lane 6 Shingle is the control roof built with asphalt shingles on the roof deck (Kosny et al. 2012). The three test roofs were constructed with different configurations of rigid fiberglass insulation, PCM, and air gap (ASV) under the metal panels, as shown in Figure 2. Figure 2 also shows how the roof assemblies were instrumented to measure the temperature distribution and heat flow through the oriented strand board (OSB) deck into and out of the attic.

Lane 4 consisted of rigid fiberglass insulation on the OSB roof deck, followed by a layer of macro-encapsulated biobased PCM and the standing-seam metal panels on top. The metal panels were placed on top of metal subpurlins that provided an air gap of about 5.1 cm over the PCM layer; the air gap was vented both at the ridge and the eave providing ASV (Miller and Kosny 2007). The fiberglass insulation is about 2.5 cm thick and has a thermal resistance of 0.76 $m^2 \cdot K/W$. The PCM was packed in arrays of plastic cells of dimensions 4.4 × 4.4 × 1.3 cm with 1.3 cm spacing (air pockets in plastic cells represented about 20% of the total volume). The nominal heat storage capacity of the PCM packed in the plastic film pouches is about 560 kJ/m^2 of the roof area. The PCM has nominal melting and freezing temperatures of about 30°C and 26°C, respec-

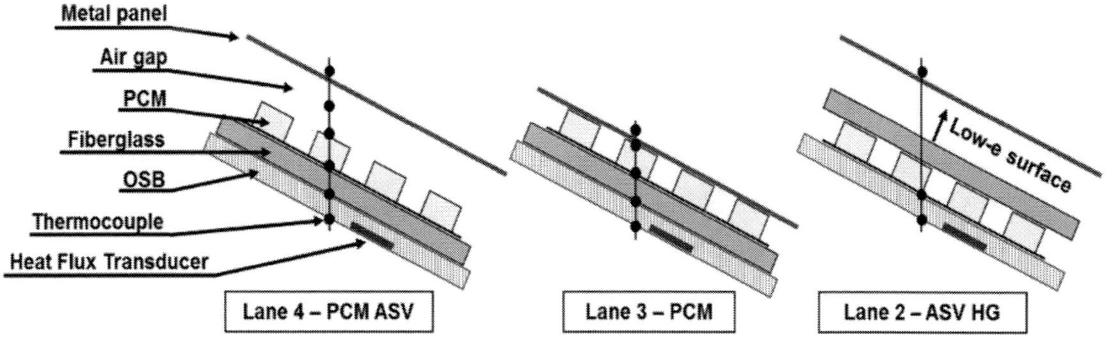

Figure 2 *Test roof configuration schematics.*

tively. The melting-phase transition enthalpy of the PCM is about 190 J/g.

Lane 3 was constructed with the same rigid fiberglass insulation on the roof deck, followed by the PCM layer, but did not contain any air gap above the PCM layer. It should be noted that the 1.3 cm spacing between the air channels still provided some ASV in this roof.

The fiberglass board in Lane 2 contained a reflective-foil facing that introduces a low-e surface which acts to reflect the incoming solar radiation and helps reduce daytime cooling loads. This roof was similar to the one studied during Phase 1 of this project (Biswas et al. 2011; Kosny et al. 2012), except for the absence of laminated photovoltaics. In addition to the thermocouples shown in Figure 2, Lane 2 was instrumented with two thermocouples, each in the air gap close to the ridge and the eave sections. The intent was to try and quantify the heat gain in the buoyant airflow generated in the air gap by natural convection (hence the nomenclature ASV HG).

ATTIC CONSTRUCTION AND INSTRUMENTATION

Figure 3 shows the typical attic construction and instrumentation on the ESRA facility. The attics are built on top of a conditioned basement. The basic roof construction consists of a weather barrier (moisture shield) on top of a 1.6 cm OSB deck. The asphalt shingles (in Lane 6, the control roof) or other test roof components are installed on top of the roof deck. The attic floor or ceiling consists of two wooden fiberboards on top of a corrugated metal sheet. No ceiling insulation was used in this study. However, during phase one of this study, R-6.7 $m^2 \cdot K/W$ ceiling insulation was installed in all attics (Biswas et al. 2011; Kosny et al. 2012). In phase three, the ceiling insulation was removed to check the energy-saving impact of the prototype roofs in retrofit applications where the original attic is poorly insulated or missing insulation.

Type T copper-constantan thermocouples are placed within the attic, roof assembly, and the ceiling, as shown in Figure 3. These are in addition to the thermocouples shown in Figure 2 and the ones added to the ASV air gap in Lane 2. Heat flux transducers (HFT), with an accuracy of ±5% and a sensitivity of 5.7 $(W/m^2)/mV$, measure the heat flows through the ceiling and the roof deck. The HFTs were calibrated using a heat flow meter apparatus while sandwiched by the same materials as in the roof deck and the ceiling, respectively. The heat flow towards the conditioned space is defined as positive (heat flow into the attic from the roof and heat flow into the conditioned basement from the attic), and vice versa.

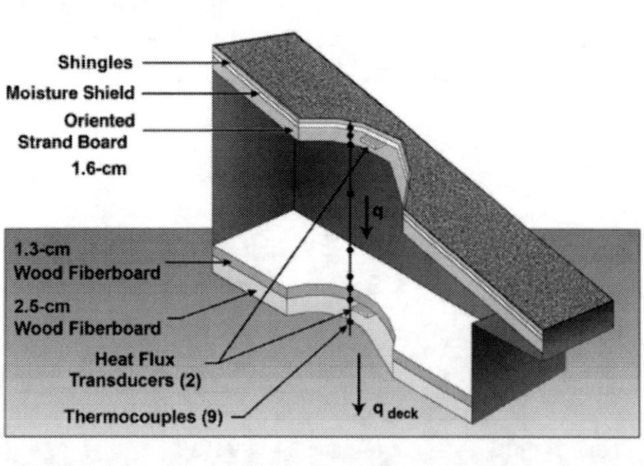

Figure 3 *Typical ESRA attic instrumentation.*

In addition to the attic instrumentation, an on-site weather station was installed that measures the solar irradiance on the sloped roofs, longwave radiation beyond 3 micrometers (μm), ambient dry bulb temperature, ambient air relative humidity, etc.

RESULTS AND DISCUSSION

In the following sections, data from the test attics are presented and discussed. The temperature and heat flux data were averaged over 15 minute periods and stored in weekly data files. Here, data collected during summer, fall, and winter of 2012–2013 are shown. Specifically, the evaluation period extends from May 30, 2012 to January 1, 2013.

Roof Surface Temperatures

Figure 4 shows the temperature variations during two typical summer days and two typical winter days. Also shown is the outdoor ambient temperature during those days. The maximum roof surface temperatures rose well above the outdoor temperature due to the daytime solar irradiance. At night, radiation losses to the sky (especially in the absence of cloud cover) lower the roof temperatures below the ambient. No significant differences were observed in the different metal panel and asphalt shingle roof temperatures. The recorded roof surface temperatures of Lane 3 (PCM) were lower than the other roofs during the winter days shown in Figure 4. It was later discovered that the Lane 3 roof thermocouple detached from the roof surface and was measuring the air temperature above the roof. Once fixed, Lane 3 roof temperatures were similar to the other roofs.

PCM Behavior

The PCM used in the present experiment was tested using a differential scanning calorimeter (DSC). The phase-change characteristics of the PCM are shown in Figure 5. The heating and cooling rate used in the DSC tests was 0.3°C per minute, and the resulting melting and freezing thresholds were observed to be 25.7°C and 27.2°C, respectively.

Figures 6 and 7 show the temperatures above and below the PCM layer during the typical summer and winter days in Lane 4. The roof design studied in Phase 1 of this project was very effective in regulating the attic temperature and reducing the ceiling heat flows, but temperature data indicated that the PCM remained frozen and was inactive throughout the winter period (Biswas et al. 2011; Kosny et al. 2012). In the current design of Lane 4, the PCM is placed on top of the fiberglass insulation.

As indicated by the temperature separation between the top and bottom of the Lane 4 PCM layer in Figures 6 and 7, the PCM did undergo both melting and freezing during both summer and winter periods, as described by Kosny et al. (2012). Figure 8 shows the weekly maximum and minimum temperatures across the PCM layer in both Lane 3 and Lane 4 roofs, compared to the phase-change onset temperatures from the DSC tests. The weekly minimum temperatures always remained below the freezing threshold temperature, and the maximum temperatures were always above the melting threshold. Thus, on a weekly basis, the PCM in both Lane 3 and Lane 4 can be expected to have been active during both summer and winter periods. It should be noted that DSC heating rate of 0.3°C per minute is higher than the temperature change rates in actual building envelopes. The data shown in Figures 6 and 7 revealed a temperature rise rate of about 0.1°C per minute at the PCM layer. At lower heating rates, the melting threshold can be expected to be lower than 25.7°C and the freezing threshold higher than 27.2°C (Castellón 2008), making the observed temperature in Lane 3 and Lane 4 even more conducive to melting and freezing of the PCM.

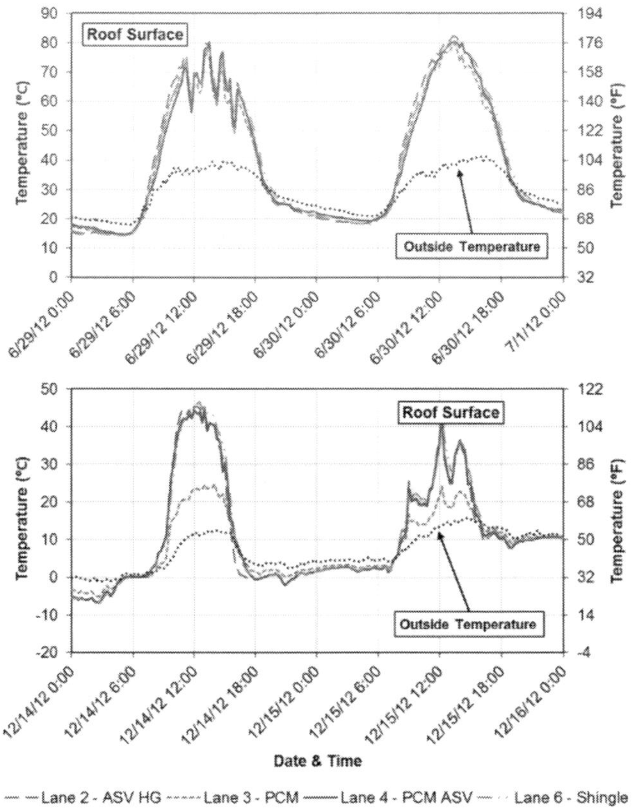

Figure 4 *Roof surface-temperature variations.*

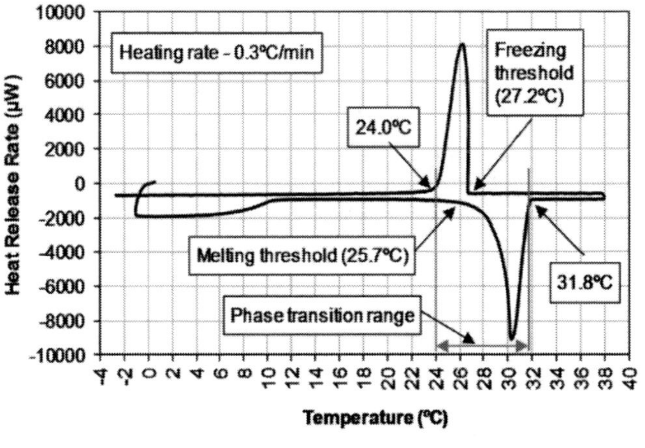

Figure 5 *Heat flow data of the PCM from differential scanning calorimetry (Kosny et al. 2012).*

Summer Performance

This section presents temperature and heat flux data from a summer week (June 27–July 3, 2012) when highest peak outdoor temperatures were observed and the overall summer

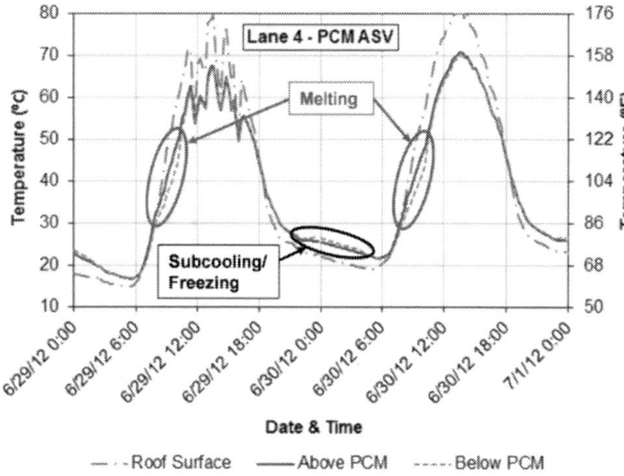

Figure 6 *Summer PCM behavior.*

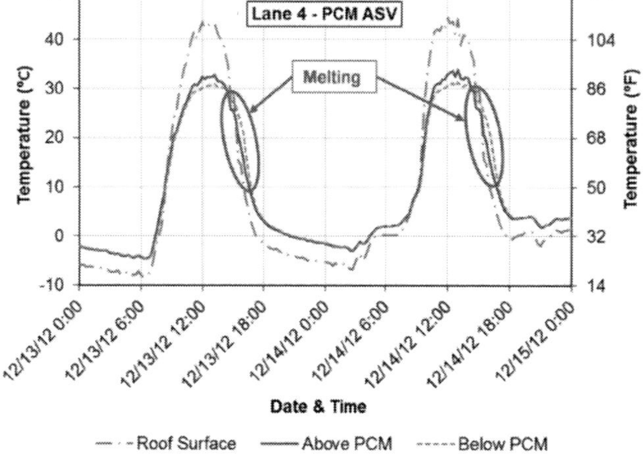

Figure 7 *Winter PCM behavior.*

period (defined as June–September 30, 2012). Figure 9 shows the variation of the roof heat flux during the summer week. Compared to the control (Lane 6), the ASV HG roof reduced the peak daytime heat gain by about 90% while PCM and PCM ASV roofs reduced the peak heat gain by 80%. Due to the melting of PCM (and the proximity of the PCM to the roof HFT) on some days, the roof HFT in the ASV HG roof actually showed a reversal of heat flow direction (negative heat flow) during the afternoons.

Figure 10 shows the attic temperatures during the summer week. All test lanes were effective in lowering the diurnal fluctuations on the attic temperature, and reduced the peak daytime temperatures. The ASV HG attic experienced the lowest peak temperatures, about 9.9°C lower than peak control attic temperatures during this week; the PCM and PCM ASV peak-attic temperatures were 7.6°C and 8.3°C lower than the control.

From a space-conditioning perspective, it is most interesting to examine the ceiling heat flux, which directly impacts the heating and cooling loads. Figure 11 shows the weekly ceiling heat-flux variations in the different attics. Again, all the test roofs were highly effective in reducing the ceiling heat gains compared to the asphalt shingle roof. During this week, on average, the peak ceiling heat gains in the different attics were 7.0 W/m^2 (ASV HG), 7.3 W/m^2 (PCM), 7.7 W/m^2 (PCM ASV), and 18.2 W/m^2 (Shingle).

To further evaluate the performance of the test roofs over the entire cooling period, the attic temperatures and the ceiling heat flows were averaged over June 1–September 30, 2012. The averaging (referred to as bin-averaging) was done over corresponding 15-minute periods of each day, and the resulting average data are shown over a 24-hour period in Figures 12 and

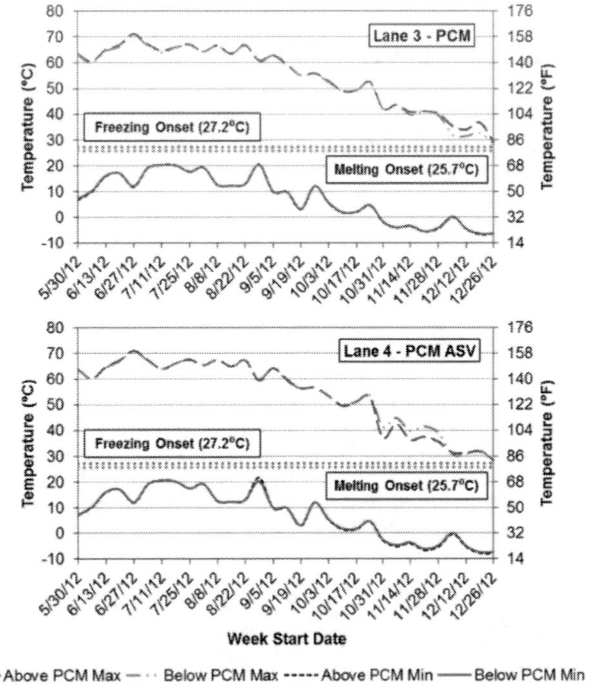

Figure 8 *Weekly maximum and minimum PCM surface temperatures.*

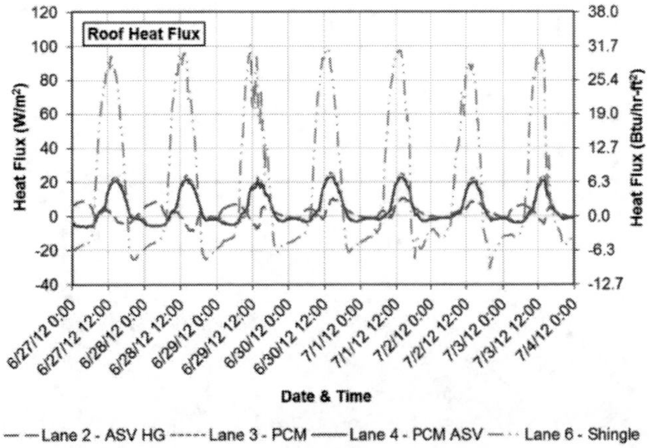

Figure 9 *Summer roof heat-flux variation.*

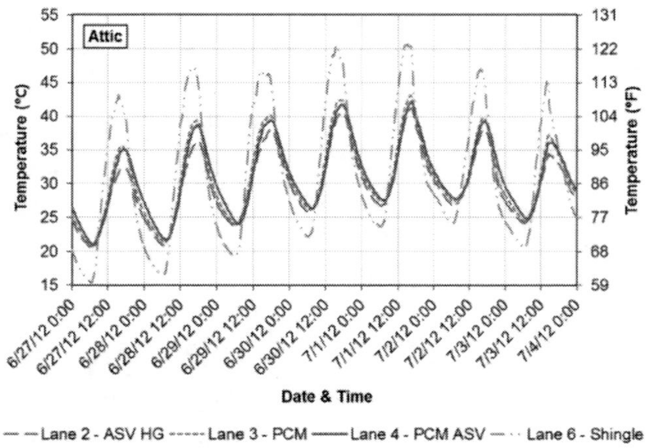

Figure 10 *Summer attic-temperature variation.*

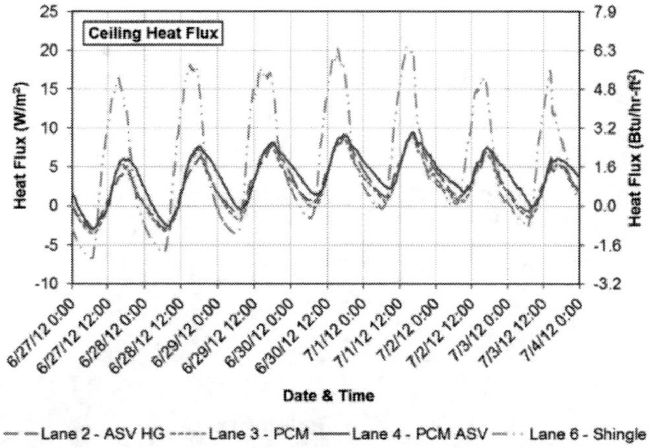

Figure 11 *Summer ceiling heat-flux variation.*

13. Bin-averaged outside temperatures are also shown in Figure 12.

The peak average-attic temperatures in the test lanes were 31.7°C (ASV HG), 33.9°C (PCM), and 33.6°C (PCM ASV), compared to 39.7°C in the asphalt shingle attic. Further, a delay of about 2–2.5 h was observed in occurrence of the peak temperature in the test attics compared to the asphalt shingle attic, which allows the benefit of peak-load shifting for utilities. Further, space-conditioning equipment have higher efficiency at lower ambient temperatures, which results in less energy usage to meet the loads with the peak shifting. The averaged ceiling heat flows showed a similar trend as the attic temperatures, with a delay of about 2–2.5 hours in the occurrence of the peak-ceiling heat flow. The peak-ceiling heat flows in the test attics were 3.6 W/m^2 (ASV HG), 4.2 W/m^2 (PCM), 5.0 W/m^2 (PCM ASV), and 12.2 W/m^2 (Shingle). The average-ceiling heat flows in the PCM ASV lane were consistently higher than the PCM lane. This is surprising, since the PCM ASV lane showed lower attic temperatures than the PCM lane between the hours of about 9:00–18:00, which includes the peak attic-temperature period. This could be the result of some uncertainty in the HFT measurements, and needs to be further investigated.

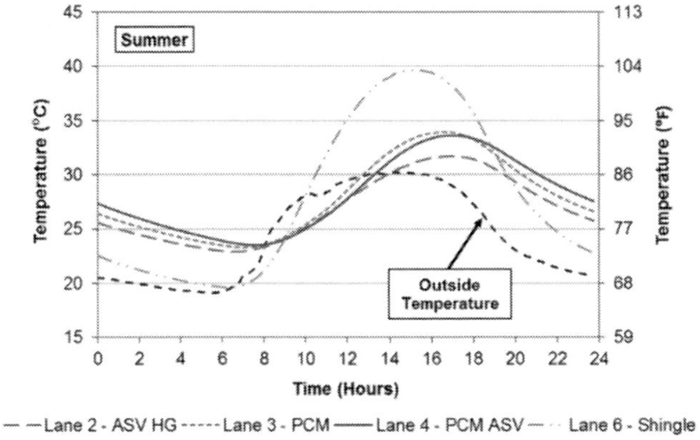

Figure 12 *Bin-averaged summer attic temperatures.*

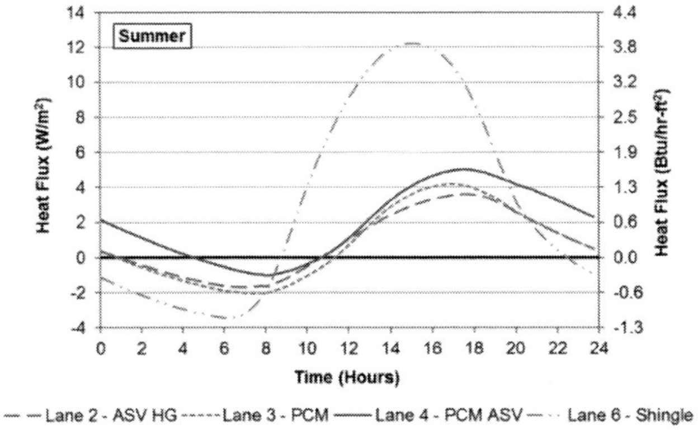

Figure 13 *Bin-averaged summer ceiling heat flows.*

Winter Performance

Here, temperature and heat flux data from a typical winter week (December 12–18, 2012) and the overall winter period (defined as November 1, 2012–January 1, 2013) are shown. Figure 14 shows the weekly variation of the roof heat flux. All test roofs (Lane 2 ASV HG, Lane 3 PCM, and Lane 4 PCM ASV) significantly reduced both the daytime heat gains and nighttime losses compared to the control asphalt shingle roof. The reductions in daytime heat gain could potentially result in a heating penalty in winter for these test roofs. However, all the attics with the dynamic test roofs remained warmer than the control attic, except during certain periods of peak solar irradiance (Figure 15). The PCM ASV roof performed the best in maintaining a warmer attic, with minimum temperatures about 6.1°C higher than the asphalt shingle attic; the minimum PCM and ASV HG attic temperatures were 4.2°C and 3.1°C higher than the control.

Figure 16 shows the weekly ceiling heat flows during the winter week. The test roofs reduced the ceiling heat loss significantly. During this week, the heat flows were predominantly out of the conditioned space. The asphalt shingle attic

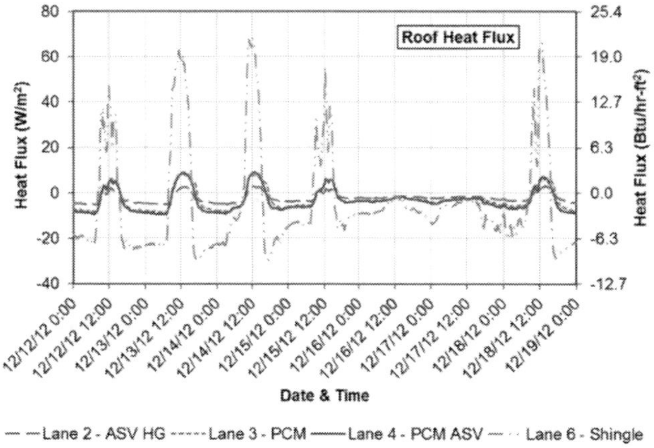

Figure 14 *Winter roof heat-flux variation.*

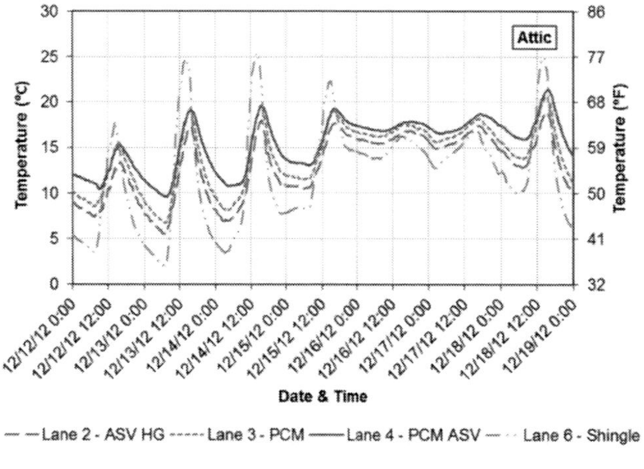

Figure 15 *Winter attic-temperature variation.*

added some heat to the conditioned space for brief daytime periods. On average, the peak nighttime ceiling losses for the different attics were 10.8 W/m^2 (ASV HG), 9.6 W/m^2 (PCM), 9.1 W/m^2 (PCM ASV), and 14.4 W/m^2 (Shingle). Note that the negative signs have been omitted, since heat loss indicates heat flow out of the conditioned space.

To evaluate the performance of the test roofs over the entire heating period, the bin-averaged attic temperatures and the ceiling heat flows from November 1, 2012–January 1, 2013 are shown in Figures 17 and 18. During winter, the PCM ASV attic stayed warmer than both the ASV HG and PCM lanes. The asphalt shingle attic had warmer daytime temperatures due to the solar heat gain, but lower nighttime temperatures than the test lanes. The minimum average-attic temperatures in the test lanes were 10.9°C (ASV HG), 12.0°C (PCM), and 14.0°C (PCM ASV), compared to 7.9°C in the asphalt shingle attic. The minimum ceiling heat flows (heat loss) were 10.7 W/m^2 (ASV HG), 9.6 W/m^2 (PCM), 9.1 W/m^2 (PCM ASV), and 14.4 W/m^2 (Shingle).

CONCLUSION AND FUTURE WORK

This article presents various metal roof configurations that were tested at the ESRA facility of ORNL during summer, fall, and winter of 2012–2013, and describes their potential for reducing the attic-generated space-conditioning loads. These roofs contained different combinations of PCM, rigid insulation, low-e surface, and above-sheathing ventilation, with standing-seam metal panels on top. Three test roofs were constructed and are referred to as the ASV HG roof, PCM roof and PCM ASV roof. These roofs were built on side by side test lanes, and compared to a control Shingle roof containing asphalt shingles on OSB roof deck. All three test roofs showed the potential for substantial energy savings compared to the asphalt shingle roof. They were highly effective in reducing the heat flows through the roof and ceiling, and in reducing the diurnal attic-temperature fluctuations. The ASV HG roof showed the best performance during the cooling period, while the PCM ASV roof performed the best during the heating season.

It should be noted that the present work does not seek to compare metal and asphalt shingle roofs *per se*. Rather, the intent was to evaluate the energy-saving potential of roofs containing rigid insulation and dynamic features like PCM, ASV and low-e surface with respect to a conventional shingle roof, which is the most common roof-type in the United States. The metal roofs in this study also possess different surface optical properties from shingle roofs, which also contributes to the thermal performance differences.

The data generated from these test lanes are useful in evaluating their performance in the field. However, these results are limited to the current test setup and the specific climate zone. In order to evaluate these roofs in different climate zones and for different construction types, numerical modeling is important. The data from these tests are invaluable for validating such energy models.

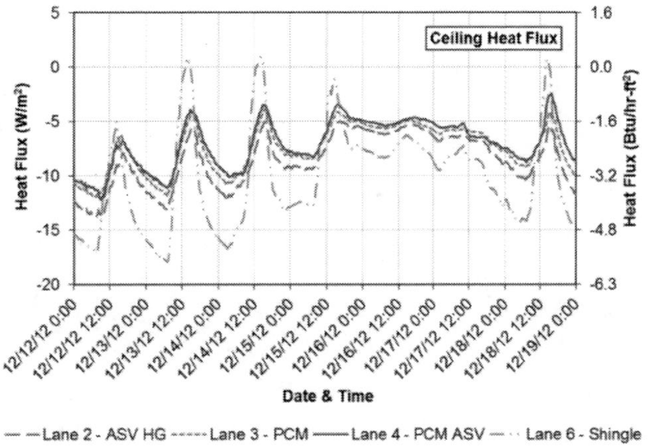

Figure 16 Winter ceiling heat-flux variation.

ACKNOWLEDGMENTS

The authors would like to thank Joe Harter and Derrick Fowler (ATAS International, Inc.), Sam Yuan (CertainTeed) and Pete Horvath (Phase Change Energy Solutions) for their contributions. The technical assistance of Phillip Childs and

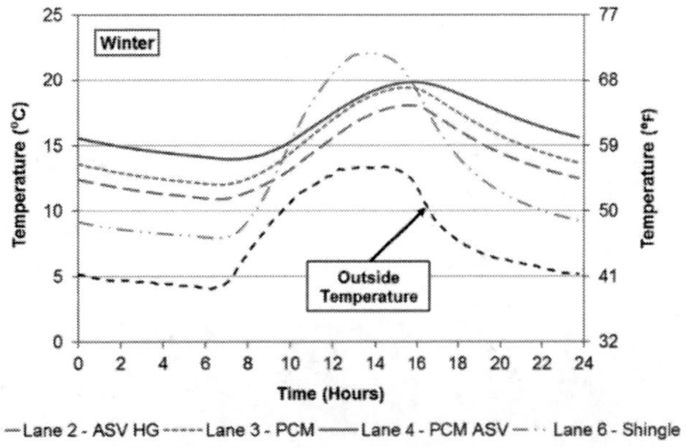

Figure 17 Bin-averaged winter attic temperatures.

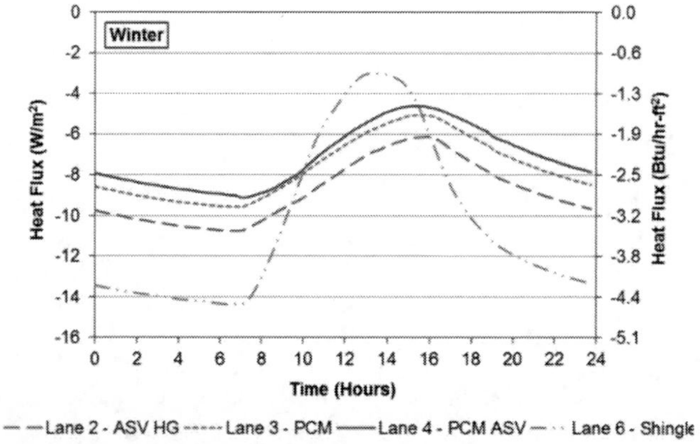

Figure 18 Bin-averaged winter ceiling heat flows.

Jerald Atchley (ORNL) in setting up the data acquisition systems is gratefully acknowledged.

REFERENCES

Akbari, H., P. Berdahl, R. Levinson, R. Wiel, A. Desjarlais, W. Miller, N. Jenkins, A. Rosenfeld, and C. Scruton. 2004. Cool colored materials for roofs. *Proceedings of the ACEEE 2004 Summer Study on Energy Efficiency in Buildings, Washington, D.C.*

Biswas, K., W. Miller, P. Childs, J. Kosny, and S. Kriner. 2011. Performance evaluation of a sustainable and energy efficient re-roofing technology using field-test data. *Proceedings of the 2011 NRCA International Roofing Symposium, National Roofing Contractors Association, Washington D.C.*

Biswas, K. and P. Childs. 2012. Performance evaluation of advanced retrofit roof technologies using field-test data—Phase two final report. ORNL/TM-2012/23, Oak Ridge National Laboratory, Oak Ridge, TN.

Castellón, C., E. Günther, H. Mehling, S. Hiebler, L. Cabeza. 2008. Determination of the enthalpy of PCM as a function of temperature using a heat-flux DSC—A study of different measurement procedures and their accuracy. *International Journal of Energy Research* 32(13):1258–65.

Dodge, F.W. 2002. Construction outlook forecast. www.fwdodge.com.

Huang, J., J. Hanford, and F. Yang. 1999. Residential heating and cooling loads component analysis. LBNL-44636, Lawrence Berkeley National Laboratory, Berkeley, CA.

Kosny, J., K. Biswas, W. Miller, P. Childs, and S. Kriner. 2011. Sustainable retrofit of residential roofs using metal roofing panels, thin-film photovoltaic laminates and PCM heat sink technology. *Journal of Building Enclosure Design*, National Institute of Building Sciences, Building Enclosure Technology and Environment Council (BETEC), Winter 2011/3.

Kosny J., K. Biswas, W. Miller, and S. Kriner. 2012. Field thermal performance of naturally ventilated solar roof with PCM heat sink. *Solar Energy* 86:2504–14.

Medina, M.A. 2010. *Performance of Attic Radiant Barriers (RBs) and Interior Radiation Control Coatings (IRCCs): A Summary of Published Research*. Reflective Insulation Manufactures Association International.

Miller, W.A. 2006. The effects of infrared-blocking pigments and deck venting on stone-coated metal residential roofs. ORNL/TM-2006/9, Oak Ridge National Laboratory, Oak Ridge, TN.

Miller, W.A. and J. Kosny. 2007. Next generation roofs and attics for residential homes. *Proceedings of the 2007 ACEEE Summer Studies on Energy Efficiency, Pebble Beach, California*.

Parker, D., D. Beal, and S. Chandra. 1995. The measured summer performance of tile roof systems and attic ventilation strategies in hot, humid climates. *Proceedings of the Thermal Performance of the Exterior Envelopes of Buildings VI, U.S. DOE/ORNL/BETEC, Clearwater, FL.*

Sharma, A., V.V. Tyagi, C.R. Chen, and D. Buddhi. 2009. Review on thermal energy storage with phase change materials and applications. Renewable and Sustainable *Energy Reviews* 13(2):318–45.

Sengoz, B. and A. Topal. 2005. Use of asphalt roofing shingle waste in HMA. *Construction and Building Materials* 19(5):337–46.

Xin, W., Z. YinPing, X. Wei, Z. RuoLang, Z. QunLi, and D. HongFa. 2009. Review on thermal performance of phase change energy storage building envelope. *Chinese Science Bulletin* 54(6):920–28.

Zalba, B., J.M. Marin, L.F. Cabeza, and H. Mehling. 2003. Review on thermal energy storage with phase change: materials, heat transfer analysis and applications. *Applied Thermal Engineering* 23:251–83.

A Home Ignition Assessment Model Applied to Structures in the Wildland-Urban Interface

Kaushik Biswas, PhD
Associate Member ASHRAE

David Werth

Narendra Gupta, PhD

ABSTRACT

The issue of exterior fire threat to buildings, from either wildfires in the wildland-urban interface or neighboring structure fires, is critically important. To address this, the Wildfire Ignition Resistant Home Design (WIRHD) program was initiated. The WIRHD program developed a tool, the WildFIRE Wizard, that will allow homeowners to estimate the external fire threat to their homes based on specific features and characteristics of the homes and yards. The software then makes recommendations to reduce the threat. The inputs include the structural and material features of the home and information about any ignition sources or flammable objects in its immediate vicinity, known as the "home ignition zone."

The tool comprises an ignition assessment model that performs explicit calculations of the radiant and convective heating of the building envelope from the potential ignition sources. This article describes a series of material ignition and flammability tests that were performed to calibrate and/or validate the ignition assessment model. The tests involved exposing test walls with different external siding types to radiant heating and/or direct flame contact. The responses of the test walls were used to determine the conditions leading to melting, ignition, or any other mode of failure of the walls. Temperature data were used to verify the model predictions of temperature rises and ignition times of the test walls.

INTRODUCTION

The wildland-urban interface (WUI) fire problem has appropriately received much attention in the last decade, both in the United States and internationally. The National Interagency Fire Center (NIFC) and the National Fire Protection Association (NFPA) provide a historical perspective on wildland fires in the U.S. in terms of economic loss, loss of life, and the fire size in acres burned (NIFC 2012; NFPA 2012). According to NFPA (2012), the ten largest wildland fires in the U.S. resulted in losses ranging from $396 million to $2.4 billion (adjusted to 2010 dollars). Approximately one hundred lives were lost to wildfire in Europe in 2007 (Lampin-Maillet et al. 2009). In Australia, the 2009 bushfires, among the worst in recent memory, resulted in widespread destruction and significant loss of life. The expanding residential development at the WUI has been widely recognized as a factor influencing forest management policies in the U.S. (Theobold and Romme 2007) and also places property and human life at risk from wildfire destruction (Reams et al. 2005). In 2000, the WUI encompassed 11% of the land area (719,156 km^2) and 38% of all housing units (44.3 million) in the contiguous U.S. (Hammer et al. 2009). Using different metrics, Theobald and Romme (2007) estimated the WUI area to be 465,614 km^2 and projected at least a 10% increase by 2030. Albini (1997) noted in an invited talk to the International Association of Fire Safety Science (IAFSS) that "Interest by the general public in matters of wildland fire safety has grown with increased exposure of affluent society to the hazards posed by building flammable structures in flammable wildland settings."

Recently, Mell et al. (2010) reviewed and summarized the challenges posed by wildfire spreading into the WUI and identified the research needs for addressing the fire problem in terms of laboratory and field experiments and fire behavior models. The article reported that "there is currently no stan-

***Kaushik Biswas** is an R&D associate (building scientist) in the Energy and Transportation Science Division at Oak Ridge National Laboratory, Oak Ridge, TN. **David Werth** is a principal scientist at and **Narendra Gupta** is retired from Savannah River National Laboratory, Aiken, SC.*

dardized method of risk assessment that can be applied to WUI communities in the US" (p. 240). The research needs identified by Mell et al. include: (1) structure exposure conditions (flame heat flux and firebrands) for a range of WUI settings (terrain, housing density, fuels, etc.), and (2) the vulnerability of a structure or building when subjected to a given exposure. The ignition threat of firebrands and their generation from burning vegetation in wildland fires has also received significant attention (Manzello et al. 2006, 2007, 2009).

There has also been significant work in modeling wildfires and their impacts in the WUI. Papadopoulos and Pavlidou (2011) reviewed various physical, quasi-physical, empirical, and quasi-empirical wildfire models and tabulated their relative merits and disadvantages. Mell et al. (2007) reported the development of a three-dimensional, fully transient, physics-based fire spread model to simulate grassland fires. The authors noted various fire behavior problems that are outside the scope of empirical and semi-empirical models and are only amenable to physical modeling. However, they also recognized the significantly higher computational resources required by the physical models. Rehm (2008) noted that application of physical or field models to compute multiple burning houses and vegetative (WUI) fires over large areas is difficult due to the range of time and spatial scales that need to be solved; time scales associated with this problem range from tens of seconds with burning grass and individual trees to hours for burning structures. The use of field models for modeling large-area WUI fires in the near term is unlikely due to the constraints on the computational resources (Rehm 2008). The key advantages of empirical models, which are based on observation and experiments, are relatively straightforward implementation and direct relation to behavior of real fires (Papadopoulos and Pavlidou 2011). Quasi-empirical models contain statistical analyses based on functional relationships between dependent and independent variables and have the added benefit of reduced influence of the data set size and personal choice (Papadopoulos and Pavlidou 2011).

In 2009, the United States Department of Homeland Security initiated the Wildfire Ignition Resistant Home Design (WIRHD) program to develop a semi-empirical software tool, the Wildfire Ignition Resistance Estimator (WildFIRE) Wizard, which evaluates fire threats to homes and provides guidelines for reducing them. The main outcome of this project is an enhanced ignition assessment model based on the structure ignition assessment model (SIAM) of Cohen (2004), which forms the core of the WildFIRE Wizard. This tool allows homeowners to graphically create a reproduction of their home and yard, including burning objects ("threats") within the home ignition zone (HIZ) that can possibly cause an ignition of the structure. It calculates temperatures and a flux-time product (FTP) from both radiative and convective heating by the threats to estimate the ignition potential of the structure based on field studies and known building material properties (Cohen 2004). One of the major challenges facing decision makers while formulating wildfire risk mitigation programs is how to change the behavior of private property owners regarding vegetation management (Reams et al. 2005). A validated WildFIRE Wizard, an easy-to-use tool needing minor computational resources, will be a major step towards overcoming this challenge.

The ignition model (enhanced SIAM) requires empirical data on the thermal and ignition properties of the external siding materials used on homes. Empirical models should be based on well-characterized, science-based, and repeatable experiments (Mell et al. 2010). This has been partially accomplished through field tests (Cohen 2004), but the full range of model calculations needed verification and validation. Therefore, a series of flammability tests of small-scale wall assemblies with different exterior sidings were conducted. These small-scale tests were specifically designed as surrogates for full-scale field testing of wildfires, which are very difficult to perform for a wide range of wildfire-structure configurations. These tests were interpreted in the context of the model—validating the way the model calculated heat flux from flames and ignition predictions or to determine more appropriate response factors. *PATRAN* (MSC 2008) simulations were also performed to get more detailed information about the thermal responses of the wall assemblies. These simulations were also verified against data from the flammability tests. Once validated, the *PATRAN* models were used as verification tools for the SIAM calculations that could not directly be validated by measurements. In the following sections, the flammability tests and the software validation using test data are presented.

EXPERIMENTAL ARRANGEMENT

The enhanced SIAM accounts for both radiant and convective heating as potential igniters of the home, and various siding samples were tested for several scenarios. The radiant tests were intended for validation of the materials' ignition characteristics, while the convective tests were done to validate the way the model converts a flame of prescribed power into a heat flux. Small-scale wall assemblies of 1.2 × 1.2 m^2 area were built and exposed to either a radiant source or a flame (or both) of prescribed intensities. The walls were constructed using wood studs and contained fiberglass cavity insulation, with a gypsum board "interior" finish. The test wall "exterior" consisted of oriented strand board (OSB) and different external claddings and was exposed to the different heat sources. Four external cladding types were tested: (1) vinyl siding, (2) cedar wood siding, (3) fiber cement board, and (4) exterior fiber cement board with a 2.54 cm extruded polystyrene (XPS) insulation board underneath.

Each test assembly was subjected to each of the following exposures, chosen to simulate various likely scenarios in the wildland-urban interface:

- Short-duration, high-intensity radiation flux: 40–50 kW/m^2 for a short time (60 s), with a pilot flame. This was designed to simulate radiation from short-exposure actively spreading forest fires (Cohen 2004).

- Direct flame contact: 100 kW flame for 60 s; the flame source was a slot burner adjacent to the wall. This exposure represented burning debris and plants next to the house.
- Long-exposure, coupled radiant heating and direct flame contact: a 25 kW/m^2 radiant source and 100 kW flame from the slot burner. This was to simulate both the radiation from a burning neighboring structure and convective heat from the burning debris next to the house. The combined radiation heat and flame exposure was continued until either the exterior cladding ignited or 15 min had elapsed with no ignition.

Thermocouples were installed at regular intervals on the sample face (Figure 1) and were sampled at 10 Hz. Type K 0.127 mm diameter thermocouples were used, with a tolerance limit of 0.75% for the temperature range of 0°C–1250°C. These were arranged along the vertical and horizontal centerlines both on the exterior surface and behind it (above the OSB layer). In the flame contact tests, two floating thermocouple arrays were also installed along the vertical centerline about 3 mm and 10 mm away from the exterior surface (Figure 1). These were installed to measure the flame temperatures adjacent to the sample surface.

The assembly test data were intended to be used in the following manner:

- Validation of the FTP ignition thresholds for wood siding for use by SIAM.
- Air and surface temperatures to validate the SIAM calculations of the convective heat flux and surface temperatures or determine a more appropriate response function.
- Interior OSB temperatures to determine the failure or ignition of internal layers (especially with cement board siding).
- External and internal surface temperatures to validate the *PATRAN* calculations for the fiber cement tests.

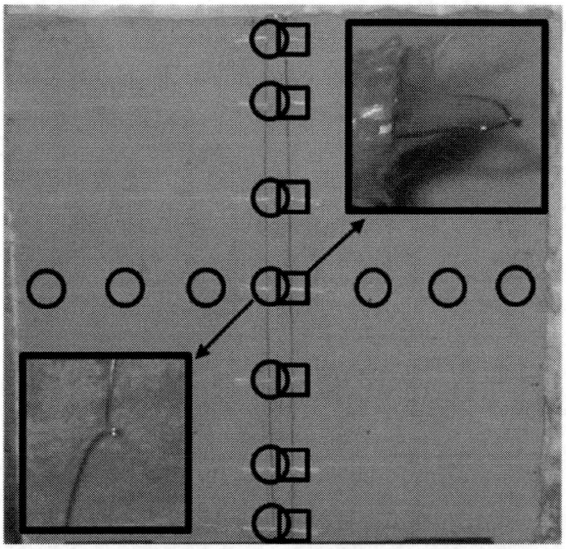

Figure 1 *Test wall sample with surface and floating thermocouples: circles = surface thermocouples; squares = floating thermocouples. Close-up photographs of each type are shown as insets.*

THE STRUCTURE IGNITION ASSESSMENT MODEL (SIAM)

This section describes the essential features of the SIAM. The user inputs details of the home and the HIZ through a graphical user interface. Once the details are obtained, the model operates by first subdividing the walls of the home into a series of panels, each characterized by its height, width, and location, and then defining a series of burning threats that can each impart radiant and convective heat to the different panels. SIAM calculates the heat flux from each threat individually and sums them to get the total heat flux on a panel. For radiation heat flux, the model uses the geometry of the flame front and the wall and their relative positions to calculate the view factor ($V_{f \to w}$) and then the radiation exchange (Q, W) using the following equation for gray, diffuse surfaces (Incropera and Dewitt 1996), using prescribed flame and wall temperatures:

$$Q = \frac{\sigma\left(T_f^4 - T_w^4\right)}{\frac{1-\varepsilon_f}{\varepsilon_f A_f} + \frac{1}{V_{f \to w} A_f} + \frac{1-\varepsilon_w}{\varepsilon_w A_w}} \qquad (1)$$

where σ is the Stefan-Boltzmann constant (5.67×10^{-8} W/m$^2 \cdot$K^4), T is the temperature, ε is the emissivity, and A is the projected area; subscripts f and w represent the flame and wall, respectively. Equation 1 is specific to a two-surface enclosure, in which the two surfaces only exchange radiation with each other, and is not an accurate representation of a fire threat facing a wall. However, this represents the worst-case scenario since all the radiant heat of the flame is assumed to be impinging on the wall. Radiation heat fluxes on the wall were calculated using the two-surface enclosure scenario, a three-surface enclosure formulation with the environment as the third surface (Modest 1993), net radiation exchange at the wall surface formulation (Modest 1993) and the original SIAM radiation heat transfer equation (Cohen 2004). The two-surface enclosure equation (Equation 1) yielded the highest heat flux impinging on the wall.

For convective heating, the model uses the wall height and flame position and length to calculate the point on the wall at which the flame or the hot air column above the flame intercepts the wall. Given the temperature difference, a modified heat transfer equation calculates the convective heat flux.

The heat fluxes produced by these calculations serve as input to two other algorithms that determine the response of the home, which depends on the siding. Vinyl absorbs heat until the melting point is reached, following which it is assumed that the siding falls away from the wall, exposing the interior layers. Wood siding absorbs heat until it ignites, and this is done according to an empirical equation (Cohen 2004). Both these calculation procedures were calibrated or validated with the current flammability tests and are described in the Test Results and Discussion section. In the tests for this research, the radiation heat flux was prescribed and, therefore, the radiation flux calculation procedure was not used for the simulations. The primary intent for the purposes of this paper was to collect data from the sample tests and apply them to the model algorithms to determine if the model can accurately reproduce the responses of the test samples.

TEST RESULTS AND DISCUSSION

Radiant Heating

The radiant heating tests involved the use of a calibrated radiant panel, shown in Figure 2, which could be adjusted to impart any desired heat flux to the samples. The radiant panel consisted of a thick metal sheet heated by gas-fired burners, and the desired heat flux was obtained by adjusting the fuel flow rate to the burners and the distance between the radiant panel and the test walls. The radiant heat flux incident on the test samples was calibrated using a panel instrumented with five radiometers (Figure 2). Heat flux calibration tests were performed both before and after the samples were tested at each desired radiant exposure (25, 40, and 50 kW/m^2). Figure 3 shows the pre- and post-test calibration plots for the 40 kW/m^2 radiant flux, which showed little loss of calibration between measurements.

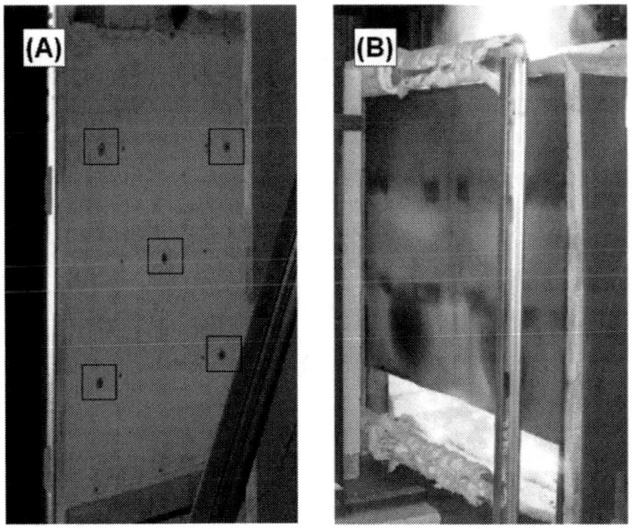

Figure 2 *(A) Calibration panel instrumented with five radiometers (indicated by □) and (B) radiant panel heated by gas-fired burners.*

For this first set of tests, the 40 and 50 kW/m^2 exposures were used, and the test walls were exposed to the radiant panel after a 15 min warm-up period. The heat flux values used as inputs to the model calculations are based on the calibration panel tests. These tests were used to validate the simulated response of the samples. Data from two radiant tests are presented: a vinyl sample and a cedar siding sample, each of which is treated differently in SIAM.

Vinyl Sample. Figure 4 shows the instrumentation of the vinyl wall sample for the radiant exposure test. Thermocouples were installed along the vertical and horizontal centerlines on the exterior vinyl and interior OSB surfaces. This test involved the application of the calibrated 40 kW/m^2 radiant heat flux to the wall sample with vinyl siding. SIAM uses an explicit heat transfer equation to calculate the vinyl temperature T, removing the siding to expose the OSB underneath when the melting temperature is exceeded. This is done according to Equation 2:

$$T_{t+1} = T_t + \left(\frac{Q'' - C}{C_p}\right)\Delta t, \text{ for } Q'' > C \qquad (2)$$

where Q'' is the imposed heat flux (kW/m^2), heat capacity C_p = 1287 J/m^2·K for vinyl, Δt is the time step, and C is a prescribed cooling rate (in kW/m^2) to account for the heat loss to the surroundings. The heat capacity was based on the

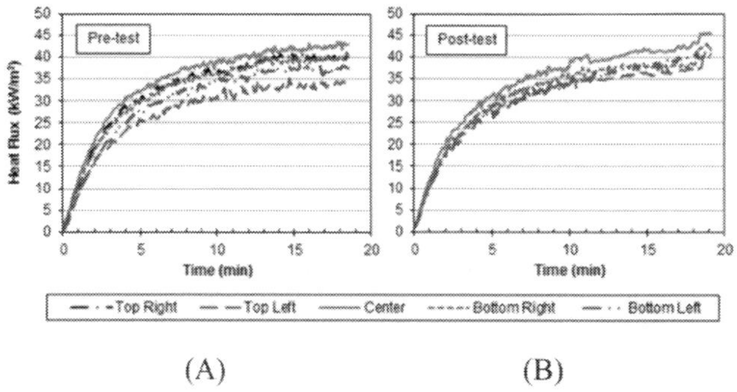

Figure 3 *(A) pre-test and (B) post-test radiant panel calibration data for the 40 kW/m^2 heat flux.*

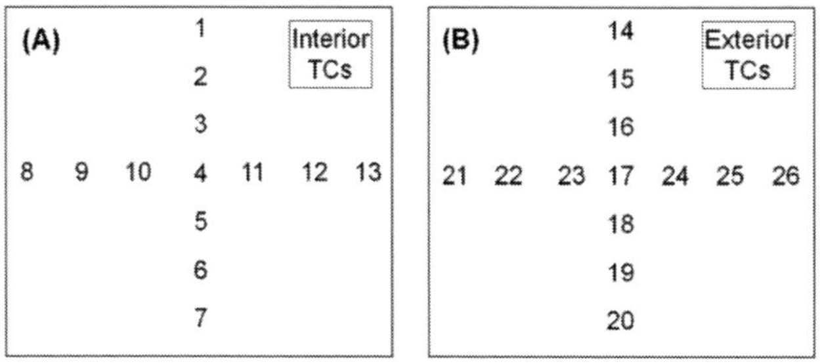

Figure 4 *Interior and exterior thermocouple (TC) locations for the vinyl radiant exposure test sample.*

measured specific heat (c_p) of vinyl according to ASTM E 1269 (2011a). The imposed heat flux (Q'') was 38.44 kW/m^2, which was the average heat flux measured from 15:00 to 15:30 min in the 40 kW/m^2 calibration tests. The cooling rate (C) represents a reduction in the rate at which the vinyl temperature increases; it was estimated according to the rate at which the actual vinyl sample heated up.

Measured interior and exterior vinyl surface temperatures were averaged to get the vinyl temperature and compared to calculated temperatures, as shown in Figure 5. The sample temperatures showed an approximately linear increase with time in response to the constant heating rate, at an average of 5.19°C/s. Different cooling rates were tested in the SIAM calculations and 31.8 kW/m^2 was chosen as the optimum cooling rate, which yielded a temperature rise rate of 5.16°C/s. Equation 2 in the ignition model was therefore recoded with the updated cooling rate, and the model was modified so that the vinyl siding temperature rise occurs only when the imposed heat flux exceeds 31.8 kW/m^2.

This test was also useful for determining the melting point of the vinyl siding, which the model uses to determine when the siding has ablated and exposed the interior OSB. The interior thermocouples (Figure 6) showed a sharp rise in temperature when the siding failed, which occurred when the exterior vinyl temperature was between 230°C and 310°C. An average value of 270°C was assigned in the SIAM module for vinyl siding melting.

Cedar Sample. This test also used a 40 kW/m^2 radiant heat panel but applied it to a cedar siding wall, which was also instrumented as shown in Figure 4. For wood siding, the model uses the FTP calculation of Cohen (2004) to establish ignition. The radiant heating is used to calculate the FTP value according to Equation 3:

$$FTP = \int (Q'' - Q''_C)^{1.828}\, dt, \text{ for } Q'' > Q''_C \qquad (3)$$

where Q''_C is the critical radiant heat flux (13.1 kW/m^2) (Cohen 2004) and Q'' = 38.44 kW/m^2 from the calibration panel tests. As the FTP values crosses 11500, ignition is more and more likely until the value of 13500 is reached, when it is a certainty (Cohen 2004). The FTP value increases until the heating stops or the ignition threshold is reached. The goal of this test was to determine if the threshold value obtained by earlier research (Cohen 2004) is accurate enough to be used in the model. With Q'' = 38.44 kW/m^2, Equation 3 predicted ignition between 32 and 37 s, going from likely to certain ignition. In the test, ignition, as indicated by the jump in temperature of thermocouples 16–18 (Figure 7), occurred at about 28 s after exposure, close to the value predicted by the model.

This test was rendered less conclusive by the fact that the pilot flame intended to ignite the sample after off-gassing began was extinguished and had to be relit. Off-gassing was observed to start after 10 s, with the siding becoming dark after 19 s. The sample ignited immediately after the pilot light was relit at 27 s, so it was concluded that this test still represents a good confirmation of the ignition threshold currently used by the model.

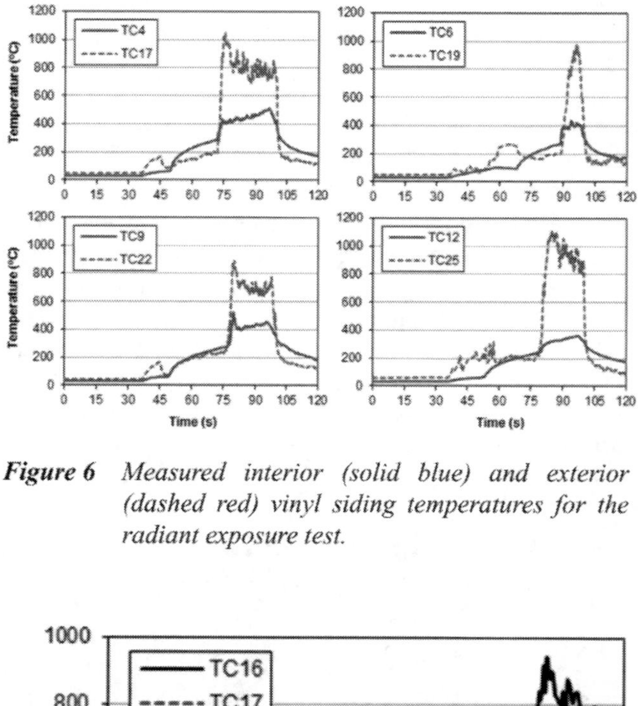

Figure 6 Measured interior (solid blue) and exterior (dashed red) vinyl siding temperatures for the radiant exposure test.

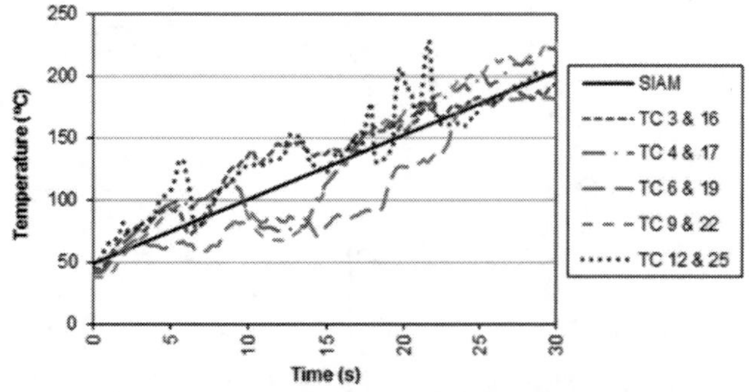

Figure 5 Measured (TC) and calculated (SIAM) temperatures of vinyl sample during the 40 kW/m^2 radiant exposure test. Measured temperatures are averages of the interior and exterior thermocouples.

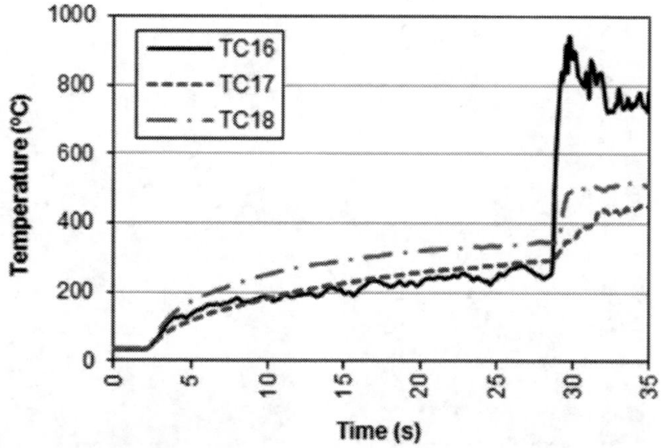

Figure 7 Measured external cedar siding temperature in response to 40 kW/m^2 radiant heating.

Convective Heating

Convective heating is imparted to the sample from either active flames or the hot air column above a flame front that is in position to allow for contact. A propane-fueled slot burner of approximately half the width of the test walls was used for the convective heating tests.

Figure 8 shows the slot burner and a test in progress (fiber cement sample). The fuel flow rate was adjusted to get the desired 100 kW flame (based on the lower heating value) output using an air calibrated rotameter. Results from these tests were used to help calibrate this process and provide some further validation of the model adjustments made in response to earlier tests.

Cedar Sample. This test comprises a 100 kW flame applied to a wall with cedar siding. The model calculates the heat flux by calculating a flame/air column temperature (T_f) and assuming a wall temperature (T_w), then applying the following heat transfer equation:

$$Q'' = h(T_f - T_w) \tag{4}$$

Here h is the heat transfer coefficient (kW/m^2·K). The model assumes a flame temperature and a hot air column above the flame tip that is relatively cooler. T_f can represent either the flame or the hot air temperature. For T_w, the model assumes a constant value of 32°C (instead of solving for the wall temperature explicitly) to maintain a conservatively large heat flux. This was necessary as, in order to simplify the process and speed up the calculations, the heat transfer module calculates and saves all heat fluxes during the burning time, which then serve as input to a separate module that calculates the thermal response of the siding. As before, the cedar ignition was estimated by calculating the FTP value. The ignition time in the experiment was 25 s after exposure, so the heat transfer coefficient (h) needed to be calibrated so that the heat flux was such that the FTP reached the critical value at the appropriate time.

Figure 9 shows the locations of the thermocouples on the cedar wall sample. The instrumentation consisted of interior and exterior surface and floating air thermocouples. Ignition was first observed at the lower edge of the sample, so the temperature reading of TC21 was used as the flame temperature. The model originally used an algorithm to determine h at each time step, which produced a series of values ranging between 20 and 40 kW/m^2·K. The simulated heat flux calculated using the above values of h resulted in FTP values that did not reach the critical FTP needed for ignition at the proper time, remaining below the threshold long past the point at which the test sample actually ignited. The model was, therefore, updated by assigning a constant value of 58.42 kW/m^2·K to h.

The heat flux was calculated using the updated value of h in Equation 4 and then the FTP was calculated by Equation 3. Figure 10 shows the evolution of the convective heat flux and resultant FTP with the original and revised val-

Figure 8 *Burner used for convective heating of the test samples.*

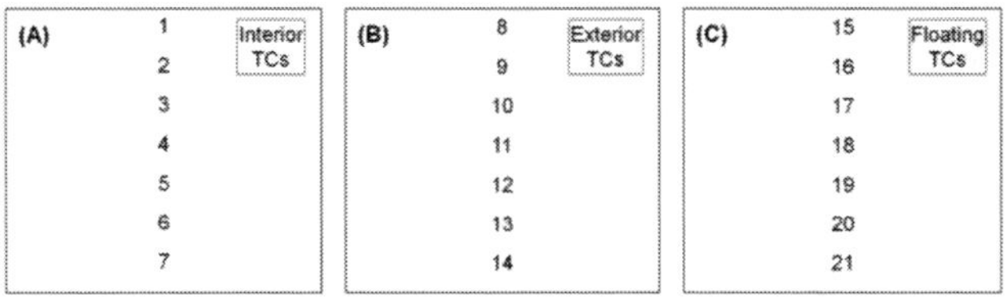

Figure 9 *Thermocouple (TC) locations for the cedar convective exposure test sample.*

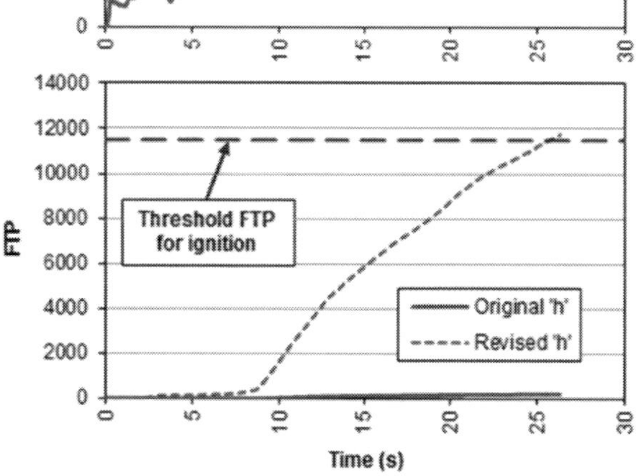

Figure 10 *Heat fluxes calculated with the original and revised* h *values (top) and the corresponding FTP values (bottom) for the cedar sample exposed to*

ues of h. The calculation yielded an FTP curve that reached 11236 at 25 s with an h of 58.42 kW/m^2·K, the same time as the actual ignition. This value was therefore selected as the optimal heat transfer coefficient in the model.

Vinyl Sample. Next, the 100 kW flame was applied to a wall sample with vinyl siding, with the thermocouple locations as shown in Figure 9, except no floating air thermocouples were used. This test was used to validate the parameters derived in previously described tests: (1) the constant value of h (58.42 kW/m^2·K) and (2) the vinyl cooling rate (C) of 31.8 kW/m^2. For this test, since no floating thermocouples were used, TC16 from the cedar flame test (Figure 9) was selected to represent the flame temperature (T_f). The focus here was on the response of the upper half of the sample. The top half was exposed to the hot gases and the fire plume, rather than direct flame, which would have caused immediate melting of the vinyl siding. The flame/hot air temperature was used in Equation 4 to get the heat flux into the sample (assuming a constant T_w = 32°C and h = 58.42 kW/m^2·K).

Next, the calculated convective heat flux was used in Equation 2 to calculate the sample temperature, and this was compared to the measured temperatures, averaged over the inner and outer probes, of the top half of the wall. The comparison of the measured and calculated temperatures is shown in Figure 11. The calculated temperature provides a good approximation of the measured interior and exterior wall temperatures, verifying the revisions to Equations 2 and 4.

Fiber Cement Sample. Many modern homes are covered with fiber cement siding, which is considered fire resistant up to very high temperatures, and the model currently considers such siding to be failure-proof. To validate this, tests were conducted on wall samples with fiber cement siding. These were also simulated with the *PATRAN* radiative transfer software, which provided a more detailed picture of heat transfer within the material. These tests of the software are also valuable because *PATRAN* has been used extensively to verify the SIAM algorithms—so comparing the *PATRAN* calculations with the experimental data from the tests provided the necessary independent check.

The actual wall specimens measured 1.2 × 1.2 m^2, but to reduce the computation time, 0.6 × 0.6 m^2 samples were simulated, with the four edges assigned adiabatic boundary conditions because of symmetry. Scoping results for symmetric radiant heat flux loading conditions showed that the end effects were minimal and two-dimensional models would yield equally good results.

Besides the imposed heat flux, the boundary conditions for the simulated heat flux loading must include convective and radiation losses to the ambient from the wall surface. For the 100 kW flame, it was not possible to estimate the heat losses from the flame to the ambient. Therefore, the temperatures recorded by the thermocouples offset from the wall were used as the flame temperature, and forced convection with 5 m/s flame velocity was used. The velocity was chosen based on measurements and simulations of a buoyant methane fire stabilized on a 1 m diameter fuel source (Tieszen et al. 2002; Xin et al. 2008). The initial temperature of the whole assembly in the *PATRAN* model was 25°C. The fiber cement board wall geometry used for the model is shown in Figure 12.

Fiber cement siding is not expected to ignite, but it may fail if it becomes too hot or allows too much heat to pass through its interior into its supporting material. These tests and the associated *PATRAN* simulations were performed to determine that *PATRAN* can accurately simulate heat transfer through the siding, a key variable in estimating the possibility of failure.

In this test, the fiber cement siding sample was exposed to the 100 kW flame. Figure 13 shows a comparison of the calculated *PATRAN* temperature at the center of the panel with the measurements from thermocouples TC11 and TC12. The thermocouple locations for the fiber cement sample were the same as those for the cedar sample (Figure 9). The calculated temperatures were within 10% of the measurements, except at the beginning of the calculations when the difference

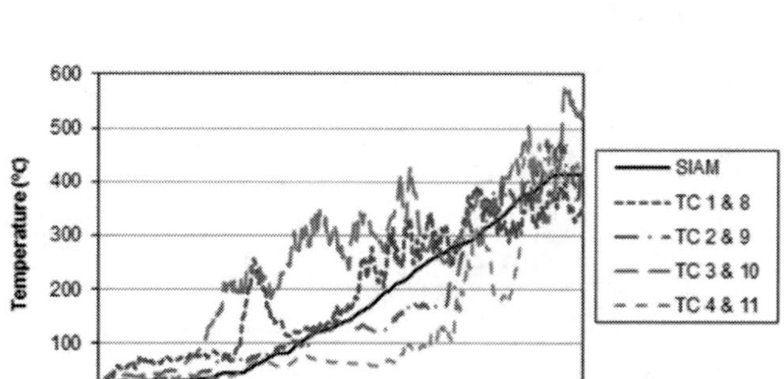

Figure 11 *Measured (TC) and calculated (SIAM) temperatures of the vinyl sample when exposed to a 100 kW flame. Measured temperatures are averages of the interior and exterior thermocouples.*

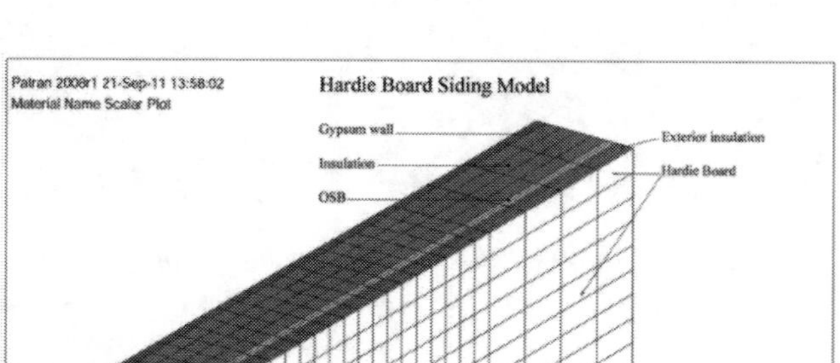

Figure 12 *Fiber cement wall thermal model geometry used in PATRAN simulations.*

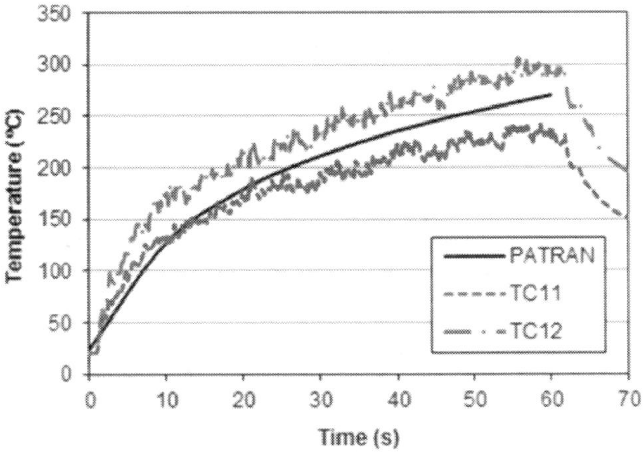

Figure 13 Measured (TC) and PATRAN calculated surface temperatures of the fiber cement sample exposed to the 100 kW flame.

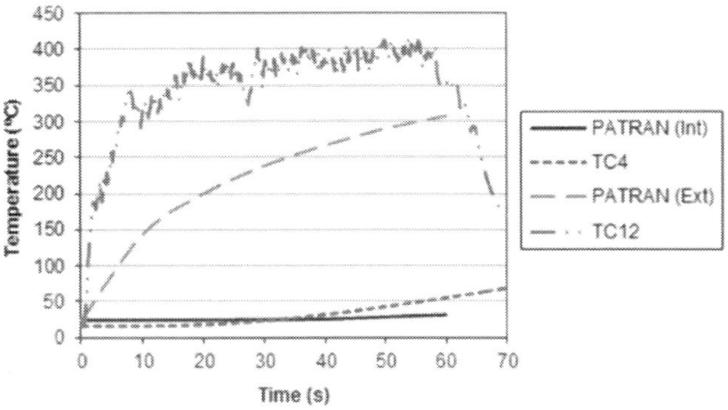

Figure 14 Measured (TC) and PATRAN calculated surface temperatures of the fiber cement sample exposed to the 100 kW flame; TC4 = internal surface and TC12 = external surface.

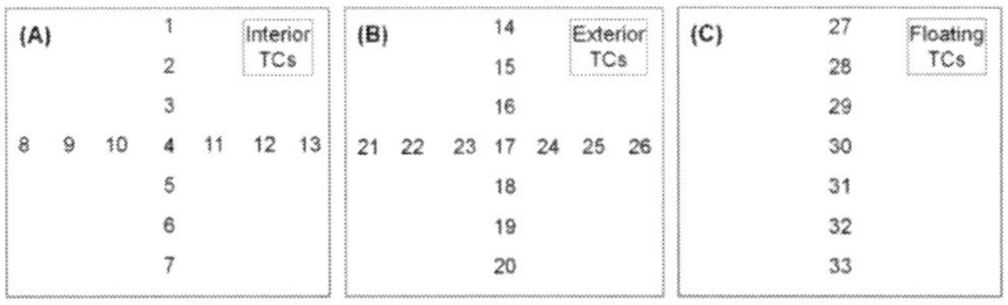

Figure 15 Thermocouple (TC) locations for the cedar sample for the combined radiant and convective exposure test.

was about 20% with the prescribed 25°C initial model temperature.

Fiber cement boards are rated non-combustible up to 750°C, according to ASTM non-combustibility test, ASTM E 136 (2011b). The measured temperatures rose to about 300°C during this short-duration convective heating test and there were no signs of ignition or other visible failure of the wall, other than some surface charring, at the end of the test.

Fiber Cement Sample with XPS Backing. This test also used the 100 kW flame heating, but here the test specimen had the XPS insulation inserted between the fiber cement siding and the OSB. Figure 14 compares the *PATRAN* calculated surface temperatures to the measured internal and external surface temperatures.

The *PATRAN* calculations of this test did not match the test data very well, with up to 50% differences between measurements and calculations. Part of the discrepancy can be attributed to measurement uncertainties. Comparing the external temperatures between this and the previous test (Figure 13), temperature differences of 100°C or more were observed for the same location. However, the test did show that the siding/XPS construction withstood the 100 kW exposure without any ignition or visible failure. It is also quite evident that the XPS temperatures did not rise enough to cause internal failure of the wall. The melting (softening) temperature of XPS is 104°C and its flash point is greater than 324°C.

Combined Radiant and Convective Heating

Cedar Sample. This test involved both a radiant panel (25 kW/m^2) and a flame (100 kW) applied to a cedar siding wall. For this test, the radiant panel was allowed a 7 min warm-up period before the test was initiated. A radiant heat flux of 27.36 kW/m^2, based on the pre- and post-test calibration data, was used for the calculations. Figure 15 shows the arrangement of thermocouples for this test.

First, the convective heat flux was calculated using Equation 4, again with the lower-edge floating temperature probe (TC33) as the flame temperature and h = 58.42 kW/m^2·K, based on prior calculations. Next, the constant radiant heat flux was added to get the total heat flux. The total heat flux was entered into the FTP equation (Equation 3) to calculate the ignition time, which was 5.8 s (when the FTP crossed the threshold value of 11,500). Figure 16 shows the calculated heat flux and resulting FTP values. Figure 17 shows the measured external cedar siding temperatures. Unlike the 40 kW/m^2 radiant heating exposure (Figure 7), no sharp temperature rises were observed to indicate wall ignition. Based on video evidence, the cedar sample ignited between 7 to 10 s after exposure. The calculated ignition time of 5.8 s is a close match to the actual ignition time, validating the revised convective heat flux calculation.

Fiber Cement Sample with XPS Backing. In this test, the fiber cement siding/XPS construction was exposed for a full 13 min to the combined 25 kW/m^2 radiation heat flux and the 100 kW flame. When the test was stopped the siding had cracked, but the combined siding/XPS construction remained stable, showing no signs that it would fail and cease protecting the home. Figure 18 shows the fiber cement/XPS sample at the conclusion of the combined radiation and convection test.

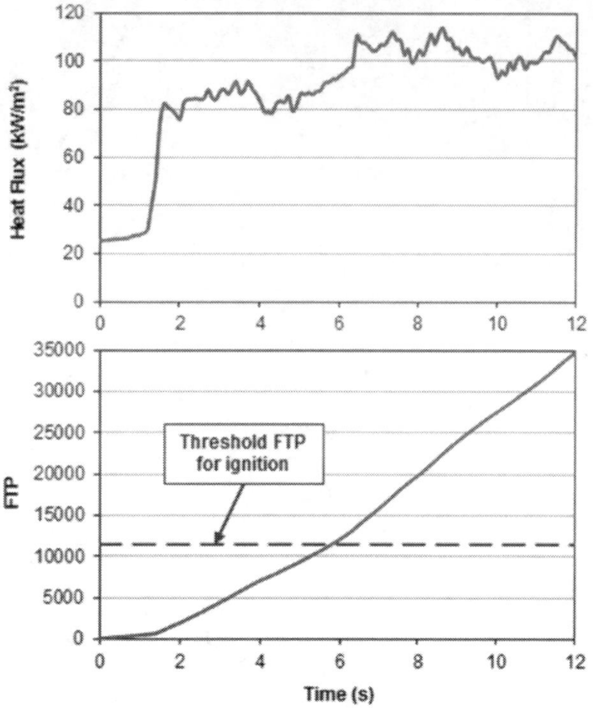

Figure 16 *Calculated incident heat flux (top) on the cedar sample for the combined radiant and convective exposure and the corresponding FTP values (bottom).*

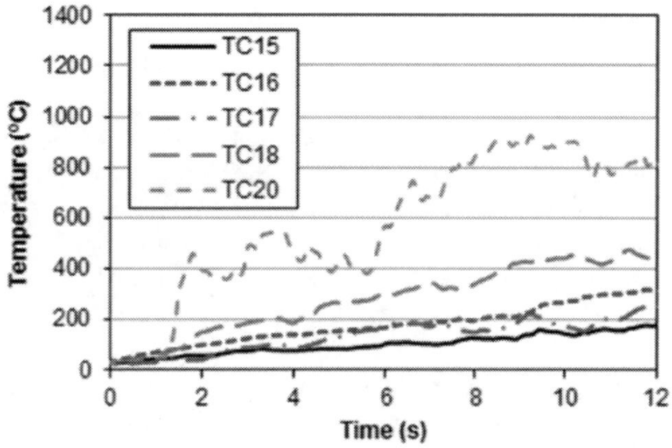

Figure 17 *Measured external cedar siding temperatures in response to combined radiant and convective heating.*

CONCLUSIONS AND FUTURE WORK

This paper describes a series of flammability tests of small-scale wall assemblies with different siding types that are commonly used in North America. These tests were designed to represent the radiant and convective heating of a building in the wildland-urban interface from common fire threats in the event of a wildfire. Data from these tests were

Figure 18 *Fiber cement board sample with XPS foam insulation at the end of the combined radiant and convective exposure test.*

utilized to validate and calibrate the calculation procedures within the structure ignition assessment model (SIAM), which is designed to allow homeowners to assess the level of ignition threat to their homes from their surroundings. The different siding types tested were cedar, vinyl, and fiber cement. The ignition threshold flux-time product embedded in the ignition model was verified by the radiant and convective tests of the wall samples with cedar siding. The convection calculation procedure was revised, with an updated constant heat transfer coefficient.

In addition to model validation, the tests provide important flammability data that can be utilized for future enhancements of the current model and also provide homeowners and community builders the knowledge to better design communities in the wildland-urban interface. Of the different sidings tested, the fiber cement siding proved to be non-combustible and the test data indicated that, for common fire threats, it could adequately protect the building exterior from failure by ignition.

The WildFIRE Wizard calculates the imposed heat flux on the structure and the response of the structure to heating. The former makes use of many estimated variables (e.g., the heat emitted from burning objects of various sizes, burning times, etc.), which can be estimated with greater confidence by performing laboratory tests. Ideally, a series of tests would be conducted with different burning objects that are typically present near a home (e.g., large shrub, shed, landscaping, etc.)

to precisely measure the time-dependent radiant heat flux. The Wizard also contains a module for calculating the risk to the home from airborne embers, mostly based on estimated probabilities. Tests exposing model homes to an ember attack would be ideal to better quantify these parameters.

ACKNOWLEDGMENTS

The authors would like to thank the Department of Homeland Security, Science and Technology Directorate, Infrastructure Protection & Disaster Management Division, for funding this research. This work was accomplished under Contract No. HSHQDC-08-X-00856 with the U.S. Department of Homeland Security. The contributions of Jack Cohen (Fire Sciences Laboratory, Missoula, MT), Jan Kosny (formerly with Oak Ridge National Laboratory, Oak Ridge, TN), Tim Smail (formerly with Savannah River National Laboratory, Aiken, SC), and Monica Philips (Savannah River National Laboratory, Aiken, SC) are gratefully acknowledged.

NOMENCLATURE

A	=	projected area, m^2
C	=	cooling rate, kW/m^2
C_p	=	heat capacity, $J/m^2 \cdot K$
c_p	=	specific heat, $J/kg \cdot K$
h	=	heat transfer coefficient, $kW/m^2 \cdot K$
Q	=	radiation exchange, W
Q''	=	heat flux, kW/m^2
Q_C''	=	critical radiant heat flux, kW/m^2
T	=	temperature, K
Δt	=	time step, s
v	=	view factor
ε	=	emissivity
σ	=	Stefan-Boltzmann constant, 5.67×10^{-8} $W/m^2 \cdot K^4$

Subscripts

f	=	flame
w	=	wall

REFERENCES

Albini, F. 1997. An overview of research on wildland fire. *Proc. Fifth International Symposium Fire Safety Sci.*, Melbourne, Australia, March 1997, pp. 59–74.

ASTM. 2011a. ASTM E 1269-11, *Standard Test Method for Determining Specific Heat Capacity by Differential Scanning Calorimetry*. West Conshohocken, PA, USA: ASTM International.

ASTM. 2011b. ASTM E 136-11, *Standard Test Method for Behavior of Materials in a Vertical Tube Furnace at 750°C*. West Conshohocken, PA, USA: ASTM International.

Cohen, J. 2004. Relating flame radiation to home ignition using modeling and experimental crown fires. *Can. J. For. Res.* 34:1616–26.

Hammer, R.B., S.I. Stewart, and V.C. Radeloff. 2009. Demographic trends, the wildland-urban interface, and wildfire management. *Soc. Nat. Resources* 22:777–82.

Incropera, F.P., and D.P. Dewitt. 1996. *Fundamentals of Heat and Mass Transfer*, 4th Ed. Wiley.

Lampin-Maillet, C., M. Jappiot, M. Long, D. Morge, and J.P. Ferrier. 2009. Characterization and mapping of dwelling types for forest fire prevention. *Comp., Env. Urban Systems* 33:224–32.

Manzello, S.L., T.G. Cleary, J.R. Shields, and J.C. Yang. 2006. Ignition of mulch and grasses by firebrands in wildland-urban interface fires. *Int. J. Wildland Fire* 15:427–31.

Manzello, S.L., A. Maranghides, and W.E. Mell. 2007. Firebrand generation from burning vegetation. *Int. J. Wildland Fire* 16:458–62.

Manzello, S.L., S-H. Park, and T.G. Cleary. 2009. Investigation on the ability of glowing firebrands deposited within crevices to ignite common building materials. *Fire Safety J.* 44:894–900.

Mell, W.E., M.A. Jenkins, J.S. Gould, and N.P. Cheney. 2007. A physics-based approach to modeling grassland fires. *Int. J. Wildland Fire* 16:1–22.

Mell, W.E., S.L. Manzello, A. Maranghides, D. Butry, and R.A. Rehm. 2010. The wildland-urban interface fire problem—Current approaches and research needs. *Int. J. Wildland Fire* 19:238–51.

Modest, M.F. 1993. *Radiative Heat Transfer*. McGraw-Hill.

MSC. 2008. *PATRAN*, Ver 2008 Rev. 1. Heat Transfer Software. MSC Software, Costa Mesa, CA.

NFPA. 2012. Quincy, MA: National Fire Protection Association. www.nfpa.org/research/fire-statistics/deadliest-and-large-loss-fires/largest-loss-wildland-fires

NIFC. 2012. Boise, ID: National Interagency Fire Center. www.nifc.gov/fireInfo/fireInfo_statistics.html

Papadopoulos, G.D., and F.-N. Pavlidou. 2011. A comparative review on wildfire simulations. *IEEE Systems J.* 5:233–43.

Reams, M.A., T.K. Haines, C.R. Renner, M.W. Wascom, and H. Kingre. 2005. Goals, obstacles and effective strategies of wildfire mitigation programs in the wildfire-urban interface. *Forest Policy Eco.* 7:818–26.

Rehm, R.G. 2008. The effects of winds from burning structures on ground-fire propagation at the wildland-urban interface. *Combustion Theory and Modeling* 12:477–96.

Theobald, D.M., and W.H. Romme. 2007. Expansion of the US Wildland-Urban Interface. *Landscape Urban Planning* 83:340–54.

Tieszen, S.R., T.J. O'Hern, R.W. Schefer, E.J. Weckman, and T.K. Blanchat. 2002. Experimental study of the flow field in and around a one meter diameter methane fire. *Combust. Flame* 129:378–91.

Xin, A., S.A. Filetyev, K. Biswas, J.P. Gore, R.G. Rehm, and H.R. Baum. 2008. Fire dynamics simulations of a one-meter diameter methane fire. *Combust. Flame* 153:499–509.

Field Testing of Nano-PCM-Enhanced Building Envelope Components in a Warm-Humid Climate

Kaushik Biswas, PhD
Associate Member ASHRAE

Jue Lu, PhD

Parviz Soroushian, PhD

ABSTRACT

The U.S. Department of Energy Building Technologies Program's goal of developing high-performance, energy-efficient buildings will require more cost-effective, durable, energy-efficient building envelopes. Forty-eight percent of the residential end-use energy consumption is spent on space heating and air conditioning. Reducing envelope-generated heating and cooling loads through application of phase-change material (PCM) enhanced envelope components can facilitate maximizing the energy efficiency of buildings. Field testing of prototype envelope components is an important step in estimating their energy benefits.

An innovative PCM (nano-PCM) was developed with PCM encapsulated with expanded graphite (interconnected) nanosheets, which is highly conducive for enhanced thermal storage and energy distribution, and is shape-stable for convenient incorporation into lightweight building components. During 2012, two test walls with cellulose cavity insulation and prototype PCM-enhanced interior wallboards were installed in a natural exposure test (NET) facility in Charleston, SC. The first test wall was divided into four sections separated by wood studs and thin layers of foam insulation. Two sections contained nano-PCM-enhanced wallboards: one was a three-layer structure in which nano-PCM was sandwiched between two gypsum boards, and the other one had PCM dispersed homogeneously throughout graphite nanosheet-enhanced gypsum board. The second test wall also contained two sections with interior PCM wallboards; one contained nano-PCM dispersed homogeneously in gypsum and the other was gypsum board containing a commercial microencapsulated PCM (MEPCM) for comparison. Each test wall contained a section covered with gypsum board on the interior side that served as control or a baseline for evaluation of the PCM wallboards. The walls were instrumented with arrays of thermocouples and heat flux transducers. This paper presents the measured performance and analysis to evaluate the energy-saving potential of the nano-PCM-enhanced building components.

INTRODUCTION

According to the 2009 Residential Energy Consumption Survey (RECS) of the U.S. Energy Information Administration (EIA), about 48% of the total residential site energy consumption is due to space heating and air conditioning (EIA 2009). The U.S. Department of Energy (DOE) Building Technologies Program's goal of developing high-performance, energy-efficient buildings will require more cost-effective, durable, energy-efficient building envelopes. Application of phase-change materials (PCMs) to building envelopes to take advantage of their high latent heat capacities in reducing the envelope-generated heating and cooling loads has received a lot of attention in the last two decades (Zhou et al. 2007).

Sharma et al. (2009) and Xin et al. (2009) reviewed the research and analysis of available thermal storage systems incorporating PCMs for building applications. Different approaches to PCM applications in building envelopes have been investigated: PCM wallboards (Stovall and Tomlinson 1995; Zhou et al. 2007), PCM mixed with loose-fill insulation (Shrestha et al. 2011), macro-packaged PCM in plastic pouches (Kosny et al. 2012), etc.

The subject of this article is an innovative PCM (nano-PCM) with the PCM encapsulated in expanded graphite

***Kaushik Biswas** is a research and development associate (building scientist) at Oak Ridge National Laboratory, Oak Ridge, TN. **Jue Lu** is a senior scientist at and **Parviz Soroushian** is the president of Technova Corporation, Lansing, MI.*

Published by ASHRAE 2013

nanosheets, which is highly conductive, for enhanced thermal storage and energy distribution and is shape-stable for convenient incorporation into lightweight building components. This nano-PCM was incorporated in gypsum boards, which are commonly used as interior wallboards on walls in residential construction. Three prototype nano-PCM-enhanced interior wallboards were produced and were installed in a natural exposure test (NET) facility in Charleston, SC, for testing. For comparison, two types of control panels were installed: gypsum board incorporating a commercial microencapsulated PCM (MEPCM) (Micronal® PCM [BASF 2013]) and standard gypsum board. In the subsequent analysis and discussion, these PCM wallboards will be referred to using the following list:

- PCM-Soaked Gypsum
- Three-Layer Structure
- Dispersed Nano-PCM
- MEPCM in Gypsum

The first two products were installed in January 2012 and the latter products were installed in July 2012. The fourth product was also used as a control. Brief descriptions of the PCM products are provided in the following section.

DESCRIPTIONS OF THE PCM PROTOTYPE SAMPLES

Following are brief descriptions of and details about the material and thermophysical characteristics of the PCM wallboards.

Materials

Gypsum powder was used for fabrication of gypsum board. A paraffin, n-heptadecane ($C_{17}H_{36}$), was used as the PCM. High thermal conductivity is a desirable characteristic of PCMs in energy storage applications, but the lower conductivity of paraffins decreases the rates of heat stored and released during the phase-change processes and limits their utility areas (Zalba et al. 2003; Sari and Karaipekli 2007). Exfoliated graphite nanoplatelets (xGnP) (Kim and Drzal 2009) and expanded graphite (interconnected) nanosheets (Drzal and Fukushima 2009) were added for increasing the thermal conductivity of the PCM wallboards for enhanced heat transfer and better utilization of PCMs.

PCM-Soaked Gypsum

This PCM wallboard is comprised of gypsum board that absorbed 20%-by-weight n-heptadecane by soaking. The board composition was 99% gypsum and 1% xGnP for enhanced thermal conductivity. The xGnP was dispersed in 1 g/L sulfonated polystyrene by tip sonication (using ultrasound energy for particle dispersion) for one hour and then mixed with the gypsum powder and water. The board was cured overnight at room temperature, followed by curing at 50°C for 48 hours. After cooling down to room temperature, the board was soaked in melted n-heptadecane, with a final PCM content of 20% by weight. Figure 1 shows the heat flow data from differential scanning calorimetry (DSC) testing of n-heptadecane. During melting, the n-heptadecane is trapped in the air voids of the gypsum matrix without weeping due to capillary action and surface tension.

Three-Layer Structure

This wallboard contained a 4 mm layer of nano-PCM sandwiched between two 5 mm layers of gypsum. The nano-PCM contains 8%-by-weight expanded graphite nanosheets and 92%-by-weight n-heptadecane. Expanded graphite nanosheets were processed following a microwave process from expandable graphite (Drzal and Fukushima 2009). The weight percentage of nano-PCM in the wallboard was about 20%. Figure 1 shows the heat flow data from DSC testing of the nano-PCM. The positive heat flows represent exothermic conditions, indicating the freezing phenomenon of the nano-PCM; negative heat flows indicate a melting process. The

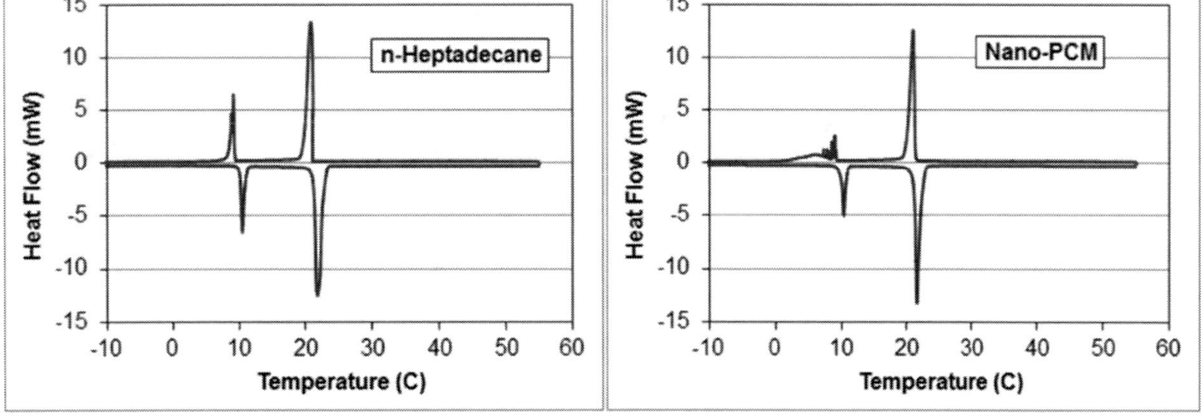

Figure 1 *Melting and freezing characteristics (DSC curve) of n-heptadecane (left) and nano-PCM (n-heptadecane with 8%-by-weight expanded graphite nanosheets) (right). The DSC heating and cooling rate was 1°C/min.*

DSC data provide information about the active temperature range of the PCM (where phase changes occur) and the respective latent heats. For design and analysis purposes, the smaller peaks on the left of the major melting and freezing peaks have been ignored.

Dispersed Nano-PCM

This board contains 20%-by-weight nano-PCM homogeneously dispersed in gypsum board. Gypsum powder was mixed with 65%-by-weight water, followed by an addition of nano-PCM. This board was cured for one week at room temperature. The DSC data for this product are shown in Figure 2.

MEPCM in Gypsum

This is gypsum board incorporating 20%-by-weight MEPCM (Micronal®). This board was intended for use as another control, in addition to regular gypsum board, to evaluate the performance of three prototype PCM wallboards (PCM-soaked gypsum, three-layer structure, and dispersed nano-PCM). This board was cured overnight at room temperature then for five hours at 50°C. Figure 2 shows the DSC data for the MEPCM in gypsum product.

Table 1 lists the thermophysical properties of the PCM wallboards. The properties were estimated using DSC data at 1°C/min heating and cooling rates. The melting and freezing onset temperatures were determined using an integration function based on a tangents method that is included in the DSC analysis software.

TEST FACILITY AND TEST WALL DETAILS

These wall tests are ongoing in a NET facility in Charleston, SC, shown in Figure 3. NET facilities expose side-by-side roof/attic and wall assemblies to natural weathering in four different humid U.S. climates. The data from these facilities help the industry develop products to avoid adverse moisture-related impacts in buildings and are essential in validating hygrothermal and energy models. NET structures are located at Oak Ridge, TN; Charleston, SC; Tacoma, WA; and Syracuse, NY. Each is temperature and humidity controlled and instrumented to measure moisture content in materials, vapor pressure, temperature, heat flux, humidity, etc. Figure 3 shows the south wall of the Charleston NET facility, representing a warm-humid climate, which houses multiple side-by-side test walls. Also shown is a weather station on the west gable end of the building.

The test walls were built using wood framing (2 × 6 studs) with oriented strand board (OSB) on the outside and PCM wallboards on the inside. The outer side of the OSB is covered with a weather-resistant barrier and vinyl siding. Figure 4 shows the installed test walls. On the left is the first wall, marked as *test wall 1*, which was built in January 2012. It was divided into four equal sections; two sections were covered with the PCM-soaked gypsum and three-layer structure PCM wallboards, and another section was covered with standard gypsum board. The second wall, *test wall 2*, was installed

Table 1. PCM Properties

	Latent Heat, J/g		Phase-Change Onset, °C	
	Melting	**Freezing**	**Melting**	**Freezing**
PCM- Soaked Gypsum	28.2	27.9	21.2	21.2
Three-Layer Structure	29.2	29.3	21.1	21.3
Dispersed Nano-PCM	26.9	26.8	21.1	21.3
MEPCM in Gypsum	18.8	18.9	18.9	20.9

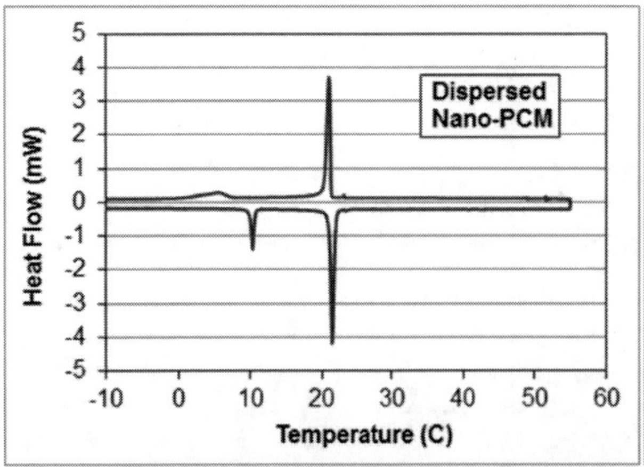

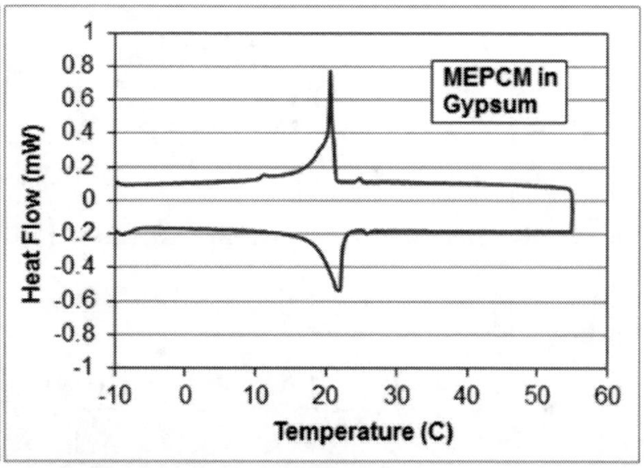

Figure 2 *Melting and freezing characteristics (DSC curves) of the dispersed nano-PCM wallboard (left) and MEPCM in gypsum wallboard (right). The DSC heating and cooling rate was 1°C/min.*

during July 2012 and contained the dispersed nano-PCM and MEPCM in gypsum wallboards, along with a taller gypsum section.

The smaller wall cavities are about 1.1 m high and 0.4 m wide, and the taller gypsum-covered cavity in test wall 2 is 2.2 m high and 0.4 m wide. Each cavity or section is separated from the others using wood studs and thin layers of foam insulation. The cavity depth is 14 cm. During construction, each wall cavity was covered with a plastic mesh and then cellulose insulation was added by cutting a slot in the mesh at the top of the cavity. The cellulose was allowed to settle under its own weight. The measured cavity volume and estimated weight of cellulose insulation added to the cavities yielded an average density of 40.9 kg/m^3. It should be noted that, due to the ad-hoc application method, the resulting density was lower than the recommended density for blown-in dense-packed cellulose, which is about 56.1 kg/m^3. The area surrounding the test cavities was filled with fiberglass insulation to prevent convection loops in empty air spaces and minimize the thermal interaction with the neighboring test wall sections. The perimeter cavities are used for routing instrumentation cables, and fiberglass batts were used to fill those perimeter cavities due to ease of installation.

Figure 3 *Charleston, SC, NET facility.*

DATA ACQUISITION SYSTEM AND INSTRUMENTATION

Figure 5 shows a typical instrumentation layout in the wall cavities. The wall has vinyl siding and a weather barrier over OSB on the exterior side of the wall, which is exposed to the atmosphere (as seen on the south wall in Figure 3). The interior side is covered with a gypsum board. Each cavity contains a thermistor and relative humidity sensor combination (T/RH sensor) on the OSB and gypsum surfaces facing the cavity, a thermistor inside the cavity (mid-depth) and on the gypsum surface facing the room interior, and heat flux transducers (HFTs) on the gypsum surface facing the cavity. Within each cavity, these sensors are located approximately in a line along both the vertical and horizontal midpoints of the cavity. In addition, a single thermistor is attached to the wall exterior (between the OSB and exterior siding) and a T/RH sensor combination is attached on the OSB surface facing the exterior (not shown in Figure 5). The T/RH sensors are indicated by the white packets seen in Figure 5.

In addition to the sensors attached to the test wall, the NET facility is equipped with sensors and instruments to monitor the local weather conditions, including temperature, humidity, solar irradiance, wind conditions, etc. These sensors are controlled and monitored by dataloggers and multiplexers. Each sensor is scanned at five-minute intervals, and the data are averaged and stored at hourly intervals. The data are downloaded on a weekly basis at Oak Ridge National Laboratory

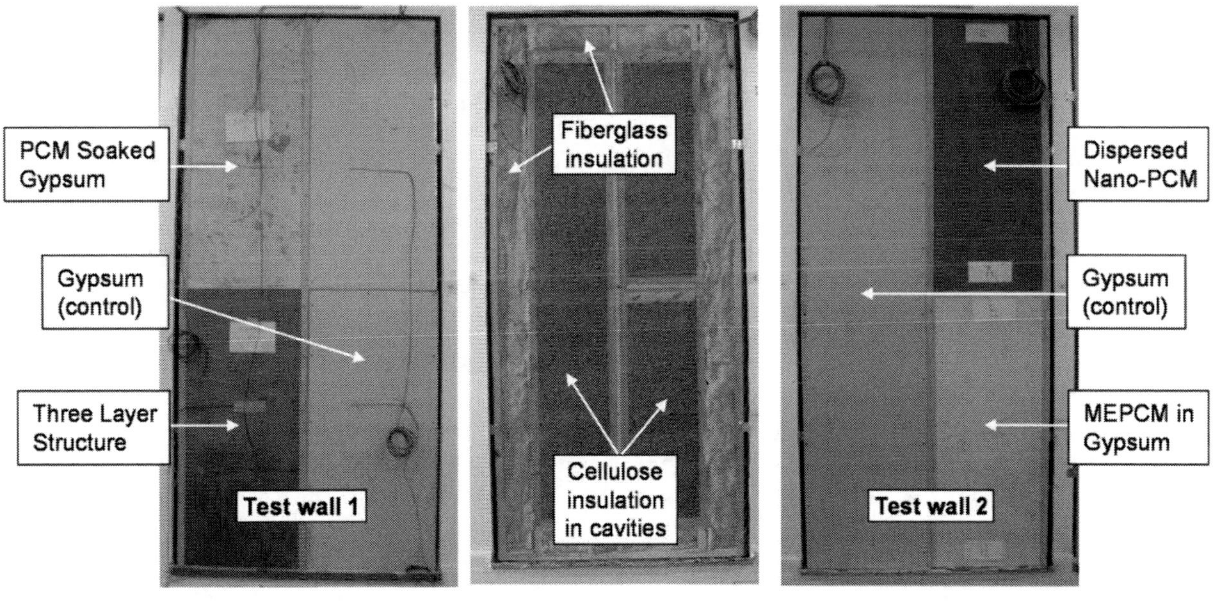

Figure 4 *Test walls with the different PCM prototype wallboards.*

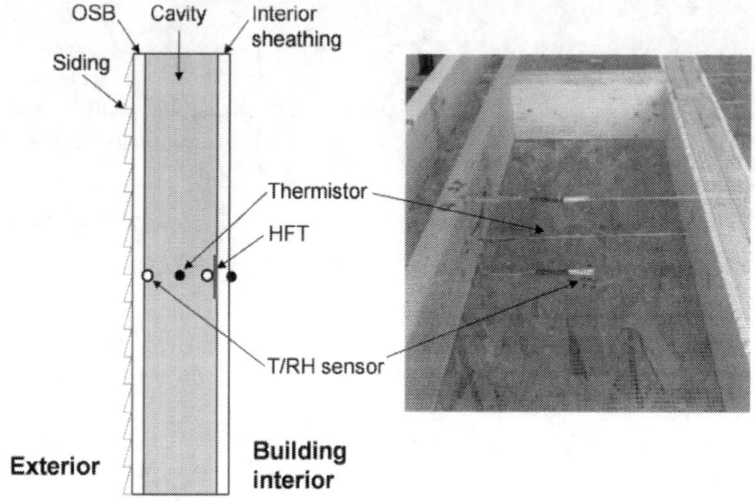

Figure 5 *Sensor placement in test wall cavities.*

(ORNL) using a dedicated computer and modem. Table 2 provides the sensor specifications. The HFTs were calibrated using a heat flowmeter apparatus (HFMA) sandwiched by insulation and regular gypsum board. The HFMA is used for thermal transmission property measurements following the ASTM C518 (2010) test method and is accurate within 1%. The HFMA consists of two independently temperature-controlled plates, both of which are equipped with heat flow sensors. The calibration constants of the HFTs used in the Charleston test walls were obtained by correlating the measured heat flows of the HFMA to the HFT voltage outputs.

RESULTS AND DISCUSSION

This section describes the temperature and heat flux data from the test walls to examine the behavior and impact of the PCM prototype wallboards. The data are compiled into weekly files containing hourly-averaged data. The period covered in this section is July–December 2012, after the installation of the second test wall. This period is deemed sufficient for the weather-related evaluation of the PCM products as it includes summer, winter, and shoulder (fall) months. The focus is on summer and winter periods, when the cooling and heating loads are at their peak.

Temperature Distribution and PCM Behavior

The temperature distributions in the different sections of the two test walls are shown in Figures 6 and 7. The hourly-averaged temperature data are shown for a typical summer week (July 18–25). The temperatures are shown at the wall exterior, the OSB interior (facing the cavity), the cavity center, the wallboard exterior (facing the cavity), and the wallboard interior (facing the room).

The room interior was maintained at about 20°C–22°C during the monitoring period. A clear trend is seen in the temperatures from the exterior to the interior. Some differences are seen in the cavity center and the wallboard exterior between the different test walls and the test sections. However, these differences are fairly insignificant and do not reveal much about the behavior and impact of the PCMs in the different wallboards.

Table 2. Installed Sensor Accuracy

	Accuracy	Sensitivity	Repeatability
10 K ohm Thermistor	± 0.2%	—	± 0.2%
Humidity Sensor	± 3.5%	—	± 0.5%
Heat Flux Transducer	± 5%	5.7 $(W/m^2)/mV$	—
Outdoor Relative Humidity Sensor	± 3%	—	—
Solar Pyranometer, Vertical	± 3%	0.2 $(kW/m^2)/mV$	—
Solar Pyranometer, Horizontal	± 3%	10 $\mu V/(W\text{-}m^2)$	—
Channel Multiplexer	± 0.1% of Full Scale Reading	—	—

To further examine the PCM behavior, the wallboard exterior and interior temperatures are shown in Figure 8, along with the corresponding melting and freezing onset temperatures. The data shown in Figure 8 were compiled from the weekly data files from July 11 to December 26, 2012.

For maximum benefit from a PCM, it needs to melt and freeze on a diurnal basis, i.e., its temperature should rise above its melting point and fall below its freezing point on a daily basis. The hourly variations of the exterior and interior surface

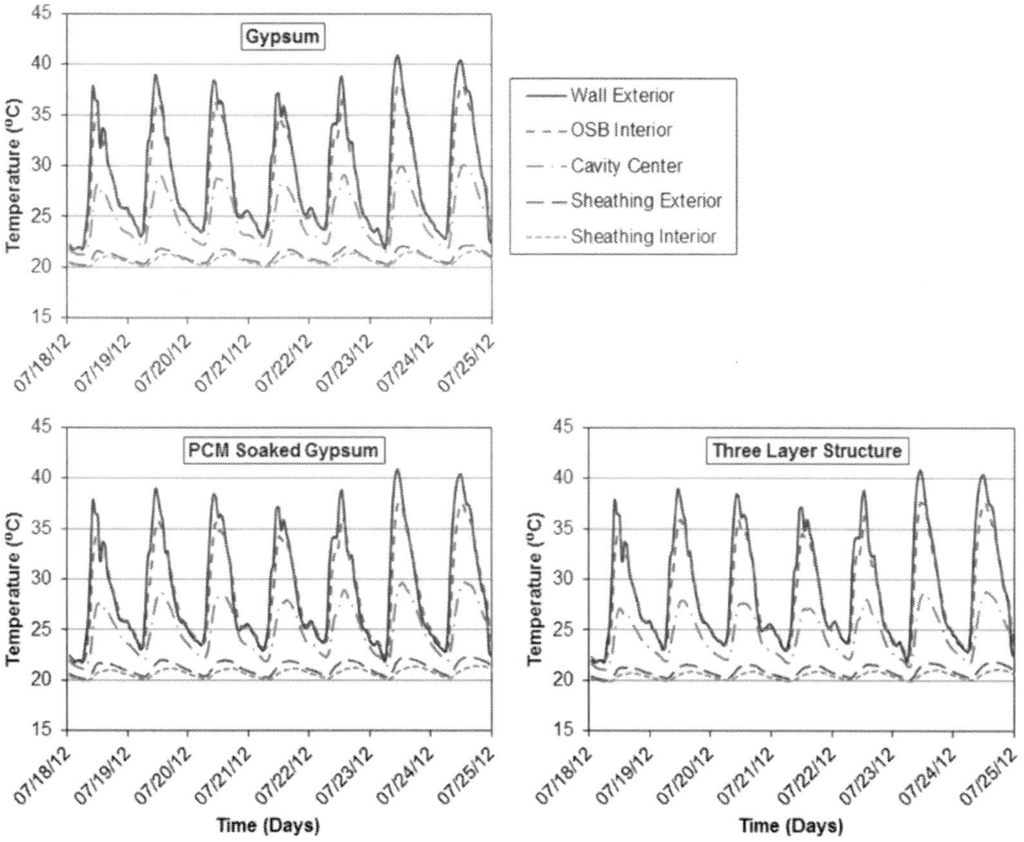

Figure 6 *Temperature distribution in different sections of test wall 1.*

temperatures compared to the phase-change onset temperatures can provide information about the effectiveness of the PCM prototype wallboards. The first plot in Figure 8 shows the behavior of the PCM-soaked gypsum wallboard. The exterior temperatures cycled above the melting onset temperature and below the freezing onset temperature almost every day during the summer months (July–September) and periodically in October, indicating frequent phase-change activity near the surface of the wallboard facing the cavity. The interior wallboard temperatures, however, remained predominantly below the melting onset temperature during this time. From November onwards, the wallboard temperatures stayed predominantly below the melting onset temperature, and the PCM can be expected to have been frozen during this time.

The second plot shows the behavior of the three-layer structure wallboard. The interior wallboard temperature remained below the melting point at all times. The exterior wallboard temperature cycled above and below the phase-change temperatures from July till early October but remained below the melting point after that. The dispersed nano-PCM wallboard had melting and freezing onset temperatures similar to those of the three-layer structure wallboard. The diurnal exterior and interior temperature oscillations were greater such that the exterior temperatures rose above and below the melting and freezing onset temperatures from July through early November. Even the interior temperatures rose above the melting point several times July through September.

The fourth plot shows the temperatures of the MEPCM in gypsum wallboard. This product has the lowest melting onset temperatures of all the wallboards, with the result that both exterior and interior wallboard temperatures were predominantly above the melting point. The wallboard temperatures cycled above and below the freezing onset temperature on a daily basis from July till early October and less frequently from October to mid-November, then remained below the freezing point after that. Thus, unlike the other PCM products, the MEPCM in gypsum wallboard can be expected to have undergone at least partial melting and freezing even during the winter period.

In summary, the temperature data indicate that the first three wallboards underwent at least partial melting and freezing between July and mid-November, especially near the exterior surface facing the wall cavities, and remained predominantly frozen thereafter. The MEPCM in gypsum wallboard, however, showed phase-change activity throughout the evaluation period.

Heat Flux Variation

In this section, data from the HFTs installed in the test cavities are shown and described. It should be noted that the

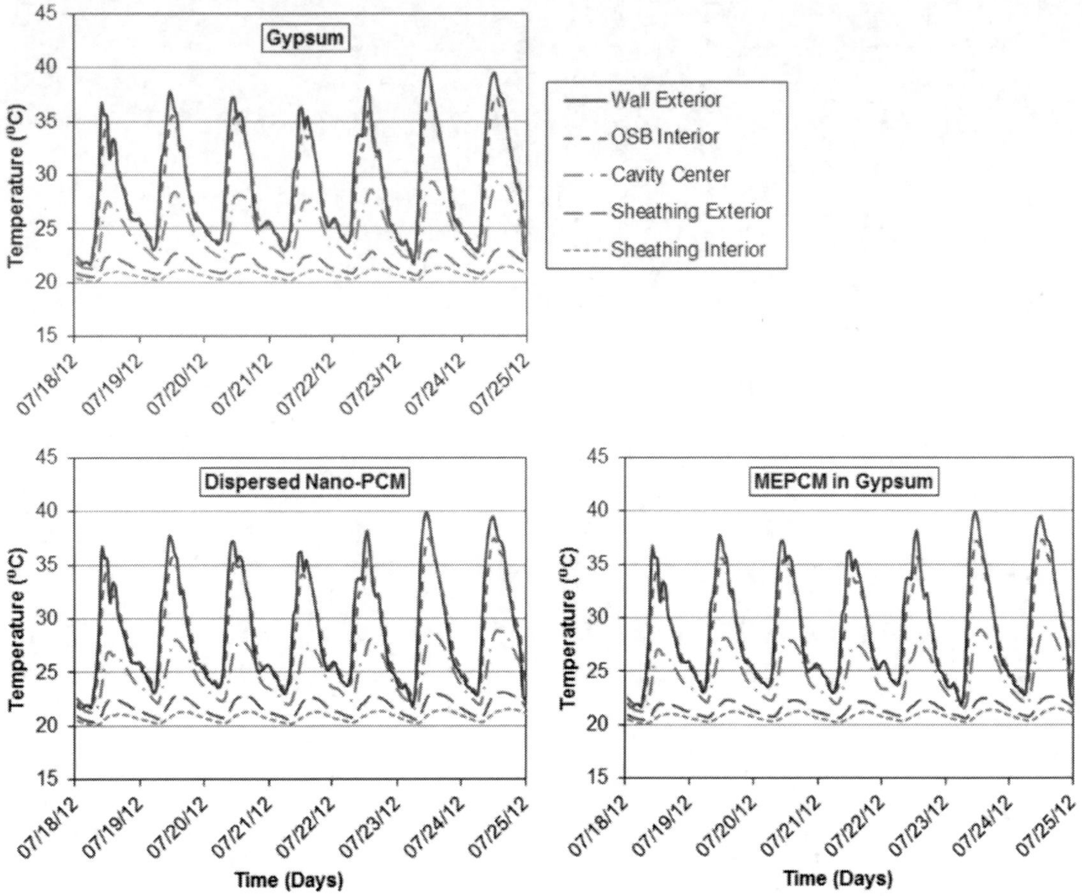

Figure 7 *Temperature distribution in different sections of test wall 2.*

HFTs measuring the heat flux were installed on the exterior surface of the wallboards (facing the cavity). From a space-conditioning perspective, the heat flux that actually enters the conditioned area is of primary interest. In the test sections, since the PCMs are contained in the wallboards, the locations of the HFTs are not ideal, as they do not fully capture the impact of the melting and freezing of PCMs on the heating and cooling loads in the conditioned area. The HFTs only capture the heat flow between the wallboard and the cavity, which may be different from the heat flow at the wallboard interior due to the latent heat storage/release within the wallboards.

Figures 9 and 10 show the hourly heat fluxes during a summer and winter week, respectively, through the different test sections in the two test walls, test wall 1 and test wall 2. The control gypsum test sections are shown for comparison. Also shown is the hourly solar irradiance on the test wall exteriors. As expected, the daytime heat fluxes show a strong correlation with the solar irradiance.

In Figure 9, the first plot shows the test sections in test wall 1. The behavior of the three-layer structure wallboard is similar to the control gypsum section, but the peak daytime heat gains in the PCM-soaked gypsum section are higher by up to 30% compared to the control. Both PCM wallboards in test wall 2 reduced the peak daytime heat gains by about 20% compared to the gypsum section, which is beneficial in reducing the cooling loads on the conditioned space.

Figure 10 shows the heat flux variations during a winter week. There was very little heat gain measured by the HFTs during this week, with substantial heat losses, especially during the night. The daytime peaks were similar between the PCM wallboards and their respective control sections. In test wall 1, both PCM wallboard sections exhibited greater peak heat losses compared to the gypsum control; conversely, in test wall 2, both PCM sections reduced the heat losses at night compared to the gypsum section.

To further investigate the impact of the PCM-enhanced wallboards, the heat flux data though the different wall sections were integrated over 30-day winter and summer periods to determine the total heat gains and losses through the different sections. The integration was performed using the trapezoidal rule:

$$\text{Total Integrated Heat Flow, } Q \ [\text{kJ/m}^2] = \sum_{n=0}^{m} 0.5 \times \Delta X_n \times (Y_n + Y_{n+1}) \quad (1)$$

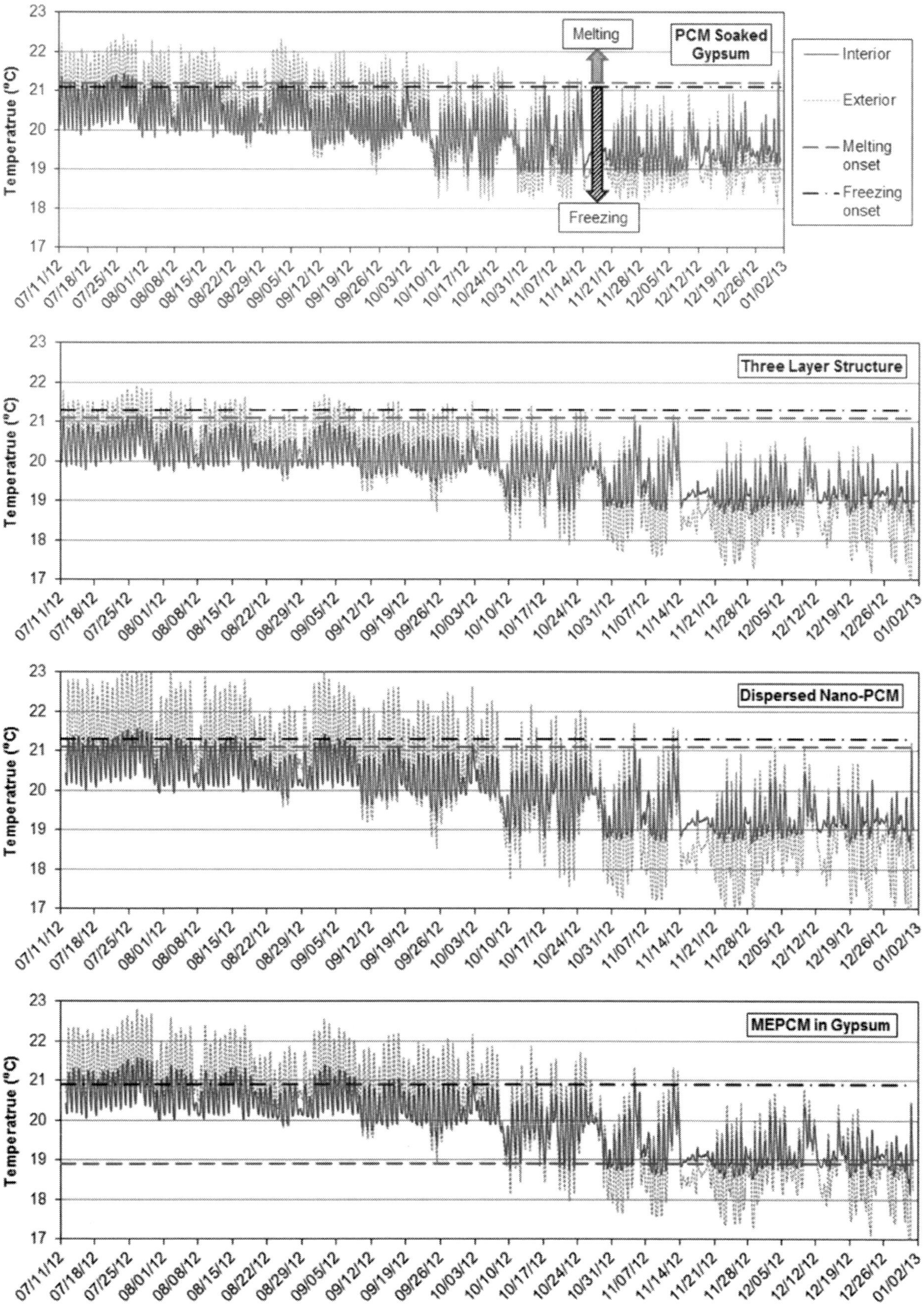

Figure 8 *Comparison of PCM-enhanced wallboard interior and exterior temperatures with their respective phase-change onset temperatures.*

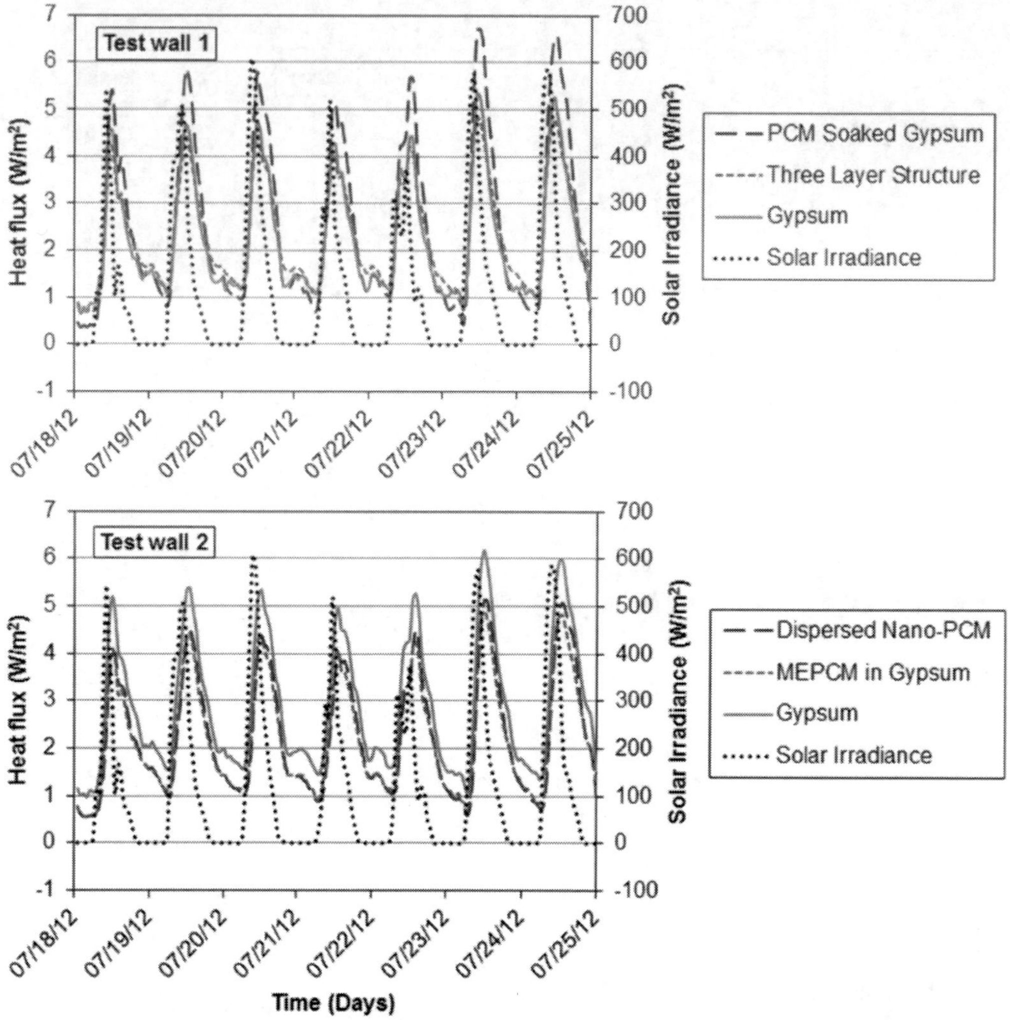

Figure 9 *Summer heat flux variation and solar irradiance.*

In Equation 1, Y_n and Y_{n+1} correspond to the heat flux during current and future time steps, ΔX_n is the time step (1 h in this case), and the subscript m denotes the final point in the series, which varies depending on the summer or winter data sets. The positive and negative heat fluxes were integrated separately to determine the heat gains and losses for each section. The integrated or total heat gains and losses and the net heat transfer during summer and winter periods are shown in Tables 3 and 4. The net heat transfer for each period was obtained by integrating the positive and negative heat fluxes together. Also shown in the tables are the percent reductions compared to the respective gypsum control sections.

Using the general uncertainty analysis method described by Coleman and Steele (1989), the uncertainty (U_Q) in the integrated heat flow (Q) can be described by:

$$U_Q^2 = \left[\left(\frac{\partial Q}{\partial Y_n}U_Y\right)^2 + \left(\frac{\partial Q}{\partial Y_{n+1}}U_Y\right)^2\right]^{1/2} \qquad (2)$$

Assuming a 5% uncertainty (U_Y) in the HFT measurements, the uncertainty of the integrated heat flows reported in Tables 3 and 4 is 7.1%.

Thirty-day periods were chosen that represent typical summer and winter conditions to analyze the data. The 30-day summer period extended from July 13 to August 11, 2012, and the winter period was December 2–31, 2012. Table 3 lists the integrated heat flows from test wall 1. Clearly, neither PCM wallboard performed as well as expected, allowing both higher heat gains during summer and higher heat losses during winter. Both PCM wallboards used in test wall 2 performed better than the gypsum control (Table 4), with about 21%–26% reductions in heat gain and heat loss during the cooling and heating seasons, respectively. The prototype dispersed nano-PCM wallboard performed better than the wallboard with the commercially available MEPCM in reducing both heating and cooling loads.

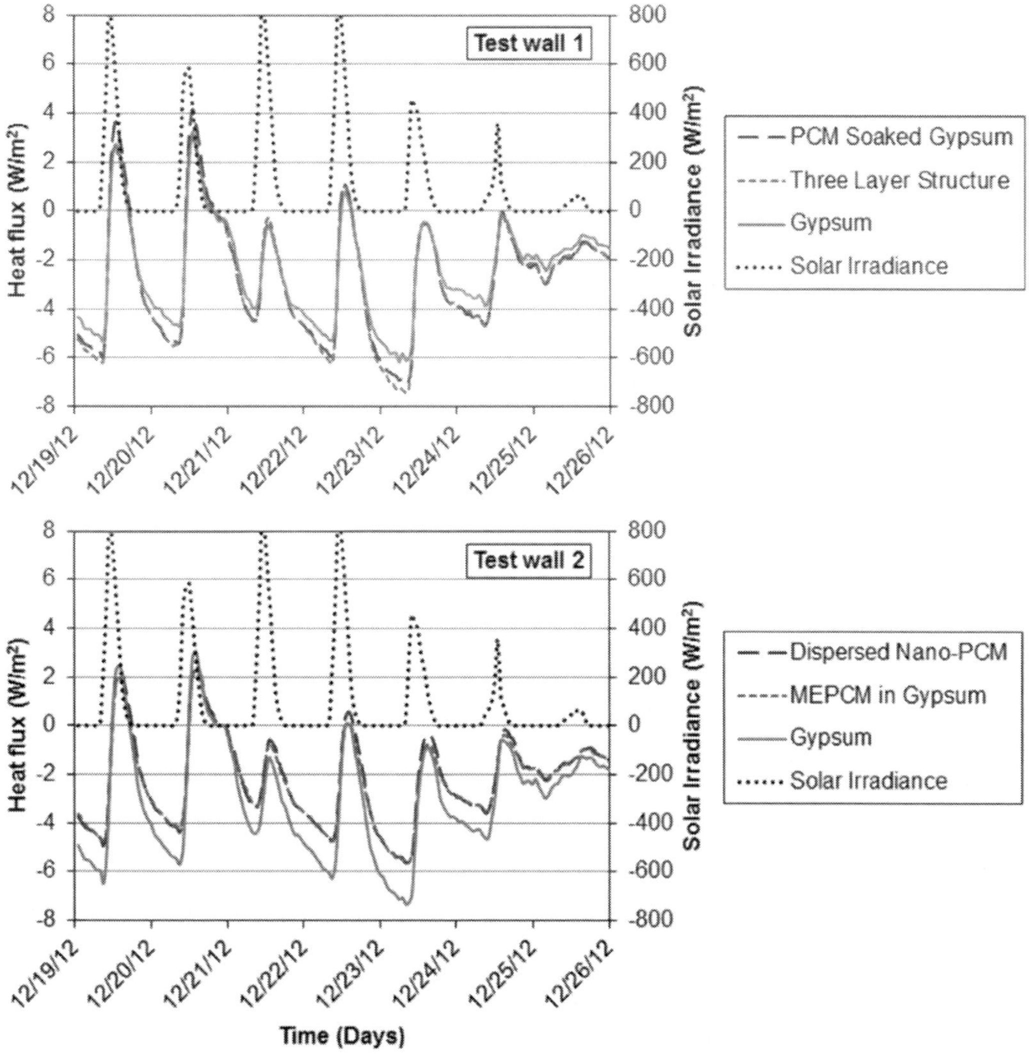

Figure 10 *Winter heat flux variation and solar irradiance.*

Table 3. Heat Flux Statistics for Test Wall 1

	Heat Gain, kJ/m^2	% Reduction	Heat Loss, kJ/m^2	% Reduction	Net, kJ/m^2	% Reduction
Summer 30-Day Period (July 13–August 11, 2012)						
PCM-Soaked Gypsum	5839.6	–9.76	–6.6		5832.9	–9.63
Three-Layer Structure	5459.5	–2.61	–0.7		5458.8	–2.60
Gypsum	5320.4		0.0		5320.4	
Winter 30-Day Period (December 2–31, 2012)						
PCM-Soaked Gypsum	633.2	–36.59	–5727.3	–17.12	–5094.1	–15.08
Three-Layer Structure	483.9	–4.38	–5780.5	–18.21	–5296.6	–19.66
Gypsum	463.5		–4890.1		–4426.5	

Table 4. Heat Flux Statistics for Test Wall 2

	Heat Gain, kJ/m^2	% Reduction	Heat Loss, kJ/m^2	% Reduction	Net, kJ/m^2	% Reduction
			Summer 30-Day Period (July 13–August 11, 2012)			
Dispersed Nano-PCM	5263.0	22.86	–0.2		5262.8	22.86
MEPCM in Gypsum	5336.2	21.78	0.0		5336.2	21.78
Gypsum	6822.3		0.0		6822.3	
			Winter 30-Day Period (December 2–31, 2012)			
Dispersed Nano-PCM	453.9	5.65	–4333.0	25.88	–3879.1	27.70
MEPCM in Gypsum	406.9	15.42	–4403.5	24.68	–3996.6	25.50
Gypsum	481.0		–5846.0		–5365.0	

It is again noted that the heat fluxes shown in Figures 9 and 10 and the integrated heat flows listed in Tables 3 and 4 are not truly representative of the heat addition to or loss from the conditioned space, due to the location of the HFTs. Further, the positive daytime heat flows measured in the PCM sections during summer may be augmented if the PCM was melting and absorbing heat at the cavity-wallboard interface. Conversely, the nighttime losses may include the heat released at the cavity-wallboard interface due to the freezing of the PCM.

FUTURE WORK

In its current form, this paper describes a work-in-progress project. Further analyses and measurements will enable a better understanding of the behavior of the PCM prototypes being studied. The next step in the process is developing a numerical model of the test walls, with appropriate thermophysical properties of the PCM prototypes, followed by validation studies using the field test data. The numerical model can provide information about the actual impact of the PCM products on the annual heating and cooling loads in the test building. Measurements of the thermal conductivity (in the fully frozen and fully melted states) and temperature-dependent specific heat of the PCM products will enable better understanding of the field data and are essential for accurate modeling of the test walls.

CONCLUSIONS

This paper describes the testing of four prototype PCM-enhanced interior wallboards in a natural exposure test facility in Charleston, SC. Two test walls were built using wood-frames with cellulose cavity insulation. The walls were divided into sections, each containing one of the PCM wallboards. The PCM wallboards were produced by combining gypsum boards with PCMs using different techniques. Based on the PCM incorporation method, the wallboards were referred to as PCM-soaked gypsum, three-layer structure, dispersed nano-PCM, and MEPCM in gypsum. Each test wall also contained a section covered with regular gypsum board that served as control to evaluate the test sections with the PCM wallboards. The walls were instrumented using temperature and heat flux sensors. The temperature data revealed that all four PCM wallboards were expected to have undergone at least partial phase-change phenomenon during the summer and shoulder months; the MEPCM in gypsum wallboard exhibited phase-change activity even during winter, unlike the PCM in the other wallboards, which remained frozen. The heat flux data indicated that the dispersed nano-PCM reduced the space-conditioning loads compared to a standard gypsum board and also performed better than the control wallboard incorporating a commercial microencapsulated PCM. However, the PCM-soaked gypsum and three-layer structure did not perform very well compared to the corresponding gypsum control section. Numerical modeling and validation studies using the test data can provide valuable information about the actual impact of the PCM wallboards on the building heating and cooling loads.

ACKNOWLEDGMENTS

The authors would like to acknowledge the funding support for this work from the Department of Energy Small Business Innovation and Strategies Program (Recovery Act) Award No. DE-SC0003309. The contributions of Jerald Atchley and Phillip Childs of ORNL in installing and instrumenting the test walls, data gathering, and troubleshooting are also gratefully acknowledged.

REFERENCES

ASTM. 2010. ASTM C518-10, *Standard Test Method for Steady-State Thermal Properties by Means of the Heat Flow Meter Apparatus*. West Conshohocken, PA: ASTM International.

Coleman, H.W. and W.G. Steele, 1989. Experimentation and Uncertainty Analysis for Engineers. John Wiley & Sons, New York.

Drzal, L., and H. Fukushima. 2009. Expanded graphite and products produced therefrom. US Patent 7,550,529 B2.

Kim, S., and L. Drzal. 2009. High latent heat storage and high thermal conductive phase change materials using exfoliated graphite nanoplatelets. *Solar Energy Materials & Solar Cells* 93:136–42.

Kosny, J., K. Biswas, W. Miller, and S. Kriner. 2012. Field thermal performance of naturally ventilated solar roof with PCM heat sink. *Solar Energy* 86:2504–14.

BASF. 2013. Micronal® phase change material, DS 5008 X, http://worldaccount.basf.com/wa/EU/Catalog/ACIndustry/pi/BASF/segment/new_building_materials

EIA. 2009. Residential Energy Consumption Survey (RECS). Washington, DC: U.S. Energy Information Administration. www.eia.gov/consumption/residential/data/2009/index.cfm?view=consumption

Sari, A., and A. Karaipekli. 2007. Thermal conductivity and latent heat thermal energy storage characteristics of paraffin/expanded graphite composite as phase change material. *Applied Thermal Engineering* 27(8-9):1271–77.

Sharma, A., V.V. Tyagi, C.R. Chen, and D. Buddhi. 2009. Review on thermal energy storage with phase change materials and applications. *Renewable and Sustainable Energy Reviews* 13(2):318–45.

Shrestha, S., W. Miller, T. Stovall, A. Desjarlais, K. Childs, W. Porter, M. Bhandari, and S. Coley. 2011. Modeling PCM-enhanced insulation system and benchmarking EnergyPlus against controlled field data. *Proceedings of Building Simulation 2011: 12th Conference of International Building Performance Simulation Association*, p 800–807.

Stovall, T.K., and J.J. Tomlinson. 1995. What are the potential benefits of including latent storage in common wallboard? *Journal of Solar Energy Engineering, Transactions of the ASME* 117(4):318–25.

Xin, W., Z. YinPing, X. Wei, Z. RuoLang, Z. QunLi, and D. HongFa. 2009. Review on thermal performance of phase change energy storage building envelope. *Chinese Science Bulletin* 54(6):920–28.

Zalba, B., J.M. Marin, L.F. Cabeza, and H. Mehling. 2003. Review on thermal energy storage with phase change: Materials, heat transfer analysis and applications. *Applied Thermal Engineering* 23:251–83.

Zhou, G., Y. Zhang, X. Wang, K. Lin, and W. Xiao. 2007. An assessment of mixed type PCM-gypsum and shape-stabilized PCM plates in a building for passive solar heating. *Solar Energy* 81:1351–60.

Blower Door Techniques for Measuring Interzonal Leakage

Erin L. Hult, PhD

Max H. Sherman, PhD
Fellow ASHRAE

Iain Walker, PhD
Fellow ASHRAE

ABSTRACT

The standard blower door test methods, such as ASTM E779, describe how to use a single blower door to determine the total leakage of a single-zone structure such as a detached single-family home. There are no standard test methods for measuring interzonal leakage in a two-zone or multi-zone building envelope such as might be encountered with an attached garage or in a multi-family building. Some practitioners have been using techniques that involve making multiple measurements with a single blower door as well as combined measurements using multiple blower doors. Even for just two zones there are dozens of combinations of one-door and two-door test protocols that could conceivably be used to determine the interzonal air tightness. We examined many of these two-zone configurations using both simulation and measured data to estimate the accuracy and precision of each technique for realistic measurement scenarios. We also considered the impact of taking measurements at a single pressure versus over multiple pressures. We compared the various techniques and evaluated them for specific uses. Some techniques work better in one leakage regime; some are more sensitive to wind and other noise; some are more suited to determining only a subset of the leakage values. This paper makes recommendations on which techniques to use or not use for various cases and provides data that could be used to develop future test methods.

INTRODUCTION

Interzonal leakage can have a negative impact on indoor air quality due to transport from an attached garage to a house, between townhouses, or between units in multi-family housing. In the house-garage case, transfer of automotive exhaust fumes into the home is of particular concern. Interzonal leakage testing methods are also used for energy-efficiency objectives to identify leakage pathways in multi-family homes or single-family homes with adjacent attic or basement zones. While a number of strategies have been used to determine interzonal leakage, currently no standard exists for this measurement. Determining leakage between adjacent zones poses a much greater challenge than determining the total leakage in, say, a single-family home. This is because typically the interzonal leakage is a small fraction of the total zone leakage; isolating the interzonal leakage effectively requires calculating a small quantity difference from larger quantities, each with some uncertainty. The accuracy necessary for measurement of interzonal leakage depends to some extent on the purpose of the measurement. While air sealing a building, it can be useful to quantify interzonal leakage along different possible leakage pathways to determine where to focus air-sealing efforts. In that case, inaccuracies of 50% or more may be acceptable. If a guideline or standard for acceptable interzonal leakage for indoor air quality purposes were being considered, higher measurement accuracy would be necessary. To address the question of whether it would be feasible to enforce such a standard, it is necessary to investigate how accurately interzonal leakage can be determined with fan pressurization methods.

The objective of this study is to identify the most accurate methods to quantify interzonal leakage using fan pressurization (blower door) testing. While infrared and smoke techniques can give a qualitative indication of interzone leakage, blower door methods provide quantitative metrics.

***Erin L. Hult** is a postdoctoral fellow, **Max H. Sherman** is a senior scientist, and **Iain Walker** is a staff scientist all at Lawrence Berkeley National Laboratory, Berkeley, California.*

Published by ASHRAE 2013

Tracer gas methods can be used to determine leakage under operational conditions but are typically more expensive and time-consuming than blower door tests. Various data collection and analysis methods were compared using both simulated data sets as well as field data. Results of the field data and simulations were used to identify the most robust methods and to quantify the uncertainty of the different methods. This information can be used to assess the trade-off between the accuracy of the methods and the time and effort required to use the methods. Additional details of the methodology can be found in the report by Hult et al. (2012).

In the field, single-zone blower door tests are often performed by taking flow-rate measurements only at a single pressure (typically 50 Pa [1.0 lb_f/ft^2]) rather than fitting a curve to measurements over a range of pressure differences. The ASTM and ISO standards to determine air leakage using fan pressurization require measurements over a range of pressure differences to improve the accuracy of extrapolation to calculate the air leakage at low operational pressure differences (ASTM 2010; ISO 2006). In this study, the uncertainty in the interzonal leakage associated with single and multiple pressure difference methods was compared using simulation and field test results.

Background

A number of studies have developed methods to determine leakage between adjacent zones (some focusing specifically on the house and attached garage scenario), but there is no existing standard for how to make this measurement. Parallels exist between interzonal leakage methods and ASTM test methods for measuring duct leakage (ASTM 2007), which also include methods to distinguish leakage to the outside from total leakage and employ more than one pressurization device. A number of strategies have been explored to use a single blower door to test interzonal leakage in buildings with two or more zones. Blasnik and Fitzgerald (1992) provide an accessible overview to the benefits of interzonal leakage testing to facilitate air sealing and describe several strategies to determine the leakage between adjacent zones using different single blower door tests.

Study Objectives

The objective of this study is to determine a) if blower door tests can be used to determine the leakage between a house and an attached garage to a degree of accuracy sufficient for developing standards or regulations and b) what is the simplest blower door test that will reliably deliver accurate interzonal leakage results. This study used simulations and field data to assess:

- single pressure difference vs. multiple pressure difference tests and
- different methods using one or two blower doors.

First, the most common testing methods are outlined. Then the methods used to generate and analyze the synthesized data are presented. A discussion of the results of the synthesized data analysis follows. Then, results are presented on the analysis of field data collected in a variety of configurations in order to examine the agreement between the different techniques in practice. Because the true interzonal leakage at the field sites is not known, the accuracy of the tests in the field cannot be directly determined. Finally, conclusions from the study are presented in the form of a decision tree on how to select a testing method.

Diagnostic Methods

Several methods using different sets of single blower door tests are outlined in this subsection, including the following:

- Three single zones method
- Conventional two-test method
- Garage 0/1 method
- LBNL IzLT method

The LBNL IzLT method was developed in this study and is described in more detail by Hult et al. (2012), where it is referred to as the *991/190 method*.

Methods that use two blower doors simultaneously include the following:

- Pressure balancing method
- Herrlin and Modera (1988) method

These methods and similar variations have been used most often, but other methods are possible (Nylund 1981; Love and Passmore 1987; Wouters et al. 1988; NYSERDA 1995; ALA 2006). Additional test methods as well as two blower door tests are not described in detail but are included in the report by Hult et al. (2012).

The blower door is used to measure the flow rate Q, where the subscripts in the equations that follow indicate the location of the door (HO for house-outside interface segment, HG for house-garage interface, and GO for garage-outside interface). P is the pressure difference across the interface indicated by the subscript. In each method, the measured pressure differences and the fan flow rate are used to calculate the leakage parameters.

Three Single Zones Method. Emmerich et al. (2003) used results from the three configurations illustrated in Figure 1 to calculate the leakage between the two zones. This is equivalent to performing three single-zone tests where the single zone is bounded by the $HO + HG$ interface (A), the $HO + GO$ interface (B), and then the $GO + HG$ interface (C):

$$Q_{HO,A} = C_{HO}P_{HO,A}^{n_{HO}} + C_{HG}P_{HG,A}^{n_{HG}} \quad (1)$$

$$Q_{HO,B} = C_{HO}P_{HO,B}^{n_{HO}} + C_{GO}P_{GO,B}^{n_{GO}} \quad (2)$$

$$Q_{GO,C} = C_{GO}P_{GO,C}^{n_{GO}} + C_{HG}P_{GO,C}^{n_{HG}} \quad (3)$$

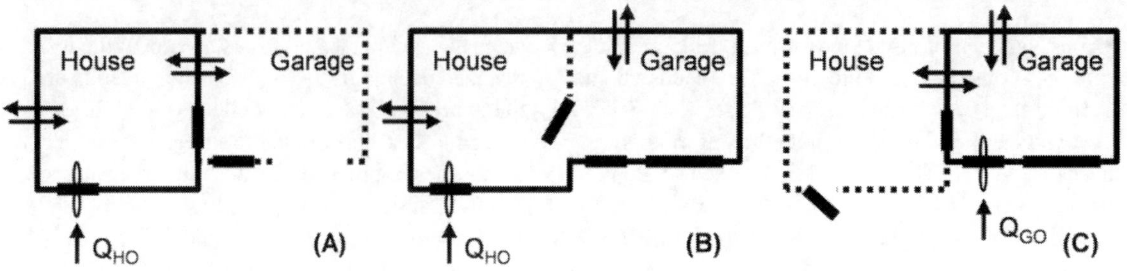

Figure 1 *Three single zones method using test configurations A, B, and C.*

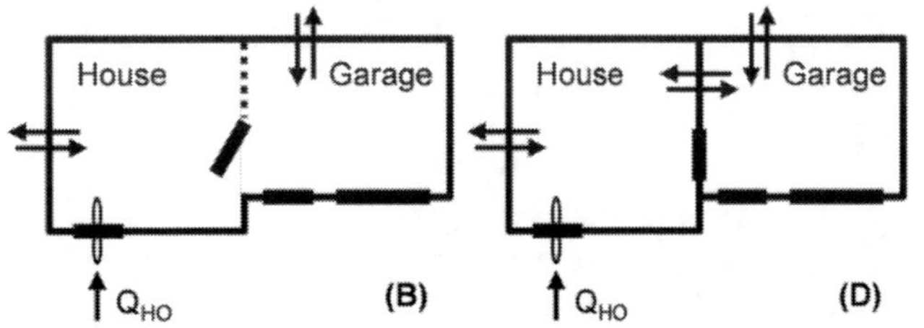

Figure 2 *Conventional two-test method, using configurations B and D.*

where C_{HO}, C_{GO}, and C_{HG} are the flow coefficients and n_{HO}, n_{GO}, and n_{HG} are the pressure exponents associated with the leakage through each envelope segment.

Similarly to determining the leakage of a single zone, this system can be solved for the coefficients using measurements at a single pressure difference if the pressure exponent is assumed; a value of $n = 0.65$ is common (Chan et al. 2012). Alternatively, the parameters C_{ij} and n_{ij} can both be fitted explicitly if measurements are taken at multiple pressure differences. Emmerich et al. (2003) took measurements at four to seven pressure differences for four houses with attached garages. Using a slightly different formulation of the equations above, they determined a value of n and C for each single-zone control volume using linear regression, from which the leakage flow through each interface could be determined. If n is not assumed, the system has three equations and six unknowns. The calculation methods used in this study are described in the Methods section below.

Conventional Two-Test Method. For determining house-garage leakage, Blasnik and Fitzgerald (1992) recommend completing a pair of tests with the blower door in the house-outside interface and the garage door closed. In the first test, the door between the house and garage is opened (B), and then in the second test the door between the house and garage is closed (D), as shown in Figure 2. In configuration D, only the house zone is directly pressurized (or depressurized) by the blower door, but due to the leakage in the *HG* interface, the garage zone will also become slightly pressurized relative to the outdoors, and both P_{HO} and P_{HG} are recorded. Blasnik and Fitzgerald performed this test pair at a single pressure ($P = 50$ Pa [1.0 lb_f/ft^2]), but the house zone pressure can also be increased over a range of pressure differences (Offermann 2009). This test method is convenient to use because although it requires two tests, the blower door only needs to be installed once.

The first test, with the door between the house and garage open (configuration B), leads to Equation 2. Then the following equations govern the air leakage in the D configuration:

$$Q_{HO,D} = C_{HO}P_{HO,D}^{n_{HO}} + C_{HG}P_{HG,D}^{n_{HG}} \tag{4}$$

$$C_{GO}P_{GO,D}^{n_{GO}} = C_{HG}P_{HG,D}^{n_{HG}} \tag{5}$$

As in the three single zones case, this system of equations can be solved either using measurements at a single test pressure, P_{HO}, or using measurements at a range of pressure difference values for P_{HO}. As for the three single zones method, for single pressure difference testing, we need to assume a pressure exponent, n, and we can solve for the flow coefficient, C_{HG}.

Garage 0/1 Method. Alternatively, the interzonal leakage can also theoretically be determined using two tests without moving the blower door using configurations A and D: the home is pressurized with the garage open and then closed, measuring both P_{HO} and P_{HG}. In practice, however, these configurations do not provide significantly different conditions to be able to determine the interzonal leakage as indicated in the Results section, because P_{HG} is typically very close to P_{HO} in configuration E.

LBNL IzLT. While the conventional two-test method is convenient because it does not require moving the blower door location, other test configuration pairs are also possible. Given that the blower door can be installed in doorways in each of the three interfaces (*HO*, *HG*, and *GO*), and the remaining interface doors can each be either open or closed, 12 unique configuration pairs can be used to determine the interzonal leakage. Hult et al. (2012) explored the 12 pairs and identified the LBNL IzLT method as the most robust pair of single blower door tests to determine interzonal leakage. Figure 3 illustrates configurations A and E used in the LBNL IzLT method. By applying control volume analysis to the house zone in A and the house and garage zones in E, the problem can be described with three equations.

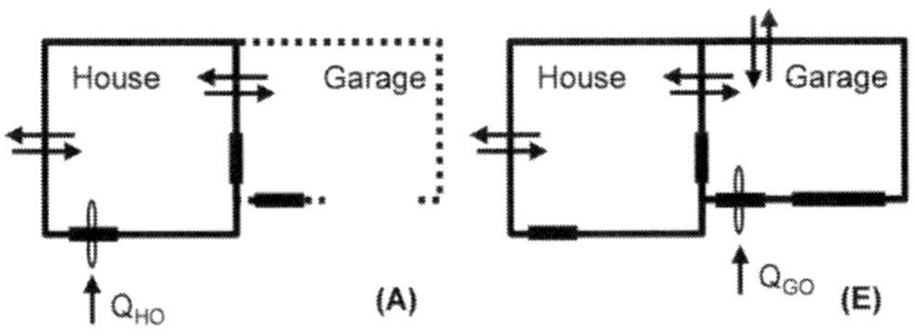

Figure 3   LBNL IzLT, using configurations A and E.

For some configurations (e.g., A and C), the blower door could be placed in either doorway to the pressurized zone. The remaining configuration, referred to in this paper as configuration F, is when the blower door is placed between the two adjacent zones and all other doors are closed.

METHODS

Two approaches were used to assess interzonal leakage measurement strategies: Monte Carlo simulations and field tests, both described in more detail in the subsections that follow. The simulations allow for a more thorough exploration of the parameter space but require some assumptions about the nature of actual leakage and measurement noise. The field tests were helpful to measure conditions such as the magnitude of leakage quantities and wind-induced fluctuations but were limited in terms of how many tests could be performed.

Monte Carlo Simulations

To determine the uncertainty in the total leakage from a single zone, it is possible to use uncertainty propagation techniques (Sherman and Palmiter 1995). However, the nonlinear system of equations makes it difficult to estimate the uncertainty in the interzonal leakage case. Instead, we used Monte Carlo simulations to mimic the collection of blower door measurements. In each simulation, a set of exact leakage parameters was selected, then the exact flow rates were calculated at a set of pressure differences for a given testing configuration. For multiple pressure difference tests, the six pressure differences used were 12.5, 25, 37.5, 50, 62.5, and 75 Pa (0.26, 0.52, 0.78, 1.04, 1.31 and 1.57 lb_f/ft^2). Then, to simulate the effects of measurement noise, randomly chosen fluctuations from a normal distribution with zero mean and a specified standard deviation were added to generate each "measured" flow rate and pressure. Using this new set of "measured" data, each calculation strategy was then used to fit the leakage parameters C_{ij} and n_{ij}. The leakage flows calculated from these fitted parameters C_{ij} and n_{ij} were then compared with the "exact" leakage rates specified initially, allowing us to assess the error resulting from the added noise and the assumptions of each calculation method.

Because the fluctuations were chosen randomly from the specified distribution, results vary to some extent between subsequent simulations, in the same way field test results at the same site may vary between repeated tests. Therefore, 500–1000 simulations were run for each set of conditions to determine the typical (median) uncertainty as well as the range (one standard deviation above and below the median). Other factors taken into account in the selection of parameters and simulation of measurements include the following:

- Difference between pressurization and depressurization leakage parameters due to valving effects (mean offset and fluctuation)
- Distribution of actual pressure exponents vs. assumed value ($n = 0.65$)
- Uncertainty in the mean pressure exponent in distribution
- Calibration errors in pressure and flow rate measurements

Additional details of the Monte Carlo simulation methods can be found in the report by Hult et al. (2012). Although beyond the scope of this report, additional considerations were addressed in the Hult et al. (2012) report, including the importance of completing pressurization and depressurization testing and the number of pressure differences tested.

Field Tests

Blower door tests were completed at seven houses in order to compare the results obtained from the various test configuration pairs. Data from House 1 was not used because at the end of the testing it was found that the attic access had blown open at some undetermined time. Field testing was completed independently from the synthesized data analysis; therefore, the two blower door methods completed at the field sites did not include the procedure described by Herrlin and Modera (1988). In addition, configuration E was not tested at House 2 or 7. At each site, the house to outdoors and house to garage pressure differences were measured using a digital manometer (Energy Conservatory DG 700) with an uncertainty of 1% of the pressure measurement or 0.2 Pa (0.004 lb_f/ft^2), whichever is greater. Blower door tests were completed using one or two blower doors from The Energy Conservatory controlled by a laptop computer. Details of the field sampling are described in the report by Hult et al. (2012). For field tests, measurements were taken at 5 Pa (0.1 lb_f/ft^2) pressure increments between 0 and 75 Pa (1.6 lb_f/ft^2), where the upper limit depended on the leakage area of the pressurized zone. Calculation procedures used pressure differences of 10 Pa (0.21 lb_f/ft^2) and higher for multiple pressure difference tests. Single pressure difference tests were at 50 Pa (1.0 lb_f/ft^2), but because 50 Pa could not always be reached the tests were also performed at 35 Pa (0.73 lb_f/ft^2).

Calculation Methods

As described in the study objectives, a number of different calculation methods were explored by Hult et al. (2012). The method with the lowest uncertainty for a range of conditions was found to be the five-parameter fitting method. In this method, pressure measurements are collected for pressurization and depressurization conditions. The pressure offset was determined from the mean pressure difference measured when the blower door was installed but off. After pressure offsets are subtracted, a set of coefficients C_{HO}, C_{HG}, C_{GO} and pressure exponents n_{HO} and n_{GO} are fit jointly to pressurization and depressurization data. A fixed-pressure exponent of n_{HG} = 0.65 is assumed. An optimization solver was used to determine the best fit values of the parameters, where the objective function minimized was the sum of the square of the difference between the measured and predicted airflow rate. This optimization was implemented by the authors in Microsoft® Excel® as well as in the statistical programming package *R*. This five-parameter fitting method was used for all multiple pressure difference tests in this study.

For the single pressure difference methods, n = 0.65 is assumed for all envelope interfaces. This results in a system of three or four linear equations (depending on the test configurations chosen) in terms of three parameters: C_{HO}, C_{HG}, and C_{GO}. For cases resulting in three equations, the system was solved to determine the three coefficients explicitly. In cases resulting in four equations, the coefficients are the least squares best fit to the data.

RESULTS

Simulation Results

Simulations were used to compare the uncertainty resulting from different test configurations and calculation methods under the same conditions. Figure 4 compares the uncertainty in the interzonal leakage using different test methods for the case when both zones are relatively tight (unvented garage, C_{GO}/C_{HO} = 0.7) and the case when the garage zone has a much larger leakage area than the house zone (vented garage, C_{GO}/C_{HO} = 8). In this comparison, it was assumed that the standard deviation of the fluctuations in the pressure was 0.5 Pa (0.01 lb_f/ft^2) and that the interzonal leakage area as a fraction of the house to outside leakage area was C_{HG}/C_{HO} = 0.05 (sensitivity to these parameters was found to be low, as discussed in the Hult et al. [2012] report).

Figure 4 summarizes the results from the synthesized data analysis. The pair of single door configurations with the lowest uncertainty was the LBNL IzLT method using multiple pressure differences. While other pairs of single blower door configurations had similar results when the leakage area of the two zones was comparable, the LBNL IzLT method was more accurate when the second zone was very leaky (the mirror image pair of C and D are recommended if the first zone has higher leakage area). The three single zones method requires an additional test, but the results were excellent when the leakage area of the two zones was comparable: for C_{GO}/C_{HO} = 0.7, the median uncertainty in Q_{HG4} was 13% of itself when multiple pressure differences were used. When the garage zone was very leaky (C_{GO}/C_{HO} = 8), the uncertainty increased but only to 32% of Q_{HG4}. When the interzonal leakage fraction C_{HG}/C_{HO} is larger, the accuracy of the methods tested tended to improve: the uncertainty of the three single zones method was less than 20% for all cases when C_{HG}/C_{HO} was greater than 0.15. The conventional two-test method used by Blasnik and Fitzgerald (1992) and Offermann (2009)

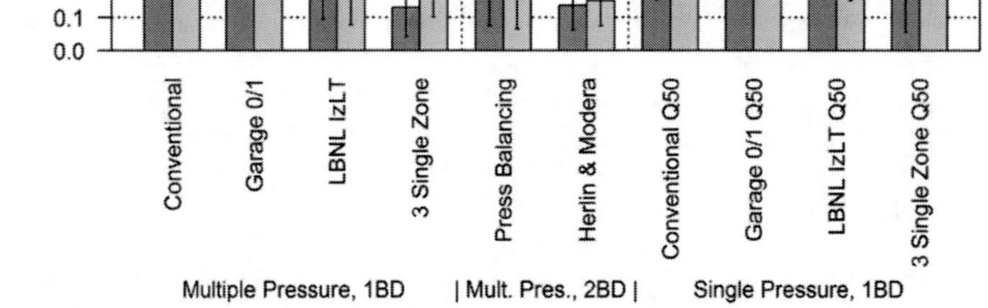

Figure 4 *Comparison of the uncertainty resulting from different measurement methods to estimate interzonal leakage using simulated data. Bars show the median uncertainty* $u(Q_{HG4})$ *in the interzonal leakage scaled by the exact interzonal leakage and error bars show one standard deviation above and below the median. Shown are single blower door tests (left) and two blower door tests (center) using multiple pressure differences. For single pressure difference methods (right), measurements were taken at 50 Pa (1.0* lb_f/ft^2*). Bars are truncated at 100% uncertainty, but several extend above.*

provided relatively consistent results if the two zones had comparable leakage areas (uncertainty about 50% of Q_{HG4} at $C_{GO}/C_{HO} = 0.7$), but when $C_{GO}/C_{HO} = 8$, the uncertainty is near 100% of Q_{HG4}. Although this uncertainty may seem small relative to the total house leakage, we found that the calculated values were often not meaningful because a good fit to the measured data could not be found (Hult et al. 2012). Results for the garage 0/1 method are also included to show that the performance was also poor and the test should be avoided.

Analysis of the simulation results suggests that the two blower door methods can be used to determine the interzonal leakage to within 20%. The method developed by Herrlin and Modera (1988) was used to determine Q_{HG4} to within 16%, regardless of C_{GO}/C_{HO} or C_{HG}/C_{HO}. This measurement routine was also largely insensitive to fluctuations in the measured quantities, making it a very robust choice if two blower doors are available for use. The pressure balancing method led to uncertainty of approximately 25% of Q_{HG4}.

Overall, the simulations suggest that using a single pressure difference (e.g., 50 Pa [1.0 lb_f/ft^2]) lead to unreliable estimates of the interzonal leakage. If C_{GO}/C_{HO} is not large (i.e., less than 3), simulation results indicate the uncertainty using the three single zones method is relatively low. However, the field test results discussed in the following subsection indicated that the three single zones method was not reliable using a single pressure difference. Under very calm conditions, fluctuations in the measured pressure may be as low as 0.1 or 0.2 Pa (0.002 or 0.004 lb_f/ft^2), which may lower the expected uncertainty slightly for single pressure difference tests compared to the results shown here, which assume fluctuations with an average magnitude of 0.5 Pa (0.01 lb_f/ft^2).

Other issues that were explored in detail by Hult et al. (2012) include additional possible testing configurations, sensitivity to fluctuations in pressure and flow rate measurements, and sensitivity to the magnitude of the interzonal leakage relative to the house leakage (C_{HG}/C_{HO}) and the relative leakage in the two zones (C_{GO}/C_{HO}). Their report also compared various calculation methods, including fitting coefficient C_{ij} and pressure exponent n_{ij} values to pressurization and depressurization data separately or jointly, fitting or specifying n_{HG} for the interzonal leakage explicitly, and fitting C_{ij} but assuming a fixed n for all leakage interfaces.

Field Testing

The results of applying different test methods in the field are summarized in Figure 5. The most consistent methods included the LBNL IzLT and three single zones methods using multiple pressure differences. Consistent with the simulation results, results from single pressure difference tests generally had higher deviations from the best estimate of the interzonal leakage. The number of tests performed with each method was relatively low, but the minimum and maximum values illustrate the wide range of deviations found for all single pressure difference methods as well as many multiple pressure difference methods. Thus, although many test methods are possible, most of them are unreliable. Although the simulations suggested that the three single zones method using a single pressure difference may have relatively low uncertainty, the field test results did not reflect this. While the median deviation for the conventional two-test method using a single pressure difference is low compared with the simula-

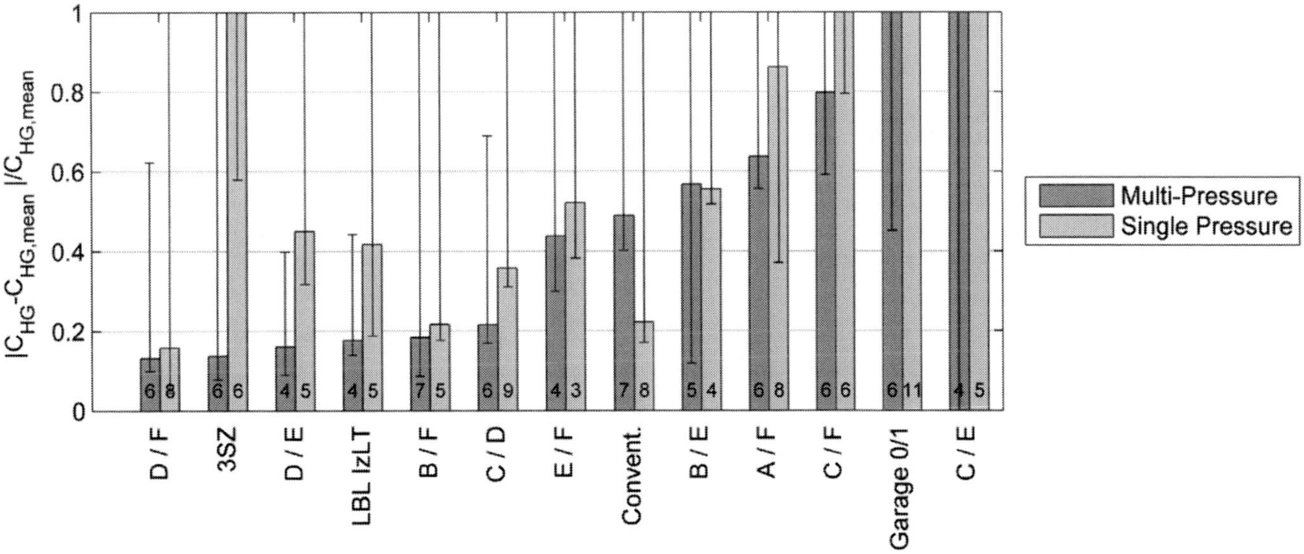

Figure 5 *Estimated errors in different methods based on field results. Bars show the median deviation from the best estimate of the interzonal leakage area for field test data from all houses. Multiple pressure methods (left bar) and single pressure methods at 50 Pa (1.0 lb_f/ft^2) and 35 Pa (0.73 lb_f/ft^2) (right bar) are shown. Error bars show the maximum and minimum deviations, and the number of tests performed is shown at the base. The best estimate $C_{HG,mean}$ is the uncertainty-weighted mean of the multiple pressure method results.*

tion result, this result is based on a limited number of tests. Overall, the agreement with simulation results is good: the three single zones and LBNL IzLT multiple pressure difference methods performed well, and the single pressure difference tests as well as the multi-point garage 0/1 methods were unreliable.

Looking at the test results at each house provides additional insight into the performance of different methods. From Figure 6, it is clear that different testing and calculation methods led to different leakage flow rates for the same house, tested on the same day. From Figure 6 and Table 1, it can be seen that the conventional two-test method performed

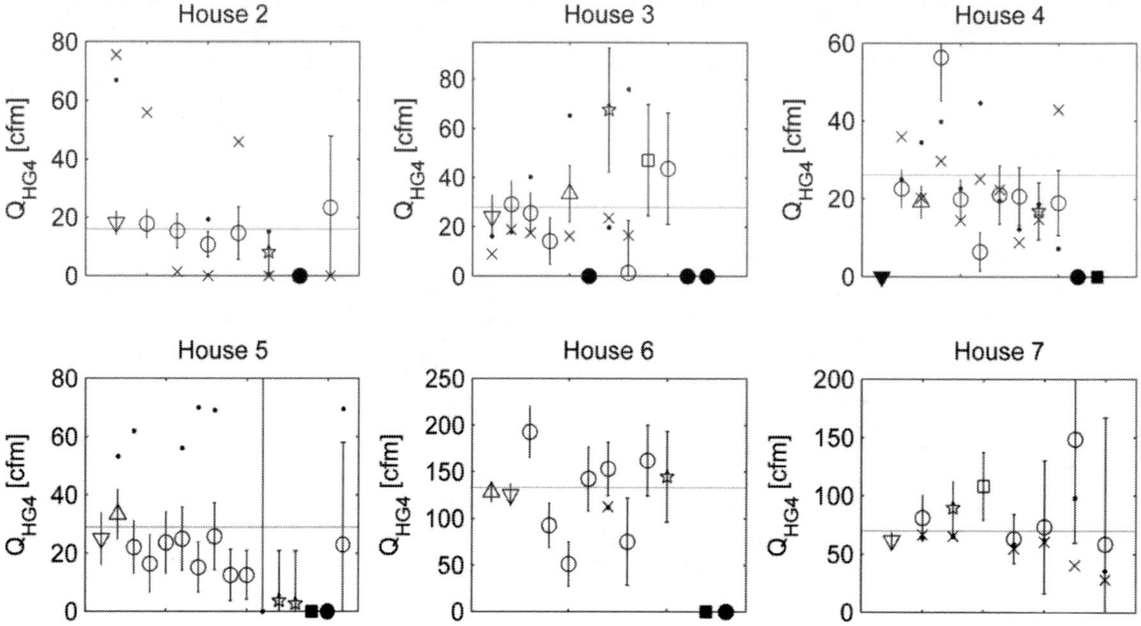

Figure 6 *Interzonal leakage at 4 Pa (0.08 lb_f/ft^2)(Q_{HG4}) from field data for houses 2–7. The symbol represents the test type. For each home, test configurations are ordered by the expected uncertainty of the multiple pressure difference tests, ∇ is the three single zones method, Δ is the LBNL IzLT method, ☆ is the conventional two-test method, □ is the garage 0/1 method, and ○ is all other multiple pressure configurations. Failed tests are plotted as filled symbols at 0. Error bars on the multiple pressure results show the uncertainty predicted by simulation results. Single pressure difference results are shown above or below the corresponding configuration using multiple pressure differences: × for tests at 50 Pa (1.0 lb_f/ft^2) and · for tests at 35 Pa (0.73 lb_f/ft^2). The dashed line shows the uncertainty-weighted mean of the multiple pressure difference results.*

Table 1. Parameters Resulting from the Field Test Configuration Set of Single Blower Door Tests with the Lowest Uncertainty as Estimated from the Simulation Results

	House 2	House 3	House 4	House 5	House 6	House 7
$u(P)$, Pa	0.61	1.42	0.46	0.49	0.35	0.78
C_{HO}, cfm/Pa^n	211	585	146	239	119	273
C_{HG}, cfm/Pa^n	7.3	9.8	9.2	10.2	52	25
C_{GO}, cfm/Pa^n	160	117	102	771	1129	166
n_{HO}	0.60	0.50	0.67	0.60	0.63	0.62
n_{GO}	0.60	0.49	0.57	0.53	0.55	0.51
Q_{GO}/Q_{HO}	0.76	0.20	0.60	2.9	8.5	0.53
Q_{HG}/Q_{HO}	0.037	0.021	0.061	0.046	0.45	0.096

Note: The configuration method used and estimated uncertainty is shown by the left-most, non-zero symbol in each subplot of Figure 6. Here, $u(P)$ is the standard deviation of the pressure difference measured when the blower door is off, to give a metric of the uncertainty in the pressure measurement.

reasonably well when the interzonal leakage is larger than 5% of the house to outside leakage (Houses 4, 6, and 7) but was further from the mean when the interzonal leakage is small (Houses 2, 3, and 5). The garage 0/1 method performed poorly, as expected. Failed tests are defined as tests in which one or more of the fitted parameter values fell at the boundary constraint value during the optimization, i.e., no good fit was found within the domain. Constraints were $0.4 < n_{ij} < 1.2$ and $0.01 < C_{ij} < 4000$.

Looking at the single pressure difference test results in Figure 6, in general there was large variability in test results. *Why were the single point results so variable?* To extrapolate from higher pressures down to find the leakage flow at 4 Pa (0.08 lb_f/ft^2), a pressure exponent of 0.65 was assumed. The pressure exponent fitted to fan pressurization measurements has been shown to vary across single-family residences in the U.S., with a mean value of 0.65 and a standard deviation of 0.057 (Walker et al. 2013). When determining the leakage flow at 4 Pa (0.08 lb_f/ft^2) from a single zone, this variability in the pressure exponent can increase the uncertainty from about 4% if n is known exactly to about 14% if $n = 0.65$ is assumed. The interzonal leakage, however, is typically small relative to the total house and garage leakage, so this uncertainty associated with the pressure exponent is compounded. In summary, if the assumption that $n = 0.65$ happens to be correct, then using a single pressure difference method may give consistent, accurate results. If the pressure exponent is not precisely 0.65, this assumption can lead to inaccurate (but repeatable) results.

It is possible that an interzonal leakage test at a single pressure will agree with test results taken at multiple pressure differences, but this was not typically the case in our field measurements. At House 7, the single pressure tests at both 50 Pa (1.0 lb_f/ft^2) and 35 Pa (0.73 lb_f/ft^2) were consistent and agree well with the multiple pressure difference results. In other cases, such as House 5, the set of single pressure tests was self-consistent but there is a substantial offset between the single and multiple pressure difference results. This was also the case at House 3, where single pressure tests clustered for the more reliable configurations (at the left of Figure 6) but were substantially lower than the leakage calculated from the multiple pressure difference tests. Large differences between tests at 50 Pa (1.0 lb_f/ft^2) and 35 Pa (0.73 lb_f/ft^2) (e.g., House 3) suggested that the pressure exponents assumed for the building envelope segments may not have been equal to the assumed pressure exponent. Table 1 indicates that the pressure exponents fitted at House 3 were indeed much lower than 0.65 ($n_{HO} = 0.50$, $n_{GO} = 0.49$). Larger-scale field testing would be needed in order to draw conclusions about how n_{GO} or n_{HG} might differ from n_{HO} on average.

Recommended Procedure

Figure 7 provides a decision tree for choosing a testing method, showing the expected uncertainty for each approach. The goal is to illustrate the best choices as well as the impact of making worse choices. This chart does not include variable wind fluctuations, which did not tend to change which tests

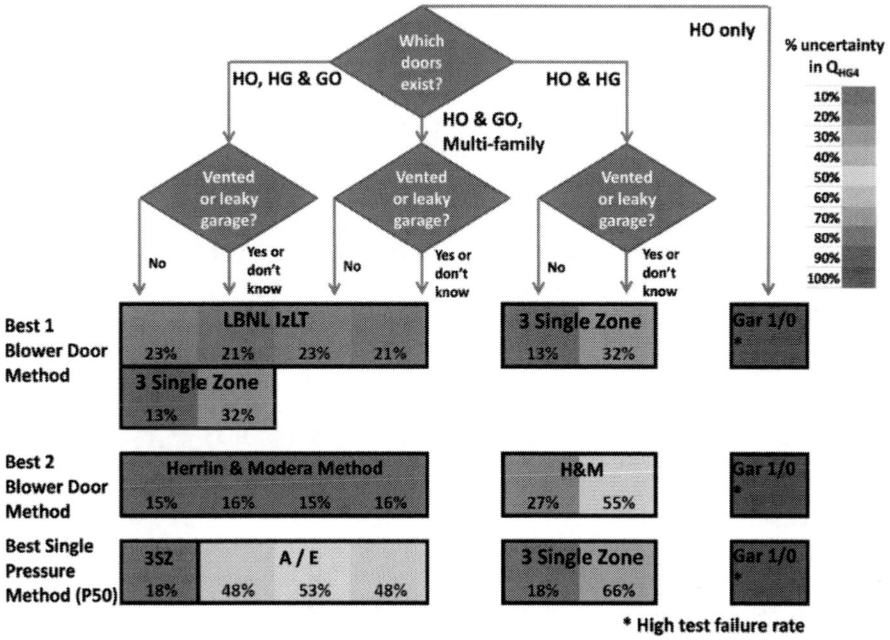

Figure 7 *Decision tree for choosing testing method. HO, HG, and GO refer to whether doors exist in the house-outdoors, house-garage, and garage-outdoors interfaces. Shading indicates the median uncertainty in* Q_{HG4} *from the simulation analysis, assuming* $u(P) = 0.5$ *Pa (0.01* lb_f/ft^2*),* $C_{HG}/C_{HO} = 0.05$*, for the unvented garage case* $C_{GO}/C_{HO} = 0.7$*, and for the vented garage case* $C_{GO}/C_{HO} = 8$*.*

performed best. Walking through the decision chart, the first question to consider in choosing a testing strategy is where it is possible to mount the blower door. Some homes will have doors in the *HO*, *GO*, and *HG* interfaces, but some homes will have only some of these doorways. If it is known that the garage is vented directly to the outdoors (through intentional or unintentional means), this can significantly increase the uncertainty of the best test options. If it is not known whether the garage zone is vented or leaky, it should be assumed that it is (this is a conservative assumption because uncertainty tends to be higher if the garage leakage area is high). If all doorways are available, the best test option is the three single zones method, followed by the LBNL IzLT method if there is a preference to run only two test configurations rather than three. If higher accuracy is desired, or if high fluctuations due to wind are expected, it may be desirable to use the two blower door method of Herrlin and Modera (1988). The uncertainties for single pressure difference methods are listed for reference to show the increased uncertainty for these tests. The chart can be followed the same way if there are *HO* and *HG* doorways or *HO* and *GO* doorways present. If there is only a doorway in the *HO* interface to mount the blower door and there is no opening between the house and garage zones, the only test configuration pair possible is the garage 0/1 method, which cannot be reliably used to determine the interzonal leakage.

In practice, it is difficult to obtain accurate results when the leakage area of one zone is much greater than the other, both because larger uncertainties tend to occur in this scenario and because it is difficult to establish high pressures in the leaky zone. In the discussion of how to determine the leakage between a house and a vented attic, Blasnik and Fitzgerald (1992) suggest temporarily sealing attic vents during testing to improve accuracy. This strategy would also be a quick and straightforward means to improve the accuracy when testing leakage between a house and vented garage, provided the vents can be temporarily sealed.

Although the house-garage scenario is considered in this paper, similar techniques using a single blower door in a series of different configurations could be applied to develop methods for testing a row of townhouses or other multi-family building configurations. For two connected units with a single shared interface, there are typically doors connecting each zone to the outside but not connecting the two units to each other. Thus, the LBNL IzLT method is likely to be the best choice if only one blower door is available.

CONCLUSION

The best single blower door methods (LBNL IzLT and three single zones) were used to determine the interzonal leakage to within 20% of its value, based on simulation results and supported by field measurements. Poor testing and calculation methods can lead to errors of up to 100% in the interzonal leakage area. The choice of analysis method can reduce uncertainty in the calculation of house-garage leakage significantly. Making the assumption that the pressure exponent for the interzonal wall is 0.65 was better than fitting for that pressure exponent, regardless of how many pressure differences were used. Additionally, the uncertainty was reduced by fitting a single set of parameters to both pressurization and depressurization data.

Two blower doors can be used simultaneously to reduce the uncertainty slightly further. The best of the measurement and analysis methods was the method developed by Herrlin and Modera (1988), which uses two blower doors simultaneously to determine the interzonal leakage to within 16%, over the range of expected conditions. When two blower doors are used simultaneously, there is a large range of combinations of pressure differences at which testing can be performed. While some two blower door methods consistently obtained accurate results, many did not give accurate results. Care should be taken to follow recommended testing procedures.

The single pressure difference approach could not reliably be used to determine interzonal leakage due to uncertainty in measured quantities and the pressure exponents in the different interfaces. If the objective is simply to identify which interzonal partitions may have high leakage flows for air-sealing purposes, using single-point testing may be sufficient. However, if the objective is to determine whether interzonal leakage is below a threshold level in a standard, higher accuracy is necessary. Analysis of field datasets confirmed that performance across test methods was consistent with the analysis of synthesized data sets.

ACKNOWLEDGMENTS

This work was supported by the Assistant Secretary for Energy Efficiency and Renewable Energy, Building Technologies Program, of the U.S. Department of Energy under Contract No. DE-AC02-05CH11231, and by the California Energy Commission Public Interest Energy Research (PIER) Program under Work for Others Contract No. 500-08-061. The authors also thank Darryl Dickerhoff and Phillip Price for their contributions to this work.

NOMENCLATURE

P_{ij} = pressure across interface indicated by subscript

Q_{ij} = flow rate (cfm) through blower door in interface indicated by subscript

C_{ij} = flow coefficient

n_{ij} = pressure exponent

Q_{HG4} = leakage flow (cfm) through the house-garage interface at 4 Pa (0.08 lb_f/ft^2)

Subscripts

HG = house-garage interface

HO = house-outside interface

GO = garage-outside interface

REFERENCES

ASTM. 2007. Standard E1554-07, *Standard Test Methods for Determining Air Leakage of Air Distribution Systems by Fan Pressurization*. West Conshohocken, PA: ASTM International.

ASTM. 2010. Standard E779-10, *Test Method for Determining Air Leakage by Fan Pressurization*. West Conshohocken, PA: ASTM International.

ALA. 2006. *American Lung Association® Health House® Builder Guidelines*. St. Paul, MN: American Lung Association. www.healthhouse.org/build/2007-2008HH builderguidelines.pdf

Blasnik, M., and J. Fitzgerald. 1992. In search of the missing leak. *Home Energy* 9(6).

Chan W.R., J. Joh, and M.H. Sherman. 2012. Analysis of air leakage measurements from residential diagnostics database. LBNL Report, LBNL-5967E, Lawrence Berkley National Laboratory, Berkeley, CA.

Emmerich, S.J., J.E. Gorfain, and C. Howard-Reed. 2003. Air and pollutant transport from attached garages to residential living spaces—Literature review and field tests. *International Journal of Ventilation* 2(3).

Herrlin, M.K., and M.P. Modera. 1988. Analysis of errors for a fan-pressurization technique for measuring inter-zonal air leakage. *Proceedings of the 9th AIVC Conference: Effective Ventilation* 1:215–32, LBL-24193.

Hult, E.L., D.J. Dickerhoff, and P.N. Price. 2012. Measurement methods to determine air leakage between adjacent zones. LBNL Report LBNL-5887E, Lawrence Berkley National Laboratory, Berkeley, CA.

ISO. 2006. Standard 9972:2006(E), *Thermal performance of buildings—Determination of air permeability of buildings—Fan pressurization method*. Geneva: International Organization for Standardization.

Love, J.A., and R.S. Passmore. 1987. Airtightness testing methods for row housing. *ASHRAE Transactions* 93(1).

Nylund, P.O. 1981. Tightness and its testing in single and terraced housing. *Air Infiltration Instrumentation and Measuring Techniques (Proceedings of the First Air Infiltration Centre Conference)*.

NYSERDA. 1995. Simplified Multizone Blower Door Techniques for Multifamily Buildings. Report 95-16, New York State Energy Research and Development Authority.

Offermann, F.J. 2009. Ventilation and indoor air quality in new homes. Report CEC-500-2009-085, California Air Resources Board and California Energy Commission, PIER Energy-Related Environmental Research Program.

Sherman, M., and L. Palmiter. 1995. Uncertainties in fan pressurization measurements. *Airflow performance of Building Envelopes, Components, and Systems, ASTM STP 1255*, M.P. Modera and K.K. Persily, eds. West Conshohocken, PA: ASTM International.

Walker, I.S., M.H. Sherman, J. Joh, and W.R. Chan. 2013. Applying large datasets to developing a better understanding of air leakage measurement in homes. *International Journal of Ventilation* 11(4):323–38.

Wouters, P., D. L'Heureux, P. Voordecker. 1988. Advanced single fan pressurization. *Proceedings of AIVC Measurement Techniques Workshop, Koge, Denmark.*

2012 IECC Performance Testing: Lessons from the Duct and Envelope Tightness (DET) Verifier Program

Mike Barcik
Associate Member ASHRAE

ABSTRACT

The 2012 International Energy Conservation Code (IECC) requires new homes and major renovations to have a pressure test of the building envelope and duct systems that are located outside of the thermal envelope. Testing is vital as leaky homes and ductwork often represent major sources of energy waste in homes. Many states will likely adopt the 2012 IECC over the next few years. However, many states currently lack the capacity to meet the testing requirements mandated by the 2012 IECC. To ensure code adoption and compliance, it is critical that states build a lasting infrastructure to offer testing services and that these services be available at a competitive price. If not, there could be a serious threat of a backlash against the testing provisions of the 2012 IECC, or perhaps the entire code.

In 2011, Georgia implemented a building energy code that requires duct and building envelope leakage testing (DET) and addresses performance testing issues left unresolved by the IECC such as who is qualified to perform the required testing, where the testing results should be recorded, how the testing requirements apply to upgrades and renovations to existing homes and duct systems, whether there should be special considerations for multifamily buildings (e.g., sampling), and what is an acceptable and effective duct sealant. This case study addresses key issues such as the appropriate experience and training required for DET professionals, acceptance of existing national testing certifications, and effective outreach strategies to recruit and train DET professionals.

INTRODUCTION

The U.S. Department of Energy has identified air leakage in building envelopes and duct systems as major sources of energy waste (PNNL and ORNL 2010). Air leakage can also be a source of comfort, durability, and indoor air quality problems (PNNL and ORNL 2010). Historically, updating building energy codes focused on increasing insulation levels and window and equipment efficiency requirements. However, the most recent versions of the International Energy Conservation Code (IECC) recognize the important contribution of air leakage to energy use and are more explicitly addressing both the best practice recommendations for reducing air leakage, as well as performance testing to set maximum thresholds for air leakage of the building envelope and duct systems.

Testing provisions were first referenced in the 2009 IECC, which required testing for ducts outside the thermal envelope and allowed for optional building envelope testing versus prescriptive air-sealing measures (ICC 2009). The 2009 IECC set criteria for both duct and envelope testing. The 2012 IECC requires both whole house air-leakage testing and the testing of duct systems when they are located outside of the thermal envelope (ICC 2012). The 2012 IECC also strengthened the passing criteria for both tests above the 2009 IECC performance thresholds. The majority of U.S. states have adopted or are at least on track for adopting the 2009 IECC and many will likely adopt the 2012 IECC over the next few years, with some states implementing the latest code as early as 2013 (see Figure 1).

***Mike Barcik** is the Director of Technical Services at Southface Energy Institute, Atlanta, GA.*

Although the 2009 IECC includes testing requirements or options, the manner in which testing is executed is unclear. Code enforcement bodies throughout the nation are struggling to address the testing criteria of both the 2009 and 2012 IECC. A primary concern is the greatly varying distribution of professionals trained in conducting whole house and duct system air-leakage testing, such as Residential Energy Services Network (RESNET) Home Energy Rating System (HERS) Raters (RESNET 2012) or Building Performance Institute (BPI) Professionals (BPI 2012). Furthermore, the level of expertise required to verify the testing criteria of the building energy code may not be as stringent as required for testing certification by existing organizations and beyond-code programs such as ENERGY STAR® or green building certification. To ensure successful code adoption and compliance, it is critical that states build lasting capacity to offer testing services that meet the code criteria, and that these services are available at a competitive price. Program administrators, government agencies and others need to plan to fill this gap in order to prepare the new construction marketplace for the full testing required by 2012 IECC implementation.

BACKGROUND

In 2010, Georgia adopted the Georgia State Minimum Standard Energy Code (IECC with Georgia State Supplements and Amendments) to be effective January 1, 2011. The Georgia energy code is based on the 2009 IECC; however, Georgia adopted amendments to the 2009 IECC that require duct and whole house building envelope air-leakage testing (GA DCA 2010). Georgia's mandated duct and envelope testing (DET) was later mirrored by the 2012 IECC. The Georgia energy code also addressed performance testing issues left unresolved by the 2009 and 2012 IECC codes. Table 1 presents a comparison of the 2009 and 2012 versions of the IECC and the 2011 Georgia energy code. For example, Georgia fully vetted the questions of who is qualified to perform the required testing (called a verifier), where the testing results should be recorded, how the testing requirements apply to upgrades and renovations to existing homes and duct systems, whether there should be special considerations for multifamily dwellings (e.g., sampling), and whether mastic should be required as a duct sealant. In answering these questions, Georgia developed a new certification program for individuals wishing to demonstrate compliance

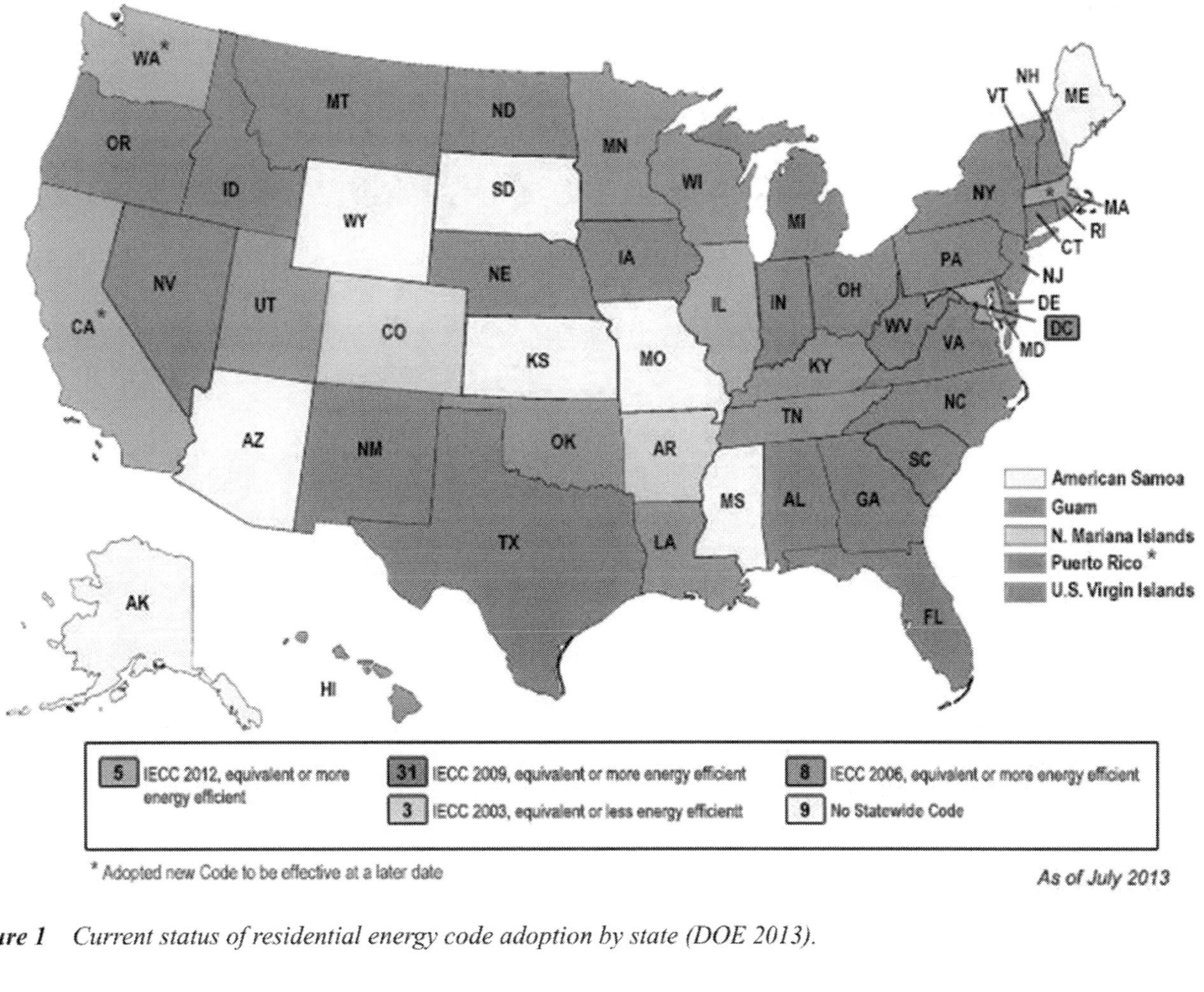

Figure 1 *Current status of residential energy code adoption by state (DOE 2013).*

with the duct and envelope testing requirements known as the DET Verifier Certification Program.

The relevance of having a successful DET Verifier Program was made evident during an attempt to roll back the Georgia energy code amendments by a small, but politically connected, faction in early 2012. The demonstration of substantial numbers of certified DET verifiers around the state prepared to offer the testing at affordable rates, plus a testing equipment loan program, helped quell this code-weakening effort.

Georgia determined that the verifiers who conduct the code-required testing must be trained and certified and that they must understand how to properly record the tightness results on an amended energy code compliance certificate. See Figure 2 for a completed example of the Georgia energy code compliance certificate template. The Atlanta-based non-profit organization Southface Energy Institute volunteered to develop a Duct and Envelope Tightness (DET) Verifier training curriculum for the Georgia Department of Community Affairs (DCA), the state agency tasked with oversight of building code adoption and interpretation. In the interest of consistent code implementation, Southface chose to provide the DET Verifier curriculum at no charge to other training entities that are deemed qualified by the Georgia DCA (Southface 2012). Additionally, Georgia recognized RESNET certified Home Energy Raters and BPI certified Building Analysts and Envelope Professionals as DET verifiers and did not require them to have additional training. Southface's DET experience, training curriculum, and supporting resources have been made available free-of-charge to assist other states in preparing for implementing the new code.

The DET Verifier Training is a one to two-day course designed to introduce the skills necessary to conduct pressure

Table 1. Duct and Envelope Tightness Requirements

Issue	2009 IECC	2011 GA Energy Code	2012 IECC
Envelope testing—single family	Optional: Blower door (BD) test or visual inspection checklist	Mandatory BD test with optional visual inspection checklist	Mandatory BD test and visual inspection checklist
Envelope testing—multifamily	Optional: BD test or visual inspection checklist	Optional – Visual Inspection or BD test. Sampling 1 in 4 units per floor or RESNET protocol	Mandatory BD test (no mention of sampling)
Envelope passing criteria	<7 ACH50 all climate zones (CZ)	<7 ACH50 all CZ (2–4)	≤3 ACH50 in CZ 3-8 ≤5 ACH50 in CZ 1-2
Clarify if envelope test required on alteration or renovation	No guidance	"When construction affects all aspects of building envelope (gut renovation)"	No guidance
Duct testing criteria at rough-in (RI) (Total)	4% – RI total no air handler 6% - RI total w/ air handler	6% - RI total w/ air handler	3% – RI total no air handler 4% – RI total with air handler
Duct Testing criteria at postconstruction—Postconstruction total (PCT) or postconstruction to outside (PCO)	12% – PCT 8% – PCO	12% – PCT 8% – PCO	4% PC (no incentive for testing at final or To Outside)
Record/display test results	Not required	On energy code certificate – template provided	On energy code certificate – no specifics on what to provide
Exempt from duct testing	Ducts and air-handler unit (AHU) entirely inside building thermal envelope	Ducts and AHU entirely inside building thermal envelope	Ducts and AHU entirely inside building thermal envelope
Duct pressure test required when modifying an existing system	No guidance	When >50% of existing duct system is modified. When AHU is changed out, test is not required but duct sealing with mastic through plenum connections is required	No guidance
Building cavities allowed as ducts	Only for returns	Not allowed for supply or returns	Not allowed for supply or returns
Duct sealing material	UL* tape, mastic, etc.	No UL* tape, only mastic and mastic tape	UL* tape, mastic, etc.
Qualified testers	No guidance	DET verifiers and RESNET and BPI certified professionals	No guidance

*UL = Underwriter's Laboratory.

Georgia Residential Energy Code Compliance Certificate*

Builder/Design Professional: Joe Builder Phone: 404-555-1234

Envelope Summary:

- List the R-Value for the following components:

Flat ceiling/roof:	R-30	Sloped/vault ceiling:	
Exterior wall:	R-13	Above grade mass wall:	
Attic kneewall:	R-18	Attic kneewall sheathing:	
Basement stud wall:		Basement continuous:	
Crawlspace stud wall:		Crawlspace continuous:	
Foundation slab:		Floors over unconditioned space:	R19
Cantilevered Floor:		Other insulation:	

- Fenestration Components:

Window U-factor:	0.35	Window SHGC:	0.28
Skylight U-factor:		Skylight SHGC:	
Glazed Door U-factor:	0.50	Opaque Door U-factor: (<50% glazed)	0.60

- Building Envelope Tightness (BET):

BET test conducted by: Dee E.T. Verifier Phone: 404-555-1111

Fan Flow at 50 Pascals= 1700 CFM_{50} Total Conditioned Volume = 16200 ft^3

ACH_{50} = CFM_{50} x 60 / Volume= 6.30 ACH_{50} (must be less than 7 ACH_{50})

Low Rise Multifamily Visual Inspection Option

(The visual inspection option may be conducted by a third-party instead of the BET test for R-2 buildings only.)

Visual inspection conducted by: ____________ Phone: ________

Mechanical Summary:

Water Heater Energy Factor: 0.58 Ef Fuel type: ☑ Gas ☐ Electric ☐ Other

Number of Heating and Cooling Systems: 2

Heating System Type:

☑ Gas: 80% AFUE ☐ Air-Source Heat Pump: ________ HSPF

☐ Other: ________ Efficiency: ________

Cooling System Type (Standard DX, Heat Pump, Geothermal, etc.): Standard DX

Cooling System Efficiency: 13 ☑ SEER ☐ EER ☐ Other

Heating/Cooling Load Calculations Performed by: Jay Manual Phone: 404-555-2222

Total Heating Load (Based on ACCA Man. J or other approved methodology): 36,777 Btu/h

Total Cooling Load (Based on ACCA Man. J or other approved methodology): 31,980 Btu/h

Cooling Sensible Load: 25,480 Btu/h Cooling Latent Load : 6,500 Btu/h

Total Air Handler CFM (based on design calculations): 1160 CFM

Duct Tightness Test Conducted by: Dee E.T. Verifier Phone: 404-555-1111

CFM_{25} per 100 ft^2 of conditioned floor area = CFM_{25} x 100 / Conditioned floor area served

If all ducts are not located within conditioned space, builder must verify that either the postconstruction duct leakage to outdoors (**PCO**) is ≤ 8%, the post construction total duct leakage (**PCT**) is ≤ 12%, or the rough-in total duct leakage (**RIT**) with air handler installed is ≤ 6%. State which method was used to conduct the duct tightness test: duct blower (**DB**), modified blower door subtraction method (**MBDS**), or automated multipoint blower door (**AMBD**).

System	Method (DB, MBDS, AMBD)	Test (PCO, PCT, RIT)	CFM_{25}	Area served (ft^2)	Result (%)
1 1st Floor	DB	PCO	44	900	4.9
2 2nd Floor	DB	PCO	65	900	7.2
3					

*Note: This permanent certificate shall be posted on or in the electrical distribution panel or air handler. Certificate shall be completed by the builder or registered design professional. Where there is more than one value for each component, certificate shall list the value covering the largest area.

Figure 2 *Sample of The Georgia Residential Energy Code Compliance Certificate source (Georgia Department of Community Affairs 2010).*

tests of the ducts and building envelope. Prior to taking the training, participants must demonstrate math proficiency through an on-line prerequisite course and must view ten-minute videos on duct testing using a duct blower and building envelope testing using a BD. In the classroom, participants then learn:

- Basics of the 2011 Georgia energy code including mandatory requirements such as air sealing and duct sealing details (2009 IECC based);
- Setup and testing protocols for BD and duct-leakage testing based on RESNET guidelines;
- Calculations for leakage testing, per the code requirements, plus how to fill in the results on the GA Energy Code Certificate; and
- Tips for successfully completing the written and in-field DET verifier certification exams, which are offered as part of the training.

The Georgia Code implementation schedule provided for approximately six months of time from adoption to statewide requirement of the DET verification implementation. To meet this ambitious schedule, Southface developed a train-the-trainer program on DET, which was targeted primarily at community college faculty and energy efficiency professionals. Southface, along with electric utility partners, private firms, and community colleges, trained over 500 verifiers in just four months. Many of the verifiers work in the housing industry in fields such as heating and air conditioning, construction management, and weatherization (see Figure 3). To help meet the statewide need for verifiers, the Georgia DCA partnered with local home builder associations to offer a loan program for BD and duct-testing equipment. The loan program charges a modest daily fee for equipment rental and is designed to allow verifiers to test the market for DET services and to serve rural areas of the state where new home construction is not strong enough to support equipment purchase by individual verifiers.

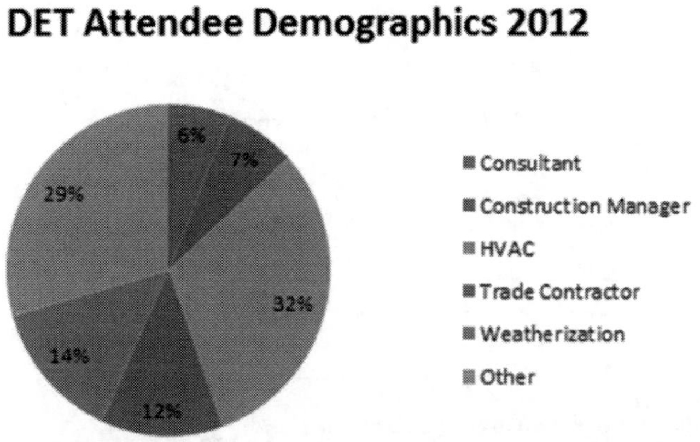

Figure 3 *Percentage of DET attendees who fell into each demographic category (Southface Energy Institute 2010).*

Lessons Learned

Several important lessons have been learned from the past two years, both in terms of training and outreach as well as in effectively implementing and maintaining the DET Verifier Program:

1. Need to educate industry professionals on ventilation. Ventilation strategies are now required by the 2012 codes (2012 IRC mandates the prescriptive ventilation tables, but not the formulas, of ASHRAE 62.2 [2010]). In section R303.4 of the *2012 International Residential Code*, intentional, whole-house ventilation is mandated for any home tighter than 5 air changes per hour at 50 Pa (ACH50) whereas in the 2012 IECC section R402.4.1.2, all homes must be <3 or 5 ACH50, depending on CZ. There is a need to teach how to determine the cubic feet per minute (CFM) of ventilation air required, the methods of ventilating a home, and how to commission and validate the actual quantity being delivered. This education is needed because intentional, whole-house ventilation has generally only been implemented by builders participating in beyond-code programs.

1. Need for the DET Verifier training to be multiday. An expanded format is necessary in order to allow additional hands-on field testing opportunities, as well as to provide more direction for a novice DET verifier if the home's envelope or duct system fails.
2. Need for one central database of certified DET verifiers. Unfortunately, Georgia does not have a central database of all DET verifiers. The Georgia DCA does keep a list of all approved DET trainers but the lack of a central database for DET verifiers makes it more challenging to identify a certified DET verifier. Alabama seems primed to have a superior approach wherein the State Licensing Board for Heating, Air Conditioning, and Refrigeration will keep a master list of all entities certified in the entire state (Alabama's testing requirement is currently scheduled to go active on January 1, 2014).
3. Need for continuing education and quality assurance (QA) on certified DET verifiers. While the opportunity still exists for Georgia to implement, unfortunately there are no formal provisions for actively maintaining the DET verifier certification. It is important to note that the DET verifier certification is for the individual and not for a company. Illinois is developing a DET verifier program that will utilize their established HERS industry group to administer QA to DET verifiers on an annual basis (IAER 2012).
4. Need for equipment loan program. the Georgia Environmental Financing Authority (state energy office). Allocated funds that placed testing equipment in a number of local home builder associations around the state. The intent is that this equipment can be rented for a modest fee and the revenue would be maintained by the local HBA. This access to equipment helps prevent spikes in testing

costs in smaller population centers and allows contractors to determine if there is a viable market for testing services prior to purchasing equipment.

5. Need for reciprocity. As other states come on board with DET verifier programs, it is important to establish reciprocity between states that at a minimum enables DET verifiers to work in housing markets that span state boundaries. Due to amendments, each individual state often has their own unique test results template (often the energy code compliance certificate) and the testing options allowed may be different. Examples for the state of Alabama and city of Chattanooga, Tennessee are presented in Figures 4 and 5. Although states may modify testing and reporting requirements, the fundamental skill set of how to conduct pressure tests remains the same.
6. Need for a discussion on certified versus third-party testing. The code is vague on this issue and always refers to the local code official. Third-party testing impacts costs, QA, and other issues. DET verifier is a certification that the contractor knows how to correctly conduct the test and the general intention is that certified DET verifiers could potentially test their own work. In other words, builders with the DET certification could conduct BD tests of the houses they have built and HVAC installers could perform duct-leakage tests of duct systems that they have installed.
7. Need for a consistent curriculum. Although the amendment was originally written in language that called upon the Georgia DCA to approve any curriculum, the agency quickly determined the value of just one training package. Southface developed the curriculum and has made an overview presentation available for general audiences to view for free. The instructional materials, including presentations, speaker notes, tests, and evaluation forms, have been shared with all DCA-approved training entities. These instructional materials have been modified for other states as typically the code and test results form has been unique to a given state.

DISCUSSION

Unfortunately, there have not been funds to fully evaluate the effectiveness of the DET program. However, the state and local chapters of the National Association of Home Builders have been supportive of the approach. In fact, these groups worked successfully to prevent repeal of the DET requirements in Georgia. This industry support indicates that the training and equipment loan programs are letting market forces meet the requirement for DET. As an active participant in a variety of housing forums in the state, Southface has anecdotal evidence that home builders have adequate access to DET verifiers and that testing services are not posing an undue financial burden.

For this paper, Southface polled four companies for data on their experience providing DET testing services. The four companies represented a variety in terms of company size, scopes of work (some performed sealing in addition to testing), and geographic and CZ areas where they offered services. Data was provided for single-family homes as well as multifamily dwellings and included projects simply being tested for minimum-code compliance as well as homes participating in beyond-code programs, such as EarthCraft House and ENERGY STAR.

Company Data Provided/Geographic Region/ Housing Type/Code vs. Beyond Code

Company A tested envelope and/or ductwork for over 1,200 single-family homes in the greater metro Atlanta area in 2012 (Climate Zones 3A and 4A). Data was made available for 944 homes of which 595 were for minimum-code compliance while 349 participated in a beyond-code program (EarthCraft or other).

Company B focused mainly on leakage-testing measurements for multifamily developments in the southern and eastern parts of the state (Climate zones 2A and 3A). Data was provided for 77 units in three different multifamily developments in three different cities. The buildings were all certified under a beyond-code program (EarthCraft) and dwelling unit floor areas ranged from 774 ft^2 to 1320 ft^2.

Company C provided testing data for 22 single-family homes in the southern portion of the state (Climate Zone 2A), with 19 of these homes tested for minimum-code compliance and three for ENERGY STAR v2.5.

Company D provided data for 55 single-family homes in some of the northernmost counties of the state (Climate Zone 4A). All homes were tested only for simple code compliance.

Envelope Testing Results

Company A released data for 936 homes that were BD tested (587 minimum code and 349 beyond code). The average building envelope leakage for code compliant homes was 4.42 ACH50 while the average for homes in beyond-code programs was 3.41 ACH50.

Company B showed that for 77 units in three different multifamily developments in three different cities under a beyond-code program (EarthCraft), the overall average ACH50 was 4.26. The data shows that, in spite of an ACH50 bias that favors large volume homes and works against small volume homes, multifamily units can still successfully pass leakage criteria, particularly when participating in a beyond-code program.

Company C provided BD results for twenty homes with an average ACH50 of 3.76. Seventeen of the homes featured spray foam rooflines and easily passed the BD test on the first attempt. The three remaining homes were conventional vented attic-style construction; two of these required retesting after not passing their initial envelope tightness test. These vented attic homes were also the only ones that required duct testing (since the spray foam houses created fully encapsulated ductwork).

Company D provided simple code compliance data for 55 single-family homes in north Georgia (Climate Zone 4A). Of the

Alabama Residential Energy Code Duct and Envelope Testing Results*

Address: 1234 Sample House Lane

Builder/Designer: Bill D. Home Phone: 222-333-4444

Envelope Summary: Building Envelope Tightness (BET)

BET test conducted by: Joe Tester Phone: 222-555-6666

Fan Flow at 50 Pascals= 1,844 CFM_{50} Total Conditioned Volume = 22,600 ft^3

ACH_{50} = CFM_{50} x 60 / Volume= 4.9 ACH_{50} (must be less than 7 ACH_{50})

Visual Inspection Option (may be conducted by an approved third-party instead of the BET test)

Visual Inspection Conducted by: ________ Phone: ________

AIR BARRIER AND INSULATION INSPECTION

	COMPONENT	CRITERIA
	Air barrier and thermal barrier	Exterior thermal envelope insulation for framed walls is installed in substantial contact and continuous alignment with building envelope air barrier. Breaks or joints in the air barrier are filled or repaired. Air-permeable insulation is not used as a sealing material.
	Ceiling/attic	Air barrier in any dropped ceiling/soffit is substantially aligned with insulation and any gaps are sealed Attic access (except unvented attic), knee wall door, or drop down stair is sealed.
	Walls	Corners and headers are insulated. Junction of foundation and sill plate is sealed.
	Windows and doors	Space between window/door jambs and framing is sealed.
	Rim joists	Rim joists are insulated and include an air barrier.
	Floors (including above garage and cantilevered floors)	Insulation is installed to maintain permanent contact with underside of subfloor decking. Air barrier is installed at any exposed edge of floor.
	Crawlspace walls	Insulation is permanently attached to walls. Exposed earth in unvented crawlspaces is covered with Class I vapor retarder with overlapping joints taped.
	Shafts, penetrations	Duct shafts, utility penetrations, knee walls and flue shafts opening to exterior or unconditioned space are sealed.
	Narrow cavities	Batts in narrow cavities are cut to fit, or narrow cavities are filled by sprayed/blown insulation.
	Garage separation	Air sealing is provided between the garage and conditioned spaces.
	Recessed lighting	Recessed light fixtures are airtight, IC rated and sealed to drywall. Exception—fixtures in conditioned space.
	Plumbing and wiring	Insulation is placed between outside and pipes. Batt insulation is cut to fit around wiring and plumbing, or sprayed/blown insulation extends behind piping and wiring.
	Shower/tub on exterior wall	Showers and tubs on exterior walls have insulation and an air barrier separating them from the exterior wall.
	Electrical/phone box on exterior wall	Air barrier extends behind boxes or air sealed type boxes are installed.
	Common wall	Air barrier is installed in common wall between dwelling units.
	HVAC register boots	HVAC register boots that penetrate building envelope are sealed to subfloor or drywall.
	Fireplace	Fireplace walls include an air barrier.

Mechanical Summary: Duct Tightness Verification (DTV)

DTV Test Conducted by: Jane Tester Phone: 777-888-9999

Tool used to conduct the duct tightness test: duct blower (**DB**), blower door subtraction method (**BDS**), or flow hood (**FH**).

Unless all ducts are located within conditioned space, builder must verify one of the following:

- Post-construction duct leakage to outdoors (**PCO**) is ≤ 8%,
- Post-construction total duct leakage (**PCT**) is ≤ 12%
- Rough-in total duct leakage (**RIT**) with air handler installed is ≤ 6%
- Rough-in total duct leakage without air handler installed (**RITnah**) is ≤ 4%

% Duct Leakage Result = CFM_{25} x 100 / Conditioned floor area served

System	Tool (DB, BDS, FH)	Test (PCO, PCT, RIT, RITnah)	CFM_{25}	Area served (ft^2)	Result (%)
1 Main	DB	PCO	165	2,300	7.2%
2					
3					

*Note: This document to be posted on or in the electrical distribution panel or on main air handling unit.

Figure 4 *Sample of The Alabama Residential Energy Code Duct and Envelope Testing Results data sheet (Alabama Power 2013).*

City of Chattanooga Residential Energy Code Compliance Certificate

HOUSE ADDRESS: ____________________

BUILDER/DESIGN PROFESSIONAL NAME: ____________________

COMPANY: ____________________ PHONE: ____________________

EMAIL: ____________________

ENVELOPE SUMMARY *(To be completed by Builder or Design Professional.)*

FLAT CEILING/ATTIC	______R-VALUE	WINDOWS:		
SLOPED/VAULTED CEILING	______R-VALUE		______SHGC	______U-FACTOR
ATTIC KNEEWALL	______R-VALUE			
EXTERIOR WALL	______R-VALUE	SKYLIGHT:		
ABOVE GRADE MASS WALL	______R-VALUE		______SHGC	______U-FACTOR
BASEMENT STUD WALL	______R-VALUE			
BASEMENT (CONTINUOUS)	______R-VALUE	OPAQUE DOOR (<50% Glazed):		
CRAWLSPACE STUD WALL	______R-VALUE		______SHGC	______U-FACTOR
CRAWLSPACE (CONTINUOUS)	______R-VALUE			
FLOOR	______R-VALUE	OTHER INSULATION COMMENTS:		
CANTILEVERED FLOOR	______R-VALUE			
FOUNDATION SLAB	______R-VALUE	____________________		

MECHANICAL SUMMARY *(To be completed by HVAC Contractor.)*

NUMBER OF HEATING AND COOLING SYSTEMS: ______

HEATING SYSTEM 1 TYPE (choose one) ☐ GAS: ______AFUE ☐ AIR-SOURCE HEAT PUMP: ______HSPF ☐ OTHER

COOLING SYSTEM 1 TYPE (Standard DX, Heat Pump, Geothermal, Etc.) ____________________

COOLING SYSTEM 1 EFFICIENCY: ______ ☐ SEER ☐ EER ☐ OTHER

HEATING SYSTEM 2 TYPE (choose one) ☐ GAS: ______AFUE ☐ AIR-SOURCE HEAT PUMP: ______HSPF ☐ OTHER

COOLING SYSTEM 2 TYPE (Standard DX, Heat Pump, Geothermal, Etc.) ____________________

COOLING SYSTEM 2 EFFICIENCY: ______ ☐ SEER ☐ EER ☐ OTHER

WATER HEATER ENERGY FACTOR: ______Ef FUEL TYPE: ☐ GAS ☐ ELEC ☐ OTHER

FIELD TESTING *(To be completed by a qualified Third-Party Verifier)*

BUILDING ENVELOPE TIGHTNESS:

FAN FLOW AT 50 PASCALS: ______ CFM_{50} TOTAL CONDITIONED VOLUME: ______ ft^3

$ACH_{50} = CFM_{50} \times 60$ / Volume = ______ ACH_{50} *(must be less than 7 ACH_{50})*

DUCT TIGHTNESS TESTING:

SYSTEM	METHOD* (PCO, PCT, RIT)	CFM_{25}	AREA SERVED (ft^2)	TEST RESULTS**	COMMENTS
1					
2					

* **PCO** (Post Construction Leakage to Outdoors) 8% max; **PCT** (Post Construction Total Leakage) 12% max; **RIT** (Rough-in w/Air Handler installed) 6% max

** CFM_{25} per 100 ft^2 of Conditioned Floor Area = $CFM_{25} \times 100$ / Conditioned Floor Area Served

BUILDING ENVELOPE TIGHTNESS / DUCT TIGHTNESS TEST CONDUCTED BY:

NAME: ____________________ SIGNATURE: ____________________

COMPANY: ____________________ QUALIFICATION: ☐ HERS ☐ BPI ☐ DET

Figure 5 *Sample of The City of Chattanooga Energy Code Compliance Certificate (City of Chattanooga Department of Public Works [Southface 2013]).*

45 homes that passed, the average BD test score was 4.7 ACH50. For the 10 homes that did not pass the BD test, the results ranged from 7.5 to 12 ACH50. Only two homes chose to retest since the county code officials chose to grant the Certificate of Occupancy without enforcing energy code performance requirements.

Duct-Leakage Testing Results

Company A performed a total of 1,617 duct-leakage tests (1,022 systems to meet minimum code requirements and 595 systems for homes in beyond-code programs). The ducts were tested either at rough-in stage (RIT) or at final stage measuring leakage to outside (PCO). The average duct leakage for minimum-code compliance was 3.9% while the average beyond-code program duct leakage was 2.9%.

Company B measured duct leakage in 77 multifamily beyond code units that averaged 2.7%.

Company C provided duct testing data for the five homes out of 22 that were conventionally vented attic-style construction. These vented attic homes were also the only ones that required duct testing (since the sprayed-foam roofline houses created fully encapsulated ductwork). Five duct systems were tested but only three passed initially; the other two required minor sealing around the boot penetrations and some touch-up around the air handler but, after this were able to pass while still on the initial visit.

Company D only leak tested 18 duct systems out of the 55 single-family homes in the northern part of the state (Climate Zone 4A). About half of the remaining duct systems did not require testing since the ductwork was inside the thermal envelope. The other half ignored the required testing, but the home still received a Certificate of Occupancy due to the jurisdiction's lack of energy code enforcement. The average passing duct-leakage test score was around 11% total leakage at final (PCT). This value is close to the noncompliant threshold of 12%. Five of 18 duct systems failed but only two chose to retest since code compliance was not being enforced.

Additional

Company A also performed air sealing and inspection services in many of their over 1, 200 single-family homes. In all cases, the need to perform BD or duct-leakage retests was fairly small (less than 2%).

Company B also evaluated 32 single-family homes that passed all envelope and duct-leakage tests. Failure rate here was estimated at less than one percent.

Company C did not perform air sealing as part of their scope of work. They did bring air-sealing materials along to help educate on how to seal top plates, penetrations, and chases. Company C did not charge for this service but saw the value of fostering good business relationships in case the homes did not pass on the first BD attempt.

Company D expressed frustration at the lack of or inconsistent code enforcement. Retests were rarely performed even if the envelope or duct system failed; only fear of liability was enough to spur some builders to pay for a retest. Other blatant lack of enforcement issues included walkout basements with no insulation on the concrete walls as required by code and frustration that certain counties were not even performing insulation inspections.

CONCLUSIONS

The 2009 IECC (2011 Georgia Energy Code) Requirements Can be Met

Overall, the data indicate that with a fairly modest effort, the performance levels for DET required by the code are easily obtainable. The thresholds arguably are too easy, especially for larger homes and floor areas served by larger duct systems. The performance thresholds in the 2012 IECC are significantly more challenging to achieve and may be especially difficult for the majority of states that have not mandated testing in the 2009 code cycle.

Beyond-Code Programs Work

The data show that beyond-code certification programs consistently produce homes that exceed code-built homes. Many beyond-code programs, such as EarthCraft House and ENERGY STAR, mandate complete compliance with the energy code as a baseline. Beyond code homes will typically exceed energy code regardless of local enforcement activity and capability.

Companies that Performed Additional Services Achieved Higher Performance

While some may question the inherent bias of testing homes where the subcontractor performed the work and later did the testing, the data indicate that the these homes achieved high performance levels. Generally the subcontractor has a vested interest in the home passing and quickly learns where the trouble spots are in the envelope and ductwork. Use of BD guided-air sealing was occasionally performed by all of the contractors.

Energy Code Enforcement Matters

Regions where energy code enforcement was spotty indicate poor performance testing results for ducts and homes. Although admittedly a small sample size, nearly 20% (10 out of 55) of the BD tested homes in counties with minimal enforcement failed the leakage test the first time out. And, duct systems that did pass were very close to the failure threshold.

SOUTHFACE ENERGY INSTITUTE RECOMMENDATIONS

In addition to the results of this limited poll, Southface offers the following recommendations based on our experience teaching over 1,500 Energy Code and DET verifier courses over the past decade, as well as our experience working directly with over 25,000 single-family and multifamily homes certified under the EarthCraft green building program and other beyond-code programs.

- Improve the IECC Table R402.4.1.1 – Air Barrier and Insulation Installation checklist. The concept of Table R402.4.1.1 is solid, but the wording is cryptic and confusing; a graphic demonstration of many of the items is recommended. An example is the provision that "corners and headers shall be insulated" without providing details or minimum R-values to the builder and code official (IECC 2012).
- There are numerous ways to report and adjust BD scores based on house size. The ACH50 approach (ACH50 = CFM50 × 60/Volume) is biased against small homes and not a sufficiently robust standard for larger homes as their much greater volumes tend to dominate the calculation. The code should move away from ACH50 and to a different metric such as envelope leakage ratio (ELR). This approach (ELR_{50} = CFM_5/square feet of building thermal envelope) does not artificially favor larger homes since the CFM50 of measured leakage is divided by the size of the home's building thermal envelope instead of the volume. This approach is more consistent across all building sizes and has been used in certain beyond-code programs for some time. Also, a similar version (tested at 75 Pa instead of 50 Pa) has been incorporated as a commercial building testing standard and adopted by the U.S. Army Core of Engineers and the commercial section of the 2012 IECC.
- Single-family homes should be individually tested for code compliance but the code should include a sampling protocol for multifamily buildings. Georgia adopted a standard of testing one in four dwelling units per floor of a multifamily building. One piece of language missing from the Georgia amendments is what to do if the tested unit fails—the implication is that the unit must be retested and the other three must also be tested. If the sampling identifies problems, then all units must be tested.
- Duct testing techniques have changed in the 2012 IECC. Postconstruction testing at final has effectively been discouraged since the threshold for passing is no different (4%) than when testing total leakage at rough-in (RIT). Rough-in testing of ducts should include a stricter passing criteria than when testing at final since additional construction activity could impact the final leakage rate. Duct testing without an air handler (AH) should be eliminated as an option since the cabinet of the AH is allowed to leak up to 20 cfm on a relatively small, 2.5-ton system (section R403.2.2.1, up to 2% of system airflow is allowed by the AH cabinet alone) (IECC 2012). Finally, certain methods of indirect duct testing should be deemed unacceptable, namely the BD subtraction method and simply adding flow-hood measurements at each grill while the house is pressurized with a BD, as these techniques do not provide the accuracy and repeatability attained by direct pressure testing.
- The code should adopt regionally appropriate ventilation strategies. For example, the code should prohibit exhaust only strategies for whole house ventilation in humid regions of the country, such as CZs 1A, 2A, and 3A, since a building experiencing continuous negative pressure increases the risk of mold or other moisture failure
- Finally, the code should provide a detailed template of the energy code certificate that clearly indicates what testing information must be provided. The IECC should mimic aspects of the Georgia, Alabama, and Chattanooga, Tennessee compliance templates. The Georgia template also requires a summary of the manual J-load calculations plus the name and contact info of the individual who performed the calculations as well as the DET verifier.

ACKNOWLEDGMENTS

Southface Energy Institute

Diana Burk, Building Energy Codes, Project Engineer
Dennis Creech, Executive Director
Rebekah Lynn Hogg, Research Intern
Bourke Reeve, Building Energy Codes, Project Manager
David Smedick, Research Fellow

Others

Mike DeWein, Manager, NE Regional Office, Codes and Energy, Institute for Building Technology and Safety
Richard Faesy, Principal, Energy Futures Group
Roxanne Greeson, Energy Policy Associate, Southeast Energy Efficiency Alliance

REFERENCES

Alabama Power. 2013. Alabama residential energy code duct and envelope testing results data sheet. http://www.alabamapower.com/business/services/hvac-training/pdf/08a-2009%20IRC%20DET%20Energy%20Code%20AL%20Test%20Results%20Formv6.pdf

ASHRAE. 2010. ASHRAE Standard 62.2, *Ventilation and Acceptable Indoor Air Quality in Low-Rise Residential Buildings*. Atlanta: ASHRAE.

Building Performance Institute (BPI). 2012. Accredited companies. www.bpi.org/companies.aspx.

DOE. 2013. Energy Efficiency and Renewable Energy. Status of state energy code adoption. www.energycodes.gov/status-state-energy-code-adoption.

Georgia Department of Community Affairs. 2010. *Georgia state minimum standard energy code (international energy conservation code with Georgia state supplements and amendments)*. www.dca.state.ga.us/development/constructioncodes/programs/codes2.asp.

Illinois Association of Energy Raters and Home Energy Professionals. 2012. *DET verifier certification*. www.ilenergyraters.org/detverifier.

International Code Council. 2009. *2009 International Energy Conservation Code*. www.shop.iccsafe.org/2012-international-energy-conservation-code-soft-cover.html.

International Code Council. 2012. *2012 International Energy Conservation Code*. shop.iccsafe.org/2012-international-energy-conservation-code-soft-cover.html.

Pacific Northwest National Laboratory and Oak Ridge National Laboratory. 2010. *Air Leakage Guide*. Washington, DC: U.S. Department of Energy, Building Technology Program.

Residential Services Network (RESNET). 2012. Certified home energy professional. www.resnet.us/dir-ectory/search.

Southface Energy Institute. 2010. Duct and envelope tightness verification training. www.southface.org/learning-center/trainings/duct-and-envelope-tightness-verification-training.

Southface Energy Institute. 2013. City of Chattanooga Residential Energy Code Compliance Certificate. http://www.southface.org/default-interior/09-chattanooga2009codecompliancecertificateairbarrierandinsulationchecklist.pdf

United States Army Core of Engineers. 2012. Air leakage test protocol for measuring air leakage in buildings. www.airbarrier.org/whole_building/index_e.php.

Characterization of Air Leakage in Residential Structures—Part 1: Joint Leakage

David Wolf, PhD **Frank Tyler**

ABSTRACT

Air will leak through a building envelope that is not well sealed. This leakage of air decreases the comfort of a residence by allowing moisture, cold drafts, and unwanted noise to enter, and air leakage can account for up to 40% of the energy used for heating and cooling in a typical residence. With nearly a mile of exterior joints in a typical residence that can leak air, knowing which joints leak the largest quantity of air allows for the most strategic placement of sealant. This two-part paper describes an extensive investigation to quantify the leakage characteristics of various types of joints and openings in a residential structure. All in all, 17 different joints/openings were characterized through both laboratory and real-house measurements using fan pressurization. Part 1 presents the methods and results associated with the air leakage of the individual joints/openings. Part 2 adapts these individual results to the whole house, including an examination of the joint leakage interdependence in the wall cavity.

INTRODUCTION

It is well known today that a properly insulated building uses less energy for heating and cooling than the same uninsulated building. And most people know that the higher the R-values of ceiling and wall insulation, the greater the energy savings. What is less known is that the air leakage of a home is also a major factor in how much energy a building uses for heating and cooling.

Air will leak through a building envelope that is not well sealed. This leakage of air decreases the comfort of a residence by allowing moisture, cold drafts, and unwanted noise to enter and may lower indoor air quality by allowing in dust and airborne pollutants. In addition, air leakage accounts for between 25% to 40% of the energy used for heating and cooling in a typical residence (EPA 2000).

To reduce air infiltration and achieve an energy-efficient building, the builder or building owner must seal gaps in the building's thermal enclosure. Installing high-quality, tightly sealed windows and doors is a good start, but it is also important to seal joints and openings in the walls, ceiling, and flooring/foundation. To properly address the negative effects of air leakage, such as wasted energy, occupant discomfort, condensation, and so on, all of the joints and openings in the building enclosure should be air-sealed. However, some builders or building owners may only have limited funds available to devote to air-sealing. With nearly a mile of exterior joints in a typical residence that can leak air, knowing which joints leak the largest quantity of air allows for the most strategic placement of sealant. As the drawing in Figure 1 depicts, there are many leakage paths through the building enclosure. This study is an investigation to quantify the leakage characteristics of various types of joints and openings in a residential structure (Part 1) and to prioritize the joints/openings in terms of the amount of air leakage per unit cost to seal it—a kind of air leakage bang-for-your-buck ranking of the joints/openings (Part 2). The air leakage results are primarily reported at a pressure difference of 50 Pa (0.2 in. water) due to its prevalence in the pre-commission testing of residential structures in the United States.

There is an abundance of information available in the literature on the air leakage characteristics of buildings. Sherman et al. (2004) and ASHRAE (2001) provide an overview

***David Wolf** is a research associate and **Frank Tyler** is a senior scientist, Owens Corning Science and Technology, LLC, Granville, OH.*

Onysko and Jones (1989) performed airtightness tests associated with vertical joints formed by waferboard sheathing. They investigated the effects of transitioning from green lumber to dried lumber (~15% moisture content) on air leakage. Siitonen (1982 and referenced by Relander et al. 2011) measured the airtightness of the joint between the wood bottom plate and a concrete slab floor of an actual house. The 2001 *ASHRAE Handbook—Fundamentals* also includes an extensive table of air leakage values, although most of the descriptions are either vague (e.g., drop ceiling, chimney, etc.) or obscure (e.g., aluminum double horizontal slider window with weather stripping). All of these microreferences will be reviewed in more detail later, where their results are compared to the results of this study.

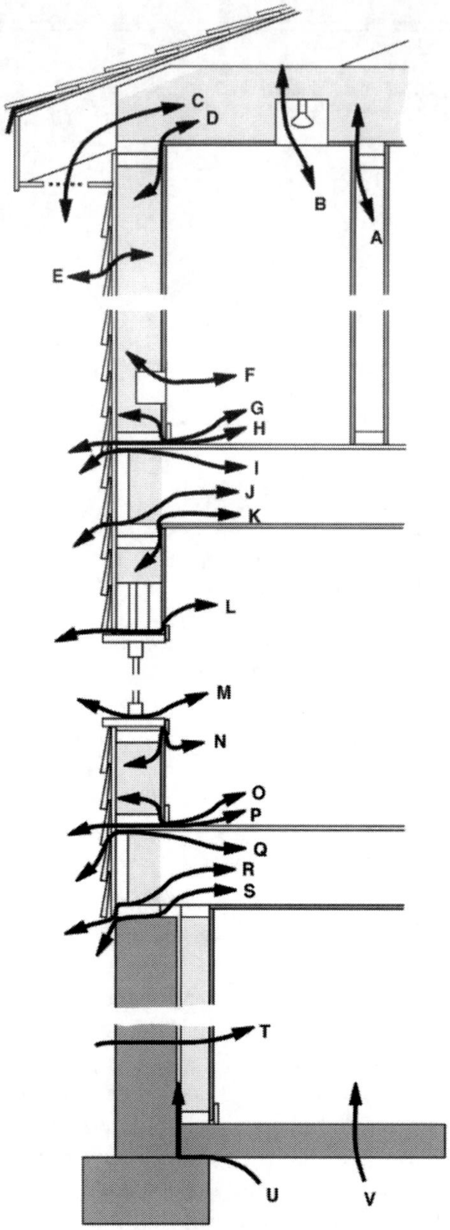

Figure 1 *Schematic of a house cross section showing the various air leakage paths.*

of many of the key papers. This information is almost all a macroview of air leakage, meaning that the information on air leakage is at the building level, not the joint level. A microview of air leakage (i.e., knowledge about the joint-level behavior) is necessary in order to understand how houses leak and where they leak from. The available literature that takes the microview is limited to a few references. But, even in those cases, only a few joints/openings have been investigated, which prevents the insightful exercise of taking the individual performance of joints/openings and integrating them into the whole-house performance (i.e., taking the microview in order to get a better appreciation for the macro view).

EXPERIMENTAL SETUP

This study involved two forms of testing. One form was to test individual components and wall assemblies in a laboratory setting. The other form was to test these same elements in a whole-house setting. The lab setting had its advantages in that it provided well-controlled conditions, ease with isolating individual joints/openings, and ease with measuring small leakage quantities. The lab's disadvantages were that the test assemblies could inadvertently be constructed to a higher quality than is typical of site-built construction due to the equipment and environmental conditions being more ideal and that the length of joints is often restricted due to practical limitations. The whole-house setting had its advantages in that it provided field-prototypic conditions, more joint length, and typical construction quality. The whole-house challenges were that testing needed to occur amidst uncontrolled conditions (e.g., wind) and more difficulty in isolating joints. In short, one form's strength was the other form's weakness, which is why both forms of testing were pursued.

Component and Assembly Testing

Airtightness of a building component or assembly was determined by exposing the specimen to a steady-state pressure difference and measuring the airflow rate needed to maintain the pressure difference per ASTM E 283 (2004). Two pressure chamber designs were used; each integrated with the same pressurization and measurement system. An 8 × 8 × 2 ft (2.4 × 2.4 × 0.61 m) chamber was used for wall testing; a 4 × 4 × 2 ft (1.2 × 1.2 × 0.61 m) chamber was used to evaluate components. Figure 2 shows a photograph and sketch of the larger chamber. The chambers were constructed with silicone sealant applied to all of the joints to prevent extraneous leakage. The flange around the perimeter of the chamber opening had a gasket present to form a good seal with the mating flange of the test plug. The two flanges, one on the chamber and one on the test plug, were clamped around the perimeter with the compressible gasket sandwiched in between.

An Infiltec Duct Leakage Tester (DL1-DM4-110) with the low-flow orifice tubes was used for the testing. This system allows for testing components with low airflow rates down to

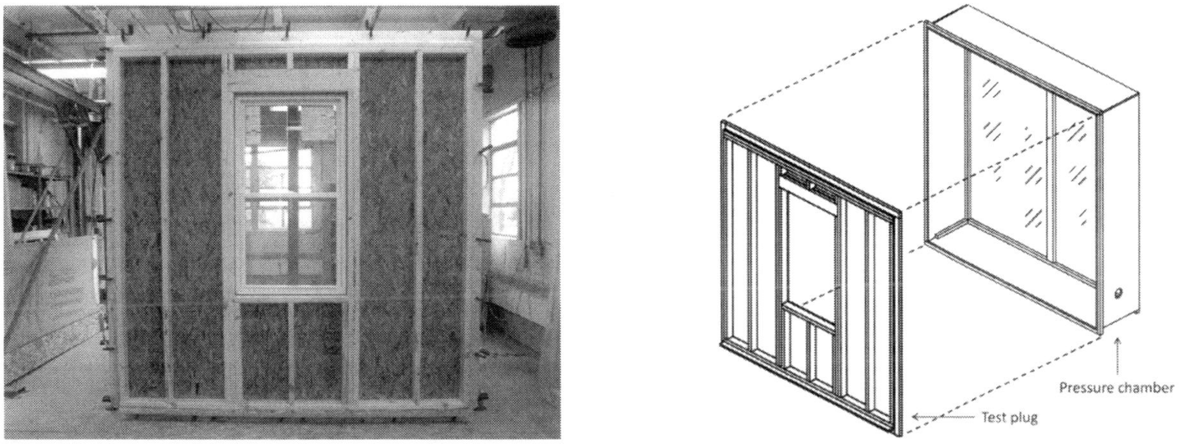

Figure 2 *(a) Photograph showing the test plug side of the larger pressure chamber. (b) Drawing showing the mating of the test plug and pressure chamber (b).*

0.33 ft^3/min (0.16 L/s), which was sufficient precision for the leakage rates that were anticipated. The Infiltec unit was used to blow air into (or out of) the test chamber and the pressure difference between the chamber and the outside was monitored and recorded along with the flow rate.

Prior to testing each assembly, a tare measurement was performed to establish the extraneous air leakage by eliminating the leakage through the test plug (typically with a plastic film). The protocol for testing the assembly involved pressurizing the chamber (mimicking infiltration) to the pressure differences of 40, 50 and 60 Pa (0.16, 0.20, and 0.24 in. water) with respect to the interior side of the wall assembly. The tare reading for each pressure was subtracted from the measurements of the test plug. The pressure-flow data were then curve-fit and interpolated to a pressure of 50 Pa (0.2 in. water) using a power-law relationship ($Q = C\Delta P^n$). A pressure difference of 50 Pa (0.2 in. water) was selected in order to relate the results to a whole-house blower door assessment, which is typically performed at 50 Pa (0.2 in. water) pressure difference (infiltration) by an energy rater on an actual house.

Test walls were constructed to represent 15 construction details of interest. The air leakage from individual joints on a test plug could be measured by sealing the other joints with removable clay. All tests were performed with the test plug mounted vertically, including ceiling assemblies, where the drywall-to-top plate joint was of interest. For the case of wall assemblies, the outer skin was 7/16 in. (11 mm) oriented strand board (OSB) (no housewrap) on 2 × 4 in. nominal wood framing (38 × 89 mm actual), and there was no interior skin (no drywall), so these laboratory measurements isolated the resistance of the joint only, excluding any upstream or downstream resistances.

Whole-House Testing

The test house was a 1400 ft^2 (130 m^2) single-story structure with a basement and crawlspace that is located on the Owens Corning Science & Technology grounds in Granville, Ohio. It was originally constructed in 1977 by a local custom home builder. During the original construction, Owens Corning personnel observed the progress and documented construction details, but did not supervise the actual construction. It was felt, therefore, that the house was of better than average quality but was still typical of many houses being built. The original sheathing on the house was 1/2 in. (12.7 mm) fiberboard. Since the predominant construction practice currently in the United States is to use 7/16 in. (11 mm) OSB, it was thought important to replace the sheathing. All of the siding and sheathing was removed from the house and replaced with 7/16 in. (11 mm) OSB and housewrap by a local contractor.

The basic measurement system consists of a computer-controlled blower door system with multichannel pressure measurement and data logging capabilities. The fan was the Minneapolis Blower Door™ System (Model 3) from The Energy Conservatory (TEC). The data logger was the Automated Performance Testing (APT) system, also from TEC, which had eight differential pressure channels with autozeroing capability and resolution of 0.1 Pa (4×10^{-4} in. water) in the range of ±400 Pa (1.6 in. water). The system is operated using software supplied by TEC (TECHLOG2). A pressure tap was located at each exterior face of the envelope – one at the base of each wall (north, south, east, and west), buried in gravel to minimize localized wind effects, and one located in the ventilated attic. Other measurements included indoor and outdoor temperatures, barometric pressure, and relative humidity. All measurements were made with the pressure difference across the house enclosure of 50 Pa (0.2 in. water). The data reduction was done per the method described in ASTM E 1827 (1996).

The whole house leakage fell within the range from 500 to 900 ft^3/min (236 to 425 L/s) at 50 Pa (0.2 in. water) throughout the sealing experiments (hereafter the unit CFM50 will be used to refer to a flow rate of ft^3/min at 50 Pa (0.2 in. water) pressure difference), which enabled orifice ring B to be used with the fan. The question that needed to be answered was whether this measurement system, using the B-ring, would provide the needed detection level for differences expected in whole house leakage, which was estimated to be 10 CFM50 (4.7 L/s). This led to the development of a means of introducing precision holes into the test house via a variable orifice plate (or VOP) concept. The VOP essentially comprised a thin metal plate with five precision holes cut in it. Using a cover plate with a single hole larger than the maximum hole size in the base plate, one could then dial in the hole size of interest, either during a test or between baseline envelope pressure checks.

The first step was to verify the system's detection abilities in the laboratory, where the variability from wind was absent. The blower door and the VOP were installed in an 8 × 8 ft (2.4 × 2.4 m) wall, which was then mated with the pressurization chamber. As Figure 3 clearly shows, differentiation down to 10 CFM50 (4.7 L/s) was easily achievable, given no wind-related variation in exterior pressures during the measurement. This led to the next step of semi-empirically establishing meteorological limitations (i.e., pressure data filters), so that the desired level of differentiation could be achieved when testing the actual house. Combining noise reduction filters and averaging techniques, a balance was derived whereby the desired precision could be attained within a practical length of time and number of repeats. Since it was already established that the blower door could effectively measure a 10 CFM50 (4.7 L/s) change in a controlled environment, it was feasible to empirically establish a data rejection protocol for the test house measurements, for $n = 5$ tests, as follows:

a. All five baseline envelope pressures within 2 Pa (8×10^{-3} in. water) of each other

b. All five baseline envelope pressures with standard deviation <1.5 Pa ($< 6 \times 10^{-3}$ in. water)

c. Fan pressure standard deviation less than 3 Pa (12×10^{-3} in. water)

d. Fan flow change of less than 7 CFM50 (3.3 L/s) over five one-minute averages

This resulted in an effective differentiation down to 10 CFM50 (4.7 L/s). These considerations for testing and data filtering resulted consistently in the ability to measure down to a 10 CFM50 (4.7 L/s) change in the test house venue. Figure 4 shows individual consecutive tests run in which the 1.75 in. (44 mm) orifice was alternately opened and closed, causing a 20 CFM50 (9.4 L/s) change in house leakage.

A total of 13 different joints were measured in the test house. The incremental air leakage for a given joint was the difference in whole-house leakage between the test-in and test-out measurements. What separated the test-in and test-out measurements was the act of sealing with a commercially-available air sealant.

RESULTS AND DISCUSSION

Component and Assembly Results

A total of 15 different joints were studied in this venue. Each of the joints/openings are described in this section, along with a summary of the results in Table 1 and a sketch/photograph in Figure 5. All of these leakage results are for the unconstrained condition, where restrictions to airflow on the interior (e.g., base trim, drywall, etc.) and exterior (e.g., a water-resistive barrier and cladding) are omitted. The potential effect of such constraints will be addressed later in this

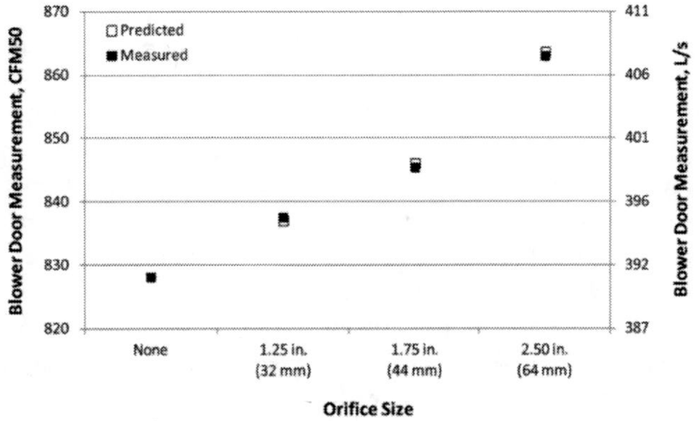

Figure 3 *Blower door results for controlled laboratory conditions and simulating the whole house leakage anticipated from the test house (828 CFM50 [391 L/s]), but where holes of a known size are introduced to determine if the measurement system is capable of differentiating the change, which this chart verifies.*

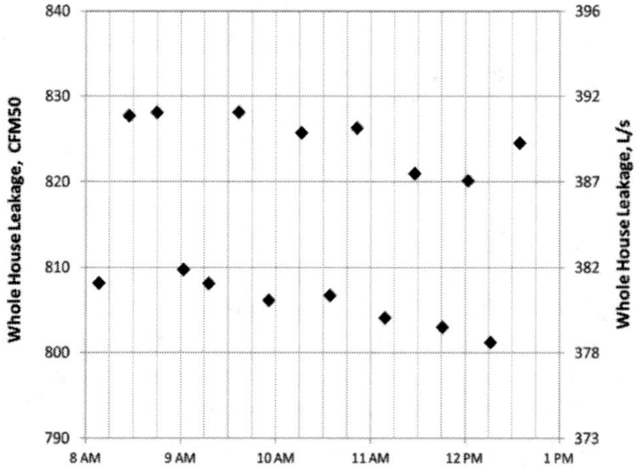

Figure 4 *Blower door results for the test house, where a 1.75 in. (44 mm) sharp-edged orifice was repeatedly opened and closed. The predicted flow through that orifice at 50 Pa (0.2 in. water) pressure difference is approximately 20 CFM50 (9.4 L/s), which is the approximate offset seen above between alternating data points.*

Table 1. Summary of Lab Leakage Data for Various Joints/Openings, Excluding AnyApplicable Upstream or Downstream Flow Resistances

Joint	Length Tested, ft	Normalized Leakage Min., CFM50/ft	Max., CFM50/ft	Ave., CFM50/ft	CFM50/ft = $C\,\Delta P^n$ *C*, CFM50/ft·Pa^n	*n*
Bottom plate-to-subfloor	104	0.016	0.306	0.113	0.00428	0.837
Plate-to-sheathing	160	0.040	0.266	0.091	0.00300	0.871
with engineered 1/8" defect	16	0.259	0.728	0.493	0.03102	0.707
Double-top plate	56	0.025	0.424	0.179	0.00700	0.829
Vertical sheathing-to-stud	64	0.046	0.380	0.129	0.00339	0.929
with engineered 1/8 in. defect	16	0.498	0.595	0.547	0.02555	0.783
Band joist-to-plate	56	0.079	0.307	0.171	0.00535	0.886
Band joist-to-subfloor	56	0.033	0.316	0.110	0.00389	0.854
Drywall-to-top plate						
top plate w/ studs, cantilever fastened	64	0.688	2.225	1.759	0.20212	0.553
top plate w/ studs, top plate fastened	64	0.340	2.187	1.168	0.09833	0.633
top plate w/ studs & hurr. clips, cantilever fastened	48	2.105	2.561	2.343	0.28394	0.539
top plate w/ studs & hurr. clips, top plate fastened	48	1.165	1.839	1.547	0.13885	0.616
Inside corner	48	0.123	0.788	0.325	0.01201	0.843
Outside corner	32	0.014	0.070	0.037	0.00147	0.822
Garage wall base	32	0.105	0.376	0.201	0.00910	0.791

Opening	Units Tested	Normalized Leakage Min., CFM50/each	Max., CFM50/each	Ave., CFM50/each	CFM50/ea = $C\,\Delta P^n$ *C*, CFM50/ea-Pa^n	*n*
Recessed light	4	6.86	10.58	9.10	1.048	0.552
Duct boot	4	5.93	8.89	7.69	0.879	0.554
Light switch	4	0.39	3.33	1.40	0.181	0.524
Electrical outlet (interior)	7	1.63	3.87	2.62	0.249	0.601
Electrical outlet (exterior)	4	0.56	4.11	2.79	0.205	0.668

paper. The results are presented as follows for each of the joints/openings in terms of the average air leakage and corresponding 95% confidence interval associated with that average.

Bottom Plate-to-Subfloor (0.113 ± 0.051 CFM50/ft [0.175 ± 0.079 L/s·m]). The leakage is shown to range from a best case of 0.016 CFM50/ft (0.025 L/s·m) to a worst case of 0.306 CFM50/ft (0.474 L/s·m) (~19X). This variability in air leakage is likely the result of the construction tolerances that are inherent to the assembly of this joint, despite these joints being constructed in a controlled environment and by the same person.

Plate-to-Sheathing (0.091 ± 0.028 CFM50/ft [0.141 ± 0.043 L/s·m]). For this testing, top plate-to-sheathing and bottom plate-to-sheathing are nominally identical in their construction, so the results are combined. Two of these results are for joints with an engineered defect, where the center stud on an 8 ft (2.4 m) wide wall was offset from the plates by 1/8 in. (3.2 mm) toward the sheathing. In that case, the average normalized leakage is shown to increase to 0.493 CFM50/ft (0.763 L/s·m). This off-setting creates a larger gap, resulting in larger leakage. The 1/8 in. (3.2 mm) gap was not chosen arbitrarily. This is an expected tolerance on wall assemblies as specified by the Gypsum Association (2010), which represents the major drywall manufacturers.

Double Top Plate (0.179 ± 0.093 CFM50/ft [0.277 ± 0.144 L/s·m]). The double top plate joint represents the leakage through the mating surface of the two plates at the top of the wall.

Vertical Sheathing-to-Stud (0.129 ± 0.074 CFM50/ft [0.200 ± 0.115 L/s·m]). The sheathing seam was located at the stud and the sheathing gap was 1/8 in. (3.2 mm) per industry recommendations (APA – The Engineered Wood Association, 2011). Note that these leakage values apply to the combined leakage on both sides of the stud. There were two additional measurements for joints where the center stud was offset from

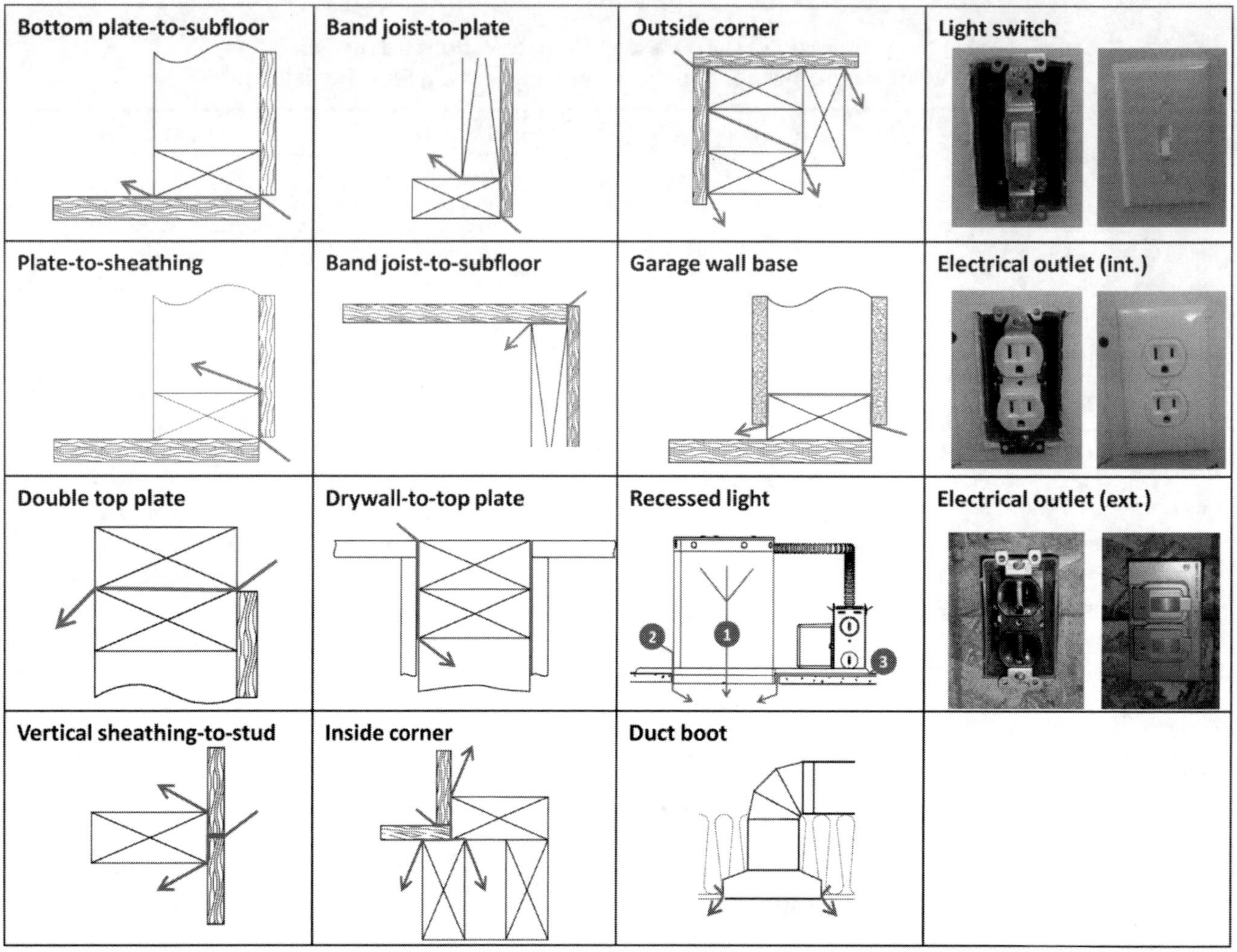

Figure 5 *Drawings/photographs of the various joints/openings studied in the laboratory testing.*

the plates by 1/8 in. (3.2 mm) toward the sheathing to simulate a construction defect, where the average normalized leakage is shown to increase to 0.547 CFM50/ft (0.857 L/s·m).

Band Joist-to-Plate (0.171 ± 0.058 CFM50/ft [0.265 ± 0.090 L/s·m]).

Band Joist-to-Subfloor (0.110 ± 0.072 CFM50/ft [0.170 ± 0.111 L/s·m]).

Drywall-to-Top Plate. Various conditions of the drywall-to-plate joint were studied, which included fastening location and the presence of hurricane clips. According to the Gypsum Association (2010), fasteners can be held back 8 in. (203 mm) from the top of the wall (i.e., the uppermost fastener is 8 in. down from the top of the wall) "to minimize the effects of truss uplift and the possibility of fastener popping in areas adjacent to wall and ceiling intersections." This is called "floating interior angles" by the Gypsum Association. For the purpose of this testing, the drywall was either top plate fastened, which meant that fasteners were placed in the top plate on 16 in. (406 mm) centers (i.e., uppermost fastener is within 2 in. of the top of the wall), or cantilever fastened, which meant that fasteners were held back 8 in. (203 mm) from the top of the wall and placed into the studs (i.e., uppermost fastener is within 8 inches of the top of the wall), mimicking the Gypsum Association recommendation. The average normalized leakage for the condition of cantilever fastening was 1.759 ± 0.711 CFM50/ft (2.73 ± 1.10 L/s·m), whereas the condition of top plate fastening was 1.168 ± 0.748 CFM50/ft (1.81 ± 1.16 L/s·m). Hurricane ties are often used to attach the roof framing to exterior wall framing. The combination of the thickness of the tie and the thickness of the nail head can create a localized offset of 3/16 in. (5 mm) between the drywall and the plates. In all cases, there were three hurricane ties used every 8 ft (2.4 m) of top plate. The average normalized leakage for the condition of cantilever fastening with hurricane clips was 2.343 ± 0.259 CFM50/ft (3.62 ± 0.40 L/s·m), whereas the condition of top plate fastening with hurricane clips was 1.547 ± 0.392 CFM50/ft (2.40 ± 0.60 L/s·m). Notable is that both of these conditions are leakier than the case where the hurricane ties were absent, which would be expected due to the larger gaps between the drywall and top plate that are created locally at the hurricane ties. All of these drywall-to-top-plate results are for the case where there is no downstream restriction once the air is inside the wall cavity, such as the drywall-to-bottom plate connection. This downstream restriction will be addressed in Part 2 of this paper.

Inside Corner (0.325 ± 0.212 CFM50/ft [0.503 ± 0.328 L/s·m]) and Outside Corner (0.037 ± 0.023 CFM50/ft [0.057 ± 0.036 L/s·m]). An inside corner is one that points toward the interior of a building and vice versa for an outside corner. It is notable that the leakage for an outside corner is roughly one-tenth the leakage for an inside corner. A possible explanation is the tortuousity of the flow path. Figure 6 shows the possible flow paths for an outside corner. Note that Joint A, which is at the sheathing intersection, would not be expected to be tight, because these two pieces of sheathing are not nailed to one another and the sheathing edges, which form this joint, are likely to be field-cut and not precise. The flow then bifurcates at Joint B and Joint C, both of which are comprised by the mating of nailed sheathing and framing. Such joints tend to be tight due to the intimate contact, and since they are the only means of leakage across the assembly, the entire joint tends to be tight, significantly retarding flow paths 1, 2, and 3. Figure 7 shows the possible flow paths for an internal corner. Joint A, which is at the sheathing intersection, would not be expected to be tight for the same reason as mentioned above for a sheathing-to-sheathing joint. Joint B similarly would not be expected to be tight, because it also is likely to be field-cut and not precise. Joint C is the mating of a 1/2 in. (13 mm) wide surface of two framing members that are not nailed to one another, which is another joint that is very vulnerable to air leakage. Collectively, the joints of the inside corner assembly would not be expected to be very restrictive to airflow path 1 in the figure.

Garage Wall Base (0.201 ± 0.124 CFM50/ft [0.311 ± 0.192 L/s·m]). This wall is unique in that it is an exterior wall (i.e., separates the conditioned living space from the unconditioned garage space) that is sheathed on the exterior side with drywall, as opposed to some form of structural sheathing, like OSB or plywood. This is significant in that drywall has far less stiffness than OSB/plywood, which adversely affects the airtightness of the joint that is formed between the drywall and the framing members. The above result is the leakage past both sheets of drywall (interior and exterior) as depicted in Figure 5.

Recessed Lights (9.10 ± 1.56 CFM50/unit [4.29 ± 0.74 L/s-unit]). Four recessed light fixtures (Juno, 6 in., IC-22 with Air-Loc® sealed housing) were tested with a standard trim piece (Juno, 24W-WH). This recessed light, like many others, has three main air leakage paths, which are shown in Figure 8. Even so-called airtight recessed lights, which are required by many building codes, can leak an appreciable amount of air at the juncture between the light housing and the mounting flange (red arrow 2 in Figure 8), as well as the mounting flange and the drywall (red arrow 3 in Figure 8). Leakage through the

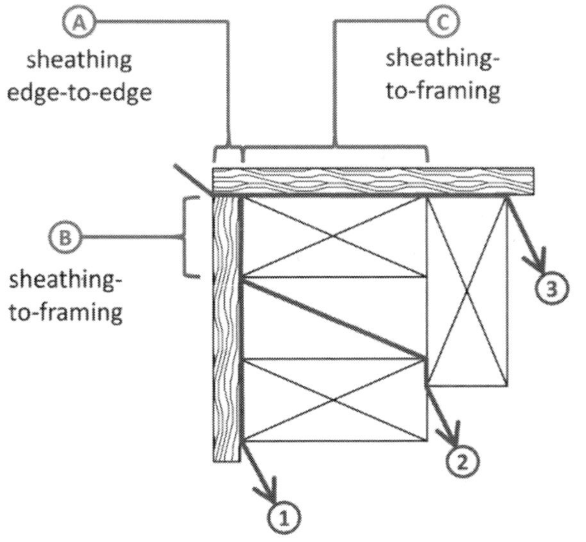

Figure 6 *Possible leakage paths at an outside corner.*

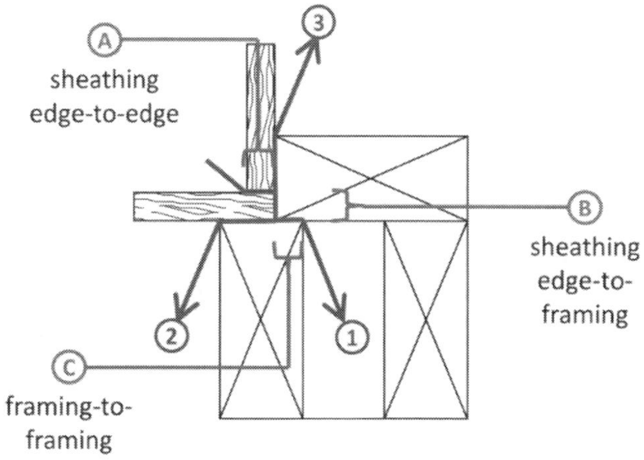

Figure 7 *Possible leakage paths at an inside corner.*

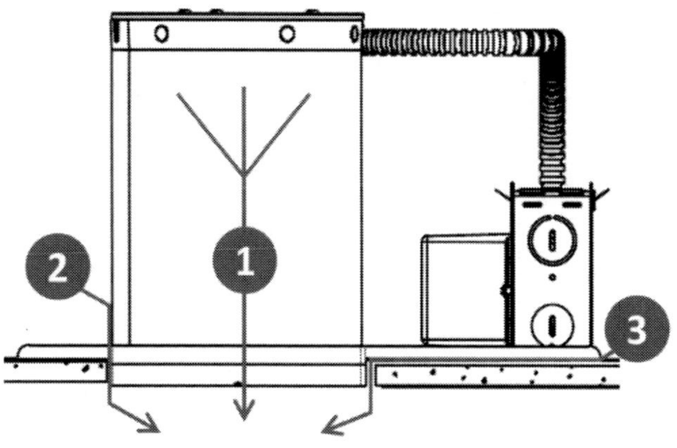

Figure 8 *Possible leakage paths associated with a recessed light: (1) leakage from the light housing, (2) leakage between the flange and housing, and (3) leakage between the flange and drywall.*

light housing (red arrow 1 in Figure 8) is fairly tight, as is required by many U.S. building codes.

Duct Boots (7.69 ± 1.31 CFM50/unit [3.63 ± 0.62 L/s-unit]). For cases where HVAC ducting is located in an unconditioned attic, the duct boots are at the ceiling and penetrate through the drywall, resulting in an air leakage path at the junction between the drywall and the perimeter of the boot. Four 6 × 10 in. (152 × 254 mm) boots, suspended from hanger brackets and mated to the drywall, were tested. The normal inlet to the boot was capped, and all of the joints on the boot itself were sealed with a silicone sealant. This was to isolate the leakage to the joint between the boot and the drywall.

Light Switch (1.40 ± 1.34 CFM50/unit [0.66 ± 0.63 L/s-unit]). Electrical light switches, positioned on the interior side of the wall cavity and penetrating the drywall, were tested with the switch cover plate installed.

Electrical Outlet, Interior (2.62 ± 0.73 CFM50/unit [1.24 ± 0.34 L/s-unit]). Electrical outlets, positioned on the interior side of the wall cavity and penetrating the drywall, were tested with the outlet cover plate installed.

Electrical Outlet, Exterior (2.79 ± 1.56 CFM50/unit [1.32 ± 0.74 L/s-unit]). Electrical outlets, positioned on the exterior side of the wall cavity and penetrating the sheathing, were tested with the manufacturer-supplied cover and gasket installed (Thomas & Betts, Red-Dot, part no. CCD).

Whole-House Results

A total of 13 different joints were studied in this venue through the incremental sealing of the joints with a spray-applied air sealant. The results from the sealing process are shown in Table 2.

Within this data set there are a few aberrant results that resulted from unique construction details of this test house, which couldn't be altered. The effect of the inside corner was too small to detect. Based on the laboratory results, which showed the inside corner to be quite leaky, this undetectable result with the test house was likely the result of a low occurrence for this joint type (16 ft [4.9 m]). The sill plate-to-foundation joint was also undetectable, which was to be expected since the house was constructed with a sill gasket between the sill plate and the foundation wall, which would be expected to cause flow resistance. The top of the band joist had a very small amount of leakage (0.17 CFM50/ft [0.26 L/s·m]), which was also to be expected since the subfloor was adhered to the joist members at the time of original construction, thereby forming a seal.

For top plate-to-drywall connection at the attic, there is just one entrance into the wall cavity, which is the top plate-to-drywall connection, but there are two exits from the wall cavity, which is at the termination of the drywall (bottom of the wall) and the penetrations of the drywall (mainly electrical outlets and switches). Table 2 lists two entries for this joint type. One indicates the case where the bottom of the drywall and all 50 penetrations/outlets were unsealed. The other indicates the case where only the 50 penetrations/outlets were unsealed. The latter obviously has less leakage associated with

Table 2. Summary of Test House Leakage Data for Various Joints/Openings, Excluding Any Applicable Upstream or Downstream Flow Resistances

Joint	Test House Leakage		
	CFM50	ft	CFM50/ft
Bottom of band joist	98	179	0.55
Top plate-to-sheathing	80	146	0.55
Outside corner	27	40	0.68
Bottom plate-to-sheathing	55	143	0.39
Top of band joist	32	184	0.17
Bottom plate-to-subfloor	7	143	0.05
Vertical sheathing seam	21	152	0.14
Sill plate-to-foundation	0	170	0.00
Inside corner	0	16	0.00
Window/Door framing-to-sheathing	32	98	0.32
Between top plates	27	155	0.17
Garage wall	18	30	0.60
Top plate-to-drywall at attic			
Bottom of drywall and 50 outlets unsealed	217	318	0.68
Bottom of drywall sealed and 50 outlets unsealed	93	318	0.29

it because the predominant path at the bottom of the wall is still sealed. This can be a realistic condition in cases where all of the trim in a house is well caulked, including caulking of the trim to a tile floor.

Comparison to Other Data in the Literature

Onysko and Jones (1989) performed airtightness tests associated with vertical joints formed by waferboard sheathing. They investigated the effects of transitioning from green lumber to dried lumber (~15% moisture content) on air leakage. Figure 9 shows their results (gray and black columns), along with the individual results obtained in this study (red columns), which shows very good overall similarity.

Siitonen (1982 and referenced by Relander et al., 2011) measured the airtightness of the joint between the wood bottom plate and a concrete slab floor of an actual house. The results were reported for a 10 Pa pressure difference. By assuming that the pressure difference exponent is 0.65, the reported leakage rate at 50 Pa (0.2 in. water) is estimated to be 0.057 CFM50/ft (0.088 L/s·m at 50 Pa) with a standard deviation of 0.046 (0.071). This compares to leakage rates obtained in this study for the bottom plate-to-subfloor that ranged from 0.016 CFM50/ft (0.025 L/s·m) to 0.306 CFM50/ft (0.474 L/s·m) with an average of 0.113 ± 0.051 CFM50/ft (0.175 ± 0.079 L/s·m), which is reasonable agreement.

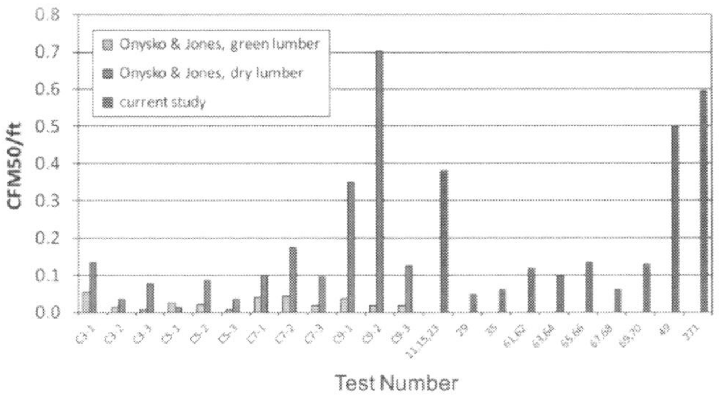

Figure 9 *Comparison of vertical sheathing joint leakage to that of Onysko and Jones (1989).*

As a part of an earlier Owens Corning study, Gavin (1985) investigated the leakage characteristics of joints located around the band joist area. Three of those joints were directly applicable to this study and are shown in Table 3, along with the results of Gavin and this present study. All of the results are expressed in terms of the effective leakage area, which was Gavin's figure-of-merit. With the exception of the band joist-to-plate connection, the agreement between the two studies is pretty good. Gavin's result for the band joist-to-plate connection is an order of magnitude larger than what was found in this study. There is no commentary by Gavin on why this joint was so much leakier than the others he studied. It is noteworthy that Gavin had no sheathing on the exterior side of the band joist when the result below was generated, whereas there was a piece of sheathing on the exterior side of the band joist in this study, which overlapped the joint in question. When Gavin added such a piece of sheathing, comparably placed to this study, the effective leakage area decreased by roughly a factor of three, which is better agreement to the present study, but still much higher.

There were several joints/openings in the 2001 *ASHRAE Handbook—Fundamentals* that were seemingly similar to this study. A comparison of some results for openings and joints is shown in Tables 4 and 5, respectively. The agreement between those data and the present data are good, with the exception of the bottom plate-to-subfloor joint, where the HoF value is one to two orders of magnitude higher.

Minimum, Maximum, and Best Estimate of Leakage Severity

Table 6 contains a summary of all the data presented here within, including both from the laboratory and test house. To the right of these measured values are columns that denote the recommended leakage values, which are broken down into a minimum, maximum and best estimate of the leakage severity, similar to how multiple measured results were handled in a previous study on the air leakage of building components (2001 *ASHRAE Handbook—Fundamentals*).

The minimum value is intended to represent the leakage severity associated with a well constructed joint. This is clearly a subjective term, but is intended to describe a condition where the construction equipment and environmental conditions are ideal, thereby enabling better-than-normal construction tolerances to be maintained. For the purposes of this testing, such conditions were routinely achieved with the assemblies tested in the laboratory, because they were constructed by highly skilled craftsmen, with precision tools, in a conditioned environment. Consequently, most of the minimum values listed in the table came from the average lab result. The lowest measured lab result (as opposed to the average lab result) was not used because it was not reproducible, even under ideal conditions.

Table 3. Comparison of Various Joint Leakages to that of Gavin (1985)

Joint	Effective Leakage Area, $ft^2/ft \times 10^5$		
	Gavin (1985)	Present Study	
		Lab	House
Bottom plate-to-subfloor	5.4	3.0 ± 1.6	1.2
Band joist-to-subfloor	3.8	2.5 ± 1.6	4.0
Band joist-to-plate	120	3.7 ± 1.5	11

Table 4. Comparison of Various Opening Leakages to that of the 2001 *ASHRAE Handbook—Fundamentals*

Opening	Effective Leakage Area, in^2 per unit			
	ASHRAE (2001)			Present Study Lab
	Min.	Best Est.	Max.	
Recessed lights	0.23	1.6	3.3	0.64 ± 0.11
Electrical outlets (no gaskets)	0.08	0.38	0.96	0.16 ± 0.05
Electrical outlets (with gaskets)	0.012	0.023	0.54	0.098 ± 0.034

Table 5. Comparison of Various Joint Leakages to that of the 2001 *ASHRAE Handbook—Fundamentals*

Joint	Effective Leakage Area, in^2/ft				
	ASHRAE (2001)			Present Study	
	Min.	Best Est.	Max.	Lab	House
Ceiling-wall	0.0075	0.070	0.12	0.033 ± 0.025	0.039
Sole plate, floor/ wall, uncaulked	0.018	0.2	0.26	0.0043 ± 0.0023	0.0017
Top plate, band joist	0.0035	0.005	0.018	0.0054 ± 0.0022	0.017

Table 6. Summary of All Air Leakage Results with Best Estimate, Minimum and Maximum Values, Excluding Any Applicable Upstream or Downstream Flow Resistances, Unless Otherwise Noted.

Joint	Lab Leakage			Test House Leakage			Recommended Leakage Values			Rationale for Best Estimate of Leakage Value
	Low, CFM50/ft	Ave., CFM50/ft	High, CFM50/ft	CFM50	ft	CFM50/ft	Best Est., CFM50/ft	Min., CFM50/ft	Max., CFM50/ft	
Bottom of band joist	0.08	0.17	0.31	98	179	0.55	0.55	0.17	0.55	1
Top plate-to-sheathing	0.04	0.09	0.73	80	146	0.55	0.55	0.09	0.73	1
Outside corner	0.01	0.04	0.07	27	40	0.68	0.68	0.04	0.68	1
Bottom plate-to-sheathing	0.04	0.09	0.73	55	143	0.39	0.39	0.09	0.73	1
Top of band joist	0.03	0.11	0.32	32	184	0.17	0.32	0.11	0.32	2
Bottom plate-to-subfloor	0.02	0.11	0.31	7	143	0.05	0.11	0.11	0.31	3
Vertical sheathing seam	0.05	0.13	0.60	21	152	0.14	0.14	0.13	0.60	4
Sill plate-to-foundation		Did not test.		0	170	0.00	0.00	0.00	0.00	5
Inside corner	0.12	0.33	0.79	0	16	0.00	0.33	0.33	0.79	6
Window/Door framing-to-sheathing		Did not test.		32	98	0.32	0.32	0.32	0.32	7
Between top plates	0.03	0.18	0.42	27	155	0.17	0.17	0.17	0.42	1
Garage wall	0.11	0.20	0.38	18	30	0.60	0.60	0.20	0.60	1
Top plate-to-drywall at attic	0.28[8]	0.53[8]	0.61[8]	217	318	0.68	0.68	0.538	0.68	1

Opening	Lab Leakage			Test House Leakage	Recommended Leakage Values			Rationale for Best Estimate of Leakage Value
	Low, CFM50/ea	Ave., CFM50/ea	High, CFM50/ea		Best Est., CFM50/ea	Min., CFM50/ea	Max., CFM50/ea	
Recessed light	6.86	9.10	10.58	Did not test.	9.10	9.10	10.58	9
Duct boot	5.93	7.69	8.89	Did not test.	7.69	7.69	8.89	9
Light switch	0.39	1.40	3.33	Did not test.	1.40	1.40	3.33	9
Electical outlet (interior)	1.63	2.62	3.87	Did not test.	2.62	2.62	3.87	9
Electrical outlet (exterior)	0.56	2.79	4.11	Did not test.	2.79	2.79	4.11	9

1 The test house value is generally considered to be more representative of typical construction tolerances than the laboratory value.

2 The band joist-to-subfloor connection in the test house had adhesive placed in the joint during the original construction. Since this is not a common building practice, the recommended value is taken from the laboratory data. Arguably, the actual value could be even higher than this, since laboratory results have generally been lower than experienced in an actual house. Also, one would expect the bottom of the band joist to have comparable leakage to the top of the band joist, which is the case for the recommended value (0.32 for the top vs. 0.55 for the bottom).

3 The average laboratory lab result is used, rather than the test house result, because the test house had 2 × 6 in. nominal (38 × 140 mm actual) framing, which would impart a greater resistance to air flow than the more typical 2 × 4 in. nominal (38 × 89 mm actual) framing, which was used in the laboratory testing. The laboratory lab assembly also used OSB for the subfloor, which is more typical of current construction than the plywood encountered in the test house.

4 The recommended value is from the test house, which had no gap between the sheathing seams, because this is perceived to be the more common situation. A potential upside of 0.60 CFM50/ft can still be realized for cases where a 1/8 in. gap is present.

5 A properly installed sill plate gasket on a nominally flat foundation should result in no leakage, which is what was experienced with the test house.

6 The quantity of inside corners in the test house was small, so resolving the effect on air leakage was difficult.

7 Only one set of results were obtained for this joint, which came from the test house. However, this joint is geometrically similar to the bottom plate-to-sheathing joint (0.39 CFM50/ft) and the top plate-to-sheathing joint (0.55 CFM50/ft), so the comparable results among all of these joints makes sense.

8 The lab leakage results for the top plate-to-drywall joint include the downstream constraint of the bottom-plate-to-drywall joint as well. This downstream resistance was inferred from the laboratory-based garage wall measurements described here within. The garage wall leakage includes two drywall-to-plate joints, one on the interior and one on the exterior. An estimate of the individual joint contribution (i.e., one drywall-to-plate joint) can be made by assuming that the interior and exterior joints leak the same, which implies an intermediate wall cavity pressure difference of 25 Pa (0.1 in. water). The flow rate associated with the overall pressure difference of 50 Pa (0.2 in. water), which includes the flow through both the interior and exterior joints, is the same as the flow through one of the joints with the pressure difference of 25 Pa (0.1 in. water), all of which are predicted by the same empirical relationship ($Q = C\,\Delta P^n$), so C can be inferred for the one drywall-to-plate joint. Also, the worst-case garage wall leakage was used in this calculation (see max normalized leakage in Table 1), rather than the average, based on the premise that lab-built wall assemblies generally have tighter tolerances than site-built construction.

9 Average results from the laboratory.

The maximum value is intended to represent the leakage severity associated with a "typically constructed joint". This is also a subjective term, but is intended to describe a condition where the construction equipment and/or environmental conditions are not ideal. Rough-cut (i.e., nonprecise) lumber and/or poor weather (e.g., cold, rain, snow) would be examples of such nonideal conditions, many of which can be found on a residential construction site. For the purposes of this testing, such conditions were routinely achieved with the test house, because it was site constructed with nonprecision tools and mixed weather. Consequently, most of the maximum values listed in the table came from the test house results.

The best estimate is the value that is believed to best represent what could be expected for typical site-built houses. It is not a measure of central tendency (i.e., not a mean or median). Rather, it is a value extracted from the overall test results, based on the judgment of the authors. The final column in the table lists the rationale for why the best estimate value was chosen.

CONCLUSION

This study was an extensive investigation to quantify the air leakage characteristics of 17 different joints/openings through both laboratory and real-house measurements using fan pressurization. This information is of greatest value when it can be applied to assessing the impact of joint leakage on actual houses and comprehending the effect of upstream and/or downstream restrictions in the wall cavity, such as drywall on the interior and cladding on the exterior. This will be addressed in Part 2 of this two-part manuscript.

ACKNOWLEDGMENTS

The authors wish to acknowledge the important contributions to this investigation by Mikael Salonvaara of Owens Corning, who consulted with them on many matters throughout the study, including experimental design, troubleshooting, and data analysis.

REFERENCES

APA - The Engineered Wood Association. 2011. Engineered wood construction guide. Form No. E30V, Tacoma: APA.

ASHRAE. 2001. *ASHRAE Handbook—Fundamentals*. Chapter 26, Ventilation and Infiltration, Table 1. Atlanta: ASHRAE.

ASTM. 1996. ASTM Standard E1827–96, *Standard Test Methods for Determining Airtightness of Buildings Using an Orifice Blower Door*. Philadelphia: ASTM.

ASTM. 2004. ASTM E283–04, *Standard Test Method for Determining Rate of Air Leakage Through Exterior Windows, Curtain Walls, and Doors Under Specified Pressure Differences Across the Specimen*. Philadelphia: ASTM.

EPA. 2000. Building envelope improvement: Air sealing. EPA 430-F-97-028. U.S. Environmental Protection Agency.

Gavin, P.M. 1985. Band Joist Component Leakage Tests. Owens Corning intra-company correspondence, March, 27 1985.

Gypsum Association. 2010. Application and finishing of gypsum panel products. Specification GA-216-2010, Section 4.1.8 & 4.9, Washington, Gypsum Association.

Onysko, D.M. and S.K. Jones. 1989. Airtightness of wall sheathing as a function of lumber drying. Proceedings of the ASHRAE/DOE/BTECC/CIBSE Conference *Thermal Performance of the Exterior Envelopes of Buildings IV Conference*: 458–474.

Relander, T.O., B. Heiskel, and J.S. Tyssedal. 2011. The influence of the joint between the basement wall and the wood-frame wall on the airtightness of wood frame houses. Energy and Buildings.

Sherman, M.H. and R. Chan. 2004. Building Airtightness: Research and Practice. LBNL-53356.

Siitonen, V. 1982. Measurement of local air tightness in buildings. Research Notes 125, Technical Research Centre of Finland, Espoo, Finland.

Characterization of Air Leakage in Residential Structures—Part 2: Whole House Leakage

David Wolf, PhD **Frank Tyler**

ABSTRACT

Air will leak through a building envelope that is not well sealed. This leakage of air decreases the comfort of a residence by allowing moisture, cold drafts, and unwanted noise to enter, and air leakage can account for up to 40% of the energy used for heating and cooling in a typical residence. With nearly a mile of exterior joints that can leak air in a typical residence, knowing which joints leak the largest quantity of air allows for the most strategic placement of sealant. This two-part paper describes an extensive investigation to quantify the leakage characteristics of various types of joints and openings in a residential structure. All-in-all, 17 different joints/openings were characterized through both laboratory and real-house measurements using fan pressurization. Part 1 presents the methods and results associated with the air leakage of the individual joints/openings. Part 2 adapts these individual results to the whole house, including an examination of the joint leakage interdependence in the wall cavity.

INTRODUCTION

To reduce air infiltration and achieve an energy-efficient building, the builder or building owner must seal gaps in the building's thermal enclosure. Installing high-quality, tightly sealed windows and doors is a good start, but it's also important to seal joints and openings in the walls, ceiling, and flooring/foundation. To properly address the negative effects of air leakage, such as wasted energy, occupant discomfort, condensation, and so on, all of the joints and openings in the building enclosure should be air sealed. However, some builders or building owners may only have limited funds available to devote to air sealing. With nearly a mile of exterior joints that can leak air in a typical residence, knowing which joints leak the largest quantity of air allows for the most strategic placement of sealant. This study is an investigation to quantify the leakage characteristics of various types of joints and openings in a residential structure (Part 1) and to prioritize the joints/openings in terms of the amount of air leakage per unit cost to seal it—a kind of air leakage "bang-for-your-buck" ranking of the joints/openings (Part 2). The air leakage results are primarily reported at a pressure difference of 50 Pa due to its prevalence in the pre-commission testing of residential structures in the United States.

RESULTS AND DISCUSSION

The extensive amount of data obtained in Part 1 is only of value if it can be applied to assessing the impact of joint leakage on actual houses. Applying these results to actual houses requires several steps, which include:

- Establishing a means by which to scale the lab/test-house results so that they could be applied to any house;
- Comprehending the effect of upstream and/or downstream restrictions, such as drywall on the interior and cladding on the exterior;
- Projecting these results to actual houses to predict the impact of individual joints on whole-house air leakage;

David Wolf *is a research associate and* ***Frank Tyler*** *is a senior scientist at Owens Corning Science and Technology, LLC, Granville, OH.*

- Incorporating the cost implications of sealing the various joints to establish a cost-versus-performance prioritization.

The following sections will address these various steps.

Joint/Opening Severity and Frequency

The results presented in Part 1 were often expressed both in terms of the total leakage for a particular joint (e.g., CFM50 or L/s) as well as the normalized leakage, where the normalization was achieved by dividing the total leakage for a joint by the relevant scaling parameter. For a joint, such as the bottom plate-to-subfloor, the relevant scaling parameter is the joint length (e.g., CFM50/ft or L/s-m). This normalization was applied to most of the leakage that was measured in this study. However, some items, such as recessed lights, duct boots, and outlets, were better normalized per unit (e.g., CFM50/unit or L/s-unit).

These normalized leakages, independent of whether the normalization was by length or unit, could be viewed as a leakage severity, which, when multiplied by the joint/opening frequency, would constitute the total leakage as shown below.

$$\text{leakage [CFM50]} = \text{severity [CFM50/ft]} \times \text{frequency [ft]} \quad (1)$$

This simple relationship allows the scaling of lab/test-house results, cast as a severity, to be applied to any house, where the frequency of joints or openings is known.

Effect of Upstream and/or Downstream Restrictions

The leakage from outside to inside associated with the joints/openings reported in Part 1 has not comprehended any upstream resistance (e.g., cladding) or downstream resistance (e.g., drywall)—they are considered unconstrained leakage values. The one exception is the results for the top plate-to-drywall connection, which necessarily included the downstream resistance at the drywall termination (i.e., bottom of the wall) and penetrations (e.g., electrical outlets).

Depending on the joint/opening, these upstream and downstream resistances can be important. The leakage with some joints/openings are exempt from this complication, such as ceiling penetrations (e.g., recessed lights, duct boots, etc.), where the pressure difference across these items is 50 Pa during the testing conducted in this study as well as what would be experienced in an actual house undergoing a blower door test. In contrast, the leakage with some other joints/openings, particularly those found in the wall cavity, can be significantly affected by the upstream cladding, the downstream drywall and trim, as well as be coupled to other leakage paths in the wall.

Quantifying any cladding-induced upstream resistance in a wall was beyond the scope of this study. Qualitatively, cladding systems that are air permeable and/or have air spaces between themselves and the sheathing are expected to have minimal impact on the results reported here, because there is believed to be ample open area for the air to flow freely. Examples of such claddings would include vinyl siding, wood siding, fiber cement siding, and full-thickness masonry that is offset from the sheathing (e.g., brick and non-veneer stone). Cladding systems that are relatively air impermeable and/or have intimate, continuous contact between themselves and the sheathing are expected to have a significant impact on the results reported here because of the flow resistance that they impart. Examples of such claddings would include stucco and stone veneer without the presence of a rainscreen.

Quantifying the drywall-induced downstream resistance was addressed in this study. Figure 1 conceptually shows a wall cavity (black rectangle), separating the interior from the exterior. The red line shows the air flow from the exterior to the interior and draws an analogy to a series-resistive electrical circuit, where the airflow rate (Q) represents the current, the pressures (P_e, P_w, P_i) represent the voltages, and the resistances to air flow (R_e, R_i) represent the resistors. With such a series-resistive flow condition, the wall cavity pressure (P_w) will approach the exterior pressure (P_e) for cases where the interior resistance is much greater than the exterior resistance ($R_i >> R_e$). The test results reported in Part 1 have assumed that there is no interior resistance ($R_i = 0$, $P_w = P_i$), when in fact there is.

Consider Figure 2, which shows various leakage paths into the wall cavity (red arrows) and out of the wall cavity (green arrows) on the first story of a two-story house (more on why this is important later). This shows an example with four inlets (there could be more) and three outlets. The inlets are: (2) top plate-to-sheathing, (3) between the double top plates, (4) bottom plate-to-sheathing, and (12) a vertical sheathing seam. The outlets are: bottom of drywall (B), electrical outlets (C), and drywall-to-top plate (D), which feeds an open floor joist cavity. Since each of the flows

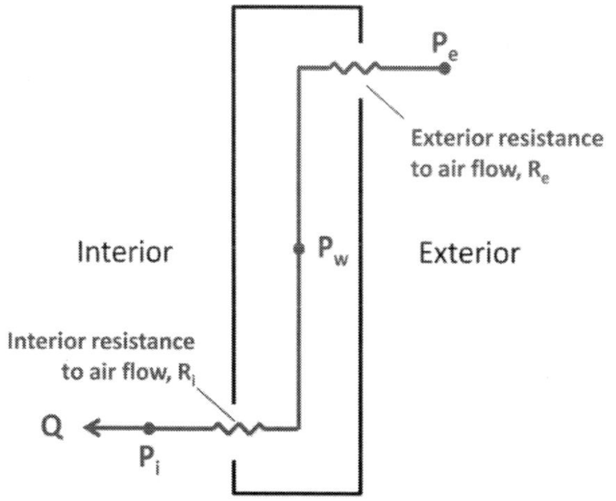

Figure 1 *Sketch of a series-resistive flow in a wall cavity, where P is pressure, R is flow resistance, and Q is flow rate for locations at the interior* (i), *exterior* (e), *and within the wall* (w).

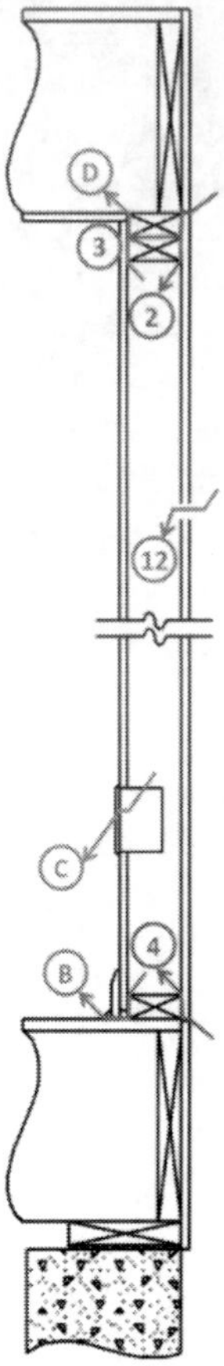

Figure 2 *Sketch showing leakage paths into and out of a wall cavity.*

depicted here has a correlated relationship to pressure difference of the form $Q = C\,\Delta P^n$, it is possible to compute the flow associated with each of the arrows in this figure, as well as calculate the wall cavity pressure. Before doing so, however, it is interesting to think about some possible outcomes.

Figure 2 shows the flow-equivalent of multiple parallel resistors on the exterior skin of the wall (red arrows) that are in series with multiple parallel resistors on the interior skin of the wall (green arrows). Like an electrical circuit, the relative distribution of these resistances not only controls the amount of air that leaks through the wall assembly but also controls the dominant leakage paths. For instance, it will be shown later that the sealing of one individual joint on the exterior side of the wall could result in an increase in leakage of the other unsealed joints on the exterior skin, even though the overall wall leakage decreases somewhat. This could happen if the resistances on the interior skin of the wall (through which the green arrows pass) are appreciable.

Table 1 is an example from a spreadsheet that shows such a set of calculations. The top of the worksheet establishes the specified length of the wall that is being calculated (8 ft [2.4 m] in this example), the specified boundary pressures (P_i being that of the interior and P_e being that of the exterior), the calculated wall cavity pressure (P_w), the calculated flow rate difference between the total inlet and outlet flows (dQ, which should be zero), the calculated wall pressure difference with respect to the interior (wall dP), and the calculated total flow through the wall per unit length of wall (Q/L). The next section of the worksheet is associated with the inlet joints on the exterior side of the wall cavity (the red arrows in Figure 2), followed by a section associated with the outlet joints on the interior side of the wall cavity (the green arrows in Figure 2). The nomenclature section describes the other parameters.

There are several notables regarding the normalized leakage values and exponents (originally reported in Part 1) shown in the spreadsheet table.

First, for those joints where a test house value was used for the normalized leakage (indicated by a "TH" in the Ref. column), the exponent came from the corresponding laboratory results. This was because the laboratory results were obtained under multiple pressure differences, thereby enabling the correlation of a pressure exponent. A pressure exponent couldn't be obtained from the test house measurements, because they were conducted primarily at 50 Pa only.

Second, the normalized leakage for the "bottom of drywall" (0.651 CFM50/ft or 1.01 L/s-m) was inferred from the laboratory-based garage wall measurements described in Part 1. The garage wall leakage includes two drywall-to-plate joints, one on the interior and one on the exterior. An estimate of the individual joint contribution (i.e., one drywall-to-plate joint) can be made by assuming that the interior and exterior joints leak the same, which implies an intermediate wall cavity pressure difference of 25 Pa. The flow rate associated with the overall pressure difference of 50 Pa, which includes the flow through both the interior and exterior joints, is the same as the flow through one of the joints with the pressure difference of 25 Pa, all of which are predicted by the same empirical relationship ($Q = C\,\Delta P^n$), so C can be inferred for the one drywall-to-plate joint. Also, the worst-case garage wall leakage was used in this calculation (see max normalized leakage in Table 1 of Part 1), rather than the average, based on the premise that lab-built wall assemblies generally have tighter tolerances than site-built construction.

Third, the electrical outlet spacing was assumed to be one for every 8 ft (2.4 m). The electrical code requires that one outlet be placed a minimum of every twelve feet apart. The closer spacing used in this study is based on general field observations, where some combination of wall layout and/or additional penetrations (e.g., light switches, telephone junctions, television junctions, internet junctions, wall light fixtures) results in more electrical-outlet-like penetrations in the drywall.

Armed with all of this information and the requirement for conservation of mass between the inlet and the outlet of the wall cavity, the spreadsheet computes the individual flows and the cavity pressure, all through iteration. Such a tool is quite handy because it then permits calculation of the leakage for a variety of permutations, where different combinations of inlet

Table 1. Calculations Showing a Balance in Flow between the Inlet and Outlet Joints of a Wall Cavity, Along with the Calculated Pressure Within the Wall Cavity

Length of Exterior Wall [ft] = 8	P_i [Pa] = 0	P_w [Pa] = 21.2	wall dP [Pa] = 21.2
	P_e [Pa] = 50	dQ [CFM] = 0.00	Q/L [CFM50/ft] = 1.21

INLET JOINTS

Joint Number	Joint Name	Ref.	Sealed? y or n	Normalized Leakage, CFM50/ft	n	C, CFM/ft·Pa^n	Q/L, CFM/ft	Joint Length, ft	Q, CFM	%Q	$Q50$, CFM50	$Q/Q50$
2	Top Plate-to-Sheathing	TH	n	0.546	0.820	0.0221	0.347	8	2.78	29%	4.37	64%
3	Between Double Top Plates	TH	n	0.170	0.829	0.0066	0.107	8	0.86	9%	1.36	63%
4	Bottom Plate-to-Sheathing	TH	n	0.387	0.820	0.0156	0.246	8	1.96	20%	3.09	64%
	Inside Corner	ALTA	n	0.325	0.843	0.0120	0.204	1.6	0.33	3%	0.52	63%
	Outside Corner	TH	n	0.680	0.822	0.0273	0.432	1.6	0.69	7%	1.09	63%
	Window/Door Framing-to-Sheathing	TH	n	0.322	0.820	0.0130	0.205	8	1.64	17%	2.58	64%
12	Vertical Sheathing Seam	TH	n	0.140	0.864	0.0048	0.087	16	1.39	14%	2.24	62%
						TOTAL FLOW INTO WALL CAVITY			9.65			

OUTLET JOINTS

Joint Letter	Joint Name	Ref.	Sealed? y or n	Normalized Leakage, CFM50/ft	n	C, CFM/ft·Pa^n	Q/L, CFM/ft	Joint Length, ft	Q, CFM	%Q
B	Bottom of Drywall	ALTA	n	0.651	0.791	0.0295	0.331	8	2.64	27%
C	Electrical Outlet	ALTA	n	2.62	0.601	0.2493	1.565	1	1.56	16%
D	Drywall-to-Top Plate	ALTA	n	1.168	0.633	0.0983	0.680	8	5.44	56%
						TOTAL FLOW OUT OF WALL CAVITY			9.65	

and outlet joints are sealed. Note from the sample calculation shown in Table 1, which assumes that all inlet and outlet joints are unsealed, the individual joints are leaking an amount that is only 63% of the amount measured when the drywall wasn't present (see column entitled *Q/Q50*).

Much emphasis is being placed on the model that generated the calculations in Table 1, so a simple experiment was designed in the laboratory to validate this model. It used the same pressure chamber that generated the laboratory leakage results discussed in Part 1. This experiment involved a wall assembly with sheathing on the exterior and drywall on the interior. The interior drywall had five locations where the air could leak from the wall cavity—the top plate-to-drywall joint, the bottom plate-to-drywall joint, and up to three electrical outlets. While measuring the flow rate and the wall cavity pressure, various combinations of these joints were sealed and unsealed. Using the above-mentioned model, the flow rates and wall cavity pressures were also calculated. Figure 3 plots the measured results versus the corresponding predicted results, one for flow rate and the other for wall cavity pressure. It can be seen that the agreement is quite good in both cases.

The leakage severity for some of the joints could be applied directly from Table 6 in Part 1, because they are not affected in any meaningful way by the flow resistance of the drywall/trim. These joints include:

- Band joist (top and bottom): The interior side of the band joist is typically connected to an open floor joist system, which is voluminous enough to impose little/no restriction on the airflow (i.e., the local pressure at the interior side of the band joist is equivalent to the house pressure).
- Sill plate-to-foundation: Although this was not shown to be a significant leakage path when some form of a sill gasket is used, any leakage that does occur would typically have little/no resistance associated with finishes because it is connected to an open floor joist system, like the band joist.
- Garage wall: The garage wall leakage was obtained with drywall present on both sides of the test assembly, so the values reported here already include all of the drywall effects. The testing did not include the effects of sealing the bottom of the drywall on one side. However, it can be shown that such an effect has minimal influence on the leakage because there is still ample area for leakage to occur through any electrical outlets.
- Duct boots: For those cases where duct boots penetrate the ceiling drywall from an unconditioned attic, there is no additional flow restriction to consider beyond the boot-drywall interface.
- Recessed lights: For those cases where recessed lights penetrate the ceiling drywall from an unconditioned attic, there is no additional flow restriction to consider beyond the recessed light-drywall interface.

Also, the leakage values in Table 6 in Part 1 that were associated with the top plate-to-drywall joint for a partition wall included the constraint of the drywall. Namely, the best estimate of 0.68 CFM50/ft (1.1 L/s-m) was obtained for the case where the bottom of the drywall was not sealed and the outlets were not sealed. For the case where the bottom of the drywall was sealed, implying leakage only through the outlets, the test house results gave a leakage value of 0.29 CFM50/ft (0.45 L/s-m).

Any other joint that is a part of a wall cavity must comprehend both the flow resistance that the drywall/trim imparts as well as the influence of the other leakage paths. These joints include:

- Plate-to-sheathing (top and bottom)
- Corners (inside and outside)
- Bottom plate-to-subfloor/slab
- Vertical sheathing seams
- Window/door framing-to-sheathing
- Between top plates
- Top plate-to-drywall at attic on an exterior wall

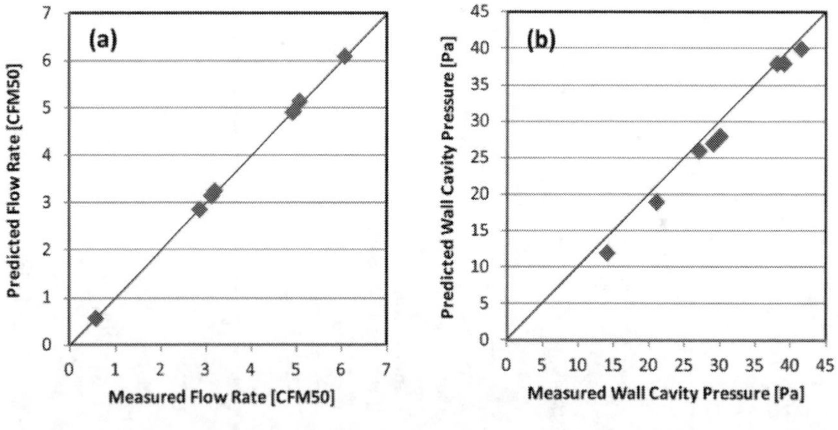

Figure 3 *Comparison of measured and predicted (a) flow rate and (b) wall cavity pressure through a wall assembly.*

The leakage severity of all of these joints was determined for two conditions—one was with maximum drywall/trim constraint and the other was with minimum drywall/trim constraint. The former is the condition where the bottom of the wall is sealed on the interior with caulking of the trim to the drywall and the trim to the floor (e.g., ceramic tile), all for aesthetic reasons by the builder (see Figure 4). The latter is the condition where the bottom of the drywall/trim is unsealed, such as when flooring treatment is hardwood or carpet, thereby precluding caulking to the floor. In either case, the spreadsheet model described earlier in this section is used to predict the overall leakage.

Figure 5 shows the effects of these constraints on the various exterior wall joints. The unconstrained condition is for the case where no drywall is present on the interior side of the wall cavity, which is the result that was presented in Part 1. The remaining conditions show the effects of: (1) whether the bottom of the drywall is sealed on the interior and (2) the location of the wall (upper story or lower story). This chart shows a significant reduction in leakage when any form of drywall constraint is applied. As would be expected, the constraint becomes greater when the bottom of the drywall is sealed, thereby allowing leakage out of the wall cavity through penetrations only (e.g., electrical outlets). Perhaps unexpected is the fact that the wall's location (lower story or upper story on a 2-story house) also affects the amount of constraint.

Figure 6 shows two identical exterior wall sections, except one has a story above it (Figure 6a) and the other has an unconditioned space/attic above it (Figure 6b). The sole difference in the exterior wall assembly between the upper story of a house, which has attic space above it, and the lower story of a two-story house, which has a floor system above it, is the leakage between the drywall and the top plate. In the former case, the leakage is into the wall cavity (see red arrow 13 in Figure 6b), whereas for the latter case, the leakage is out of the wall cavity (see green arrow D in Figure 6a). This results in different leakage behavior between the first floor and second floor exterior walls, with the first floor wall enabling more leakage due to the additional outlet at the top plate, which feeds the floor system, all else being equal.

The main take-away from Figure 5 is that the leakage associated with any joint in the wall cavity will vary, depending on the downstream conditions (drywall/trim treatment) and the leakage of other joints sharing the same wall cavity. This prevents the use of single leakage severity value for these types of joints, and it requires the type of calculations described above to obtain a realistic estimate.

Projection to Whole-House Leakage with Cost Impact

With the frequency of the joints/openings obtained from the actual house or set of plans, the whole house leakage associated with those joints/openings can be readily determined with Equation 1. Furthermore, knowledge of the volume of the house allows the results to be expressed in terms of air changes per hour at 50 Pa (ACH50), which is a common metric used to express whole-house airtightness.

Two different houses were analyzed—a one-story and a two-story house—both constructed on a slab. The basic characteristics of these houses are listed in Table 2. The floor plan associated with each of these houses was used to obtain the various joint lengths (i.e., the frequency term in Equation 1). The normalized leakage rates (i.e., the severity term in Equation 1) were obtained either from Table 6 of Part 1 (joints/openings not associated with a wall cavity) or Figure 5 (joints/openings constrained by drywall). Similarly, a cost can be associated with the sealing of each joint, which enables the calculation of a ratio of the leakage reduction achieved by sealing an individual joint to the corresponding incremental cost to seal it. This can be interpreted as a bang-for-your-buck

Table 2. Floor Area and Volume of Two Houses Used to Estimate Whole-House Leakage Values

House Type	Floor Area, ft^2	Volume, ft^3
One-story, slab	2926	26,334
Two-story, slab	3984	35,856

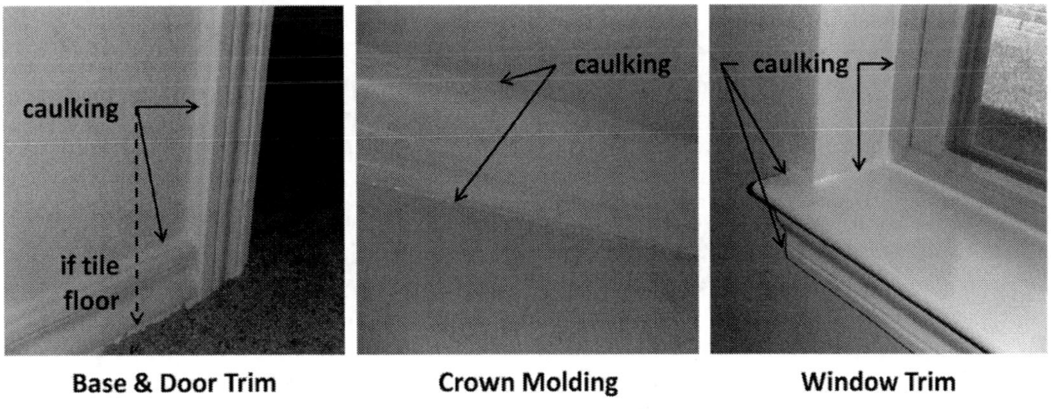

Figure 4 Several photographs that show the extent of caulking that is done by a large production builder for aesthetic reasons.

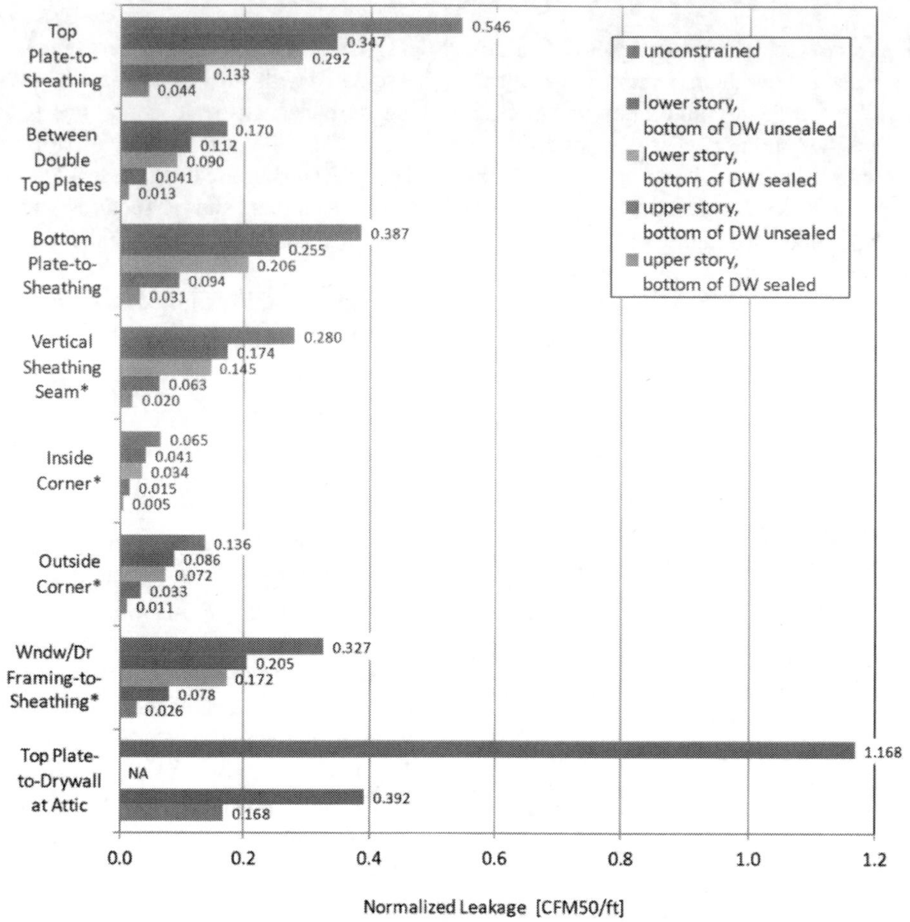

Figure 5 *Normalized leakage associated with wall joints, where the normalizing length is the length of the wall, not necessarily the length of the joint (e.g., corner length ≠ wall length). The unconstrained condition is for the case where no drywall is present on the interior side of the wall cavity. Drywall is present for the remaining conditions, which constrains the leakage, albeit by differing amounts depending on the location of the wall (upper story or lower story) and whether the bottom of the drywall (DW) is sealed on the interior.*
**Note: Inside and outside corners are assumed to be present at 8 ft per every 40 ft of wall section; a window/door joint is assumed to be present at every 16 ft of perimeter (one 3 ft × 5 ft window) per every 16 ft of wall section; vertical sheathing joint is assumed to present at 16 ft per every 8 ft of wall section.*

metric. Clearly, joints that have a large amount of leakage and/or don't require much material to seal will rise to the top of the list as being the most cost effective to seal. It should be noted that this cost assumes easy access to the joint, such as would be the case with new construction. For existing houses, access to certain joints may be obstructed (e.g., wall joints), which would increase the cost.

Figure 7 shows a chart of incremental performance improvement for the one-story house on the *y*-axis (ACH50 for a minimal drywall constraint) as a function of the installed cost on the *x*-axis. Since the cost of sealants varies with the type and method of application, the cost units have been omitted. Figure 8 shows a comparable chart of incremental performance improvement for the two-story house. The order with which the joints and openings are displayed in these figures is based on their individual contribution to air leakage (i.e., the "bang," if sealed) per unit of cost to seal that joint (i.e., "for your buck.") Using Figure 7 as an example, the four biggest bang-for-your-buck joints to seal would be recessed lights, duct boots, garage wall, and top-plate-to-drywall at attic. Also note that the top-plate-to-drywall at attic joint gives the single biggest opportunity for lowering the blower door result.

Note that both of these charts display a non-linear behavior, where the curve is steep initially (large reduction in air leakage per unit cost), and relatively flat finally (small reduction in air leakage per unit cost). Recall that the main objective of this study was to prioritize the joints/openings in terms of the amount of air leakage per unit cost to seal it—a kind of air leakage bang-for-your-buck ranking of the joints/openings. If every joint in the house leaked comparably and required the same amount of sealant to address it, then the relationship in

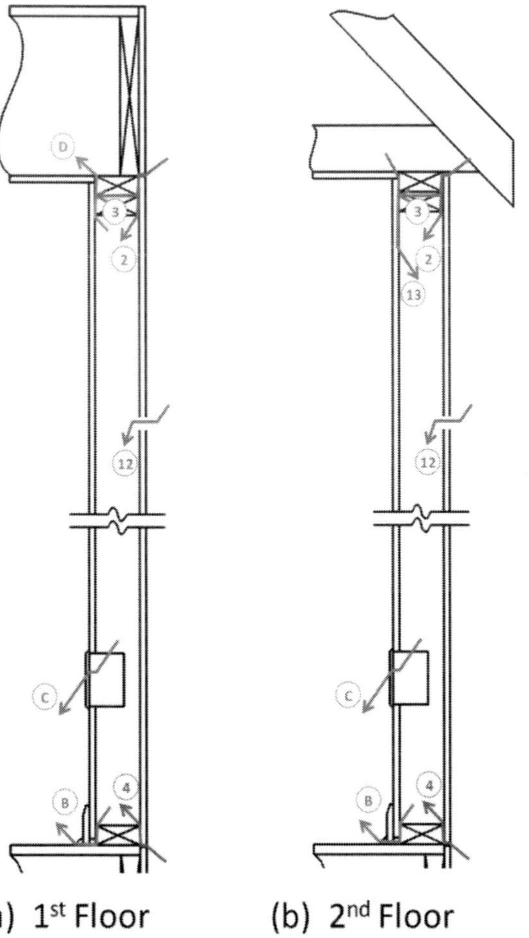

Figure 6 *Schematics showing the inlet (red arrows) and outlet (green arrows) leakage paths through the exteriors walls on the 1st floor and 2nd floor of a two-story building.*

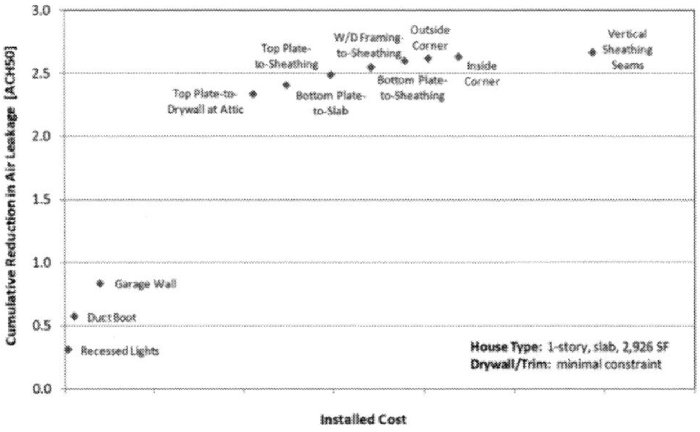

Figure 7 *Chart of cumulative reduction in air leakage as a function of installed cost (arbitrary) for a one-story house.*

Figures 7 and 8 would be linear, but that's clearly not the case. Indeed, some joints/openings are far more important than others from a blower door perspective.

CONCLUSION

This study was an extensive investigation to quantify the air leakage characteristics of various types of joints and openings in a residential structure and to prioritize the joints/openings in terms of the amount of air leakage per unit cost to seal it—a kind of air leakage bang-for-your-buck ranking of the joints/openings. The air leakage in all cases was assessed under the blower door test pressure difference across the building enclosure of 50 Pa. A summary of the results are provided in Table 3. These rankings were bucketed into the three categories of (1) most effective joints to seal, (2) moderately effective joints to seal, and (3) least effective joints to seal.

The most effective joints to seal included the following:

- Top Plate-to-Attic: This joint was shown to leak in the range of 0.29 to 0.68 CFM50 per foot of joint. This range is driven by differences in how well the drywall is sealed to the interior finishes. For an average-sized house, there can be upwards of 500 ft of this joint, and the size of the gap between the drywall and framing at this location can be relatively large, due to misalignment in framing (stud-to-plate) or the presence of top plate-to-rafter ties (hurricane ties), which can create a localized offset of 3/16 in. to 1/4 in. between the drywall and plate. The cumulative air leakage for this joint is significant, resulting in normalized results in the range of 0.29 to 1.6 ACH50.
- Recessed Lights: Even so-called airtight recessed lights, which are mandated by many building codes, can leak an appreciable amount of air at the juncture between the light housing and the mounting flange, as well as the mounting flange and the drywall. Test results showed an

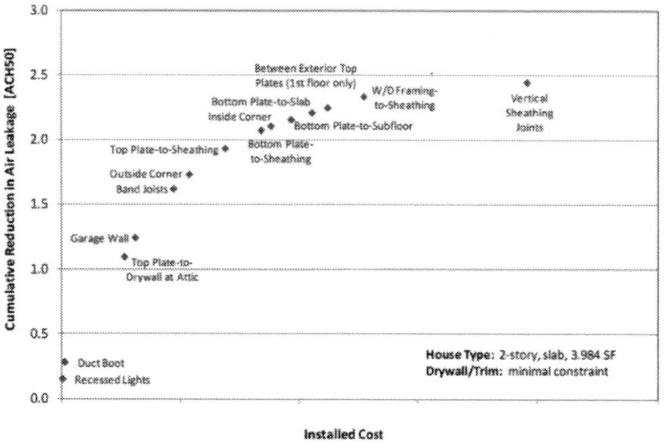

Figure 8 *Chart of cumulative reduction in air leakage as a function of installed cost (arbitrary) for a two-story house.*

Table 3. Summary of Per Unit Leakage Rates and Projected Whole-House Leakage for the Various Studied in this Investigation

Joint/Opening	CFM50[1]		ACH50[1]
Recessed light	9.1	per light	0.15 to 0.31
Duct boot	7.7	per boot	0.13 to 0.26
Top plate-to-attic	0.29 to 0.68	per foot	0.29 to 1.6
Garage-house common wall	0.60	per foot	0.14 to 0.26
Band joist (top and bottom)	0.86	per foot	0.37 to 0.42
Sheathing-to-plate (top and bottom)	0.074 to 0.62	per foot	0.040 to 0.38
Bottom plate-to-subfloor	0 to 0.11	per foot	0 to 0.11
Corners (interior pointing)	0.024 to 0.21	per foot	0.0021 to 0.032
Corners (exterior pointing)	0.054 to 0.45	per foot	0.0069 to 0.11
Window/door framing-to-sheathing	0.031 to 0.11	per foot	0.020 to 0.10
Between exterior top plates	0.10 to 0.11	per foot	0.033 to 0.046
Vertical sheathing joints	0.010 to 0.090	per foot	0.011 to 0.11
Sill plate-to-foundation[2]	0 to 0.030	per foot	0 to 0.025

[1] Assumes all other joints in the wall cavity are sealed.
[2] Assumes the presence of a sill gasket.

average leakage of 9.1 CFM50 per light, which could amount to upwards of 0.15 ACH50, depending on the number of lights penetrating the ceiling into the unconditioned attic.

- Duct Boots: For cases where the HVAC ductwork is located in the attic, there will likely be duct boots that penetrate the ceiling drywall to supply or return air to the living space. The leakage at the interface between the duct boot and the drywall contributes 7.7 CFM50 per boot, which could amount to upwards of 0.13 ACH50, depending on the number of duct boots present.
- Band Joist: This joint was shown to leak an average amount of 0.86 CFM50 per foot of band joist (includes the leakage from both the upper and lower joints), which results in approximately 0.4 ACH50 for the whole-house result. It is notable that the band joist joints are the only wall joints that are not meaningfully constrained by drywall, which makes them important contributors to air leakage. The air leakage associated with all other exterior wall joints must negotiate the joint on the exterior skin of the wall (the sheathing layer) and the interior skin of the wall (the drywall layer). Once the air passes the exterior skin associated with the band joist, it typically encounters a large, open space in the floor system, where it travels relatively unimpeded.
- Garage-to-House Common Wall: This wall is unique in that it is an exterior wall (i.e., separates the conditioned living space from the unconditioned garage space) that is sheathed on the exterior side with drywall, as opposed to some form of structural sheathing, like oriented-strand board or plywood. This is significant in that drywall has far less stiffness than OSB/plywood, which adversely affects the airtightness of the joint that is formed between the drywall and the framing members. This joint was shown to leak an average amount of 0.6 CFM50 per foot of joint, which results in approximately 0.14 to 0.26 ACH50 for the whole-house result. While this is fairly high leakage on a per foot basis, its implication on the whole-house leakage is not as large because there are typically not many feet associated with a garage wall. However, such a joint could be very important from the standpoint of indoor air quality, since it connects the living space to an area where harmful gases can be found (e.g., combustion products, solvents).

The moderately effective joints to seal included the following:

- Top and Bottom Plate-to-Sheathing: This joint was shown to leak an amount that ranged from 0.074 to 0.62 CFM50 per foot of joint (includes the leakage from both the top and bottom joints), which results in approximately 0.040 to 0.38 ACH50 for the whole-house result. This range is driven by differences in how well the drywall is sealed to the interior finishes. This matters because of what was stated above—namely, the air leakage associated with most exterior wall joints must negotiate the joint on the exterior skin of the wall (the sheathing layer) and the interior skin of the wall (the drywall layer). Once the air passes the exterior skin associated with the plate-to-sheathing joint, it must also pass interior skin through drywall penetrations (electrical outlets and switches, plumbing fixtures, etc.) and terminations (bottom of the wall, windows, doors, etc.). However, these terminations can occasionally (and unintentionally) be well sealed with caulk by the painter or finish carpenter for aesthetic reasons. Such sealing, however, is of questionable durability, since such caulks are almost always low performing (i.e., low flexibility, resulting in cracking).
- Bottom Plate-to-Subfloor: Numerous independent measurements in this investigation showed the leakage associated with this joint to be relatively small (0.1 CFM50/ft on average and up to 0.1 ACH50 for the whole house). It is interesting to note that many builders and building code officials expect/require this joint to be sealed, often

ignoring far more important joints. It is possible that this joint can be quite large/leaky in some localized cases where the wall is constructed on the floor, tilted up into place, and potentially have some construction debris (wood chips, fasteners, etc.) lodged between the bottom plate and the subfloor. While this should be uncommon or very localized, the joint should obviously be sealed in this case (i.e., "If you see daylight, seal it!").

The least effective joints to seal included the following:

- Corners, Window/Door Framing-to-Sheathing, and Between Exterior Top Plates: These are three very different types of joints, but the thing that they have in common are they are not significant contributors to whole-house leakage (ACH50) for one or more of the following reasons: (1) small amount of overall joint length, (2) not a large amount of leakage, and (3) the leakage is constrained to some extent by the drywall, which was mentioned above. However, the leakage from any one of these joints could still cause thermal or acoustic comfort issues.
- Bottom Plate-to-Slab: This study did not make measurements for the case where the bottom plate mated with a slab, so this least-effective categorization is an extension of the low leakage comments above for the bottom plate-to-subfloor joint as well as the reality that the bottom plate-to-slab connection typically has two advantages related to airtightness. One is that the slab is often very smooth because of the finish flooring requirements for uniformity (e.g., ceramic tile, wood laminate flooring). The other is that this joint is often addressed with an inexpensive sill gasket, which is sandwiched between the bottom plate and the smooth slab for the very purpose of air sealing.

ACKNOWLEDGMENTS

The authors wish to acknowledge the important contributions to this investigation by Mikael Salonvaara of Owens Corning, who consulted with them on many matters throughout the study, including experimental design, troubleshooting, and data analysis.

NOMENCLATURE

C = proportionality constant in the relation $Q = C\,\Delta P^n$. It is calculated by taking the normalized leakage and dividing by the pressure difference of 50 Pa to the exponent n. The pressure difference of 50 Pa is used here because this was the pressure difference associated with the normalized leakage.

Joint length = appropriate joint length associated with the length of wall section specified above

n = measured exponent to the pressure difference across the joint in the relation $Q = C\,\Delta P^n$. This also was reported in Part 1 for a particular joint.

Normalized leakage = leakage value reported in Part 1 for a particular joint (also referred to previously as the leakage severity)

P = pressure

Q = leakage associated with the particular joint, when a part of the overall wall assembly. It is the product of Q/L and joint length.

$\%Q$ = fraction of the total flow into or out of the wall cavity associated with that particular joint

$Q50$ = flow associated with a particular joint if the pressure difference across the joint were 50 Pa, which is the equivalent of having no drywall present for an exterior joint

$Q/Q50$ = ratio of the joint leakage as a part of the wall assembly (Q) to the joint leakage in isolation and unobstructed by a downstream resistance, such as drywall ($Q50$)

Q/L = leakage per unit length associated with the particular joint when a part of the overall wall assembly. It uses the relation $Q = C\,\Delta P^n$ to compute the flow, where the pressure difference is with respect to the wall cavity. pressure (P_w).

Ref = indicates whether the leakage data used in that row came from the laboratory results of this study (denoted ALTA) or the test house results (denoted TH)

Sealed? = specifies whether the particular joint should be considered sealed (indicated with a "y") or not (indicated with a "n"). If it is sealed, then no flow associated with that joint is included in the calculation. If it is unsealed, then the flow is calculated based on the relation $Q = C\,\Delta P^n$.

Subscripts

i = interior

e = exterior

w = wall

REFERENCES

Wolf, D.H. and Tyler, F.S. 2013. Characterization of Air Leakage in Residential Structures—Part 1: Joint Leakage. Thermal Performance of the Exterior Envelopes of Whole Buildings XII International Conference.

Thermal Performance Evaluation of Attic Radiant Barrier Systems Using the Large Scale Climate Simulator (LSCS)

Som Shrestha, PhD
Member ASHRAE

William Miller, PhD
Member ASHRAE

Andre Desjarlais

ABSTRACT

Application of radiant barriers and low-emittance surface coatings in residential building attics can significantly reduce conditioning loads from heat flow through attic floors. The roofing industry has been developing and using various radiant barrier systems and low-emittance (low-e) surface coatings to increase energy efficiency in buildings; however, minimal data are available that quantify the effectiveness of these technologies.

This study evaluates performance of various attic radiant barrier systems under simulated summer daytime conditions and nighttime or low solar gain daytime winter conditions using the large scale climate simulator (LSCS). The four attic configurations that were evaluated are 1) no radiant barrier (control), 2) perforated low-e foil laminated oriented strand board (OSB) deck, 3) low-e foil stapled on rafters, and 4) liquid applied low-emittance coating on roof deck and rafters. All test attics used nominal R_{US} 13 h·ft²·°F/Btu (R_{SI} 2.29 m²·K/W) fiberglass batting insulation on the attic floor. Results indicate that the three systems with radiant barriers had heat flows through the attic floor during summer daytime condition that were 32.8%, 49.8%, and 19.1% lower than the control, respectively.

INTRODUCTION

Among the various modes of heat transfer, radiation is the predominant mode of heat transfer in typical building attics, particularly during summer months. Therefore, the use of radiant barriers or low-emittance (low-e) surface coatings in residential building attics can significantly reduce conditioning loads from heat flow through attic floors. In order to determine the effectiveness of various radiant barrier systems, the Building Envelope Research (BER) group within the Building Technologies Research and Integration Center (BTRIC) at Oak Ridge National Laboratory (ORNL) conducted a series of steady-state guarded hot box evaluations in the large scale climate simulator (LSCS) from May to August 2011. Results from this study are presented in this paper.

DESCRIPTION OF THE TEST APPARATUS

The large scale climate simulator (Figure 1) is a facility that is capable of testing whole roof and attic systems under either steady-state or transient conditions. The climate chamber of the LSCS can expose specimen to ambient temperatures ranging from –40°F (–40°C) to 150°F (65.6°C). The roof surface can be maintained at a higher temperature than the climate chamber air temperature because it can be heated with infrared lamps that can be controlled to simulate solar heat gain. The lower portion of the LSCS contains both a guard and a metering chamber, each of which can be controlled independently from 45°F (7.2°C) to 150°F (65.6°C). Both upper and lower chambers have controlled humidity and other capabilities that were not utilized in this set of experiments. Figure 2 shows the test attic in the LSCS environmental chamber.

The LSCS can be operated as an environmental chamber (heat flows through the ceiling are measured with heat flux transducers attached to it) or as a guarded horizontal hot box test facility. Guarded hot box tests are conducted in accordance with ASTM C1363-05 *Standard Test Method for Thermal Performance of Building Materials and Envelope*

***Som Shrestha, William Miller,** and **Andre Desjarlais** are affiliated with the Building Technologies Research and Integration Center, Oak Ridge National Laboratory, Oak Ridge, TN.*

Published by ASHRAE 2013

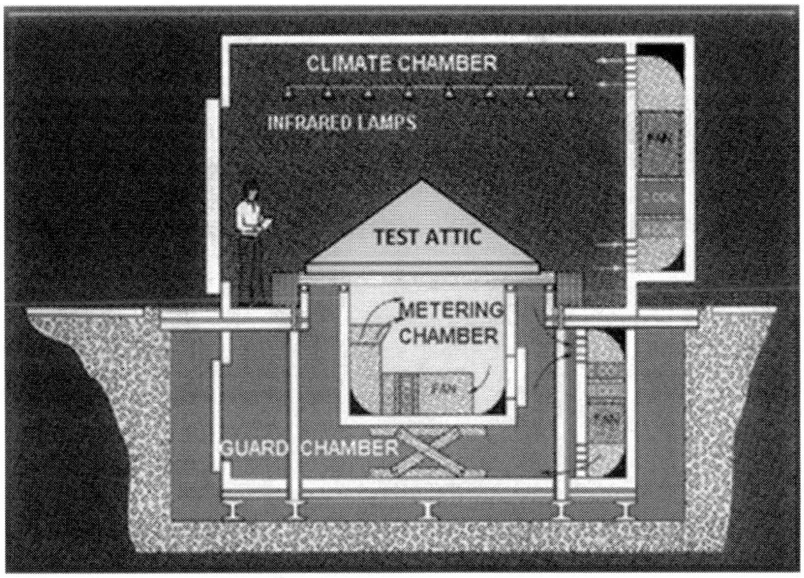

Figure 1 *Sketch of the large scale climate simulator (LSCS).*

Figure 2 *The test attic in the LSCS environmental chamber.*

Assemblies by Means of a Hot Box Apparatus or in accordance with ASTM C1373-03 *Standard Practice for Determination of Thermal Resistance of Attic Insulation Systems Under Simulated Winter Conditions*. The chamber is designed to test the thermal performance of 12 ft, 9 in by 12 ft, 9 in (3.9 m by 3.9 m) roof sections and all controls and data acquisitions are automatic and programmable. Details of the LSCS are described by Huntley (1989).

The metering chamber serves as a guarded hot box. The chamber is manually raised so that it seals against the ceiling, and provides a measurement of the total heat flow through the 8 ft (2.44 m) by 8 ft (2.44 m) central area of the attic floor. The surrounding guard chamber temperature is maintained close to the metering chamber temperature to minimize heat flows across the metering chamber walls. Heat flow through the attic floor can then be calculated from an energy balance of the metering chamber (Wilkes et al., 1995). The tests described in this paper were conducted in this manner.

Prior to the test, the metering chamber was calibrated with a series of experiments conducted on a homogeneous panel made of 4 in. (0.102 m) of expanded polystyrene (EPS) foam (with known R-value) which was painted on both sides with white latex paint. The calculated bias from this calibration test was used to adjust the collected experimental data. The precision and bias test procedure for the LSCS is described by Wilkes et al. (1995).

DESCRIPTION OF THE TEST ATTIC

The test attic module was 12 ft, 8 in. (3.86 m) by 12 ft, 8 in. (3.86 m) with 6:12 roof slope (Figure 2). Nominal 2 by 4 (actual 1.5 in. [38 mm] by 3.5 in. [89 mm]) rafters and ceiling joists were spaced at 24 in. (0.61 m) on center. Nominal R_{US} 13 h·ft^2·°F/Btu (R_{SI} 2.29 m^2·K/W) fiberglass batting insulation was installed between the joists. The insulation was 3.5 in. (89 mm) thick; therefore, the top of the attic floor insulation was level with the top edge of the ceiling joists. A ½ in. (13 mm) gypsum board was used on the ceiling. The roof consisted of nominal ¾ in. (19 mm) oriented strand board (OSB) and 15 lb (6.8 kg) roofing felt. Ridge vents and eaves made by drilling holes were used to achieve a net free ventilation area of 1:150 (total unobstructed cross-sectional area of ventilation openings of 1 ft^2 for every 150 ft^2 [0.093 m^2 for every 13.94 m^2] of attic floor area).

The performance of the following four attic configurations was evaluated:

Attic 1: ordinary OSB without radiant barrier (control)

Attic 2: perforated low-e foil laminated on OSB (TechShield) (Figure 3)

Attic 3: low-e foil stapled on rafters (Figure 4)

Attic 4: liquid applied low-e coating on roof deck and rafters (Figure 5)

The thermal emittance of the OSB, radiant barrier materials, and the OSB with low-e coating were measured using Devices & Services emissometer model AE which was operated in accordance with ASTM C1371 *Standard Test Method for Determination of Emittance of Materials Near Room Temperature Using Portable Emissometers* or Devices and Services Technical Note 79-17 (transient method). Table 1 shows the measured emittances. The foil stapled on rafters had a low-e surface on both sides. The specimen with liquid applied low-e on OSB was cured at 410°F (210°C) for about 30 minutes prior to measuring the thermal emittance.

Figure 3 *Test attic with low-e foil laminated on OSB.*

Table 1. Measured Thermal Emittance of Materials

Material	Measured Average Thermal Emittance, ε	Measurement Method
OSB	0.89	Transient
Perforated low-e foil laminated on OSB	0.03	Transient
Low-e foil stapled on rafters	0.02	ASTM C1371
Low-e coated on OSB	0.23	Transient
Rafters	0.87	Transient

Figure 4 *Test attic with low-e foil stapled on rafters.*

Figure 5 *Test attic with liquid applied low-e coating on roof deck and rafters.*

TEST DESCRIPTION

The tests were conducted in laboratory environments simulating summer daytime and winter nighttime or low solar gain daytime conditions for a cooling dominant climate zone. For the summer daytime condition, the climate chamber air temperature around and above the test attic was set to 100°F (37.8°C) and the IR lamps were used to maintain roof exterior surface temperature at 140°F (60.0°C). For the winter night condition, the climate chamber air temperature was set to 32°F (0°C) and the infrared lights were turned off (the roof exterior surface temperature was not controlled). These conditions were selected based on AtticSim (attic simulation software developed at ORNL, Wilkes 1991) simulation result using TMY3 weather file for Austin, TX. Metering chamber temperature (measured at about 3 in. (76 mm) below the attic floor) and the guard chamber temperatures were set at 70°F (21.1°C) in both cases to simulated indoor conditions.

In total; 5 thermocouples were used to measure the climate chamber air temperature, 20 thermocouples were used to measure the roof exterior surface temperature, 13 thermocouples were used at various heights to measure the attic air temperature, 5 thermocouples were used to measure the attic floor temperature on top of the insulation, and 5 thermocouples were used to measure the temperature of the gypsum board surface facing the metering chamber. Averages of the measured temperatures are presented in Table 2. As can be seen from the table, the climate chamber air temperature was maintained within ± 0.5°F (± 0.3°C) of the setpoint tempera-

Table 2. Average of the Measured Temperatures

Measurement Location	Attic Configuration	Summer Day Condition	Winter Night Condition
Climate Chamber Air Temperature, °F (°C)	Attic 1	99.8 (37.7)	32.0 (0.0)
	Attic 2	100.1 (37.8)	31.9 (–0.1)
	Attic 3	99.5 (37.5)	32.1 (0.1)
	Attic 4	99.9 (37.7)	32.1 (0.0)
Roof Exterior Surface Temperature, °F (°C)	Attic 1	140.5 (60.3)	32.2 (0.1)
	Attic 2	140.6 (60.3)	32.0 (0.0)
	Attic 3	139.6 (59.8)	32.2 (0.1)
	Attic 4	139.5 (59.7)	32.0 (0.0)
Attic Air Temperature, °F (°C)	Attic 1	120.2 (49.0)	33.9 (1.1)
	Attic 2	111.1 (44.0)	35.0 (1.7)
	Attic 3	105.6 (40.9)	38.5 (3.6)
	Attic 4	114.1 (45.6)	34.6 (1.4)
Attic Floor Temperature, °F (°C)	Attic 1	121.3 (49.6)	37.1 (2.8)
	Attic 2	105.7 (40.9)	39.2 (4.0)
	Attic 3	98.1 (36.7)	41.9 (5.5)
	Attic 4	110.7 (43.7)	38.3 (3.5)
Gypsum Board Surface Temperature Towards the Metering Chamber, °F (°C)	Attic 1	74.6 (23.7)	68.1 (20.0)
	Attic 2	73.2 (22.9)	68.2 (20.1)
	Attic 3	72.6 (22.6)	68.4 (20.2)
	Attic 4	73.7 (23.2)	68.2 (20.1)
Metering Chamber Air Temperature, °F (°C)	Attic 1	70.0 (21.1)	69.9 (21.1)
	Attic 2	70.0 (21.1)	70.0 (21.1)
	Attic 3	70.1 (21.1)	70.0 (21.1)
	Attic 4	70.0 (21.1)	70.0 (21.1)
Gable Exterior Surface Temperature, °F (°C)	Attic 1	105.0 (40.5)	32.8 (0.4)
	Attic 2	102.1 (38.9)	32.6 (0.3)
	Attic 3	100.8 (38.2)	33.9 (1.0)
	Attic 4	102.6 (39.2)	32.8 (0.4)

ture, the roof exterior surface temperature was maintained within ± 0.6°F (± 0.33°C) of the setpoint temperature, and the metering chamber air temperature was maintained within ±0.1°F (± 0.06°C) of the setpoint temperature.

RESULTS AND DISCUSSION

The rate of heat flow to or from the metering chamber through the attic floor was calculated by conducting an energy balance calculation in the metering chamber. The rate of heat flow was converted to heat flux by dividing the effective area of the ceiling exposed to the metering chamber, which is 64 ft^2 (5.95 m^2).

Energy in and out of the metering chamber, including the heat flow through the metering chamber walls due to the negligibly small imbalance between the metering and guard chamber air temperatures, was measured in accordance with ASTM C1363. After the steady-state conditions were reached, data from each sensor were recorded for 18 hours at 5 minute intervals. Time constant of the LSCS is approximately 3 hours; therefore, 18 hours period is equivalent to six time-constants of the LSCS. Energy balance calculations on the metering chamber were performed for each five-minute interval, and these values were averaged over the 18-hour period to calculate an average heat flow through the attic floor. Heat flows through the metering chamber walls were estimated using 32 differential thermocouples across the walls.

Figure 6 shows the hourly cooling load during summer daytime condition and heating load during winter condition to the metering chamber due to the heat flow through the attic floor. The error bars on the plot show the two standard deviations of the calculated average hourly values over the same period. The calculated heating and cooling loads, heat flux, and 95% reproducibility intervals are shown in Table 3.

The key measure of the thermal performance of the radiant barrier is the percentage reduction in heat flow to conditioned space caused by the use of the radiant barrier. From Table 3, it can be seen that the test attic with perforated low-e foil laminated OSB deck, low-e foil stapled on rafters, and liquid applied low-e coating reduced the cooling load from the attic floor compared to the load from the attic without a radiant barrier on roof deck by 32.8% ± 4.9%, 49.8% ± 3.2%, and 19.1% ± 3.7%, respectively, during summer daytime conditions and by 8%, 10%, and 6%, respectively, during winter conditions. However, while the result is statistically significant with better than 95% confidence for summer condition, the result is not statistically significant at the same level for winter condition.

It should be noted that the attic with low-e foil stapled on rafters had 3.5 in. (88.9 mm)-high air space between the foil and roof sheathing. Air movement in these cavities takes away some of the heat penetrated through roof sheathings (similar to the case with above-sheathing ventilation design, as described in Miller et al. 2007), reducing heat flow to attic.

It should also be noted that, due to the limitations in the LSCS capability to control radiation heat flux from infrared light source, the exterior roof surface temperature was held fixed in all attic configurations for a given climate condition. If these tests were conducted in an actual building exposed to natural conditions, the exterior roof surface temperatures would not be the same for all attics. It is expected that for similar environmental conditions, attic construction, and all other building parameters; the exterior roof surface temperature would be higher in an attic with low-e foil laminated on OSB or liquid applied low-e coating on roof deck than in an attic without radiant barrier. The natural convection currents in attics with low-e foil stapled on rafters could feasibly drop the exterior roof surface temperature as compared to the case with foil faced OSB or liquid applied low-e coating on roof deck.

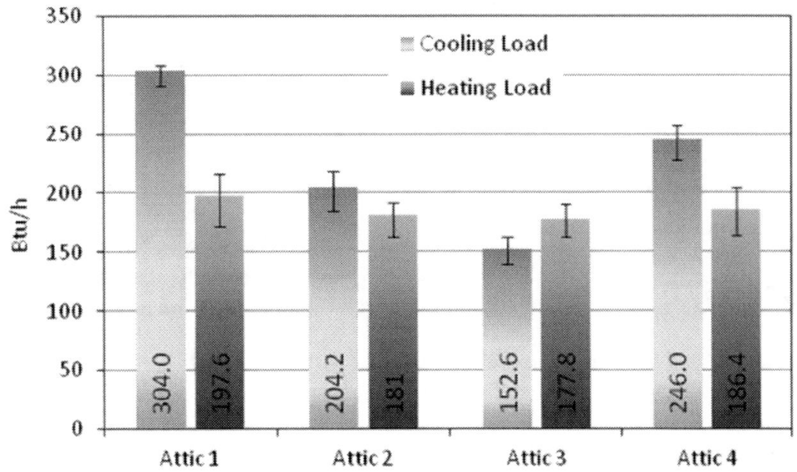

Figure 6 *Load to the metering chamber due to the heat flow through the attic floor.*

Table 3. Heating and Cooling Loads, 95% Reproducibility Intervals, and Heat Flux

Attic	Cooling Load		Heating Load		95% Reproducibility Intervals		Heat Flux, Summer		Heat Flux, Winter	
	Btu/h	W	Btu/h	W	Cooling Load, Btu/h (W)	Heating Load, Btu/h (W)	Btu/h·ft²	W/m²	Btu/h·ft²	W/m²
1 (Control)	304.0	89.0	197.6	57.9	± 4.4 (1.3)	± 18.2 (5.3)	4.8	15.0	–3.1	–9.7
2 (TechShield)	204.2	59.8	181.0	53.0	± 14.8 (4.3)	± 10.9 (3.2)	3.2	10.1	–2.8	–8.9
3 (Foil Stapled on Rafters)	152.6	44.7	177.8	52.1	± 9.4 (2.8)	± 12.4 (3.6)	2.4	7.5	–2.8	–8.8
4 (Liquid Applied)	246.0	72.1	186.4	54.6	± 10.6 (4.3)	± 17.8 (5.2)	3.8	12.1	–2.9	–9.2

SUMMARY

Implementation of radiant barriers is becoming more prominent in building codes, like California's Title 24. The US Department of Energy (DOE) conducted elaborate tests for radiant barrier in attics at the large scale climate simulator at Oak Ridge National Laboratory. The purpose of this testing was to evaluate the performance differences in new construction and retrofit applications, as well as the performance differences between the radiant barrier sheet and the radiant barrier paint, also known as an interior radiant control coating (IRCC).

Results show that the use of perforated low-e foil laminated OSB deck, low-e foil stapled on rafters, and liquid applied low-e coating resulted in a decrease in heat flow to the conditioned space by 32.8% ± 4.9%, 49.8% ± 3.2%, and 19.1% ± 3.7%, respectively, when compared to the load from attic without a radiant barrier during laboratory conditions representative of summer daytime temperatures and irradiance. During winter conditions, a 6% to 10% reduction in heat loss through the ceiling was observed, but the results were not statistically significant at 95% confidence. The test attic had nominal R_{US} 13 h·ft^2·°F/Btu (R_{SI} 2.29 m^2·K/W) fiberglass batting insulation on the floor. The heat flow through the attic floor decreases with higher insulation R-values or during milder weather conditions. Therefore, for attics with higher insulation R-values the potential savings due to the application of radiant barrier systems will be lower than the results shown in this study.

The DOE was hopeful that the IRCC would offer better performance since spraying on IRCC paint is easier than installing a sheet radiant barrier once the house is built, but the IRCCs on the market today do not perform as well as a sheet radiant barrier.

In new construction, the radiant barrier laminated to the OSB sheathing is certainly the easiest and least expensive to install, but going to the added labor in installing a radiant barrier under rafters can result in greater savings. This may be for a variety of reasons, like the higher emissivity of the OSB, a lack of space between the barrier and the sheathing to vent, and the presence of the rafters below the barrier to reduce the coverage of the low emittance surface.

The most efficient method was the radiant barrier stapled on rafters. This particular experiment did not have HVAC ducts in the attic; homes with attic ducts could see even greater savings in the summer. Radiant barrier stapled on the rafters is the preferred method of installing a retrofit radiant barrier by the US DOE, ENERGY STAR®, California Title 24, ASTM International, RIMA International, and others.

FUTURE WORK

The experimental data can be used to calibrate an attic model developed in any building energy simulation software. The calibrated model then can be used to run an annual simulation with local weather data and calculate annual energy savings potential of the various radiant barrier systems and attic insulation level for various climate zones.

For this study, the exterior roof surface temperature was held fixed in all attic configurations for a given climate condition, which would not be exactly the same if the test was conducted at natural environmental condition. Therefore, further experimental study should be carried out in natural exposure buildings.

ACKNOWLEDGMENTS

This work was supported by the Building Technologies Program, U.S. Department of Energy. The author would like to acknowledge the contributions of Jerald Atchley who supervised the operation of the LSCS; R&D Services, who constructed the attic modules; and industry supporters, who provided materials to build the attic.

DISCLAIMER

REFERENCES

ASTM C1363. 2005. *Standard Test Method for Thermal Performance of Building Materials and Envelope Assemblies by Means of a Hot Box Apparatus*, American Society of Testing and Materials, West Conshohocken, PA, 2005.

ASTM C1371. 2010. *Standard Test Method for Determination of Emittance of Materials Near Room Temperature Using Portable Emissometers*, American Society of Testing and Materials, West Conshohocken, PA, 2005.

ASTM C1373. 2003. *Standard Practice for Determination of Thermal Resistance of Attic Insulation Systems Under Simulated Winter Conditions*, American Society of Testing and Materials, West Conshohocken, PA, 2005.

Devices and Services Technical Note 79-17 Emissivity Measurements for In-Place Surfaces and for Materials with Low Thermal Conductivity http://www.devicesandservices.com/TechNotes/TN79-17.pdf

Miller, W., M. Keyhani, T. Stovall, and A. Youngquist. 2007. Natural Convection Heat Transfer in Roofs with Above

Sheathing Ventilation. Proceedings of the Thermal Performance of the Exterior Envelopes of Buildings X. December 2–7, 2007 Clearwater, Florida.

Huntley, W. R. 1989. Design Description of the Large Scale Climate Simulator, ORNL/TM-10675, Oak Ridge National Laboratory, Oak Ridge, TN, 1989.

Wilkes, K.E., and P.W. Childs. 1993. Thermal Performance of Clean Horizontal Radiant Barriers under Winter Conditions – Laboratory Measurements and Mathematical Modeling. *ASHRAE Transactions* 1993, Vol. 99, Pt. 1.

Wilkes, K.E., T.W. Petrie and P.W. Childs. 1995. Precision and Bias of the Large Scale Climate Simulator in the Guarded Hot Box and Cold Box Modes. Proceedings of the 21st International Thermal Conductivity Conference. Oak Ridge National Laboratory, Oak Ridge, TN.

Wilkes, K. E. 1991. *Thermal Model of Attic Systems with Radiant Barriers*. ORNL/CON-262. Oak Ridge, TN: Oak Ridge National Laboratory.

Important Factors Affecting the Risk of Mold Growth in Well-Insulated Wood Frame Walls in Northern European Climates

S. Olof Mundt-Petersen **Lars-Erik Harderup, PhD** **Jesper Arfvidsson, PhD**

ABSTRACT

Due to increased awareness of climate change and energy costs, well-insulated buildings have become more common. Furthermore, greater interest has been shown in using wood in building to produce more carbon dioxide-efficient houses. However, thicker thermal insulation in walls increases the risk of high relative humidity (RH) levels and the risk of mold-related damage in wood frame houses.

This paper presents important factors affecting the risk of mold growth in well-insulated wood frame walls in Northern European climates. Recent findings regarding important factors are first briefly summarized. The paper continues with a parametric study in which moisture-critical positions in traditional Swedish wood frame designs are investigated by using hygrothermal modeling. Traditional North European walls with insulation thicknesses of 220 mm are compared to walls with thicker thermal insulation and alternative designs. The influences of the different wall designs in relation to the risk of mold growth are compared using isopleth and a visual mold chart.

It has been found that there is a higher risk of moisture-related damage in thicker insulated walls. However, this risk could be reduced by choosing more suitable designs in which well-ventilated air gaps behind the cladding and exterior vapor-permeable moisture resistant thermal insulation boards are of great importance.

INTRODUCTION

Background

Interest in the use of wood frame constructions has increased as greater attention has been given to building more carbon dioxide-efficient (CO_2-efficient) houses (Dodoo et al. 2012). In Northern European countries there is also a tradition of building wooden houses and timber is readily available (Björk et al. 2009; Björk et al. 2003). Furthermore, increased awareness of climate changes and energy costs has made well-insulated (U-factor < 0.2 $W/m^2 \cdot K$) walls more common. However, thicker insulation increases the risk of high relative humidity (RH) levels in parts of the construction (Nevander and Elmarsson 1991; Hägerstedt 2012). Since wood and other organic materials has low mold growth resistance compared to other building materials (Nielsen et al. 2004), this property increases the risk of mold damage in well-insulated wood constructions.

State of the Art

Recent findings regarding important factors when building well-insulated houses that are not covered in this parametric study are briefly summarized below.

Besides the variations of mold growth sensitivity in different materials there is also general agreement among researchers around the world that temperature (T), relative humidity, and duration are the main factors affecting mold growth (Viitanen et al. 2010; Sedlbauer 2001; Clarke et al. 1999; Hens 1999). Several mold growth models have also been invented based on these findings. However, these models are limited and have a number of different theories and models regarding the reduction of the amount of mold growth during

***S. Olof Mundt-Petersen** is a PhD student, **Lars-Erik Harderup** is an associate professor, and **Jesper Arfvidsson** is a professor in the Department of Building Physics at Lund University, Lund, Sweden.*

non-favorable mold growth climate conditions. Furthermore, different critical levels with regard to the effects of duration are presented and none of the studied mold growth models include the influence of short time climate variations (Viitanden and Ojanen 2007; Sedlbauer 2001; Isaksson et al. 2010; Pierzyk et al. 2011; Cornick et al. 2003).

It is well-established that rendered undrained wood frame walls, without a drainage and ventilated air gap behind the cladding with outer EPS insulation, so-called ETICS or EIFS walls, are a risk design solution that cannot handle the influence of wind driven rain (WRD) penetrating the facade (Brown et al. 1997; Quirouette and Arch 1997; Künzel 2008). A great number of instances of moisture damage have also been found in Swedish wood frame houses with ETICS walls (Jansson 2011). The influence of a drainage and ventilated air gap behind the cladding (Straube et al. 2004; Piñon et al. 2004; Salonvaara et al. 2007) is further analyzed and discussed in the paper.

Recent Swedish field and laboratory studies showed that the risk of mold growth damage is highly dependent on the prevailing weather conditions during the on-site production phase until the building is weather-proof. Mold growth has been seen in wooden houses that have been exposed only once to rain during the on-site erection of the wooden building elements, even in cases where this has been completed in one day (Olsson et al. 2011; Olsson et al. 2012).

The need for an interior vapor barrier and its location in the building envelope in cold climates has been thoroughly investigated. Possible leakages can be found by blower-door tests. To further improve the moisture safety an exhaust ventilation can be used to prevent interior humid air from penetrating and damaging the wall by a high relative humidity (Vinha 2007; Hagentoft 2001). When an additional interior waterproof membrane is installed for wet rooms, it is important to ensure that the interior waterproof membrane, closest to the interior surface, has a high vapor resistance in relation to the ordinary vapor barrier (Jansson and Samuelson 2005).

Aim

The aim of this study is to give recommendations and show important factors that need to be taken into account during design process and construction in order to build moisture-safe well-insulated wooden houses without risk of sick building syndrome (SBS) and moisture damage. The intention is to present relevant factors by showing examples of different designs, evaluated by using one-dimensional calculations and the Folos 2D visual mold chart (Mundt-Petersen et al. 2012), as elements of the quantitative part of a moisture-safety design process (Hägerstedt and Harderup 2011b).

Limitations

This case study has been carried out as a parametric study and only gives general recommendations based on a southern Swedish standard climate. All calculations were made in one-dimension and the influence of the wooden studs and other thermal bridges as well as possible influence of convection was negligible. The study only considers traditional external wood frame walls and the designs have only been evaluated from a quantitative aspect where professional workmanship was assumed. The definition of RH_{crit} that was used was limited to show the conditions when mold growth on wood based materials is possible.

METHOD

The investigation was carried out as a parametric study with calculations made on differently designed wood frame constructions with the one-dimensional transient heat and moisture calculation tool WUFI 5.0. The materials used were retrieved from the WUFI 5.0 material data base (WUFI 2009a). The one-dimensional calculation tool has been previously blindly verified, using similar walls and material data, with good results (Mundt-Petersen and Harderup 2013).

A reference wood frame wall, designed as a traditional Swedish wall, was modeled. The wall and its boundary conditions are henceforth referred to as the *reference case*.

Initially, the most moisture-critical position (pos) in the wall was established in the reference case. The design and surrounding conditions of the reference case were then varied. The effect of the changes in the most moisture-critical positions were then analyzed and compared to the reference case. Possible measures in order to achieve moisture-safe wooden designs were also studied.

All the different conditions and investigated designs are described separately in connection with the results of each specific case. The variations of conditions and designs that are presented are based on personal experience, previous calculations, and possible changes that could be made in some of the construction systems used by Swedish timber house manufacturing companies (Hägerstedt 2012). Calculation results from the fifth year after construction were used in order to reach quasi stationary conditions and avoid construction moisture and other initial conditions influencing the calculations.

The Folos 2D visual mold chart and isopleth charts were used to evaluate and compare results in different designs, positions within the walls, and surrounding conditions to a moisture-critical limit regarding mold growth (Mundt-Petersen et al. 2012). The chart is briefly described in the analysis of the reference case. The moisture-critical limit used was the LIM I limit (Sedlbauer 2001).

RESULTS AND ANALYSIS

Reference Case and Most Moisture-Critical Position

The reference case intends to simulate a traditional Swedish wood frame wall with mineral wool insulation (λ = 0.037 W/mK). A horizontal cross-section and simplified one-dimensional calculation model are shown in Figure 1. Since a one-dimensional calculation tool was used, layers with mixed materials have been simplified. Wooden studs in the insulation layer and battens in the air gap behind the cladding were disregarded.

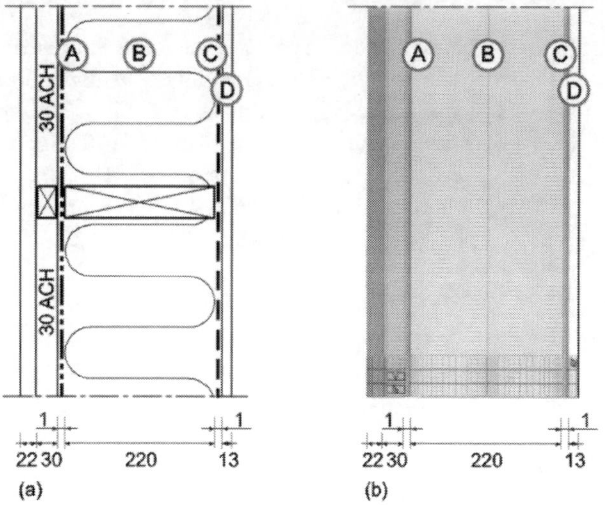

Wall from the outside:

22 mm Wood panel - Spruce radial[1,2]. Including paint[3], Sd = 1 m

30 mm Air gap[1], (including 30x45 mm battens in real wall) 30 ACH[4], 1 % driving rain behind the panel[5]

1 mm Weather resistive barrier[1], Sd = 0,2 m

220 mm Mineral wool[1, 6, 7], (including 220x45 mm studs in real wall)

1 mm Vapor barrier[1], Sd = 50 m

13 mm Gypsum board[1, 8]

(c)

Figure 1 *Reference case: (a) cross-sectional top view of an ordinary Swedish wood frame wall, (b) simplified one-dimensional calculation model, and (c) material specifications, where superscripts indicate the following references: 1. WUFI 2009a; 2. Vik 1996; 3. Nevander and Elmarsson 1994; 4. Hägerstedt and Harderup 2011a; 5. ASHRAE 2009; 6. IEA Annex 24 1996; 7. Paroc 2002; 8. Krus 1996.*

The most moisture-critical Swedish outdoor climate, from the city of Lund, was used during the entire study (Hägerstedt 2012). Lund is located in the far south of Sweden and had an average annual temperature of 9.2°C and average annual relative humidity of 81%. The climate data used showed a significantly higher driving rain load on the south-facing facade (450 mm/a) compared to the north-facing facade (100 mm/a) (WUFI 2009b).

The indoor climate was based on EN 13788 with an initial indoor climate of 20°C and an RH that varies dependent on exterior vapor content using moisture load level 2, (i.e., 4 g/m^3 is added to the exterior vapor content for temperatures below 0°C [SS-EN 2001; WUFI 2009b]).

The wall in the reference case was oriented towards the north and the influence of driving rain was taken into account by adding 1% of the driving rain load in the air gap behind the cladding (ASHRAE 2009). Other parameters and boundary conditions were set to simulate as real conditions as possible (Hägerstedt 2012).

Most Moisture-Critical Position. To find the most moisture-critical area in the wall, climate conditions in four different positions, A–D in Figure 1, were calculated and plotted in an isopleth chart with a moisture-critical limit, RH_{crit}, as shown in Figure 2. The RH_{crit} line is defined as the limit above which mold growth is possible and varies depending on the specific material. In this study the LIM I curve referring to biodegradable materials, such as wood, was chosen (Sedlbauer 2001).

The isopleth chart, in Figure 2, shows that the most moisture-critical conditions occur in position A, in the outer part of the wall. Further on, this study focuses on position A and the exterior part of the wall.

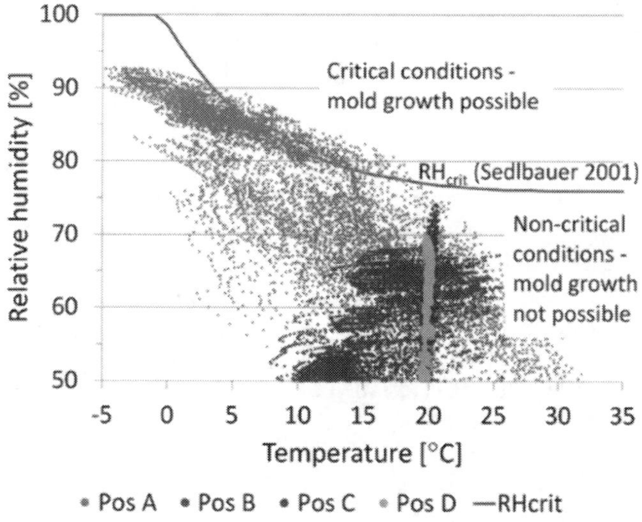

Figure 2 *Calculated climate conditions in position A (turquoise), B (brown), C (purple), and D (green) compared to RH_{crit} (red) (Sedlbauer 2001).*

It should be mentioned that the wooden studs shown in Figure 1 create thermal bridges that increase the temperature in the exterior parts. This reduces the RH and decreases the risk of mold growth in the wood. In this one-dimensional study the influence of the studs has been neglected, as shown in Figure 1, and the results in Figure 2 should be seen as a worst case situation. Two-dimensional studies of thermal bridges show that the temperature approximately increases with 0.5°C to 1°C, which reduces the relative humidity with 2.5% to 5% in the outer part of the stud (Forsberg 2011; Olsson 2011). Since the negative effect from thermal bridges on energy use is evident, the thermal bridges normally are eliminated by using lightweight studs or layers with separated studs, which make the one-dimensional worst case calculations appropriate.

Analyzing the Calculated Results Using the Folos 2D Visual Mold Chart. The Folos 2D visual mold chart, in Figure 3, shows the conditions in the most moisture-critical position A, based on information from the isopleth chart in Figure 2. The Folos 2D visual mold chart visualizes temperature (yellow) on the right *y*-axis and RH (turquoise), RH_{crit} (red), and the RH > RH_{crit} difference (light brown) on the left *y*-axis. The time presented on the *x*-axis indicates the conditions at any specific time, and of particular interest are the periods when RH > RH_{crit}. The hourly RH isopleth dots (turquoise) for one year (8760 h), as shown in Figure 2, resulted in a line when presented over time in the Folos 2D visual mold chart in Figure 3.

Critical conditions occur, and mold growth is possible, when the RH is above the RH_{crit} line. The RH_{crit} line is defined by the temperature that, at any specific time, exceeds the RH_{crit} limit as shown in Figure 2, (i.e., the chosen RH_{crit} line from Figure 2 is converted over time by using the actual temperature at each point in time). This means that critical conditions depend on the prevailing RH and temperature, where a high temperature gives a lower RH_{crit} line and vice versa. The RH_{crit} line in this study is based on the LIM I curve invented by Sedlbauer (2001). Based on the findings in the previous State of the Art subsection, it might take months or longer before mold growth occurs if the RH is only slightly above RH_{crit}, especially at low temperatures. Single short periods, even with RH high above RH_{crit}, do not cause damage since mold needs more than a few hours above the RH_{crit} limit to start to germinate.

Depending on the moisture resistance of different materials and code regulations in different countries, another mold growth limit could be used by choosing another appropriate RH_{crit} curve. A further description of the Folos 2D visual mold chart and how it can be used can be found in a separate paper (Mundt-Petersen et al. 2012).

The most moisture-critical conditions occur in the outer part of the wall, close to the air gap, particularly in September, October, and November. The critical conditions also depend on the moisture sensitivity of the material in this position and thermal bridges may decrease the risk of mold growth.

To limit the RH, and the critical conditions, there are generally two types of measures that can be taken: measures to increase the temperature and measures to reduce the amount of vapor content, as shown further on in the study.

Protection Against Mold Growth by Exterior Mold-Resistant Insulation Boards

By attaching exterior vapor permeable insulation boards to the outside of the wooden studs, as shown in Figure 4 and 5, the surrounding temperature on the exterior side of the studs will increase (Maref et al. 2011). The RH and the risk of mold growth on the studs will then decrease (Hägerstedt and Harderup 2011a, 2011b). Exterior insulation boards must be located on the outside of the wood studs, between the studs and the weather resistive barrier, and must be made of moisture-resistant materials so they are not damaged by the high RH that occurs in position A.

The influence of increased total insulation thickness on the conditions in position A was studied. Furthermore, the required minimum thickness of the exterior insulation board

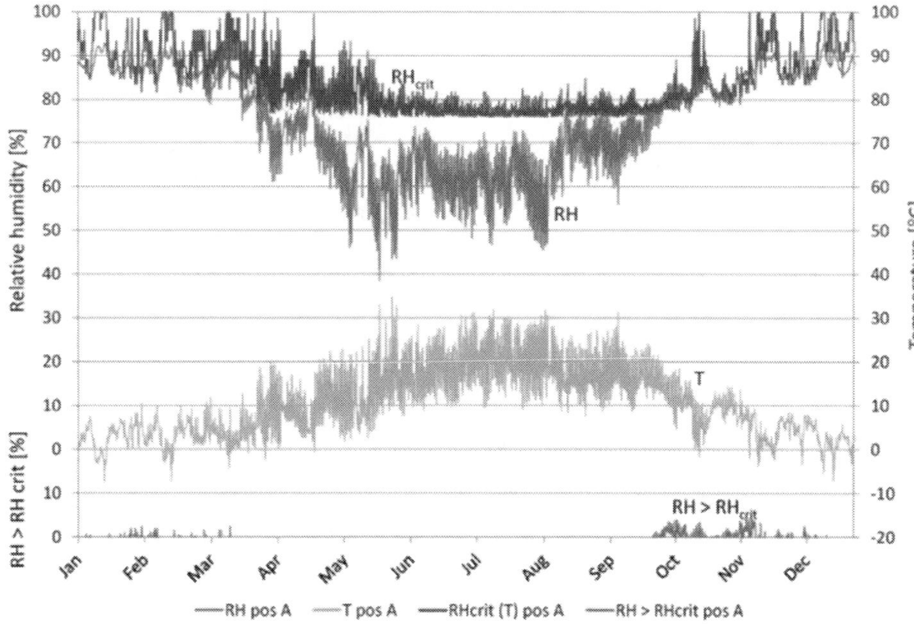

Figure 3 *RH compared to RH_{crit} in position A. RH (turquoise), RH_{crit} (red), temperature (yellow), RH > RH_{crit} (light brown).*

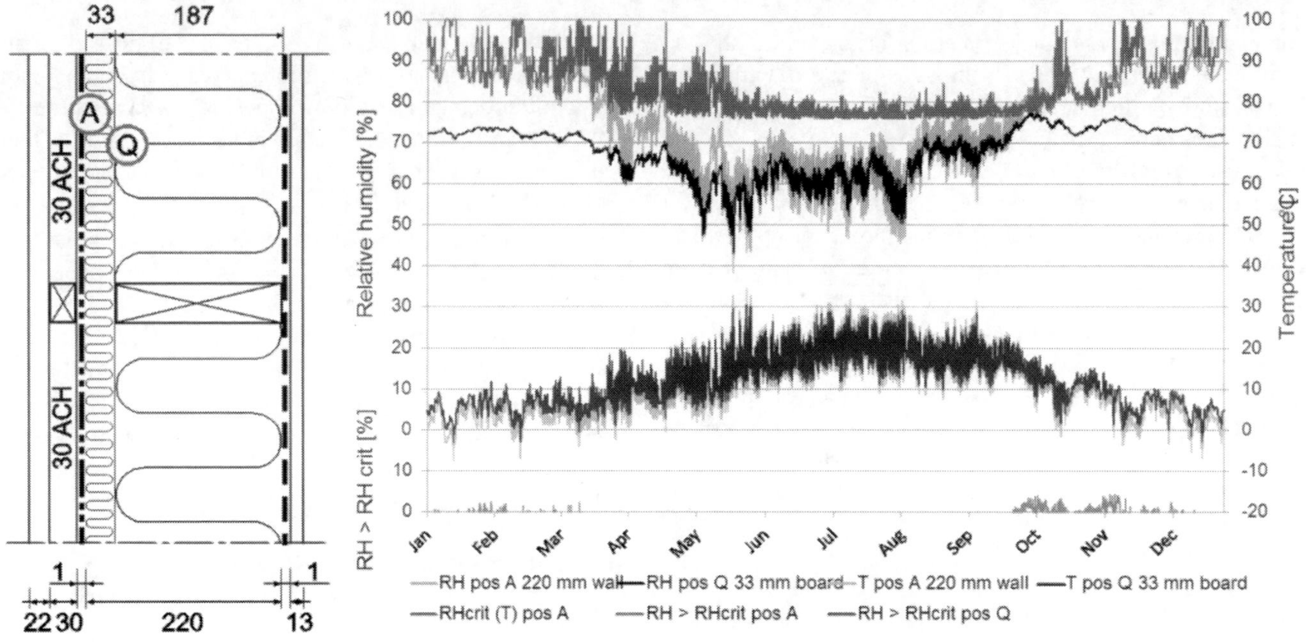

Figure 4 *RH in position A (turquoise) and position Q (black) compared to RH_{crit} for a wall with a total insulation thickness of 220 mm, as shown in the cross sectional top view drawing. Temperatures in position A (yellow) and temperatures in position Q (dark blue). RH_{crit} is dependent on T in position A (red), RH > RH_{crit} in position A (light brown), RH > RH_{crit} in position Q (purple) is constant below the critical limit and not shown in the figure.*

on the outside of the wooden studs, in order to avoid mold growth on the studs, was also studied. The minimum thickness of the exterior insulation board was calculated in relation to the total insulation thickness and was determined through iterative calculations. Position Q was located between the exterior insulation board and the mixed layer of studs and insulation. The thermal conductivity (λ = 0.037 W/m·K) was the same in both the exterior insulation board and the layer with mixed insulation and studs.

RH in Positions A and Q with a Total Insulation Thickness of 220 mm and 33 mm Exterior Insulation Board. The required minimum thickness of the exterior insulation board for a wall with a total of 220 mm insulation, as shown in Figure 4, was studied. Calculation results from both position A and position Q are shown together in the same chart.

By iteration it was found that the exterior insulation board needs to be 33 mm thick in order to avoid critical conditions in position Q if the wall has a total insulation thickness of 220 mm. A similar situation to such iteration in a real case study was made by Hägerstedt and Harderup (2011b). Critical conditions in position A are equal to those in the reference case. RH_{crit} in position Q is not shown when the two cases are compared in order to limit the number of plots.

RH in Positions A and Q with a Total Insulation Thickness of 420 mm and 52 mm Exterior Insulation Board. In thicker insulated walls the conditions in position A change and there is a need for a thicker exterior insulation board. The conditions in position A and required minimum thickness of the exterior insulation board for a wall with a total of 420 mm insulation, as shown in Figure 5, were studied. Calculation results from both position A and position Q are shown together in the same chart.

By iteration it was found that the exterior insulation board needs to be at least 52 mm thick in order to avoid critical conditions in position Q if the wall has a total insulation thickness of 420 mm. By comparing position A in Figures 4 and 5 it can be stated that the occurrence of critical conditions increased in the wall with thicker insulation, which also was shown in Table 1.

Summarized results from several calculations show that higher thermal resistance increases the moisture conditions (RH) in position A. Walls with a higher thermal resistance also require a thicker exterior insulation board to avoid critical conditions in position Q, see Table 1.

Calculations when using cellulose fiber insulation and an exterior moisture-resistant mineral insulation board generally show similar results to those presented in Table 1 (Hägerstedt 2012).

Well-Ventilated Air Gap Behind the Cladding

Several of studies discuss the importance of a well-ventilated air gap behind the cladding in wood frame walls. A high airflow in the air gap will vent out moisture that has penetrated the cladding and will contribute to an increased dry out of a wood frame wall (Salonvaara et al. 2007; Straube et al. 2004;

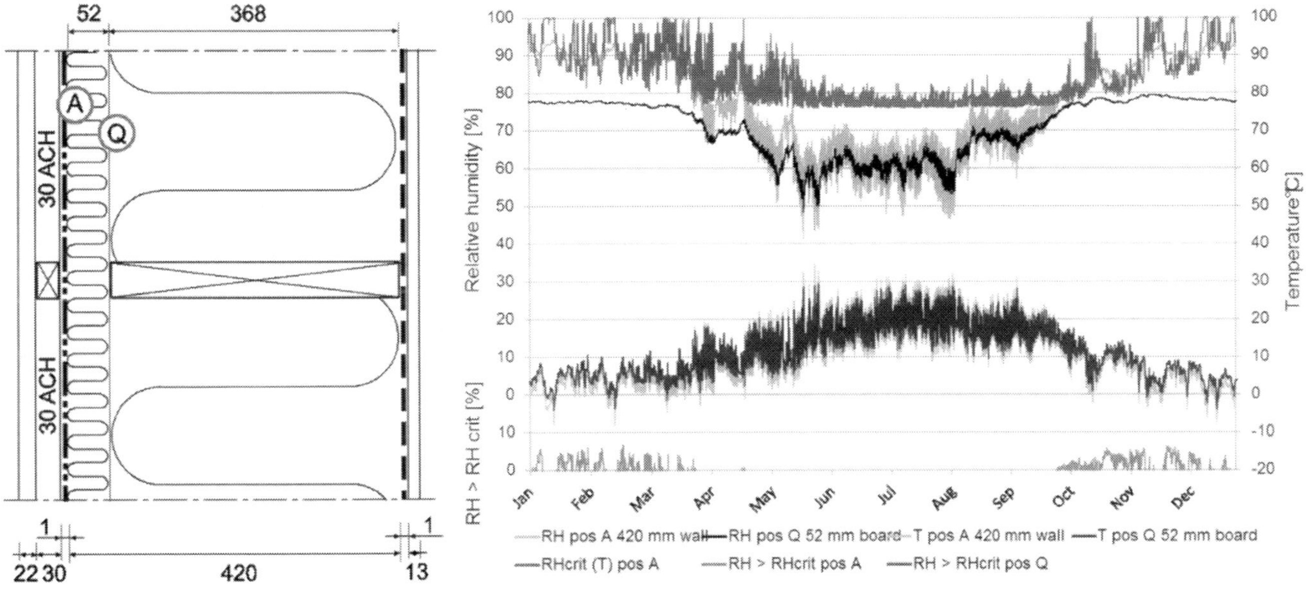

Figure 5 *RH in position A (turquoise) and position Q (black) compared to RH_{crit} for a wall with a total insulation thickness of 420 mm, as shown in the cross sectional top view drawing. Temperatures in position A (yellow) and temperatures in position Q (dark blue). RH_{crit} is dependent on temperature (T) in position A (red); $RH > RH_{crit}$ in position A (light brown); $RH > RH_{crit}$ in position Q (purple) is constant below the critical limit and not shown in the figure.*

Table 1. Minimum Thicknesses of Exterior Insulation Board and the Highest RH above RH_{crit} in Position A Depending on the Total Insulation Thickness of the Wall

Total Insulation Thickness, mm	Thickness of Exterior Insulation Board, mm $RH < RH_{crit}$ Position Q	Highest $RH > RH_{crit}$ (%) in position A Inside the Weather Resistive Barrier
220	33	4.21
270	39	4.61
320	45	5.15
370	49	6.15
420	52	6.71
470	55	7.04
520	59	7.52

Piñon et al. 2004; Hägerstedt and Harderup 2010). The influence on critical conditions in position A was also studied when there was a lower airflow in the air gap behind the cladding than in the reference case. Calculation results from position A with 30 ach compared to 1 ach are shown together in the same chart in Figure 6.

A low airflow in the air gap behind the cladding has a negative influence and the critical conditions occur more frequently and during other periods of the year in position A. In a well-ventilated air gap with vertical battens and wide openings at the bottom and top there is normally a higher air change rate than 30 ach (Falk 2013; Tichy and Murray 2007). Normally, a ventilation rate more than 30 ach does not further improve the conditions in the wall since there is no more moisture available to dry out (Hägerstedt and Harderup 2010).

The Need for Well-Ventilated Air Gaps in Well-Insulated Walls. Differences in the critical conditions in positions A and Q were studied in walls with increased insulation in combination with a low and high air change rate in the air gap. Four cases, as shown in Figure 7, were compared in isopleth charts for positions A and Q: Two cases with 1 or 30 ach in the air gap and a total insulation thickness of 220 mm and two cases with 1 and 30 ach in the air gap and total insulation thickness of 420 mm.

As previously observed, comparing the cases with 1 and 30 ach in the 220 mm insulated wall showed that a low airflow in the air gap behind the cladding had a negative influence on

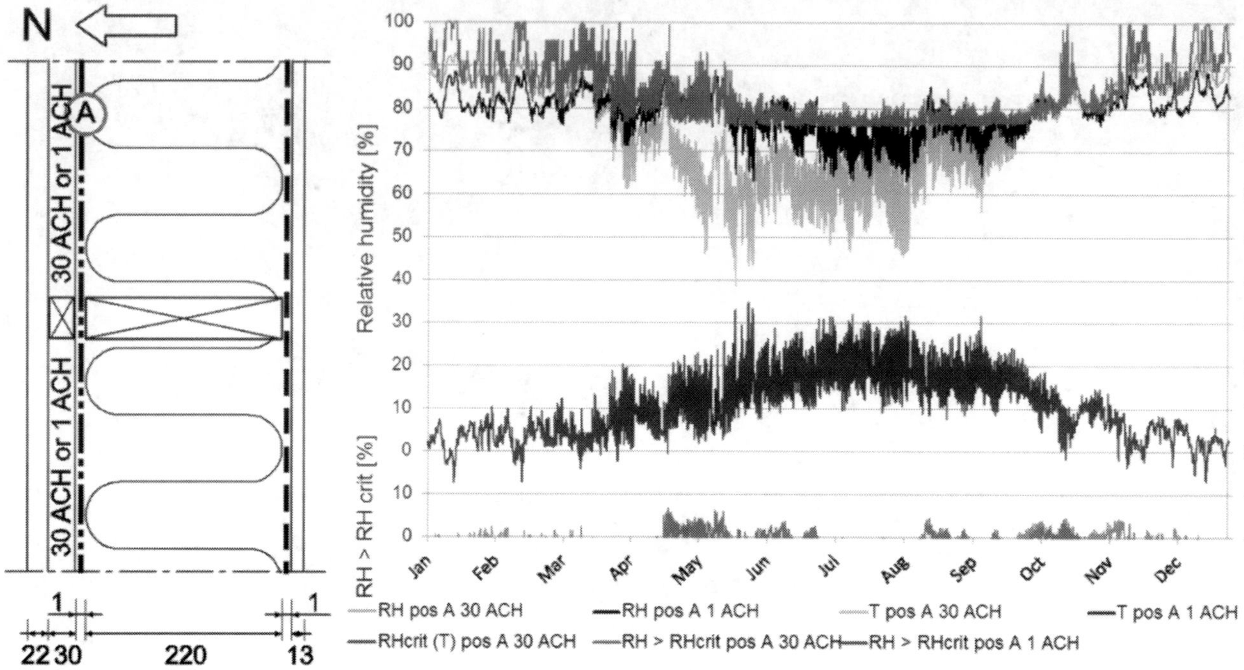

Figure 6 *RH in position A compared to* RH_{crit} *in a wall with 220 mm insulation, as shown in the cross sectional top view drawing, oriented towards the north with an air change rate of 30 ach (turquoise) and 1 ach (black) in the air gap. The temperature at 30 ach (yellow) is hidden behind the temperature at 1 ach (dark blue).* RH_{crit} *(red) depends on temperature (T) for 30 ach, RH >* RH_{crit} *30 ach (light brown), RH >* RH_{crit} *1 ach (purple).*

the critical conditions in position A. The low airflow of 1 ach also has a negative influence in position Q. The isopleth charts show that the critical conditions in position A and Q increased with thicker insulation. However, comparing positions A and Q at 1 and 30 ach in the 220 mm and the 420 mm insulated wall showed that there was a higher negative influence on the critical conditions in the wall with 420 mm insulation. I.e., a low airflow in the air gap had a higher negative influence when the walls were well-insulated. Results in position Q also showed that the critical conditions in the wall could be handled by a high air change rate in the air gap and an exterior moisture resistant insulation board.

Influence of Driving Rain. Driving rain is caused by rain and wind acting at the same time. A driving rain load varies depending on the exterior climate conditions and the facade orientation. As mentioned above, the climate data used showed a significantly higher driving rain load on the south-facing facade (450 mm/a) compared to the north-facing facade (100 mm/a) (WUFI 2009b). The influence of a higher amount of driving rain on a facade oriented towards the south and different airflows of 1 or 30 ach in the air gap, as shown in Figure 8, were studied in position A. The results in Figure 8 can be compared to those for the same wall oriented towards the north, as shown in Figure 6.

If there is a low air change rate, the negative impact of driving rain in position A is significantly higher, as shown in Figure 8, when compared to cases with a lower driving rain load, as shown in Figure 6. A sufficiently high airflow can also handle a high amount of penetrating driving rain. It is therefore very important to establish a sufficient airflow in the air gap behind the cladding if there is a high amount of driving rain. Comparing Figure 6 and 8 also shows that the colder north directed walls have slightly higher moisture critical conditions.

Influence of Different Cladding Materials. The ability to store moisture in different cladding materials also affects the risk of mold damage. Critical conditions in position A in a wood frame wall with an exterior brick facade (solid brick masonry, including mortar joints) (WUFI 2009a; IBP 1994) with a high moisture storage capacity, as shown in Figure 9, were therefore studied. The results in Figure 9 can also be compared to the same type of frame with a wood facade as shown in Figure 6.

The results confirm pervious results by Straube and Finch (2009). The critical conditions in position A are higher, longer, and also occur during other periods when a brick facade, as shown in Figure 9, was compared to the reference case. A low air change rate in the air gap behind the brick facade in Figure 9 created significantly more critical conditions compared to the case with a wood facade, as shown in Figure 6. The increase in occurrence of critical conditions was due to moisture storage capacity in the brick facade. However, a well-ventilated air gap resulted in significantly better conditions in position A when there was a brick facade with a high moisture storage capacity.

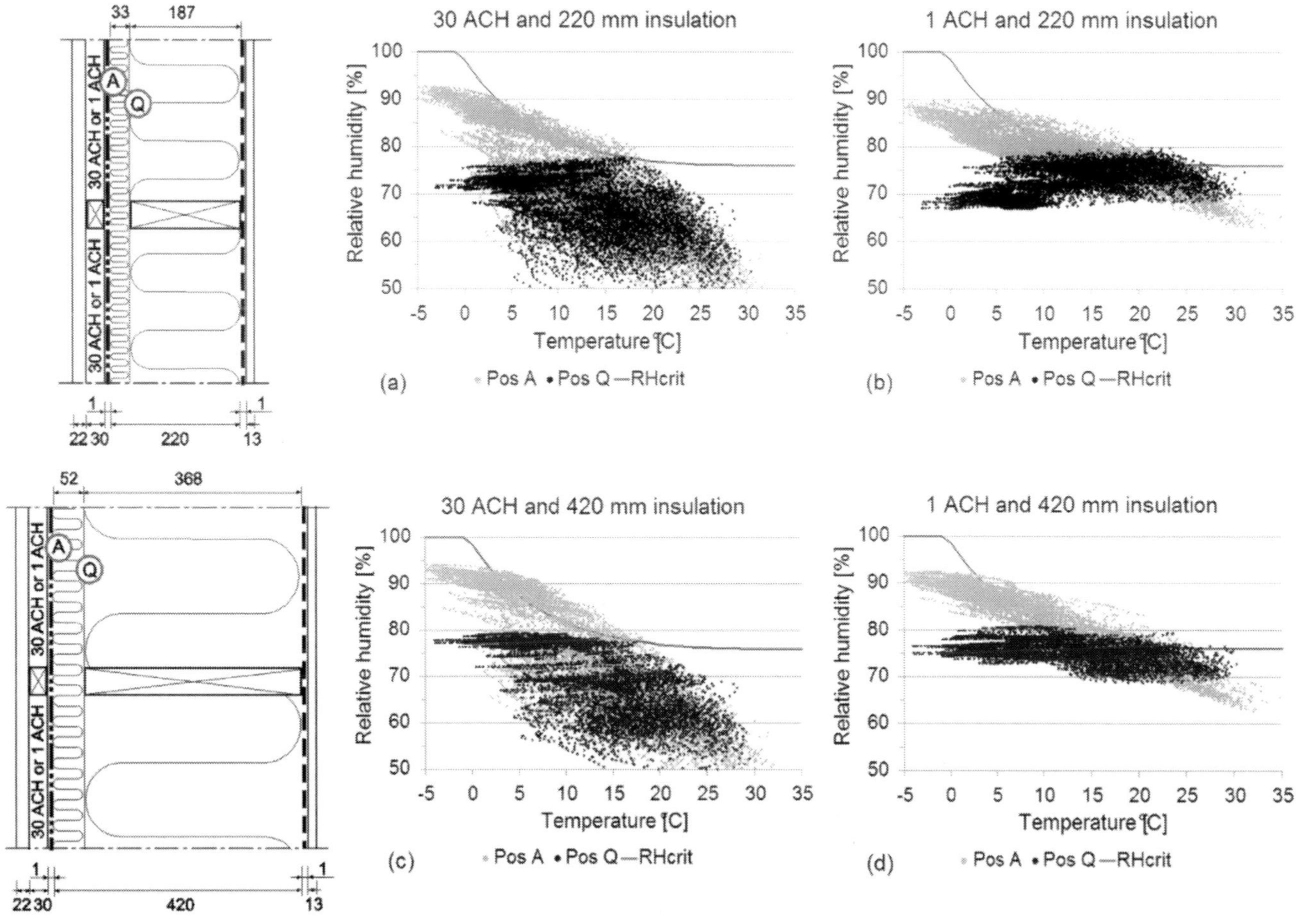

Figure 7 *RH in positions A and Q compared to RH_{crit} (red) for two different insulation thicknesses and two different air change rates in the air gap behind the cladding, as shown in the cross sectional top view drawings. (a) RH 220 mm insulation and 30 ach, position A (turquoise) and position Q (black). (b) RH 220 mm insulation and 1 ach, position A (turquoise) and position Q (black). (c) RH 420 mm insulation and 30 ach, position A (yellow) and position Q (dark blue). (d) RH 420 mm insulation and 1 ach, position A (yellow) and position Q (dark blue).*

Possibilities for Drying out Water Caused by Leakages and Initial Moisture from the Construction Phase

Leakages into walls from rain or driving rain are probably more common than expected. Furthermore, it must also be possible to dry out initial construction moisture. In order to estimate the drying out potential in different constructions, the influence of a leakage of 1% of the driving rain (ASHRAE 2009) located 152 mm from the air gap, close to the middle of the construction, was studied. Four cases in positions A and Q, as shown in Figure 10, were compared in isopleth charts.

Two of the designs had insulation thicknesses of 220 and 420 mm insulation respectively. Two designs had an exterior mold-resistant insulation board of vapor-permeable mineral wool (λ = 0.037 W/mK). The other two designs had exterior mold-resistant insulation of rather vapor-tight expanded polystyrene (EPS) and insulation board with the thermal conductivity 0.037 W/mK and a density of 30 kg/m^3 (WUFI 2009a; Achtziger and Cammerer 1984).

All cases with leakages in Figure 10 showed a higher RH in positions A and Q compared to situations without leakages, as shown in Figure 7 (a) and (c). However, there is a significant difference in high RH in position Q between the cases with vapor-permeable mineral wool and an EPS vapor-tight exterior insulation board. In the case with a total thicker insulation of 420 mm, there is also a significantly higher RH in position Q compared to the case with thinner insulation of 220 mm.

To handle initial construction moisture and possible leakages, the exterior mold-resistant board must be vapor-permeable.

CONCLUSIONS

The study assessed a number of important factors that must be taken into account if moisture-safe well-insulation

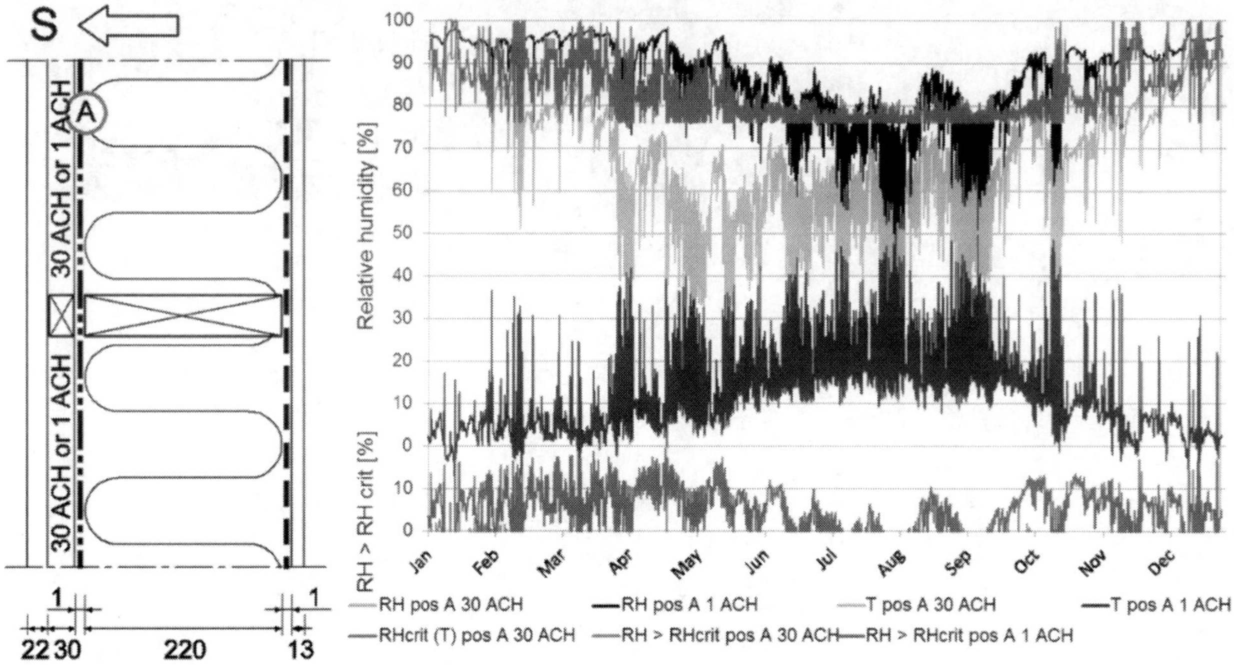

Figure 8 *RH in position A in a wall with 220 mm insulation, as shown in the cross sectional top view drawing, oriented towards the south with an air change rate of 30 ach (turquoise) and 1 ach (black) in the air gap compared to* RH_{crit}*. The temperature at 30 ach (yellow) is hidden behind the temperature at 1 ach (dark blue).* RH_{crit} *(red) depends on temperature (T) for 30 ach,* $RH > RH_{crit}$ *30 ach (light brown),* $RH > RH_{crit}$ *1 ach (purple).*

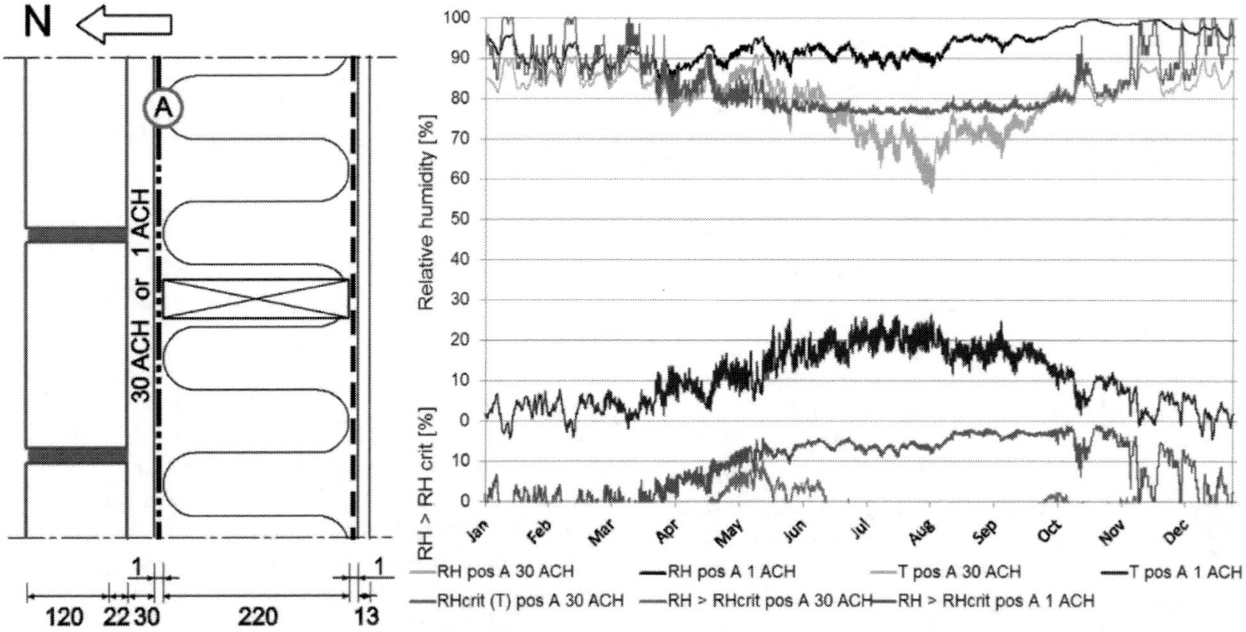

Figure 9 *RH in position A in a wall with 220 mm insulation, as shown in the cross sectional top view drawing, a north oriented brick façade and air change rate of 30 ach (turquoise) and 1 ach (black) in the air gap compared to* RH_{crit}*. The temperature at 30 ach (yellow) is hidden behind the temperature at 1 ach (dark blue).* RH_{crit} *(red) depends on temperature (T) for 30 ach,* $RH > RH_{crit}$ *30 ach (light brown),* $RH > RH_{crit}$ *1 ach (purple).*

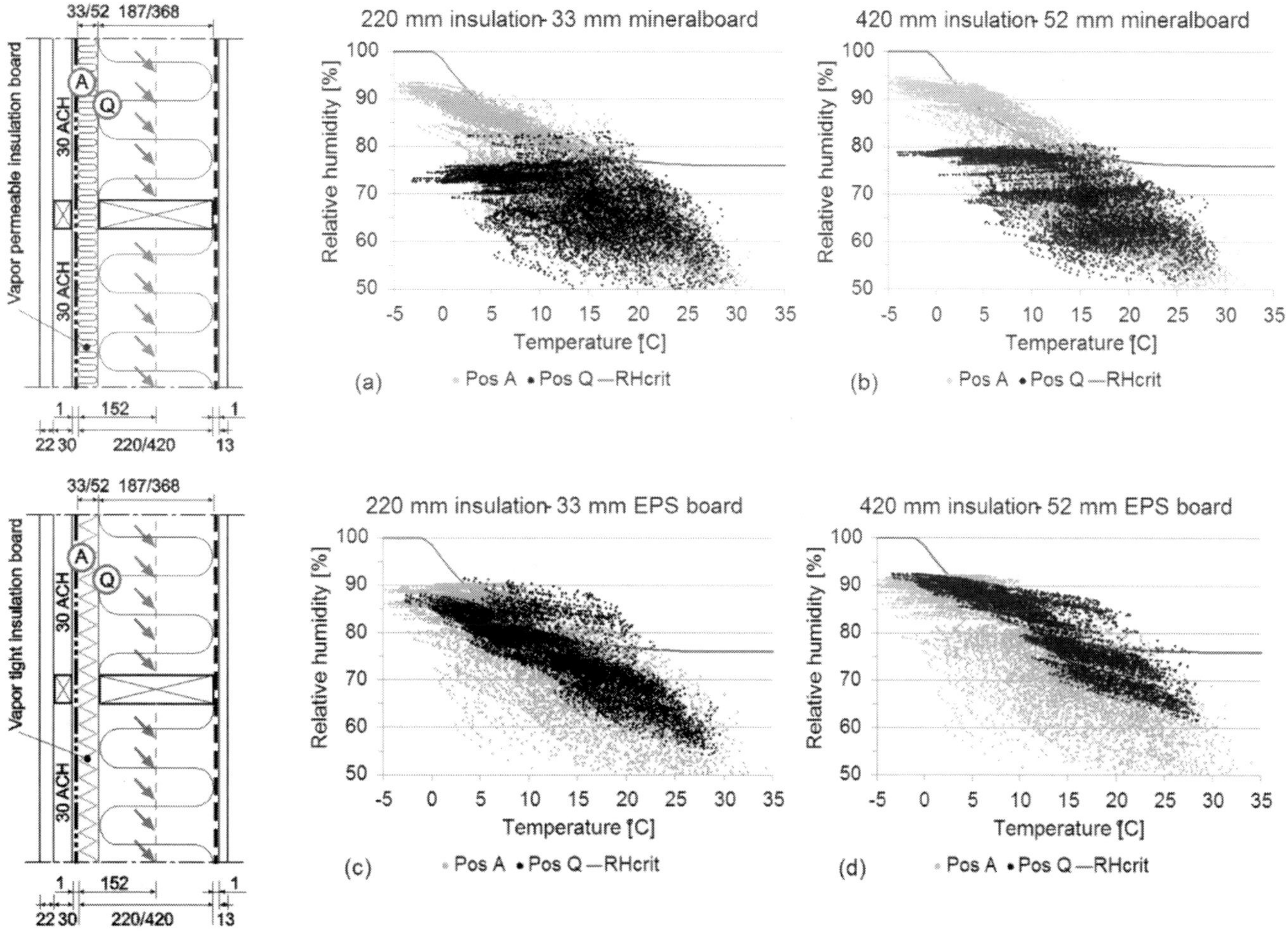

Figure 10 *RH in positions A and Q compared to* RH_{crit} *(red) for walls subject to a leakage with two different materials used for the exterior insulation board and two different insulation thicknesses, as shown in the cross sectional top view drawings. (a) RH 220 mm–33 mm mineral insulation board, position A (turquoise) and position Q (black). (b) RH 420 mm–52 mm mineral insulation board, position A (yellow), and position Q (dark blue). (c) RH 220 mm–33 mm EPS insulation board, position A (turquoise), and position Q (black). (d) RH 420 mm–52 mm EPS insulation board, position A (yellow), and position Q (dark blue).*

walls are to be built. The results were based on a great number of hygrothermal calculations. The main conclusion was that there is a higher risk of moisture damage in well-insulation wood frame walls. However, this risk can be reduced by implementing suitable designs in which a well-ventilated air gap behind the cladding and the addition of a vapor-permeable exterior moisture-resistant insulation board with high density are of great importance. Further important factors are listed below:

1. Moisture-critical conditions in wood frame walls mainly occur in the exterior part of the wall, based on hygrothermal modeling and Northern European climate used in this study.
2. The occurrences of moisture-critical conditions increase as the thickness of thermal insulation in walls is increased.
3. Studs and other organic materials in the exterior part of a wood frame wall could be protected from moisture-critical conditions by using an exterior moisture-resistant thermal insulation board on the outside of the studs.
   a. The required thickness of the exterior moisture-resistant thermal insulation board varies depending on the total thermal insulation of the entire wall.
4. There is a great need for a well-ventilated air gap behind the cladding in order to build moisture-safe well-insulated walls.
   a. The need increases with higher thermal resistance of the wall.

b. It is especially important when there are high amounts of driving rain
c. It is especially important when the facade material has a high moisture storage capacity, such as a brick facade.

5. Exterior moisture-resistant insulation boards must be vapor permeable to allow initial construction moisture and water from possible leakages to dry out.

ACKNOWLEDGMENTS

This study was supported by Vinnova (the Swedish Government Agency for Innovation Systems) and their partnership, with representatives of the timber industry, which initiated the research projects "Framtidens trähus" (Wood frame buildings of the future) and "ECO2 – Carbon-efficient timber constructions."

NOMENCLATURE

RH_{crit}	=	limit for possible mold growth in wooden materials
$RH > RH_{crit}$	=	critical conditions showing that mold growth is possible
T	=	Temperature
Pos	=	Position

REFERENCES

Achtziger, J. and Cammerer, J. 1984. Influence of moisture content on the thermal conductivity of building and insulation materials. Report BI 5-800883-4. Institute for heat protection, München, Germany. In German.

ASHRAE. 2009. BSR/ ASHRAE Standard 160P, *Criteria for Moisture-Control Design Analysis in Buildings*. Atlanta: ASHRAE.

BBR. 2011. Swedish Building Regulations 2012. Karlskrona: The Swedish National Board of Housing, Building and Planning. In Swedish.

Björk, C., Kallstenius, P., and Reppen, L. 2003. Old house design 1880–2000, 5th Ed. Stockholm: Edita Ljunglöfs. In Swedish.

Björk, C., Nordling, L., and Reppen, L. 2009. Single family house design—Swedish single family house architecture 1890–2010. Värnamo: Fälth & Hässler. In Swedish.

Brown, W., Ullett, J., Karagiozis, A., and Tonyan, T. 1997. Barrier EIFS clad walls: Results from a moisture engineering study *Journal of Building Physics* 20(3):206–26.

Clarke, J.A., Johnstone, C.M., Kelly, N.J., Mclean, R.C., Anderson, J.A., and Rowan N.J. 1999. A technique for prediction of the conditions lading to mould growth in buildings, *Building and Environment* 34(4):515–521.

Dodoo, A., Gustavsson, L., and Sathre, R. 2012. Effect of thermal mass on life cycle primary energy balance of a concrete- and wood-frame building. *Applied Energy* 92:462–72.

Falk, J., and Sandin, K. 2013. Ventilated rainscreen cladding: Measurements of cavity air velocities, estimation of air change rates and evaluation of driving forces. *Building and Environment* 59:164–76.

Forsberg, T. 2011, Moisture rearrangements in exterior walls. Bachelor thesis. Report 2011:14. KTH Royal Institute of Technology. In Swedish.

Hagentoft, C. E. 2001. *Introduction to Building Physics*. 1st Ed. Göteborg. Studentlitteratur AB.

Hägerstedt, S.O. 2012. Moisture safe wood constructions - guidelines for wall design. Report TVBH-3052. Lund University. In Swedish.

Hägerstedt, S.O., and Harderup, L. E. 2010. Importance of a proper applied airflow in the façade air gap when moisture and temperature are calculated in wood framed walls. Proceedings, 5th International BUILDAIR-Symposium "Building and Ductwork Air-tightness" Copenhagen, Denmark.

Hägerstedt, S. O., and Harderup, L. E. 2011a. Comparison of measured and calculated temperature and relative humidity with varied and constant air flow in the façade air gap. Proceedings of the 9th Nordic Symposium on Building Physics - NSB 2011, Tampere, Finland, 1:147–54.

Hägerstedt, S. O., and Harderup, L. E. 2011b. Control of moisture safety design by comparison between calculations and measurements in passive house walls made of wood. Proceedings of the 12th International Conference on Durability of Building Materials and Components - XII DBMC, Porto, Portugal, 1:25–32.

Hens, H.L.S.C. 1999. Fungal defacement in buildings: A performance related approach. *International Journal of Heating, International Journal of HVAC & R Research Ventilation, Air-Conditioning and Refrigerating Research* 5(3):256–280.

IBP. 1994. IBP-Report 35. In German.

IEA Annex 24. 1996. Heat, air and moisture transport in insulated envelope parts, Material properties. International Energy Agency, E*Nergy Conservation in Building Community System*. 3:3.

Isaksson, T., Thelandersson, S., Ekstrand-Tobin, A., and Johansson, P. 2010. Critical conditions for onset of mould growth under varying climate conditions. *Building and Environment* 45(7):1712–21.

Jansson, A. 2011. External plaster stud walls 2011 - Experience from studies that SP has performed. Report

2011:61. SP Technical Research Institute of Sweden. In Swedish.

Jansson, A., and Samuelson, I. 2005. Double sealing layers in the external walls of wet rooms with internal tile cladding. Report 2005:20. SP Technical Research Institute of Sweden. In Swedish.

Krus, M. 1996. Moisture transport and storage coefficients of porous mineral building materials - Theoretical principles and new test methods. Doctoral thesis. Fraunhofer Institute for Building Physics. University of Stuttgart.

Künzel, H. M. 2008. Influence of rain water leakage on the hygrothermal performance of exterior insulation system. Proceedings of the 8th Nordic Symposium on Building Physics - NSB 2008, Copenhagen, Denmark, 1:253–60.

Maref, W., Rousseau, M. Z., Armstrong, M. M., Lee, W., Leroux, M., and Nicholls, M. 2011. Evaluating the Effects of Two Energy Retrofit Strategies for Housing on the Wetting and Drying Potential of Wall Assemblies: Summary Report for Year 2007–08 Phase of the Study. RR-315. NRC-CNRC Institute for research in construction. National research council Canada.

Mundt-Petersen, S.O. 2012. Literature study / State-of-the-art - Mould and moisture safety in constructions. Report TVBH-3053. Lund University, Sweden.

Mundt-Petersen, S.O. and Harderup L. E. 2013. Validation of a 1D transient heat and moisture calculation tool under real conditions. Proceedings - Thermal Performance of the Exterior Envelopes of Whole Buildings XII - International Conference, Florida, USA.

Mundt-Petersen, S.O., Wallentén, P., Toratti, T., and Heikkinen, J. 2012. Moisture risk evaluation and determination of required measures to avoid mould damage using the Folos 2D visual mould chart. Thermophysics 2012 - Conference proceedings, 17th International Meeting of Thermophysical Society, Podkylava, Slovakia 134–41.

Nevander, L.E., and Elmarsson, B. 1991. Moisture safety design of timber constructions. Report R38:1991, Lund University, In Swedish.

Nevander, L.E., and Elmarsson, B. 2007. *Moisture Handbook—Practice And Theory*, 3rd Ed. Chapter 92.2. Mölnycke: AB Svensk byggtjänst. Elanders Infologistics Väst AB. In Swedish.

Nielsen, K. F., Holm, G., Uttrup, L. P., and Nielsen P. A. 2004. Mould growth on building materials under low water activities - Influence of humidity and temperature on fungal growth and secondary metabolism. *International Biodeterioration & Biodegradation* 54(4):325–36.

Olsson, L. 2011. Laboratory investigation of timber frame walls with various weather barriers. Report 2011:56. SP Technical Research Institute of Sweden. In Swedish.

Olsson, L., Mjörnell, K. 2012. Laboratory investigation of sills and studs exposed to rain. Proceedings of the 5th International Building Physics Conference - IBPC 2012, Kyoto, Japan, 155–61.

Olsson, L., Mjörnell, K., and Johansson, P. 2011. Moisture and mould in prefabricated timber framed constructions during production until enclosure of the house. Proceedings of the 9th Nordic Symposium on Building Physics - NSB 2011, Tampere, Finland, 1:339–46.

Paroc. 2002. Construction paper - Material information. Skövde: Paroc Ltd. www.stenull.paroc.se/produktdat/pdf_down/Pinfo.pdf. In Swedish.

Pietrzyk, K., Samuelson, I., and Johansson, P. 2011. Modelling reliability of structure with respect to incipient mould growth. Proceedings of the 9th Nordic Symposium on Building Physics - NSB 2011, Tampere, Finland, 2:875–82.

Piñon J.P., Burnett, E.F.P., Davidovic, D., and Srebric, J. 2004. The airflow characteristics of ventilated cavities in screen-type enclosure wall systems (RP-1091). Proceedings - Thermal Performance of the Exterior Envelopes of Whole Buildings X - International Conference, Florida, USA.

Quirouette, R., and Arch, B. 1997. Review of the NRCC and USG study of the EIFS barrier walls for wood framed houses in Wilmington, North Carolina. *Journal of Building Physics* 20(4):305–58.

Salonvaara, M., Karagiozis, A. N., Pazera, M., and Miller, W. 2007. Air cavities behind claddings - what have we learned? Proceedings - Thermal Performance of the Exterior Envelopes of Whole Buildings X - International Conference, Florida, USA. XI:A.

Sedlbauer, K. 2001. Prediction of mould fungus formation on the surface of and inside building components. Doctoral thesis. Fraunhofer Institute for Building Physics. University of Stuttgart.

SS-EN. 2001. SS-EN 13788:2001 - *Hygrothermal performance of building components and elements - Internal surface temperature to avoid critical surface humidity and interstitial condensation - Calculation methods*. Swedish Standards Institute.

Straube, J., and Finch, G. 2009. Ventilated wall claddings: Review, field performance, and hygrothermal modeling. Research report - 0906. Building science press.

Straube, J.F., van Straaten, R., and Burnett E. 2004. Field studies of ventilation drying. Proceedings - Thermal Performance of the Exterior Envelopes of Whole Buildings X - International Conference, Florida, USA.

Tichy, R., and Murray, C. 2006. Developing innovative wall systems that improve hygrothermal performance of residential buildings. Technical report FC26-02NT41498. Washington State University.

Viitanen, H., and Ojanen, T. 2007. Improved model to predict mold growth in building materials. Proceedings - Thermal Performance of the Exterior Envelopes of Whole Buildings X - International Conference, Florida, US. V:A.

Viitanen, H., Vinha, J., Salminen, K., Ojanen, T., Peuhkuri R., Paajanen, L., and Lähdesmäki, K. 2010. Moisture and biodeterioration risk of building materials and structures. *Journal of Building Physics* 33(3):201–24.

Vik, T.A. 1996. Determination of suitable wet properties of wood and calculation the moisture balance using determined characteristic values. Diploma thesis. Norwegian University of Science and Technology. In German.

Vinha, J. 2007. Hygrothermal performance of timber-framed external walls in Finnish climatic conditions: A method for determining the sufficient water vapour resistance of the interior lining of a wall assembly. Doctoral thesis publication 658. Tampere University of technology.

WUFI. 2009a. WUFI 5.0 Material data base - Generic materials and Fraunhofer-IBP - Holzkirchen, Germany. Release: 5.0.1.521.DB.24.70. Fraunhofer Institute for Building Physics, Holzkirchen, Germany.

WUFI. 2009b. WUFI 5.0 Climate data base - Lund, Sweden, Europe. Release: 5.0.1.521.DB.24.70. Fraunhofer Institute for Building Physics, Holzkirchen, Germany.

Validation of a One-Dimensional Transient Heat and Moisture Calculation Tool under Real Conditions

S. Olof Mundt-Petersen **Lars-Erik Harderup, PhD**

ABSTRACT

Stricter Swedish building regulations require a moisture-safety design process to be carried out before houses are built. In order to predict moisture-critical conditions, a properly verified, user-friendly and reliable calculation tool that could be used in this design phase, is required.

This paper initially presents a blind method that could be used in order to verify heat and moisture calculation tools in a reliable manner. Furthermore, general results and findings from blind validations using a transient heat and moisture calculation tool are summarized and presented. The comparisons include measurements and calculations of temperature and relative humidity and were carried out in northern European climates.

In general, the results show a good correlation between measured and blindly-calculated values. Comparisons show that the studied tool can be used during the design phase to predict moisture risks. However, factors such as the influence of impaired-temperature measurements on relative humidity have to be taken into account. There is also a need for outdoor climate-boundary conditions that take into account critical periods. Measurements and calculations also established that the most moisture-critical conditions in general occurred in the exterior part of the frame, behind the air gap.

INTRODUCTION

Background

Several studies show that mold and moisture-related damage is linked to high costs. As much as 30% of single-family houses and 15% of other buildings in Sweden have moisture-related damage (Boverket 2009). Individuals, as well as insurance companies and the national economy, are affected by these costs.

At the same time, the risk of moisture-related damage generally increases in new houses. The awareness of climate changes, increased energy costs, and new energy demands have resulted in new houses being more well-insulated (BBR 2011). Besides the positive effect of reduced energy needs, thicker insulation also results in a building envelope in which critical parts more often become exposed to higher levels of relative humidity (RH). Higher relative humidity leads to the increased probability of occurrences of mold growth (Nevander and Elmarsson 1991; Hägerstedt and Arfvidsson 2010; Hägerstedt 2012).

As a consequence, recent Swedish building regulations have stipulated stricter requirements when predicting the risk of mold and moisture damage in order to reduce this kind of problem. Moisture conditions that create odors, unhealthy indoor climates, and mold growth that affect the health of a building's occupants are forbidden. Furthermore, it is strictly recommended that these factors should be taken into account and verified before a house is built (BBR 2011).

To minimize the risk of moisture damage, a reliable and validated user-friendly moisture calculation tool is needed to predict the hygrothermal performance. A tool that could be used to make it possible to check and compare different designs before a house is built (Boverket 2009; Mjörnell et al. 2012). From experience, it is known that the construction

S. Olof Mundt-Petersen *is a doctoral student and* ***Lars-Erik Harderup*** *is an associate professor in the Department of Building Physics, Lund University, Lund Sweden.*

industry today only uses one-dimensional and fast calculation tools, if any tool is used at all, in the moisture-safety design process.

Today, there are several tools available that could be used to predict the hygrothermal performance and mold risks in constructions. Unfortunately there are only two calculation tools, WUFI and DELPHIN, that are commercially available and could be seen as user-friendly. Neither of these calculation tools seems to have been properly verified by independent researchers using blind methods, i.e., verified without knowing the measurement results before making comparisons with unadjusted calculated results (Mundt-Petersen 2012). Experiences from the northern European building trade also show that only one-dimensional methods are used. Furthermore, the only studies found with explicit blind comparisons focus on moisture control by allowing a high airflow in the air gap between the cladding and exterior mold-resistant insulation boards (Hägerstedt and Arfvidsson 2010; Hägerstedt and Harderup 2011a; 2011b). Only a few studies in northern European climates are found in which it is possible to compare measured and calculated values, such as Nore (2009); Geving and Holme (2010); Geving et al. (2011); Hägerstedt and Arfvidsson (2010); Hägerstedt and Harderup (2011a; 2011b). Most of the other verifications that were found could not be seen as independent as they were needed during the software development process or carried out by the developers, such as Künzel (1995); Krus (1996); Künzel et al. (2002); Künzel et al. (2004); Zirkelbach et al. (2011). There is also a lack of real-life field studies in houses where people are actually living (Nore 2009; Salonvaara et al. 2010; Geving and Holme 2010; Zirkelbach et al. 2011). It can therefore be established that there is no independent and blind verified user-friendly moisture calculation tool. There is also a lack of verified calculation tools in which real field conditions and northern European climate conditions have been studied (Mundt-Petersen 2012).

In order to determinate an independent blindly verified moisture calculation tool, measurements were carried out in a Swedish research project with starting points at the end of 2008 and in early 2009. The project comprised 148 measuring positions, of which 85 positions were in walls and 63 positions in roofs and attics, and measured temperature (*T*), relative humidity (RH) and moisture content (MC) in five different houses located in four different parts of Sweden over a period of approximately three years (Framtidens trähus 2012).

Aim

The aim of the entire project was to validate a hygrothermal tool to be used in the design phase to avoid mold and moisture damage and evaluate the risk of damage in studied building envelopes. However, the aim of this paper was to summarize the results of comparisons between measurements and blind calculations of temperature and relative humidity carried out in wood-frame walls during the period 2008 to 2011 in northern European climates.

The study aimed to show whether it might be possible to use the one-dimensional transient heat and moisture calculation tool WUFI 5.0 as a tool during the design phase in order to predict and evaluate the risk of mold growth and moisture damage. It was also intended to analyze under what conditions the calculation tool can be used and to show important factors that highly affect the correlation between measured and calculated values.

Furthermore, this paper aims to present a method that describes how calculation tools could be reliably verified under real conditions by using blind calculations.

Since the Folos 2-D visual mold chart was used in the comparison between calculated and measured values, the risk of mold growth was simultaneously evaluated in the studied positions in the walls (Mundt-Petersen et al. 2012).

Limitations

This paper presents a limited number of examples with general results from the entire investigation of 85 different positions. Only comparisons between measured and calculated temperature and relative humidity are shown. Complete comparisons between measured and calculated values in all the studied positions are presented in five separate reports (Mundt-Petersen 2013a; 2013b; 2013c; 2013d; 2013e).

The study is limited to an investigation of wood-frame walls with well-ventilated air gaps behind the cladding and an interior, well functioning vapor barrier. A detailed description of studied designs, calculation models and used material data are not given, but could be found in Mundt-Petersen (2013a; 2013b; 2013c; 2013d; 2013e; 2013f). Measurements and; calculations were carried out in northern European climates.

The study does not in detail analyze analytical solutions and physical and numerical models in the investigated calculation tool. Influence of possible measurement errors is discussed by Sandberg et al. (2011) and Mundt-Petersen (2013f). No detailed analysis of material data in the calculations models was made.

All calculations were made in one-dimension since this is the only tools that is widely used in the construction industry. The influence of wood studs and other thermal bridges were therefore deemed as neglectable in the calculations. Furthermore, possible influence of convection was neglected. The walls were also assumed to have been the results of perfect workmanship when compared to the drawings and calculation models used.

The definition of RH_{crit} that was used was limited to showing the conditions when mold growth on wood based materials are possible.

METHOD

All calculations were made blind, i.e., made without knowing the results of any measurements. Blind comparisons could also be called single-blind. All calculations and comparisons in this study were made single-blind and were carried out by independent organizations and testers, although this is not specifically highlighted in each specific case. Blind calculations are equivalent to situations when the designers carry out the heat and moisture calculations before a house is built. Comparisons with calculated values, made after the measured values had been received, were evaluated using the Folos 2-D visual mold chart.

Blind Comparisons Between Measured and Calculated Values. Initially, measuring sensors for temperature, relative humidity and MC were mounted at different depths and locations in the walls during the construction phase. The position of each sensor was well documented in drawings and photos. All the construction phases were monitored to establish any possible deviations between the drawings and the real conditions in the built wall. Measurements were started as soon as possible, sometimes before the houses were occupied, and were carried out using a wireless Protimeter Hygro Trac system (Sandberg 2011; Mundt-Petersen 2013f; GE Sensing 1996). Hourly measurements of temperature, relative humidity, and MC for each specific position were then separately stored by a measurement collector, inaccessible to the persons involved in evaluating the calculation tool.

Over a period of three years, when the measurements were carried out, calculation models of each studied position were made. The calculation models were based on drawings and photos from the construction phase with the intention of reflecting as real conditions as possible. However, since the calculation tool being evaluated was one-dimensional, the possible influence of beams, sills and studs was not included in the calculations.

In 2012 calculations were carried out for each studied position for the period 2008 to 2011, without knowing the measured results. The calculations were made using the indoor and outdoor climate-boundary conditions collected from indoor measurements and closely located outdoor climate stations (SMHI 2012).

After the blind calculations had been completed and sent to the measurement collector, the previously inaccessible measurements were retrieved. Comparisons between the measurements and the calculated temperature and relative humidity were then made over time using the Folos 2-D visual mold chart.

Note that it was possible to make adjustments to the calculation models to achieve better correlation, or even a perfect match, between the measured and blindly-calculated values in almost all the studied positions. However, this was not a part of the blind verification presented in this paper.

The Need for Blind Comparison. Software used as a tool to predict mold and moisture-related damage before a house is built to be blindly verified and without being influenced by the program developer. In this case, this was fulfilled by carrying out the calculations before the results of the measurement were known, i.e., they were blind calculations, and they were then compared to the measurements. Blind validations are reliable since intentional or unintentional adjustments of calculated results, to obtain better correlations to the measured values, are impossible. There are also other positive effects since the blind calculations are similar to the situation the designer has to deal with before a house is built. This provides important information about how the user perceives the tool. Many poor calculation results are due to inaccurate models or incorrect boundary conditions—such as unrealistic climate data, false material data, or inaccurate surface resistance—created by the user. Errors like these could be avoided by providing better default values, clearer instruction manuals, or stricter parameter limit values in the software.

Notice that the method used is not double-blind since it is possible to guess the measured results before the calculations are made. This is also possible during the design phase. It should also be mentioned that nonblind comparisons might be needed in the early development phase of equations and calculation tools.

Folos 2-D Visual Mold Chart. The Folos 2-D visual mold chart was used to evaluate and compare results between measured and calculated values in different designs, positions within the walls, and surrounding climates. Besides showing the measured and calculated values and the correlation between measured and calculated values, the Folos 2-D visual mold chart also indicates the risk of mold growth. By using the Folos 2-D visual mold chart, it is also possible to establish measures to be taken and to compare different designs in order to reduce the risk of mold growth (Mundt-Petersen et al. 2012).

The chart is briefly described together with the general results to make it easier to understand when it is used in context. The moisture-critical limit used was the LIM I limit (Sedlbauer 2001).

MATERIALS

The Studied Houses. Measurements and calculations were carried out in five different wood-frame houses. The measurement and calculation positions were located at different depths and locations in the walls, which had five different designs and faced different directions. The studied designs are shown together with examples of comparisons in the results section. The houses were located in four different towns in Sweden (as shown in Figure 1) each with different climate conditions.

Choice of Evaluated Calculation Tool. The criteria for choosing the evaluated tool were that it had to be user-friendly and available to the Swedish timber industry, i.e., a commercial tool (Boverket 2009). The most user-friendly and commercial software seems to be WUFI, and that is why it was chosen for the blind evaluation. The chosen tool was also specifically mentioned as a possible tool in previous investigations (Boverket 2009).

The materials and material data used in the calculation models were chosen from the WUFI 5.0 material data base (WUFI 2009). All calculation models, including the used materials and material data, are presented together with the description of the studied positions in separate reports. In these, the positions and the results from comparisons between

Figure 1 *Locations of the studied houses in Sweden.*

measurements and calculations over the investigated period are presented (Mundt-Petersen 2013a; 2013b; 2013c; 2013d; 2013e; 2013f). Detailed boundary conditions and material data are given by Mundt-Petersen (2013f).

The Climate-Boundary Conditions and Periods with Impaired Climate-Boundary Data. To achieve better agreement between measured and calculated values, use of the standard outdoor climate in the calculation tool was avoided. Real measured climate data from climate stations closely located to the studied houses was collected during the measuring period and then used in the calculations (SMHI 2012).

Periods with flawed climate data or a lack of climate data have been replaced and are shown in the Folos 2-D visual mold chart together with the other results.

RESULTS, ANALYSIS, AND DISCUSSION

This section presents examples of general results and findings that could be established from 85 positions at different depths in walls with different designs, locations and orientations in five different houses (Mundt-Petersen 2013a; 2013b; 2013c; 2013d; 2013e). The investigated design is shown together with the results in each figure, in which the studied positions are marked with a red cross.

General Results and Description of the Folos 2-D Visual Mold Chart

In general, there was a good correlation between the measured and blindly-calculated values in most of the studied positions, as shown in the examples in Figures 2, 3, and 6, i.e., the calculated values can be used in further analysis to predict the risk of mold growth in a reliable manner. However, there are also differences between measured and calculated values in many positions. These are presented and analyzed below. Possible factors influencing differences are also discussed.

Description of the Folos 2-D Visual Mold Chart. The Folos 2-D visual mold chart (in Figures 2 to 10) shows calculated temperature (yellow) and measured temperature (dark blue) on the right *y*-axis. The calculated relative humidity (turquoise), measured relative humidity (black), RH_{crit} dependent on the calculated temperature (red), and the RH > RH_{crit} difference for calculated values (light brown) and measured values (purple) are shown on the left *y*-axis. The time presented on the *x*-axis indicates the conditions at any specific time, and of particular interest are the periods when RH > RH_{crit}.

Critical conditions occur, and mold growth is possible, when the RH is above the RH_{crit} line. The RH_{crit} line is defined by the temperature that, at any specific time, exceeds the RH_{crit} limit LIM I invented by Sedlbauer (2001), i.e., the chosen RH_{crit} line is converted over time by using the actual temperature at each point in time. This means that critical conditions depend on the prevailing RH and temperature, where a high temperature gives a lower RH_{crit} line and vice versa.

However, depending on the moisture resistance of different materials and legislation in different countries, other mold growth limits could be used by choosing another appropriate RH_{crit} curve. A further description of the Folos 2-D visual mold chart and how it can be used can be found in a separate paper (Mundt-Petersen et al. 2012).

General Results. Examples of general results are shown in Figures 2, 3, and 6.

Impaired Climate-Boundary Data. Periods with impaired or lack of climate-boundary data have been replaced. Those periods are shown as dots or a line (green) at the top of the Folos 2-D visual mold chart. Vertical lines indicate a short period with lack of climate-boundary data and dots creating a horizontal line indicate longer periods that lack climate-boundary data. Lower values of dots or longer vertical lines indicate the lack of outdoor temperature or relative humidity data, which has a higher influence on the calculation results than the higher values or shorter lines that indicate lack of other less important climate-boundary conditions.

Influence of Temperature on the Relative Humidity

Differences between measured and blindly-calculated relative humidity depend in many cases on differences between measured and calculated temperatures, i.e., the vapor content is the same but different measured and calculated temperatures give different vapor contents at saturation. Furthermore, this creates differences between the measured

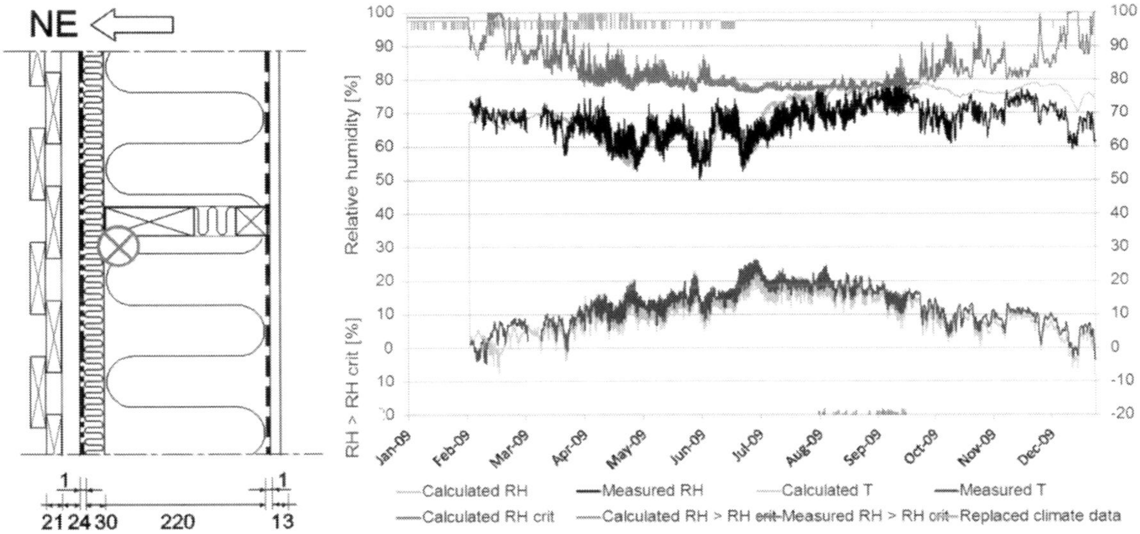

Figure 2 *Example of comparisons between measured and calculated RH and temperature behind a mold-resistant facade insulation board in the exterior part of a wall. Calculated RH (turquoise) and measured RH (black). Calculated temperature (yellow) and measured temperature (dark blue).The RH_{crit} derived from the calculated temperature (red). Calculated RH > RH_{crit} (light brown) and measured RH > RH_{crit} (purple). Periods with replaced climate-boundary data (green).*

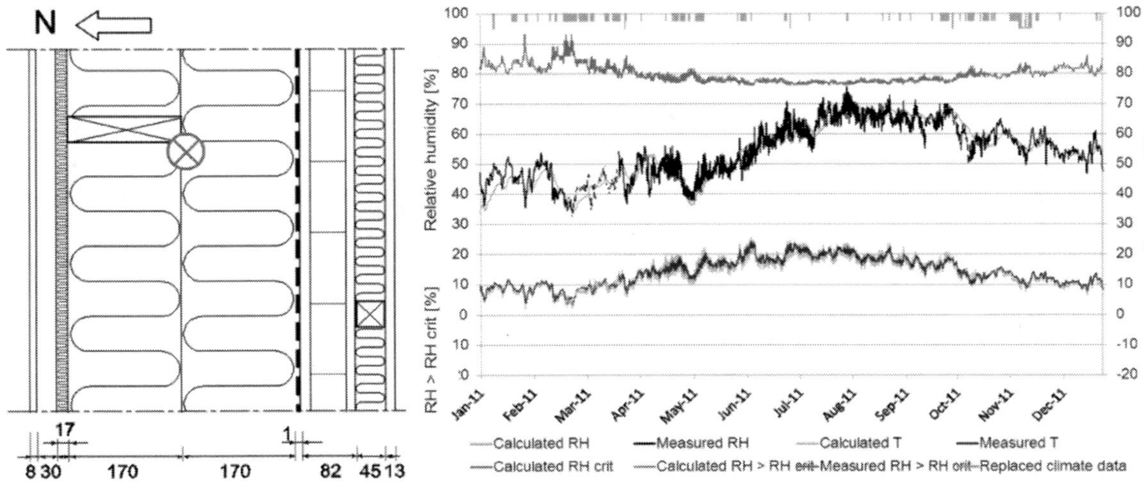

Figure 3 *Example of comparisons between measured and calculated RH and temperature in the middle of a wall. Calculated RH (turquoise) and measured RH (black). Calculated temperature (yellow) and measured temperature (dark blue). RH_{crit} derived from the calculated temperature (red). Calculated RH > RH_{crit} (light brown) and measured RH > RH_{crit} (purple). Periods with replaced climate-boundary data (green).*

and calculated relative humidity. This particular effect of temperature on relative humidity can be found in all the studied designs and houses. In most studied positions, a higher measured temperature creating a lower relative humidity, as shown in Figures 4 and 5, were found.

Differences Between Measured and Calculated Values During Cold Periods. Differences, depending on temperature, between measured and blindly-calculated relative humidity mainly occur during the colder periods in the outer part of the studied walls. As expected, differences tend to be greater in a colder climate, as shown in the example in Figure 4.

There may be several reasons for differences between measured and blindly-calculated temperature in the exterior part of studied walls during colder periods.

One factor was the thermal bridges due to studs and beams that are disregarded in the one-dimensional calculations. Studs and beams located close to the measuring sensor in the exterior part of the wall cause higher temperatures. This also explains why this effect becomes more obvious in colder climates. Two-dimensional studies of thermal bridges in a central Sweden climate show that the temperature increases by approximately

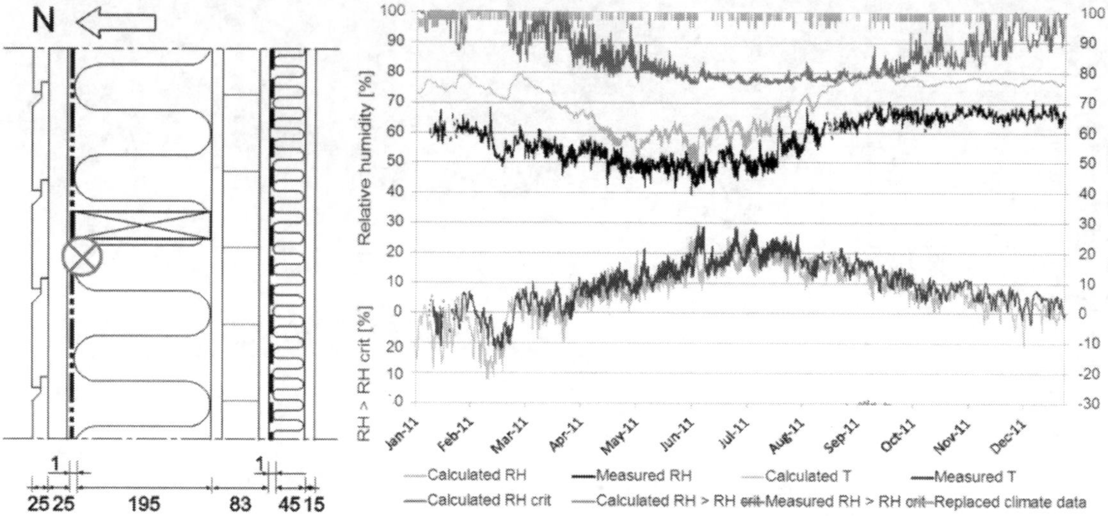

Figure 4 *Example of comparisons between measured and calculated RH and temperature in the exterior part of a wall located in the northern part of Sweden. Calculated RH (turquoise) and measured RH (black). Calculated temperature (yellow) and measured temperature (dark blue). The RH_{crit} derived from the calculated temperature (red). Calculated RH > RH_{crit} (light brown) and measured RH > RH_{crit} (purple). Periods with replaced climate-boundary data (green).*

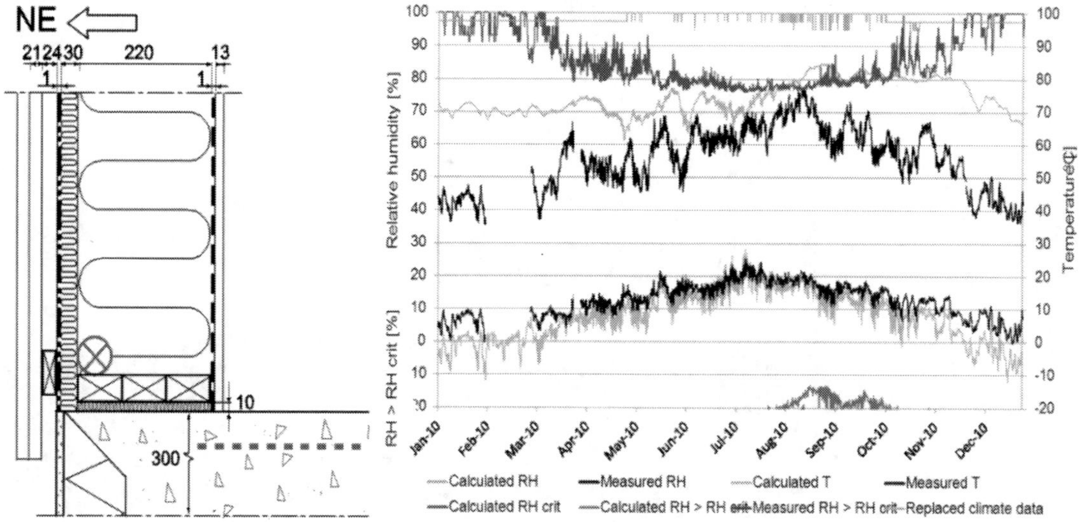

Figure 5 *Example of comparisons between measured and calculated RH and temperature, where measurements were carried out on the topside of the sill behind a mold-resistant facade insulation board in the exterior part of a wall. Calculated RH (turquoise) and measured RH (black). Calculated temperature (yellow) and measured temperature (dark blue). RH_{crit} derived from the calculated temperature (red). Calculated RH > RH_{crit} (light brown) and measured RH > RH_{crit} (purple). Periods with replaced climate-boundary data (green).*

0.5°C to 1°C. This reduces the relative humidity by 2.5% to 5% in the outer part of the stud (Forsberg 2011; Olsson 2011).

The sensor thickness of approximately 50 mm may also affect the temperature. Its size reduces the thickness of the surrounding thermal insulation which creates a higher temperature in the studied position, especially when the outdoor temperature is low. The temperature may also depend on the specific location of the temperature sensor within the larger measuring sensor. Furthermore, the sensor might create heat during measurement processing.

Another factor might be that differences between measured and calculated temperatures have a greater influence on the relative humidity of the lower vapor content that occurs in a cold climate.

Differences between the measured and calculated values may also depend on incorrect or delayed measured relative

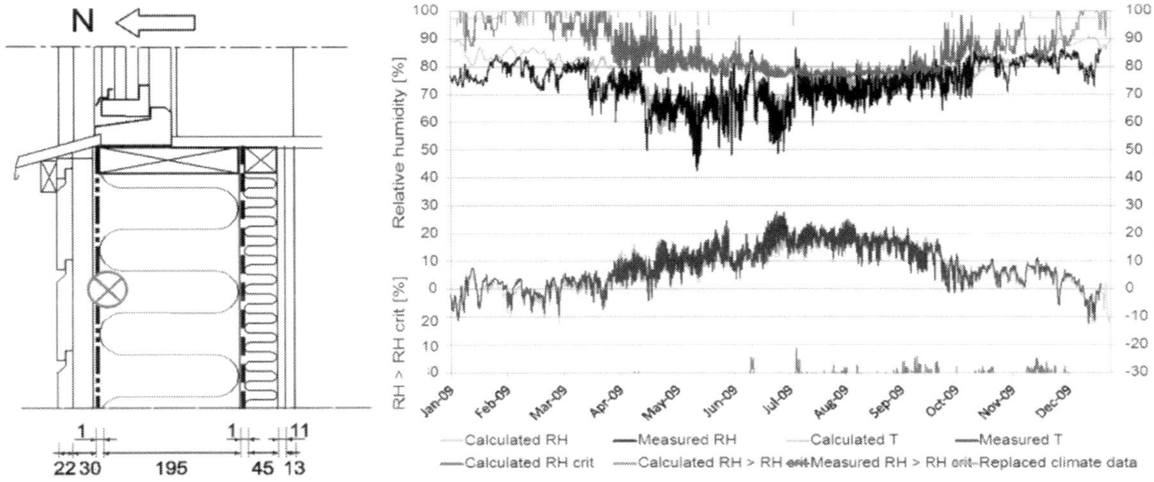

Figure 6 *Example of comparisons between measured and calculated RH and temperature in the exterior part of a wall. Calculated RH (turquoise) and measured RH (black). Calculated temperature (yellow) and measured temperature (dark blue).* RH_{crit} *derived from the calculated temperature (red). Calculated RH >* RH_{crit} *(light brown) and measured RH >* RH_{crit} *(purple). Periods with replaced climate-boundary data (green).*

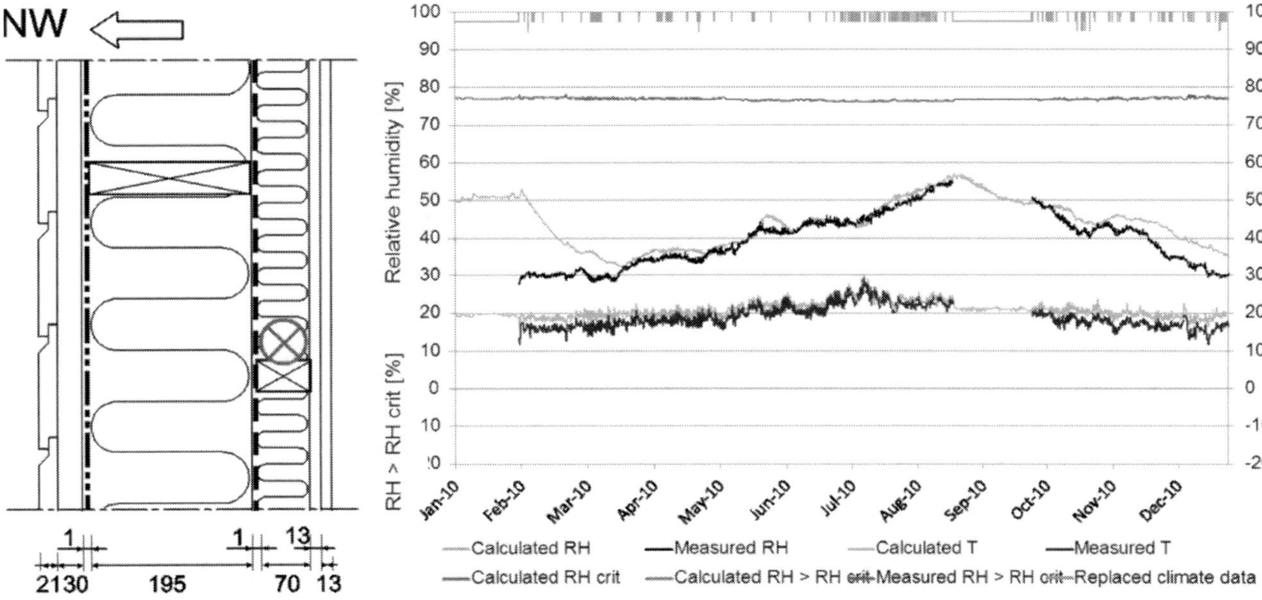

Figure 7 *Example of comparisons between measured and calculated RH and temperature in the installation layer on the inside of the vapor barrier in the interior part of a wall. Calculated RH (turquoise) and measured RH (black). Calculated temperature (yellow) and measured temperature (dark blue).* RH_{crit} *derived from the calculated temperature derived from the calculated temperature (red). Calculated RH >* RH_{crit} *(light brown) and measured RH >* RH_{crit} *(purple). Periods with replaced climate-boundary data (green).*

humidity. This becomes more evident in the winter when the outdoor relative humidity is high.

Influence of Under-Floor Heating Close to the Sill. A number of sensors were mounted at different depths in the wall on top of the sill, as shown in Figure 5.

In houses with under-floor heating, as shown in the slab in Figure 5, a positive effect was generated as a higher temperature reduced the relative humidity in this area. Higher temperatures also improve the drying-out process and make the sill warmer. It is important to observe that this positive effect requires vapor-permeable insulation that allows the drying-out process to take place.

However, the differences between measured and calculated values clearly show that it is not possible to use one-dimensional

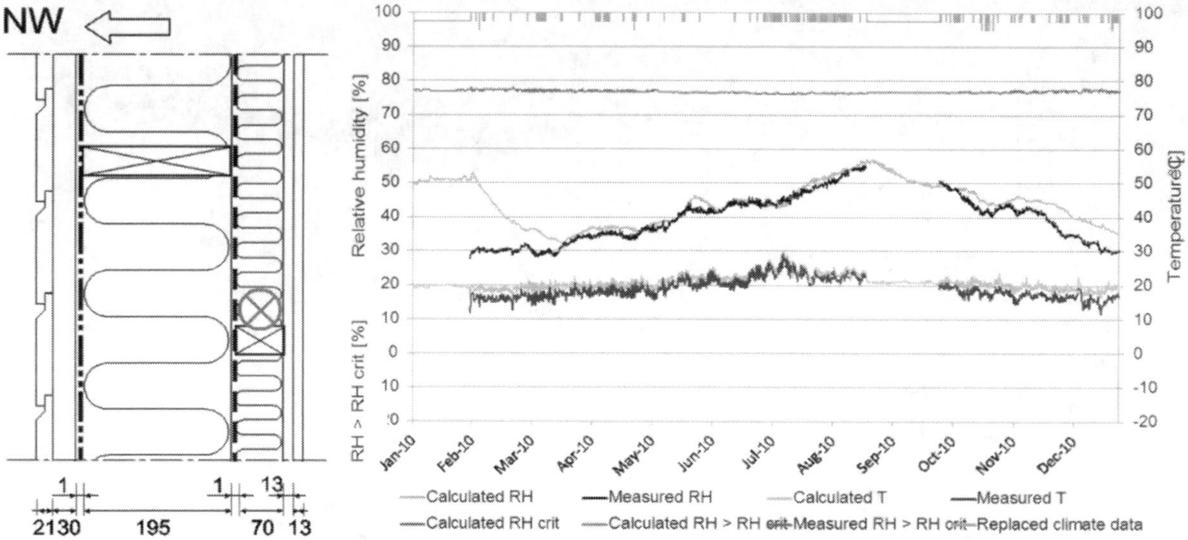

Figure 8 *Example of comparisons between measured and calculated RH and temperature in the installation layer between the vapor barrier and the interior waterproof membrane in the interior part of a bathroom wall over a period of three years. Calculated RH (turquoise) and measured RH (black). Calculated temperature (yellow) and measured temperature (dark blue).* RH_{crit} *derived from the calculated temperature (red). Calculated RH >* RH_{crit} *(light brown) and measured RH >* RH_{crit} *(purple). Periods with replaced climate-boundary data (green).*

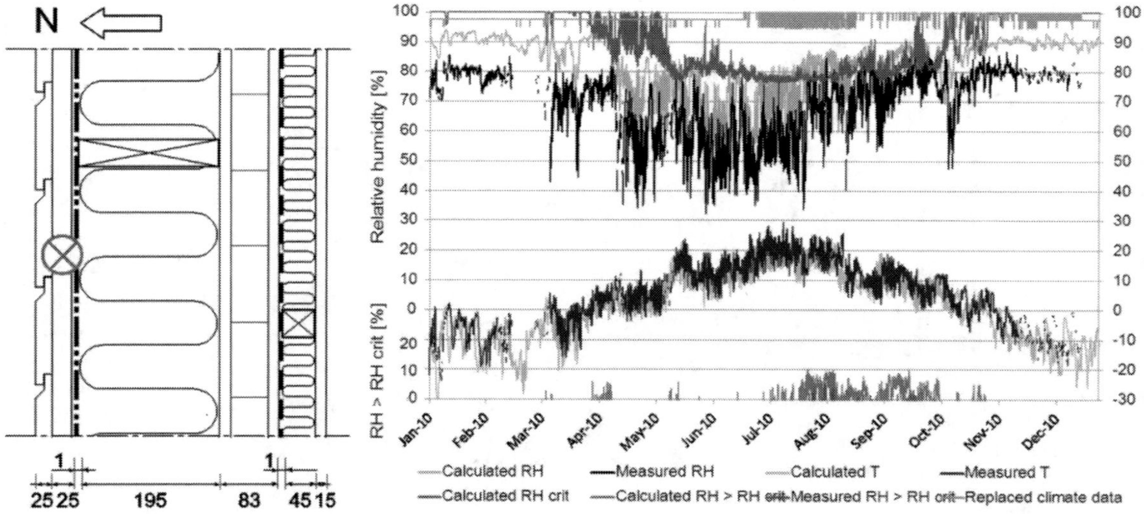

Figure 9 *Example of comparisons between measured and calculated RH and temperature in the air gap behind the cladding. Calculated RH (turquoise) and measured RH (black). Calculated temperature (yellow) and measured temperature (dark blue).* RH_{crit} *derived from the calculated temperature (red). Calculated RH >* RH_{crit} *(light brown) and measured RH >* RH_{crit} *(purple). Periods with replaced climate-boundary data (green).*

calculation tools in all situations and that heating sources need to be included in the evaluation of a design.

Most Moisture-Critical Positions. By using the Folos 2-D visual mold chart for the studied positions at different depths it could be established by both measurements and calculations that the most moisture-critical conditions occurred in the outer part of the wall. The critical conditions mainly occurred during longer periods at the end of the summer or in early autumn when there was high vapor content and low temperature that created a high relative humidity, as shown in Figure 6. Higher temperatures deeper within the wall reduced the relative humidity and furthermore the risk of mold growth. An exterior mold-resistant facade insulation board on the outside of the studs behind the air gap, as shown in Figure 2, which increases the temperature where there are organic mold-sensitive wood beams, can therefore be used as moisture protection. However, when the relative humidity is high at low outdoor temperatures below 0°C, there are no

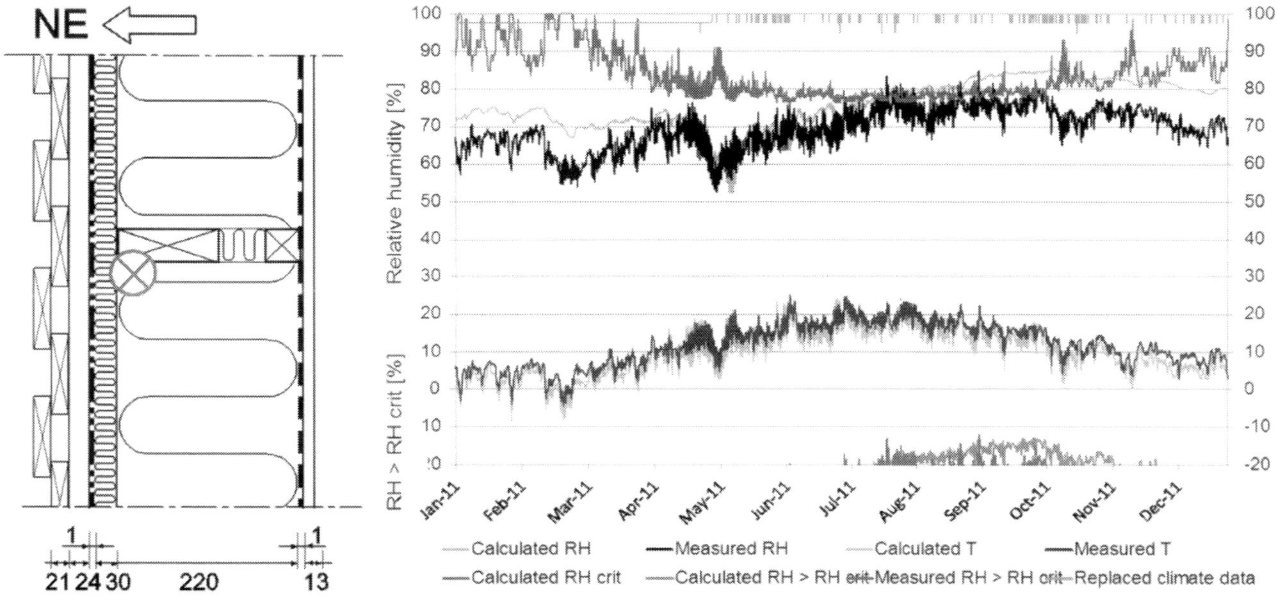

Figure 10 *Example of comparisons between measured and calculated RH and temperature behind a mold-resistant facade insulation board in the exterior part of a wall. Calculated RH (turquoise) and measured RH (black). Calculated temperature (yellow) and measured temperature (dark blue).* RH_{crit} *derived from the calculated temperature (red). Calculated RH >* RH_{crit} *(light brown) and measured RH >* RH_{crit} *(purple). Periods with replaced climate-boundary data (green).*

moisture-critical conditions. Such nonmoisture-critical conditions occur during the winter period in the studied house in northern Sweden, as shown in Figure 4. Thinner walls also have higher temperatures on the exterior parts, which reduce the relative humidity and the risk of mold growth.

Influence of Vapor Content on the Relative Humidity

In some cases the differences between the measured and blindly-calculated values is difficult to explain. They occurred during the winter, when the relative humidity was low, in the inner part of some of the studied walls, mainly in the installation layer (in which cables and sockets are located) behind the vapor barrier, as shown in Figure 7. It should not have been possible to record that both the measured temperature and relative humidity were below the calculated values at the same time. The reason may have been the influence of the moisture capacity of the stud in the installation layer or flaws in the measurements or calculation models. The location of the measured indoor climate-boundary conditions, remote from the studied wall, might also have created the anomalous result.

However, this is not as common, nor does it have as big an influence as the differences in temperatures have on the relative humidity.

Drying-Out in Installation Layers in Bathrooms between Two Vapor-Tight Membranes

In four positions in two of the studied designs, sensors were mounted in the installation layer in bathrooms, between the vapor barrier and the interior waterproof membrane (blue dashed line) as shown in Figure 8. Measurements in these positions indicated that the speed of the drying-out process was faster here than predicted by the calculations, as shown in Figure 8. A faster rate of drying out could also be observed in the not presented MC measurements.

The faster drying-out process may depend on several factors. The vapor barrier, outside of the installation layer, probably have bad joints and overlaps. Vertical or horizontal vapor transport, which cannot be taken into account in the one-dimensional calculations, may also speed up the drying-out process. The vapor barrier might also have been damaged or have a lower vapor resistance in reality than the vapor barrier used in the calculation model.

Correlation between Measured and Calculated Values in Air Gaps behind the Cladding and on the Exterior Facade

Several positions in the air gap behind the cladding were studied in all five houses. In the house located in the north of Sweden, comparisons between measured and calculated values were also carried out on the outside of the façade surface.

In general, all comparisons between measured and calculated values in the air gap showed a significantly lower measured relative humidity of approximately 10% to 15% compared to calculated values. The lower measured relative humidity is a result of a higher temperature and occurs during the entire studied period (as shown in Figure 9).

Differences between measurements and calculations on the cladding surface in the house located in the north of Sweden were similar to the corresponding differences in the air gap. However, the comparisons in the wall oriented towards the south had bigger differences than the other studied directions. Those differences were probably created by the influence of solar radiation that could be better simulated in the calculation.

Amplitude Variations in Temperature and Relative Humidity

Where moisture-critical conditions are concerned, the amplitude might be of interest as mold growth is dependent on the duration of the critical conditions (Viitanden and Ojanen 2007; Isaksson et al. 2010). The amplitudes in different positions, in Figures 2 to 8, show that there were greater amplitudes in the measured temperature and relative humidity than in the blindly-calculated values in the construction. Close to the inside of the wall there were low amplitudes in both the measured and calculated values (as shown in Figures 7 and 8). In the middle of the wall, the amplitudes are slightly larger, mainly in the measured values (as shown in Figure 3). Closer to the air gap, the amplitudes, mainly of the measured values, become significantly greater, (as shown in Figures 2, 4, 5, and 6). In the air gap and on the outside of the facade, the amplitudes are mainly the same when the measured and calculated values are compared, as shown in Figure 9. The amplitudes of both the measured and calculated values were also lower during the cold periods of the year (as shown in Figures 2, 4, 5 and 6). Comparisons (not presented here) also show that there were larger amplitudes in the calculated and measured values in positions orientated towards the south (Mundt-Petersen 2013b; 2013d).

Variations in amplitudes during different periods of the year and for different orientations of the studied positions depend on differences in temperature. There were greater variations in temperature during the summer and these created larger amplitudes during those periods. Furthermore, the positions orientated towards the south were affected more by the heat variations created by solar radiation.

The studied positions in the exterior part of the wall were more affected by the variations in the outdoor climate than the positions closer to the interior side of the wall, which were thermally influenced by the more stable indoor climate.

The reason for different amplitudes between measured and blindly-calculated values in the exterior part of the studied walls may have depended on the difference in heat and moisture capacity in the actual materials compared to the materials in the calculation model. The specific measurement sensors were also protected by a plastic shell and not directly exposed to the surrounding material. This material may include a volume of air that is more quickly affected by temperature changes than the surrounding materials.

Influence of Variations in Climate-Boundary Conditions between Different Years

The outdoor surrounding climate affects the conditions in the exterior part of a wall. Significant variations between different years were found by comparing the same positions over longer periods. By comparing the relative humidity in Figures 2 and 10, for the same position during two different years (2009 and 2011), it was found that there were significantly different behaviors.

The conclusions are that the different outdoor climates during different years created major variations in the studied positions. The variations affected temperature, relative humidity and duration, all of which affect the risk of mold growth (Viitanden and Ojanen 2007; Isaksson et al. 2010). Outdoor climate variations between different years especially affected the most moisture-critical positions in the exterior part of the wall.

It may therefore be questionable whether mean or standard climate-boundary conditions should be allowed in the moisture-safety design process for external walls as they may reduce or disregard the influence of moisture-critical periods. Climates that do not include critical outdoor periods may result in negative consequences when the risk of mold growth is investigated before a house is built. There is therefore a need for reliable outdoor climate-boundary conditions including the effect of periods with moisture-critical climate or other safety factors that could be used during the moisture-safety design phase.

The variations between different years also show that correct climate-boundary conditions must be used during validation of transient mold growth models and of transient heat and moisture calculation tools.

The influence of periods with impaired climate-boundary conditions (as shown in green in Figures 2 to 10) that affect the correlation between measured and calculated values could also be found in some positions. However, as long as the period with impaired climate data not is too long, it does not seem to have any major influence on the comparison between measured and calculated values.

CONCLUSIONS

Although there are a number of differences between the measured and blindly-calculated values shown above, it must be stated that most of the 85 blind comparisons show such a good correlation, as shown in Figures 2, 3 and 6, that they could be used to predict the risk of mold growth in further analysis. It may therefore be established that WUFI 5.0 can be used as a tool to predict moisture risks in wood-frame walls with a well-ventilated air gap and an interior vapor barrier. However, there are factors and observations that affect the results and these need to be taken into account if the tool is to be used for moisture safety purposes. These include:

1. Differences in temperature can have a great effect on the relative humidity.

   a. In the studied positions the measured temperature was in general higher than calculated values, which make the measured relative humidity lower than the calculated values.
2. A correct calculation model must be created in the calculation tool. It is essential that reliable outdoor climate-boundary conditions are used.
   a. This is necessary because outdoor climates have a great influence on the exterior moisture-critical part of wood-frame walls.
   b. Mean or standard outdoor climate-boundary conditions without extremes must not be used.
   c. It should be investigated whether it is possible to use other safety limits combined with mean or standard climates.
   d. Solar radiation has a higher influence on the hygrothermal conditions in the outer part of a wall in reality compared to calculated results.
3. One-dimensional models cannot be used in all situations and influence of heating sources need to be considered in the hygrothermal calculations.
4. The amplitudes of the measured values were higher than the calculated values, especially in moisture-critical positions in the exterior part of the wall.
5. Note that previous studies also show that a correct airflow in the air gap was required in the calculation model in order to obtain correct results (Hägerstedt and Arfvidsson 2010). It must therefore be possible to obtain these flows in the finished house.

Furthermore, it was found that the most moisture-critical positions generally occurred in the exterior part of the wall, in organic material on the inside of the air gap. Under-floor heating had a positive influence on the studied walls with vapor permeable materials on the outside.

Comparisons between measured and calculated values over time must also be made using real climate-boundary conditions as there are significant variations in annual climate that affect the results between different years.

ACKNOWLEDGMENTS

This study was supported by Vinnova (the Swedish Government Agency for Innovation Systems) and their partnership, with representatives of the timber industry, which initiated the research projects "Framtidens trähus" (Wood frame buildings of the future) and "ECO2 – Carbon-efficient timber constructions".

NOMENCLATURE

RH	=	relative humidity
RH_{crit}	=	limit for possible mold growth in wooden materials
$RH > RH_{crit}$	=	critical conditions showing that mold growth is possible
T	=	temperature

REFERENCES

BBR. 2011. *Swedish Building Regulations 2012* (In Swedish). Karlskrona: The Swedish National Board of Housing, Building and Planning.

Boverket. 2009. The standard of our houses—Report on the Swedish governments commissions regarding the technical standard of Swedish buildings (In Swedish). Karlskrona: The Swedish National Board of Housing, Building and Planning.

Forsberg, T. 2011. Moisture rearrangements in exterior walls, Report 2011:14 (Report 2011:14). Bachelor thesis. KTH Royal Institute of Technology.

Trähus, F. 2012. Project information (In Swedish). www.framtidenstrahus.se.

Sensing, G.E. 2006. Protimeter HygroTrac – Wireless environmental monitoring. www.veronics.com/products/Relative_humidity-transmitter/Hygrotrac.pdf.

Geving, S. and J. Holme. 2010. The drying potential and risk for mold growth in compact wood frame roofs with built-in-moisture. *Journal of Building Physics* 33(3):249–69.

Geving, S., Kvalvik, M., and Martinsen, E. 2011. Rehabilitation of basement walls with moisture problems by the use of vapour open exterior thermal insulation. *Proceedings of the 9th Nordic Symposium on Building Physics—NSB 2011, Tampere, Finland,* 1:323–30.

Hägerstedt, S.O. 2012. Moisture safe wood constructions—guidelines for wall design, Report TVBH-3052 (In Swedish). Lund University.

Hägerstedt, S.O. and J. Arfvidsson. 2010. Comparison of field measurements and calculations of relative humidity and temperature in wood framed walls. *Thermophysics 2010 – Conference proceedings, 15th International Meeting of Thermophysical Society, Valtice, Czech Republic,* 93–101.

Hägerstedt, S.O. and L.-E. Harderup. 2011a. Comparison of measured and calculated temperature and relative humidity with varied and constant air flow in the facade air gap. *Proceedings of the 9th Nordic Symposium on Building Physics – NSB 2011, Tampere, Finland,* 1:147–54.

Hägerstedt, S.O. and L.-E. Harderup. 2011b. Control of moisture safety design by comparison between calculations and measurements in passive house walls made of wood. *Proceedings of the 12th International Conference on Durability of Building Materials and Components – XII DBMC, Porto, Portugal,* 1:25–32.

Isaksson, T., S. Thelandersson, A. Ekstrand-Tobin, and Johansson, P. 2010. Critical conditions for onset of mold growth under varying climate conditions. *Building and Environment* 45(7):1712–21.

Krus, M. 1996. Moisture transport and storage coefficients of porous mineral building materials – Theoretical principles and new test methods. Doctoral thesis, Fraunhofer Institute for Building Physics. University of Stuttgart.

Künzel, H.M. 1995. Simultaneous heat and moisture transport in building components, One- and two-dimensional calculation using simple parameters. Doctoral thesis, Fraunhofer Institute of Building Physics. University of Stuttgart.

Künzel, H.M., H. Künzel, and A. Holm. 2004. Rain protection of stucco facades. *Proceedings of the Thermal Performance of the Exterior Envelopes of Whole Buildings IX – International Conference, Florida, USA*.

Künzel, H.M., T. Schmidt, and A. Holm. 2002. Exterior surface temperature of different wall constructions – comparison of numerical simulation and experiment. *Proceedings of the 11th Symposium for Building Physics, Dresden, Germany*, 1:441–49.

Mjörnell, K., J. Arfvidsson, and E. Sikander, 2012. A method for including moisture safety in the building process. *Indoor and Built Environment* 21(4):583–94.

Mundt-Petersen, S.O. 2012. Literature study/State-of-the-art —Mould and moisture safety in constructions, Report TVBH-3053. Lund University, Sweden.

Mundt-Petersen, S.O. 2013a. Comparison of hygrothermal measurements and calculations in a single-family wooden house on the west coast of Sweden, Report TVBH-3054. Lund University. In press.

Mundt-Petersen, S.O. 2013b. Comparison of hygrothermal measurements and calculations in a multi-family wooden house on the north-eastern coast of Sweden, Report TVBH-3055. Lund University. In press.

Mundt-Petersen, S.O. 2013c. Comparison of hygrothermal measurements and calculations in a single-family wooden house in the Swedish town of Växjö, Report TVBH-3056. Lund University. In press.

Mundt-Petersen, S.O. 2013d. Comparison of hygrothermal measurements and calculations in a multi-family wooden house in the Swedish town of Växjö, Report TVBH-3057. Lund University. In press.

Mundt-Petersen, S.O. 2013e. Comparison of hygrothermal measurements and calculations in a single-family wooden house in the Swedish municipality of Upplands-Bro, Report TVBH-3058. Lund University. In press.

Mundt-Petersen, S.O. 2013f. Moisture safety in wood frame walls – Blind evaluation of the hygrothermal calculation tool WUFI 5.0 using field measurements and determination of factors affecting the moisture safety. Licentiate thesis, Report TVBH-3059. Lund University.

Mundt-Petersen, S. O., P. Wallentén, T. Toratti, and J. Heikkinen, 2012. Moisture risk evaluation and determination of required measures to avoid mold damage using the Folos 2D visual mold chart. *Thermophysics 2012—Conference proceedings, 17th International Meeting of Thermophysical Society, Podkylava, Slovakia* 134–41.

Nevander, L.E. and B. Elmarsson. 1991. Moisture safety design of timber constructions, Report R38:1991 (In Swedish), Lund University.

Nore, K. 2009. Hygrothermal performance of ventilated wooden cladding. Doctoral theses, Report NTNU 2009:31. Norwegian University of Science and Technology.

Olsson, L. 2011. Laboratory investigation of timber frame walls with various weather barriers, Report 2011:56 (In Swedish). SP Technical Research Institute of Sweden.

Salonvaara, M., M. Pazera, and A. Karagiozis. 2010. Impact of weather on predicting drying characteristics of spray-applied cellulose insulation. *Proceedings of the Thermal Performance of the Exterior Envelopes of Whole Buildings XI—International Conference, Florida, USA*, I:B.

Sandberg, K., Pousette, A., and Dahlquist, S. 2011. Wireless in-situ measurements of moisture content and temperature in timber constructions. *Proceedings of the 12th International Conference on Durability of Building Materials and Components – XII DBMC, Porto, Portugal*, 1:191-98.

Sedlbauer, K. 2001. Prediction of mold fungus formation on the surface of and inside building components. Doctoral thesis, Fraunhofer Institute for Building Physics, University of Stuttgart.

SMHI. 2012. Climate data statistics (In Swedish). Swedish Meteorological and Hydrological Institute. www.smhi.se/Professionella-tjanster/Professionella-tjanster/statistik-och-data/bestall-vaderstatistik-1.1629.

Viitanen, H., and T. Ojanen. 2007. Improved model to predict mold growth in building materials. *Proceedings of the Thermal Performance of the Exterior Envelopes of Whole Buildings X – International Conference, Florida, USA*, V:A.

WUFI. 2009. WUFI 5.0, Material data base. Release: 5.0.1.521.DB.24.70. Holzkirchen, Germany: Fraunhofer Institute for Building Physics.

Zirkelbach, D., B. Schafaczek, and H.M. Künzel. 2011. Thermal performance degradation of foam insulation in inverted roofs due to moisture accumulation. *Proceedings of the 12th International Conference on Durability of Building Materials and Components – XII DBMC, Porto, Portugal*, 1:529–36.

Interior Insulation Retrofit of a Brick Wall Using Vacuum Insulation Panels: Design of a Laboratory Study to Determine the Hygrothermal Effect on Existing Structure and Wooden Beam Ends

Pär Johansson Berit Time, PhD Stig Geving, PhD Bjørn Petter Jelle, PhD

Angela Sasic Kalagasidis, PhD Carl-Eric Hagentoft, PhD Egil Rognvik

ABSTRACT

The increasing demand on energy-efficient buildings requires energy retrofitting measures in the existing building stock. Conventional thermal insulation materials, such as mineral wool and expanded polystyrene (EPS), demand a thick layer of insulation to reach the energy targets. Vacuum insulation panels (VIPs) are a novel thermal insulation component with 5–10 times lower thermal conductivity than the conventional insulation materials, depending on the VIP ageing state. The thermal transmittance of the building envelope can thereby be substantially reduced using a limited additional insulation thickness. Previous research has shown that interior energy retrofitting of exterior walls may increase the moisture content of the walls and increases the risk of freeze-thaw damages at the surface. This study analyzes the hygrothermal consequences on a 250 mm (9.8 in.) thick brick wall retrofitted with 20 mm (0.8 in.) interior VIP (glued directly on the plastered brick wall). Hygrothermal simulations in WUFI 2D are used to study the hygrothermal effects by different material properties and boundary conditions. Apart from the material properties, the amount of driving rain available at the surface is the most important influential parameter in the simulations. The conclusions from this study are used to plan a measurement study in a climate simulator where driving rain and solar radiation will be simulated.

INTRODUCTION

One of the main challenges in the building sector is to reduce the energy use in the existing building stock. In Sweden, 47% of the building stock is from before 1960. These buildings are the ones with the highest average thermal transmittance of the exterior wall, 0.58 W/(m^2·K) (R-10 ft^2·°F·h/Btu), compared to 0.44 W/(m^2·K) (R-13 ft^2·°F·h/Btu) for all the existing buildings in Sweden (Boverket 2009). In Oslo, the capital of Norway, there are 3000 brick buildings built between 1800 and 1900. Their technical characteristics and guidelines to how they should be maintained are described by Bjørberg et al. (2011). Many of these buildings are protected for their architectural qualities. The Swedish board of housing and planning, Boverket, estimates that 41% of the buildings in the Swedish building stock are not suitable for exterior retrofitting of the façade and 28% are considered dubious (Boverket 2010). The situation in Norway is somewhat similar, though there are no reports available analyzing the building stock with regard to suitability for interior or exterior insulation. However, it is known that about 100 000 multifamily buildings were built before 1945, with an average thermal transmittance of the exterior wall of 0.9 W/m^2·K (R-6.3 ft^2·°F·h/Btu) (Thyholt et al. 2009). The façades of many of these buildings are likely protected for their architectural qualities. Hence, a majority of the buildings will not be possible to retrofit on the exterior and other energy efficiency measures have to be used instead.

Retrofitting of an exterior wall changes the hygrothermal performance of the wall, which could lead to damages and, in the worst case, building failure. When adding insulation to the exterior of the wall, the existing structure is kept in a warm and dry condition, which is beneficial from a building physics point of view. Listed buildings whose features and aesthetics are to be preserved (see Figure 1a for an example of

***Pär Johansson** is a doctoral student, **Angela Sasic Kalagasidis** is an assistant professor, and **Carl-Eric Hagentoft** is a professor at Chalmers University of Technology, Sweden. **Berit Time** is a senior scientist at SINTEF Building and Infrastructure, Norway. **Stig Geving** is a professor at Norwegian University of Science and Technology, Norway. **Bjørn Petter Jelle** is a professor at Norwegian University of Science and Technology and a senior scientist at SINTEF Building and Infrastructure. **Egil Rognvik** is an engineer at SINTEF Building and Infrastructure.*

(a)

(b)

Figure 1 *(a) An old industrial building considered to have a façade with special aesthetic values that should be preserved and (b) a connection between a wooden beam and brick wall.*

such a building) cannot be insulated on the exterior. In these cases interior insulation is the only option to reduce the heat losses through the exterior walls. Interior insulation of walls is a topic that has been investigated continuously during the years. Straube et al. (2011) studied a number of brick buildings in the U.S. to which interior insulation had been added and concluded that rainwater has to be treated properly and that there is a risk that wooden beam ends in the walls (see Figure 1b) may be damaged. Künzel (1998) showed that exterior walls insulated on the interior need to be combined with rain protection measures to avoid moisture-induced damages.

The thickness of the additional interior insulation layer is an important limitation because the rentable floor area will be reduced after the retrofit. Materials with a lower thermal conductivity make it possible to reduce the thermal transmittance while minimizing the rentable floor area reduction. Vacuum insulation panels (VIPs) are a novel thermal insulation component with 5–10 times lower thermal conductivity than the conventional insulation materials (Baetens et al. 2010) such as mineral wool and expanded polystyrene (EPS), depending on the VIP ageing state. Sveipe et al. (2011) showed that the VIP may represent a condensation risk when used on the exterior. They investigated timber frame walls with 100 mm (3.9 in.) mineral wool and 20–30 mm (0.8–1.2 in.) VIP in laboratory experiments. The walls lacked an interior vapor barrier. The exterior climate was –18°C (–0.4°F) with 60% relative humidity and the interior climate was 20°C (68°F) with 50%–60% relative humidity, i.e., there was a moisture excess of 8–10 g/m^3 (100–130 lb/ft^3). With these boundary conditions, the vapor in the air could reach the saturation vapor content and condense on the cold surfaces. However, with a vapor barrier on the interior this problem could have been avoided but a new moisture problem could arise where moisture is trapped between the interior vapor barrier and the exterior VIP.

The decreasing heat resistance of the VIP during many years of use, due to diffusion of water vapor and air through the enclosing VIP laminate (Baetens et al. 2010; Wegger et al. 2011), also represents a calculation challenge when studying VIPs. Pfluger et al. (2008) used hygrothermal simulations, laboratory investigations, and field tests to investigate interior insulation retrofit of walls using VIPs. The results showed that interior insulation with VIPs is possible without causing moisture-induced damages as long as the attachment details are designed to minimize air leakage from the interior and the façade is protected from driving rain.

This paper presents one- and two-dimensional simulations performed in WUFI 2D (Fraunhofer IBP 2010) with the purpose of preparing an experimental benchmark in a laboratory environment where a brick wall with wooden beam ends is insulated with VIPs on the interior. The goal is to propose design conditions for a measurement procedure and for the design of retrofit solutions to minimize the risk of damages to the existing structure. Special focus is pointed at investigating the influence of the retrofitting on the wooden beam ends and practical limitations when using VIPs. Wooden beam ends are studied because this is a known risk area when insulating brick buildings on the interior. The study is part of a research project run in cooperation with Chalmers University of Technology in Gothenburg, Sweden, and the Norwegian University of Science and Technology and SINTEF Building and Infrastructure in Trondheim, Norway. A laboratory investigation where a brick wall was exposed to varying conditions in a climate simulator following the results of the simulations presented in this paper was performed in January to June 2013.

INTERIOR RETROFIT RISKS

The brick walls that are targeted for this study are normally 1.5 bricks thick—380 mm (15 in.)—which is a common wall thickness in buildings from the late 19th century (Kvande and Edvardsen 2007). The walls often have a wooden beam inserted around 200 mm (8 in.) into the brick (see Figure 2). When the wall is insulated on the interior, the beam end will be in a colder environment. Rasmussen (2010) used numerical simulations to find the change in temperature of the wooden beams when retrofitted on the interior and concluded that the temperature was substantially reduced. With 95 mm (3.7 in.) mineral wool, an interior temperature of 20°C (68°F), and 0°C (32°F) at the exterior, the temperature could be reduced from 10°C to 7°C (50°F to 45°F) after the wall is insulated. If the moisture load is constant, the relative humidity at the coldest point will increase, thus increasing risk of mold growth and rot in the beam.

Brick walls are capillary active, which means they will absorb water from driving rain until the surface layer is fully covered by a water film and the water runs down along the façade. The influence of driving rain on a brick wall insulated on the interior with 200 mm (8 in.) mineral wool was investigated numerically by Morelli et al. (2010). The influence by the amount of water available for capillary absorption was studied using the coupled heat and moisture program DELPHIN (Nicolai et al. 2013). A rain shielding factor between 0.1 and 0.5 was investigated in the study. When the façade was fully protected and no capillary absorption took place the factor was 0; if all the rain was available for capillary absorption the factor was 1. The driving rain showed to have a major influence on the performance of the wall where the relative humidity at the beam end was studied in detail. The additional

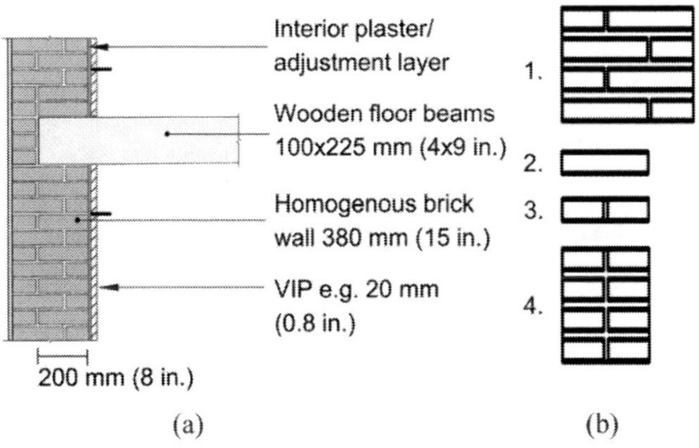

Figure 2 *(a) Schematics of a brick wall with a wooden floor beam inserted in the brick wall (© 2013 SINTEF) where the black horizontal lines through the insulation layer mark the 300 mm (12 in.) gap in the insulation described by Morelli et al. (2010) and (b) the simulated cases of brick and mortar investigated in this paper.*

thermal insulation reduced the temperature in the wall, thus elevating the relative humidity in the wooden parts, leading to risk for moisture damages. It was shown that the high moisture content in the wooden beam end could be reduced by enlarging the thermal bridge around the wooden beam by removing 300 mm (12 in.) of the insulation above and below the wooden beam, as shown by the horizontal black lines in Figure 2. Using this measure, the relative humidity of the beam end was decreasing from its starting value of 85%. However, with a rain shielding factor of 0.5 instead of 0.1, the relative humidity increased to more than 95%.

The movement of water through brick masonry has many important consequences in building construction and has therefore been studied by a number of authors, e.g., Hall (1977) and Brocken (1998). While the majority of these studies involved water suction experiments from a free water surface, large-scale experiments where water suction in brick walls is studied during a real or artificial rain load, such as presented by Abuku et al. (2009) and Piaia et al. (2013), are rare. To the knowledge of the authors, similar studies for brick masonry are not readily available.

Also, bricks that are not freeze-thaw resistant can be damaged if too-high moisture levels occur in the wall during freeze-thaw cycles. The water in the capillaries of the brick expands when it freezes, and after a number of cycles the structure starts to disintegrate. Mensinga et al. (2010) studied the risk of frost damage on bricks subjected to freeze-thaw cycles in a retrofitted wall. A risk assessment methodology based on frost dilatometry was developed where the critical level of moisture saturation for freeze-thaw damage could be found. They experienced that the critical degree of saturation (defined as the ratio of the moisture content over the moisture content when all accessible pores are filled with water) varied from 0.25 to 0.8 for two different investigated types of brick. This information is important when deciding on the retrofit design of the brick wall to ensure a safe moisture state in the brick.

When using VIPs it is essential to ensure airtight connections between the panels to avoid any convective moisture transfer into the wall from the interior of the building. Old brick walls can be damaged internally with cracks in the brickwork and mortar, which could mean the moist indoor air can be transferred into the colder parts of the construction and eventually condense. It is also important to make sure the moisture content in the wall close to the VIPs is not too high, since it has not yet been entirely evaluated how the panels can handle a longer exposure to a moist environment. Brunner et al. (2008) studied the formation of soluble aluminum oxide and transparent aluminum oxide in the enclosing VIP laminate by high heat and moisture load. In accelerated ageing experiments representative of 25 years of normal exposure, aluminum oxides did not form. If aluminum oxides formed in the VIP laminate, the diffusion of water vapor and air through the laminate would increase the thermal conductivity to around half the thermal conductivity of conventional insulation materials.

HYGROTHERMAL SIMULATIONS

In this study one- and two-dimensional hygrothermal simulations are used to investigate the hygrothermal performance of a masonry wall based on the following parameters:

- Climate: temperature, relative humidity, solar radiation, and driving rain
- Climate load at the location of the wall
- Wall thickness
- Type of brick
- Type of mortar
- Internal thermal insulation

The investigation starts with an analysis of the climate at the location of the building. The climates of Bergen (west coast of Norway) and Gothenburg (west coast of Sweden) are compared since these are the cities with the highest driving rain loads in Norway and Sweden. Figure 3 shows the temperature and relative humidity in Bergen and Gothenburg.

As can be seen, the climates in Bergen and Gothenburg are very similar, with average yearly temperatures of 8.1°C and 8.8°C (46.6°F and 47.8°F), respectively. The average relative humidity is 80.6% in Bergen and 74.5% in Gothenburg. The annual rainfall in Bergen is 2 421 mm (95.3 in.) and in Gothenburg is 1 074 mm (42.3 in.). Both cities are located close by the sea, which means a large portion of driving rain hits the façades of the buildings.

The measurement study should be based on real climate data to simulate the behavior of a future full-scale retrofitting measure. However, the climate simulator has some limitations and there is a time limitation for how long each test sequence can be run. The climate simulator generates a controlled dynamic climate condition on both sides of the brick wall. On the exterior side of the chamber, the climate can be set to vary between –20°C and +80°C (–4°F and +176°F) with a relative humidity of 20%–95%. The rain hitting the test wall is produced by a set of nozzles able to produce 15–35 μm (590–1380 μin.) large droplets with a rain intensity of 10–100 mm/h (0.39–3.9 in./h). Nine metal halide global lamps produce an artificial solar radiation with a wavelength distribution similar to that of natural solar radiation. The maximum solar intensity is 1000 W/m^2 (3412 Btu/h), which can be controlled continuously between 50% and 100%. On the interior side of the wall the temperature can be varied between +5°C and +50°C (41°F and 122°F) with a relative humidity of 20%–95% (Angelantoni Industries S.p.A., n.d.).

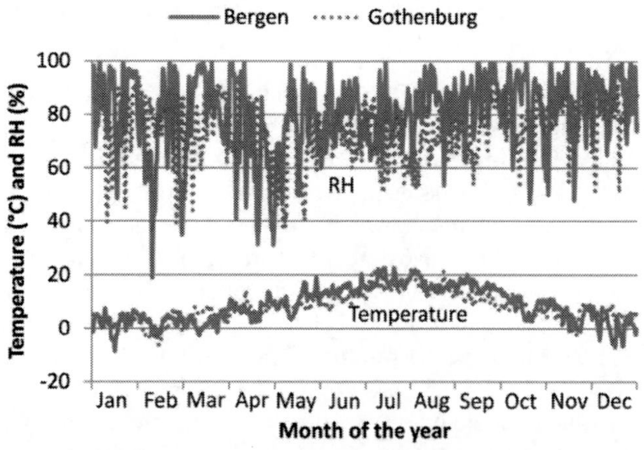

Figure 3 *Temperature and relative humidity in Bergen (solid line) compared to Gothenburg (dotted line) starting with January based on data from WUFI 2D (Fraunhofer IBP 2010).*

Climate Load on the Location of the Wall

The moisture conditions in 380 mm (15 in.) brick walls in Bergen and Gothenburg with the material data of extruded brick and coarse lime cement mortar defined in Table 1 were simulated with two-dimensional heat and moisture transfer. The sorption isotherms of the bricks included in this study are presented in Figure 4. The thermal conductivity and specific heat capacity of the brick and mortar are 0.6 W/(m·K)

Table 1. Material Data for the Hygrothermal Simulations of Four Brick Types and Six Mortar Types based on Data from WUFI 2D (Fraunhofer IBP 2010)*

Material	Density, kg/m^3	Porosity	Water Vapor Diffusion Resistance Factor	Liquid Water Transfer Coefficient, m^2/s
Masonry brick	1900	0.24	10	1.7
Extruded brick	1650	0.41	9.5	70
Hand-formed brick	1725	0.38	17	200
Historical brick	1800	0.31	15	300
Cement (C) type S mortar	1885	0.50	15.5	0.00075
Cement (C) type N mortar	1885	0.50	14.8	0.0015
Coarse lime cement (LC) mortar	1910	0.25	46	0.87
Fine lime cement (LC) mortar	1880	0.28	50	0.31
Coarse hydraulic lime (L) mortar	1830	0.27	20	0.9
Fine hydraulic lime (L) mortar	1700	0.35	14.8	0.33

* The sorption isotherms of the bricks included in this study are presented in Figure 4.

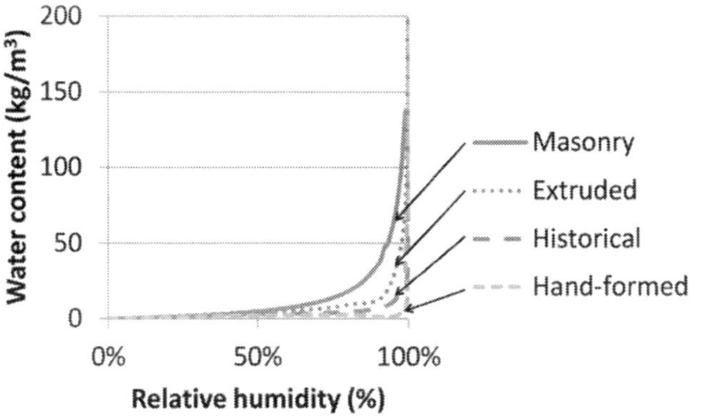

Figure 4 *Sorption isotherms at 23°C of the four types of brick included in the study based on data from WUFI 2D (Fraunhofer IBP 2010).*

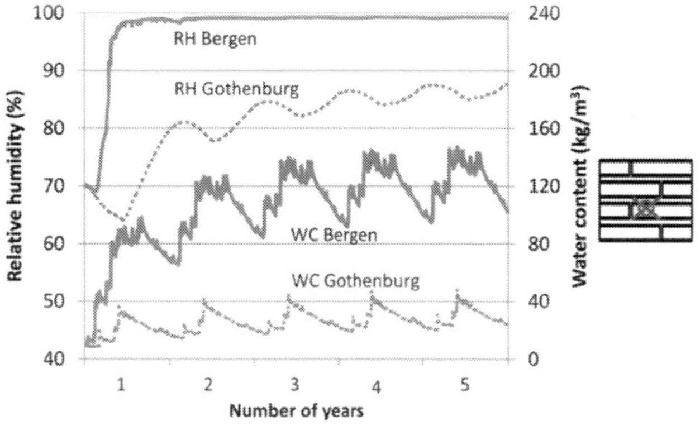

Figure 5 *Relative humidity in the middle of a 380 mm (15 in.) thick homogenous brick wall (marked by a cross with circle in the right figure) and total water content in the wall, which is made of extruded brick and coarse lime cement mortar. The wall is facing south in both Bergen and Gothenburg with a 0.3 rainwater absorption factor.*

(0.35 Btu/[h·ft·°F]) and 850 J/(kg·K) (0.2 Btu/[lb·°F]), respectively. The initial conditions of the materials were 15°C (59°F) and 70% relative humidity. The sizes of the bricks were 250 × 120 × 62 mm (9.8 × 4.7 × 2.4 in.) and the thicknesses of the mortar joints between the bricks were 10–12 mm (0.39–0.47 in.) (see number 1 in Figure 2). On the interior, the climate is defined by EN 15026 (CEN 2007) with a normal moisture load. The interior heat transfer coefficient, *h*, is 8 W/m^2·K (R-0.7 ft^2·°F·h/Btu), and the interior diffusion equivalent air layer thickness, *sd*, is 3 m (1.1 perm). The *sd* value describes the resistance to moisture transfer by the surface material in terms of equivalent air thickness. The rainwater hitting the façade, available for capillary absorption, is determined by the rainwater absorption factor. When the façade is fully protected and no capillary absorption takes place the factor is 0, and if all the rain is available for capillary absorption the factor is 1. In WUFI 2D, a rainwater absorption factor of 0.7 is considered to be adequate for normal walls (Fraunhofer IBP 2010). The dominant wind direction for both Bergen and Gothenburg is south, which therefore was chosen as the direction the wall is facing. The wall is considered to be protected from most of the rain with a rain water absorption factor of 0.3. The exterior heat transfer coefficient is 25 W/m^2·K (R-0.2 ft^2·°F·h/Btu). The results are shown in Figure 5.

In Bergen, the resulting relative humidity in the middle of the wall is close to 100% after only 1 year of simulation. Also, the water content in the middle of the wall increases rapidly during the first years of the simulation, stabilizing during years 4 and 5. The moisture is evenly distributed in the wall. The maximum water content is 147 kg/m^3 (9 lb/ft^3) in December of year 4, while the saturation moisture content is 370 kg/m^3 (23 lb/ft^3). At the surface of the brick, where there could be risk of freeze-thaw damages, the moisture content is significantly higher, peaking at 298 kg/m^3 (19 lb/ft^3). The degree of saturation becomes 0.4 at maximum, and the number of freeze-thaw cycles during the simulated year is 16. However, the critical degree of saturation is not known for the brick used in the simulations. As mentioned previously, measurements have shown that the critical degree of saturation can be in the range of 0.25–0.8 (Mensinga et al. 2010). Investigation of the critical degree of saturation for the specific brick used in the wall to be retrofitted has to be determined to conclude the risk for freeze-thaw damages.

In Gothenburg the brick wall is also be exposed to driving rain; however, the rain load is not as high as it is in Bergen. The relative humidity peaks at 88% in the middle of the wall in August of year 5. The water content in the middle of the wall is only 48 kg/m^3 (3 lb/ft^3) at maximum while it is 211 kg/m^3 (13 lb/ft^3) at the exterior surface of the brick wall. The moisture content is highest in the exterior part of the wall. The degree of saturation becomes 0.13, which should be below the critical degree of saturation for most brick types. Although, the number of freeze-thaw cycles in a year in Gothenburg is 35, which is much higher than that in Bergen. The combination of many freeze-thaw cycles and unprotected brick façades could explain why many brick buildings in Gothenburg have freeze-thaw damage. If the wall is not protected from driving rain, the moisture content in the middle of the wall will be very high, even in its original state.

Wall Thickness

One-dimensional hygrothermal simulations are used to estimate how long of a testing period the full wall thickness would require to notice the moisture accumulation in the wall. The hypothesis is that the same hygrothermal mechanisms will take place in the wall independent of the thickness. Only the time scale will be influenced by a reduced thickness in the laboratory study. Figure 6 shows the relative humidity 60 mm (2.4 in.) from the interior of the wall in two brick walls of three different thicknesses. In this part of the study it is the influence of the bricks only that is of interest; therefore, no mortar was simulated between the bricks (see number 2 in

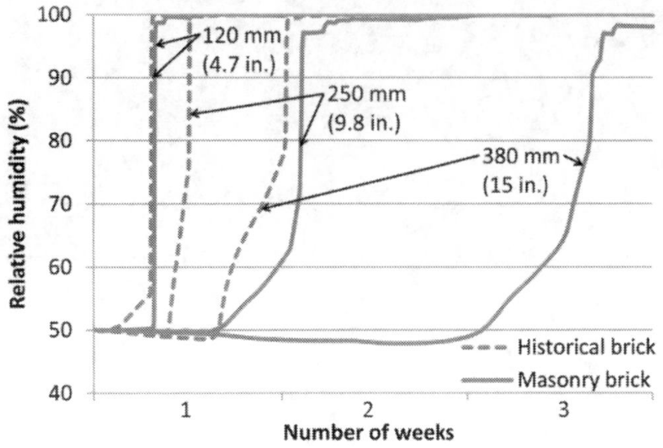

Figure 6 *Relative humidity 60 mm (2.4 in.) from the interior surface of the two brick walls with varying thicknesses composed of homogenous historical bricks and masonry bricks as defined in Table 1 and during the climate cycle defined in Table 2.*

Figure 2). The walls are composed of homogenous historical bricks and masonry bricks and are defined in Table 1. The initial conditions of the materials are 20°C (68°F) and 50% relative humidity. The interior climate is 20°C (68°F) and 40% relative humidity. The interior heat transfer coefficient is 8 $W/m^2 \cdot K$ (R-0.7 $ft^2 \cdot °F \cdot h/Btu$) and the interior *sd* value is 0.2 m (16 perm). The exterior climate is defined in Table 2, which is based on the HAMSTAD Benchmark Project (Hagentoft et al. 2004) where the moisture accumulation in a brick wall was studied during a climate cycle of heavy rainfall with drying periods in between. Note that this is a fictitious climate where the exterior relative humidity is fixed at 90%. The hours with rainfall are limited, which means the drying during these hours is limited compared to the wetting by the rain and the drying during the hours without rain. The exterior heat transfer coefficient is 25 $W/m^2 \cdot K$ (R-0.2 $ft^2 \cdot °F \cdot h/Btu$) with a wind-dependent vapor resistance. The wall is facing the dominant wind direction towards the south with a rain water absorption factor of 1; i.e., all rain is available for capillary absorption.

The time before the wall reaches saturation 60 mm (2.4 in.) from the interior surface of the brick wall is different depending on the thickness of the wall and the type of bricks in the wall. For the 120 mm (4.7 in.) thick historical brick wall it takes 53 hours, for the 250 mm (9.8 in.) thick wall it takes 87 hours, and for the 380 mm (15 in.) thick wall it takes 173 hours. It takes a longer time for the masonry brick wall to reach saturation. For the 120 mm (4.7 in.) thick wall it takes 64 hours, for the 250 mm (9.8 in.) thick wall it takes 218 hours, and for the 380 mm (15 in.) thick wall it takes 578 hours. The relation is not linear between the thickness and moisture accumulation. For the thinnest historical brick wall the moisture condensates after 30% of the time for the thickest wall, while it takes 11% of the time for the masonry

Table 2. Exterior Climate for Wetting of Brick Wall in a 120 Hour Cycle*

Duration, h	Rain, mm	Sun, W/m^2	Exterior Temperature, °C	Exterior Relative Humidity, %
6	0	0	10	90
16	0	0	–2	90
27	0	0	10	90
6	2	0	10	90
7	0	0	10	90
4	2.5	0	10	90
6	0	0	10	90
7	0	1000	10	90
5	0	0	10	90
4	2	0	10	90
7	0	1000	10	90
3	3	0	10	90
8	0	1000	10	90
4	2	0	10	90
10	0	0	10	90

* Based on the climate used in the HAMSTAD Benchmark Project (Hagentoft et al. 2004).

brick wall; the thickness is 32% of the thickest wall. In the middle wall, 250 mm (9.8 in.), the moisture condensates after 50% of the time for the historical brick, while it takes 38% of the time for the masonry brick wall; the thickness is 66% of the thickest wall. Despite the differences in the time of wetting, the moisture levels in the walls are at the same level at the end of the simulation and thus it is concluded that a thinner wall can be used in the experiment but the time scale has to be decided based on the permeability of the bricks.

Type of Brick

In previous research, it was found that the type of brick is very important to the test results; modern bricks are less capillary active than the hand-formed bricks manufactured many years ago (Hammett 1997). The main difference between the four bricks given in Table 1 is the liquid water transfer coefficient, which varies from 1.7 to 300 m^2/s (0.2 to 27.9 ft^2/s). For the masonry and extruded bricks, the liquid transfer is lower than for the hand-formed and historical bricks. The sorption isotherms of the bricks included in this study are presented in Figure 4.

Figure 7 shows the results from one-dimensional simulations of the moisture accumulation in a 250 mm (9.8 in.)

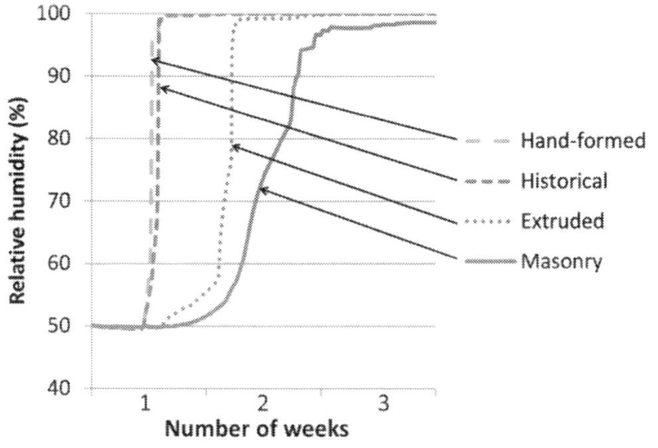

Figure 7 *Relative humidity 60 mm (2.4 in.) from the interior surface of the 250 mm (9.8 in.) brick wall with one of the brick types and fine lime cement mortar from Table 1 during the climate cycle defined in Table 2.*

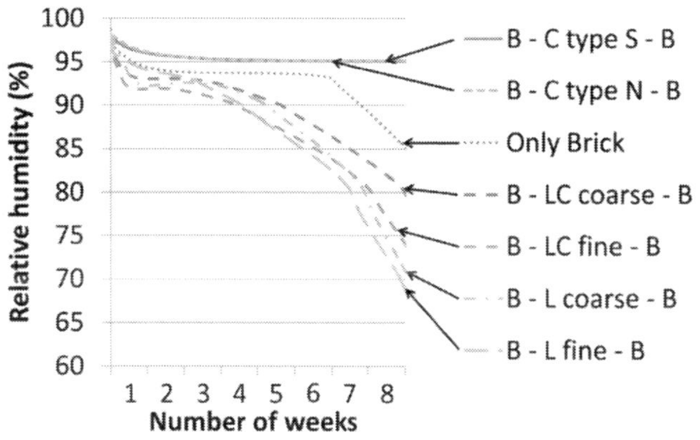

Figure 8 *Relative humidity in the middle of the mortar in a 250 mm (9.8 in.) extruded brick (B) wall with varying mortar properties placed indoors in Trondheim, Norway, starting on October 1 based on data from the WUFI database (Fraunhofer IBP 2010). For comparison, a wall with only brick is simulated under the same conditions.*

brick wall with one of the four different brick types in Table 1. There is a layer of fine lime cement mortar in the middle of the wall to simulate the real case of a wall with two bricks placed next to each other (see number 3 in Figure 2). The initial conditions and boundary conditions are the same as in the previous section.

The difference between the bricks is clear, and the four types of brick tested can be divided into two main groups where the historical and hand-formed bricks are similar while the extruded and masonry bricks are less similar. The time before the walls are saturated with moisture at the point 60 mm (2.4 in.) from the interior differs from 99 and 108 hours for historical and hand-formed bricks to 217 hours for extruded brick and 588 hours for masonry. These differences confirm that it is important to investigate the properties of the specific type of bricks used in the study to be able to predict the results of the laboratory study.

Type of Mortar

The type of mortar also influences the moisture properties of the wall. To investigate the time required for drying the wall during the laboratory investigations, a wall was placed indoors in the laboratory environment and the drying period was investigated. The material data for the six different mortars are given in Table 1. The main difference between the mortars is the liquid water transfer coefficient, which is substantially larger for the lime cement mortars and hydraulic lime mortars compared to the pure cement mortar. In the early 20th century, lime mortar was used in masonry buildings. Pure lime mortar requires a longer time for curing than a mixture of lime and cement. The difference between cement types S and N is in the final strength. Type S has higher final strength than type N, which is the type normally recommended for masonry work (BIA 2006). Figure 8 shows the drying time for a wall dependent on the type of mortar used in a 250 mm (9.8 in.) thick wall with the extruded brick from Table 1 and a 10–12 mm (0.39–0.47 in.) thick mortar layer between the bricks with two-dimensional heat and moisture transfer (see number 4 in Figure 2). The starting condition is a fully saturated mortar at 20°C (68°F) and brick with 80% relative humidity. The interior climate is 20°C (68°F) and 40% relative humidity. The interior heat transfer coefficient is 8 $W/m^2 \cdot K$ (R-0.7 $ft^2 \cdot °F \cdot h/Btu$) and there is no surface coating applied. The wall is fully exposed to the indoor climate of the laboratory in Trondheim, Norway, to simulate the drying during the period starting on October 1 with a normal moisture load defined by EN 15026 (CEN 2007). For comparison, a wall of only brick is simulated in the same conditions.

Pure cement mortar (C) gives a very slow drying process, and after 8 weeks the relative humidity in the wall has only been reduced by 3 percentage points, to 95%. For the other mortar types, the drying is much faster. The fastest is for the fine hydraulic lime mortar (L), where the relative humidity is reduced to 70% after 8 weeks of drying. The coarse and fine types of the mortars have different properties as well, but the difference in relative humidity is not as large after 8 weeks as when comparing the lime cement mortar (LC) and hydraulic lime mortar (L). The wall with only brick has a slow drying during the first 6 weeks, but thereafter it dries quickly, down to 83% after 8 weeks. The influence of the mortar is that it shortens the drying process of the wall, which could be explained by a higher moisture transport through the thinner capillaries of the mortar than through the coarser brick capillaries. This could also have an influence during the wetting phase of the walls with different thicknesses, described previously. However, as Figure 8 shows, the difference is limited between the wall with only brick and the different mortars. For the forthcoming laboratory study, a mixture of lime and cement will be used as mortar.

Internal Thermal Insulation

The thermal conductivity of pristine VIP is below 5 mW/(m·K) (R-29 per in.) but increases, due to diffusion of water vapor and air through the enclosing VIP laminate, to around 7 or 8 mW/(m·K) (R-18-21 per in.) in 25 years (Brunner and Simmler 2008). The VIPs investigated in this study are pristine, which is why the lower thermal conductivity of 5 mW/(m·K) (R-29 per in.) was chosen.

The VIPs have a very high vapor resistance compared to, for example, mineral wool insulation. On the other hand, the gaps between the panels can be around 2 to 4 mm (0.08 to 0.16 in.), which could allow vapor transfer. To investigate this, four cases are studied more in detail:

1. Original brick wall; $h = 8$ W/ m^2·K (R-0.7 ft^2·°F·h/Btu), $sd = 0.2$ m (16 perm)
2. Vapor barrier on the interior; $h = 8$ W/ m^2·K (R-0.7 ft^2·°F·h/Btu), $sd = 1500$ m (0.002 perm)
3. Vapor-open insulation with a heat resistance equivalent to 20 mm (0.8 in.) VIP; $h = 0.242$ W/m^2·K (R-235 ft^2·°F·h/Btu), $sd = 0.2$ m (16 perm)
4. Interior insulation with 20 mm (0.8 in.) VIP; $h = 0.242$ W/m^2·K (R-235 ft^2·°F·h/Btu), $sd = 1500$ m (0.002 perm)

The original wall is 250 mm (9.8 in.) thick with the historical brick from Table 1 and a 10 mm (0.39 in.) thick layer of fine lime cement mortar between the bricks as shown in number 3 in Figure 2. The initial conditions of the materials are 20°C (68°F) and 50% relative humidity. The interior climate is 20°C (68°F) and 40% relative humidity. The exterior climate is defined in Table 2. The exterior heat transfer coefficient is 25 W/m^2·K (R-0.2 ft^2·°F·h/Btu) with a wind-dependent vapor resistance. The wall is facing south with a rainwater absorption factor of 0.1, i.e., a small amount of rain is available for capillary absorption. WUFI suggests 0.7 for most wall arrangements where the wall is fully exposed to the climate loads (Fraunhofer IBP 2010). In this case, the wall got fully saturated after a very short time period and no difference could be seen for the insulation cases. Therefore, a lower rainwater absorption factor was chosen for these simulations where 10% of the rain is available for capillary absorption and the remaining water runs down the façade or splashes off at impact. The results from the one-dimensional simulations of the four cases are shown in Figure 9.

As can be seen in Figure 9, adding a vapor barrier on the original brick wall does not change the total moisture content of the wall compared to the case without additional insulation. However, at the location 60 mm (2.4 in.) from the interior surface, the relative humidity is higher in the case of adding a vapor barrier than that without a vapor barrier, which indicates a lack of drying towards the interior space. This vapor transport is blocked when adding the vapor barrier.

In the two cases with thermal insulation installed on the interior surface of the brick wall, the moisture accumulation rate is higher than that of the original wall. With a vapor-open insulation, the moisture accumulation rate is slightly lower than for the case with VIP, which has a very high vapor resistance. Also, the relative humidity is slightly higher for the wall with VIP than for the wall with vapor-open insulation. The main part of the vapor and water flow in the wall is caused by the rain events and the indoor moisture content is of minor importance for the conditions studied here.

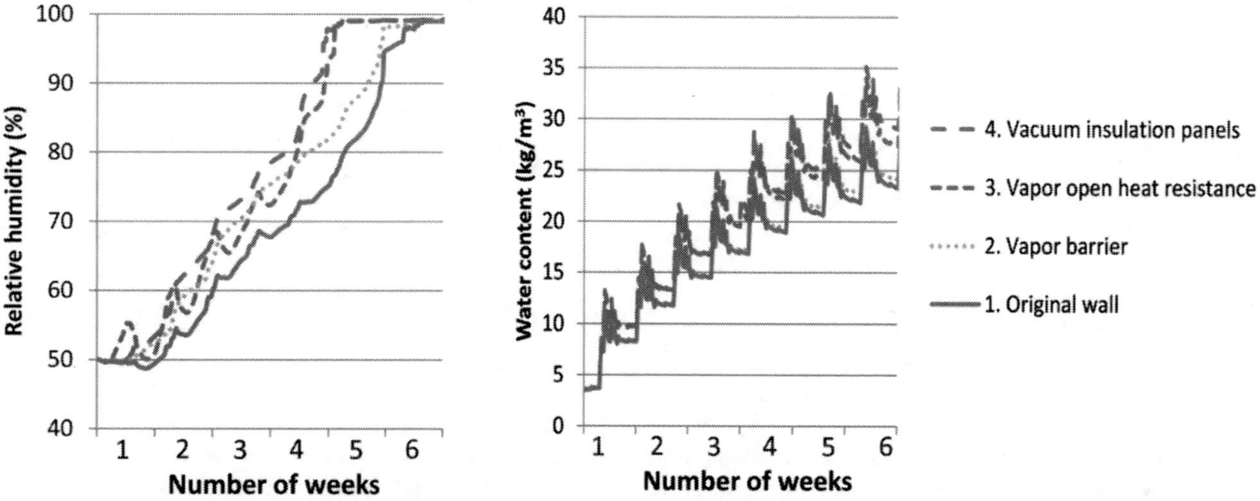

Figure 9 *Simulation of the relative humidity 60 mm (2.4 in.) from the interior surface and the total water content in the 250 mm (9.8 in.) wall composed of homogenous historical bricks and fine lime cement mortar between the bricks with varying interior boundary conditions during the climate cycle defined in Table 2.*

CONCLUSIONS

One- and two-dimensional hygrothermal simulations were used to investigate the relative humidity in a 250 mm (9.8 in.) thick brick wall. The purpose is to prepare an experimental benchmark in a laboratory environment where a brick wall with wooden beam ends is insulated with vacuum insulation panels (VIPs) on the interior. More specifically, the purpose is to estimate the influence of different brick and mortar types on the wetting and drying of the wall, to estimate the time scale of wetting, to estimate the rain load, and to investigate a possible reduction of the wall thickness from 380 mm (15 in.) to 250 mm (9.8 in.).

Two climate loads were investigated, where the Bergen, Norway, climate gave the highest moisture accumulation rate compared to the Gothenburg, Sweden, climate. The wall could be at risk of suffering freeze-thaw damage on the exterior surface in Bergen since the degree of saturation is close to the critical degree of saturation for some brick types. In Gothenburg, the wall did not reach as high a degree of saturation, but the number of freeze-thaw cycles was substantially higher.

The thickness of the wall influences the moisture accumulation rate when investigating the bricks 60 mm (2.4 in.) from the interior surface. The moisture accumulation rate does not decrease linearly with increasing thickness but has a more exponential relationship.

The moisture content in the wall is highly influenced by the properties of the brick. Two main groups of bricks were investigated, where hand-formed and historical bricks had similar properties while masonry and extruded bricks had similar properties. The time before the walls are saturated with moisture at the point 60 mm (2.4 in.) from the interior surface differed with a factor of 6 between the least and most permeable bricks.

The type of mortar influenced the drying of the wall during the investigated drying in a laboratory in the Trondheim, Norway, climate. The mortar that gave the lowest drying rate was the pure cement mortar, while the mixture of lime and cement gave a lower drying rate than pure lime mortar.

The properties of the interior insulation material were found to have a lesser influence on the moisture accumulation rate. The rain load is the dominating factor determining the vapor and water transport in the wall. Having the possibility of inward drying lowers the moisture accumulation rate slightly.

Based on the results from the numerical analysis, a 250 mm (9.8 in.) brick wall is proposed where the brick properties should resemble those of the brick type used in the early 20th century, i.e., a brick that is highly capillary active with a high liquid transport coefficient. The difference between using lime mortar and a lime cement mortar was not substantial; thus, lime cement mortar can be used in the laboratory tests. The rain load in the laboratory experiment should be limited so that the effect of the added interior VIP on the brick wall and wooden beam ends is visible. With the results from the ongoing laboratory measurements, the influence of driving rain and solar radiation on a brick wall with interior VIP and wooden beam ends can be evaluated. The results can be used to refine the hygrothermal simulations so that more advanced simulations can be performed to draw more general conclusions of the applicability of VIPs in interior retrofitting of existing brick buildings. Using VIP, drying towards the interior is more limited and the only moisture transfer between the interior climate and the wall will be through the wooden beam ends and through possible cracks and air gaps between the VIP and at the beams. If the surface at the intermediate floor-to-wall attachment is left in its original state, moisture can also be transferred into the wall from here. In future studies these matters will be investigated numerically and compared to the results of the laboratory test.

ACKNOWLEDGMENTS

The work is supported by the Swedish Research Council FORMAS, the Lars Hierta Memorial Foundation, the Research Council of Norway, and several partners through the Research Centre on Zero Emission Buildings (www.ZEB.no).

REFERENCES

Abuku, M., Blocken, B., and Roels, S. 2009. Moisture response of building facades to wind-driven rain: Field measurements compared with numerical simulations. *Journal of Wind Engineering and Industrial Aerodynamics* 97(5-6):197–207.

Angelantoni Industries S.p.A. n.d. Double room test equipment for building walls thermal performance testing Mod. UC2x18 RT/SR. Massa Martana, Italy: Angelantoni Industries S.p.A.

Baetens, R., Jelle, B.P., Thue, J.V., Tenpierik, M.J., Grynning, S., Uvsløkk, S., and Gustavsen, A. 2010. Vacuum insulation panels for building applications: A review and beyond. *Energy and Buildings* 42(2):147–72.

BIA. 2006. *Mortars for Brickwork—Selection and Quality Assurance*. Reston, Virginia, USA: The Brick Industry Association.

Bjørberg, S., Lirhus, E.H., and Skattum, P. 2011. Vedlikeholdsveiledning: Murgårdsbebyggelsen i Oslo (Maintenance manual: Masonry buildings in Oslo). Oslo, Norway: Multiconsult AS.

Boverket. 2009. Så mår våra hus – redovisning av regeringsuppdrag beträffande byggnaders tekniska utformning m.m. (The state of our buildings – report of governmental mission on the technical design of buildings etc.). Karlskrona, Sweden: Boverket.

Boverket. 2010. Energi i bebyggelsen - tekniska egenskaper och beräkningar – resultat från projektet BETSI (Energy in the built environment – technical properties and calculations – results from the BETSI study). Karlskrona, Sweden: Boverket.

Brocken, H.J.P. 1998. Moisture transport in brick masonry: The grey area between bricks. Dissertation, Building Physics and Systems, Technische Universiteit Eindhoven, Eindhoven, The Netherlands.

Brunner, S., and Simmler, H. 2008. In situ performance assessment of vacuum insulation panels in a flat roof construction. *Vacuum* 82(7):700–707.

Brunner, S., Tharian, P.J., Simmler, H., and Ghazi Wakili, K. 2008. Focused ion beam (FIB) etching to investigate aluminium-coated polymer laminates subjected to heat and moisture loads. Surface and Coatings Technology, 202(24):6054-6063.

CEN. 2007. EN 15026:2007, *Hygrothermal Performance of Building Components and Elements – Assessment of Moisture Transfer by Numerical Simulation*. Brussels: European Committee for Standardization.

Hagentoft, C.-E., Kalagasidis, A.S., Adl-Zarrabi, B., Roels, S., Carmeliet, J., Hens, H., Grunewald, J., Funk, M., Becker, R., Shamir, D., Adan, O., Brocken, H., Kumaran, K., and Djebbar, R. 2004. Assessment method of numerical prediction models for combined heat, air and moisture transfer in building components: Benchmarks for one-dimensional cases. *Journal of Thermal Envelope and Building Science*, 27(4):327–52. <Benchmark files are available for download from www.byggnadsteknologi.se/downloads.html>

Hall, C. 1977. Water movement in porous building materials—I. Unsaturated flow theory and its applications. *Building and Environment* 12(2):117–25.

Hammett, M. 1997. *Resisting Rain Penetration With Facing Brickwork*. London, UK: The Brick Development Association.

Fraunhofer IBP. 2010. WUFI 2D (Version 3.3.2) [Computer Program]. Holzkirchen, Germany: Fraunhofer IBP.

Künzel, H. 1998. Effect of interior and exterior insulation on the hygrothermal behaviour of exposed walls. *Materials and Structures* 31(2):99–103.

Kvande, T., and Edvardsen, K. I. 2007. Eldre yttervegger av mur og betong. Metoder og materialer. (Old exterior brick and concrete walls. Methods and materials). Building Research Design Guide 723.308. Oslo, Norway: SINTEF Byggforsk.

Mensinga, P., Straube, J., and Schumacher, C. 2010. Assessing the freeze-thaw resistance of clay brick for interior insulation retrofit projects. *Proceedings of the Thermal Performance of the Exterior Envelopes of Whole Buildings XI*, Clearwater Beach, FL, USA, December 5–9.

Morelli, M., Scheffler, G.A., Nielsen, T.R., and Svendsen, S. 2010. Internal insulation of masonry walls with wooden floor beams in northern humid climate. *Proceedings of the Thermal Performance of the Exterior Envelopes of Whole Buildings XI*, Clearwater Beach, FL, USA, December 5–9.

Nicolai, A., J. Grunewald, and H. Fechner. 2013. DELPHIN: Simulation program for the calculation of coupled heat, moisture, air, pollutant, and salt transport. Bauklimatik-Dresden, Germany.

Piaia, J.C.Z., Cheriaf, M., Rocha, J.C., and Mustelier, N.L. 2013. Measurements of water penetration and leakage in masonry wall: Experimental results and numerical simulation. *Building and Environment* 61:18–26.

Pfluger, R., Hasper, W., Schulze Darup, B., Sommer, D., and Forstner, M. 2008. Wissenschaftlige Analyse eines auf vorgefertigten Vakuum-Paneel-Verbundplatten beruhenden Innendämmsystems. Abschlussbericht (Scientific analysis of an interior insulation system based on prefabricated vacuum insulation composite panels. Final report). Neumarkt, Germany: VARIOTEC GmbH & Co.KG.

Rasmussen, T.V. 2010. Post-insulation of existing buildings constructed between 1850 and 1920. *Proceedings of the Thermal Performance of the Exterior Envelopes of Whole Buildings XI*, Clearwater Beach, FL, USA, December 5–9.

Straube, J., Ueno, K., and Schumacher, C. 2011. Internal insulation of masonry walls: Final measure guideline. Prepared for U.S. Department of Energy, Office of Energy Efficiency and Renewable Energy.

Sveipe, E., Jelle, B.P., Wegger, E., Uvsløkk, S., Grynning, S., Thue, J.V., Time, B., and Gustavsen, A. 2011. Improving thermal insulation of timber frame walls by retrofitting with vacuum insulation panels—Experimental and theoretical investigations. *Journal of Building Physics* 35(2): 168–88.

Thyholt, M., Pettersen, T.D., Haavik, T., and Wachenfeldt, B.J. 2009. Energy Analysis of the Norwegian Dwelling Stock. Subtask A—Internal working document. IEA SHC Task 37 Advanced Housing Renovation by Solar and Conservation.

Wegger, E., Jelle, B.P., Sveipe, E., Grynning, S., Gustavsen, A., Baetens, R., and Thue, J.V. 2011. Aging effects on thermal properties and service life of vacuum insulation panels. *Journal of Building Physics* 35(2):128–67.

Selection Tool for Water-Resistive Barriers with Suitable Vapor Permeability

Marcus Jablonka **Achilles Karagiozis, PhD** **John Straube, PhD**
Associate Member ASHRAE

ABSTRACT

Over the past several decades, advances in the application of polymers brought new polymeric water-resistive barriers with a wide range of vapor permeability to market, commonly referred to as "breathable" housewraps. However, at this time there is virtually no guidance available regarding the selection of the optimum vapor permeability of such membranes under specific conditions.

This paper evaluates the impact of various water-resistive barriers with a large range of vapor permeability on the hygrothermal performance of different wall assemblies. The information enables designers to select products with the most suitable vapor permeability for particular geographical locations and construction conditions. Variations in boundary conditions included climatic conditions (seven climatic locations), cladding type (three-coat stucco, manufactured stone, cement board, brick), and type of WRB (low versus high vapor permeability) deployed. The results for the performance of the wall systems are presented in form of a mold index.

INTRODUCTION

The primary function of a water-resistive barrier (WRB) in a building enclosure system is to serve as a second line of defense and shed water that penetrates the cladding. Even though the building enclosure may be designed properly by design professionals, experience shows that defects created during the construction process or those occurring during the service life of the structure may allow water to enter the wall assembly. Hence, for a wall assembly to function well, it should be designed to permit drainage on the surface of the water-resistive barrier and—particularly important for wood frame construction—drying of any excess moisture. Therefore, the water-resistive barrier is required to be vapor permeable in order to allow for outward diffusion of water vapor. The moisture balance of the building material adjacent to the water-resistive barrier will be strongly affected by the water vapor flow caused by thermal drive, which may vary depending on the moisture content, temperature of outdoor air, and solar radiation. A reverse thermal gradient may cause inward vapor diffusion into the wall cavity. For this reason water-resistive barriers need to be evaluated in regards to their effect on the performance of a wall assembly (Jablonka 2011).

Different types of water-resistive barriers may be incorporated into the wall assembly, and—depending on climatic conditions and the type of sheathing and cladding material used—may have different effects on its performance and durability. The available variety and characteristics of such membrane products have changed significantly over the past several decades. Advances in the application of polymers brought a large variety of "breathable" housewraps with a wide range of vapor permeability to market. However, the large variety of performance characteristics of such membrane products has resulted in uncertainty among design professionals and the construction industry at large regarding the in-service performance of various types of water-resistive barriers and the selection of the optimum vapor permeability under specific conditions. This paper describes a research project that evaluates the impact of various water-resistive

***Marcus Jablonka** is director at Cosella-Dorken Products, Beamsville, ON, Canada. **Achilles Karagiozis** is a distinguished research engineer at Oak Ridge National Laboratory, Oak Ridge, TN, USA. **John Straube** is a principal at Building Science Consulting, Waterloo, ON, Canada.*

barriers with a large range of vapor permeability on the hygrothermal performance of different wall assemblies—information vitally important for proper product selection. For the convenience of designers the results of this research project have been summarized in a software selection tool.

OBJECTIVE

The objective of this research project was to understand the performance of different water-resistive barriers with various vapor permeabilities in different climates and cladding applications in building wall assemblies. The sensitivity to different types of water ingress (location in the wall assembly) was examined as a function of water-resistive barrier and climate.

SCOPE

Variations in boundary conditions included climatic conditions (seven climatic locations), cladding type (brick, adhered manufactured stone veneer, cement-board, three-coat stucco), and type of WRB (low versus high vapor permeability) deployed.

The research approach was structured into several phases: In the first phase the water vapor permeability of different water-resistive barriers were determined as per ASTM E96-00. Subsequently a subassembly laboratory test was designed to simulate performance of a small component of a wall system during operation under controlled conditions to predict the performance of large-scale assemblies and to validate the performance simulation tool. A variation of the hygrothermal loads was performed to allow gapping between perfectly built wall assemblies and walls with realistic imperfections (workmanship issues). A number of parameters were varied to understand the sensitivity of the results to the different types of substrate, cladding, and climatic locations. In the next phase simulations were carried out with a hygrothermal computer model (WUFI 5, Karagiozis et al. 2001). The sheathing moisture content, temperature, and relative humidities were plotted against time for comparison and analysis, and presented as an index of moisture performance. In the final phase the results were embedded into a software selection tool, allowing an architect to select a specific climate zone, cladding type, water-resistive barrier with a particular perm rating, and wall orientation. Results are being presented as a function of moisture performance index (mold index).

WATER VAPOR PERMEANCE TESTING

Material property and subassembly tests were performed to support and strengthen the computer simulations. Table 1 shows a summary of the water-resistive barriers that were tested and their dry cup and wet cup vapor permeance values determined using ASTM E96-00 Method A and Method B (Straube et al. 2010). The test dishes were sealed with aluminum tape to ensure that the only vapor movement observed was through the test specimen. The standard requires a minimum sample size of 3000 mm^2. Samples of 16,200 mm^2 were used to ensure that the test results are not influenced by local variations in vapor permeance of the sample. The temperature was controlled to 23°C as specified for Method B. The relative humidity was kept constant at 50%. Further details of this part of the laboratory testing are described in Straube et al. (2010).

The wet cup (Method B) testing did result in higher average permeance values than the dry cup (Method A) testing, as anticipated. The greatest increase in vapor permeance occurred with the WRB B water-resistive barrier which nearly doubled in vapor permeance between the dry cup and wet cup tests. The wet cup vapor permeance test is more appropriate for determining the drying performance of walls as the cladding in many climates is more often between 50% and 100% relative humidity (wet cup conditions) than between 0% and 50% relative humidity.

SUBASSEMBLY TESTING

Testing the vapor permeance according to ASTM E96 demonstrates how the water-resistive barrier performs as an individual material, but it is also important to understand how the water-resistive barrier performs in combination with OSB or exterior gypsum sheathing, which more closely simulates a wall assembly. A subassembly laboratory study was undertaken to more clearly understand the drying ability of water-resistive barriers in combination with OSB or exterior grade gypsum sheathing. The subsystem testing was designed to simulate performance of a small component of a wall system during operation under controlled conditions to predict the performance of large-scale assemblies, and to validate the performance simulation model.

Two different types of polymeric, vapor permeable water-resistive barriers and #15 asphalt impregnated building paper were tested in the subassembly test. Twenty-seven subassembly test samples were made using three different water-resistive barriers installed on either OSB or exterior grade gypsum sheathing as shown in the testing matrix in Table 2. The differences between interior and exterior wetting are shown in Figure 1.

Table 1. Summary of Vapor Permeance as per ASTM E96 (Dry Cup Versus Wet Cup Method)

Water-Resistive Barrier	Method A	Method B
	Dry Cup, ng/Pa·m^2·s	Wet Cup, ng/Pa·m^2·s
WRB A	12,284 (214 perms)	13,812 (241 perms)
WRB B	804 (14 perms)	1597 (28 perms)
WRB C	3444 (60 perms)	3737 (65 perms)

Square samples measuring 330 × 330 mm (13 × 13 in.) were cut from sheets of OSB and exterior grade gypsum sheathing. The edges of the samples were wrapped with foil tape to create a 305 × 305 mm (12 × 12 in.) active test area.

Four layers of moisture storage media were installed between the sheathing and the water-resistive barrier to simulate exterior wetting or installed on the opposite side of the sheathing from the water-resistive barrier to simulate wetting on the interior, in the stud cavity as shown in the schematic in Figure 1. A 0.125 in. ID tube was installed to provide water to the moisture storage media.

Three layers of 6 mil poly were installed on the interior surface of the sheathing and sealed with aluminum foil tape to the edges of the sample. For interior wetting, the water storage media was visible through the poly to inspect the storage media for saturation.

Figures 2 and 3 show the exterior and interior surfaces of a subassembly testing sample. Figure 3 shows the wetting storage media on the interior surface through the multiple layers of polyethylene sheet.

Table 2. Testing Matrix—Number of Subassembly Samples of Each Construction

	OSB—Wetting		Gypsum Sheathing—Wetting	
	Interior	Exterior	Interior	Exterior
WRB A	4	1	3	1
WRB B	4	1	3	1
Building Paper	4	1	3	1

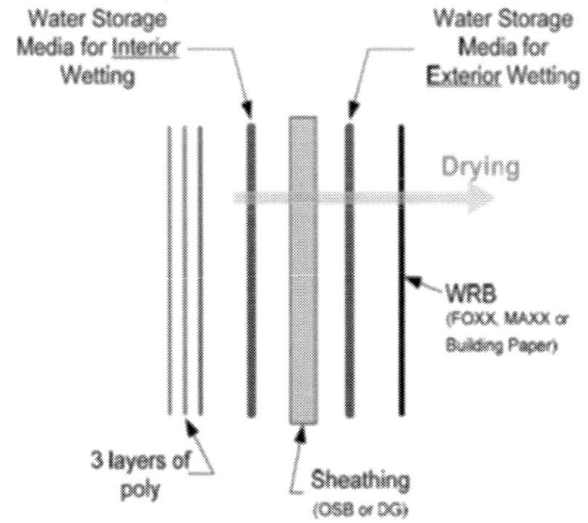

Figure 1 *Subassembly testing sample schematic.*

Procedure

The test for each sample began by adding 100 mL (5 doses of 20 mL) by syringe to the water storage media. Each dose of water was injected over approximately 10 seconds followed by 20 seconds of wait time for water to redistribute into water storage media before adding more water. The samples were held at a constant angle for all water injections.

The samples were weighed before and after any water was added to determine the actual mass of water added by syringe. Each subsequent day, the sample was weighed to

Figure 2 *Exterior surface of subassembly test sample with exterior wetting.*

Figure 3 *Interior surface of subassembly test sample with interior wetting.*

determine the mass of water lost, and the same amount of water lost was added. The total mass lost and water added were graphed in Figure 4 to show the rates of both wetting and drying. If the two rates (line slopes) for wetting and drying were similar, the sample was determined to be at equilibrium and an effective permeance could be calculated.

If no water had been lost from the previous day, 20 more mL of water was added each subsequent day until "ponding" of water was observed in the water storage media through the polyethylene. This occurred when the storage media was saturated and water was not being absorbed into the substrate quickly enough.

Once the rates of wetting and drying were calculated, repeatability was determined by increasing the amount of water injected into the water storage media for two subsequent days at a rate of twice the daily loss. This dosage was increased to determine if the effective system permeance would change when a higher volume of water was added to the sample. Following the two days of increased loading, only the amount of water lost was added back to the sample.

Boundary Conditions

The testing was conducted in a constant climate room (CCR) where the RH and temperature can be tightly controlled. The climate controls for the room were set at 23°C (±1°C) and 50% rh (±2%).

To determine the vapor pressure gradient it was assumed that the relative humidity in the sample during the testing and continuous daily wetting of the water storage media was 100%. These subassembly tests help simulate real wetting conditions, but in wall systems there are usually temperature gradients across a wall that will affect the drying rate. These tests were run with no temperature gradient across the subassembly system, which is the worst-case scenario for drying performance. If a temperature gradient was added, the drying rates would increase, but the ratio of drying amounts would remain the same.

Results

Water-Resistive Barrier A. The dry cup vapor permeance of the WRB A water-resistive barrier according to ASTM E96, Method A is 214 perms (12,284 ng/Pa·s·m^2), and the wet cup vapor permeance according to ASTM E96, Method B is 241 perms (13,812 ng/Pa·s·m^2). This was the highest vapor permeance water-resistive barrier in the subassembly testing.

- OSB substrate: OSB has a vapor permeance of approximately 1 to 2 US perms (57 to 115 ng/Pa·s·m^2), although the vapor permeance will change with RH as well as age of the OSB.

  Figure 5 shows a typical data set from samples of OSB with interior wetting. Very little mass was lost over the duration of the test, and the water storage media became saturated with water and unable to store more. At the beginning of the test, the OSB substrate adsorbed moisture from the constant climate room that was at 50% relative humidity and 23°C and gained mass, resulting in negative mass lost on the graph, until equilibrium was reached.

  The average mass lost for all OSB samples with interior wetting and WRB A was 1.4 grams per day, which is an effective sample vapor permeance of 2.2 US perms (126 ng/Pa·s·m^2).

  Simulating an exterior wetting with the water storage media between the OSB and WRB A resulted in a loss of 40.6 g/day or effective sample permeance of 89 US perms (5107 ng/Pa·s·m^2).

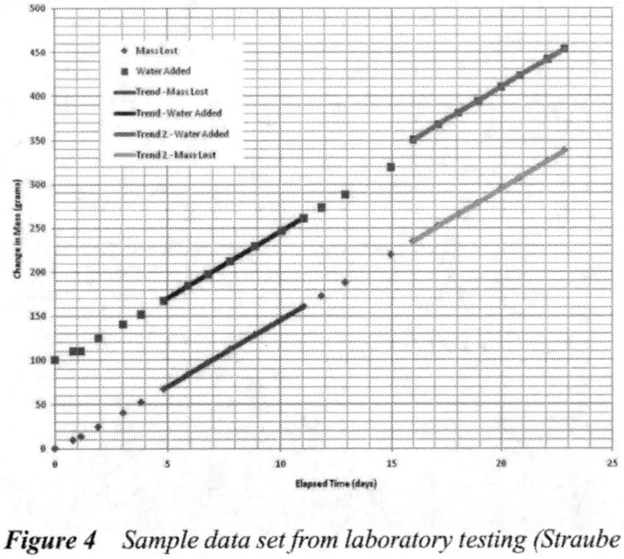

Figure 4 *Sample data set from laboratory testing (Straube et al. 2010).*

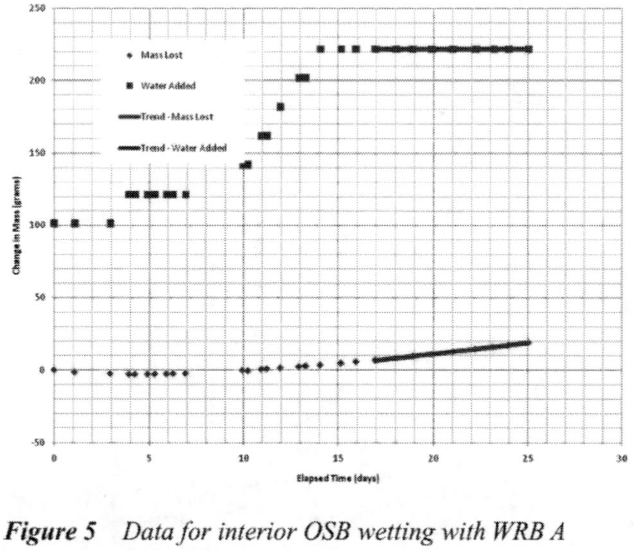

Figure 5 *Data for interior OSB wetting with WRB A (Straube et al. 2010).*

- Exterior grade gypsum substrate: Moisture from interior wetting of exterior grade gypsum sheathing was able to dry much more quickly than the OSB case. The vapor permeance of the exterior grade gypsum product used in the test is 23 perms (1300 ng/Pa·s·m^2), which is a value reported by the manufacturer. The average mass lost for all samples was 37.1 g/day. The effective sample permeance for wetting on the interior of the exterior grade gypsum sheathing with WRB A is 56 perms (3240 ng/Pa·s·m^2).

  Exterior wetting of the exterior grade gypsum sheathing was approximately the same as the exterior wetting of OSB, resulting in a daily mass loss of 52.2 g, and an effective sample vapor permeance of 92 US perms (5317 ng/Pa·s·m^2).

  The interior wetting of the exterior grade gypsum subassembly with WRB A dried more slowly than the exterior wetting of both OSB and exterior grade gypsum sheathing, indicating that with the higher vapor permeance WRB A, the exterior grade gypsum sheathing may limit vapor diffusion drying. WRB A is approximately ten times more vapor permeable than exterior grade gypsum sheathing.

Water-Resistive Barrier B. The dry cup vapor permeance of WRB B according to ASTM E96 Method A is 14 US perms (804 ng/Pa·s·m^2) and the wet cup vapor permeance according to ASTM E96 Method B is 28 US perms (1597 ng/Pa·s·m^2). WRB B has a much greater vapor permeance than OSB which is 1 to 2 perms (57 to 115 ng/Pa·s·m^2), and WRB B has approximately the same vapor permeance as exterior grade gypsum sheathing which is 23 perms (1300 ng/Pa·s·m^2) when WRB B is in a high RH environment (simulated by Method B wet cup ASTM E96).

- OSB substrate: Subassembly testing of OSB and WRB B performed similarly to the WRB A. The vapor permeance of the OSB was the controlling force in drying from interior wetting. The mass lost during interior wetting was 1.7 g/day, which is an effective sample permeance of 1.7 perms (96 ng/Pa·s·m^2).

  Exterior wetting of the OSB with WRB B resulted in 16.1 g/day or an effective sample permeance of 24.8 perms (1423 ng/Pa·s·m^2).
- Exterior grade gypsum substrate: Interior wetting of the exterior grade gypsum sheathing with WRB B resulted in a mass loss of 15.0 g/day, which is an effective sample permeance of 23 perms (1327 ng/Pa·s·m^2).

  Exterior wetting between the exterior grade gypsum substrate and WRB B resulted in a mass loss of 16.1 g/day and an effective sample permeance of 24.8 perms (1423 ng/Pa·s·m^2).

  The effective permeance for the exterior and interior wetting of exterior grade gypsum sheathing, and the exterior wetting of OSB with WRB B all have similar mass loss and effective sample vapor permeances. This means that in the subassembly test of interior wetting on exterior grade gypsum sheathing, the limiting factor for drying was the WRB B, not the exterior grade gypsum sheathing. The vapor permeances in all cases were lower with WRB B than with WRB A, but in the case of interior wetting of OSB, the difference is insignificant.

Building Paper (#15 Felt). The range of properties for building paper can vary significantly. According to *ASHRAE Handbook—Fundamentals* the vapor permeance of different thickness #15 felt ranges from 0.5 to 40 perms (28 to 2300 ng/Pa·s·m^2) over a range of relative humidities. For these tests, the relative humidities during wetting events were quite high, so the permeances were likely at the higher end of the range at that time. As the subassemblies dried out the permeance would drop again to a low range depending on the relative humidities.

- OSB substrate: Interior wetting of OSB with building paper resulted in an average mass loss of 0.5 g/day or an effective sample permeance of 0.8 perms (45 ng/Pa·s·m^2). This is the slowest drying of all interior OSB wetting.

  Exterior wetting of the OSB with building paper resulted in a mass loss of 12.0 g/day or an effective sample permeance of 18.6 perms (1070 ng/Pa·s·m^2). This is the lowest of all exterior wetting on OSB subassembly tests.
- Exterior grade gypsum substrate: Interior wetting of the exterior grade gypsum sheathing with building paper resulted in a mass loss of 9.3g/day, which is an effective sample permeance of 13.8 perms (794 ng/Pa·s·m^2). Exterior wetting between the exterior grade gypsum sheathing and building paper resulted in a mass loss of 10.7 g/day and an effective sample permeance of 15.8 perms (908 ng/Pa·s·m^2).

Summary Subassembly Testing

Table 3 shows a summary of the subassembly testing results and the effective sample vapor permeances.

The following conclusions can be drawn from Table 3:

- If water were to enter on the interior of the OSB, it would dry twice as quickly with WRB B as with building paper and three times as quickly with WRB A as with building paper, although the rate of drying is quite slow since the permeance in all cases is controlled by the absorptivity and vapor permeance of the OSB.
- If water were to enter on the interior of exterior grade gypsum sheathing, it would dry 1.5 times as quickly with WRB B instead of building paper, and would dry 2.5 times as quickly with WRB A as with building paper.
- If water were on the exterior of the OSB sheathing between the sheathing and the water-resistive barrier, it would dry 1.3 times more quickly with WRB B than with building paper, and 3.3 times more quickly with WRB A than with building paper.

Table 3. Summary of Subassembly Test Results and Effective Sample Permeances

	Mass Loss, g/day	Effective Sample Permeance	
		US Perms	ng/Pa·s· m^2
WRB A			
Interior wetting on OSB	1.4	2.2	126
Exterior wetting on OSB	40.6	89	5108
Interior wetting on gypsum sheathing	37.1	56.4	3240
Exterior wetting on gypsum sheathing	52.2	92.6	5317
WRB B			
Interior wetting on OSB	1.1	1.7	96
Exterior wetting on OSB	16.1	24.8	1423
Interior wetting on gypsum sheathing	15	23.1	1327
Exterior wetting on gypsum sheathing	16.1	24.8	1423
Building Paper			
Interior wetting on OSB	0.5	0.8	45
Exterior wetting on OSB	12	18.6	1070
Interior wetting on gypsum sheathing	9.3	13.8	794
Exterior wetting on gypsum sheathing	10.7	15.8	908

If water were on the exterior of the exterior grade gypsum sheathing between the sheathing and the water-resistive barrier, the water would dry 1.5 times more quickly with WRB B than with building paper, and almost 5 times more quickly with WRB A than with building paper.

These tests were conducted without a temperature gradient across the subassembly sample. Using a temperature gradient would increase the vapor pressure equally for all samples, and should increase the drying rate equally.

In all cases, the effective permeance of the samples with WRB A had a higher effective vapor permeance than the similar tests with WRB B, and the WRB B samples had a higher effective vapor permeance than similar tests with building paper.

In all cases of interior wetting of the OSB, with no temperature gradient, the drying was very slow and controlled by the OSB sheathing. The type of water-resistive barrier did not significantly affect the drying rate of the OSB.

Comparing the exterior wetting of OSB and exterior grade gypsum sheathing with WRB B, the resulting effective vapor permeance is very close to the ASTM E96 wet cup value of 28 perms (1597 ng/Pa·s·m2) determined in phase one of this study. Exterior wetting beneath the WRB A resulted in effective vapor permeances of approximately 1/3 of the ASTM E96 wet cup value. WRB A has a high vapor permeance and may not have been maintaining 100% rh between the sheathing and the water-resistive barrier.

SIMULATIONS WITH HYGROTHERMAL MODEL

The thermal and hygric behavior of building enclosure components are closely interrelated and therefore have to be investigated together. Permanently increased moisture content in a building enclosure component may result in moisture damages and mold growth. Increased moisture content in building components favors heat losses, and thermal conditions affect moisture transport.

To investigate how the water-vapor permeability of a water-resistive barrier affects the performance of a building enclosure under various climatic conditions and in conjunction with different sheathing materials and cladding materials, hygrothermal simulations were performed with WUFI 5. This software model allows the one-dimensional investigation of the hygrothermal performance of building components including effects like built in moisture, driving rain, solar radiation, long-wave emission, capillary transport, and summer condensation (Künzel and Karagiozis, ASTM Manual 40 2001).

The following climatic locations were chosen for the hygrothermal simulations (climate zone references as per the International Energy Conservation Code zoning):

- Miami, FL (Climate Zone 1)
- New Orleans, LA (Climate Zone 2)
- Atlanta, GA (Climate Zone 3)
- San Francisco, CA (Climate Zone 3)
- Baltimore, Maryland (Climate Zone 4)
- Portland, OR (Climate Zone 4)
- Seattle, WA (Climate Zone 4)
- Chicago, IL (Climate Zone 5)
- Minneapolis, Minnesota (Climate Zone 6)
- Fairbanks, Alaska (Climate Zone 7)

Two different exterior sheathing materials were considered for the hygrothermal simulations: Oriented strand board (OSB) and exterior grade gypsum board.

The following cladding materials were investigated:

- Brick
- Adhered manufactured stone veneer
- Cementitious stucco
- Cementitious siding

Six different water-resistive barrier scenarios were investigated:

- Building paper
- Low-perm membrane (1 perm)
- WRB A

- WRB B
- WRB C
- Combination of WRB C with ventilated rainscreen membrane (Product D)

In regards to the inner wall construction IRC code requirements were deployed, (e.g., a vapor retarder on the warm side of the insulation was only used where required by building code). The building paper and the low-perm membrane case were used for comparison. The last case (combination of WRB C with ventilated rainscreen D) comprised a highly water-vapor permeable water-resistive barrier with an impermeable, three-dimensional rainscreen HDPE membrane (dimple sheet) which was ventilated on the front and backside. This scenario was chosen to investigate the beneficial effect of combining drying potential for any moisture within the wall cavity (via vapor diffusion through the permeable water-resistive barrier into the ventilated cavity outside) and an impermeable layer outboard of the ventilated airspace in order to prevent inward moisture movement from absorptive cladding due to solar drive. The beneficial effect of a ventilated vapor impermeable rainscreen product has been evaluated and discussed in detail by Straube et al (2009) and by Jablonka et al. (2010).

The key properties for water-resistive barriers that were measured earlier on in the project were used in the analysis. The heat (conduction) and moisture transport (vapor diffusion and capillary conduction) were deployed in the simulations in one-hour time steps. Hourly indoor and outdoor climatic conditions as per ASHRAE Standard 160-2009 were used and the assumption was applied that 1% of the precipitation water that hits the cladding would leak through or enter behind the cladding. Additional analysis was performed where 0.75% water penetration took place between the water-resistive barrier and sheathing board, and 0.5% of water penetration between the sheathing board and insulation were also included into the parametric. Analysis was performed in one-hour steps for a two-year period for the selected wall enclosure systems. The interior conditions were calculated from the exterior weather file used in the hygrothermal simulation by applying the intermediate method from ASHRAE Standard 160-2009.

Material property inputs were taken from the WUFI North American database with the exception of the adhered manufactured stone veneer where measured data was used. The WUFI database for North America includes the data from NRC (Kumaran 2001).

Connection of Model and Subsystem Testing

As with all modeling activities, it is important to capture the subsystem effects. When these subsystem effects are properly captured, the accuracy of the predictions are expected to be higher. Prior to the execution of the hygrothermal analysis, the basic properties of the water-resistive barriers were measured (Straube et al. 2010). Then a series of subsystem tests were performed as described in the previous sections. The 1D hygrothermal model (WUFI 5) was used to validate the drying performance of the various experimental results. Good agreement was found between the model predictions and the measured drying rates for the various laboratory subsystems tested. The validation provided the necessary confidence in the results to engage in the comprehensive hygrothermal parametric analysis.

For the simulation cases that included the ventilated rainscreen D, two air cavities were included into the WUFI 5 model. The air exchange rates within the two air cavities were calculated based on the procedure developed within ASHRAE TRP 1091 (Burnett et al. 2004) which was also detailed in *Journal of ASTM International* by Karagiozis and Kuenzel (2009). The flow equations with the entrance and exit pressure drops were used, as well as the flow resistance along the length of the ventilated rainscreen membrane (product D). The effect of stack pressures, wind pressures, and moisture concentration gradients are combined to produce a net force for driving airflow in each of the two cavities that the ventilated rainscreen D creates. These mass flows were converted into cavity air changes per hour (Karagiozis and Kuenzel 2009) and allowed to calculate the dynamic impact of exterior and interior boundary conditions. Input files were created for the hygrothermal analysis using WUFI. The WUFI model has been extensively validated for many different cladding cavity configurations and has shown excellent agreement with measured field data.

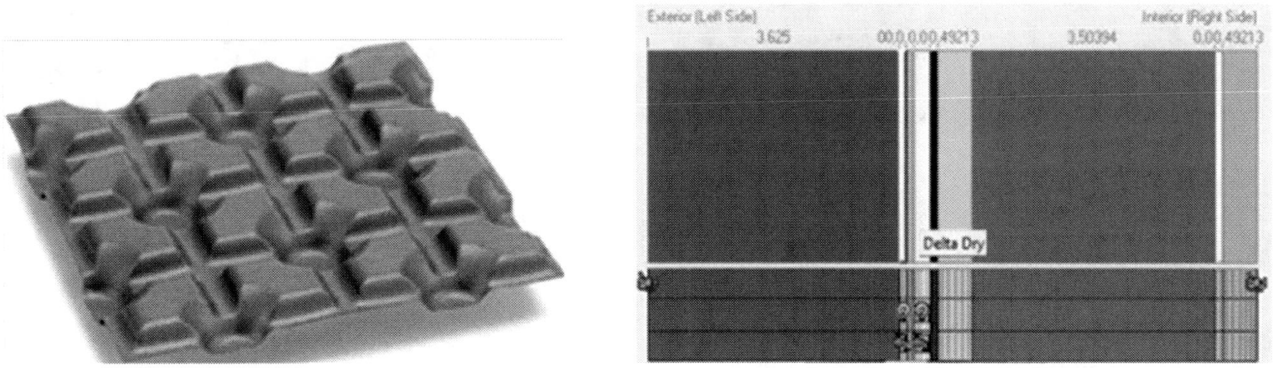

Figure 6 *Ventilated rainscreen D, configuration within wall assembly.*

Three surfaces were selected as the critical layers for detailed analysis: The exterior surface of the exterior sheathing board (P1), the interior surface of the exterior sheathing board (P2), and the interior side of the insulation layer in the wall cavity (P3). Selection points both on the interior and exterior side of the wall cavity ensured that the climatic effects were captured in the performance analysis.

The maximum mold growth index for each point was investigated. The mold growth index, as described in depth by Viitanen (Viitanen 2010) makes it possible to analyse the critical conditions needed for the start of mold growth and to measure the progress of mold growth.

First, all points were checked against a reference value. The average of the maximum mold growth index of these three points was calculated, and this value was then used as a performance indicator for ranking. The lowest value has the highest ranking, representing the best wall performance in regards to moisture management.

SOFTWARE SELECTION TOOL

Wall assemblies with various cladding types in different climates demand the proper selection of water-resistive barriers for optimal performance. The results of the measurements and hygrothermal simulations with WUFI 5 described in this paper were summarized in a software tool that helps the designer simply and effortlessly choose the most suitable water-resistive barrier for a particular wall assembly configuration. Furthermore, it provides a design recommendation for optimal wall performance based on a performance ranking by water-resistive barrier type. A damage function is used for guidance to provide insight on performance of these different water-resistive barriers.

The software tool allows for a step-by-step selection process as shown in Figure 7.

The user gets prompted to select the location from a list or map. A total number of 13 locations are available for selection by the user. These locations were carefully selected in order to cover all climatic zones of the US; the user would choose the location that would be closest to the location of interest within the same climate zone. In the next step the user gets prompted to choose the type of exterior sheathing (OSB or exterior grade gypsum board). Furthermore the user can choose from one of six options for water-resistive barrier with a wide range of water vapor permeability. The last step in the selection process allows the user to decide between four different cladding options: Brick, adhered manufactured stone veneer, cementitious stucco, and cementitious cladding.

The performance results for the selected wall assembly for that particular climatic location are then presented in a summary screen as shown in Figure 8.

The summary screen provides three gauges with a range from 1 to 6. In this way the designer is provided with a simple overview of how well a particular wall assembly would function under the chosen conditions.

The first gauge on the left provides the maximum mold growth index for the assumed best case scenario that 1% of the precipitation water that hits the façade will leak through the cladding but remain on the outside of the water-resistive barrier. The second gauge in the middle provides the maximum mold growth index for the assumed case that 0.75% of that water enters between the water-resistive barrier and the sheathing board. The third gauge on the right presents the maximum mold growth index for the assumed case that 0.5% of that water enters into the wall cavity (backside of the exterior sheathing material). By presenting all three gauges in one view the designer gets a quick impression how well a wall assembly would perform under optimal versus suboptimal (more realistic) conditions (e.g., missing flashing, penetrations in water-resistive barrier).

If desired, the designer can pull up detailed graphs for temperature, relative humidity, and moisture content for each of the three scenarios. The graph in Figure 9 shows a typical example for temperatures measured in the three sensor locations P1, P2, and P3 for the simulated two-year period.

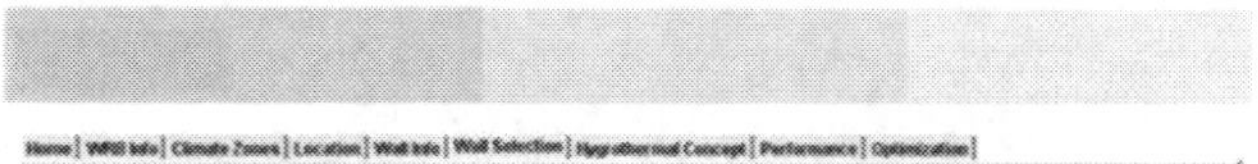

Figure 7 *Software tool for simple selection of most suitable water-resistive barrier.*

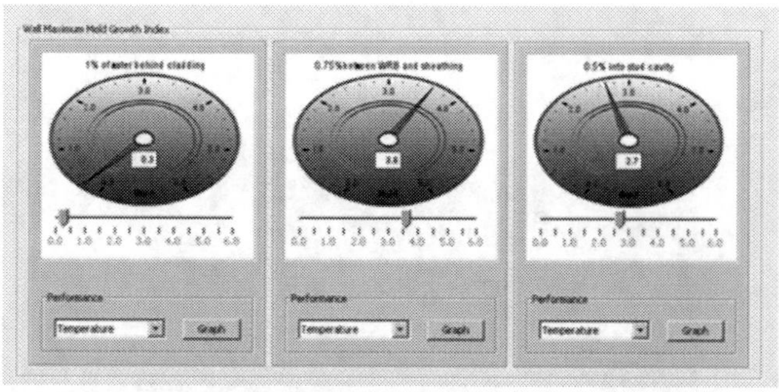

Figure 8 *Summary screen for wall performance under selected conditions.*

For simplicity a performance overview is provided to the designer as shown in Figure 10.

The performance overview chart shows the six different of water-resistive barrier scenarios that were analyzed. Lowest mold growth index (optimum hygrothermal performance) is achieved in the center of the diagram. The further the red line moves to the outside of the diagram, the more mold growth has to be expected under the chosen circumstances. The list below the diagram shown in Figure 10 provides a performance ranking of the different water-resistive barrier scenarios for that particular climate zone and the chosen sheathing board and cladding type.

It is apparent that the vapor permeability of the water-resistive barrier has an influence on the average mold growth index, as it affects the drying of interior moisture via diffusion.

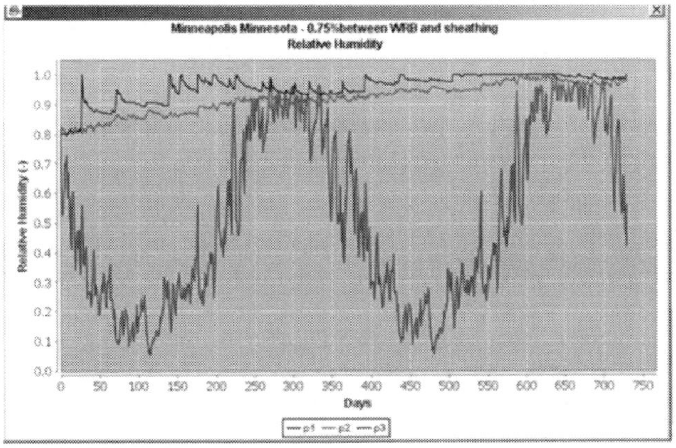

Figure 9 Relative Humidity predicted over 2 years in locations P1, P2, and P3 for selected conditions.

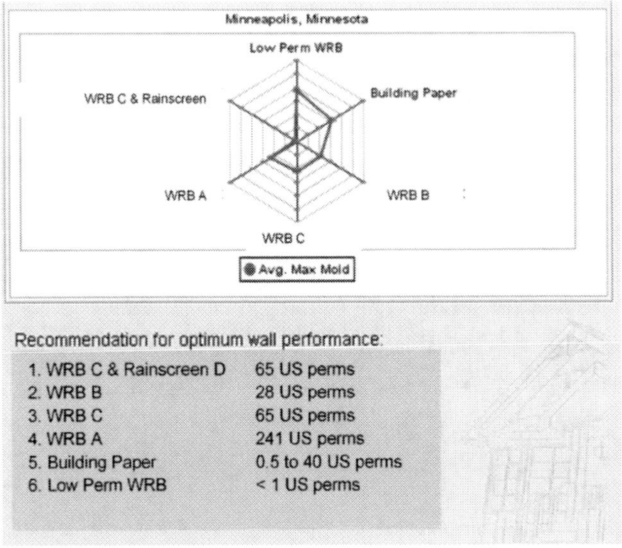

Figure 10 Performance overview and performance based ranking for various WRB options: Smaller values (center of diagram) indicate better performance.

Figure 11 and Figure 12 show the average mold growth index for Chicago for adhered manufactured stone veneer and cementitious siding with OSB as exterior sheathing material.

Figures 13 and 14 show the relative humidity in the three different sensor locations for the same location and cladding materials (1% of precipitation leaked behind cladding).

Figure 13 shows that in the Chicago climate, the use of a highly vapor permeable water-resistive barrier used behind adhered manufactured stone veneer would lead to elevated

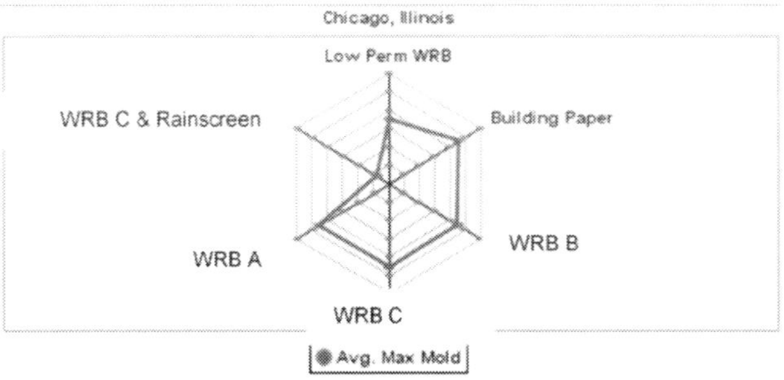

Figure 11 Average mold growth index Chicago, adhered manufactured stone veneer, OSB as exterior sheathing.

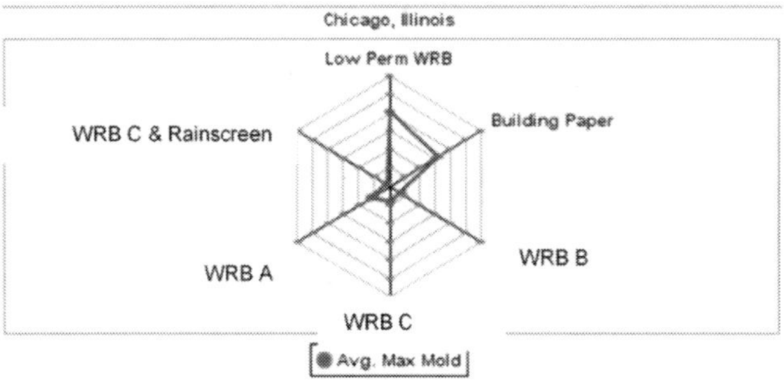

Figure 12 Average mold growth index Chicago, cementitious siding, OSB as exterior sheathing.

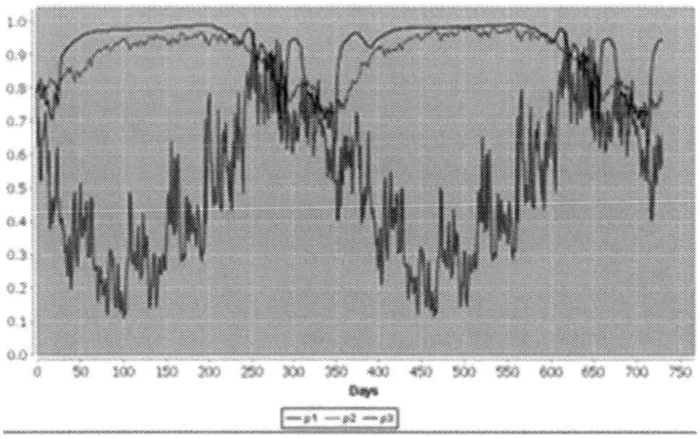

Figure 13 Relative humidity Chicago, adhered manufactured stone veneer, OSB and WRB C (65 US perms).

levels of relative humidity for long time periods per year (between 90% and 100% rh). The same water-resistive barrier used behind cementitious siding would lead to significantly lower RH values for most of the year (between 70% and 90%). The elevated levels can be explained with solar moisture drive from highly absorptive claddings into the wall cavity. A highly vapor-permeable membrane would allow moisture from the inside of the cavity to easily diffuse to the outside, but in case of reverse vapor pressure differential moisture can also easily diffuse inwards and elevate the moisture levels inside the wall assembly.

The reverse moisture flow can be prevented by using a ventilated, vapor impermeable rainscreen outboard of a highly vapor permeable water-resistive barrier as shown in Figures 15 and 16 (1% of precipitation leaked behind cladding).

The graphs show that by utilizing a highly vapor permeable water-resistive barrier in conjunction with an impermeable rainscreen product on the outside, the relative humidity levels drop noticeably.

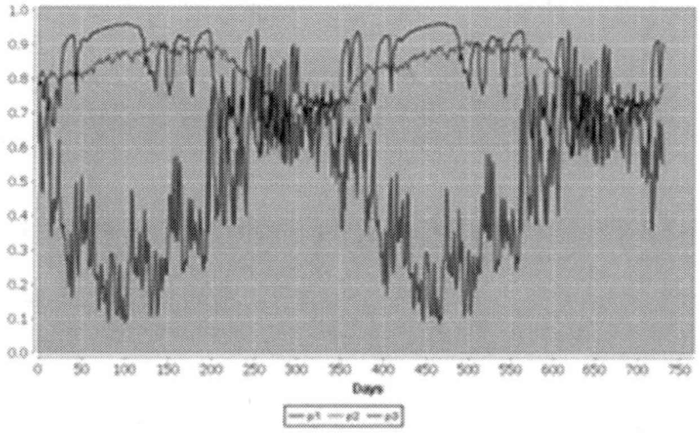

Figure 14 *Relative humidity Chicago, cementitious siding, OSB and WRB C (65 US perms).*

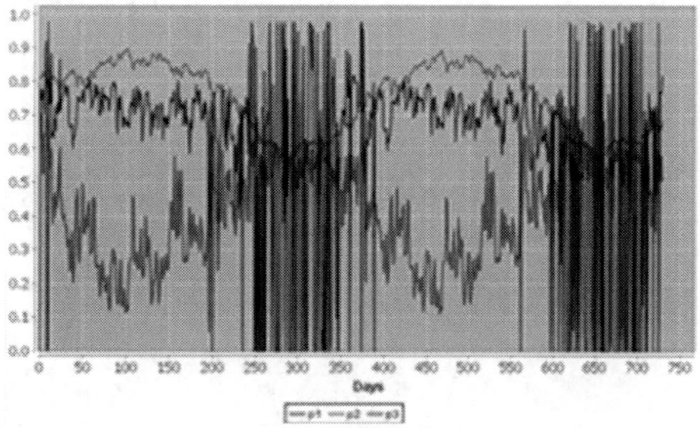

Figure 15 *Relative humidity Chicago, adhered manufactured stone veneer, OSB, and WRB C (65 perms) with rainscreen.*

Similar results can be seen for other climate zones. Optimum wall performance in any climate zone can generally be achieved by combining a highly vapor permeable water-resistive barrier with an impermeable, ventilated rainscreen material. This combination allows for quick drying of moisture from within the wall assembly to the outside, while moisture from the outside (i.e., stored in absorptive cladding material like adhered manufactured veneer) cannot migrate inward.

CONCLUSIONS

This paper evaluates the impact of various water-resistive barriers with a large range of vapor permeability on the hygrothermal performance of different wall assemblies. A simple-to-use tool has been developed after it was validated with laboratory results.

The results from this tool enable designers to select products with the most suitable vapor permeability for a particular geographical location under specific construction conditions. Variations in boundary conditions included climatic conditions (from seven climatic zones), cladding type (three-coat stucco, manufactured stone, cement board, brick), and type of water-resistive barrier (low versus high vapor permeability) deployed. The results for the performance of the wall systems are presented in form of a mold index. The approach presented includes three possible wetting locations and examines the best drying potential as a function of a range of vapor permeability of the water-resistive barriers.

REFERENCES

ANSI/ASHRAE Standard 160-2009, *Criteria for Moisture-Control Design Analysis in Buildings*, ASHRAE.

ASHRAE. *ASHRAE Handbook—Fundamentals* (2009): Heat, Air and Moisture Control in Building Assemblies—Material Properties, Chapter 26.

ASTM E96-00, *Standard Test Methods for Water Vapor Transmission of Materials*.

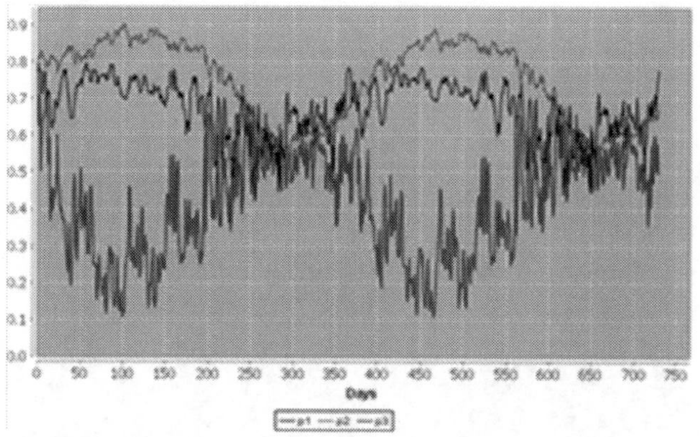

Figure 16 *Relative humidity Chicago, cementitious siding, OSB and WRB C (65 perms) with rainscreen.*

Burnett, E., Straube, J., and Karagiozis, A. (2004): Synthesis Report and Guidelines, ASHRAE TRP-1091, Report No. 12.

Carmeliet, J., Karagiozis, A., and Derome, D. (2007): Cyclic Temperature-Gradient-Driven Moisture Transport in Walls with Wetted Masonry Cladding, Thermal Performance of the Exterior Envelopes of Buildings X, *Proceedings of ASHRAE THERM X*, Clearwater, FL, December 2007.

Jablonka, M. (2011): DELTA-iSelect, an easy to use software tool to select WRBs based on durability analysis, *IEA Annex 55, Proceedings of 4th Meeting*, San Antonio, Oct. 14–16, 2011.

Jablonka, M., Karagiozis, A., and Straube, J. (2010): Innovative Passive Ventilation Water-resistive Barriers—How do they work?, *ASHRAE Transactions* (2010, Vol. 116, Part 2).

Karagiozis, A., and Kuenzel, H. (2001): WUFI-ORNL/IBP-A North American Hygrothermal Model, *Proceedings for the Performance of Exterior Envelopes of Whole Buildings VIII*; Integration of Building Envelopes, Clearwater, FL, December 2001, pp. 547–554.

Karagiozis, A., and Kuenzel, H. (2009): The Effect of Air Cavity Convection on the Wetting and Drying Behavior of Wood-Frame Walls Using a Multi-Physics Approach, *Journal of ASTM International*, Vol. 6, No. 10, Paper ID JAI101455.

Künzel, H.M., Karagiozis, A.N. and Holm, A. (2001): *Moisture Analysis and Condensation Control in Building Envelopes*, Manual 40, West Conshohocken, Pa., American Society for Testing and Materials, Chapter 9.

Kumaran, K. (2001): Hygrothermal Material Properties, ASHRAE TRP-1018.

Straube, J., Schumacher, C., Smegal, J., and Jablonka, M. (2009): Adhered Veneers and Inward Vapor Drives: Significance, Problems and Solutions, *Journal of Building Enclosure Design*, Summer 2009.

Straube, J., Smegal, J., and Schumacher, C. (2010): Water-resistive barrier laboratory testing – ASTM E96, Building Science Consulting Inc., Lab report, February 16, 2010.

Straube, J., Smegal, J., Schumacher, C. (2010a): Water-resistive barrier sub-assembly laboratory testing, Building Science Consulting Inc., Lab report, March 24, 2010.

Viitanen, H. (2010): Moisture conditions and biodeterioration risk of building materials and structure, *Journal of Building Physics* January 2010 vol. 33 no. 3 201–224.

Prediction of Phase Change Material (PCM) Degradation

Elizabeth Gao **Bao Zhang** **L.D. Stephenson, PhD**
Veera Boddu, PhD **Jonathan Trovillion**

ABSTRACT

Phase change materials (PCMs) are now well-known as potential additives to building insulation to provide a thermal mass effect to help conserve energy and maintain comfortable temperatures. There are many PCMs available, with phase change temperatures covering a wide range of applications; however, the performance envelope of most PCMs over time is not well known. For example, PCMs may lose energy storage capacity at temperatures at which phase transitions occur or may change due to chemical and physical transformations over many cycles of use. In this paper we are simulating long–term use of selected PCMs by laboratory testing in a thermal cycling chamber where temperatures are varied over the PCM samples' transition temperature ranges every 90 minutes. PCM samples being tested include: (1) encapsulated octodecane, (2) un-encapsulated mixtures of palm-oil and soy oil, and (3) samples of core materials excised from PCM panels. PCMs samples are removed from thermal cycling at two week intervals and differential scanning calorimetry (DSC) analysis is conducted to determine any degradation in thermophysical properties including melting point, heat of fusion, specific heat capacity, melting onset, and melting peak transition range. All PCM samples tested will be subjected to 5,400 cycles during the one-year long experiment to simulate 20 years of use in buildings. This research is expected to result in an accelerated aging methodology that allows reliable prediction of long-term PCM performance.

INTRODUCTION

Phase change materials (PCMs) are materials with a high capacity to absorb and release energy in the form of latent heat over a short temperature range at ambient conditions (Mondal 2008). The phase changes involved may be solid-liquid or solid-solid. Incorporation of the PCMs and their performance in building cooling and heating applications have been reviewed by several researchers (Sharma at al. 2009; Tyagi and Buddhi 2007; Farid et al. 2004; Kosny et al. 2012). These materials and concepts are being evaluated with renewed interest due to the increased costs associated with energy and the environment. In particular, applications of PCMs for building envelopes for heating and cooling are considered very promising applications as the market for cooling of buildings is expanding rapidly and free cooling systems can be retrofitted in existing buildings. The design of a thermal storage system within this narrow temperature range must be founded on reliable and high resolution material data. The accurate determination of the PCM's heat storage capability as a function of temperature is crucial. Another very important factor in the design of free cooling applications is the heat transfer between storage material and ambient air. There have been several research articles on simulation and modeling of the PCM based materials; often the stability of the PCMs over extended periods of time and the effect of thermal cycling have not been extensively studied. Experimental measurements on PCMs are needed to confirm calculations. In this paper we report experimental measurements on changes in the latent heat, melting and freezing points, heat of fusion, specific heat capacity (C_p), melting and freezing onset, and peak transition ranges of four commercial PCMs to assess their stability over 5,400 thermal cycles representing a duration of 20 years.

***Elizabeth Gao** is a graduate student contractor to U.S. Army Engineer Research and Development Center Construction Engineering Research Laboratory (ERDC-CERL) through the University of Illinois at Urbana-Champaign, Urbana, IL. **Bao Zhang** is a post-graduate from Oak Ridge Institute for Science and Education (ORISE), Oak Ridge, TN. **L.D. Stephenson, Veera Boddu**, and **Jonathan Trovillion** are staff research scientists at US Army ERDC-CERL.*

Published by ASHRAE 2013

EXPERIMENTAL

Materials

In this research, four commercially available PCM materials were selected. These materials represent different chemical natures, including two encapsulated octadecane mixtures with different melting and freezing points (designated as MA and MB in this paper, respectively), one un-encapsulated mixture of palm oil and soy oil (designated as POSO), and one sample of paraffin wax extracted from a PCM panel (designated as PWAX). The micro-encapsulated PCMs (Figure 1a and 1b), have the consistency of free-flowing powders and are generally mixed in with loose fill insulation. The POSO samples were extracted from one of the polymeric pouches of a PCM blanket (Figure 1c). In practice, the PWAX materials are used as the core materials of a PCM wall or ceiling panel with the wax and elastomer binder sandwiched in between two aluminum sheets (Figure 1d). The purpose of the elastomer binder is to hold the paraffin in place to prevent sagging and dripping as it melts. For this research, the PCM samples were extracted from one of the panels and subjected to thermal cycling.

Sample Preparation

Samples of the four types of PCM materials (MA, MB, POSO, and PWAX) were placed in aluminum pans for DSC measurements, which were then hermetically sealed. Three DSC pans were prepared for each material. In order to reduce the variation in measurements of the heat storage capacity with respect to temperature, all samples weighed approximately between 5 mg and 6 mg. The DSC pans and lids were obtained from TA instrument, Inc. (New Castle, DE). Baseline DSC measurements for each sample were conducted prior to thermal cycling.

Differential Scanning Calorimetry (DSC)

The thermal characterization of samples including melting and freezing points, heat of fusion, specific heat capacity (C_p), melting and freezing onset, and peak transition range were determined by the DSC (Model Q20 from TA instrument, Inc., New Castle, DE), which is shown in Figure 2. All DSC measurements were carried out at constant heating and cooling rate of 1°C min^{-1} and were repeated three times for each sample. The beginning and ending temperatures of the DSC scans depended on the physical properties of each PCM being evaluated in order to ensure complete melting and solidification.

Thermal Cycling Chamber and Thermal Cycling Profile

The DSC pans containing PCM samples were placed in Model EC127 temperature test chambers (Sun Electronic Systems, Inc., Titusville, FL), as shown in Figure 3 (left), where thermal cycling was performed under controlled thermal and humidity conditions. The thermal cycle begins at –10°C and the heating rate proceeds at 3.5°C min^{-1} up to 60°C. The samples are held at 60°C for 25 minutes and then cooled back to –10°C with a cooling rate of 3.5°C min^{-1}. The samples are held at –10°C for another 25 minutes, and the thermal cycle is repeated, as shown in Figure 3 (right). Each complete cycle requires 90 minutes. The thermal cycling chamber was outfitted with a small battery operated thermometer to provide an independent verification of the chamber temperatures. Samples were taken out of thermal cycling chamber at two week intervals, and they were subjected to DSC measurements to determine any changes in thermal properties. Each thermal cycle represents approximately one day of actual use in a building envelope. Since PCMs used in building envelopes would likely only undergo phase changes during 75% of a year, it is reasonable to assume that PCM materials would undergo 270 complete phase change cycles per year. As the duration of each cycle in the test chamber was 90 minutes, one year of use can be simulated in these cham-

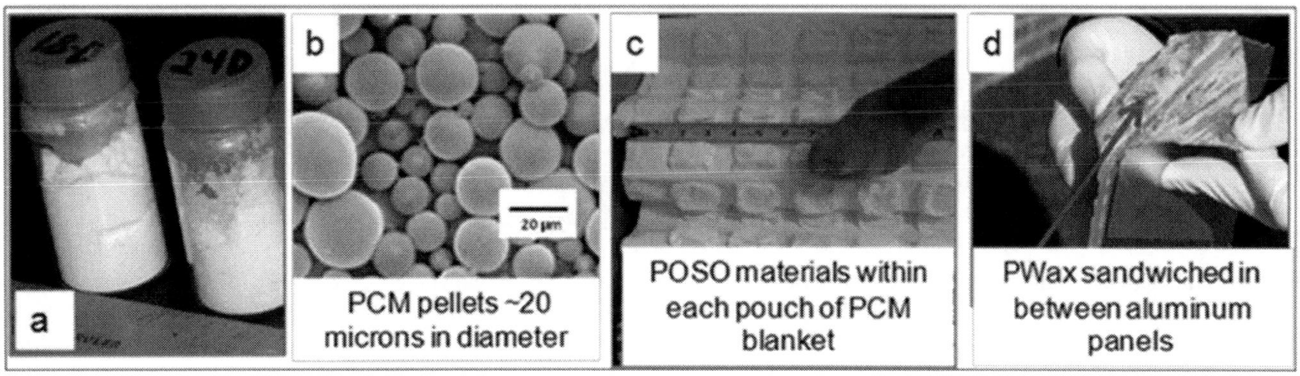

Figure 1 *(a) Vials containing micro-encapsulated octadecane mixtures with two different melting and freezing points; (b) Scanning electron micrograph of microencapsulated PCM pellets shown in (a); (c) blanket with pouches of mixture of palm-oil and soy oil; (d) paraffin wax with elastomer binder sandwiched in between aluminum panels.*

Figure 2 *Differential scanning calorimetry (DSC) model Q20 from TA instrument, Inc. with inset showing the sample and reference pans.*

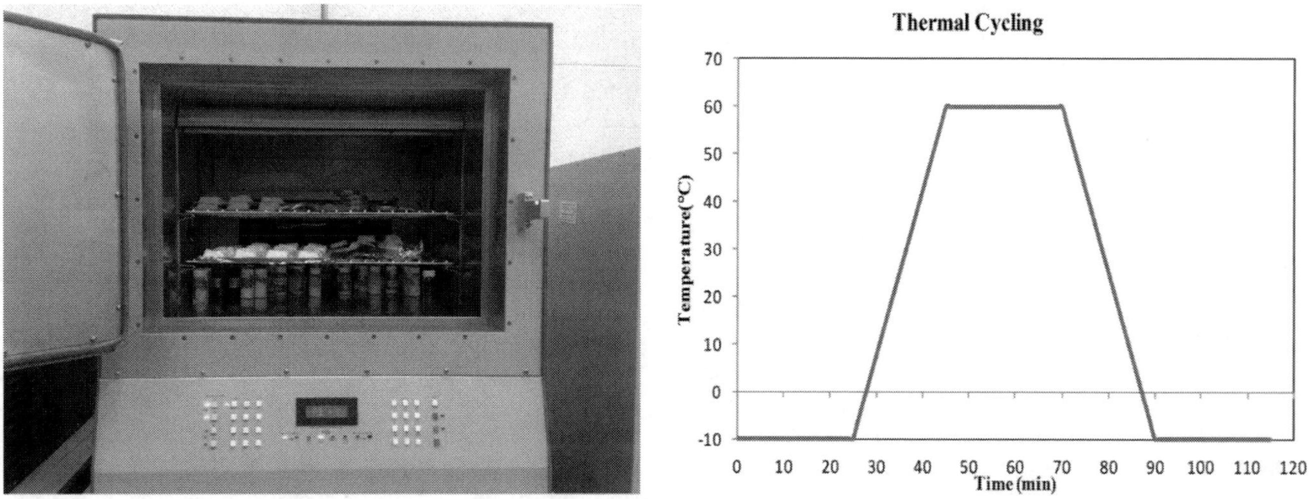

Figure 3 *Left: EC127 thermal cycling chamber (Sun Electronic Systems). Right: Thermal cycling profile used for the thermal cycling experiments.*

bers in 17 days. All the PCM samples are subjected to 5,400 thermal cycles during this one-year long research effort to simulate 20 years of use in building envelopes.

RESULTS

Changes in latent heat capacity were determined for each sample using DSC. The results of melting point on-set, average melting point, freezing point, and latent heats for all the four PCMs at different cycles are included in Tables 1–4 and the results are also presented in Figures 4–7.The same DSC pan sample was periodically removed from the chamber (after approximately 200 cycles), analyzed with the DSC and then replaced into the test chamber.

As can be seen in Tables 1–4,the measured melting point temperatures and latent heat values for baseline samples were generally in accordance with the manufacturer's stated values. Thus far, the samples have been exposed to 848 heating/cooling cycles, which represents about 3.1 years of PCM use, based on 5,400 cycles for 20 years. It is readily apparent that all PCM materials tend to lose latent heat capacity as they undergo thermal cycling.

PCM samples comprised of micro- encapsulated octadecane mixture samples with low melting point (T_m ~ 18°C), desig-

Table 1. Differential Scanning Calorimetry Data (Melting and Freezing) for PCM Comprised of Paraffin Wax and Binder Elastomer

Samples	Cycles	Melting			Freezing		
		Onset (°C) Average	Melting Point (°C) Average	Latent Heat (J/g) Average	Onset (°C) Average	Freezing Point (°C) Average	Latent Heat (J/g) Average
PWAX 1*	0	16.56	22.27	68.50	21.20	16.82	68.22
	212	17.37	23.11	64.29	20.54	17.58	67.95
	424	18.75	24.17	57.73	21.57	18.31	64.10
PWAX 2	0	16.94	22.47	72.31	20.43	17.74	76.29
	212	18.32	23.17	69.05	20.56	17.45	76.15
	424	18.72	23.97	67.17	21.27	18.13	74.63
	636	19.40	24.53	62.76	21.94	18.71	70.44
	848	19.24	24.87	56.48	20.56	18.72	67.18
PWAX 3	0	16.40	22.36	69.82	21.97	16.87	74.16
	212	17.27	23.21	67.71	20.91	17.26	74.80
	424	18.22	24.19	61.46	21.68	18.02	70.74
	636	19.40	24.53	62.76	21.94	18.71	70.44
	848	19.11	24.81	50.07	25.12	18.57	61.71

* PWAX: mixture of paraffin wax and elastomer binder

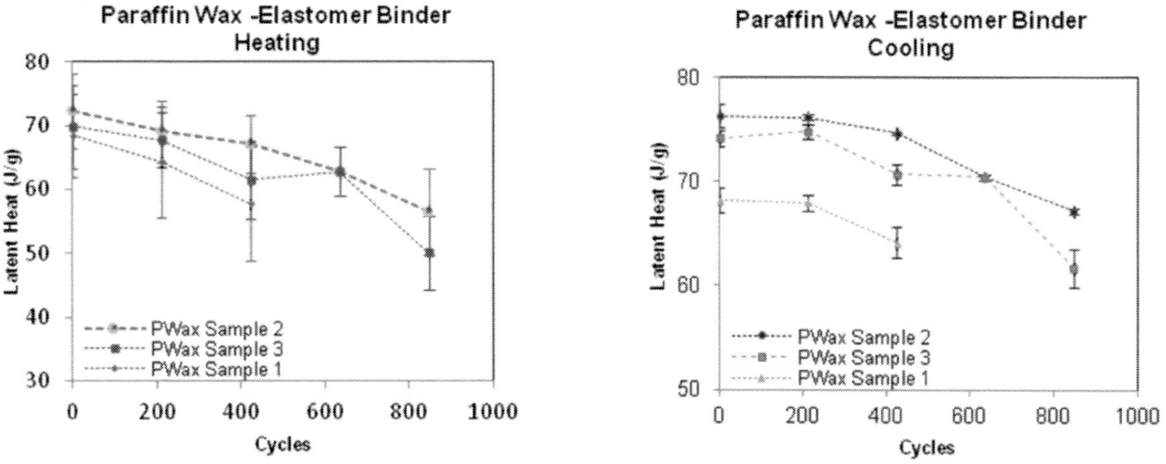

Figure 4 *Latent heat versus number of cycles for PCM comprised of paraffin wax and elastomer binder.*

nated as MA, showed the greatest decrease of latent heat for both heating and cooling curves, at rates of about –0.040 J/g/cycle. PCM samples comprised of micro-encapsulated octadecane mixture samples with high melting point (T_m ~ 24°C), designated as MB, showed the least degradation in terms of loss of latent heat capacity in both heating and cooling curves, at rates of about –0.10 J/g/cycle. PCM samples comprised of palm oil and soy oils (POSO) showed latent heat degradation rates for both heating and cooling curves of –0.30 J/g/cycle and –0.020 J/g/cycle, respectively, except for one sample which showed anomalous behavior, with approximately five times higher degradation rates. PCM samples comprised of paraffin wax and binder (POSO), showed latent heat degradation rates of –0.025 J/g/cycle and –0.014 J/g/cycle, for heating and cooling curves, respectively.

DISCUSSION

Thermal performance models of the building envelopes that include the dynamics of phase change materials must depend on the materials properties of the PCMs employed;

Table 2. Differential Scanning Calorimetry Data (Melting and Freezing) for PCM Comprised of Palm-Oil and Soy-Oil Mixture

Samples	Cycles	Melting			Freezing		
		Onset (°C) Average	Melting Point (°C) Average	Latent Heat (J/g) Average	Onset (°C) Average	Freezing Point (°C) Average	Latent Heat (J/g) Average
POSO 1*	0	22.46	26.37	130.55	22.19	21.04	131.25
	212	23.01	24.61	92.85	22.33	21.24	97.44
	424	22.71	24.70	63.43	22.41	21.49	69.09
	636	22.40	24.42	45.78	22.23	21.56	47.11
POSO 2	0	22.87	26.46	136.30	22.01	21.23	137.25
	212	24.36	26.37	136.47	21.81	21.39	137.70
	424	24.30	26.16	130.93	21.67	20.99	133.83
	636	24.11	26.04	117.27	21.56	21.06	125.13
	848	24.13	25.93	110.33	21.55	21.07	120.67
POSO 3	0	22.81	26.59	136.03	22.04	21.12	137.43
	212	24.12	26.22	136.93	21.82	21.00	138.53
	424	24.16	26.08	130.83	21.73	21.11	136.77
	636	24.05	25.96	124.03	21.65	21.16	135.47
	848	23.20	25.90	119.90	21.55	21.18	133.60

* POSO: PCM containing palm-oil and soy-oil mixture

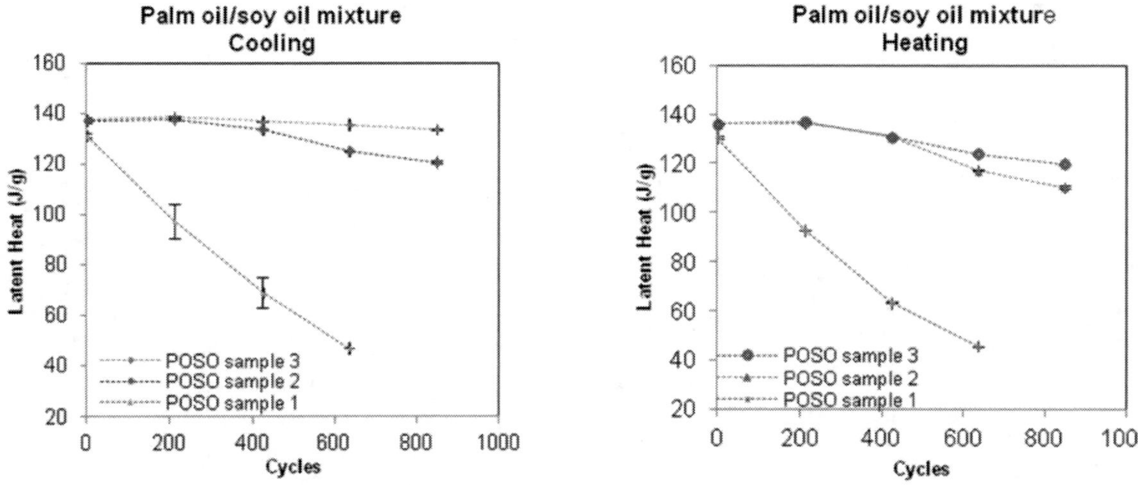

Figure 5 *Latent heat versus number of cycles for PCM comprised of palm-oil and soy-oil mixture* T_m *~ 25°C).*

however, if those properties change over time, the long term performance expectations and service life may be reduced. For example, if the transition temperatures drift or if latent heat values have degraded over time, the expected savings in energy due to installation of the PCM will not be achieved.

Obviously, all PCM products should exhibit high stability of the thermal properties over time (i.e., withstand thousands of cycles without degradation). Builders that use PCMs with unknown or unreliable thermal performance are taking some risk, especially if cyclic testing does not indicate long-term stability. We have presented an experimental approach to evaluate long-term thermal stability of PCMs. The laboratory-based cyclic testing described in this work provides a means of screening and comparing candidate PCM materials performance in a simulated relevant environment. The information obtained from these studies is expected to assist in selection of the optimum PCM for a given application.

Table 3. Differential Scanning Calorimetry Data (Melting and Freezing) for PCM Comprised of Micro-Encapsulated Octadecane Mixtures with Low Melting Point (18°C)

Samples	Cycles	Melting			Freezing		
		Onset (°C) Average	Melting Point (°C) Average	Latent Heat (J/g) Average	Onset (°C) Average	Freezing Point (°C) Average	Latent Heat (J/g) Average
MA 1*	0	14.39	17.21	164.88	14.43	12.95	167.75
	212	14.22	17.25	160.13	14.49	13.25	157.33
MA 2	0	14.40	17.21	170.70	14.54	12.97	175.65
	212	14.15	17.34	162.60	14.74	13.09	166.93
	424	13.60	17.14	153.47	14.82	13.74	156.43
	636	13.28	17.07	144.27	14.88	13.77	146.60
	848	13.30	17.05	136.17	14.95	13.77	138.37
MA 3	0	14.49	17.29	174.25	14.60	13.18	177.05
	212	14.53	17.19	169.93	14.73	13.60	172.93
	424	14.32	17.17	162.30	14.74	13.67	165.73
	636	13.52	17.07	156.17	14.80	13.76	158.63
	848	13.12	16.99	148.30	14.86	13.84	146.80

* MA: encapsulated octadecane mixture sample with low melting point
* Sample MA1 failed the test after 424 cycles because the pan and lid fell apart.

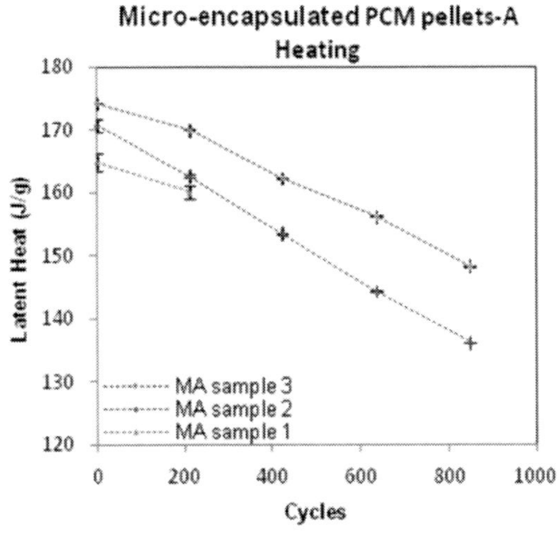

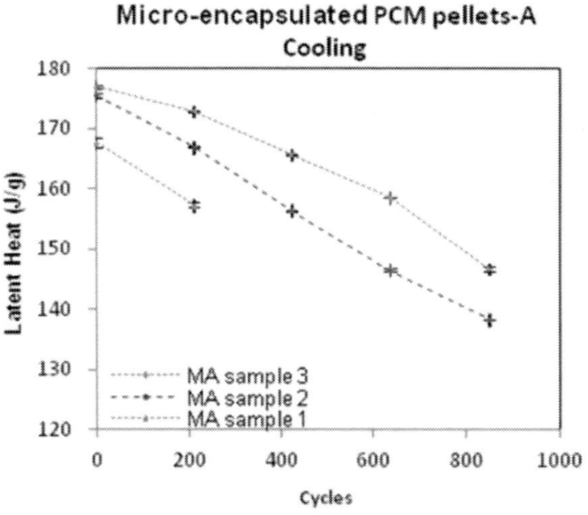

Figure 6 *Latent heat versus number of cycles for PCM comprised of micro-encapsulated encapsulated octadecane mixtures with low melting point (18°C).*

Although European standards for quality assurance of PCM performance (e.g., RAL-GZ-839), have been established, most US manufacturers rely on their own internal standards for quality assurance, and some US PCM manufacturers maintain limited data validating the cyclic performance of their products. Currently, one of the ASTM committees is working on developing a US standard for quality control/quality assurance of PCMs (Private Communication, 30 May 2013). In the meantime, most US manufacturers have focused on moisture transmission testing, based on the procedures of ASTM E96 (Method B) *Standard Test Methods for Water Vapor Transmission of Materials*, and fire resistance testing, as per ASTM E84-09 *Standard Test Method for Surface Burning Characteristics of Building Materials* results, rather than on tests simulating long-term thermal performance of their products. The industrial quality assurance standards used in Europe are well thought out, and US manufacturers would be well-advised to give them due consideration.

Table 4. Differential Scanning Calorimetry Data (Melting and Freezing) for PCM Comprised Micro-Encapsulated Octadecane Mixtures with High Melting Point (24°C)

Samples	Cycles	Melting			Freezing		
		Onset (°C) Average	Melting Point (°C) Average	Latent Heat (J/g) Average	Onset (°C) Average	Freezing Point (°C) Average	Latent Heat (J/g) Average
MB1	0	20.03	22.40	125.30	21.75	20.46	122.47
	212	19.33	22.54	121.88	22.29	20.78	120.60
	424	19.26	22.78	118.30	22.37	20.96	117.38
	636	19.53	22.89	118.15	22.52	21.40	115.80
MB2	0	19.84	22.20	127.28	21.91	20.68	123.08
	212	19.39	22.39	123.63	22.29	20.69	123.17
	424	19.22	22.68	120.10	22.34	20.85	118.67
	636	19.49	22.82	120.07	22.45	21.43	117.30
	848	19.72	22.94	118.13	22.59	21.64	114.47
MB 3	0	19.94	22.32	127.63	21.88	20.56	123.80
	212	18.99	22.68	120.83	22.26	20.44	121.77
	424	19.16	22.77	119.80	22.34	20.94	119.67
	636	19.48	22.88	120.67	22.45	21.37	119.10
	848	19.67	22.96	120.30	22.57	21.57	117.27

* MB: micro-encapsulated octadecane mixture sample with high melting point

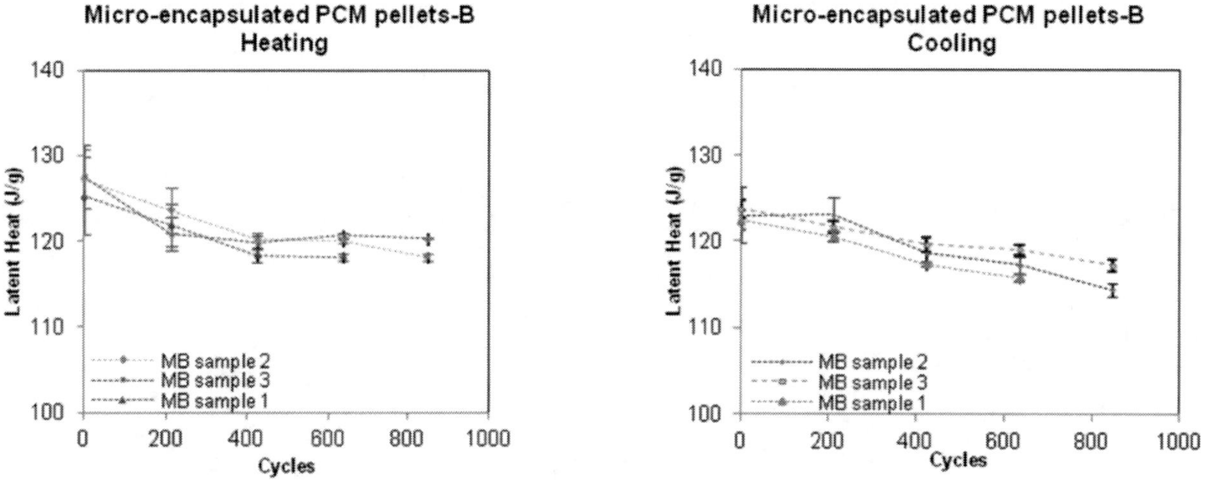

Figure 7 *Latent heat versus number of cycles for PCM comprised of micro-encapsulated octadecane mixtures sample with high melting point (24°C).*

The motivation for evaluating change in performance of phase change materials as they are subjected to repeated cycling stems from the concern that the materials may undergo physical and chemical transformations, as well as changes in crystal structure. The materials may also absorb moisture, leading to chemical decomposition and degradation of the PCM material and degradation of its thermal properties. Such physical/chemical transformations are evident in all four PCM materials, however, the extent and rate of decrease in latent heat varies for different materials. The PCMs evaluated in this investigation have so far shown only slight degradation of latent heat during 848 repeated heating and cooling cycles, with no appreciable change in transition temperatures. It is also possible that flame retardants and other volatile components of the PCMs are released into the surroundings, contributing to the loss of latent heat. In the case of the PWAX samples, it is possible that the elastomer binder had reacted with the paraffin wax, and contributed to latent heat loss. Also, other impurities in the PCM mixtures can interfere with bond breakage and formation, or (possibly by reacting with the primary PCM) causing this degradation over time.

While the PCMs in these experiments have only been subjected to 848 thermal cycles, simulating 3.1 year of use, this investigation is on-going and eventually the PCMs will be subjected to 5,400 cycles, in order to simulate 20 years of use. In future research, large samples of the PCMs will be collected in vials and will be subjected to thermal cycling, according to the same schedule as described for the samples in the DSC pans. The vial samples will allow visual examination of the PCM, bulk mass loss analysis and use of Fourier Transform Infrared (FTIR) spectroscopy to determine any chemical changes that may have occurred. The use of large samples in the vials will also circumvent the problems of loss of sample due to hermetic seal breaching during repeated cycles. However, it is noted that the process of evaluating the same sample each time, ensures that an accurate history of performance is attributed to changes in that particular sample. Both methods need to be evaluated in future experiments and compared. The anomalous behavior observed for sample 1 of the POSO samples, perhaps due to the breach of the hermitic seal during DSC measurements, will be further investigated. Overall, the maximum decrease in latent heat for each of the four materials over the 848 phase change cycles was as follows: (a) for PWAX (paraffin and elastomer binder), the decrease in latent heat was 22%–25% during heating, and ~15% during cooling; (b) for POSO (palm oil and soy oil mixture), the decrease in latent heat was 13%–19% during heating and 3%–12% during cooling; (c) for MA (low transition temperature microencapsulated octadecane mixtures samples, the decrease in latent heat during both heating and cooling was 15%–20%; for MB (high transition temperature microencapsulated octadecane mixture samples), the decrease in latent heat during both heating and cooling was 6%–7%. Both high temperature low temperature micro-encapsulated samples (MA and MB) exhibited the least variability in latent heat data.

CONCLUSIONS

Thermal cycling experiments are being conducted in order to establish a standard protocol for evaluation of the long-term performance of phase change materials, in a relatively short term tests. Thus far, samples of four types of PCM have been exposed to 848 heating/cooling cycles, which represents about 3.1 years of PCM use, based on 5,400 cycles for 20 years. The results have indicated that baseline latent heat values were generally in accordance with the manufacturer's stated values. All PCM materials evaluated tended to lose latent heat storage capacity as they underwent thermal cycling. PCM samples comprised of micro- encapsulated octadecane mixture samples with high melting point (T_m ~ 24°C), designated as MB, showed the least degradation in terms of loss of latent heat for both heating and cooling curves, at rates of about –0.10 J/g/cycle. PCM samples comprised of micro- encapsulated octadecane mixtures samples with low melting point (T_m ~ 18°C), designated as MA, showed the greatest loss of latent heat for both heating and cooling curves, at rates of about –0.040 J/g/cycle. Thermal cycling experiments are continuing to expose the four PCM materials to a total of 5,400 cycles.

ACKNOWLDGMENTS

This research was supported in part by an appointment to the Postgraduate Research Participation Program at the US Army Engineer Research and Development Center-Construction Engineering Research laboratory administered by the Oak Ridge Institute for Science and Education through an interagency agreement between the U.S. Department of Energy and ERDC-CERL.

REFERENCES

ASTM E96-95 (Method B), *Standard Test Methods for Water Vapor Transmission of Materials*, ASTM International, West Conshohocken, PA, 19428–2959, 1995.

ASTM E84-09, *Standard Method for Surface Burning Characteristics of Building Materials*, ASTM International, West Conshohocken, PA, 19428–2959, 2009.

Borderon, Julien, Joseph Virgone, Richard Cantin, and Frederic Kuznik, 2011, Full-scale study of a building equipped with a multi-layer rack latent heat thermal energy storage system, *HVAC Research*, 17(4): pp. 566–576.

Fang, Yuan and Mario A. Medina, 2009, Proposed modifications for models of heat transfer problems involving partially melted phase change processes, *Journal of ASTM International*, 6(9): pp. 1–20.

Farid, Mohammed, M., Amar M. Khudhair, Siddique Ali K. Razack, and Said Al-Hallaj, 2004, A review on phase change energy storage: materials and applications, *Energy Conversion and Management*, 45: pp. 1597–1615.

Gunther, E., S. Hiebler, and H. Mehling, Determination of the heat storage capacity of PCM and PCM-objects as a function of temperature. http://intraweb.stockton.edu/eyos/energy_studies/content/docs/final_papers/11b-2.pdf

Janda, Kathryn, B. and Yael Parag, 2013, A middle-out approach for improving energy performance in buildings, *Building Research and Information*, 41(1):39–50.

Kosny, Jan, Elizabeth Kossecka, Andrzej Brzezinski, Akhan Tleoubaev, and David Yarbrough, 2012, Dynamic thermal performance analysis of fiber insulations containing bio-based phase change materials (PCM), *Energy and Buildings*, 52: pp. 122–131.

Mondal, S., 2008, Phase change materials for smart textiles – An overview, *Applied Thermal Engineering*, 28: pp. 1536–1550.

Petersen, T. P., J. Kalb, W.K. Njoroge, D. Wamwangi, and M. Wuttig, 2001, Mechanical stresses upon crystallization in phase change materials, *Applied Physics Letters*, 79(22): pp. 3597–3599.

Sharma, Atul, V.V. Tyagi, C.R. Chen, and D. Buddhi, 2009, Review on thermal energy storage with phase change materials and applications, *Renewable and Sustainable Energy Reviews*, 13: pp. 318–345.

Som, Shrestha, William Miller, Therese Stovall, Andre Desjarlais, Kenneth Childs, Wallace Porter, Mahabir Bhandari, and Steven Coley, 2011, Modeling PCM-enhanced insulation system and benchmarking energyplus against controlled field data, Proceedings of Building Simulation.

Tyagi, V. V. and Buddhi, D. 2007. PCM thermal storage in buildings: A state of art, *Renewable and Sustainable Energy Reviews*, 11: pp. 1146–1166.

RAL-GZ 896, Phase Change Material Quality Assurance, German Institute for Quality Assurance and Certification, E.V., Stuttgart, Germany, September 2009.

Private Communication with Dr. Kaushik Biswas (Oak Ridge National Laboratory), 30 May 2013.

Highly Insulated, Ventilated, Wood-Framed Attics in Cool Marine Climates

Patrick Roppel, PEng **Neil Norris** **Mark Lawton, PEng**
Member ASHRAE

ABSTRACT

Attic ventilation is a firmly established requirement for residential wood-framed buildings. Attic ventilation is cited to be a benefit for moisture control, reducing cooling loads in hot climates, minimizing ice dams in cold climates, and extending the service life of roof materials by reducing surface temperatures. There is an ongoing debate on the real significance of some of these benefits and mandatory venting requirements.

Despite plenty of attention to moisture control and ventilation of attics, there is growing evidence that buildings, which are built to code in cool marine climates, have mold growth in attic spaces. Problems are being reported in marine climates in both Europe and North America.

The objective of this paper is to demonstrate how cold attic spaces, which are subject to high outdoor moisture levels and are built to code (ventilation area and distribution and ceiling airtightness) can still have moisture problems and to identify the contributing factors leading to mold growth. This paper covers a study in British Columbia, Canada of four attics and a control roof at the same location but not over conditioned space. Testing and measurements included long-term monitoring, air leakage and venting area characterization using dual blower door fans, smoke testing, and tracer gas testing to measure air transfer rates.

INTRODUCTION

The connection between attic ventilation and the hygrothermal performance of attics of wood-framed sloped roofs has been studied for many years, at least back to the 1930s (Rose and TenWolde 1999, 2002). Most building codes stipulate specific prescriptive requirements for the ventilation of wood-frame roofs.

Originally the identified purpose of attic ventilation appears to be to minimize condensation and moisture collection in attics (Rose and TenWolde 2002) which can lead to biological degradation of wood based material, corrosion of metals, and, most relevant to this paper, mold growth on wood surfaces.

The lack of ventilation is routinely blamed for a variety of problems and failures. In addition to moisture control, attic ventilation is also cited to benefit summer cooling of the attic air and reducing cooling loads in hot climates, minimizing ice dams in cold climates, and extending the service life of roof materials by reducing material temperatures. The debate is ongoing on the real significance of some of these benefits and the mandatory requirement for venting for all types of roof construction has come into question (Parker and Sherwin 1998; Rose 2001; Rose and TenWolde 2002; Tobiasson et al. 2001; Lstiburek 2006).

Most Canadian research into attic moisture problems has identified that the leading cause of moisture troubled attics and high sheathing moisture content, with associated mold growth, is the transfer of moist indoor air into the attic space from high humidity indoor spaces (Forgues 1985; BLP 1991; Sheltair 1997). Attic ventilation is intended to reduce the potential for problems by diluting interior air moisture sources and in theory to provide some drying of moisture in the attic. The general assumption in Canada is that if the ceiling is reasonably airtight and you have the code required attic ventilation then there is little potential for moisture collection.

***Patrick Roppel** is a principal and building science specialist, **Neil Norris** is a building science consultant, and **Mark Lawton** is a principal and senior building science specialist in the Buildings, Technology, and Energy Division, Morrison Hershfield Limited, Vancouver, BC, Canada.*

In contrast, the University of Alberta (Forest and Berg 1993) predicted through a validated computer model that generally the 1:300 code requirement provides too much ventilation in Canadian marine climates (Vancouver, Halifax, St. John's) and the dominant moisture source for the attic is the outdoor air. Furthermore, they predicted that the ceiling airtightness had little impact on the sheathing moisture contents and virtually did not affect the quantity of moisture deposited by condensation. They suggested that sheathing moisture content can be reduced by sealing the attic or at least substantially reducing vent area by not installing any vents and relying only on the background leakage of the attic envelope.

There is growing evidence that in cool marine climates that buildings seemingly built to code are experiencing high incidences of moisture problems and mold growth in attics. In the coastal climate of British Columbia we are seeing problems related to mold growth in attics in many recently constructed wood-frame buildings. This problem is not unique to British Columbia's climate and construction practice. There is similar experience in Europe. Surveys are showing that as many as 60% to 80% of the single family houses in the Gothenburg region of Sweden, also a cool marine climate, are showing significant mold growth (Arfvidsson and Harderup 2005; Hagentoft et al. 2008; Hagentoft and Sasic Kalagasidis 2010; Hagentoft 2011). Coincidently, the frequency of reported attic moisture problems has increased as insulation levels in attics has steadily increased to address energy efficiency goals in both jurisdictions.

There are several factors leading to potential issues for highly insulated wood-framed attics constructed in cool maritime climates. The drying capacity of outdoor air during the winter is low because of constant wet conditions and the lack of sunshine hours limits the drying benefits from solar exposure. The ability to dry out materials is dependent on the temperature and moisture capacity of the attic air. High levels of insulation limit the heat supplied to the attic from the conditioned space. There is even visual evidence that attic ventilation might increase moisture and mold growth in some attics in cool marine climates. Figure 1 illustrates a staining pattern seen in many recent reviews of attics in the Lower Mainland of British Columbia. Mold growth is occurring at soffit vent locations, exactly where the ventilation rate should be the highest.

The hypothesis for these conditions is that the sheathing and framing can become colder than the surrounding air temperature and lead to wetting due to:

a. Cooling of the roof surface by radiation during clear nights, and
b. The thermal mass of the wood, relative to rising air temperatures in the morning due to solar exposure.

When the outdoor air is saturated, condensation or frost can form on the sheathing and framing in the attic even without a significant moisture contribution from inside the house. The theory is that the wood in the attic will pick up moisture as temperatures steadily drop in the winter, and average relative humidity rises in the attic space to levels that any significant wetting event from condensation or frost will result in conditions optimum for mold growth (i.e., relative humidity 90% to 96%).

There is a growing body of anecdotal information indicating that:

1. Attic ventilation may *not* be helpful in controlling moisture in, and mold growth on, the wood sheathing and framing in well insulated attics located in cool maritime climates.
2. In some cases, attic ventilation may actually *increase* moisture in, and mold growth on, the wood sheathing and framing in well insulated attics located in cool maritime climates.

This paper summarizes a research study carried out in British Columbia, Canada, where comprehensive testing and measurements were conducted of attics in buildings that were seemingly built to code requirements and good practice with respect to attic ventilation area, venting area distribution, and ceiling airtightness, but still exhibited staining from mold growth.

SIGNIFICANCE OF MOLD IN ATTICS

One major reason for this study is the observation of surface mold growth on wood roof sheathings of wood frame buildings that seemingly meet code requirements with respect to ceiling air tightness and attic venting area. A perfectly valid question is "Is surface mold in the attic a problem?"

Some individuals believe that the presence of mold anywhere in a building is not acceptable and measures must be taken to eliminate visible staining. Others note that mold located in attic spaces is unlikely to affect exposure in the occupied space because air transfer between the indoor space and attic is predominately in the direction of the indoor space to the attic space,[1] and any transfer in the opposite direction should be insignificant. Tracer gas testing done in this study supports this point.

Figure 1 *Staining (mold) at soffit vents in a recently constructed wood-frame project.*

Additionally, the dominant molds that have been sampled in many attics are surface molds that are abundant in our outdoor environment.[2] Discovery of only common outdoor molds helps alleviate concerns for some individuals. However, the fact that more harmful molds are not identified doesn't preclude their existence and cannot be relied upon to alleviate health concerns of occupants.

Another way of looking at the issue is to simply focus on marketability and property values. Many occupants and potential buyers naturally have concerns with any mold within attic spaces that are identified by building inspections. The visible presence of mold affects marketability, and ultimately property values, even if experts conclude there is no real health risk or cause for concern of accelerated material degradation.

The visible presence of mold in buildings without identifiable deficiencies or clear solutions presents a challenge to industry.

STUDY OVERVIEW

Study Location and Units

The study units are in a townhouse development near the ocean and surrounded by mountains. This is an area noted for persistent mist and fog. Attic surveys had identified many units in the complex that had surface mold growth especially near soffit vents. Each of the four units was selected from a larger pool of potential volunteers with this condition. All the buildings and units are of similar design and construction but completed in separate phases in 2004 and 2005.

The selection of study units was based on the apparent indoor humidity levels (signs of past condensation on windows), unit orientation (north-south or east-west), apparent venting area (venting provided on two sides or three), and level of observed staining at soffits. Table 1 summarizes the conditions and occupancy of the test units recorded during the visual review of the units. Figure 2 shows the site plan and location of each of the selected study units in the complex. Figure 3 shows the elevation drawings for the study units. Figures 4 and 5 show the north elevation of Unit 4 and the east elevation of Unit 1 for context.

Table 1. Summary of Conditions and Occupancy of Study Units

Unit	Indoor Conditions	Occupants	Staining
1	56% rh, DP 14.7°C, 25°C	1 child, 2 adults	very light
2	49.6% rh, DP 12.7°C, 24°C	2 children, 2 adults	light
3	46.5% rh, DP 12.1°C, 24°C	2 adults	severe
4	48% rh, DP 12.14°C, 23.9°C	2 children, 2 adults	heavier

The test (control) roof assembly shown in Figure 6 was selected because this sloped roof is not subject to moisture sources from an occupied space and has similar weather exposure as the test units. The wood sheathing tested in this assembly allowed us to evaluate the conditions of the roof sheathings in an attic with abundant ventilation. The test roof assembly included monitoring of plywood samples that were installed directly below the roof underlayment and at the underside of R-5 extruded rigid insulation that was attached to the underside of the roof structure. Figure 7 shows the plywood samples for the test roof assembly. This side by side comparison allowed us to also evaluate the impact that condensation due to night-sky radiation and rain-water absorption through the asphalt singles had on the wetting of the plywood samples.

Attic Construction and Ventilation

The attic construction and ventilation (shown in Figure 8) is typical of sloped roofs for wood-framed buildings in coastal BC. As required by code, the venting area is provided at the top of the attic near the ridge line and at the bottom of the attic space at the soffits at opposite ends. The vents at the ridge are square low profile vents and baffles are installed at the soffits. Unit 4 is the end of the building and is part of the hip roof end.

Figure 2 Site plan and orientation of test units.

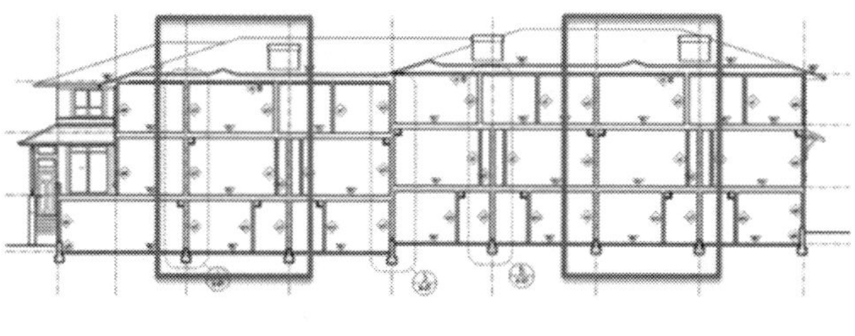

Figure 3 Building elevation.

1. This is because both stack effect (i.e., heat rises and cold air falls) and wind pressures (suction on the roof) create driving forces for air transfer in the direction of the indoor space to the attic

2. These same molds are also not considered pathogenic for humans, which means the molds do not cause infections. However, some people are allergic to the spores of these molds similar to being allergic to plants.

Figure 4 End Unit 4.

Figure 5 East elevation of Unit 1.

Figure 6 Test (control) roof assembly with Unit 2 in the background.

Dryer, exhaust, and fresh air ducts, as well as plumbing pipes penetrate through the attic ceiling and up through the roof. The dryer and exhaust ducts are insulated with fiberglass insulation wrapped in a polyethylene bag. The ceiling penetrations are sealed at the ceiling air barrier and to the metal ducts at the roof level with contractors tape. The attics are accessed through hatches that friction fit to a wood framed opening and batt insulation wrapped in polyethylene above the drywall hatch cover. Drywall with taped joints separates the attics between units.

INVESTIGATION AND MEASUREMENT

This section describes the procedures of the investigation and measurement program conducted for this study to provide a complete understanding of the contributing factors leading to localized staining at soffits in wood-frame attics in coastal BC. Results, Analysis, and Discussion follow in later sections.

Building Characterization

The air leakage area of the ceiling interface between the indoor and attic space and the attic venting area was deter-

Figure 7 Control assembly sensors.

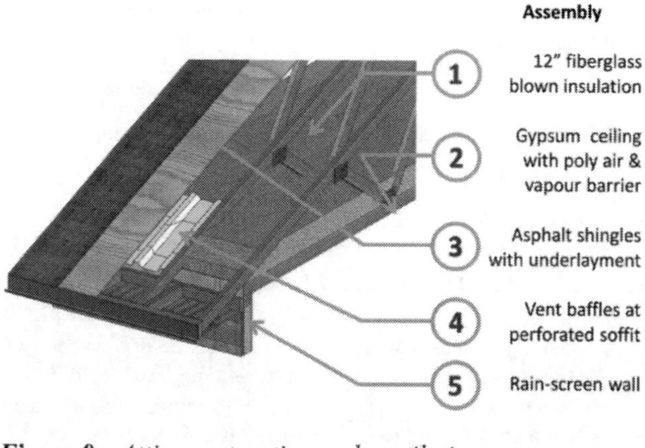

Figure 8 Attic construction and ventilation

mined using a two fan depressurization method. This method is similar to what was done in the past in previous Canadian attic research studies (BLP 1991; Sheltair 1997).

The primary purpose of conducting the air leakage and attic venting area measurements are twofold:

1. Provide an empirical basis to compare the relative airtightness of the attic ceiling to other buildings for a standard pressure differential, and
2. Confirm that the venting area meets or exceeds the code requirement of 1:300 venting area of the insulated ceiling area.

Two calibrated fans were required, one in the attic hatch and another in the main entry door. The fans were connected so that the fan speeds automatically adjusted until a target pressure difference was achieved in both the attic and indoor spaces.

The first step was to pressurize the attic with respect to the outdoors. All the windows and doors were opened and the pressure difference was checked to ensure that there was no pressure difference between the indoors and outdoors. In the second step, both the attic and indoor space were pressurized to equal amounts with respect to the outdoors. This in theory yielded no flow across the attic-to-indoor interface, which allowed the attic air exchange rate to be determined. Then the flow across the attic-to-indoor interface was determined from the results of the first test. Figures 9 and 10 illustrate these procedures and calculations to determine the airtightness of the attic ceiling and venting area.

A pressure difference of 40 Pa was maintained for three of the units and generally the fluctuations in the air leakage area readings were fairly stable, less than 1.5%, indicating reliable measurements.[3] Unit 4, however, has soffit vents on three sides, the flow rates were significantly higher than in the other units, and a pressure difference of 40 Pa was not achieved. A pressure difference of only 15 Pa was ultimately achieved but there was a lot of fluctuation in the readings. The fluctuation was so significant that the calculated attic-to-indoor interface value is unreliable because the fluctuation in the attic venting area was greater than the resolution of the ceiling airtightness area being calculated. However, the testing did confirm that the attic venting area at 15 Pa was much greater than required by code as will be discussed below.

Smoke Test

Smoke testing was completed to visually inspect significant air leakage paths. This was done by filling the attic full of smoke by both positively pressurizing the attic with the fan in the attic hatch and negatively pressurizing the indoor air with the house fan. Smoke was visible coming through the fresh air grill in two units (Units 2 and 3) and through the seal of the blower door to the attic hatch in all the units. The leakage at the attic hatch during the testing is not representative of actual conditions, but likely a source of some leakage from the indoor to attic space during normal conditions. The leakage at the fresh air grill was a result of the tape not being well adhered at the connection from the polyethylene insulation

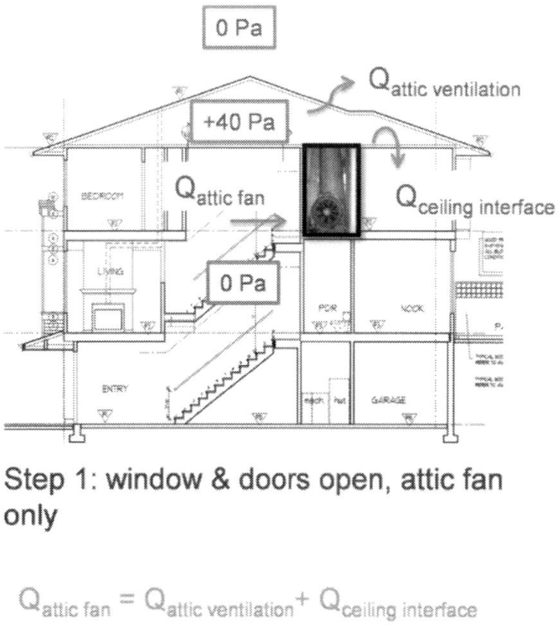

Figure 9 Building characterization step one, window and doors open, attic fan only.

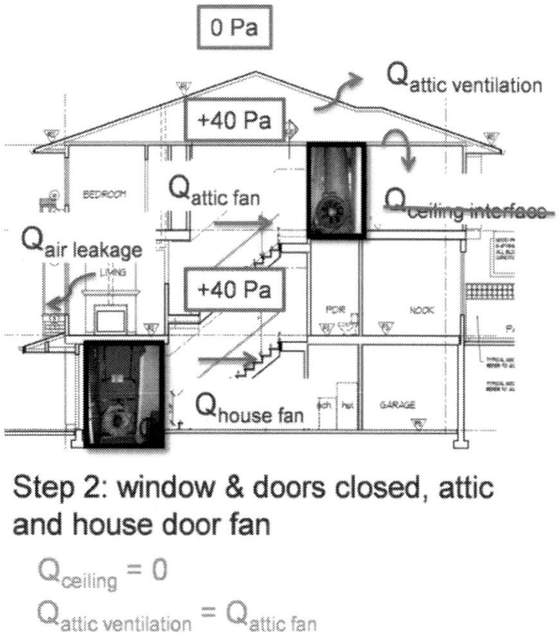

Figure 10 Building characterization step two, window and doors closed, attic and house door fan.

3. A pressure differential less than 50 or 75 Pa was selected due to the limitations associated with the two fan setup, available power, the size of the venting areas, and ultimately the fan capacity (maximum 6300 cfm).

bag to the flange of the grill. The connections were sealed before the tracer gas sampling was completed in December as described below.

Tracer Gas Testing

The venting and ceiling air leakage areas provide an indication of the rates but are measured at specific pressure differentials. Tracer gas testing provides a means to determine average air transfer rates over a specific period of time.

Two key questions about attic performance were answered using the tracer gas sampling:

1. What is the attic air ventilation rate for real-life conditions?
2. What is the transfer rate of indoor air into the attic space?

The sampling was completed for two 1-week periods, December 8 to December 15, 2011 and February 20 to February 27, 2012. The air transfer rates calculated for these two periods represent average rates over these one week periods.

The tracer gas testing was done using a perflourocarbon tracer (PFT) method developed by Brookhaven National Laboratories (BNL). This method provides a convenient and practical method to measure average air transfer rates in multi-zone buildings (Dietz et al. 1985). Small passive sources and samplers are left in place for a period of time then sent to BNL to be analyzed and to determine the average transfer rates for a given period. PFTs are good tracers because they are very stable, not susceptible to oxidation in the atmosphere, and are present in the atmosphere at low levels. Low enough levels that small amounts of PFTs released in a building provide clear signals that can be easily and reliably absorbed by passive charcoal samplers. This method works on the steady-state assumption that over several days the average concentrations of the tracer vapour in a zone is equal to the emission rate of the tracer source divided by the air leakage or infiltration rate. Knowing the rate from deployed passive PFT sources and measuring the average concentration with passive samplers provides a means to calculate the air transfer rates.

Three types of PFTs were deployed in test units: on the second floor living room/kitchen, third floor hallway, and into the attic. A small fan ran continuously in the attic to ensure the air was well mixed during the testing. No additional fans were added to the indoor space and the CATS (samplers) were placed at return air ducts to provide average mixing in the zone. Figure 11 illustrates the deployment of the sources and samplers.

Long Term Monitoring

Sensors were installed in the units at the beginning of September and the end of September for the control roof. Readings were recorded every 15 minutes.

For each unit the following measurements were made:

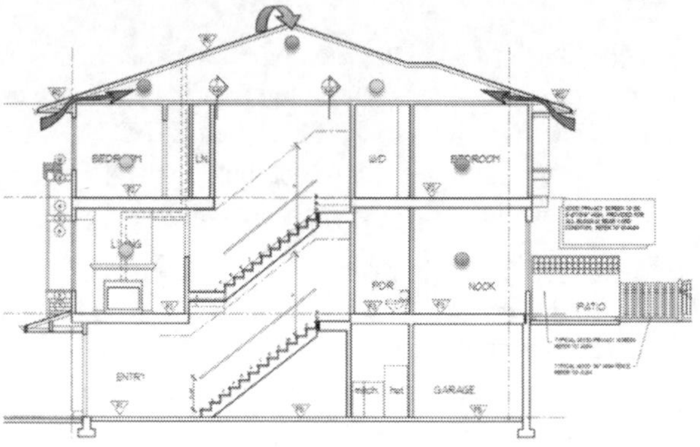

PFT - perfluorocarbon tracer (source)
CATS – capillary absorption tube sampler (sink)

Figure 11 Tracer gas testing experiential program.

a. Moisture content and temperature of the roof sheathing at three locations, at the vent baffles at the soffits and at the framing bay beside the baffles at the soffit
b. Moisture content and temperature of top chord of truss at one location
c. Condensation detection on sheathing at a baffle location
d. Attic air temperature and relatively humidity
e. Indoor air temperature and relatively humidity

At the control assembly the following measurements were made:

a. Moisture content and temperature of the plywood sample installed at the underside of the roof underlayment
b. Moisture content and temperature of a plywood sample with R5 rigid insulation installed between the underside of the roof structure and plywood sample
c. Outdoor air temperature and relatively humidity

Weather data was obtained from local environmental stations as needed for analysis.

RESULTS

A summary of results of the testing and measurements is presented in this section. Analysis and discussion considering how the test and measurements are interconnected is presented in the next section.

Building Characterization and Smoke Testing

The attic venting area, including both intentional and unintentional openings, is presented in Table 2 for the measured pressure differential and is compared to the applicable required venting area required by code for the construction of the test attics (1:300 per insulated ceiling area).

The measured venting areas are higher than the average areas reported in the BLP (1991) and Sheltair (1997) studies and exceed the building code requirement by as much as three times.

The measured air leakage area for the attic ceiling, calculated normalized leakage area (NLA), and observations of the smoke test are summarized in Table 3. These values were derived using equation 43, Chapter 16 Ventilation and Infiltration, of the *ASHRAE Handbook—Fundamentals* to convert to a 10 Pa pressure differential basis.

Comparison of the NLA values in Table 3 to the reported values in past Canadian studies (BLP 1991; Sheltair 1997; NRCan 1997) suggests that the ceiling airtightness of the units in this current study can be considered to have at least average airtightness levels, by Canadian standards, and a convincing argument can be made to classify the ceiling to attic interface as airtight.

The NLA values for Units 1 and 2 are lower than all the leakage areas of the ceiling to attic interface for all the eight units measured in the Sheltair study, which is interesting because that study included three R-2000 houses.[4] Unit 4 with a NLA of 4, which is a less reliable value and likely lower in reality because of the difficulties identified in the previous section, is even lower than the measured R-2000 houses in Langley that were part of the Sheltair study.

Comparison of the reliable data for Units 1 to 3, with NLAs 1.6 to 2.2, to the 1997 NLA Survey for whole building airtightness by National Resources Canada summarized in Table 4 further supports the argument that the study units have at least average airtightness at the ceiling to attic interface. The measured NLAs for the ceiling to attic interface of Units 1 to 3 are close to the average NLAs for whole buildings but not for the R-2000 buildings. However, the ceiling interface NLAs for the study units are lower than all the R-2000 measurements in the Sheltair study, and built in the same period; and all R-2000 houses must meet a NLA of 0.7 for the entire house. Further recognizing that the normalized air leakage for the ceiling interface of a row townhouse with only 600 to 750 ft^2 area (55 to 70 m^2) and an attic hatch is likely higher than the overall whole building normalized area for detached homes, then there is a convincing argument that the ceiling interfaces are airtight for the study units.[5]

Table 2. Measured Attic Venting Area Compared to 1:300 Venting Area

Unit	Measured Venting Area	Insulated Area, m^2 (ft^2)	Required Area (1/300 per insulated area)	% Measured Area/ Required Area
1	2450 cm^2 @ 40 Pa	60 (642)	1900 cm^2	129%
2	2435 cm^2 @ 40 Pa	68 (728)	2160 cm^2	113%
3	3900 cm^2 @ 40 Pa	57 (614)	1990 cm^2	196%
4	7530 cm^2 @ 15 Pa	60 (642)	2315 cm^2	325%

Table 3. Measured and NLA Attic Ceiling Leakage Area and Smoke Test Observations

Unit	Measured Leakage Area	Calculated Normalized Leakage Area* cm^2/m^2 @ 10 Pa	Smoke Test Observations
1	110 cm^2 @ 40 Pa	1.6	Smoke at hatch
2	110 cm^2 @ 40 Pa	1.4	Smoke at hatch and the fresh air vent in the bedroom closet
3	160 cm^2 @ 40 Pa	2.2	Smoke at hatch and the fresh air vent in the washer/dryer closet
4	300 cm^2 @ 15 Pa	4	Smoke at hatch (less than others)

*Air leakage through surface(s), such as a ceiling surface, is commonly presented as NLA in previous Canadian studies and programs, therefore is continued in this paper. ΔP = 10 Pa, CD, 2 = 1, CD, 2 = 0.611, and n = 0.65 was utilized to calculate the NLA values.

Tracer Gas Testing

The weekly average flow rates determined by tracer gas testing with comparison to the flow rates determined by fan testing at static pressure differentials are presented in Tables 5 and 6. In interpreting the results, it is important to keep in mind the objectives of the testing. The objective of the fan-testing is to provide an estimate of the attic venting and ceiling to attic air leakage areas, while the objective of the tracer gas testing is to provide the average air transfer rates over a specific period of time. The purpose of comparing the flow rates is to show the relative difference and provide a general indication of the forces driving airflow compared to standard assumed pressure differentials.

The large difference between the attic venting rates suggests that on average there was not much wind pressure to induce ventilation during the tracer gas testing. Moreover, there were very few thermal forces to drive airflow through the attic. The average temperature difference between the attic and

Table 4. 1997 NLA Survey by National Resources Canada

Region	1981–1990	1991–1997	R-2000
BC	2.8	1.9	0.7
National	2.3	1.4	0.6

4. Certified R-2000 houses must have airtightness testing to confirm that a level of airtightness of at least 1.5 ach at 50 Pa or a NLA of 0.7 cm2/m^2 (1.0 in2/100 ft^2) is achieved.

5. This statement is largely based on recognizing the absolute leakage areas relative to the ceiling to wall area for the row townhouse, but it is also supported by the findings in the reference studies for detached houses (BLP 1991; Sheltair 1997).

Table 5. Comparison Between Measured Attic Ventilation Rates and Venting Rates Derived from Fan Testing

	Tracer Gas Testing, ACH (m^3/h)		
Unit	Round 1 (December 8 to 15)	Round 2 (February 20 to 27)	Derived from Fan Testing, ACH (m^3/h) @ 4 Pa
1	1.2 (69.9)	1.0 (61.1)	8.3 (488)
2	1.3 (75.7)	1.2 (73.4)	8.3 (485)
3	2.6 (112.8)	3.7 (91.9)	18.2 (778)
4	4.1 (102.4)	2.1 (123.0)	58.7 (1475)

Table 6. Comparison Between Measured Indoor to Attic Air Transfer Rates and Leakage Rates Derived from Fan Testing

	Tracer Gas Testing, CFM (m^3/h)		
Unit	Round 1 (December 8 to 15)	Round 2 (February 20 to 27)	Derived from Fan Testing, CFM (m^3/h) @ 4 Pa
1	16.5 (28.0)	11.4 (19.3)	26.2 (44.5)
2	16.4 (27.8)	8.5 (14.4)	26.2 (44.5)
3	18.8 (32.0)	11.7 (19.9)	38.1 (64.7)
4	18.5 (31.4)	12.6 (21.4)	82.7 (140.5)

outdoor air was 1°C to 2°C during the tracer gas testing periods, with periods during the nights where the attic air was below the exterior air temperature. The average wind speed, irrespective of direction, at a nearby weather station during the tracer gas testing was 0.6 m/s for the first period and 1.5 m/s for the second period. The test units are sheltered from wind.

To demonstrate the order of magnitude flow rates derived by fan testing at a "standard" 4 Pa pressure differential, as presented in Table 5, we assumed an inlet area of half the total venting area. In reality, the inlet and outlet area are not likely equal and the airflows are much more complex than this simple extrapolation. The flow will be governed by the ratio of the inlet to outlet areas,[6] the pressure distribution due to varying wind direction, and shelter provided by the adjacent row housing and woodland.

The average pressure difference due to wind was probably less than 0.5 Pa when accounting for the low wind speeds at the buildings during the tracer gas testing periods. This estimate is based on rough estimates for the pressure coefficients and shelter factors outlined in Chapter 16 of the 2009 *ASHRAE Handbook – Fundamentals*. The flow rates determined by tracer gas testing and venting areas determined by fan testing appear to be aligned when the low wind speeds, shelter, pressure coefficients due to roof orientation and wind direction, and reduced inlet areas based on the venting area distribution are considered concurrently.

The attic ventilation rates measured by the tracer gases are in a range of 1 to 5 ach, which represents 50% of the measured values in the BLP field study (1991). The other reported ranges were 10% within 5 to 10 ach, 30% within 10 to 15 ach, and 10% greater than 15 ach. The sampling period was one hour for the BLP study.

The ventilation rates are also within the range of 0 to 7 ach measured for one of the research houses at the University of Alberta (Forest and Walker 1993) that had no intentional venting area added to the roof assembly. This is in contrast to a much wider range of 0 to 50 ach for the attic with a venting area provided to meet the code requirement of 1:300 for the other research house. Given the wind exposure, the measured ventilation rates for this study are within the range of ventilation rates measured for the attic with a venting area per 1:300 for the University of Alberta study.

There was a temperature differential to drive air from the indoors to the attic, albeit less than the common assumption of 4 Pa. The average temperature difference between the indoor and outdoor air was between 13°C and 20°C for all the units during the two tracer gas testing periods. This temperature difference corresponds to a pressure difference, due to stack, between 1 to 2 Pa for a neutral plane level at 0.75 of the total building height.

The lower flow rates for the tracer gas testing compared to the fan testing can be fully explained by stack effect for the measured temperature differences with a neutral plane level in the range of 0.7 to 0.85 of the total building height. A NPL level in this range is consistent with NPL data for houses with exhaust systems, fresh air intakes and a chimney through the roof (NRCC 1995; ASRHAE 2009).

For BLP study (1991), the indoor to attic air transfer rate was determined by tracer gas testing for eight of the 20 units. The flow rate ranged from 2 to 85 cfm (3 to 144 m^3/h) and the average rate was 31 cfm (52 m^3h). The measured rates for the all the units in this current study are much lower than the average BLP measured rates. The rates are however higher, roughly double, than the 5 to 7 cfm (12 to 15 m^3/h) measured for two research houses at the University of Alberta during extremely cold weather (1993).

Monitoring

This sub-section presents observations and analysis specific to the monitoring of the indoor, outdoor, attic, and control spaces. Discussion of the monitoring in context to the other measurements, building characterization and air transfer rates, is presented later in the paper.

First, we note that there were many difficulties with the monitoring equipment during the winter of this study and con-

6. Distribution of the intentional venting area in the attic is approximately 25% at the roof ridge and the remaining 75% distributed between the soffits at each end of the roof.

sequently there are holes in the data during various periods of the study. Though this was disappointing and made the analysis more difficult, there was fortunately enough data to test our hypothesis and meet the objectives of this study.

Monitoring continued the winter of 2012/2013 for Units 1 and 2 and the control roofs. The ongoing monitoring is exploring the impact of changes to the venting area. Unit 1 had an additional button vent installed at the ridge and Unit 2 had the soffit venting restricted by filling the free area with plastic bags filled with semi-rigid insulation.

At the time of writing this paper in January of 2013, the monitoring from the winter 2012/2013 was not comprehensively analyzed. We note that the general trends continue, but Unit 1 appears to have larger swings in moisture content and Unit 2 does not dry out as quickly after wetting events.

Many of the previous Canadian studies indicated that moisture problems in attics were generally not evident without the presence of high indoor humidity (BLP 1991; Sheltair 1997). The units that were part of this study do not follow this trend. The measured indoor humidity during the heating season of 2011 to 2012, in all the units, is considered within normal operating conditions; higher moisture levels are often assumed for the design of building envelope assemblies. Table 7 summarizes the indoor conditions measured for the test units. A ΔVP of approximately 800 Pa is considered high, 550 Pa moderate, and 250 low (Roppel et al 2009; ISO 13788-01).

The difference in vapour pressure between the indoor and outdoor air, ΔVP, is a useful metric to categorize indoor moisture levels, since indoor relative humidity is variable depending on the outdoor conditions and indoor operating temperature (Roppel et al. 2009). Comparing the ΔVP for the test units shows that the indoor moisture levels in the study units ranged from low to moderately high during the monitored heating season (December 1, 2011 to March 15, 2012).

Long-Term Trends

The staining and occurrence of high sheathing moisture levels are correlated, but the highest moisture for the longest duration did not necessarily coincide with the most visible staining. For example, the heaviest staining was observed at Unit 1 at the east baffle, Unit 2 at the east baffle, and Unit 3 at the west baffle, but some of the highest moisture levels were recorded at Units 1, 2, and 3 at the east non-baffle location.[7] Figure 12 summarizes the duration (hours) of elevated sheathing moisture levels and Table 8 summarizes observations of staining at the sensor locations at the start of the monitoring period.

The correlation between moisture thresholds and staining provides a benchmark for the conditions favourable for mold growth. However, mold research tells us that exposure time, temperature, and RH must be considered concurrently when evaluating the risk of mold growth in attics (Haukka et al 1999; Viitanen et al 2007; Clarke et al 1996; Sedlbauer 2001). Moreover, analysis of the field monitoring data and visual observation of staining demonstrates the need to consider these factors together. An interesting finding was that the control assembly had visible staining after 2 years, but only spent 2 to 3 weeks above 25% MC. The control also was not subject to conditions above 28% MC and had more staining than wetter locations in the attics. However, the control assembly spent more time above 20% MC and near 25% MC than compared to the attics. These differences are explainable using a predictor, such as mold index (Haukka et al 1999; Viitanen et al. 2007), which accounts for exposure time, temperature, and RH concurrently, and will be discussed in future work.

Two significant moisture spikes occurred that affected the long-term sheathing moisture levels of the east soffit for Units 1, 2, and 3 during the winter of 2011/2012. Spikes in the moisture levels occurred on December 10 to 12th, 2011 and January 12, 2012. These events are discussed further in the next sub-section, diurnal wetting.

This winter (2012/13) moisture levels spiked to 35% MC in both the monitored units during a 10 day period, in the middle of January, when there was several consecutive cold clear nights. Unit 1 (additional vent) spiked less times than Unit 2 (blocked soffit vents) and dried out more quickly. Moisture levels in Unit 1 dropped to similar moisture levels as the control assembly in approximately 1 week period after wetting events. In contrast, the moisture levels in Unit 2 remain elevated around 28% MC at the end of January.

7. The sensors at Unit 3 west soffit, the location with the most visible staining, malfunctioned during critical wetting periods

Table 7. Measured Indoor Conditions

Unit	Average Temp., °C	Average RH, %	95% Percentile Dewpoint Temp.,°C	Average ΔVP, Pa	95% Percentile ΔVP, Pa	99% Percentile ΔVP, Pa
Outdoors	**6.9**	**84.4**	**8.5**	**N/A**	**N/A**	**N/A**
1*	24.1	30.7	9.3	250	600	750
2	22.1	40.7	10.7	250	450	550
3	19.1	37.5	7.1	0	200	300
4	23.3	32.8	9.0	100	300	500

*Unit 1 was only analyzed for December 2011 and January 2012

The west and south sheathing benefit more from solar heating than the north and east elevations, but all the elevations experience similar night sky cooling. Night sky cooling and solar heating are important factors for the elevated sheathing moisture levels, but exposure to outdoor air and moisture appears to be the critical reason for the elevated sheathing moisture levels. This statement is drawn from the observation that the control assemblies have moisture levels near 25% MC, regardless of thermal resistance outboard the plywood sheathing, and the control assemblies are strongly linked to the outdoor air and do not have any moisture source from the indoor air. The attic sheathing takes longer to reach elevated moisture levels than the control assembly sheathings, but eventually reaches similar levels as the sheathings absorb moisture with decreasing outdoor temperatures, ventilation, and higher relative humidity. Exterior air will pass the non-baffle locations, similar to the baffle locations but likely at different rates, because airflow is not greatly restricted by the insulation. The sheathing at the soffits will pick up the most moisture because this is the entry point of outdoor air and the coolest surface temperatures. Spikes in the sheathing moisture levels observed in the monitoring over a period of a day are presented in the next section.

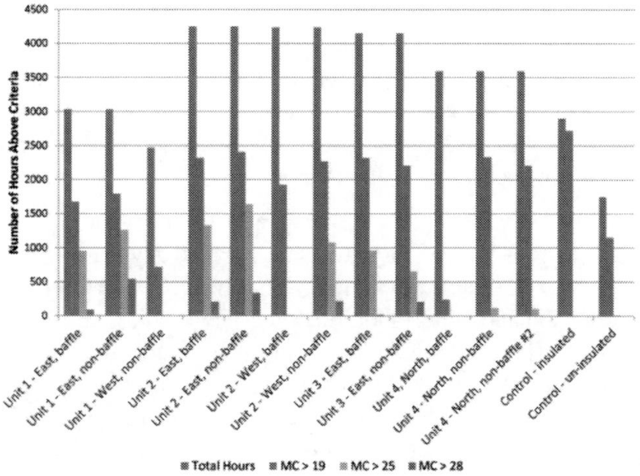

Figure 12 Duration of elevated sheathing moisture levels.

Another important observation showing that outdoor air is the principal source of wetting of the attic sheathing is the attic air has essentially the same overall moisture level as the outdoor air. Figures 13 and 14 show the attic air moisture content for Unit 1 compared to the outdoor air moisture content. The time series graph in Figure 13 shows that the attic air follows the same trend as the outdoor air. The scatter plot in Figure 14 shows that on average the attic and outdoor air are at the same moisture levels, with the attic air being slightly drier than the outdoor air at high moisture content levels.

Note that the times where the attic air moisture content levels are higher than the outdoor moisture levels coincide with decreasing moisture content levels in the sheathing and elevated sheathing temperatures (i.e., moisture is driven out of/into the wood to/from the attic air depending on the relative difference in vapour pressures).

Diurnal Wetting

Sharp spikes in the sheathing moisture content occurred in all the units at the same time and appeared to happen when moisture was deposited by condensation or frost as a result of surfaces cooled by night-sky radiation that was subsequently exposed to warmer humid outdoor air as temperatures rose in the daytime. Figure 15 illustrates these conditions for Unit 1 on December 12, 2011. Review of the weather records for the night before this wetting event showed periods of clear skies during the night, and this was also the case for the night before the significant wetting period on January 12, 2012 and in January 2013.

ANALYSIS AND DISCUSSION

The average attic ventilation rates measured in this study are low despite the abundant venting area. Nevertheless, these low rates appear to be sufficient enough to dilute any moisture transfer from the interior air to the attic space. The ceiling between the indoor and attic space appears to be relatively airtight, compared to other studies and expected values, and the transfer of air from the indoor to attic space does not appear to be a significant contributing factor to the moisture problems observed in the attics. Moreover, the average attic air moisture content is very close to the moisture con-

Table 8. Staining Pattern Observed at Sensor Locations

Unit	Moderate to Heavy – Field Area Spotty or Covered	Light – Localized at Fasteners	No Visible Staining
1	East: baffle	East: non-baffle, West: baffle and non-baffle	East: truss West: truss
2	East: baffle	East: non-baffle West: baffle and non-baffle	East: truss West: truss
3	West: baffle	East: baffle and non-baffle West: baffle and non-baffle	East: truss West: truss
4	North: non-baffle	North: baffle	South: baffle and non-baffle South: truss

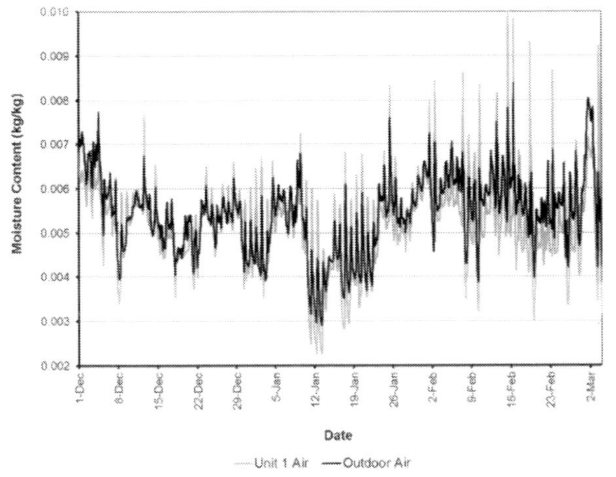

Figure 13 Unit 1 attic and outdoor air moisture levels per time.

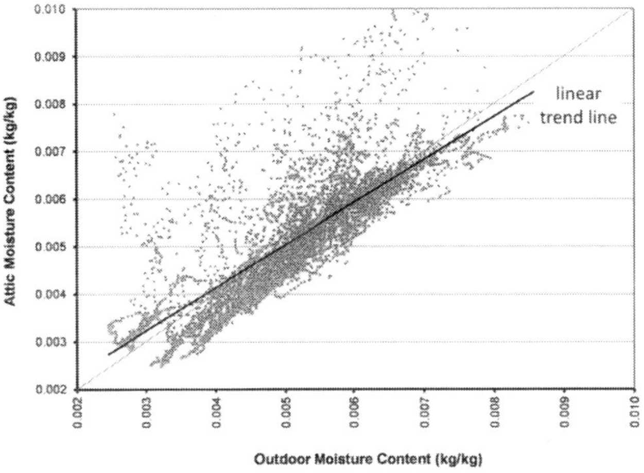

Figure 14 Unit 1 attic versus outdoor air moisture levels.

tent of the exterior air. Though the background moisture content of the attic sheathing appears to be largely dependent on the exterior air moisture levels, there are diurnal cycles due to daytime solar gains and nighttime radiative losses that result in differences of the sheathing moisture MC for the various locations. The fact that the moisture levels of the plywood sheathings in the control assemblies also reached elevated levels, up to 25% MC, suggests that higher ventilation rates will not significantly decrease the moisture levels in the attic and will not alleviate the occurrence of staining.

The testing and measurements completed for this study provided sufficient information to demonstrate how attics with ample venting area, subject to high outdoor moisture levels, can lead to mold growth in ventilated attics. A follow-up study is in progress and a summary of this work follows.

FOLLOW-UP STUDY

Visible mold in attics spaces for good and common construction practice without identifiable deficiencies or clear solutions presents a real issue for industry. This paper highlights this problem and provides an understanding of the contributing factors leading to visible mold in code compliant buildings. A follow-up study is focused on identifying solutions that industry can implement to respond to this issue, for both new construction and existing buildings. Included in the follow-up study is an evaluation of the impact of various parameters using a heat-air-moisture model that is validated and calibrated to the measured data presented in this paper.

The primary objectives of the follow-up study are to:

- develop strategies to reduce the likelihood of staining occurring and/or wetting in wood-frame attics,
- identify solutions that have a high probability of being successful if implemented by industry,
- identify any risks with the various alternatives, and
- help guide decisions for any contemplated changes to the building code.

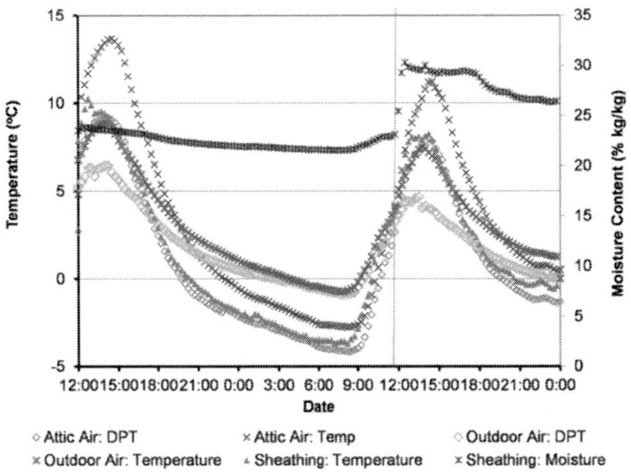

Figure 15 Unit 1 sheathing moisture spike after night-sky cooling on December 12, 2012.

Strategies and related solutions that are being explored, appropriate for the British Columbia costal climate, are as follows:

- **Strategy 1.** Apply treatments and coatings to create surfaces that are unfavourable for mold growth for a broad range of environmental conditions. For example, a treated wood sheathing surface that can be repeatedly exposed to condensation, long periods of high relative humidity and dynamic roof temperatures without growing mold.
- **Strategy 2.** Provide insulating boards and mold resistant sheathing outboard ventilated spaces. Liquid water is controlled by the insulating board by potentially two mechanisms, depending on the roof type. The principal mechanism of controlling exposure to liquid water, for all roof types, is keeping the roof sheathing temperature warm enough to eliminate wetting by night-sky radia-

tion. A secondary mechanism of controlling exposure to liquid water, for sloped roofs with asphalt shingles, is the insulating board will provide an effective capillary break between the sheathing and the shingles. The roof sheathing still needs mold resistance greater than unprotected wood because of the exposure to high humidity for long periods of time, but does not need to meet the same requirements as the treatments for Strategy 1 because exposure of the sheathing to liquid water will be minimized. Mold resistance of the sheathing can be provided by material selection, treatment, and/or coatings.

- **Strategy 3.** The optimum solution from a technical and durability perspective is to provide all the roof insulation outboard the roof sheathing and keep the roof structure warm and dry.
- **Strategy 4.** Insulate the underside of the roof sheathing with foam insulation to stop mold spores from getting in contact with the roof sheathing, while in service, and limit the available oxygen and moisture. Although mold will not likely grow on roof sheathings covered with foam insulation, the spray foam also provides a barrier and eliminates the air paths between the wood sheathing and interior space.

Other possible strategies that were considered but are not being followed up include:

- **Strategy 5.** Provide a mechanical system that controls airflow into the attic space and only ventilates when there is not the potential to add moisture to the attic space. This is an engineered strategy that will require more in-depth study; a calibrated heat-air-moisture modeling, field testing, and/or lab testing; before it is practical to implement in standard building practice.
- **Strategy 6.** Hide the mold growth. There are many types of roof constructions that follow this strategy that have similarities to the roofs covered by the other strategies. Examples are cathedral ceilings with ventilated cavities, low-sloped roof ventilated below the roof sheathing, and exterior insulated roofs with ventilated decking. The common theme is that inspection of the roof sheathing is more difficult and as a result is inspected with significantly less frequency than attic spaces with hatches. Consequently, these roofs are likely to circumvent unnecessary alarm of mold growth within roof structures. However, mold growth is still likely to occur.

A few of these strategies are already done in practice but a lot less frequently than conventional sloped ventilated wood-framed roofs, since generally all these solutions come at additional cost. Follow-up work is focused on identifying the cost-benefit of each solution for several scenarios; manufacturer and site solutions, low-slope and sloped roofs, and new and existing buildings.

Work also continues on validating a "whole building" heat-air-transfer model and evaluating the relative impact of ceiling airtightness, attic ventilation rates, sheathing thermal resistance, roof colour, and insulation levels on the moisture levels in ventilated attics. This effort is being supplemented through continued monitoring at the building site identified in this study, with different venting areas, and another location with low-sloped roofs with differing sheathing thermal resistance.

SUMMARY AND CONCLUSIONS

The attics in this study and the measured data are distinctive in the context of previous Canadian studies into the connection of attic ventilation and the hygrothermal performance of wood-framed attics because:

1. The attic construction with regard to controlling the heat-air-moisture flows represent current good practice
2. Venting areas and distribution are per or exceed code
3. A reasonable level of airtightness of the attic ceiling has been achieved
4. All ducts and plumbing that penetrate through the attic are brought up to the roof sheathing and are sealed, with no indication that they are contributing to higher moisture loads in the attic space
5. The indoor moisture loads are principally low to moderate levels
6. Despite all the above, the attics are getting wet, leading to localized staining on the plywood sheathing near the soffits

The implication is that the provision of venting area and an airtight ceiling alone is not enough to control moisture, to limit mold growth, in insulated attics of wood-framed sloped roofs in marine climates, similar to Vancouver's climatic zone. More ventilation will not solve the problem for attics constructed similar to the ones in this study and experience has shown us that less ventilation can lead to problems if an airtight ceiling is not achieved in practice.

ACKNOWLEDGMENTS

Morrison Hershfield collaborated with several industry partners, in addition to HPO-BC Housing (our primary client), to meet the objectives of this study. We would like to thank these partners that provided technical expertise and time to specific aspects of the study:

- FP Innovations (FPI)—Technical expertise on the durability and performance of wood products and treatments in building construction
- Polygon Construction Management—Construction practicality and costs of solutions
- University of British Columbia's School of Population and Public Heath—Expert opinion of health risks related to exposure of molds in buildings

REFERENCES

ASHRAE. 2009. *ASHRAE Handbook—Fundamentals*. Atlanta: ASHRAE.

BLP. 1991. Survey of Moisture Levels in Attics, Final Report submitted by Buchan, Lawton, Parent Ltd. to Canada Mortgage and Housing Corporation (CMHC), Ottawa.

Arfvidsson, J., and L.E. Harderup. 2005. "Moisture Safety in Attics Ventilated by Outdoor Air", *Proceedings of 7th Nordic Symposium on Building Physics*, Reykjavik, Iceland.

Clarke, J.A., C.M. Johnstone, N.J. Kelly, R.C. McLean, J.A. Anderson, N.J. Rowan, and J.E. Smith. 1996. "Development of a technique for the prediction/ alleviation of conditions leading to mould growth in houses". Scottish Homes Research Programme 1995–6. Final Report.

Dietz, R.N., T.W. D'Ottavio, and R.W. Goodrich. 1985. "Multizone Infiltration Measurements in Homes and Buildings Using a Passive Perfluorocarbon Tracer Method", *ASHRAE Transactions* 91 (2). 1761–1776. ASHRAE. Atlanta, GA.

Forest, T.W., and I.S. Walker. 1993. "Attic Ventilation and Moisture", Department of Mechanical Engineering University of Alberta, Final Report, Canada Mortgage and Housing Corporation (CMHC), Ottawa.

Forest, T.W., and K. Berg. 1993. "Simulations of Attic Ventilation and Moisture", Department of Mechanical Engineering University of Alberta, Final Report, Canada Mortgage and Housing Corporation (CMHC), Ottawa.

Forgues, Y.E. 1985. "The Ventilation of Insulated Roofs", Technical Information Section, Division of Building Research, National Research Council Canada, Ottawa.

Hagentoft, C. E, and A.S. Kalagasidis. 2010. "Mold Growth Control in Cold Attics through Adaptive Ventilation. Validation by Field Measurements", *Proceedings of the XI International Conference on the Performance of the Exterior Envelopes of Whole Buildings*, Clearwater, FL.

Hagentoft, C. E, A.S. Kalagasidis. S.F. Nilsson, M. Thorin.2008. "Mould growth control in cold attics through adaptive ventilation", *Proceedings of 8th Nordic Symposium of Building Physics*, Copenhagen, Denmark, 2008.

Hagentoft, C. E. 2011. "Probabilistic Analysis of Hygrothermal Conditions and Mould Growth Potential in Cold Attics. Impact of Weather, Building System and Construction Design Characteristics", *Proceedings of International Conference on Durability of Building Materials and Components*, Porto, Portugal.

Hukka, A., and H. Viitanen. 1999. "A mathematical model of mold growth on wooden material". *Wood Science and Technology* 33(6):475–85.

ISO 13788:2001 (E) "Hygrothermal performance of building components and building elements – Internal surface temperature to avoid critical surface humidity and interstitial condensation – Calculation methods", Geneva, Switzerland.

Kumaran, M.K. 1996. Final Report, Volume 3, "Material Properties", International Energy Agency Annex 24 Report, KU Leaven, Belgium.

Lstiburek, J. 2006. "Understanding Attic Ventilation", *ASHRAE Journal* 48 (4):36–45, ASHRAE. Atlanta, GA.

Hutcheon, N., and G. Handegord. 1995. "NRC-CNRC Building Science for a Cold Climate". National Research Council of Canada. Ottawa.

Parker, D.S., and J.R. Sherwin. 1998. "Comparative Summer Attic Thermal Performance of Six Roof Constructions." FSEC-PF-337-98, Florida Solar Energy Center, Cocoa, FL.

Roppel, P., M. Lawton, and W.C. Brown, 2009. "Setting Realistic Design Indoor Conditions for Residential Buildings by Vapour Pressure Difference", *Journal of ASTM International*, Vol. 6, No. 9, West Conshohocken, PA. Paper available on www.astm.org.

Rose, W.B. 2001. "Measured Summer Values of Sheathing and Single Temperatures for Residential Attics and Cathedral Ceilings", *Proceedings of the VII International Conference on the Performance of the Exterior Envelopes of Whole Buildings*, Clearwater, FL.

Rose, W.B., and A.A. TenWolde. 2002. "Venting of Attics and Cathedral Ceilings:, *ASHRAE Journal* 44 (10): 26 – 33, ASHRAE. Atlanta, GA.

Sedlbauer, K. 2001. "Prediction of Mould Fungus Formation on the Surface of/and Inside Building Components". PhD thesis, University of Stuttgart, Fraunhofer Institute for Building Physics, Stuttgart, Germany.

Sheltair Scientific Ltd. 1997. "Attic Ventilation and Moisture Control Strategies", Final Report, Canada Mortgage and Housing Corporation (CMHC), Ottawa.

TenWolde, A., and W.B. Rose. 1999. "Issues Related to Venting of Attics and Cathedral Ceilings", *ASHRAE Transactions* 105 (1). ASHRAE. Atlanta, GA.

Tobiasson, W.N., J.S. Buska, and A.R. Greatorex. 2001. "Guidelines for Ventilating Attics and Cathedral Ceilings to Avoid Icings at Their Eaves", *Proceedings of the VII International Conference on the Performance of the Exterior Envelopes of Whole Buildings*, Clearwater, FL.

Viitanen, H, and T. Ojanen. 2007. "Improved Model to Predict Mold Growth in Building Materials", Proceedings *of the X International Conference on the Performance of the Exterior Envelopes of Whole Buildings*, Clearwater, FL.

Window Related Thermal Bridges

Benjamin Barnes
Member ASHRAE

Axy Pagán-Vázquez
Member ASHRAE

Richard Liesen, PhD
Member ASHRAE

Justine Yu

Nicholas Alexander

ABSTRACT

Window thermal performance in the US is rated by the National Fenestration Rating Council's (NFRC) Standard 100, which requires estimates of a window frame's effective conductivity. Meanwhile, most energy estimation methods require effective conductivities to be derived for the rest of the wall. These two U-factors, when weighted by their respective areas, however, do not capture the thermal bridging caused by a window installation; flashing, lintels, masonry sills, additional structural framing, and the window frame's interaction with these components can all contribute to thermal bridging. The magnitude of this additional "as installed" thermal bridge is estimated as the difference between area weighted U–factors and a 2D or 3D simulation capturing "as installed" details. Some thermal bridge catalogs now report such values. These "Ψ -values," however, could be different for different windows installed with the same sill detail. This work estimates the sensitivity of a sill detail's "as installed" Ψ -value to the properties of the window frame installed. Six different window frames with progressively lower frame U-factors, from high performance wood to low performance aluminum, are simulated as they would be installed in two common commercial wall constructions with two sill detail strategies. Simulations use the finite-element heat transfer software THERM. Resulting knowledge of the impact of frame characteristics on the Ψ -value suggests that it is not always possible to use the Ψ -value from a catalog if a different window is chosen than that used for the original calculation, especially when combining well-detailed sills with lower performing windows.

BACKGROUND

ISO Standard 10211 defines a thermal bridge as "part of the building envelope where the otherwise uniform thermal resistance is significantly changed by full or partial penetration of the building envelope by materials with a different thermal conductivity, and/or a change in thickness of the fabric, and/or a difference between internal and external areas, such as occur at wall/floor/ceiling junctions" (2007). The calculation method used to determine the increase in thermal transmittance attributable to a thermal bridge is dependent upon the type of bridge encountered. For repeating thermal bridges, like studs or brick ties, the additional heat flow is taken into account in the calculation of the U-values of the surfaces containing the thermal bridges.

For non-repeating thermal bridges, which encompass the window-related bridges discussed in this paper, the heat flow attributable to the thermal bridge is calculated and added separately. For reasons described in Morrison Hershfield (2011), the additional heat transfer should be quantified as the linear thermal transmittance or Ψ -value. The Ψ -value has units of W/(m·K) and expresses the additional heat transfer caused by a feature per unit of length per degree of temperature difference at steady or quasi-steady state.

Window related thermal bridges are particularly interesting because they exhibit all three conditions described in the definition: materials of different resistance, change in the thickness of the building fabric, and a difference between internal and external surface areas. They also occur where the uniform thermal resistance —the building clear wall—meets

***Benjamin Barnes**, **Axy Pagán-Vázquez**, **Richard Liesen**, and **Justine Yu** are researchers at US Army Corps of Engineers, Engineer Research and Development Center, Construction Engineering Research Laboratory, Champaign, IL. **Nicholas Alexander** is a construction representative for US Army Corps of Engineers, USAF Academy, CO.*

Published by ASHRAE 2013

an entirely different type of surface—the window—which is quantified in the US using the area-weighted approach according to NFRC Standard 100 (NFRC 2010). The building clear wall, if determined correctly, captures 2D effects such as repeating structural elements. If the correct component U-factors are used with the correct areas, the window U-factor will capture the 2D thermal effects within the window itself. Estimating a whole wall performance by area-weighting these two resulting U-factors, however, fails to capture the substantial thermal impacts of masonry sills and lintels, flashing, other structural elements related to window installation, and the interaction of these features with the window frame itself.

This window-related thermal bridge is shown for a sill detail as the red shaded area in Figure 1. Figure 1 (b) shows a badly bridged detail and a correspondingly greater red area, while (c) shows a detail that maintains the thermal control layer and has a correspondingly smaller red area. This net additional heat flux, normalized to the temperature difference and the length of the detail, becomes the Ψ-value and can be used as a correction to the assumption pictured in Figure 1(a). The aim of this work is to determine whether this Ψ-value is sensitive to the window chosen, or if it can be considered a property of the sill detail.

Window Thermal Bridge Issues

Schild and Blom (2010) reviewed and compiled the experiences of several European Union (EU) member states' attempts to provide guidance on thermal bridges. Their conclusion was that thermal bridge regulation and guidance should be accomplished via catalogs of recommended details, complete with quantified linear thermal transmittance values. It is also useful to have poorly performing details quantified to facilitate justifying changes or retrofits. Such catalogs free designers from the time consuming and difficult process of performing their own 2D heat transfer simulations for every detail. For windows it is conceivable that a reasonable number of sill, jamb and head details could be quantified and could represent most situations. The catalog detail could even include a specific window position in the wall, which has a well established effect on the Ψ-value, as discussed in Cappelletti et al (2011). If, however, the Ψ-value of the sill, jamb, and head bridges depend strongly on the window frame cross-section interacting with them, beyond simply the frame's position in the wall, the number of required detail permutations expands greatly, and the feasibility of a go-to catalog is decreased. This work aims to explore when it is necessary to compute the Ψ-value of a unique combination of window cross section and installation detail and when it can be taken as independent of the window frame installed and used

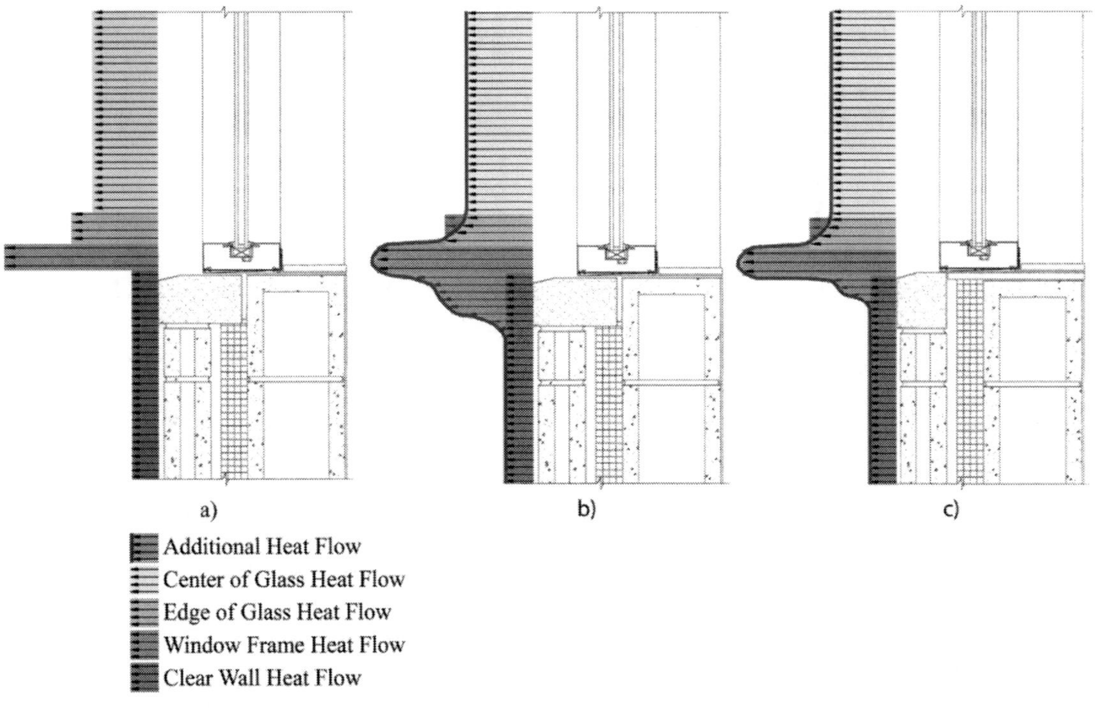

Figure 1 *(a) A common model for estimating whole wall U-factor is to area weight the 1D effects over the window center, edge of glass, frame, and the clear wall. (b) The real heat transfer, shown by the blue line, of a thermally bridged detail. (c) The real heat transfer, shown by the blue line, of an unbridged window detail. Typically, the three window U-factors would be area averaged and modeled as one, uniform U-factor prior to area-weighting with the wall, but they are shown separately here for clarity.*

as a widely applicable catalog detail. Similarly, this will clarify applicability of existing catalogs where the user wishes to specify a different window than the one shown in the catalog detail.

To accomplish this, we will combine multiple wall types and window types across the spectrum of construction techniques and window frame thermal performance. Window sills are used for this work because the window-wall coupling characteristics are thought to be representative for other details; that is, the implications concerning the design of a Ψ-value catalog are expected to be relevant to jamb and head details as well.

Ψ-VALUE CALCULATIONS FOR WINDOWS

Most energy estimation methods require a whole wall U-factor estimated by area-weighting the window U-factor and the clear-wall U-factor. This area-weighting approximates conduction as 1D over each area and implicitly assumes that the two elements meet at an adiabatic plane. These two terms alone fail to capture the thermal impacts of masonry sills and lintels, other structural elements, and the 2D interaction of these features with the window frame itself. A third term, the Ψ-value, makes up for this deficiency and is calculated as the difference between a calculation that includes these complexities and a simple area-weighted approach. For a given wall cross section, these three terms can be expressed as the heat transfer per length of the wall, or thermal coupling coefficient, as in ISO (2007):

$$L_{2D} = U_{clear-wall}h_{clear-wall} + U_{window}h_{window} + \Psi_{sill} \quad (1)$$

where:

L_{2D} = thermal coupling coefficient of wall cross section in W/(m·K)

$U_{clear-wall}$ = U-factor of the wall where no thermal bridges occur in W/(m²·K)

$h_{clear-wall}$ = height of clear wall in section

U_{window} = U-factor of the window, typically from area-weighting U-factors from 2D simulations of the frame, edge-of-glass, and center-of-glass regions in W/(m²·K) (Btu/h·sf·°F)

h_{window} = height of window included in section

Ψ = linear thermal transmittance of a bridge in W/(m·K)

Equation 1 (and this paper) only examines a section containing the sill. To include the full effect on the wall, one would need to include the bridges associated with the jambs and head as well. As discussed above, this paper focuses on sills in order to explore in greater depth the effects of window choice on window-related Ψ-value. The sill thermal bridge heat flow can be calculated by 1) simulating the window frame including 150 mm of glass, following NFRC 100 guidelines, to obtain U_{window}, 2) simulating the clear wall, not including any sill details such as masonry sills and flashing, to obtain $U_{clear-wall}$, 3) simulating the assembly including all sill details and the window to obtain the thermal coupling coefficient, L_{2D}, and finally 4) solving Equation 1 for Ψ_{sill}.

ASSEMBLIES SIMULATED

Two wall types common to commercial construction were chosen for investigation, including block construction with brick veneer and light gage steel construction with a brick veneer. Both wall types were simulated with high performing (unbridged) and typical (bridged) details. The four resulting combinations are shown in Figures 1 and 2. The bridged details are taken from recent construction documents. To create the unbridged details, the two bridged details are modified to better maintain the thermal control layer as recommended in Lstiburek (2011). The four details are later combined with each of six windows. These windows, shown in Figure 3, represent the performance range of modern window frames, from a very high performance wood window to a typical aluminum curtain wall with a minimal thermal break. All windows were inspired by commercially available windows but are not identical to the actual products. Weep holes, curtain wall bolts, and other fasteners are ignored.

SIMULATION METHODS

THERM 6.3 is used to perform all heat transfer simulations (Finlayson et al. 1998). WINDOW 6.3 is used to create the glazing units and estimate their center-of-glass properties before importing into THERM (Mitchell et al. 2001). Unless otherwise stated, simulations were carried out in accordance with NFRC 100 guidelines, including boundary conditions (NFRC 2010). Though boundary conditions from ISO Standard 6946 (2007) are a more typical choice, this work attempts to discover any relationship between the window and the window-related Ψ-value, and since the window U-factor would normally come from NFRC 100 certification, the wall boundary conditions were taken from this standard to keep from having adjacent boundary conditions differ.

For light gage steel stud walls, the effective conductivity of the stud cavity was determined from a simulation of a plan-view cross section, which accounts for the effect of the vertical studs. The horizontal sill stud is the only steel stud that is explicitly modeled in the section view. The boundary conditions for all simulations were as follows:

1. Internal, non-glazing surfaces were assigned appropriate emissivities based on their materials and were modeled as radiatively communicating with a 21.0°C (69.8°F) room surface and with each other. Where the material was likely to have a higher internal surface temperature (e.g., in the case of the interior wall surface or the interior of a wood window), it was assigned a convection coefficient of 2.44 W/(m²·K) (0.430 Btu/h·ft²·°F). In the case of the Aluminum window, where a colder surface is expected, 3.0 W/(m²·K) (0.528 Btu/h·ft²·°F) is assigned. Air temperature is 21.0°C (69.8°F).

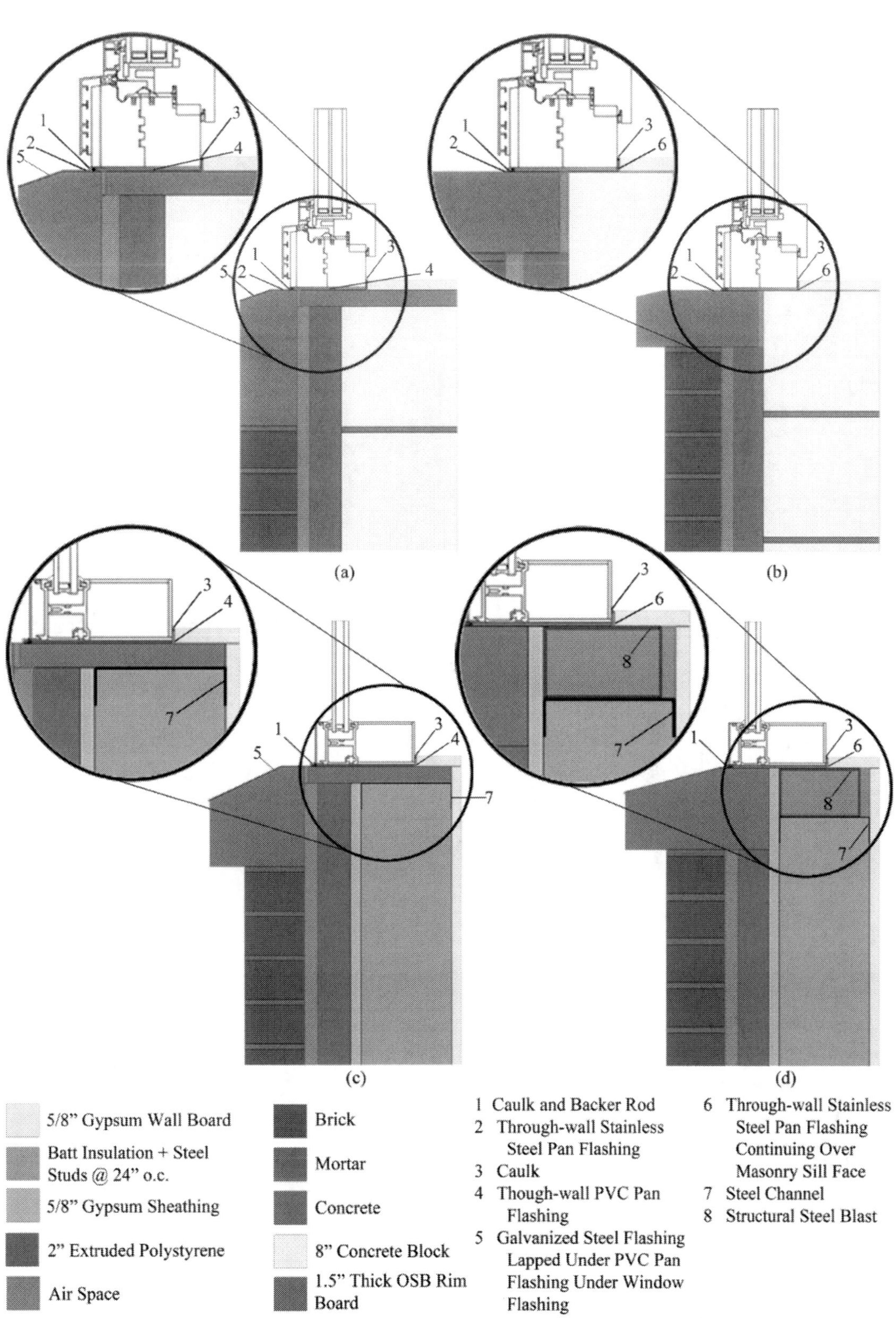

Figure 2 *The four details, made up of two wall types each with bridged and unbridged sills: a) unbridged CMU wall; b) bridged CMU wall; c) unbridged steel stud wall; d) bridged steel stud wall. Note that the four details shown here were studied with all windows, for a total of 24 combinations, not just the four combinations shown.*

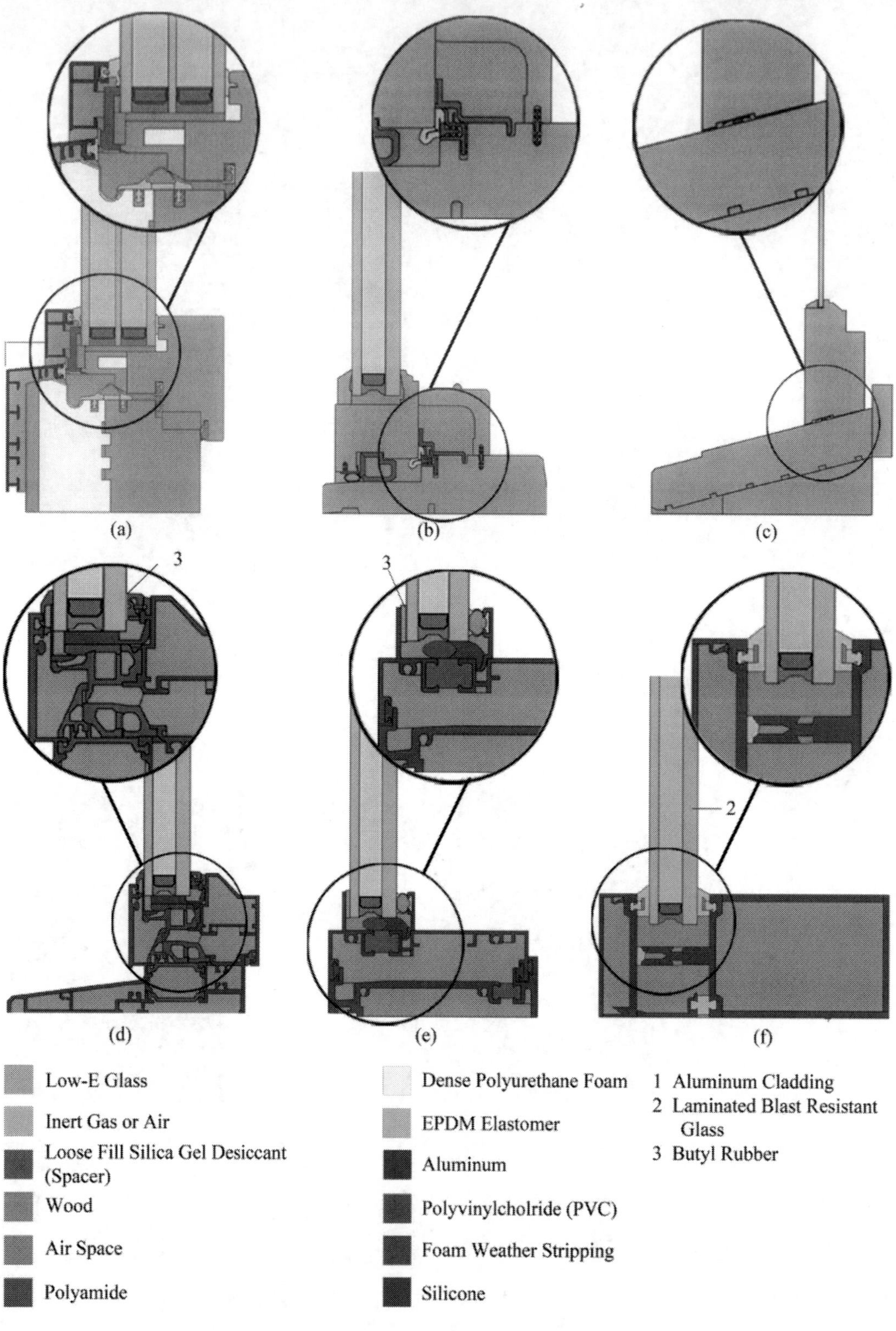

Figure 3 *The six windows examined attempt to capture the range of modern windows, from very high performing triple glazed wood to typical aluminum curtain wall, with progressively lower frame U-factors (note, overall window U-factor does not decrease in the same order, largely due to the single-pane glazing for the low performance wood).*

2. Internal glazing surfaces take their boundary condition from an automatic THERM computation, which uses the center-of-glass temperature computed by WINDOW. Radiation is handled the same as with other internal surfaces.
3. External Surfaces are assigned a convection coefficient of 26 (m^2·K) (4.6 Btu/h·ft^2·°F), which corresponds to a 5.5 m/s (12 mph) wind. Radiation is modeled as communication with a black body at –18°C (–0.40°F) and a view factor of 1. Air temperature is also taken to be –18°C (–0.40°F).
4. Adiabatic surfaces are used on the bottom of windows when they are modeled to determine their component U-values.The glass is extended high enough that it can effectively be cut at an adiabat as well. Walls also extend down far enough that they are terminated at an adiabatic boundary condition. Clear-walls simulations to determine $U_{clear\text{-}wall}$ are also terminated at their top and bottom at an adiabat.

Each of the 24 resulting combinations of wall type, sill detail, and window is then simulated in three different configurations:

1. Baseline thermal bridge: In this case, sill details are in place, but the surfaces where a window frame would touch the wall are given adiabatic boundary conditions. The window itself is not included, but the adiabat must be of the correct depth. Any trim that would block the inside surface of the window is not included, as in Figure 4(a). The thermal coupling coefficient predicted by the simulation is input as L_{2D} in Equation 1, and $U_{window}h_{window}$ is input as zero. The Ψ-value solved for is identical for any window of the same depth. It estimates the thermal bridge effect of all masonry sills, additional framing and flashing, but excludes the effect of thermal interaction with the window frame and any effect of trim blocking the inside surface of the window frame.
2. Window installed without trim: In this case, the window is simulated attached to the wall with the appropriate shim gap and sealant and allowed to interact with the wall conductively and, to a limited extent on the interior sill, radiatively. Trim is still not included. See Figure 4(b). The thermal coupling coefficient from the simulation is again used as L_{2D} in Equation 1, but $U_{window}h_{window}$ is no longer zero. Any change in the thermal bridge effect from Case 1 is attributed to thermal coupling with the window.
3. Full assembly: In this case, the window is simulated as in Case 2, but realistic interior trim is included, as in Figure 4(c). All trim modeled is 16 mm (5/8 in.) gypsum wall board. Computation of thermal bridge effect is similar to Case 2. Any change in the thermal bridge effect from Case 2 is attributable to the effects of covering part of the interior of the window frame and, to a lesser extent, part of the interior wall.

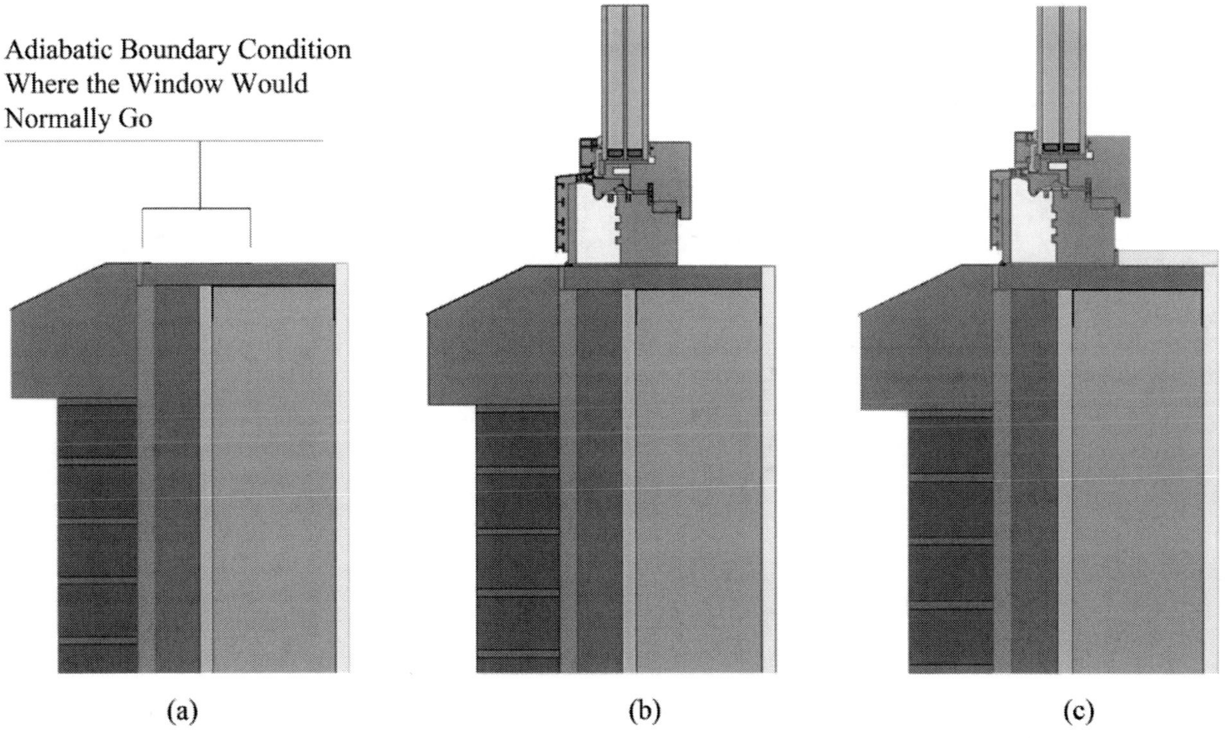

Figure 4 *The different thermal bridge cases including a) the baseline thermal bridge, b) the thermal bridge including window coupling effects, and c) the thermal bridge including the effects of interior trim on the wall and window.*

RESULTS

Tables 1 and 2 show the results from the clear-wall and window simulations, and Table 3 shows the Ψ-values for each window-wall combination for the bridged and unbridged case. Figure 5 displays the primary results in terms of L_{2D} from Equation 1, color coded to match Figure 1. Note that the top bar is a function of the window only, excluding any trim effects, while the bottom bar is a function of the wall only, excluding any sill details. The middle bar encompasses trim effects, masonry sills, additional framing, flashing, and all interactions between the wall and the window. The relationship between this quantity and the window in the assembly is the focus of this study.

Table 1. Clear-Wall U-Factors of the Walls Analyzed

	Description	U-Factor, W/(m²·K) (Btu/h·ft²·°F)
Wall A	8 in. CMU with 2 in. rigid insulation and brick veneer	0.382 (0.0673)
Wall B	6 in. stud, 24 in. o.c. with batt insulation, 2 in. rigid insulation and brick veneer	0.247 (0.0435)

Table 2. Window Properties (Note Progression of Frame U-Factors)

	Description	Center of Glass Height	Edge of Glass Height	Frame Height	Center of Glass U-Factor	Edge of Glass U-Factor	Frame U-Factor	Whole Window U-Factor[1]
		mm (in.)			W/(m²·K) (Btu/h·ft²·°F)			
1	Wood, high performance	86.5 (3.41)	63.5 (2.5)	135 (5.32)	1.08 (0.190)	1.20 (0.211)	0.81 (0.143)	1.00 (0.176)
2	Wood, medium performance	86.5 (3.41)	63.5 (2.5)	73.6 (2.90)	1.93 (0.340)	2.41 (0.424)	1.68 (0.296)	1.96 (0.345)
3	Wood, low performance	86.5 (3.41)	63.5 (2.5)	98.7 (3.89)	6.06 (1.07)	5.93 (1.04)	2.67 (0.470)	5.11 (0.900)
4	Aluminum, high performance	86.5 (3.41)	63.5 (2.5)	76.3 (3.00)	1.91 (0.336)	2.14 (0.377)	4.62 (0.834)	2.53 (0.446)
5	Aluminum, medium performance	86.5 (3.41)	63.5 (2.5)	65.9 (2.59)	1.92 (0.338)	2.31 (0.407)	6.20 (1.09)	2.79 (0.491)
6	Aluminum, low performance	86.5 (3.41)	63.5 (2.5)	68.1 (2.68)	1.87 (0.330)	2.03 (0.358)	8.03 (1.41)	3.09 (0.544)

1. Window U-factors are computed for rough comparison only, assuming a window 1200mm by 1500mm with jamb and head U-factors equivalent to sill U-factors.

Table 3. Resulting Ψ-Values for All Assemblies

		With Bridged Sill Details, W/(m·K) (Btu/h·ft·°F)		With Unbridged Sill Details, W/(m·K) (Btu/h·ft²·°F)	
		CMU Wall	Steel Stud Wall	CMU Wall	Steel Stud Wall
1	Wood, high performance	0.297 (0.172)	0.346 (0.200)	0.0263 (0.0152)	0.0332 (0.0192)
2	Wood, medium Performance	0.306 (0.177)	0.362 (0.209)	0.0256 (0.0148)	0.0359 (0.0207)
3	Wood, low performance	0.258 (0.149)	0.317 (0.183)	0.00756 (0.00437)	0.0192 (0.0111)
4	Aluminum, high performance	0.257 (0.149)	0.324 (0.187)	0.0162 (0.00938)	0.0204 (0.0118)
5	Aluminum, medium performance	0.306 (0.177)	0.347 (0.201)	0.0809 (0.0468)	0.0725 (0.0419)
6	Aluminum, low performance	0.307 (0.177)	0.305 (0.176)	0.0785 (0.0453)	0.0799 (0.0462)

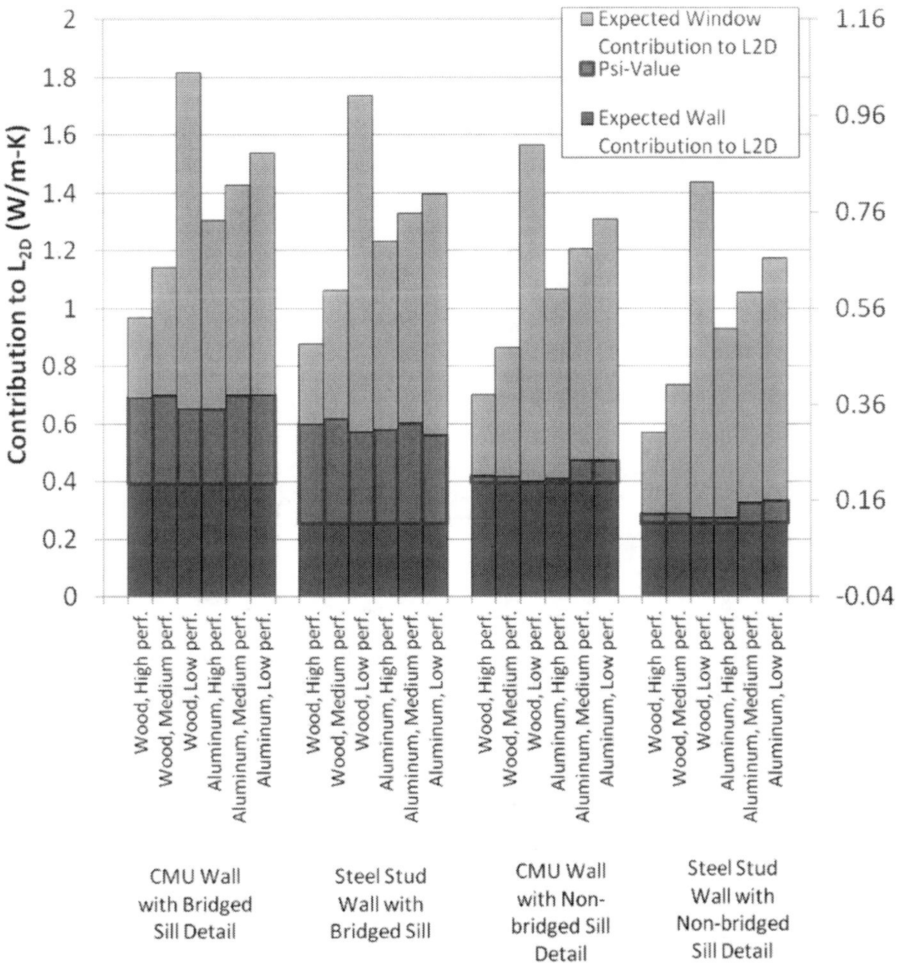

Figure 5 *The thermal coupling coefficient of each assembly. The contributions are separated into those from the wall, the contribution from the window, and the thermal bridge effect. Colors match those of Figure 1. Note that the expected contribution for the wall is based on the height of the wall used in the assemblies and the expected contribution for the window is also for only that height of window included in the assembly (see Table 2).*

Figure 6 shows the thermal bridge effects broken down into components for each of the window-wall combinations for the bridged details. Figure 7 shows it for unbridged details. The window interaction is the change in thermal bridging with the addition of the window. The trim contribution is the reduction in thermal bridging caused by the addition of 16 mm (5/8 in.) interior gypsum trim. This progression is described above and shown in Figure 4.

The window interaction effect can be broken into conduction and radiation effects. If we limit our discussion to conduction heat transfer, it is clear that the insertion of any real window in place of the baseline case will not decrease heat flux. It can only increase thermal cross section—that is, the possible heat transfer paths at every point through the wall. For example, heat transfers from the masonry sill into the highly conductive window frame through most of the depth of the window; at the spacer, the reverse happens, effectively using the masonry to bridge the spacer.

Radiative interaction, however, does not always increase the heat flow. The interior sill surfaces can partially "see" the window frame and glass. In the baseline case, which captures a detail thermally decoupled from the window, all they can see is a 23.0°C (73.4°F) surface, which could tend to cause more room heat to radiatively transfer to the surface without the window.

In some cases, this effect dominates the conductive coupling effect, causing the overall coupling effect to be negative; this is especially true for the low-performance wood window, which gives the interior sill a generous view of cold single pane glass. It is also evident in severely bridged walls, where a substantial bridging element has a large view factor to the window. This example shows that the presence of a window can actually reduce the calculated thermal bridge effect, but this is likely only true in particularly strange cases. The total heat flow, of course, increases over the baseline case with the addition of a window, even with the negative impact to Ψ_{sill};

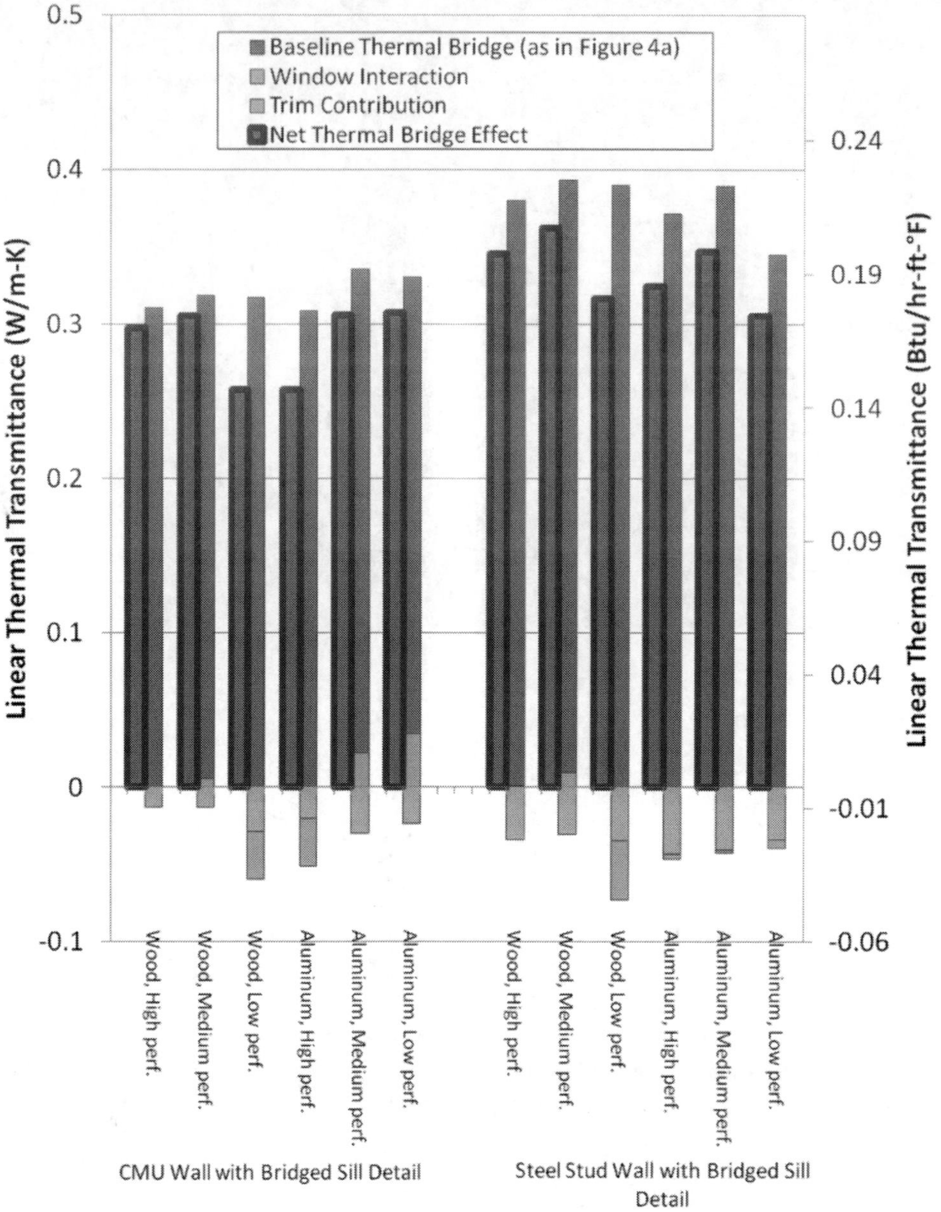

Figure 6 *The net ψ_{sill} from Figure 5 for every window on the bridged details, broken into components determined from the series of simulations described in Figure 4.*

it is obvious from Figure 5 that a poor window cannot make up for its high U-factor with its small reduction in Ψ_{sill}.

Trim, on the other hand, always decreases the calculated Ψ_{sill}. In part, this is attributable to insulating the inside wall beyond the baseline case. The greater contribution, however, is insulating a part of the inside edge of the window frame, which would otherwise exhibit very cold temperatures and substantial heat flux. This latter effect does not occur by any actual thermal bridging reduction, but rather because of an unaccounted for reduction in the heat flow from the interior window frame surface, making the window behave differently than is assumed during the NFRC window simulation and, subsequently, in the application of Equation 1.

The magnitudes of the coupling and trim effects will be discussed separately for the bridged and unbridged cases.

Bridged Details

For all windows on the bridged details, the net thermal bridge effect is severe, mostly owing to the bridging inherent in the details themselves, not window interaction or trim. For the CMU wall, the Ψ_{sill} values are equivalent to adding between 0.67–0.80 m (2.2–2.6 ft) to the clear-wall height. For the steel stud wall, it is 1.3–1.5 m (4.1–4.8 ft). Table 4 shows

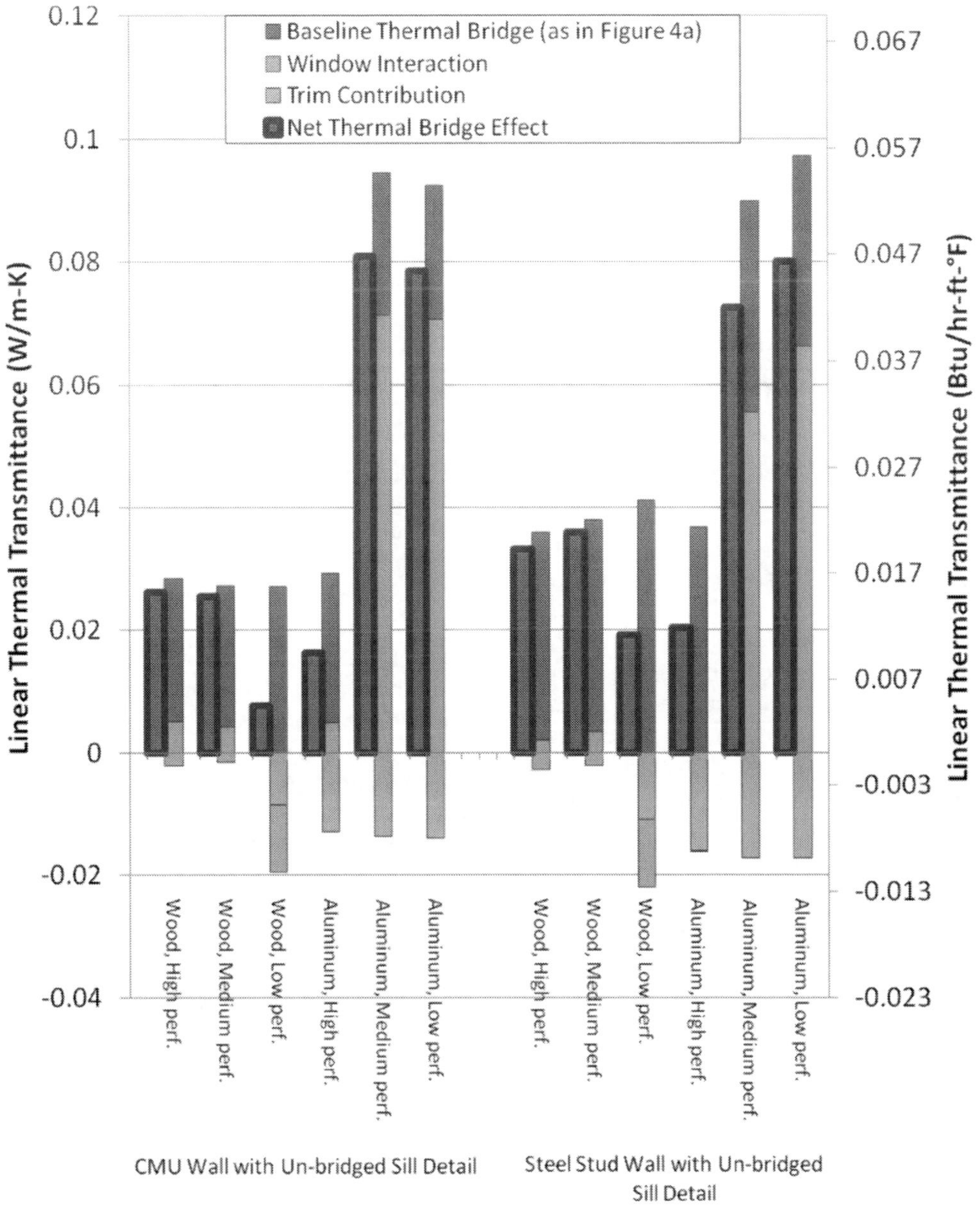

Figure 7 *The net Ψ -value from Figure 5 for every window on the unbridged details, broken into components determined from the series of simulations described in Figure 4.*

Table 4. Statistics for Bridged Details Across All Six Windows

	Average Ψ_{sill}, W/(m·K) (Btu/h·ft·°F)	Standard Deviation of Ψ_{sill}, W/(m·K) (Btu/h·ft·°F)
CMU Wall	0.288 (0.167)	0.0242 (0.0140)
Steel Stud Wall	0.334 (0.193)	0.021 (0.0124)

the average and standard deviation of the Ψ_{sill} for the bridged details across all six windows. The thermal bridge is dominated by the effects of the masonry sills, steel reinforcement, and flashing; trim and window interaction play a minor part. This indicates that a catalog containing these details could purport, with reasonable accuracy, to be relevant for any window installed in the insulation plane. With thermal bridges this severe, temperature gradient through the wall does not encourage heat flow to short-circuit through the window; there is no continuous thermal control layer to short-circuit around.

Unbridged Details

The range in the unbridged CMU wall details of Ψ_{sill} is equivalent to the additional heat flow caused by between 0.02–0.21 m (0.065–0.69 ft) of additional clear wall. For the steel stud wall, the equivalent additional clear wall would be 0.08–0.32 m (0.26–1.1 ft). Table 5 shows the average and standard deviation of the Ψ_{sill} for the unbridged details across all six windows. Though the absolute deviation from the baseline thermal bridge is small, the relative deviation is quite large. This can be substantial if one is attempting to minimize thermal bridges. The baseline Ψ_{sill}'s for the unbridged are on the same order of the Ψ-value cutoff for consideration in the passive house standard, which is 0.01 W/(m·K) (0.006 Btu/h·ft·°F). Below this value, a thermal bridge does not have to be considered during calculations for Passive House certification. For designers attempting this level of performance, deviations seen here can mean a great deal.

The window interaction impact increases the thermal bridge effect in all of the unbridged cases, except for that of the low performance wood window. This is likely due to minimal conductive coupling and strong radiative coupling due to the single pane glass. In all other wood cases, and in the high performance aluminum case, the interaction is small. In the lower performing aluminum windows, however, the interaction dominates the thermal bridge effect. This effect is not seen for the bridged details because they have temperature gradients that do not tend to encourage heat flux through the windows; in the unbridged case, however, a conductive path through a window can short circuit an otherwise well maintained thermal control layer.

CONCLUSIONS

It is important to remember that the primary window-related increase in heat flux over the clear wall is captured in the U-factor of the window chosen, not in the thermal bridge effect. For all but the highest performing window in this study, a mere 250 mm or less of window frame and glass exhibits heat flux greater than over 1 m of clear-wall height and also greater than any thermal bridge effect, even in the badly bridged details (see Figure 5). That is, when one looks at the whole picture, it is obvious that the minor reductions in thermal bridging evident with some windows do not justify choosing that window over with a higher overall U-factor.

Table 5. Ψ_{sill} Statistics for Unbridged Details Across All Six Windows

	Average Ψ_{sill}, W/(m·K) (Btu/h·ft·°F)	Standard Deviation of Ψ_{sill}, W/(m·K) (Btu/h·ft·°F)
CMU Wall	0.0392 (0.0226)	0.0321 (0.0186)
Steel Stud Wall	0.0435 (0.0251)	0.0263 (0.00152)

For badly bridged details (e.g., those with masonry sills and metal flashing passing through the insulation plane) the window frame choice does not strongly change the Ψ-value of the thermal bridge. The additional thermal cross section provided by allowing the window to interact conductively does not greatly increase heat flux when the existing thermal path is already quite conductive. This means that such details can be presented in a catalog for use with nearly any window.

For details that maintain the thermal control layer, however, different windows in the same sill detail give rise to relatively large Ψ-value differences, especially relative to Ψ-values of details that meet Passive House level performance. In fact, the baseline thermal bridge can be magnified by a factor of up to 3.6 by choosing a poorly performing window. This alone makes it difficult to publish a Ψ-value for a particular detail, since it is really a property of the sill detail in combination with the window frame. The trim effects are also quite large in the unbridged cases for window frames that would otherwise have a very cold inside surface. As such, a catalog entry for high performing details would need caveats limiting its use to situations in which the details are combined with high performance windows.

REFERENCES

Cappelletti F., Gasparella A., Romagnoni P., Baggio P. 2011. Analysis of the influence of installation thermal bridges on windows performance: The case of clay block walls. 2011. University IUAV of Venezia, Venezia, Italy

Finlayson, E., Mitchell R., Arasteh D., Huizenga C. and Curcija D. 1998. THERM 2.0: Program description. A PC program for analyzing the two-dimensional heat transfer through building products. University of California, Berkeley, CA, USA.

International Organization for Standardization (ISO). ISO 10211. 2007. *Thermal bridges in building construction –Heat flows and surface temperatures– Detailed calculations*. Geneva, Switzerland: ISO.

International Organization for Standardization (ISO). ISO 14683. 2007. *Thermal bridges in building construction –Linear thermal transmittance – Simplified methods and default values*. Geneva, Switzerland: ISO.

International Organization for Standardization (ISO). ISO 6946. 2007. *Building Components and building elements –Thermal resistance and thermal transmittance– Calculation method*. Geneva, Switzerland: ISO.

Lstiburek, Joseph. 2011. BSI-062 Thermal Bridges Redux. http://www.buildingscience.com/documents/insights/bsi062-thermal-bridges-redux.

Mitchell R., Kohler C., Arasteh D., Huizenga C., Yu T., and Curcija D. 2001. *WINDOW 5.0 User Manual*. Lawrence Berkeley National Laboratory, Berkeley, CA, USA.

Morrison Hershfield. 2011. Thermal Performance of Building Envelope Details for Mid- and High-Rise Buildings (1365-RP). Vancouver, BC, Canada

National Fenestration Rating Council (NFRC). 2010. NFRC 100: *Procedure for determining fenestration product U-factors*. Silver Spring, MD: NFRC.

O'Leary, Thomás. 2012. Thermal Bridge Analysis and Design. High Performance Building Envelope Workshop, March 6–8, 2012, San Antonio, TX, USA.

Schild, Peter G., and Peter Blom. 2010. Good practice guidance on thermal bridges & construction details, Part I: Principles. EPBD, http://buildup.eu/system/files/content/P188_Thermal_bridge_guidance_principles_ASIEPI-WP4.pdf.

An Assessment of Envelope Measures in Mild-Climate Deep Energy Retrofits

Iain Walker, PhD
Fellow ASHRAE

Brennan Less

ABSTRACT

Energy end uses and interior comfort conditions were monitored in 11 deep energy retrofits (DERs) in a mild marine climate. Two broad categories of DER envelope were identified: first, bringing homes up to current code levels of insulation and airtightness, and second, enhanced retrofits that go beyond these code requirements. The efficacy of envelope measures in DERs was difficult to determine, due to the intermingled effects of enclosure improvements, HVAC system upgrades, and changes in interior comfort conditions. While energy reductions in these project homes could not be assigned to specific improvements, the combined effects of changes in enclosure, HVAC system, and comfort led to average heating energy reductions of 76% (12,937 kWh) in the five DERs with pre-retrofit data, or 80% (5.9 kWh/ft^2) when normalized by floor area. Overall, net site energy reductions averaged 58% (15,966 kWh; n=5*), and DERs with code-style envelopes achieved average net site energy reductions of 65% (18,923 kWh;* n=4*). In some homes, the heating energy reductions were actually larger than the whole-house reductions that were achieved, which suggests that substantial additional energy uses were added to the home during the retrofit that offset some heating savings.*

Heating system operation and energy use was shown to vary inconsistently with outdoor conditions, suggesting that most DERs were not thermostatically controlled and that occupants were engaged in managing the indoor environmental conditions. Indoor temperatures maintained in these DERs were highly variable, and no project home consistently provided conditions within the ASHRAE Standard 55-2010 heating season comfort zone. Thermal comfort and heating system operation had large impacts on performance and were found to depend upon the occupant activities, so DERs should be designed with the occupants' needs and patterns of consumption in mind. Beyond-code building envelopes were not found to be strictly necessary for the achievement of deep energy savings in existing uninsulated homes in mild marine climates, provided that other energy end uses were comprehensively reduced. We recommend that mild-climate DERs pursue envelopes in compliance with the 2012 International Energy Conservation Code *(*IECC*) and pair these with high-efficiency, off-the-shelf HVAC equipment. Enhanced building envelopes should be considered in cases where very low heating energy use (<1,000 kWh/year or <0.5 kWh/ft^2·year) and enhanced thermal comfort (ASHRAE Standard 55-2010) are required, as well as in those situations where substantial energy uses are added to the home, such as decorative lighting, cooling, or smart-home audio/visual and communication equipment.*

INTRODUCTION

A deep energy retrofit (DER) is a home energy upgrade aimed at energy reductions above and beyond those achieved in traditional weatherization or home performance programs. These ambitious projects take existing inefficient homes and transform them into very energy-efficient, comfortable, and potentially low-energy homes. Often sustainability, historic preservation, and occupant comfort, health, and safety are intertwined with the energy reduction goals. The energy reductions are typically achieved using a combination of building enclosure air sealing, additional insulation, window replacement, HVAC and domestic hot water system upgrades, lighting and appliance replacement, and sometimes the addition of renewable energy technologies, such as solar photo-

Iain Walker *is a scientist and* ***Brennan Less*** *is a senior research associate in the Residential Building Systems Group at Lawrence Berkeley Laboratory, Berkeley, CA.*

Published by ASHRAE 2013

voltaics or solar hot water. These building upgrades are often combined in varying degrees with occupant conservation efforts.

While the exact definition of a DER is not yet clear, most working in the field consider energy reductions of 50% to 90% to be readily achievable with existing technologies, materials, and construction practices (Henderson and Mattock 2007; Wigington 2010). Published definitions of DERs range from 30% to 75% of annual energy use compared with a pre-retrofit baseline (ACI 2010a; Chandra et al. 2012). Given advances in minimum building codes and the general DER objective to make significant changes in energy use, we believe that the most appropriate DER definition should be on the high end of this scale, at the 70% level.

In this paper, we summarize the building envelope improvements, heating energy use, and thermal comfort in eleven case study homes in Northern California that have targeted energy reductions of 70% or more. All project homes were located in the San Francisco Bay Area and Sacramento regions. The retrofit measures and performance of each DER are documented in detail in Less et al. (2012). Each home was retrofitted by the homeowner prior to joining the research project, so the study had no influence on the retrofit measures taken. The project goals, strategies used, and results achieved represent actual results of the homeowners', designers', and contractors' approaches to a high-performance retrofit.

These DER case study homes were equipped with wireless energy monitoring equipment, providing one-minute resolution on each electrical and gas energy end use, as well as indoor temperature and relative humidity. This live data stream was made available to the home occupants and research team via a web application. Pre-retrofit utility billing data were available in only 5 of 11 projects. These pre-retrofit usages were weather-adjusted by heating degree-day regression for direct comparison to the monitoring year.

Project and Enclosure Descriptions

Table 1 provides basic descriptions for the 11 DER homes in this study, and retrofit measures are broadly summarized in Table 2. Significant diversity was observed between homes, which is likely to be true of DERs as they reach a wider audience because of the variability in existing home construction, location, and occupancy. The breadth of available paths to deep energy savings in even one climate zone is evident in Table 2. All DERs were constrained by the existing site, building, equipment, and fuel types, and all projects uncovered unforeseen obstacles during construction. This required flexibility throughout the process. Furthermore, homeowners and developers had different goals and reasons for performing the retrofits, which led them to different solutions. No particular technological or behavioral solution was required for success; rather, solutions had to be evaluated on a project-by-project basis for appropriateness and acceptability to occupants. The wide range of approaches is reflected in the range of Home Energy Rating System (HERS) indices (RESNET 2006) between projects for the retrofitted homes, ranging from 25 to 86. Despite this large range, these indices imply that these homes are all better than the HERS reference home. Pre-retrofit HERS indices were not calculated, because project teams were not able to consistently provide data on pre-retrofit conditions.

Insulation. Nearly all pre-retrofit homes were uninsulated structures, with occasional small amounts of insulation in the attic. Post-retrofit insulation levels varied from approximately code-compliant (termed *code-style*) to either 50% or 100% better than code (termed *highly* and *super* insulated, respectively). The code-style homes were consistent with the prescriptive requirements of the *California Building Energy Code* (Title 24) (CEC 2008).

Table 1. Project Summaries

ID	Location	Year Built / Retrofitted	HDD_{65}	Floor Area Pre / Post	Occupants Pre / Post	HERS
P1	Berkeley, CA	1904 / 2008	2909	960 / 1630	2 / 4	72
P2	Palo Alto, CA	1936 / 2008	2563	2780 / 2780	NA / 2	55
P3	Sonoma, CA	1958 / 2010	2844	1937 / 2357	NA / 1	25
P4	Petaluma, CA	1940 / 2010	2844	1540 / 2510	2 / 2	36
P5	Point Reyes Station, CA	1920 / 2010	3770	800 / 905	NA / 3	86
P6-N	Davis, CA	1932 / 2011	2702	1179 / 1462	NA / 4	28
P6-S	Davis, CA	1934 / 2011	2702	1496 / 1496	NA / 4	37
P7	San Mateo, CA	1910 / 2011	3042	3288 / 3288	2 / 2	76
P8	Oakland, CA	1915 / 2008	2909	1440 / 1627	NA / 4	33
P9	Folsom, CA	1998 / 2006	2702	3114 / 3114	NA / 4	72
P10	Pacifica, CA	1934 / 2008	3770	1503 / 1706	2 / 2	25

Table 2. Project Retrofit Comparison

Project ID	P1	P2	P3	P4	P5	$P6_N$	$P6_S$	P7	P8	P9	P10
Building Enclosure											
Super Insulated (100% > CA Title 24)			X			X	X				
Highly Insulated (50% > CA Title 24)	X				X						
Insulated (Meets CA Title 24)		X		X				X	X	X	X
All Triple-Pane Glazing			X								
All Double-Pane Glazing	X	X		X	X	X	X			X	X
Airtightness: Passive House Standard (<0.6 ACH_{50})			X								
Airtightness: <3 ACH_{50} (recommended)	X		X		X					X	
Airtightness: ENERGY STAR® Ver. 3 (<5 ACH_{50})	X		X		X					X	
HVAC											
Heat/Energy Recovery Ventilation	X	X	X								
Electric Resistance Heating	X				X						
Heat Pump Heating and Cooling		X	X								
Air Conditioner with Evaporative Cooling										X	
Heat and Domestic Hot Water Combisystem		X	X						X		X
Night Ventilation Cooling				X		X	X			X	
Domestic Hot Water											
Electric Resistance					X						
Heat Pump		X									
On-Demand Condensing Natural Gas	X		X	X		X	X	X			
Atmospheric Draft Natural Gas Tank										X	
Solar Thermal Tank with Condensing Natural Gas Backup			X			X	X		X		X
User Behavior											
24/7 Electric Baseload <200 Watts	X			X	X	X	X	X		X	X
24/7 Electric Baseload >200 Watts		X	X						X		
Solar Technology											
Photovoltaics		X	X	X		X	X		X		X
Solar Thermal			X			X	X		X		X

A variety of methods were used to insulate above-grade walls, and multiple assemblies were often employed on a single home due to varying envelope constructions. Three homes exclusively drilled and filled using dense packed cellulose (P2, P4, and P8). A number of other projects exposed wall framing cavities from either the interior or exterior, then filled them with cellulose, fiberglass, or spray foam insulation. Often these homes also added continuous exterior foam insulation of thickness ranging from 1 to 5 in. (P1, P3, and P5). P3 provided the most exterior foam insulation, with those assemblies constructed from 2 × 4 framing, adding 5 in. of foam, and those using new 2 × 6 framing, adding 2.5 in. P1 used a mixed approach, resulting from the lifting of the home and rebuilding of the ground floor, using 2 × 6 framing. This 2 × 6 frame was insulated with blown cellulose, and the existing 2 × 4 framed wall on the second floor was insulated with blown cellulose; a continuous layer of exterior foam was added on the second floor only. Mixed assemblies such as this were seen on P1, P3, P7, and P10. P7 used a unique "house within a house" approach that sought to fully insulate a portion of the first floor with respect to inside and outside using a fiberglass insulation blow-in blanket system (BIBS),[1] providing a central zone that could be kept more comfortable while the rest of the home was not directly conditioned and whose temperature was allowed to vary over a wide range (50°F–75°F). The two P6 homes are notable for their construction of an additional stud wall to the interior, which was spaced 1/2 in. from the existing framed wall. Both assemblies were filled with sprayed cellulose insulation. P9 was a newer home that already had fiberglass batts in all walls. Finally, some homes left portions of their above-grade walls uninsulated (P2, P7, and P8), which was usually due to existing historical/decorative elements that could not to be disturbed. The overall R-values for all these alternatives are summarized in Table 3.

Attic or roof insulation was upgraded in all homes. Six projects used the ceiling plane as the thermal barrier, and five projects moved the thermal barrier to the slope of the roof, producing an unvented attic. For those homes that created unvented attic assemblies, two approaches were used: (1) spray polyurethane foam (SPF) insulation on the undersides of rafters (P2, P8, and P10), and (2) a combination of blown insulation in rafters and continuous foam insulation on either the underside of rafters (P7) or on top of roof decking (P3). Both approaches create conditioned space in the attic, which can reduce distribution losses for attic HVAC systems, though two homes used this approach with no mechanical services in the attic whatsoever. SPF installed beneath the roof deck tended to produce the lowest R-values in the range of only 25 to 30, usually due to a combination of the project team's decision to install only one lift of SPF and the manufacturer's depth limits on single-lift applications (CPI 2012). High R-values were achieved in attics that added blown cellulose insulation to the attic floor, ranging between approximately 40 and 60. These were the lowest-cost attic retrofits by a large margin. P9 used this approach to bury HVAC ductwork in the attic to reduce distribution losses.

Foundations were a mixture of slab-on-grade, crawlspace, and basement types, with a number of homes combining types. Slabs were treated in Passive House (PH) style homes (P1 and P3) with both perimeter foam insulation and continuous foam on top of the slab. P3 consisted of two structures joined together during the renovation, with the slabs at slightly different heights. A continuously level floor and sufficient insulation were achieved using solely expanded polystyrene (EPS) foam on one slab and a mixture of foam and aeorgel on the other. P1 installed a new perimeter foundation and slab during the renovation. The other slab homes (P4 and P9) added no insulation. A number of homes insulated the framed floor

1. BIBS involves attaching a vapor- and air-permeable material to the rough framing members, with insulation then blown into the resulting cavity.

Table 3. Summary of Above-Grade Wall Insulation Methods

ID	Wall Insulation Method	R-Value Cavity Insulation*
P1	First floor: 5.5 in. dense pack (dp) cellulose	19
	Second floor: 3.5 in. dp cellulose, 2 in. polyiso exterior	23
P2	3.5 in. dp cellulose	13
P3	1: 3.5 in. dp fiberglass, 5 in. EPS exterior	38
	2: 5.5 in. dp fiberglass, 2.5 in. EPS exterior	33
P4	1: 5.5 in. dp cellulose	19
	2: 3.5 in. dp cellulose	13
P5	3.5 in. cellulose, 1 in. XPS exterior	18
P6-N	7 in. cellulose	25
P6-S	7 in. cellulose	25
P7	Rear zone: 5.5 in. BIBS, 1 in. polyiso	23
	Upstairs: 3.5 in. blown fiberglass	13
	Downstairs: None	—
P8	3.5 in. dp cellulose	13
P9	Fiberglass batts, improved installation and air sealed in kitchen and under stairs, insulated attic knee wall	13
P10	3.5 in. low-density SPF	13
	5.5 in. low-density SPF in garden room	19

* Values do not account for thermal bridging in framing members.

of the basement or crawlspace, using either SPF (P2 and P8), fiberglass batts (P8), foam board beneath floor joists (P7), or dense packed cellulose in floor joists using oriented strand board as an air barrier beneath the joists (P5). The two P6 homes created sealed crawlspaces, installing 2 in. of extruded polystyrene (XPS) foam board on the interior of new poured foundation stem walls and 6 in. of SPF in the rim joists. Neither home directly conditioned, dehumidified, or vented the crawlspaces, in contradiction to current best practices (Dastur et al. 2005). Several potential problems were observed with the sealed crawlspace homes. Cupping of hardwood floors was observed in one P6 house after sealing of the crawlspace, suggesting a potential moisture issue. During the removal of monitoring equipment at P5, a puddle of water was observed on top of the ground vapor barrier installed as part of the sealed crawlspace, and the seal of the vapor barrier to the foundation wall had failed in numerous locations. The puddle of water was most likely due to water intrusion from beneath, as the puddle was in the lowest part of the crawlspace and moisture was observed trapped between the two vapor barrier layers throughout the crawlspace. Failure to follow best practices, including improper use of sealants and mechanical attachment at the ground-to-wall interface, likely contributed to this moisture issue.

Windows. Most homes began with single-pane windows with either wood or steel frames. Windows were replaced in most homes; the other homes rehabilitated their windows through air sealing and use of storm windows. Most new windows were double-pane gas-filled units with U-factors ranging from 0.29 to 0.34 and SHGCs from 0.18 to 0.35. P3 used triple-pane windows imported from Europe with a U-factor of 0.125 and an SHGC of 0.53. In the P6 homes, one structure replaced all windows with high-performance[2] units, and the other structure's windows were rehabbed, adding new weather-stripping and an additional glass pane to the wood frame. Windows in P2 were historically significant, so a custom interior double-pane low-e storm window was installed that attached magnetically to the existing steel frames. Unfortunately, a number of these storm windows were observed to be uninstalled or broken. P7 used a mixed approach, with new windows in the "house within a house" zone and existing windows elsewhere. A number of homes that replaced windows left intact individual units that were particularly large and costly or otherwise unavailable (P4, P8, and P10). A complete discussion of existing window retrofit options is provided by Baker (2012).

Airtightness. Enclosure airtightness averaged 4.8 ACH_{50} but varied from relatively loose (10.8 ACH_{50}) to extremely airtight (0.48 ACH_{50}). Envelope airtightness for each project is summarized in Table 4. Consistent with the PH emphasis on superinsulation and airtightness, the projects inspired by the PH standard were substantially more airtight. Average airtightness in PH style projects was 1.3 ACH_{50}, whereas other DERs averaged 6.3 ACH_{50}. PH style projects approached air sealing systematically and methodically, with a specific goal in hand. All other project homes did not necessarily prioritize airtightness, and their results leave much room for cost-effective improvement. The only airtight non-PH style home was P9 at 2.4 ACH_{50}, which was constructed using modern materials in the 1990s. Despite the touted air-sealing capabilities of spray foam insulation, those homes that used SPF did not achieve good airtightness. They averaged 7.0 ACH_{50}, making them amongst the leakiest homes in the sample. Only two projects were assessed for air leakage prior to beginning the retrofit, P7 and P9, and they achieved reductions of 37% and 35%, respectively.

Average levels in this study can be compared to those summarized in Table 5 from the DER literature (ACI 2010b; BA-PIRC 2012; Berges and Metcalf 2013; Blanchard et al. 2012; Christian et al. 2011; McIlvaine 2010; Keesee 2012; Less 2012; McIlvaine et al. 2010; Neuhauser 2012; Osser et al. 2012; Chandra et al. 2012). Airtightness levels in this project were similar to those measured in DERs located in hot-dry, mixed-humid, and marine climates, whereas hot-humid DERs have been more leaky and cold-climate DERs have been much more airtight, on average. Cold-climate DERs have achieved

2. Compliant with current ENERGY STAR South-Central regional requirement (U-factor <0.35 and SHGC <0.3) (EPA 2011).

Table 4. Summary of Blower Door Airtightness Testing

Project ID	CFM_{50}	ACH_{50}	CFM_{50}/ft^2, Surface Area	CFM/ft^2 Floor Area	Effective Leakage Area, in^2	nACH	Specific Leakage Area (ELA/ft^2 Floor Area)
P1	271	1.1	0.063	0.166	10.3	0.05	0.00004
P2	2260	5.7	0.325	0.588	124.6	0.27	0.00031
P3	151	0.4	0.019	0.064	8.3	0.02	0.00002
P4	1983	5.4	0.322	0.790	110.0	0.26	0.00028
P5	292	2.4	0.097	0.323	14.0	0.10	0.00011
P6-N	991	5.1	0.222	0.678	49.4	0.18	0.00023
P6-S	1114	5.6	0.247	0.745	55.9	0.20	0.00026
P7	5336	10.8	0.790	1.623	300.6	0.72	0.00064
P8	2397	9.3	0.476	1.474	130.6	0.63	0.00056
P9	1227	2.4	0.183	0.394	69.8	0.14	0.00016
P10	1455	6.1	0.288	0.853	75.4	0.28	0.00031
Average	1588	4.9	0.276	0.700	86.3	0.26	0.00027

Table 5. Summary of Average Airtightness Levels (ACH_{50}) in DER Literature by DOE Building America Climate Zone

DOE Building America Climate Zone	Pre-Retrofit Mean (n)	Post-Retrofit Mean (n)	% Mean Reduction
Cold	20.7 (14)	3.3 (32)	80% (15)
Hot-Dry	15.3 (3)	5.2 (6)	65% (3)
Hot-Humid	24.7 (3)	7.0 (12)	59% (3)
Marine	6.6 (1)	5.1 (7)	35% (1)
Mixed-Humid	16.7 (2)	5.3 (3)	62% (2)
All	19.8 (23)	4.5 (60)	72% (24)

tighter homes because airtightness was a project priority due to the extreme climate. Many participated in the National Grid Deep Retrofit Pilot project, which specified approximately 1.5 ACH_{50} as a goal (Neuhauser 2012). Hot-humid homes likely achieved leakier post-retrofit results due to the definition by programs operating in that region of a DER as 30% or greater energy savings, which is the lowest published value in the U.S. Air leakage reductions averaged 72% from the pre-retrofit baseline. Chan and Sherman (2013) provide a summary of standard airtightness retrofits in the U.S. for Weatherization Assistance Program (WAP) retrofits and non-WAP retrofits, which achieved average percent reductions in ACH_{50} of 30% (n=13,093) and 20% (n=9,999), respectively (Chan and Sherman 2013). DERs have regularly doubled and tripled these average reductions. Ninety-fifth percentile reductions reported by Chan and Sherman were 47% and 61%, respectively, which makes these DERs amongst the most aggressive airtightening efforts in the U.S.

Some general similarities existed between observations from this project and the reports in the literature. As in our study, Eldenkamp and DuClos (2013) found that the major difficulty in getting to lower airtightness levels was details at interfaces. The construction and air-sealing details around eaves, rim joists, plumbing stacks, chases, and top interior partitions can be difficult to address depending on the construction details of each particular home. In other words, some homes are easier to seal than others. Also, it was evident that DERs targeting aggressive leakage reductions were able to achieve impressive results, consistent with high-performance new homes. Very airtight homes (<3 ACH_{50}) were nearly solely achieved in DERs that completely redid the exterior finish of the above-grade walls, including new continuous air barriers and exterior foam insulation.

New construction literature and standards are also relevant for comparison. The 2009 *International Energy Conservation Code* (*IECC*) target (7 ACH_{50}) was met by all but two homes in this study, and the 2012 *IECC* target (3 ACH_{50}) was met by only four homes in this study (see Table 4) (ICC 2009, 2012). On average, DER project homes were more leaky than California single-family homes built between 2001 and 2011 (mean 3.9) (Chan 2012). Similarly, Offermann (2009) reported a median airtightness of 4.8 ACH_{50} in 106 new California homes (ranging from 3.6 to approximately 8 ACH_{50}).

Thermal Distribution. The thermal distribution systems installed in these DERs included a range of solutions. Five homes used no thermal distribution, providing either point-source (P3, P6-N, and P6-S) or electric baseboard (P1 and P5) heat. Four homes used full forced-air systems, with three of the four installing new ductwork throughout (P2, P4, and P7). Effort was made to include ducts in conditioned spaces, but only P4 was able to fully achieve this. Both P2 and P7 created conditioned attics but insulated them at the framed floor so that the upstairs systems were inside conditioned space and the downstairs systems were outside. P9 buried ducting in several inches of attic insulation to achieve lower distribution losses. For P2, the post-retrofit duct air leakage to outside was indistinguishable from zero. Duct systems were tested before and after retrofit at P7, and it was revealed that total supply and return leakages to outside were reduced by 25% (from 115 to 86 cfm) and 46% (from 124 to 67 cfm), respectively. Total leakage of the duct system in P9 was reduced 61% (from 103 to 40 cfm). Testing of post-retrofit leakage to outside for the system revealed only 10 cfm of supply and 15 cfm of return leakage. Two homes used hydronic radiant heat, with in-floor heat used in P10 and wall radiators in P8. Thermal distribution in P8 was particularly troubling, as the solar storage tank with integrated boiler was located in an unconditioned, detached garage. This required an underground supply line to the unconditioned crawlspace, where a manifold system created lots of water pipe surface area; they used R-2 pipe wrap insulation.

Durability. DERs provide an opportunity to increase the durability of building envelopes through correction of moisture management issues, such as roof leaks, improper or missing flashing, door pans, etc. When the building control layers for air, heat, and moisture are exposed, significant improvements are possible. Rain screen systems were used in P1, P3, and P5, where exterior cladding was installed over a ventilated air space, which allows the building assembly to drain. Fully adhered weather-resistive barriers and meticulously detailed flashing can also be incorporated into project design. The P3 remodel took this strategy the furthest, using a REMOTE/PERSIST system, with a fully adhered membrane tying together the foundation, walls, and roof (Benesh 2009). The P4 remodelers installed new roof overhangs to protect the walls and the foundation from rain and for solar control. The North P6 home added awnings over each window opening, which, in addition to providing shade, provide protection from rain. Most projects installed gutters to remove bulk rainwater, but P8 built a rainwater harvesting system to retain water for other uses. Even in those homes that did not provide rain-screen systems, new cladding was common. Some projects did not incorporate these kinds of improvements, which may have been a lost opportunity for improved building durability.

Envelope and Energy Performance

Heating Energy Savings. Post-retrofit heating energy consumptions are presented in Table 6 (whole house) and Table 7 (per ft^2); these are compared with estimated pre-retrofit consumptions for those five homes where usage was known. The estimated pre-retrofit heating energy consumptions were obtained from billing data that was adjusted for post-retrofit weather conditions. Post-retrofit heating consumptions were determined from submetered data or by weather regression in P2 due to use of a combined heating and hot-water system. Data are not included for P8 and P10, because both homes used solar combined heating and hot-water systems, which were not disaggregated by our submetering. Homes using forced air include air-handler energy use. Absolute heating energy reductions averaged 12,937 kWh (76%), with reductions per square foot averaging 5.9 kWh/ft^2 (80%). Not surprisingly, those homes with relatively little pre-retrofit heating energy usage did not achieve dramatic savings (P1 and P4) despite substantial envelope improvements, nearly to the Passive House standard in the case of P1. These heating savings can in general be attributed to three sources: (1) envelope improvements, (2) HVAC equipment changes, and (3) changes in indoor temperature. In a number of cases, floor area was added to the homes (see Table 1), so heating energy reductions had to overcome these increases.

Table 6. Summary of Post-Retrofit Heating Energy Consumptions and Estimated Heating Energy Reductions, per House

Project ID	Pre-Retrofit Heating Consumption, kWh	Post-Retrofit Heating Consumption, kWh	Heating Energy Reduction, kWh	Heating Energy Reduction, %
P1	8,939	2,182	6,757	76%
P2	27,641	3,205	24,436	88%
P3	NA	576	NA	NA
P4	8,348	2,834	5,514	66%
P5	NA	415	NA	NA
P6-N	NA	489	NA	NA
P6-S	NA	1,781	NA	NA
P7	19,324	688	18,636	96%
P9	18,013	8,668	9,345	52%
Average	16,453	2,315	12,938	76%

Table 7. Summary of Post-Retrofit Heating Energy Consumptions and Estimated Heating Energy Reductions, per Square Foot Floor Area

Project ID	Pre-Retrofit Heating Consumption, kWh/ft^2	Post-Retrofit Heating Consumption, kWh/ft^2	Heating Energy Reduction, kWh/ft^2	Heating Energy Reduction, %
P1	9.3	1.3	8.0	86%
P2	9.9	1.2	8.8	88%
P3	NA	0.2	NA	NA
P4	5.4	1.1	4.3	79%
P5	NA	0.5	NA	NA
6-N	NA	0.3	NA	NA
P6-S	NA	1.2	NA	NA
P7	5.9	0.2	5.7	96%
P9	5.8	2.8	3.0	52%
Average	7.3	1.0	6.0	80%

When annual heating energy consumptions are compared between groups of highly or super insulated and code-style DERs (see Table 2), major differences are observed. The superior insulated homes averaged 1089 kWh (0.7 kWh/ft^2) in annual heating energy, whereas code-style homes averaged 3753 kWh (1.3 kWh/ft^2). The extreme examples in these groups can illustrate the true variability. P3, the first certified Passive House retrofit in the U.S., used only 576 kWh (0.24 kWh/ft^2), whereas code-style P9 used 7954 kWh (2.55 kWh/ft^2).[3] The difference between these two is greater than the total annual consumption of three of the project homes (P4, P5, and P6-N). Yet, as discussed in Less et al. (2012), low post-retrofit energy use did not always lead to large absolute energy reductions. In fact, P9 was the largest single reducer of net-source energy, despite having the highest post-retrofit heating energy use of any project home. This indicates that possibly more than one metric needs to be used when assessing DERs: percent savings, final energy use, energy use per square foot, code compliance as a target, etc. It also shows that other things significantly contribute to mild-climate energy performance and that a holistic approach to DER assessment is crucial rather than a simplistic focus on space-conditioning energy reductions.

Comparing Heating and Total Reductions. In order to assess the overall contributions of envelope and HVAC improvements to DER energy savings, the estimated heating energy savings are compared with total net site energy savings and envelope insulation designations in Table 8. The ratios of estimated heating energy savings to total net site savings varied from just below 50% to over 150%, with an average of 91%. Overall, heating energy reductions averaged 76%, whereas total net site reductions were only 58% for the same homes. Notably, four of the five homes in Table 8 had code-style envelopes, and they achieved an average net site energy reduction of 65%, which suggests that beyond-code envelopes are not strictly required in mild-climate DERs.

Those homes that achieved heating energy savings greater than total net site energy savings increased energy usage in other end-use areas, which offset their impressive heating reductions. For example, P1—the only highly insulated home with pre-retrofit data—reduced heating energy by 76%, but its total savings were only 31%. P1 nearly doubled its finished floor area (from 960 to 1630 ft^2) during the renovation by creating a full-height, livable ground floor, and its occupancy doubled from two to four. The fact that 76% heating energy reduction was achievable while nearly doubling the floor area is a testament to the value of the Passive House style approach used in this DER. Yet, increases in hot-water energy use and in other electrical end uses wiped out more than half of these savings.[4] Similarly, P2 reduced estimated heating energy by 88%, but its total net site savings were only 61%. No expansion of floor area occurred in P2, but it had the highest electrical baseload of any project home (562 W continuous power, estimated at 4926 kWh annually) and the highest combined usage of lights, appliances, and plugs (8875 kWh/yr). While baseload and miscellaneous electrical usage could not be assessed in the pre-retrofit home, P2 most likely substantially increased its miscellaneous power demand during renovation due to installation of mechanical cooling, large communication and A/V systems (see Figure 1), decorative lighting, etc. Nevertheless, P2's total net site reductions were the largest of any project home.[5]

Table 8. Summary Comparison of Total Net Site and Heating Energy Reductions and Insulation Designation

Project ID	Net Site Energy Reduction, kWh (%)	Estimated Heating Energy Reduction, kWh	Proportion of Heating-to-Total Reduction, %	Insulation Designation (from Table 2)
P1	4,138 (31%)	6,757	163%	Highly Insulated
P2	24,492 (61%)	24,436	100%	Code-style
P4	8,955 (74%)	5,514	62%	Code-style
P7	22,653 (72%)	18,636	82%	Code-style
P9	19,592 (54%)	9,345	48%	Code-style
Average	15,966 (58%)	12,938	91%	NA

These two projects highlight the importance of including all energy end uses in analyses of DERs as well as an awareness of how increases in other end uses can offset heating energy savings. This offsetting can be challenging, as many DERs include the addition of much needed or wanted services, such as additional appliances, enhanced lighting, and smart-home features. Yet these additions can offset costly heating and cooling energy savings. If no other changes are made to the home, a strictly heating and cooling DER is possible, though it is not necessarily the most cost-effective or valuable to the occupants, who often reap substantial benefits in aesthetics and convenience from non-space-conditioning DER measures.

In contrast to these two homes, P4, P7, and P9 (all DERs with code-style envelopes) most likely reduced most other household energy uses as part of their DER, resulting in heat-

3. Some of this is certainly attributable to use of a mini-split heat pump in P3 versus a gas furnace in P9.

4. Net-source energy performance in P1 was quite bad (consumption increased by 12%) due to changes from natural gas for heating to electric resistance baseboard heat. When combined with increased miscellaneous electrical uses (home office, two additional occupants, etc.), this shift to electric heat led to source energy increases, despite the aggressive Passive House approach.

5. The net-source energy performance in P2 was degraded similarly to P1 through a switch to an all-electric home, where previously gas was used for cooking, heating, hot water, and clothes drying. As a result, P2's net site savings of 61% were reduced to only a 7% reduction in net-source energy.

ing energy savings that were only fractions of total net reduction. P9 was particularly successful at targeting energy reductions equally towards heating and other uses such as cooling, lighting, and miscellaneous electrical loads. In homes that have already achieved insulation levels near code, non-heating end-uses provide what may be the most cost-effective energy reductions in the home. Lighting retrofits and appliance upgrades are almost certainly less costly than super insulation of walls or extreme airtightness.

A holistic assessment of whole-house energy use is essential in determining the impacts of DER projects. The examples above clearly illustrate the distortions that can be introduced by offsetting space-conditioning savings with increases in other end uses. Enhanced whole-house savings were demonstrated in homes that did not increase other uses, showing a very effective mixed approach to deep savings in residences.

Heating System Response to Weather Conditions. We explored the thermal responses of the DERs by using submetered heating system energy data and site-specific weather data to create linear regression models. Daily average energy use and weather data were used to assess the variability of heating system energy use with indoor-outdoor temperature difference and with outdoor temperature alone. Homes P1, P2, P4, and P9 showed relatively consistent correlation between heating consumption and indoor-outdoor temperature difference, all of which had coefficients of determination (R^2) greater than 0.4. Figure 2 shows the relationship between daily average heating power consumption and daily average indoor-outdoor temperature difference for P9. A number of other homes had highly erratic relationships between weather conditions and heating system energy use; these homes were characterized by R^2 values less than 0.3 (P3, P5, P6-N, P6-S, P7, and P10). These less consistent results are illustrated in Figure 3 for project P6-N. Relationships were also poor in these erratic homes on a monthly basis, with an average R^2 of 0.52 versus 0.76 in the other projects. These results show that factors other than temperature difference can dominate heating consumption in low-load homes in mild climates. These

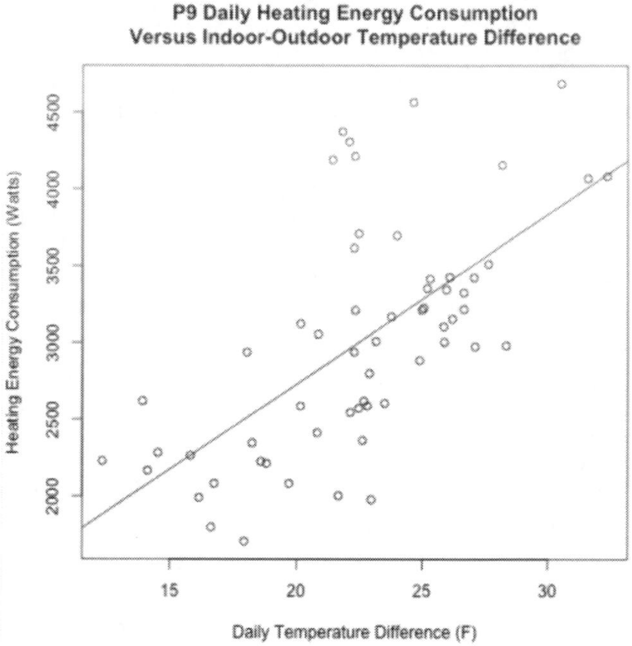

Figure 2 *P9 comparison of hourly and daily heating energy methods.*

Figure 1 *P2 audio/visual and data equipment in basement cabinet.*

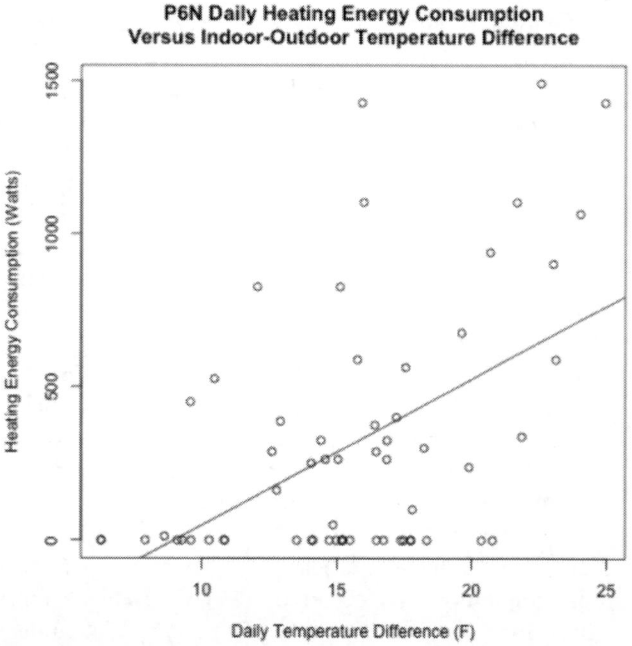

Figure 3 *P6 North, illustration of erratic daily plot.*

other factors include occupant temperature control, solar gain, and internal gains. This makes the prediction of heating energy use in such low-energy homes challenging, because weather data cannot be used exclusively to successfully predict usage.

In addition, many of these DER project homes are not strictly controlled by a thermostat; rather, the heat is controlled by the occupants on an as-needed basis. This is illustrated in Figure 3 on those days where substantial temperature differences existed but average daily heating power varied from 0 to 1500 W. So, on equally cold days, sometimes the system was operated regularly and on other such days not at all. Occupant behavior, comfort, and preference drive this phenomenon. It may be that this behavior is the result of particularly engaged building occupants who elected to invest in deeply retrofitting their home for environmental reasons. Yet, even without occupant manipulation, the variation in internal loads and solar gains can have similar impacts, particularly in homes where the heating load is comparable to these gains. For example, it is recommended in Passive Houses that the winter heating load not exceed 3.17 Btu/hr·ft^2·°F (Schnieders 2003), which can be met in a 1000 ft^2 home by less than 1 kW of miscellaneous electrical equipment or solar gain.

The balance point temperature[6] of a home is an indicator of building envelope performance that is independent of the heating system but is still dependent on occupant temperature control as well as solar and internal gains. In an attempt to determine balance point temperature, heating system consumption was also compared directly to outdoor temperature. A mix of well-behaved and erratic results was obtained. Yet, even for the best fitted model (P4), a single standard deviation confidence interval for the balance point of 57°F was 46 to 70 degrees. All other models produced even poorer results. This suggests it may not be possible to generate valid balance points for low-energy homes from weather regressions.

Despite the limitations of weather regression in DERs, the heating energy regressions still demonstrated an obvious difference between homes that were highly or super insulated and code-type homes. Highly or super insulated homes had an average heating energy use per degree of daily average outdoor temperature of –24 W/°F (–0.016 W/°F·ft^2), compared with –85 W/°F (–0.035 W/°F·ft^2) in code-style homes. These suggest that enhanced building enclosures in mild-climate DERs still provided reduced heating loads—approximately 50% lower than code-style DERs by square foot of floor area.

Thermal Comfort

Among the many non-energy benefits of DERs, improvements in thermal comfort may be a primary driver for homeowners investing in improvements. The hourly temperature profiles for the months of December and January are compared for each DER in Figure 4. Thermal comfort is highly variable in these DER homes, with indoor average hourly temperatures varying by up to 10°F between homes. The heating season thermostat setpoint used in all Building America simulation assessments is 71°F and is indicated for reference as a bold dotted line in Figure 4 (Engebrecht and Hendron 2010). All DER project homes, even those providing the warmest and most consistent indoor conditions (P1 and P3), did not quite meet the Building America setpoint requirement, which is based upon the "optimum seasonal temperature for human comfort as defined in ASHRAE Standard 55-2004" (Engebrecht and Hendron 2010, p. 49).[7] In fact, only P1, P2, and P3 provided any average hour within the Standard 55-2010 acceptable range of thermal comfort (ASHRAE 2010). The lower temperatures maintained in these DERs may be the result of occupant thermal preferences, efforts to reduce heating energy consumption, or both. It is also possible that lower setpoints were maintained due to reductions in drafts due to airtightening and a better radiant environment due to improved windows and wall insulation.

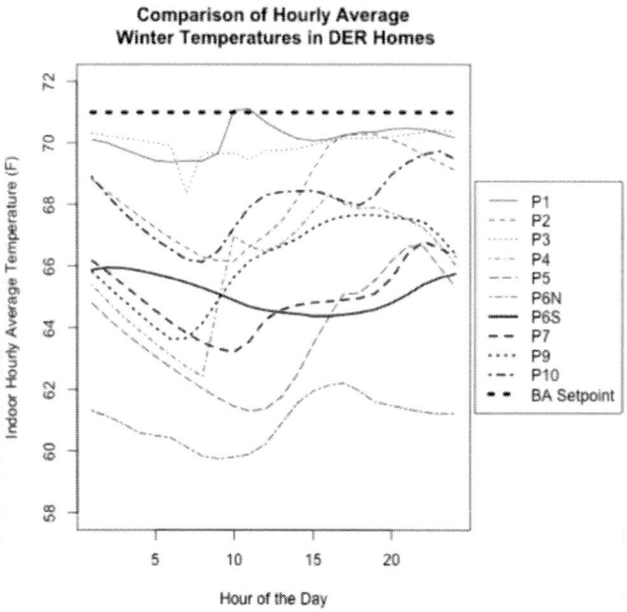

Figure 4 *Indoor hourly temperature profiles in DER homes for December and January.*

6. By balance point temperature we are referring to the outdoor temperature at which heating input is required to maintain indoor conditions. This is not the same as the balance point temperature of a heat pump, with capacity that varies with outdoor temperature.

7. It is not clear that ASHRAE Standard 55 stipulates an "optimum" temperature; rather, a range of acceptable temperatures are provided that will result in an 80% acceptance level by building occupants.

Thermal comfort surveys were not distributed to occupants, but comfort complaints were not reported, with the exception of P10 and P7. Occupants in P10 complained that the heating system was not able to properly recover to a comfortable temperature after a nighttime set-back in their living room. In an energy-efficient home, the energy savings that can be obtained via setback are reduced and may not be worth the comfort trade-off when equipment is sized to meet the very small steady-state load. Research in super insulated cold-climate homes has shown that daytime setbacks may not be beneficial and may increase energy use (CARB 2010). Occupants in P7, while not directly admitting discomfort, did mention the need to heat to higher temperatures during visits by guests.

The impacts of building envelope and interior temperature can be compared by examining interior temperatures in DERs that used very little heating energy—P3, P5, P6-N, and P7. P3 used only 576 kWh for annual heating, and the other three homes averaged 531 kWh. P3 maintained average hourly winter temperatures at 70°F with very little variability from hour to hour. It was by far the most airtight and best insulated home in this research. Indoor conditions were much less consistent and cooler in the other three homes. Average hourly temperatures for P5, P6-N, and P7 were 64°F, 61°F, and 65°F, respectively.

Comparing these four homes allows us to understand some of the variables involved in DER thermal comfort. Similar to P3, P5 was built in a Passive House style, but it used electric resistance heaters. As a result, P5 required much more energy to produce equivalent indoor temperatures to those in P3, which used a high-efficiency ductless heat pump. Had P5 used such a heat pump, its interior conditions could have been greatly improved without sacrificing energy performance. P6-N was a super insulated home, but it lacked substantial airtightness (5.1 ACH_{50}) and its occupants were willing to accept very cool winter temperatures. With increased airtightness, much warmer conditions could likely have been achieved with little heating energy increase. P6-S did achieve better comfort (average of 65°F), but its heating energy was a little over 3.5 times the North house (1781 kWh). Airtightness was clearly overlooked in these super insulated homes to the detriment of indoor comfort. Finally, a home like P7, with an envelope only partially brought to code, achieved very low heating energy by using its "house within a house" approach, providing comfort in only one of three house zones. The P7 temperatures in Figure 4 are for the insulated kitchen zone, whereas the other house zones were substantially colder on average (55°F and 60°F in the non-kitchen first floor and second floor zones, respectively). This temperature manipulation was occurring alongside envelope improvements and high-efficiency HVAC, leading to heating energy reductions of nearly 19,000 kWh.

Similar heating energy usages were attained by these four homes, but significantly different indoor comfort conditions were necessary to do this. Both P3 and P6-N were super insulated and P5 was highly insulated, yet indoor comfort conditions were widely variable. Enhanced insulation did not necessarily lead to higher interior temperatures, though it was clearly related to very low heating energy use. DERs that are targeting consistent, high heating season temperatures and very low heating energy use will need to employ very high-performance envelopes (insulation and airtightness) and HVAC equipment. Code-style homes can still be comfortable and have highly successful whole-house energy reductions (P2, P4, and P9), but their heating energy use will remain high, which will need to be accounted for by reductions elsewhere.

DER Building Envelope Recommendations

The type of thermal envelope that is targeted in a mild-climate DER project will depend upon the project goals, the comfort demands of the occupants, and the existing condition and energy use of the home. This research demonstrates that enhanced insulation and high levels of airtightness are not necessarily required to achieve deep energy reductions. In the authors' opinion, most existing uninsulated homes in mild climates can achieve deep energy reductions by getting to new building code levels of envelope performance, provided that all energy end uses are comprehensively addressed. Enhanced building envelopes can contribute to increased thermal comfort and result in very low heating energy usage (<1000 kWh/yr or <0.5 kWh/ft^2·yr). Furthermore, an enhanced envelope allows leeway for the potential increases in select other end uses that can occur in DER projects—decorative lighting, cooling, and miscellaneous electric loads, for example—as well as increases in floor area and/or comfort. The occupant should always be considered directly in project planning because of the substantial impact that behavior and comfort requirements have been shown to have on heating and whole-house energy use, as well as HVAC system operation and indoor environmental conditions.

It is our recommendation for mild-climate DERs to pursue building envelopes compliant with current energy codes for the region, namely the 2012 *IECC*. The 2012 *IECC* airtightness requirement of 3 ACH_{50} should be targeted by all DERs. This target may be excessively difficult to meet in homes where neither the interior nor exterior claddings are removed. In such cases, the ENERGY STAR Version 3 target of 5 ACH_{50} for Climate Zone 3 can be used (EPA 2011). Insulation in excess of energy code requirements should be considered when other retrofit activities (such as re-siding) make it easier and cheaper to do so. For example, if exterior siding must be replaced during a DER, then continuous exterior insulation should be considered along with the required new sheathing and weather-resistive barrier.

High-efficiency, off-the-shelf HVAC equipment should be combined with these code-compliant envelopes to achieve the best results. Electric resistance heaters as primary heat should be avoided, even in super insulated projects, particularly for source energy and carbon emissions reasons, as discussed in footnotes 4 and 5.

CONCLUSION

Assessing the impacts of building envelope improvements in DERs is complicated by the intermingled effects of enclosure improvements, HVAC system upgrades, and changes in interior comfort conditions. Yet, beyond-code building envelope assemblies have been shown to be unnecessary to achieve deep energy reductions in existing uninsulated homes in a mild marine climate, provided that other energy end uses are comprehensively addressed. In this research, DERs with code-style envelopes averaged a 65% net site energy reduction. Nevertheless, the benefits of an enhanced beyond-code envelope include (1) the potential to increase indoor comfort conditions, (2) the ability to maintain enhanced comfort while minimizing heating energy use (<1000 kWh/yr or <0.5 kWh/ft^2·yr), and (3) to compensate for energy usage increases in other end-use categories, which has been observed to occur in many owner-conducted DERs. The building occupants have been shown to have substantial impact on indoor comfort conditions, heating system operation, and miscellaneous energy use, all of which can dramatically affect DER performance. While beyond-code envelope improvements can play a key role in deep energy reductions, we recommend that mild-climate DERs target envelope compliance with the 2012 *IECC* code requirements.

Ultimately, the decision to pursue a DER building envelope beyond prescriptive energy code requirements is at the discretion of the project designer and the occupant and by no means is it required in order for deep reductions to be realized. In our opinion, it is more important that deep retrofit strategies be tailored toward the needs and wants of the occupant and to current patterns of energy use. If high levels of thermal comfort are required and if substantial additional energy usages are to be added as part of the project, then an enhanced envelope may be prudent. On the other hand, a much less disruptive and lower-cost retrofit may be possible if comfort requirements are somewhat lower or if comprehensive energy reductions are achieved across all end uses.

REFERENCES

ACI. 2010a. What is the Thousand Home Challenge (THC). Thousand Home Challenge. Moon Township, PA: Affordable Comfort, Inc. Retrieved September 17, 2013, from http://thousandhomechallenge.com/sites/thousandhomechallenge.com/files/user-files/thc_description_2012.pdf

ACI. 2010b. Case Studies. Thousand Home Challenge. Moon Township, PA: Affordable Comfort, Inc. Retrieved April 2, 2013, from http://thousandhomechallenge.com/case-studies

ASHRAE. 2010. ANSI/ASHRAE Standard 55-2010, *Thermal Environmental Conditions for Human Occupancy.* Atlanta: ASHRAE.

BA-PIRC. 2012. Building America Partnership for Improved Residential Construction—Case Studies. Cocoa, FL: Florida Solar Energy Center of the University of Central Florida. Retrieved April 2, 2013, from www.ba-pirc.org/casestud/

Baker, P. 2012. *Measure Guideline: Wood Window Repair, Rehabilitation, and Replacement.* Washington, DC: U.S. Department of Energy. http://apps1.eere.energy.gov/buildings/publications/pdfs/building_america/measure_guide_wood_windows.pdf

Benesh, I. 2009. *REMOTE: A Manual.* Cold Climate Housing Research Center. Fairbanks, AK. http://cchrc.org/docs/best_practices/REMOTE_Manual.pdf

Berges, M., and M. Metcalf. 2013. Lessons learned on energy-efficient affordable housing: Practical insights on combining deep energy retrofits with affordable housing for 12 Cleveland homes. *Journal of Light Construction* (February), pp. 55–63.

Blanchard, J., E. Giever, S. Widder, and M. Baechler. 2012. Actual and Estimated Energy Savings Comparison for Deep Energy Retrofits in the Pacific Northwest. PNNL-21870, Pacific Northwest National Laboratory, Richland, WA.

CARB. 2010. *Point-Source Heating Systems in Cold-Climate Homes: Wisdom Way Solar Village.* Consortium for Advanced Residential Buildings.

CEC. 2008. *2008 Building Energy Efficiency Standards for Residential and Nonresidential Buildings-Regulations/Standards.* No. CEC-400-2008-001-CMF, California Energy Commission, Sacramento. www.energy.ca.gov/2008publications/CEC-400-2008-001/CEC-400-2008-001-CMF.PDF

Chan, W. 2012. Re: Air leakage database numbers. Email correspondence with the author, May 24.

Chan, W.R., and M.H. Sherman. 2013. Improving building envelope and duct airtightness of U.S. dwellings—The current state of energy retrofits. Presented at the Tight-Vent Workshop on Building and Ductwork Airtightness, Washington, DC.

Chandra, S., S. Widder, G. Parker, S. Sande, J. Blanchard, D. Stroer, et al. 2012. Pilot Residential Deep Energy Retrofits and the PNNL Lab Homes. PNNL-21116, Pacific Northwest National Laboratory, Richland, WA.

Christian, J., T. Gehl, P. Boudreaux, and J. Munk. 2011. Electric Utility Research Pathways to Deep Retrofit. Presented at the Denver Building America Meeting, Denver, CO, August 9. http://apps1.eere.energy.gov/buildings/publications/pdfs/building_america/ns/eemtg082011_b11_utility_deep_retrofit.pdf

CPI. 2012. *Guidance on Best Practices for the Installation of Spray Polyurethane Foam.* Spray Foam Coalition of the Center for the Polyurethanes Industry. http://polyurethane.americanchemistry.com/Spray-Foam-Coalition/Guidance-on-Best-Practices-for-the-Installation-of-Spray-Polyurethane-Foam.pdf

Dastur, C., B. Davis, and B. Warren. 2005. *Closed Crawl Spaces—An Introduction to Design, Construction and Performance.* Raleigh, NC: Advanced Energy.

www.advancedenergy.org/buildings/knowledge_library/crawl_spaces/pdfs/Closed%20Crawl%20Spaces_An%20Introduction%20for%20the%20Southeast.pdfClosed%20Crawl%20Spaces_An%20Introduction%20for%20the%20Southeast.pdf

Eldenkamp, P., and M. DuClos. 2013. Deep energy retrofits. *Home Energy Magazine*. January/February. www.homeenergy.org/show/article/id/1835/nav/retrofit

Engebrecht, C., and B. Hendron. 2010. *Building America Simulation Protocols*. No. DOE/GO-102010-3141, U.S. Department of Energy, Washington, DC. www.nrel.gov/docs/fy11osti/49246.pdf

EPA. 2011. *ENERGY STAR Qualified Homes, Version 3 (Rev. 3) Inspection Checklists For National Program Requirements*. Washington, DC: U.S. Environmental Protection Agency. www.energystar.gov/ia/partners/bldrs_lenders_raters/downloads/InspectionChecklists.pdf?104f-c645

Henderson, S., and C. Mattock. 2007. *Approaching Net-Zero Energy in Existing Housing*. Ottawa, Ontario: Canada Mortgage and Housing Corporation.

ICC. 2009. *International Energy Conservation Code*. Washington, DC: International Code Council.

ICC. 2012. *International Energy Conservation Code*. Washington, DC: International Code Council.

Keesee, M. 2012. Deep energy retrofits: Six real world examples and lessons learned. *2012 ACEEE Summer Study for Energy Efficiency in Buildings-Fueling Our Future with Efficiency* 1:141–52). www.aceee.org/files/proceedings/2012/data/papers/0193-000006.pdf

Less, B. 2012. *Indoor Air Quality in 24 California Residences Designed as High Performance Green Homes*. Master's thesis, Department of Architecture, University of California, Berkeley, Berkeley, CA. http://escholarship.org/uc/item/25x5j8w6

Less, B., J. Fisher, and I. Walker. 2012. Deep Energy Retrofits—11 California Case Studies. No. LBNL-6166E, Lawrence Berkeley National Laboratory, Berkeley, CA. http://eetd.lbl.gov/publications/deep-energy-retrofits-eleven-california-case-studies

McIlvaine, J. 2010. Deep Energy Retrofits: Hot Humid Climate Case Study. Presented at the Residential Buildings Energy Efficiency Meeting, Denver, CO, July.

McIlvaine, J., K. Sutherland, K. Schleith, and S. Chandra. 2010. Exploring Cost-Effective High Performance Residential Retrofits for Affordable Housing in Hot Humid Climate. No. FSEC-PF-448-10, Florida Solar Energy Center, Cocoa, FL.

Neuhauser, K. 2012. National Grid Deep Energy Retrofit Pilot. No. KNDJ-0-40337-00, National Renewable Energy Laboratory, Golden, CO. www.nrel.gov/docs/fy12osti/53684.pdf

Offermann, F. 2009. *Ventilation and Indoor Air Quality in New Homes*. No. CEC-500-2009-085, California Energy Commission, Sacramento. www.energy.ca.gov/2009publications/CEC-500-2009-085/CEC-500-2009-085.PDF

Osser, R., K. Neuhauser, and K. Ueno. 2012. Proven Performance of Seven Cold Climate Deep Retrofit Homes. No. KNDJ-0-40337-00, Building Science Corporation, Somerville, MA. Retrieved from www.nrel.gov/docs/fy12osti/54205.pdf

RESNET. 2006. 2006 Mortgage Industry National Home Energy Rating Systems Standards. Oceanside, CA: Residential Energy Services Network. www.resnet.us/standards/RESNET_Mortgage_Industry_National_HERS_Standards.pdf

Schnieders, J. 2003. CEPHEUS—Measurement results from more than 100 dwelling units in Passive Houses. *ECEEE 2003 Summer Study—Time To Turn Down Energy Demand*, p. 341. www.eceee.org/conference_proceedings/eceee/2003c/Panel_2/2047schnieders/paper

Wigington, L. 2010. Staged Approaches for Deep Energy Reductions in Existing Homes. *Proceedings of ACEEE Summer Study on Energy Efficiency in Buildings*. Washington, DC: American Council for an Energy-Efficient Economy. http://aceee.org/files/proceedings/2010/data/papers/2299.pdf

Modeling of Vacuum Insulating Glazing

Robert Hart, PE
Student Member ASHRAE

D. Charlie Curcija, PhD
Member ASHRAE

ABSTRACT

Windows with thermal resistance of $R = 10$ h·ft²·°F/Btu ($U = 0.10$ Btu/h·ft²·°F) or greater across the U.S. building stock could significantly reduce energy consumption (Arasteh 2006). Vacuum insulating glazing (VIG) is a promising technology that is capable of meeting the rigorous thermal performance requirements, but there is currently no simple method of evaluating its performance for research and development or certification.

In this work, the thermal transmittance (U-factor) of VIG is calculated with three methods: (1) analytical analysis using published correlations for modeling vacuum and pillar array conduction heat transfer; radiation heat transfer is solved explicitly using parallel plate theory; (2) two-dimensional finite element method computational heat transfer (2D FEM) where the pillar and vacuum space are replaced with an effective solid since 2D FEM cannot handle point thermal bridges such as pillars; and 3) three-dimensional finite volume numerical heat transfer modeling where pillar array conduction heat transfer and radiation heat transfer between surfaces in vacuum space are modeled to account for point thermal bridges such as pillars. The center-of-glass and edge-of-glass models from each method are compared to demonstrate a strong correlation between the methods and confirm applicability of representing vacuum space and pillar arrays using 2D FEM, the standard method of modeling and certifying windows and glazing.

INTRODUCTION

It is estimated that the implementation of "zero energy" windows with thermal resistance of $R = 10$ h·ft^2·°F/Btu ($U = 0.10$ Btu/h·ft^2·°F) performance across the U.S. building stock could reduce energy consumption $300 billion in the 20 years following these performance improvements (Arasteh 2006). Vacuum insulating glazing (VIG) for residential and commercial use is a promising technology that is capable of meeting the rigorous R-10 thermal performance requirements. While some VIG technology has been available in the Japanese and other markets as early as 1996, it has achieved minimal market penetration in the U.S. Some drivers for its low adoption include high costs, design features that restrict use in very hot or cold environments, and insulating performance ($U = 1.2$ W/m^2·K center-of-glass [COG] by NSG) no greater than traditional triple glazing. New VIG products with claimed insulating performance down to $U = 0.4$ W/m^2·K center-of-glass (far better than standard triple glazing), as well as the claimed ability to be installed in all U.S. climate regions, are being developed (Russo 2012; ENoB 2013; Jelle 2012; Eversealed 2013).

The COG thermal performance of VIG is well understood and the process for evaluation by analytical method is outlined in this work. The diversity of designs and complexities of edge-of-glass (EOG) heat flow though does not provoke an analytical model; therefore the three-dimensional finite volume computational method (3D FVM) is typically used for evaluation. 3D FVM models require trained and experienced users of specialized software and significant setup and evaluation effort. Such effort precludes the widespread use of these models in the fenestration industry and therefore the certification of windows using this technology. A simple

***Robert Hart** is a scientific engineering associate and **D. Charlie Curcija** is a deputy group leader in the Windows & Envelope Materials Group, Lawrence Berkeley National Laboratory, Berkeley, CA.*

Published by ASHRAE 2013

and accurate method of evaluating whole window VIG performance using the U.S. fenestration industry standard two-dimensional finite element computational method (2D FEM) is proposed and evaluated in this work.

ANALYSIS

The analysis done in this study involves modeling procedures only. The thermal transmittance (U-factor) of VIG is calculated with three methods: (1) analytical analysis using published correlations for modeling vacuum and pillar array conduction heat transfer; radiation heat transfer is solved explicitly using parallel plate theory; (2) 2D FEM where the pillar and vacuum space are replaced with an effective solid since 2D FEM cannot handle point thermal bridges such as pillars; and (3) 3D FVM where pillar array conduction heat transfer and radiation heat transfer between surfaces in vacuum space are modeled to account for point thermal bridges such as pillars.

Analytical Method

The COG U-factor of VIG with support pillars is calculated analytically as a function of the surface resistances, glass resistance, and the vacuum gap resistance. The fundamental equations are presented here for completeness; the works of Collins (1991) and Corruccini (1959) should be reviewed for more detailed explanation of the methods.

$$U\text{-}factor = \frac{1}{R_o + 2 \cdot R_{glass} + R_{gap} + R_i} \quad (1)$$

where:

- R_o = exterior surface resistance, $m^2 \cdot K/W$
- R_i = interior surface resistance, $m^2 \cdot K/W$
- R_{glass} = glass pane resistance, $m^2 \cdot K/W$
- R_{gap} = vacuum gap resistance, $m^2 \cdot K/W$

The interior and exterior surface resistances are boundary conditions defined as a combination of convection and radiation resistance and listed in Table 3. The glass pane resistance is a function of total glass thickness and conductivity.

$$R_{glass} = \frac{t_{gl}}{k_{gl}} \quad (2)$$

where:

- t_{gl} = glass thickness, m
- k_{gl} = glass conductivity, $W/m \cdot K$

The vacuum gap resistance is a function of the low-pressure gap conductance, the radiation conductance between glass panes, and the conductance of the support pillars between glass. The conductance of low-pressure gas is calculated using formula by Corruccini (1959). Equation 6 gives exact conductance of a pillar array, but pillar conductivity does not play a substantial role in overall heat transfer when the conductivity is equal to or higher than the conductivity of the glass pane. Therefore, a good approximation for common materials used in this technology can be made by ignoring the ratio of conductivities in the denominator of Equation 6.

$$R_{gap} = \frac{1}{C_{cond} + C_{rad} + C_{pa}} \quad (3)$$

$$C_{cond} = \left[\frac{\alpha_1 \cdot \alpha_2}{\alpha_2 + \alpha_1(1-\alpha_2)}\right] \cdot \left[\frac{\gamma+1}{\gamma-1}\right] \cdot \left[\frac{R}{4 \cdot \pi \cdot M \cdot (T_{1c} + T_{2c})}\right]^{\frac{1}{2}} \cdot P \quad (4)$$

$$C_{rad} = \frac{\sigma}{\varepsilon_1^{-1} + \varepsilon_2^{-1} - 1} \cdot \left[\frac{T_1^4 - T_2^4}{T_1 - T_2}\right] \quad (5)$$

$$C_{pa} = \frac{2 \cdot \kappa_{gl} \cdot a}{l^2 \cdot \left(1 + \frac{2 \cdot \kappa_{gl} \cdot h}{\kappa_p \cdot \pi \cdot a}\right)} \quad (6)$$

where:

- C_{cond} = conductance of low pressure gas between glass, $W/m^2 \cdot K$
- C_{rad} = radiation conductance between glass panes, $W/m^2 \cdot K$
- C_{pa} = conductance of pillar array between glass panes, $W/m^2 \cdot K$
- α_1, α_2 = accommodation coefficients of the gas molecules. Dependent on the temperature, surface conditions, etc. For the present configuration and conditions, $\alpha_1, \alpha_2 \cong 0.5$. If conservative, 1.0 could be used.
- γ = specific heat ratio, $\gamma_{air} = 1.40$
- R = universal gas constant, $R = 8{,}314.462175$ J/mol·K
- M = molecular weight, $M_{air} = 28.97$ mol/g
- T_{1c} = exterior temperature, K
- T_{2c} = interior temperature, K
- P = gas pressure, N/m^2
- σ = Stefan-Boltzmann constant 5.67×10^{-8} $W/m^2 \cdot K^4$
- ε_1 = emissivity of the first facing glass surface, –
- ε_2 = emissivity of the second facing glass surface, –
- T_1 = temperature of the first facing glass surface, K
- T_2 = temperature of the second facing glass surface, K
- a = pillar radius, m
- h = pillar height, m
- k_p = pillar conductivity, W/m·K
- l = pillar spacing, m

Two-Dimensional Finite Element Method

2D FEM is used to solve the conductive heat-transfer equation for two-dimensional analysis. The quadrilateral mesh is automatically generated. Refinement is performed in accordance with section 6.3.2b of ISO 15099 (ISO 2003). The energy error norm is less than six percent in all cases, where it has been shown that energy error norm of 10% or less correlates to an error of less than one percent in the total thermal transmittance of the object(s) being modeled. More information on the thermal simulation program algorithms can be found in Finlayson (1998) and Carli (2006).

Center-of-glass. Modeled COG geometry for 2D FEM makes use of symmetry as shown in Figure 1. 2D FEM cannot handle point thermal bridges such as pillars, which are properly modeled only in three-dimensional space, therefore, the pillars and vacuum space are replaced with an effective solid whose properties are calculated for each configuration by Equation 3.

Edge-of-glass. Modeled EOG geometry for 2D FEM analysis makes use of symmetry as shown in Figure 2. The vacuum space from the edge-of-glass to the first pillar is replaced with an effective solid whose properties are calculated for each configuration by Equation 3 with the conductance of the pillars set to zero. Starting at first pillar, the vacuum space is modeled similar to the COG, with an effective solid whose properties are calculated for each configuration by Equation 3. The influence of edge conduction in the geometries modeled becomes negligible after the third pillar, as is shown later.

Due to the relative high conductivity of the edge compared to COG in current VIG design, an exposed edge region would exhibit very high heat transfer rates that are difficult to measure in real systems due to condensation and frost build-up. For these reasons the edge is modeled as covered by a 12.5 mm square wood block on the interior and exterior sides of the VIG. This technique effectively reduces heat transfer through the edge region to a magnitude in line with a VIG installed into a window frame. A similar approach is taken by Simko (1996).

Three-Dimensional Finite Volume Method

3D FVM is used to solve the coupled heat and fluid-flow equations for three-dimensional analysis. Conduction, convection, and radiation are simulated numerically. Three-dimensional analysis is necessary to account for the non-continuous pillar shapes between glass. Because the Navier-Stokes fluid equations typically used by commercial solvers are invalid at the low pressures seen in VIGs, two alternative methods are used to approximate the gap conductance: (1) The "solid" model, where the vacuum gap is defined as a solid with a combined fluid and radiation conductivity solved by the analytical methods of Equation 3; (2) The "radiation" model, where the vacuum gap is defined as a fictitious fluid and radiation is solved for explicitly. In both methods, pillars are modeled with their actual geometry.

The semi-implicit method for pressure-linked equations consistent (SIMPLEC) was used to model the interaction between pressure and velocity. The energy and momentum variables at cell faces were found by using the

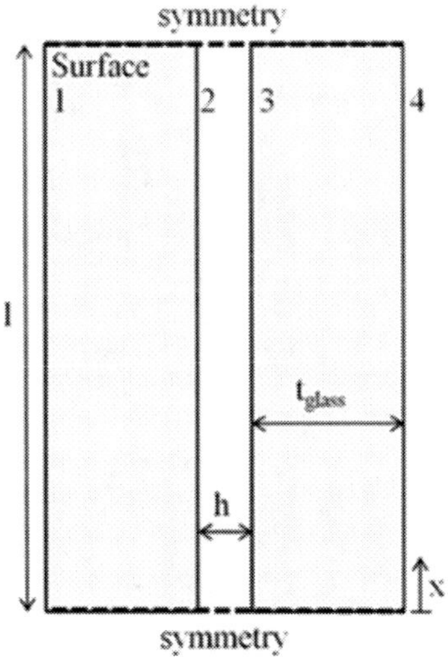

Figure 1 COG model geometry for 2D FEM analysis (not to scale).

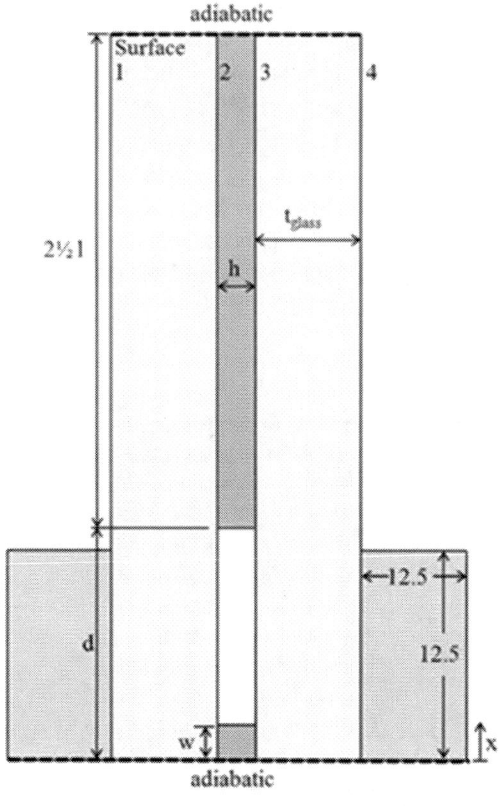

Figure 2 EOG model geometry for 2D FEM analysis (not to scale).

Quadratic Upstream Interpolation for Convective Kinetics (QUICK) scheme. The commercial software used solves central differences to approximate diffusion terms and relies on the PREssure STaggering Option scheme (PRESTO) to find the pressure values at the cell faces. Convergence is determined by checking the residuals for energy and ensuring that they are lower than 1E-13. Radiation heat transfer is included in the simulations through use of Surface to Surface (S2S) radiation model, which calculates the energy exchange between surfaces taking into account their size, separation distance, and orientation using a geometric function called *view factor*. The S2S model ignores absorption, emission, and scattering phenomena in the vacuum cavity. The internal cavity walls were assumed to be diffuse gray. The meshing program automatically generated meshes. Mesh refinement was done at pillars and edge seals to ensure acceptable mesh density and orthogonal quality.

Center-of-glass. Modeled COG geometry used for 3D FVM analysis makes use of symmetry as shown in Figure 3. This geometry is referred to as a *¼ pillar model* as it utilizes symmetry around ¼ of a pillar as shown in section (B). Additional model geometries were tested, including a full pillar and a 3 × 3 grid of pillars. The variation between the three models in all cases was less than 0.50 percent.

Edge-of-glass. Modeled EOG geometry used for 3D FVM analysis makes use of symmetry as shown in Figure 4. This model uses the same ¼ pillar symmetry as the COG model but extends three and one-half pillars from the edge. Additional models geometries were generated, including single pillar and 5 pillar models. The influence of edge conduction in the geometries modeled becomes negligible after the third pillar. As in 2D FEM, an exposed edge region would exhibit very high heat transfer rates, so the edge model is covered by a 12.5 mm square wood block on the interior and exterior sides of the EGU.

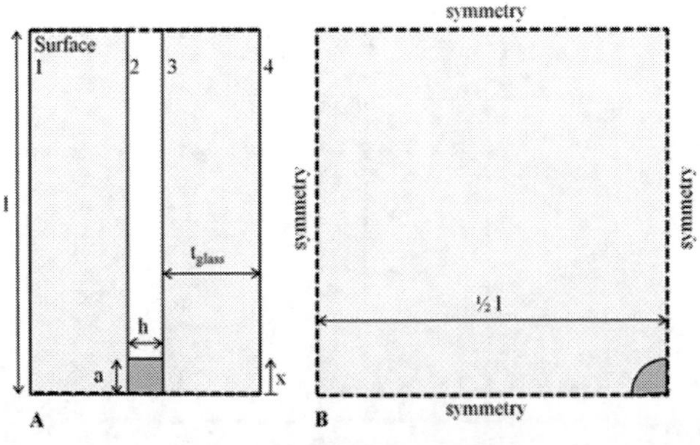

Figure 3 COG model geometry for 3D FVM analysis (not to scale).

Model Parameters

Table 1 lists the model dimension ranges used for this study. The minimum (min) and maximum (max) are selected to represent the range of expected design. Thermal performance sensitivities are determined by varying material properties and model dimensions. Table 2 displays the material properties for each component used in the numerical simulations. Base material data is obtained from NFRC 101 (2010). Min and max values are selected based on available materials and to properly demonstrate sensitivities.

Boundary conditions used in the 3D FVM and 2D FEM simulations are shown in Table 3. The outdoor and indoor boundary conditions use a fixed combined convection and radiation coefficient. Fixed boundary conditions are used, as opposed to temperature dependent conditions such as ASHRAE, to simplify modeling. The methods outlined are valid for fixed and variable boundary conditions.

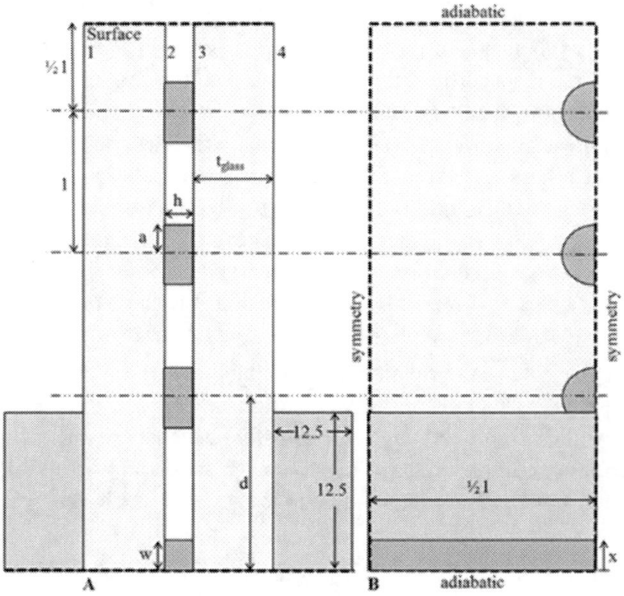

Figure 4 EOG model geometry for 3D FVM analysis (not to scale).

Table 1. Model Dimensions

Dimension	Variable	Base (mm)	Min (mm)	Max (mm)
Pillar radius	a	0.5	0	1
Vacuum gap	h	0.13	0.05	1
Glass thickness	t_{glass}	5.7	2.7	7.2
Pillar spacing	l	50.8	25	125
Pillar to edge of glass	d	25	10	50
Edge seal thickness	w	7.2	0	10.5

RESULTS AND DISCUSSION

Center-of-Glass (COG)

Figure 5 shows the temperature of each glass surface at a specified distance, x, from the center of a pillar (see Figure 3). To account for the localized pillar effects, this model is performed using 3D FVM. The external surfaces 1 and 4 show approximately one Kelvin temperature changes over the length of the model. This means that using an area weighted equivalent conductance (2D FEM) has little impact on external surface temperatures. The internal surfaces 2 and 3 reach 95% of their total temperature change at 5.4 and 6.1 mm, respectively, from the pillar center, meaning the thermal bridging effect of the pillar is a radius of approximately 6.5 mm for the configurations used in this study. A similar plot of the 2D FEM model shows constant temperatures along x since Equation 3 averages pillar heat transfer over the entire vacuum volume.

Table 2. Conductivity and Emissivity of Modeled Materials

Component	Conductivity (W/m·K) Base	Min	Max	Emissivity Base	Min	Max
Outside glass	1	–	–	(1) 0.84	(1) —	(1) —
				(2) 0.02	(2) 0.02	(2) 0.84
Inside glass	1	–	–	(3) 0.84	(3) —	(3) —
				(4) 0.84	(4) —	(4) —
Pillar	16	–	–	0.9	–	–
Edge seal	1	0.01	100	0.9	–	–
Wood	0.14	–	–	0.9	–	–

Table 3. Boundary Conditions Used in the Simulations

Boundary Condition	Value	Units
Indoor temperature	293.15	K
Outdoor temperature	255.15	K
Density of heat flow rate of incident solar radiation on surface	0	W/m²
Combined convection and radiation surface coefficient of heat transfer for the indoor section (surface 4)	7.6	W/m²·K
Combined convection and radiation surface coefficient of heat transfer for the outdoor section (surface 1)	30.0	W/m²·K
Density of heat flow rate at lines of symmetry	0	W/m²

Sensitivities to changes in variables. Analytical, 2D FEM, and 3D FVM models are compared for twenty-six cases of COG U-factor. The 3D FVM radiation method is used as the base for comparison since the model is calculated on the most fundamental basis and has been shown in previous studies to agree well with measurements (Simko 1996). Summarized sensitivity results are presented in Table 4. These results show the four models solve for total thermal transmittance within ~0.01 W/m^2·K, or 2 percent, of each other for all parameter variations in this study. The results for the range of parameters chosen confirm that 2D FEM is an accurate alternative to 3D FVM.

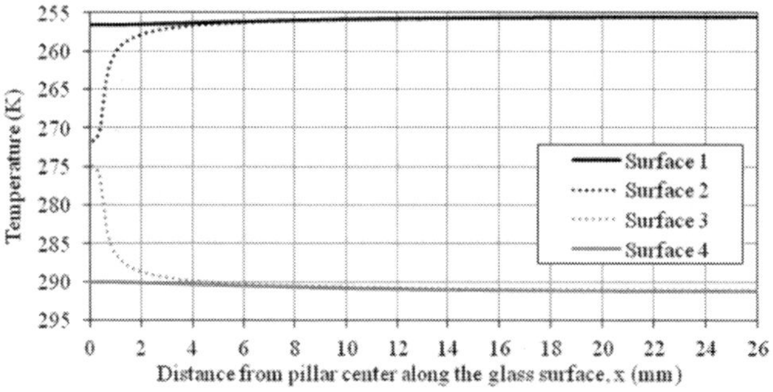

Figure 5 *COG surface temperature distribution from 3D FVM.*

Table 4. COG Thermal Transmittance (U-Factor) Using Four Solution Methods

Parameter	3D FVM "Radiation" (W/m²·K) Min U	Max U	3D FVM "Solid" (W/m²·K) Average ΔU	Average Δ%	2D FEM (W/m²·K) Average ΔU	Average Δ%	Analytical (W/m²·K) Average ΔU	Average Δ%
Vacuum Gap	0.422	0.449	–0.002	–0.6%	–0.010	–2.3%	–0.010	–2.3%
Pillar Radius	0.097	0.766	–0.001	–0.2%	–0.005	–1.1%	–0.005	–1.1%
Pillar Spacing	0.154	1.313	–0.001	–0.2%	–0.001	–1.4%	–0.001	–1.4%
Surface 2 Emissivity	0.446	2.215	–0.020	–1.3%	–0.026	–2.1%	–0.026	–2.1%
Vacuum Pressure	0.441	0.866	–0.001	–0.2%	–0.009	–1.7%	–0.009	–1.7%

The 3D FVM radiation model accounts for localized radiation effects within the radius of pillar thermal bridging while the 3D FVM solid, 2D FEM, and analytical models do not. This results in a slight divergence of the solved thermal transmittance when surface emissivity is increased, as can be seen in Figure 6. The best agreement between 3D FVM and 2D FEM is therefore at low surface 2 emissivity.

Edge-of-Glass (EOG)

NFRC 100 considers all glazed vision areas within 63.5 mm of any part of the frame EOG (NFRC 2010). In order for this assumption to be accurate, and not overestimate the whole window U-factor, the penetration length of thermal effects from frame and end of glazing must be less than 63.5 mm. To determine the penetration length of the EOG effects into the VIG unit, the EOG was modeled with 3D FVM to six pillars in length. The results showed the temperature change on surfaces 1 and 4 at a distance halfway between pillars 3 and 4 is less than 1 percent different from the temperature between pillars 5 and 6. This shows that the edge effect has stabilized by the third pillar and a 3.5 pillar model is used for the remainder of the EOG calculations.

For all combinations of edge seal conductance, pillar spacing, and distance to first pillar, the external surfaces 1 and 4 reach 95% of their total temperature change at approximately 50 mm from the edge, meaning the thermal bridging effect of the edge is within the NFRC EOG for the configurations used in this study. The localized thermal effects of pillars cannot be solved for with a 2D FEM model. The surface temperatures though do closely match those from 3D FVM models at locations outside of pillar effects. Figure 7 compares the EOG glass surface temperatures of each glass surface at a specified distance, *x*, from the EOG (see Figures 2 and 4) of 2D FEM and 3D FVM models on a unit with surface 2 emissivity of 0.02.

Sensitivities to changes in variables. Since the goal of this study is to account for total heat flow, the EOG distance in this study is longer than 63.5 mm and the reported U-factors are based on the total interior length of the EOG model (includes heat flow though the wood block). Therefore, these

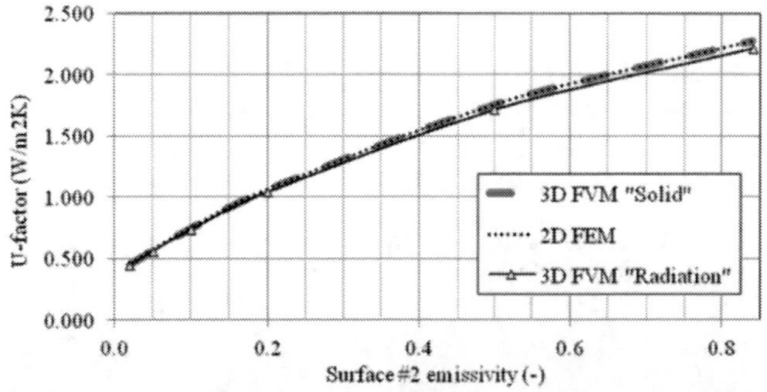

Figure 6 *COG sensitivity to surface 2 emissivity by modeling method.*

models are not directly comparable to typically reported EOG values in standards.

2D FEM, 3D FVM radiation and 3D FVM solid model results are compared for eleven cases of EOG U-factor. Due to the complexity of the edge seal, an entirely analytical solution for the EOG is not attempted. As in the COG calculations, 3D FVM radiation method is used as the base for comparison since the model is calculated on the most fundamental basis.

When pillars are located close to the EOG, they can become a localized conduction path to the frame that is ignored in the area weighted (2D FEM) calculations. The total thermal transmittance though shows negligible divergence between 2D FEM and 3D FVM. Summarized results are presented in Table 5. These results show the three models solve for total thermal transmittance within 0.01 $W/m^2 \cdot K$, or less than 1.3%, of each other for the three parameter variations in this study. The results for the range of parameters chosen confirm that 2D FEM is an accurate alternative to 3D FVM.

CONCLUSION

The center-of-glass (COG) thermal performance of VIG is well understood and simple analytical evaluation has been shown in previous work, and again in this study, to match 3D FEM approximations within 2.3% of total thermal transmittance. The diversity of designs and complexities of EOG heat flow though does not allow for a simple analytical model; therefore difficult and time consuming 3D FVM models are typically used for evaluation. The 2D FEM alternative to such models outlined in this work requires far less computational effort and produces similar total thermal transmittance approximations, within 1.3% of 3D FVM. Concerns regarding EOG length and localized pillar conductance are shown to have little impact on total performance and therefore can be neglected in 2D FEM. This alternative approach will allow for the widespread use of these models in the fenestration industry and therefore the certification of windows using this technology.

ACKNOWLEDGMENTS

This work was supported by the assistant secretary for Energy Efficiency and Renewable Energy, Building Technologies Program, of the U.S. Department of Energy under Contract no. DE-AC02-05CH11231.

REFERENCES

Arasteh, D., S. Selkowitz, J. Apte, and M. LaFrance. Zero Energy Windows. ACEEE Summer Study on Energy Efficiency in Buildings. 2006

Carli. 2006. Conrad 5 & Viewer 5 Technical and Programming Documentation. Carli, Inc. Technical Report. June 20, 2006.

Collins, R.E., and Robinson, S.J. 1991. Evacuated Glazing. *Solar Energy*. Vol. 47, No. 1, pp. 27–38.

Collins, R.E., and Fischer-Cripps, A.C. 1991. Design of Support Pillar Arrays in Flat Evacuated Windows. Aust. *J. Phys.*

Corruccini, R.J. 1959. Gaseous Heat Conduction at Low Pressures and Temperatures. *Vacuum*. Vol. 7–8, pp.19–29

ENoB (Research for Energy Optimized Buildings). Next Step: Vacuum insulation glass. Retrieved April, 17 2013. http://www.enob.info/en/new-technologies/projects /details/next-step-vacuum-insulation-glass/

Eversealed Windows. Technically Speaking Retrieved April 17, 2013. http://www.eversealedwindows.com/

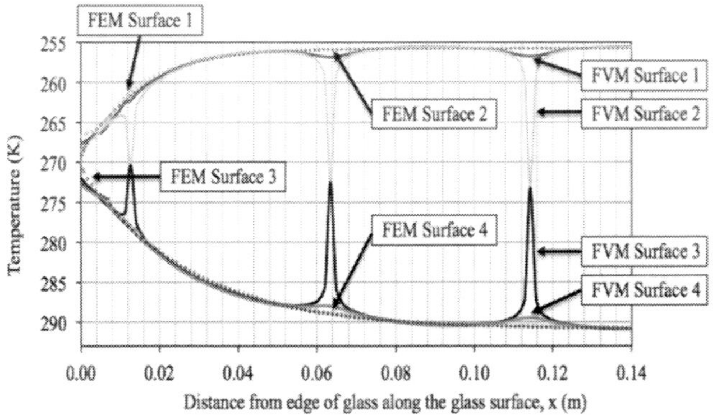

Figure 7 *EOG surface temperature distribution.*

Finlayson, E., R. Mitchell, D. Arasteh, C. Huizenga, and D. Curcija. 1998. *THERM 2.0*: Program description. A PC program for analyzing the two-dimensional heat transfer through building products. Berkeley, CA: University of California.

ISO 15099. 2003. *Thermal Performance o Windows, Doors and Shading Devices—Detailed Calculations*. Geneva, Switzerland: International Organization for Standardization.

Jelle, B.P., A. Hynd, A. Gustavsen, D. Arasteh, H. Goudey, and R. Hart. 2012. Fenestration of today and tomorrow: A state-of-the-art review and future research opportunities. *Solar Energy Materials and Solar Cells*, Vol. 96, pp 1–28.

National Fenestration Rating Council. 2010. NFRC-100-2010E0A1. *Procedure for Determining Fenestration Product U-factors*.

National Fenestration Rating Council. 2010. NFRC-101-2010E2A7. *Procedure for Determining Thermophysical Properties of Materials For Use in NFRC-Approved Software Programs*.

Russo, A. May 1, 2012. New Glass Technologies Promise More Energy Savings. *Window and Door*. Retrieved April, 17 2013. http://www.windowanddoor.com/article/may-2012 /new-glass-technologies-promise-more-energy-savings

Simko, T.M. 1996. Heat Transfer Processes and Stresses in Vacuum Glazing. PhD thesis. The University of Sydney.

Table 5. Table 5. EOG Thermal Transmittance (U-Factor) Using Three Solution Methods

	3D FVM Radiation Model (W/m²·K)		3D FVM Solid Model (W/m²·K)		2D FEM (W/m²·K)	
Parameter	**Min *U***	**Max *U***	**Average Δ*U***	**Average % difference**	**Average Δ*U***	**Average % difference**
Glass Thickness, t_{glass}	0.835	0.988	–0.001	–0.1%	0.003	0.3%
Pillar from EOG, *d*	0.871	0.998	–0.001	–0.1%	0.004	0.5%
Edge Seal Conductance	0.652	0.963	–0.001	–0.1%	0.010	1.3%

Time Step Considerations When Simulating Dynamic Behavior of High-Performance Homes

Paulo Cesar Tabares-Velasco, PhD
Associate Member ASHRAE

ABSTRACT

Building energy simulations, especially those concerning pre-cooling strategies and cooling/heating peak demand management, require careful analysis and detailed understanding of building characteristics. Accurate modeling of the building's thermal response and material properties for thermally massive walls or advanced materials like phase change materials (PCMs) are critically important. However, one input variable that many times is not correctly assessed is time-step dependency. This study is part of an ongoing research effort that has led to the verification and validation of the EnergyPlus PCM and Conduction Finite Difference (CondFD) algorithms in EnergyPlus as well as the simplification of enthalpy linearization for PCMs. This paper demonstrates the importance of using the correct time step depending on the simulation objectives and technology analysed. Using field data of different pre-cooling strategies, this analysis shows and explains the time-step dependency shown in some cases. It also suggests appropriate time-step values depending on the application and HVAC controls.

INTRODUCTION

Residential building air-conditioning and electric heating play a major role in driving peak demand. Together with residential lighting, they can account for up to 40% of total peak load (Koomey and Brown 2002). Thus, a natural place to start addressing peak demand utility challenges is in the residential sector. Researchers have made significant efforts to retrofit houses with energy efficiency measures that focus on utilizing existing thermal mass (distributed thermal storage) or hot water storage (concentrated thermal storage) in residential building along with strategies such as pre-cooling (Xu et al. 2004; Keeney and Braun 1997; Henze et al. 2005; Henze et al. 2007; Sparn et al. 2012). However, using building energy simulations, especially those concerning (1) pre-cooling strategies, (2) cooling/heating peak demand management, and (3) advance envelope assemblies requires careful analysis and detailed understanding of building envelope and load profile characteristics. The analyses of these technologies and strategies are more sensitive to the time step selected than the analyses of energy efficiency technologies that only consider annual energy use. Thus, accurate modeling of the building thermal response and material properties for thermally massive walls or advanced materials like phase-change materials (PCMs) are critically important.

However, one input variable that many times is not correctly assessed is time-step dependency. Time steps are series of discrete bins of time used in time marching algorithms to solve transient problems. In each time step, boundary conditions and dynamic input variables are held constant (EnergyPlus 2012a). Typically, the shorter the time step, the more accurate the solution is at the expense of computational resources and runtime. For this reason, some programs have the capabilities to have variable time steps (Crawley et al. 2005; Sahlin et al. 2005; Sousa 2012).

A previous study analysed different variables that affect the runtime in EnergyPlus, the DOE's building energy simulation program (Hong et al. 2008). Depending on the objective or design stage, the authors recommend time steps from 10 min to 60 min. Their study found important energy differences depending on time step. In addition, EnergyPlus contin-

***Paulo Cesar Tabares-Velasco** is a research engineer at National Renewable Energy Laboratory, Golden, CO.*

Published by ASHRAE 2013

uous testing done for every new release is done using different test suites. For building envelope, the tests are ASHRAE 140 (ASHRAE 140 2007) and ASHRAE 1052-RP Toolkit (Rees et al. 2002). In each case, the EnergyPlus developing team uses time steps of 10 and 6, min respectively (Henninger and Witte 2012a; Henninger and Witte 2012b). Other studies have conducted time-step-dependency related research. For example, Cuilla et al. (2010) reviewed some of the previous studies looking at numerical stability of transfer functions and the influence of the time step in conduction heat transfer and thermal inertia. Their studies expressed concern related to reliability of CTFs for walls with high thermal inertia. Dos Santos and Mendes (2004) performed time-step-sensitivity analysis concluding that 1 h time steps can lead to important errors in indoor air temperature and the calculation of conduction loads. However, they also pointed out that linear interpolation of outdoor variables for sub-hourly time steps introduces uncertainty and ended up recommending a 1.5 min time step. Subsequent studies for heat and moisture transfer in soil coupled with a simple lumped building room showed that the results were not very sensitive to time step (Dos Santos and Mendes 2006).

Garde et al. (2001) compared experimental results in a residential house with results using 1 h and 1 min time steps. Overall, their results showed that short time steps can more accurately predict energy consumption for low part-load conditions than hourly time step. Other studies have looked at differential time scale solutions to allow the use of larger time steps for some building component than the building domain time step (e.g., 5 versus 60 min) (Bourgeois and Reinhart 2007). Other nonenvelope inter-program comparative testing for diagnosing errors has used 5 min time steps (Beausoleil-Morrison et al. 2006) and co-simulation studies that have addresses this issue (Treka et al. 2009). Finally, other studies have looked into the effects of the Fourier and Biot numbers on the accuracy of modeling conduction heat transfer in opaque surfaces. In these studies, analytical or experimental data was compared to the results from finite difference algorithms (Pupeikis et al. 2010; Hensen and Nakhi 1994; Waters and Wright 1985).

Among building energy simulation programs, many of them default the time step with 60 min and or use variable time step integration schemes (Crawley et al. 2005). Many users may not change this default time step when running simulations. For example, BEopt, NREL residential building energy optimization software, uses 1 h time steps. However, the impacts of time step on annual, peak, and hourly results as well as optimization procedure have not been fully explored. In fact, the EnergyPlus *Input and Output Reference* considers 1 h as a "long" time step that should only be used sparingly due to problems with accuracy and the introduction of more lag which leads to a dampened dynamic response when compared to a shorter time step (EnergyPlus 2012b). Given the fact that EnergyPlus is generalized building energy simulation software, it is not completely clear how large the impact would be for residential buildings, which are more envelope/weather dependent. Thus, the objective of this study is to analyse and show the importance of using the correct time step when electric peak demand is the key design variable using a simple analytical test and minute average field data. This study is part of a larger project that has an overall goal to improve the accuracy of energy analysis methods for residential buildings (Polly et al. 2011).

TECHNICAL APPROACH

Building Simulation Considerations

This study uses EnergyPlus Version 8.0 to analyse the impact of time step on the dynamic behaviour of high-performance homes. For this particular case, the conduction heat transfer (CTF) algorithm was selected in EnergyPlus from the different conduction heat transfer algorithms. This selection was done as CTFs are the default and most-used conduction heat transfer algorithms in EnergyPlus. CTFs are also very powerful as they relate the current surface heat flux and temperature values to previous surface heat flux and temperature values (ASHRAE 2009). CTFs assume constant thermal properties, thus any simulations for the analysis of PCMs in EnergyPlus need to use the conduction finite difference (CondFD) model. However, the selection of time step with CondFD is limited up to 3 min, avoiding any problems with time-step dependence as shown in previous studies (EnergyPlus 2012a; Tabares-Velasco et al. 2012).

According to the literature, prediction of variables can be reported in two main ways (EnergyPlus 2012a):

- For state variables, such as temperature, wind speed, and heat flux, their predictions are averages over the time step.
- For summed variables, such as heating and/or cooling energy use, their predictions are aggregated totals over the time step.

There are two time steps used in EnergyPlus for calculating the zone temperature: a zone and a system time step. The zone time step is constant and user specified (or default). It is used to update/calculate envelope-related loads and internal loads to the zone. The system time step is variable and can go from one minute up to the zone time step. Zone loads from HVAC system response, infiltration, and air mixing are updated at the system time step.

Users typically have the option to select the reporting frequency or how often a variable will be reported in the output file. This can be as long as the simulation runtime or as short as the selected time step. Because EnergyPlus has two time steps, detailed frequency is used for the system time step and for variables such as cooling energy. In addition, time step frequency is used for the zone time step and for variables such as surface heat flux, surface temperature, or weather data.

Following analytical solutions used in a previous case study to validate EnergyPlus conduction finite difference model (Tabares-Velasco and Griffith 2012), this study initially uses three simple cases that have analytical solutions (wall subjected to sudden constant heat flux change, wall subjected to a step-up temperature boundary condition, wall with a periodic surface temperature), the ASHRAE Standard 140 Case 600 model, and a model of an actual retrofitted house in Sacramento (Booten and Tabares-Velasco 2012). However, this publication will only focus on one of the analytical solutions (wall with a constant heat flux), a retrofitted house, and the same house with higher thermal mass, as the other two analytical solutions did not uncover any additional information concerning conduction heat flux than the heat flux test.

Analytical Solution

This case has a wall with constant heat flux (Carslaw and Jaeger 1986). The wall is made of 5.08 cm (2 in.) of drywall with an initial homogeneous temperature of 26.5°C (80°F). The outside surface is suddenly exposed to a constant heat flux of 150 W/m^2 (47.5 $Btu/h \cdot ft^2$) while inside surface temperature is kept at 26.5°C (80°F). This problem was selected as it allows the analysis to focus on the conduction heat transfer process through a wall. However, it lacks the actual boundary conditions observed in homes, so extrapolating from it is not suitable.

Retrofitted House Description

The simulated house is an all-electric, 1980s era ranch house located in Sacramento, CA. It has slab-on-grade construction and a total floor area of 161 m^2 (1732 ft^2). The house was retrofitted through the Sacramento Municipal Utility District's (SMUD) deep energy retrofit demonstration program. Figure 1 shows the model implemented in EnergyPlus. The model was originally developed from BEopt but was further expanded in EnergyPlus to allow more customization of the ventilation, wall insulation asymmetry, occupancy schedules, and individual characteristics for items such as ceiling areas (Sparn et al. 2012).

The actual house was instrumented as part of a larger study and the data collected were used to validate hourly simulation predictions (Sparn et al. 2012, Booten and Tabares-Velasco 2012). However, all comparisons done in the previous study were done using 1 h average data in contrast with this study where 1 min data is used. The unoccupied home had simulated occupancy where heaters and lighting was used to simulate occupant's behaviour. Appliance usage was based on the standard occupants assumptions defined in the Building America House Simulation Protocols (Hendron and Engebrecht 2010). Some energy efficiency and renewable energy measures are listed next and more information about the retrofit measures can be found in the literature (Sparn et al. 2012):

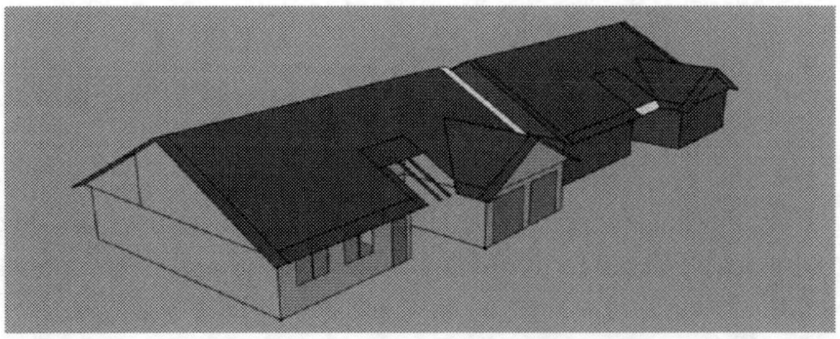

Figure 1 3D model used in energy simulations.

1. Attic floor has R-42 blown-in cellulose and the roof has a radiant barrier.
2. Windows were vinyl frame, dual pane Argon filled, low-e with a U-factor = 1.62 $W/m^2 \cdot K$ (0.285 $Btu/ft^2 \cdot h \cdot R$) and solar heat gain coefficient (SHGC) = 0.0205.
3. West wall had R-15 blown-in cellulose insulation while the rest of walls had R-11.
4. Blower door infiltration results show an infiltration rate of 3.6 ach @ 50 Pa.
5. A programmable thermostat used to analyse different precooling strategies.
6. House has a heat pump rated at SEER 16/EER 13 HSPF 9.75.

From the different cooling strategies tested during the monitoring period, only one is assessed in this study: a pre-cooling strategy with three different setpoints. This strategy consist in setting the thermostat equal to 21.1°C (70°F) between 10 a.m. and 4 p.m., 26.0°C (79°F) between 4 p.m. and 8 p.m., and 24.1°C (75.4°F) at all other times. The idea of this approach was to provide additional cooling during the morning and early afternoon to attempt to reduce the cooling load during afternoon. It is selected for this study as it had many features that allow the building temperature to float over a couple of hours a day as well as having a drastic indoor air temperature change.

The actual test for the analyzed pre-cooling strategy lasted two weeks. For purposes of eliminating any issues with initial conditions, the house was modeled for several days before the day used for time step analysis to allow for thermal and other transient simulation effects to diminish (Ellis 1998).

RESULTS

Analytical Solution

Figure 2 shows the exterior surface heat flux. Figure 3 shows predicted temperature of the surface subject to the heat flux. Both figures show the analytical solution (Eqn) and EnergyPlus solutions. Actual analyses considered 1, 5, 10, 15, 20, 30, and 60 min time steps. However, in the following graphs only 1, 5, 30, and 60 min time steps are shown to avoid overcrowding the graphs. Results are reported every zone time step. Figure 2(a) shows the typical scatter plots with straight lines while Figure 2(b) shows the actual discrete/step-up nature of the results. As expressed earlier, EnergyPlus

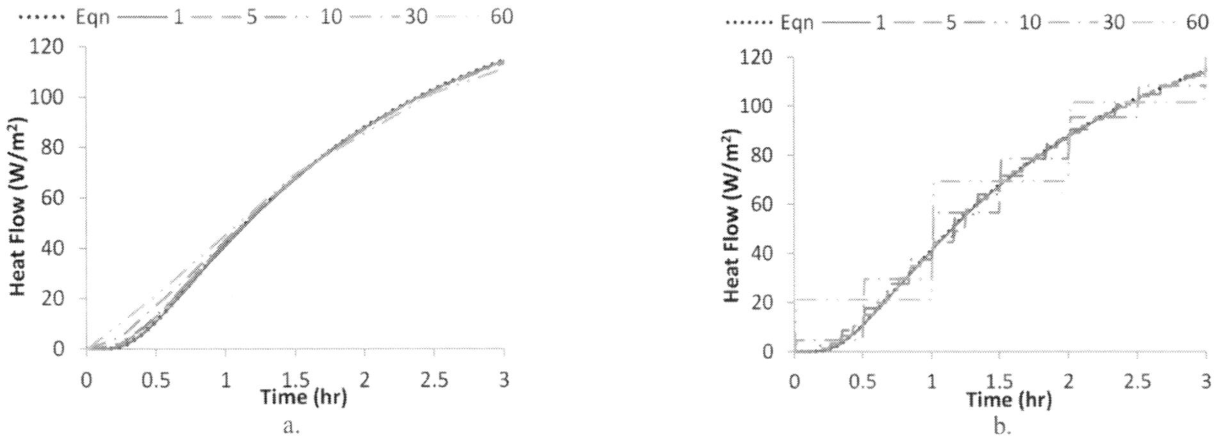

Figure 2 *Exterior surface heat flux for a simple wall subjected to a rapid heat flux. Data shown represents solution for analytical solution (Eqn) and simulation results using multiples times from 1 min time step (1) to 60 min time step (60) for (a) interpolation curve and (b) actual step curve.*

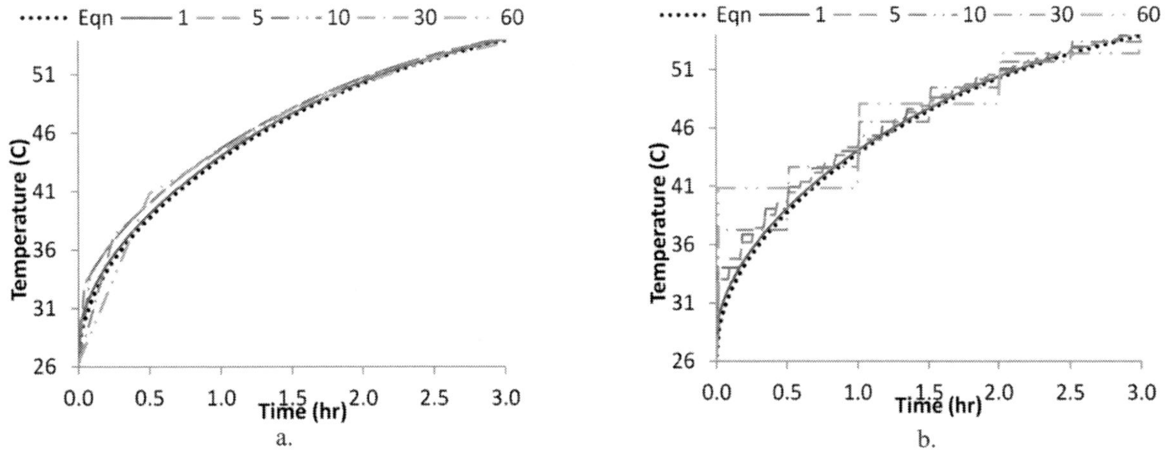

Figure 3 *Interior surface temperature for simple wall subjected to a rapid heat flux change.*

reports state variables, such as temperature and heat flux as averages over the time step, thus the need to plot the data as in Figure 2(b) and 3(b).

Figure 2(b) shows that using a 60 min time step totally missed the initial dynamic behaviour of the wall. In fact, only the time steps below 15 react fast enough to catch up with the actual phenomena. However, plotting results with different frequency can be visually misleading as shown in Figure 2(a).

Table 1 shows the results for the first two hours as well as the percentage difference with respect to the analytical solution. The highest differences are observed for the largest time step for the first hour, but tend to decrease with time as it is shown in Figure 3(b) where the 1 h time step overestimates the inner surface temperature and exterior heat flux by about ¾ of the first hour.

The higher difference in the first hour compared to the subsequent ones is mainly due to the slow response of CTFs

Table 1. Average Exterior Surface Heat Flux for the First Two Hours

Time Step	1st hour		2nd hour	
	Average	Difference (%)	Average	Difference (%)
Analytical	14.7		67.1	
1 min	14.6	–1%	66.7	0.6%
5 min	15.5	–7%	66.9	0.3%
10 min	15.6	–8%	66.9	0.2%
15 min	15.9	–10%	67.0	0.0%
20 m	16.2	–12%	67.2	–0.2%
30 min	17.1	–18%	67.6	–0.8%
60 min	21.1	–46%	69.4	–3.4%

and numerical error. The only continuous and smooth curve is the analytical solution, while the simulation solutions are discrete curves that assume average heat flux within a time step (discrete jumps between time steps) as shown in Figure 2(b) and Figure 3(b). Thus the higher differences at larger timesteps are due to a discretization error that is exacerbated at the beginning of the sudden change in the boundary condition problem, when the wall is in equilibrium before being subjected to a heat flux. In other words, during the first hour, the 60 min result overestimated the heat flux about 60% of the first hour, while in the second hour it is overestimated for half of the time but is underpredicted by a similar amount in the second half. Using 1 min time steps have negligible differences with the analytical solution.

As explained earlier, this problem was selected as it has an analytical solution that allows comparison between CTFs and the actual heat flux through the envelope. However, it lacks the typical boundary conditions observed in homes and the temperature values shown in Figure 3 are outside of the typical range for an interior surface, so extrapolating from it is not suitable.

Retrofitted House

Figures 4, 5, and 6 show the cooling electric power, living zone temperature, and the internal loads imposed in the test house. The same nomenclature is used here as in previous figures. Figures 4 and 5 are shown with values reported at the (a) zone time step and (b) system time-step frequency. Using system frequency shows a more rapid response to changes since EneryPlus is using the variable system time step to manage the rapid zone load changes but smaller time steps still show faster response to the dynamic of the house.

Field test data shown are minute averages and are shown for reference only. They are not to necessary determine which time step gets closer to the actual data, as other factors could be affecting the final results of the simulations such as air thermocouple location for the living zone temperature. Figure 4 shows the difference regarding how A/C equipment is typically modeled in BEopt and in other building simulation programs with a part-load ratio (PLR). PLR assumes A/C runs continuously when cooling is needed (from 9 a.m. to 3 p.m. in this example) instead of the on/off cycle that the actual equipment went through.

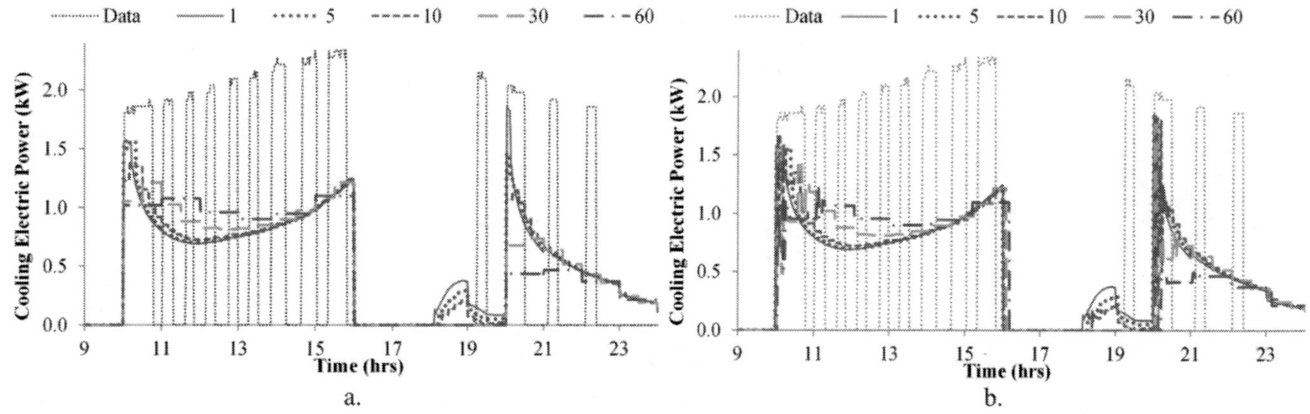

Figure 4 On-site electric cooling power plotted using (a) time-step frequency and (b) detailed frequency.

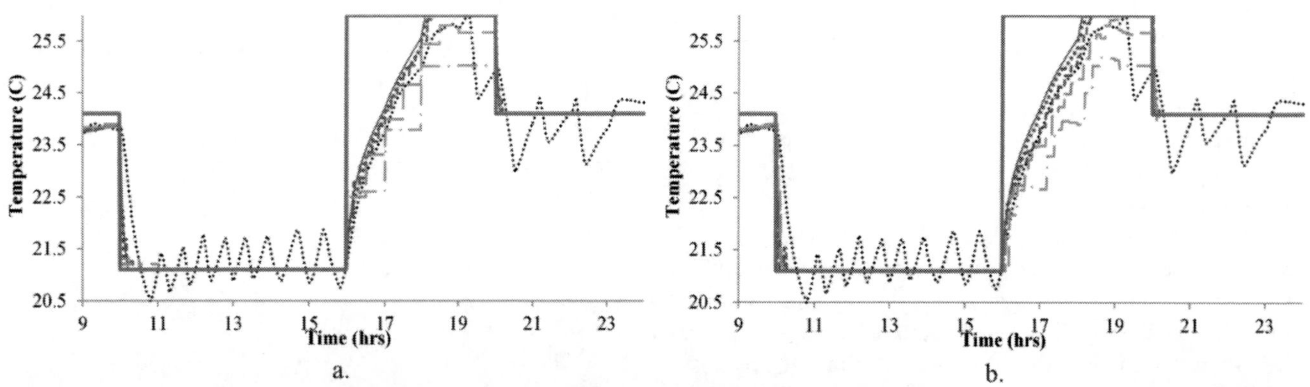

Figure 5 Living zone air temperature plotted using (a) zone time step frequency and (b) system time-step frequency. (Continuous thick blue line represents thermostat setting.)

Figure 6 shows the simulated internal loads that were distributed in the house. Location and loads profiles were based on Building America House Simulation Protocols (Hendron and Engebrecht 2010, Sparn et al. 2012). The large internal gain around 18–19 h reflects cooking range among other interior loads that are shifted less than one hour in the simulations with respect with the cooling electric data. This is probably due to perfect mixing assumption in the model for the entire living zone (all bedrooms, kitchen, and living room).

Measured living room temperature in Figure 5 shows how the actual living indoor temperature varies while the A/C is cycling on/off. This effect was not captured in any of the simulations due to part-load ratio assumption used in BEopt. The 30 and 60 min solutions had the slowest response to the setback from 21°C to 26°C. In fact the 30 min and 60 min results did not reach the temperature setpoint, while the other solutions with smaller time step did. While in this case living room temperature data shows closer agreement with the 30 min results, actual house temperature varied depending on the location and room: west side rooms were 1–2°C (2–4°F) warmer in the afternoon than the east oriented rooms.

Figure 4 supports the previous statements that larger time steps (20–60 min) tend to damp some of the heat transfer or, at around hour 18, completely ignoring the dynamic thermal behavior of the house. More importantly, the solutions with the larger time steps also underestimate the afternoon peak cooling electric demand. This confirms that using 30 and 60 min time steps tends to underestimate the peak cooling late in the afternoon. Figure 7 shows the same variables as Figure 4 but reporting hourly average cooling electric power instead of every time step. Larger time steps continue underestimating the afternoon peak load even when all solutions are hourly averages. Reporting data hourly as in Figure 7 also tends to underestimate overall peak cooling demand (time = 18–20 h) no matter the time step used and ignore/damps higher frequency thermal behavior of the house. Thus peak cooling load analysis should use time step lower than 60 min (preferably 5 min or lower) and report/output variables with a time step frequency as in Figure 4.

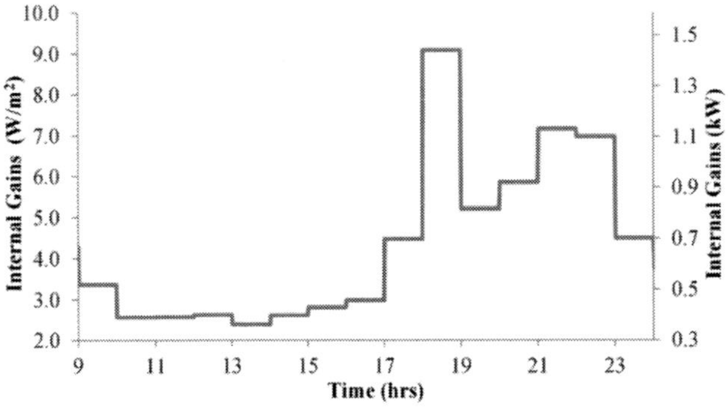

Figure 6 *Total internal heat gains.*

Tables 2 and 3 summarize the results shown in Figure 4 and Figure 7 for the zone and system time step. Time step selection is not critical when daily or longer periods are being analyzed, as the difference between results with different time steps is less than 4%. However, it is very important to report data with system time step when analyzing peak cooling load. Predicted peak cooling load in the afternoon was very sensitive to time step and frequency (zone or system) because of averaging of system time step over the zone time step. For this particular case, reporting values with the zone time step and selecting a 60 min time step can underestimate the afternoon cooling peak demand by 75%. Using the system time step will underestimate the afternoon cooling peak demand by 46%. It is only when results are reported using system time step with time steps lower than 10 min that results are close to the measured afternoon cooling peak demand.

Figure 8 shows the interior surface heat flux for the south wall. This figure is only shown in zone time step since EnergyPlus keeps surface temperature and calculated heat fluxes constant during the time step. Notice in Figure 8 how the larger time steps lead to lag in the heat transfer between the air and the wall. In the late afternoon, larger time steps also completely missed the dynamic behaviour of the wall reacting to the change in setpoints and increase in internal loads. Unfortunately, no heat flux data was measured in any wall in the house.

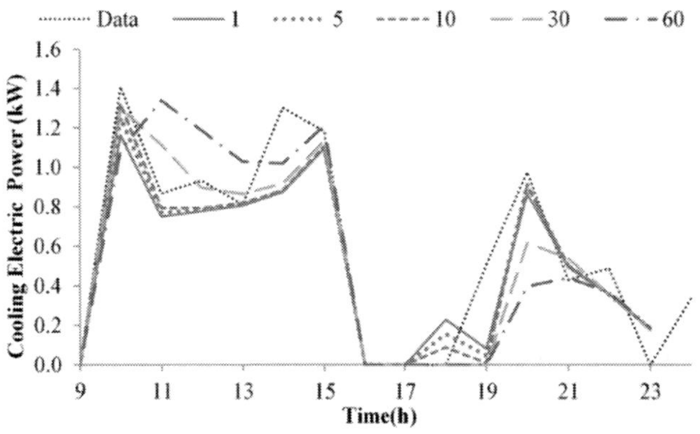

Figure 7 *On-site electric cooling power, hourly output.*

Table 2. On-Site Cooling Energy Use and Afternoon Peak Demand Using Zone Time Step

Time Step	Data	1 min	5 min	10 min	15 min	20 min	30 min	60 min
24 h Cooling Energy (kWh)	9.56	8.08	8.02	8.00	7.98	8.01	7.96	7.78
Difference (%)			0.8	0.9	1.1	0.8	1.5	3.6
Afternoon Peak Load (kW)	2.2	1.8	1.5	1.1	1.0	0.9	0.7	0.5
Difference (%)			21	39	46	51	61	75

Figure 9 shows the slab interior heat fluxes. For this variable, two heat flux meters were installed in a bedroom closet under the carpet. There is a large variation in the heat flux in the afternoon due to both the change in setpoint and the solar gain. There are also shorter oscillations due to the on/off A/C cycling. As with other figures, field data is shown only as a reference, as in this case two heat flux meters are not representative of the entire slab heat transfer which is calculated by the model. However, all simulations follow the same trend as the field data, but the heat flux magnitudes are shifted.

Both Figure 8 and Figure 9 show how larger time steps affect the heat flux and lead to an overall lag in the heat transfer between the slab/wall and the indoor air. In particular, the predicted heat flux from the south wall using larger time steps completely misses the afternoon peak. Floor and south wall heat flux peaks in the afternoon can be underestimated by more than 50% for cases with an hourly time step.

Figure 10 shows the total internal convective heat gain rate (kW) for the living zone. This figure also explains the differences observed previously in cooling energy use. Calculated south wall convection coefficients are not shown here, due to space limitations but for this particular afternoon and precooling strategy, the south wall convection coefficient changed from 1.2 to 3.5 W/m^2·K (0.21–0.62 Btu/h·ft^2·R) for the 1 min time step while the 60 min solution only varies from 1.2 to 2 W/m^2·K (0.21–0.35 Btu/h·ft^2·R) with a two hour delay in which the heat transfer coefficient changes. This further explains the slow response seen in the 60 min solution in Figure 8.

Finally, to fully understand the impact of time step, Figure 11 shows the interior surface temperature for the south wall. Unfortunately, wall surface temperatures were not measured either, thus there is no data to compare. The difference between the 1 min and 60 min solutions for the floor and south wall can be as large as 2°C (4°F).

Table 3. On-site Cooling Energy Use and Afternoon Peak Demand Using System Time Step

Time Step	Data	1 min	5 min	10 min	15 min	20 min	30 min	60 min
24 h Cooling Energy (kWh)	9.56	8.08	8.02	8.00	7.98	8.01	7.96	7.78
Difference (%)			0.8	0.9	1.1	0.8	1.5	3.6
Afternoon Peak Load (kW)	2.2	1.85	1.85	1.83	1.81	1.8	1.76	1
Difference (%)			0.3	1	2.5	2.7	4.8	46

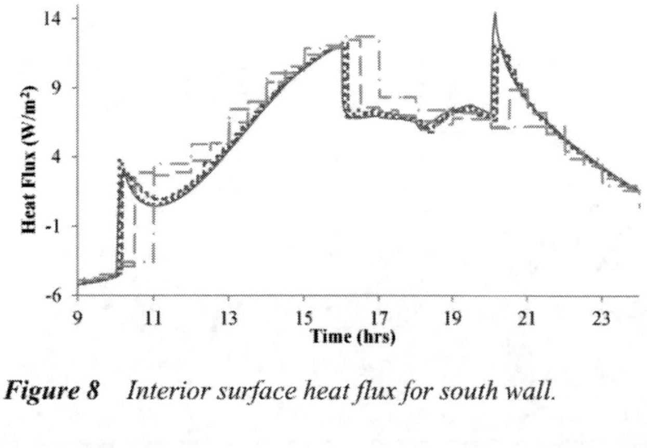

Figure 8 *Interior surface heat flux for south wall.*

Retrofitted House with High Thermal Mass

The same retrofitted house was analysed but in a scenario where the internal thermal mass is increased by using PCMs or installing large furniture. This was done by increasing the thickness of the drywall from 0.013 m (½ in.) to 0.10 m (4 in.). Figure 12 shows the cooling electric power for the retrofitted house (light) and high thermal mass (heavy) using a 1 and 60 min time step reported with the system time step. Morning cooling load increase and thermal behaviour is

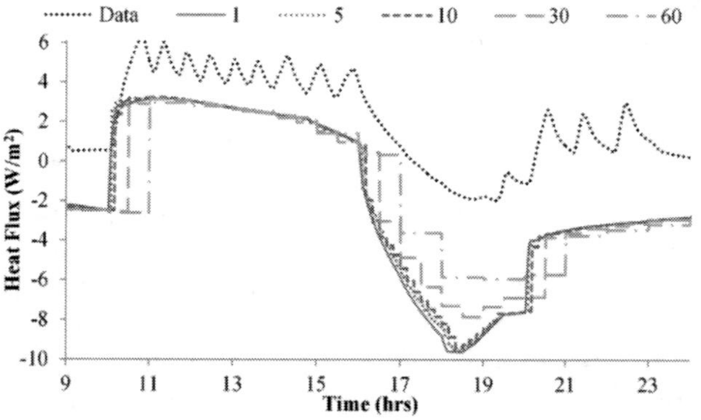

Figure 9 *Slab interior heat flux.*

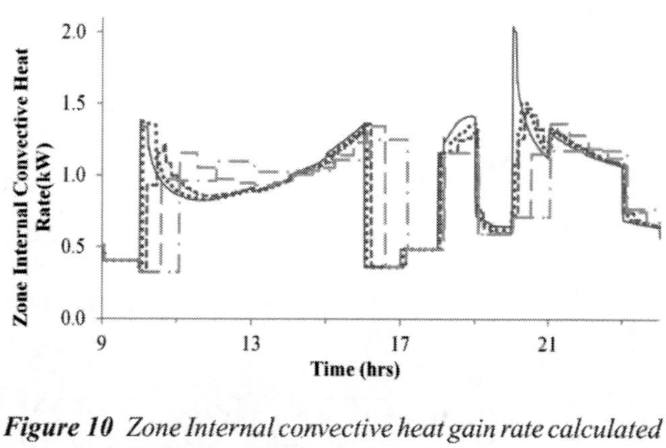

Figure 10 *Zone Internal convective heat gain rate calculated with system time step.*

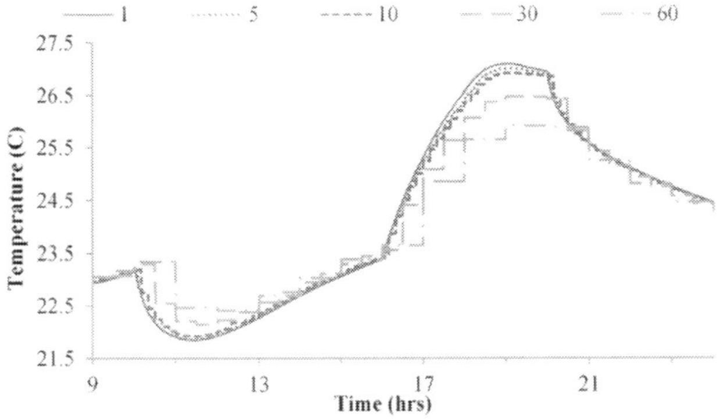

Figure 11 *South wall interior surface temperature.*

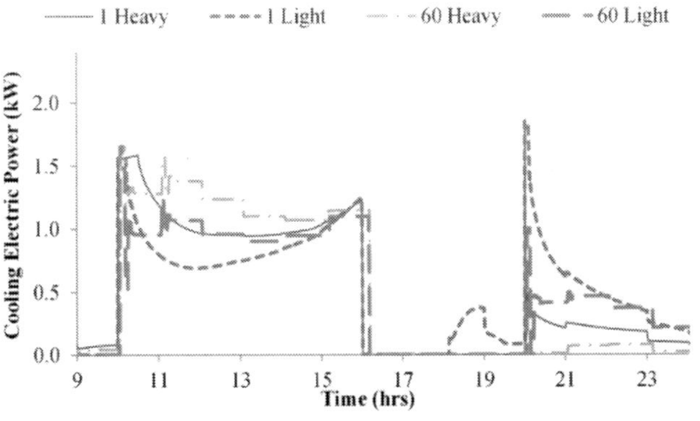

Figure 12 *On-site electric cooling power for a light and heavy retrofitted house calculated with 1 and 60 min time step.*

well predicted with both time steps due to the added thermal mass that needs to cool down. However, the 60 min time steps oscillates slightly at 10 a.m. when the A/C starts running. The most important differences are in the afternoon, the 60 min model predicts almost zero cooling energy use for the high thermal mass, while the 1 min time step does not, with a short spike, and an almost constant load of 0.4 kW. More important, this example shows that users should always compare technologies or designs using the same time step.

DISCUSSION FROM TIME STEP ANALYSIS

Results presented here show that if peak air-conditioning load is an important factor to consider, users should use time step of 10 min or less using the system time-step frequency. This will ensure that all important transient phenomena are well calculated at the expense of increased runtime. The same recommendation might apply when TOU rates are used. However, if the interest of a study is only annual energy consumption, then users can freely select hourly simulations to speed up the results. The time-step selection is an important variable to consider since it affects all heat transfer processes in the building whenever fast changes take place inside the building due to HVAC controls (set back) or internal gains profiles as shown in this study. Missing this dynamic behaviour of the building can ultimately lead to incorrect predictions of the cooling load. This conclusion is consistent with results found in Garde et al. (2001) and Dos Santos and Mendes (2004) that suggest time steps of only a few minutes. Although not explored here, this preliminary research also agrees with Cuilla et al. (2010) that weather data interpolation for subhourly values impacts the subhourly results, but a definitive conclusion requires further analysis. For this particular case, the runtime for the 1 min time step was 5 min, while the runtime for the 60 min case was 37 s. The runtime difference between 60 and 1 min cases is a factor of around 8.

CONCLUSIONS

Building energy simulations, especially those concerning pre-cooling strategies and cooling/heating peak demand management, require careful analysis and detailed understanding of building characteristics. Depending on the simulation objectives and technology analysed, this paper shows the importance of using the correct time step and reporting frequency (when variable time steps are allowed). By examining a simple case with an analytical solution and field data from a lab home using a pre-cooling strategy this study shows and explains the time-step dependency of building energy simulations with rapid temperature oscillations. Overall, this study found that for these particular cases, users should use time-steps lower than 10 min to ensure the thermal dynamic behaviour of residential buildings are fully modeled for peak load scenarios. The results for this particular pre-cooling strategy showed that using a 60 min time step could underestimate the peak load up to 70%. However, if annual energy consumption is the main interest, users can use larger time steps.

ACKNOWLEDGMENTS

This work was supported by the U.S. Department of Energy under Contract No. DE-AC36-08-GO28308 with the National Renewable Energy Laboratory.

REFERENCES

ASHRAE (2004). ANSI/ASHRAE Standard 140-2004. *Standard Method of Test for the Evaluation of Building Energy Analysis Computer Programs*. Atlanta, GA, ASHRAE: 151.

Beausoleil-Morrison, Griffith B., Vesanen T., Lerson S., and Weber A. 2006. A case study demonstrating the utility of inter-program comparative testing for diagnosing errors in building simulation programs, Sim 2006, 181–188.

Booten, C., and Tabares-Velasco, P.C. 2012. Using EnergyPlus to Simulate the Dynamic Response of a Residential Building to Advanced Cooling Strategies, *Proceedings: The Second International Conference on Building Energy and Environment*, Boulder, CO, August 2012: 894–902.

Bourgeois, D., and Reinhart, C. 2007. Differential time scale solutions for dynamic boundary conditions within whole-building energy simulation, *Proceedings: Building Simulation* 2007:1354–1360.

California Energy Commission 2008. 2008 *Building Energy Efficiency Standards for Residential and Commercial Buildings*. California Code of Regulations, Title 24, Part 1 CEC-400-2008-001-CMF.

Carslaw, H.S., and Jaeger, J.C., 1986. *Conduction of Heat in Solids*. Oxford: Oxford University Press.

Crawley, D.B., Hand, J.W., Kummert, M., and Griffith, B.T. 2005. Contrasting the Capabilities of Building Energy Simulation Programs. Version 1.0 (http://gundog.lbl.gov/dirpubs/2005/05_compare.pdf) EnergyPlus, 2012a. *EnergyPlus Engineering Reference: The Reference to EnergyPlus Calculations*. Berkeley, CA: Ernest Orlando Lawrence Berkeley National Laboratory.

Ciulla, G., Brano, V., and Orioli, A. 2010. A criterion for the assessment of the reliability of ASHRAE conduction transfer function coefficients, *Energy and Buildings* 42:1426–1436.

Dos Santos, G.H., and N. Mendes. 2004. Analysis of numerical methods and simulation time step effects on the prediction of building thermal performance, *Applied Thermal Engineering* 24: 1129–1142.

Dos Santos, G.H., and Mendes, N. 2006. Simultaneous heat and moisture transfer in soils combined with building simulation. *Energy and Buildings* 38:303–314.

Ellis, P.G. 1998. Development and Validation of the Unvented Trombe Wall Model in EnergyPlus. Master's Thesis, University of Illinois at Urbana-Champaign.

EnergyPlus, 2012b. *Input Output Reference: The Encyclopedic Reference to EnergyPlus Input And Output*. Berkeley, CA: Ernest Orlando Lawrence Berkeley National Laboratory.

Garde, F. Boyer, H., Lucas, F., and Brau, J. 2001. Multiple Model Approach of a Residential Heat Pump for Integration in a Building Thermal Simulation Code and Comparison with Experimental Results. *ASHRAE Transactions* vol. 107 (1):14–25.

Hendron, R., and Engebrecht, C. 2010. Building America Research Benchmark Definition NREL/TP-550-47246, Golden, CO. National Renewable Energy Laboratory.

Henninger, R.H., and Witte, M. 2012a. EnergyPlus Testing with ASHRAE 1052-RP Toolkit – Building Fabric Analytical Tests. October 2012. http://apps1.eere.energy.gov/buildings/energyplus/pdfs/energyplus_ashrae_1052rp_envelope.pdf

Henninger, R.H., and Witte, M. 2012b. EnergyPlus Testing with Building Thermal Envelope and Fabric Load Tests from ANSI/ASHRAE Standard 140-2011. October 2012. http://apps1.eere.energy.gov/buildings/energyplus/pdfs/energyplus_ashrae_140_envelope.pdf

Hensen J.L.M., and Nakhi, A.1994. Fourier and Biot numbers and the accuracy of conduction modeling *Proceeding of BEP* 94 "Facing the future" 6–8 April: 247–256.

Henze, G., et al. 2005. Experimental analysis of model-based predictive optimal control for active and passive building thermal storage inventory" *HVACR Research*, vol. 11, no. 2, pp. 189–213.

Henze, G., et al. 2007. Sensitivity analysis of optimal building thermal mass control, *Journal of Solar Energy Engineering*, vol. 39, no. 2, pp 221–235.

Hong, T., Buhl, F., and Haves, P. 2008. EnergyPlus Run Time Analysis. Ernest Orlando Lawrence Berkeley National Laboratory.

Keeney, K., and Braun, J. 1997. Application of Building Precooling to Reduce Peak Cooling Requirements", *ASHRAE Transactions*, vol. 103, no. 1, pp. 463–469.

Koomey, J., and Brown R. 2002. The Role of Building Technologies in Reducing and Controlling Peak Electricity Demand, Energy Analysis Department, Environmental Energy Technologies Division, Ernest Orlando Lawrence Berkeley National Laboratory, LBNL-49947.

Polly, B., Kruis, N., and Robert, D., 2011. Assessing and improving the accuracy of energy analysis methods for residential buildings. DOE/GO-102011-3243. Golden, CO.

Pupeikis D, Stankevičius, V., and Burlingis, A. 2010. The effect of the Fourier number on calculation of an unsteady heat transfer of building walls, *Journal of Civil Engineering and Management* 16(2):298–305.

Rees, S.J., Spitler, J.D., and Xiao, D., 2002. An analytical verification test suite for building fabric models in whole building energy simulation programs. *ASHRAE Transactions*, 108 (1), 11.

Sahlin P. et al. 2003. Will Equation-based Building Simulation Make it? Experiences from the Introduction of IDA Indoor Climate and Energy, Eight International IBPSA Conference, Eindhoven, Netherlands, August 11–14: 1147–1154.

Sparn B., K. Hudon, L. Earle, C. Booten, P.C. Tabares-Velasco, G. Barker, and E. Hancock. 2012. Greenbuilt Retrofit Test House Final Report, NREL Technical Report, NREL/TP-5500-54009.

Sousa J. 2012. Energy Simulation Software for Buildings: Review and Comparison, Information Technology for Energy Applications 2012, Lisbon, Portugal, September 6–7:57–68.

Tabares-Velasco, P.C., and Griffith, B. 2012. Diagnostic Test Cases for Verifying Surface Heat Transfer Algorithms and Boundary Conditions in Building Energy Simulation Programs, *Journal of Building Performance Simulation*, 5 (5):329–346.

Tabares-Velasco, P.C., Christensen, C., and Bianchi, M. 2012. Verification and Validation of EnergyPlus Phase Change Material Model for Opaque Wall Assemblies, *Building and Environment*, 54:186–196.

Treka, M., Hensen, J.L.M., and Wetter, M. 2009. Co-simulation of innovative integrated HVAC systems in buildings, *Journal of Building Performance Simulation*, 2:3, 209–230.

Waters, J.R., and A.J. Wright. 1985. Criteria for the distribution of nodes in multilayer walls in finite difference modelling, *Building and Environment* 20 (3): 151–162.

Xu P, et al. (2004). Peak Demand Reduction from Pre-Cooling with Zone Temperature Reset in an Office Building, *Proceedings of 2004 ACEEE Summer Study on Energy Efficiency in Buildings*, Pacific Grove, CA, August 23–27, 2004.

Design Implication of Glazing Ratio Restrictions

Medgar Marceau, PE
Member ASHRAE

Stéphane Hoffman, PE, PEng
Associate Member ASHRAE

ABSTRACT

The 2009 and 2012 International Energy Conservation Codes set the trend for significant improvements to the thermal performance of the building envelope. These new codes challenge building owners who desire large expanses of vision glass to utilize higher performance technologies. The implication of glazing ratio on glazing system U-factors and spandrel panel design will be presented, including a comparison of prescriptive versus performance-based approach to code compliance. Through a case study of a high-rise residential tower in the Pacific Northwest, this paper will present solutions involving both traditional glazing technology as well as emerging technologies that provide solutions that comply with one of the strictest energy codes in the nation. The paper is based on an analysis undertaken to quantify the impact of the glazing ratio on the design of a glazing system. Looking beyond code compliance, the effect of thermal bridging will be demonstrated using the latest 3D thermal modeling. The implications of raising the bar when comparing a code-compliant baseline building to the proposed building and its impact on LEED projects will be discussed. This will be of interest to designers looking to understand the implications of the restriction imposed on glazing ratio as well to the manufacturer of glazing systems looking to provide systems that meet these new requirements

INTRODUCTION

This paper examines how the maximum allowable glazing ratio mandated by current energy codes affects the design and construction of buildings. The scope of this paper is commercial buildings, which includes multi-unit residential buildings over three stories. The scope is limited to the thermal performance of the building envelope. The case study is based on an analysis undertaken during the design of a multi-family high-rise residential building in western Washington state. Energy codes stipulate requirements for energy-efficient design and construction of commercial and residential buildings. In the United States there is no nation-wide energy code, except for federal buildings (CFR 10 Part 433 2012). So, code adoption and enforcement generally occur at the state and local level. Some states have adopted ANSI/ASHRAE/IESNA Standard 90.1-2007, *Energy Standard for Buildings except Low-Rise Residential Buildings* (ASHRAE 90.1-2007). However, most states have adopted the *2009 International Energy Conservation Code* (2009 IECC). A few states, including California, Oregon, and Washington, have their own "home-grown" energy codes (DOE 2013). Every three years energy codes are revised to continuously improve the energy efficiency of new and renovated buildings. Consequently, stricter energy conservation measures are adopted with each code cycle. Since this paper's case-study building was permitted under the *2009 Washington State Energy Code* (WSEC), the paper will also discuss some of the requirements of the 2009 WSEC. At the time of writing, the state of Washington is in the process of switching to the IECC with state amendments.

Demonstrating Compliance with Energy Codes

Energy codes typically have three options for demonstrating compliance: prescriptive, performance, and trade-off. Prescriptive requirements are specified minimum performance requirements in the code. For example, in western Washington

***Medgar Marceau** is a building science engineer and **Stéphane Hoffman** is a senior building science consultant at Morrison Hershfield, Bellevue, WA.*

(which is in climate zone Marine 4), exterior steel-framed walls enclosing a residential occupancy (other than single-family) are required to have fiberglass batt insulation with a minimum thermal resistance (R-value) of 3.3 $m^2 \cdot K/W$ (19 $h \cdot ft^2 \cdot °F/Btu$) plus continuous insulation with a minimum thermal resistance of 1.5 $m^2 \cdot K/W$ (8.5 $h \cdot ft^2 \cdot °F/Btu$) without thermal bridges other than fasteners (WSEC 2009). Using the performance option, compliance is demonstrated through whole-building computer simulation to show that a proposed building has an annual energy performance that is less than or equal to the annual energy performance of the standard reference design over a typical meteorological year. For example, if the proposed heating ventilation and air-conditioning (HVAC) system is much more energy efficient then the code-mandated minimums, energy simulation could be used to show that the expected energy savings from the HVAC can make up for the poorer thermal performance of a less-than-code-compliant wall. In ASHRAE 90.1-2007 and 2009 IECC, annual energy performance is based on predicted annual energy cost. Hence this method is also called the *energy cost budget method*. In WSEC 2009, annual energy performance is based on annual energy consumption. ASHRAE 90.1-2007, 2009 IECC, and 2009 WSEC have a third option for demonstrating compliance of the building envelope called, respectively, "building envelope trade-off option," "total UA alternative," and "component performance building envelope option." UA is the overall rate of heat transfer through the building envelope per unit of time induced by a unit temperature difference between the environments on each side of the envelope. It is equal to the sum of the weighted product of thermal transmittances (that is, the U-factor) and area of roofs, opaque wall areas, fenestration areas, etc. All of these third options are trade-off options, which can be thought of as an intermediate path between the prescriptive and performance paths. Using the trade-off option, buildings whose design heat loss rate is less than or equal to the target heat loss rate will be considered in compliance. The 2009 IECC and the 2009 WSEC require accounting for conductive heat loss and solar heat gain (refer to 2009 IECC, *Section 402.1.4, Total UA Alternative* and 2009 WSEC, *Section 1330, Component Performance Building Envelope Option*). However, the procedure in ASHRAE 90.1-2007 is more complex because it also requires accounting for climate, lighting, thermal mass, and building orientation (refer to ASHRAE 90.1, *Section 5.6, Building Envelope Trade-Off Option* and *Normative Appendix C*).

Maximum Allowable Fenestration Area: 40%

With most states having implemented the 2009 IECC during the global financial crisis of 2007–2012, the so-called Great Recession (IMF 2012), relatively few buildings were permitted under this new code, and hence few designers have had to deal with the challenges of complying with the new thermal performance requirements of the building envelope. One of these new requirements is a 40% limit on fenestration area. For those buildings that have been permitted under the new code (and where the new code is actually enforced), this new limit has had a significant impact on the design of so-called "glass" buildings.

Under the prescriptive building envelope requirements of ASHRAE 90.1-2007, Section 5.5.4.2 Fenestration Area states, "the total vertical fenestration areas shall be less than 40% of the gross wall area" and "the total skylight area shall be less than 5% of the gross roof area". Fenestration is defined as "all areas (including the frames) in the building envelope that let in light, including windows, plastic panels, clerestories, skylights, doors that are more than one-half glass, and glass block walls."

Similar requirements exist in the 2009 IECC. Under the prescriptive building envelope requirements, Section 502.3.1 Maximum Area states that the vertical fenestration area shall not exceed 40% of the gross wall area and that skylights shall not exceed 3% of the gross roof area. Fenestration is defined as "skylights, roof windows, vertical windows (fixed or moveable), opaque doors, glazed doors, glazed block, and combination opaque/glazed doors."

And there are similar requirements in the 2009 WSEC, although there is a subtle but importance difference. Under the prescriptive building envelope requirements, Section 1323.1 states that the percentage of *total* glazing relative to the gross exterior wall areas shall not be greater than 40% for the *vertical glazing and overhead glazing*. Although the requirement is specified in terms of glazing and not fenestration, the definition of glazing in the 2009 WSEC is similar to the definition of fenestration in ASHRAE 90.1-2007 except that the 2009 WSEC does not specifically mention plastic panels and the threshold of glass in doors is not specified. However, the requirement is in terms of *total* glazing, which includes skylights. Skylights are included in the glazing area even though they are not counted as part of the gross exterior wall area. This has the effect of further limiting vertical glazing areas on buildings when there is also a significant skylight area. One important exception to Section 1323 is that glazing on "the display side of street-level of retail" can be excluded from the glazing area calculation.

In all three of these codes, the limitation on fenestration area—or glazing area as it is called in the 2009 WSEC—essentially comes down to a limit on the area of *vision* glass. Fenestration area is all areas (including frame) in the building envelope that let in light. One can still design an all-glass building using the prescriptive path as long as no more than 40% of the gross wall area is vision glass and at least 60% is the cladding for the opaque portions of the envelope, such as spandrel glazing. The spandrel glass need not be opaque as long as it is the cladding on an opaque wall (such as in a curtain wall shadow box). For example, Figure 1 shows an all-glass building that has a combination of reflective and transparent vision glass (50% of the gross wall area) and reflective and opaque spandrel glass (also 50% of the gross wall area). Under the 2007/2009 prescriptive requirements, if this building envelope were designed today, it would not comply with current

Figure 1 *This building has a 100% glazed wall, but 50% of the wall area is vision glass (reflective and transparent) and 50% of the area is spandrel glass (reflective and opaque).*

energy codes. Yet, this building is closer to current typical building designs, which often have vision glazing areas in excess of 60%. With the new codes, buildings taking the performance or trade-off paths to compliance with the aim of increasing the area of vision glass more than 40% must now demonstrate that the effective U-factor of the opaque wall area—be it spandrel glass or some other opaque cladding systems—meets or exceeds the overall U-factor of the prescriptive approach. This is made even more challenging with the additional requirements to also account for heat loss due to thermal bridging. The amount of thermal bridging that must be accounted for varies by code. For example, the 2009 IECC states "the UA calculation shall be done using a method consistent with the *ASHRAE Handbook—Fundamentals* and shall include the thermal bridging effects of framing materials". The 2009 WSEC takes it even further when defining continuous insulation as "insulation that is continuous across all structural members without thermal bridges other than fasteners and service openings." As we will demonstrate in this case study, it can be difficult to significantly exceed the 40% limit on the vision glazing area using glazing systems that are commonly available on the market today.

IMPACT OF GLAZING ON ENERGY

Why does the energy code address thermal performance of glazing? Glazing performance is directly related to both heating and cooling loads in buildings. The thermal performance of glazing units and their framing system drive the majority of heat loss through the building enclosure. Furthermore, typical glazing assemblies significantly under-perform with respect to heat loss when compared to typical opaque wall assemblies, especially when considering the implementation of the codes' requirement for continuous insulation. Solar gain through vision areas is one of the three main sources of heat gains (the other two are lighting and people). Solar gains are a major component of cooling loads. Unmanaged heat gains can have a significant impact on the design of commercial buildings; therefore, unlike in residential building, the code specifies maximum solar heat gain coefficients (SHGC). For example, in the prescriptive residential requirements of 2009 IECC Section 402.1.1, there is no SHGC requirement in Climate Zones 4 to 8, and it is 0.3 in Climate Zones 1 to 3; whereas in the prescriptive commercial requirements, the SHGC requirements are more stringent in Climate Zones 1 to 3, and there are requirements in all climate zones. Although commonly available technologies such has low-emissivity coatings can mitigate solar heat gains, the impact on cooling loads is still significant when large amounts of glazing are used.

There is an optimum fenestration area that minimizes the energy consumption in buildings but it is building- and envelope-specific. And depending on occupancy, it may or may not be climate-specific. Determining the optimum area requires whole-building energy simulation (commonly called energy modeling). The key to finding the optimum is to realize that there are trade-offs to be made between synergistic but competing design objectives. According to Johnson (1983), in buildings with high internal heat loads, such as in typical commercial buildings, the relationships are:

1. Decrease window area or its solar transmission and cooling energy use is decreased; and
2. Increase window area or its daylight transmission and lighting energy use and associated heat gains are decreased.

The trade-offs apply to the building's perimeter, and are actually independent of climate. In both Chicago and Houston, the optimum fenestration area is 15% of the gross wall area for double-pane clear windows and daylighting controls. These windows have thermal transmittance (U-factor) 3.41 $W/m^2 \cdot K$ (0.60 $Btu/h \cdot ft^2 \cdot °F$), solar heat gain coefficient (SHGC) 0.6, and visible transmittance (VT) 0.63. The U-factor of fenestration products includes the effects of framing but not the interface between framing and adjacent constructions. In the same cities, the optimum fenestration area jumps to 45% for triple-pane, clear low-e windows with overhang and daylighting controls. These windows have U-1.14 $W/m^2 \cdot K$ (0.20 $Btu/h \cdot ft^2 \cdot °F$), SHGC 0.22, VT 0.37). However, it should be noted that these percentages were determined from energy simulation that ignored heat loss at thermal bridges. Therefore, we caution designers in applying these results blindly to all buildings. It would be instructive to reproduce the referenced study while also taking heat loss at thermal bridges into account.

The objectives of placing a limit on the maximum fenestration area are to minimize heat loss in winter and heat gains in summer. On building projects that are targeting a larger fenestration area, the limit will encourage designers to use new technologies to minimize heat loss and solar heat gain to trade-off the additional heat exchange though large fenestration areas.

As our case study will demonstrate, buildings designed with fenestration areas greater than 40% will require higher performing glazing in better thermally designed framing systems to achieve glazing U-factors significantly below the 2009 WSEC mandated value of 2.27 W/m^2·K (0.40 Btu/h·ft^2·°F). And designers will need to consider opaque wall assemblies that have much more thermal resistance than what is common today. However, the real challenge will be in achieving higher U-factors for spandrel glazing in all-glass buildings. As we will demonstrate in this case study, the inherent thermal bridging of the framing system can significantly impact the performance of the opaque wall assembly in these buildings, especially when the three-dimensional (3D) heat loss between the vision glazing and spandrel glazing is considered.

CASE STUDY ON MAXIMIZING FENESTRATION AREA

The case-study building is a multi-family high-rise residential building located in western Washington. It has an all-glass custom-designed curtain wall, five levels of below-grade parking, two levels of retail space, and 19 residential floors as shown in Figure 2. The gross wall area is 7786 m^2 (83,809 ft^2). As Figure 2 shows, the design goal was to maximize vision glazing areas (upwards of 80%) to take advantage of abundant views of the region's natural beauty of forests, lakes and sound, mountain ranges, and snow-capped volcanoes, and to appeal to consumer demand for dramatic floor-to-ceiling windows.

Starting with a code-matching building that has 40% vision glazing and opaque walls with R-3.3 m^2·K/W (R-19 h·ft^2·°F/Btu) batt insulation and R-1.5 (R-8.5) continuous insulation, the target heat loss rate (UA) is 7908 W/K (14,991 Btu/h·°F). Of this, 1502 W/K (2841 Btu/h·°F) is for the opaque envelope, and 6406 W/K (12,150 Btu/h·°F) is for fenestration. The relative UA of the fenestration compared to the opaque portion of the envelope shows that 81% of the heat loss is through the fenestration (6406/7908 × 100% = 81%). That leaves a very small percentage of the opaque wall heat loss (19%) that can be used to offset additional fenestration heat loss from a larger fenestration area. And even a better fenestration U-factor will not change the fact that most of the heat loss is through the fenestration.

At this point the developer was confident that a fenestration U-factor of 1.99 W/m^2·K (0.35 Btu/h·ft^2·°F) would be attainable with the custom-designed curtain wall. This U-factor is a weighted average that includes the vision glazing and the framing for the vision glazing. Note that this glazing system already has a much better U-factor than the code maximum U-factor of 2.27 W/m^2·K (0.40 Btu/h·ft^2·°F). By incrementally increasing the thermal performance of the glazed spandrel assembly (expressed as 1/U, or the overall effective R-value), the authors calculated the maximum allowable fenestration area for a given opaque wall thermal performance as shown in Figure 3. For example, to get to a 50% fenestration area, the glazed spandrel assembly would have to have an over-

Figure 2 *The case-study building is an all-glass high rise-residential building.*

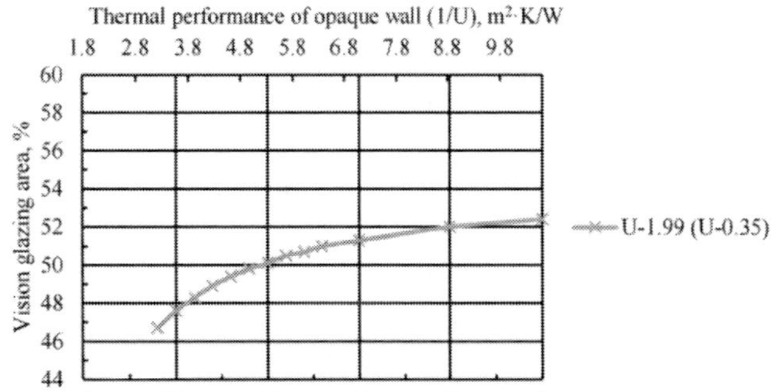

Figure 3 *For a given fenestration U-factor of 1.99 W/m^2·K (0.35 Btu/h·ft^2·°F), the maximum allowable fenestration area increases as the overall effective R-value of the opaque wall (that is, the glazed spandrel assembly) increases but it has diminishing returns after about 51% of the gross wall area.*

all effective R-value of about R-5.3 $m^2 \cdot K/W$ (R-30 $h \cdot ft^2 \cdot °F/Btu$). With 50% fenestration, the proposed UA is 7894 W/K (14,964 Btu/h·°F). Of this, 904 W/K (1714 Btu/h·°F) is for the opaque envelope, and 6990 W/K (13,250 Btu/h·°F) is for fenestration. Thus, the relative UA of the fenestration compared to the opaque portion of the envelope shows that now 89% of the heat loss is through the fenestration (compared to 81% for a code-matching building). However, the architect felt that 50% glazing would not meet their design goal for large expanses of floor-to-ceiling vision glass, so we explored other options for increasing the fenestration area.

Taking it One Step Further

The next step was to look at how improving the fenestration U-factor could help increase the amount of allowable vision glass. Using the same incremental procedure described above, the authors calculated the maximum allowable fenestration area for fenestration U-factors ranging from 1.76 to 1.99 $W/m^2 \cdot K$ (0.31 to 0.35 $Btu/h \cdot ft^2 \cdot °F$) in 0.06 $W/m^2 \cdot K$ (0.01 $Btu/h \cdot ft^2 \cdot °F$) increments. The resulting required thermal performance of the glazed spandrel assembly for a given fenestration U-factor is shown in Figure 4. The results show that very significant improvements to thermal performance of the fenestration system would be required to increase the maximum allowable fenestration area. But even with a U-factor of 1.76 $W/m^2 \cdot K$ (0.31 $Btu/h \cdot ft^2 \cdot °F$) and an overall effective R-value of R-5.3 $m^2 \cdot K/W$ (R-30 $h \cdot ft^2 \cdot °F/Btu$), the maximum allowable fenestration area is only about 57%. But based on the project's budget and technical constraints, the design team determined that a fenestration U-factor of 0.35 and overall effective R-value of the glazed spandrel assembly of R-33 was the most realistic choice. This allowed the project to have a fenestration area of 51%—a far cry from the architect's original vision of upwards of 80% vision glazing. The glazed spandrel assembly is shown in Figure 5. Two-dimensional (2D) thermal modeling (THERM 2012) was used to determine the U-factor of the curtain wall. The curtain wall is thermally broken. The thermal model accounts for 2D thermal bridging through mullions and steel studs in the wall. In order to maximize the amount of floor-to-ceiling vision glass, the designer chose to orient the spandrel panels vertically instead of horizontally as shown in Figure 6.

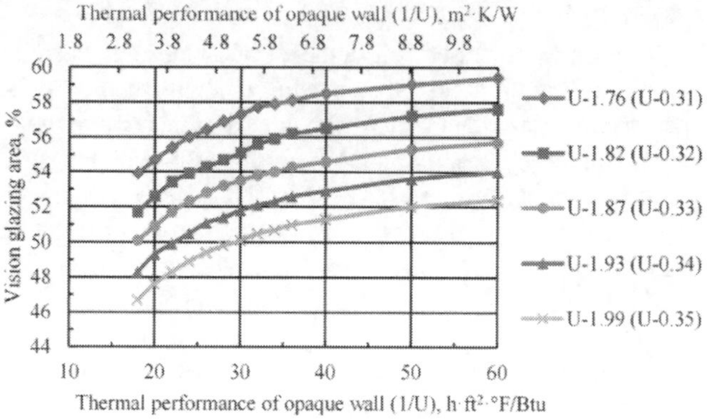

Figure 4 *Maximum allowable fenestration area for a given thermal performance of the opaque wall (that is, the glazed spandrel assembly) assuming fenestration U-factors in the range of 1.99 to 1.76 $W/m^2 \cdot K$ (0.35 to 0.31 $Btu/h \cdot ft^2 \cdot °F$).*

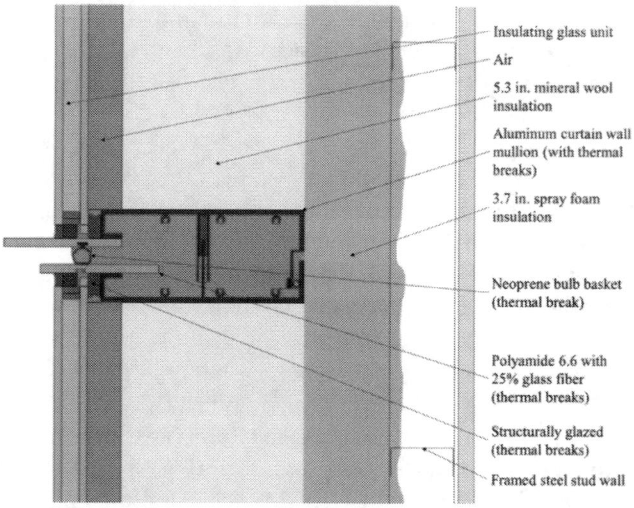

Figure 5 *The glazed spandrel assembly has a 2D overall effective R-value (1/U) of 5.8 $m^2 \cdot K/W$ (33 $h \cdot ft^2 \cdot °F/Btu$). The figure shows a horizontal cross-section through a vertical mullion.*

ENERGY CODES AND SUSTAINABILITY RATING SYSTEMS

Designing for higher glazing ratio under the new code gets significantly more challenging for project pursuing sustainability ratings. To earn more than three points under LEED Energy and Atmosphere Credit 1 to optimize energy performance (LEED 2012) the design team must demonstrate through whole-building energy simulation that the energy cost performance of their proposed design exceeds the baseline design. The baseline design is a building meeting ASHRAE Standard 90.1-2007. The challenge lies in the fact that the bar has been raised across all building systems including energy using systems, energy conversion equipment, and building envelope components. The baseline design is already an energy-efficient building. Where once the energy performance savings from HVAC and lighting systems could be counted upon to make up for shortfalls in the performance of glazing systems, now the new baseline has started to close the gap on the performance expectation of these systems. This means that unless a project makes use of unusual measures for these systems or incorporates some aspect of site generated energy, there are no longer sufficient energy savings to allow for a significant increase in glazing area. Going forward on most projects the building envelope will have to be at least code neutral from an energy performance perspective in order for any energy savings from the HVAC and lighting systems to be used to demonstrate above code performance.

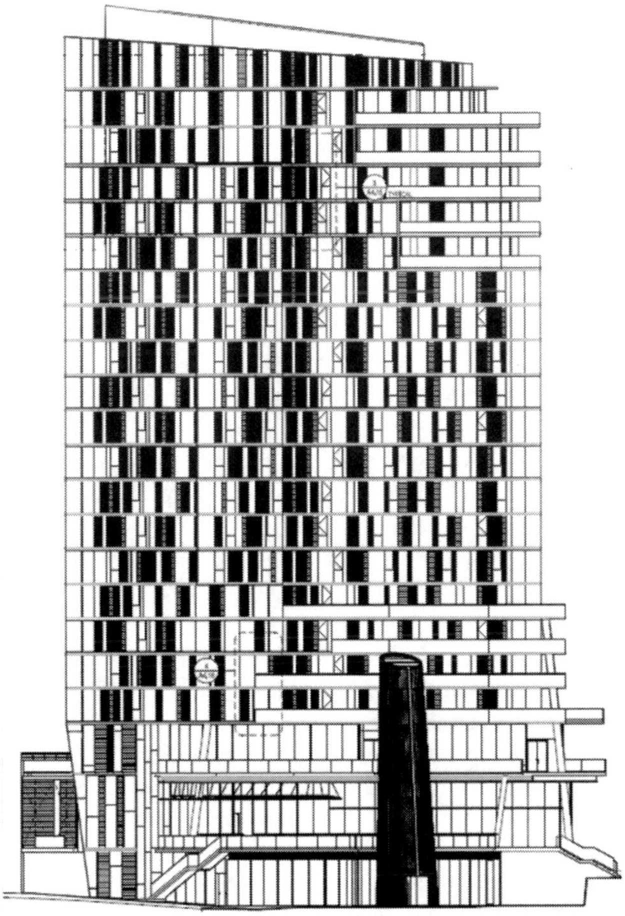

Figure 6 *The final design will consist of a combination of vertical vision and opaque glazing to meet the calculated maximum fenestration area of 51%.*

In a second case-study building, again a high-rise residential building located in the same jurisdiction as the first case study, this time the goal was to achieve a LEED Gold rating. It is a window-wall system with a combination of metal and glass spandrel panels. The building has four levels of below-grade parking, one level of retail space at grade, 11 residential floors, and roof-level amenity spaces. Initially starting with a glazing ratio in excess of 60%, the design was revised to a code-matching ratio of 40% vision glazing. The fixed glazed portions of the window wall were better than the code prescriptive requirement: they had a U-factor of 1.82 W/m^2·K (0.32 Btu/h·ft^2·°F) based on a thermally broken aluminum framing system and low-e glazing with warm edge spacers and argon filled cavity. The operable windows did not meet the 2009 WSEC requirements: they had a U-factor of 2.38 W/m^2·K (0.42 Btu/h·ft^2·°F). The opaque spandrel areas had a better than code U-factor of 0.27 W/m^2·K (0.048 Btu/h·ft^2·°F) based on an insulated backpan design with 100 cm (4 in.) of spray foam insulation applied across the interior face of the system. The U-factor of insulated slab-edge covers was 0.50 W/m^2·K (0.089 Btu/h·ft^2·°F). The mechanical and lighting systems included such high performance features as high-efficiency through-wall heat pumps in each unit, high-efficiency domestic hot water heaters, domestic hot water heat recovery, and occupancy sensor-controlled lighting in parking garage and stairwell. Despite all of these energy conservation measures, whole-building energy simulation demonstrated that the energy cost performance of their proposed design bettered the baseline design by only 17.2% (worth three EAc1 points). Therefore, the design achieved the Energy and Atmosphere Prerequisite Credit 2 for minimum energy performance of 10% improvement above the baseline (that is, two EAc1 points) and one additional EAc1 point. But to achieve LEED Gold the project was going to have to look at securing credits from the other categories. As it was, the project was already over budget, and the developer conceded and accepted that a LEED Silver rating was a more achievable goal given the project constraints.

ENERGY CODES AND REALITY

The first case study above was based on overall effective R-values and accounted for thermal bridging as required by the energy code. Standard practice in North America to account for thermal bridging within the building envelope is to consider thermal bridging within an assembly, for example a steel stud wall, but to ignore thermal bridging at architectural and structural details—including interfaces—where walls, windows, floors, and roofs come together. Whole-building energy modeling procedures for performance-based compliance in energy codes and standards are either silent on thermal bridges relating to details and transitions (such as slab edges, shelf angles, and sheet metal flashings), they allow these thermal bridges to be ignored through partial or full exemptions, or the procedures reduce the apparent significance of thermal bridges through oversimplification. The reasons for these omissions appear to be based on:

1. The belief that details do not have a significant impact on the overall building envelope performance and on whole building energy use because they comprise a small area compared to the total envelope area.
2. Past experience that shows it would take too much effort to quantify all thermal bridges, which often have complex three dimensional (3D) heat flow paths.
3. The lack of comprehensive thermal transmittance data for standard details.

However, recent work accounting for 3D heat flow through details (Morrison Hershfield Ltd 2011) has shown that the overall performance of many common wall assemblies is much less than what is currently assumed by many practitioners. Irrespective of the small areas of highly conductive materials that bypass thermal insulation, the effect on overall energy consumption is significant, and simple changes to assembly design may be more effective at reducing energy use than adding more insulation. In addition, accounting for these details is now easier because straightforward procedures to quantify the impact of common details have been developed and thermal transmittance data for standard details are now readily available in a catalogue published by ASHRAE (Morrison Hershfield LTD 2011). Realistic expectations of building envelope performance are necessary to make informed decisions related to building energy efficiency.

3D heat loss through curtain wall systems is very significant and should not be ignored. For example, using the results of 3D thermal modeling, such as that shown in Figure 7, installed insulation with a nominal R-value of R-5.8 (R-33) results in an assembly with an overall effective R-value of about R-1.6 (R-9). In fact, due to heat loss through exposed vertical and horizontal mullions in vision areas down to the opaque spandrel areas, at the intersection of vertical and horizontal mullions, and at curtain wall anchors, there is a diminishing return on the effectiveness of installed insulation as shown by the results in Figure 8. With conventional materials and a typical good thermally broken curtain wall system, the overall effective R-value will be in the range of R-0.9 to R-1.6 (R-5 to R-9). However, with new materials such as vacuum insulated panels—about R-7 (R-40) per inch—and non-metal curtain wall framing, the authors are beginning to see these limitations be exceeded.

THE NEED FOR ALTERNATIVE ASSEMBLIES

As demonstrated by the two case studies, glazing ratios in the range of 50% (that is, ten percentage points above the code prescriptive maximum of 40%) can be achieved using the current nominally thermally-broken frames and good quality dual glazed units combined with highly insulated opaque wall assemblies. However, those projects seeking even higher glazing areas or those seeking energy savings exceeding the code baseline will have to turn to higher performing assemblies.

There are existing high performance technologies on the market that have been used on projects where extreme climates or extreme interior environmental conditions dictated a high performance system. These include such technologies as high performance thermal breaks for aluminum frames, triple glazed units including the use of suspended films, as well as double low-e coatings in and on glazing units. Combined, these technologies can deliver a U-factor below 1.70 W/m^2·K (0.30 Btu/h·ft^2·°F). To date the cost premiums associated with these assemblies has limited their applications. With the demands of the new energy code we should start to see increasing demand and a general trend towards a greater commoditization of existing high performance technologies

In addition, there are a number of promising emerging technologies that are being adapted to significantly improve the thermal performance of glazing systems both in terms of vision and opaque assemblies. Some of these include alternate framing material such as fiberglass: traditionally reserved for the low-rise residential market now being developed for more high-rise commercial construction. Likewise there are alternate glazing assemblies such as translucent fiberglass or polycarbonate panels insulated with nanogels to provide increased daylighting without compromising thermal performance. Vacuum insulating technologies are being applied to both vision glazing assemblies as well as spandrel assemblies to significantly improve their thermal performance. Electrochromic and photochromic glass can be used to significantly reduce solar heat gain beyond what can be

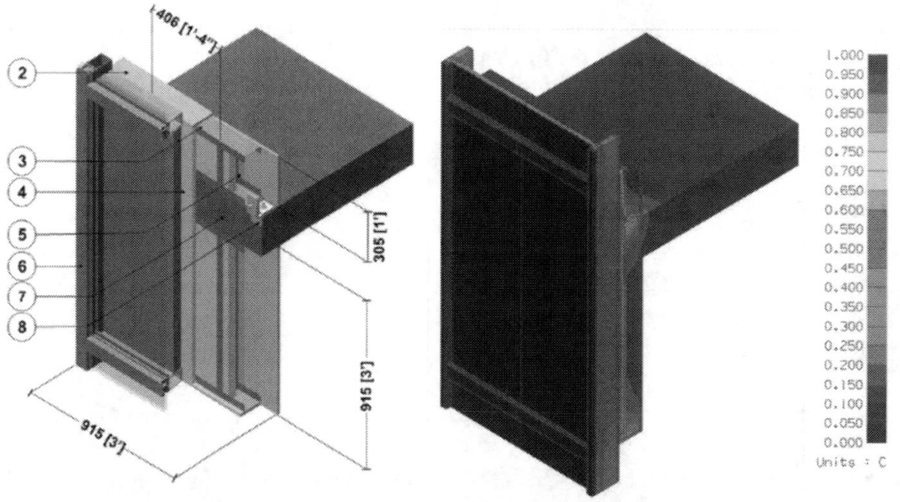

Figure 7 *On the left, a typical opaque spandrel area of curtain wall with insulation in the backpan and spray foam insulation applied to the inside face of the backpan (through the steel framed wall). On the right, the same detail showing the typical pattern of temperature distribution (normalized temperature index) as a result of three-dimensional heat loss (Morrison Hershfield LTD 2011).*

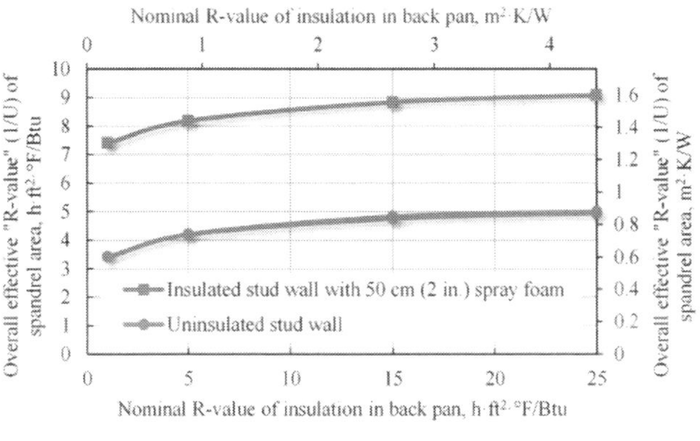

Figure 8 *Overall effective R-value (1/U) of spandrel areas of curtain wall with a range of insulation in the backpan and either no insulation in the stud wall cavity or 2 in. of spray foam insulation in the stud wall cavity applied to the inside surface of the backpan (Morrison Hershfield LTD 2011).*

achieved with fixed shading devices. Lastly, innovative integrated photovoltaic systems are generating electricity on-site that can significantly impact the energy cost budget for a project. Again, the demands of the new energy code should drive a general trend towards an accelerated commercialization of these emerging technologies

Returning to our original case study demonstrates the impact of some of these technologies. Assuming the project budget could have afforded a triple glazed system with high performance thermal breaks U-factor of 1.42 $W/m^2 \cdot K$ (0.25 $Btu/h \cdot ft^2 \cdot °F$) compared to the code maximum U-factor of 2.27 $W/m^2 \cdot K$ (0.40 $Btu/h \cdot ft^2 \cdot °F$), we determined a maximum allowable glazing ratio of 72% would have been attainable with the same R-5.8 (R-33) spandrel assembly. This demonstrates that improving the thermal performance of the glazing system has the single biggest impact for increasing the glazing ratio above the prescriptive requirement.

Alternatively, using the case study's glazing assembly with a U-factor of 1.99 W/m^2 K (0.35 $Btu/h \cdot ft^2 \cdot °F$) and introducing a vacuum insulated spandrel panel in the glazing pocket of the spandrel area would have replaced the backpan insulation and stud cavity insulation. Alternatively it could have raised the opaque wall value to R-12.3 $m^2 \cdot K/W$ (R-70 $Btu/h \cdot ft^2 \cdot °F$) allowing a glazing ratio of 54%. This demonstrates that the super insulated spandrel assembly developed for the case study can be more efficiently achieved; however, even with highly insulating materials, increasing the thermal performance of the opaque wall has limited impact in terms of increasing the glazing ratios.

Finally, one must consider what is the goal in increasing the glazing ratio? If increase daylighting potential is the goal, there may be more efficient ways to achieve it. Combining the case study's glazing assembly with a U-factor of 1.99 $W/m^2 \cdot K$ (0.35 $Btu/h \cdot ft^2 \cdot °F$) in a 60/40 split with a translucent insulated fiberglass panels insulated with nanogel infill with a U-factor of 0.11 $W/m^2 \cdot K$ (0.02 $Btu/h \cdot ft^2 \cdot °F$) would have allowed an 82% glazing ratio with the same opaque wall R-value of 5.8 $m^2 \cdot K/W$(33 $h \cdot ft^2 \cdot °F/Btu$).

CONCLUSION: THE FUTURE OF ENERGY CODES

As challenging as the maximum fenestration area is, it will be even more challenging once the new 2010/2012 codes are adopted. Both ASHRAE 90.1-2010 and the 2012 IECC have decreased the maximum fenestration area to 30% of the gross wall area. As states adopt these codes, getting beyond 30% will be even more challenging than getting past 40% with the current code. Therefore, it will be even harder to meet code if one wants to also exceed maximum fenestration areas using commonly available systems that are on the market today. We expect that these stricter requirements in the energy code will eventually trickle down to LEED. One of the proposed changes for the next version of LEED for New Construction is to reference ASHRAE Standard 90.1-2010 as the baseline design (LEED 2013). With a maximum fenestration area of 30% in the baseline design, the total energy budget that the design team has to work with becomes even less. Whether it is to meet code or to earn points under Energy and Atmosphere to optimize energy performance, exceeding 30% fenestration area will require even higher performing envelope systems or greater energy savings in other areas to trade-off.

The focus on continuously improving the energy efficiency of new building through energy codes will drive a demand for higher performance glazing and better thermal performing frames. It will also provide an incentive for emerging technologies and other innovative applications that can help improve the thermal performance of glazing systems. At the same time, it will also bring a closer examination of the justifications for increased applications of vision glass on projects. Increasingly designers will be required to demonstrate that increasing the fenestration area will add value to a project. The 2012 IECC will have an exception to the glazing ratio for buildings where 50% or more of the floor plate benefits from daylighting. In turn this will drive a need to address the benefits and impacts of increased glazing early in the conceptual design phase of the projects. Early collaboration between the design architects, mechanical engineers, and building envelope consultants will be even more crucial on these projects.

REFERENCES

ASHRAE. 2007. *ANSI/ASHRAE/IESNA Standard 90.1-2007, Energy Standard for Buildings except Low-Rise Residential Buildings*. Atlanta: ASHRAE.

ASHRAE. 2010. *ANSI/ASHRAE/IESNA Standard 90.1-2010, Energy Standard for Buildings except Low-Rise Residential Buildings*. Atlanta: ASHRAE.

CFR 10 Part 433. 2012. Code of federal regulations, Energy efficiency standards for the design and construction of new federal commercial and multi-family high rise residential buildings, Title 10, CFR Part 433. Washington: U.S. National Archives and Records Administration. http://www.ecfr.gov/. Retrieved 19 February 2013.

DOE. 2013. Status of state energy code adoption maps, U.S. Department of Energy Building Energy Codes Program, http://www.energycodes.gov/status-state-energy-code-adoption. Retrieved 21 February 2013.

IECC. 2009. *International Energy Conservation Code*. Washington: International Code Council.

IECC. 2012. *International Energy Conservation Code*. Washington: International Code Council.

IMF. 2012. *World economic outlook: a survey by the staff of the International Monetary Fund*. Washington: International Monetary Fund. http://www.imf.org/external/pubs/ft/weo/2012/01/pdf/text.pdf. Retrieved 19 February 2013.

Johnson, R., R. Sullivan, S. Nozaki, S, Selkowitz, C. Conner, and D. Arasteh. 1983. *Building Envelope Thermal and Daylighting Analysis in Support of Recommendations to Update ASHRAE/IES Standard 90.1*. Richland, WA: Battelle Pacific Northwest Laboratories.

LBNL. 2012. THERM Finite Element Simulator, Version 6.3.45. Regents of the University of California. Developed and maintained by Lawrence Berkeley National Laboratory, Berkely, CA.

LEED. 2012. *LEED 2009 for New Construction and Major Renovations*. Washington: U.S. Green Building Council.

LEED. 2013. *LEED v4 for New Construction–draft*. Washington: United Stated Green Building Council. http://new.usgbc.org/credits/new-construction/v4-draft. Retrieved 2013 February 21.

Morrison Hershfield Ltd. 2011. *Thermal Performance of Building Envelope Construction Details for Mid- and High-Rise Buildings*. ASHRAE RP-1365. Atlanta: ASHRAE.

WSEC. 2009. *2009 Washington State Energy Code*. Olympia: Washington State Building Code Council.

Thermal Bridges of Modern Windows—Experience with Minimum Acceptable Surface Temperature

Ernst Jan de Place Hansen, PhD **Eva B. Møller, PhD** **Anker Nielsen, PhD** *Member ASHRAE* **Erik Brandt**

ABSTRACT

With its focus on reduced energy consumption, contemporary housing construction requires a highly insulated and airtight building envelope with as few thermal bridges as possible. Windows must be carefully designed, as thermal bridges can lead to surface condensation or mold growth, even if the window has an U-factor of 1 W/(m^2·K) or lower.

This paper describes the development of modern, energy efficient Danish windows with reduced thermal bridges. It focuses on materials, geometry, and sealing of window panes based on a literature review. Examples of modern windows are presented. Experience with the minimum acceptable surface temperature regarding surface condensation or mold growth, implemented in the Danish Building Regulations in 2010 (Danish Enterprise and Construction Authority 2010), and the calculation method for this temperature based on international standards is discussed.

The introduction of the minimum acceptable surface temperature has been an important driver for the development of new window solutions in Denmark, increasing the inner-surface temperature at the sealing of window panes. However, it will not stop complaints from consumers, as this temperature is calculated under standardized conditions. Increasing requirements for airtightness increases the importance of sufficient ventilation in order to prevent problems with condensation or mold growth.

INTRODUCTION

Contemporary housing construction focuses on reducing energy consumption. This means that the building envelope has to be highly insulated and airtight with as few thermal bridges as possible. The surface temperature becomes lower at thermal bridges than at surrounding surfaces. Indoor air that gets in contact with the surface at the thermal bridge will be cooled down resulting in a higher relative humidity. This means that the relative humidity is higher at thermal bridges than at other surfaces.

Windows are normally the coldest area in a room and previously the presence of dew on windows was used as an indication that the room was insufficiently ventilated. The pane of traditional double-glazed windows or coupled sashes was previously the coldest area of a window. However, the corners of the sash and the edge of the glazing are normally the coldest parts of windows with low-e glazing (Brandt 2013).

To minimize the thermal bridges, windows must be carefully designed. Otherwise it will lead to moisture-related problems such as condensation, degradation, or mold growth even if the window has a total U-factor of 1 W/(m^2·K) or lower.

Condensation occurs when the surface temperature is lower than the dew-point temperature, which is given by the moisture content in the ambient air. Even at higher temperatures, there might still be a risk of moisture problems, as molds are known to grow at a relative humidity as low as 75%, depending on the temperature and material (Sedlbauer 2001). Therefore, to prevent mold growth, a critical surface temperature may not only be based on the risk of condensation (100% rh) but also based on a lower moisture level, e.g., the minimum acceptable surface temperature should be higher than the dew-point temperature.

The present requirement for U-factors for windows means that some of the problems that previously were

***Ernst Jan de Place Hansen, Eva B. Møller,** and **Erik Brandt** are senior researchers and **Anker Nielsen** is a professor in the Department of Construction and Health at the Danish Building Research Institute/Aalborg University, Copenhagen, Denmark.*

common are not found in contemporary constructions (Brandt 2013). For instance, condensation is very rare on the center of modern glazing with a U-factor of 1 W/(m^2·K) or lower. However, windows should not be characterized by energy performance alone. Acceptable moisture levels must be ensured as well. The paper discusses the minimum acceptable surface temperature as a parameter and the method for calculating this based on the building legislation and experience gained with windows in Denmark. The minimum acceptable surface temperature approach is relevant for a major part of Europe situated in the same warm temperate, fully humid climate zone, defined by the Köppen-Geiger climate classification (Kottek et al. 2006).

The development of modern, energy-efficient windows in Denmark is described, highlighting the necessary choices that had to be made concerning materials and design. Examples of modern windows are shown. Then the energy and moisture-related requirements for windows in the Danish Building Regulations (DBR) are described, as these have been an important driver for the development of windows. The moisture-related requirements are based on an international standard for calculating the minimum acceptable surface temperature for mold growth or surface condensation, which is briefly described (ISO 2012b). Examples of using the standard in a Danish context are given as a background for discussing the method itself and its practical consequences. The choice of level for the minimum acceptable surface temperature is also discussed.

DEVELOPMENT OF ENERGY-EFFICIENT WINDOWS

A window is a complex construction where different parts all contribute to the total U-factor: the glazing, the sash/frame construction, and the edge sealing. Since 2000, research conducted at the Technical University of Denmark, financed by the Danish Energy Agency, has aimed at inspiring and helping the Danish window industry and it has been an important background for the development of modern, energy-efficient windows. Development was necessary to comply with the gradually-tightening requirements for the energy performance of windows and recently also the requirement concerning the avoidance of surface condensation as presented in the section "Requirements for Windows in the DBR."

U-factor for Glazing

As glazing constitutes the major part of a window area, the first efforts at reducing the U-factor of windows focused on reducing the U-factor of the glazing by increasing the number of layers of glazing, by filling the space between panes with gas instead of air or by applying low-emission coating. Table 1 shows some typical U-factors for different types of glazing (Nielsen 1991).

While condensation on the inside of the glazing was common on windows with a high U-factor, improved U-factors of the glazing have resulted in condensation on the outside instead. This condensation can occur on clear nights with humid air (above 85% rh), usually in the spring and autumn, and in locations where the window has an unobstructed view of the sky. Since the outbound heat transport through the window is small, the outer pane thus becomes colder than the surrounding air, and condensation will form. The condensation starts to form at the bottom of the pane (the coldest part), and in some cases it spreads over the whole pane (Persson et al. 1994).

Table 1. U-factors for Different Types of Glazing, All with 4 mm Glass and 12 mm Distance between the Layers of Glazing, Except when Noted (Nielsen 1991)

Type of Glazing	Type of Glass	U-factor (Center) W/(m^2·K)
Double glazing	Clear pane	2.9
Double Low-e glazing	Same + pane with a Low-e coating	1.9
Aerogel 20	Two clear panes with 24 mm distance, space filled with Aerogel	0.6
Aerogel 30	Same with 34 mm distance	0.4

Materials for Sash/Frame Construction

Improvement of the U-factor of glazing resulting in U-factors below 1 W/(m^2·K) meant that it became important to reduce the U-factors of the sash/frame construction. Development of new constructions has taken place for several years involving different kinds of materials: wood (Finland, Sweden, and Denmark), aluminum (e.g., Denmark), polyvinyl chloride (PVC) (Germany), or glass-fiber reinforced polyester (GFRP) (USA and Canada) (BYG DTU 2009a; 2009b). Or a combination, e.g., where a traditional inner wood part is combined with a weather-resistant outer part made of aluminum or PVC as shown in Figure 1.

Constructions made of aluminum, PVC or GFRP can be produced with a lot of cavities. Thereby the thermal conductivity can be reduced depending on the number and size of the cavities and whether the cavities are empty or filled with insulating material, etc. GFRP is stronger than PVC and this avoids the use of steel as reinforcement. Aluminum is strong and light, but it has a very high thermal conductivity (see Table 2). This makes it difficult to make sash/frame profiles of aluminum with a low U-factor without thermal interruption and the only real solution is to use another material, e.g., GFRP.

Design of Sash/Frame Construction

Reduction of the U-factor of the sash/frame construction often resulted in very large dimensions of the sash/frame part of the window. This in turn reduced the amount of solar radiation coming through the window and eventually reduced the

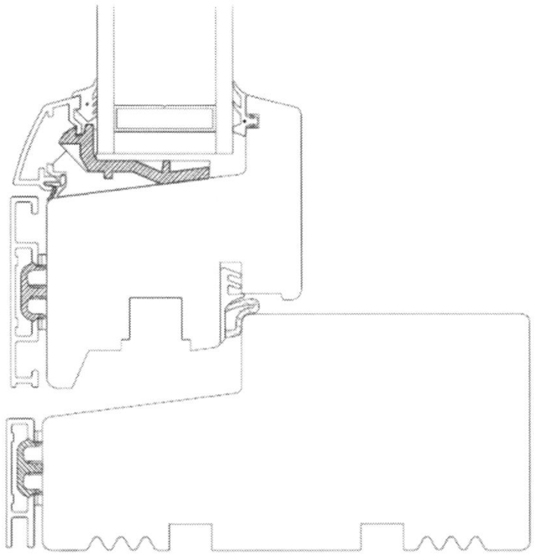

Figure 1 *Wood sash/frame construction with an outer cover of aluminum (BYG DTU 2001). External side to the left.*

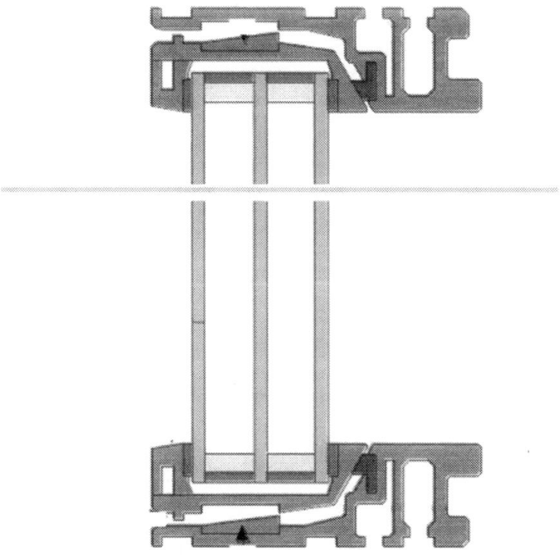

Figure 2 *Detail from a window with a sash/frame construction of GFRP, designed to optimize the glass area of the window,* $U_w = 0.78\ W/(m^2{\cdot}K)$, $E_{ref} = 27\ kWh/m^2$ *(Kragh and Svendsen 2009). External side to the left.*

Table 2. Thermal Conductivity of Different Materials Used for Windows

Aluminum	Stainless Steel	PVC	GFRP
220 W/(m·K)	15 W/(m·K)	0.2 W/(m·K)	0.2–0.4 W/(m·K)

energy performance of the window, expressed by the net-energy gain, E_{ref}.

Therefore, attempts were made to develop windows with a reduced sash/frame ratio without increasing the thermal bridge at the edge sealing (BYG DTU 2009b; Kragh and Svendsen 2009). Such solutions are primarily relevant for dwellings and not for offices, etc., where there is a risk of overheating the building. Figure 2 shows a solution with a sash/frame construction made of GFRP with better energy performance than traditional Danish products with a net-energy gain of –60 to –40 kWh/m^2 (Kragh and Svendsen 2009; Applefield et al. 2010).

For a simpler solution based on traditional sash/frame constructions, it is suggested to place the sash construction on the external side of the frame construction (BYG DTU 2009b). Solutions with a sash construction on the external side of the frame construction (as shown in Figure 3) are produced for the Danish market by a number of manufacturers.

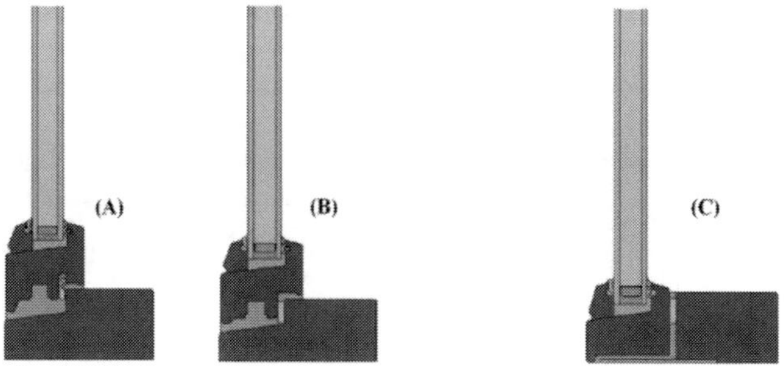

Figure 3 *A traditional design of (A) a wood sash/frame construction, (B) a slimmer profile and (C) a design where the sash is placed on the external side of the frame (BYG DTU 2009b). External side to the left.*

Design of the Edge Sealing/Construction

Concurrently with the development of new designs of the sash/frame construction, work has been done to reduce the linear-thermal transmittance (or equivalent thermal conductivity) of the edge construction between the panes and the sash. The equivalent thermal conductivity of different edge constructions is given in Table 3. The edge constructions consist of a spacer that provides the construction with a specified stiffness and sealing materials (see Figure 4).

The total U-factor of the window can be reduced by up to 10% by using a spacer with a low equivalent thermal conductivity instead of using a profile made of aluminum. The improvement of a specific window depends on the size of the window, the design of the sash/frame construction, the type of glazing, etc.

Apart from improving the U-factor of the window, the choice of material for the spacer also affects the risk of condensation, as it affects the inner-surface temperature. The lowest inner-surface temperature of a traditional double-glazed window with a wood sash/frame construction, 15 mm distance between layers, and a spacer made of aluminum, is 8.2°C (BYG DTU 2009a), when the external and internal

Table 3. Equivalent Thermal Conductivity of Edge Constructions Depending on the Material Used for the Spacer in Windows with Double Glazing and 15 mm Distance between Layers (BYG DTU 2009a)

0.4 mm Aluminum Profile	0.2 mm Stainless Steel Profile	1 mm PVC Profile	Massive Polyisobutylene
2.6 W/m·K	0.7 W/m·K	0.5 W/m·K	0.3 W/m·K

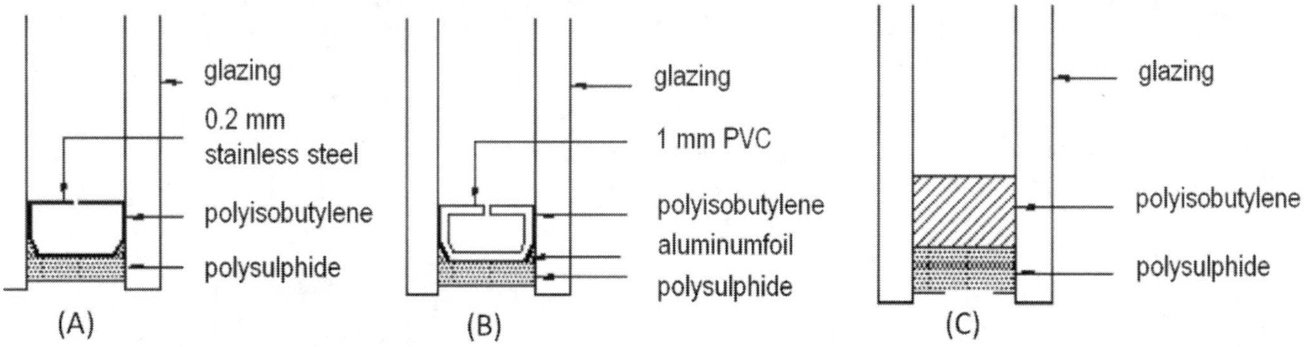

Figure 4 *Design of edge constructions for a window with double glazing. The spacer made of 0.4 mm aluminum, (A) 0.2 mm stainless steel or (B) 1 mm PVC is attached to the glass with a 0.25 mm polyisobutylene sealant and 3 mm polysulphide sealant in the bottom for extra mechanical stability. (C) Solution with a spacer of polyisobutylene (BYG DTU 2001; 2009a).*

Table 4. Lowest Inner-surface Temperature for Windows with a Wood Sash/Frame Construction and 12 mm Distance between Layers of Glazing, External and Internal Temperature –10°C and 20°C (Nielsen 1991)

Type of Glazing	θ_{si} Aluminum Spacer	θ_{si} Stainless Steel Spacer
Double glazing	4.3°C	7.0°C
Triple glazing	7.1°C	10.0°C
Low-e double glazing	5.2°C	8.7°C
Low-e triple glazing	8.8°C	11.7°C

temperature is 0°C and 20°C respectively. Lowest means that the inner-surface temperature refers to the location on inner-glass surface closest to the dew-point temperature of the indoor air. If the air change is 0.5 h^{-1} and the inhabitants produce 9.5 kg moisture per 24 h, the equivalent of 4–5 people, there is a risk of condensation 57 h/yr. By using a spacer made of stainless steel or PVC instead, there is a risk of condensation of only about 17 h and 11 h/y (BYG DTU 2001; 2009a). The number of hours with a risk of condensation should be as low as possible, but the result of condensation expressed as discoloring as well as rot and mold growth will of course depend on whether condensation takes place often, each time for a short period, or seldom, but for longer periods.

Table 4 shows the effect of using a spacer made of aluminum or stainless steel on the inner-surface temperature for traditional windows with a wood sash/frame construction and a 12 mm distance between the layers of glazing (Nielsen 1991).

The choice of spacer not only depends on its thermal conductivity. It is important that the spacer is tight and stays tight for the window's service life to prevent moisture from entering the construction or punctured glazing with resulting reduced insulating capacity. This is the reason for using a double-sealed unit, as shown in Figure 4. Other properties that are relevant: stiffness, ability to adhere to glass, that the spacer is easy to work on, and cost (Nielsen 1991). Altogether, this has for a long time made aluminum the best choice.

The durability of spacers made of aluminum is known to be good, while the durability of spacers made of PVC is unknown. The experience of spacers made of materials with a low-thermal conductivity has been gained over about 8 years. Today all windows for new buildings in Denmark have warm spacers made of stainless steel, polycarbonate, or a combination of stainless steel and polypropylene.

Examples of Modern Danish Windows

Table 5 shows examples of modern, openable Danish windows and their data that refer to the U-factor, the net-energy gain E_{ref}, and the lowest inner-surface temperature θ_{si} at 20°C (internal) and 0°C (external). Windows with similar properties for the U-factor and net-energy gain from other Danish manufacturers and from the USA, Canada, Germany, etc. can also be found. Windows on the Danish market seem to be the only ones with specifications for the inner-surface temperature. In the USA, the condensation resistance (CR) calculated according to the National Fenestration Rating Council is widely used by manufacturers as a relative indicator of the ability to resist the formation of condensation at a specific set of environmental conditions, however it is not mandatory (NFRC 2010).

Table 5. Examples of Modern, Openable Windows on the Danish Market. Layers of Glazing, Sash/Frame Material, U-factor, E_{ref}, Lowest Inner-surface Temperature (θ_{si}), and CR According to (NFRC 2010)

Type	Layers of Glazing	Sash/Frame, Material	U-factor W/(m²·K)	E_{ref} (2) kWh/m²	θ_{si} °C	CR
A1	2	Wood with aluminum cover (1)	1.36	–29	11.7	48
A2	3	Same	0.77	3	14.1	63
A3	3	Wood/GFRP	1.01	–17.2	13.5	59
B1 (3)	2	Wood/aluminum	1.6	–40	7.2	25
B2 (4)	2	Wood/aluminum with thermal interruption	1.4	–24	9.5	37
B3 (5)	2	Wood/GFRP	1.3	–17	10.3	41
C1	2	PVC	0.82	8	14.0	63
D1	3	Wood/PUR/alu	0.87	24	14.1	63

Notes: (1): Separated by a composite profile, (2): Existing requirement for net-energy gain in the DBR –33 kWh/m², (3): Marketed in 1985–2008, (4): Marketed since 2009, (5): Marketed since 2011.

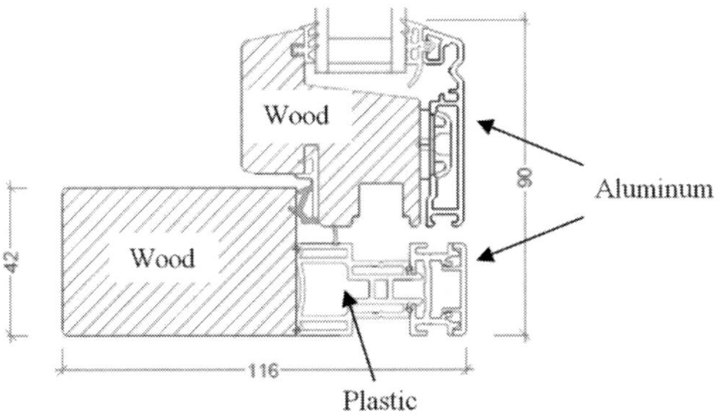

Figure 5 *Design of window type A1. External side to the right. Measures in mm.*

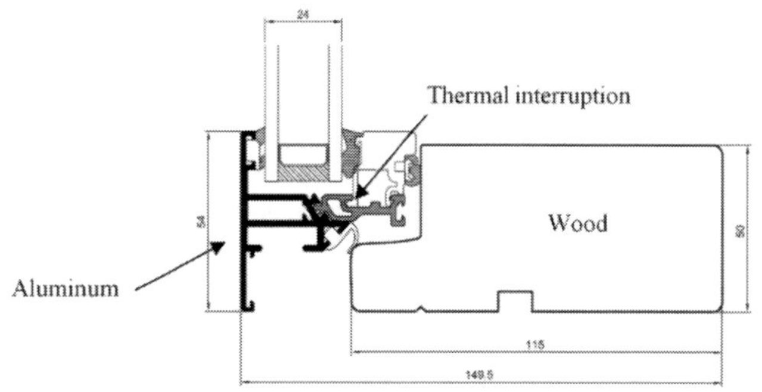

Figure 6 *Design of window type B1. External side to the left. Measures in mm.*

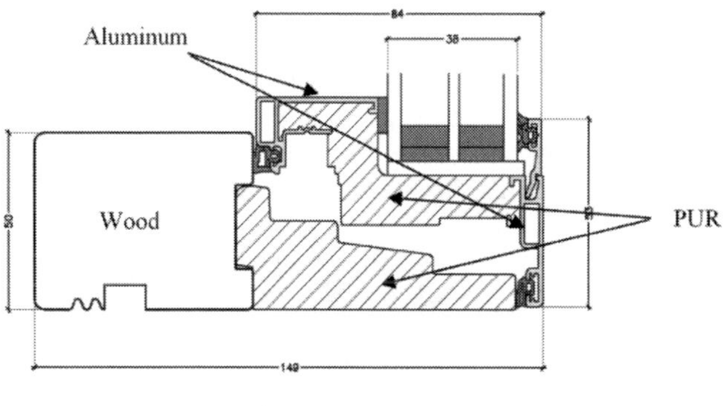

Figure 7 *Design of window type D1. External side to the right. Measures in mm.*

With one exception, (B1), all the windows in Table 5 have been marketed in Denmark since 2008 and later, and are produced with a warm spacer made of stainless steel, polycarbonate, or similar materials. Types A1–A3 are designed with a traditional mutual placement of the sash and frame construction, as shown in Figure 5. Types B1–B3 are designed with the sash construction placed on the external side of the frame construction, as shown in Figure 6, giving the optimal amount of solar energy.

Types C1 and D1 are examples of windows where the sash/frame construction is a combination of wood at the internal side and plastic (PVC or polyurethane [PUR]) at the external side, covered by a thin aluminium profile, as shown in Figure 7.

Type B1 resulted in condensation on the inner side of the window in many cases in Denmark, although the lowest inner-surface temperature of this window type is higher than 7°C, recommended since 2003 by the Association of Danish Window Manufacturers (2003). This shows that the recommendation was obviously too low. Type B1 also had a relatively high U-factor, which together with the low inner-surface temperature probably is the reason for removing it from the market.

Table 5 also shows CR for these windows. The calculation method described by the NFRC is based on 2D heat transfer simulation of the cross sections of the whole window and calculations are made separately for the frame, edge-of glazing and the center-of-glazing (2010). The resulting CR for the whole window is determined from the lower of these ratings. The values for Danish windows shown in Table 5 are the result of a simplified calculation based on the inner-surface temperature (θ_{si}). θ_{si} is adjusted as NFRC (2010) uses 21°C and –18°C as internal and external temperatures while θ_{si} is based on 20°C and 0°C.

As the inner-surface temperature in Table 5 represents the coldest spot on the window, a calculation of CR based on this value would underestimate CR compared with American windows, where the calculation is based on the conditions for the window as a whole. Instead a weighted calculation where 25% of the window (the lower part) is represented by the inner-surface temperature given in Table 5 and the other 75% of the window is represented by a 1°C higher temperature according to (BYG DTU 2001). It is assumed that the inner-surface temperature given in Table 5 represents the worst case, i.e., no separate calculations are made for frame, edge-of-glazing, or center-of-glazing.

The values for CR calculated for the Danish windows are comparable with values for windows in the USA, according to the NFRC Certified Products Directory (search.nfrc.org). As the calculation of CR for Danish windows is based on simplified conditions, no further discussion on specific differences in CR for windows with two or three layers of glazing is made in this paper.

REQUIREMENTS FOR WINDOWS IN THE DBR

Since 1966, minimum requirements for windows in new buildings were expressed by U-factor in the DBR (Ministry of Housing 1966) and they were gradually tightened from 3 W/(m^2·K) in 1966 to 1.5 W/(m^2·K) in 2008. It was also prescribed that thermal bridges in windows must be insignificant. A requirement for U-factor at the replacement of windows was introduced in 2006 (Danish Enterprise and Construction Authority 2006).

Although a window is considered to be energy efficient if it has a low U-factor, that is not sufficient to describe a window's energy performance, as it can have a very low U-factor, while more or less neglecting another important property, e.g., the amount of solar radiation through the glazing. Therefore, the minimum requirement for windows both in new buildings and at replacement has since 2010 been expressed by the net-energy gain through the window in the heating season: –33 kWh/m^2 per year for vertically-oriented windows (Danish Enterprise and Construction Authority 2010). The net-energy gain E_{ref} is based on the U-factor U_w and the solar energy transmittance of the window g_w expressed as

$$E_{ref} = I \cdot g_w - G \cdot U_w = 196.4 \cdot g_w - 90.36 \cdot U_w \qquad (1)$$

The solar heat gain I and the number of degree hours G during the heating season are determined on the basis of the Design Reference Year DRY (Lund 2001). The solar heat gain through windows depends on the orientation of the windows. A single-family house with the following distribution of windows is used as a point of reference: north 26%, south 31%, east/west 33%. The calculation uses an opening reference window 1.23 × 1.48 m.

The Danish standard DS 418 (Danish Standard Association 2011) on the calculation of the heat loss from buildings has been part of the Danish construction legislation since the first national Building Regulations was implemented (Ministry of Housing 1961) and it specifies how the U-factor for windows is calculated. In the newest version of the standard, it is stated that the calculation of the U-factor only takes place in instances where the manufacturer has not declared a value according to harmonized product standards (Danish Standard Association 2011). DS 418 shows a graph with U-factors for wood sashes and frames depending on the thickness of the sash or frame. Reference is made to EN ISO 10077-2 for a more detailed calculation of the U-factor for the sash/frame construction (2012a).

Moisture-Related Requirements

Together with the introduction of net-energy gain in 2010 in DBR, a requirement for the minimum acceptable surface temperature on the window frames was introduced for windows at replacement (9.3°C) (Danish Enterprise and Construction Authority 2010). For a calculation of the surface temperature, DBR10 refers to EN ISO 10077-2, although the standard only specifies the calculation of U-factors for the sash/frame construction and the linear-thermal transmittance for the spacer (2012a). But the surface temperature can at least be calculated when the U-factor is known.

Since 2008, DBR has stipulated that building structures and materials must not, on moving in, have a moisture content that is liable to increase the risk of mold growth (Danish Enterprise and Construction Authority 2008). This requirement minimizes the risk of moving into overly-damp buildings, both in new buildings and renovation projects and derives from the number of cases of mold growth in buildings in Denmark. The moisture content of a material that is in equilibrium at 75% rh is normally taken as the critical surface humidity regardless of surface quality, temperature, and duration of moisture load, as knowledge of the specific level of the critical moisture content for different materials is rather scarce (de Place Hansen 2012; Nevander and Elmarsson 2006; Sedlbauer 2001).

Then, when the indoor relative humidity based on knowledge about moisture production, air change, etc., is known in a specific case, it is possible to calculate the minimum inner-surface temperature to avoid critical situations, as described in the following section.

METHODS TO CALCULATE MINIMUM ACCEPTABLE SURFACE TEMPERATURE

A simplified and easily-accessible method for evaluation of temperature and moisture conditions in and around a building element is described in EN ISO 13788, Section 5 (2012b). Using this method, the risk of surface mold growth or surface condensation can be calculated. The method is described briefly in this section, followed by some examples in the next section.

For each month the following steps are to be followed:

1. Define the external air temperature;
2. Define the external humidity;
3. Define the internal temperature in accordance with national practice;
4. Calculate the internal relative humidity;
5. Calculate the minimum acceptable saturation humidity or vapor pressure;
6. Determine the minimum acceptable surface temperature, $\theta_{Si,min}$; and
7. Calculate the minimum temperature factor, $f_{Rsi,min}$ from the minimum acceptable surface temperature, the assumed internal temperature and external temperature.

Then it is possible to determine the month with the highest risk of mold growth or surface condensation, expressed by the highest required value of $f_{Rsi,min}$.

The standard describes different ways of calculating internal relative humidity (step 4) designing for avoidance of mold growth, based on internal humidity classes, a constant relative humidity, or a fixed or variable air-change rate and fixed moisture supply. The standard operates with 80% rh as the limit for mold growth unless more specific information is available. Another calculation based on constant external relative humidity (95% rh) and the mean annual minimum on a daily basis is to be used for design to prevent surface condensation on low thermal-inertia elements such as windows.

Internal Humidity Classes

It is generally considered that the moisture supply or internal humidity load, i.e., the difference between moisture content (in g/m^3) in the indoor and the outdoor air tends to be relatively constant in a house in the colder part of the heating season (external temperature below 0°C), while it decreases with increasing external temperature due to increased ventilation. EN ISO 13788 defines five humidity classes to describe the internal humidity load for different types of buildings, based on data from buildings in Western Europe (maritime climates), see Figure 8 (2012b).

The placement of dwellings with normal occupancy in humidity class 2 is in accordance with findings reported in (Geving et al. 2008; Künzel 2006; Francisco and Rose 2010; Brandt 2013).

Fixed or Variable Air-Change Rate and Fixed Moisture Supply

Instead of using humidity classes, a constant internal relative humidity or a fixed or variable air-change rate (h^{-1}) and fixed moisture supply can be used. This is especially relevant in mechanically ventilated buildings. The DBR stipulates an air-change rate of 0.3 l/s/m^3 (equaling 0.5 h^{-1} in a building with a normal ceiling height of 2.5 m). The size of the moisture production should be set according to the use of the building and the number of people using it; in a single-family house with three or four people 7.0–9.5 kg per 24 hours is appropriate.

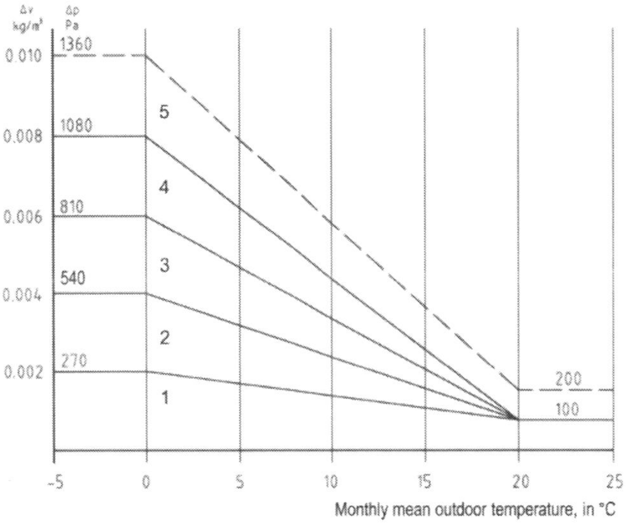

Figure 8 *Variation of internal humidity classes with external temperature. Maritime climates (Western Europe). Humidity class 1: unoccupied buildings, storage of dry goods, class 2: offices, dwellings with normal occupancy and ventilation, class 3: dwellings with unknown occupancy, class 4: sports halls, kitchens, canteens, and class 5: special buildings, e.g. laundry, brewery and swimming pool (ISO 2012b).*

Constant External Humidity and Mean Annual Minimum Temperatures

The standard EN ISO 13788 also prescribes how to calculate the risk of surface condensation on a lightweight construction (2012b). In this case, the temperature shifts relatively fast and therefore the mean annual minimum temperature on a daily basis should be used as external temperature and 95% rh as external humidity.

EXAMPLES OF CALCULATION OF MINIMUM ACCEPTABLE SURFACE TEMPERATURE

A calculation based on internal humidity classes is made for the Danish Test Reference Year TRY (Andersen et al. 1982), with an internal humidity load corresponding with the upper limit of humidity class 3 and 75% rh as critical surface humidity for mold growth during the winter period (Brandt 2013). Table 6 shows that the highest value of the minimum temperature factor $f_{Rsi,min}$ is found in January. Comparing the corresponding minimum acceptable temperature $\theta_{si,min}$ with the inner-surface temperature of a window at a specific location, e.g., at the edge of the glazing, it can be evaluated whether there is a risk of mold growth.

Table 6. Calculation of the Minimum Acceptable Surface Temperature and the Minimum Temperature Factor According to EN ISO 13788 (2012b) for Buildings in Humidity Class 3 (Upper Limit) Based on TRY (Andersen et al. 1982) and 75% rh as Critical Moisture Level (Mold Growth), Internal Temperature: 20°C. (Brandt 2013)

Month	θ_e °C	φ_e	p_e Pa	Δp Pa	p_i Pa	$p_{sat}(\theta_{si})$ Pa	$\theta_{si,min}$ °C	$f_{Rsi,min}$
Jan	–0.6	0.94	546	810	1356	1808	15.9	0.802
Feb	–1.1	0.91	507	810	1317	1756	15.5	0.785
Mar	2.6	0.91	670	718	1388	1850	16.3	0.786
Apr	6.6	0.82	799	576	1375	1833	16.1	0.711
May	10.6	0.78	996	434	1430	1907	16.8	0.655
Jun	15.7	0.67	1194	253	1447	1929	16.9	0.289
Jul	16.4	0.74	1380	228	1607	2143	18.6	0.614
Aug	16.7	0.71	1349	217	1566	2088	18.2	0.453
Sep	13.7	0.85	1332	324	1656	2207	19,1	0.854
Oct	9.2	0.87	1012	483	1495	1994	17.5	0.765
Nov	5	0.91	793	633	1426	1901	16.7	0.781
Dec	1.6	0.88	603	753	1356	1808	15.9	0.778

Table 7. The Minimum Acceptable Surface Temperature Based on TRY (Andersen et al. 1982) at the Upper Limit of Different Internal Humidity Classes Calculated According to EN ISO 13788 (2012)

Critical RH	Minimum Acceptable Surface Temperature, $\theta_{si,min}$			
Φ_{si}	Internal Humidity Class			
	1	2	3	4
75% rh	8.2 °C	12.5 °C	15.9 °C	18.8 °C
100% rh	4.1 °C	8.2 °C	11.5 °C	14.3 °C

The example in Table 6 has been expanded in Table 7, which shows the minimum acceptable surface temperature for mold growth (75% rh) and surface condensation (100% rh), calculated for the upper limit of the different internal humidity classes shown in Figure 8. Table 7 is based on average monthly values from TRY (Andersen et al. 1982) for temperature (–0.6°C) and relative humidity (94% rh) for January, which is the critical month in Denmark regardless of humidity class. Internal temperature is 20°C.

Table 8 compares the level of the minimum acceptable surface temperature using internal humidity classes and the other preconditions according to EN ISO 13788 (2012b). For 'internal humidity classes', the upper limit of humidity class 3 is used as internal moisture load, like in Tables 6 and 7. For 'fixed or variable air-change rate and fixed moisture supply', the air-change rate is $0.25 + 0.04\ \theta_e\ h^{-1}$ and the moisture production is 0.4 kg per 24 hours, representing 4-5 persons. The chosen conditions are estimated to be comparable.

Table 8. Calculation of the Minimum Acceptable Surface Temperature Using Different Preconditions According to EN ISO 13788 (2012b). Climate Data for January in Denmark (Andersen et al. 1982), Internal Temperature: 20°C.

Method	Preconditions	$\theta_{si,min}$ 75% rh	$\theta_{si,min}$ 100% rh
Internal humidity classes	Upper limit class 3	15.9°C	11.5°C
Constant internal RH	50% rh	13.6°C	9.3°C
Fixed air-change rate and moisture production	$0.25 + 0.04\theta_e\ h^{-1}$, 0.4 kg/h	17.2°C	12.7°C
Fixed external rh	95% rh	16.6°C	12.2°C

In all four cases, results based on 75% rh (mold growth) and 100% rh (condensation) are shown, although the first three methods are aimed at mold growth and the last one at surface condensation. Only the results for January are shown, like in Table 7, as this is the critical month.

Apart from the case of constant internal RH, Table 8 shows that the different methods give comparable results and that 9.3°C used as critical surface temperature to prevent surface condensation in the DBR is too low. The results for a specific building will, of course, depend on the actual moisture conditions. Note that 9.3°C is based on 0°C as external temperature, while the results of the calculation based on internal humidity classes and fixed external RH in Table 8 are based on –0.6°C as external temperature, representing January. The temperature 9.3°C corresponds to 9.0°C at –0.6°C as external temperature.

DISCUSSION

Minimum Acceptable Surface Temperature—Based on a Daily Changing Temperature

In preparing the requirements for DBR10, the Technical University of Denmark recommended a value of 12°C to avoid surface condensation, which is comparable with the values for 100% rh in Table 8 (Laustsen and Svendsen 2008). Calculations based on the upper limit of humidity class 3 (ISO 2012b) as internal humidity load, 95% rh as external relative humidity, 20°C as internal temperature, and daily minimum external temperature based on the Danish DRY (Lund 2001) gave a temperature factor with a maximum value of 0.65 corresponding to 12.9°C as the minimum acceptable surface temperature. As this is an extreme situation that occurs just once a year, 12°C was recommended, corresponding to a temperature factor of 0.6.

EN ISO 13788 prescribes the use of a mean annual minimum temperature on a daily basis and 95% rh as external relative humidity, implying that only one single calculation is to be made, representing the whole year (2012b). By using daily minimum temperatures and calculating the temperature factor for each day, it is not only possible to calculate whether there is a risk of surface condensation in a specific case or not, but also the number of days when there is a risk. This makes it possible to compare different window solutions to see which of them gives the highest risk of surface condensation. However, in combination with the daily minimum temperature, Laustsen and Svendsen used 95% rh as external humidity (2008). It can be argued that such preconditions are overdone to be on the safe side, and thus exaggerating the risk of condensation of specific window solutions.

BYG DTU describes an even more detailed calculation of the risk of surface condensation using the external temperature in the heating season on a time basis, and also using a program to calculate the U-factor as a function of the external temperature with a fixed internal temperature and another program to calculate the inner-surface temperature as a function of the external temperature (2009a). The calculations show that the minimum temperature factor, f_{Rsi} is nondependent on an external (θ_e) and internal (θ_i) temperature between −10°C and 20°C. This means that it is possible to calculate the inner-surface temperature (θ_{si}) hour by hour by using a reference year for the external temperature:

$$\theta_{si} = \theta_i \cdot f_{Rsi} + (1 - f_{Rsi}) \cdot \theta_e \quad (2)$$

Minimum Acceptable Surface Temperature—Based on a Fixed Air-Change Rate

When the minimum acceptable surface temperature became a requirement in DBR10, the value was set at 9.3°C, but only at replacement of windows (Danish Enterprise and Construction Authority 2010). DBR10 also introduced the concept of moisture safety in relation to energy savings to highlight that care has to be taken when renovation or replacement is carried out to save energy. In some cases, renovation is not advisable, as it is not possible to renovate without a high risk of causing moisture-related problems.

Since 2003, the Association of Danish Window Manufacturers had recommended that windows should have an inner-surface temperature of at least 7°C (Association of Danish Window Manufacturers 2003) which was increased to 8.5°C in 2008 and 9.3°C in 2010. The Association argued that 12°C corresponds to 60% rh indoors, which is a very high level during winter, indicating that it represents a situation with insufficient ventilation (Association of Danish Window Manufacturers 2009). Figure 9 shows how the internal relative humidity changes over the year depending on the humidity class.

The Association suggested that calculations should be based on 0.5 h^{-1} as the fixed air-change rate, as required in DBR10 and a moisture production of 10l per 24 hours, representing 4–5 persons. This is in accordance with EN ISO 13788, which prescribes the use of humidity classes, a constant relative humidity or a known moisture supply and ventilation rate (2012b). In that case, an 80 m^2 house, which is very small for 4–5 persons, results in 9.3°C as the minimum inner-surface temperature, indicating that 9.3°C should be on the safe side.

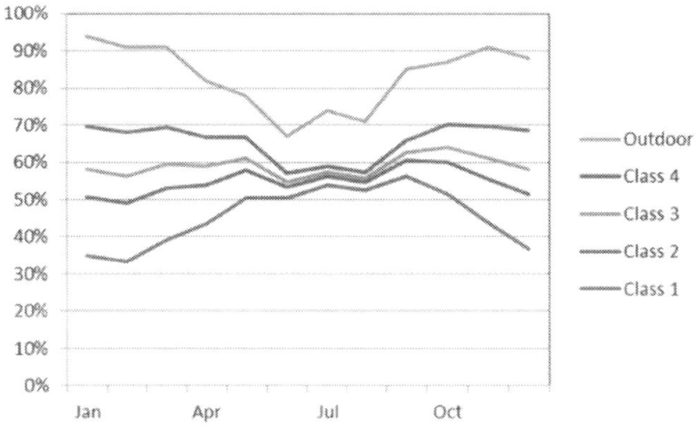

Figure 9 *Internal relative humidity in internal humidity classes 1–4 (upper limit) (simplified) and in external air during the year. Based on EN ISO 13788 (2012) and Danish Test Reference Year (TRY) (Andersen et al. 1982). Internal temperature 20°C, except June and September (22°C), and July and August (23°C).*

The U-factor of a window is calculated for the window as a whole, according to EN ISO 10077-1 (2006) and DS 418 (2011), which means that the inner-surface temperature is also calculated for the window as a whole. However, the lower spacer becomes 0.2°C–1.0°C colder than the rest of the window due to internal convection, which means that some kind of safety margin is necessary, which speaks for a higher value than 9.3°C to be used (BYG DTU 2001).

It can also be argued that the use of 0°C as external temperature is too high. The inner-surface temperature at the spacer is reduced from 8.2°C to 5.3°C for a window with wood sash/frame construction, double glazing, 15 mm distance between glazing panes and a traditional spacer made of aluminum if the external temperature is reduced from 0°C to −5°C (BYG DTU 2009a) resulting in moisture-related problems.

The risk of condensation is very dependent on the air-change rate and often people do not operate the fresh air valves correctly, closing them during winter to avoid a draft. A 50% reduction of the air-change rate can increase 50–100 times the number of hours per year with a risk of condensation (BYG DTU 2001), depending on the materials used for the sash/frame construction. Table 8 shows a case with a fixed moisture supply (0.4 kg/h) and a variable air-change rate (0.25 + 0.04 θ_e h^{-1}), resulting in 12.7°C as minimum acceptable surface temperature (surface condensation). With 0.5 h^{-1} as fixed air-change rate, the temperature becomes 6.5°C showing the effect of sufficient ventilation all year round.

Minimum Acceptable Surface Temperature vs. Amount of Ventilation

Problems with specific types of wood/aluminum windows, e.g., type B1 in Table 5 and a similar window with a 7.5°C inner-surface temperature, demonstrate that it was relevant to introduce the minimum acceptable surface temperature in the Danish building legislation concurrently with the U-factor or net-energy gain as a parameter for windows. It also demonstrates that the recommendation by the Danish window industry (7.0°C and later 8.5°C) was too low. The manufacturer of window B1 has since improved the construction by inserting a thermal interruption between the internal wood part and the external aluminum part (type B2) or by exchanging aluminum with a composite material (type B3). The inner-surface temperature for these solutions is 9.5°C and 10.3°C, respectively. They apply to the requirements in DBR10 and are markedly higher than 7.2°C, but they are still lower than 12°C as given by Table 8 and recommended by Laustsen and Svendsen (2008).

Other window manufacturers have improved their products as well, in some cases introducing materials with low-thermal conductivity such as GFRP (see Table 5), acknowledging that the thermal bridge at the edge construction becomes too big when using aluminum, which results in problems with surface condensation. This shows that the industry responds to new requirements. Since 2008, warm spacers made of stainless steel, polycarbonate, or a combination of stainless steel and polypropylene have in most cases been standard for new windows in Denmark and since 2010 for all new windows (Association of Danish Window Manufacturers 2013). In some cases, spacers of aluminum are still used at replacement. It remains to be seen how durable the warm spacers are compared with the traditional aluminum solution.

During the preparation of DBR10, introducing a requirement for the inner-surface temperature, the Danish window industry argued that internal climate problems should not only be solved by setting requirements to windows, but by focusing also on the importance of sufficient heating and ventilating (Association of Danish Window Manufacturers 2009). Sufficient heating becomes more important as radiators are replaced by floor heating.

Ensuring sufficient ventilation by opening the windows from time to time is also important, not least when the building is new and construction moisture needs to be removed or when the windows in an old building are replaced. However, in many cases surface condensation was detected, although the requirements for air change in the building regulations were followed obliging people to remove water from the inside of windows every morning. As buildings are becoming more air tight to reduce energy loss, this problem might increase.

On the other hand, DBR10 prescribes that new multi-family houses must have mechanical ventilation with heat recovery. This might ensure sufficient ventilation, as the building user will no longer experience cold air coming through fresh-air valves and therefore do not block them. Furthermore, the system does not rely on the occasional opening of windows. However, internal humidity may still be high if the internal moisture production is higher than foreseen in the humidity classes of EN ISO 13788 (2012). In the future, the problem might therefore increasingly be limited to single-family houses.

Level of Minimum Acceptable Surface Temperature—Political Considerations

By choosing 9.3°C as the minimum acceptable surface temperature, the Danish authorities were accused of listening too much to the window manufacturers and letting the consumers down, being aware that some windows with a sash/frame construction and thermal interruption on the market in 2009 were not able to meet a 12°C requirement recommended by (Laustsen and Svendsen 2008). If this is true, the minimum acceptable surface temperature is not only a matter of protecting the building user, but also a political question of protecting national interests. However, the authorities have had this problem before and solved it by indicating future requirements years before they became effective.

Concerning energy performance, DBR has, since 2006, included voluntary classes (low-energy classes). At present, low-energy classes indicate the minimum requirements for 2015 and 2020, leaving time for the industry and clients to develop and promote building components that comply with these voluntary classes. The introduction of net-energy gain and minimum acceptable surface temperature for windows instead of U-factor has had the same effect. As a result the implementation of warm spacers has led to a number of products that comply with what could be expected to be requirements in 2020. However, it remains to be seen how much the requirement for the minimum acceptable surface temperature will be raised. Other countries could learn from this experience as well. In 2015, the requirement for net-energy gain is expected to be tightened to –17 kWh/m^2 per year and the requirement for the minimum acceptable surface temperature is to be reconsidered.

How Different Internal Humidity Classes Can Be Taken into Account

The relationship between the minimum acceptable surface temperature at the upper limit of the different internal humidity classes and the acceptable relative humidity is shown in Figure 10, as an extension of the results in Table 7. Figure 10 shows that if the acceptable RH level increases, the acceptable minimum surface temperature decreases. For example, a building in humidity class 3: If 75% rh can be accepted at the inner surface, then the window should be designed to ensure an inner-surface temperature that is at least 16°C. However, if 85% rh can be accepted, 14°C is acceptable as inner-surface temperature.

From Figure 10 it can be seen that the present requirement for the inner-surface temperature (9.0°C, based on –0.6°C as external temperature) fits well with the upper limit for internal

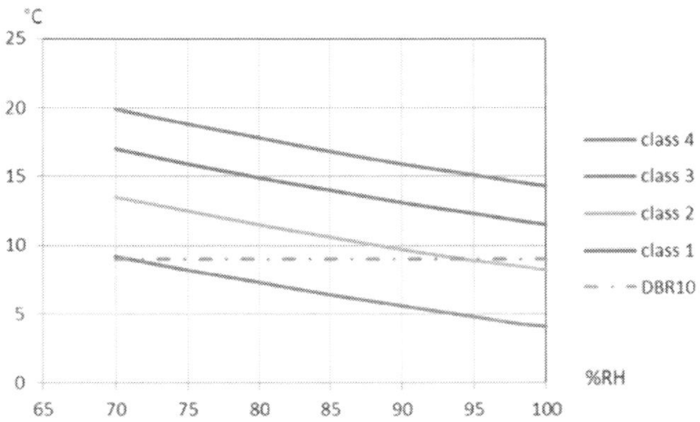

Figure 10 *Minimum acceptable inner surface temperature, $\theta_{si'min}$ depending on the acceptable RH at internal humidity classes 1, 2, 3, and 4 (upper limit). External air temperature and relative humidity based on the Danish Test Reference Year TRY (Andersen et al, 1982). DBR10: The present requirement for inner-surface temperature in the Danish Building Regulations (Danish Enterprise and Construction Authority 2010) adjusted to –0.6°C as external temperature.*

humidity class 2 representing dwellings with normal occupancy and ventilation. However, this indicates that dwellings with unknown occupancy, represented by internal humidity class 3, will have problems with surface condensation using windows with 9.0°C as inner-surface temperature, highlighting the need of raising the requirement.

Suitability of the Method

Depending on whether it is a question of evaluating the risk of surface condensation or mold growth for a specific window, or whether it is a question of comparing different designs of windows, it can be argued how detailed the calculations should be.

The calculation for the risk of mold growth according to EN ISO 13788 is a static calculation based on monthly mean values, not taking into account the daily variations and the heat capacity (2012b). The calculation of the risk of surface condensation is based on a single case where the external temperature is an average, taken over several years, of the lowest mean daily temperature of each year.

Another possibility is to make a calculation based on daily values for external temperatures and relative humidity. Or even data for each hour (BYG DTU 2009a). Then the number of days or hours where the acceptable surface temperature is lower than the inner-surface temperature for the window can be calculated. However, as long as the criteria for mold growth is not known in detail, i.e., for how many days or weeks the moisture conditions should be above a certain level to be critical, it is difficult to decide whether a day-to-day or month-to-month calculation is the more relevant.

Requirements Based on Risk of Surface Condensation or Mold Growth

The requirement of the inner-surface temperature in the DBR relates to the risk of surface condensation. As the examples in Tables 7 and 8 show, a requirement related to the risk of mold growth (75% or 80% rh) would become a problem for most of the windows available on the market today. However, most of the sash/frames of modern windows have some kind of surface treatment and are relative easy to clean. It is easy to see whether cleaning is necessary compared with a wall where mold growth can occur behind a closet or in other places that are difficult to access.

Condensation on windows is not only a question of good technical solutions in window design, but it is very much influenced by the moisture production and the airflow at the internal side of the window, i.e., height/width ratio of the window, how close the internal side of the window is to the internal side of the wall, whether there is a curtain, a radiator, and/or a window sill, etc.

Part of this can be influenced by the architect and engineer that designs the building, but the behavior of the user is decisive. The user controls internal temperature, moisture load (for instance plants in the window), ventilation (opening of windows), and use of curtains. Replacement of old windows with new, tight, energy-efficient windows does not necessarily solve the problem with surface condensation if behavior patterns do not change.

Simulations of condensation on single glazing based on typical variations in internal and external climate in Norway (given in Figure 11) show how different conditions can be and this explains why some people complain of moisture problems while others have no problems.

However, as the development of windows has shown, it is possible to design windows with an inner-surface temperature higher than 13.2°C, which is the minimum acceptable surface temperature using the upper limit of humidity class 2 as internal moisture load, (see Table 6) representing dwellings with normal occupancy and ventilation. Therefore, in cases where the air-change rate complies with the requirements in the DBR and the internal moisture load is as expected, it should not be accepted that drying the windows on a daily/regular basis is necessary to avoid mold growth. Of course, if the internal moisture load is high, it is reasonable to require that people air rooms more often as a precaution.

CONCLUSION

The introduction of a minimum acceptable surface temperature in the DBR has been an important driver for the development of new window solutions, especially for the design of the sealing system for window panes as this is the critical point in relation to surface condensation and mold growth. The requirement is not strict enough to get rid of surface condensation. It can be stricter without ruining the window market as a number of windows are developed with an

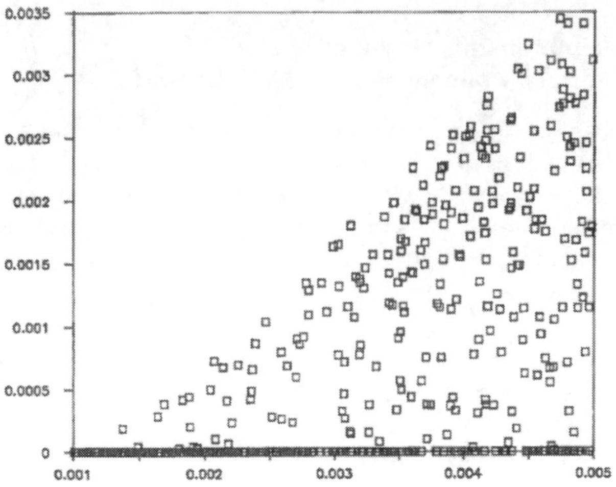

Figure 11 *Simulation of condensation on single glazing (500 cases). Difference in moisture concentration (kg/m^3) between the surface and the indoor air (y-axis) versus extra moisture content (kg/m^3) in the indoor air (Nielsen 1988). Values above zero on the y-axis are cases with condensation and cases with high values are cases with high condensation.*

inner-surface temperature high enough to avoid mold growth in dwellings with normal occupancy and ventilation.

Indications in the building legislation of the future requirements to the inner-surface temperature are recommended based on the experience from low-energy classes. Introduction of requirements in countries with similar problems is recommended as well.

More detailed calculations of the minimum acceptable surface temperature than described by the international standard are possible, but as long as the criteria for mold growth is not known in detail the present methods seems suitable.

Information on the minimum acceptable surface temperature gives the consumer a possibility to ask for this property for specific windows giving the opportunity to select between different windows. But the introduction of the minimum acceptable surface temperature does not stop complaints from consumers as this temperature is measured or calculated under specific conditions. It cannot take into account whether windows are mounted in the inner or outer part of the wall or the presence of curtains in front of the window during night. In both cases, the risk of surface condensation increases as the internal heat resistance is increased.

Increasing requirements for airtightness increases the importance of sufficient ventilation in order to avoid problems with condensation or mold growth, as the requirement for a minimum acceptable surface temperature in the building legislation is based on a certain internal humidity load.

NOMENCLATURE

E = energy

G = degree hours in the heating season based on 20°C as internal temperature

I = sunlight corrected for the variation of g value with angle of entry

U = thermal transmission coefficient

f = temperature factor

g = solar energy transmittance

p = water vapor pressure

θ = temperature

φ = relative humidity

Subscripts

e = external (outdoor)

i = internal (indoor)

ref = net-energy gain

si = internal surface

sat = saturation

w = window

REFERENCES

Andersen, B., S. Eidorff, L. Hallgreen, H. Lund, E. Pedersen, S. Rosenørn, and O. Valbjørn. 1982. Weather data for plumbing and energy. SBi Report 135, Danish Test Reference Year TRY (author's translation from Danish). Hørsholm: Danish Building Research Institute.

Applefield, D., C.S. Hansen, and S. Svendsen. 2010. Development of a slim window frame made of glass fibre reinforced polyester. *Energy and Buildings* 41(10):1918–25.

Association of Danish Window Manufacturers. 2003. *Technical Conditions for DWV* (Danish Window Verification), 5th ed., addendum 3. (author's translation from Danish). Aarhus.

Association of Danish Window Manufacturers. 2009. Note to the Danish Enterprise and Construction Authority and the Danish Energy Agency (author's translation from Danish). Aarhus.

Association of Danish Window Manufacturers. 2013. Johny Jensen, personal communication with the author, January 21.

Brandt, E., ed. 2013. Moisture in buildings (author's translation from Danish). *SBi Directions 224*, 2nd ed. Copenhagen: Danish Building Research Institute.

BYG DTU. 2001. Energy properties of panes and windows—Part 11: Edge constructions with reduced thermal bridge, Report U-010 (author's translation from Danish). Lyngby: Technical University of Denmark www.vinduesvidensystem.dk/Kompendier/Kompendium11-vidensystem.pdf

BYG DTU. 2009a. Energy properties of panes and windows—Part 4: Development of energy efficient panes and windows, Report U-004 (author's translation from Danish). Lyngby: Technical University of Denmark www.vinduesvidensystem.dk/Kompendier/Kompendium4-vidensystem.pdf

BYG DTU. 2009b. Energy properties of panes and windows—Part 9: Overview of possibilities for developing better panes and windows, Report U-008 (author's translation from Danish). Lyngby: Technical University of Denmark. www.vinduesvidensystem.dk/Kompendier/Kompendium9-vidensystem.pdf

Danish Enterprise and Construction Authority. 2006. *Building Regulation 1995,* Addendum 12 to (author's translation from Danish). Copenhagen.

Danish Enterprise and Construction Authority. 2008. Building Regulations 2008. Copenhagen.

Danish Enterprise and Construction Authority. 2010. Building Regulations 2010. Copenhagen.

de Place Hansen, E.J., ed. 2012. Guidelines on building regulations 2010. *SBi Guidelines 230*, 2nd ed. Hørsholm: Danish Building Research Institute.

Danish Standard Association. 2011. DS 418, *Calculation of Heat Loss from Buildings*, 7th ed. Copenhagen: Danish Standard Association.

ISO. 2006. EN ISO 10077-1, *Thermal performance of Windows, Doors and Shutters—Calculation of Thermal Transmittance—Part 1: General*. Brussels: European Committee for Standardization.

ISO. 2012a. EN ISO 10077-2, *Thermal Performance of Windows, Doors and Shutters—Calculation of Thermal Transmittance—Part 2: Numerical Method for Frames.* Brussels: European Committee for Standardization.

ISO. 2012b. EN ISO 13788, *Hygrothermal Performance of Building Components and Building Elements—Internal Surface Temperature to Avoid Critical Surface Humidity and Interstitial Condensation—Calculation Methods.* Brussels: European Committee for Standardization.

Francisco, P.W., and W.B. Rose. 2010. Temperature and Humidity Measurements in 71 Homes Participating in an IAQ Improvement Program. *Proceedings of the Thermal Performance of the Exterior Envelopes of Whole Buildings XI*, Clearwater Beach, Florida Dec 5–9, 2010.

Geving, S., J. Holme, and J.A. Jenssen. 2008. Indoor air humidity in Norwegian houses, Report R-189. C. Rode, ed. *Proceedings of the 8th Symposium on Building Physics in the Nordic Countries*, Lyngby: Technical University of Denmark.

Kottek, M., J. Grieser, C. Beck, B. Rudolf, and F. Rubel. 2006. World Map of Köpppen-Geiger Climate Classification Updated. *Meteorologische Zeitschrift* 15:259–63.

Kragh, J. and S. Svendsen. 2009. Development of new types of energy windows of composite materials. *Design Proposals for Profile Systems,* Report BYG DTU R-203, (author's translation from Danish). Lyngby: Technical University of Denmark. http://www.byg.dtu.dk/~/media/Institutter/Byg/publikationer/byg_rapporter/byg_r203.ashx

Künzel, H.M. 2006. *Indoor Air Humidity in Dwellings—Limits for the Assessment of Protection Against Moisture* (author's translation from German). Stuttgart: Fraunhofer Institut für Bauphysik.

Laustsen, J.B. and S. Svendsen. 2008. *Analysis of Energy Requirements for Windows in the Energy Labeling System and the Danish Building Regulations for 2010, 2015 and 2020,* Report BYG DTU SR-08-06 (author's translation from Danish). Lyngby: Technical University of Denmark. www.byg.dtu.dk/~/media/institutter/byg/publikationer/byg_rapporter/byg_sr0806.ashx

Lund, H. 2001. Design reference years and test reference years in Europe, Turkey and Israel, Electronic CIB Publication No. 262. Rotterdam: CIB General Secretariat.

Ministry of Housing. 1961. *Building Regulations for market towns and the country 1961* (author's translation from Danish). Copenhagen.

Ministry of Housing. 1966. Building Regulations for market towns and the country 1966 (author's translation from Danish). Copenhagen.

Nevander, L.E. and B. Elmarsson. 2006. Moisture handbook. *Practice and Theory* (author's translation from Swedish). Stockholm: Svensk Byggetjänst.

NFRC. 2010. Procedure for determining fenestration product condensation resistance values, NFRC 500-2010. Greenbelt, MD: National Fenestration Rating Council.

Nielsen, A. 1988. Prediction of risk for surface condensation. *Proceedings of Healthy Buildings, Volume 2: Planning, Physics, and Climate Technology for Healthier Buildings*, Stockholm: Sweeden, Swedish Council for Building Research, 325–333.

Nielsen, A. 1991. Analysis of heat loss and temperature conditions for spacers in sealed glazing, BYGGEFORSK Report O3411(author's translation from Norwegian). Trondheim: Norwegian Building Institute.

Persson, A., B. Jonsson, P.-O. Marklund, P. Wickman, and S.-O. Eriksson. 1994. Condensation on windows—causes, steps, dimensioning (author's translation from Swedish). *NUTEK Project 'Window now.'* Stockholm: Sweeden.

Sedlbauer, K. 2001. Prediction of Mould Fungus Formation on the Surface of and Inside Building Components. Stuttgart: Fraunhofer Institute for Building Physics.

Case-Study Analysis on Hygrothermal Performance of ETICS on Concrete Wall after Low-Budget Energy Renovation

Simo Ilomets **Targo Kalamees, PhD**

ABSTRACT

Hygrothermal performance of mineral wool and expanded polystyrene external thermal insulation composite system (ETICS/EIFS) was studied in field conditions and by computer simulations with the Delphin program. Temperature and relative humidity were measured in a additionally insulated large panel concrete element wall from September 2011 to January 2013 in the cold climate of Estonia. Measurements of indoor climate, air leakage, and thermal bridges through building fabric were conducted before and after the low-budget energy renovation pilot project of an apartment building.

Results showed that indoor climate and thermal comfort were not improved a lot in all aspects and the same was true for airtightness. Unacceptable thermal bridges remained at external wall/balcony, external wall/foundation wall junction, and also around the windows because they stayed in their original position. Built-in moisture of the whole wall dried out during the first heating season. Better agreement was found between measured and results if the convective moisture flow in addition to diffusion was taken into account in simulations. Average measured thermal transmittance during the winter was U_{wall} *0.17 W/(m²·K) in the case of graphite enhanced EPS and 0.19 W/(m²·K) with mineral wool. Long-term durability of ETICS in cold climate and under wind driven rain loads need further investigation.*

INTRODUCTION

It is estimated that there are 200 million dwellings and 25 billion m^2 of useful floor space in the EU27, Switzerland, and Norway (Dol and Haffner 2010; Economidou et al. 2011). In Europe, especially in Eastern Europe, a large number of buildings were constructed after the second World War. About 70% of the residential building stock is over 30 years old and about 35% are more than 50 years old (Balaras 2005). A general characteristic of dwellings in Eastern Europe constructed after the second World War and before the energy crisis in the 1970s is that buildings are generally poorly insulated compared to today's requirements for energy performance. Apartment buildings constructed in the late 1970s and 1980s typically have a much higher energy consumption. The energy-renovation measures of old apartment buildings are under discussion internationally (Hagentoft 2010; Silva et al. 2013; Nemry et al. 2010; Dall'O' et al. 2012; Uihlein and Eder 2010) as well as in Estonia (Kalamees et al. 2009).

In Estonia 65% of people live in apartment buildings. Nearly half of the apartment buildings constructed between 1961 and 1990 are composed of large panel prefabricated concrete elements. The condition of concrete façades is mainly satisfactory, except the low frost resistance and corrosion of reinforcement when the carbonation depth has reached close to reinforcement; steel connections of balconies and awnings have turned out to be problematic in some cases (Kalamees et al. 2009; Ilomets et al. 2011). In combination with the high indoor humidity load and thermal bridges, mould can grow at internal surfaces building envelope. Thermal transmittance of original external walls is 0.7–1.0 W/(m^2·K), being 3–5 times higher than considered reasonable today. Therefore, energy-renovation and better understanding of renovation solutions of apartment building is needed.

***Simo Ilomets** is a PhD student and chair of Building Physics & Energy Efficiency, Tallinn University of Technology, Tallinn, Estonia. **Targo Kalamees** is a professor and chair of Building Physics & Energy Efficiency, Tallinn University of Technology.*

The aim of the study was to explore hygrothermal performance of the existing concrete wall insulated with ETICS and to get data for calibrating a hygrothermal computer model for ETICS to permit applying the model to other climatic conditions and other structural dimensions. Other outcomes from energy-renovation pilot project are discussed briefly.

METHODS

An energy-renovation pilot project "Healthy and Cost-Saving Home" was started in spring 2010 in co-operation with two financing institutions together with a ministry, an energy company, a local municipality, and a university. The global purpose was to carry out an example renovation of a typical apartment building to demonstrate energy saving and improve indoor climate as well to motivate people to renovate their buildings. The renovation works lasted from summer to the end of 2011.

The case-study building is a five-story block of flats (Figure 1) constructed during 1966 with prefabricated large panel concrete elements. That type of construction was typical in Estonia during the period between 1961–90. Approximately half of apartment buildings in Estonia are composed of prefabricated concrete large panel elements.

Renovation Solution

Renovation solution was selected based on the energy and economical calculations:

- Improvement of building envelope:
    - External walls: +15 cm graphite enhanced expanded polystyrene (EPS) in the external thermal insulation composite system (ETICS): $U_{external\ walls} = 0.17$ W/(m^2·K);
    - Basement walls: +10 cm EPS covered with drained and ventilated cladding: $U_{basementwall} = 0.36$ W/(m^2·K);
    - Roof: +30 cm EPS above the roof $U_{roof} = 0.11$ W/(m^2·K);
    - Replacing of 33% of the old windows: $U_{old\ window} = 1.8$ W/(m^2·K), $U_{new\ window} = 1.0$ W/(m^2·K); because only 1/3 of the windows were replaced with new; all windows were left at their original position: inboard of the exterior concrete core of the external wall;
- Ventilation system: new exhaust ventilation with heat recovery (exhaust air [water to water] heat pump);
- Heating system: new hydronic radiator heating with thermostats.

Measurements

Measurements on the case-study building concentrated on the indoor climate, energy performance, and hygrothermal performance of the building envelope. During the study the following measurements were taken:

- The values of indoor temperature (t) and relative humidity (RH) conditions were measured with small data loggers at 1 h intervals over a 2 year period;
- Surface temperature of thermal bridges were measured by using infrared image camera "FLIR ThermaCam E320" (EN 13187) with minimum indoor and outdoor temperature difference at least 20 K. Tabulated surface emissivities from FLIR ThermaCam E320 user's manual was used;
- The air leakages of the building fabric of the apartment were measured with the standardized fan pressurization method (EN 13829), using "Minneapolis Blower Door Model 4" equipment; hygrothermal performance of east facing external wall (see Figure 1 and Figure 2), additionally insulated with ETICS, was studied using temperature and relative humidity sensors and heat flux plates.

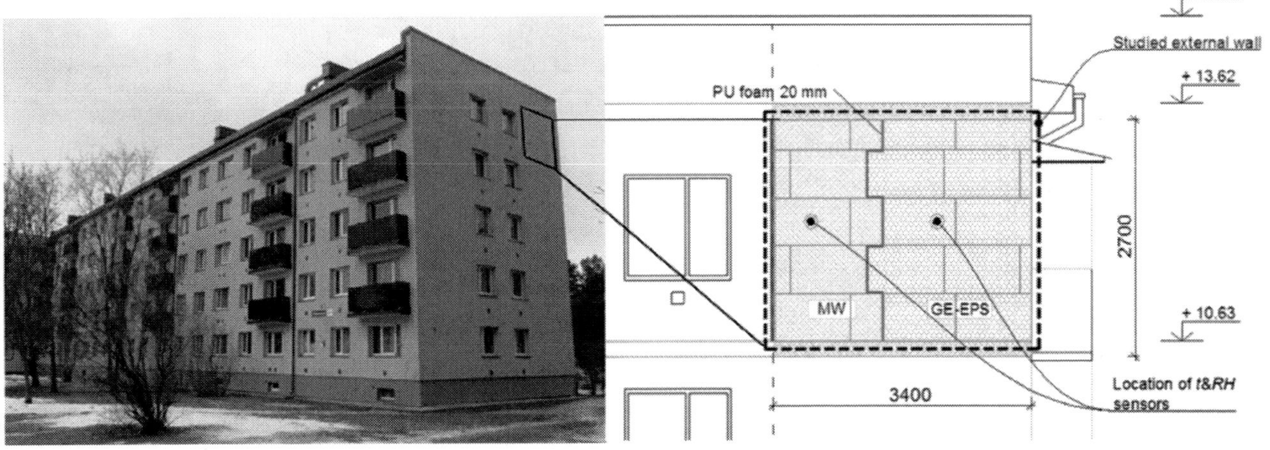

Figure 1 *View to the external wall with measurement areas of ETICS.*

Hygrothermal performance of face sealed, non-drainable external thermal insulation composite system (ETICS) (also called exterior insulation and finish systems, EIFS) with two different insulation materials was compared in two test-walls: mineral wool (MW) and graphite enhanced expanded polystyrene (EPS), see Figure 1 and Figure 2. All the other components (adhesive, base coat, reinforcement [glass fiber mesh], mineral finishing coat, paint coating) of two test-walls were the same. Test-walls were located exterior to the same room so that both test-walls experienced the same climatic conditions. Two walls were separated from each other with polyurethane (PU) foam insulation to minimize wall-direction moisture movement. The location of the test-walls at the upper corner of the wall was selected based on it having the highest driving rain intensity.

Simulations

The temperature and humidity measurement results from test-walls were compared with a hygrothermal simulation model Delphin 5.7.4 (Nicolai 2008):

- to validate the simulation model for future simulations with different initial and climatic conditions as well as with different dimensions of the building envelope;
- to understand better the performance of the external thermal insulation composite system.

Modified material properties were used from the Delphin's database. Table 1 shows the properties of materials used as compared to measured and simulated results when the best correlation was obtained. The dependency of hygrother-

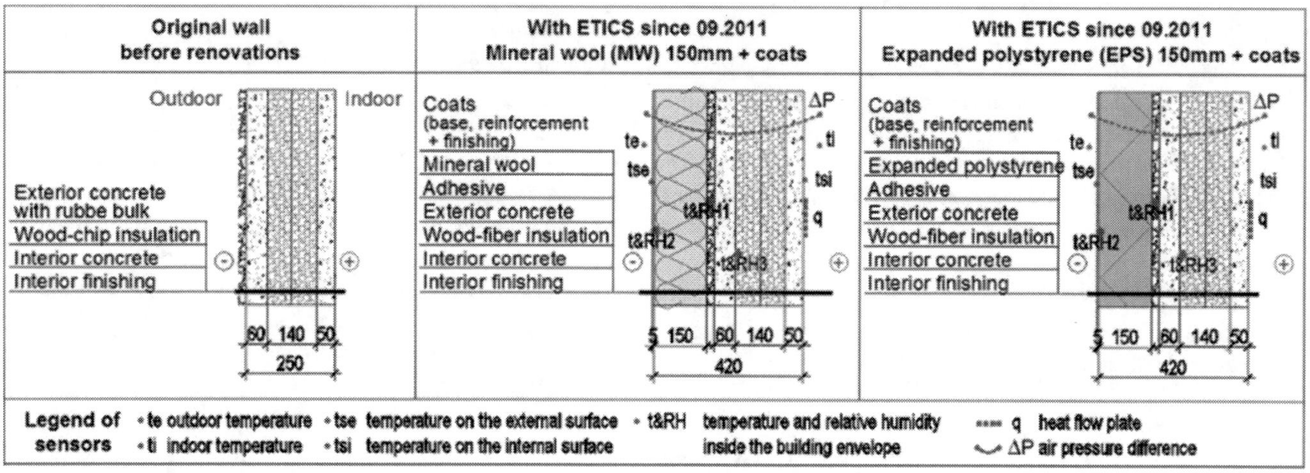

Figure 2 *Test-wall schematics and measurement points (red dots).*

Table 1. Main Hygrothermal Properties of Materials Used in the Simulation

	Concrete	Wood-Cement Chip Board	Adhesive Mortar	EPS	MW	Exterior Coating
Bulk density ρ, kg/m^3	2320	500	700	35	75	1270
Porosity *f*, m^3/m^3	0.14	0.93	0.73	0.94	0.92	0.50
Specific heat capacity *c*, J/(kg·K)	850	1470	945	1500	840	960
Thermal conductivity* λ, W/(m·K)	1.5	0.12	0.19	0.035	0.038	1.0
Water vapour diffusion resistance factor* μ, —	19	3.8	15	15	2	10
Liquid water conductivity* *k*, kg/(m·s·Pa)	$44 \cdot 10^{-12}$	$16 \cdot 10^{-9}$	$3.2 \cdot 10^{-9}$	0	0	$0.27 \cdot 10^{-6}$
Air permeability Kg, s	$1 \cdot 10^{-6}$	$7 \cdot 10^{-3}$	$1 \cdot 10^{-5}$	$1 \cdot 10^{-6}$	$1 \cdot 10^{-2}$	$1 \cdot 10^{-6}$
Built-in moisture *w*, kg/m^3 / rh, %	48 / 65	27 / 60	200 / 100	0.6 / 60	3.1 / 60	300

*Dry material

mal properties on the environmental conditions was taken into account: water vapour permeability, liquid water conductivity, thermal conductivity dependent on water content of a material (example of concrete is shown in Figure 3 [left]). Cross-section of simulated wall is presented in Figure 3 (right).

Thermal bridges before and after the renovation were evaluated by using a thermal camera and two-dimensional heat-transfer simulation software THERM 6.3 according to standard EN ISO 10211. Thermal bridges were assessed using the temperature factor (EVS-EN ISO 13788). The limit value for temperature factor in a dwelling with high or unknown humidity loads is considered to be ≥0.8 (Kalamees 2006).

RESULTS

Technical Condition of External Walls

External walls of the analysed building were composed of large-panel prefabricated concrete elements. Fourteen centimeters of wood-cement chip board (thermal conductivity λ ~0.12 W/[m·K]) was used for insulation between two layers of concrete and external surface is covered with rubble bulk. The main problems related to external walls are thermal bridges, corrosion of reinforcement, frost resistance, and large thermal transmittance. The layer between insulation and external concrete core is not ventilated but slightly drained (small tubes installed into the intersection of movement joints). The external core can become wet due to vapour condensation and wind-driven rain. This means that frost damages of concrete and corrosion of reinforcement can occur. Structures and technical condition of a building are shown in Figure 4–6.

Indoor Climate Conditions

Temperature and relative humidity were measured in four apartments before and after the renovation. There was significant difference ($P < 0.001$) in temperature measurement results, see Figure 7 left. A new adjusted heating system has even temperatures during a whole heating season after the renovation while there was a clear overheating before. There was no significant ($P = 0.3$) difference in RH before and after the renovation. Moisture excess (the difference between vapour content of the indoor and outdoor air), Δv, g/m^3 was calculated from indoor and outdoor climate, Figure 7 (right). Somewhat surprisingly, moisture excess was not lower but even higher (significant difference [$P < 0.001$]) after the renovation, especially during very cold days with outdoor air temperature lower than –10°C. Still, the level of moisture excess being up to ~3 g/m^3 on average and up to ~4 g/m^3 at 90% level are lower than in these types of buildings in general.

Thermal Bridges

Thermographic measurements before the renovation (see Figure 8) show very low surface temperatures (temperature factors f_{Rsi} ~0.6–0.7) that cannot be accepted in the cold climate of Estonia. In addition to lower average value of the temperature factor, the range of measurement results is wider before the renovation. After renovation and installation of additional thermal insulation, all thermal bridges were decreased. Thermal bridges that ended up being lower than the limit value of 0.8 after the renovation were at the external wall/window, external wall/balcony, and external wall/foundation wall junction.

Because before the renovation already 2/3 of windows were replaced, it was decided to change only remaining 1/3 windows. Therefore all windows were left at their original position (see Figure 9 left). That solution is comfortable from the point of view of inhabitant living conditions during the renovation (less work inside the apartment), but it is unreasonable in terms of energy efficiency, hygrothermal performance, and esthetics since serious thermal bridges remain at the junction of window/external wall.

If windows remain at their original position, linear thermal transmittance of external wall/window increases, see Figure 9. For that reason, this junction remains a domi-

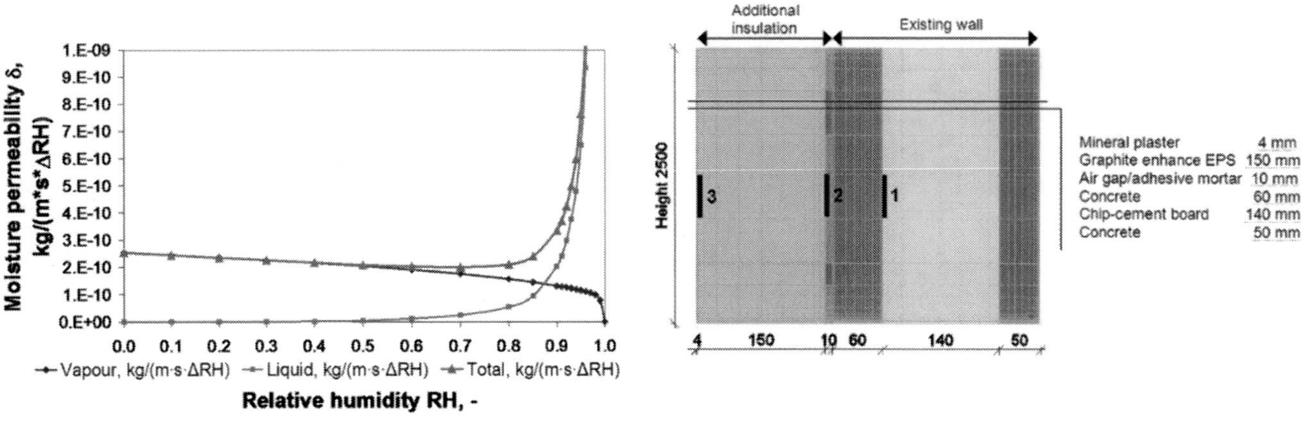

Figure 3 *Moisture permeability of concrete as liquid and vapour at different RHs (left). Example of wall section in simulation tool Delphin with measurement points of temperature and RH 1, 2, 3 (right).*

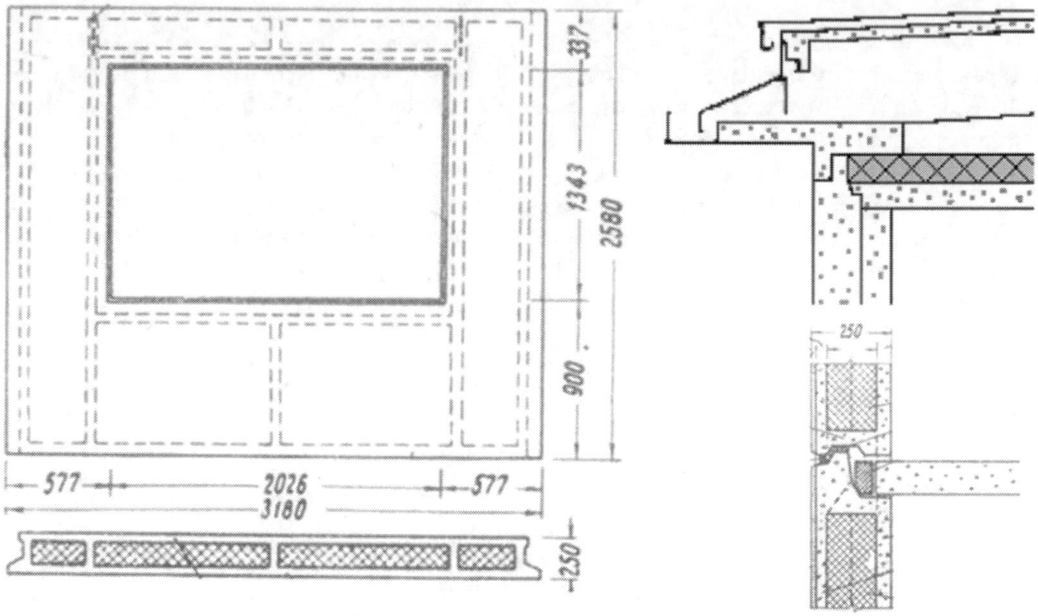

Figure 4 *Original design drawing of a large panel prefabricated concrete element (left). External wall/roof junction (right top) and external wall/inserted ceiling (right bottom).*

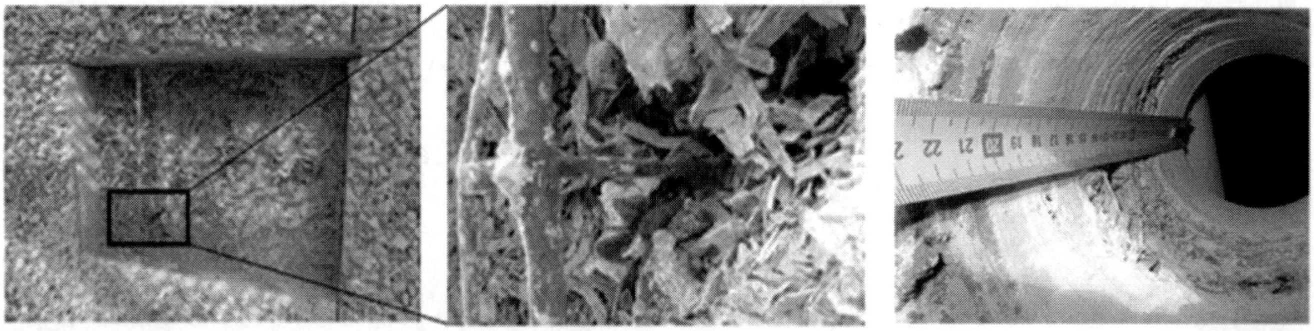

Figure 5 *Reinforcement steel between layers of concrete has started to corrode (left and middle). In principle, this steel should be cast into expanded clay concrete, but in reality it is often not so. Cross-section of a concrete wall element (right).*

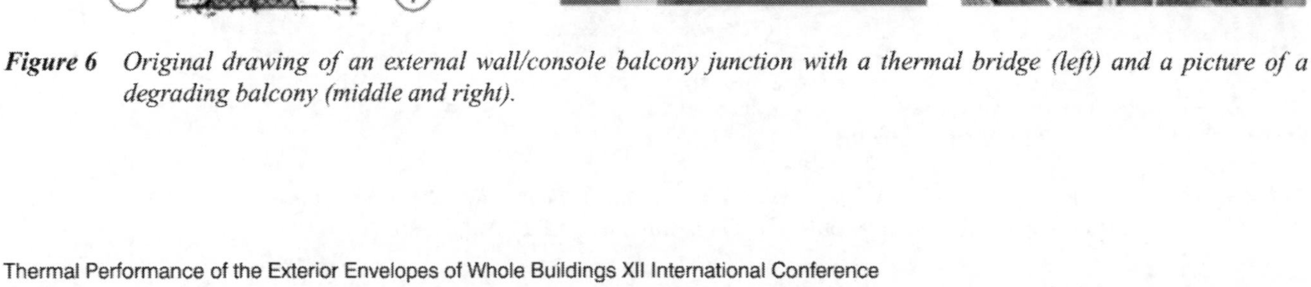

Figure 6 *Original drawing of an external wall/console balcony junction with a thermal bridge (left) and a picture of a degrading balcony (middle and right).*

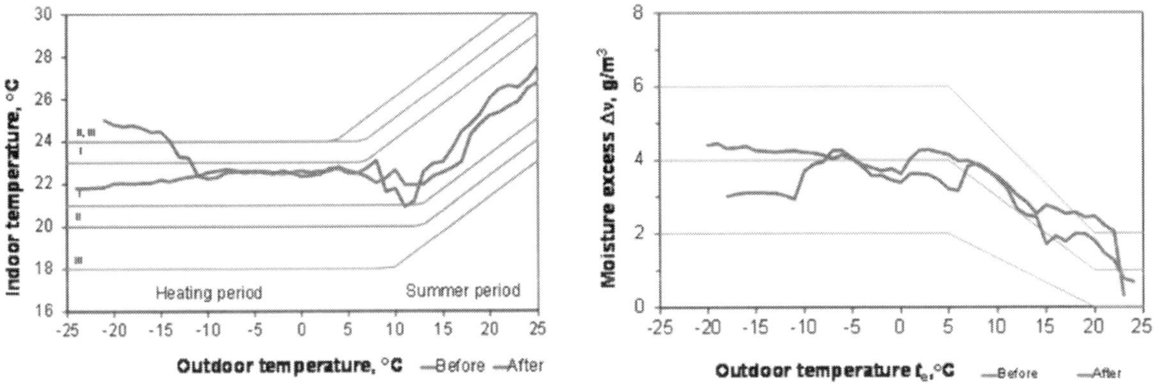

Figure 7 *Measurement results of indoor air temperature (left) and moisture excess (right) depending on the outdoor air's temperature before and after the renovation.*

Corner of external walls	External walls/ inserted ceiling	External walls/ windows (jamb insulated with 5...20 mm insulation)
		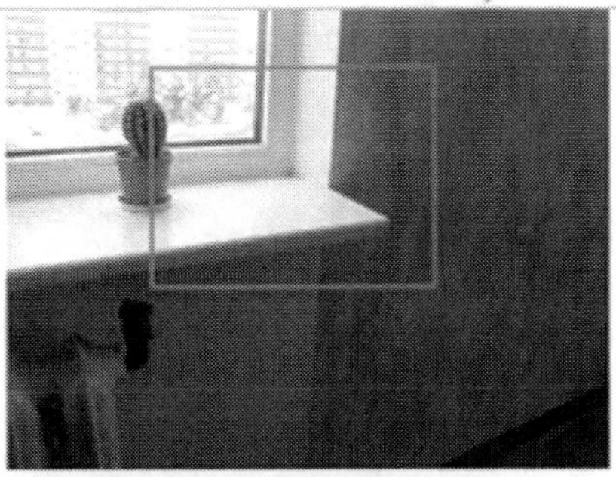
	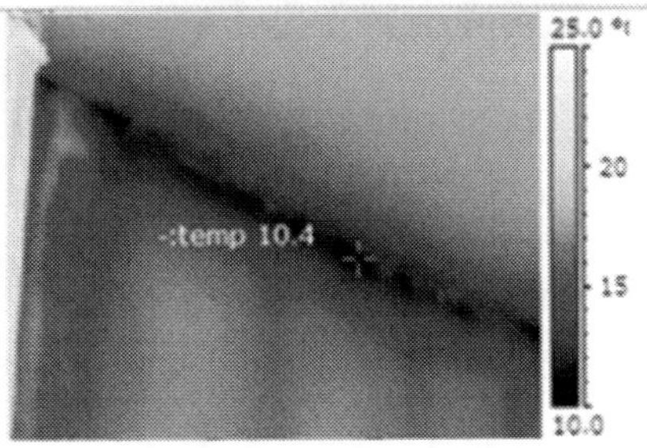	
Before $\overline{f_{Rsi}}$ 0.77; σ 0.09	Before $\overline{f_{Rsi}}$ 0.66; σ 0.12)	Before $\overline{f_{Rsi}}$ 0.74; σ 0.08
After $\overline{f_{Rsi}}$ 0.93; σ 0.03	After $\overline{f_{Rsi}}$ 0.87; σ 0.08)	After $\overline{f_{Rsi}}$ 0.78; σ 0.09

Figure 8 *Measured range of temperature factor f_{Rsi}, of thermal bridges with average and standard deviation σ. Pictures were taken before the renovation.*

nant thermal bridge in terms of heat loss. Results are very sensitive about the thickness of insulation on a window's jamb (cheek) (window return).

It is not possible to install a sufficient layer of insulation (>30 mm) to the window jambs because of the shape of the concrete element that windows are attached to, see Figure 10. To avoid serious thermal bridges around the windows, all the windows should be placed to the external side of the existing concrete element that moves them more in-line with the additional exterior insulation.

Hygrothermal Performance of External Walls

Temperature and RH were measured in both test-walls: MW and EPS external thermal insulation composite systems

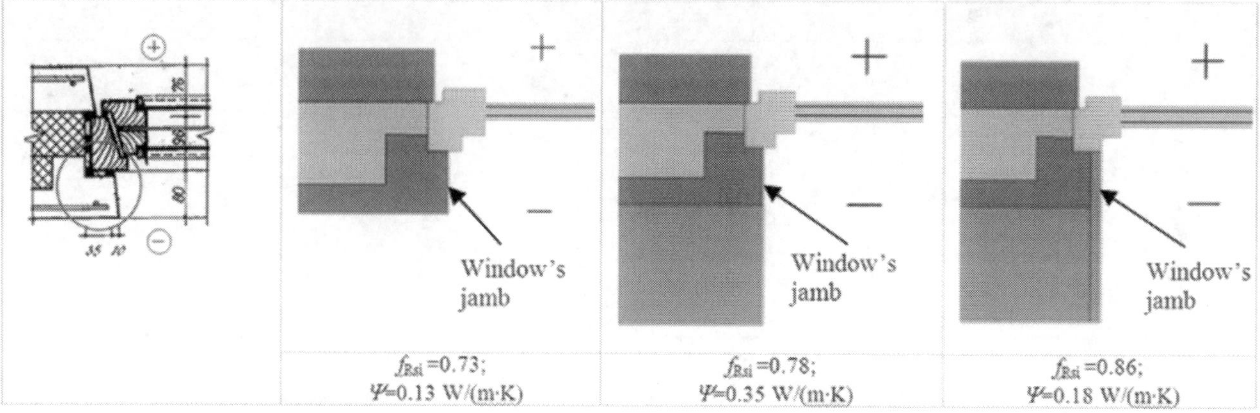

Figure 9 *Calculation results of temperature factor f_{Rsi} and linear thermal transmittance ψ of the external wall/window junction. Inside area of original drawing (left) the red circle shows the proportion of the jamb of a window covered.*

Figure 10 *Picture of an old window (left) and of new windows placed at their original position. Most of the jamb is covered and only a thin layer of insulation (5...20 mm) can be attached to the jamb.*

(ETICS). Measurements lasted for one and a half heating periods from September 2011 until December 2012. Hygrothermal performance of both walls was also simulated by Delphin 5.7.4.

In general, agreement of temperature, vapour pressure, and RH between the measurements and calculations was good, except RH between mineral wool insulation and plaster (measuring point 3, see Figure 14) where the measured value was higher. Results presented in Figure 12–Figure 14 of the cold period fit also with steady-state distributions of temperature, vapour pressure, and RH. After first calculations, there was also a mismatch about the drying out period. According to the measurements, moisture in the wall (existing moisture in the external core and additional moisture from adhesive mortar) dried out during within 3–4 months, mainly by diffusion and with minor convection. We ended up with acceptable agreement after adding small air gaps between measurement points 1 and 2 with air pressure difference at outdoor climate conditions in cross-section 2, see Figure 3 right. The effect of moisture movement without convection can be seen in Figure 11. If in fact moisture in measuring points 1 and 2 dried out during half of heating season it was much slower by diffusion only according to calculations. There is also some difference in the beginning of the second winter.

Calculated and measured temperatures of both walls match quite well. In the case of MW, calculated temperatures at measurement point 1 and 2 are up to 1°C lower than measured (Figure 12) and up to 1°C higher than measured in the case of EPS (Figure 15). Measured temperatures of two test-walls are similar, maximum 0.5°C higher in the case of EPS.

Also, vapour content/vapour pressures were calculated from the measured temperature and RH in both test-walls and calculated with software. Results for MW are shown in Figure 13 and for EPS Figure 16, respectively. Vapour pressure starts to increase right after the beginning of the heating season in the second half of October 2011 but then starts to decrease. Measurement point nr 2 is not graphically presented. It behaves as point nr 1 but its level is somewhat lower.

Results of measured and calculated RH are presented in Figure 14 for MW and in Figure 17 for EPS. As it

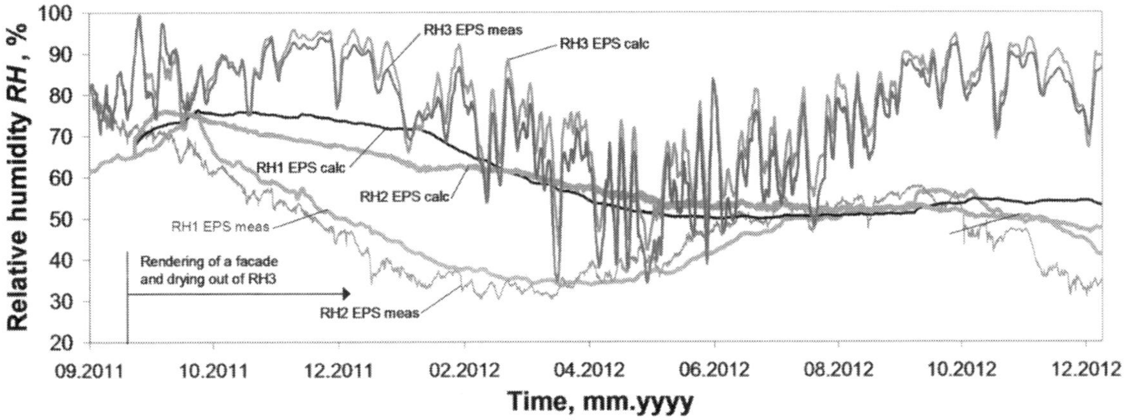

Figure 11 *Comparison of measured (meas) RH and calculated (calc) RH moisture movement only by diffusion of ETICS with graphite enhanced EPS.*

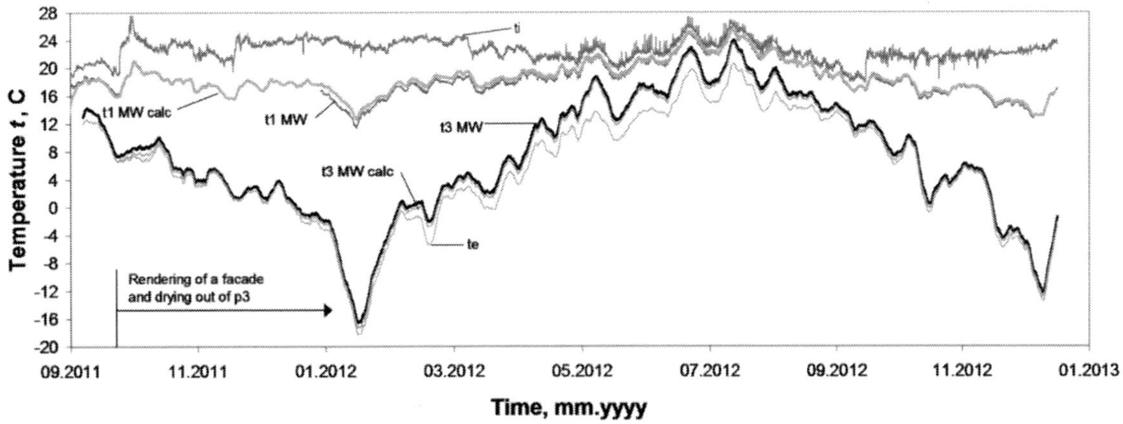

Figure 12 *Measured and calculated temperature (t) in the wall additionally insulated with MW. Location of measurement points see in Figure 2.*

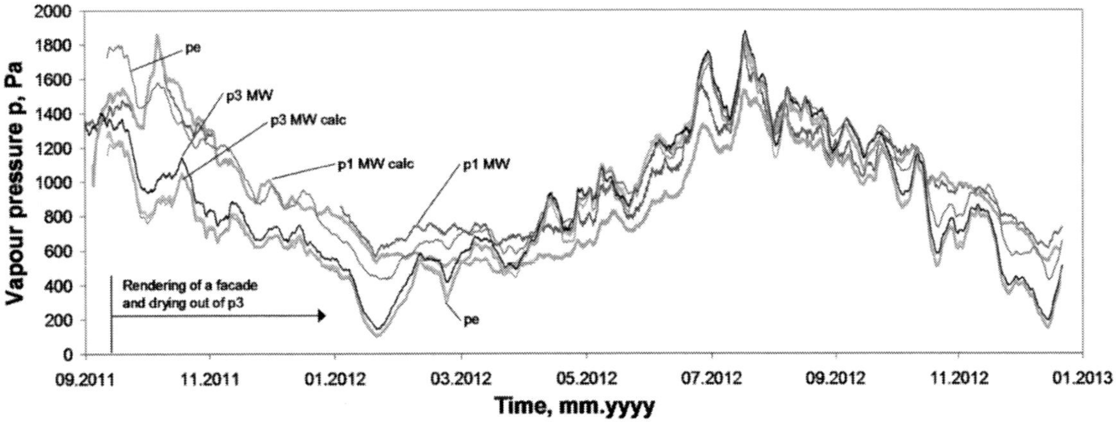

Figure 13 *Measured and calculated vapour pressure (p) in the wall additionally insulated with MW.*

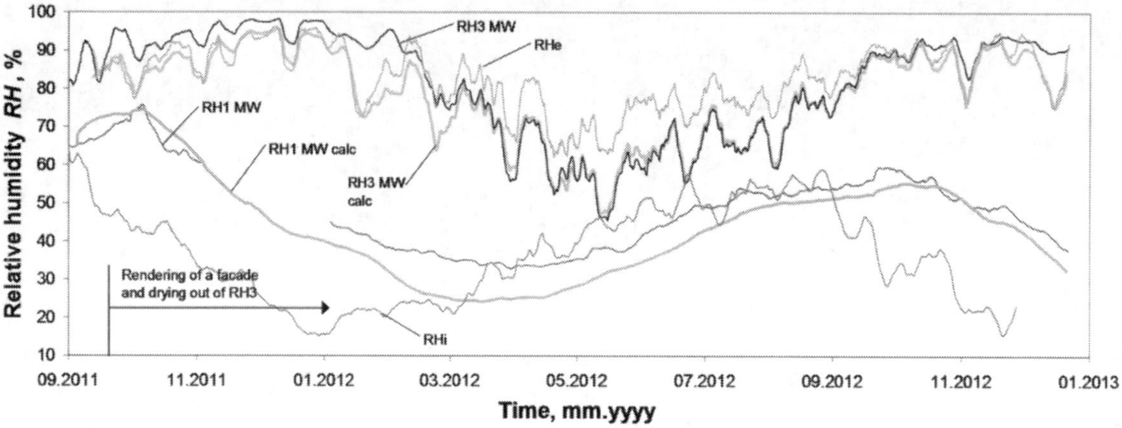

Figure 14 *Measured and calculated RH in the wall additionally insulated with MW.*

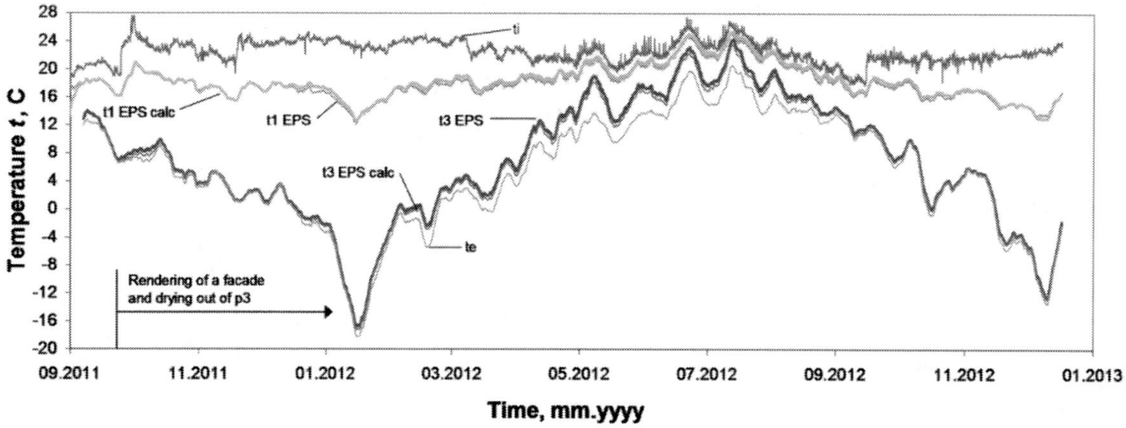

Figure 15 *Measured and calculated temperature in the wall additionally insulated with expanded polystyrene (EPS). Location of measurement points see in Figure 2.*

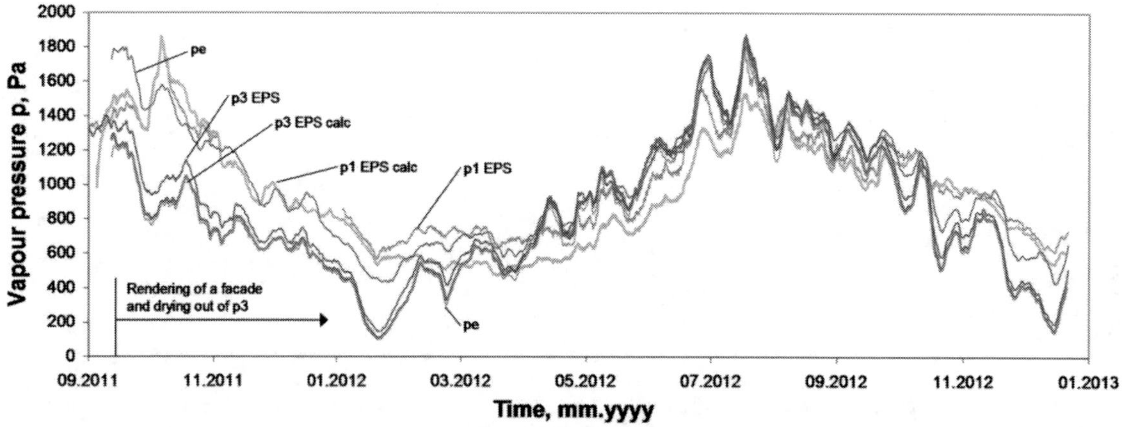

Figure 16 *Measured and calculated vapour pressure in the wall additionally insulated with EPS.*

appears, measured RH at point 3 with MW is somewhat higher than calculated during winter period. In spring measured RH at point 1 is somewhat higher than calculated with both insulation materials. Correlation in summer period is very good, especially in the case of MW. To the contrary, agreement with EPS is better at winter time. Results of measuring point 2 follow the results of point 1 but at 10%–15% lower level. That is a result of higher vapour pressure at point 1 compared to point 2 at almost same temperatures.

Results of thermal transmittance of walls are presented in Figure 18. It can be seen that the calculated thermal transmittance of MW is substantially higher than the rest of three lines. Fluctuation of measurement results appears probably because solar radiation affects temperature of the external surface of the wall. Average thermal transmittances of winter months are 0.17 W/(m^2·K) for EPS and 0.19 W/(m^2·K) for MW.

One of the important research questions was the drying out period of the external core of the original external wall. Analysis of time for drying out period at different initial moisture conditions (from 80% to 100% rh) is shown in Figure 19.

DISCUSSION

Indoor Climate and Thermal Comfort

Indoor climate measurements before and after the retrofit showed small improvement at air temperatures, where no overheating appears at very cold period because of better adjustment of a new heating system.

Natural passive stack ventilation was changed into mechanical exhaust ventilation with exhaust air heat pump. Recovered heat was used for domestic hot water and heating. The air change did not improve after the renovations. It can be seen from moisture excess (Figure 7 right). There are several reasons that are more or less related to each other. First, the

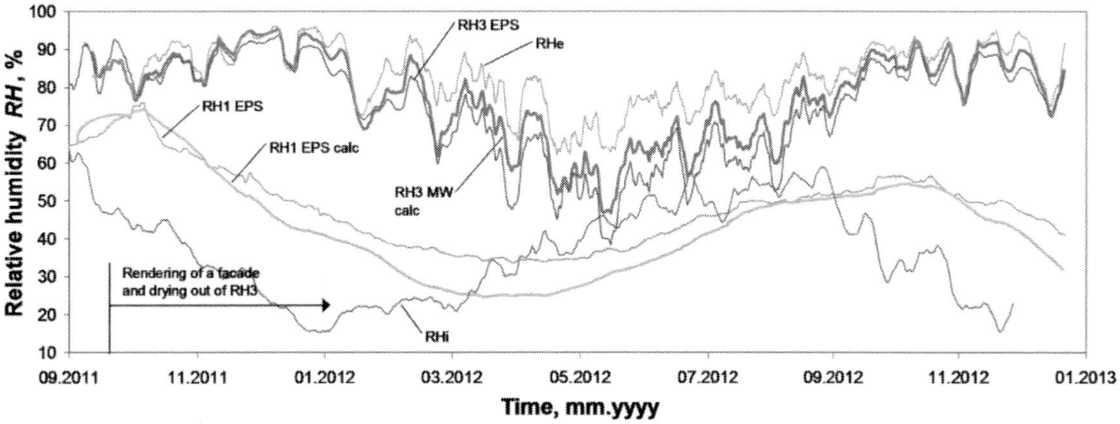

Figure 17 *Measured and calculated RH in the wall additionally insulated with EPS.*

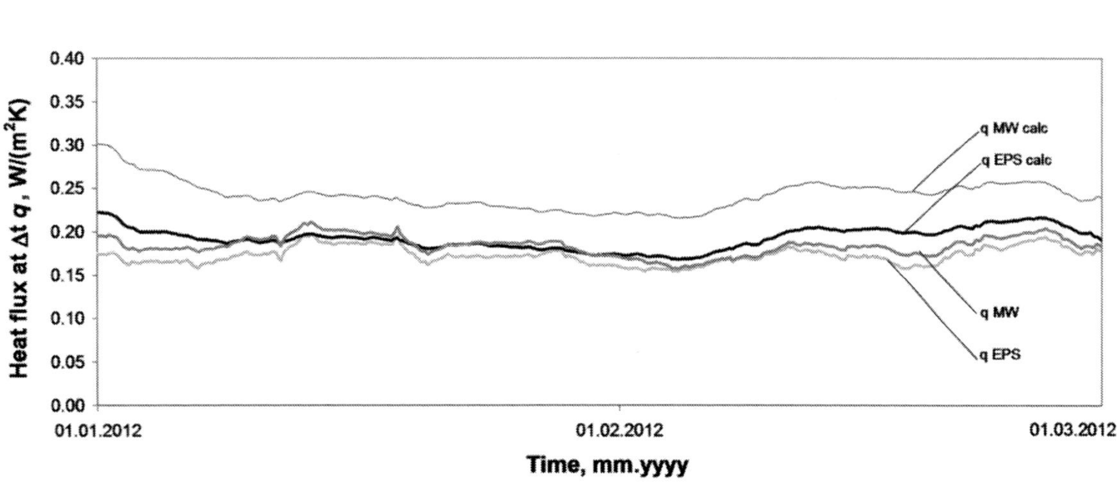

Figure 18 *Measured and calculated thermal transmittances in the wall additionally insulated with EPS and MW. Lines represent weekly average values.*

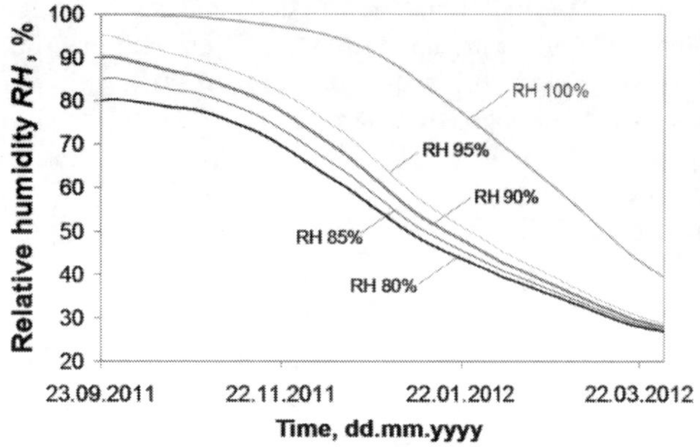

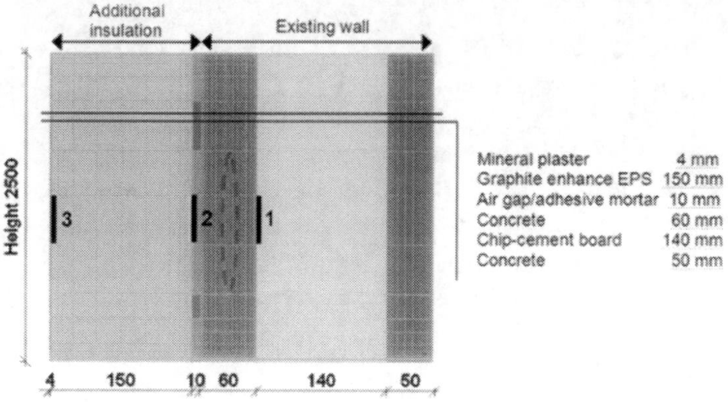

Figure 19 *Moisture drying out time from the external layer of concrete between measuring points 1 and 2 (right) with different initial moisture.*

exhaust fan was set to minimum speed soon after the installation due to energy saving, lower sound level, and better thermal comfort. Also, building fabric became more airtight with new windows and additional insulation that decreased infiltration. The indoor moisture excess 2–3 g/m^3 after the renovation indicates average indoor humidity load level. This means that moisture production/living density had been relatively low.

There was a problem with low floor temperatures on the first floor. The floor was not insulated from the bottom side (cellar's ceiling) because of lack of finance, low height of cellar, and impracticability of workmanship (pipes, lighting, cables, etc. are attached to the cellar's ceiling).

Thermal Bridges

During typical low-budget energy renovation, only old wooden windows are replaced and previously replaced windows remain in the same location. Since the thickness of additional insulation into walls is typically 150–200mm, windows become located 200–270 mm inside the surface of the façade instead of 50–70mm originally. This is a problem because of esthetics and architecture, reduced day lighting, energy efficiency, and mainly thermal bridges around the windows. Negative impact is greater in the case of small and narrow windows. Generally, some insulation is recommended to be placed at window's jamb, but in practice this is complicated and often only 5–20mm is achievable. If the temperature factor f_{Rsi} does not change much, then the linear thermal transmittance Ψ is very sensitive to the jamb's insulation. A remarkable impact of a window's position in the wall and sensitivity on the jamb's insulation has been also found by Hens and Carmeliet (2002). Heat loss through thermal bridges around the windows compared to the rest of a wall is often at a similar scale. In the future, moving of all the windows towards outside is advisable. Also, most of the windows installed about 10 years ago can be replaced again since selective covering was not yet common then and remarkable development of windows has taken place during last years.

Another thermal bridge that remained after the renovation is the external wall/balcony junction. It could be eliminated by insulating a balcony slab from the bottom and top. Technical condition of concrete balconies and awning is sometimes problematic and steel elements, including reinforcement mesh, is sometimes corroded. Since after additional insulation the original 1 m width of a balcony becomes only about 0.8 m, removal of all existing balconies should be considered. New, durable, and wider balconies containing minor thermal bridges would be a sustainable investment.

Hygrothermal Performance of ETICS on a Concrete Element

One of the main goal of this study was to analyse hygrothermal performance of an external wall insulated additionally with mineral wool and expanded polystyrene ETICS. Initial simulation results considering only diffusion showed remarkable mismatch during the first heating season. Although the adhesive mortar/air gap was sealed with PU foam at the basement wall and roof, there was probably some air leakage. Air leakage between the insulation and original wall (measurement point 2) is possible because of rubble bulk covered with particles of different sizes. This in contact with adhesive mortar did not assure totally airtight connection. The air pressure difference created by stack effect enabled some air movement along the wall. Calculations show that a narrow 0.5–1 mm opening below and above the wall is able to transport more moisture than by diffusion. Cracks with such a small width could originate from shrinkage of monolitization concrete connecting pre-fabricated elements or those could be in the elements produced from three layers. Considerable impact of air gaps in the ETICS has been also found by Olson and Taylor (2009) and Sedlbauer and Krus

(2002). In Sedlbauer and Krus (2002), a drying out time of up to 8 years is stated.

The impact of air convection between the additional thermal insulation and original wall is small enough not to appear in the results of thermal transmittance of the wall in practice, see Figure 18. Small impact of air convection to thermal transmittance is also found by Hens and Carmeliet (2002) where 2 mm air gap increased thermal transmittance 12% at ETICS on masonry. Higher calculated thermal transmittance might be caused by air leakage through MW insulation. Slightly higher thermal transmittance in case of MW can be explained with lower thermal conductivity of EPS, having declared value $\lambda_D = 0.032$ W/(m·K) compared to $\lambda_D = 0.038$ W/(m·K) in the case of MW. The addition of 150 mm of additional EPS/MW insulation contributes more than 2/3 to the total thermal resistance of the wall.

Measurements and calculations showed that moisture in the analysed wall dried out in half of the heating season (see Figure 11 and Figure 19), approximately 3–4 months. If dried only by diffusion, it would have taken about a year for most of the moisture and the rest of it would have dried out during the second year. Slow drying out by diffusion from ETICS on masonry is also declared by Hens and Carmeliet (2002). In the analyzed case with convection totally wet, an external layer would have dried out from most of the moisture also within 3–4 months but in that case, large moisture flux through the wall and possible condensation behind the plaster would be a threat. In that case there would probably be some drying out toward indoors as found in work made by Hens and Carmeliet (2002).

Different aspects of hygrothermal performance of ETICS have been studied for decades. There should be separate viewpoint between ETICS on concrete/masonry and wooden structures. Although it has been found to be a functioning solution in principle (Karagiozis and Kumaran 1997; Holm and Künzel 1999; Kvande 2008; Künzel, H.M. and Zirkelbach, D. 2008) it is very sensitive to driving-rain-caused water leakage from cracks in finishing layer (Bomberg et al. 1997; Krus et al. 2008). Already, 1% of driving rain leakage can be crucial (Künzel 1998). Performance of ETICS has also been studied by Balocco at al. (2008). Solid computational analyses are done by Vares et al. (2012) and Desjarlais (2001). Literature review is conducted by Cheple and Huelman (2000).

According to Zwayer (1995), performance of ETICS walls depends on the design and materials choice, connections between plaster and other materials in joints, workmanship, technology, and quality. In the cold climate of Finland, Pakkala (2011) has found poor frost resistance of ETICS to be a problem and testing methods and criteria according to ETAG 004 worked out in central Europe might not be proper.

Tested walls had also minor driving rain loads since it was sheltered by a neighboring higher building and because of other factors of the surrounding environment. Performance of ETICS walls under high driving rain loads and in the case of higher buildings must be handled as another field of research that was not the aim of this paper.

Also, durability and factors impacting it in addition to climate loads (content and type of plaster, vapour permeability, water uptake coefficient, adhesion, cracking, frost resistance, etc.) is a field of the future research. During maintenance of ETICS, one should also consider a possible layer of painting that might be added later.

General Success of Current Energy-Renovation Pilot Project

The main progress of the current energy renovation pilot project was its better design and workmanship quality. It must be noticed that since it was a pilot project, workmanship under strict supervision was probably somewhat better than usual. From now on, similar renovations can be started with existing know-how about success and problems. In addition to lower heating bills, renovation should lead also to better durability, longer service life of structures, and better indoor climate and aesthetics. These all have also a positive impact on the value of real estate. One should also move on towards the next more progressive steps by modernizing older housing stock.

Similar low-cost renovation described in this paper is wide-spread in Estonia and in neighboring countries today. Often renovation is conducted in many stages without a holistic approach. Main renovation solutions employed are typical: additional external insulation and changing of old wooden windows. Together with the change of membrane waterproofing, an additional thermal insulation is added to the roof. The renovation of balconies and awnings depends on damages.

Overall success of this example project can acquire satisfactory evaluation avowals. A novel solution in this pilot project was holistic approach and an air heat pump for heat recovery of the new mechanical ventilation. Attitudes of inhabitants to the renovation varied both before and after renovation. In the current pilot project, people were somewhat more tolerant since approximately half of the whole investment came from sponsors and subsidizers. Still, there were complaints about designed solutions, execution of workmanship, and some unfamiliar changes.

CONCLUSIONS

This case-study analysis studied measured and calculated hygrothermal performance of a mineral wool (MW) and graphite enhanced expanded polystyrene (EPS) external thermal insulation composite systems (ETICS/EIFS). Also, indoor climate, air leakage rate and thermal bridges were measured before and after the low-budget energy renovation of an existing concrete large panel elements apartment building.

Indoor climate measurements showed no major changes in moisture excess, mostly because mechanical ventilation airflows were decreased by the operator. Overheating that existed before was avoided after the renovation with a new adjusted heating system. Airtightness of the building fabric improved only in apartments where old leaky windows were replaced. Additional external insulation solves the serious problem of thermal bridges, except at external wall/window junction if windows remain at their original position and also at external wall/balcony and external wall/foundation wall junction. Therefore, windows should be attached to the external side of the existing façade.

Measured hygrothermal performance of both walls was good and correlation existed with the calculated temperature, vapour pressure, and RH distributions. Built-in moisture of the whole wall (measuring spots 1 and 2 on both sides of the external layer of concrete) dried out during the first heating season, achieving RH values below 50%. Moisture in thin plastering (measuring spot 3 between insulation and plaster) dried out within few days but stayed quite high, especially in the case of mineral wool, being over 90%. It means that the content and properties of the external finishing plaster must be carefully designed to achieve low enough vapour permeability and low water uptake. This becomes even more crucial at a greater vapour flow due to material with higher vapour permeability instead of concrete. Average measured thermal transmittance U during the winter was 0.17 W/(m^2·K) in the case of graphite enhanced EPS and 0.19 W/(m^{2}K) for mineral wool. This is close to the expected value calculated from a producer's data and allows us to make positive conclusions about thermal performance. Hygric performance was also satisfactory and moisture dried out during the first heating season. Measurements and calculations showed that there had been some convection of moist air in the wall. A drying out process by diffusion only would have been much slower.

Short term hygrothermal performance of ETICS is normal when proper design and workmanship is performed, in particular, related to joints. Additional insulation lowers the RH inside the existing wall that stops the corrosion process and frost damages. Long term performance and durability of ETICS in cold climate and under driving rain loads needs to be further investigated with special attention to content and properties of render, frost resistance, cracking, staining, maintenance, and repainting.

ACKNOWLEDGMENTS

The research has been conducted as part of the IUT1–15 project "Nearly-zero energy solutions and their implementation on deep renovation of buildings."

REFERENCES

Balaras, C.A., Droutsa, K., Dascalaki, E., Kontoyiannidis, S. 2005. Deterioration of European apartment buildings. *Energy and Buildings*, 37 (5), pp. 515–527.

Balocco, C. / Grazzini, G. / Cavalera, A. Transient analysis of an external building cladding, *Energy and Buildings*, 40 (7), p.1273–1277, Jan 2008.

Bomberg, M. / Lstiburek, J. / Nabhan, F. Long-Term, Hygrothermal Performance of Exterior Insulation and Finish Systems (EIFS), *Journal of Building Physics*, 20 (3), p.227–248, Jan 1997.

Cheple, M. and Huelman, P. (2000). Literature Review of Exterior Insulation Finish Systems and Stucco Finishes, Report MNDC/RP B80-0130, University of Minnesota.

Desjarlais, A.O., Karagiozis, A.N., and Aoki-Kramer, M. (2001): Wall Moisture Problems in Seattle. Buildings VIII proceedings, ASHRAE, 8p.

Dol, K.; Haffner, M. 2010. Housing Statistics in the European Union 2010. OTB Research Institute for the Built Environment, Delft University of Technology.

Dall'O', G., Galante, A., Pasetti, G. A methodology for evaluating the potential energy savings of retrofitting residential building stocks. *Sustainable Cities and Society*. 4-1, October 2012, Pages 12–21.

Economidou, M., Atanasiu, B., Despret, C., Maio, J., Nolte, I., and Rapf, O. (2011). European buildings under the microscope: A country-by-country review of the energy performance of European buildings. Buildings Performance Institute. Europe 2011.

EN ISO 10211 *Thermal bridges in building construction - Heat flows and surface temperatures - Detailed calculations*. Part 1:2000 and Part 2:2001.

EN ISO 13788 *Hygrothermal performance of building components and building elements - Internal surface temperature to avoid critical surface humidity and interstitial condensation - Calculation methods.*

EN 13187 *Thermal performance of buildings - Qualitative detection of thermal irregularities in building envelopes - Infrared method.*

EN 13829 *Thermal performance of buildings - Determination of air permeability of buildings - Fan pressurization method.*

Hagentoft, C. E. 2010. Importance of a correct overall performance assessment – Probability assessment of performance and cost. Large scale national implementation plans for building airtightness assessment: a must for 2020. We should start now to be ready in 2020. June 14–15, 2010 Brussels, Belgium.

Hens, H. and J. Carmeliet, (2002), Performance prediction for masonry walls with EIFS using calculation procedures and laboratory testing, *Journal of Thermal Envelope and Building Science*, 25(03).

Holm, A. and Künzel, H.M, (1999), Combined effect of temperature and humidity of the deterioration process of insulation materials in ETICS, Building Physics in the Nordic Country, August, Gothenburg.

Ilomets, S.; Kalamees, T.; Agasild, T.; Õiger, K.; Raado, L. M. (2011). Durability of concrete and brick facades of apartment buildings built between 1960–90 in Estonia. In proceedings: XII DBMC International Conference on Durability of Building Materials and Components, Portugal, Porto, 2011, 1171–1178.

Holopainen, R., Tuomaala, P. (2011) Effect of energy renovation on thermal sensation and comfort during heating season. 9th Nordic Symposium on Building Physics - NSB 2011 p. 1069–1076.

Kalamees T; Õiger K; Kõiv T. A., et al. 2009. Technical condition and service life of Estonian apartment buildings, built with prefabricated concrete elements. Tallinn University of Technology (in Estonian).

Kalamees, T., 2006. Hygrothermal Criteria for Design and Simulation of Buildings. Doctoral theses, Tallinn University of Technology, Tallinn: TTU Press.

Karagiozis, A. and Kumaran, K. (1997). Drying Potential of EIFS Walls: Innovative Vapor Control Strategies. STP 1339, American Society for Testing and Materials (ASTM), West Conshohocken.

Krus, M., Hofbauer, W. and Lengsfeld, K. Microbial Growth on ETICS as a Result of the New Building Technology? Building Physics Symposium. Leuven, Belgium 2008-10-29 - 2008-10-31.

Kvande, B. Bergheim Einar Thue, Jan Vincent. Durability of external thermal insulation composite systems (ETICS) with rendering. Building Physics Symposium Leuven, Belgium 2008.

Künzel, H.M.: Drying of Masonry with Exterior Insulation. Proceedings of the Fifth International Masonry Conference. British Masonry Society, No. 8, Stoke-on-Trent 1998, pp 245–250.

Künzel, H.M.; Zirkelbach, D. Influence of rain water leakage on the hygrothermal performance of exterior insulation systems. In Rode, C.; Technical University of Denmark -DTU-, Lyngby: 8th Nordic Symposium on Building Physics in the Nordic Countries 2008. Proceedings. Vol.1: Copenhagen, June 16–18, 2008 Lyngby, 2008 (DTU Byg Report 189).

Nemry, F., Uihlein, A., Colodel, C.M., Wetzel, C., Braune, A., Wittstock, B., Hasan, I., Kreißig, J., Gallon, N., Niemeier, S., Frech, Y. Options to reduce the environmental impacts of residential buildings in the European Union-Potential and costs. Energy and Buildings Volume 42, Issue 7, July 2010, 976–984.

Nicolai, A. 2008. Modeling and Numerical Simulation of Salt Transport and Phase Transitions in Unsaturated Porous Building Materials, Dissertation, Syracuse University.

Olson P.E., E.K., Taylor, J.A. 2009. Retrofitting barrier exterior insulation and finish system: Methodology and performance assessment. ASTM Special Technical Publication 1493 STP, pp. 229–237

Sedlbauer, K. and Krus, M., (2002), Information and Technology Transfer from IBP: Mold Growth on ETICS (EIFS) as a Result of "Bad Workmanship"?, *Journal of Thermal Envelope and Building Science*, vol. 26, no. 2, pp. 117–121.

Silva, P.C.P., Almeida, M., Bragança, L., Mesquita, V. Development of prefabricated retrofit module towards nearly zero energy buildings. *Energy and Buildings*, Volume 56, January 2013, 115–125.

Uihlein, A., Eder, P. Policy options towards an energy efficient residential building stock in the EU-27. 2010. *Energy and Buildings* 42 (6), pp. 791–798

Vares Sirje, Pulakka Sakari, Toratti Tomi, Fulop Ludovic, Hradil Petr, Vesikari Erkki, et al. 2012. Sustainable refurbishment of exterior walls and building facades. Final report, Part B – General refurbishment concepts. VTT TECHNOLOGY 33. ISBN 978-951-38-7849-8, ISSN 2242-122X. VTT. Espoo Finland. 244 p. + App. 30 p.

Roof and Attic Design Guidelines for New and Retrofit Construction of Homes in Hot and Cold Climates

William Miller, PhD
Member ASHRAE

Andre Desjarlais

Marc LaFrance

ABSTRACT

Guidelines for improving the energy efficiency of roofs and attics are documented from research conducted for the Department of Energy (DOE) Building Technologies Office. The results of combined analytical and experimental studies were used to benchmark computer tools, which in turn, were used to simulate homes in hot and cold climates. Simulations of adding floor and roof insulation, above-deck ventilation, radiant barriers, cool-color shingles, and metal or tile roofs; sealing the attic floor and the duct system; and making the attic a semi-conditioned space (sealed attic) were performed to compute the cost of energy savings. Results are prioritized to help building owners make an informed economic decision when contemplating roof and attic retrofits. Sealing the attic floor is a top retrofit option. The approach of making the attic a semi-conditioned space (sealed attic) and exploiting a new prototype roof assembly—an insulated and ventilated roof—are good options for retrofit work but have paybacks ranging from 15 to 25 years. A new sealed attic concept is proposed, and computations show its simple payback is about 10 to 12 years in hot and cold climates; its first cost is significantly reduced from that of a spray foam approach. For new construction the best option is to keep the ducts out of the attic, make sure the attic floor is sealed to limit whole-house air leakage, and add at least the code level of insulation to the ceiling.

INTRODUCTION

Several conservation measures are described for improving the energy efficiency of new and retrofit roofs and attics. We define the attic to broadly include the roof as well as the space between the roof and the finished ceiling. The attic often includes air-conditioning ducts and a heating, ventilating, and air-conditioning (HVAC) system. In the United States (mostly the south), single-family homes with slab foundations rose from 26% of homes built in 1971 up to 70% of homes completed in 2011 (CB 2011). These statistics imply a growing trend to place HVAC and air distribution systems in the attic. In the colder northern climates, basements dominate about 80% of the residential market, and a smaller percentage of HVAC systems are placed in attics (CB 2011). As land demand and prices have risen, however, the percentage of two-story homes built in the northeast has jumped from 44% in 1971 up to 81% of homes built in 2011.

Dodge (2002) reported that in about 85% of U.S. homes, existing worn-out roofs have been replaced with asphalt shingles, as shown in Table 1. The consensus shows that retrofit work predominates over new construction as homeowners upgrade their home. Therefore, the paper first focuses on retrofit options that reflect the present market; it then addresses options for roof and attic systems.

Demographics provided by the U.S. Census Bureau (2009A, 2009B) reveal that roughly half of the existing 112 million homes in the country were built from about 1960 to 1979. ASHRAE 90-1980 was the code of practice; however, few states measured or enforced code compliance because of the complexity and or expense of the task. Too often a state's adoption of the current IECC code lags the code of practice by several years (U.S. DOE 2010).

William Miller *is a building scientist and* ***Andre Desjarlais*** *manages the Building Envelope program at Oak Ridge National Laboratory, Oak Ridge, TN.* ***Marc LaFrance*** *serves as an analyst on building energy and policy at the International Energy Agency, Paris, France.*

Published by ASHRAE 2013

Several roofing contractors with businesses in hot climates were surveyed, and almost all contractors commented that the homes they worked in had at best R_{US}-10 (R_{SI}-1.7)[1] on the attic floor because they could see the ceiling joists in the attics. The observation is very important because it controls the level of savings for retrofit applications.

Table 2 shows current ASHRAE 90.2 (ASHRAE 2007) practice and ASHRAE 90 (ASHRAE 1980) practice. The current ASHRAE 90.2 practice was used to estimate savings for new construction. ASHRAE 90-1980 was selected as the base for cold climates; however, R_{US}-10 (R_{SI}-1.7) was assumed for hot climates.

The best retrofit option for your attic depends on climate, attic geometry, duct arrangement, amount of ceiling insulation, air leakage, and thermostat setting used to comfort condition the home. If the ceiling currently has less than code insulation, savings will be greater and payback period will be shorter. If the ductwork has recently been repaired, sealed, and insulated, the savings for other measures will be less. Savings are also affected by the run time of the HVAC unit.

The breakdown of fuels used for cooling and heating is also very interesting, with electricity being the fuel of choice for comfort cooling. However, natural gas and bottled LP gas are used in about 56% of all American homes. Even in New York state about 3.8 million homes use natural gas, while some 2.2 million homeowners in New York utilize fuel oil to heat their homes (EIA 2005). A SEER 13 heat pump and a 0.85 energy efficient gas furnace are assumed when converting saved cooling and heating energy into costs using EIA 2012 state average fuel prices. Electricity rates are based on EIA (2012a) data, while comfort heat is based on the prices for natural gas in EIA (2012b), as shown in Table 2.

LEAKY DUCTS IN ATTICS

The convenience of the attic space is very appealing to builders, and all too often they install the HVAC unit and ducts in the attic. Parker et al. (1993) simulated the effects of ducts on space conditioning Florida homes and observed that air leakage and heat transfer to the duct were major contributors to the peak electrical burden on the utility. Energy costs are also increased if the attic floor leaks air to or from the conditioned space. Walker (1998), in his bibliography of duct leakage, provides average leakage rates of 10% to 20% of fan flow, as shown in Table 3.

The impacts of supply and return leaks can be different because the origin of the leak can affect airflow patterns in the attic. However, the computer simulations are based on overall energy balances for the attic and attic air; therefore, equal supply and return leakages were used in simulations.

The computer tool AtticSim (Version 11)[2] with input from Energy Plus was exercised to estimate potential savings.

[1] R_{US} represents the thermal resistance of ceiling insulation and has units: (h·ft²·°F)/Btu; R_{SI} units: (m²·°C)/W.

Table 1. Number of Squares (100 ft²) of New and Retrofit Roof Products Installed in the United States

	Asphalt Shingles	Metal	Concrete	Clay	Wood Shakes	Other
New	29,955,734	1,705,073	1,965,500	323,763	280,821	461,263
Reroof	115,054,533	9,550,899	1,657,307	892,926	6,445,277	4,400,195
Total	145,010.267	11,255,972	3,622,807	1,216,689	6,726,098	4,861,458

Table 2. Cities Representing the ASHRAE Standard 90.2 Climate Zones Selected for Simulating Retrofit and New Construction[a]

				ASHRAE Standard 90.2 New Construction		ASHRAE Standard 90-1980 Retrofit Construction		Fuel Costs	
				Ceiling	Duct[b]	Ceiling	Duct[c]	Electricity (EIA 2012a)	NG (EIA 2012b)
Zone	City, State	HDD_{65}	CDD_{65}	R-Value, (h·ft²·°F)/Btu		R-Value, (h·ft²·°F)/Btu		¢ per kWh	$/1000 CF
1	Miami, FL	222	9368	30	8	20	5.5	12.52	21.69
2	Austin, TX	1481	7435	30	8	20	5.5	12.88	13.79
3	Atlanta, GA	2614	4814	30	8	20	5.5	9.55	18.5
4	Baltimore, MD	4731	3598	38	8	21	5.5	14.57	16.05
6	Minneapolis, MN	7787	2513	49	8	29	5.5	9.66	11.3

[a]Retrofit cases assumed the ceiling and duct levels of insulation listed under the column ASHRAE 90-1980 to compute the cost of energy savings.
[b]Inspected duct system assumes 4% air leakage.
[c]Existing duct system assumed 10% and 20% air leakage. ASHRAE 90-1980 does not address leakage.

The HERS BESTEST model home (NREL 1995a) was used to simulate a house having an attic footprint of 1550 ft^2 (144 m^2). The roof slope was set at 18° and a dark heat-absorbing shingle roof having solar reflectance of 0.10 was assumed for hot and cold climates using TMY2 (NREL 1995b) weather data. A supply and return duct placed in the attic was modeled by AtticSim. Energy Plus does not compute the radiation coefficients between the duct and seven attic surfaces. However, we used Energy Plus to estimate the hourly run times for a 2-ton heat pump certified with a seasonal energy efficiency ratio (SEER) of 13. The hourly indoor air temperature for the house, the run time for the HVAC, and the flow rate of air in the ducts were computed by Energy Plus and read by AtticSim to better estimate the roof and attic load as coupled to the building envelope. The procedures developed by Fairey et al. (2004) were used to de-rate the SEER and HSPF for climatic impacts. Savings are computed using as base a conventional stick-built roof and attic with 1:300 opening at the soffit and ridge.

Figure 1 illustrates why duct renovations should be a priority in a building program. The turquoise bars represent energy use for a roof and attic where the roof is dark and the attic is poorly insulated and has 20% leaky ducts and a leaky attic floor. The dark blue bars are for the same roof and attic but with a 10% leaky duct. The orange bars show energy use where the practitioner repaired the leaks in the attic and sealed and rewrapped the ducts in R_{US}-8 (R_{SI}-1.4) insulation. The green bars show the benefits of sealing the attic floor and moving ducts into the conditioned space. Often it is not feasible to remove ducts from the attic and totally eliminate duct losses, but the simulations vividly show the magnitude of the losses from the duct system. A new prototype roof studied by Miller et al. (2011) is also shown (black color bars) having the same dark shingle roof but with the attic floor sealed and the ducts in the conditioned space.

Figure 1 is of keen interest because it shows that adding ceiling insulation (view *x*-axis) reduces the energy bill for all roof and attic assemblies whether with or without a duct system. But the heat transfer tends to level off and there are diminishing returns for adding floor insulation above about R_{US}-19 (R_{SI}-3.3)[3] because losses from the ducts predominate. Therefore, if all one does is put more insulation down and ignore the HVAC system, then hundreds of dollars' worth of energy is still being lost from the HVAC in the attic. Frankly, adding only insulation still leaves a poorly performing home that has higher-than-necessary utility costs month after month.

The green bars (Figure 1) show energy savings if the ducts are in the conditioned space. Add to that a new roof design (Miller et al. 2011), and the annual roof and attic load drops from 45 MBtu/year down to 4.5 MBtu/year, about a 40 MBtu annual savings in energy in Austin, TX (Figure 1 viewed at R_{US}-30 [R_{SI}-5.3] level of ceiling insulation). The simple payback for this system is about 10 to 15 years, depending on the climate, the initial level of floor insulation, and the temperature settings on the thermostat. That 40 MBtu translates into about 4000 kWh, which in turn for electricity at 10¢/kWh, yields annual savings of $400. Therefore, implementing a single strategy is good, but more can be accomplished. Of course, renovations are more costly and most people want an affordable roof replacement. The initial cost and the simple payback are the major hurdles for market acceptance.

Table 3. Survey of Homes Checked for Duct Leakage

Literature Cited in Walker (1998)	Supply Duct Leakage	Return Duct Leakage
Cummings et al. (ACEEE 1990) 91 homes surveyed	10%	10%
Jump et al. (ACEEE 1996) 24 homes surveyed	17%	16%
Modera et al. (CEC 1995 report) 100 homes (new construction)	8%	7%

Source: Data is courtesy of Walker (1998).

PRIORITIZING RETROFITS

The attics of the homes used to estimate the above savings, which represent those built between 1960 and 1979, have at best R_{US}-5.5 (R_{SI}-0.97) insulation wrapped around leaky ducts operating in the attic. Air leakage of the ductwork

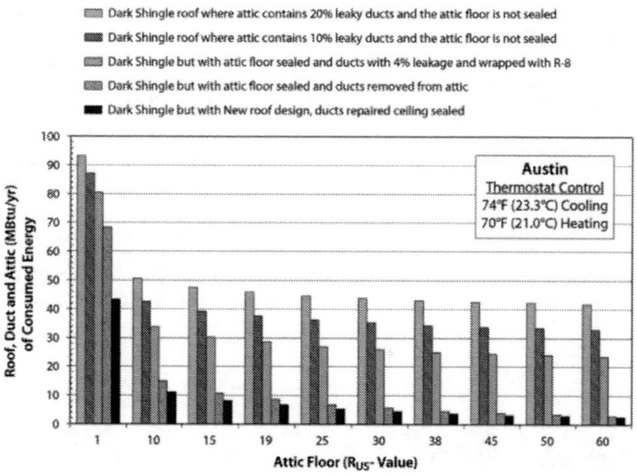

Figure 1 *Comparison of energy impacts of leaky ducts in attic space, sealing attic floors, insulating attic floors, and totally eliminating energy losses from HVAC ducts in unconditioned attics.*

---

2. AtticSim version 11 originated from the ASTM C1340 (2004) protocol. Version 11 contains revisions for above sheathing ventilation benchmarked by Miller et al. (2007).

3. Climate Zones 1 through 3 require by ASHRAE Standard 90.2-2007 code R_{US}-30 (R_{SI}-5.3).

is unknown; however, based on measurements made for a sampling of homes (Table 3), simulations for retrofit options assumed air losses of 4%, 10%, and 20% of supply airflow. Two different thermostat duty cycles for controlling the HVAC were implemented for the savings computations. The one described herein as "conservation mode" is for those interested in energy conservation (home at 68°F [20°C] in the winter and 78°F [25.6°C] in the summer). The other thermostat control is for comfort control (70°F [21°C] in the winter and 74°F [23.3°C] in the summer), herein referred to as comfort control. These conditions were used to compute a range for simple paybacks for implementing a given retrofit practice.

Seal the Attic Floor

The most important retrofit option, whether the duct system is in the attic or not, is to seal the attic floor. The ceiling is part of the home's perimeter, and air leakage is the major source of heat gains and losses to the home. d'Ambrosio et al. (2012) cites ceiling leakage as contributing to about 18% of the typical leakage in a home. Savings were estimated more conservatively based on 65 cfm (0.03 m^3/s) of air leakage crossing the ceiling, which contributes about 5% to the whole house leakage assumed at 5 ach by IECC 2012 code. Sealing yields annual savings ranging from just \$31–39 in Miami to \$126–132 in Atlanta. The cost of the renovation is about \$0.40 per square foot, and the savings payback for the renovation is about 16–22 years for a home in Miami, 8–9 years in Austin, 5 years in Atlanta, and 4 years in Baltimore, as shown in Table 4.

Sealing the attic floor saves money (Table 4); however, if the intent is to make the attic a semi-conditioned space, then sealing the attic floor will have only secondary effects on building load. Sealing the attic floor in conjunction with repair of a duct system operating in the attic is a viable option because both are part of the perimeter and the repairs reduce the overall house leak rate, which, in turn, reduces the building load.

Repair the Ducts in the Attic

Table 5 shows the savings for repairing only the ductwork. Again, these are based on a pre-1980 home built to the historical code (ASHRAE 1980) in cold climates. For hot climates, the ceiling of the control assumed R_{US}-10 (R_{SI}-1.8). As noted, a SEER 13 air conditioner was assumed for comfort conditioning. Repairing the ductwork saves \$46 for 10% leaky ducts with the air conditioner set for energy conservation in Miami but increases to \$138 if the duct leakage is as high as 20% and thermostat controls are for occupant comfort. Therefore, savings can vary significantly depending on duct leakage and thermostat setpoint. In Miami the base home has computed heat gains from the roof and attic that cost the homeowner about \$230 per year with the air conditioner set for energy conservation.

The EIA (2010) publishes typical residential electric and gas heating bills for various U.S. cities and for Miami the roof and attic contributes to about 13% of the simulated whole house electrical bill. For Austin a homeowner would pay \$457 per year for a poorly performing duct system in the attic, and the roof and attic is about 20% of the home's total use of gas and electricity. For Austin the electricity attributed to a conventionally designed roof and attic is about 25% of the total kilowatt-hour draw stated by the EIA (2010) for the Texas census region. Therefore, the consumption rate is within reason, considering the whole building load.

In Baltimore's colder climate, the pre-1980 home's roof and attic is roughly 30% of the total energy consumed by the home. Therefore, as the heating load increases, so does the cost for having a duct system in the unconditioned attic, and the payback for duct renovation becomes more and more enticing. The simple payback can be as low as 1 year for a home with 20% leaky ducts and thermostat set for comfort control. It is also seen that cold climate design of roof systems and attics is even more important than hot climate design because on a national basis, residential homes use 4.3 quads of total site energy for heating as compared to 0.64 quads for cooling (EIA 2013).

Radiant Barriers

Stovall et al. (2010) documented the performance of radiant barrier systems in limiting the radiation heat transfer occurring in the attic. Shrestha et al. (2013) showed it most effective when attached to the underside of the roof rafters so that the foil acts as a dual-acting radiation shield and as a channel for directing buoyancy-driven hot air in from the soffit and out the ridge vent, as in Figure 2. The radiant barrier foil costs about \$0.15 per square foot of material, and the labor to attach the foil directly to the underside of the roof rafters is estimated at \$0.65 per square foot[4] of coverage for a 4:12 pitch roof.

Table 4. Cost of Energy Attributed to the Base Roof, Attic, and Leaky Duct and the Subsequent Monetary Savings in Energy and Payback for Sealing the Attic Floor Having Footprint of 1550 ft^2 (144 m^2)[a]

	Miami	Austin	Atlanta	Baltimore
Base 1980 roof and attic	\$230–398	\$457–652	\$714–991	\$925–1287
Retrofit savings[b]	\$31–39	\$74–75	\$126–132	\$155–162
Payback[b] (years)	16–20	8–9	≈5	≈4

[a]All data ranges include values for attics containing leaky and poorly insulated ducts with 10% and 20% leakage.
[b]Sealing the attic floor eliminates 0.26 air changes per hour between the home and the attic.

Table 5. Cost of Energy Savings and Simple Payback for Repair of Ductwork Simulated with 10% and 20% Leakage in the Unconditioned Attic Having Footprint of 1550 ft² (144 m²)[a]

	Miami	Austin	Atlanta	Baltimore
Base 1980 roof and attic	$230–398	$457–652	$714–991	$925–1287
Retrofit savings[b]	$46–138	$85–216	$120–348	$169–470
Payback (years)[b]	4.5–56	3–30	2–21	1–15

[a]All data ranges include values for repairing ducts of 10% and 20% initial leakage.
[b]Repair includes replacing R_{US}-5.5 with R_{US}-8 insulation and sealing duct to 4% of flow leakage. Indoor air handler is assumed to be in conditioned space.

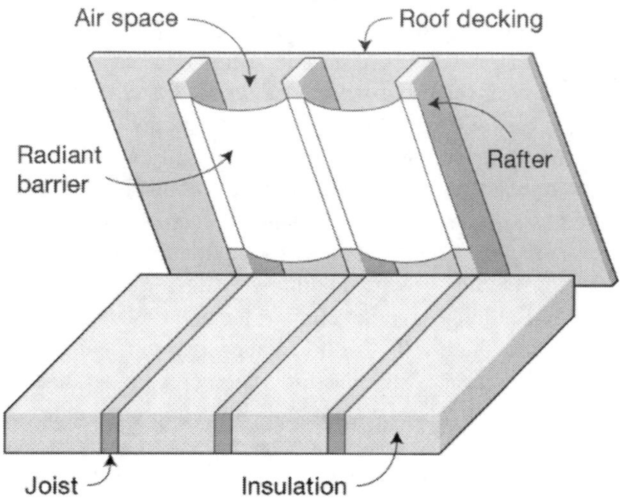

Figure 2 *Radiant barrier is installed to the underside of the roof rafters.*

Therefore, retrofitting the attic with a radiant barrier would cost about $0.80 per square foot of attic footprint.

Another alternative is to spray paint the attic interior with low emittance paints. The application qualifies as an internal radiation control coating (IRCC) if the paint has a thermal emittance less than or equal to 0.25. The cost of material and labor is comparable to the foil application.

Simulations for hot climates indicate that a radiant barrier can pay for itself in 20–25 years. Without leaky ducts in the attic, the payback increases to about 38–50 years. In cold climates the radiant barrier is not an effective measure.

Adding Insulation to the Attic

Adding insulation to meet IECC code (R_{US}-30 [R_{SI}-5.3]) has an incremental material and labor charge of $1.04 per square foot (RS Means 2011) for adding R_{US}-20 (R_{SI}-3.5) insulation in Miami, Austin, and Atlanta. Code for Baltimore is R_{US}-38 (R_{SI}-6.7), and the additional insulation needed to meet code costs $1.38 per square foot. The savings (Table 6) are estimated with no ductwork in the attic and also with 20% leaky ducts in the attic. In hot climates adding R_{US}-20 (R_{SI}-3.5) to a ceiling already covered with R_{US}-10 (R_{SI}-1.8) and with no ducts in the attic yields annual savings ranging from $64 in Miami to $238 in Baltimore (Table 6).

However, if the attic contains 20% leaky ducts, the savings are reduced as compared to having no ducts in the attic. This occurs because the leaky ducts actually condition the attic space, which reduces the heat transfer crossing the floor insulation. So, adding insulation to an attic that does not include ductwork has a larger savings than an attic that contains leaky ductwork. Savings also increase as the climate incurs more and more days of required comfort heating. Simple payback is 10–12 years in Atlanta and 9–11 years in Baltimore.

[4] Estimates for labor based on costs provided by local distributors conducting attic renovations.

Cool-Color Shingles

Dark roofing can be formulated to reflect like a highly reflective white roof in the near infrared (NIR) portion of the solar spectrum. ORNL and LBNL collaborated on a cool roof study for the California Energy Commission and worked with pigment manufacturers to identify and characterize pigments with optical properties suitable for cool-color roof products (Levinson 2004a, 2004b).

Simulations for a home with cool-color shingles (solar reflectance of 0.25) on the roof in the hot climates of Miami and Austin show a reduction in the cooling load but also a slight increase in heating load, which results in a net annual savings of only $6–9 in Austin and an actual penalty in Atlanta (Table 7). The simulation was conducted with only R_{US}-10 (R_{SI}-1.8) on the attic floor to view the benefits of the cool roof shingle. Shingles with solar reflectance of 0.40 and 0.70 (white roof) were also simulated (Table 7).

In Austin and Atlanta the net savings for a 0.40 solar reflective shingle is respectively $13–18 and –$11–10, assuming R_{US}-10 (R_{SI}-1.8) on the attic floor. The cost premium for cool-color shingles according to local distributors installing the product is about $50 per 100 ft² (Mullenax 2010). Therefore, the simple payback for a commercially available 0.25 solar reflective shingle is on the order of 50 years.

However, in many existing homes, especially in the southwestern United States, the roof pitch is less than 4 in 12, making it difficult to renovate the attic. For example, retrofit-

Table 6. Cost of Saved Energy and Payback for Adding Code-Level of Insulation to the Attic Floor Having Footprint of 1550 ft² (144 m²)[a]

	Miami		Austin		Atlanta		Baltimore	
	No Duct	20% Leaky Duct	No Duct	20% Leaky Duct	No Duct	20% Leaky Duct	No Duct	20% Leaky Duct
Base Roof and Attic	$160	$400	$286	$652	$445	$990	$562	$1,288
Retrofit Savings[b]	$64	$52	$110	$88	$163	$135	$238	$196
Payback[b] (Years)	25	30	15	18	10	12	9	11

[a]Thermostat controlled for occupancy comfort. All data ranges include values for attics with no duct and attic with ducts of 20% leakage.
[b]R_{US}-20 (R_{SI}-3.5) insulation added to attic floor for hot climates.

Table 7. Cost of Saved Energy for Replacing Dark Shingles with 25% and 40% Solar Reflective Cool-Color Products[a]

	Miami	Austin	Atlanta	Baltimore
Base roof and attic	$230–398	$457–652	$714–991	$925–1287
Retrofit savings for SR25[b] shingle[c]	$9–12.5	$6.3–9	$(5)–(4.5)[d]	$(3.7)–(2.6)
Retrofit savings for SR40 shingle[c]	$18–25	$13–18	$(11)–(10)	$(8)–(6)
Retrofit savings for SR70 shingle[c]	$36–50	$24–34	$(23)–(22)	$(17)–(13)

[a]Attic footprint is 1550 ft² (144 m²). All data ranges include values for attics containing ducts with 10% and 20% leakage.
[b]SR, solar reflectance.
[c]Homeowner replaces shingles and makes no other improvements.
[d]Values in parentheses imply a cost penalty.

ting a home in El Centro, CA with 0.25 reflective cool-color shingles drops roof and attic cooling energy by 6.4% of that used by a base roof with 0.05 dark shingles if the attic has only R_{US}-19 (R_{SI}-3.3) insulation on the attic floor. California Title 24 code calls for R_{US}-38 (R_{SI}-6.7). If an attic space cannot be modified, then the cool-color roof becomes a good option, with simple payback of about 10 years in El Centro (Miller and Kosny 2008).

SYSTEMS APPROACH TO HOME RETROFITS

Attic renovations show that incremental changes for some retrofits yield marginal savings and also that losses from ducts in attics predominate. A question one might ask is whether a combination of practices such as the insulated and ventilated roof deck reported by Miller et al. (2011) provides enough improvement to compensate for an attic with poorly performing, leaky ducts.

Ventilating the roof deck above the sheathing was proven to effectively reduce heat transfer crossing the roof deck, as shown in Figure 3. The ventilation also redirects solar-driven moisture out the ridge vent instead of it crossing the roof deck (Miller et al. 2006). Field data collected for an insulated and ventilated roof deck (R_{US}-8 [R_{SI}-1.4]) measured almost an 85% drop in the peak day heat transfer crossing the ceiling. Ventilation occurring above the sheathing reduced shingle temperature 16°F (9°C) and dropped the attic air temperature to about the outdoor air temperature. The design improved the durability of the shingles by keeping them cooler, and it reduced heat flows crossing into the conditioned space, thereby reducing energy consumed for comfort conditioning. The free ventilation flow occurring during the day helped dry the sheathing and limited moisture intrusion into the attic, as shown in Figure 3.

Annual performances of the insulated and ventilated roof deck and a sealed attic were cast into the scenario used in Figure 1 to show whether either system compensates for an attic with poorly performing, leaky ducts, as shown in Figure 4. The bars in the pale green color represent consumed energy for an insulated and ventilated roof deck (new design) installed on a poorly insulated attic that has a leaky attic floor and 10% leaky ducts. The black bars represent a sealed attic system that exploits the lost duct energy to make a semi-conditioned space in the attic. Insulation for the sealed attic is in the rafters, not the attic floor. Hence, the floor of the sealed attic was simulated without insulation.

The new design cannot compensate for the losses incurred by the leaky ductwork or for the air leakage crossing the attic floor. Repairing the leaky ducts (turquoise and navy blue bars) by sealing them to 4% leakage and wrapping them with R_{US}-8 (R_{SI}-1.4) insulation yields slightly more savings (orange bars) than simply installing a new roof design. The sealed attic (black bars) shows better performance (Figure 4); however, the initial cost for these retrofits is important because homeowners want the system at an affordable price. Repairing 140 ft of ductwork attached to a two-ton heat pump is estimated at $1.67 per square foot for an attic of 1550 ft^2 (144 m^2). Retrofitting a new roof design costs \$1.36 per square foot. Sealing the attic with code level of closed cell foam (R_{US}-30 [R_{SI}-5.3]) costs \$3.86 per square foot. Total first costs are as follows:

	Repair 10% Leaky Ducts on 2-Ton Heat Pump	Installation of New Roof and Attic Design	Seal Attic
Initial cost	\$2600	\$2200	\$6300
Simple payback	25 years	25 years	16 years

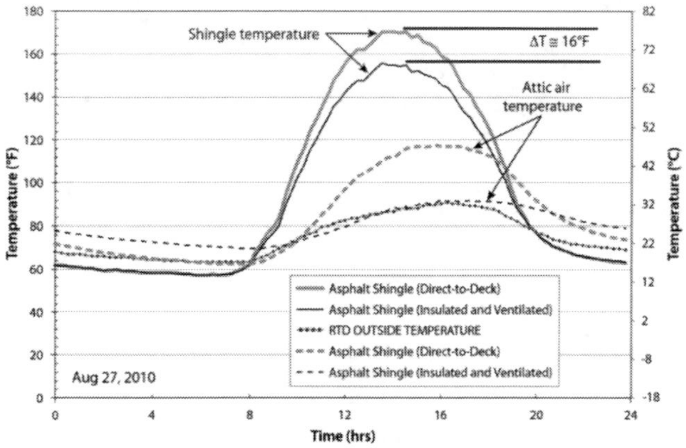

Figure 3 *Shingle and attic air temperatures for insulated and ventilated roof deck and direct-to-deck shingle roof.*

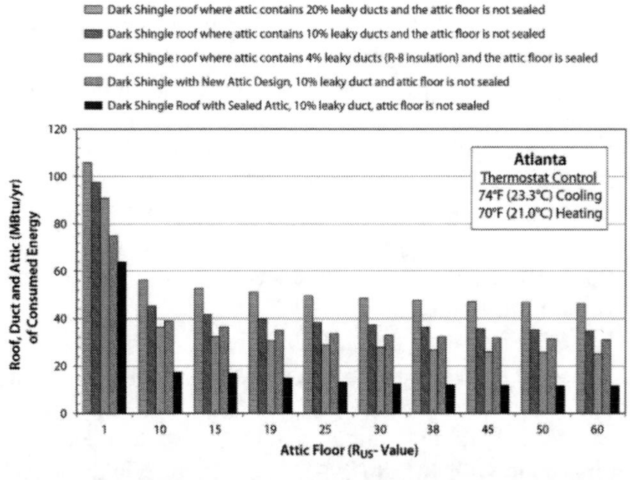

Figure 4 *Energy savings from retrofitting an insulated and ventilated roof deck onto an attic having leaky ducts and a poorly sealed attic floor. Results compared to sealed attic and to improving ducts and also removing ducts.*

The insulated and ventilated roof design is about \$400 less expensive than renovating the ducts and about \$4000 less than sealing the attic.

Converting Attics into Semi-conditioned Space. Sealing the roof deck and gables of an attic with spray polyurethane foam insulation (Figure 5) has gained popularity among many builders and code officials, especially in hot and very humid climates. Sealing recaptures the energy losses from a leaky air distribution system, which can cost \$240 in Miami and upwards of \$725 in Baltimore. The amount of applied foam should at least match the code requirements for the attic floor. Otherwise the system may actually increase heat flows into the conditioned space, especially if there are no ducts in the attic. In fact, the procedure should not be considered unless there is a leaky duct in the attic.

Sealing the attic and eliminating exterior vents is done to exclude moist outside air from the inside, but the sealing must be done properly or a dark attic can become a damp breeding ground for mold and mildew. The foam is sprayed between the roof rafters and over the gable ends of the attic to literally seal the attic from the outdoor ambient, as shown in Figure 5. However, this requires that the insulation be removed from the attic floor because the design makes the attic part of the conditioned space. The roof deck must be sealed to protect against moisture intrusion because the leaky duct will cause pressure imbalances between the attic and the conditioned space, especially if the air handler is in the attic.

Recent data collected from a test facility in Charleston, SC (Miller et al. 2013) revealed very interesting summertime trends in attic humidity for a sealed attic as compared to the conventionally ventilated attics. Asphalt shingles and a semi-permeable underlayment (15# felt paper) protected the OSB roof deck, and the attic was sealed by adhering open-cell spray foam to the deck's underside. Summer field data showed drops in relative humidity as the attic air warmed for all ventilated attics field tested by Miller

Figure 5 *Sealing the roof deck of an attic.*

et al. (2013). He observed that the larger the vent openings in the attic, the higher was the humidity content in the attic air, which peaks around 9 a.m. for the vented attics, as shown in Figure 6.

However, the opposite was true in the sealed attic. Peaks in measured relative humidity from two different sensors[5] in the sealed attic showed values in excess of 80%–90%, and occasionally saturated air (i.e., 100%) occurred around solar noon. In other words, the moisture content in the sealed attic was consistently 80%–100% rh from solar noon to around 6 p.m. for the seven contiguous days shown in Figure 6. Further, the trend was observed throughout the hot summer months.

The temperature and relative humidity data for the sealed attic and the indoor and outdoor ambient were converted to partial pressures to reveal the driving potentials for mass transfer in the sealed attic, as shown in Figure 7. The horizontal irradiance measured at the Charleston site is also included on a secondary y-axis of Figure 7. The partial pressure of attic air for the sealed attic lags the peak in solar irradiance by about 2 hours (view blue dashed line versus the black and orange lines in Figure 7). The high moisture content in the sealed attic does not come from the outdoor ambient or the indoor ambient because both have partial pressures during peak irradiance that are less than that observed in the sealed attic air, Figure 7.

Field measurements imply that some of the moisture from a previous rainstorm migrates to the underside of the shingles and underlayment. Irradiance drives moisture from an earlier rainstorm into and through the OSB deck and open-cell foam.[6] As a result, the partial pressure of the attic air and the partial pressure of water at the foam-to-attic air interface occasionally exceed the saturation pressure of water vapor in the attic air, Figure 7. If moisture is coming from the open-cell foam as driven by irradiance, then the partial pressure of water vapor at the foam-to-attic air interface would be higher than the partial pressure of water vapor in the attic, and Figure 7 data shows that this is indeed the case around solar noon. On these particularly hot days, relative humidity measurements showed super saturation (i.e., partial pressure of water vapor exceeded saturated pressure). At this condition, water vapor is in equilibrium with liquid water and therefore all interior attic surfaces are wet! Without a leaky duct, the sealed attic approach actually exacerbates humidity control in attics exposed in hot, humid climates. However, the presence of leaky ducts does not necessarily provide adequate protection from attic moisture. Colon (2011) field tested open-cell spray foam in a home in the hot muggy climate of south Florida. The roof deck was protected by an impermeable underlayment and the air-handler unit and ductwork were contained in the attic, yet moisture levels in the conditioned space increased above that measured *a priori* sealing the attic.

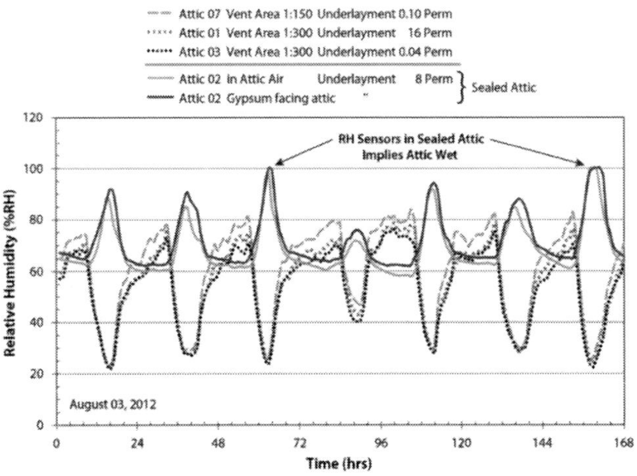

Figure 6 *Relative humidity measured in ventilated and sealed attics field tested in Charleston, SC.*

[5] Relative humidity measured by thermoset polymer capacitive type sensor. Calibrations in controlled environmental chamber showed the sensor ±0.5 rh at 90% rh, and the sensor's output is fairly linear with increase in RH.

Prototype Sealed Attic Design

Implementing a single retrofit option (adding floor insulation) cannot compensate for the losses incurred by leaky ducts or for air leakage crossing the attic floor. Sealing an attic into a semi-conditioned zone makes good practical sense because the losses from the ducts are recaptured and used to make the attic a semi-conditioned space. The duct becomes part of the conditioned space and losses no longer predominate. These results and those for the insulated and ventilated roof deck have led us to envision another sealed attic design that uses less-expensive insulations and less labor, thereby improving both affordability and simple payback.

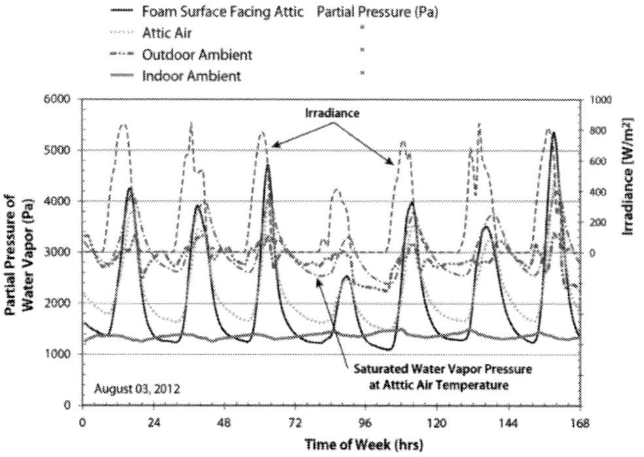

Figure 7 *Partial pressure of water vapor at various locations in and around the sealed attic.*

[6] Personal communication with Joe Lstiburek of Building Science Corporation.

The prototype design is applicable to new and retrofit work and makes use of roof deck ventilation, which has not been tried in sealed attic assemblies. Blown fiberglass insulation is blown into a spacer fastened between rafters, as shown in Figure 8a. The black liner (Figure 8a) serves as a vent spacer and trough and as the sealing plane (air barrier) for the cathedral design; it is critical to the design. The blown fiberglass insulation is held in place by the black liner whose underside may also have a low-e foil facing into the attic for reducing the radiation heat transfer to the attic floor. The black liner could be either retrofit or new construction, whereas the depicted fiberglass blanket (Figure 8b) is for new construction where the silver lining serves as the air barrier and as a low-e surface and ventilation cavity. A key difference between an attic sealed with spray foam and the proposed sealed attic approach is the inclusion of the inclined air space fashioned between the sheathing and the fiberglass insulation. Heat and moisture driven by irradiance through the permeable felt paper and OSB roof deck will be carried away by free ventilation inside the inclined air space. Each adjacent pair of rafters serves as channels for free ventilation from soffit to ridge, and the inclined cavity is sealed by the spacer (black liner) much like drywall that seals the classical cathedral ceiling. For new construction the fiberglass could feasibly be blanket insulation with a foil facing the underside of the sheathing; it was assumed draped over rafters from the exterior and cut normal to each rafter or truss so that the blanket is not compressed by the sheathing. The R-value of the insulation (whether blanket or blown fiberglass) was preliminarily set at R-38 (i.e., two R-19 blankets). The gable ends of the home are also insulated with R-19 fiberglass blankets.

Computations for Austin, TX show that a conventional attic with R-30 insulation on the attic floor consumes about 3.96 kWh of power per square foot of attic footprint, as shown in Table 8. As compared to the conventional stick-built roof and attic, the proposed sealed roof design reduces energy consumption of the sealed attic from 3.96 kWh down to 2.85 kWh, a 28% reduction in Austin, TX. Including some minor repair to the ducts to drop leakage from 20% to 10% would boost energy savings to 50% of the conventionally built roof and attic.

A 26% energy reduction is estimated for Baltimore, MD; duct renovation increases the energy savings to 43%. For Minneapolis, MN a 22.5% drop in energy compared to the base is computed, and inclusion of some duct repair increases savings to 41% of the control. These results with inclusion of some duct repair exceed a 35% reduction compared to an attic built to IECC 2009 code. The duct repair is a compromise between complete repairs for best heat pump delivery performance and tolerating some leakage to support both the semi-conditioned space and the weather-induced and occupant-generated latent load potentially making its way into the attic.

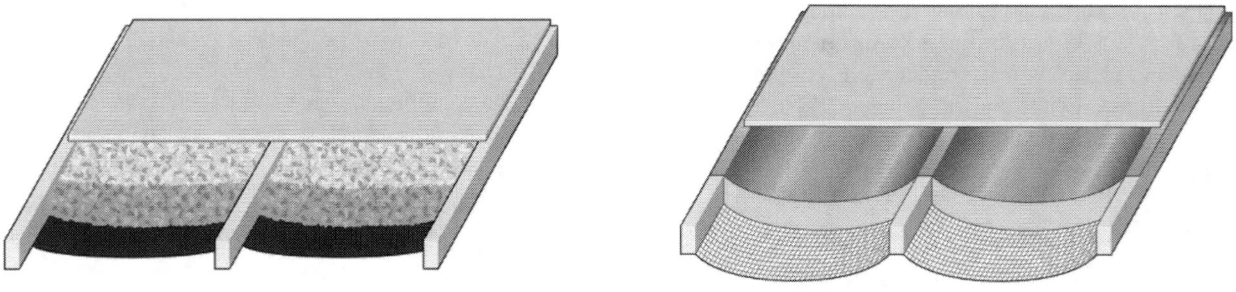

a) Retrofit Practice

b) New Construction

Figure 8 *Retrofit and new construction making the attic a semi-conditioned space.*

Table 8. Energy Attributed to the Roof and Attic and to the Duct System in the Attic

	IECC 2009	Conventional[a] Ventilated[b] Attic	Sealed Attic Prototype R-38 Fiberglass Insulation and Spacer: Duck Leakage 20%	Sealed Attic Prototype R-38 Fiberglass Insulation and Spacer: Duct Leakage 10%	Conventional[a] Ventilated[b] Attic	Sealed Attic Prototype R-38 Fiberglass Insulation and Spacer: Duck Leakage 20%	Sealed Attic Prototype R-38 Fiberglass Insulation and Spacer: Duct Leakage 10%
City, State	R-Value	Electricity (kWh/ft²)			Energy Cost ($/ft²)		
Austin, TX	30	3.96	2.85	1.98	0.511	0.367	0.255
Baltimore, MD	38	2.87	2.12	1.63	0.418	0.309	0.237
Minneapolis, MN	49	5.25	4.07	3.09	0.507	0.393	0.299

[a] Conventional attic equipped with soffit and ridge vents (1:300 area ratio).
[b] Conventional attic assumes IECC 2009 code insulation on attic floor. Duct with 20% leakage included in attic.

Cost and Simple Payback

For the same R-value, the material and labor charges for sealing an attic are about the same using closed-cell as compared to open-cell foam insulations (Faulkner 2010). Closed-cell insulation has a thermal resistance of about R_{US}-6 (R_{SI}-1.1), whereas open-cell polyurethane is lower at R_{US}-3.5 (R_{SI}-0.6). RS Means material charges for closed-cell foam and blanket insulation are listed in Table 9. The labor charge is based on data gleaned from various contractors and is less than RS Means; Means is based on organized labor rates. Contractor labor charges for foam, however, were comparable to RS Means.

The total cost for foam material and labor ranges from \$3.86 per square foot in Austin to \$6.3 per square foot in Minneapolis as applied to a 1550 ft^2 home with a 4:12 pitch. Table 9 costs for spray foam include R_{US}-10 (R_{SI}-1.74) of foam sprayed onto the gable walls each of 513 ft^2 (47.2 m^2). Initial material and labor cost for Austin is \$7900, and in Minneapolis to seal the roof to R_{US}-49 (R_{SI}-8.5) would cost about \$11,800. The application is very expensive and the proposed off-the-shelf materials (except for the spacer) can substantially reduce costs especially in the colder climates. The new sealed attic design costs about \$3900 for Austin, which includes duct repair. The repaired ducts have bladders installed on the supply and return to control air circulation to maintain the semi-conditioned space. The technique will also roughly set a minimal pressure drop across the attic floor to limit air and moisture movement across it. Simple payback for the proposed sealed attic with inclusion of duct renovation is 10 years in Austin to 12 years in Minneapolis. Note that the new design shows superior performance to the spray foam attic and is estimated to cost \$8000 less than the spray foam application for a footprint of 1550 ft^2 (144 m^2), Table 9.

NEW CONSTRUCTION

Figure 9 illustrates that new construction should always keep the ducts out of the attic. The black bars represent energy use for a roof and attic where the roof is dark and the attic has code level of insulation, but the attic contains an inspected duct with 4% leakage and is well insulated with R_{US}-8 (R_{SI}-1.4) insulation. The home's roof and attic total load is about 15 MBtu/year more than all other systems that do not contain ductwork in the attic. This translates to about \$100 to \$150 cost due to heat gains and losses in the conditioned air. So, comparing the attic containing inspected ducts with the newly designed insulated and ventilated roof deck (Table 10) will yield savings incurring simple paybacks as low as 5–10 years (Table 10). Material and labor assumed the added cost of the profiled and foil-faced EPS insert and the labor to install it.

But the key point of Figure 9 is seen by the prototypes against an attic containing no duct, the attic floor sealed, and with code level of insulation (Figure 9 blue bars with white cross hatching). Incremental savings are very small because a properly constructed roof and attic lose very little heat and there is not that much improvement for a new roof design. Hence for new construction, keep the ducts out of the attic, seal the attic floor, and add insulation to at least code level for the climate zone.

Assuming IECC 2012 code of R_{US}-38 (R_{SI}-6.7) and new housing starts at about 700,000,[7] new housing would save annually about 0.2 quads of energy. In comparison, a proactive roof maintenance program that retrofits sealed attics (Figure 4) in 50% of the existing homes would yield annual savings of 2 quads.

CONCLUSION

Building heat transfer gains and losses are low-grade energy; therefore, retrofitting a single component is not effective in substantially reducing energy use attributed to the roof and attic. For example, a radiant barrier foil or the addition of R_{US}-20 (R_{SI}-3.5) insulation to an attic with R_{US}-10 (R_{SI}-1.8) insulation saves about the same amount of money, but both

[7] The U.S. Census Bureau and the Department of Housing and Urban Development jointly announced the following new residential construction statistics for April 2013.

Table 9. RS Means Material Costs for Closed-Cell Spray Foam and for Fiberglass Insulation[a]

	IECC 2009	Sealed Attic Costs, \$			Energy Savings,[b] \$	
		Spray Foam[c]	R-38 Fiberglass and Spacer[d]		R-38 Fiberglass and Spacer[d]	
			No Duct Repair	Duct Repair	No Duct Repair	Duct Repair
City, State	RUS Insulation	Material and Labor, \$/ft²			Attic Footprint of 1550 ft²	
Austin, TX	30	5.08	0.83	2.53	222	396
Baltimore, MD	38	6.16	0.83	2.53	169	280
Minneapolis, MN	49	7.65	0.83	2.53	177	323

[a] Cost of energy savings as compared to conventional roof and attic design. The attic contains 20% leaky ducts.
[b] Savings for sealed attic based on conventional roof and attic with IECC code insulation on floor.
[c] Spray foam costs based on R-value needed in roof rafters to meet IECC Code.
[d] Fiberglass blankets each R-19 in roof rafters plus R-19 on gable walls.
[e] Duct repair does not include new wrapped insulation. Repair assumes only leakage reduced from 20 to 10% of fan flow.

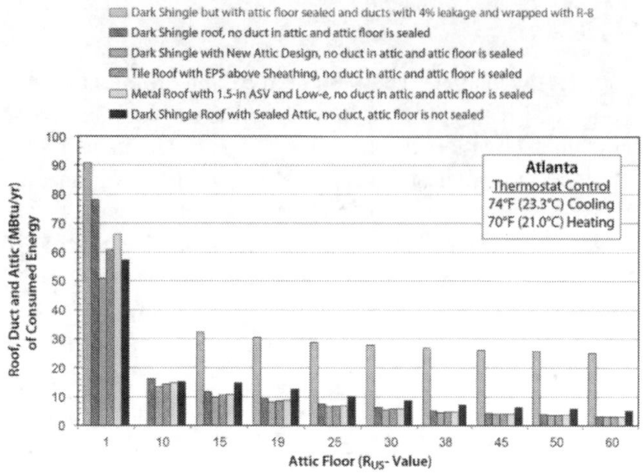

Figure 9 *New roof designs applied to new construction. Design described in Table 10.*

Table 10. Simulated Yearly Cost ($) of Energy Attributed to Roofs and Attics for New Homes as Compared to a IECC 2012 Compliant Home but with Inspected Ducts in the Attic

	New House[a] Compliant with IECC 2012	New Design 1[b]	New Design 2[b]	New Design 3[b]
Location	Roof and Attic Energy Cost	Annual Homeowner Savings from the New Designs Reducing Whole-House Energy Use		
Miami	100	80	70	80
Austin	190	140	130	135
Atlanta	170	115	110	115
Baltimore	860	760	750	750

[a]IECC-compliant home has ducts in attic. Ducts wrapped with R_{US}-8 (R_{SI}-1.4) insulation; duct leakage 4%.
[b]The new designs assume the duct system is in the conditioned space.
Design 1: Foil-faced 1 in. (0.0254 m) EPS insulation fitted between roof rafters; ventilation from soffit and attic to the inclined air space; radiant barrier in attic, two low-e surfaces in air space.
Design 2: Cool-color metal roof offset-mounted from roof deck 1½ in. (0.038 m); low-e surface in inclined air space covers non-breathable membrane; radiant barrier in attic.
Design 3: Peel-and-stick vapor barrier, 1¼ in. (0.032 m) EPS insulation above deck with high-profile tile; spray foam adhered EPS foam and tile to roof deck (no fasteners required).

retrofits have long payback periods. Savings for installing cool-color shingles (solar reflectance of 0.25 or 0.40) yields meager results if this is the only retrofit made to the house. Results show that a systems approach must be used to achieve a cost-effective repair of the roof and attic. Installing a ventilated and insulated roof deck with no other effective repair saves energy but does not compensate for the losses incurred by leaky ducts in an unconditioned attic.

The sealed attic approach and the insulated and ventilated roof assembly are very competitive to one another. Both approaches have paybacks ranging from 10 to 20 years with payback improving as the climate becomes more heating dominant. However, the attic moisture management reported herein led to a conceptual design that includes an inclined air space in the roof deck and controlled recirculation of air in the attic supply and return ducts. Simple payback for the prototype sealed attic design is about 10–12 years in hot and cold climates, and its initial cost is about half that spent for a conventional spray foam installation.

For new construction, the best option is to keep the ducts out of the attic, make sure the attic floor is sealed, and add at least code level of insulation to the ceiling. The advanced attic systems save some energy, but the total load attributed to the roof and attic is so low that the incremental savings are too low to be cost effective. However, even with inspected and sealed ducts having 4% leakage and wrapped in R_{US}-8 (R_{SI}-1.4), the homeowners pay about $100 to $150 more to comfort condition their home. In this scenario, a ventilated and insulated roof deck design with ducts not in the attic would yield simple payback in less than 10 years.

ACKNOWLEDGMENTS

Funding for this project was provided by the Building Technology Program of the U.S. Department of Energy. Oak Ridge National Laboratory is managed by UT-Battelle, LLC, for the U.S. Department of Energy under contract DE-AC05-00OR22725. The submitted manuscript has been authored by a contractor of the U.S. Government under contract DE-AC05-00OR22725. Accordingly, the U.S. Government retains a nonexclusive, royalty-free license to publish or reproduce the published form of this contribution, or allow others to do so, for U.S. Government purposes.

REFERENCES

ASHRAE. 1980A. ASHRAE 90-1980, *Energy Efficient Design of Low-Rise Residential Buildings*. Atlanta: ASHRAE.

ASHRAE. 2007. ASHRAE 90.2-2007, *Energy Efficient Design of Low-Rise Residential Buildings*. Atlanta: ASHRAE.

ASTM. 2004. *Standard Practice for Estimation of Heat Gain or Loss through Ceilings under Attics Containing Radiant Barriers by Use of a Computer Program*. Standard C 1340-04. West Conshohocken, PA: American Society for Testing and Materials.

Colon, C. 2011. New Construction Builders Challenge: Sealed Attic and High Efficiency HVAC in Central Florida: A Year in Review, Florida Solar Energy Center, FSEC-PF-454-11.

Cummings, J.B., Toole, J.J., and Moyer, N.A. 1990. Duct Leakage Impacts on Airtightness, Infiltration, and Peak Electrical Demand in Florida Homes. Professional Paper, Florida Solar Energy Center, Cocoa, FL, FSEC-PF-212-90.

d'Ambrosio Alfano, F.R., Dell'Isola, M., Ficca, G., and Tassini, F. Experimental analysis of air tightness in Mediterranean buildings using the fan pressurization method. *Building and Environment*, 53(2012):16-25.

Dodge, F.W. 2002. Construction Outlook Forecast. www.fwdodge.com. F.W. Dodge Market Analysis Group, Lexington, MA.

EIA. 2005. Washington, D.C. Total Households by Space Heating Fuels Used. Table SH2. 2005.

EIA. 2010. Washington, D.C.: Energy Information Administration. Table 5a Residential Average Monthly Bill by Census Division and State. web site: http://www.eia.gov/electricity/sales_revenue_price/html/table5_a.html

EIA. 2012a. Washington, D.C.: Energy Information Administration. Electric Power Monthly, web site: http://www.eia.gov/electricity/monthly/epm_table_grapher.cfm?t=epmt_5_06_b.

EIA. 2012b. Washington, D.C.: Energy Information Administration. Natural Gas Prices, web site: http://www.eia.gov/dnav/ng/ng_pri_sum_a_EPG0_PRS_DMcf_m.htm

EIA. 2013. Washington, D.C.: Energy Information Administration. Table CE3.1 Household Site End-Use Consumption in the U.S., Totals and Averages, 2013 web site: http://www.eia.gov/dnav/ng/ng_pri_sum_a_EPG0_PEU_DMcf_a.htm

Fairey, P., Wilcox, B., Parker, D., and Lombardi, M. 2004. Climatic impacts on heating seasonal performance factor (HSPF) and seasonal energy efficiency ratio (SEER) for air-source heat pumps. *ASHRAE Transactions*, July 2004.

Faulkner, D., and Ferrante, B. 2010. Private communication with Polyfoam Corporation, a subsidiary of 3M Company.

IECC. 2009. *International Energy Conservation Code.*

IECC. 2012. *International Energy Conservation Code.*

Jump, D., Walker, I.S., and Modera, M.P. 1996. Field measurements of efficiency and duct retrofit effectiveness in residential forced air distribution systems. *Proc. ACEEE Summer Study*, 1:147-157.

Levinson, R., Berdahl, P., and Akbari, H. 2004a. Solar Spectral Optical Properties of Pigments, Part II: Survey of Common Colorants.

Levinson, R., Berdahl, P., and Akbari, H. 2004b. Solar Spectral Optical Properties of Pigments, Part I: Model for Deriving Scattering and Absorption Coefficients from Transmittance and Reflectance Measurements.

Miller, W.A., Wilson, P., and Karagiozis, A. 2006. The impact of above-sheathing ventilation on the thermal and moisture performance of steep-slope residential roofs and attics. Paper presented at the 15th Symposium on Hot and Humid Climates, Orlando, FL, July.

Miller, W. A., Keyhani, M., Stovall, T., and Youngquist, A. 2007. Natural Convection Heat Transfer in Roofs with Above-Sheathing Ventilation, in Thermal Performance of the Exterior Envelopes of Buildings, X, *Proceedings of ASHRAE THERM X*, Clearwater, FL, Dec. 2007.

Miller, W.A., and Kosny, J. 2008. Next generation roofs and attics for homes. ACEEE Summer Study on Energy Efficiency in Buildings, *Proceedings of American Council for an Energy Efficient Economy*, Asilomar Conference Center in Pacific Grove, CA, Aug. 2008.

Miller, W.A., Graves, R., and Atherton, S. 2011. A prototype roof deck designed to self-regulate deck temperature and reduce heat transfer. NRCA International Roofing Symposium, Washington D.C., Sep 2011.

Miller, W.A., Biswas, K., Kehrer, M., Desjarlais, A., and Atherton, S. 2013. Analytical and Field Study of the Effects of Ventilation on Thermal Performance and Moisture Control in Residential Attics. ORNL/TM-2013/38, Oak Ridge National Laboratory.

Modera, M.P., and Wilcox, B. 1995. Treatment of Residential Duct Leakage in Title-24 Energy Efficiency Standards, CEC contract report.

Mullenax, A. 2010. Private communication with Shelter Distributors, Lavergne, TN.

NREL/TP-472-7332a. 1995. Home Energy Rating System Building Energy Simulation Test (HERS BESTEST), Volume 1, used with EnergyPlus: DOE's Next Generation Simulation Program.

NREL. 1995b. TMY2s. Typical meteorological years derived from the 1961–1990 national solar radiation database. Data Compact Disk. Golden, CO: National Renewable Energy Laboratory.

Parker, D., Fairey, P., and Gu, L. 1993. Simulation of the effects of duct leakage and heat transfer on residential space cooling energy use. *Energy and Buildings*, 20(2):97-113.

Shrestha, S.S., Miller, W.A. and Desjarlais, A.O. 2013. "Thermal Performance of Attic Radiant Barrier Systems Using the Large Scale Climate Simulator (LSCS)," to be published in Thermal Performance of the Exterior Envelopes of Buildings, XII, *Proceedings of ASHRAE THERM X*, Clearwater, FL, Dec. 2013.

Stovall, T., Shrestha, S., Arimilli, R.V., Yarbrough, D.W., and Pearson, T. 2010. Analysis in Support of the radiant Barrier Face Sheet 2010 Update, in Thermal Performance of the Exterior Envelopes of Buildings, XI, *Proceedings of ASHRAE THERM X*, Clearwater, FL, Dec. 2010.

RS Means. 2011. Building Construction Cost Data 2012, Catalog #60012.

U.S. Census Bureau. 2009A. Tenure by House Heating Fuel - Universe: Occupied Housing Units. Table B25117.

U.S. Census Bureau. 2009B. http://factfinder.census.gov/servlet/STTable?_bm=y&-geo_id=01000US&-qr_name=ACS_2009_5YR_G00_S2504&-ds_name=ACS_2009_5YR_G00

U.S. Census Bureau. 2012. Characteristics of New Housing. http://www.census.gov/construction/chars/completed.html

U.S. Department of Energy. 2010. Measuring State Energy Code Compliance. Prepared by Pacific Northwest National Laboratory under contract DE-AC05-76RL01830.

Walker, I.S. 1998. Technical background for default values used for forced air systems in proposed ASHRAE standard

152P. *ASHRAE Trans.*, 104(1); presented at ASHRAE TC 6.3 Symposium, January 1998. LBNL 40588.

The Trade-Off between Solar Reflectance and Above-Sheathing Ventilation for Metal Roofs on Residential and Commercial Buildings

Scott Kriner

William Miller, PhD
Member ASHRAE

Andre Desjarlais, PE

ABSTRACT

An alternative to white and cool-color roofs that meets prescriptive requirements for steep-slope (residential and non-residential) and low-slope (non-residential) roofing has been documented. Roofs fitted with an inclined air space above the sheathing (herein termed above-sheathing ventilation or ASV) performed as well as, if not better than, high-reflectance, high-emittance roofs fastened directly to the deck. Field measurements demonstrated the benefit of roofs designed with ASV. A computer tool was benchmarked against the field data. Testing and benchmarks were conducted at roofs inclined at 18.34°; the roof span from soffit to ridge was 18.7 ft (5.7 m). The tool was then exercised to compute the solar reflectance needed by a roof equipped with ASV to exhibit the same annual cooling load as that for a direct-to-deck cool-color roof. A painted metal roof with an air space height of 0.75 in. (0.019 m) and spanning 18.7 ft (5.7 m) up the roof incline of 18.34° needed only a 0.10 solar reflectance to exhibit the same annual cooling load as a direct-to-deck cool-color metal roof (solar reflectance of 0.25). This held for all eight ASHRAE climate zones complying with ASHRAE/IES Standard 90.1 (2007a). A dark heat-absorbing roof fitted with 1.5 in. (0.038 m) air space spanning 18.7 ft (5.7 m) and inclined at 18.34° was shown to have a seasonal cooling load equivalent to that of a conventional direct-to-deck cool-color metal roof. Computations for retrofit application based on ASHRAE/IES Standard 90.1 (1980) showed that ASV air spaces of either 0.75 or 1.5 in. (0.019 or 0.038 m) would permit black roofs to have annual cooling loads equivalent to the direct-to-deck cool roof. Results are encouraging, and a parametric study of roof slope and ASV aspect ratio is needed for developing guidelines applicable to all steep- and low-slope roof applications.

INTRODUCTION

In moderate and hot climates, a roof surface with high solar reflectance and high thermal emittance was shown by Akbari et al. (2004) and Parker and Sherwin (1998) to reduce the exterior temperature and produce savings in comfort cooling. Akbari and Levinson (2008), in a compilation of cool roof studies conducted for non-residential low-slope buildings, observed summertime daily air-conditioning savings ranging from 10% to 30%, though some reported data showed values as low as 2% and as high as 40%. Their findings clearly show that cool roofs can be a viable strategy for reducing energy consumption; therefore, many U.S. states have implemented prescriptive requirements for cool roofs in their energy codes based on ASHRAE Standard/IES Standard 90.1 (2007a), ASHRAE Standard 90.2 (2007b), or the International Energy Conservation Code (2009).

California's building code regulations, referred to here as 2008 Title 24, specify either prescriptive requirements or a whole-building performance approach for demonstrating the energy-efficiency compliance of buildings in California (CEC 2008). The performance approach allows the building owner to simulate the energy use of a proposed building with an approved whole-building model. Alternatively, the prescriptive approach requires that each building component meet or surpass the respective component requirements

***Scott Kriner** is the technical director for the Metal Construction Association, Macungie, PA. **William Miller** and **Andre Desjarlais** work in the Building Envelope Program of Oak Ridge National Laboratory under the directive of DOE Buildings Technology.*

Published by ASHRAE 2013

in 2008 Title 24, which prescribes the initial solar reflectance (SR) of steep-slope residential and non-residential cool roofs. Aged solar reflectance must equal 0.15 and emittance must be greater than or equal to 0.75 for residential roofing.

While industry agrees with many of the cool roof guidelines, there is concern among roof manufacturers that the strong emphasis on cool roofs ignores other viable construction practices such as above-sheathing ventilation (ASV). Therefore, this paper demonstrates the benefits of ASV through both field and computer verification. Field tests were conducted on the Envelope Systems Research Apparatus (ESRA). The data were used to benchmark the computer tool AtticSim (Version 13).[1] The computer tool uses overall energy balances to calculate the heat flows and temperatures on the external and internal surfaces of the attic. The tool can predict the temperature difference across the air space of an ASV system and was validated against field data. The validated code was exercised to simulate roofs having a pitch of 18.34°, the same pitch as validated against the ESRA field measurements. AtticSim simulated the performance of roofs installed directly to the deck as compared to roofs fitted with ASV. Results of the simulations are described to show that roofs equipped with ASV can meet existing prescriptions for cool roofs.

ABOVE-SHEATHING VENTILATION–FIELD DATA AND COMPUTER BENCHMARKS

Several steep-slope roofs and attics were installed on top of the ESRA, located on the campus of the Buildings Technology Research Integration Center at Oak Ridge National Laboratory. Stone-coated metal and standing-seam metal roofs were equipped with ridge and soffit vents for ventilating the attic. The ratio of the vent opening area to attic floor area was 1 to 300. Miller (2006) field tested stone-coated metal products with shake profiles to observe and quantify the effect of ASV (Figure 1a). Stone-coated metal is made of 26-gauge Zincalume (pre-primed 55% Al-Zn hot-dip-coated steel). It is coated with a layer of stone chips sealed by an acrylic base coat and over-glazed.

Standing-seam metal roofs (green color in Figure 1a) were also field tested to document the effects of offset metal roofs as compared to metal fastened directly to the roof deck. The metal panels snap-lock together and have a straight pan profile. Panels are 24 gauge galvanized steel painted with a polyvinylidene fluoride (PVDF) finish that contains infrared-reflective paint pigments. Solar reflectance of the painted metal is about 0.30 and is warranted for 35 years.

The distance (L) from the soffit to ridge is 18.7 ft (5.7 m) for the roofs pictured in Figure 1a, and the inclination of the roofs is 18.43°. The double-batten arrangement (Figure 1b) provides an air space of height (W) equal to 0.75 in. (0.019 m). Therefore, the aspect ratio (L/W) for the stone-coated metal roof is about 300:1. Air space heights of 0.75 and 4.0 in. (0.019 and 0.1 m) were field tested in the standing-seam metal roofs (Figure 1a) yielding aspect ratios of 300:1 and 56:1, respectively.

Stone-Coated Metal Field Measurements

Miller, Wilson, and Karagiozis (2006) field tested dark-gray stone-coated metal shakes having a solar reflectance of 0.08 and thermal emittance of 0.89 (abbreviated SR08E90) and light-gray metal shakes (SR26E90). Both were tested on identical double-batten constructions (air space of ¾ in. [0.019 m], Figure 1b). The dark-gray metal shake and the control shingle have almost identical solar reflectance and thermal emittance, yet the peak heat flow crossing the roof deck from the dark-gray shake was just 70% of the heat flow crossing the roof deck of the control

[1] AtticSim Version 13 originated from the ASTM C1340 (2004) protocol. Version 13 contains revisions for above-sheathing ventilation benchmarked by Miller et al. (2007) and attic ventilation as developed by Ken Childs.

Figure 1 *(a) Roofs offset-mounted from the roof deck for ASV or fastened directly to the roof deck on the Envelope Systems Research Apparatus. (b) Stone-coated metal offset from the roof deck using double battens.*

shingle (Figure 2). The 30% lower heat flow was caused in part by the thermal mass of the double battens, the thermal resistance of the air space, and the low emittance (0.35) of the bare metal underside of the stone-coated roof shake.

Increasing solar reflectance from 0.08 to 0.26 caused the heat flow crossing the roof deck of the light-gray shake to be less than the heat flow crossing the deck of the dark-gray shake (Figure 2). Miller, Wilson, and Karagiozis (2006) also determined that the heat flow diverted away from the attic by ASV of the hotter dark-gray shake is more than double that removed by the light-gray shake.

The hotter dark-gray roof causes greater buoyancy-induced airflows; therefore, the ventilation scheme is somewhat self-regulating. The darker the roof, the hotter the roof, so there is greater buoyancy to carry heat away from the attic. For the data shown in Figure 2, the surface temperature of the asphalt shingle peaks at 165.6°F (74.2°C). Buoyancy-driven airflow in the inclined air space cooled the dark-gray stone-coated metal shake 10°F (5.6°C) below that of the shingle; the dark shake had a peak temperature of 155.5°F (68.6°C). The addition of cool-color pigments further dropped the surface temperature of the light-gray shake to 151.3°F (66.3°C). Temperature profiles measured from the soffit to ridge at four discrete cross sections of the roof deck for the light-gray shake reveal the benefit of the inclined air space (Figure 3). Heat flow crossing the roof deck was measured with heat flux transducers (HFTs) embedded into the underside of the OSB deck (Figure 3). The HFTs show a linear but slight increase in heat flow, which is also reflected in the temperature trends observed in the oriented strand board (OSB) and inclined air space. The temperature difference from air to underside of OSB increases with length (Figure 3). This causes the increase in heat transfer observed in the HFTs from soffit to ridge.

The air space temperature was 124°F (51.1°C) after 4 ft (1.2 m) of travel up the roof (Figure 3). However, after another 12 ft (3.6 m) of travel,[2] its temperature rose to 138°F (58.9°C). Slight increases are observed in the top side temperature of the OSB (Figure 3); however, the thermal buoyancy in the air space keeps the OSB cooler than that of the direct-nailed shingle roof. The OSB for the shingle roof was 138°F (58.9°C) at a 10 ft (3.05 m) station. In contrast, the OSB for the light-gray shake was 116°F (46.7°C) at a similar location.

STANDING-SEAM PAINTED METAL FIELD MEASUREMENTS

Field tests (Figure 4) revealed that the control asphalt shingle (solar reflectance of 0.09 and thermal emittance of 0.89, abbreviated SR09E89), had heat flows crossing the deck similar to those of a metal roof applied directly to the deck. The cool-color painted metal (solar reflectance of 0.28 and thermal emittance of 0.81, abbreviated SR28E81) had similar deck heat flows to the shingle roof. The results are due in part to the higher thermal resistance of the shingle and its higher thermal mass. Shingles overlap and the added bulk causes the shingle to perform better than painted metal if both systems have the same solar reflectance and thermal emittance.

Two painted metal roofs were offset about 0.75 in. (0.019 m) from the roof deck. One of the two roofs had its underside painted with conventional backer[3] paint. The other metal roof was painted with a low-emittance coating for im-

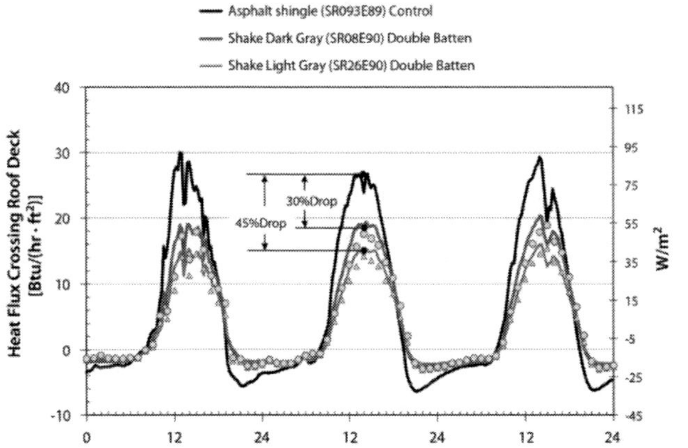

Figure 2 *Heat flow measured crossing the roof deck of a direct-nailed shingle roof and stone-coated metal roofs with and without cool-color pigments. Stone-coated metal roofs were installed on 1½ × 1½ in. (0.038 × 0.038 m) double battens. The open symbols represent AtticSim benchmarks against the field data for the dark-gray shakes (open circles) and light-gray shakes (open triangles).*

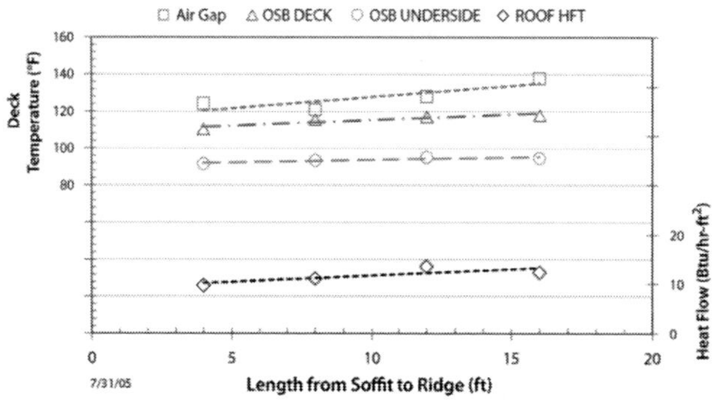

Figure 3 *Temperatures in the inclined air space, the oriented strand board (OSB) facing the metal shake, the OSB facing into the attic, and the roof deck heat flux are measured at discrete locations from soffit to ridge of the light-gray stone-coated metal shake. Measurements made at 12:00 PM EDST.*

2. Measurement location is within 2 ft (0.61 m) of ridge vent; total roof length is about 18.7 ft (5.7 m).

proved radiation control. Field tests showed that the painted metal roof with 0.75 in. (0.019 m) air space and backer paint on its underside had almost the same deck heat flow as observed for a direct-to-deck metal roof that had an EPS[4] insert (Figure 4). The finding implies that the 0.75 in. (0.019 m) air space is roughly equivalent to a thermal resistance of R_{US}-1 (R_{SI}-0.18). Adding the low-e undercoat dropped daytime heat flows below that of the roof with backer paint (Figure 4) and resulted in deck heat flows almost identical to another metal roof with 2 in. (0.51 m) of air space (not shown in Figure 4). ASV air spaces of 2 in. (0.51 m) and 4 in. (0.10 m) were also compared to the cool metal with EPS insulation inserted between the pan and the deck. Field data showed that both air gaps reduced deck heat flows below that observed for the metal roof with EPS insert having R_{US}-1 (R_{SI}-0.18) thermal resistance.

AtticSim Benchmarks to Stone-Coated Metal and Standing-Seam Metal Field Data

Wilkes (1991), Petrie et al. (2004), and Miller et al. (2007) all demonstrated AtticSim's ability to accurately predict the surface temperature and ceiling heat flow for attic assemblies with direct-nailed roof products. The AtticSim model can account for different insulation R-values and the effect of air-conditioning ducts placed in the attic, as reported by Petrie et al. (1998) and described in ASTM C 1340 (2004). Salient features of AtticSim are provided by Wilkes (1991). Miller et al. (2007) modified the code for inclusion of ASV.

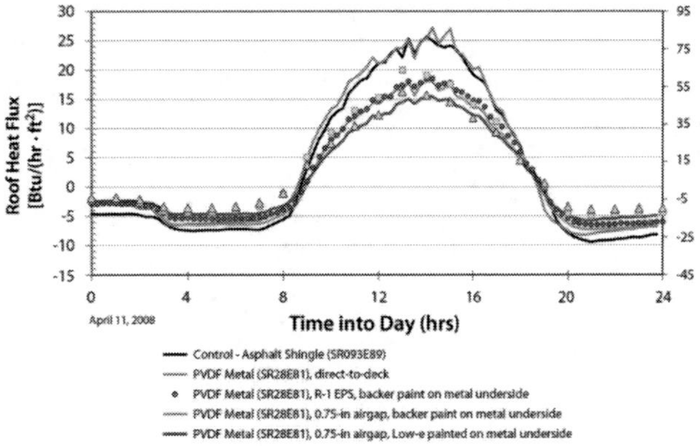

Figure 4 *Heat flow crossing the roof deck of cool-color standing-seam painted metal roofs. Open symbols (circle and triangle) represent AtticSim benchmarks against the PVDF metal roofs with 0.75 in. (0.019 m) air space.*

3. Prepainted metal roofing uses backer paint on the interior surface of metal panels to provide additional corrosion resistance, scratch resistance, and aesthetics where the underside is seen in open purlin structures.
4. EPS is expanded polystyrene insulation that is R_{US}-1 for its thickness of ¼ in. (0.071 m).

He describes tracer gas tests that were used to benchmark AtticSim's ability to predict the natural convection airflow in the ASV inclined air space. Herein, the code was also benchmarked against the stone-coated metal data (Figure 2), the standing seal metal data (Figure 4), and ASHRAE (2005) data for a closed air space. The code was then exercised to predict the benefit of opening the air space in terms of the effective R-value (shown later in Table 1).

Stone-Coated Metal. Weather data[5] from a station adjacent the ESRA was read by the AtticSim code and used to simulate the dark-gray and light-gray stone-coated metal roofs (refer back to Figure 2). The code accounted for the effects of ASV and predicted the heat transfer crossing the roof deck of the dark-gray shake within ±15% of measures made with HFTs embedded in the roof deck (view open circles with yellow highlight). AtticSim also predicted the heat flux crossing the roof deck of the light-gray shake and closely followed the trends in measured heat flux. These results are surprisingly good because the ASV algorithm does not account for the thermal mass of the double battens or for the radiation view factors among all battens and roof deck. It simply computes the radiation exchange between two parallel plates of finite dimension.

The AtticSim algorithm for ASV assumes an inclined air space (gap height of W) that runs from the soffit to the ridge (length of L). Note, however, that the aspect ratios used by experimentalists to derive convective heat transfer correlations from which the convection heat transfer coefficient is computed and used to calculate heat transfer crossing the air space in AtticSim are only of the order 50 to 1. AtticSim uses the correlation by Azevedo (1984) for an open inclined channel with aspect ratios ranging from 10:1 to 22:1. For a closed air space with the warmer surface below the colder surface, the ASV algorithm uses Holland's (1976) inclined closed cavity correlation. Holland's correlation was developed from an experimental facility with aspect ratio of 48:1. During summer exposure, the roof cover is hotter than the roof deck, and a lighter air layer hugs the underside of the roof cover as it travels up the roof; it is naturally above the denser air layer in contact with the cooler bottom surface of the air space (i.e., OSB deck). Arnold et al. (1974) and ElSherbiny et al. (1984) offer a correlation that is based on an adaptation of heat transfer in a vertical closed cavity useful for aspect ratios exceeding 5. Hence the benchmarks reported herein are very important and verify the applicability of the multiplicity of correlations used in the ASV algorithm within AtticSim because the correlations are based on aspect ratios different than found in roof applications.

Standing-Seam Metal. AtticSim also predicted the heat transfer crossing the roof deck of standing-seam metal roofs fitted with a 0.75 in. (0.019 m) air space. Here the in-

5. A local weather station provided outdoor temperature, relative humidity, wind speed and direction, irradiance, and night-sky radiation for use in validating the AtticSim code.

clined channel was clear of any obstructions inside the cavity, and therefore AtticSim's predictions are well within 10% of heat flux measurements depicted in Figure 4. The open squares with yellow highlight (Figure 4) represent the heat flow crossing the OSB deck of the PVDF metal having backer paint on its underside and 0.75 in. (0.019 m) of air space. Note that the open circles with brown highlight represent the heat flow through the roof deck where R_{US}-1 (R_{SI}-0.18) EPS insulation was sandwiched between the metal and OSB. The field data and AtticSim predictions are very consistent and prove the code is capable of accurately calculating the heat transfer crossing roof decks equipped with ASV. The triangle symbols highlighted in yellow represent simulations results for a PVDF metal roof with 0.75 in. (0.019 m) of air space but with the underside of the metal painted with a low-emittance paint (Figure 4). The results show AtticSim's ability to capture the physics of the combined convection and radiation heat transfer occurring in an ASV inclined air space.

The AtticSim code was also validated for a case having an aspect ratio of 55:1 (e.g., standing-seam metal roof with inclined air space of 4 in. [0.032 m]). The standing-seam metal roof was equipped with thermocouples measuring temperatures of the air space and roof deck at several discrete locations up the roof (similar to that shown in Figure 3). The thermocouples in the air space (four discrete locations) were averaged and plotted against AtticSim estimates. AtticSim is designed to compute overall heat balances for the various surfaces of the attic. Results in Figure 5 prove AtticSim is capable of accurately predicting the average air temperature in an ASV inclined air space.

***ASHRAE Handbook—Fundamentals* Data.** The *ASHRAE Handbook—Fundamentals* (2005) lists data for the thermal resistance of an inclined and closed air space. The data accounts for the effects of aspect ratio, slope, and convection and radiation heat transfer occurring within the air space. ASV results computed by AtticSim for a closed cavity are in excellent agreement with ASHRAE data (Table 1).

For the case of a painted metal roof with 0.75 in. (0.019 m) of open air space and a span of 18.7 ft (5.7 m), AtticSim predicts R_{US}-0.91 (R_{SI}-0.16) for a 0.75 in. (0.019 m) air space spanning 18.7 ft (5.7 m). The Table 1 result is within 10% of the experimentally derived result in Figure 4 for the SR28E81 metal roof with R_{US}-1 (R_{SI}-0.17) of expanded polystyrene insulation between the sheathing and metal pan. Increasing the air space from 0.75 in. (0.019 m) to 4 in. (0.10 m) caused a 40% jump in the thermal resistance of the open air space. However, including a low-emittance surface (E = 0.05) caused the air space resistance to jump by a factor of 6 as compared to the open cavity with effective emittance of 0.82 (Table 1). ASHRAE (2005) documents similar effects of emittance in an inclined but closed air space.

METHODOLOGY USED FOR SR AND ASV TRADE-OFF

AtticSim and Energy Plus were coupled through parallel simulations to compute the heat transfer through ceilings under attics. Simulations were made for the ASHRAE climate zones and for Sacramento, California (Table 2). An attic of 1539 ft^2 (143 m^2)[6] with a roof pitch of 18.34° was modeled with cool-color stone-coated metal, standing-seam metal, and asphalt shingle; solar reflectance of the cool roofs was set at 0.25 and 0.40. Simulations assumed supply and return air-conditioning ducts (cylindrical metal) installed in the attic. The supply duct surface area was set at 304 ft^2 (28.7 m^2). The surface area of the return duct exposed in the unconditioned attic was assumed to be 176 ft^2 (16.4 m^2). Energy Plus estimated the hourly indoor air temperature and hourly run times for a SEER 13 air conditioner as it cooled the home; heating assumed an 85% efficient gas furnace. The Home Energy Rating System (HERS) Building Energy Simulation Test (BESTEST) served as the simulated home (NREL 1995). It is used as a standard for evaluating the credibility of software used by HERS to predict energy use in homes.

Energy Plus computed hourly values of the mass flow rate of air in the duct system and the bulk air temperature entering the ducts for all climates listed in Table 2. Hourly values of duct flow rate, duct air leakage, bulk air temperature, percentage "on" time of the HVAC, and indoor air temperature were written to auxiliary files and read by AtticSim for computing the roof and attic load as coupled to the home simulated in the various ASHRAE climate zones. Low-slope simulations assumed the HVAC in the conditioned space.

The roof heat transfer for metal roofs with ASV was computed for the cooling and heating season and compared with data for a painted cool-color metal roof with a prescrip-

[6] Footprint of 57 ft by 27 ft wide (17.4 m by 8.2 m).

Table 1. R-Value for an Inclined Air Space Computed by AtticSim and Compared to ASHRAE (2005) for the Case of a Closed Air Space. AtticSim Also Computed the Thermal Resistance for an Open Cavity

Air Space for 4-in-12 Pitch Roof	ASHRAE[1] (2005) (Closed Air Space)	AtticSim Simulation[1] (Closed Air Space)	AtticSim Simulation (Open Air Space)	AtticSim Simulation (Low-ε Surface in Open Air Space)
0.75 in. (0.019 m)	0.70	0.68	0.91	5.32
4.00 in. (0.10 m)	0.74	0.77	1.30	8.74

[1]An effective emittance of 0.82 was assumed with a mean temperature of 133°F (56.1°C) having 11°F (6.1°C) temperature gradient for heat flows moving downward across the air space.

tive requirement of 0.25 solar reflectance. The base case roof was fastened directly to the roof deck.

The ceiling insulation was set to comply with ASHRAE/IES Standard 90.1 for low-slope commercial roofing (ASHRAE 2007a) and with ASHRAE Standard 90.2 for steep-slope commercial and residential homes (ASHRAE 2007b). For retrofit practice, the ceiling and duct insulation were assumed to comply with thermal resistance levels published in ASHRAE 90 (1980), Table 2. Air leakage of the ductwork is unknown; however, for demonstration purposes, simulations assumed air losses of 10% of supply airflow for existing homes built to ASHRAE 90 (1980) code and 4% leakage for inspected, well-installed ductwork (Cummings et al. 1990; CEC 2008).

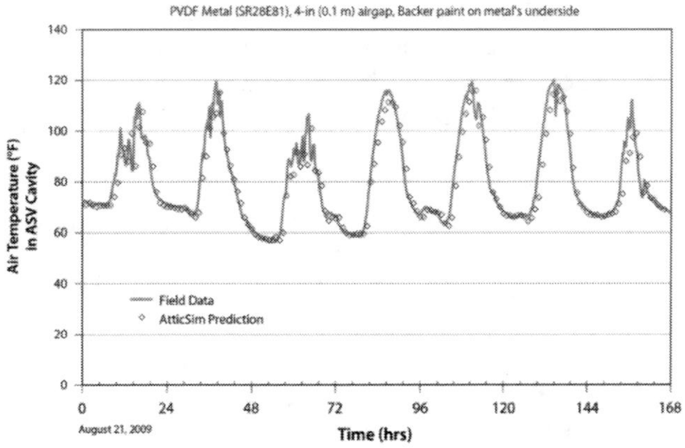

Figure 5 *AtticSim's prediction of the average air temperature in an inclined cavity of a standing-seam metal roof offset from the roof deck by 4 in. (0.032 m). Aspect ratio (L/W) of 48:1.*

TASKS REQUIRED TO JUDGE THE SR TRADE-OFF WITH ASV

Simulations were completed to judge the trade-off between solar reflectance and ASV for both low- and steep-slope roof assemblies. The ceiling heat transfer was computed over the cooling season for direct-to-deck cool roofs and for roofs equipped with ASV. Aspect ratio for the ASV roofs was 240:1 for a 0.75 in. (0.019 m) air gap and about 115:1 for a 1.5 in. (0.038 m) air space. Attic ventilation was set to 1:300. Computations assumed weathered reflectance values consistent with the 2008 Title 24 guidelines. However, PVDF painted metal and stone-coated metal roofs are very fade resistant and lose less than 10% of their original solar reflectance even after 30 years of weathering (Berdahl et al. 2006, Miller et al. 2010). AtticSim was exercised for the following tasks:

Task 1: Determine the solar reflectance needed for a steep-slope (18.34°) roof with ASV (min ¾ in. of continuous air space) to match load of an SR-0.25 (0.15 weathered) cool roof

Task 2: Determine the solar reflectance needed in a steep-slope (18.34°) roof with ASV (min ¾ in. of continuous air space) to match load of an SR-0.40 roof (0.25 weathered) fastened directly to deck

Task 3: Determine the solar reflectance needed in a low-slope (9.5°) roof with ASV to match load of an SR-0.65 (0.50 weathered) cool roof fastened directly to deck

Table 2. Trade-Off of ASV with Solar Reflectance for Low- and Steep-Slope Roofs in ASHRAE Climate Zones and California Climate Zone 12

ASHRAE Climate Zones					ASHRAE Standard 90.2 Residential Code				ASHRAE/IES Standard 90.1 Commercial Code			
					ASHRAE Standard 90.2		ASHRAE Standard 90-80		ASHRAE/IES Standard 90.1 New		ASHRAE Standard 90-80	
Zone	City	State	HDD_{65}	CDD_{65}	Ceiling	Duct[1]	Ceiling	Duct[2]	Ceiling	Duct	Ceiling	Duct
1	Miami	FL	222	9368	30	8	20	5.5	15	NA	10	NA
2	Austin	TX	1481	7435	30	8	20	5.5	20	NA	10	NA
3	Atlanta	GA	2614	4814	30	8	20	5.5	20	NA	10	NA
4	Baltimore	MD	4731	3598	38	8	21	5.5	20	NA	11.8	NA
5	Chicago	IL	6139	2895	43	8	24	5.5	20	NA	13.3	NA
6	Minneapolis	MN	7787	2513	49	8	29	5.5	20	NA	16.4	NA
7	Fargo	ND	10052	1332	49	8	30	5.5	20	NA	16.7	NA
8	Fairbanks	AK	2697	1040	52	8	33	5.5	20	NA	19.6	NA
12	Sacramento	CA	2697	1202	38	6	20	5.5	20	NA	10	NA

[1]Inspected duct system "iiDuct" having 4% air leakage.

[2]Existing duct system "kkDuct" assumed 10% air leakage. ASHRAE 90-80 does not address leakage.

Three simulations per climate zone were needed (Figure 6) to develop the solar reflectance labels in Figures 7 and 8. As an example, simulations were conducted for a given climate zone (say Atlanta, which represents ASHRAE Climate Zone 3). Cumulative cooling and heating loads were written to an external file for the case of a roof fastened directly to the roof deck. A second run was made for an SR-0.25 roof fitted with ASV; its cooling and heating loads were also written to the external file. The third run read the data from the external file and used an interpolating scheme in AtticSim to adjust the solar reflectance of the roof with ASV until its annual cooling load matched the annual cooling load of the cool

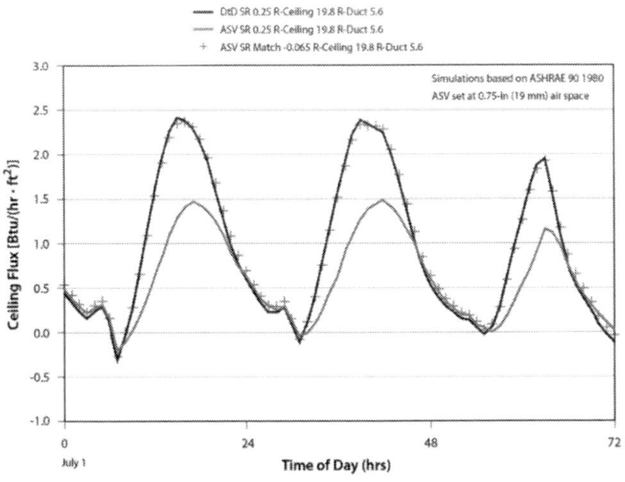

Figure 6 *Simulations showing (on the line defined by the + symbols) the match of ceiling load between a direct-to-deck cool metal roof and a roof fitted with ASV. AtticSim matched the loads by adjusting the solar reflectance of the ASV roof.*

metal roof fastened directly to the deck. These are the solar reflectance labels shown in Figures 7 and 8.

A check of the procedure is shown in Figure 6 for three contiguous July days. The direct-to-deck cool metal roof having solar reflectance of 0.25 has a peak-day heat flow crossing the attic floor that is about 40% higher than the load for a cool roof fitted with ASV (Figure 6). By adjusting the solar reflectance of the roof with ASV, its seasonal load is made about equal to the load for the direct-to-deck base roof. (See control roof black line comparison to the ASV roof with adjusted SR defined by + symbols in Figure 6.)

Task 1. AtticSim/Energy Plus computed the heat flux crossing the ceiling below a cool metal roof that was attached directly to the deck (Figure 7). The solar reflectance of a cool metal roof fitted with ASV was modified until its seasonal cooling load matched the computed load for the conventional direct-to-deck cool roof assembly. New construction (Figure 7) and retrofit practice (Figure 8) were compared to the base, a cool-color painted metal attached directly to the deck.

A painted metal roof with 0.75 in. (0.019 m) air space spanning 14.2 ft (4.3 m) from soffit to ridge needed only a 0.10 solar reflectance to have the same annual cooling load as a direct-to-deck cool-color metal roof based on ASHRAE Standard 90.2 (2007) code. Increasing the air space to 1.5 in. (0.038 m) for the same span caused the required solar reflectance to drop to almost the equivalent of a black roof. In other words, a roof fitted with 1.5 in. (0.038 m) of ASV spanning 14.2 ft (4.3 m) can be black and still have a seasonal cooling load within 0.05% of the conventionally constructed cool-color metal roof. The finding was also observed for stone-coated metal roofing. The stone-coated metal with an air space equal to or exceeding 0.75 in. (0.019 m) can be black and still have a seasonal cooling load that is less than that of a cool metal (SR-0.25) roof fastened directly to the roof deck.

Computations based on existing construction using ASHRAE Standard 90 (1980) were very eye opening

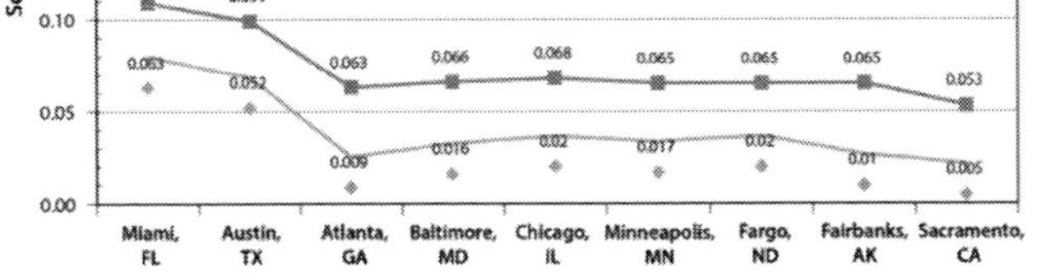

Figure 7 *Solar reflectance needed for a standing-seam metal roof with ASV (new product on new construction) to match the cooling load of a cool roof fastened directly to the roof deck.*

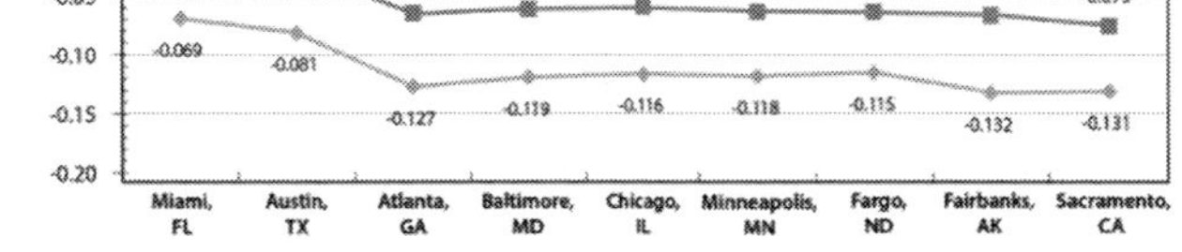

Figure 8 *Solar reflectance needed for a standing-seam metal roof with ASV (new product on existing construction) to match the cooling load of a cool roof fastened directly to the roof deck.*

(Figure 8). For several simulations, iterations yielded a negative solar reflectance to match loads, which is unrealistic because the lowest achievable solar reflectance is zero for a perfectly black absorber. Therefore, the negative values imply that the ASV roof can be black and still perform better than a direct-to-deck cool metal roof having a solar reflectance of 0.25.

For retrofit work, the offset-mounted metal roof could be black and still have the same annual cooling load as the cool metal direct-to-deck base case! The data suggest that offset-mounted standing-seam and stone-coated metal products can qualify as cool roof systems.

Task 2 is a continuation of the work in Task 1 with the exception that the solar reflectance of the cool roof is increased from 0.25 to 0.40. Pending legislation has suggested increasing the solar reflectance of roof products to as high as 0.40. Levinson et al. (2006) indicated that coated steel and glazed clay-tile products painted with cool pigments can achieve near-infrared solar reflectance of up to 0.50 and 0.75, respectively, resulting in a solar reflectance limit of about 0.50. We conducted the computations based on ASHRAE Standard 90.2 (2007) code to document the solar reflectance needed in an ASV metal roof to match the cooling load for a 0.40 solar reflective roof of conventional direct-to-deck construction (Figure 9). Again the roof slope is 18.34°, and the roof span from soffit to ridge is 14.2 ft (4.3 m).

Simulations for the hot climates surrounding Miami and Austin resulted in the highest solar reflectance needed by an offset-mounted roof to have the same seasonal cooling load as the 0.40 high-reflectance, direct-to-deck metal roof. Atlanta has 4814 CDD_{65} but also sees 2614 HDD_{65}, double that of Austin and 11 times that of Miami. From Atlanta on through the colder climates such as Baltimore, Chicago, and Minneapolis, the solar reflectance of a painted roof offset 0.75 in. (0.019 m) needs to be about 0.26 to have the same cooling load as the direct-to-deck base roof with high solar reflectance of 0.40. The stone-coated metal roof needs a solar reflectance of only 0.11 in Atlanta and the colder climates.

A standing-seam cool roof (0.25 SR) that has a 1.5 in. (0.038 m) air space spanning 14.2 ft (4.3 m) has about the same annual cooling load as the direct-to-deck, 0.40 high-reflectance roof assembly (Figure 9). The stone-coated metal with an air space of 1.5 in. (0.038 m) needs a solar reflectance of only about 0.1 in hot climates and 0.05 in cold climates to match the cooling load for a 0.40-SR cool roof that is fastened directly to the deck.

Task 3. AtticSim/Energy Plus computed the heat flux for a low-slope (9.5°), SR-0.65 (0.50 weathered) roof and compared it to the load for a roof fitted with a 2 in. (0.051 m) air space spanning 14.2 ft (4.3 m). Simulations investigated all climate zones listed in Table 2. The code determined the heat flux for the low-slope roof with 0.65 SR (0.50 weathered) and matched its seasonal cooling load to the offset-mounted assembly by altering the solar reflectance of the ASV system. Results for ASHRAE/IES Standard 90.1 (2007) and ASHRAE Standard 90 (1980) are listed respectively in Tables 3 and 4.

The ASV roof must have a solar reflectance greater than 0.55 for all climates zones (Table 3) in order to have the same

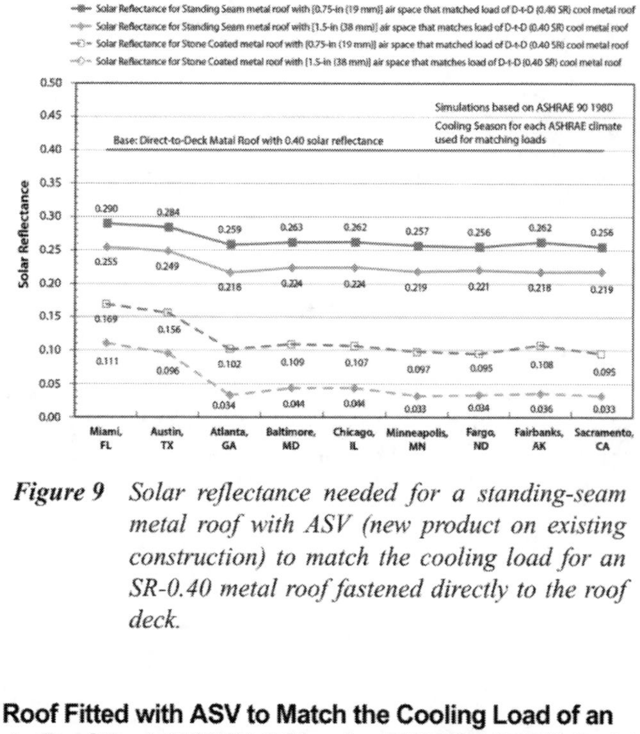

Figure 9 *Solar reflectance needed for a standing-seam metal roof with ASV (new product on existing construction) to match the cooling load for an SR-0.40 metal roof fastened directly to the roof deck.*

Table 3. Solar Reflectance Needed for a Painted Metal Roof Fitted with ASV to Match the Cooling Load of an SR-0.65 (SR-0.55 Aged) Cool White Roof Mounted Directly to Roof Deck (ASHRAE Standard/IES 90.1 [2007] Code)

Task 1C	Miami, FL	Austin, TX	Atlanta, GA	Baltimore, MD	Chicago, IL	Minneapolis, MN	Fargo, ND	Fairbanks, AK	Sacrament, CO
HDD_{65}	222	1481	2614	4731	6139	7787	10052	13940	2697
CDD_{65}	9368	7435	4814	3598	2895	2513	1332	1040	1202
R_{US} Ceiling	15	20	20	20	20	20	20	20	20
Direct-to-Deck	Solar reflectance needed by roof with inclined air space set at 2 in. (50.8 mm)								
New Product (SR 0.65)	0.577	0.572	0.546	0.543	0.539	0.535	0.526	0.547	0.529
Aged Product (SR 0.55)	0.399	0.392	0.356	0.356	0.351	0.347	0.340	0.354	0.337

Table 4. **Solar Reflectance Needed for a Painted Metal Roof with ASV to Match the Cooling Load of an SR-0.65 (SR-0.55 Aged) Cool White Roof Mounted Directly to the Roof Deck (ASHRAE 90 [1980] Code)**

Task 1C	Miami, FL	Austin, TX	Atlanta, GA	Baltimore, MD	Chicago, IL	Minneapolis, MN	Fargo, ND	Fairbanks, AK	Sacrament, CO
HDD_{65}	222	1481	2614	4731	6139	7787	10052	13940	2697
CDD_{65}	9368	7435	4814	3598	2895	2513	1332	1040	1202
R_{US} Ceiling	10	10	10	11.8	13.3	16.4	16.7	19.6	10
Direct-to-Deck	Solar reflectance needed by roof with inclined air space set at 2 in. (50.8 mm)								
New Product (SR 0.65)	0.565	0.557	0.537	0.536	0.536	0.535	0.526	0.547	0.529
Aged Product (SR 0.55)	0.386	0.375	0.346	0.347	0.347	0.347	0.340	0.354	0.333

cooling load as new product fastened directly to the roof deck and having a roof solar reflectance of 0.65. For aged product, the solar reflectance needed to match loads is less than the prescriptive code of 0.55. As an example, in Miami a 0.40 solar reflectance is needed for a roof with 2 in. (0.051 m) of air space to match the load of an SR-0.55 (aged) roof that is fastened directly to the roof deck. In Chicago, an SR-0.35 roof with 2 in. (0.051 m) of air space has the same cooling load as the 0.55 direct-to-deck assembly.

Generally the results for existing construction based on the ASHRAE Standard 90 (1980) code are very similar to those computed for the new 2007 code (Table 3 compared to Table 4). However, the annual loads are very different between the two cases because, for example, Chicago called for R_{US}-20 (R_{SI}-3.5) ceiling insulation in 2007 while in 1980 ASHRAE specified R_{US}-13.3 (R_{SI}-2.3).

CONCLUSION

Field tests were conducted to examine the effects of cool-color pigments on standing-seam metal roofs and stone-coated metal roofs attached directly to the roof deck or fitted with ASV. Results show that light-gray stone-coated metal shakes with ASV reduced the peak day heat transfer penetrating the roof deck by about 45% compared to the heat penetrating the deck of an attic covered with an asphalt shingle roof.

Field data also indicated that the measured summertime deck heat flows for an asphalt shingle roof having solar reflectance of 0.09 and thermal emittance of 0.89 was about the same as that measured for a cool-color standing-seam metal roof having a higher solar reflectance of 0.28 and thermal emittance of 0.85. Both roof assemblies were attached directly to the deck. Therefore AtticSim/Energy Plus simulations based on equal cool roof reflectance for the two systems revealed that the asphalt shingle yielded better performance if both roofs were direct-to-deck applications. As result, simulations were conducted using a painted metal fixed directly to the roof deck as the base to best show the trade-off between solar reflectance and ASV.

Simulations to determine the solar reflectance and ASV trade-off were conducted at the same roof slope (18.34°) as field tested on the ESRA and benchmarked by AtticSim against the field data. It is expected that the steeper the pitch of the roof, the greater will be the buoyancy-driven convection within the air space, and, therefore, the trade-off between solar reflectance and ASV will be more and more weighted toward ASV as roof slope increases. However, additional simulations conducted at steeper roof pitch and different roof spans (implying different aspect ratios) are needed to qualify potential code requirements for ASV as applied to steep- and low-slope roofs.

A standing-seam metal roof with 0.75 in. (0.019 m) air space with span of 14.2 ft (4.3 m) needs only a 0.10 solar reflectance to have the same annual cooling load as a direct-to-deck cool-color metal roof. A standing-seam metal roof fitted with 1.5 in. (0.038 m) of ASV or a stone-coated metal fitted with 0.75 in. (0.019 m) of air space (spanning 14.2 ft [4.3 m]) can both be black and still have a seasonal cooling load within 0.05% of the conventionally constructed cool-color metal roof (solar reflectance of 0.25). Computations for retrofit application based on ASHRAE Standard 90 (1980) showed ASV air spaces of 0.75 and 1.5 in. (0.019 and 0.038 m) would permit black roofs to have cooling loads equivalent to the direct-to-deck cool roof.

The hot climates of Miami and Austin require solar reflectance of 0.30 and 0.29 respectively for an ASV standing-seam metal roof (0.75 in. [0.019 m] air space spanning 14.2 ft [4.3 m]) to have the same seasonal cooling load as the 0.40 solar reflective direct-to-deck metal roof. For climates colder than Atlanta, the required solar reflectance for the ASV assemblies remain relatively level. Therefore, a cool-color roof (solar reflectance of 0.25) fitted with at least 0.75 in. (0.019 m) of ASV performs as well as a 0.40 reflective roof attached directly to the roof deck. The stone-coated metal with an air space of 1.5 in. (0.038 m) need only have a solar reflectance of about 0.10 to match the cooling load for a 0.40 solar reflective cool roof that is fastened directly to the deck.

A low-slope painted metal roof with 2.0 in. (0.051 m) of air space must have a solar reflectance slightly above 0.55 for all climate zones to match 0.65 direct to deck. Roof span was span was 14.2 ft (4.3 m). For aged product the solar reflectance needed to match load is less than the prescriptive code of 0.55. The results for existing construction based on the

ASHRAE Standard 90 (1980) code are very similar to those computed for the current 2007 code.

ACKNOWLEDGMENTS

The U.S. Department of Energy funded work under the direction of Marc LaFrance of the Buildings Program. The Metal Construction Association and its affiliate members provided the stone-coated shake and S mission roofs. Metro Roof Products constructed the attic assemblies and provided valuable assistance in installing the roofs on the steep-slope assemblies. Custom-Bilt Metals provided the cool-color standing-seam metal roofs.

ABBREVIATIONS, NOMENCLATURE, AND SUBSCRIPTS

air	=	air within inclined channel
ASV	=	above-sheathing ventilation
C_{DD}	=	cooling degree days based on 65°F
E	=	thermal emittance
EPS	=	expanded polystyrene insulation
H_{DD}	=	heating degree days based on 65°F
HFT	=	heat flux transducer
IR	=	infrared spectrum
L	=	length of roof from soffit to ridge
L/W	=	aspect ratio
PVDF	=	polyvinylidene fluoride
R_{SI}	=	thermal resistance ($m^2 \cdot °C/W$)
R_{US}	=	thermal resistance ($h \cdot ft^2 \cdot °F/Btu$)
sky	=	sky temperature
solar	=	irradiance
SR	=	solar reflectance
T	=	temperature
W	=	height of inclined air space

REFERENCES

Akbari, H., and R. Levinson. 2008. Evolution of cool-roof standards in the U.S. *Advances in Building Energy Research* 2:1–32.

Akbari, H., P. Berdahl, R. Levinson, R. Wiel, A. Desjarlais, W. Miller, N. Jenkins, A. Rosenfeld, and C. Scruton. 2004. Cool colored materials for roofs. ACEEE Summer Study on Energy Efficiency in Buildings, *Proceedings of American Council for an Energy Efficient Economy*, Pacific Grove, CA, August.

American Society for Testing and Materials (ASTM). 2004. *Standard C 1340-04, Standard Practice for Estimation of Heat Gain or Loss through Ceilings under Attics Containing Radiant Barriers by Use of a Computer Program.* West Conshohocken, PA: American Society for Testing and Materials.

ASHRAE. 2007. ASHRAE/IES Standard 90.1-2007, *Energy Standard for Buildings Except Low-Rise Residential Buildings.*

Arnold, J.N., P.N. Bonaparte, I. Canton, and D.K. Edwards. 1974. *Proceedings of the 1974 Heat Transfer and Fluid Mechanics Institute.* Stanford, CA: Stanford University Press.

ASHRAE. 2007. ASHRAE 90.2-2007, *Energy Efficient Design of Low-Rise Residential Buildings.*

ASHRAE. 2005. Thermal resistances of plane air spaces. Chapter 22.2, Table 2. *ASHRAE Handbook—Fundamentals.* Atlanta: ASHRAE.

Azevedo, L.F., and E.M. Sparrow. 1985. Natural convection in open-ended inclined channels. *Journal of Heat Transfer* 107:893–901.

Berdahl, P., H. Akbari, R. Levinson, and W.A. Miller. 2006. Weathering of roofing materials–an overview. *Construction and Building Materials* 22:423–433.

California Energy Commission (CEC). *2008 Energy Efficiency Standards for Residential and Nonresidential Buildings.* P400-03-001F. Sacramento: California Energy Commission.

Cummings, J.B., J.J. Toole, and N.A. Moyer. 1990. Duct leakage impacts on airtightness, infiltration, and peak electrical demand in Florida homes. Professional Paper FSEC-PF-212-90, Florida Solar Energy Center, Cocoa, FL.

ElSherbiny, S.M., G.D. Raithby, and K.G.T. Hollands. 1984. Heat transfer by natural convection across vertical and inclined cavities. *Journal of Heat Transfer* 104:96–102.

International Energy Conservation Code (IECC), ISBN: 978-1-58001-742-8. 2009.

Hollands, K.G.T., T.E. Unny, G.D. Raithby, and L. Konicek. 1976. Free convection heat transfer across inclined air layers. *Journal of Heat Transfer* May:189–193.

Levinson, R., P. Berdahl, H. Akbari, W. Miller, I. Joedicke, J. Reilly, Y. Suzuki, and M. Vondran. 2006. Methods of creating solar-reflective nonwhite surfaces and their application to residential roofing materials. *Solar Energy Material, Solar Cells* 91:304–314.

Miller, W., M-D Cheng, J. New, R. Levinson, H. Akbari, P. Berdahl, J. New, A. Desjarlais, Y.J. Juang, E. Erdem, R. Levinson, and H. Akbari. 2010. Natural exposure testing in California. Task 2.5.5 milestone report to be submitted to *Atmospheric Environment*.

Miller, W.A., M. Keyhani, T. Stovall, and A. Youngquist. 2007. Natural convection heat transfer in roofs with above-sheathing ventilation. *Thermal Performance of the Exterior Envelopes of Buildings X.* Atlanta: ASHRAE.

Miller, W.A. 2006. The effects of infrared-blocking pigments and deck venting on stone-coated metal residential roofs. Report ORNL/TM-2006/9. Oak Ridge, TN: Oak Ridge National Laboratory.

Miller, W.A., P. Wilson, and A. Karagiozis. 2006. The impact of above-sheathing ventilation on the thermal and moisture performance of steep-slope residential roofs and attics. 15th Symposium on Hot and Humid Climates, Orlando, FL, July.

Miller, W.A., K.T. Loyle, A.O. Desjarlais, H. Akbari, R. Levenson, P. Berdahl, S. Kriner, S. Weil, and R.G. Scichili. 2004. Special IR reflective pigments make a dark roof reflect almost like a white roof. *Thermal Performance of the Exterior Envelopes of Buildings, IX, Proceedings of ASHRAE THERM IX,* Clearwater, FL, December.

NREL/TP-472-7332a. 1995. *Home Energy Rating System Building Energy Simulation Test (HERS BESTEST), Volume 1,* used with "Energy Plus: DOE's Next Generation Simulation Program."

Parker, D.S., and J.R. Sherwin. 1998. Comparative summer attic thermal performance of six roof constructions. *ASHRAE Transactions* 104(2):1084–1092.

Petrie, T.W., T.K. Stovall., K.E. Wilkes, and A.O. Desjarlais. 2004. Comparison of cathedralized attics to conventional attics: where and when do cathedralized attics save energy and operating costs? *Thermal Performance of the Exterior Envelopes of Buildings, IX, Proceedings of ASHRAE THERM VIII,* Clearwater, FL, December.

Petrie, T.W., K.E. Wilkes, P.W. Childs, and J.E. Christian. 1998. Effect of radiant barriers and attic ventilation on residential attics and attic duct systems: new tools for measuring and modeling. *ASHRAE Transactions* 104: 1175–1192.

Wilkes, K.E. 1991. Thermal model of attic systems with radiant barriers. Report ORNL/CON-262. Oak Ridge, TN: Oak Ridge National Laboratory.

The Equivalent Thermal Resistance of Tile Roofs with and without Batten Systems

Rick Olson

William Miller, PhD
Member ASHRAE

Russell Graves

ABSTRACT

A clay roof, several concrete tile roofs, and one conventional direct-nailed shingle roof were installed on a fully instrumented attic test facility operating in east Tennessee's climate. The roof tiles were attached to batten and counter-batten systems and also fastened directly to the deck to observe the benefits of the venting occurring between the roof's sheathing and the profiled tile. Roof, attic, and deck temperatures and heat flows were recorded for each of the tile roofs and for the attic cavity covered with a conventionally pigmented and direct-nailed asphalt shingle roof. The data were used to benchmark a computer tool for simulation of roofs and attics and the tool used to develop an approach for computing an equivalent seasonal R-value for sub-tile venting. The approach computed the heat flux through the ceiling of the direct-nailed shingle roof and also the flux crossing the ceiling of a ventilated tile roof. The load for the shingle roof and attic was made equivalent to the load crossing the ceiling of the tile roof and attic by modifying the amount of ceiling insulation for the shingle assembly until both attics had the same annual load. The added R-value equates to the effective annual thermal resistance for the tile roof having different combinations of surface radiation properties (cool color pigments) and or building constructions. Therefore the direct nailed shingle roof served as a control for estimating the equivalent thermal resistance of the air space. Simulations were benchmarked to data in the ASHRAE Handbook—Fundamentals *for the thermal resistance of inclined and closed air spaces.*

INTRODUCTION

The venting of the underside of roof tile provides thermal benefits for comfort cooling. Residential roof tests by Beal and Chandra (1995) demonstrated a 45% reduction in the daytime heat flux penetrating a counter-batten tile roof as compared to a direct nailed shingle roof. Parker et al. (2002) measured an 8% reduction in annual cooling load for a home with terra cotta barrel-shaped tile as compared to an adjacent, identical footprint home with asphalt shingle roof. These reported energy savings are in part attributed to the venting that occurs on the underside of the roof tile, although it is difficult to quantify this benefit.

The reduced heat flow occurs because of a thermally driven airflow within the air gap formed between the tile and the roof deck. The thickness of the air space depends on the construction practice whether laid direct-to-deck, fastened to battens or attached to double-battens. For double-batten construction, wood furring strips are laid vertically (soffit-to-ridge) against the roof deck and a second batten running parallel to the roof's ridge is laid horizontally across the vertical counter-battens (Figure 1). The bottom surface of the inclined channel is formed by the sheathing (oriented strand board [OSB] deck) and 30# felt paper placed on top of the sheathing. The underside of the roof tiles establish the upper surface of the inclined channel, and the tile overlaps are designed to be air porous to allow pressure equalization and reduce the wind uplift on the tiles. The airflow in the inclined channel is driven more by buoyancy than by wind-driven forces because building codes require fire stops be used to block any path for potential burning embers to enter the inclined channel. The air gap also provides an improvement in the insulating effect of the roof system. However, measuring and correctly describing the

***Rick Olson** is the technical director for the Tile Roof Institute, Eugene, Oregon. **William Miller** serves as a building scientist at the Oak Ridge National Laboratory, Oak Ridge, TN. **Russell Graves** is an undergraduate student studying Mechanical Engineering at the University of Tennessee, Knoxville, TN.*

Published by ASHRAE 2013

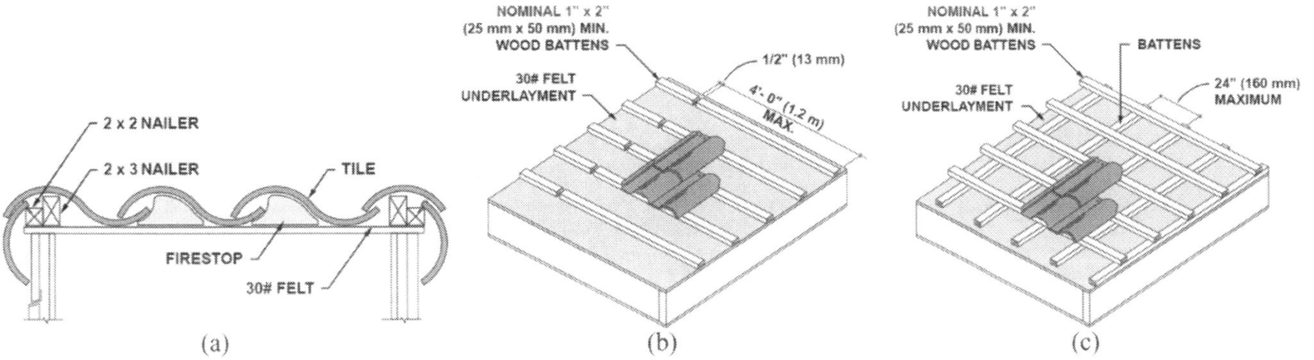

Figure 1 *The attachment of tile (a) direct-to-deck, (b) mounted on battens, and (c) mounted on double-battens.*

heat flow within the cavity of a tile roof is a key hurdle for predicting the roof's thermal performance.

IECC (2012) and ASHRAE 90.2 (2007) do not directly address sub-tile ventilation. Title 24 of the California Energy Commission (CEC 2008, Section 152) specifies that steep-slope roof products with a density of 5 pounds per square foot or more shall have a minimum aged solar reflectance of 0.15 and a minimum thermal emittance of 0.75 in all California climates. The CEC allows several exceptions for the solar reflectance and emittance requirements, and two of the exceptions are pertinent to work described herein:

1. Exception granted if insulation with a thermal resistance of at least 0.85 (h·ft^2·°F)/Btu [0.15 (m^2·°C)/W] or at least a ¾ in. (0.02 m) air space is added to the roof deck; or
2. R_{US}-3 (R_{SI}-0.53) or greater roof deck insulation is placed above the vented attic (exception applicable for climate zones 10, 11, 13 and 14).

A combined experimental and analytical study therefore documented the effective annual thermal resistance of the inclined and opened air space formed by clay and concrete tile applied direct-to-deck, applied on battens, and applied on double batten systems. The batten and double batten assembly were studied to determine added benefit for elevating tile off the roof deck and to also assess systems that would provide a free flow path between the tile and the roof deck (see Pennington comment on blockage in air space made at CEC workshop [Peters, 2006]). Tile roofs were selected having high-, medium-, and flat-profiles.

TILE ROOF AND ATTIC TEST STAND

The experimental portion of the study included testing of attic assemblies on the Envelope Systems Research Apparatus (ESRA). All attics were equipped with heat flux transducers (HFTs) embedded in the roof deck and in the attic floor. A Fox 670 heat flux apparatus was used to calibrate each HFT in a guard made of the same material used in construction to correct for shunting effects (i.e., distortion due to three-dimensional heat flow, ASTM [2006]). Salient features of the ESRA and instrumentation details of the roof and attic assemblies are provided by Miller (2006). A commercially available asphalt shingle with a solar reflectance of 0.093 and thermal emittance of 0.89 (SR10E89) served as the control for comparing the thermal performances of the prototype roof systems. Solar reflectance (SR) was measured using the ASTM protocol C 1549 (2009). Thermal emittance (TE) was measured using the ASTM protocol C 1371 (2004).

The field data were used to benchmark the attic simulation code AtticSim (ASTM C 1340 2004). Ceiling insulation in the test assemblies was purposely set low at R_{US}-5 (h·ft^2·°F)/Btu (R_{SI}-0.9 [m^2·K]/W) to help reduce the experimental uncertainty in measured heat flux. Hence, the discussions of field data will focus on heat crossing the roof deck of the attic, and simulations will provide results for energy consumption of attics with code-compliant levels of insulation.

CONCRETE TILE FIELD DATA

Miller and Kosny (2008) assessed the effects of cool color roofs, thermal mass, and placement of batten and double-batten systems under tile roofs. Miller and Kosny observed that medium-profile tiles exhibited less heat transfer across the roof deck as compared to a conventional shingle roof. Most important, though, is the results among the medium-profile tile; the one with conventional color pigments (0.10 solar reflectance) and on double battens (1.5 in [0.038 m]) had measured deck heat transfer very similar to that of the cool color tile (0.40 solar reflectance) attached directly to the deck. Therefore, for the two tiles, the nominal 1.5 in (0.038 m) inclined air space is as effective as is the benefit derived from cool color roofs. The performance of the air space agrees with similar work by Beal and Chandra (1995), who field-tested two identical medium-profile concrete tile roofs; one tile roof was direct-nailed and the other was offset mounted about 1½ in. (0.038 m) above the deck. Beal and Chandra (1995) measured an 11% reduction in the daytime heat flux penetrating the concrete tile roof on double battens compared with the adjacent direct-nailed tile roof.

Miller et al. (2005) field tested flat concrete tile (SR13E83) and a medium-profile tile (SR10E93) that had very similar solar reflectance and thermal emittance to the control, an asphalt shingle (SR10E89) roof (Figure 2). The heat transfer through the roof and ceiling of the attic with the flat concrete roof and the medium-profile tile roof were about half that measured for the asphalt shingle roof (Figure 2). The reduction is due to the buoyancy effects occurring in the inclined air channel that dissipates heat away from the deck. The flat concrete tiles are attached to batten and counter-batten strips, which form a vent cavity that is about 1½ in. (0.038 m) deep. The medium-profile tile is on battens and also forms its own half-cylindrical channel of about 0.5 in. (0.0127 m) radius.

Bulk air and sheathing temperatures for the flat concrete tile (heat flux depicted in Figure 2) were measured along the length of the tile roof from soffit to ridge (Figure 3). Data is displayed for a hot August day and also for a cold day in January to view the differences between hot and cold weather.

August (8/17/2004) field data: At 8 AM the air in the inclined cavity and the sheathing temperatures are about the same from the soffit to the ridge vent. As the day progresses,

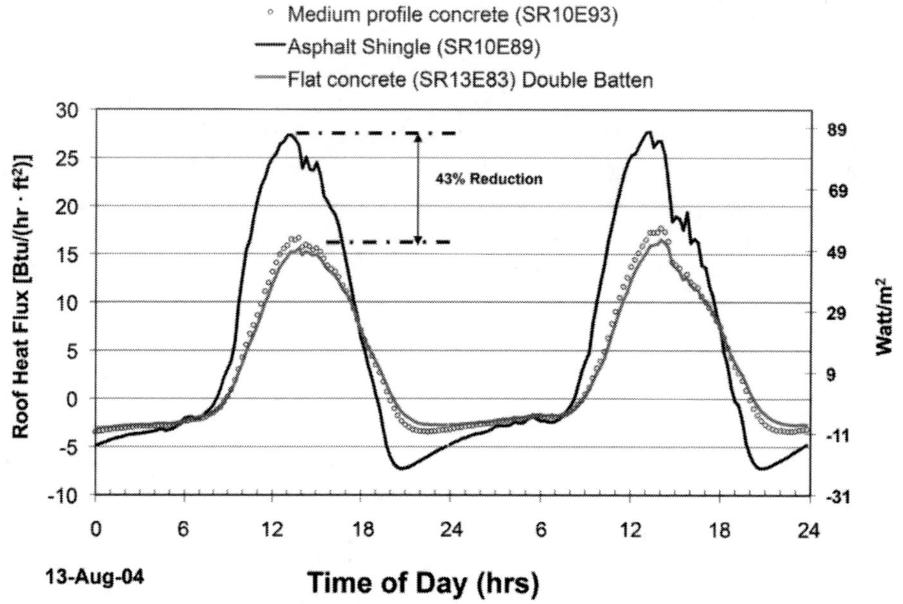

Figure 2 *Heat penetrating the roof of each attic assembly being field tested on the ESRA (Miller et al. 2005).*

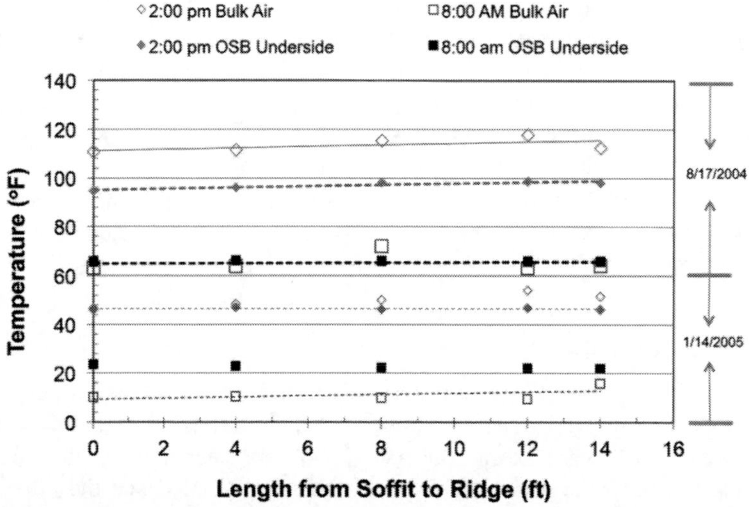

Figure 3 *Local temperature measures for a flat concrete tile made along the length of the inclined air space at 8:00 a.m. and at solar noon (2:00 p.m.).*

irradiance heats the roof and there is observed a linear increase in the sheathing's underside temperature; it being about 94.5°F (34.7°C) near the soffit and increasing to 98°F (36.6°C) just upstream the ridge vent. The linear profile is expected because the irradiance has no preferred location on the south-facing roof deck, implying a constant solar flux. The bulk air temperature in the inclined air space is observed to be roughly 20°F (10.5°C) warmer than the sheathing (Figure 3). Its profile dips as the air nears the ridge vent. The tiles are designed porous to alleviate wind uplift and therefore opposing winds at or around the ridge are possibly causing the slight drop. The bulk air temperature is about 30°F (16.7°C) warmer than the outdoor air.

January (1/14/2005) field data: At 8 a.m. the bulk air temperature measured near the soffit is 13°F (7.2°C) less than the sheathing's underside temperature (Figure 3). This temperature difference lessens as one moves from the measurement station at the soffit to the station at the ridge. Night sky radiation causes the tile roof to chill below the outdoor air temperature, and therefore air in the inclined air space is cooled below the outdoor air temperature and is believed to flow from ridge to soffit as the tile cools the air in the channel. At solar noon there is only a slight increase in the bulk air temperature occurring around 12 feet (3.6 m) from the soffit. However, the irradiance has caused a flow reversal in the channel whereby flow is not from soffit to ridge simply because of the air temperatures as affected by the tile roof.

ALGORITHM FOR SUB-TILE VENTILATION

Miller et al. (2007) previously benchmarked AtticSim against the field data for stone-coated metal roofs fixed to double-battens. The tool has also been validated in this study against the tile field experiments (Figures 2 and 3) and predicted the ceiling heat flow within ±10% of the field measure. The tool contains overall moisture balances but is not designed to predict moisture content of wood surfaces.

The airflow in the inclined air space can be driven by both buoyancy and wind; however, the air space is obstructed at the soffit to resist burning embers landing on the structure and to resist flame impingement (Quarles and TenWolde 2004). Therefore, the flow phenomenon is driven more by natural convection rather than by forced convection because the fire code calls for an obstruction which for tile roofs takes the form of fire stops[1] which still allows for outdoor makeup air. Blocking the eave also helps alleviate wind uplift on the offset-mounted roof. Hence, by fire code and wind design, the formulation of the algorithm is based solely on natural convection flow fields within the air space of the offset mounted roof system.

Hollands et al. (1976); Arnold, et al. (1974); Azevedo and Sparrow (1985), and, most recently, Brinkworth (2000) studied the air flows occurring in inclined air channels. During winter exposure, the warmer roof deck is positioned below the cooler roof cover and heat flows upwards through the roof, much as in the problem studied by Hollands et al. (1976) for a closed cavity. Here, a denser air layer near the roof's underside overlies lighter air adjacent to the roof deck. The density inversion causes the heat transfer across the air channel to switch from conduction to single-cell convection to Bénard cell convection, depending on the magnitude of the Ra_S number. For Ra_S numbers of less than $1708/\cos(\theta)$, there is no naturally induced airflow within the cavity, and the heat transfer occurs exclusively by conduction. However, as the flow increases as a result of buoyancy, the heat transfer within the channel can switch to Bénard cell convection.

During summer exposure, the roof cover is hotter than the roof deck, and Bénard cell convection does not occur within a closed inclined air space because the lighter air layer is naturally above the denser air layer. Arnold et al. (1974) offers a correlation which is based on an adaptation of heat transfer in a vertical closed cavity; it being

$$Nu(\theta) = 1 + [Nu_{90^\circ} - 1]\sin(\theta) \qquad (1)$$

where Nu_{90° is estimated from the correlations by ElSherbiny et al. 1984, and θ is the angle of inclination the roof makes with the horizontal plane.

The correlation by Azevedo and Sparrow (1985) is used for the case of an open inclined air space. Their correlation was developed from experimental data where both walls were heated, the top wall heated with the bottom wall unheated, and also for the case of the top wall unheated and the bottom wall heated. Several inclinations and channel heights were studied and used to derive the following correlation which is programmed in an algorithm for open channels:

$$Nu = 0.673\left[\frac{S}{H}Ra_S\right]^{0.25}$$

where

S = the height of the air space

H = the length of the roof from soffit to ridge

Ra_S = the Rayleigh number for an inclined air space of S height.

The algorithm in AtticSim was benchmarked against the *ASHRAE Handbook—Fundamentals* (1989) data for closed inclined air spaces, Table 1. Two effective emittance values were investigated to review the computed R-value. If the thermal emittance of the two facing surfaces are high at 0.90 then radiation heat transfer dominates over convention in the closed cell. However, if one of the two surfaces has a low emittance (i.e., 0.05 to 0.25) then convection heat transfer dominates.

The computed R-value by AtticSim for a closed air space is within 1.5% of ASHRAE measures (Table 1) if there are no low-emittance surfaces. The inclusion of a low-e surface showed an error of 6.7% for a 0.75 in. (0.019 m) air space and an error of 12% for the larger air space of 1.5 in (0.038 m). The

[1] Tile industry uses the term *bird stop* for keeping rodents and birds from entering and nesting in the inclined air space.

mean air temperature in the air space and the temperature change (ΔT) across the air space were gleaned from summer field data collected at solar noon. AtticSim and ASHRAE (Yarbrough and Desjarlais 1991) algorithms fixed these temperatures and computed the thermal resistance of the air space. Note that changing the air temperature and the temperature drop across the air space changes the value of the thermal resistance.

If one of the surfaces has low-emittance, then the thermal resistance is seen to increase by almost a factor of 5 over the case of the closed channel with high emittance surfaces, Table 1.

The results imply that inclusion of a reflective underlayment or painting the underside of the tile with low-emittance paint will enhance performance of tile roofs. Therefore simulations were conducted with and without a low emittance surface in the air space.

Climatic Simulation Results

The AtticSim code was used in conjunction with Energy Plus to compute the equivalent annual thermal resistance of the inclined air space under tile roofs. The standardized HERS BESTEST home (NREL 1995) clad with clay and concrete tile roofs was simulated for several cities each in a unique ASHRAE climate zone, Table 2. Sacramento, CA was included but with ceiling insulation set to Title 24 building code. Energy Plus estimated the hourly run times for the HERS BESTEST home (NREL 1995) conditioned by an air-conditioner certified with a Seasonal Energy Efficiency Ratio (SEER) of 13 and heated by an 85%. efficient gas furnace.

Table 1. Benchmark of the Effective Thermal Resistance Computed by the AtticSim Code as Compared to *ASHRAE Handbook—Fundamentals* (1989) Data

	ε_{eff}	Air Space ΔT, °F (°C)	Mean Air Temperature, °F (°C)	Thermal Resistance[2] (°F · ft^2 ·h)/Btu ASHRAE[1]	AtticSim	% Error
0.75 in. (0.019 m)	0.7456	13.7 (7.61)	130.4 (54.7)	0.762	0.752	1.22%
Air space	0.0497	33.3 (18.5)	123.3 (50.7)	2.976	2.775	6.76%
1.5 in. (0.038 m)	0.7456	14.7 (8.17)	130.0 (54.4)	0.814	0.818	0.49%
Air space	0.0497	37.7 (20.9)	121.8 (49.9)	3.191	3.588	12.44%

1 Yarbrough and Desjarlais (1991) and ASHRAE Handbook—Fundamentals, (1989) Chapter 22.2, Table 2.
2 R_{SI} [(m^2 ·K)/W] = R_{US}/5.678

Table 2. Simulations Conducted in the ASHRAE Climate Zones to Estimate the Benefit of Sub-Tile Ventilation. IECC 2009 Code Assumed for Ceiling R-value

ASHRAE Zone	City	State	HDD_{65}	CDD_{65}	Ceiling R-value [h · ft^2 · °F)/Btu]
1	Miami	FL	222	9368	30
2	Austin	TX	1481	7435	30
3	Atlanta	GA	2614	4814	30
4	Baltimore	MD	4731	3598	38
5	Chicago	IL	6139	2895	43
6	Minneapolis	MN	7787	2513	49
7	Fargo	ND	10052	1332	49
8	Fairbanks	AK	13940	1040	52
Title 24	Sacramento	CA	2697	1202	38

AtticSim used these outputs from Energy Plus to better estimate the roof and attic load as coupled to the building envelope. The attic floor was modeled assuming no air leakage crossing the attic floor. Ducts for the air conditioner were assumed in the conditioned space.

The TRI and its affiliates provided specifications for several tiles with high-, medium-, and flat profile, see Appendix A. Members computed the volume of the air space made by tiles applied direct-to-deck and derived an effective height of the tile's air space by dividing the underside volume by the footprint for a square (100 square feet [9.3 m^2]) of tile. Elevations adjustments were made for batten and double-batten systems, but the volume of wood within the footprint was subtracted from the total volume in determining the effective height of the air space. Simulations used the effective air space and results for all tile with the tile attached directly to the deck, the tile placed on battens, and the tile fastened to double batten assemblies is provided in Appendix B.

A procedure was formulated that matched the annual load crossing the ceiling of a tile roof and attic as compared to the load computed for a direct-nailed shingle roof (Figure 4). As example, an asphalt shingle roof with solar reflectance of 0.10 and thermal emittance of 0.90 was simulated for the climate of Miami, Florida; annual load crossing the ceiling was 4.25 MBtu/yr (4483 MJ/yr) assuming R_{US}–30 (R_{SI} 5.3) of insulation on the attic floor. The same home but with a tile roof placed on double battens needed R_{US}–25.3 (R_{SI} 4.5) to match the load of the home with a shingle roof.

Therefore, the equivalent annual R-value of the inclined air space is estimated at R_{US}–4.7 = [R_{US}–30 – R_{US}–25.3]. The computed R-value is termed an equivalent annual value because it accounts for the diurnal and seasonal changes in mean air temperature and temperature change across the inclined air space, which in turn affects the temperatures and heat flows within the attic. Results imply what the added level of ceiling insulation needs to be for the control shingle roof to have the same annual load as the tile roof.

The field data for summer shown back in Figure 3 reveals that the open air space helps drop the heat flow crossing the roof deck which, in turn, drops the underside temperature of the deck and the radiation and convection heat transfer crossing from the deck to the attic floor. As the surface temperatures of the roof and attic vary so does the radiation and convection heat transfer coefficients; however, by comparing the ceiling load for the attic with tile roof to the base shingle roof, the effect of the air space on attic temperatures and subsequently the heat flows is accounted for in prediction of the equivalent annual thermal resistance. Therefore the procedure does not yield a true academic measure of thermal resistance for the air space, rather we computed an effective annual resistance needed to match load to the base shingle roof assembly.

Direct-to-Deck ε_{eff} = 0.82 Simulation Results

AtticSim/Energy Plus simulations for the concrete and clay tile fastened directly to the deck show best results in the hottest US climates (Figure 5). The tiles with high profile (Espana and Barcelona 900) yielded the highest thermal resistance of all tile. In Miami, Florida (ASHRAE zone 1) these tiles yielded an annual equivalent R_{US}–2.7, which implies that for a conventional shingle roof to have the same annual ceiling

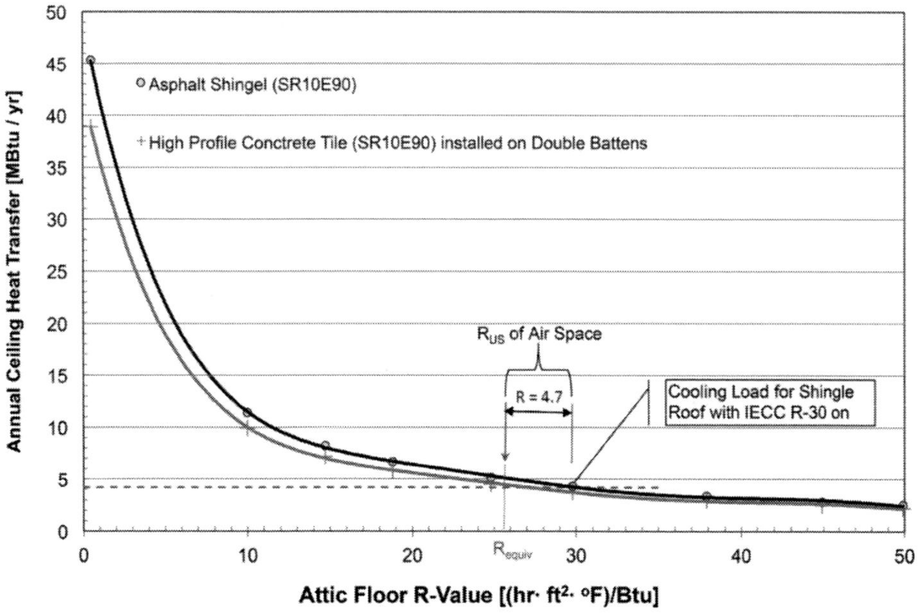

Figure 4 *Annual load crossing the ceiling of home simulated for Miami, Florida with shingle roof and with concrete tile roof placed on double battens. The R-value of ceiling insulation needed to match load for shingle roof is used to estimate the equivalent R-value of the inclined air space.*

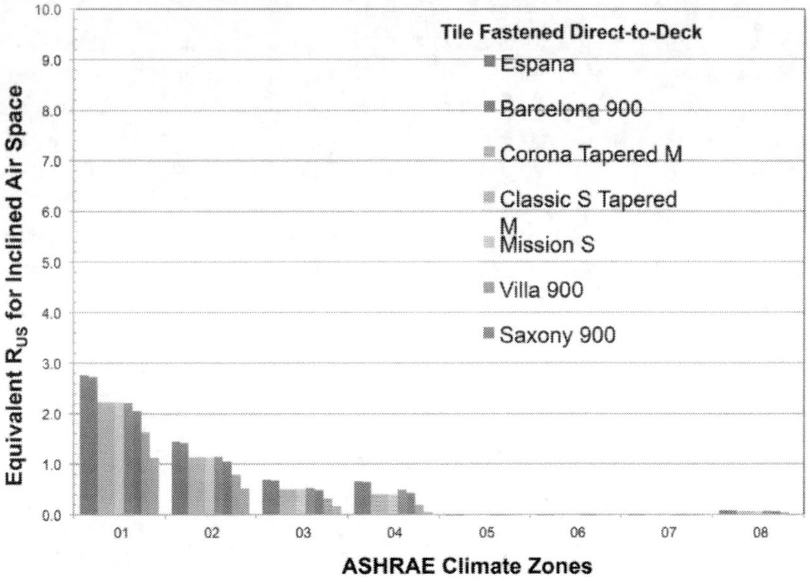

Figure 5 *The annual equivalent thermal resistance for DtD tile roofs;* $\varepsilon_{eff} = 0.82$.

load, it would need R_{US}–2.7 more insulation above ASHRAE 90.2 (2007) code requirement. Or said differently, the inclined air space formed by the Barcelona 900 and Espana tile with code level of insulation on the attic floor represents an equivalent annual R_{US} of 2.7 [(h·ft^2·°F)/Btu]; R_{SI} of ([0.48 m^2·°C]/W).

The high profile clay tiles, Corona Tapered M, Classic S Tapered M, and Mission S) all yielded an equivalent thermal resistance of about R_{US}–2.2 (R_{SI}–0.39) for exposure in Miami, Florida. In Austin, Texas (Zone 2) the equivalent thermal resistance drops to about R_{US}–1.1 (R_{SI}–0.19). The medium profile tiles (Villa 900) show results very similar to the clay tile roofs. The Saxony 900, Madera, and Eagle FT tile, because of their flat profile, had the lowest computed equivalent thermal resistance measures (Figure 5).

In climate zone 3 (Atlanta, Georgia), the R-value is of the order 0.7 to 0.2, and in the moderate climate of Baltimore, Maryland (zone 4) the equivalent annual R-value is about 0.2. The more predominantly heating climates of Chicago, Minneapolis, Fargo, and Fairbanks showed that the inclusion of the air space had no impact on thermal performance (Figure 5).

Direct-to-deck ε_{eff} = 0.22 Simulation Results

Additional AtticSim/Energy Plus simulations were made for the case of the tile's underside spray painted with a low-emittance paint having a thermal emittance of 0.22. The concrete and clay tile were again fastened directly to the deck (Figure 6). The low emittance paint increased the effective annual thermal resistance of all tile assemblies (Figure 6). Comparing Figure 5 to 6 reveals that the low emittance paint reduces the radiation heat transfer crossing the air space and therefore improves the thermal resistance. Even the cold climates show some resistance value for the air space; of the order R_{US}–0.5 (R_{SI}–0.09).

For ASHRAE zone 3 (Atlanta, Georgia) the high-profile concrete and clay tiles (Espana, Barcelona 900, Corona Tapered, Classic S, and Mission S) all showed an almost tripling of the effective annual thermal resistance with the addition of the low emittance paint. The clay tiles all jumped from R_{US}–0.5 (R_{SI}–0.09) to R_{US}–1.6 (R_{SI}–0.28).

Double Batten Simulation Results

AtticSim/Energy Plus simulations for Austin, Baltimore and Minneapolis were made with the tile offset mounted on battens and double battens. Figure 7 provides results for double batten applications with the air space having an effective thermal emittance of 0.05, 0.21, and 0.82, respectively. Appendix B lists results for flat-, medium-, and high-profile tile attached direct-to-deck, and on batten and double batten applications.

An underlayment is typically designed to provide hygrothermal protection for the roof deck. Some also offer a low-emittance foil. The foil's emittance is about 0.05 and if used in the air space would further reduce the radiation heat transfer crossing the air space. Because it is positioned facing up, the low-emittance surface is expected to degrade as fine dusts settle and tarnish the foil. Dust accumulation over time was shown to reduce the effectiveness of foil placed on an attic floor (Fairey, Swami, and Beal 1988). However, simulations were made to document feasible gains in performance. More work is needed to document the sustained effectiveness of the foil when used as an underlayment for tile roof applications.

In the hot climate of Austin, the effective annual thermal resistance of the Barcelona and Espana concrete tile applied on double battens increases from R_{US} 1.54 (R_{SI}–0.27)

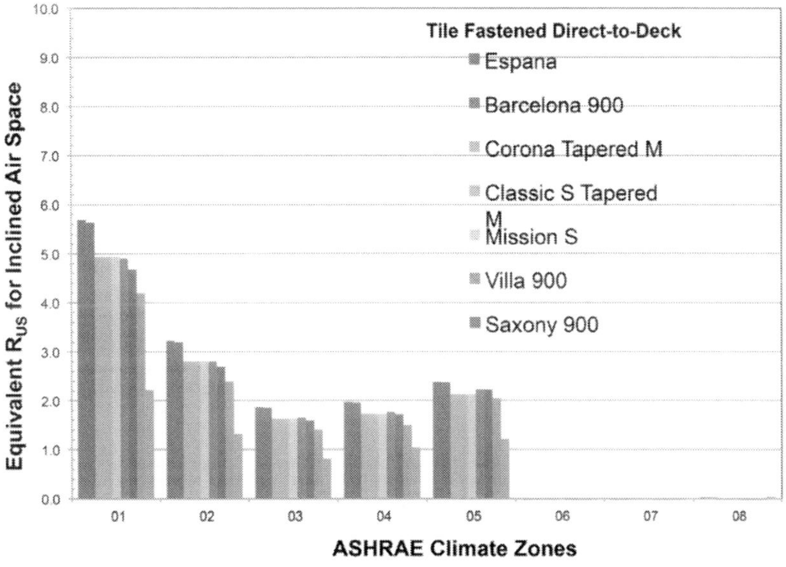

Figure 6 *The annual equivalent thermal resistance for direct-to-deck tile roofs;* $\varepsilon_{eff} = 0.21$.

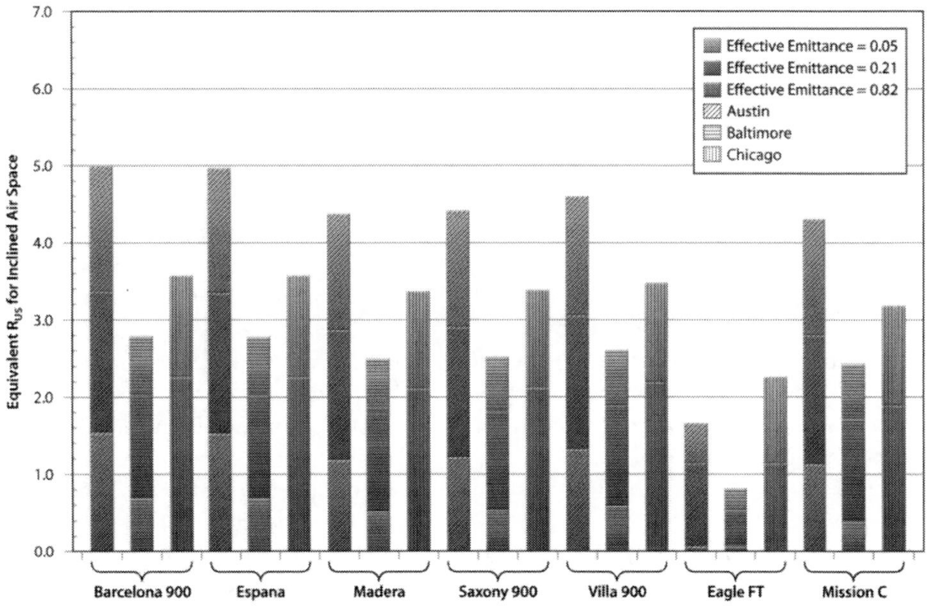

Figure 7 *Simulations for Atlanta, Georgia showing the annual equivalent thermal resistance for tile roofs offset mounted on double battens fastened to the deck.*

(air space $\varepsilon_{eff} = 0.82$) to R_{US} 3.4 (R_{SI}–0.60) (air space $\varepsilon_{eff} = 0.21$) up to a thermal resistance of R_{US}–5 (R_{SI}–0.88) if air space $\varepsilon_{eff} = 0.05$. The Mission C clay tile by practice is fastened direct-to-deck but is also shown in Figure 8 to view its performance against the concrete tiles on double battens. The Mission C yields an effective thermal resistance of R_{US}–4.3 (R_{SI}–0.76) which is just 14% less than that of the best performing concrete tile on double battens. Baltimore is more moderate in climate and maximum effective annual thermal resistance was about R_{US}–1.5 (R_{SI}–0.26) for the case having an $\varepsilon_{eff} = 0.05$. In Minneapolis, the effective annual thermal resistance is only R_{US}–0.60 (R_{SI}–0.1).

Miller (2011) developed design guidelines for roofs and attics and showed that the inclusion of the foil in an inclined air space would yield annual savings of about \$80 for simulation in Austin but in the colder climates of Baltimore and Minneapolis savings is only about \$10 annually. However, closer inspection shows that the inclusion of a foil in the inclined air space would increase the temperature on the underside of the roof deck during the cold evening hours as

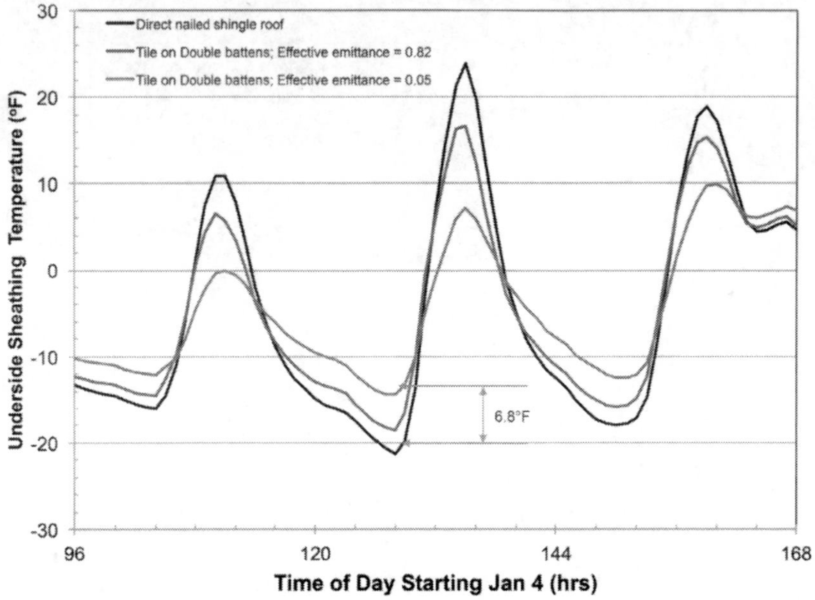

Figure 8 *Simulations for Minneapolis, Minnesota, showing the temperature of the OSB deck facing into the attic for tile roofs on double battens with and without a low emittance surface and also for a direct nailed shingle roof. Data depicted for winter exposure from January 4 through 6.*

compared to a direct nailed shingle roof (Figure 8). All roofs have about the same exterior roof temperature after a cold winter evening. However, at about 8 a.m. the tile roof with low-e foil shows its deck temperature (facing into the attic) to be 7°F (3.9°C) warmer than that computed for the direct nailed shingle roof.

The foil retards the radiation loss to the night sky and keeps the attic and subsequent surfaces warmer than the direct nailed base case. The warmer deck implies that the ventilated attic with tile on double battens has less potential for condensate to form on the underside of the deck as compared to the conventional ventilated and direct nailed roof and attic.

CONCLUSIONS

An inclined air space in or above the roof deck provides thermal buffering and is an excellent design strategy for reducing roof heat gains and losses in hot climates. Field tests of flat concrete tile and medium-profile concrete tile having almost the same solar reflectance and thermal emittance as a conventional shingle roof demonstrated peak day heat flows penetrating the roof deck that were 43% less than that measured for the control shingle roof. Hygrothermal performance also improves because the natural ventilation occurring in the air space during the day helps dry the roof. Thermal resistance of the air space increases if the effective thermal emittance of the air space is reduced by adding a foil (low-emittance surface) or low-emittance paint. Dropping the effective emittance drops the radiation heat transfer crossing the air space and at night causes the temperature of the sheathing's underside to not drop as much as compared to the case without the foil. Therefore, natural ventilation helps dry the roof during the day and the foil surface drops the potential for condensation on the underside of the sheathing during the winter's coldest evening hours.

Field temperatures collected within the air space from soffit to ridge show the heat transfer to be constant flux at the roof deck. At solar noon the bulk air temperature is about 30°F (16.7°C) warmer than the outdoor air and the natural convection flow is transitioning from laminar to turbulent flow as flow proceeds toward the ridge.

A procedure was formulated and validated for predicting the effective annual thermal resistance of the air space. First, an algorithm for combined convection and radiation in a closed air space was benchmarked against published data (*ASHRAE Handbook—Fundamentals* 1989). Second a computer tool, AtticSim with support from Energy Plus, modeled the HERS BESTTEST standard simulation and applied the code to several climates across the U.S. Finally, results were compared to a control shingle roof having the same solar reflectance and thermal emittance as the concrete and clay tile roofs. Fixing the annual load for the control shingle with code level of insulation on the ceiling and comparing the ceiling thermal resistance needed to yield the same load for a tile roof yields the annual effective thermal resistance of the inclined air space.

Tiles with a high profile yielded the highest thermal resistance of all tile. Offset mounting of all tiles improves the thermal resistance; however, best results were observed only in hot to mild climates. The more predominantly cold climates showed no thermal advantage for inclusion of the air space.

However, adding a low emittance surface to the air space causes a significant jump in the effective annual thermal resistance. In the hot climate of Austin, the effective annual thermal resistance of the high profile tile applied on double battens increases from R_{US} 1.54 (R_{SI}–0.27) (air space $\varepsilon_{eff} = 0.82$) to R_{US} 3.4 (R_{SI}–0.60) (air space $\varepsilon_{eff} = 0.21$) up to a thermal resistance of R_{US}–5 (RSI–0.88) if (air space $\varepsilon_{eff} = 0.05$).

ACKNOWLEDGMENTS

Funding for this project was provided by the Tile Roofing Institute under a DOE User Agreement, UR-07-580 and by the Department of Energy Building Technology program. Oak Ridge National Laboratory is managed by UT-Battelle, LLC, for the US Department of Energy under contract DE-AC05-00OR22725. The submitted manuscript has been authored by a contractor of the U.S. Government under contract DE-AC05-00OR22725.

NOMENCLATURE

CDD_{65} = cooling degree days with 65°F (18.3°C) base

HDD_{65} = heating degree days with 65°F (18.3°C) base

R_{SI} = thermal resistance in units of $m^2 \cdot K/W$

R_{US} = thermal resistance in units of $h \cdot ft^2 \cdot °F/Btu$

Nu = Nusselt number

Ra = Rayleigh number

REFERENCES

Arnold, J.N., P.N. Bonaparte, I. Canton and D.K. Edwards. 1974. Proceedings of the 1974 Heat Transfer and Fluid Mechanics Institute, Stanford University Press. Stanford, CA 1974.

ASHRAE. 1989. Thermal Resistances of Plane Air Spaces, Chapter 22.2, Table 2. *ASHRAE Handbook—Fundamentals*. Atlanta: ASHRAE.

ASHRAE. 2007. ASHRAE 90.2-2007: *Energy Efficient Design of Low-Rise Residential Buildings*. Atlanta: ASHRAE.

ASTM. American Society for Testing and Materials 2009. Designation C 1549-09: *Standard Test Method for Determination of Solar Reflectance Near Ambient Temperature Using a Portable Solar Reflectometer*. American Society for Testing and Materials, West Conshohocken, PA

ASTM. 2006. *Test Method for Steady-State Heat Flux Measurements and Thermal Transmission Properties by Means of the Heat Flow Meter Apparatus*. Standard C 518. West Conshohocken, Pennsylvania: American Society for Testing and Materials.

ASTM. 2004. Designation C 1371-04a: *Standard Test Method for Determination of Emittance of Materials Near Room Temperature Using Portable Emissometers*. American Society for Testing and Materials, West Conshohocken, PA

ASTM. 2004. ASTM Standard C 1340-04, *Standard Practice for Estimation of Heat Gain or Loss Through Ceilings Under Attics Containing Radiant Barriers by Use of a Computer Program*. West Conshohocken, PA: American Society for Testing and Materials.

Azevedo, L.F. and Sparrow, E.M. 1985. Natural Convection in Open-Ended Inclined Channels, *Journal of Heat Transfer*, 1985. Vol. 107, P. 893–901.

Beal, D. and S. Chandra. 1995. The Measured Summer Performance of Tile Roof Systems and Attic Ventilation Strategies in Hot Humid Climates. In *Thermal Performance of the Exterior Envelopes of Buildings* VI, 753–760. Atlanta: ASHRAE.

Brinkworth, B.J. 2000. A procedure for the routine calculation of laminar free and mixed convection in inclined ducts, *International Journal of Heat and Fluid Flow* 21:456–462.

CEC. California Energy Commission. 2008 *Energy Efficiency Standards for Residential and Nonresidential Buildings*. P400-2008-001-CMF. Sacramento: California Energy Commission.

ElSherbiny, S.M., G.D. Raithby and K.G.T. Hollands, Heat Transfer by Natural Convection across Vertical and Inclined Cavities, *J. Heat Transfer*, Vol. 104, pp. 96–102, 1984.

Fairey, P., M. Swami, D. Beal. 1988. RBS Technology: Task 3 Report. Florida Solar Energy Center. Publication Number: FSEC-CR-211-88.

Hollands, K.G.T., T.E. Unny, G.D. Raithby, and L. Konicek. 1976. Free convection heat transfer across inclined air layers, *Journal of Heat Transfer* May: 189–193.

International Energy Conservation Code (IECC), 2012.

Miller, W. A. 2011. Roof and Attic Design Guidelines for Hot Climates, DOE Deliverable to BTO TM Marc LaFrance.

Miller, W.A. and J., Kosny. 2008. Next Generation Roofs and Attics for Homes, in ACEEE Summer Study on Energy Efficiency in Buildings, proceedings of American Council for an Energy Efficient Economy, Asilomar Conference Center in Pacific Grove, CA., Aug. 2008.

Miller, W. A., M. Keyhani, T. Stovall and A. Youngquist. 2007. Natural Convection Heat Transfer in Roofs with Above-Sheathing Ventilation, in *Thermal Performance of the Exterior Envelopes of Buildings X*. Atlanta: ASHRAE.

Miller, W.A. 2006. The Effects of Infrared-Blocking Pigments and Deck Venting on Stone-Coated Metal Residential Roofs. ORNL/TM-2006/9. Oak Ridge, Tennesse: Oak Ridge National Laboratory.

Miller, W.A., W.M. MacDonald, A.O. Desjarlais, J.A. Atchley, M. Keyhani, R. Olson, and J. Vandewater. 2005. Experimental analysis of the natural convection effects observed within the closed cavity of tile roofs. Presented at Cool Roofs: Cutting Through the Glare, RCI Foundation Conference, Atlanta, May 12–13.

NREL/TP-472-7332a. 1995. Home Energy Rating System Building Energy Simulation Test (HERS BESTEST), Volume 1, used with EnergyPlus: DOE's Next Generation Simulation Program.

Parker, D.S., Sonne, J. K., Sherwin, J. R. 2002. Comparative Evaluation of the Impact of Roofing Systems on Residential Cooling Energy Demand in Florida, in ACEEE Summer Study on Energy Efficiency in Buildings, proceedings of American Council for an Energy Efficient Economy, Asilomar Conference Center in Pacific Grove, CA., Aug. 2002.

Quarles, S.L. and A. TenWolde. 2004. Attic and crawlspace ventilation: Implications for homes located in the Urban-Wildland Interface, Proceedings of the Wood-frame Housing Durability and Disaster Issues Conference, Forest Products Society, Las Vegas, October 2004.

Yarbrough, D.W. and Desjarlais, A.O. 1991. Prediction of the Thermal Performance of Single and Multi-Airspace reflective Insulation Materials, *Insulation Materials: Testing and Applications*, 2nd Volume, ASTM STP 1116,Graves and D.C. Wysocki, Eds. American Society for Testing and Materials, Philadelphia.

APPENDIX A

Table A1. Clay and Concrete Tile Specifications for Air Space with Tile Attached Directly to the Roof Deckand also Offset Mounted from the Roof Deck on Battens and Double Batten Assemblies

Manufacturer	Tile Profile	Material	Roof Application	Volume of Air Space for 100 ft^2 of Tile Roof		Average Thickness of Air Space of Tile Roof	
				in.^{3}	m^3	in.	m
MCA Corona Tapered M	High	Clay	Direct-to-Deck	21398	0.351	1.49	0.0378
MCA Classic S	High	Clay	Direct-to-Deck	24746	0.406	1.72	0.0437
MCA Mission S	High	Clay	Direct-to-Deck	25828	0.423	1.79	0.0455
Boral Saxony 900	Flat	Concrete	Direct-to-Deck	16499	0.270	1.15	0.0291
	Flat	Concrete	Battens	20082	0.329	1.39	0.0354
	Flat	Concrete	Double Battens	23771	0.390	1.65	0.0419
Boral Barcelona 900	High	Concrete	Direct-to-Deck	34652	0.568	2.41	0.0611
	High	Concrete	Battens	38670	0.634	2.69	0.0682
	High	Concrete	Double Battens	44047	0.722	3.06	0.0777
Boral Standard Villa	Medium	Concrete	Direct-to-Deck	20088	0.329	1.40	0.0354
	Medium	Concrete	Battens	24195	0.396	1.68	0.0427
	Medium	Concrete	Double Battens	29052	0.476	2.02	0.0512
Boral Espana	High	Concrete	Direct-to-Deck	35948	0.589	2.50	0.0634
	High	Concrete	Battens	36727	0.602	2.55	0.0648
	High	Concrete	Double Battens	42788	0.701	2.97	0.0755
Boral Madera	Flat	Concrete	Direct-to-Deck	7833	0.128	0.54	0.0138
	Flat	Concrete	Battens	17133	0.281	1.19	0.0302
	Flat	Concrete	Double Battens	22440	0.368	1.56	0.0396
Eagle	Flat	Concrete	Direct-to-Deck	16313	0.267	1.13	0.0288
	Flat	Concrete	Battens	26342	0.432	1.83	0.0465
	Flat	Concrete	Double Battens	31427	0.515	2.18	0.0554

APPENDIX B

The Barcelona 900™ is an extruded interlocking concrete roof tile that complies with ASTM C 1492. The tile is a popular selection among homeowners because it reflects the classic style of historic Spanish architecture. It has distinct hue combinations that create an aesthetically pleasing and rustic look. The tile is made of concrete. About 85 tiles are needed for coverage of 100 square feet (9.3 m^2) of roof coverage. Tile weight is reported at 900 lb (431 kg) for 100 square feet (9.3 m^2) of coverage. Tables B1, B2, and B3 list the equivalent annual thermal resistance of the tile for direct-to-deck, batten and double batten applications simulated for the 8 ASHRAE climatic zones (Table 2) and also for Sacramento, CA assuming Title 24 code requirements.

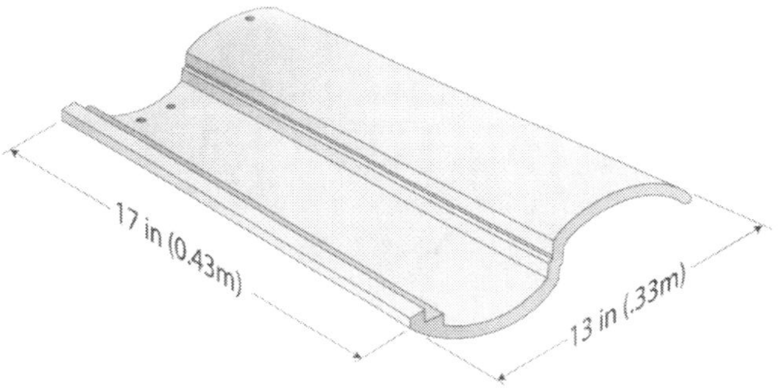

Figure B1 Barcelona 900 Concrete Tile.

Table B1. Barcelona 900 Effective Annual Thermal Resistance [R_{US}] of Inclined Air Space Yielding Same Ceiling Load as Direct-to-Deck Asphalt Shingle Roof. Air Space Effective Emittance = 0.82

Roof Configuration	Air Space Height, in. (mm)	ASHRAE Climate Zones								Sacramento, CA
		1	2	3	4	5	6	7	8	Title 24
Direct-to-Deck	2.406 (61.1)	2.72	1.42	0.68	0.29	0.08	0.05	0.00	0.17	0.52
Battens	2.69 (68.2)	2.84	1.47	0.71	0.30	0.09	0.05	0.00	0.16	0.55
Double Battens	3.058 (77.7)	2.97	1.54	0.74	0.32	0.09	0.05	–0.01	0.16	0.58

R_{US} units = [(h · ft² · °F)/Btu]; R_{SI} [(m² · K)/W] = R_{US} ÷ 5.678263
A solar reflectance of 0.10 and thermal emittance of 0.90 was assumed for control shingle and for Barcelona 900 simulations.

Table B2. Barcelona 900 Effective Annual Thermal Resistance [R_{US}] of Inclined Air Space Yielding Same Ceiling Load as Direct-to-Deck Asphalt Shingle Roof. Air Space Effective Emittance = 0.22

Roof Configuration	Air Space Height, in. (mm)	ASHRAE Climate Zones								Sacramento, CA
		1	2	3	4	5	6	7	8	Title 24
Direct-to-Deck	2.406 (61.1)	5.66	3.20	1.86	0.99	0.51	0.36	0.23	0.35	2.10
Battens	2.69 (68.2)	5.83	3.29	1.91	1.01	0.52	0.36	0.23	0.35	2.15
Double Battens	3.058 (77.7)	6.03	3.38	1.95	1.03	0.53	0.36	0.23	0.34	2.21

R_{US} units = [(h · ft² · °F) / Btu]; R_{SI} [(m² · K)/W] = R_{US} ÷ 5.678263
A solar reflectance of 0.10 and thermal emittance of 0.90 was assumed for control shingle and for Barcelona 900 simulations.

Table B3. Barcelona 900 Effective Annual Thermal Resistance [R_{US}] of Inclined Air Space Yielding Same Ceiling Load as Direct-to-Deck Asphalt Shingle Roof. Air Space Effective Emittance = 0.05

Roof Configuration	Air Space Height, in. (mm)	ASHRAE Climate Zones								Sacramento, CA
		1	2	3	4	5	6	7	8	Title 24
Direct-to-Deck	2.406 (61.1)	8.54	4.79	2.81	1.50	0.80	0.56	0.37	0.42	3.29
Battens	2.69 (68.2)	8.75	4.89	2.87	1.53	0.81	0.57	0.37	0.42	3.35
Double Battens	3.058 (77.7)	9.02	5.02	2.93	1.56	0.82	0.57	0.37	0.42	3.42

R_{US} units = [(h · ft² · °F) / Btu]; R_{SI} [(m² · K)/W] = R_{US} ÷ 5.678263
A solar reflectance of 0.10 and thermal emittance of 0.90 was assumed for control shingle and for Barcelona 900 simulations.

The Villa 900 concrete tile has a moderate tile profile whose shape and style of tile is often seen on buildings in Italy and Southern France. About 85 tiles will cover 100 square feet (9.3 m^2) of roof. Tile weight is about 920 lbs (417.2 kg) per 100 square feet (9.3 m^2) of coverage. Tables B4, B5 and B6 list the equivalent annual thermal resistance of the tile for direct-to-deck, batten and double batten applications simulated for the eight ASHRAE climatic zones (Table 2) and also for Sacramento, California assuming Title 24 code requirements.

The MCA Corona and Classic S Tapered M tiles are a two-piece pan and cover units. The clay tiles are extruded and fire cured and comply with ASTM C 1167. The tile meets all freeze/thaw and intrusive salt environments. About 150 tiles are needed for 100 square feet (9.3 m^2) of roof coverage. Tile weight is reported at 1050 lb (476.2 kg)

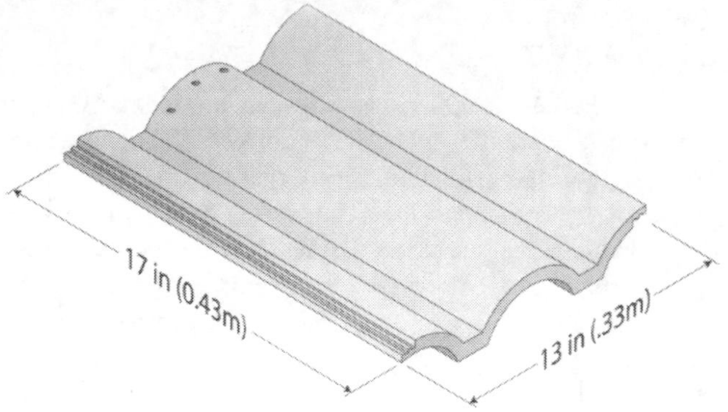

Figure B2 Villa 900 Concrete Tile.

Table B4. Villa 900 Effective Annual Thermal Resistance [R_{US}] of Inclined Air Space Yielding Same Ceiling Load as Direct-to-Deck Asphalt Shingle Roof. Air Space Effective Emittance = 0.82

Roof Configuration	Air Space Height, in. (mm)	ASHRAE Climate Zones								Sacramento, CA
		1	2	3	4	5	6	7	8	Title 24
Direct-to-Deck	1.4 (35.6)	2.21	1.14	0.53	0.23	0.06	0.05	0.00	0.17	0.37
Battens	1.68 (42.7)	2.37	1.23	0.58	0.25	0.07	0.05	0.00	0.17	0.42
Double Battens	2.02 (51.3)	2.55	1.32	0.63	0.27	0.08	0.05	0.00	0.17	0.47

R_{US} units = [(h · ft² · °F) / Btu]; R_{SI} [(m² · K)/W] = R_{US} ÷ 5.678263
A solar reflectance of 0.10 and thermal emittance of 0.90 was assumed for control shingle and for Villa 900 simulations.

Table B5. Villa 900 Effective Annual Thermal Resistance [R_{US}] of Inclined Air Space Yielding Same Ceiling Load as Direct-to-Deck Asphalt Shingle Roof. Air Space Effective Emittance = 0.22

Roof Configuration	Air Space Height, in. (mm)	ASHRAE Climate Zones								Sacramento, CA
		1	2	3	4	5	6	7	8	Title 24
Direct-to-Deck	1.4 (35.6)	4.91	2.81	1.66	0.89	0.47	0.34	0.22	0.34	1.87
Battens	1.68 (42.7)	5.15	2.94	1.72	0.92	0.49	0.34	0.22	0.35	1.94
Double Battens	2.02 (51.3)	5.40	3.07	1.79	0.95	0.50	0.35	0.22	0.35	2.02

R_{US} units = [(h · ft² · °F) / Btu]; R_{SI} [(m² · K)/W] = R_{US} ÷ 5.678263
A solar reflectance of 0.10 and thermal emittance of 0.90 was assumed for control shingle and for Villa 900 simulations.

Table B6. Villa 900 Effective Annual Thermal Resistance [R_{US}] of Inclined Air Space Yielding Same Ceiling Load as Direct-to-Deck Asphalt Shingle Roof. Air Space Effective Emittance = 0.05

Roof Configuration	Air Space Height, in. (mm)	ASHRAE Climate Zones								Sacramento, CA
		1	2	3	4	5	6	7	8	Title 24
Direct-to-Deck	1.4 (35.6)	7.61	4.32	2.55	1.37	0.75	0.52	0.35	0.40	2.99
Battens	1.68 (42.7)	7.90	4.46	2.63	1.41	0.77	0.54	0.36	0.41	3.08
Double Battens	2.02 (51.3)	8.21	4.62	2.72	1.45	0.79	0.55	0.36	0.42	3.18

R_{US} units = [(h · ft² · °F) / Btu]; R_{SI} [(m² · K)/W] = R_{US} ÷ 5.678263
A solar reflectance of 0.10 and thermal emittance of 0.90 was assumed for control shingle and for Villa 900 simulations.

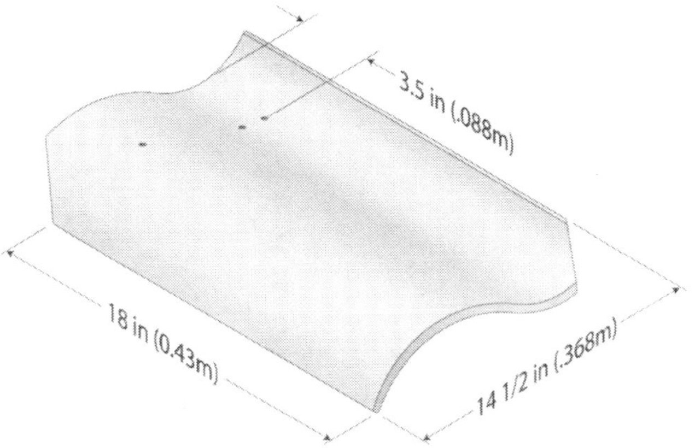

Figure B3 MCA Corona tapered M clay tile.

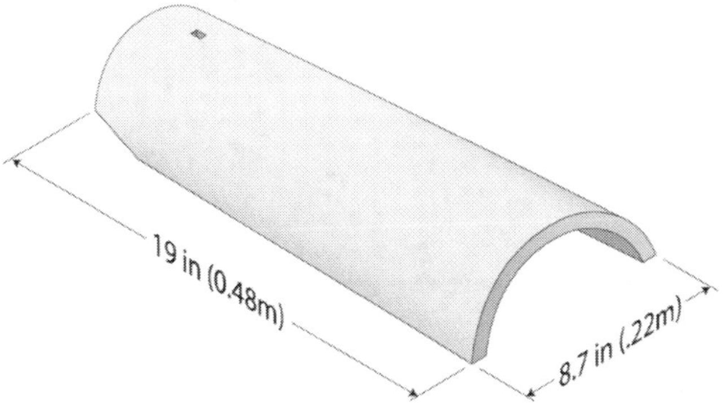

Figure B4 MCA Classic S tapered M clay tile.

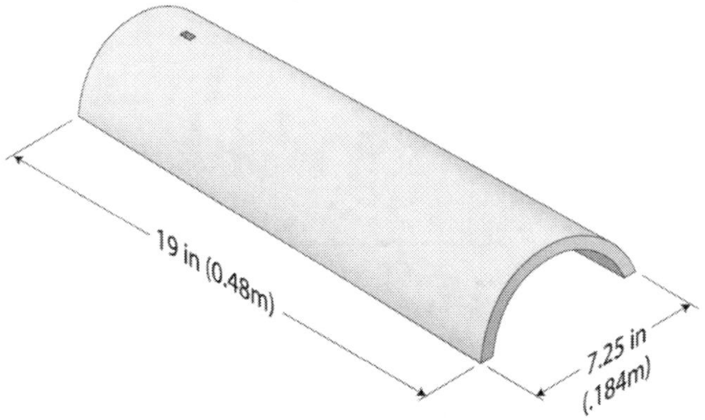

Figure B5 MCA Mission S clay tile.

for 100 square feet (9.3 m^2) of coverage. The Mission S style is a one-piece tile with weight of 780 lb (353.7 kg) for 100 square feet (9.3 m^2) of coverage. Tables B7, B8, and B9 list the equivalent annual thermal resistance of the tile for direct-to-deck applications simulated for the eight ASHRAE climatic zones (Table 2) and also for Sacramento, California, assuming Title 24 code requirements.

Table B7. MCA Effective Annual Thermal Resistance [R_{US}] of Inclined Air Space Yielding Same Ceiling Load as Direct-to-Deck Asphalt Shingle Roof. Air Space Effective Emittance = 0.82

Roof Configuration	Air Space Height, in. (mm)	ASHRAE Climate Zones								Sacramento, CA
		1	2	3	4	5	6	7	8	Title 24
Corona Tapered M	1.49 (37.8)	2.23	1.13	0.50	0.21	0.04	0.03	–0.01	0.17	0.29
Classic S Tapered M	2.24 (56.8)	2.36	1.20	0.54	0.23	0.05	0.04	–0.01	0.17	0.33
Mission S	2.99 (75.9)	2.40	1.22	0.55	0.23	0.05	0.04	–0.01	0.17	0.34

R_{US} units = [(h · ft^2 · °F) / Btu]; R_{SI} [(m^2 · K)/W] = R_{US} ÷ 5.678263
MCA clay tile fastened directly to the roof deck without batten system.
A solar reflectance of 0.10 and thermal emittance of 0.90 was assumed for control shingle and for Corona Tapered clay tile simulations.

Table B8. MCA Effective Annual Thermal Resistance [R_{US}] of Inclined Air Space Yielding Same Ceiling Load as Direct-to-Deck Asphalt Shingle Roof. Air Space Effective Emittance = 0.22

Roof Configuration	Air Space Height, in. (mm)	ASHRAE Climate Zones								Sacramento, CA
		1	2	3	4	5	6	7	8	Title 24
Corona Tapered M	1.49 (37.8)	4.95	2.82	1.64	0.87	0.46	0.33	0.21	0.34	1.81
Classic S Tapered M	2.24 (56.8)	4.95	2.82	1.64	0.87	0.46	0.33	0.21	0.34	1.81
Mission S	2.99 (75.9)	4.95	2.82	1.64	0.87	0.46	0.33	0.21	0.34	1.81

R_{US} units = [(h · ft^2 · °F) / Btu]; R_{SI} [(m^2 · K)/W] = R_{US} ÷ 5.678263
MCA clay tile fastened directly to the roof deck without batten system.
A solar reflectance of 0.10 and thermal emittance of 0.90 was assumed for control shingle and for Corona Tapered clay tile simulations.

Table B9. MCA Clay Tile Effective Annual Thermal Resistance [R_{US}] of Inclined Air Space Yielding Same Ceiling Load as Direct-to-Deck Asphalt Shingle Roof. Air Space Effective Emittance = 0.05

Roof Configuration	Air Space Height, in. (mm)	ASHRAE Climate Zones								Sacramento, CA
		1	2	3	4	5	6	7	8	Title 24
Corona Tapered M	1.49 (37.8)	7.67	4.33	2.54	1.36	0.74	0.52	0.34	0.40	2.95
Classic S Tapered M	2.24 (56.8)	7.90	4.45	2.61	1.39	0.76	0.53	0.35	0.41	3.03
Mission S	2.99 (75.9)	7.96	4.48	2.63	1.40	0.76	0.53	0.35	0.41	3.05

R_{US} units = [(h · ft^2 · °F) / Btu]; R_{SI} [(m^2 · K)/W] = R_{US} ÷ 5.678263
MCA clay tile fastened directly to the roof deck without batten system.
A solar reflectance of 0.10 and thermal emittance of 0.90 was assumed for control shingle and for MCA clay tile simulations.

Influence of Building Design on Energy Benefit of Thermal Mass Compared to Prescriptive U-Factors

Cheryl M. Saldanha, PE

Joseph P. Piñon, PE
Associate Member ASHRAE

ABSTRACT

The building industry is continually moving toward higher insulation levels and continuous insulation. Many building codes and standards recognize that thermally massive buildings need less insulation because they can store and gradually release heat compared to lightweight metal or wood-framed buildings. This relative reduction in required minimum insulation values for a massive building compared to a lightweight building varies by climate, but in general, ASHRAE Standard 90.1 allows about a 30% reduction in insulation levels. California's energy code allows up to an 86% reduction in insulation levels for "mass heavy" walls in certain climate zones.

However, most state building codes and energy standards do not currently reflect whether the thermal mass is located on the interior or exterior side of the insulation in the wall system. Also, the building type, geometry, window-to-wall ratio, and internal loads are not considered with respect to thermal mass in the prescriptive tables of ASHRAE 90.1 and various state energy codes. Our paper discusses the results of a series of whole-building energy simulations that we completed using the EnergyPlus computer program to study the effect of thermal mass relative to these parameters.

Our results show that the increased U-factors in the prescriptive tables in ASHRAE 90.1 and various state energy codes are justified in terms of accounting for the benefit of thermal mass. However, in all cases, having insulation on the exterior side of the thermal mass outperforms cases where the thermal mass is on the exterior side of the insulation. We show the most important parameters to maximize the benefit of thermal mass with regards to energy efficiency and demonstrate how the prescriptive tables in ASHRAE 90.1 and most state energy codes do not currently account for these benefits or parameters.

INTRODUCTION

The prescriptive U-factor tables in ASHRAE 90.1-2010, *Energy Standard for Buildings Except Low-Rise Residential Buildings*, the 2009 *International Energy Conservation Code* (IECC), and various state energy codes allow a significant increase (about 30%) in allowable U-factor when the exterior walls incorporate thermal mass such as concrete or masonry. The 2010 California Energy Code (CEC) divides the state up into 16 climate zones (compared to ASHRAE's 8 zones for the entire United States), each with varying benefits in the prescriptive tables for having thermally massive walls. California even defines a mass heavy category for concrete walls about 10 in. wide and thicker, with even larger allowed increases in U-factors (an allowable increase of up to 86% compared to lightweight framed walls). However, with some exceptions, these codes and standards do not currently dictate whether the thermal mass should be located on the interior or exterior side of the insulation in the wall system. Also, they do not relate the benefit of thermal mass to many other variables, such as window-to-wall ratio or internal loads (e.g., lighting).

We have seen some exceptions to the code trends described above. The residential sections of the 2009 *International Energy Conservation Code* (2009 IECC), Table R402.1.1, do require about 25% more insulation for mass walls where insulation is placed on the interior side of

***Cheryl M. Saldanha** is a staff II engineer and **Joseph P. Piñon** is a senior project manager with Simpson Gumpertz & Heger Inc., San Francisco, CA.*

the mass wall. This indicates that the code assumes that thermal mass has more of an energy benefit when located inboard of the insulation. The 1989 Model Energy Code as reported by Christian (1991) had separate U-factor tables to distinguish the benefit of having the insulation on the exterior of the mass, however, that distinction was not incorporated into ASHRAE 90.1. The *2010 Oregon Energy Efficiency Specialty Code*, Table 502.1.3, allows the U-factor to be 30% higher if the glazing area for the building is 15% instead of 30%, but only if the insulation is on the exterior side of the concrete, masonry, or integral (e.g., between wythes). Oregon also allows up to a 38% increase in U-factor for mass walls that have the insulation on the exterior side compared to mass walls with insulation on the interior side.

In contrast, most previous research has demonstrated the better energy performance when the mass is located on the interior side of the insulation so that it can interact more directly thermally with the interior conditioned air (Burch et al. 1984, Christian 1991; Kossecka and Kosny 1998; Marceau and VanGeem 2007; Zhu et al. 2008).

Objectives and Scope

To evaluate the relative benefit of thermal mass given variables such as location of insulation, glazing area, climate, building height, and internal loads, we conducted a parametric study using a series of energy models. The objectives of this study were as follows:

1. Verify whether the allowable increases in U-factor in the prescriptive tables of ASHRAE 90.1-2010 and 2010 CEC are justified for various climates, locations of insulation, amount of glazing area, building height, and internal loads.
2. Determine if the accuracy of the prescriptive tables can be improved to account for the relative benefit of thermal mass given variables such as location of insulation and glazing area.
3. Provide information to designers and code officials on the most important parameters to maximize the benefit of thermal mass with regards to energy efficiency and identify how the prescriptive tables in ASHRAE 90.1-2010 and most state energy codes do not currently account for these benefits or parameters.

In our study, we did not attempt to optimize the use of thermal mass in any way, such as mechanical ventilation strategies to cool or heat the mass when the outside temperatures are favorable (Christian 1991).

MODEL SETUP

Software

We used the EnergyPlus version 7.2 computer program to conduct a series of whole building energy simulations. EnergyPlus is a whole building energy simulation program, developed by the United States Department of Energy (DOE). The energy simulation takes into account the building's three-dimensional geometry, enclosure (wall, window, and roof) construction and materials, and estimates of internal loads such as lighting and space conditioning. EnergyPlus has been validated based on various test standards, such as ASHRAE Standard 140 and ASHRAE Research Projects 865 and 1052.

Reference Building Model

We used the EnergyPlus commercial reference building model for a medium office building as a basis for our comparison. The medium office is three stories, has a floor area of 53,628 square feet (4982 square meters) and a floor-to-floor height is 9 ft (2.74 m). The commercial reference building models were developed by the US DOE in conjunction with three of the national laboratories: the National Renewable Energy Laboratory, Pacific Northwest National Laboratory, and Lawrence Berkeley National Laboratory. The models were created to represent typical buildings in the United States for the purpose of having a benchmark to assess the benefit of various energy efficiency design strategies.

The reference building model inputs were derived from ASHRAE Standards 90.1-2004, 62.1-2004, and 62-1999 for new construction (guidelines for energy efficiency and ventilation). As described in the following sections, we modified some of the model parameters to assess the effect of thermal mass in buildings. We also revised the insulation levels and glazing performance characteristics to comply with ASHRAE Standard 90.1-2010 or the 2010 CEC requirements.

Model Setup—Wall Assemblies

We started with two basic exterior wall assemblies: a massive and lightweight assembly. The massive assemblies have a 6 in. thick concrete back-up wall and the lightweight assemblies have a gypsum-sheathed light gauge steel-framed back-up wall.

We then modified the insulation location within those assemblies. For the massive wall assemblies, we modeled the buildings with insulation at either the exterior or interior side of the back-up wall. For the lightweight wall assemblies, we modeled insulation in the stud cavity only, however, we adjusted the conductivity so that the insulated stud cavity included the code required amount of continuous insulation (i.e., the steady-state U-factor was equivalent to that of the massive wall). The insulation amount varied in all cases to create wall assemblies that are compliant with prescriptive ASHRAE 90.1-2010 or the 2010 CEC requirements for mass walls.

The exterior wall cladding in all cases was modeled as 5/16 in. (8 mm) cement board siding. Although this cladding is not necessarily typical of office buildings, we selected this cladding because it is relatively lightweight and radiates solar heat gain toward the interior with a minimal time delay. This also keeps factors such as surface emissivity constant between all of our iterations.

Table 1 summarizes the basic wall assemblies that were used in our EnergyPlus simulations. Materials are listed from exterior to interior. Insulation thicknesses are not listed because, as described later in this paper, we varied the thicknesses to meet prescriptive code requirements based on the climate zone location.

MODEL ITERATIONS

We applied the three wall assembly types listed in Table 1 to the medium office building to determine the relative impact of thermal mass position on building energy consumption. We specifically analyzed the effect of climate, glazing, building height, and interior loads. The following sections describe our setup of the iterations.

Effect of Climate

We performed annual simulations using time-varying exterior conditions, including temperature and solar radiation, based on typical meteorological year weather data for the following climate zones:

- ASHRAE Climate Zone 5A (e.g., Chicago, Illinois)
- ASHRAE Climate Zone 2A (e.g., Houston, Texas)
- California Climate Zone 3 (e.g., San Francisco, California)
- California Climate Zone 9 (e.g., Los Angeles inland region or Pasadena, California)

All other building geometries, components, and systems were unchanged between model iterations. We selected these climate zones to encompass the cold and warm climate extremes in the continental United States, as well as some moderate climates.

Table 2 summarizes the model iterations performed to determine the effect of climate and insulation position on en-

Table 1. Basic Wall Assemblies

Massive Wall—Exterior Insulation	Massive Wall—Interior Insulation	Lightweight Wall
• Cement board siding	• 5/16 in. (8 mm) Cement board siding	• Cement board siding
• Continuous Insulation	• 6 in. (150 mm) thick concrete wall	• 1/2 in. (13 mm) gypsum board
• 6 in. (150 mm) thick concrete wall	• Fiberglass insulation between steel studs at 16 in. (406 mm) o.c. (includes effect of thermal bridging of the studs)	• Fiberglass insulation between steel studs at 16 in. (406 mm) o.c. (conductivity reduced to include Code required amount of continuous insulation)
• 1 in. (25.4 mm) thick empty stud wall	• 1/2 in. (12.7 mm) gypsum board	• 1/2 in. (12.7 mm) gypsum board
• 1/2 in. (12.7 mm) gypsum board		

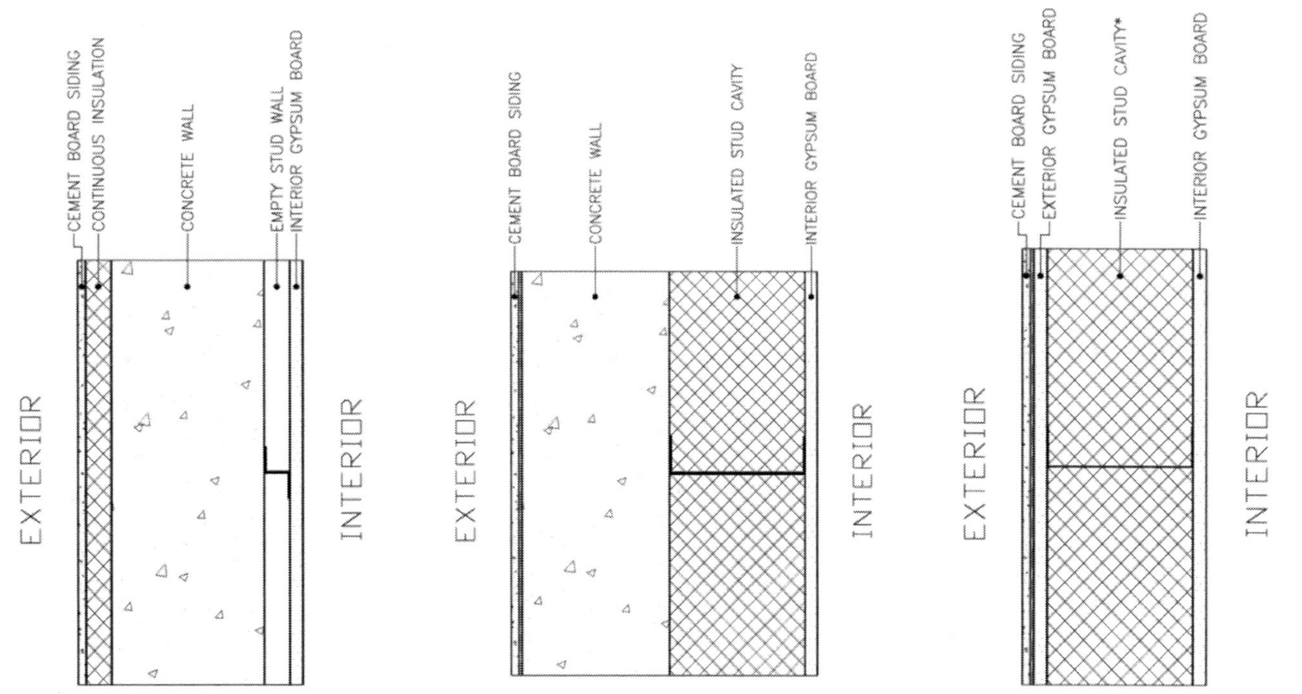

ergy consumption in buildings. Also included in Table 2 are the prescriptive maximum U-factors for the various wall assemblies.

For each wall assembly listed in Table 1, we calculated the insulation layer conductivity required to achieve the U-factors listed in Table 2.

Effect of Glazing

We also considered the effect of increasing both the glazing area and glazing performance. The medium office reference building has a 33% window-to-wall area ratio. For each climate zone, we increased the window-to-wall area ratio to 40% and 50%, representing a 21% and 52% respective increase in glazing. We chose 40% window-to-wall area ratio as a threshold for comparison because 40% is typically the current maximum glazing area allowed by most energy codes on a commercial building where you can still use the prescriptive tables to show envelope compliance. We also performed the same analysis with a 20%, 50%, and 60% glazing area ratio to see if any trends emerged.

In addition to the amount of glazing, we reviewed the effect of using better windows with lower U-factors for the case with 33% window-to-wall area ratio. Table 3 summarizes the glazing performance parameters used in our models that represent prescriptive U-factors for double glazed units and achievable U-factors for triple glazed units in a thermally broken aluminum frame. The U-factor of 0.55 is the maximum allowed in Climate Zone 5, but exceeds the requirement in Climate Zones 2 and 3 where non-thermally broken aluminum frames are allowed.

Effect of Building Height

We also considered the effect of building height on energy use. We increased the multiplier for thermal zones at the middle floor of the medium office building to 10. This effectively models a building with twelve stories instead of three. We wanted to determine the effect of the roof-to-wall area ratio on energy consumption and how that impacts the relative benefit of thermal mass compared to a lightweight wall.

Effect of Internal Loads

We reduced the electrical equipment loads and lighting by half to see if the effectiveness of thermal mass is also similarly reduced. Reducing lighting and electrical equipment loads reduces the internal heat generated by the equipment. This reduced internal heat load decreases cooling demand in summer but increases heating demand during winter.

MODEL RESULTS AND DISCUSSION

We calculated the annual and peak energy use for each model case. We considered the lightweight wall models to be the baseline cases. We then compared the massive wall model results to the lightweight model results to assess the relative change. We summarize and discuss the results in the next sections.

Effect of Climate

Tables 4 and 5 summarize the results of our analysis of the effect of climate zone on buildings with mass and lightweight wall assemblies.

Table 2. Prescriptive Code Requirements

Climate Zone	Massive Wall—Interior or Exterior Insulation		Lightweight Wall	
	Prescriptive Effective R-Value	Prescriptive U-Factor	Prescriptive Effective R-Value	Prescriptive U-Factor
ASHRAE CZ 5A	11.1 h·ft^2·°F/Btu	0.09 Btu/h·ft^2·°F	15.6 h·ft^2·°F/Btu	0.064 Btu/h·ft^2·°F
	2.0 m^2·K/W	0.511 W/m^2·K	2.8 m^2·K/W	0.363 W/m^2·K
ASHRAE CZ 2A	6.6 h·ft^2·°F/Btu	0.151 Btu/h·ft^2·°F	8.1 h·ft^2·°F/Btu	0.124 Btu/h·ft^2·°F
	1.2 m^2·K/W	0.857 W/m^2·K	1.4 m^2·K/W	0.704 W/m^2·K
CA CZ 3	3.7 h·ft^2·°F/Btu	0.268 Btu/h·ft^2·°F	12.2 h·ft^2·°F/Btu	0.082 Btu/h·ft^2·°F
	0.7 m^2·K/W	1.522 W/m^2·K	2.1 m^2·K/W	0.466 W/m^2·K
CA CZ 9	2.3 h·ft^2·°F/Btu	0.440 Btu/h·ft^2·°F	16.1 h·ft^2·°F/Btu	0.062 Btu/h·ft^2·°F
	0.4 m^2·K/W	2.498 W/m^2·K	2.8 m^2·K/W	0.352 W/m^2·K

Table 3. Glazing Performance Inputs

Climate Zone	Double Glazed		Triple Glazed		Prescriptive SHGC
	U-factor, Btu/h·ft^2·°F	U-factor, W/m^2·K	U-factor, Btu/h·ft^2·°F	U-factor, W/m^2·K	
ASHRAE CZ 5A					0.40
ASHRAE CZ 2A	0.55	3.12	0.39	2.21	0.25
California CZ 3					0.41

Table 4. Annual Heating & Cooling Energy Use and Percentage Difference Compared to Baseline

Climate	Wall Type	Heating MJ/m²	Heating kBtu/ft²	% Difference in Heating	Cooling MJ/m²	Cooling kBtu/ft²	% Difference in Cooling
ASHRAE CZ 5A: Chicago, IL	Lightweight	231	20	—	55	5	—
	Mass—Exterior Insulation	197	17	−14.5%	50	4	−9.1%
	Mass—Interior Insulation	221	19	−4.1%	51	5	−7.8%
ASHRAE CZ 2A: Houston, TX	Lightweight	63	6	—	147	13	—
	Mass—Exterior Insulation	45	4	−28.5%	135	12	−8.2%
	Mass—Interior Insulation	51	4	−19.6%	138	12	−6.3%
CA CZ 3: San Francisco, CA	Lightweight	80	7	—	46	4	—
	Mass—Exterior Insulation	52	5	−34.7%	40	3	−13.0%
	Mass—Interior Insulation	57	5	−28.5%	39	3	−13.5%
CA CZ 9: Los Angeles Inland, CA	Lightweight	28	3	—	79	7	—
	Mass—(No Insulation Required)	13	1	−53.8%	70	6	−10.9%

Table 5. Total Building Annual Energy Use and Percentage Difference Compared to Baseline

Climate	Wall Type	Total Building Energy MJ/m²	Total Building Energy kBtu/ft²	% Difference in Total Building Energy
ASHRAE CZ 5A: Chicago, IL	Lightweight	676	60	—
	Mass—Exterior Insulation	692	61	−5.9%
	Mass—Interior Insulation	651	57	−2.2%
ASHRAE CZ 2A: Houston, TX	Lightweight	592	52	—
	Mass—Exterior Insulation	615	54	−5.2%
	Mass—Interior Insulation	583	51	−3.7%
CA CZ 3: San Francisco, CA	Lightweight	498	44	—
	Mass—Exterior Insulation	528	47	−6.7%
	Mass—Interior Insulation	493	43	−5.8%
CA CZ 9: Los Angeles Inland, CA	Lightweight	485	43	—
	Mass—(No Insulation Required)	510	45	−5.0%

The results presented in Tables 4 and 5 show that massive walls perform better than lightweight walls in all climates. The massive outperform the lightweight cases even with U-Factors that are 40% higher (allowed by ASHRAE prescriptive code tables) or 600% higher (allowed by California prescriptive code tables), than the lightweight wall U-factors.

The results in Table 5 show that positioning the mass so that the insulation is on the exterior side of the mass significantly outperforms the cases where the insulation is placed on the interior side of the mass. For example, in Climate Zone 5A, the building with insulation at the exterior side of the massive wall uses 3.7% less energy on an annual basis than the same building with insulation positioned at the interior side of the massive wall.

For all four climate zones analyzed, a building with insulation at the exterior side of the thermally massive backup wall uses approximately 1% to 4% less energy annually than buildings where the insulation is positioned at the interior side, as shown in Table 5. When the insulation is inboard of the thermally massive wall, the mass is able to absorb and store heat from solar radiation. However, during cold winter days, it is disconnected from the interior and so stays colder longer. This delays the benefit of solar radiation when it is cold and heating is called for by the HVAC systems, resulting in higher heat losses during heating periods. Thermal mass on the exterior side of the insulation can also act as a cold sink. For example, if the outside temperature is 75°F (24°C), but the thermal mass is 60°F (16°C) due to night cooling, if you are heating the building interior to 70°F (21°C), you can actually lose heat through the exterior walls (to the thermal mass), even though it is warmer outside than inside.

Tables 4 and 5 demonstrate that the code allowed insulation reductions for massive walled assemblies are justified. However, it also shows that the code should consider allowing even higher reductions for mass walls when the insulation is placed on the exterior side of the thermal mass.

One interesting point to note from Tables 4 and 5 is that for Climate Zone 9, the massive wall assembly meets the code prescribed U-factor without any insulation. Therefore, the total annual energy use is the same for both the massive wall cases.

We also separated out the annual heating and cooling loads and compared the massive wall buildings with the lightweight wall buildings in Table 5. The results show similar trends, where the cases with exterior insulation outperform the interior insulation cases, however the difference between the two is an order of magnitude greater than when comparing total annual energy use for the same cases. This is because of the effect of large lighting and equipment loads, which stay constant between all of the cases we modeled.

Looking specifically at annual heating loads in the colder climate zone (CZ 5A), the exterior insulation case uses 10% less heating energy annually than the interior insulation case. In the warmer climate zone, this difference in heating is less pronounced (only a 6% difference).

When considering annual cooling loads in the colder climate zone, the exterior insulation cases only use about 2% less cooling energy annually, while in a warmer climate zone there is almost no difference. These results indicate that location of the insulation primarily affects the efficiency of the thermal mass for heating loads versus cooling loads.

Peak Loads

Table 6 shows the peak heating and cooling loads and the percentage compared to the baseline cases.

The results presented in Table 6 summarize how massive walls can reduce peak heating or cooling demand in the three climates studied. Massive walls lower peak heating or cooling demand by up to 15% compared to lightweight walls.

In all cases, mass on the interior side of the insulation outperforms mass on the exterior side by about a 5% additional reduction in peak loads. This benefit of reducing peak demand of mass walls compared to lightweight walls is not currently accounted for in the code. Reducing peak loads results in smaller physical equipment sizes, generally lower installation and maintenance costs, and higher efficiencies.

Additionally, since mechanical engineers use peak load calculations to size mechanical equipment, they may be oversizing the equipment if not considering the effect of building thermal mass.

Table 6. Peak Heating or Cooling Demand and Percentage Difference Compared to Baseline

Climate	Wall Type	Electricity Peak Demand, kW	% Reduction of Peak Demand
ASHRAE CZ 5A: Chicago, IL	Lightweight	383	—
	Mass – Exterior Insulation	355	–7.4%
	Mass – Interior Insulation	376	–1.8%
ASHRAE CZ 2A: Houston, TX	Lightweight	293	—
	Mass – Exterior Insulation	263	–10.4%
	Mass – Interior Insulation	274	–6.6%
CA CZ 3: San Francisco, CA	Lightweight	262	—
	Mass – Exterior Insulation	223	–15.0%
	Mass – Interior Insulation	234	–11.0%
CA CZ 9: Los Angeles Inland, CA	Lightweight	111	—
	Mass – (No Insulation Required)	101	–9.5%

EFFECT OF GLAZING

Glazing Area—Total Annual Energy Results

Figures 1–3 summarize the results of our analysis of the effect of glazing area on buildings with mass versus lightweight wall assemblies. The figures show the total annual energy use according to the window-to-wall area ratio for each climate zone.

The results from Figures 1–3 show the following four trends when increasing the window-to-wall area ratio:

- The exterior insulation cases consistently outperform the interior insulation cases for mass walled buildings in all climate zones for all simulated glazing percentages.
- Increasing the glazing area results in a roughly linear increase in annual energy use, though the code prescriptive tables are only allowed currently for cases up to 40% glazing area.
- The difference in annual energy use between the lightweight and massive walled buildings decreases as the glazing area increases.
- The mass wall with higher glazing areas outperforms the lightweight wall with lower glazing areas. For example, in Climate Zone 5A, a building with insulation on the exterior of mass walls with 60% glazed area outperforms a building with 20% glazed area with lightweight walls.

The first trend was previously discussed in Section 4.1, and holds true even at varying glazing area amounts.

The second trend emerges when comparing the total annual energy use to the window-to-wall area ratio. We see that for each wall type, the annual energy use increases linearly as the glazing area is increased. For example, in Climate Zone 2A the lightweight walled building with 40% glazing uses approximately 7.7% more energy than the massed wall (exterior insulation) with 33% glazing and about

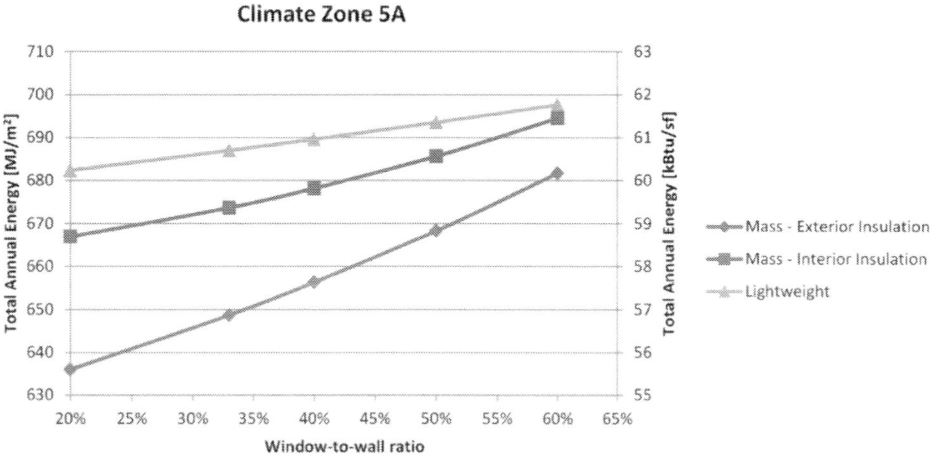

Figure 1 *Effect of glazing area on total annual energy use compared to baseline (Climate Zone 5A).*

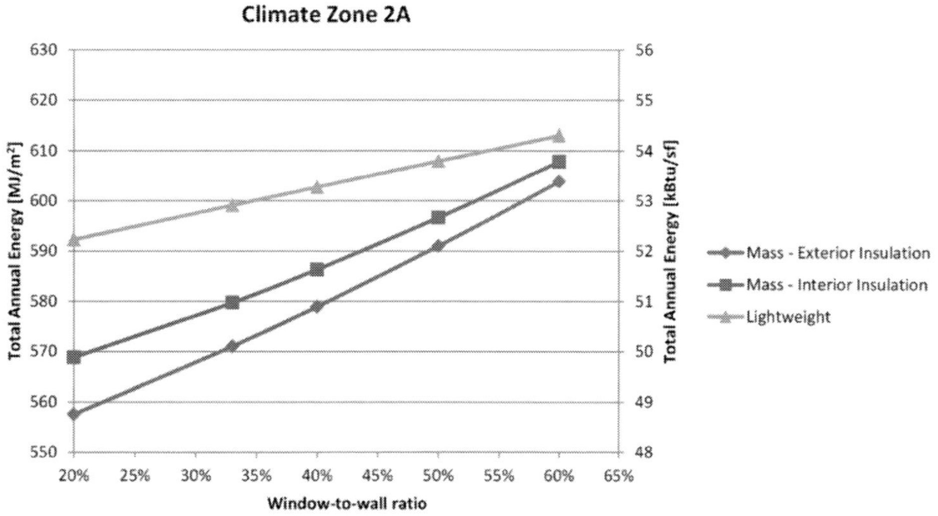

Figure 2 *Effect of glazing area on total annual energy use compared to baseline (Climate Zone 2A).*

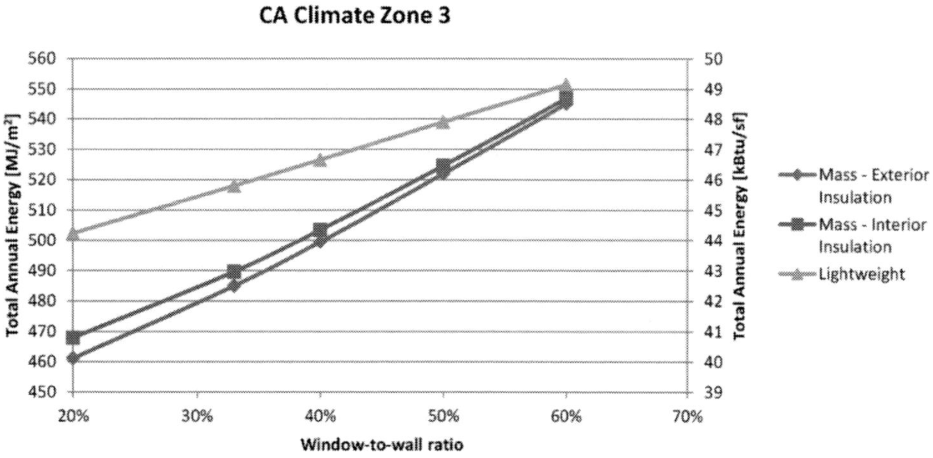

Figure 3 *Effect of glazing area on total annual energy use compared to baseline (CA Climate Zone 3).*

8.7% more energy if the window area of the massed wall building is decreased to 20%. Yet all of those cases are allowed and considered equal by the prescriptive tables in the code.

The third trend is that as glazing area increases, the total energy use converges between the three cases: lightweight, massive with exterior insulation, and massive with interior insulation. For example, in Climate Zone 3, the three cases converge at about 60% glazing area. In the colder Climate Zone 5A, the convergence occurs at a glazing area greater than 60%. At 60% glazing, the difference between the massive wall with exterior insulation and lightweight wall is still about 3%.

The fourth trend is that a mass wall with higher glazing areas can use less energy than a lightweight wall with lower glazing areas. For example, in the Climate Zone 5A, the mass wall with exterior insulation and 60% glazing uses less energy than the massive wall with interior insulation and 40% glazing. It is also uses less energy than the lightweight wall case with only 20% glazing. This indicates that the thermally massive walls can in essence be used to offset the higher glazing areas. However, according to current codes a mass wall with 60% glazing area does not comply with the prescriptive compliance path of the code, while a lightweight wall with 20% glazing does comply.

Glazing Area—Heating and Cooling Energy Results

Figures 4 and 5 show the percentage reduction in annual heating and cooling loads of the mass walled buildings compared to the lightweight walled buildings.

The figures show that for heating and cooling loads, a building in Climate Zone 3 has the greatest reduction in heating (10% to 50%) and cooling (4% to 19%) energy use, at all glazing area ratios. A building in Climate Zone 2A uses from 9% to 40% less heating energy than the baseline, which makes sense because it is a warmer climate. Conversely, a building in Climate Zone 5A uses from 2% to 12% less cooling energy than the baseline case, because it is a colder climate.

The magnitude of energy savings when comparing only the percentage change in heating or cooling loads exceeds the total energy load, because the total includes lighting and equipment loads which remain constant between cases.

Looking at the breakdown by climate, it is apparent that Climate Zone 3, the mild climate, shows the greatest heating and cooling load percent reductions for the mass walls compared to lightweight walls. For example, for heating loads, the reduction compared to the lightweight case ranges from 10% to 40% and for cooling loads ranges from 2% to 15%. However, this is because the heating and cooling loads are very small in a mild climate. For example, a 2.0 MJ/m^2 (0.18 kBtu/ft^2) reduction in Climate Zone 5A, where the heating load is 200.7 MJ/m^2 (17.7 kBtu/ft^2), results in a modest percentage change compared to Climate Zone 3 where the heating load is only about 60.2 MJ/m^2 (5.3 kBtu/ft^2).

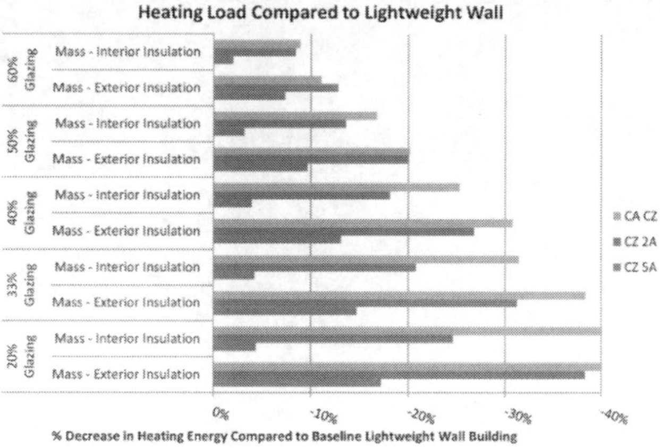

Figure 4 *Percentage difference in annual heating load of the mass walled building compared to baseline, given different glazing areas.*

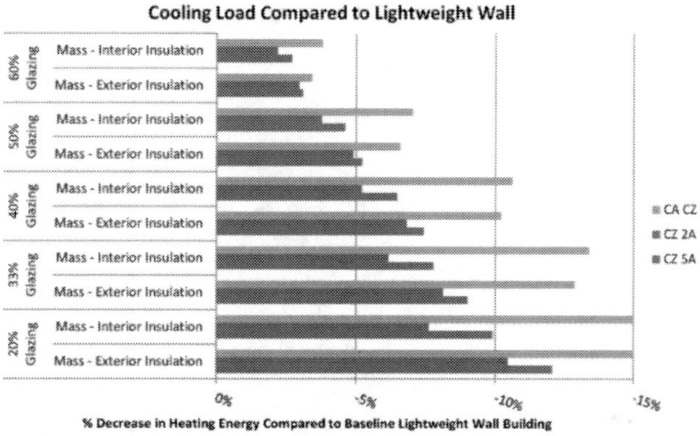

Figure 5 *Percentage difference in annual cooling load of the mass walled building compared to baseline, given different glazing areas.*

Glazing Performance—Total Annual Energy Results

Figures 6 through 8 show the effect of including triple glazed windows compared to double glazed (lower U-factor for windows per Table 3) in the building in Climate Zones 5A, 3A, and CA Climate Zone 3, with a 33% window-to-wall area ratio.

The results show that with the exception of Climate Zone 5A, the lightweight walled building benefits most from the upgrade to triple glazed windows. However, the massed wall buildings still out-perform the lightweight walled building. Also, insulation placed on the exterior side of the mass out-performs the case with insulation on the interior side of the mass, most significantly in the colder Climate Zone 5A. The heating and cooling energy data shows that this is because the better windows reduce heating loads by a larger amount than cooling loads (given the same solar heat gain coefficients), so that magnifies the benefit of mass on the interior side in heating dominated climates. The results also show that the better win-

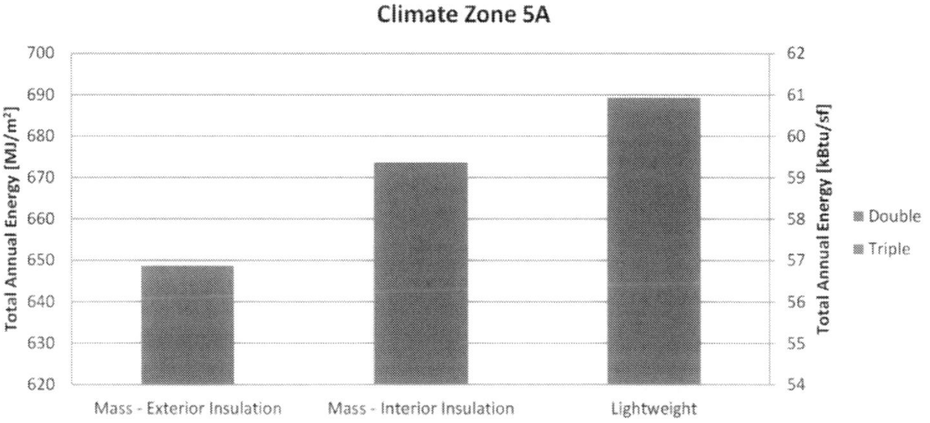

Figure 6 *Effect of including triple glazed windows on total annual energy use compared to double glazed (Climate Zone 5A).*

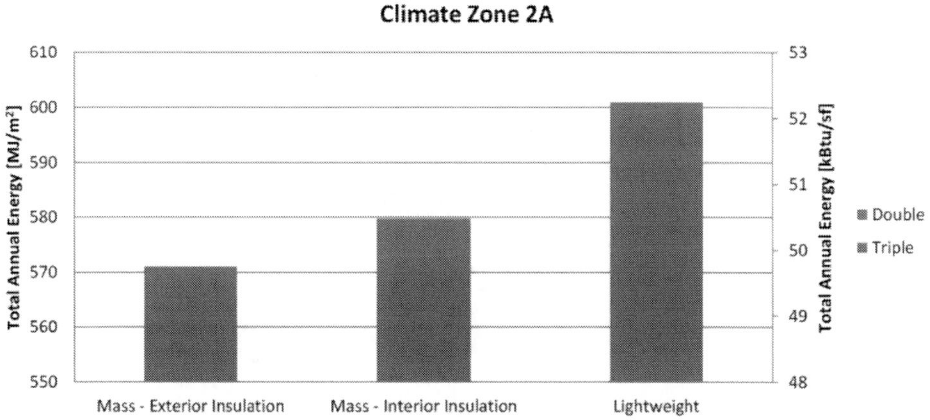

Figure 7 *Effect of including triple glazed windows on total annual energy use compared to double glazed (Climate Zone 2A).*

dows could offset the increased energy use from having a larger glazed area. For example, comparing Figures 1 and 6 shows that in Climate Zone 5A, a massed wall building with insulation on the exterior side of the mass, and with a 33% glazed area, outperforms the same building with 20% glazed area if the windows are improved to triple glazed. The benefit of using better windows combined with a thermal mass strategy is not currently accounted for in the prescriptive tables of the code.

Effect of Building Height

Figure 9 summarizes the results of our analysis of the effect of building height on buildings with mass and lightweight wall assemblies in various climates. The figures show the percentage difference in total energy for the mass wall building compared to the lightweight wall buildings, according to climate zone and building height. The results show less than a 1% relative difference due to building height between a building with mass wall and lightweight wall assemblies.

When comparing the results of a 3 story building to a 12 story building, some trends discussed in earlier sections are similarly true. For example, the exterior insulation cases outperform the interior insulation cases for massive walled buildings.

While the 12 story building cases all use more energy than the 3 story cases, we can compare each to their respective lightweight cases to extract some trends. For all climate zones, the 12 story building percent reduction in energy for the massive cases are about 1% better than for the 3 story building cases. This slight difference can be attributed to the increased wall-to-roof area for the taller building. This amplifies the benefit of the thermal mass in the walls, and at the same time diminishes the effect of the roofing assembly.

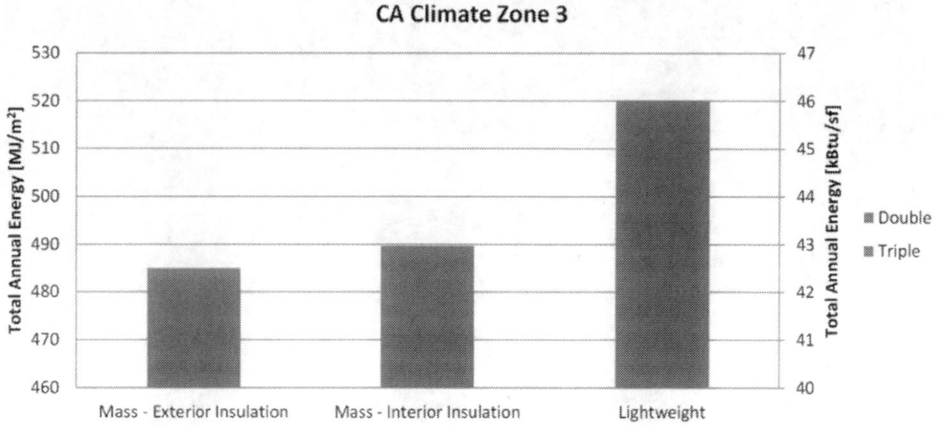

Figure 8 *Effect of including triple glazed windows on total annual energy use compared to double glazed (CA Climate Zone 3).*

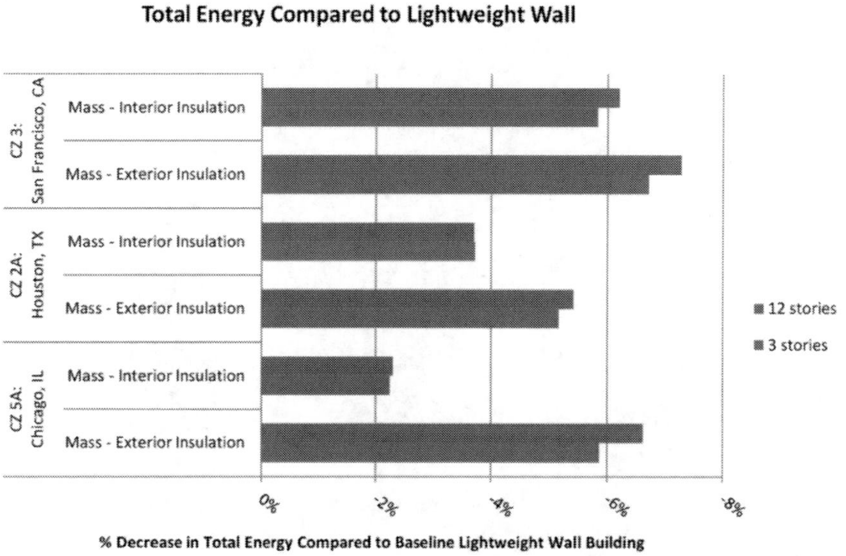

Figure 9 *Effect of building height on relative benefit of mass walled buildings compared to baseline.*

Effect of Internal Loads

Figures 10 to 12 summarize the results of our analysis of the effect of internal loads on buildings with mass and lightweight wall assemblies. From this figure, we see that the reduction in internal loads has a greater effect in the warmer climate zone.

Figures 11 and 12 show the annual heating and cooling energy results.

The heating energy increases by 24% to 39%, with the larger differences seen in the colder climate zone.

The cooling energy decreases by 18% to 24%, with the larger differences seen in the warmer climate zone.

We see that the total energy savings is largest in the warmer climate zone when reducing lighting and equipment loads by 50%. This is attributed to the reduced cooling load in these climates resulting from the reduced radiant heat energy from smaller lighting and equipment loads. When we separate out and analyze the annual heating and cooling loads, we see that, as anticipated, the heating loads increase, while the cooling loads decrease.

One note to add here is the difference in the cost of different energy sources. Typically a building uses natural gas for heating, and internal loads such as lighting and equipment use electricity. Since natural gas is currently cheaper than electricity, the heating cost will be much lower than the cost for internal loads from equipment. Therefore, reducing the internal loads by 50% will result in a significant cost savings, perhaps enough to offset the increased heating required during the winter.

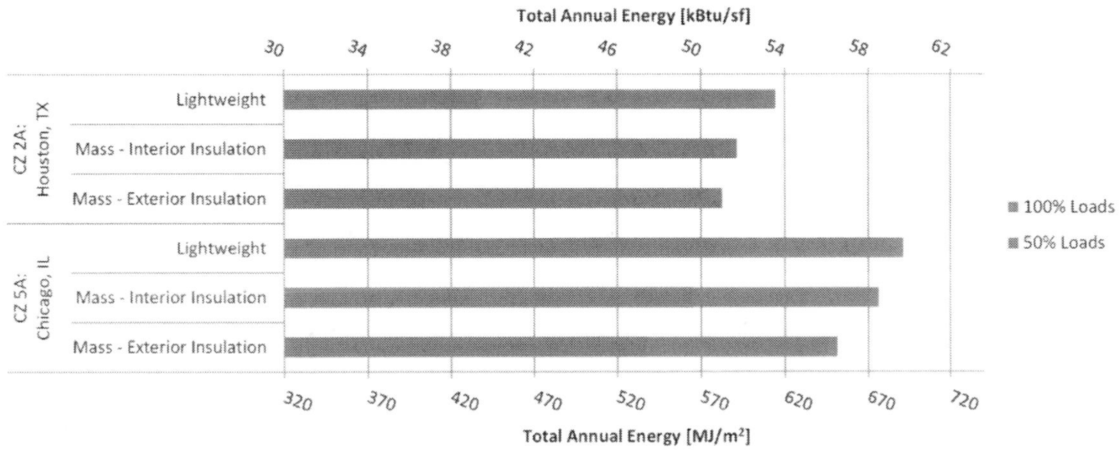

Figure 10 *Total annual energy at reduced internal loads compared to baseline.*

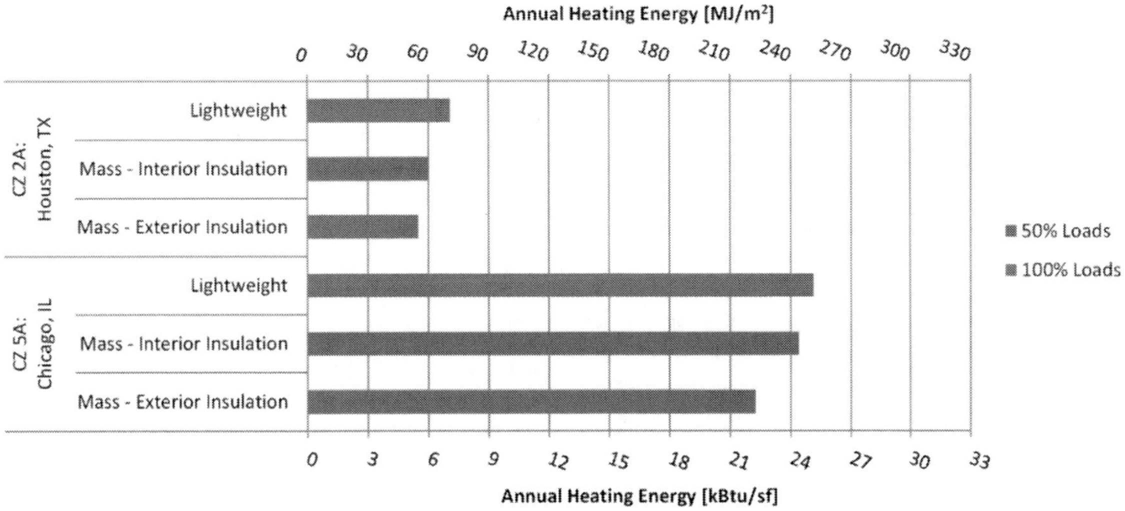

Figure 11 *Total heating energy at reduced internal loads compared to baseline.*

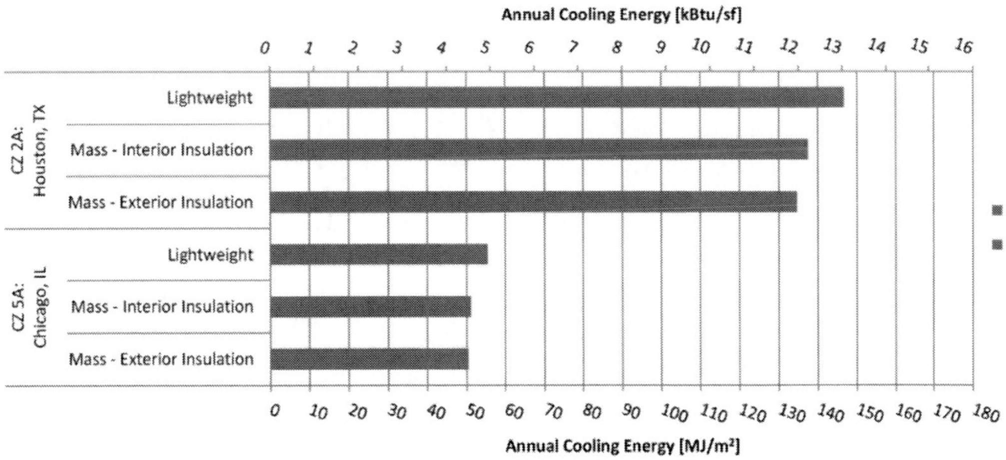

Figure 12 *Total cooling energy at reduced internal loads compared to baseline.*

CONCLUSIONS AND RECOMMENDATIONS

Based on our analysis, we conclude the following:

- From a total heating/cooling energy standpoint, buildings with mass walls outperform those with lightweight walls, despite the higher U-factors allowed by code for mass walls compared to lightweight walls. The allowed higher U-factors are justified for the all the cases and variations we studied, even for the California climate zone 9 where insulation was not even required in the mass walls.
- Positioning insulation at the exterior of the massive wall results in greater energy savings than insulation at the interior side in all the cases and climate zones we studied. The effect is stronger in heating climates, up to 10% reduced heating load and almost 4% reduction in total building energy loads compared to cases where the mass is positioned on the exterior side of the insulation. We recommend that future editions of ASHRAE 90.1 and international and state energy codes provide a larger benefit (increased U-factor) for mass on the interior side of the wall insulation.
- Our study confirmed that mass walled buildings also reduce peak loads by up to 15%, the benefit for mass on the interior side of the insulation being about 5% higher reduction in peak loads compared to mass on the exterior side of insulation. This reduction in peak loads is not accounted for in the prescriptive method of ASHRAE 90.1-2010 and state energy codes, but we recommend that it be considered for incorporation, particularly for the interior mass case. In addition, mechanical engineers may oversize equipment if their load calculations do not consider the effects of thermal mass and the location of that mass with respect to the insulation.
- The difference in annual energy use between the lightweight and massive walled buildings decreases to the point of convergence as the glazing area increases, however this does not occur until reaching much larger glazing areas than the 40% maximum allowed by current codes for prescriptive compliance.
- Mass walled buildings with higher glazing areas significantly outperform lightweight walled buildings with lower glazing areas, indicating that the mass can be used as an energy efficient strategy to offset high glazing areas. We recommend that future editions of ASHRAE 90.1 or state energy codes allow a higher glazed area for buildings with mass walls, or alternatively lower the maximum glazed area for lightweight walls that can be used with the prescriptive method of compliance.
- Lightweight walled buildings appear to benefit more from windows with a lower U-factor than mass walled buildings as a percent reduction in total energy, but mass walled buildings still out-perform lightweight walled buildings (only the relative benefit of having the mass decreases). Also, windows with a lower U-factor benefit all buildings in all climates, and therefore the use of a better window can offset a higher glazed area. This benefit is not currently captured by the prescriptive tables in the code.
- Taller buildings have an increased wall-to-roof area ratio and therefore slightly increase the positive effect of the mass walls by reducing whole building energy use by up to an additional 1% compared to a lightweight walled building.
- Reducing internal loads (lighting and equipment) reduces the demands on the cooling system in the summer, while increasing the heating system demands in the winter. The effect on energy savings will therefore depend on the climate zone, with greater reductions in cooling dominated climates. Since the cost for heating is typically much lower than the cost for lighting and equipment, reducing internal loads may result in enough savings to offset the increased cost of heating during the winter.

REFERENCES

ASHRAE. (2010). Standard 90.1-2010, *Energy Standard for Buildings Except Low-Rise Residential Buildings.*

ASHRAE. (2011). Standard 140, *Standard Method of Test for the Evaluation of Building Energy Analysis Computer Programs.*

ASHRAE. (2004). Standard 62.1, *Ventilation for Acceptable Indoor Air Quality.*

ASHRAE Research Projects 865 and 1052

Burch, D.M., W.L. Johns, T. Jacobsen, G.N. Walton, and C.P. Reeve. 1984-a. The effect of thermal mass on night temperature setback savings, *ASHRAE Transactions*, Vol. 90, Part 2.

Christian, J.E. 1991. Thermal mass credits relating to building energy standards, *ASHRAE Transactions*, Vol. 97, Part 2.

California Building Standards Commission (2010). 2010 *California Energy Code*, Title 24, Part 6.

International Code Council (ICC). (2009). 2009 *International Energy Conservation Code.*

Kossecka E., J. Kosny. 1998. The effect of insulation and mass distribution in exterior walls on the dynamic thermal performance of whole buildings. *Conference Proceedings: Thermal Performance of the Exterior Envelopes of Buildings VII*, December 6–10, Clearwater Beach, Florida, 721–731.

Marceau, M.L and M.G. VanGeem. 2007. Modeling energy performance of concrete buildings for LEED-NC Version 2.2: Energy and Atmosphere, Credit 1. Portland Cement Association R&D Serial No. 2880a.

State of Oregon. 2010. *Oregon Energy Efficiency Specialty Code.* International Code Council, 2010.

United States Department of Energy. (2012) Commercial Reference Building Models http://www1.eere.energy.gov/buildings/commercial/ref_buildings.html [Date accessed: September 2012].

United States Department of Energy. 2012. EnergyPlus Version 7.2.0.006.

Zhu, L., R. Hurt, D. Correia, and R. Boehm. 2008. Detailed energy saving performance analyses on thermal mass walls demonstrated in a zero energy house. *Energy and Buildings* 41, 303–310.

Effects of Overhangs on the Wind-Driven Rain Wetting of a Low-Rise Building

S. Sepehr M. Foroushani **Hua Ge, PhD, PEng** *Member ASHRAE* **David Naylor, PhD, PEng**

ABSTRACT

Wind-driven rain is known to be a major source of moisture loads on building envelopes and is responsible for numerous cases of building envelope failures. One of the classic solutions for preventing the building envelope from being extensively exposed to wind-driven rain is the use of overhangs. Previous studies have shown that the introduction of the overhang can change both the amount and the pattern of the wind-driven rain wetting of the facade significantly. In this work, the effects of overhangs on the distribution of wind-driven rain on the windward facade of a low-rise building are discussed based on the results of numerical simulations. It is shown that roof overhangs significantly reduce the amount of wind-driven rain deposition on the upper half of the facade. Wind speed is found to have a large influence on the protection that the overhang provides, while the intensity of rainfall does not have a significant effect. A physical explanation of the effects of overhangs is also presented.

INTRODUCTION

It was a rush of water leaks in multi-unit residential buildings in southern British Columbia during the early 1990s that drew attentions to the causes of such failures in buildings in Canada. Accordingly, a field survey conducted by the Building Envelope Research Consortium (BERC) confirmed that, rather than moisture from construction or condensation due to air leakage, rain penetration was the main source of the problem (Hazleden and Morris 1999). The field survey recognized the major concern to be poor design details and practice, not material quality or problems in maintenance (Hershfield 1996). Morris has mentioned "underestimation of wind-driven rain (WDR) and extended wetting" to be one of the causes of the problems in building envelopes, specifically in the regions with damp coastal weather (2006). One important finding was that the walls facing the prevailing wind directions had more problems in terms of moisture penetration (Hazleden and Morris 1999). In Europe too, WDR studies were first conducted in the regions with similar conditions in terms of climate. It is now well established that among different sources of moisture in buildings, WDR is the most critical, especially in climates like the costal climate of British Columbia, Canada.

Overhangs have been traditionally used for purposes including protection against rain. Nonetheless, the protective effect of roof overhangs against WDR wetting of building facades and its relation with the overhang configuration (sizing and geometry) are not clear to the designer. Consequently, design practice has been generally based on rule of thumb and is governed by the architectural expression. It is the objective of the present work to provide insight into the effectiveness of overhangs.

WDR studies usually consist of two main tasks: quantifying rain loads on the building facade and investigating the hygrothermal response of the envelope. There are different methods for quantification of WDR loads, namely, measurement, semi-empirical methods, and numerical simulation based on computational fluid dynamics (CFD). Due to its importance in the durability, evaluation of building envelopes and the proper design of envelopes, there have been quite a number of studies on quantifying WDR on building facades. However, it remains a difficult task to predict the distribution

S. Sepehr M. Foroushani is a graduate student and **David Naylor** is a professor in the department of Mechanical and Industrial Engineering, Ryerson University, Toronto and **Hua Ge** is an assistant professor in the department of Building, Civil, and Environmental Engineering, Concordia University, Montreal, Canada.

of WDR on buildings accurately, especially with the consideration of design details such as overhangs.

So far, measurements have been the primary tool of quantifying WDR. Nonetheless, as discussed by Blocken and Carmeliet, establishing a systematic experimental method for WDR quantification seems to be very difficult (2007). Different problems have been enumerated for WDR measurements, including that they can easily suffer from large error (Blocken and Carmeliet 2006), that they are time consuming and expensive, and that measurements on a particular building site would have restricted application in other cases (Blocken and Carmeliet 2007). Semi-empirical relationships correlating the quantity of WDR to the climatic parameters such as wind speed and direction and rainfall intensity have been developed. These correlations are generally simple and easy-to-use relations, but they can provide only rough estimations of the exposure of the building to WDR. As noted by Blocken and Carmeliet, with the drawbacks of measurements and semi-empirical methods on one hand, and the vast developments in CFD techniques and the computational power available to the engineer on the other hand, researchers have resorted to CFD-based numerical modeling techniques for quantifying WDR loads (2007).

A detailed review of WDR research is given by Blocken and Carmeliet (2004), where the pioneering work of Choi (1991) has been recognized as the starting point of the numerical research on WDR conducted in the past twenty years. In Choi's method, which is the basis of the present study, the raindrop trajectories are calculated based on a 3-D steady-state RANS (Reynolds-Averaged Navier Stokes) solution of the wind flow around the building using the standard k-ε turbulence model (1991; 1993). The spatial distribution of WDR on buildings is then determined based on the computed raindrop trajectories. This technique, which was originally applicable to steady wind and rain conditions only, has been extended by Blocken and Carmeliet (2000a, 2000b). In this extended method, a new weighted time-averaging technique allows the determination of both the spatial and temporal distribution of WDR.

Blocken and Carmeleit (2002) have studied a low-rise building of complex geometry under a few on-site recorded rain events. They also have made a "full-scale experimental verification" of the numerical method. As related to the subject of the present study, they have mentioned that for fixed wind speed and raindrop diameter the shelter effect of the overhang increases as the overhang width increases. In addition, the overhang has been reported to be less effective in sheltering the facade as the raindrop diameter decreases.

In the present work, based on the technique developed by Choi, the effectiveness of different overhangs on the wind-driven wetting of a low-rise building is investigated (1991; 1993). The effects of wind and rain conditions on the performance of the overhang are also examined. The local effects of the overhang and physical explanations for these effects have been the focus of discussion.

BUILDING GEOMETRY

A low-rise cubic building has been studied in this work. There are two reasons for choosing this simple geometry. First, there are published numerical studies in the WDR literature that provide a basis for validation of the present numerical simulations (Blocken 2004). Second, the simplicity of the building, and consequently the flow field around it, makes it easier and less time consuming to examine a number of different wind and rain conditions. In addition, it is expected that the effects of overhangs would be clearer and easier to assess on a simple building like this. The tested overhangs vary in width (OH), but they are all 15 cm thick and 10 m long.

Schematics of the cubic building, without and with an overhang, are shown in Figure 1. The building is subjected to undisturbed wind approaching normal to the windward facade with different speeds. An exponential profile, as described by Equation 1, is set for the approaching wind speed. Different rainfall intensities are studied under each wind condition. Three rectangular overhangs widths of 30 cm, 60 cm, and 90 cm are tested.

$$U(y) = U_H\left(\frac{y}{H}\right)^{\alpha} \quad (1)$$

In Equation 1, U is the velocity in x direction, y is the elevation from the ground, H is some reference height at which the reference velocity U_H is determined, and α is a profile exponent depending on the terrain type. A value of $\alpha = 0.15$, which represents an open terrain with little obstruction (Wieringa et al. 2001), and a reference height of $H = 10$ m are used in all the cases presented in the current study.

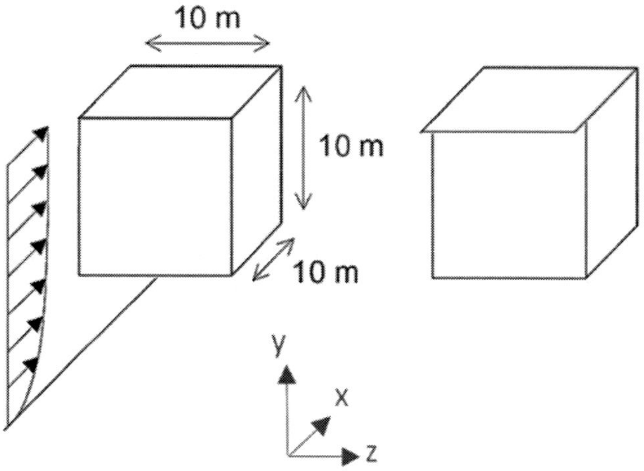

Figure 1 *LEFT: Cubic building subject to undisturbed exponential wind (Blocken 2004). RIGHT: The building with added overhang.*

Inlet turbulence kinetic energy and a modified dissipation rate are set at the inlet boundary based on the work of Blocken (2004) as

$$k = \frac{(u^*_{ABL})^2}{\sqrt{C\mu}} \tag{2}$$

and

$$\varepsilon(y) = \frac{(u^*_{ABL})^3}{\kappa(y + y_0)} \tag{3}$$

In these equations, y_0 is the aerodynamic roughness length and u^*_{ABL} is the friction velocity. Blokcen has computed different values for u^*_{ABL} based on measurements of the wind velocity profile (2004). A representative value of 0.69 m/s has been used in the current study. $C_\mu = 0.09$ is one of the constants of the k-ε model (Wilcox 1994). The more frequently used value of 0.42 has been used here for the Von Karman constant κ. Table 1 describes the boundary conditions of the present numerical simulations.

NUMERICAL SIMUILATION OF WDR

The commercial CFD code ANSYS FLUENT is used to solve the wind flow using the realizable k-ε turbulence model and standard wall functions. Standard interpolation for pressure and second order upwind discretization schemes for the momentum and turbulence equations are used. The SIMPLE method is selected for pressure-velocity coupling.

Figure 2 depicts a schematic of the computational domain with the building at the center of the bottom boundary. Information on the size and discretization of the domain are summarized in Table 2. A grid convergence study has been performed using three different grids based on the procedure proposed by the *Journal of Fluids Engineering* (ASME 2008). The area-averaged values of the pressure coefficient of the building have been used as the monitored field variable. The grid convergence index (GCI) has been shown to be substantially small (GCI ~5%).

The discrete phase model (DPM) module of FLUENT is utilized to perform the Lagrangian particle tracking and to compute the raindrop trajectories for a range of raindrop diameters. Trajectory data extracted from FLUENT for each diameter are further processed using a MATLAB code developed to calculate specific catch ratios. Catch ratio calculations are performed for 30 × 30 cm rectangular cells on the facade surface. These cells are the numerical representation of WDR gages. Finally, as discussed in the next section, these specific catch ratios are multiplied by the proper weighting function and integrated to yield the (total) catch ratio values.

The current approach is validated by comparison to the results of an earlier numerical study carried out by Blocken (2004). Details of the numerical solution are further discussed in a recent paper of Mohaddes Foroushani et al. (2012).

CATCH RATIO

Catch ratio is one of the main parameters used to describe the WDR loads on building facades quantitatively. By definition (Blocken 2004), catch ratio is the ratio of the WDR intensity on vertical surfaces (R_{wdr}) to the intensity of rainfall on the horizontal surface (the horizontal rainfall intensity, R_h), as given in Equation 4.

$$\eta(d, t) = \frac{R_{wdr}(d, t)}{R_h(d, t)} \tag{4}$$

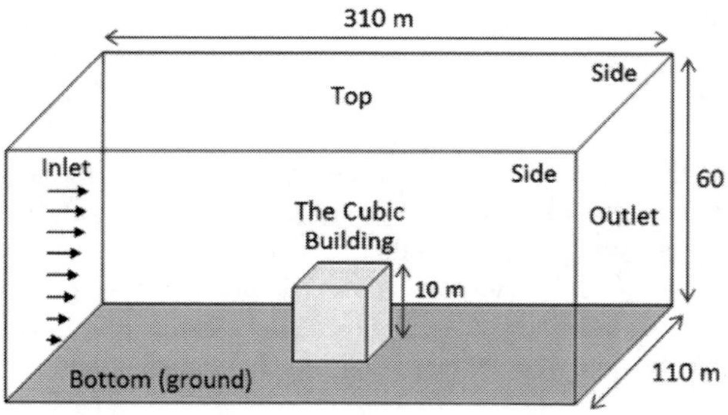

Figure 2 *The computational domain (not scaled).*

Table 1. Summary of the Boundary Conditions

Boundary		Inlet	Sides and Top	Bottom	Outlet	Building
Condition	Momentum	Normal to boundary U_{10} = 2, 5, 10 m/s α_p = 0.15	Symmetry	No slip, impermeable	Zero gauge pressure for backflow	No slip, impermeable
	Turbulence	k: Equation 2 : Equation 3		Roughness height (K_s = 0.03 m)	Backflow: k_{bf} = 1 m²/s² $_{bf}$ = 1 m²/s³ (default values)	Roughness height (K_s = 0)
DPM		—	—	Trap	—	Trap

In the case of a constant-intensity rainfall event and under steady-state wind flow conditions, the definition of the specific catch ratio may be rewritten as

$$\eta_d = \frac{A_h(d)}{A_f(d)} \tag{5}$$

In this equation, A_h is the area of a reference horizontal surface, R_{wdr} is the horizontal component of the rainfall flux vector (deposited on the vertical surface of the facade) and A_f is the wetted area on the building facade (Blocken 2004).

Integrated over the range of raindrop diameters that are likely to occur in a rainfall event, the parameter usually referred to as catch ratio is defined as

$$\eta = \int \eta_d f(d) dd \tag{6}$$

where the weighting function is the probability density function of each raindrop size (Best 1950).

RESULTS AND DISCUSSION

Three main parameters that are suspected to affect the effectiveness of the overhang are the overhang size, the wind speed, and the rainfall intensity. In this section, the effects of each factor are examined while the other two are fixed. Three rectangular overhangs with widths of 30, 60, and 90 cm are tested. Wind speeds of 2, 5 and 10 m/s representing low, moderate, and high-speed winds and rainfall intensities of 2, 5, and 10 mm/h are examined. Wind is approaching the windward facade in the normal direction in all the cases.

As mentioned previously, under steady-state wind and rain conditions, the calculation of catch ratio is simplified to calculating the ratio of areas from which raindrops are falling to the areas of the facade that are wetted by WDR. Therefore, raindrop trajectories and their deposition patterns on the facade play a key role in catch-ratio calculations. Hence, an understanding of the raindrop trajectories provides insight into the effects of overhangs too.

Two forces, namely gravity and drag, affect WDR drops. The balance between these forces determines the raindrop trajectory. Small raindrops (d < 2 mm) tend to be mainly affected by the wind drag force, more strictly follow the wind flow patterns, and thus have highly curved trajectories. On the other hand, large raindrops are gravity driven, are less influenced by the flow field, and tend to have straight trajectories. Their behavior has smaller deviation from free fall and their motion can be projected as vertically free falling under gravity at terminal (constant) velocity and horizontally being swept, without slip, by wind. Figure 3 shows the trajectories of small (d = 0.5 mm) and large (d = 3 mm) raindrops in side and top views. These graphs are generated for U_{10} = 5 m/s, but the trajectories follow similar patterns under different wind speeds.

The effect of overhangs on WDR deposition on buildings can be attributed to two major effects (Mohaddes Foroushani et al. 2012). The first is the direct protection of the facade from rain exposure. This effect is expected to be dominant in high rainfall intensities where large (gravity-driven) raindrops are more likely to appear along with lower speed winds. The second is the indirect effect, which is due to the disturbance introduced by the overhang in the wind flow pattern. This disturbance is pronounced especially in the standing upstream vortex, the high vorticity regions around the corners and edges, and the wake above the building roof. The indirect effect is expected to be more significant under low rainfall intensities where small (drag-driven) raindrops hold a greater share. The actual protecting influence of the overhang will be a combina-

Table 2. Computational Domain Size and Discretization

Domain Length (x)			310 m
Domain Height (y)			60 m
Domain Width (z)			110 m
Surface Mesh	Size	Domain sides	3 m
		Building walls	0.3 m
	Type		Quadrangle
Volume Mesh	Number of CVs		~6 × 10^5
	Type		Tetrahedral

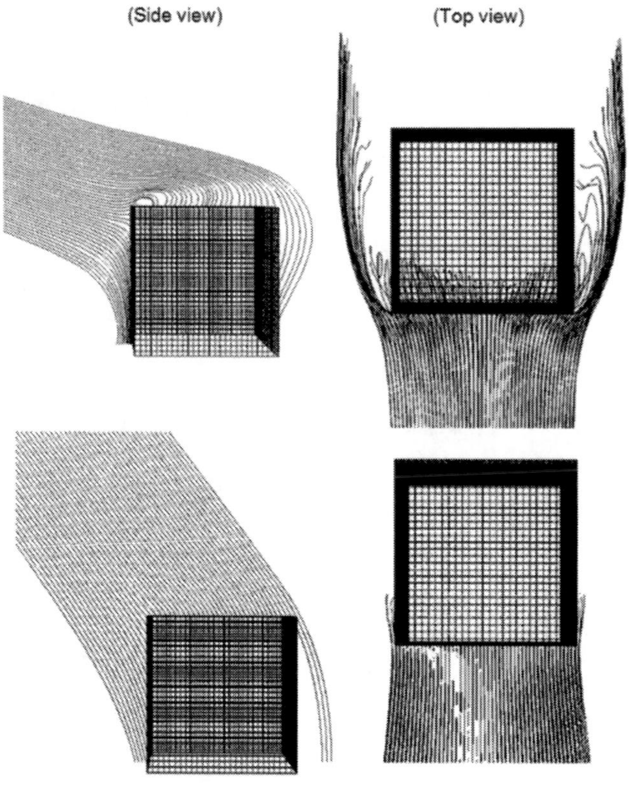

Figure 3 *Raindrop trajectories (U_{10} = 5 m/s), TOP: d = 0.5 mm, BOTTOM: d = 3 mm.*

tion of these two effects. Both these effects can be seen in Figure 4. Note that in addition to the few raindrops approaching the upper edge that are directly blocked by the overhang, others are also influenced by the presence of the overhang; the overhang has created a dead zone where raindrops have highly curved trajectories and approach the ground almost free falling. This is the indirect effect of the overhang.

Figure 5 shows the deposition patterns of 0.5 and 3 mm raindrops on buildings without an overhang and with a 60 cm overhang. Wind speed is fixed at $U_{10} = 5$ m/s. It can be seen that the small drag-dominant raindrops are driven towards the edges and corners. Therefore, corners and edges are subject to a higher concentration of WDR. The large gravity-dominant raindrops, on the other hand, are deposited on the facade more uniformly. There is, nonetheless, a higher concentration of WDR deposition at the upper edge of the facade. It can also be seen that the overhang almost completely protects the facade from smaller raindrops; only a few of these raindrops make it to the lower half of the facade near the edges and the upper half remain completely dry. On the other hand, only a small area at the top of the facade is protected from large raindrops by the direct shading effect that the overhang provides. The lower parts of the facade are almost not influenced at all.

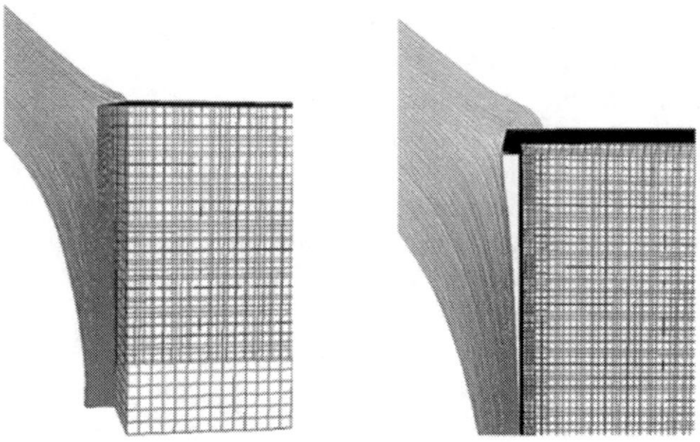

Figure 4 *Trajectories of 0.5 mm raindrops ($U_{10} = 2$ m/s) LEFT: No overhang, RIGHT: with 30 cm overhang.*

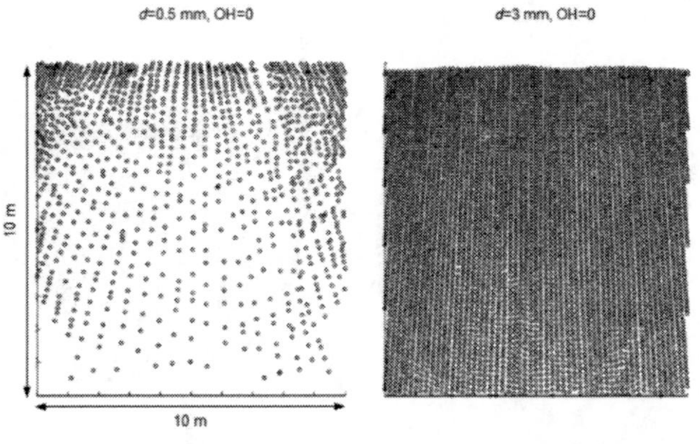

Figure 5 *Distribution of raindrops on the windward facade with no overhang ($U_{10} = 5$ m/s).*

Figure 6 shows the distribution of specific catch ratio on the facade for the same two raindrop sizes (0.5 and 3 mm), a fixed wind speed of $U_{10} = 5$ m/s, without and with a 60 cm overhang. It is observed that the introduction of overhangs not only changes the value of catch ratio at individual points, but also leads to changes in the wetting pattern of the facade (the configuration of the contour lines). The observations made based on the raindrop distribution are confirmed by these contours. Almost 40% of the facade is protected from 0.5 mm raindrops, for which no contour line is shown. Less than 10% of the facade is protected from 3 mm raindrop. The remainder of the upper half is also significantly, though not completely, protected. Specific catch ratio values ($d = 3$ mm) are reduced as much as 50% by introducing the 60 cm overhang.

Catch ratio contours for a wind speed of $U_{10} = 5$ m/s and a rainfall intensity of $R_h = 5$ mm/h are shown in Figure 7. Buildings with different overhang widths are compared to the building without an overhang. It can be seen that, under these wind and rain conditions, the wider overhangs provide much better protection at the upper half of the facade. A narrow strip beneath the upper edge of the facade has been fully protected from WDR after an overhang has been introduced. The height of this dry strip increases from almost 0.5 to around 2 m as the overhang width increases from 30 to 90 cm. On the other hand, the overhang has almost no effect on the lower half of the facade.

Figure 8 shows the distribution of total catch ratio on the windward facade under rainfalls with different intensities. These results are for a wind speed of $U_{10} = 5$ m/s. The results suggest that, with or without overhang, the rainfall intensity does not have a significant effect on wind-driven wetting of the facade. In addition, the performance of the overhang, at least at such moderate wind speeds, seems to be independent of the rainfall intensity. Under all the examined rainfall intensities, the top 15% region of the facade remains completely dry when the 60 cm is added to the building.

Finally, the effect of wind speed can be seen in the contours of Figure 9. In these contours, the rainfall intensity and the overhang width are fixed at 5 mm/h and 60 cm respectively. Wind speed seems to play a significantly larger role than the intensity of rainfall.

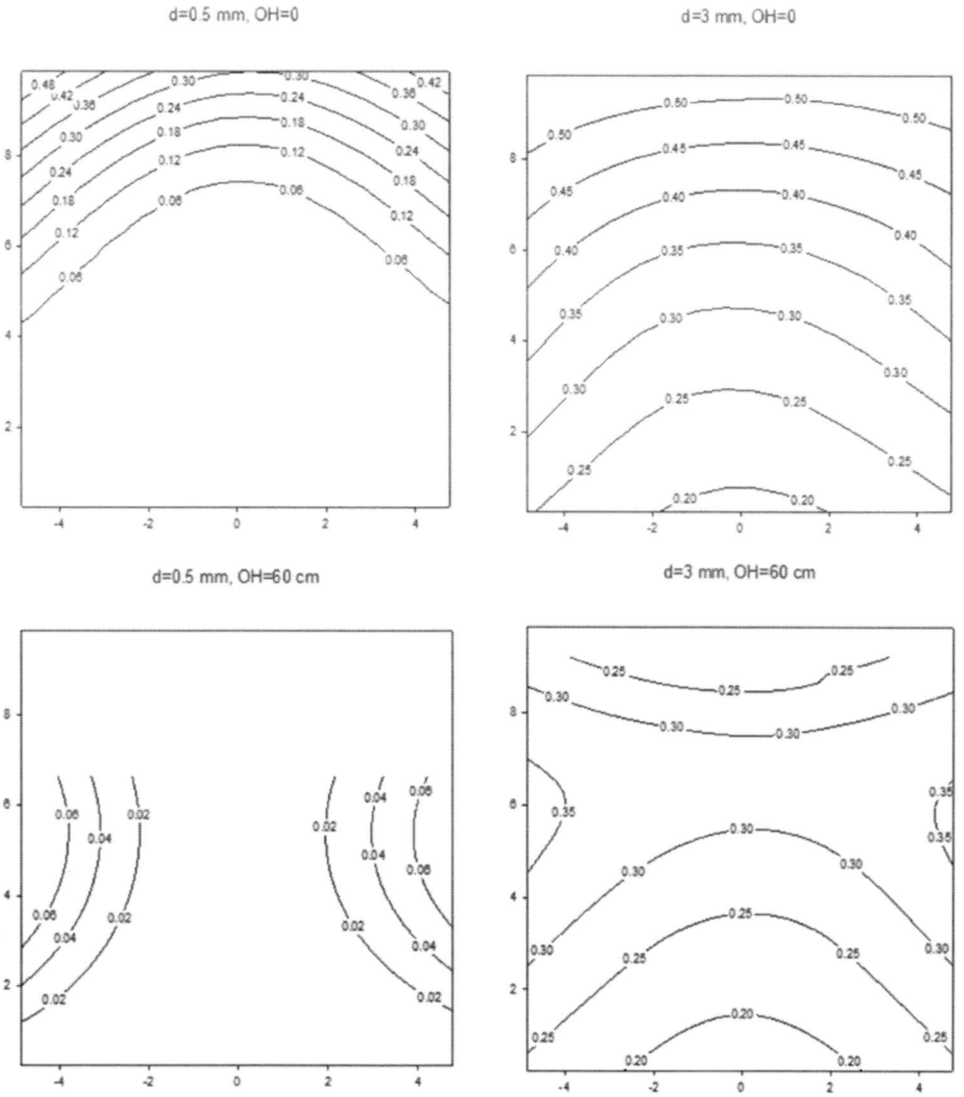

Figure 6 *Specific catch ratio contours on the windward facade* (U_{10} = *5 m/s*).

The entire facade is exposed to considerably larger WDR loads under the high-speed wind of U_{10} = 10 m/s when compared to the low and moderate-speed winds. The effectiveness of the overhang also varies under different wind speeds. Based on these results, a 60 cm overhang reduces the WDR deposition on the upper half of the facade under low, moderate, and high wind speeds. Yet, in terms of relative protection with respect to the case without an overhang, the 60 cm overhang appears to be much more effective in the low-wind speed of U_{10} = 2 m/s. The percentage of the facade area fully protected by the 60 cm overhang reduces from approximately 25% under the U_{10} = 2 m/s wind to approximately 15% and 10% under 5 m/s and 10 m/s wind speeds respectively.

The slight deviation from fully symmetric wetting patterns that can be observed in the presented contours may be attributed to asymmetry in the flow field (especially for high wind speeds), numerical errors in the CFD solution of the wind flow, and the numerical errors associated with the catch ratio calculations and contour generation. Numerical tests at different grid levels have revealed that this asymmetric behavior is pronounced for coarser girds. It is therefore assumed to be mainly due to the numerical errors of the CFD solution.

CONCLUSION

The effects of roof overhangs on the wind-driven wetting of the windward facade of a low-rise building have been examined using numerical simulation. Three overhang widths were tested. The effects of wind speed and rainfall intensity on the performance of the overhangs are also studied. It is observed that the introduction of the overhang changes both the magnitude and the pattern of the WDR deposition on the facade. All the tested overhangs helped protect the upper half of the facade from WDR wetting under all the studied wind and rain conditions. The lower half of the facade, however, appears to be

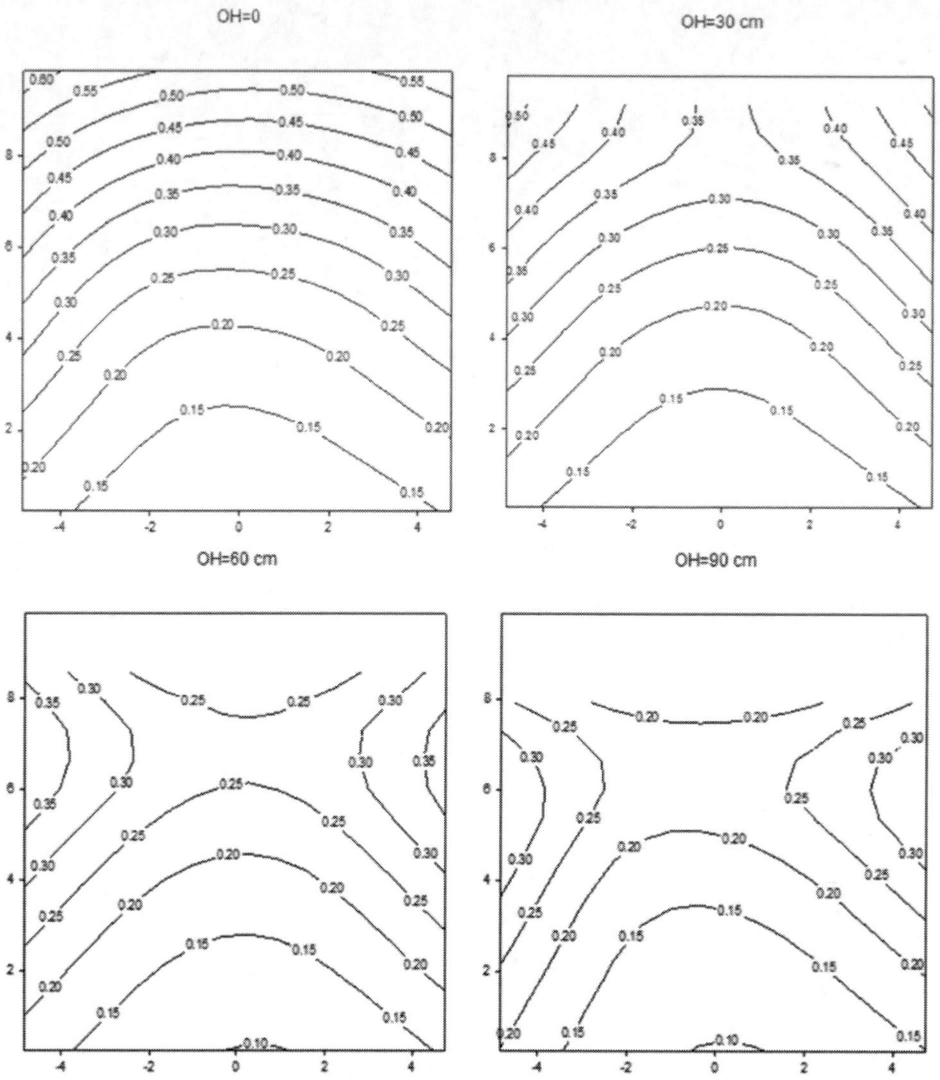

Figure 7 *Catch ratio contours on the windward facade with different overhangs (U_{10}=5 m/s, R_h=5 mm/h).*

almost not affected at all by the overhang. It is also shown that, compared to rainfall intensity, wind speed has a much larger influence on the WDR loads as well as the protection that the overhang can provide from these loads. The overhangs have significantly better protecting performances under low-speed winds.

FUTURE WORK

This work is part of a project that is intended to provide guidelines and recommendations on the design of overhangs that are effective in protecting building facades from large WDR loads. In order to achieve this, a number of other factors and cases must be further investigated in the future among which are different overhang shapes, (specifically pitched overhangs), realistic building geometries, and different wind angles. In addition, as is essential for any numerical study, results are to be compared with and validated by measured data collected from a number of test buildings.

ACKNOWLEDGMENTS

The authors would like to acknowledge the financial support of NSERC NEWBluidS Network, FP Innovations, and Ryerson University.

NOMENCLATURE

A_h	= reference horizontal area, m^2
A_f	= wetted area on facade, m^2
d	= raindrop diameter, mm
f	= raindrop diameter PDF (probability density function), 1/mm
H	= reference height, m
K_s	= surface roughness height, m
k	= turbulence kinetic energy, m^2/s^2
OH	= overhang width, cm
R_h	= rainfall intensity on the horizontal surface, mm/h

R_{wdr} = WDR intensity, mm/hr
t = time, s
U = wind speed, m/s
U_H = reference wind speed, m/s
u^*_{ABL} = friction velocity, m/s
y = vertical coordinate/ elevation from the ground, m
y_0 = aerodynamic roughness length, m
α = wind speed profile exponent, —
ε = turbulence dissipation rate, m^2/s^3
η = catch ratio, —
κ = von Karman constant, —

REFERENCES

ASME. 2008. Procedure for estimation and reporting of uncertainty due to discretization in CFD applications (Announcement). *Journal of Fluids Engineering* 130(7).

Best, A.C. 1950. The size distribution of raindrops. *Quarterly Journal of the Royal Meteorological Society* 76(327):16–36.

Blocken, B. and J. Carmeliet. 2000a. Driving rain on building envelopes—I: Numerical estimation and full-scale experimental verification. *Journal of Thermal Envelope and Building Science* 24(1):61–85.

Blocken, B., and J. Carmeliet. 2000b. Driving rain on building envelopes—II: Representative experimental data for driving rain estimation. *Journal of Thermal Envelope and Building Science* 24(2):89–110.

Blocken, B. and J. Carmeliet. 2002. Spatial and temporal distribution of driving rain on a low-rise building. *Civil Engineering* 5(5):441–62.

Blocken, B. 2004. Wind-driven rain on buildings. Doctoral. thesis, Katholieke Universiteit Leuven, Lueven, Belgium.

Blocken, B. and J. Carmeliet. 2004. A review of wind-driven rain research in building science. *Journal of Wind Engineering and Industrial Aerodynamics* 92(13):1079–1130.

Blocken, B. and J. Carmeliet. 2006. On the accuracy of wind-driven rain measurements on buildings. *Building and Environment* 41(12):1798–1810.

Blocken, B. and J. Carmeliet. 2007. Validation of CFD simulations of wind-driven rain on a low-rise building facade. *Building and Environment* 42(7):2530–2548.

Choi, E.C.C. 1991. Numerical simulation of wind-driven-rain falling onto a 2-D building. *Asia Pacific Conference on Computational Mechanics, Hong Kong*, 1721–28.

Choi, E.C.C. 1993. Simulation of wind-driven-rain around a building. *Journal of Wind Engineering and Industrial Aerodynamics* 46(47): 721-729.

Hazleden, D.G. and P.I. Morris. 1999. Designing for durable wood construction*: The 4 Ds, 8th International Conference On Durability Of Building Materials And Components*, Vancouver, Canada.

Foroushani, M.S.S., D. Naylor, and H. Ge. 2012. Numerical investigation of the effects of different overhang configurations on the wind-driven rain wetting of building facades. *Proceedings of ASME 2012 International Mechanical Engineering Congress & Exposition* (IMECE2012), Houston, Texas.

Morris, P.I. 2006. Damp coastal weather field issues and improving performance. *In Proceedings: Wood Protection.*

Hershfield, M. 1996. Survey of building envelope failures in the coastal climate of British Columbia. Research report for Canadian Mortgage and Housing Corporation, Ottawa, Canada.

Wieringa, J., A.G. Davenport, C.S.B. Grimoond, and T. R. Oke. 2001. New revision of davenport roughness classification. *The 3rd European and African Conference on Wind Engineering, Eindhoven, Netherlands.*

Wilcox, D. C. 1994. *Turbulence modeling for CFD*. California: DCW Industries.

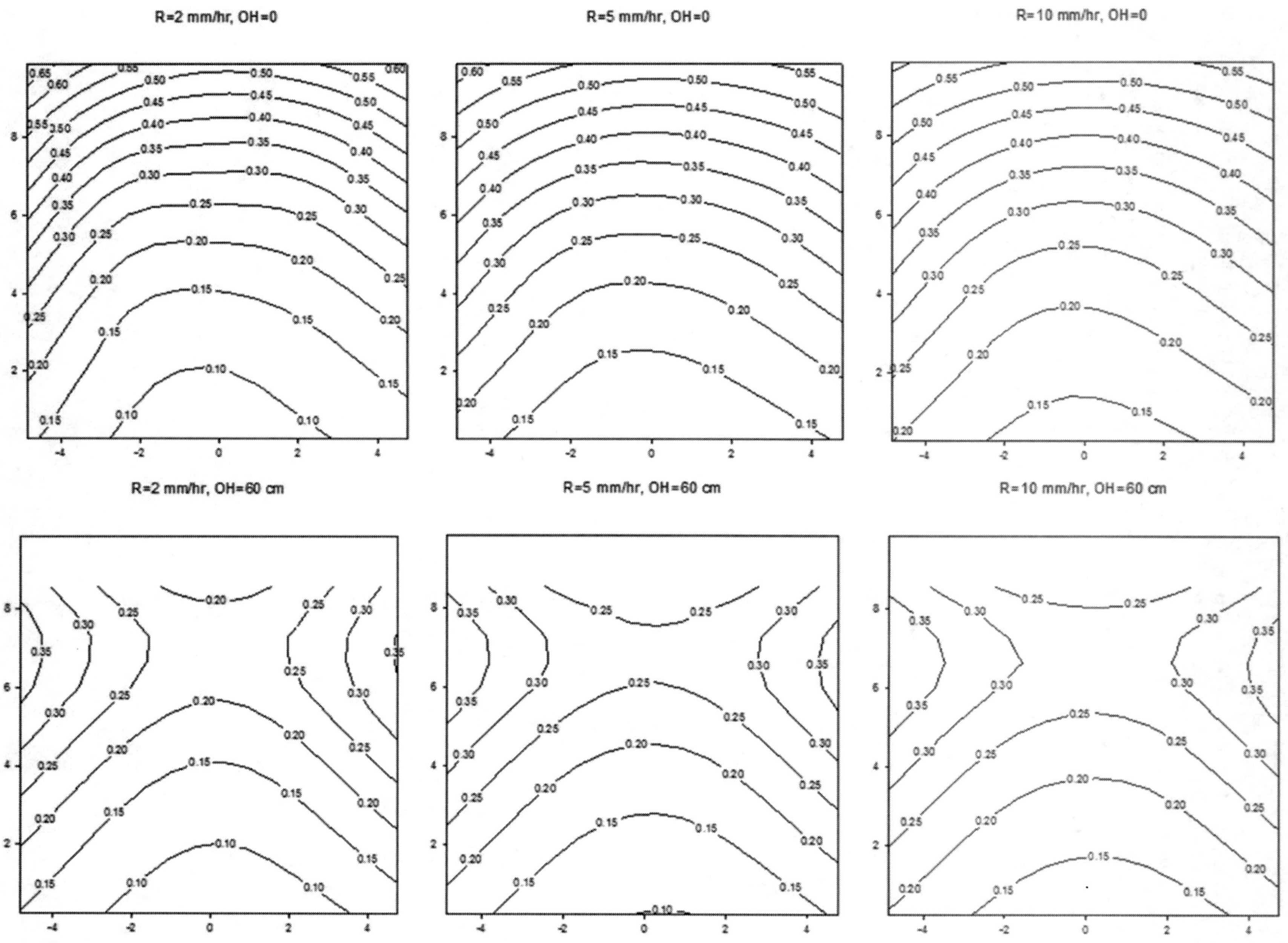

Figure 8 *Catch ratio contours on the windward facade under different rainfall intensities, without (TOP) and with (BOTTOM) 60 cm overhang* (U_{10}*=5 m/s).*

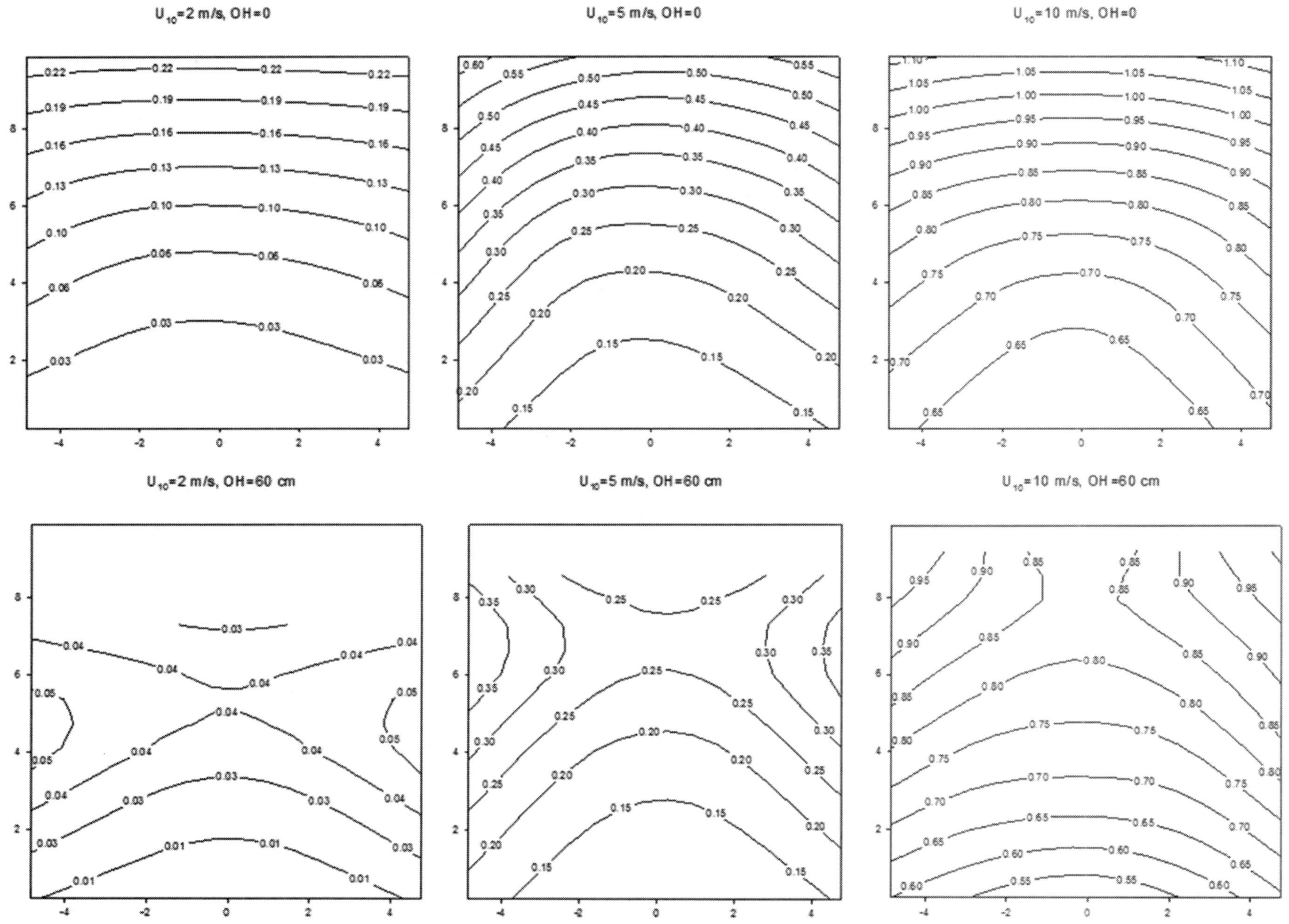

Figure 9 *Catch ratio contours on the windward facade under different wind speeds, without (TOP) and with (BOTTOM) 60 cm overhang (R_h=5 mm/h).*

Permeance of Steel Roof Decks and Effect on Hygrothermal Performance of Roofing Systems

Joseph P. Piñon, PE
Associate Member ASHRAE

Raymond W. LaTona, PhD, PE
Member ASHRAE

ABSTRACT

The majority of commercial roofs consist of corrugated steel roof decks, with or without concrete fill. Steel decks used with concrete fill typically contain perforations along the ribs (flutes) to allow drying of the concrete, termed "vented decks." Steel roof decks without concrete fill are typically "unvented," but still have seams that are only tack welded or mechanically fastened together. Decks without concrete fill have additional penetrations due to fasteners used to attach insulation and roofing membrane layers. Despite being an integral part on the majority of all commercial roofs, surprisingly little research has been performed to study the air and vapor permeance of steel roof deck construction. Designers are forced to estimate the effect on the air and vapor permeance of steel roof decks that results from unsealed laps and fastener penetrations, as well as intentional vents (perforations). The quantification of these effects is needed for use as inputs into hygrothermal models.

This paper presents the results of a series of laboratory experiments designed to quantify the air and vapor permeance of several steel deck constructions, including vented and non-vented decks and including the effect on permeance when the metal deck is filled with concrete. This paper provides values for designers to use in hygrothermal models for both design and investigation of roof systems containing metal deck.

INTRODUCTION

Description of Typical Steel Roof Deck Construction

The majority of commercial roofs consist of corrugated steel roof decks, with or without concrete fill. Steel decks used with concrete fill typically contain perforations along the ribs (flutes) to allow drying of the concrete, termed "vented decks." Steel roof decks without concrete fill are typically "unvented," but still have seams that are only tack welded or mechanically fastened together. Decks without concrete fill have additional penetrations due to fasteners used to attach insulation and roofing membrane layers. Virtually all roof decks have various penetrations including HVAC ducts, drains, pipe and conduit penetrations, roof hatches, skylights, smoke vents, etc. These penetrations also create gaps that affect the permeability of the in-situ deck.

In the United States, most roof decks are 1.5 in. deep and are characterized as Type A, narrow rib; Type F, intermediate rib; or Type B, wide rib decks. Standard decks have a six-inch spacing of the ribs, and the variation from one type to another is in the geometry of the ribs as shown in Figure 1. Typical deck units are 24 in., 30 in., or 36 in. wide and generally come in lengths up to about 42 ft. In constructing the roof deck, the deck units are lapped at the end joints and generally have nested side laps. Although the field of the steel deck is impermeable, the side and end laps contain gaps that permit water vapor and air to pass through the deck; consequently, the vapor permeability of the deck needs to be evaluated as it relates to the performance of the overall in-situ roof deck system. Side laps are generally welded or mechanically fastened using sheet-metal screws, button punches, or by crimping (Figure 2). The ends of the

***Joseph P. Piñon** is a senior project manager and **Raymond W. LaTona** is a senior principal at Simpson Gumpertz & Heger Inc. San Francisco, CA.*

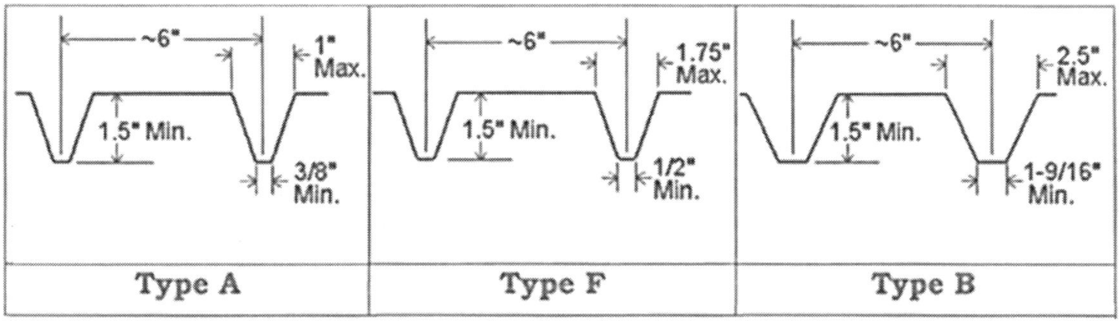

Figure 1 *Common types of steel roof deck used in the United States.*

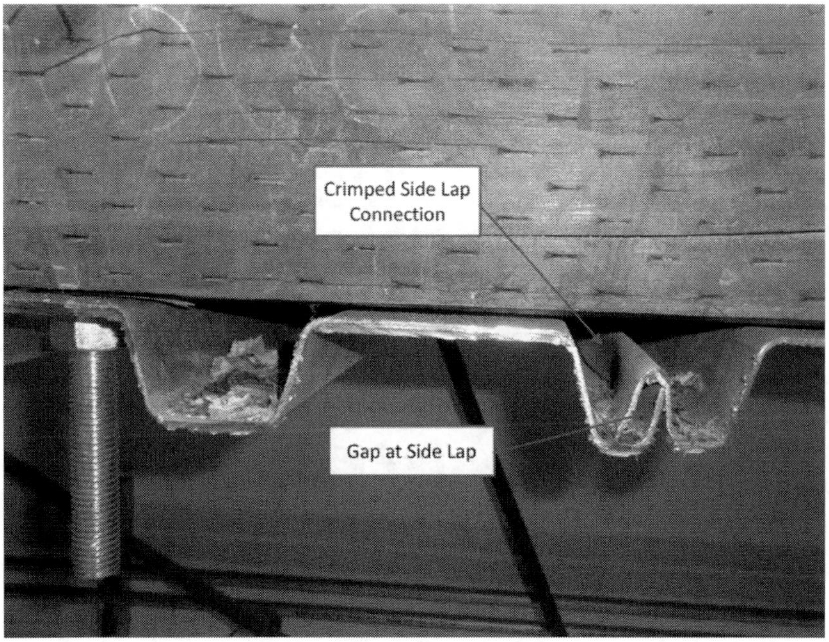

Figure 2 *Common types of steel roof deck used in the United States.*

deck units are generally overlapped and welded or mechanically fastened using self-drilling, self-tapping fasteners or pneumatically driven fasteners. Mechanical fasteners are mis-driven sometimes and removed, leaving holes, and sometimes welding leaves burn holes. These methods of joining the deck units often leave substantial gaps and openings (Figure 3).

In addition to the gaps at the deck unit laps, large gaps or holes in the deck often occur at the deck penetrations. Figures 4 and 5 show some typical penetrations that occur in steel roof decks. At drains, skylights, and perimeter walls, there are gaps equal to the cross sectional area of several ribs. In addition, there are often additional gaps at locations where there is a change in slope of the roof deck or where there is a change in direction of the roof deck (Figure 6).

Overview of Typical Roofing Systems Installed over Steel Deck

Figure 7 shows the typical commercial roofing systems that are installed without a vapor retarder over steel deck either with or without concrete fill. The typical systems consist of a roofing membrane over a gypsum or perlite based cover board, over multiple layers of insulation. When a vapor retarder is specified, it is typically installed over the concrete or over the steel deck (for non-concrete decks), often with a gypsum based cover board under the vapor retarder to provide support and fire protection to the assembly.

When the decks are filled with concrete, the insulation, cover board, and roofing membrane layers are typically adhered with adhesive or asphalt. For steel decks without concrete fill, the layers are typically mechanically fastened,

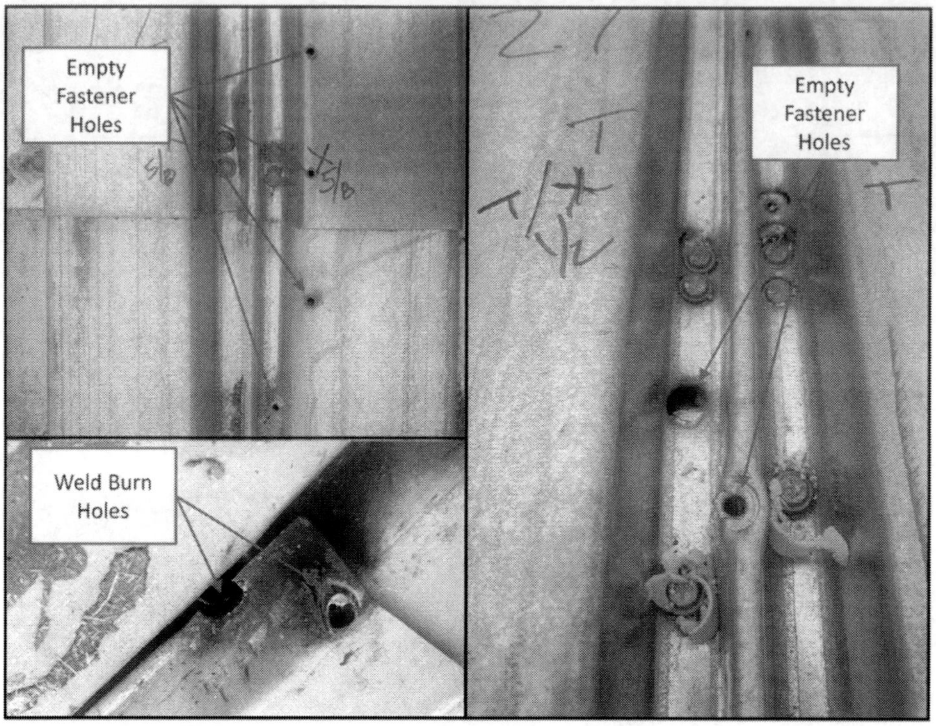

Figure 3 *Typical gaps and holes in steel deck side and end laps.*

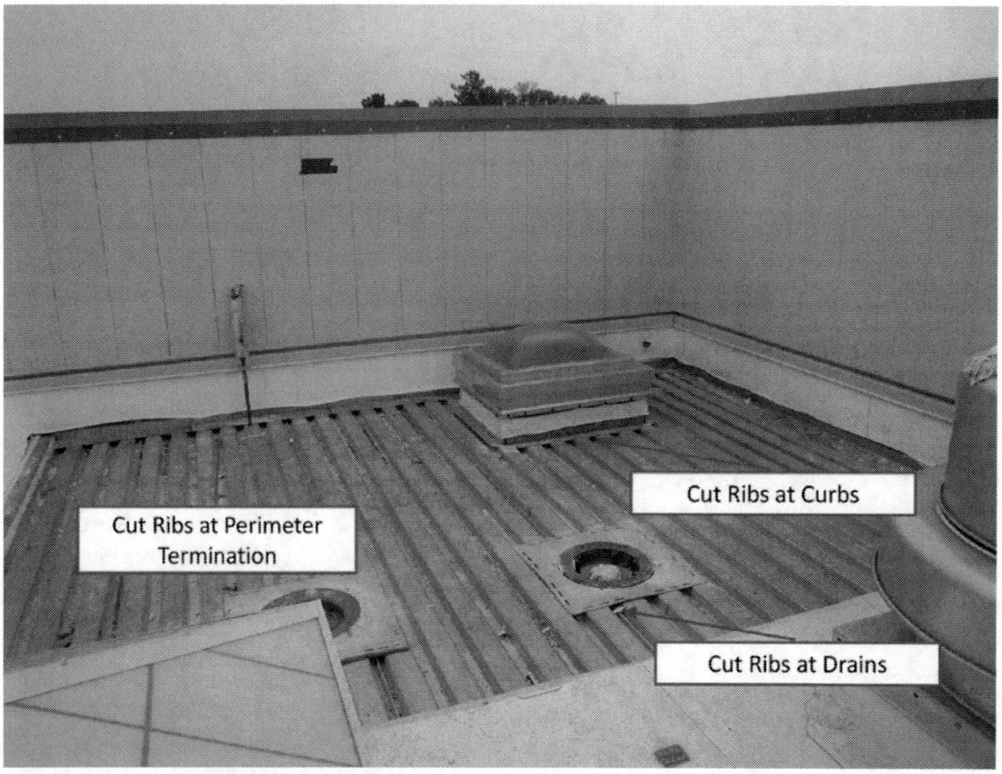

Figure 4 *Typical penetrations in a steel roof deck including a skylight hatch, roof drains, and perimeter terminations.*

Figure 5 *Large gap around pipe penetrations.*

Figure 6 *Large gap at change in steel deck orientation*

except for the roofing membrane which can either be mechanically fastened or fully adhered to the cover board.

Overview of Previous Research

Despite being an integral part on the majority of all commercial roofs, surprisingly little research has been performed to study the air and vapor permeance of steel roof deck construction. Seiffert (1970) performed ASTM E-96 test of aluminum foil that showed that the effective permeance of the foil increased by a factor of 4,000 when perforated with holes representing as little as 0.22% of the surface area of the foil. In a research study that included perforated metal foil radiant barriers, Wilkes (1991), compared theoretical calculations using a parallel path analogy with unpublished experimental results from colleagues to estimate the effect of holes on a low permeance vapor retarder to be described by the following equation:

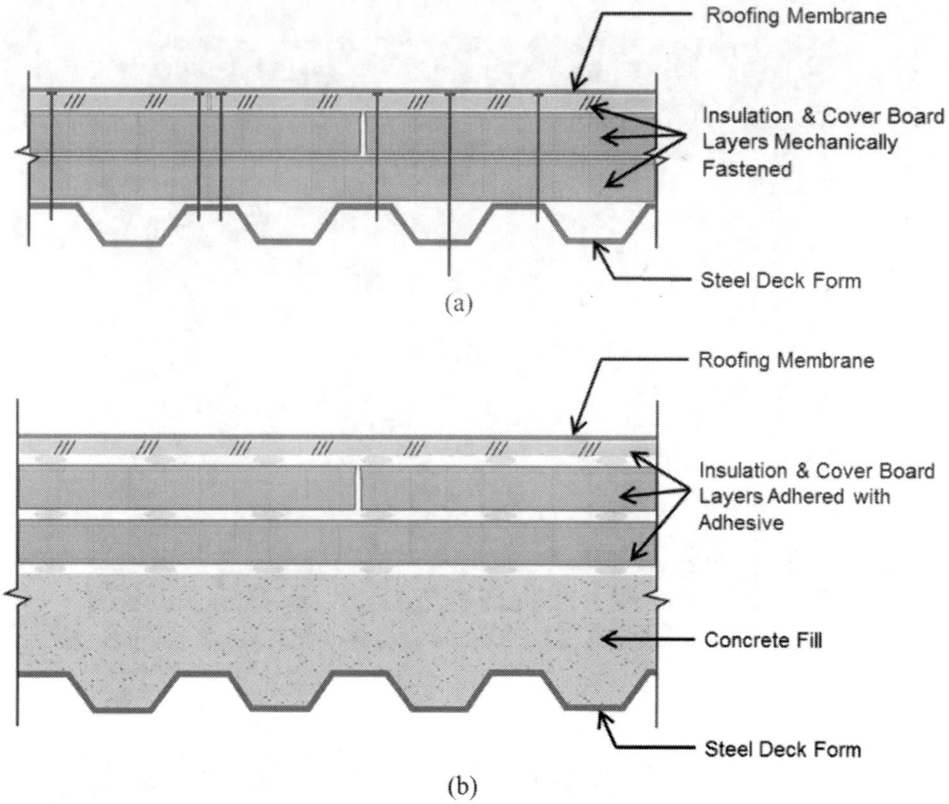

Figure 7 *Typical roof assemblies over steel deck: (a) mechanically fastened with no concrete fill and; (b) adhered over concrete filled deck.*

$$P = 56,200 \times \left(\frac{A_{holes}}{A_{totals}}\right) \quad (1)$$

Later in this paper we compare both Wilkes's estimate and Seiffert's test results to our test results.

Other than Wilkes and Seiffert's papers, the only other information available on the effect of holes in vapor retarders is anecdotal evidence or differing opinions on whether a small percentage area holes can significantly increase the permeance of a vapor retarder. The Steel Deck Institute (2008), for example, has issued a technical bulletin claiming that the 0.5% to 1.5% area of holes available on "vented" (perforated) steel deck has an insignificant effect on the permeance of the deck in terms of the drying (to the interior) of concrete that is cast in the metal deck.

Most building scientist and building enclosure designers agree that if there is airflow through the holes, then the effective permeance of the steel deck, or vapor retarder membrane, is significantly increased to the point where it can be ignored, due to the ability of the airflow through small gaps to both bypass the low permeance surface and to transport significant amounts of water vapor compared to diffusion.

Example Theoretical Calculations of Effective Permeance for Steel Roof Deck

Since diffusion of water vapor is analogous to conduction of heat, one can consider the diffusion of water vapor across a steel deck with holes (or other vapor retarder membrane with holes) to be similar to parallel path heat transfer that results from thermal bridging of a steel stud (or other highly conductive element) across an insulated cavity. The electrical circuit analogy often used in heat transfer calculations is shown in Figure 8.

Similar to heat flow, the vapor flow will take the path of least resistance. Since the vapor resistance of the steel without holes is close to infinity, the path through the steel goes to zero. The resulting prediction is that it takes only a small percentage area of holes to increase the effective permeance of the steel significantly. This effect is similar to a thermal bridge that can cause a much larger amount of heat loss than its area would suggest (more heat will flow along the path of least resistance). This effect is also how certain housewrap materials achieve a higher permeance by perforating a non-permeable material with micro-sized holes.

The parallel path analogy also shows that the resulting predicted effective permeance is highly dependent on the permeance of the hole. In Table 1 we show the effective permeance calculated with and without concrete fill to demonstrate that the predicted effective permeance using this parallel path analogy assuming concrete fills the holes can be more than 100 times lower than if only air is assumed to fill the holes. Furthermore, in practice, concrete fill will also block airflow that would further decrease the effective permeance of the steel deck.

Others have performed a similar calculation to determine the effect that airflow through the holes has on the effective permeance of a low permeance material and concluded that it only takes a small amount of airflow to significantly increase the effective permeance of the material (Straube and Burnett 1995).

In addition to the steel deck layer, a typical roofing system will have multiple other layers, such as insulation, cover board, and a low permeance roofing membrane (Figure 7). For analyzing thermal bridging, ASHRAE (2009) provides two methods based on the parallel path analogy. These two methods are termed the "parallel path" and "isothermal planes" methods. In Figure 9 we illustrate these two paths in terms of vapor diffusion resistance for a typical steel deck roofing system without concrete fill. According to ASHRAE, the parallel path method tends to under-predict the amount of thermal resistance, while the isothermal planes method tends to over-predict the amount of thermal resistance. In terms of flow of water vapor, these two methods provide significantly different predictions,

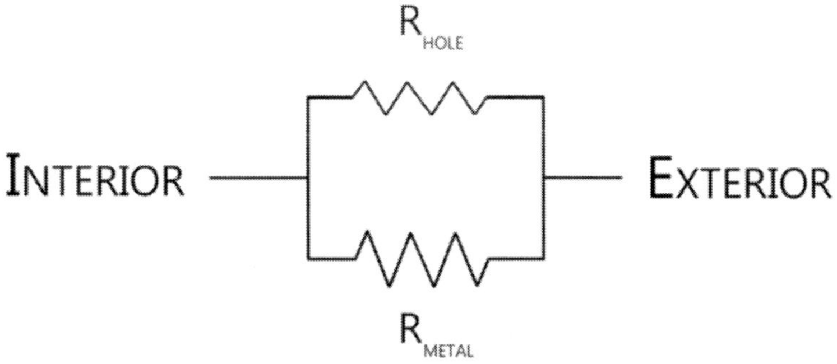

Figure 8 *Example of a parallel path analogy for vapor diffusion through a steel deck with holes.*

Table 1. Parallel Path Effective Permeance Calculation for 18 Gauge Steel Deck with Holes with and without Concrete Fill

% Area of Holes	No Concrete (assume holes in deck are filled with 18 gauge thick air layer) [Perm]	Deck Filled with 3 Different Types of Concrete (assume holes in deck are filled with 18 gauge thick layer of concrete) [Perm]		
		1.4 Perm-In. Concrete	0.716 Perm-In. Concrete	0.519 Perm-In. Concrete
0.00%	0.0	0.00	0.00	0.00
0.04%	1.0	0.01	0.01	0.00
0.06%	1.5	0.02	0.01	0.01
0.22%	5.8	0.06	0.03	0.02
0.46%	12	0.13	0.06	0.05
0.50%	13	0.14	0.07	0.05
1.0%	26	0.27	0.14	0.10
1.5%	40	0.41	0.21	0.15
2%	53	0.54	0.28	0.20
5%	132	1.4	0.69	0.50
10%	264	2.7	1.4	1.0

particularly for cases where there is another low permeance material in the flow path, as is the typical case with most roofing membranes. Table 2 shows how different predictions from these two methods can be. In general, in the isothermal planes method, it only takes a small percentage of holes to make the steel deck ineffective as a vapor retarder.

Later in this report we compare our experimental results to the predictions of these two methods and discuss how our results indicate that the isothermal planes method appears to be the more accurate analogy.

Given the above uncertainties in actual performance of the steel deck, designers are forced to estimate the effect on the air and vapor permeance of steel roof decks that results from unsealed laps and fastener penetrations, as well as intentional vents (perforations). The quantification of these effects is needed for use as inputs into hygrothermal models.

Objective

The objective of this work is to determine realistic ranges of values that can be used for the equivalent vapor permeance of steel roof decks with and without concrete fill.

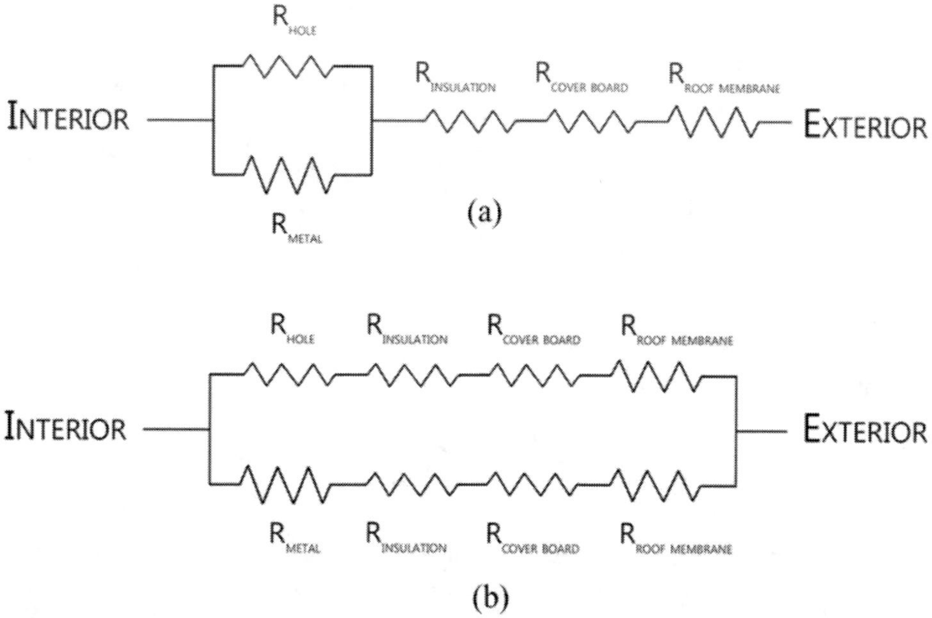

Figure 9 *(a) Isothermal planes method analogy and (b) parallel path method analogy.*

Table 2. Comparison of Predictions on Effective Permeance of Steel Deck in a Roof System for Parallel Path and Isothermal Planes Methods Shown in Figure 9

% Area of Holes	Predicted Effective Permeance of Steel Deck Layer with Holes when Installed in a Roof System [Perm, IP Units]	
	Parallel Path Method	Isothermal Planes Method
0%	0	0
0.05%	0.00012	1.32
0.1%	0.00015	2.64
0.5%	0.00033	13.2
1%	0.00056	26.4
5%	0.00249	132
10%	0.00515	264

Scope

We conducted research and laboratory testing to quantify the permeability for several deck configurations. In order to keep the specimen size as given in ASTM E 96, we utilized flat sheet metal rather than corrugated steel deck. In some regards, the use of flat sheet-metal work creates a condition with lower permeability than would be achieved with a corrugated deck due to the tighter nesting of the laps. In order to study the effect of the varying gaps and holes that can occur in actual construction (Figures 2 and 3), we tested specimens with varying number of fasteners, fastener holes, and varying width of gaps at the laps.

The size of the specimens relative to the potential paths of moisture migration through the specimen is different than occurs in standard construction. Consequently, in order to determine useful permeability for actual construction, we had to determine the equivalent vapor flow per feature of the deck such as per fastener, per hole, and per length of lap. We then calculated the permeability of in-situ decks given typical insulation fastening patterns and deck panel sizes.

In this paper we focus primarily on the effect of holes in steel deck given diffusion only (no induced airflow) for the following reasons. First, it is well documented that a small amount of airflow can transport an order of magnitude more moisture than vapor diffusion, thereby bypassing the vapor resistance of the material. Second, a paper by Kan and Piñon (2005) showed that the concept of equivalent permeance is of limited use in predicting the in-situ performance of walls due to airflow. Therefore, where there is air leakage through a steel deck roof system, the potential hygrothermal effects on the roof system are more than just making the vapor resistance of the steel deck negligible.

LABORATORY TESTING

Experimental Setup

We measured the water vapor permeance using the wet cup method in accordance with ASTM E 96 *Standard Test Methods for Water Vapor Transmission of Materials*. We conducted 17 separate tests; the configuration of the specimens for each is described in Table 3. Specimen 1 had a flat sheet of 18 gauge sheet-metal set into a specimen pan with the perimeter sealed in the same fashion as used for the remaining specimens. This specimen was used to confirm the specimen preparation was appropriate, and the test confirmed that this condition resulted in zero perms.

We selected the remaining specimen configurations described in Table 3 to simulate actual field conditions such as mechanical fasteners used to attach insulation and/or membranes to steel decks, varying thickness of the gap within the side and end laps (e.g., Figure 2), and holes left from fasteners installed incorrectly and removed and/or reroofing situations or from weld burn holes (e.g., Figure 3). In addition, we also tested configurations with a vapor permeable air barrier over the sheet metal specimen to get a sense of the

Table 3. Description of Test Specimens

Specimen Number	Condition
1	Sealed steel deck
2	2 insulation fasteners
3	1 screw hole from insulation fastener, fastener removed
4	Lap, 2 sheet metal screws
5	Lap, 2 sheet metal screws, covered with Air Barrier
6	12 screw holes from insulation fasteners, fasteners removed
7	25 screw holes from insulation fasteners, fasteners removed
8	25 screw holes from insulation fasteners, fasteners removed, covered with Air Barrier
9	2 screw holes from insulation fasteners, fasteners removed
10	3 screw holes from insulation fasteners, fasteners removed
11	3 screw holes from insulation fasteners, fasteners removed, covered with Air Barrier
12	Lap, 2 sheet metal screws 1/32 in. gap
13	Lap, 2 sheet metal screws 1/16 in. gap
14	Lap, 2 sheet metal screws 1/8 in. gap
15	Lap, two sheet metal screws 1/32 in. gap, covered with Air Barrier
16	Lap, two sheet metal screws 1/16 in. gap, covered with Air Barrier
17	Lap, two sheet metal screws 1/8 in. gap, covered with Air Barrier

permeance where the steel deck may be filled with concrete (no airflow). Examples of the test specimens are shown in Figures 10 and 11.

RESULTS

General

To relate these experimental test results to behavior in actual steel deck construction, it is necessary to calculate the effective permeability for a representative steel deck area in an actual construction based on the test values. The ratio of the lap length to the area of the deck in the test specimens is different than that ratio is in actual construction. We chose to use the area of a single deck unit as a representative steel deck area for our calculations. We use deck panels of 20, 30, and 40 ft. (6.1, 9.1, and 12.2 m) in length and 36 in. (0.91 m) wide. We used the calculations described below to compute the permeability in the actual construction based on the results obtained in our tests.

A perm is defined as

$$1 \text{ perm} = 1 \text{ grain/h per ft}^2 \text{ per in. Hg}$$
$$(57.2 \times 10^{-12} \text{ kg/s per m}^2 \text{ per Pa})$$

To calculate the permeability for actual construction from the tests results, we must calculate the number of grains/h per in. Hg that pass through a representative area of the actual steel

Figure 10 *Example of two test specimens.*

Figure 11 *Example of a test specimen with a lap joint and feeler gauge to measure lap gap.*

deck construction and then divide by that representative area to obtain grains/h per ft^2 per in. Hg to obtain perms.

We recognize that the permeability for the 12 in. × 12 in. test specimens is the number of grains/h per in. Hg for the 1 ft^2 specimen which is the number of grains passing through the gap at the lap since the steel deck itself is impervious.

We perform the following calculation:

Let w = width of deck unit

ℓ = length of deck unit

g_ℓ = number of grains that pass through a unit length of lap

Then for the actual steel deck construction:

$$\text{Perms}_{lap} = (w+\ell)g_\ell \div w\ell \quad (2)$$

The calculations for the permeability for fasteners only and for empty fastener holes only are accomplished in a similar fashion. We recognize that the specimen size for these tests was 0.5 ft^2.

For fasteners let

n_f = number of fasteners in the test specimen

n_{fa} = number of fasteners in the actual construction

g_f = number of grains that pass through a fastener location = $\text{Perm}_{specimen} \times 0.5 / n_f$,

Then,

$$\text{Perm}_{fasteners} = n_{fa}g_f \div w\ell \quad (3)$$

For empty fastener holes let

n_h = number of fastener holes in the test specimen

n_{ha} = number of fastener holes in the actual construction

g_h = number of grains that pass through a fastener hole location = $\text{Perm}_{specimen} \times 0.5/n_h$

Then,

$$\text{Perm}_{holes} = n_{ha}g_h \div w\ell \quad (4)$$

The permeability for a steel deck including the effects of the laps, fasteners, and empty screw holes can be expressed as:

$$\text{Perm} = [(w+\ell)g_\ell + n_{fa}g_f + n_{ha}g_h] \div w\ell \quad (5)$$

Calculated Results

Table 4 shows the results of the laboratory tests. The columns under the light grey headings define the specimen sizes and configurations as well as the permeability determined from the tests under ASTM E-96. The shaded columns show values determined from the calculations we made to convert the permeability values from our tests to the values for the units contained in the right most column.

Table 5 shows the permeability of in-situ deck taking into account the permeability resulting from the laps only. In this Table, the data from Table 4 are utilized to calculate the data in Table 4 consistent with Equation 2.

Table 6 shows the permeability of in-situ deck taking into account only the permeability resulting from fasteners only. In this table, the data from Table 4 are utilized to calculate the data in Table 6 consistent with Equation 3.

Table 7 shows the permeability of in-situ deck taking into account only the permeability resulting from empty fastener holes. The data from Table 4 are utilized to calculate the data in Table 7 consistent with Equation 4.

Table 8 shows the permeability of in-situ deck taking into account the permeability resulting from the gaps at laps, the presence of screw fasteners, and the presence of empty fastener holes. The data from Table 4 are utilized to calculate the data in Table 8 consistent with Equation 5.

Graphs of Results of Calculation

Figure 12 is a plot of our calculated results showing the effect on permeability resulting from varying the dimension of the gap at the deck laps for deck panels that are 20, 30, and 40 feet long.

Figure 13 shows the permeability of a deck with holes where the area of the holes is expressed as a percentage of the deck area. On Figure 13 we also compare the predictions from the parallel path method and test results from the previous research discussed in the introduction section of this paper.

DISCUSSION

Area of Gaps and Holes and the Effect on Permeability

It is well known that un-perforated steel deck is impervious, and our tests on Specimen 1 confirm that this is true. In real construction, decks without perforations are nonexistent. Since deck units come in finite sizes smaller than a typical roof area, end laps and side laps affect the permeability. Our study confirms that larger gaps at laps create higher permeability than tighter laps. In addition, decks are attached to the substructure with welds or mechanical fasteners which penetrate the deck and often leave holes in the deck as shown in Figure 3. These holes add to the deck gaps and increase permeability. The majority of the lap length for a steel deck is the side laps, which often have gaps that are wider than the gaps at end laps. Generally, end laps are fastened to the substrate at 12 or 6 in. on center, while side laps are fastened at wider intervals such as 18 or 24 in. Consequently, the gaps at side laps are generally greater than at end laps.

Laboratory tests show that fasteners penetrating the steel deck to attach insulation or membranes have a relatively small effect on the permeability of steel deck. Even when as many as 16 screws per 4 × 4 insulation board are used, the screw density is one screw per square foot, which raises the permeability of the deck to approximately 0.02 perms. On the other hand, the tests showed that empty screw holes (fastener removed) have a greater effect on permeability. One empty screw hole per sq ft (0.093 m^2) raises the permeability of the deck to approximately 0.246 perms which is in the order of 12 times as much as the case when screws are present. Furthermore, the tests show that when the percentage of the deck area that has holes is relatively small, the permeability is raised significantly to the point where a hole area of approxi-

Table 4. Laboratory Test Results

Laboratory Specimen Dimensions and Test Results						Calculated Values	
Specimen Number	Specimen Description	Lap Length [in]	Lap Gap [in]	Air Barrier Present (Y/N)	ASTM E-96 Permeability [Perms - IP Units] (Perms - SI Units)	Calculated Permeability	Calculated Units
1	Solid deck	NA	NA	N	0 (0)	0.00	perms
2	2 fasteners	NA	NA	N	0.08 (4.58E-12)	0.02 1.06E-15	g_f = grains H_2O/h per in. Hg per fastener g_f = kg/s per Pa per fastener
3	1 screw hole	NA	NA	N	0.34 (1.95E-11)	0.17 9.02E-15	g_h = grains H_2O/h per in. Hg per hole g_h = kg/s per Pa per hole
4	2 holes	NA	NA	N	1.18 (6.77E-11)	0.30 1.57E-14	g_h = grains H_2O/h per in. Hg per hole g_h = kg/s per Pa per hole
5	3 holes	NA	NA	N	1.80 (1.03E-10)	0.30 1.59E-14	g_h = grains H_2O/h per in. Hg per hole g_h = kg/s per Pa per hole
6	12 holes	NA	NA	N	6.33 (3.62E-10)	0.26 1.40E-14	g_h = grains H_2O/h per in. Hg per hole g_h = kg/s per Pa per hole
7	25 holes	NA	NA	N	10.31 (5.90E-10)	0.21 1.09E-14	g_h = grains H_2O/h per in. Hg per hole g_h = kg/s per Pa per hole
8	3 holes, Air Barrier	NA	NA	Y	0.42 (2.40E-11)	0.07 3.70E-15	g_h = grains H_2O/h per in. Hg per hole g_h = kg/s per Pa per hole
9	25 holes, Air Barrier	NA	NA	Y	6.36 (3.64E-10)	0.13 6.75E-15	g_h = grains H_2O/h per in. Hg per hole g_h = kg/s per Pa per hole
10	Lap: 0.013 in. Gap	12	0.013	N	0.71 (4.06E-11)	0.71 1.24E-13	g_l = grains H_2O/h per in. Hg per ft of lap g_l = kg/s per Pa per m of lap
11	Lap: 1/32 in. Gap	12	1/32	N	1.15 (6.59E-11)	1.38 2.41E-13	g_l = grains H_2O/h per in. Hg per ft of lap g_l = kg/s per Pa per m of lap
12	Lap: 1/16 in. Gap	12	1/16	N	1.70 (9.74E-11)	2.04 3.57E-13	g_l = grains H_2O/h per in. Hg per ft of lap g_l = kg/s per Pa per m of lap
13	Lap: 1/8 in. Gap	12	1/8	N	5.05 (2.89E-10)	6.06 1.06E-12	g_l = grains H_2O/h per in. Hg per ft of lap g_l = kg/s per Pa per m of lap
14	Lap: 0.013 in. Gap, Air Barrier	12	0.013	Y	0.49 (2.80E-11)	0.49 8.55E-14	g_l = grains H_2O/h per in. Hg per ft of lap g_l = kg/s per Pa per m of lap
15	Lap: 1/32 in. Gap, Air Barrier	12	1/32	Y	2.76 (1.58E-10)	3.31 5.79E-13	g_l = grains H_2O/h per in. Hg per ft of lap g_l = kg/s per Pa per m of lap
16	Lap: 1/16 in. Gap, Air Barrier	12	1/16	Y	0.85 (4.83E-11)	1.01 1.77E-13	g_l = grains H_2O/h per in. Hg per ft of lap g_l = kg/s per Pa per m of lap
17	Lap: 1/8 in. Gap, Air Barrier	12	1/8	Y	1.48 (8.44E-11)	1.77 3.09E-13	g_l = grains H_2O/h per in. Hg per ft of lap g_l = kg/s per Pa per m of lap

Table 5. Permeability of In-situ Deck Panels for Laps Only

Lap Gap [in]	Panel Length [ft]	Panel Width [ft]	Length of Lap per Panel [ft]	Calculated Permeability per ft (m) of Lap (g_l)		Grains per Panel	Kilograms per Panel	Permeability of Assembly for Lap Only ($Perm_{lap}$)	
(a)	(b)	(c)	(d)	(e)	(f)	(g)	(h)	(i)	(j)
0.013	20	3	24.1	0.71	1.24E-13	17.11	9.10E-11	0.285	1.63E-11
0.013	30	3	34.1	0.71	1.24E-13	24.21	1.29E-10	0.269	1.54E-11
0.013	40	3	44.1	0.71	1.24E-13	31.31	1.67E-10	0.261	1.49E-11
1/32	20	3	24.1	1.38	2.41E-13	33.29	1.77E-10	0.555	3.18E-11
1/32	30	3	34.1	1.38	2.41E-13	47.10	2.51E-10	0.523	3.00E-11
1/32	40	3	44.1	1.38	2.41E-13	60.91	3.24E-10	0.508	2.91E-11
1/16	20	3	24.1	2.04	3.57E-13	49.25	2.62E-10	0.821	4.70E-11
1/16	30	3	34.1	2.04	3.57E-13	69.69	3.71E-10	0.774	4.44E-11
1/16	40	3	44.1	2.04	3.57E-13	90.12	4.80E-10	0.751	4.30E-11
1/8	20	3	24.1	6.06	1.06E-12	146.13	7.78E-10	2.436	1.40E-10
1/8	30	3	34.1	6.06	1.06E-12	206.77	1.10E-09	2.297	1.32E-10
1/8	40	3	44.1	6.06	1.06E-12	267.40	1.42E-09	2.228	1.28E-10

(d) Length of lap per panel includes an additional 1.1 ft (0.10 m) of lap to account for the added length along the profile of the corrugations compared to the steel deck coverage width at an end lap (refer to Figure 1 for typical dimensions of corrugations)
(e) IP units: g_l = Perms × Area ÷ Length = Grains H_2O/h per in. Hg per ft of Lap
(f) SI units: g_l = Perms × Area ÷ Length = kg/s per Pa per m of Lap
(g) IP units: Grains per panel = (Grains H_2O/h per in. Hg per ft of Lap) × (Length of Lap per panel)
(h) SI units: kg per panel = (kg/s per Pa per m of Lap)*(Length of Lap per panel)
(i) IP units: $Perm_{lap}$ = Grains per panel / (Panel Area) = Perms (grains H_2O/h per ft^2 per in. Hg)
(j) SI units: $Perm_{lap}$ = kg per panel / (Panel Area) = Perms (kg/s per m^2 per Pa)

Table 6. Permeability of In-situ Deck Panels for Fasteners Only

Number of Fasteners per 4 × 4 ft Insulation Board	Number of Fasteners per ft^2	Fasteners per 30 × 3 ft Deck Panel	Calculated Permeability per Fastener (g_f)		Grains per Panel	Kilograms per Panel	Permeability of Assembly for Fasteners Only ($Perm_{fasteners}$)	
(a)	(b)	(c)	(d)	(e)	(f)	(g)	(h)	(i)
4	0.25	22.5	0.020	1.06E-15	0.45	2.39E-14	0.005	2.86E-15
5	0.3125	28.1	0.020	1.06E-15	0.56	2.98E-14	0.006	3.57E-15
6	0.375	33.8	0.020	1.06E-15	0.68	3.58E-14	0.008	4.28E-15
8	0.5	45.0	0.020	1.06E-15	0.90	4.77E-14	0.010	5.71E-15
9	0.5625	50.6	0.020	1.06E-15	1.01	5.37E-14	0.011	6.42E-15
11	0.6875	61.9	0.020	1.06E-15	1.24	6.56E-14	0.014	7.85E-15
12	0.75	67.5	0.020	1.06E-15	1.35	7.16E-14	0.015	8.57E-15
14	0.875	78.8	0.020	1.06E-15	1.58	8.36E-14	0.018	9.99E-15
16	1	90.0	0.020	1.06E-15	1.80	9.55E-14	0.020	1.14E-14

(d) IP units: g_f = Grains H_2O/h per in. Hg per fastener
(e) SI units: g_f = kg/s per Pa per fastener
(f) IP units: Grains per panel = (Grains H_2O/h per in. Hg per fastener)*(Number of Fasteners/30 ft Panel)
(g) SI units: kg per panel = (kg/s per Pa per fastener)*(Number of Fasteners/9.14 m Panel)
(h) IP units: $Perm_{fasteners}$ = Grains per panel/(Panel Area) = Perms (grains H_2O/h per ft^2 per in. Hg)
(i) SI units: $Perm_{fasteners}$ = kg per panel/(Panel Area) = Perms (kg/s per m^2 per Pa)

Table 7. Permeability of In-situ Deck Panels for Empty Fastener Holes Only

Number of Holes per 4 × 4 ft Insulation Board	Number of Holes per ft^2	Holes per 30 × 3 ft Deck Panel	Calculated Permeability per Hole (g_h)		Grains per Panel	Kilogram per Panel	Permeability of Assembly for Holes Only ($Perm_{holes}$)	
(a)	(b)	(c)	(d)	(e)	(f)	(g)	(h)	(i)
4	0.25	22.5	0.297	8.92E-13	6.65	3.48E-11	0.074	4.17E-12
5	0.3125	28.1	0.293	8.81E-13	8.19	4.24E-11	0.091	5.07E-12
6	0.375	33.8	0.288	8.71E-13	9.68	4.94E-11	0.108	5.91E-12
8	0.5	45.0	0.280	8.50E-13	12.53	6.22E-11	0.139	7.44E-12
9	0.5625	50.6	0.277	8.40E-13	13.88	6.78E-11	0.154	8.11E-12
11	0.6875	61.9	0.267	8.19E-13	16.44	7.78E-11	0.183	9.30E-12
12	0.75	67.5	0.264	8.09E-13	17.65	8.20E-11	0.196	9.81E-12
14	0.875	78.8	0.255	7.88E-13	19.92	8.92E-11	0.221	1.07E-11
16	1	90.0	0.246	7.68E-13	22.00	9.44E-11	0.244	1.13E-11

(d) IP units: g_h = Grains H_2O/h per in. Hg per hole
(e) SI units: g_h = kg/s per Pa per hole
(f) IP units: Grains per panel = (Grains H_2O/h per in. Hg per hole) × (Holes/30 ft Panel)
(g) SI units: kg per panel = (kg/s per Pa per hole) × (Number of holes/9.14 m Panel)
(h) IP units: $Perm_{holes}$ = Grains per panel/(Panel Area) = Perms in IP units (grains H_2O/h per ft^2 per in. Hg)
(i) SI units: $Perm_{holes}$ = kg per panel/(Panel Area) = Perms in SI units (kg/s per m^2 per Pa)

Table 8. Effective Permeance of In-situ Deck Panels for Laps, Fasteners and Fastener Holes

Lap Gap in.	Permeance of Assembly for Lap Only (Perms)		Permeance of Assembly for Laps + 8 Fasteners per Insulation Board (Perms)		Permeance of Assembly for Laps + 8 Fasteners & 8 Holes per Insulation Board (Perms)	
(a)	(b)	(c)	(b)	(c)	(b)	(c)
0.013	0.27	1.54E-11	0.28	1.60E-11	0.42	2.40E-11
1/32	0.52	2.99E-11	0.53	3.05E-11	0.67	3.85E-11
1/16	0.77	4.43E-11	0.78	4.49E-11	0.92	5.29E-11
1/8	2.30	1.31E-10	2.31	1.32E-10	2.45	1.40E-10

(b) IP units: Perms (grains H_2O/h per ft^2 per in. Hg)
(c) SI units: Perms (kg/s per m^2 per Pa)

mately 0.02% of the total deck area raises the permeability above one perm, rendering the steel deck ineffective as a vapor retarder.

The presence of an air barrier reduces the permeability associated with laps and screw holes to a value that is approximately 1/2 the permeance when the air barrier is not present. In our tests we did not induce an air pressure difference across the sample, therefore, we expect an air barrier in an in-situ deck to have a much greater effect, since as we discussed in the introduction, it is well documented in the literature that even a small amount of airflow through small holes can transport significantly more moisture than diffusion alone. However, the tests with the air barrier demonstrate that the large increase in permeance due to the holes is not all due to air exchange through the holes. Our data with air barrier included also suggests that when concrete fill is used on the steel deck, the effect of laps and holes is diminished; however, decks that are made for concrete placement generally have many holes (perforations) that are present to allow the concrete to dry out from below.

Deck Penetrations and the Effect on Permeability

Figure 13 shows that holes that represent a very small percentage of the deck area can increase the permeance of the steel deck system significantly. From a practical standpoint, this means the deck cannot function as a vapor retarder. The holes studied in our tests are quite small, but in actual construction, deck penetrations and terminations

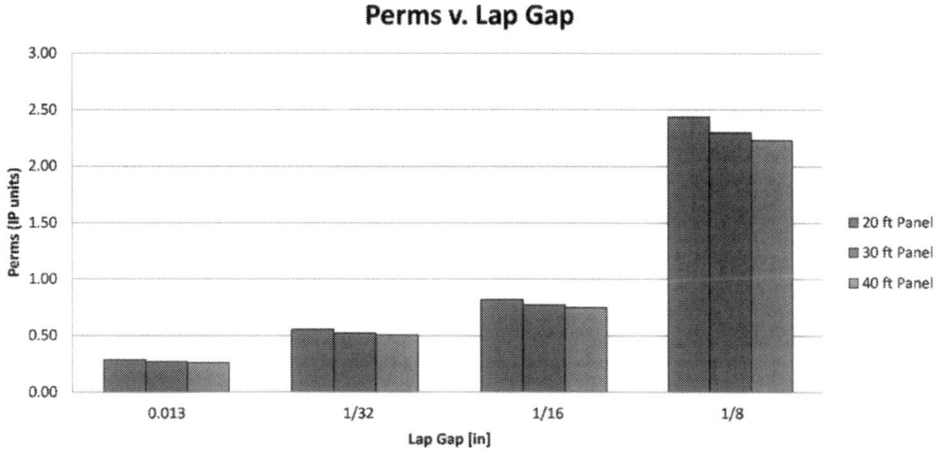

Figure 12 *Effect of gap dimension on deck permeance.*

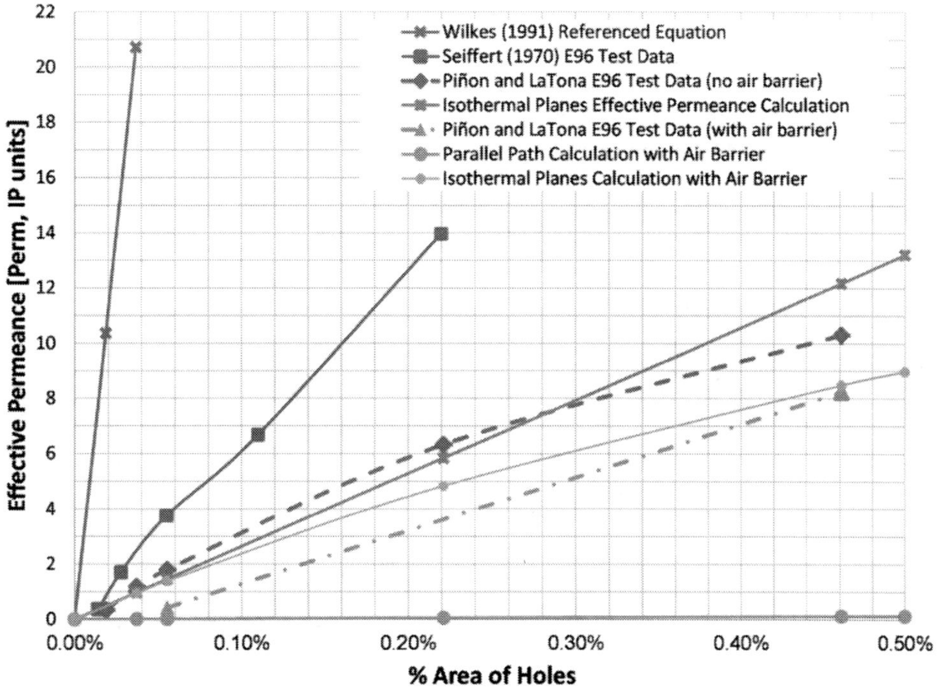

Figure 13 *Effect of percentage area of holes on deck permeance.*

can lead to a much larger percentage of holes in the deck area (Figures 4 and 5).

Figure 13 also shows how our data compares to the previous test data discussed in the introduction and that an estimate using a simple isothermal planes analogy (refer to Figure 8) matches our data relatively closely for area of holes less than 0.30%. At greater hole areas the simple isothermal planes analogy appears to over-predict the effective permeance. However, note the predicted and measured effective permeance of the deck with 0.30% area of holes is already about 8 perms, meaning that the steel deck already is too permeable to function as a vapor retarder.

How these Data Are Used in Hygrothermal Models

Our results reinforce the previous published data that it takes only a small percentage of holes to render a vapor retarder ineffective. However, there is still a question as to whether the effect of the holes would diminish as you add more layers to the exterior side of the steel deck as occurs in a roofing system and as predicted by the parallel path method calculation results shown in Table 2 and graphically shown in Figure 9. Our results with the air barrier membrane (one additional layer to the exterior of the steel) suggests that the effective permeance of a steel deck with holes in a roof system is more similar to that predicted by the isothermal planes method, meaning that a small percentage of holes will still render the steel deck ineffective as a vapor retarder.

Furthermore, we have performed extensive hygrothermal modeling that indicates that once the effective permeance of the steel deck is raised above 1 perm, it is ineffective as a vapor retarder for roofing systems where the roofing membranes are typically between 0.2 and 0.05 perms or even lower for multi-ply asphalt applied built-up roofing systems.

Based on these results and analysis, for steel decks without concrete fill, we conclude that the steel deck can be safely ignored in hygrothermal models in terms of both its diffusion or airflow resistance. If the hygrothermal analysis shows a vapor retarder and/or interior air barrier is needed, then the designer needs to specify a vapor retarder/air barrier system and not rely on the steel deck for this purpose.

For steel decks with concrete fill, the situation is not as clear cut. Table 1 shows that even a thin layer of concrete filling the holes in the steel deck can reduce the effect of the holes significantly. For concrete, the amount of permeance added by the holes would also be limited in practice by the rate of moisture redistribution with the adjacent areas of concrete that is not over a hole. However, the estimates shown Table 1 show that a steel deck "vented" with 1.5% area of perforations, can potentially increase its effective permeance from impermeable to 0.40 perms. These results are likely conservative since they assume the holes are completely filled with concrete. However, they provide a starting point for designers.

We have run hygrothermal models to show that an increase in the permeance of the steel deck from impermeable to 0.40 perms can assist the drying of the concrete to the interior. However, for most concrete roof decks in most climates there is still a significant drive of moisture from the concrete into the roofing system with or without a steel deck. Also, the concrete itself has a good amount of resistance to diffusion, the typical thicknesses used to fill steel roof decks results in a permeance of between 0.1 to 0.4 perms.

Therefore, our contention is that for concrete filled steel decks, the amount of moisture in the concrete is of more hygrothermal significance than knowing the exact effective permeance of the steel deck assembly. Until more experimental information is available on the benefit vented decks might provide to drying of concrete, we recommend that designers use an effective permeance for the steel deck assembly between 0.05 and 1.0 perms depending on the percentage of venting.

CONCLUSIONS

We conclude the following from our experimental results and analysis:

- The vapor resistance of a steel roof deck without concrete fill can be ignored in terms of its effect on the diffusion or airflow across a typical roofing system. The amount of gaps, holes, and unsealed seams render it ineffective as a vapor retarder.
- For concrete filled steel decks, we estimate the effective permeance to be between 0.05 and 1 perms depending on the area of venting (perforations) and other gaps and holes.
- In general, it only takes a small percentage area of holes (less than 0.1%) to render a vapor retarder ineffective, however, more research is needed if this same finding holds true for roof and wall systems that include multiple layers of building materials on the exterior side of the steel deck or vapor retarder.

RECOMMENDATIONS

We recommend that additional research and testing be performed on both wall and roof systems consisting of multiple layers on the exterior side of a vapor retarder to find out the relationship between percent area of holes and the effective permeance of the vapor retarder. If the finding that only a small percentage area of holes renders the vapor retarder ineffective, then that has significant implications on how we design, specify, and install vapor and air retarder membranes..

REFERENCES

ASHRAE. 2009. *ASHRAE Handbook-Fundamentals*. Atlanta: ASHRAE.

Kan, L., and Piñon, J., 2005. Predicting and Modeling the Effects of Wall Cladding Ventilation on Sun-Driven Condensation. Third International Building Physics Conference (ICBP3), Montreal

Seiffert, K. 1970. *Damp Diffusion and Buildings*. Amsterdam: Elsevier Publishing Company Ltd.

Steel Deck Institute (SDI). 2008. Venting of Composite Steel Floor Deck. Position Statement. Nov. 2008.

Straube, J.F., and Burnett, E.F.P. 1995. Vents, Ventilation Drying, and Pressure Moderation. Building Engineering Group, University of Waterloo. Ottawa: Canada Housing and Mortgage Corp. (CMHC).

Wilkes, K. 1991. *Analysis of Annual Thermal and Moisture Performance of Radiant Barrier Systems*, Oak Ridge National Laboratory. ORNL/CON-319.

Can VIP be an Alternative Solution for Energy Retrofit for Enhancing the Thermal Performance of Wood-Frame Walls?

Hamed H. Saber, PhD **Wahid Maref, PhD** *Member ASHRAE* **Ganapathy Gnanamurugan** **Mike Nicholls**

ABSTRACT

Field monitoring of the dynamic heat transmission characteristics of residential 2 × 6 wood-frame wall systems that had been retrofitted using vacuum insulation panels (VIPs) with glass fiber cores and extruded polystyrene foam (XPS) panels were undertaken in 2011–2012 at the Field Exposure of Walls Facility (FEWF) of the NRC Construction. The main objective of this research was to evaluate the steady-state and transient thermal performance of three wall assemblies (4 ft × 6 ft), two of which incorporated VIPs within an XPS tongue and groove (T&G) configuration and VIPs within an XPS clip-on (C-O) configuration, and a third assembly incorporating only XPS. The three wall assemblies were installed in the FEWF for one year cycle of exposure to outdoor natural weather conditions. The NRC's hygrothermal model, hygIRC-C, was validated against the measured data. Results showed that the model predictions were in good agreement with the experimental data. Given that the VIP could be punctured during the installation process (e.g., inadvertent use of fasteners in wall assembly) or may fail during normal operating conditions or due to imperfection during the manufacturing process, the numerical model was used to conduct parametric analyses in order to predict the thermal resistance (R-value) in cases where one or more VIPs failed.

In the final phase of this study, the model was used to predict the yearly cumulative heat losses from the respective wall systems. It is important to point out that the aging effect and the effect of the thermal bridging due to the envelope (i.e., skin) of the VIPs are not accounted for in this study. However, sensitivity analysis of the thickness and thermal conductivity of the VIP envelope was conducted in this paper to investigate the effect of these parameters on the effective R-values of VIP. Results showed that the yearly cumulative heating loads for the XPS retrofit wall specimen was 69.9% and 78.8%, respectively, greater than that obtained for the T&G VIP retrofit wall specimen and the C-O VIP retrofit wall specimen. Furthermore, the results showed that the effect of the furred-airspace assembly in the C-O VIP retrofit wall specimen when the emissivity of all surfaces that bounded the airspace was assumed to be 0.9 resulted in a reduction in the yearly heating load by 5% compared to the T&G VIP retrofit wall specimen without furred-airspace.

INTRODUCTION

Increasingly, home builders are turning towards a variety of construction and retrofit methods to improve thermal performance whilst reducing the operating cost of buildings. The energy crisis in the 1970s had a major effect on building technology related to energy performance. Thermal insulation of buildings became the key element to minimize heat losses, hence improving the energy performance. Adding more thermal insulation on the exterior wall and increasing the airtightness of the assembly was an evident solution adopted by many practitioners to achieve reduced energy usage in homes. In order to achieve better airtightness, high levels of envelope insulation will certainly be required with wall thermal resistances (R-values) of more than 3.33 $m^2 \cdot K/W$ (18.9 $ft^2 \cdot h \cdot °F/Btu$) and roof R-values of more than 5.0 $m^2 \cdot K/W$ (28.4 $ft^2 \cdot h \cdot °F/Btu$)

***Hamed H. Saber** is a research officer, **Wahid Maref** is a senior research officer, and **Ganapathy Gnanamurugan** and **Mike Nicholls** are technical officers at NRC Construction, National Research Council Canada, Ottawa, Ontario.*

(Maref et al. 2011, 2012a). Normally, the air leakage and airtightness across the wall systems depends on the type of insulation materials, air-barrier systems, and the pressure differential across the wall systems (Maref et al. 2011, 2012a; Elmahdy et al. 2009, 2010; Saber et al. 2010a, 2012a).

To achieve high levels of energy performance with existing wall systems and materials, the wall systems would need to be considerably thicker and this may be less acceptable to consumers because of the loss of usable floor area, greater weight, increase in costs due to transportation, and time of construction, as well as new challenges for the structure associated with greatly increased wall thickness. To maintain reasonable envelope thickness while having high thermal performance, a promising recent innovation in building technology was investigated within the context of its application using Vacuum Insulation Panel (VIP) systems. Due to high R-values per unit thickness of VIPs, their potential is considered to be a promising means of insulating the building envelope to high levels of thermal resistance. By considering the center of the panel, VIPs are of interest owing to their exceptional insulating R-value, up to 10.6 $m^2 \cdot K/W$ (60 $ft^2 \cdot h \cdot °F/Btu$) per inch or even higher (Ogden and Kendrick 2005; Maref et al. 2011, 2012a). This concept is gaining interest and is being considered by architects and building practitioners for new and retrofit construction (Mukhopadhyaya 2010). In the case of accounting for the thermal bridging due the envelope (i.e., skin) of the VIP, the effective R-value might be lower than that at the center of the panel.

Despite their high R-values at the center of the panels, VIPs have been slow to make inroads into construction because of three drawbacks: cost, the need to protect them against puncturing, and the absence of long-term performance data or procedures to predict their long-term performance. Recently, Canadian, European, and Chinese research efforts have mainly focused on the core materials and the envelopes of the VIPs themselves (Cremers 2005; Mukhopadhyaya 2006; Mukhopadhyaya et al. 2008, 2009; Simmler and Brunner 2005a, 2005b; Brunner and Simmler 2008; Stahl et al. 2012; Schwab et al. 2005; Ghazi Wakili et al. 2004; Xiaobo et al. 2013). Also, research has now started to investigate how to integrate VIP insulation in building construction in such a way that the thermal advantages are fully used. Preliminary efforts have been published in IEA Annex 39 on High Performance Thermal Insulation (IEA Annex 39 2005).

The VIP technology can be used in retrofitting the existing homes, for example, either at the interior or at the exterior of the wall assemblies (Binz and Steinke 2005; Rogatzki 2009). It can also be used in new construction such as wood-frame construction in double stud walls (Maref 2011; Haavi et al. 2012). Most of the VIP activities in Canada are still in the research and development phase with some demonstration projects that have been recently completed (Maref et al. 2011, 2012a). Germany and Switzerland are the first two countries where VIP technology has been introduced to the building construction sector, through research efforts such as those of the IEA Annex 39 (2005).

Cremers (2005) pointed out that since VIP technology is considerably more costly than established insulating technologies, in the future measures such as improved production capacity, efficient manufacturing techniques, and alternative and less expensive core material of the VIP will likely contribute in making VIP insulation more cost-effective for building envelopes. Several prototype projects have been completed or are being carried out in Europe using VIP technology (IEA Annex 39 2005). Further technology developments in the area of insulation technology and recent global concerns about climate change phenomena showed that there is room and need for improvement in the insulation technology for building envelopes. Hence, the National Research Council (NRC) of Canada and other organizations and laboratories around the world have established research programs and studies that focused on the applicability of high performance VIPs in building envelope construction.

OBJECTIVE

Based on PERD (Program on Energy Research and Development) research efforts on "Next Generation Building Envelope Building Systems" by NRC, Natural Resources Canada (NRCan), and Canada Mortgage and Housing Corporation (CMHC), it appears that VIP-based insulation systems are capable of reducing building energy consumption in Canada (Maref et al. 2011, 2012a). The main objective of this paper is to use a numerical model in order to evaluate the dynamic heat transmission characteristics of three mid-scale 2 × 6 wood-frame wall assemblies, two of which incorporated VIPs within an XPS tongue and groove (T&G) configuration and VIPs within an XPS clip-on (C-O) configuration, and a third assembly incorporating only XPS, in the Field Exposure of Walls Facility (FEWF) of NRC Construction, located in Ottawa for one year cycle of exposure to outdoor natural weather conditions. The numerical model was validated against the measured data. The measured data was obtained from field monitoring of the dynamic heat transmission characteristics of these wall systems that were tested in period between May 2011 and May 2012 in the FEWF. Given the fragile nature of VIPs (they could be punctured during the installation process or may fail during normal operating conditions), the numerical model was used to conduct parametric analyses to predict the R-value in cases where one or more VIPs failed. Also, the model was used to investigate the reduction in R-value of walls incorporating VIPs when one or more VIPs were replaced by XPS layer of the same thickness as the VIP. Finally, the model was used to predict the yearly cumulative heat losses from the respective wall systems.

DESCRIPTION OF RETROFITTED WALL SPECIMENS

Three wall specimens (4 ft × 6 ft) conforming to typical residential wood-frame construction were installed side-by-side in the FEWF. The different material layers and the dimensions of the wall specimens are shown in Figures 1 and 2. The backup wall for all three retrofit strategies consists of interior drywall (1/2 in. thick), polyethylene air barrier (6 mil), 2 × 6 wood-frame with friction-fit glass fibre batt insulations, OSB (7/16 in. thick), and Tyvek sheathing membrane. The backup wall was retrofitted by adding different types of external insulations. The first wall (W1) was retrofitted by adding an XPS layer (2 in. thick) between the sheathing membrane and vinyl siding. The other two walls were retrofitted with vacuum insulation panels (VIPs) using two concepts as described below. No information is available about the type of the material of the VIP envelope (i.e., skin) and the degree of vacuum inside the VIP.

In the second retrofit concept, each VIP having nominal thickness of 15 mm (5 panels in total) was sandwiched between an exterior XPS board (1 in. thick) and an interior XPS board (5/8 in. thick). Note that VIPs of thickness of 20 mm and thicker up to 40 mm are common in Germany and Switzerland. To protect the VIP, a hollow piece of XPS of the same thickness as the VIP was cut and the VIP panel was placed inside the opening such that the VIP would be protected by the XPS surround. The effect of thermal bridging of the XPS surround on the effective R-value of the wall assembly was accounted for by the numerical model. The assembly consisting of the VIP and XPS surround was placed between the exterior and interior XPS layers to form a tongue and groove (T&G) VIP "sandwich," as shown in Figure 3. A thin glue layer was applied on the surfaces of the VIP to ensure good contact between the VIP and the XPS layers. Five T&G VIP sandwiches were assembled (friction-fit at the tongue and groove interfaces, see Figure 3c). Then the entire assembly of T&G VIPs was placed vertically between the sheathing and smart board for the retrofitted wall specimen. The smart board used in study, called *waferboard*, is a class of mat-formed structural panels made predominantly of wood wafers.

In the third retrofit concept (see Figure 4), the VIP sandwich was similar to the VIP sandwich described in the second concept but without the tongue and groove assembly. In this concept, five VIP sandwiches were assembled using clips. Hence, the retrofitted wall specimen was called clip-on (C-O) VIP. To ensure good contact, a foamed polychloroprene (neoprene) gasket was inserted between the VIP sandwiches as shown in Figure 4. The C-O VIP assembly was stacked vertically against the surface of the sheathing membrane. Vertical furring strips (16 inch o.c. and 5/8 inch thick) were attached to metal clips, which supported the C-O VIP assembly and provided the nailing surface to which the smart board (waferboard) was attached. Detailed descriptions for the instrumentation (i.e., thermocouples, heat flux transducers, pressure sensors, and RH sensors) of the reference wall (W1), the C-O VIP (W2), and the T&G VIP (W3) are available in Maref et al. (2012b). As a part of the test protocol, all thermocouples used in the three test specimens were calibrated. Also, all heat flux transducers were calibrated according to the ASTM C-1130 *Standard Practice for Calibrating Thin Heat Flux Transducers* (ASTM 2009). The uncertainties of temperature and heat flux measurements were ± 0.1°C and ±5%, respectively.

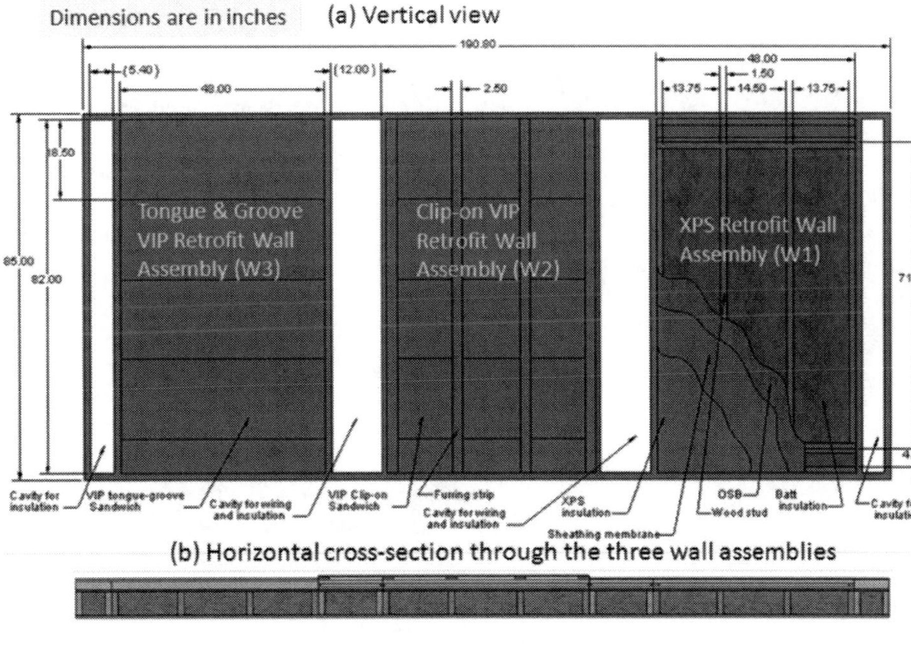

Figure 1 *Schematic of three residential 2 in. × 6 in. wood-frame wall test specimens installed side by side in the Field Exposure of Walls Facility (FEWF) of NRC Construction.*

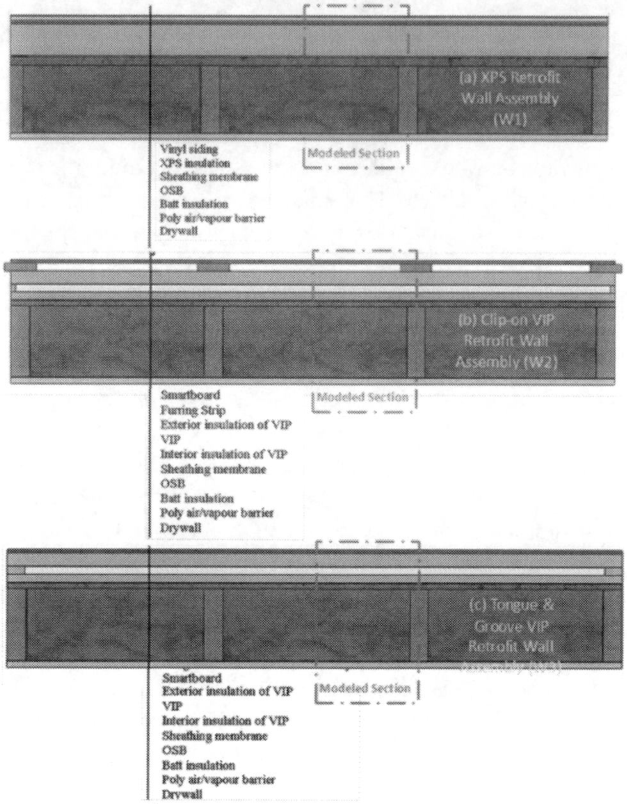

Figure 2 *Horizontal cross-section through different wall systems.*

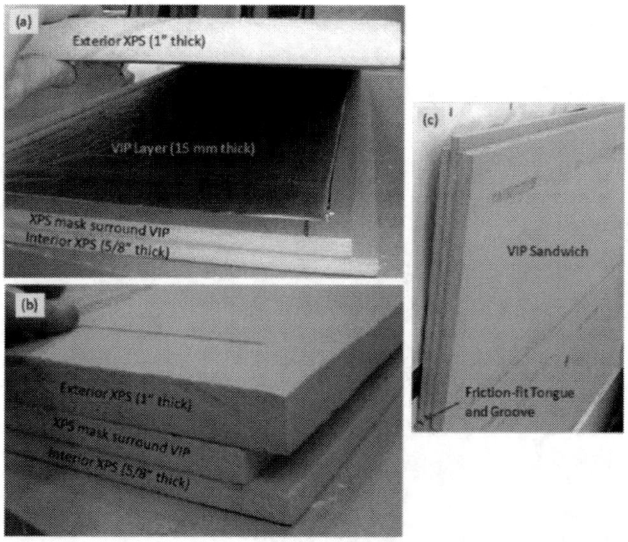

Figure 3 *Tongue and groove (T&G) VIP assembly (W3).*

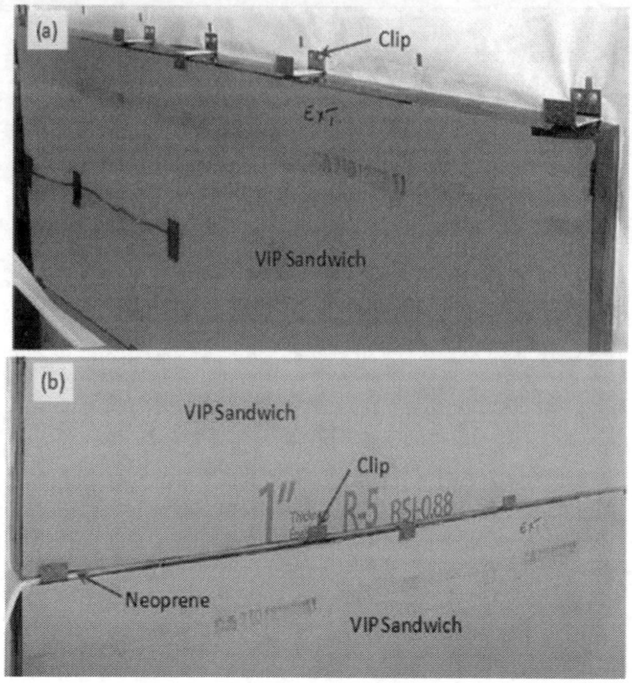

Figure 4 *Clip-on (C-O) VIP assembly (W2).*

MODEL DESCRIPTION AND VALIDATION

The numerical model that was used in this study simultaneously solves the highly nonlinear 2D and 3D heat, air, and moisture (HAM) equations. This model has been extensively validated in a number of other projects and has been used in several related studies to assess the thermal and hygrothermal performance of wall and roofing systems (Elmahdy et al. 2009, 2010; Saber and Swinton 2010; Saber and Laouadi 2011; Saber et al. 2010 a, 2010b, 2010c, 2011a, 2011b, 2011c, 2012 a, 2012b, 2012c, 2012d; Armstrong et al. 2011a, 2011b; Saber and Maref 2012; Saber 2012, 2013). The 3D version of this model was used to conduct numerical simulations for different full-scale wall assemblies incorporating, or not, penetrations representative of a window installation, such that the effective R-value of the assemblies could be predicted, taking into consideration air leakage across the assembly. The predicted R-values for these walls, which incorporated different types of spray polyurethane foams or glass fiber insulation, were in good agreement (within ±5%) with the measured R-values that were obtained from testing in a guarded hot box (GHB) (Elmahdy et al. 2009, 2010; Saber et al. 2010a, 2012a).

The present model was also validated against GHB test results (Saber et al. 2011a) and then used to conduct numerical simulations to investigate the effect of foil emissivity on the effective thermal resistance of different wall systems with foil bonded to different types of thermal insulations placed in furred assemblies, in which the foil was adjacent to the air-space (Saber and Swinton 2010; Saber and Maref 2012; Saber et al. 2012b). Also, the present model was validated and used to assess the thermal performance of insulated concrete form (ICF) wall systems when placed in the FEWF and subjected to yearly periods of Canadian climate (Saber et al. 2010c, 2011c; Armstrong et al. 2011a, 2011b). In these studies (Elmahdy et al. 2009, 2010; Saber and Swinton 2010; Saber et al. 2010a, 2010c, 2011a, 2011c, 2012a, 2012b; Saber and Maref 2012; Armstrong et al. 2011a, 2011b), no moisture transport was accounted for in predicting the thermal performance of different types of walls.

In instances where the model was used to account for moisture transport across wall assemblies, this model predicted the drying rate of a number of wall assemblies subjected to different outdoor and indoor boundary conditions (Saber et al. 2010b) in which there was a significant vapour drive across the wall. The results showed that there was overall agreement between the results derived from the present model and the hygIRC-2D model, a model that had been previously developed and validated at the NRC (Maref et al. 2002). The present model predictions were in good agreement with the experimental measurements of the drying and drying rate of the assembly in respect to the shape of the curve and the length of time predicted for drying. As well, the predicted average moisture content of the different wall assemblies over the test periods were in good agreement, all being within ±5% of those measured (Saber et al. 2010b). Furthermore, in respect to the prediction of the hygrothermal performance of roofing systems, the present model was used to investigate the moisture accumulation and energy performance of reflective (white) and non-reflective (black) roofing systems that were subjected to different climatic conditions of North America (Saber et al. 2011b, 2012d).

Having previously validated the present model against experimental data of several tests undertaken in controlled laboratory conditions, a subsequent and important step in this study was to validate the present model against field measurements. Thereafter, the general parameters affecting the thermal response of wall specimens is discussed, information is provided regarding assumptions, and initial and boundary conditions that were used in conducting the numerical simulations. Finally, the estimated yearly cumulative heating and cooling loads by the model due to the respective wall specimens are discussed.

GENERAL PARAMETERS AFFECTING THE THERMAL RESPONSE OF WALL SPECIMENS

The thermal response of the wall specimens depends on both the thermal properties of all material layers of the wall specimens and the outdoor and indoor conditions. All thermal properties of the different layers of which the wall assemblies were composed, with the exception of the vacuum insulation panels (VIPs), were taken from *ASHRAE Handbook of Fundamentals* (ASHRAE 2009). However, the properties of the VIPs had been previously characterized in another project (see references Maref et al. [2011, 2012a] for more details). The thermal conductivity of the VIP layer was measured at different temperatures. Two specimens of VIP having a nominal thickness of 15 mm and length and width of 1200 mm and 455 mm, respectively, were used to measure the thermal conductivity at three mean temperatures. The measured density of the VIP samples was 261.2 kg/m^3. It is important to point out that the type of the VIP that was used in this project is proprietary product. The details about the composition of this type of VIP are not available (such as degree of vacuum inside the VIP, type of fiber core material inside the VIP, and type of material and its thickness of the VIP envelope).

The test method used to measure the thermal conductivity of the VIP was ASTM C-518 (ASTM 2007). The use of heat flowmeter apparatus according to the ASTM C-518 test method when there are thermal bridges present in the assembly may yield results that are unrepresentative of the assembly (see ASTM [2007] for more details). According to the ASTM C-518 test method (ASTM 2007), the thermocouples embedded in the surfaces of the upper and lower plates (12 × 12 in. ([305 × 305 mm]) of the heat flowmeter measure the temperature drop across the specimen and the heat flux transducer (HFT) embedded in each plate measures the heat flow through the specimen. The size of HFT in the NRC's heat flowmeter is 6 × 6 in. (152 × 152 mm). The uncertainty of the measurements in this test method is ±2%. As indicated earlier, the size of the VIP sample is 1200 × 455 mm, which is larger than the size of upper and bottom plates of the heat flowmeter (305 × 305 mm). As such, the measured R-value of the VIP using the heat flowmeter is not the effective R-value because the effect of thermal bridging due to the VIP edges is not accounted for. The limitations of using the test method of ASTM C-518 (ASTM 2007) for measuring the R-values of samples with thermal bridging are recently addressed by Saber (Saber 2012; Saber at al. 2012c). Alternatively, the ASTM C-1363 test method using the GHB can be used to determine the effective thermal resistance of samples with thermal bridges (ASTM 2006).

In this paper, the values of the thermal resistance and the corresponding thermal conductivity of the VIP are applicable to the center core of the VIP only. These values are applicable to whole VIP only in one case when the thermal conductivity of VIP envelope is the same as the thermal conductivity of the center core of the VIP. Figure 5 shows the dependence of the thermal conductivity of the center core of the VIP samples on the temperature. The measured thermal conductivity of VIP layer, λ in W/(m·K), as a function of temperature, T in °C, that was used in the numerical simulation is given as:

$$\lambda = a + b \times T,\ a = 0.002054,\ b = 7.03088 \times 10^{-6} \qquad (1)$$

The above correlation of λ is in good agreement with all measured values at different temperatures (within ±1.6%). The corresponding average R-value of the center core of the VIP sample (15 mm thick) is 7.03 m^2·K/W (39.9 ft^2·h·°F/Btu).

To quantify the effect of the thermal bridging due to the envelope of the VIP on its effective R-value, 2D numerical simulations were conducted for VIPs of 15 mm thick, different heights (H) of 400 mm and 800 mm, and different envelope thicknesses (δ_{skin}) of 0.1 mm and 0.2 mm. The 2D numerical simulations represent the case of using VIP of a large width. In these simulations, the thermal conductivity of the VIP core was taken equal to the measured value using the ASTM C-518 test method (0.00214 W/[m·K]), whereas a wide range was considered for the thermal conductivity of the VIP envelope (ranging from 0.00214–10.0 W/[m·K]). Figure 6 shows the dependence of the effective R-values of the VIP (R_{VIP}) of different H and δ_{skin} on the thermal conductivity of its envelope (λ_{skin}). As shown in this figure, the parameters H, λ_{skin}, and δ_{skin} have a significant effect on the effective R-value of the VIP. For example, for VIP of δ_{skin} = 0.2 mm and H = 400 mm, the R_{VIP} decreases by 33% (from 39.9 ft²·h·°F/Btu to 30.0 ft²·h·°F/Btu) as λ_{skin} increases from 0.00214 W/(m·K) to 1.0 W/(m·K). At λ_{skin} = 1.0 W/(m·K) and H = 400 mm, increasing δ_{skin} from 0.1 mm to 0.2 mm resulted in a decrease in the R_{VIP} by 13% (from 33.8 ft²·h·°F/Btu to 30 ft²·h·°F/Btu). Furthermore, at λ_{skin} = 1.0 W/(m·K) and δ_{skin} = 0.1 mm, increasing H from 400 mm to 800 mm resulted in an increase of R_{VIP} by 7.4% (from 33.8 ft²·h·°F/Btu to 36.3 ft²·h·°F/Btu). In summary, the results shown in Figure 6 clearly indicated that the thermal bridging due the envelope of the VIP can have a significant effect on its effective R-value. This effect, however, was neglected in this study due to lack of information about the envelope of the VIP. Also, due to lack of information about the change of the degree of vacuum inside the VIP with time, the aging effect on the effective R-value was not accounted for in this study.

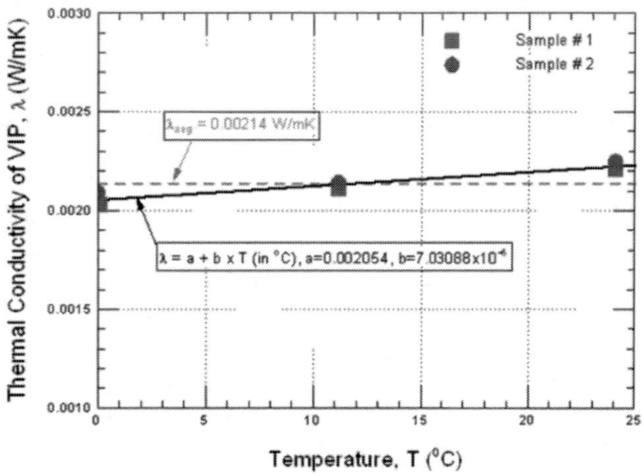

Figure 5 *Measured thermal conductivity at the center core of the VIP samples (15 mm thick).*

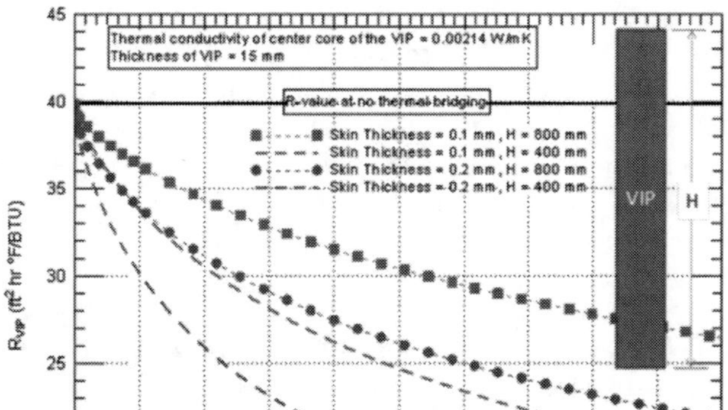

Figure 6 *Effect of thermal bridging due to VIP envelope on its effective R-value.*

The thermal conductivity and R-value of the VIP were also measured using the ASTM C-518 test method (ASTM 2007) when the VIP failed. These measurements were conducted in order to investigate the reduction in R-value of a retrofitted wall specimen incorporating VIPs for which one or more VIPs failed. The measured values of thermal conductivity and R-value when the VIP failed were 0.0257 W/(m·K) and 3.32 ft²·h·°F/Btu, respectively. It is important to point out that the thermal conductivity of the VIP when it failed is approximately equal to the thermal conductivity of still air. Furthermore, when the VIP failed, its thermal conductivity is significantly increased by a factor of ~12 (increased from 0.00214 W/[m·K] to 0.0257 W/[m·K]); and this resulted in a significant reduction in its R-value (from 39.9 ft²·h·°F/Btu to 3.32 ft²·h·°F/Btu). Consequently, consideration and care should be given in handling and installing VIPs in the wall systems so as to minimize the risk of puncturing them. The present model was used to predict the steady-state and transient thermal performance of different wall systems (W1, W2, and W3).

ASSUMPTIONS

It was assumed that all material layers are in perfect contact (i.e., the interfacial thermal resistances between all material layers were neglected). Due to lack of information about the type of VIP that was used in this project, the effects of aging and thermal bridging due to the envelope of the VIP were neglected. Note that the constructive thermal bridges in the different wall systems (W1, W2, and W3) were obviously considered in this study. The emissivity of all surfaces that bounded the airspaces (i.e., airspaces between XPS layer and vinyl siding for W1 and airspaces between vertical furring, C-O VIP assembly, and waferboard for W2) was taken equal to 0.9 (ASHRAE 2009). The effect of heat transfer by conduction, convection, and radiation inside these airspaces on the thermal performance of W1 and W2 are accounted for.

INITIAL AND BOUNDARY CONDITIONS

The initial temperature in all material layers of specimens W1, W2, and W3 was assumed uniform and equal to 10.0°C. Since this initial temperature was not the same as in the test, it was anticipated that the predicted dynamic response of the different wall specimens in the first period of the test (say first 24–48 h) would be different from that obtained in the test itself. Due to symmetry, only one module of each wall system was modeled, which is formed from a vertical plane passing through the middle of the stud and a vertical plane passing through the middle of the stud cavity (see the green dashed boxes shown in Figure 2a, b, c). The boundary conditions on these vertical planes are adiabatic due to symmetry. The boundary conditions on the top and bottom surfaces of one module of the wall systems are adiabatic (i.e., no edge heat losses).

In this study, two types of numerical simulations were conducted:

- Steady-state numerical simulations to determine the R-values of different wall specimens. In this case, the outdoor surface of the vinyl siding for W1 and smart board (waferboard) for W2 and W3 were subjected to convective boundary conditions with an air temperature of –35°C and heat transfer coefficient of 34.0 W/($m^2 \cdot K$) (ASTM 2006). Similarly, the indoor surface of the gypsum board for all wall systems was subjected to a convective boundary condition with constant air temperature and heat transfer coefficient of 21.0°C and 8.29 W/($m^2 \cdot K$), respectively (ASTM 2006). The boundary conditions on the outdoor and indoor surfaces of the wall specimens are similar to the case of measuring the R-value using the guarded hot box (GHB) in accordance with ASTM C-1363 test method (ASTM 2006).
- Transient numerical simulations were conducted in order to validate the present model and then determine the yearly heating and cooling loads for different wall specimens. In this case, the outdoor surface of the vinyl siding for W1 and smart board (waferboard) for W2 and W3 were subjected to temperature boundary conditions. Similarly, the indoor surface of the gypsum board for all wall systems was subjected to a temperature boundary condition. The temperatures on the outdoor and indoor surfaces of different wall specimens (changed with time) were taken equal to that measured on these surfaces (see reference Maref et al. [2012b] and Saber et al. [2012e] for more details). As an example, these temperatures are shown in Figure 7 for wall W1.

THERMAL RESISTANCES OF DIFFERENT WALL SPECIMENS

As indicated earlier, the present model was extensively validated in previous studies (Elmahdy et al. 2009, 2010; Saber 2012a, 2010a; Saber 2012; Saber et al. 2012c) by comparing the predicted R-values against the measured R-values obtained using two standard test methods: (a) guarded hot box in accordance with ASTM C-1363 (ASTM 2006), and (b) heat flowmeter in accordance of ASTM C-518 (ASTM 2007). Results showed that the predictions of the present model were in good agreement with the experimental data (within the uncertainties of the ASTM C-1363 test method [ASTM 2006] and ASTM C-518 test method [ASTM 2007]). After gaining confidence in the present model, it was used to predict the R-values of the three residential 2 × 6 wood-frame wall specimens that were retrofitted using thermal insulations of XPS without VIP (W1, W2, and W3).

All R-values presented in this paper for different wall specimens are the surface-to-surface R-values. Figure 8 shows a comparison between the R-values for different wall specimens. As shown in this figure, the XPS retrofit wall assembly (W1) resulted in the lowest R-value (29.6 $ft^2 \cdot h \cdot °F/Btu$); whereas the C-O VIP retrofit wall assembly (W2) resulted in

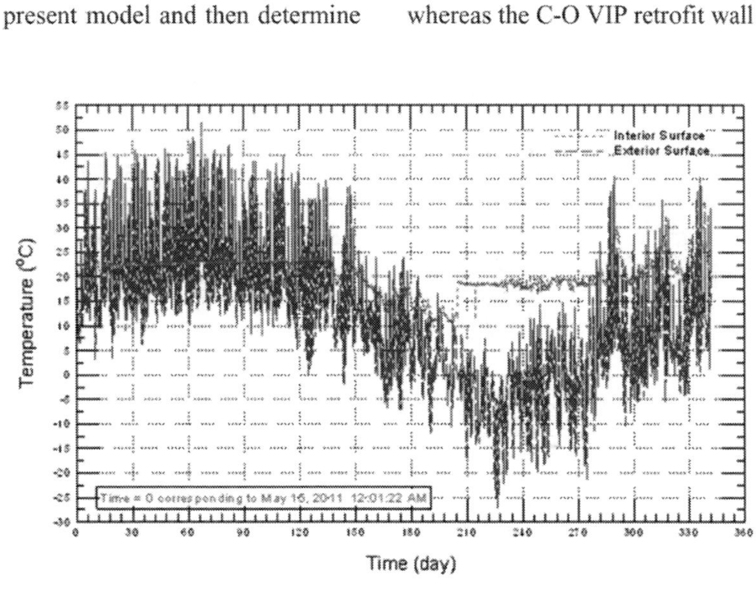

Figure 7 *Temperature measurements on the indoor and outdoor surfaces of the XPS retrofit wall assembly (W1).*

the highest R-value (55.4 $ft^2 \cdot h \cdot °F/Btu$). The R-value of the T&G VIP retrofit wall assembly, W3 (without furred-air-space) was 53.8 $ft^2 \cdot h \cdot °F/Btu$, which is lower than that for W2 by 1.58 $ft^2 \cdot h \cdot °F/Btu$. This means that the furred-air-space in the C-O VIP retrofit wall assembly (W2) having surface emissivity of all surfaces bounded in the airspace of 0.9 (ASHRAE 2009) contributed to the R-value by a value of 1.58 $ft^2 \cdot h \cdot °F/Btu$. However, installing foil with low emissivity on the surfaces of the furred-airspace would result in further enhancement of the thermal performance of the C-O VIP retrofit wall assembly.

As explained in Maref et al. (2012b), the test results of the C-O VIP retrofit wall specimen (W2) showed that one of the VIPs failed during normal operation (at time = 202 day). At time = 271 day, this VIP was subsequently replaced by a new VIP component once it became apparent that the expected thermal performance of the wall was not being achieved. The R-value of the failed VIP was measured using the ASTM C-518 test method (ASTM 2007) and found to reduce the R-value from 39.9 $ft^2 \cdot h \cdot °F/Btu$ to 3.32 $ft^2 \cdot h \cdot °F/Btu$ (a factor of ~12). Consequently, because there is always a risk of puncturing VIPs either during the installation process or over the course of its in-service use, a parametric study was conducted to predict the R-value of the wall assembly when one or more VIPs is punctured or failed by other means. Furthermore, because the VIP is more expensive than an XPS panel, this parametric analysis was conducted to determine the R-value when one or more VIPs is replaced by an XPS panel of the same dimension as that of the VIP. In these analyses, six cases were considered. These cases are shown in Figure 9a through Figure 9f for the T&G VIP wall specimen (W3). In Figure 9f, the case-VI represents the situation when the five VIPs in Figure 9a either failed or replaced by an XPS layer.

Comparisons of R-values of the T&G VIP wall specimen (W3) for case-I through case-VI (see Figure 9) when the VIP failed or replaced by an XPS layer are shown in Figure 10. This approach is similar to the approach by Nussbaumer et

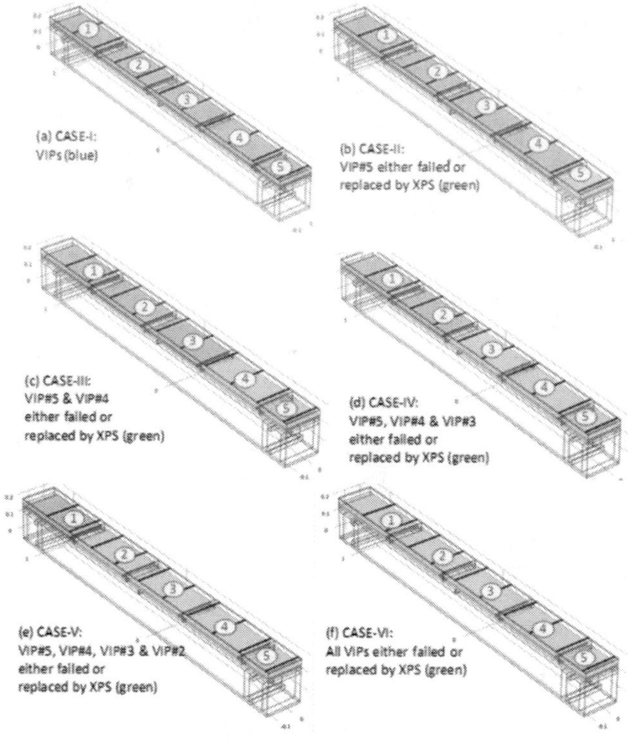

Figure 9 *Schematics of different cases when VIP either failed or replaced by XPS panel (see the green dashed boxes in Figures 2a, 2b, and 2c).*

Figure 8 *Comparisons between the R-values of different wall specimens without considering the effect of thermal bridging due to the envelope of the VIPs.*

Figure 10 *Comparisons between R-values for case I through case VI (see Figure 9) of T&G VIP wall specimen (W3) without considering the effect of thermal bridging due to the envelope of the VIPs.*

al. (2006). Also, Table 1 shows the reduction in the R-values when the VIP failed or replaced by an XPS layer.

It is important to indicate that when all VIPs failed (case-VI, Figure 10), the R-value of the T&G VIP wall specimen (R = 30.139 ft²·h·°F/Btu) is only 1.8% higher than that for the reference wall (XPS retrofit wall specimen, R = 29.608 ft²·h·°F/Btu, see Figure 8). Therefore, consideration must be given when handling and installing VIPs for retrofitting wall specimens to minimize the risk of puncturing the VIP.

As indicated earlier, the measured thermal conductivity and R-value when the VIP failed were 0.0257 W/(m·K) and 3.32 ft²·h·°F/Btu, respectively. On the other hand, the R-value of an XPS layer (thermal conductivity = 0.029 W/[m·K] [ASHRAE 2009]) of the same thickness as VIP (15 mm thick) is 2.94 ft²·h·°F/Btu, which is approximately equal to the R-value of the VIP when it failed. As such, the reduction in the R-value when replacing VIP by XPS layer is approximately the same as when the VIP failed (see Figure 10). For example, for the situation when the VIP is replaced by XPS layer, Figure 10 and Table 1 show that the R-values are reduced by 8.3%, 25.1%, 41.8%, 58.6%, and 76.2% for case-II, case-III, case-IV, case-V, and case-VI, respectively, which are approximately the same as when the VIP failed. Hence, unless the VIP panel itself can be replaced, there is no benefit with respect to thermal performance of the overall assembly if replacing it with XPS panel. Note the effect of the thermal bridging due the envelope of VIP on the total R-values of the wall systems was not accounted for. The above comparisons between the performance of different wall specimens may change in the case of accounting for the effect of thermal bridging due to the VIP envelope.

Table 1. Comparisons Between R-Values for Case I through Case VI (see Figure 9) of T&G VIP Wall Specimen (W3)

	R-Value of Wall Assembly		Reduction % in R-Value Due to	
Case	Having Failed VIPs (ft²·h·°F/Btu)	Having VIPs Replaced by XPS (ft²·h·°F/Btu)	Failed VIPs	Replacing VIP by XPS
I	53.784	53.784	0%	0%
II	49.545	49.653	8.50%	8.30%
III	42.761	42.982	25.80%	25.10%
IV	37.641	37.923	42.90%	41.80%
V	33.593	33.906	60.10%	58.60%
VI	30.139	30.468	78.50%	76.20%

During the test period, it was noted that the VIP#2 in the C-O VIP wall specimen had failed at time = 202 day (Maref et al. 2012b; Saber et al. 2012e). Numerical simulations were conducted to predict the R-value for this situation for both the T&G VIP and the C-O VIP wall specimens. For T&G VIP wall specimen (R = 53.784 ft²·h·°F/Btu), Figure 11 shows that the R-value was reduced to R = 45.876 ft²·h·°F/Btu in the case where VIP#2 only failed and R = 30.468 ft²·h·°F/Btu when all VIPs were replaced by XPS layers (a reduction of 17.2% and 76.2%, respectively). Similarly, for the C-O VIP wall specimen, Figure 12 shows that the R-value (R = 55.361 ft²·h·°F/Btu) was reduced by 16.5% and 75.5% in the cases when VIP#2 failed (R = 47.503 ft²·h·°F/Btu) and when all VIPs were replaced by XPS layers (R = 31.545 ft²·h·°F/Btu), respectively.

DYNAMIC HEAT TRANSMISSION IN DIFFERENT WALL SPECIMENS

In order to validate the present model, and hence use it to determine the yearly heating and cooling loads of different wall specimens, transient numerical simulations were conducted for the three wall specimens (W1, W2, and W3). In these simulations, the initial temperature in all material layers of the wall specimens was assumed uniform and equal to 10.0°C. Because this initial temperature was not the same as that was recorded in the test, it was anticipated that the predicted dynamic response of the different wall specimens during the first period of the test (say the first 24–48 h) would be different than that obtained from the measurements. For the purpose of validating the present model, the indoor and outdoor surfaces for all wall

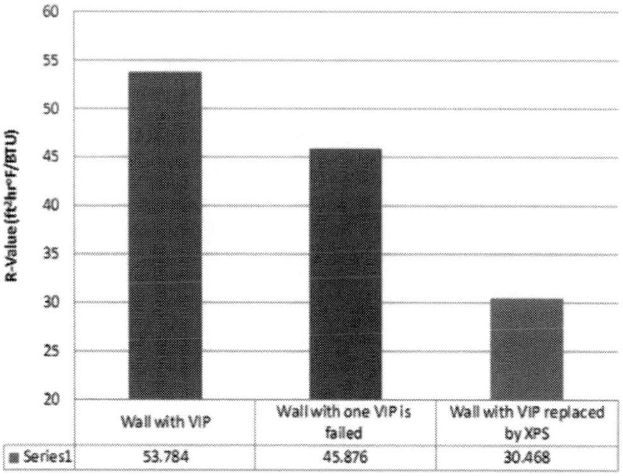

Figure 11 Comparisons between R-values for case I, when VIP#2 failed, and all VIP were replaced by XPS (see Figure 9) of T&G VIP wall specimen (W3).

systems were subjected to temperature boundary conditions. As an example, the temperatures on the outdoor and indoor surfaces of wall specimen W1 (changed with time) are shown in Figure 7. A description of all instrumentation and measurements are available in Maref et al. (2012b). In each wall system, three heat flux transducers (HFTs) were used to measure the heat flux at the middle (mid-height and mid-width) of each wall at three interfaces, namely (Maref et al. 2012b; Saber et al. 2012e):

- HFT1 at the cladding: XPS interface (cladding type: vinyl siding for W1, smart board for W2 and W3);

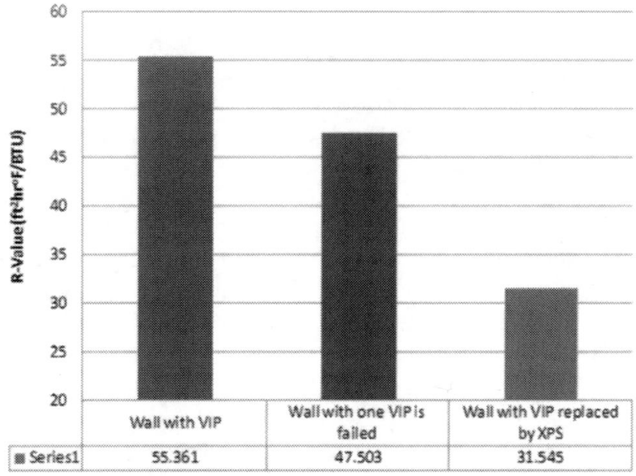

Figure 12 *Comparisons between R-values for case I, when VIP#2 failed, and all VIP were replaced by XPS (see Figure 9) of C-O VIP wall specimen (W2).*

- HFT2 at XPS: OSB interface; and
- HFT3 at polyethylene lapped air barrier: gypsum interface.

As an example, Figure 13 shows comparison between the measured and the predicted values of heat flux heat flux at the XPS/OSB interface during the test period for the wall specimen (W1). As shown in this figure, the predicted heat flux is in good agreement with the measurements. Furthermore, the detail comparisons between the predicted heat fluxes and measured heat fluxes at different layer interfaces as described above for the three wall specimens (W1, W2, and W3) are available elsewhere (Saber et al. 2012e). The results showed that both predicted and measured heat fluxes were in good agreements (see Saber et al. [2012e] for more details).

YEARLY CUMULATIVE HEATING AND COOLING LOADS

After the present model was validated, it was used to predict the yearly cumulative heating and cooling loads for different wall specimens and subjected to the climatic condition of Ottawa in the period between May 2011 and May 2012. As an example, Figure 14 shows the predicted heat flux on the indoor surface of the gypsum board for wall specimen W1. In this figure, positive values of heat flux represent the case when cooling loads are needed whereas negative values of heat flux, shown in this figure, represent the case when heating loads are needed. Note that the effect of thermal mass of all wall layers was accounted for. The negative and positive heat fluxes shown in this figure were used to calculate the cumulative energy loss (i.e., yearly

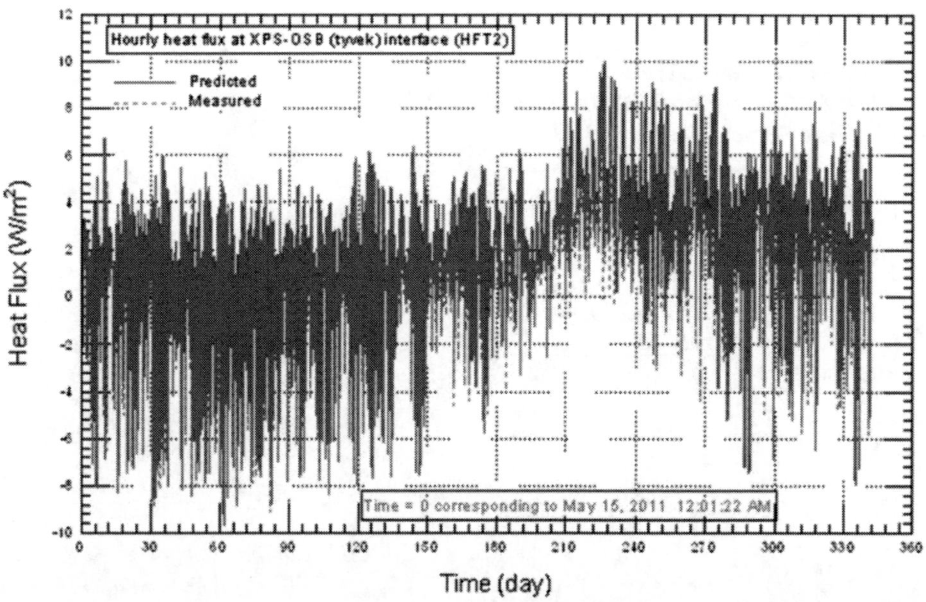

Figure 13 *Comparisons between predicted and measured heat fluxes at XPS/OSB interface for XPS retrofit wall specimen (W1).*

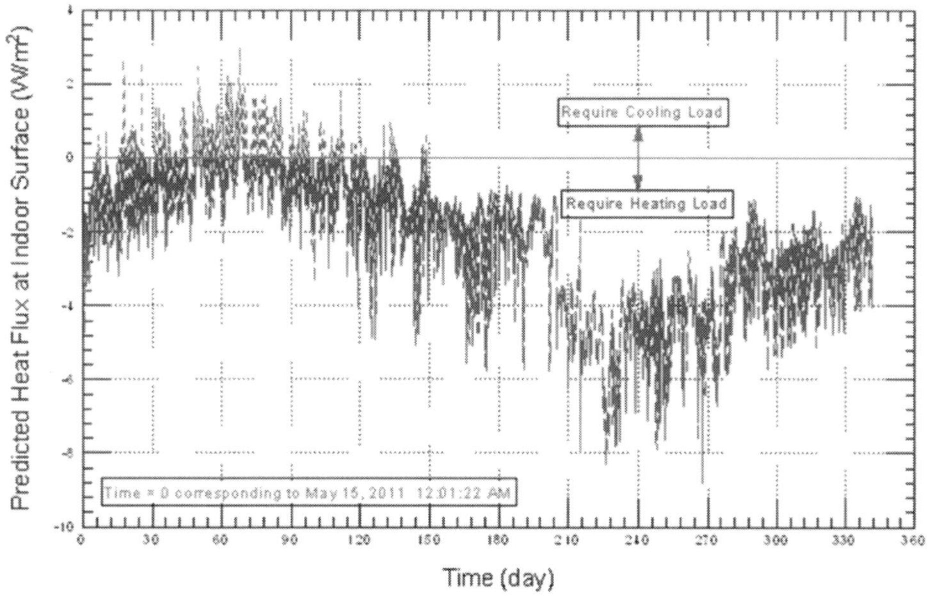

Figure 14 *Predicted heat flux at the indoor surface for XPS retrofit wall specimen (W1).*

heating load) and cumulative energy gain (i.e., yearly cooling load), respectively. The results obtained are shown in Figure 15. As shown in this figure, the yearly cooling loads were approximately the same for the different wall specimens given that over the test period which occurred over the summer season, the outdoor temperature was close to that of the indoor temperature. However, the yearly heating load for the XPS retrofit wall specimen (744 W·day/m^2) was 69.9% and 78.8%, respectively, higher than that for the T&G VIP retrofit wall specimen (438 W·day/m^2), and the C-O VIP retrofit wall specimen (416 W·day/m^2).

By comparing the wall specimens that were retrofitted with VIPs, the effect of the furred-airspace assembly in the C-O VIP retrofit wall specimen when the emissivity of all surfaces bounding the airspace was 0.9 (ASHRAE 2009) resulted in a reduction in the yearly heating load by 5.0% compared to the T&G VIP retrofit wall specimen without furred-airspace assembly.

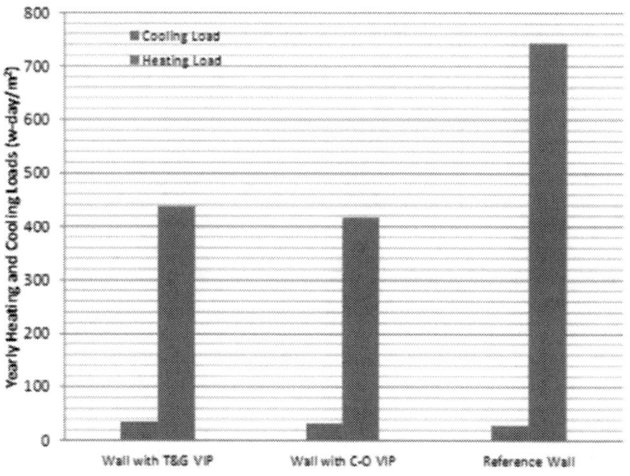

Figure 15 *Comparisons of the predicted yearly cumulative heating and cooling loads of different wall specimens.*

CONCLUDING REMARKS

To maintain reasonable envelope thickness while having high thermal performance, a promising recent innovation in building technology was investigated within the context of its application using vacuum insulation panel (VIP) systems. This concept is gaining interest and is being considered by architects and building practitioners for new and retrofit construction. In this study, steady-state and transient numerical simulations were conducted to predict the thermal performance of residential 2 × 6 wood-frame wall specimens that were retrofitted using vacuum insulation panels (VIPs) (W2 and W3) and extruded polystyrene foam, XPS (W1). Numerical simulations were conducted in order to investigate the effect of the thermal bridging due to the envelope of the VIP on its effective R-value. Depending on the VIP size, and the thickness of VIP envelope and its thermal conductivity, the results showed that the thermal bridging due the VIP envelope can have a significant effect on its effective R-value. This effect, however, was neglected in this study due to lack of information about the envelope of the VIP.

In order to determine the R-values of different wall specimens, steady-state numerical simulations were conducted using the same outdoor and indoor conditions as described in the standard test method using guarded hot box (GHB) (in accordance of the ASTM C-1363 [ASTM 2006]). Results showed that the XPS retrofit wall specimen resulted in the lowest R-value (29.6 ft^2·h·°F/Btu) while the C-O VIP

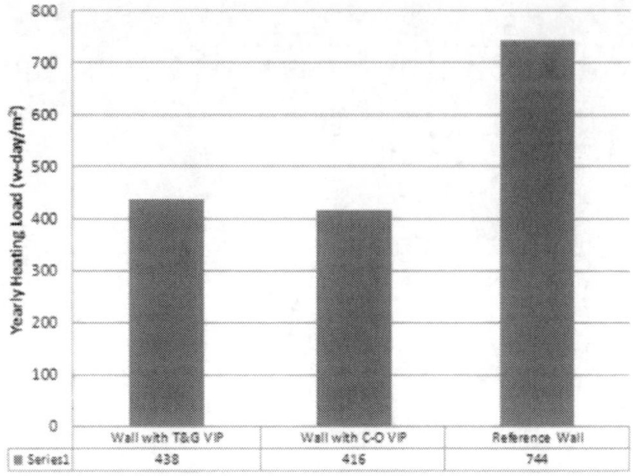

Figure 16 *Comparisons of yearly cumulative heating load of different wall specimens.*

retrofit wall specimen resulted in the highest R-value (55.4 $ft^2 \cdot h \cdot °F/Btu$). The R-value of the T&G VIP retrofit wall specimen (without furred-airspace) was 53.8 $ft^2 \cdot h \cdot °F/Btu$, which was lower than that for C-O VIP retrofit wall specimen (with furred-airspace) by 1.58 $ft^2 \cdot h \cdot °F/Btu$. Furthermore, because there is a risk of the VIP to fail either during the installation process or during in-service conditions, a parametric study was conducted to predict the R-value when one or more VIPs failed.

In the second part of this study, transient numerical simulations were conducted in order to validate the present model and then determine the yearly heating and cooling loads for different wall specimens. The numerical results of heat fluxes were compared with the measured values of heat fluxes. The results showed that the comparison between the present model predictions and experimental data were in good agreement. After the present model was validated, it was used to determine the yearly heating loads for different wall specimens. Results showed that the yearly heating loads for the XPS retrofit wall specimen (744 $W \cdot day/m^2$) was 69.9% and 78.8%, respectively, greater than that for the T&G VIP retrofit wall specimen (438 $W \cdot day/m^2$) and the C-O VIP retrofit wall specimen (416 $W \cdot day/m^2$). For the wall specimens that were retrofitted with VIPs, the effect of the furred-airspace in the C-O VIP retrofit wall specimen when the emissivity of all surfaces that bounded the airspace was 0.9 resulted in a reduction in the yearly heating load by 5.0% compared to the T&G VIP retrofit wall specimen without furred-airspace. It is important to mention that the conclusion of this paper may change when the effect of thermal bridging due to the VIP envelope is significant. In general, research study on VIPs should continue in order to improve thermal resistance by reducing, for example, thermal bridging due to the VIP envelope and improving the way that it is manufactured. By considering the center of the panel, VIPs are of interest owing to their exceptional insulating R-value, up to 10.6 $m^2 \cdot K/W$ (60 $ft^2 \cdot h \cdot °F/Btu$) per inch or even higher.

ACKNOWLEDGMENTS

The authors wish to thank Mr. S. Plescia from the Canada Mortgage and Housing Corporation (CMHC) and Mr. A. Parekh from the Natural Resources Canada (NRCan) for providing guidance and funding for this project. Our thanks are also extended to P. Mukhopadhyaya and D. van Reenen for their support.

REFERENCES

Armstrong, M., H.H. Saber, W. Maref, M.Z. Rousseau, G. Ganapathy, G., and M.C. Swinton. 2011a. Field energy performance of an insulating concrete form (ICF) wall. 13th CCBST conference, The 13th Canadian Conference on Building Science and Technology, Winnipeg, Manitoba May 10 – 13, 2011.

Armstrong, M., W. Maref, H.H. Saber, M.Z. Rousseau, G. Ganapathy, and M.C. Swinton. 2011b. The impact of the thermal mass on field energy performance of insulating concrete form (ICF) wall. International Workshop on Whole Building Testing, Evaluation and Modelling for Energy Assessment, Copenhagen, Denmark, May 18, 2011, pp. 1–12.

ASTM Designation: C 1130-07. 2009. *Standard Practice for Calibrating Thin Heat Flux Transducers*. Annual Book of ASTM Standards, sec 4, Construction, vol. 04.06, Thermal Insulation; Building & Environment Acoustic, pp. 577–584.

ASTM Designation: C-518-04. 2007. *Steady-State Thermal Transmission Properties by Means of the Heat Flow Meter Apparatus*. Section 4, Volume 04.06, Thermal Insulation, 2007 Book of ASTM Standards.

ASHRAE. 2009. *ASHRAE Handbook—Fundamentals* (SI), Chapter 26, Atlanta: ASHRAE.

ASTM. 2006. ASTM C-1363, *Standard Test Method for the Thermal Performance of Building Assemblies by Means of a Hot Box Apparatus*. 2006 Annual Book of ASTM Standards 04.06:717-59.

Binz, A., and G. Steinke. 2005. Applications of vacuum insulation in the building sector. *7th International Vacuum Insulation Symposium Proceedings*, EMPA, Duebendorf, Zurich, Switzerland, September 28–29, 2005, pp. 43–48.

Brunner, S., and H. Simmler. 2008. In situ performance assessment of vacuum insulation panels in a flat roof construction. *Vacuum* 82:700–707.

Cremers, J.M., 2005. Typology of applications for opaque and translucent VIP in the building envelope and their potential for temporary thermal insulation. *7th International Vacuum Insulation Symposium Proceedings*, pp. 189–196, EMPA, Duebendorf, Zurich, Switzerland, September 28–29, 2005.

Elmahdy, A.H., M. Maref, H.H. Saber, M.C. Swinton, and R. Glazer. 2010. Assessment of the energy rating of insulated wall assemblies a step towards building energy labeling. 10th International Conference for Enhanced Building Operations (ICEBO2010), Kuwait, October 2010.

Elmahdy, A.H., W. Maref, M.C. Swinton, H.H. Saber, and R. Glazer. 2009. Development of energy ratings for insulated wall assemblies. Building Envelope Symposium, San Diego, California, October 26, 2009, 21–30.

Haavi, T., B.P. Jelle, and A. Gustavsen. 2012. Vacuum insulation panels in wood frame wall constructions with different stud profiles. *Journal of Building Physics* 36: 212–226.

Ghazi Wakili, K., R. Bundi, and B. Binder. 2004. Effective thermal conductivity of vacuum insulation panels. *Building Research and Information* 32:293–299.

HiPTI-High Performance Thermal Insulation, IEA /ECBCS Annex 39. 2005. Vacuum insulation in the building sector: systems and applications. Subtask A and B.

Maref, W., H.H. Saber, M.M. Armstrong, R. Glazer, G. Ganapathy, M. Nicholls, H. Elmahdy, and M.C. Swinton. 2012a. Integration of vacuum insulation panels into Canadian wood frame walls, Report I: Performance assessment in the laboratory, Internal Research Report, Building Envelope Engineering Materials Program, Construction Portfolio, National Research Council of Canada, Ottawa, Canada.

Maref, W., H.H. Saber, G. Ganapathy, and M. Nicholls 2012b. Field energy performance and hygrothermal evaluation of different strategies of energy retrofit for residential wood-frame wall systems field using VIP, Part I—Field monitoring and thermal performance, Internal Research Report, Construction Portfolio, National Research Council of Canada, Ottawa.

Maref, W. 2011. Program of Energy Research and Development – PERD. Internal Communication.

Maref, W., H.H. Saber, R. Glazer, M.M. Armstrong, M. Nicholls, H. Elmahdy, and M.C. Swinton. 2011. Energy performance of highly insulated wood-frame wall systems using a VIP. 10th International Vacuum Insulation Symposium (Ottawa, Ontario 2011-09-15), 68–76.

Maref, W., M.K. Kumaran, M.A. Lacasse, M.C. Swinton, and D. van Reenen. 2002. Laboratory measurements and benchmarking of an advanced hygrothermal model. *Proceedings of the 12th International Heat Transfer Conference*, Grenoble, France, August 18, 2002, pp. 117–122.

Mukhopadhyaya, P. 2010. High-performance thermal insulation—better than ever. *Solplan Review* (153), pp. 1–3.

Mukhopadhyaya, P., 2006. High performance vacuum insulation panel-research update from Canada. *Global Insulation Magazine*, pp. 9–15, NRCC-48705.

Mukhopadhyaya, P., M.K. Kumaran, N. Normandin, D. van Reenen, and J.C. Lackey. 2009. Fiber-powder composite as core material for vacuum insulation panel. 9th International Vacuum Insulation Symposium, London, UK, September 17–18, 2009, pp. 1–9.

Mukhopadhyaya, P., M.K. Kumaran, N. Normandin, D. van Reenen, and J.C. Lackey. 2008. High performance vacuum insulation panel: Development of alternative core materials. *Journal of Cold Regions Engineering* 22(4): 103–123.

Nussbaumer, T., K. Ghazi Wakili, and Ch. Tanner. 2006. Experimental and numerical investigation of the thermal performance of a protected vacuum-insulation system applied to a concrete wall. *Applied Energy* 83:841–855.

Ogden, R., and C. Kendrick. 2005. VIP cladding panels for buildings: Applications and conceptual solutions. *7th International Vacuum Insulation Symposium Proceedings*, pp. 153–160, EMPA, Duebendorf, Zurich, Switzerland, September 28–29, 2005.

Rogatzki, P. 2009. Pre-fabricated cavity walls-the potential for vacuum insulated panels. 9th International Vacuum Insulation Symposium, September 17–18, 2009, pp. 1–26.

Saber, H.H. 2013. Thermal performance of wall assemblies with low emissivity. *Journal of Building Physics* 36(3):308–329.

Saber, H.H. 2012. Investigation of thermal performance of reflective insulations for different applications. *Journal of Building and Environment* 52:32–44.

Saber, H.H., and W. Maref. 2012. Effect of furring orientation on thermal response of wall systems with low emissivity material and furred-airspace. The Building Enclosure Science & Technology (BEST3) Conference, Atlanta, Georgia, USA, April 2–4, 2012.

Saber, H.H., W. Maref, H. Elmahdy, M.C. Swinton, and R. Glazer. 2012a. 3D heat and air transport model for predicting the thermal resistances of insulated wall assemblies. *Journal of Building Performance Simulation* 5(2): 75–91, (NRCC-53583).

Saber, H.H., W. Maref, and M.C. Swinton. 2012b. Thermal response of basement wall systems with low emissivity material and furred airspace. *Journal of Building Physics* 35(2):353–371, (NRCC-53962).

Saber, H.H., W. Maref, G. Sherrer, and M.C. Swinton. 2012c. Numerical modelling and experimental investigations of thermal performance of reflective insulations. *Journal of Building Physics* 36(2):163–177.

Saber, H.H., M.C. Swinton, P. Kalinger, and R.M. Paroli. 2012d. Long-term hygrothermal performance of white and black roofs in North American climates. *Journal of Building and Environment*, 50:141–154.

Saber, H.H., W. Maref, and G. Ganapathy. 2012e. Field energy performance and hygrothermal evaluation of different strategies of energy retrofit for residential wood-frame wall systems using VIP, Part – II: Model benchmarking and numerical simulations. Internal Research Report, National Research Council of Canada, Ottawa.

Saber, H.H., and A. Laouadi, 2011. Convective heat transfer in hemispherical cavities with planar inner surfaces (1415-RP). *Journal of ASHRAE Transactions*, Volume 117, Part 2, 2011.

Saber, H.H., W. Maref, M.C. Swinton, and C. St-Onge. 2011a. Thermal analysis of above-grade wall assembly with low emissivity materials and furred-airspace. *Journal of Building and Environment*, 46(7):1403–1414.

Saber, H.H., M.C. Swinton, P. Kalinger, and R.M. Paroli. 2011b. Hygrothermal simulations of cool reflective and conventional roofs. 2011 NRCA International Roofing Symposium, Emerging Technologies and Roof System Performance, Washington D.C., USA, Sept. 7–9, 2011.

Saber, H.H., W. Maref, M.M. Armstrong, M.C. Swinton, M.Z. Rousseau, and G. Ganapathy. 2011c. Numerical simulations to predict the thermal response of insulating concrete form (ICF) wall in cold climate. Research Report-310, NRC Institute for Research in Construction.

Saber, H.H., and M.C. Swinton, 2010. Determining through numerical modeling the effective thermal resistance of a foundation wall system with low emissivity material and furred – airspace. International Conference on Building Envelope Systems and Technologies, ICBEST 2010, Vancouver, Canada, June 27–30, 2010.

Saber, H.H., W. Maref, A.H. Elmahdy, M.C. Swinton, and R. Glazer. 2010a. 3D thermal model for predicting the thermal resistances of spray polyurethane foam wall assemblies. Building XI conference, Clearwater, Florida, December 2010.

Saber, H.H., W. Maref, M.A. Lacasse, M.C. Swinton, and K. Kumaran. 2010b. Benchmarking of hygrothermal model against measurements of drying of full-scale wall assemblies. International Conference on Building Envelope Systems and Technologies, ICBEST 2010, Vancouver, Canada, June 27–30, 2010.

Saber, H.H., W. Maref, M.M. Armstrong, M.C. Swinton, M.Z. Rousseau, and G. Ganapathy. 2010c. Benchmarking 3D thermal model against field measurement on the thermal response of an insulating concrete form (ICF) wall in cold climate. Eleventh International Conference on Thermal Performance of the Exterior Envelopes of Whole Buildings XI, Clearwater, FL, USA, December 4–9, 2010.

Schwab, H., U. Heinemann, A. Beck, H.P. Ebert, and J. Fricke. 2005. Dependence of thermal conductivity on water content in vacuum insulation panels with fumed silica kernels. *Journal of Thermal Envelope and Building Science* 28:319–326.

Simmler, H., and S. Brunner. 2005a. Aging and service life of VIP in buildings. *7th International Vacuum Insulation Symposium Proceedings*, pp. 15–22, EMPA, Duebendorf, Zurich, Switzerland, September 28–29, 2005.

Simmler, H., and S. Brunner. 2005b. Vacuum insulation panels for building application basic properties, aging mechanisms and service life. *Energy and Buildings* 37: 1122–1131.

Stahl, Th., S. Brunner, M. Zimmermann, and K. Ghazi Wakili. 2012. Thermo-hygric properties of a newly developed aerogel based insulation rendering for both exterior and interior applications. Short communication, *Energy and Buildings* 44:114–117.

Xiaobo, D., G. Yimin, B. Chonggao, H. Yongnian, and X. Zhen'gang. 2013. Optimization of glass fiber based core materials for vacuum insulation panels with laminated aluminum foils as envelopes. *Vacuum* 97:55–59.

Validating Wind-Driven Rain Module in HAM-Tools

Sam Siassi **Hua Ge, PhD, PE** *Member ASHRAE* **Miljana Horvat, PhD**

ABSTRACT

This paper is a continuation of previous research carried out to improve the hygrothermal analysis capabilities of the readily available HAM-Tools building simulation software. HAM-Tools is an open source library of Simulink models specifically constructed for thermal system analysis in building physics including one-dimensional heat, air, and moisture (HAM) analysis through building envelope components and ventilated spaces. The previous study was intended to improve the program by adding a wind-driven rain module using a semi-empirical model from ASHRAE 160P in the hygrothermal analysis. Comparison to WUFI simulation results was used to verify the wind-driven rain module in HAM-Tools. However, the previous study did not account for the difference in moisture transport mechanism and calculation between the two programs. HAM-Tools uses suction pressure differential as the driving force and the hydraulic conductivity while WUFI uses moisture content difference as the driving force and liquid diffusivity for liquid transports. In this study, the verification procedure for the wind-driven rain module was corrected. Simulation results using HAM-Tools were compared to results from WUFI simulation for different weather data and wall assemblies and to field measurements taken under coastal climate of British Columbia. The comparison with WUFI results and field measurements indicated that the newly designed HAM-Tools wind-driven rain module is performing properly.

INTRODUCTION

Wind-driven rain (WDR) is one of the most important boundary conditions and the main moisture source that affects the hygrothermal performance and durability of building envelopes (Kumaran and Sanders 2008). Wind-driven rain can penetrate through the building facade, which potentially damages the building envelope and reduces its performance and service life. Therefore, it is important to have a proper tool that takes into account heat, air, and moisture variations through the building envelope to be able to analyze the moisture that is being transported through the building envelope structures. This study is a continuation of the previous work that attempted to investigate the effect of microclimatic conditions on the hygrothermal performance of building envelopes using HAM-Tools, an open source one-dimensional hygrothermal performance analysis tool developed using Simulink. A wind-driven rain module was added into the program using a semi-empirical model from ASHRAE 160P 2009 (Wu, Ge, and Horvat 2012). From this model, the amount of rain is calculated using Equation 1.

$$R_{wdr} = F_E \cdot F_D \cdot F_L \cdot U \cdot \cos\theta \cdot r_h \tag{1}$$

where R_{wdr} is the wind-driven rain in mm/h, U is wind speed in m/s, r_h is the unobstructed horizontal rainfall intensity in mm/h, and θ is the angle between wind direction and perpendicular to façade. The exposure factor F_E is obtained based on the height of the building and the terrain surrounding it. The rain deposition factor F_D is based on the slope of the roof and if the wall material is subject to rain runoff. F_L is the empirical constant. To create the WDR module, Equation 1 was used to calculate the WDR intensity. The completed model is shown in Figure 1. Since HAM-Tools has defined south to be zero degrees and increase by moving counter clockwise, section A changes this to define north to be zero degree and counts

***Sam Siassi** is a Master's student and **Miljana Horvat** is an associate professor in the Department of Architectural Science at Ryerson University, Toronto, Ontario, Canada. **Hua Ge** is an assistant professor in Building, Civil and Environmental Engineering at Concordia University, Montreal, Quebec, Canada.*

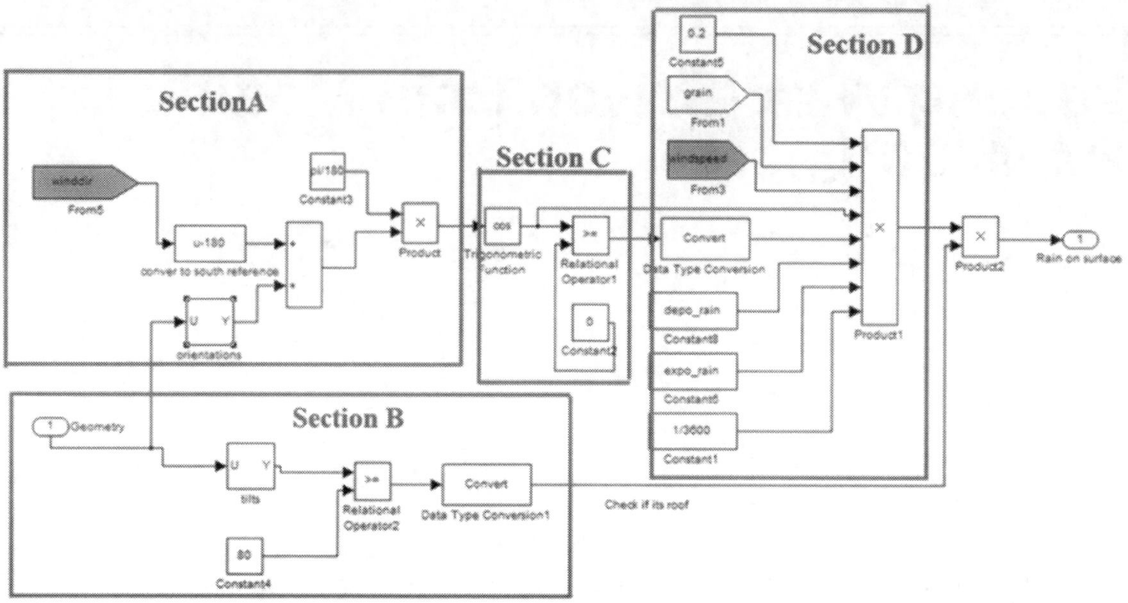

Figure 1 *Added WDR module to HAM-Tools (Wu, Ge, and Horvat 2012).*

clockwise. This ensures that the weather data that is input in HAM-Tools is consistent with the normal meteorological data. Section B verifies if the surface chosen in the geometry is a roof, while section C ensures that the wall orientation faces the direction of the wind blowing by setting the WDR intensity to zero if wind is not hitting the surface of the wall. Finally, section D is the multiplication of all these and the parameters defined in Equation 1. Since currently the time step used in HAM-Tool's calculations is in seconds, it is necessary to convert the result of the WDR module to be consistent with the software. For example, if the weather data is presented in hourly data, then a factor of 1/3600 is introduced.

The developed module was then verified by comparing the results with WUFI, a widely accepted, commercial hygrothermal simulation tool. However, HAM-Tools uses suction pressure differential as the driving force and the hydraulic conductivity while WUFI uses moisture content difference as the driving force and liquid diffusivity for liquid transports. Due to the difficulties in converting material properties from one program to the other, material with the closest property match was selected for the analysis in the previous study, which contributed to discrepancies in moisture content between these two programs. To further improve and verify the wind-driven rain module in HAM-Tools, materials with known retention curve, which allows the conversion from liquid diffusivity to hydraulic conductivity, were carefully selected for the analysis. Simulation results from HAM-Tools were compared to WUFI results using different wall constructions and weather data and to field measurements collected under the coastal climate of British Columbia, a climate with more instance of wind-driven rain and higher amount of rainfall. This paper presents the improvement and verification of the wind-driven rain module in HAM-Tools.

METHODOLOGY

Material Properties

To compare simulation results from HAM-tools to WUFI, the material properties of wall assemblies need to be consistent. Since HAM-Tools has a very limited material database, materials from WUFI's database are used to construct the simulations. Most of the material properties can either be directly input from one program to the other, or can be easily converted. Table 1 summarizes the necessary conversions.

The main difference between these two programs is the driving force and material properties to account for liquid transport. WUFI uses the gradient of water content and liquid diffusivity while HAM-Tools uses suction pressure difference across the material thickness and hydraulic conductivity. To convert liquid diffusivity for materials from WUFI's database to hydraulic conductivity required by HAM-Tools, the material's retention curve is needed.

Water Transport Conversion

The liquid diffusivity in WUFI is governed by Equation 2 (Fraunhofer IBP 2010).

$$g_w = D_W \cdot grad(w) \qquad (2)$$

where g_w is the liquid moisture flux in kg/m^{2}s, D_w is liquid transport coefficient in m^2/s, and w is moisture content in kg/m^3. WUFI uses tabulated data to provide the transport coefficient as a function of water content in the material. It applies to two

Table 1. Material Property Conversions for HAM-Tools and WUFI

WUFI Input	Conversion from WUFI to HAM-Tools	Ham-Tools Input
Thermal conductivity	Same	Lambda_dry: Thermal conductivity of dry material
Water vapour diffusion resistance factor, $\mu_{(@RH)}$	Delta_p_RH = $\delta_0/\mu_{(@RH)}$ where $\delta_0 = 2\cdot10^{-7}\cdot T^{0.81}/P$	Delta_p_RH: Vapor permeability
Temperature dependent thermal conductivity supplement	Same	Lambda_T: Thermal conductivity factor dependent on temperature
Water absorption coefficient	Same	WAC
Free water saturation	Same	W_capillary: Capillary moisture content by volume
Moisture storage function (@RH)	Slope_sorption_ksi = ∇(moisture storage function (@RH))	Slope_sorption_ksi: Slope of the sorption isotherm at specific RH
Liquid diffusion resistance factor	$K = \frac{\rho_{dry}\cdot D_y}{\frac{\partial P_{capillary}}{\partial w}}$ $P_{suc} = \frac{\ln(RH)\cdot R\cdot T\cdot\rho_w}{M_w}$	Hyd_Cond_K = Hydraulic Conductivity
Thermal conductivity	Heat resistance = (1/thermal conductivity)·thickness	Heat resistance
μ	Vapour resistance = $1/(\delta_0/\mu)$·thickness	Vapour resistance

kinds of moisture transports: the suction, which is when free water is presented at the material surface (such as rain), and redistribution, which is the transport of moisture without the presence of free water. The transport redistribution coefficient is approximated one tenth of the suction coefficient.

In HAM-Tools, the liquid transport uses the water retention curve and is based on the suction pressure difference across the material thickness. It is governed by Darcy's law, shown in Equation 3.

$$g_w = K\cdot\frac{\partial P_{suc}}{\partial x} \tag{3}$$

where K is the hydraulic conductivity in kg/Pa·m·s, P_{suc} is the suction pressure in Pa, and $\partial/\partial x$ is the differentiation term in the x direction. The relationship between these two means of water transports can be calculated using Equation 4 (Carmeliet and Roels 2000).

$$K = -\frac{\rho_{dry}\cdot D_y}{\frac{\partial P_{capillary}}{\partial w}} \tag{4}$$

where ρ_{dry} is the density of the material in kg/m^3. The relationship between moisture content and suction pressure is known as the retention curve. Thus, the differentiating term is the differentiation of the retention curve that obtains the moisture storage capacity of the material. Finding the relationship between relative humidity and suction (at high RH) allows for calculating the retention curve. Equation 5 describes this relationship (ASTM-Standard 2009).

$$\ln(\varphi) = \frac{P_{suc}M_w}{RT\rho_w} \tag{5}$$

where M_w is the molar weight of water and R is the ideal gas constant. It is possible to find the suction pressure at a specific relative humidity from the Equation 5. By using sorption isotherm curve of a material, the calculated relative humidity can then be used to find the moisture content of a material at the specific suction pressure. Subsequently, the retention curve can be obtained by correlating the moisture content and the suction pressure. Finally, this retention curve is used in Equation 4 to obtain the relationship between hydraulic conductivity and liquid diffusivity of a material.

If the sorption isotherm of a material in WUFI's database has enough details (moisture content values at high relative humidity), then the retention curve can be calculated. The retention curve is normally obtained through laboratory measurements using pressure plate technique, which uses overpressure to force water out of a saturated specimen (Janz and Johannesson 2001).

BOUNDARY CONDITIONS AND WALL CONSTRUCTIONS

In order to verify and validate the model, the boundary conditions have to be defined. This includes outdoor weather data and indoor climate data. The wall constructions are described below as well.

Exterior Weather Data

The weather data collected from a weather station located in downtown Toronto is used for the verification process. Since the wind-driven rain module is heavily dependent on wind speed, wind direction, and rain (Kumaran and Sanders 2008), it is important to verify the wind-driven rain module under a different climate with higher rainfall. Moreover, if the temperatures reach below freezing point, ice can form at high moisture levels in the pores of building materials. This affects the liquid transport by limiting capillary suction in building materials (Künzel 1995). Thus, it is important to choose weather data that not only has a high amount of rain, but also temperatures that mostly do not fall below freezing point. Therefore, the second set of weather data is obtained from an on-site weather station in Burnaby, BC, Canada, where there is higher amount of rainfall and only a few days that the temperature reaches below freezing point (Environment Canada 2012). This is an ideal climate for verification purposes for the wind-driven rain module. The simulations are conducted on the walls facing the prevailing wind-driven rain, which is east and south-east for Toronto and Burnaby locations, respectively. The driving rain diagram is shown in Figure 2. It should be noted that the duration of the Toronto weather data is for the whole year of 1984, while Burnaby data is from Beginning of January till end of June of 2008.

Interior Climate Data

The interior climate is set according to EN 15026 standard from International Association for Science and Technology of Building Maintenance and Monument Preservation (WTA) Guideline 6-2-01/E. The moisture load is set to normal use and initial conditions of the wall assemblies are set at 20°C. Both HAM-Tools and WUFI have an option to select the EN 15026 standard for the indoor conditions.

Wall Construction

The first wall assembly constructed in both programs is chosen to be a typical North American wood frame residential wall construction. Detail of the wall assembly is shown in Table 2. This assembly is tested against Toronto weather data.

The second wall represents the wall assembly tested in the test facility in Burnaby. The details of the wall can be found in Table 3.

The wall construction in Table 3 is used with Burnaby weather for both verification and validation purposes.

RESULTS AND ANALYSIS

Verification Using Toronto Weather Data

First, the weather data from Toronto (January 1 to December 31, 1984) with the wall assembly described in Table 2 was used. In order to achieve a base comparison, the wind-driven rain module was turned off. The results are shown in Figure 3 with wind-driven rain module off. The figure also shows the relative difference with WUFI's results as the reference. It should be noted that in order to keep the simulations consistent between the two programs, the numerical mesh for HAM-Tools was changed to correspond to that used in WUFI for all the simulations.

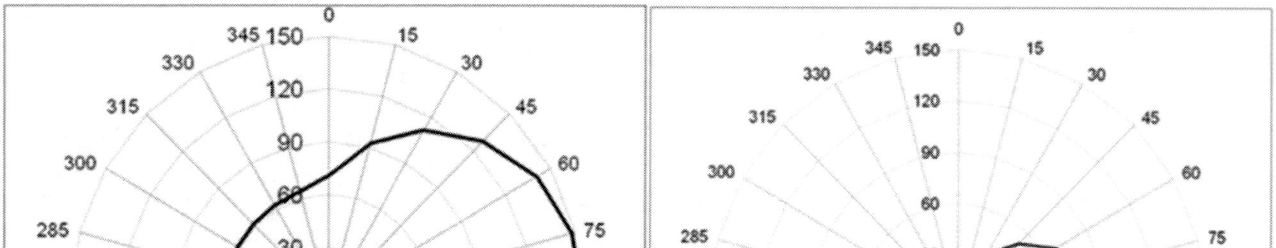

Figure 2 *(Left) Wind-driven rain rosette for Burnaby for four months from January to April 2008 (mm). (Right) Annual wind-driven rain rosette for Toronto using year 1984 weather data (mm).*

Table 2. Simulated Wall Construction

Material (Outdoor to Indoor)	Thickness, mm	Thickness, in.
Wood siding	20	0.787
Air space	19	0.984
60 minute building paper	0.1	0.00394
OSB sheathing	12.5	0.492
Fiberglass insulation	89	3.504
Polyethylene vapour retarder	0.15	0.00591
Gypsum board	12.5	0.492

Table 3. Field Experiment Wall Assembly

Material (Outdoor to Indoor)	Thickness, mm	Thickness, in.
Red matt clay brick	90	3.543
Air space	25	0.984
Spun bonded polyolefine (SBP) membrane	0.2	0.00787
Plywood	12.7	0.5
Low density fiberglass batt insulation	140	5.512
Polyethylene membrane	1	0.0393701
Gypsum board	12.5	0.492

The relative difference in moisture content is shown to be 10% or less between the two programs. This difference between the two programs should be considered when analyzing the results with the WDR module activated.

The simulation is then performed with the WDR module turned on. The result is shown in Figure 4.

When the wind-driven rain is on, the same difference can be observed when the module was off. However, just considering the spikes that show the presence of rain, it can be seen that the relative difference of these spikes is 10% or less. The reason for this effect is due to the difference in liquid diffusivity for suction and distribution that WUFI uses. When rain is present, the suction transport coefficient in WUFI is used in the calculation. This is the same value that is used to calculate the hydraulic conductivity for HAM-Tools. However, WUFI uses the redistribution transport coefficient when the rain stops and there is no presence of free water. The redistribution coefficient is generally one tenth of the suction coefficient. Therefore, HAM-Tools results tend to dry out faster during the rainy periods.

The same simulation is also performed with Burnaby weather data and wall construction described in Table 3. The moisture content of plywood, the fourth layer from exterior is used for this comparison. The result is shown in Figure 5.

The relative difference in the plywood between the two simulations is mostly around 2% to 4%, with the exception of the very end of the simulation that shows around 8% difference. Similar to previous comparison of wood siding, this difference can be explained from the slight variation in vapour diffusion and vapour permeability values between the two programs.

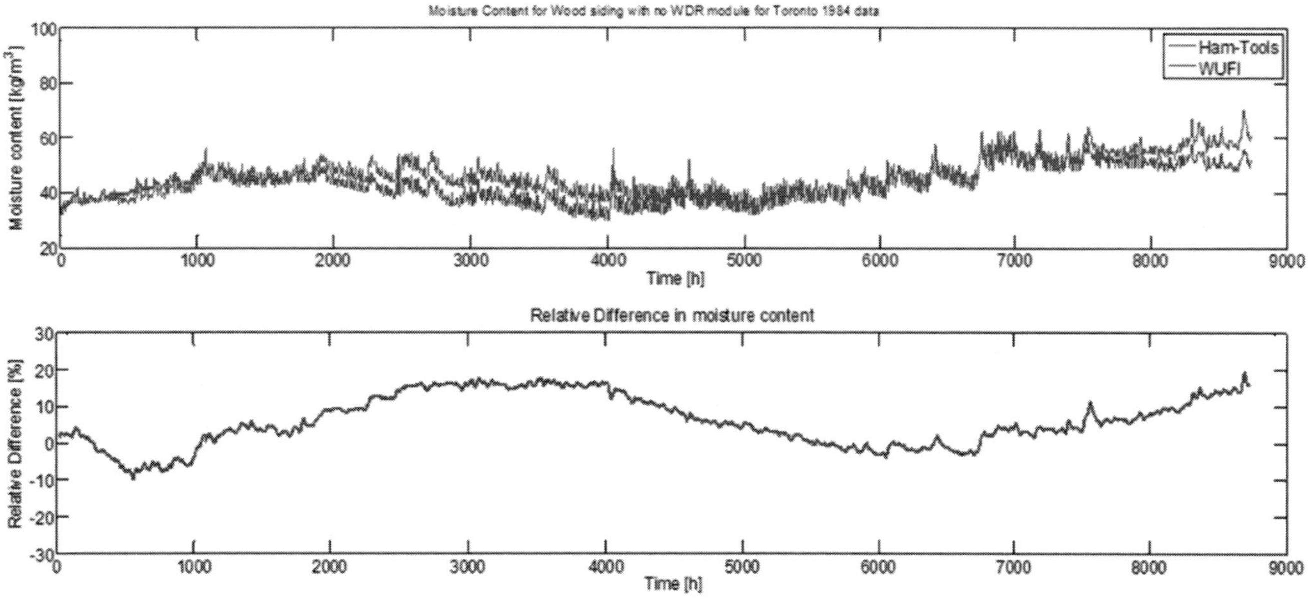

Figure 3 *(Top) Comparison of moisture content in wood siding between HAM-Tools and WUFI with wind-driven rain off. (Bottom) The resulting relative difference in % between these two programs. WUFI results are used as reference.*

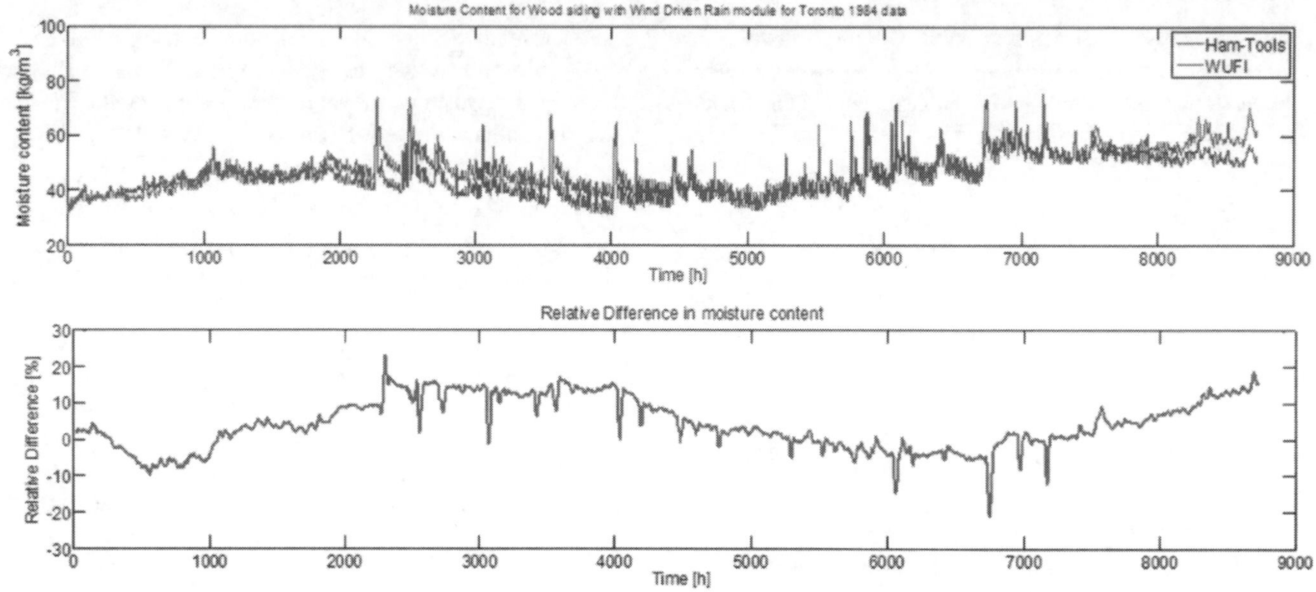

Figure 4 *(Top) Comparison of moisture content in wood siding between HAM-Tools and WUFI with the wind-driven rain module on. (Bottom) The resulting relative difference in % between these two programs. WUFI results are used as the reference.*

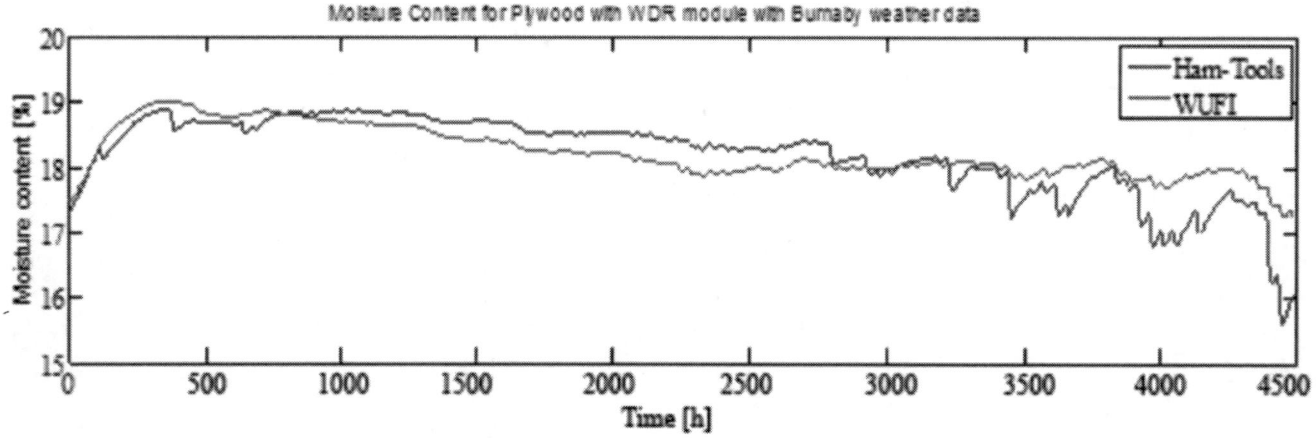

Figure 5 *Comparison of moisture content in plywood between HAM-Tools and WUFI with the wind-driven rain module on.*

MODEL VALIDATION

A field measurement experiment was conducted on the test facility in Burnaby using the wall assembly described in Table 3 and the Burnaby weather data in 2008 (Simpson and Ge 2010). The field measurements' initial moisture content of each layer is shown in Table 4.

It should be noted that there were no sensors present in the insulation and gypsum board to obtain initial relative humidity for those layers, and so they were assumed to have the same value as the interior conditions (Simpson and Ge 2010).

Table 4. Initial Relative Humidity of Wall Components in Burnaby Test Hut

Wall Component	Relative Humidity, %
Brick veneer	91.5
Plywood	87
Insulation	50
Gypsum board	50

One of HAM-Tools input data is initial relative humidity. This value controls the initial moisture content of all the layers in HAM-Tools. The influence of initial conditions will diminish for simulations over a long period, such as a couple of years. For verification section, the initial relative humidity of all materials in both WUFI and HAM-Tools were kept the same. However, for this particular case, only a few months' measurements and on-site weather data were collected. Therefore, it is important to set the accurate initial conditions. In order to demonstrate this effect, a simple comparison is performed on the plywood for half a year using WUFI. For the first setup, the initial moisture content of all the layers is set at 87% rh. In the second setup, the initial moisture content of plywood is kept at 87% while all the other layers are changed to be at 40% rh. The results are shown in Figure 6.

This result shows that having different initial moisture content in the surrounding layers directly affects the moisture content in the studied layer. This affects the result at the very beginning of the simulation by increasing or decreasing the moisture content of the surrounding layer. As a result, to better reflect the reality, different initial relative humidity inputs for each layer were added in HAM-Tools and the variable for every single node within the construction block was changed accordingly. The result was compared to WUFI and is shown in Figure 7.

With correct initial moisture content, HAM-Tools was then tested and compared with field measurements. Figure 8 shows the simulated and measured results.

The simulated result shows the same trend as the measured data with a relative difference of around 5% and a maximum difference of around 10%. The differences may be

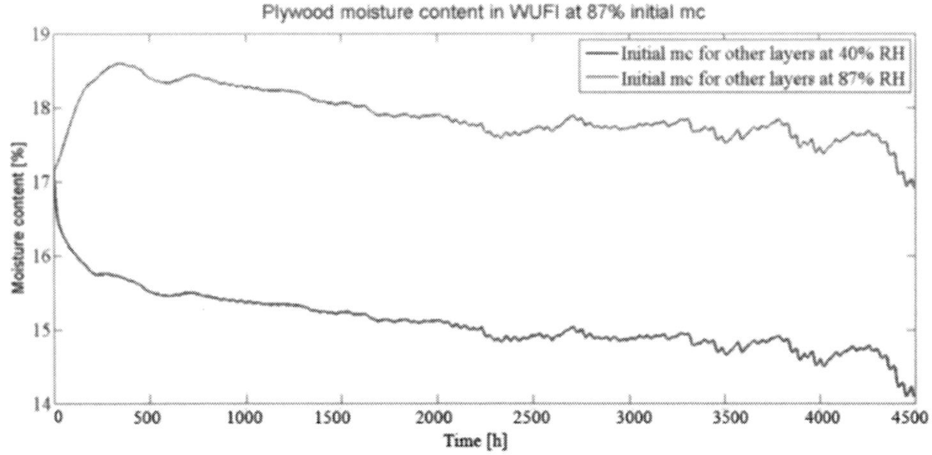

Figure 6 *Comparison of plywood's moisture content with different initial conditions.*

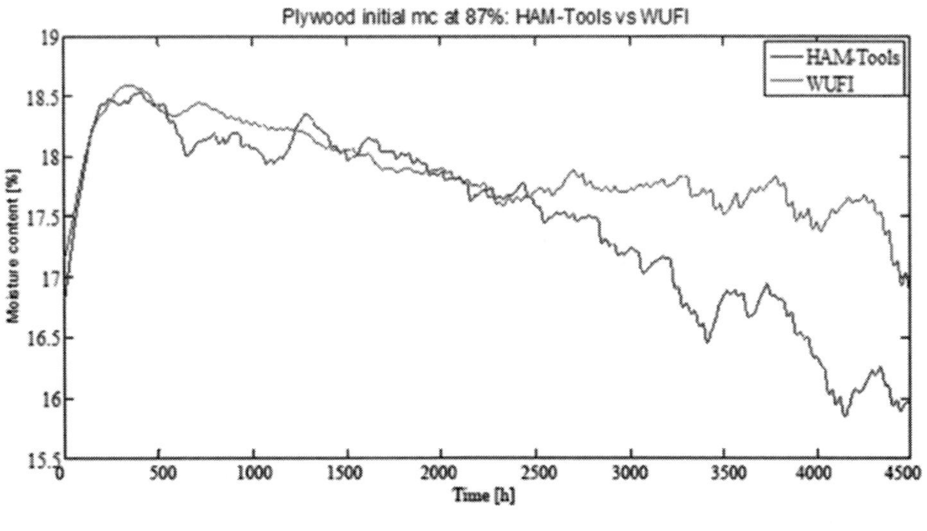

Figure 7 *Plywood's moisture content comparison between HAM-Tools vs WUFI with different initial moisture content at each layer.*

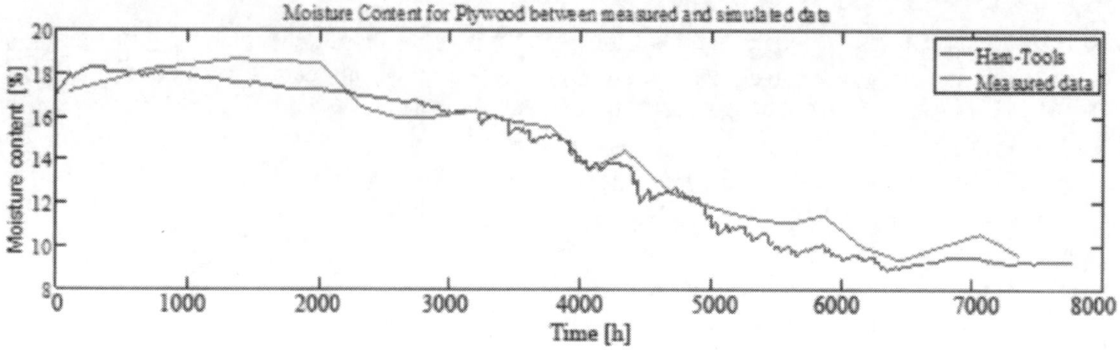

Figure 8 *Comparison of moisture content of plywood between simulated (HAM-Tools) and measured data.*

attributed to the initial moisture content of the materials, discrepancy in material properties, and measurement errors. Thus, with a more accurate measurements, the program may obtain better results in comparison with measured data. It should be noted that there were no sensors present to measure the moisture content in the layers behind the plywood (batt insulation and gypsum board) since it was not needed for the study (Simpson and Ge 2010). The assumed initial moisture content of the insulation slightly affects the outcome of the comparison. Similarly, the liquid and vapour diffusivity of each material were not directly measured. Instead, WUFI's material database was used to provide these values.

CONCLUSIONS

The previous study to create a wind-driven rain module for HAM-Tools required proper verification with other simulation programs and validation with measured data. A widely accepted commercial program, WUFI, is used for the verification. Because of the very limited material database in HAM-Tools, materials from WUFI's database were selected. Importance was placed on proper conversion of these material properties from WUFI to the HAM-Tools. The most significant one was the water transport properties, in which HAM-Tools uses hydraulic conductivity while WUFI uses liquid diffusivity. Moreover, the consistency for the weather data between each program and actual data had to be considered.

First, the conversion formula between hydraulic conductivity and liquid diffusivity was found. Then proper materials were chosen from WUFI's database that included enough information for the conversion to be applied. Using these materials, a wall assembly was chosen and used to validate the module by comparing the results with WUFI. The results showed a 10% relative difference in the exterior layer when the WDR module was off due to vapour diffusion. When the module was on, the relative difference for the spikes were less than 10% as well. This can be explained since HAM-Tools uses the same water transport coefficient (hydraulic conductivity) for both wetting and drying periods while WUFI has a smaller value for drying periods. The comparison between these two programs was also made for a wall assembly tested under Burnaby weather conditions. The result showed an average relative difference of 4% in the moisture content of plywood.

Finally, the module was compared with field measurements carried out on a building envelope test facility. The comparison showed a relative difference of around 5% and a maximum relative difference of around 10%. This difference can be due to the discrepancy in material properties, initial moisture content, effect of air movement presented in the air cavity, and measurement errors. In general, the simulation results from HAM-Tools agree well with simulation results from WUFI and field measurements. Thus, it can be concluded that the newly designed HAM-Tools wind-driven rain module is performing properly.

NOMENCLATURE

D_w	=	liquid transport coefficient, m^2/s
F_D	=	rain deposition factor
F_E	=	rain exposure factor
F_L	=	empirical constant (0.2), $kg{\cdot}s/(m^3{\cdot}mm)$
g_w	=	liquid moisture flux (or density of moisture flow rate), kg/m^2s
K	=	hydraulic conductivity, kg/Pa·m·s
M_w	=	molar weight of water (0.018, Kg/mol
P	=	atmospheric pressure, Pa
P_{suc}	=	suction pressure, Pa
R	=	ideal gas constant (8.314), j/(mol·k)
R_{wdr}	=	wind-driven rain intensity, mm/h
T	=	thermodynamic temperature, K
T'	=	absolute temperature, K
w	=	liquid moisture flux, kg/m^3
$\partial/\partial x$	=	differentiating in the x direction, m
$\frac{\partial P_{capillary}}{\partial w}$	=	suction isotherm or Hydraulic conductivity
θ	=	direction of wind, °
ρ_w	=	density of water (1,000), kg/m^3

REFERENCES

ASTM-Standard. (2009). http://www.astm.org/Standards/C1699.htm. Designation C1699.

Carmeliet, J., and Roels, S. (2000). *Determination of the Isothermal Moisture Transport Properties of Porous Building Materials.* Catholic University of Leuven, Department of Civil Engineering, Heverlee, Belgium.

Environment Canada. (2012). *National Climate Data and Information Archive.* Retrieved 9 2012, from www.climate.weather.gc.ca

Fraunhofer IBP. (2010). Retrieved 05 20, 2012, from WUFI: http://www.wufi-pro.com/.

Hansen, M., Houvenghel, G., Janz, M., Krus, M., and Stromdahl, K. (1999). *Interlaboratory Comparison of the Measurement of Retention Curves.* Technical University of Denmark, Department of Structural Engineering and Materials, Denmark.

Janz, M., and Johannesson, B. (2001, Apr). Measurement of the Moisture Storage Capacity Using Sorption Balance and Pressure Extractors. *Journal of Building Physics*, 24: 316–334.

Kalagasidis, A. (2004). *HAM-Tools: An Integrated Simulation Tool for Heat, Air and Moisture Transfer Analyses in Building Physics.* Chalmers University of Technology, Department of Building Technology, Building Physics Division, Gothenburg, Sweden.

Kalagasidis, A.S., Weitzmann, P., Nielsen, T. R., Peuhkuri, R., Hagentoft, C.E., and Rode, C. (2007). The international building physics toolbox in simulink. *Energy and Buildings*, 39 (6), 665–674.

Kumaran, K., and Sanders, C. (2008). *Annex 41, Whole Building Heat, Air, Moisture Response - Boundary Conditions and Whole Building HAM Analysis.* International Energy Agency.

Künzel, H.M. (1995). Simultaneous Heat and Moisture Transport in Building Components—One- and two-dimentional calculation using simple parameters. *Fraunhofer IRB.*

Rode, C., Ruut, P., and Woloszyn, M. (2008). Simulation tests in whole building heat and moisture transfer. *Proceeding of the third International Building Physics/Engineering Conference*, (pp. 527–534). Montreal, Canada.

Ruut, P., and Rode, C. (2005). Common Exercise 1 – Case 0A and 0B Revised. *IEA, ANNEX 41, Task 1, Modeling Common Exercise.*

Simpson, Y., and Ge, H. (2010). *Field evaluation of ventilation wetting and drying of rainscreen walls in coastal British Columbia.* Concordia University, Department of Building, Civil and Environmental Engineering, Montreal, Quebec.

Tariku, F., Kumaran, K., and Fazio, P. (2010). Integrated analysis of whole building heat, air and moisture transfer. *Int. J. Heat Mass Transfer.*

Wu, R., Ge, H., and Horvat, M. (2012). Wind-driven rain module for HAM-Tools (International Building Physics Toolbox) according to ASHRAE 160P-design and verification. *Proceedings of 5th International Building Physics conference.* Kyoto, Japan.

Hygrothermal Performance Assessment of Vented and Ventilated Wall Systems: An Experimental Study

Fitsum Tariku, PhD
Member ASHRAE

Ying Simpson

ABSTRACT

Based on analysis of the drying and wetting potentials of a particular local climate, designers choose wall systems with or without an air gap between a sheathing membrane and a cladding layer. In addition to the capillary break that the air gap provides, thereby reducing the moisture transfer from wet cladding to the interior of the wall, the airspace will add the thermal resistance of the wall system and reduce the heat flow across the wall system. These moisture and thermal performances are straightforward to understand if the air in the air cavity is assumed to be a "still air." In this paper, an experimental study is undertaken to understand the impact of airflow through an air cavity on the moisture and thermal performance of wall systems. To achieve this objective three test panels are instrumented and monitored in the field-experimental setting: one with no air gap, another one with an air gap but restricted airflow, and the third one with an air gap and open for airflow. The second and third wall systems have the same air gap width but different top flashing designs creating vented and ventilated wall systems. For the wall systems' orientation and boundary conditions considered in this study, the wall with no air gap accumulates relatively high moisture content on the sheathing board, stud, and bottom plate and also has high moisture content changes in a year cycle when compared to the vented and ventilated wall systems. In general, the hygrothermal performances of vented and ventilated wall systems are comparable. During the winter period when relatively high moisture accumulation occurs, the upper section of the ventilated wall system shows slightly lower moisture content compared to that of the vented wall system. The temperature readings of the sheathing boards in the vented and ventilated wall systems are slightly warmer than that of the wall with no air gap for 85.5% and 73% of the time (based on hourly data of a year), respectively. For the balance of a period of time, the sheathing boards in the walls with an air cavity are slightly cooler than that of the wall with an air gap. Although the low temperature on the sheathing board, which is caused by solar radiation-induced airflow, is beneficial during a cooling season, the air gap and the associated airflow may reduce the heat gain that may be obtained from solar radiation during the heating season. The implications of air cavity and flashing design (airflow rate) on the heating and cooling load calculations of different orientations, wall configurations, and climate require further investigation.

INTRODUCTION

New materials and building envelope designs are continuously being introduced in the construction industry with the objective of increasing the energy efficiency, durability, and safety of buildings. It is crucial to evaluate the performance of the new approaches in relation to the performance of existing building code accepted solution. One of the three ways of investigation of newly proposed solutions is through a field experimental study, where the other two are computer modeling and laboratory evaluation. Field experiment has the advantage of assessing performance in real operating conditions as it enables the capture of a range of hygrothermal loads and wall systems' responses to real environmental exposures. In wet climatic regions where a wind-driven rain load is significant, a water absorbing exterior layer holds and transfers moisture to moisture sensitive layers such as wood based sheathing boards

Fitsum Tariku *is a director and* ***Ying Simpson*** *is a research analyst of the Building Science Centre of Excellence at the British Columbia Institute of Technology (BCIT), Burnaby, BC, Canada.*

and frames. If the moisture accumulation on these sensitive materials is significant and lasts for an extended period of time, the durability of the materials will be compromised, and subsequently, the building will fail (Morrison Hershfield 1998; Rousseau and Dalgliesh 2004). To reduce the moisture transfer from the exterior to the interior part of a wall, the National Building Code of Canada (2005) prescribes a mandatory 10 mm (3/8 in.) air gap requirement between cladding and sheathing membrane materials. The air gap will act as a capillary break and eliminates liquid moisture transfer. In this type of wall design, which is referred to as a *rain-screen* wall design, the moisture transfer from the exterior to the interior part of the wall is limited to the maximum moisture storage capacity of the air in the air gap, and the moisture transfer mechanism is limited to vapour flow by diffusion and convection. Although the function and advantage of the air gap as a capillary break is demonstrated in a number of publications (Hazleden and Morris 2002; Straube et al. 2004; Shi and Burnett 2006; Bassett and McNeil 2005; Tariku and Ge 2010; Simpson 2010), the amount and effectiveness of cavity airflow on the moisture management potential of wall systems are active research topics. It has been shown in a laboratory setting (Burnett et al. 2004) and parametric analysis (Onysko 2004) that as the airflow in the cavity increases the drying potential of the wall system increases. In this paper, the effect of promoting cavity airflow in a mild and wet climate is investigated in a field experiment setting. To achieve this objective three test panels (one with no air gap, another one with an air gap but restricted airflow, and the third with an air gap and open for airflow) are instrumented and monitored in a field exposure test facility for fifteen months. The second and third wall systems have the same air gap width but different top flashing designs creating a vented and ventilated wall system. The hygrothermal responses of the various layers of the three wood-frame wall systems are presented and discussed in the sections below.

FIELD EXPERIMENT

The first objective of this experimental study is to understand the hygrothermal impact of an air gap between the sheathing membrane and cladding, and the associated top flashing designs that lead to vented and ventilated wall designs in a coastal climate; and the second objective is to gather experimental data of such wall system performance in "real" operating conditions for benchmarking of hygrothermal models. To achieve these objectives one conventional wall system (Figure 1a), a wall with nonair gap between the sheathing board and cladding, and another two wall systems with similar air gap width but different top flashing designs are considered (Figure 1). The flashing designs are intended to create a ventilated (Figure 1b) and a vented (Figure 1c) wall system. In the ventilated wall system, the clearance between the cladding exterior surface and the flashing is 12.7 mm (1/2 in.), which essentially results in a wall system which is open at both bottom and top and allows air movement through the air cavity. Whereas in the vented wall system, the clearance is only 1.0 mm (1/16 in.) and the airflow through the air cavity is expected to be insignificant, thus in this paper this wall system is referred to as a vented wall system (open at the bottom but nearly closed at the top).

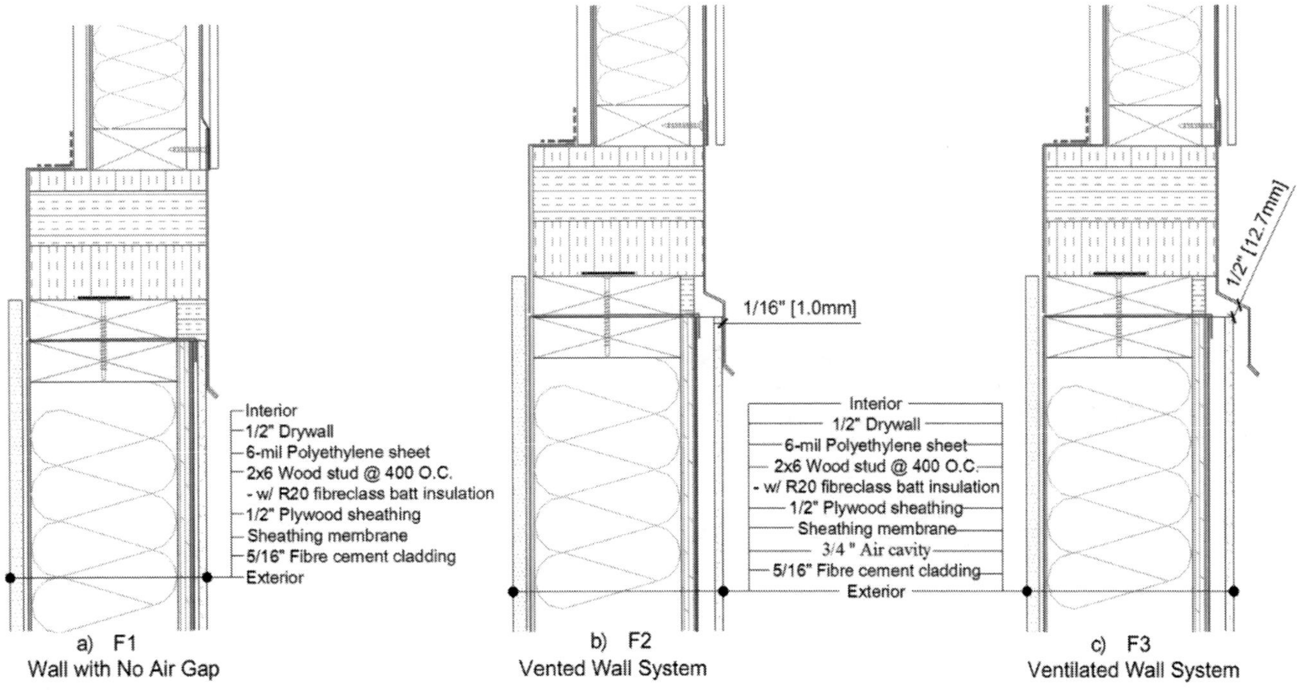

Figure 1 *Top flashing configurations and descriptions of wall assemblies for test walls F1 (no air gap), F2 (vented), and F3 (ventilated).*

The walls are instrumented with thermocouples, humidity probes, and moisture pins to measure the temperature, relative humidity, and moisture at various layers of the test panels. The three test panels are installed side by side on the southeast orientation of BCIT's Building Envelope Test Facility (BETF), a direction where the local wind-driven rain load is predominant, to ensure similar indoor and outdoor climatic loads on both the interior and the exterior sides of the test panels. The BETF is a 13.4 by 8.5 m (44 by 28 ft) two-storey facility, Figure 2, positioned in a relatively open site to allow higher wind-driven rain and solar radiation exposures. The outdoor climatic conditions, including temperature, relative humidity, wind speed and direction, and global horizontal solar radiation are measured with a weather station that is mounted on the roof of the facility. The actual wind-driven rain load and solar radiation that the samples are subjected to are measured from the wall mounted wind-driven rain gauge and pyranometer as shown in Figure 2. The indoor environmental conditions, namely temperature and relative humidity, are controlled and maintained with forced-air heating/cooling and humidification systems. Dehumidification of the indoor air is relayed on the air-conditioning system. The indoor temperature is controlled by a thermostat, which is set at 21°C. Humidistat works with the humidification system to maintain the indoor relative humidity at 35%.

Test Panels Description

The test panels are built as a regular 2 × 6 (38 by 140 mm [1.5 by 5.5 in.]) wood-frame wall system. Each test panel is 2235 mm (88 in.) high, 1219 mm (48 in.) wide, and has three cavity spaces between four studs. The center and side cavity spaces are 368 and 349 mm (14 1/2 and 13 3/4 in.) wide, respectively. The configuration and materials used to construct the test panels are shown in Figure 1. Fiber cement board is used as an exterior layer (cladding), pressure treated plywood strip (19 mm [3/4 in.] thick and 51 mm [2 in.] wide) as furring, spun bonded olefin as a sheathing membrane, plywood as a sheathing board, 6 mil polyethylene sheet as a vapour and air barrier material, and gypsum board as an interior finishing layer. The wood frame is made of spruce, and the stud cavity is filled with R-20 (RSI-3.5) fiberglass insulation. The conventional wall system is built in the same way as that as shown in Figure 1 except that there is no furring and air cavity between the sheathing membrane and cladding (Figure 1).

Sensor Layout and Installation

The layouts of moisture pins, thermocouples, and relative humidity sensors are shown in Figure 3. To assess the moisture and temperature variations along the length of the sheathing board, three pairs of moisture pins are inserted into the mid-thickness of the plywood sheathing and thermocouple sensors are installed in the interior surface of the plywood at lower, middle, and upper sections (305, 1118, and 1930 mm [12 in., 44 in., and 76 in.] from the bottom plate respectively). To avoid edge effects on the measurement, sensors are installed in the middle bay along the symmetry line. Just 127 mm (5 in.) above the location of the moisture pins, two 51 mm (2 in.) long strips of moisture detection tapes are installed in both interior and exterior surfaces of the plywood. The detection tapes are intended to alarm unintended

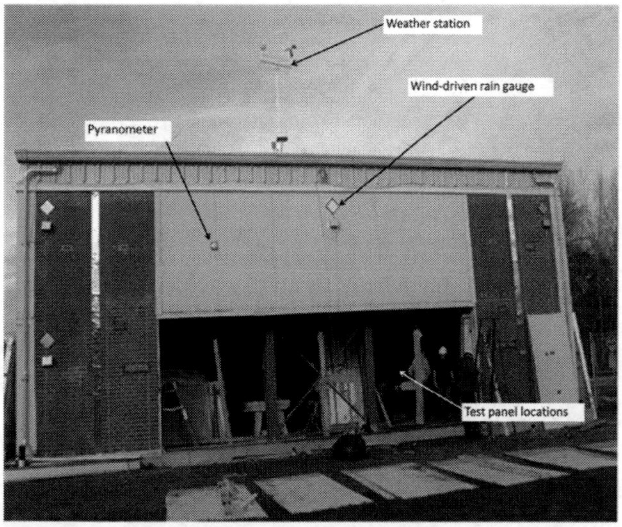

Figure 2 *Weather station on the roof of BETF with a pyranometer and five wind-driven rain gauges on the south east façade.*

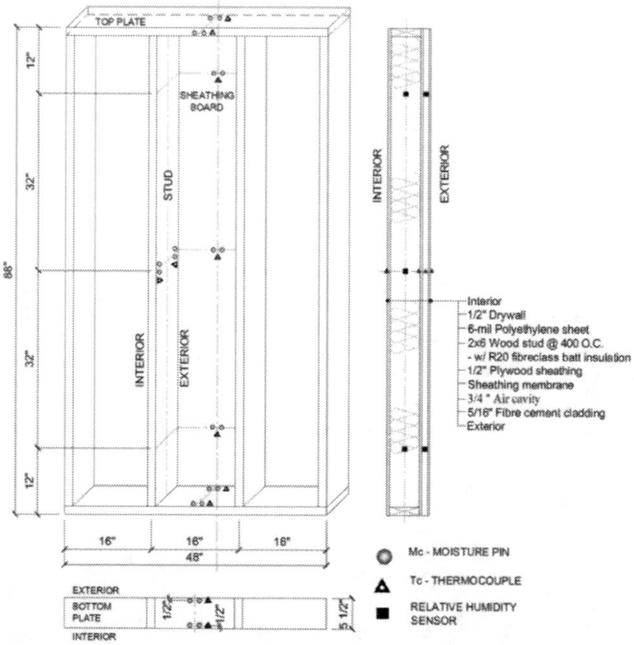

Figure 3 *Sensor locations and test panel configuration.*

water leakage to the test panels. The moisture pins are manufactured from 6 mm (1/4 in.) diameter stainless steel wire. The wire is coated with electrical insulating paint except for the tip area, which enables the measurement of moisture content at a desired depth. Here, the moisture content of the plywood at mid-thickness is sought, and therefore, the pins are inserted 6 mm (1/4 in.) from the interior surface of the plywood. To restrict possible water seepage into the plywood through the pin-plywood contact area, epoxy is applied around the moisture pin and plywood surface. In addition, the moisture pin tip surface area, which is soldered with instrumentation wire, is coated with epoxy to avoid possible short circuiting between the two pin heads through the insulation material in the event that the insulation near the moisture pin heads get wet. The moisture pins are inserted 25 mm (1 in.) apart and the wires are connected to 5M Ohm resistor in series. The electrical resistance between the end surfaces of the two pins is measured and translated to moisture content using a moisture pin calibration curve. The calibration curve is developed in house by measuring the equilibrium moisture contents of wood samples (plywood and spruce) at 30%, 50%, 70%, and 90% relative humidity and the corresponding electrical resistance. The temperatures at various locations of the test panels are measured using Type T (copper and constantan) thermocouple wires. The temperature readings are accurate to ±0.3°C as per the calibration done using an isothermal bath, which has an accuracy of ±0.1°C. The temperature and moisture content inside each wall insulation space are measured using three Vaisala HMP50 RHT sensors that are located at the same height as the lower, middle, and upper moisture pins. Like the moisture pins, the RHT sensors are located in the center plane of the middle bay and 70 mm (2 3/4 in.) deep from the interior surface of the insulation. The RHT sensors are newly acquired and have an accuracy of ±3% within the manufacturer's calibration range of 0%–90% and ±5% between 90% and 98%. Using an Agilent 34980A data-acquisition unit, the moisture content, temperature, and relative humidity measurements are collected every five minutes, whereas the indoor and outdoor climatic conditions are recorded every minute using a Campbell Scientific CR10X data-acquisition unit.

Test Panel Fabrication and Installation

The test panels are fabricated and instrumented in the lab before installing them on the field exposure test facility for monitoring. All the materials are newly acquired and kept inside the test facility for about 30 days. The exposure of these materials to the nearly constant indoor temperature of 21°C and 35% relative humidity helps to attain uniform distribution of initial moisture content in the wood-based materials (spruce and plywood), cladding, and drywall.

The construction and instrumentation series are as follows: first, the wood frames are constructed and the plywood is attached. Since the temperature, moisture, and air pressure gradients normal to the test panel will be higher than in the lateral directions during testing, the instrumentation wires are routed from the center bay to the side of the test panels through two 25 mm (1 in.) diameter holes on the studs and connected to the terminal block. As recommended in Straube et al. (2002), the locations on the studs where the strands of wires pass through are sealed with polyurethane foam to restrict air, vapour, and heat movement across the wire routing openings. The sheathing membrane is cut to the same dimension as the sheathing board plane area and stapled on the exterior surface of the sheathing board with 305 mm (12 in.) spacing. The test panel is flipped, and the three stud spaces are filled with R-20 (RSI-3.5) fiber glass insulation. Then, a polyethylene sheet is placed on top of the insulation and wrapped around the side studs and top and bottom plates and overlapped 25 mm (1 in.) on the sheathing membrane at the back of the test panel. The polyethylene sheet edge is taped onto the sheathing membrane using tuck tape, which forms a test panel with controlled test areas and no moisture and airflow along the lateral directions. The polyethylene sheet is stapled to the studs and plates with approximately 305 mm (12 in.) spacing. After the four 51 × 19 mm (2 × 3/4 in.) pressure-treated wood straps are placed on top of the sheathing membrane and nailed onto the studs with 305 mm (12 in.) spacing, the test panels are installed side by side in the southeast orientation of the BETF. As the polyethylene sheet wraps around the side of the test panels to avoid lateral air and moisture transfer between the panels, 51 mm (2 in.) rigid insulation (extruded polystyrene) is installed between the test panels to thermally isolate them. Finally, the test panels' interior and exterior are closed with gypsum boards and fiber cement claddings, respectively.

EXPERIMENTAL RESULTS AND DISCUSSION

In this paper, the results from fifteen and a half months of monitoring starting July 1, 2011 to November 15, 2012 are presented and discussed. The period prior to July 1, 2011 is considered as a conditioning period. At the time of installation, the moisture contents of the sheathing boards (plywood) in all three test walls are 9%, whereas the moisture contents in the wood-frames (studs and top and bottom plates) are 12%. In this section, the wall system with no air cavity is referred to as *F1*, the vented wall system as *F2*, and the ventilated wall system as *F3*. The hygrothermal responses of the respective wall layers, including sheathing board, wood-stud, bottom plate, and cladding are discussed. The transient moisture content profiles of the sheathing boards in the walls F1, F2, and F3 are shown in Figures 4, 5, and 6, respectively. The measured wind-driven rain and solar radiation loads on the cladding surfaces are superimposed on the moisture content plots of the respective figures. Figure 4 shows the moisture content of the plywood in the F1 wall system at the lower, middle, and upper locations. The wind-driven load on the cladding is more frequent during the period from October to May, a period which can be characterized as a wet period. During this period the solar radiation is also relatively low. The months between July and September can be characterized as a dry period as the solar radiation is relatively high while the wind-driven rain events are relatively low. The moisture content in the plywood increases to the highest level in the fourth week of March, 2012 and the lowest level at the end of August. The continued drying of the sheathing board in April to June in spite of substantial rain events occurring shows that during this period, the drying potential of the outdoor climatic condition (solar radiation and temperature) is higher than its wetting potential (rain).

In general, the moisture content variation across the height of the sheathing board is minimal, as can be seen in the figure. Compared to the middle and upper sections, the bottom section of the plywood seems to have slightly higher drying and wetting rates during the corresponding drying and wetting periods. The relatively higher drying rate leads the lower section to have moisture content (MC) of 6% moisture content while the middle and the upper sections have 8% MC. Following the drying period, the lower section also shows steep moisture uptake during the fall season and attains about the same moisture content as the upper and middle sections of the plywood. For this conventional wall system, the highest moisture content that is measured on the sheathing board during the monitoring period is 18.5% MC, which is close to the upper limit of the Canadian Building Code requirement (i.e., 19% MC).

Figure 5 shows the moisture content profiles of the lower, middle, and upper sections of the sheathing board of the vented rain screen wall system (F2). As the figure shows, the moisture content across the height of the sheathing board is about the same. In this wall system, the moisture content on the plywood varies from about 6% MC (first week of September 2012) to the highest value of about 14% MC (first week of February 2012). The moisture profile on the sheathing board in the ventilated wall system (F3) is shown in Figure 6. The moisture content on the sheathing board varies from 6% MC in the first week of September to the highest moisture content of 13% MC in the first week of February. The lowest and the highest moisture content measurements are made at the middle and lower sections of the sheathing board, respectively. Like F1, in this wall system the upper and middle sections of the sheathing board have very similar moisture content, but the lower section of the sheathing board has slightly higher moisture content in comparison to the other two sections, which is a reverse situation to wall F1. During the high moisture accumulation period (February and March) the moisture content difference between the lower section and the upper and middle sections is about 2%. The moisture content difference in these sections reduces below 1% during the drying periods.

To compare the hygrothermal performance of the three wall systems, the moisture content of the lower, middle, and upper sections of the respective sheathing boards are plotted in Figure 7 to Figure 9. As it can be seen in Figure 7 (lower section), approximately from the beginning of November 2011 to the end of April 2012, the sheathing board in F1 has a relatively high moisture content when compared to that of F2 and F3 wall systems. During this period, the moisture content difference between F1 and the other two walls (F2 and F3) can be as high as 7%. In the other six-month period

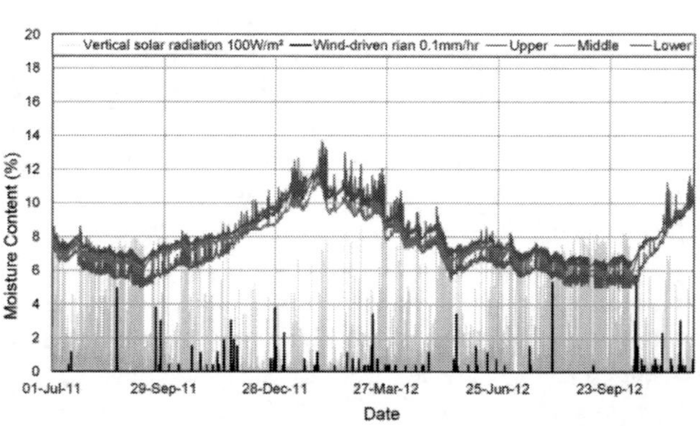

Figure 5 Hourly average moisture content in plywood sheathing of F2, a vented wall.

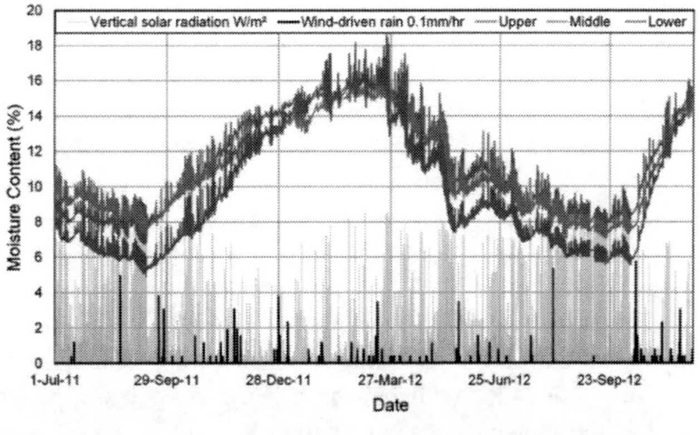

Figure 4 Hourly average moisture content in plywood sheathing of F1, a wall with no air gap.

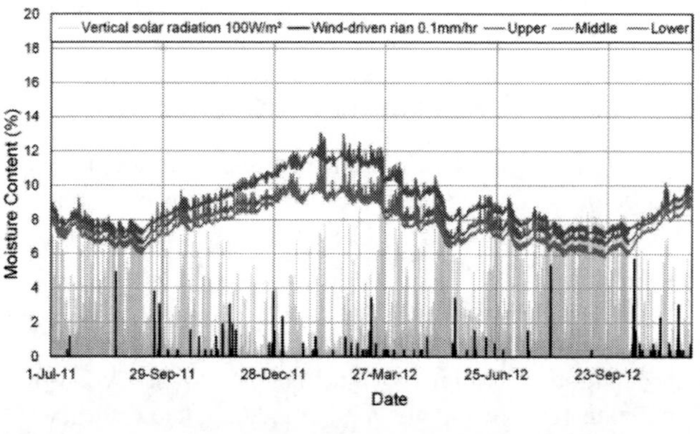

Figure 6 Hourly average moisture content in plywood sheathing of F3, a ventilated wall.

(May to October), the moisture contents in the sheathing boards are comparable and are within a 2% MC difference. Within a year cycle, the amplitude of moisture content change in F1 is 13% whereas in F2 and F3 they are 7% and 6%, respectively. Given that the three walls are exposed to similar indoor and outdoor climatic conditions and significant moisture content differences on the sheathing boards are observed while the frequencies of rain events are higher (October to April), the substantial hygrothermal performance difference between the wall systems must be associated with the presence (F2 and F3) and absence (F1) of an air gap and the role it plays as a capillary break.

In comparison to F2 and F3, the rate of moisture accumulation after the drying period is higher in F1, Figure 7. While its moisture content continues to increase for the third week of March before it starts drying, the sheathing board moisture contents in F2 and F3 reach to their maximum level in the first week of February (six weeks prior to F1's peak time) and maintain about the same level of moisture content before start drying at the end of March. The relatively stable moisture content readings observed on the F2 and F3 sheathing boards in February and March, in spite of the rain events and winter cold temperature, is believed to be the result of moisture removal from the sheathing board and cladding by the airflow through the air gap. The airflow is believed to be induced by the increases in solar radiation in February and March (compared to November to January).

As shown in Figure 8 and Figure 9, the moisture content at the middle and upper sections in F1 are consistently higher than that of the corresponding F2 and F3 wall systems. In the middle section, Figure 8, the amplitudes of sheathing board moisture content change in a year cycle (difference of maximum minus minimum values) are about 10% MC, 7% MC, and 5.5% for the wall systems of F1, F2, and F3, respectively. In the upper section, Figure 9, the moisture content changes in a year cycle in F3 is relatively low (5%: varies from 6.5% to 11.5%) compared to F2 (8.3%: varies from 5.3% to 13.6%) and F1 (12%: varies from 6.5% to 18.5%).

In general, the hygrothermal performance of vented (F2) and ventilated wall (F3) systems are comparable. The moisture content of the lower section of the ventilated wall system (F3) seems to be higher by 1% compared to the vented wall system, Figure 7, which may be attributed to the enhanced airflow in F3 that brings in moist outdoor air into the air cavity. In the two wall systems the moisture contents at the middle section of the respective sheathing boards are nearly the same during most of the monitoring period, Figure 8. Like the lower section of the sheathing board, the upper section of F3 has a slightly (about 1%) higher moisture content during the spring and summer periods, Figure 9. However, during the winter period when relatively high moisture accumulation occurred, the upper section of F3 shows slightly lower (1% to 2%) moisture content compared to that of the vented wall system (F2), Figure 9, which leads to a moisture content profile with relatively low amplitude

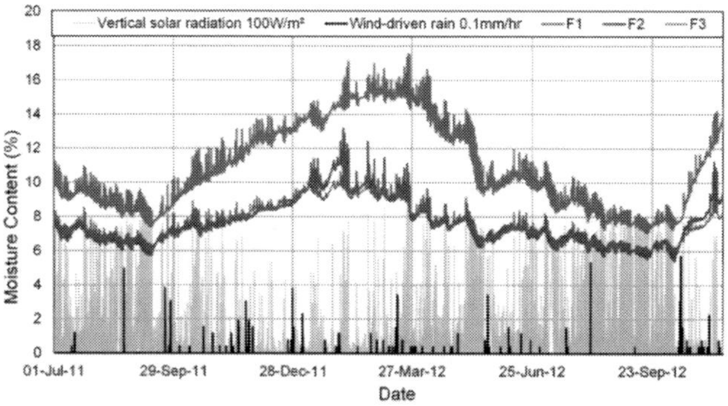

Figure 8 *Comparison of hourly average moisture content at middle part of plywood sheathing between walls: F1, wall with no air gap; F2, vented wall; and F3, ventilated wall.*

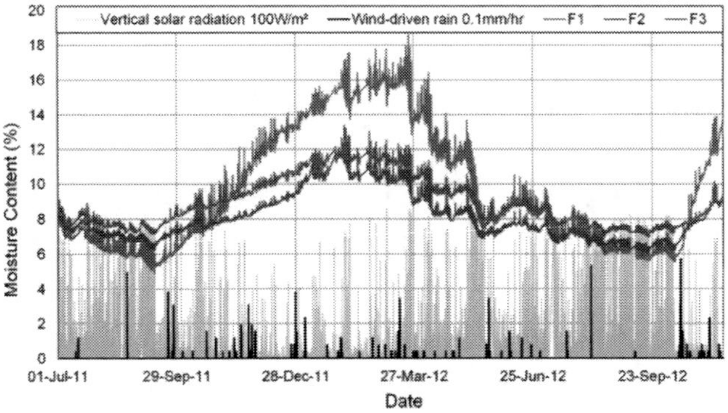

Figure 7 *Comparison of hourly average moisture content at lower part of plywood sheathing between walls: F1, wall with no air gap; F2, vented wall; and F3, ventilated wall.*

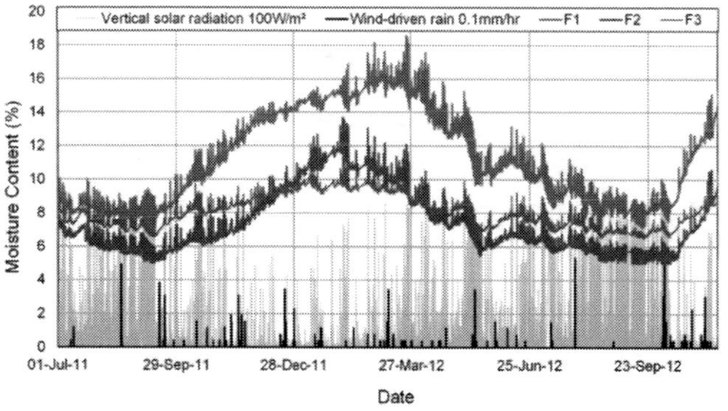

Figure 9 *Comparison of hourly average moisture content at upper part of plywood sheathing between walls: F1, wall with no air gap; F2, vented wall and F3, ventilated wall.*

between the dry and wet instances. The enhanced moisture management potential of F3 at the critical time when the high moisture accumulation occurs is believed to be the result of solar radiation-induced airflow in the wall F3 air cavity enhanced by the open flashing design. For the entire monitoring period, the sheathing boards in both wall systems are under 14% moisture content and under 12% for most of the monitoring period.

The moisture contents in the studs of the three wall systems at the mid height and 12.7 mm (1/2 in.) from the outside surface of the studs are shown in Figure 10. As the sheathing board, the moisture content of the stud on the wall with no air cavity shows higher moisture accumulation compared to that of vented and ventilated wall systems. The maximum moisture contents of the respective wall systems are about 16% (F1) and 12% (F2 and F3). The moisture content magnitudes and profiles of F2 and F3 studs are similar. The difference in moisture content between walls with and without an air gap is lower during the summer period and increases during the fall and winter periods to about 4% difference in the third week of March. The moisture content profiles at the middle and interior locations of the studs are nearly stable at about 10% MC during the majority of the monitoring period (figure not shown here because of space limitation). It is also observed that when solar radiation is significant in the spring and summer periods, the interior part of the stud seems to show a slight increase in moisture content, which is believed to be a consequence of moisture release from the sheathing board to the interior cavity and moisture drives from the exterior to the interior part of the studs as they are going through a drying process.

Similar to the interior part of the stud, the moisture contents at the interior parts of the top and bottom plates are stable (about 10%) for the majority of the time and show a slight increase in moisture content when solar radiation is high. Figure 11 shows the moisture content profiles of the bottom plates of the three walls at a location 12.7 mm (1/2 in.) from the outside surface of the plate. As it can be seen in the figure, the moisture content at the bottom plate of the vented wall (F2) is lower (by about 1%) when compared with that of the ventilated wall (F3). The difference is consistent throughout the monitoring period (wet and dry seasons). During the moisture uptake period, the moisture content of the bottom plate of the wall with no air gap (F1) increases from about 8.5% to 14% while that of F2 increases from 8.5% to only 10.5%. During the drying period, the bottom plate in F1, however, manages to dry fast and reach to the same level of moisture content as that of F2 (8%) in the fourth week of September, which is a decrease of 5.5% for F1 and 2.5% for F2 from their respective peak moisture contents. In general, the moisture accumulation and drying processes of the bottom plate in the wall with no air cavity (F1) seem to be faster in the respective wet and dry seasons.

Figure 12 shows the inside surface temperatures of the sheathing boards at the mid-height of the respective three wall systems. The lowest and the highest temperature read-

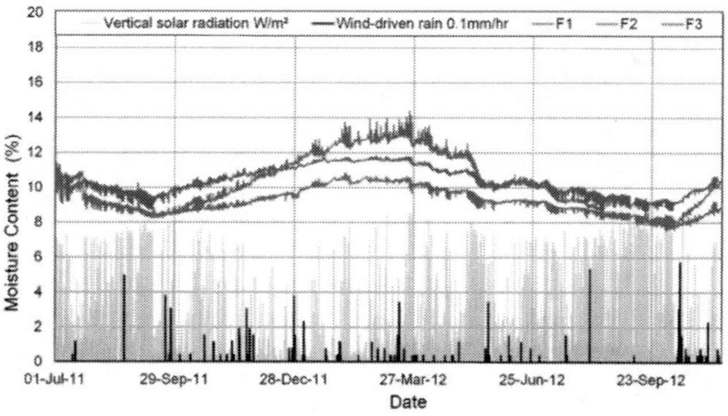

Figure 11 *Comparison of hourly average moisture content in bottom plates, 12.7 mm (1/2 in.) near exterior edge, between walls: F1, wall with no air gap; F2, vented wall; and F3, ventilated wall.*

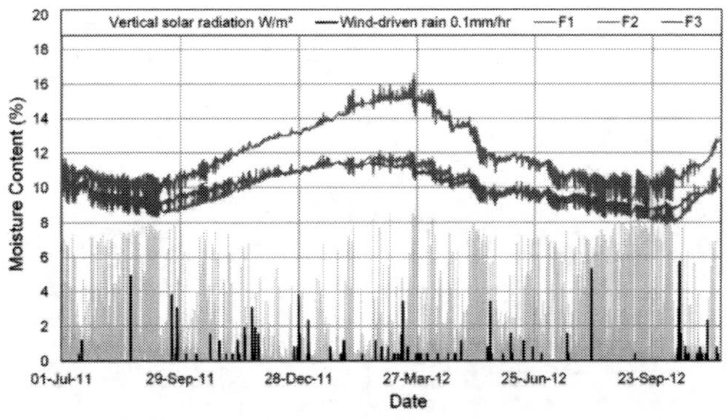

Figure 10 *Comparison of hourly average moisture content at the middle height of stud, 12.7 mm (1/2 in.) near the exterior edge, between walls: F1, wall with no air gap; F2, vented wall; and F3, ventilated wall.*

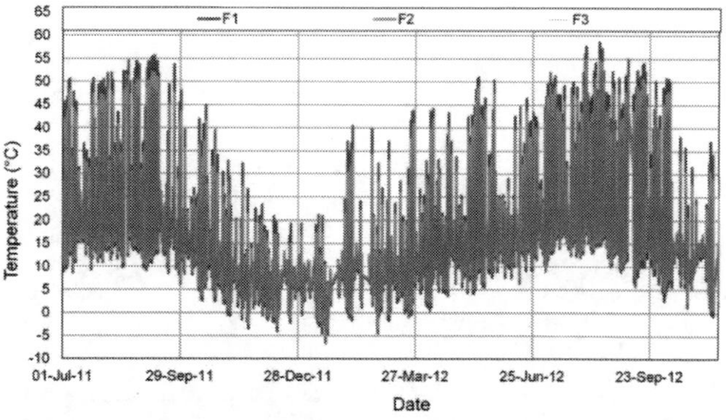

Figure 12 *Comparison of hourly average temperature at the middle part of the sheathing board, inside surface, between walls: F1, wall with no air gap; F2, vented wall; and F3, ventilated wall.*

ings on the sheathing board, which are –6.3°C on January 17, 2012 and 58.5°C on August 15, 2012, are made on the wall with no air gap (F1). Although the temperature profiles of the three walls' sheathing boards are similar, noticeable differences are observed during the entire monitoring period. It seems that the sheathing board temperature in F1 rises slightly above F2's and F3's when the solar radiation is present, otherwise its temperature is slightly lower than the other two. Figure 13 shows the cumulative probability distribution of the temperature difference between the sheathing boards using the wall with no air gap (F1) sheathing board temperature as a reference. Positive values in the x-axis indicate that the sheathing board in the wall with no air gap is warmer in comparison to that of vented and ventilated wall systems, and the negative values indicate the opposite. The temperature difference between F1 and F2 sheathing boards can vary from –3°C to 7°C, whereas the difference between F1 and F3 can range from –2°C to 13.5°C. The sheathing board in the wall with no air gap is warmer than the vented and ventilated wall systems for about 14.5% and 27% of the time (hourly data of a year). These temperature differences can be as high as 7°C in the case of the vented wall system and 13.5°C in the case of the ventilated wall system. Based on the measured solar radiation data, the 14.5 and 27 percentiles solar radiation corresponds to 150 W/m^2 and 50 W/m^2, respectively. This suggests that a stronger solar radiation (above 150 W/m^2) is required to have air movement in the vented cavity, whereas in the ventilated cavity just 50 W/m^2 may start air movement. At times when the solar radiation is between 0 to 50 W/m^2, which is about 85% of the time, the presence of the air gap in F2 and F3 leads to a slightly warmer sheathing board (to the maximum of 2°C to 3°C) as compared to wall F1.

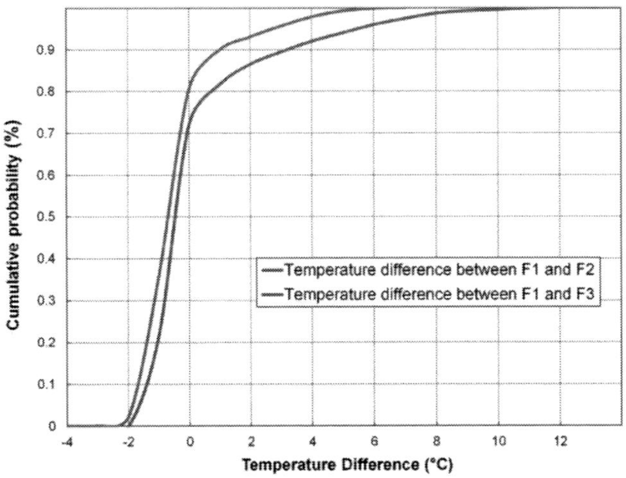

Figure 13 *Cumulative probability distribution of temperature differences between the sheathing boards, using the sheathing board temperature of F1 (the wall with no air gap) as a reference.*

The temperature difference within the three wall systems can be explained as follows: the solar radiation heats the cladding and then the air in the F2 and F3 cavities, which creates a situation where fresh air will be drawn into the air cavity and the warm air leaves the cavity, cooling the sheathing board and cladding, a phenomena which won't happen in F1. The temperature differences in the sheathing boards of the vented and ventilated wall systems are related to the amount of airflow through the air cavity, which depends on the type of top flashing configuration. The more open the flashing design, the higher solar radiation induced airflow through an air cavity. The air movement through the air cavity at nighttime or when solar radiation is minimal is expected to be none or significantly less than the solar driven airflow rate. During these times, the sheathing boards in the walls with an air cavity seem to have a slightly higher temperature, which is because of the added thermal resistance obtained from the air in the cavity without being short-circuited by cross airflow (a situation near still air condition).

In addition to the sheathing board, the airflow through the air cavity in walls F2 and F3 has an impact on the temperature of the fiber cement cladding. The cladding temperature difference between the wall with no air gap and the vented wall ranges from –3°C to 6°C, whereas between the wall with no air gap and the ventilated wall system the difference is –4°C to 9.5°C. For over 95% of the time the cladding on the wall with no air gap has a higher temperature than either of the other wall systems' cladding. For 20% of the time, the temperature difference is greater than 1°C.

CONCLUSION

Here, the hygrothermal performances of ventilated and vented wall systems are evaluated in comparison with a wall system with no air gap between sheathing membrane and cladding. The three test panels are simultaneously tested in a field experimental setting. The test panels are exposed to the same environmental loads on both interior and exterior sides of the test panels. For the wall systems, orientation, and boundary conditions considered in this study, the wall with no air gap accumulates relatively high moisture content on the sheathing board, stud, and bottom plate and also has high moisture content changes in a year cycle compared to the vented and ventilated wall systems. In this experiment, the moisture content of the sheathing boards in the vented and ventilated wall systems are under 14%, while the moisture content of the sheathing board in the wall with no air gap reaches to 18.5%. The air gap creates a capillary break and reduces the rate as well as the amount of moisture accumulations on the sheathing board. In general, the hygrothermal performance of vented and ventilated wall systems are comparable, but with some slight differences. The upper and bottom sections of the sheathing board in the ventilated wall system seem to have slightly higher moisture content than that of the corresponding sections of the vented wall system. However, during the winter period when relatively high moisture accumulation occurs, the upper section of the ventilated wall system shows slightly lower (1% to 2%) moisture content compared to that of the vented wall system. The better moisture management potential of the ventilated wall system at the critical time when the high moisture accumulation occurs is related to the enhanced (solar radiation induced) airflow that resulted from the open flashing design.

The air gap in the vented and ventilated wall systems adds additional thermal resistance and keeps the respective sheathing boards slightly warmer than that of the wall with no air gap for 85.5% and 73% of the time (based on hourly data of a year), respectively. The temperature difference can be as high as 2°C to 3°C. For the balance of a period of time, the sheathing boards in the walls with an air cavity are slightly cooler than that of the wall with an air gap. The temperature difference can be as high as 7°C and 13.5°C. Although the low sheathing board temperature, which is caused by solar radiation-induced airflow, is beneficiary during the cooling season, the air gap and the associated airflow may reduce the heat gain that may be obtained from solar radiation during the heating season. The implications of air cavity and flashing design (airflow rate) on the heating and cooling load calculations of different orientations, wall configurations, and climate require further investigation.

ACKNOWLEDGMENTS

The authors are grateful for the financial support provided by the Natural Sciences and Engineering Research Council of Canada (NSERC) and the School of Construction and the Environment at the British Columbia Institute of Technology (BCIT), and the technical support provided by Mr. Stephen Roy.

REFERENCE

Bassett, M.R., and McNeil. (2005). Drained and vented cavity walls—measured ventilation rates. *IRHACE Conference*. Institute of Refrigeration, Heating and Air Conditioning Engineers in., Nelson, New Zealand.

Burnett, E., Straube, J., and Karagiozis, A. (2004). Synthesis report and guidelines. ASHRAE 1091 Report #12. Final Report.

Hazleden, D., and Morris, P. (2002). The Influence of Design on Drying of Wood-Frame Walls Under Controlled Conditions. *Thermal Performance of Building Envelopes VIII 2002*. Clearwater, FL.

Morrison Hershfield Limited. (1998). Survey of building envelope failures in the coastal climate of British Columbia, Canadian Housing Information Center (CHIC), Technical Series 98-102.

National Building Code of Canada (2005): http://www.nationalcodes.ca/nbc/index_e.shtmlU

Onysko, M.D. (2000). Drying of walls with ventilation: a parametric analysis. Research report-Housing Technology Series, Canada Mortgage and Housing Corporation.

Rousseau, M.Z., and Dalgliesh, W.A. (2004). Selected Findings of An IRC Study of The Wetting and Drying Potentials of Wood-frame Walls Exposed to Different Climates. CIB 2004 World Building Congress, Toronto, ON, Canada.

Shi, X., and E. Burnett. (2006). Ventilation drying in walls with vapor impermeable claddings. *Proceedings of the 3rd International Conference in Building Physics and Building Engineering*. pp. 395-402. Montreal, Canada.

Simpson, Y. (2010). Field Evaluation of Ventilation Wetting and Drying of Rain screen Walls in Coastal British Columbia. M.A.Sc. Thesis (Building Engineering), Concordia University, Montreal. QC, Canada.

Straube, J., VanStraaten, R., Burnett, E., and Schumacher, C. (2004). Review of literature and Theory. ASHRAE 1091-Development of Design Strategies for Rainscreen and Sheathing Membrane Performance in Wood Frame Walls. Report#1. Final Report.

Straube, J., Onysko, M.D., and Schumacher, C. (2002). Methodology and Design of Field Experiments for Monitoring the Hygrothermal Performance of Wood Frame Enclosures. *Journal of Thermal Envelope and Building Science*. 26(2):123-151.

Tariku, F., and Ge, H. (2010). Moisture Response of Sheathing Board in Conventional and Rain-Screen Wall Systems with Shiplap Cladding. *Journal of Testing and Evaluation*, 39:381-88.

Natural Convection in Below Grade Open Cell Concrete Block Walls

Merle F. McBride, Ph.D., P.E.
Life Member ASHRAE

ABSTRACT

A series of eight calibrated hot box experimental tests were completed to determine the impact of natural convection in below grade open cell concrete block walls. The thermal boundary conditions associated with the exterior soil were included by using a tapered layer of insulation. Once the tests were completed with the open cells they were then filled with sand to eliminate any convection affects and retested. The key challenge was to investigate the 38% over-prediction of the R-value of walls insulated only on the top half by the arc-length method. The results were intended to be used to further develop a simplified analytical model that can estimate the energy savings with various insulation strategies on the interior or exterior at full depth or partial depths. Key findings were that the heat transfer in the hollow cores is extremely complex and driven primarily by natural convection but also includes radiation and conduction through the webs and shell faces of the concrete blocks. Natural convection was identified as the major driving force. The results did not lead to any new modeling procedures or correlations that would improve the arc length model. Future modeling improvements will need to focus on a thermal network that simultaneously accounts for all of the heat transfer mechanisms.

INTRODUCTION

The thermal performance of below grade masonry block walls is complex and influenced by the surrounding soil which makes it a two dimensional heat transfer problem with natural convection and radiation within the block cavities and vertical heat transfer through the concrete webs and face shells. The thermal boundary conditions associated with the exterior soil were experimentally investigated by using a tapered layer of insulation. The four cases of interior insulation tested in a calibrated hot box were: (1) no insulation, (2), R-5 (R-0.9) full wall, (3) R-11 (R-1.9) full wall and (4) R-11 (R-1.9) on just the top half of the wall. Once the four open cell tests were completed the cells were then filled with sand to eliminate convection affects and retested.

The results of these tests were initially analyzed using a one-dimensional steady-state, concentric arc heat-flow path, "series-parallel" (isothermal planes) procedure (Shipp-1983), see Figure 1. The average difference between the measured and predicted wall thermal resistances was 4% with sand filled cavities and 6% with open cavities except when the insulation only covered the top half of the interior surface. For that case the predicted thermal resistance was 38% greater than the measured value.

It was concluded that the simplified model was adequate except for the case when the wall cavities are empty and the wall was insulated only on the top half.

DESCRIPTION OF THE THERMAL TESTS

The thermal performance of below grade masonry block walls was investigated through a series of eight steady-state calibrated hot box (CHB) tests for a variety of constructions in a 9 ft (2.7 m) high by 14 ft (4.3 m) wide test frame, see Figure 2. Heat flux transducers and thermocouples were installed on the interior and exterior surfaces of the masonry block wall to define six thermal nodes. All of the tests used an interior air

***Dr. McBride** is a Senior Research Associate at the Owens Corning Center of Science and Technology, Granville, OH.*

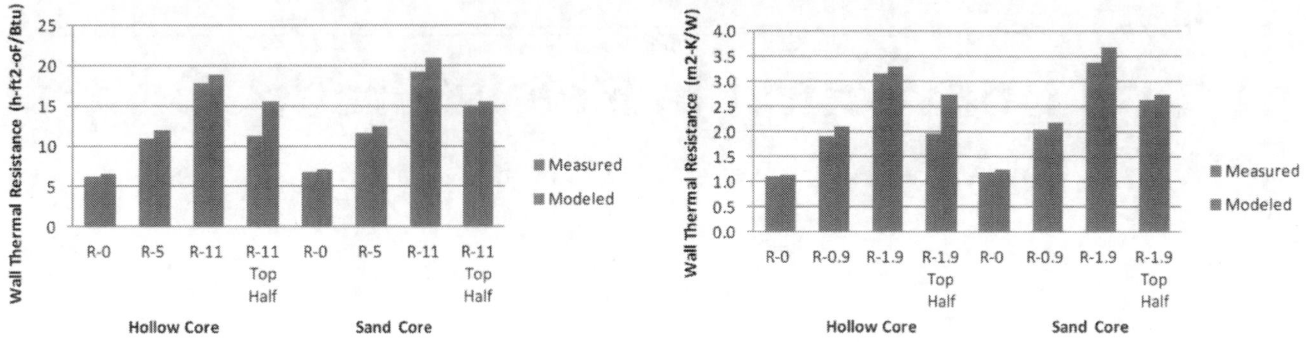

Figure 1 *Measured vs modeled wall R-values: (left) I-P and (right) SI.*

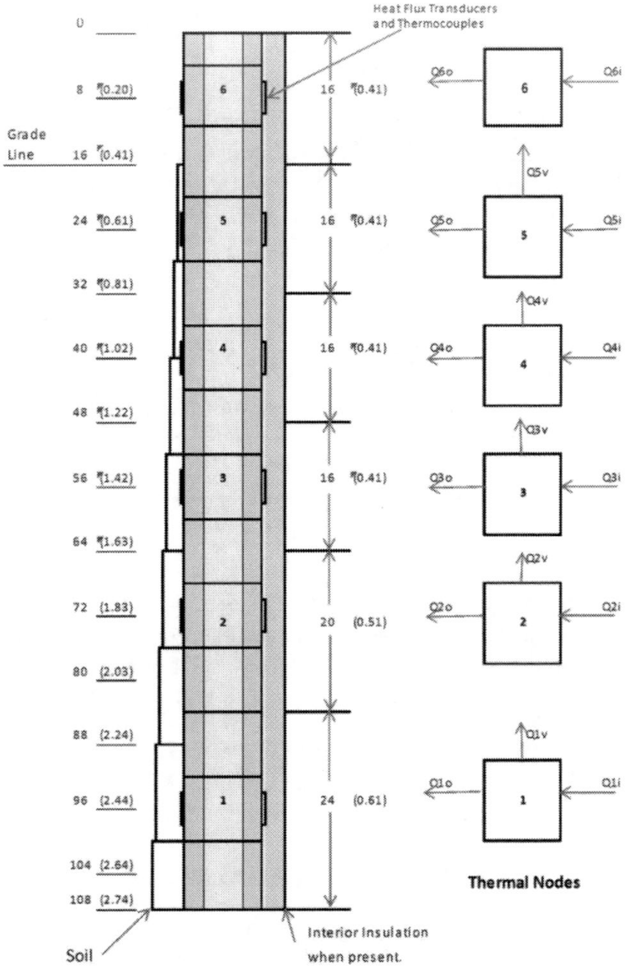

Figure 2 *Wall construction and six thermal nodes, in. (m).*

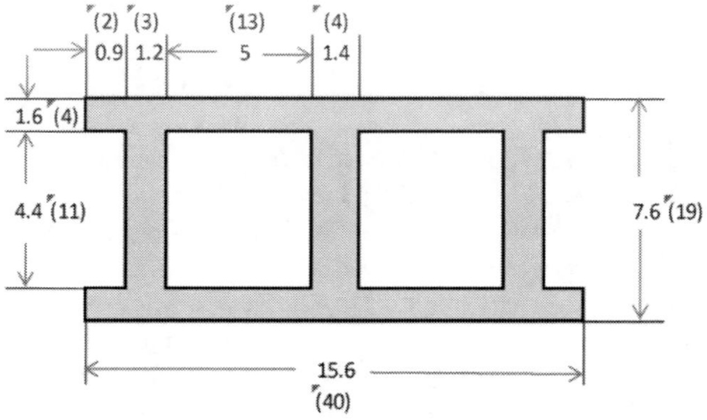

Figure 3 *Masonry block dimensions, in. (cm).*

temperature of 75°F (24°C) and an exterior air temperature of 25°F (–3.9°C).

The masonry block characteristics were: two cores, nominal dimensions of 8 in. (0.20 m) × 8 in. (0.20 m) by 16 in. (0.41 m) (Figure 3), 36 lb (16 kg), and a thermal conductivity of the concrete equal to 0.88 Btu/h·ft·°F (1.5 W/m·K). The thermal conductivity of the sand was 0.13 Btu/h·ft·°F (0.22 W/m·K).

The interior wall insulation configurations were: (1) no insulation, (2) 1 in. (0.03 m) wood furring strips with R-4.8 (R-0.84) insulation plus 0.5 in. (0.01 m) drywall, (3) 24 in. (0.61 m) on center 2 × 4 wood studs, R-11 (R-1.9) insulation full wall plus 0.5 in. (0.03 m) drywall and (4) 24 in. (0.61 m) on center 2 × 4 wood studs full wall, R-11 (R-1.9) insulation on just the top half of the wall plus 0.5 in. (0.01 m) drywall on the entire wall.

The thermal impact of the soil below grade was addressed by installing layers of insulation to the exterior portion of the wall beginning with R-0.8 (R-0.14) for the first foot (0.30 m) below grade and ending with R-15 (R-2.6) at the footer. The thermal conductivity of the soil modeled was 0.78 Btu/h·ft·°F (1.3 W/m·K).

MEASURED RESULTS

The measured results for all eight of the calibrated hot box tests are shown in Tables 1–4. Each table contains the results for the cavity with air and the corresponding results when the cavity was filled with sand. A control volume was defined for each of the thermal nodes and an energy balance was completed which determined the vertical heat flow exiting each node. The vertical heat flow in cavities with air includes natural convection, radiation plus vertical heat transfer through the webs and face shells. The vertical heat flow with sand in cavities includes conduction due to the sand, radiation but to a much smaller extent and vertical heat transfer through the webs and face shells. The difference between these two vertical heat flows at each node illustrates the vertical heat flow due to natural convection and radiation, see Figures 4–7.

Cavities with sand in the cores reduce the radiation but does not totally eliminate it. The density of sand depends upon the particle size and how tightly the particles are packed. Typical sand densities range from 90 to 112 lb/ft^3 (1442 to 1,794 kg/m^3) while the density of SiO_2 is 165 lb/ft^3 (2,643 kg/m^3)which leads to void volumes of 32 to 45%. Another consideration associated with these particular CHB tests was that the sand was installed in the wall cavities after the wall was built. Sand flowing from the top of the wall to the bottom may not have completely filled in below all of the internal webs leaving voids that would allow some radiation to proceed uninhibited.

ANALYSIS

Natural convection in vertical enclosures can be shown by graphing the Grashof (Gr) number on the horizontal axis and Ub/k on the vertical axis, see Figure 8.

The solid line (Kreith) shows pure conduction up to Gr = 6×10^3 and then laminar natural convection begins. There is a transition at Gr = 10^5 to fully turbulent flow. Empirical formulas (Jakob) were presented by (Karlekar and Desmond) for $20{,}000 < Gr < 200{,}000$ as Equation 1.

$$k_e = 0.18k\,Gr_\delta^{1/4}(\delta/L)^{1/9} \quad (1)$$

where

k_e = effective thermal conductivity, Btu/h·ft·°F (W/m·K)

k = thermal conductivity of air, Btu/h·ft·°F (W/m·K)

Gr = Grashof Number, dimensionless

δ = thickness of the air space, ft (m)

L = height of the air space, ft (m)

For $200{,}000 < Gr < 10 \times 10^6$ use Equation 2.

$$k_e = 0.065k\,Gr_\delta^{1/3}(\delta/L)^{1/9} \quad (2)$$

Results for these two formulas agree very closely with the solid line. It is important to note that these equations apply to vertical enclosures with interior and exterior surfaces that are of uniform thermal transmittance.

A typical application would be above grade masonry block walls. For example the results of tests conducted in a guarded hot box on masonry block walls are shown in Figure 8 (Van Geem). The Gr number was calculated using the interior and exterior air temperatures as the temperature difference and their average as the mean temperature. These results are consistent with the line for fully turbulent flow ($Gr > 1.0 \times 10^5$) and Equation 2.

Analysis of the experimental data began by defining six thermal nodes that were aligned with the heat flux transducers and thermocouples installed on the walls, Figure 2. All of the results for each node are presented in Tables 1A–4A in I-P

Table 1A. No Interior Insulation (I-P)

Node	Height in.	Cavity—Hollow Q_o Btu/h	T_o °F	Q_{vert} Btu/h	Q_i Btu/h	T_i °F	Gr	Cavity—Sand Q_o Btu/h	T_o °F	Q_{vert} Btu/h	Q_i Btu/h	T_i °F
6	16	415.2	45.5	0	274.6	70.8	3.40E+06	368.6	42.2	0	313.3	70.1
5	16	214.7	54.5	140.6	222.4	69.2	1.93E+06	205.1	53.9	55.3	230.2	69.4
4	16	122.5	59.7	132.8	154.0	70.1	1.34E+06	127.0	60.4	30.1	140.9	70.9
3	16	89.1	62.4	101.4	116.0	70.6	1.03E+06	96.7	64.8	16.2	97.6	71.6
2	20	99.5	65.0	74.5	127.3	70.8	7.29E+05	110.6	66.4	15.3	98.0	72.1
1	24	72.8	66.9	46.7	119.5	70.4	4.33E+05	74.9	67.1	27.9	102.8	71.4

Table 2A. R-5 Interior Insulation (I-P)

Node	Height in.	Cavity—Hollow Q_o Btu/h	T_o °F	Q_{vert} Btu/h	Q_i Btu/h	T_i °F	Gr	Cavity—Sand Q_o Btu/h	T_o °F	Q_{vert} Btu/h	Q_i Btu/h	T_i °F
6	16	226.5	33.3	0	117.0	46.4	2.01E+06	170.8	31.1	0	136.7	42.2
5	16	125.1	40.8	109.4	98.7	48.6	1.15E+06	110.9	38.1	34.0	112.9	45.8
4	16	75.1	44.0	135.8	88.7	50.8	9.81E+05	80.8	45.4	32.1	89.9	51.3
3	16	56.2	46.1	122.2	80.6	52.1	8.53E+05	69.3	50.1	23.0	73.9	54.8
2	20	61.6	49.5	97.8	101.8	53.1	5.08E+05	82.3	54.4	18.3	82.9	58.0
1	24	48.9	51.5	57.6	106.5	54.2	3.69E+05	64	55.8	17.8	81.8	59.3

Table 3A. R-11 Interior Insulation (I-P)

Node	Height in.	Cavity—Hollow Q_o Btu/h	T_o °F	Q_{vert} Btu/h	Q_i Btu/h	T_i °F	Gr	Cavity—Sand Q_o Btu/h	T_o °F	Q_{vert} Btu/h	Q_i Btu/h	T_i °F
6	16	131	29.5	0	68.4	39.6	1.60E+06	82.3	30.2	0	69.9	36.3
5	16	78.9	34.8	62.5	60.0	41.4	1.01E+06	55.9	42.3	12.4	64.1	39.8
4	16	47.6	38.5	81.2	54.5	43.3	7.30E+05	43.9	42.2	4.2	48.3	45.8
3	16	36.6	40.5	74.3	49.6	44.4	5.86E+05	41.9	48.2	0	41.2	50.7
2	20	39.3	43.4	61.4	64.9	45.7	3.43E+05	50.7	52.3	0.5	47.9	55.0
1	24	30.2	45.4	35.7	65.9	46.9	2.21E+05	37.6	54.9	3.2	40.8	57.0

Table 4A. R-11 Top Half Interior Insulation (I-P)

Node	Height in.	Cavity—Hollow Q_o Btu/h	T_o °F	Q_{vert} Btu/h	Q_i Btu/h	T_i °F	Gr	Cavity—Sand Q_o Btu/h	T_o °F	Q_{vert} Btu/h	Q_i Btu/h	T_i °F
6	16	189.7	34.5	0	40.4	44.2	1.49E+06	114.8	31.7	0	70.8	39.4
5	16	115.9	41.2	149.2	38.7	48.0	9.99E+05	70.5	41.9	44.0	53.7	41.7
4	16	70.7	45.1	226.3	34.2	50.4	7.68E+05	57	44.9	60.8	31.2	48.4
3	16	58.4	49.3	262.8	165.2	58.4	1.26E+06	62.1	55.2	86.7	126.5	61.7
2	20	66.7	53.2	156.1	144.2	59.0	7.90E+05	75.3	59.5	22.3	72.3	64.4
1	24	50.7	55.0	78.6	129.3	59.1	5.58E+05	51.6	60.5	25.3	76.9	64.1

Table 1B. No Interior Insulation (SI)

Node	Height m	Cavity—Hollow Q_o W	T_o °C	Q_{vert} W	Q_i W	T_i °C	Gr	Cavity—Sand Q_o W	T_o °C	Q_{vert} W	Q_iQ_i W	T_i °C
6	0.41	121.7	7.5	0.0	80.5	21.6	3.40E+06	108.0	5.7	0.0	91.8	21.2
5	0.41	62.9	12.5	41.2	65.2	20.7	1.93E+06	60.1	12.2	16.2	67.5	20.8
4	0.41	35.9	15.4	38.9	45.1	21.2	1.34E+06	37.2	15.8	8.8	41.3	21.6
3	0.41	26.1	16.9	29.7	34.0	21.4	1.03E+06	28.3	18.2	4.7	28.6	22.0
2	0.51	29.2	18.3	21.8	37.3	21.6	7.29E+05	32.4	19.1	4.5	28.7	22.3
1	0.61	21.3	19.4	13.7	35.0	21.3	4.33E+05	22.0	19.5	8.2	30.1	21.9

Table 2B. R-0.9 Interior Insulation (SI)

Node	Height m	Cavity—Hollow Q_o W	T_o °C	Q_{vert} W	Q_i W	T_i °C	Gr	Cavity—Sand Q_o W	T_o °C	Q_{vert} W	Q_i W	T_i °C
6	0.41	66.4	0.7	0.0	34.3	8.0	2.01E+06	50.1	–0.5	0.0	40.1	5.7
5	0.41	36.7	4.9	32.1	28.9	9.2	1.15E+06	32.5	3.4	10.0	33.1	7.7
4	0.41	22.0	6.7	39.8	26.0	10.4	9.81E+05	23.7	7.4	9.4	26.3	10.7
3	0.41	16.5	7.8	35.8	23.6	11.2	8.53E+05	20.3	10.1	6.7	21.7	12.7
2	0.51	18.1	9.7	28.7	29.8	11.7	5.08E+05	24.1	12.4	5.4	24.3	14.4
1	0.61	14.3	10.8	16.9	31.2	12.3	3.69E+05	18.8	13.2	5.2	24.0	15.2

Table 3B. R-1.9 Interior Insulation (SI)

Node	Height m	Cavity—Hollow Q_o W	T_o °C	Q_{vert} W	Q_i W	T_i °C	Gr	Cavity—Sand Q_o W	T_o °C	Q_{vert} W	Q_i W	T_i °C
6	0.41	38.4	−1.4	0.0	20.0	4.2	1.60E+06	24.1	−1.0	0.0	20.5	2.4
5	0.41	23.1	1.6	18.3	17.6	5.2	1.01E+06	16.4	5.7	3.6	18.8	4.3
4	0.41	14.0	3.6	23.8	16.0	6.3	7.30E+05	12.9	5.7	1.2	14.2	7.7
3	0.41	10.7	4.7	21.8	14.5	6.9	5.86E+05	12.3	9.0	0.0	12.1	10.4
2	0.51	11.5	6.3	18.0	19.0	7.6	3.43E+05	14.9	11.3	0.1	14.0	12.8
1	0.61	8.9	7.4	10.5	19.3	8.3	2.21E+05	11.0	12.7	0.9	12.0	13.9

Table 4B. R-1.9 Top Half Interior Insulation (SI)

Node	Height m	Cavity—Hollow Q_o W	T_o °C	Q_{vert} W	Q_i W	T_i °C	Gr	Cavity—Sand Q_o W	T_o °C	Q_{vert} W	Q_i W	T_i °C
6	0.41	55.6	1.4	0.0	11.8	6.8	1.49E+06	33.6	−0.2	0.0	20.7	4.1
5	0.41	34.0	5.1	43.7	11.3	8.9	9.99E+05	20.7	5.5	12.9	15.7	5.4
4	0.41	20.7	7.3	66.3	10.0	10.2	7.68E+05	16.7	7.2	17.8	9.1	9.1
3	0.41	17.1	9.6	77.0	48.4	14.7	1.26E+06	18.2	12.9	25.4	37.1	16.5
2	0.51	19.5	11.8	45.7	42.3	15.0	7.90E+05	22.1	15.3	6.5	21.2	18.0
1	0.61	14.9	12.8	23.0	37.9	15.1	5.58E+05	15.1	15.8	7.4	22.5	17.8

units and Tables 1B–4B in SI units.The experimental data for each node includes the measured heat flux and temperature at the outside and inside surfaces of the block walls.

A control volume was defined around each thermal node and an energy balance was completed. Beginning with node 1 there was a net vertical heat flow (Q_{vert}). The Grashof number was also calculated for each node and is included in the table. The next step in the analysis was to calculate *Ub*/*k* for each node using Equation 3.

$$U = \frac{Q_{vert}}{A(T_i - T_o)} \tag{3}$$

where

Q_{vert} = vertical heat flow, Btu/h (W)
A = area of the thermal node, ft^2 (m^2)
T_i = inside surface temperature of the face shell, °F (°C)
T_o = outside surface temperature of the face shell, °F (°C)

In addition to the following variables:

b = thickness of the cavity, 4.4 in (0.11 m)
k = thermal conductivity of air, Btu/h·ft·°F (W/m·K)

where the thermal conductivity is evaluated at the mean temperature of T_i and T_o.

The analysis for node 2 accounted for the vertical energy flow from node 1 and the net result was an increase in the vertical heat flow from node 2. Analysis of the subsequent nodes exhibited a similar trend through node 4 which had the largest vertical heat flow of any of the nodes for those walls with a constant layer of interior insulation, Figures 4–6. The exception to this was the wall which only had the top half of it insulated, Figure 7 in which node 3 had the largest vertical heat flow. Node 3 was physically located just below the bottom edge of the insulation that had been installed on the top half of the wall. The insulation on the top half of the wall kept the block faces cooler than those on the bottom half of the wall which further enhanced the vertical convection. Proceeding further up the wall the temperature difference between the inside and outside faces of the block begins to decrease which causes the natural convection to decrease.

Figures 4–7 also show the vertical heat flow for the walls which had sand in the cavities. In all cases the vertical heat flow through the walls with cavities that were filled with sand were lower than the vertical heat flow through the walls with cavities that were hollow. Clearly the sand altered the vertical natural convection and radiation heat transfer.

When the results for the four tests for below grade masonry walls with cavities that were hollow are added to Figure 8 substantial departure from the above grade masonry walls was exhibited as expected.

DISCUSSION

The analysis of vertical heat flow due to natural convection in below grade basement walls with cavities that are

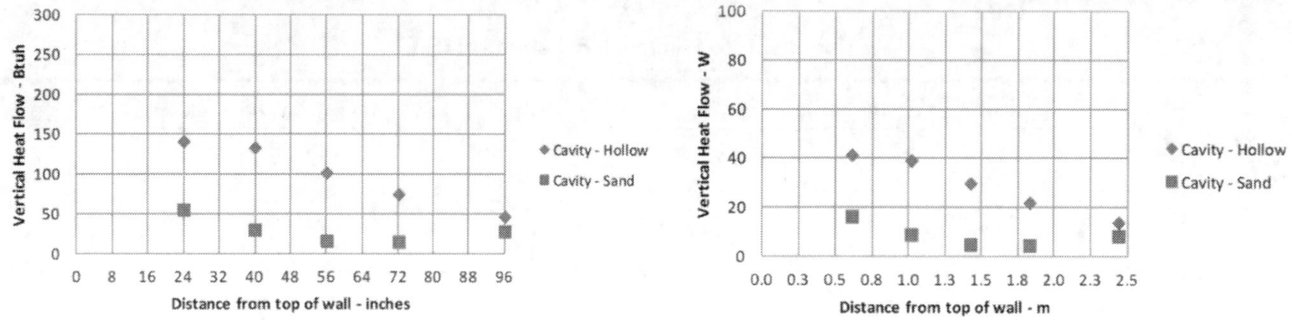

Figure 4 *Vertical heat flow—no interior insulation: (left) I-P and (right) SI.*

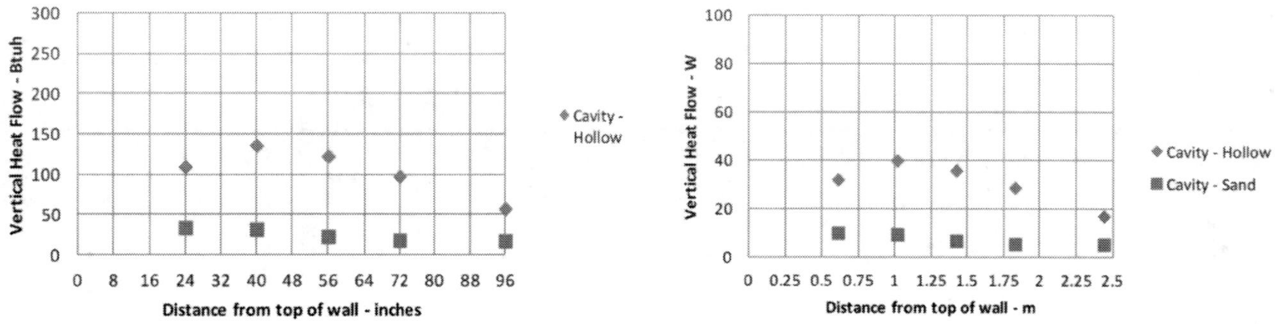

Figure 5 *Vertical heat flow—(left) interior R-5 and (right) interior R-09.*

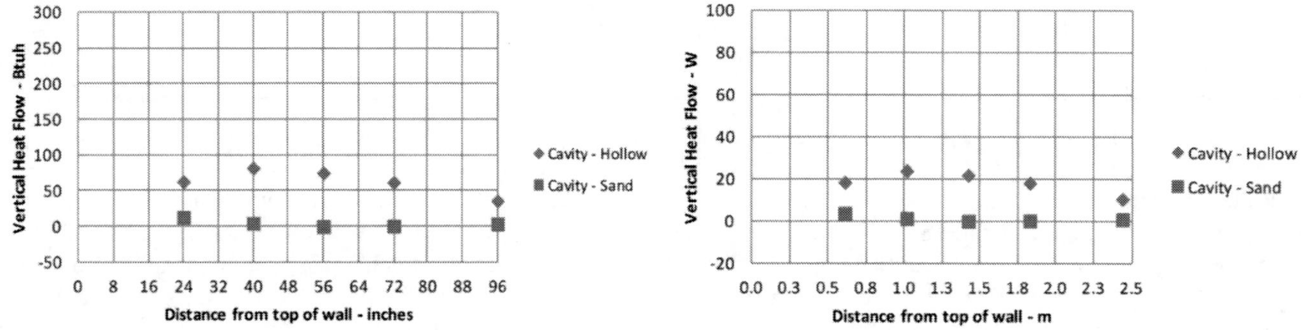

Figure 6 *Vertical heat flow—(left) interior R-11 and (right) interior R-1.9.*

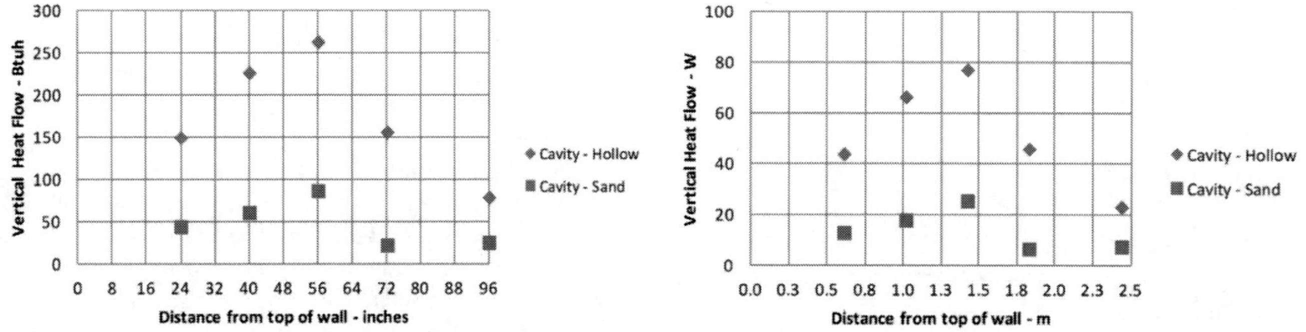

Figure 7 *Vertical heat flow on top—(left) interior R-11 on top half and (right) interior R-1.9.*

hollow has been shown to be very complex. A series of six thermal nodes were analyzed using experimental data for eight test walls. The results exhibited similar trends in that the vertical heat flow due to natural convection increased from the bottom to the top of the walls. The natural convection was shown to be significantly increased due to the impact of the surrounding soil on the exterior. As insulation was added to the interior surface of the wall the natural convection increased

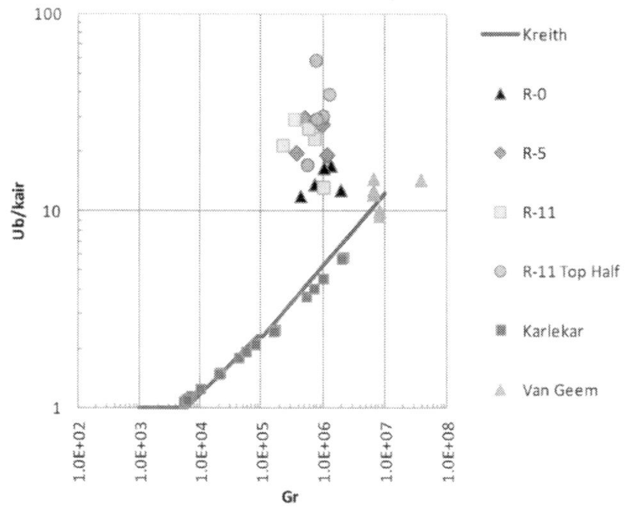

Figure 8 *Vertical air spaces.*

even further. Finally, when only the top half of the interior wall was insulated the natural convection was the highest. For this case only the arc-length method should not be used. Comparisons to prior experimental data quantified the magnitudes of the increased vertical heat flow. The complexity and interactions of the various heat flow mechanisms did not reveal any simplifications that could be made to further improve the arc-length modeling procedures. Any modeling improvements would require a more complex approach which would include a thermal network that coupled all of the nodes identified in this analysis.

CONCLUSION

The primary result of this analysis was the quantification of the vertical heat flow due to natural convection within cavities that were hollow. There was some expectation that the experimental data would exhibit some consistent trends that could be used to develop a simplified model or correlation method that would improve the arc-length calculation procedure for the case with open cavity walls insulated only on the top half. Fundamentally, the scatter is representative of the complex nature of the multiple heat transfer mechanisms that exist with below grade masonry walls. In order to truly model this complexity of vertical heat flow a thermal network would need to be created that coupled the six nodes. Furthermore, there are too few data points in this study to fully develop a predictive model that would be applicable to the broad range of temperature conditions that occur with a typical basement over the course of a year.

NOMENCLATURE

b	=	thickness of an air space, in. (m)
g	=	acceleration due to gravity, ft/sec^2 (m/sec^2)
Gr	=	Grashof Number, dimensionless
k	=	thermal conductivity, Btu/h·ft·°F (W/m·K)
L	=	clearance in enclosed space, ft (m)
Q	=	heat flow, Btu/h (W)
R	=	thermal resistance, h·ft^2·°F/Btu (m^2·K/W)
T	=	temperature, °F (°C)
U	=	thermal transmission, Btu/h·ft^2·°F (W/m·K)

Greek

δ	=	thickness of an air space, ft (m)
β	=	coefficient of volumetric expansion, 1/°F (1/°C)
ρ	=	density of air, lb/ft^3 (kg/m^3)
μ	=	viscosity of fluid, lb/ft·sec (Pa·s)

Subscripts

i	=	inside
e	=	effective
o	=	outside
v	=	vertical

REFERENCES

ASHRAE. 2009. *2009 ASHRAE Handbook—Fundamentals*. Atlanta: American Society of Heating Refrigerating and Air-Conditioning Engineers, Inc.

Jakob, M. 1958. *Heat Transfer*. Wiley, New York, NY.

Karlekar, B.V. and Desmond, R.M. 1977. Engineering Heat Transfer, West Publishing Company, St. Paul. MN.

Kreith, F. 1967. Principles of Heat Transfer, International Textbook Company, Sranton, PA.

McAdams, W.H. 1954. Heat Transmission, McGraw-Hill Book Company, New York, NY.

Shipp, P.H. 1983. Natural Convection within Masonry Block Basement Walls. ASHRAE Transactions, 89 (1B): 405-419.

Van Geem, M.G. 1985. Thermal Transmittance of Concrete Block Walls with Core Insulation. ASHRAE Transactions, 85 (2): 17-35.

ASHRAE 1478: Measuring Airtightness of Mid- and High-Rise Non-Residential Buildings

Terry Brennan
Member ASHRAE

Wagdy Anis
Member ASHRAE

Gary Nelson
Member ASHRAE

Collin Olson, PhD

ABSTRACT

ASHRAE 1478 is a research project designed to measure enclosure airtightness of mid- and high-rise buildings in the United States. Data were collected from 16 non-residential buildings in climate zones 2–7 constructed since the year 2000. The dataset includes buildings with no particular attention to making the building airtight, buildings where some attention was given to airtightness, and buildings where extensive attention was paid to airtightness. Some of the buildings were designed to be sustainable (e.g., receiving LEED status). Buildings ranged from four to fourteen stories. A fan pressure testing protocol based on ASTM E779 was developed by the project team. A number of issues in using E779 to test large building were identified, discussed, and addressed. Building airtightness was reported in CFM per square foot of above grade enclosure at a 75 pascal induced shell pressure difference and CFM per square foot of complete enclosure (including slab and below grade conditioned space walls). The results range from 0.06 cfm 75/ft^2 to 0.75 cfm 75/ft^2 of complete enclosure (1.10 m^3/h/m^2 to 13.7 m^3/h/m^2). Major air leakage sites were identified in the course of testing. Air leakage through HVAC related penetrations was measured in a subset of the buildings. Factors that are associated with the most airtight enclosures include air-barrier continuity detailed in construction documents and precast concrete panel construction. Damper air leakage turned out to be a significant portion of the total enclosure air leakage in some of the buildings. The significance of air leakage by HVAC systems is reviewed in relation to building air tightness.

INTRODUCTION

The air leakage characteristics of commercial and institutional building enclosures is important to design engineers, IAQ consultants, researchers, and forensic investigators. Unintentional air leakage through the building enclosure may result in indoor air quality issues, condensation problems in the enclosure, excessive energy consumption, and, in snow country, ice dams (Brennan 1997; Anis 2001; Brennan 2002; Ask 2003; Fennel and Haehnel 2005). Emmerich and Persily (2005) reported fan pressure test results of over 200 US non-residential buildings, only 11 of which were for buildings greater than 3 stories. ASHRAE 1478 "Measuring Airtightness of Mid- and High-Rise Non-residential Buildings" is a research project intended to provide airtightness data for buildings greater than 3 stories, that were designed and constructed during or after the year 2000. This project conducted fan pressurization testing on 16 buildings located in Climate Zones 2–7 as defined by the IECC Climate Zone map (IECC 2012).

METHODS

Test Procedure

Tests were conducted in accordance with ASTM E779-2010 *Standard Test Method for Determining Air Leakage Rate by Fan Pressurization*. A test protocol based on E779 that addresses issues specific to testing large buildings was developed for the project.

Terry Brennan is a principal with Camroden Associates and a principal researcher and contributor to the project. Wagdy Anis is a principal with Wiss, Janney, Elstner, Associates Inc. and principal researcher for ASHRAE 1478 RP, Gary Nelson is president of the Energy Conservatory and supplier of testing hardware and software for the project. Collin Olson is a physicist with the Energy Conservatory, developer of the TECLOG software used in the testing of the project.

The test protocol first required selecting a date when the fewest occupants would occupy the building. It was vital to address security issues and arrange for building personnel authorized to place HVAC in the desired test mode, provide interior access, and to reset circuit breakers. Emergency contacts were also collected.

Portable test fans to test the buildings were used unless compelling reasons stipulated the use of the HVAC system. Pressure taps were placed on all four sides of the buildings as shown in Figure 1 of E779. The measurements of these taps were averaged. Interior pressure differences between the test control location and locations in the building that were most likely to have reduced distribution of interior pressure (e.g., between bottom floor lobby and top floor of building or lobby and the end of a wing. In accordance with E779, single zone condition was achieved if the difference between the highest and lowest interior pressure was less than 10% of the pressure difference induced across the enclosure (taken to be the numerical average of all four ground floor enclosure taps).The interior measurements were monitored so single zone conditions could be maintained and documented. Doors could remain closed if a single zone condition was achieved. If portions of the building had an ambiguous air barrier, the decision made regarding these spaces was documented. Finally, HVAC penetrations were identified and either closed, sealed, or left open, depending on their typical state and other stipulations of the test method.

The test itself involved:

- Collecting baseline and induced envelope pressure and test fan flow data using a central computer
- Analyzing data using the methods in E779 while still in the test building with the test equipment still installed; identifying problems with test data, correcting problems in the building test conditions (e.g. doors blowing open) and collecting additional data to correct the problem.
- Reporting the data in terms of:
    - CFM (m^3/h) at 50 and 75 pascals induced envelope pressure
    - CFM/ft^2 (m^3/h per m^2) above grade enclosure at 50 and 75 pascals, and
    - CFM/ft^2 (m^3/h per m^2) of total enclosure at 50 and 75 pascals.

Identifying the largest air leakage sites in the building and details that contributed to the air leaks was completed during the testing.

Air leakage through HVAC related penetrations was estimated during the test, as well. The building was initially tested with all HVAC related penetrations sealed with masking and any associated dampers closed. After those tests were completed, the test equipment was set to automatically induce 75 pascals indoor-outdoor pressure. The masking was removed from the HVAC penetrations one at a time and the airflow required to maintain the pressure difference at 75 pascals was recorded. Increases in airflow are attributed to air leakage through the unmasked penetration.

Buildings Tested

Table 1 summarizes the properties of the buildings tested. The buildings in the data set are all owner-occupied buildings. Many of them were designed for the owner by the design team. Buildings 2, 3, 4, 5, 6, 7, 8, 9, 10, 11, 14, and 15 are green labeled buildings. Buildings 4, 5, 6, 10, 13, and 14 included an envelope consultant on the design team and had air barriers specified in the construction documents.

Leakage data for all HVAC related openings was collected in 11 of the test buildings. It was collected for smoke evacuators only in Building 1 and kitchen exhaust only in Building 3 and 4.

RESULTS

Table 1 summarizes average air leakage results of the 16 buildings tested. The test results for the buildings with all HVAC penetrations in dampers closed, openings sealed mode are listed in Table 2. Figure 1 plots the results of depressurization and pressurization tests and the average of the two for each building. The mean of the averaged test results is 0.29 cfm 75/ft^2 with a standard deviation of 0.20 cfm 75/ft^2 (5.3 $m^3/h/m^2$ with a standard deviation of 3.6 $m^3/h/m^2$).

The air leakage rate of buildings designated a green building is 0.32 cfm75/ft^2 (5.8 $m^3/h/m^2$). The average leakage rate of all other buildings in the data set is 0.22 cfm 75/ft^2 (4.2 $m^3/h/m^2$).

The air leakage rates for buildings with an air barrier specified and an envelope expert consulted average 0.13 cfm 75/ft^2 total enclosure (2.4 $m^3/h/m^2$). If buildings with air barriers and air barrier consultants included in the design of the buildings are removed from the data set, the mean of the remaining buildings is 0.39 cfm75/ft^2 total enclosure (7.1 $m^3/h/m^2$).

The major air leakage sites were identified at each building. The most frequently encountered large leakage sites were:

- The intersection of roofs and walls
- Soffits and overhangs
- Through portions of buildings that join either non-conditioned or semi-conditioned space (e.g., mechanical rooms, garages, basements, loading docks).
- Coiling doors, primarily at the top of the door.

HVAC related penetration leakage was assessed in 14 of the test buildings. In some of the buildings, the total HVAC related leakage was minimal, while in others it was significant. The average increase in HVAC related CFM 75 airflow across the fourteen buildings was 27% and ranged from 2% to 51%.

DISCUSSION

Biases due to diversity and size limited the analysis of the results. It proved difficult to find owners of commercial rental

Table 1. Properties and Air Leakage of the Building Data Set

Building	Function	Height, number of stories	Floor Area, ft^2 (m^2)	Total Building Enclosure Area, ft^2 (m^2)	CFM 75/ft^2 total enclosure area ($m^3/h/m^2$)
1	Institutional	4	129441 (12025)	135501 (12588)	0.36 (6.63)
2	Institutional	1–5	152333 (14152)	197999 (18394)	0.59 (10.81)
3	Corporate	5	166636 (15480)	193918 (18015)	0.14 (2.47)
4	Institutional	4	330000 (30657)	260800 (24228)	0.13 (2.44)
5	Mixed-use	6	180000 (16722)	147690 (13720)	0.22 (4.01)
6	Corporate	4	109933 (10213)	108751 (10103)	0.06 (1.04)
7	Corporate	14	476162 (44235)	169876 (15781)	0.31 (5.60)
8	Institutional	5	51563 (4790)	78457 (7289)	0.30 (5.48)
9	Corporate	4	85004 (7897)	87428 (8122)	0.57 (10.35)
10	Corporate	4	121825 (11318)	109691 (10190)	0.10 (1.79)
11	Institutional	5	123314 (11456)	126955 (11794)	0.43 (7.86)
12	Institutional	4	141472 (13143)	122269 (11359)	0.15 (2.80)
13	Mixed-use	4	54540 (5067)	77952 (7242)	0.10 (1.87)
14	Institutional	6	214656 (19942)	193824 (18006)	0.18 (3.24)
15	Institutional	4	57286 (5322)	80685 (7496)	0.75 (13.69)
16	Institutional	9	362201 (33648)	210184 (19526)	0.27 (4.99)

properties who were willing to allow the testing to occur. It is suspected that the data set is biased towards tighter buildings. The small dataset may not be representative of the entire set of mid- and high-rise buildings constructed since 2000. However, there are some conclusions that can be made. First, as seen in Figure 1, most of the results are within one standard deviation. However, there is still a large range in air leakage. In addition, pressurization air leakage is always slightly greater than depressurization results. It is believed this is because some air leakage sites are leakier when the building is pressurized (e.g., entry doors that open outward are drawn against weatherstripping in depressurization mode and are pressed to the limits of the latch during pressurization mode. Essentially the same test result was found whether using induced enclosure pressures at the top of the building or on the ground floor.

The eight tightest buildings tested meet the Army Corps of Engineers air leakage requirement of 0.25 cfm 75/ft^2 (4.6 $m^3/h/m^2$). Six of these, Buildings 4, 5, 6, 10, 13 and 14 had special attention to airtightness during design and construction.

The difference in airtightness for green buildings implies that a green building designation is counter-productive to achieving high airtightness. However, there are many other goals when working towards a green label. In addition, only four of the buildings in the data set are not considered environmentally friendly. Results may not be statistically significant.

The air leakage sites identified during the building test are almost all due to detailing at corners, penetrations, and joints. These problems can be addressed by design and installation and are much more difficult to locate and repair at a later time. It is clear that addressing airtightness early is important.

There is a large amount of air leakage associated with HVAC related penetrations. The largest air leaks were through undampered exhaust fans serving kitchens, toilets, elevator shafts, and dampers that did not cycle to the closed position when instructed to do so by the control system. These problems can be fixed, but must be located first. Leakage through fans that operate continuously (e.g., electrical room exhaust) has little effect on energy use. Leakage through fans that almost never run (e.g., smoke evacuation fans) behave as any other enclosure leakage site.

SUMMARY

There is a large range in building air tightness. However, with appropriate detailing and HVAC control, a tighter building can be achieved. This knowledge must be incorporated into all of our buildings to ensure energy efficiency and human comfort.

REFERENCES

Anis, W. 2001. The Impact of Air-tightness on System Design, *ASHRAE Journal* Vol. 43, No. 12.

Ask, A. 2003. Ventilation and Air Leakage, *ASHRAE Journal* Vol. 45, No. 11.

ASTM. 2010. E779-2010 *Standard Test Method for Determining Air Leakage Rate by Fan Pressurization*. West

Table 2. Results of Fan Pressure Tests

HVAC penetration unmasking results. The columns from left to right are: Building ID number, IECC 2012 climate zone, fan pressure test mode (depressurization, pressurization, or the average of the depressurization and pressurization results), airflow in CFM required to induce a 75 pascal envelope pressure difference, the 95% confidence interval, test result in CFM 75 per square foot of above grade enclosure area, CFM 75 per square foot of total enclosure area (all six sides of the enclosure), airflow in cubic meters per hour required to induce a 75 pascal envelope pressure difference, test result in m^3/h 75 per square meter of above grade enclosure area, in m^3/h 75 per square meter of total enclosure area (all six sides of the enclosure), the air exchange rate needed to induce a 50 pascal pressure difference between inside and outside, flow coefficient, the flow exponent, and r squared.

Building ID	Climate	Test Mode	CFM75	95% CI (±%)	CFM75/ft² above grade encl	CFM75/ft² total encl	m³/h	m³/h/m² above grade encl	m³/h/m² total encl	ACH50	C	n	r²
1	2B	2-dep	49121	0.6	0.48	0.36	81983	8.7	6.6	1.24	4479	0.56	0.9996
2	5A	1-dep	112195	0.2	0.76	0.57	187253	13.8	10.4	2.01	6501	0.66	1.0000
2	5A	2-Press	121958	6.9	0.82	0.62	203548	15.0	11.3	2.16	6405	0.68	0.9966
2	5A	ave 1-2	117077	3.5	0.79	0.59	195401	14.4	10.8	2.08	6453	0.67	0.9983
3	6A	1-dep	25537	1.4	0.18	0.13	42621	3.2	2.4	0.43	1807	0.61	0.9983
3	6A	2-Press	26893	2.6	0.19	0.14	44884	3.4	2.5	0.46	2346	0.57	0.9878
3	6A	ave 1-2	26215	1.5	0.18	0.14	43753	3.3	2.5	0.45	2076	0.59	0.9980
4	6A	1-dep	33301	2.8	0.19	0.13	55579	3.5	2.3	0.36	3157	0.55	0.9967
4	6A	2-Press	36183	1.6	0.21	0.14	60389	3.8	2.5	0.40	3582	0.54	0.9981
4	6A	ave 1-2	34742	1.6	0.20	0.13	57984	3.7	2.4	0.38	3370	0.54	0.9974
5	4C	1-dep	31286	1.4	0.27	0.21	52216	4.9	3.9	0.61	2151	0.62	0.9980
5	4C	2-Press	33552	1.4	0.29	0.23	55998	5.2	4.2	0.66	2510	0.60	0.9985
5	4C	ave 1-2	32419	1.0	0.28	0.22	54107	5.0	4.0	0.63	2330	0.61	0.9982
6	3A	1-Press	6440	2.1	0.08	0.06	10748	1.5	1.1	0.21	601	0.55	0.9938
6	3A	2-Dep	5889	3.0	0.07	0.05	9829	1.3	1.0	0.19	453	0.59	0.9932
6	3A	ave 1-2	6165	1.8	0.08	0.06	10289	1.4	1.0	0.20	527	0.57	0.9935
7	5A	1-Press	52604	1.0	0.40	0.31	87796	7.3	5.7	0.26	3462	0.63	0.9946
7	5A	2-Dep	51442	3.4	0.39	0.30	85857	7.1	5.5	0.26	3565	0.62	0.9905
7	5A	ave 1-2	52023	1.8	0.39	0.31	86826	7.2	5.6	0.26	3513	0.62	0.9926
8	4A	1-Press	25330	2.7	0.36	0.32	42276	6.6	5.9	1.14	1347	0.68	0.9766
8	4A	2-Dep	21649	1.6	0.31	0.28	36132	5.7	5.0	1.00	1457	0.63	0.9964
8	4A	ave 1-2	23490	1.6	0.34	0.30	39204	6.1	5.5	1.07	1402	0.65	0.9865
9	4A	1-dep	48605	4.2	0.77	0.56	81122	14.0	10.2	1.97	3920	0.58	0.9922
9	4A	2-Press	50384	2.1	0.79	0.58	84091	14.5	10.5	2.03	3968	0.59	0.9920
9	4A	ave 1-2	49495	2.3	0.78	0.57	82606	14.3	10.4	2.00	3944	0.59	0.9921
10	6A	1-dep	9840	1.6	0.12	0.09	16423	2.2	1.6	0.24	692	0.62	0.9978
10	6A	2- Press	11609	1.5	0.14	0.11	19375	2.5	1.9	0.28	699	0.65	0.9979
10	6A	ave 1-2	10725	1.1	0.13	0.10	17899	2.3	1.8	0.27	696	0.63	0.9979
11	4A	1-dep	53350	1.3	0.53	0.42	89041	9.7	7.7	1.34	4653	0.57	0.9967
11	4A	2-Press	55729	1.4	0.55	0.44	93012	10.1	8.0	1.38	4237	0.60	0.9960

Table 2. *(continued)*Results of Fan Pressure Tests

Build-ing ID	Climate	Test Mode	CFM75	95% CI (±%)	$CFM75/ft^2$ above grade encl	$CFM75/ft^2$ total encl	m^3/h	$m^3/h/m^2$ above grade encl	$m^3/h/m^2$ total encl	ACH50	C	n	r^2
11	4A	ave 1-2	54540	1.0	0.54	0.43	91026	9.9	7.9	1.36	4445	0.58	0.9964
12	7A	1-dep	18640	1.0	0.25	0.15	31110	4.5	2.8	0.49	1290	0.62	0.9993
12	7A	2-Press	18818	0.5	0.25	0.15	31407	4.5	2.8	0.51	1597	0.57	0.9998
12	7A	ave 1-2	18729	0.6	0.25	0.15	31259	4.5	2.8	0.50	1443	0.60	0.9995
13	5A	1-Press	8361	1.1	0.16	0.11	13955	2.9	2.0	0.55	616	0.60	0.9996
13	5A	2-Dep	7568	0.6	0.14	0.10	12631	2.6	1.8	0.50	543	0.61	0.9999
13	5A	ave 1-2	7965	0.6	0.15	0.10	13293	2.7	1.9	0.52	579	0.61	0.9998
14	5A	1-dep	32870	1.7	0.21	0.17	54860	3.9	3.1	0.49	2385	0.61	0.9984
14	5A	2-Press	35839	0.9	0.23	0.18	59815	4.3	3.4	0.53	2316	0.63	0.9998
14	5A	ave 1-2	34355	1.0	0.22	0.18	57338	4.1	3.2	0.51	2350	0.62	0.9991
15	5A	1-dep	58233	1.1	1.08	0.72	97191	19.7	13.2	3.58	5505	0.55	0.9982
15	5A	2-Press	62536	2.1	1.16	0.78	104373	21.2	14.2	3.77	4773	0.60	0.9969
15	5A	ave 1-2	60385	1.2	1.12	0.75	100782	20.5	13.7	3.68	5139	0.57	0.9976
16	5A	1-Press	62088	1.6	0.39	0.30	103625	7.1	5.4	0.61	3724	0.65	0.9966
16	5A	2-Dep	52563	1.8	0.33	0.25	87728	6.0	4.6	0.52	3791	0.61	0.9947
16	5A	ave 1-2	57326	1.2	0.36	0.27	95676	6.6	5.0	0.56	3757	0.63	0.9957
Mean						0.29			5.32				

Table 2. *(continued)* Results of Fan Pressure Tests

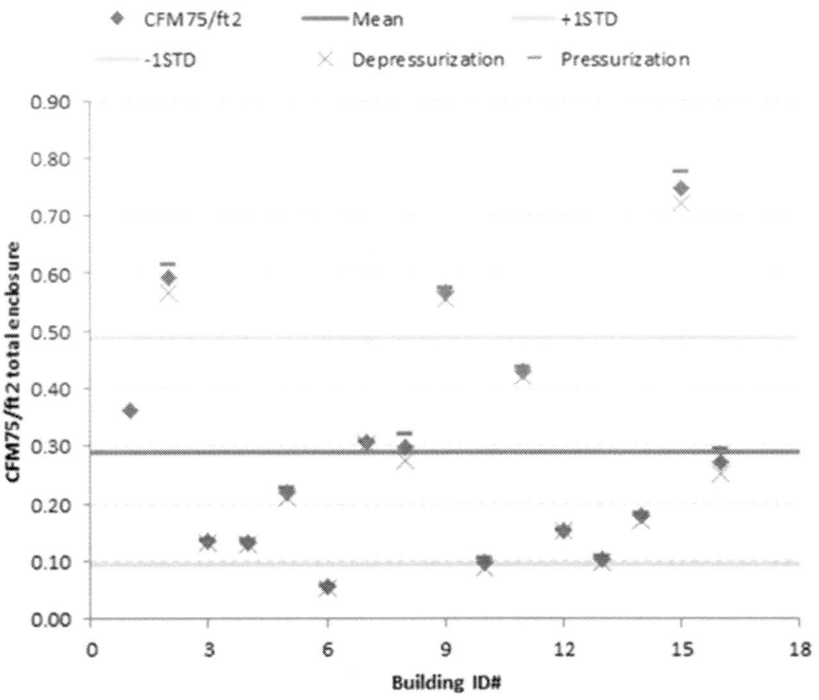

Figure 1 *The depressurization, pressurization, and average of the two test results is plotted for each building.*

Conshohocken, PA: American Society for Testing and Materials.

Brennan, T. et al. 1997. Unintended consequences of planned and unplanned airflows in commercial buildings. Proceedings of ASHRAE Healthy Buildings/IAQ 97.

Brennan, T. et al. 2002. Unplanned airflows and moisture problems. *ASHRAE Journal* Vol. 44, No. 11.

Emmerich, S.J. and Persily, A.K. (2005). Air-tightness of Commercial Buildings in the U.S. AIVC Conference, Brussels, Belgium

Fennell, H.C. and J. Haehnel (2005). Setting Airtightness Standards. *ASHRAE Journal*, Vol. 47, No. 9.

IECC. 2012. 2012 International Energy Conservation Code. International Code Council, Inc. 4051 West Flossmoor Road, Country Club Hills, IL 60478-5795.

Impact of Cool Roofing Membranes on the Hygrothermal Performance of Low-Sloped Roof Structures in Timber Construction

Christoph Buxbaum, PhD **Wolfgang Gallent** **Simon Paulitsch** **Oskar Pankratz**

ABSTRACT

This paper will present an analysis regarding the hygrothermal performance and furthermore durability of non-ventilated light-weight flat roofs with cool roofing membranes in the Central European climate.

The main focus is the assessment of the drying potential of selected roof variants to predict the potential risks of mold and especially wood decay fungi growth inside the constructions. Several numerical simulations were carried out to investigate the influence of different roofing membranes and especially cool roofing products on the overall hygrothermal conditions inside the constructions.

Additionally, roof assemblies with cool roofing membranes were investigated at the Austrian BSRTU Building Science – Research & Test Unit outdoor test facility (www.building-science.at) of the Carinthia University of Applied Sciences (CUAS). In-situ measurement results, taken at this experimental set-up will be discussed.

INTRODUCTION

This paper presents first results of a research project on the durability of non-ventilated, highly-insulated timber-framed low-sloped roofs in the climate of Central Europe. The primary aim of the work was to investigate the influence of different colored roof membranes on their hygrothermal response. The calculations were performed for light-grey, dark-grey, and especially white colored (cool roofing) membranes, basically looking to their drying potential due to solar-driven diffusion. It is important to note, that usually in Austria, according to code OENORM B8110-2, avoiding interstitial condensation is evaluated using the dew-point method (Glaser scheme). However, the method as applied in the code, only considers very simplified steady-state boundary conditions. Neither solar radiation and undercooling, nor wetting processes and hygroscopic sorption, nor liquid transport are taken into account. Therefore, the scope of this study is to predict the long-term durability as realistically as possible using transient hygrothermal calculations.

INVESTIGATED CASES

In the frame of the research, the drying potential of prefabricated timber low-sloped roof constructions, which are used more and more for larger commercial and production, but also residential buildings in Austria, was analyzed. These roofs are typically prefabricated and consist of 400 mm (16 in.) high timber beams faced with OSB decking at both sides. The space in between the beams is filled with blown-in cellulose insulation. Based on this base case, three variants with different colored waterproofing membranes were analyzed (Table 1 and Figure 2). All analyses assumed the roof structures cases 1 to 3 to be airtight and the insulation correctly installed, so no convective vapor flow could develop, affecting moisture buildup.

SIMULATIONS

The simulations were carried out with the WUFI® (Wärme und Feuchte instationär [Transient Heat and Moisture]) software, developed at the Fraunhofer Institute for Building Physics (Kuenzel 1994) in Holzkirchen/Germany and validated against outdoor measurements and laboratory tests.

Christoph Buxbaum *is an architect and professor for energy efficient and sustainable building design and* ***Wolfgang Gallent*** *and* ***Simon Paulitsch*** *are research engineers at the Department of Architecture, Carinthia University of Applied Sciences, Spittal an der Drau, Austria.* ***Oskar Pankratz*** *is an architect and consultant for building physics in Haidershofen, Austria.*

The analysis was conducted under the assumption that the roof structures had good airtightness and insulation installation, and thus without convective vapor flow affecting moisture buildup.

Default Program Settings

The surface film coefficient at the inside surface was set equal to 8 W/(m^2·K). Exterior climate data and surface temperatures, measured at three roof sections with different colored membranes (white, light grey, dark grey) at the Building Science Research & Test Unit (BSRTU) outdoor test facility were chosen as exterior boundary conditions.

The room climate varied as a sine curve between 20°C and 40% relative humidity in winter and 22°C and 60% relative humidity in summer. As initial conditions for the materials 20°C and an RH of 80% were chosen (Karagiozis 1998). Such high moisture content (i.e., 80%) may be due to rain during transport and mounting of large roofs structures or even due to flank diffusion over adjacent masonry (Kuenzel 1996).

The material parameters required for each material were taken from respective product spec sheets, approved due to third party laboratory report (Table 2). The hygrothermal behaviour was simulated over a period of five years, starting on October 1st.

RESULTS

The results concern the hygric behaviour of the assemblies. Moisture tolerance, in fact, is a critical performance in building envelope design. Hence, for analyzing the drying potential, total water content in the whole construction was calculated. Further, the water content in the different layers and especially the timber based ones was analyzed to assess the possibility of wood decay. According to Austrian codes the critical limit for the moisture ratio in timber is 20 mass percent (M%) although it is well known that wood detoriation mainly occurs at higher levels and long exposure time. In addition, the relative humidity and temperatures in the interfaces between cellulose and the exterior OSB decking and the interior vapor retarder was calculated to assess the risk of mold growth based on the exposure time. In the graphs, results for the whole period of five years are shown.

Total Water Content (TWC)

First, simulations were carried out to investigate total water content for case 1, 2, and 3. Figure 2 shows the results (TWC) for the five year period. All cases show a decreasing TWC. Anyhow, the white colored cool roofing membrane applied in case 1 has the most significant influence on remaining moisture in the roof. Total water content (TWC) in case 1 with cool roofing membrane, levels off at a seasonal maximum of ~3,8 kg/m^2 after about 5 years. Case 2, with a light-

Table 1. Cases (Layers from Outside to Inside)

Case 1	Case 2	Case 3
Roof Membrane; Type: Plastic (Polyolefin); Permeance = 0.1426 perm; (s_d – Value = 23 m)		
Color: white (cool roofing)	Color: light grey	Color: dark grey
Solar Reflectance: ~ 0,78 (*)	Solar Reflectance: ~ 0,65 (*)	Solar Reflectance: ~ 0,50 (*)
Oriented Strand Board (OSB) 15 mm (5/8 in.)		
Cellulose Blown-in Insulation 400 mm (~16 in.)		
Code approved Vapor Retarder, Permeance = 0.0252 perm (s_d – Value = 130 m) (**)		
Oriented Strand Board (OSB) 15 mm (5/8 in.)		

* Solar reflectance measured according to ASTM E903: Material data from used products (spec sheet) were inserted in simulation. Note, that these values are representing initial conditions, ignoring weathering effects. The influence of soiling (decrease of solar reflectance) during the course of the years is part of the research work at the CUAS outdoor test facility.
** Vapor retarder with permanent s_d value. No smart vapor retarder with variable s_d value.

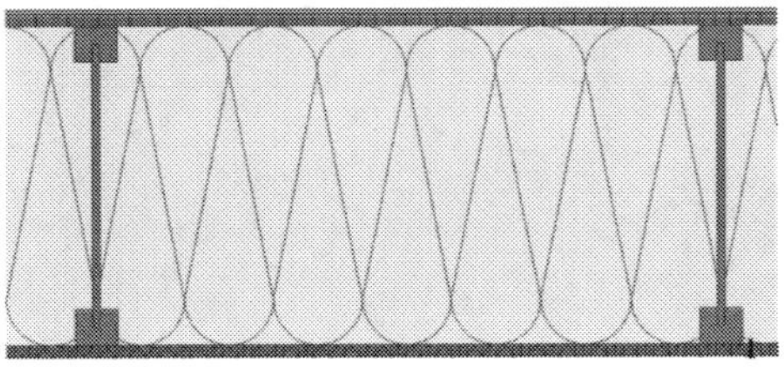

Figure 1 *Schematic draft of investigated roof assemblies, cases 1 to 3.*

Table 2. Material Properties

Material	Bulk Density, kg/m^3	Porosity, m^3/m^3	Heat Capacity, kJ/kg·K	Heat Conductivity Dry, W/m·K	Diffusion Resistance Factor Dry μ [–]
Oriented strand board (OSB)	615	0,6	1880	0,101	∅ 175
Cellulose	50	0,95	2000	0,038	1,8

* Material data taken from respective product spec sheets, approved by third party laboratory report.

grey membrane is showing an improved and even faster drying. The total water content levels off at a seasonal maximum of ~ 3,6 kg/m². Case 3 with the dark-grey colored waterproofing membrane shows the fastest drying rate of all variants. Higher temperatures at the exterior surface, due to a higher short wave absorptivity of the membrane, in fact increase the inwards vapor pressure gradient accelerating drying that way. The total water content levels off at a seasonal maximum of ~ 3,55 kg/m² after five years.

Clearly inward drying capacity should be considered as an important design factor. (Powell and Robinson, 1971; Straube, 2001; Desjarlais, 1995)

Water Content (WC) Exterior OSB Decking

The next step in this parametric study was to predict the moisture response of the wooden components and especially the exterior OSB decking. In Figure 3 one can observe that the moisture content of the OSB for all cases is exceeding 20 M% during the first wintertime and then decreasing over the years. In cases 2 and 3, the water content decreases faster during the first five years and levels off at about ~135 kg/m³ (Case 2) and ~ 120 kg/m³ (Case 3), respectively, during the winter time in the fifth year and therefore below or rather slightly above the critical limit of 20 M%.

Only the moisture content of the exterior OSB in case 1 with the white colored cool roofing membrane remains at a higher humidity level during all five winter periods increasing beyond 20 M%. This effect is quite crucial for the practical application since the unscheduled moisture within the construction will be trapped inside the roof construction for a longer time due to poor solar-driven diffusion. This potential higher humidity level during long periods would promote the risk of mold or even decay fungi growth and hence a potential premature construction failure.

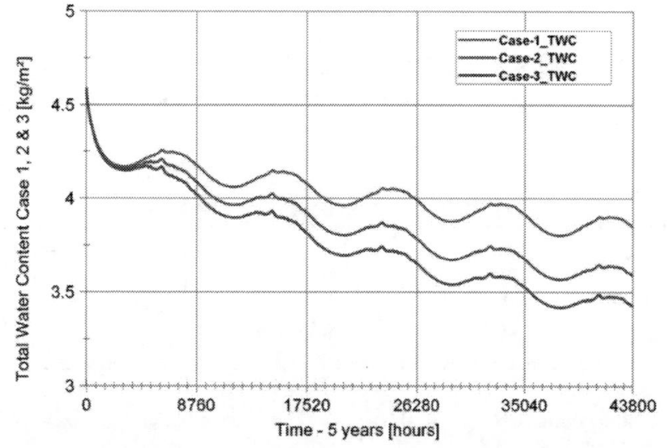

Figure 2 *Calculated total water content (TWC) of cases 1, 2, and 3.*

Relative Humidity on the Interface to Cellulose

In addition, the relative humidity on the interface between external OSB decking and the cellulose insulation was calculated to assess the risk of mold growth (Figure 4). The simulations were started with an initial moisture content of 80% rh. This higher moisture content at the start is useful to predict the construction's drying. The simulation is predicting a higher RH of ~95% during the first winter for all constructions due to the higher initial moisture conditions. In principle, mold growth is possible at RH of >80% and >5°C. In Figure 4 it is also well to see, that case 2 with a light-grey colored membrane and case 3 with a dark-colored membrane are showing the best drying. The RH at the interface external OSB cellulose is therefore continuously decreasing during the time of investigation and remains below the critical border of 95% rh after the fifth year. In addition, in case 1 with the cool roofing membrane the RH remains above 95% rh every wintertime.

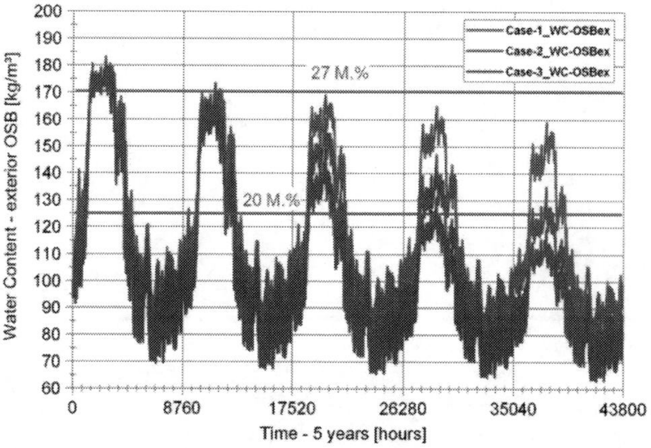

Figure 3 *Calculated moisture content (by mass) of the external OSB, cases 1, 2, and 3.*

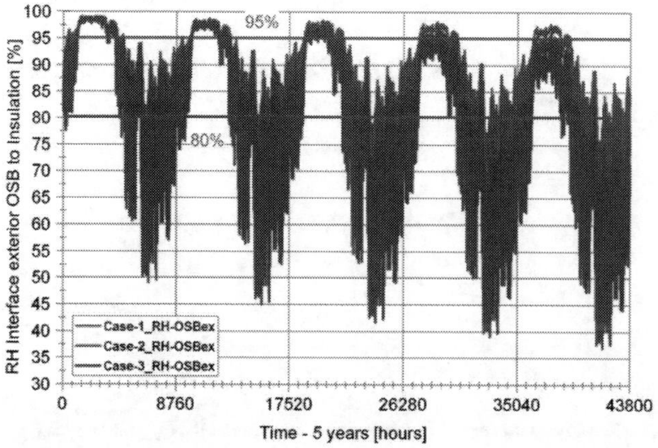

Figure 4 *Relative humidity at the external OSB-celluose interface, cases 1, 2, and 3.*

Cases 2 and 3 are showing a decreasing gradient, but nevertheless 80% rh are exceeded during some weeks within wintertime when the temperatures (Figure 6) are reduced. It has to be noted that blown-in cellulose is blended with fungicides like boric acid, boric pentahydrate, etc. by default. In cases of a direct contact between wooden surfaces and cellulose, these additives should limit the germination of mold growth under moisture conditions between 80% and 90%. The effectiveness of the above mentioned fungicidal additives in wooden constructions has been investigated in a laboratory test at the Carinthia University of Applied Sciences and in-situ measurement results from practice confirm this theory. Due to the lower temperatures during this time of the year and the fungicidal influence of cellulose, mold germination should be reduced or even limited, if cases 2 and 3 are executed with dry materials.

Furthermore, the calculation of the relative humidity at the interface between the OSB board and the cellulose in case 1 with the cool roofing membrane indicates a potential risk of mold growth during the whole period of investigation. Figure 4 illustrates that the RH is exceeding 80 % for several months during the first five years with a seasonal maximum of 90% to 95% rh. Mold growth is therefore likely to occur.

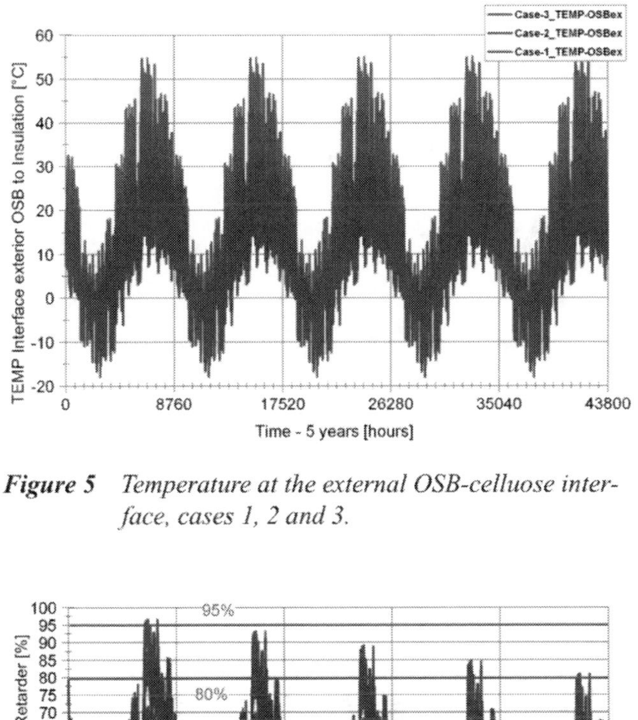

Figure 5 *Temperature at the external OSB-celluose interface, cases 1, 2 and 3.*

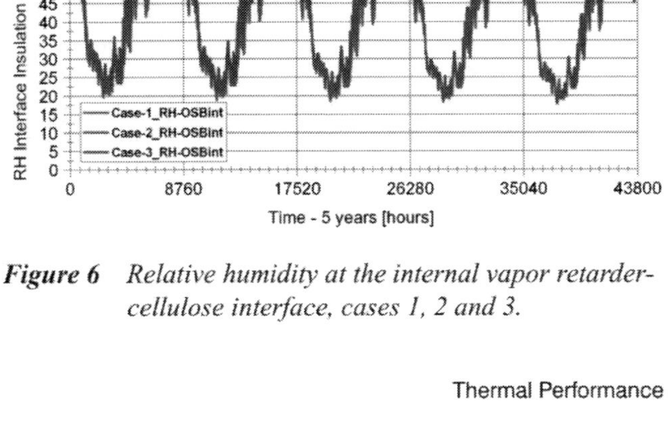

Figure 6 *Relative humidity at the internal vapor retarder-cellulose interface, cases 1, 2 and 3.*

Provided that good workmanship was done, only the roof constructions case 2 and 3 are showing a relatively good hygric behavior. Under these circumstances moisture related problems like mold and wood decay fungi growth at the exterior OSB boards should be avoided in these cases.

In a second step, the relative humidity at the interface between cellulose and the interior vapor retarder was calculated. It is well to see that in variants 2 and 3 the RH at the interface between cellulose and internal vapor retarder is generally exceeding 80% rh during the first and second summer due to the assumed higher initial moisture conditions (Figure 6). The higher short wave absorptivity of the light-grey and dark-grey colored membranes is improving the solar driven inwards diffusion, and hence a higher RH at the interface of the vapor retarder foil and the cellulose is occurring. Case 1 with the cool roofing membrane is showing a continuously decreased gradient and is remaining below 80% rh without the risk of mold germination, because the lower surface temperatures are reducing inward diffusion drying.

IN-SITU MEASUREMENTS AT BSRTU OUTDOOR TEST FACILITY

In 2010, several low sloped roof construction test assemblies were installed at the CUAS outdoor test facility, built in South Austria. One part of the roof section was used to investigate the hygrothermal performance of selected roof systems executed with different waterproofing membranes. For the roof assemblies a construction system similar to the previous calculated cases 1, 2, and 3 respectively, with OSB deckings on both sides of the timber I-beams and vapor retarder foils was chosen. All roof components are equipped with special measurement systems. The measurements started on December 30, 2010 and are still ongoing. The weather data is recorded using two separate weather stations and also the indoor conditions are monitored continuously. The slope of the investigated roof sections is 2° northwards.

Results—In-Situ Measurements

In this section measured results for the period of January 2011 to January 2012 are presented. On the interfaces between the external OSB decking and the internal vapor retarder foil to the cellulose insulation layer capacitive humidity and temperature sensors were installed to measure the hourly values for relative humidity and temperature (Figure 10). On the internal and external roof surfaces additional temperature sensors were applied and the climate data in- and outside the building was monitored continuously. In addition, optical radiation sensors were installed to verify the climate data which were used during the previous simulations.

Figures 11 and 12 are providing an overview about the measured data concerning outside climate and the solar radiation on the roof constructions.

The measured relative humidity and temperature (°C) values within the roof constructions are presented in Figures 13, 14, and 15.

In case 1 with the cool roofing membrane the RH at the interface external OSB and cellulose is varying between

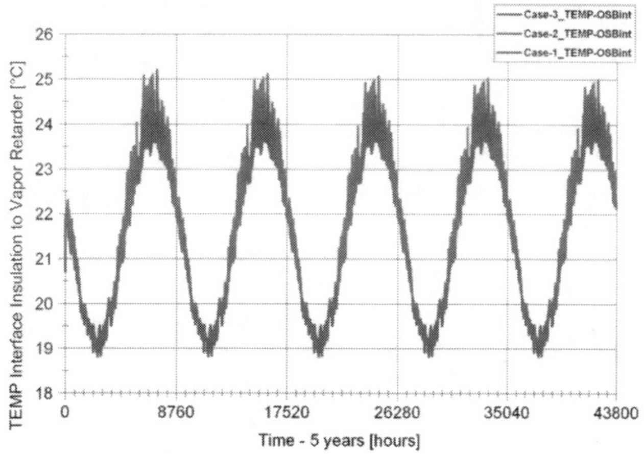

Figure 7 *Temperature at the internal vapor retarder-cellulose interface, cases 1, 2, and 3.*

Figure 8 *In-situ measurements at BSRTU Outdoor Test Facility.*

Figure 9 *Test assemblies for in-situ measurements.*

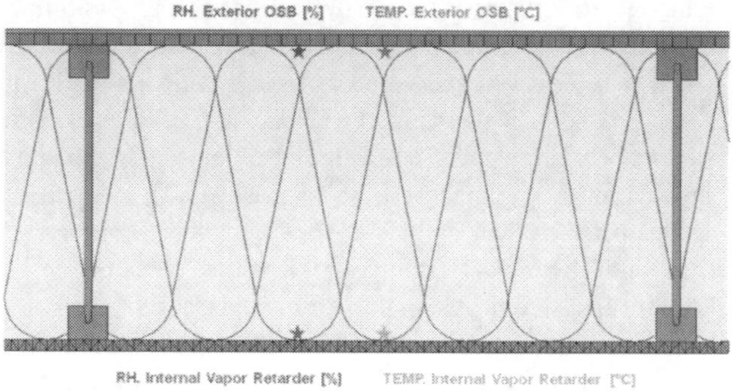

Figure 10 *Positioning of sensors in test assemblies.*

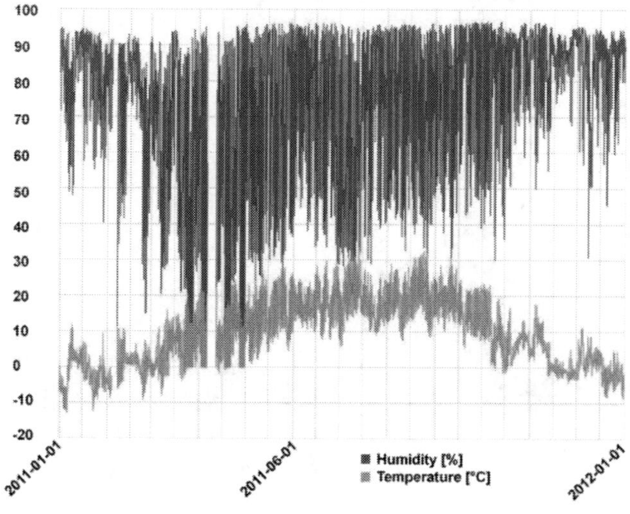

Figure 11 *Measured exterior climate (relative humidity, temperature).*

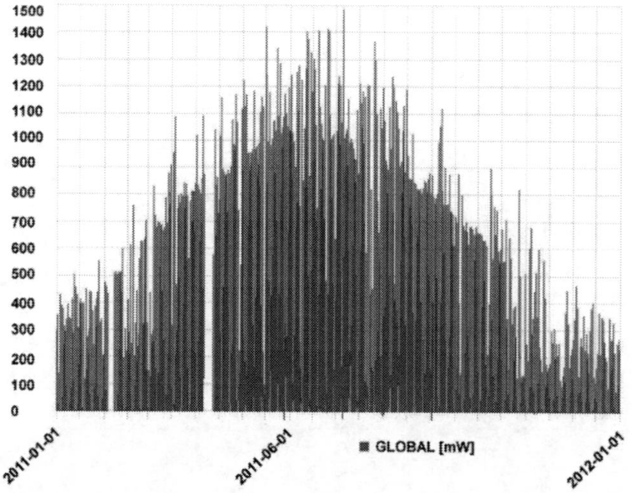

Figure 12 *Measured solar radiation on roof surfaces.*

~62% rh in summer and ~100% rh during wintertime (Figure 13). It is also well to see, that the RH is oscillating at a higher rate between ~70% and ~ 80% rh during the period from spring to fall. Due to the higher temperatures during this time, a certain risk of mold growth cannot be disregarded, but until now no distinctive mold or wood decay fungi growth was detected during probing. Nevertheless it's important to note that a higher water content of the exterior insulation was detected. Hence it's necessary to wait, to see how the future moisture development will proceed.

In addition, the RH at the interface between cellulose and internal vapor retarder is increasing up to ~75% RH in summer and up to ~30% RH in winter. Due to the white

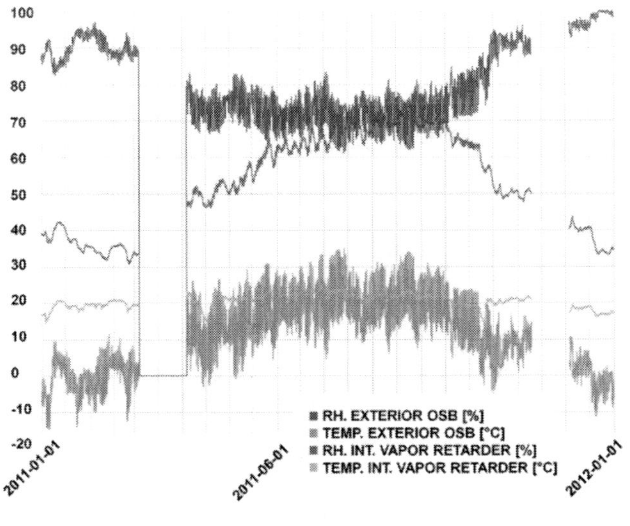

***Figure 13** Relative humidity and temperatures in test assembly 1 with white roofing membrane.*

roofing membrane lower surface temperatures are generated and hence a reduced solar-driven inward diffusions is occurring.

Figure 14 is presenting the relative humidity and temperature values within case 2, executed with a light-grey colored roofing membrane.

The RH value below the exterior OSB decking is leveling between ~50% rh in summer up to ~95% rh during a few days in winter. Although the critical limit for mold germination, 80% rh, was exceeded, no mold or wood decay fungi growth was detected during inspections until now. That's reasonable because temperatures at this time are quite low and it is assumed that the fungicidal additives in the cellulose are reducing the possibility of fungi germination on wooden surfaces, provided that a proper contact between cellulose and timber is given.

According to that, also the relative humidity on the interface between cellulose and the internal vapor retarder was measured. The RH is leveling between ~25% rh in winter and ~78% rh during summertime. The higher RH gradient is founded in the higher solar driven diffusion due to the darker membrane surface. Nevertheless, the critical limit of 80% rh is not exceeded, hence mold related problems are to be neglected.

Figure 15 is presenting the relative humidity and temperature values within case 3, executed with a dark-grey colored roofing membrane.

The RH value below the exterior OSB decking is leveling between ~45% rh in summer up to ~92% rh during a few days in winter. Analogical to case 2, the critical limit for mold germination, 80% rh was exceeded, but no mold or wood decay fungi growth was detected during probing until now.

According to that, the relative humidity on the interface between cellulose and the internal vapor retarder

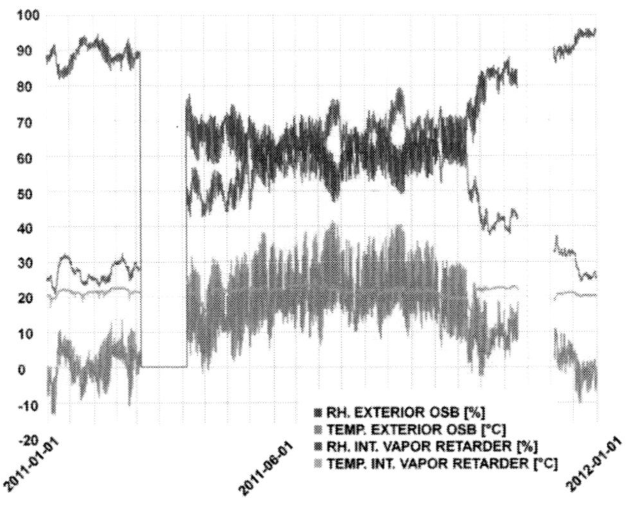

***Figure 14** Relative humidity and temperatures in test assembly 2 with light-grey roofing membrane.*

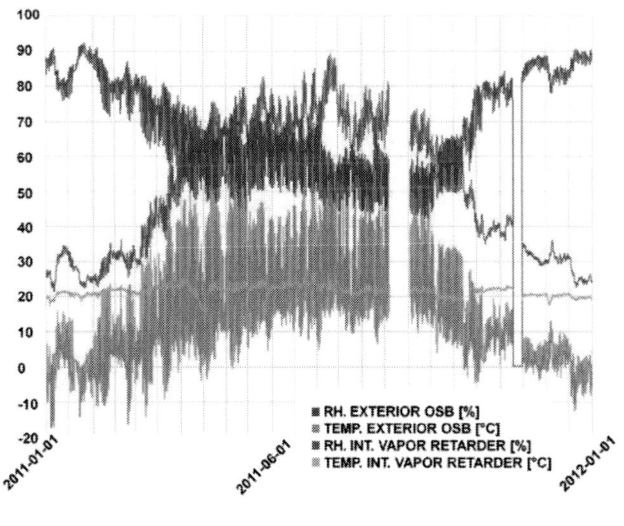

***Figure 15** Relative humidity and temperatures in test assembly 3 with dark-grey roofing membrane.*

was measured. The RH is leveling between ~22% rh in winter and ~88% rh during summertime. The higher RH gradient is founded in the higher solar driven diffusion due to the dark membrane surface. Although the critical limit of 80% rh was exceeded during higher temperatures, no mold related problems have been found.

DISCUSSION AND CONCLUSIONS

This paper presented initial results concerning the hygrothermal behavior of non-ventilated, highly-insulated timber-frame flat roofs with different waterproofing membranes in the climate of Austria. The investigations indicated that different colored waterproofing membranes are seriously influencing the external roof surface temperatures due to varying short wave absorptivity.

The varying surface temperatures are strongly influencing the moisture migration and accumulation and therefore the durability performance of low-sloped roof constructions. Darker membranes increase the temperatures in the exterior part of the roof assemblies and therefore optimize the solar-driven diffusion to the interior. Light-colored and especially white cool roofing membranes are reducing this effect, hence a drying of the roof construction is limited and moisture related problems are likely to occur.

Summing up, it is mentioned that aspects of thermal insulation and moisture control need to be investigated in common. A varying coloring of waterproofing roof membranes has to be considered as a critical factor, influencing the drying of unventilated low sloped roofs and modern computer based simulation models may be used to analyze the hygrothermal performance within the early design process.

REFERENCES

OENORM B8110-2, Thermal insulation in building construction – Part 2: Water vapour diffusion and protection against condensation, Austrian Standards Institute, 2003.

Künzel, H.M. 1994. Verfahren zur ein- und zweidimensionalen Berechnung des gekoppelten Wärme- und Feuchtetransportes in Bauteilen mit einfachen Kennwerten, PhD thesis, University Stuttgart.

Karagiozis, A. 1998. Applied moisture engineering, page 241, *Conference Proceedings Thermal Performance of the Exterior Envelopes of Buildings VII*, Clearwater Beach, FL.

Kuenzel, H.M. 1996, Tauwasserschaeden im Dach aufgrund von Diffusion durch angrenzendes Mauerwerk, wksb 41/1996, Heft 37, Seite 34–36.

WTA – Wissenschaftlich Technische Arbeitsgemeinschaft für Bauwerkserhaltung und Denkmalpflege, 2002, Merkblatt 6-2-01/D "Simulation waerme- und feuchtetechnischer Prozesse."

Powell, F.J., and Robinson, H.E. 1971, The effect of moisture on the heat transfer performance of insulated flat-roof constructions, *Building Sciences Series* 37, Gaithersburg, National Bureau of Standards.

Straube, J. 2001. The influence of low-permeance vapor barriers on roof and wall performance, *Conference Proceedings Performance of Exterior Envelopes of Whole Buildings VIII*, Clearwater Beach, Florida.

Desjarlais, A. 1995, Self-drying roofs: What?! No dripping!, *Conference Proceedings Thermal Performance of the Exterior Envelopes of Buildings VI*, Clearwater Beach, FL, page 763.

Weiß, B., Wagenführ, A. and Kruse, A. 2000. Beschreibung und Bestimmung von Bauholzpilzen.

Hens, Hugo. 2010, Applied Building Physics—Boundary Conditions, *Building Performance and Material Properties*.

Hens, Hugo. 2012, *Performance Based Building Design 2: From Timber-Framed Construction to Partition Walls*.

Hens, Hugo. 1975, Theoretical and experimental study on the hygrothermal behaviour of building and insulation materials during condensation and drying with application on low-sloped roofs, doctoral thesis, K.U.Leuven.

Hens, Hugo. 1987, Envelope solutions for residential construction: low-sloped roofs, Final report RD Energy, Services of the Ministry of Scientific Policy.

A Hygrothermal Risk Analysis Applied to Residential Unvented Attics

Simon Pallin, PhD

Manfred Kehrer
Member ASHRAE

William A. Miller, PhD
Member ASHRAE

ABSTRACT

A residential building, constructed with an unvented attic, is a common roof assembly in the United States. The expected hygrothermal performance and service life of the roof are difficult to estimate due to a number of varying parameters. Typical parameters expected to vary are the climate, direction, and slope of the roof as well as the radiation properties of the surface material. Furthermore, influential parameters are indoor moisture excess, air leakages through the attic floor, and leakages from air-handling unit and ventilation ducts. In addition, the type of building materials such as the insulation material and closed or open cell spray polyurethane foam will influence the future performance of the roof.

A development of a simulation model of the roof assembly will enable a risk and sensitivity analysis, in which the most important varying parameters on the hygrothermal performance can be determined. The model is designed to perform probabilistic simulations using mathematical and hygrothermal calculation tools. The varying input parameters can be chosen from existing measurements, simulations, or standards. An analysis is applied to determine the risk of consequences, such as mold growth, rot, or energy demand of the HVAC unit. Furthermore, the future performance of the roof can be simulated in different climates to facilitate the design of an efficient and reliable roof construction with the most suitable technical solution and to determine the most appropriate building materials for a given climate.

INTRODUCTION

A pitched roof is a very common residential building construction in which the attic space can be either ventilated with exterior air or constructed as unvented. The difference between the two alternatives is mainly the location of the thermal resistance. In a ventilated attic, the greater part of the insulation is located in the ceiling plane. As for the unvented attic, the thermal resistance is instead located inside or directly underneath the roof. Subsequently, the attic space in an unvented attic is embraced by the thermal envelope, thus making it a better environment for an HVAC system compared to a vented attic space. An unvented attic also has less energy loss from the air distribution system in comparison with a ventilated attic (Rudd 2005). The more leaky the ductwork is, the greater the energy losses of the HVAC unit (Russell et al. 2007)—which, in a ventilated attic, become very significant. Air leakages from the indoor environment or the duct system may lead to mold growth on the interior surface of the roof (Hagentoft et al. 2008) or result in ice dam creation (Lstiburek 2006). Consequently, there are advantages with an unvented attic compared with a vented attic, at least in terms of energy efficiency. In concerns of moisture safety, though, the performance of an unvented attic is not that obvious; there are other properties of the roof construction and the indoor and outdoor environment that must be taken into account.

There are usually a large number of wood-based materials in a residential pitched roof construction—rafters, ceiling joists, and roof deck sheathing. Unfortunately, a wood surface is a favorable environment for mold growth, depending on available nutrients, humidity, and temperature (Hukka and Viitanen 1999), and rot. Therefore, it is of great concern to prevent critical levels of relative humidity (RH) inside the

Simon Pallin *is a postdoctoral research associate and* ***Manfred Kehrer*** *and* ***William Miller*** *are senior research engineers in Building Envelope Research at Oak Ridge National Laboratory, Oak Ridge, TN.*

attic, which otherwise offers a potential for moisture build up in the wood materials, especially the roof sheathing (Straube et al. 2010). Naturally, air leakages from the air distribution system, leakages between the indoor and attic spaces, as well as leakages to and from the outside environment will influence the RH in the attic. An unvented attic is constructed to prevent leakages to the outdoor environment; hence leakages from the interior and the air distribution system will dominate the unintended air leakages into the attic space. In addition to these air leakages, the water vapor content of the attic air will depend on the indoor and outdoor water vapor content of the air, the dehumidifying effect from the HVAC system, and the indoor moisture generation.

An unvented attic hosting an HVAC system must be considered as a very complex hygrothermal system. The future performance and expected service life of the roof construction will vary due to a number of influencing parameters, referred to as *variables*. These variables are important to include when analyzing the future conditions of an unvented attic in terms of energy efficiency, moisture safety, and durability. However, each parameter must be expected to vary differently, as should their importance on the attic performance; hence, the most decisive parameters can be identified. A range of influencing parameters is as follows:

- Indoor heat and moisture production
- Hygrothermal material properties
- Natural and forced unintended air leakage
- Features of the HVAC system, i.e., dehumidifying/humidifying effect, airflow rate, etc.
- Geometrical variations of the building components
- Outdoor climate
- Orientation and location of the building and slope of the roof
- Workmanship
- User behavior, i.e., HVAC setpoint temperatures, airing, maintenance, etc.

The purpose of this study is to develop a reliable hygrothermal simulation model of an unvented attic, which in the future should facilitate the design of an efficient and reliable roof construction with the most suitable technical solution, and to determine the most appropriate building materials for a given climate. This study focuses on determining the importance of the most decisive parameters and analyzing their sensitivity on the overall hygrothermal performance of the unvented attic. A risk analysis is executed based on simulations with a number of varying parameters, creating different plausible scenarios of realistic attic and roof constructions. The performance indicators of interest in this study are the risk of rot in the wood-based roof sheathing and the risk of mold on roof rafters and ceiling joists. In addition, the impact from the varying parameters on the energy demand of the air distribution system is investigated.

THE SIMULATION MODEL

The unvented attic was simulated with a numerical model created in MATLAB®, which is a software for mathematical computations (MathWorks 2013). This model was then interacted with WUFI1D, which is a validated hygrothermal calculation tool (Künzel 1995). The numerical MATLAB model is designed to calculate the heat and water vapor transfer through the building components and intermediate air volumes, in exception of the outer roof construction, which is calculated in WUFI1D. The advantages of separating the hygrothermal system between two different tools are many, though there are also some important drawbacks that must be considered.

WUFI1D allows realistic calculation of the transient hygrothermal behavior of building components and contains a wide range of building material properties. This tool is also developed to include all the transfer coefficients that are expected at an exterior surface with relation to the exterior climate and incident solar radiation. A disadvantage with WUFI1D is that it is lacking the capability to include radiation heat exchange in intermediate air spaces or at an interior surface and it is not capable of calculating indoor boundary conditions. In attics, the radiation heat exchange between the interior surfaces may have great influence on the attic environment (Hagentoft et al. 2008). Instead, this radiative interaction and the heat and water vapor transfer that exist below the roof construction, such as the HVAC system, natural and forced convection, thermal conductivity, and water vapor transfer due to convection and diffusion, are simulated in the numerical MATLAB model. Consequently, the numerical tool is designed to simulate the performance of the complete building envelope and inner environment, except for the roof construction, whose hygrothermal performance is instead simulated in WUFI1D. However, since WUFI1D is a one-dimensional hygrothermal tool, the complete attic model requires two WUFI models: one left and one right roof construction. This approach requires an iterative process between these two WUFI1D models and the mathematical MATLAB model, as shown in Figure 1. An iterative simulation process is essential to enable the two WUFI models and the mathematical model to represent a complete and realistic system of the attic space and the adjacent roof construction, which consequently includes the coupling of the simulated elements of WUFI and the attic environment. A thorough description of the mathematical MATLAB model is given in the appendix.

There is a large number of material, geometrical, and system features that must be implemented into a simulation model. This information is obtained either by measurements, from previous simulations, or from standards. All input data for the simulations are intended to represent realistic and true values and not design values, excepting the varying parameters, for which upper values may be taken from design criteria.

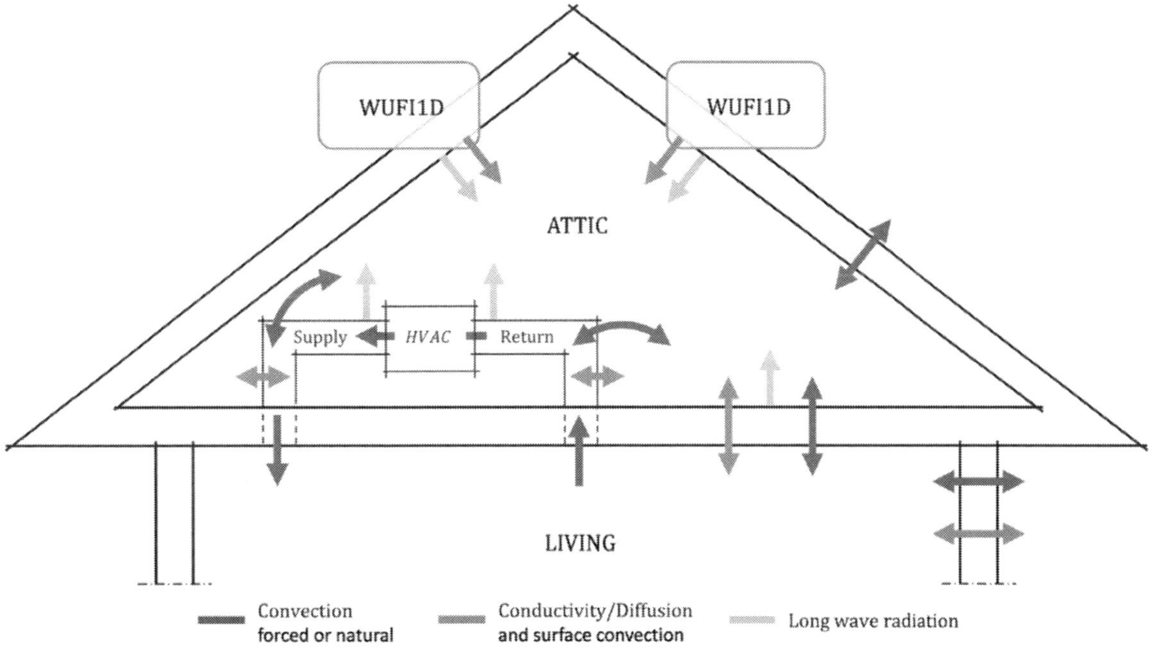

Figure 1 *The unvented attic, hosting an HVAC system, must be considered as a very complex hygrothermal system with a large number of interacting mechanisms. The arrows depict the location and direction of both the heat and water vapor transfer of convection, conductivity/diffusion, and longwave radiation.*

SIMULATION INPUT DATA

There is an extensive amount of information needed if we are expecting a realistic result from running simulations of the unvented attic. This information is used as inputs for the simulation model, of which the most essential are defined in this section.

The air distribution system, applied to the simulation model, is designed as a whole-house air distribution system with intermittent supply and with an exterior air intake on the return side of the HVAC system. The electrical efficiency of the HVAC unit is used to determine the thermal load in the attic. Assuming 90% efficiency results in 10% excess of heat, based on the power needed by the HVAC unit. The airflow rate is determined by the Total Ventilation Rate Method of ANSI/ASHRAE Standard 62.2 (2012), Equations 4.2 and 4.8. The rate of outdoor air exchange in the HVAC system is set to a minimum of 10%, as shown in Table 1. If this rate of fresh air doesn't meet the required minimum outdoor air exchange (ASHRAE 2012) in Equation 4.1, instead the rest is assumed to be fulfilled by infiltration through the building envelope.

The cycling of the HVAC system depends on the indoor temperature and the setpoint temperatures of the thermostat. The setpoint temperatures are usually chosen based on the desired indoor comfort temperatures of the occupants in a building and may vary greatly. The simulation model is designed to optimize the HVAC unit to a 50% on and off cycle at a small span of setpoint temperatures, 70°F/74°F (21.1°C/23.3°C). The cycling is adjusted with the supply airflow rate,

Table 1. Some of the Information Needed to Perform Simulations of the Unvented Attic

Simulation Input Data	
Geometric	
Dimensions of building	33 × 49 ft (10 × 15 m)
Indoor height	8 ft (2.4 m)
Roof slope	20°
Number of bedrooms	4
Number of floors	2
Fenestration area	15% of wall area
HVAC	
Electrical efficiency	90%
Duct length (Ø 10 in./0.25 m)	230 ft/2000 ft^2 of FFA* (70 m/610 m^2 of FFA)
Ratio of supply vs. return ducts	2
Supply duct insulation	R-5 ($ft^2 \cdot h \cdot °F$)/Btu (R=0.88 [$m^2 \cdot K$]/W)
Duct insulation emissivity, ε	0.10
Ratio of fresh air	≥10%
Supply temperature—Cooling	55°F (12.8°C)
Supply temperature—Heating	105°F (40.6°C)

* FFA = finished floor area. The presented ratio is an empirical value based on experience, thus subjectively chosen.

though the rate never falls below the required minimum airflow rate (ASHRAE 2012).

In order to determine the attic temperature, the indoor temperature and the cycling of the HVAC system must be estimated. The indoor thermal conditions mainly depend on thermostat setpoint temperatures, outdoor temperature together with the U-factor of the building envelope, indoor thermal inertia, fenestration area coupled with incident solar radiation, and indoor thermal load. These parameters must be established, with values are based on standards and handbooks:

- The U-factors of the walls, slab on grade, windows, and doors in the simulation model agree with the equivalent U-factors required by the *International Energy Conservation Code* (ICC 2011), Table R402.1.3.
- The indoor thermal inertia from furniture and contents is based on the values presented in ANSI/ASHRAE Standard 90.2 (2007), Section 8.8.2. The thermal inertia inside the attic space is defined by the wooden joists and rafters.
- The indoor thermal load from lights, people, and equipment coincides with Equations 30 and 31 of *ASHRAE Handbook—Fundamentals* (2005), whereas the distribution of the thermal load during the day is based on a Daily Internal Heat Gain Profile (ASHRAE 2007), Table 8.8.1.

With respect to water vapor transfer, the moisture buffering capacity of the materials inside the attic and the indoor space are equivalent with the defined materials with heat capacity. All other material properties are provided from the WUFI material database.

The roof construction, simulated in WUFI1D, is assumed to have an upper covering with shingles, applied on an oriented strand board (OSB) sheathing and with an underlying insulation material of a spray polyurethane foam (SPF). The shingles are assumed to be watertight and have a vapor permeance of 0.3 perm (sd = 10 m). A possible mechanism of solar-driven moisture between the laps of the shingles (Rudd 2005) is not taken into account in this study because of the lack of knowledge and data describing this possible phenomenon.

VARYING PARAMETERS

The varying parameters are chosen based on an importance sampling scheme (Vose 2008), in which the chosen values mostly represent the extreme tail of the parameters' variability. Six different input parameters are selected to vary in the simulation model due to their assumed high influence on the hygrothermal performance of the attic and roof construction. The varying parameters of this study are the following:

- Thermostat setpoint temperatures
- Outdoor climate
- Vapor permeance of the rigid spray foam insulation
- Air leakage rate from supply and return ducts
- Airtightness of ceiling floor
- Indoor moisture production

The thermostat setpoint temperatures are chosen to vary between 70°F/74°F (21.1°C/23.3°C) and 68°F/78°F (20°C/25.6°C), thus representing an assumed small and wide range of, according to the occupants, acceptable indoor temperatures. Practically, the simulation model determines the HVAC cycling based on 70°F/74°F and then applies the same supply airflow rate on 68°/78°F setpoints. This approach will determine the impact on the hygrothermal performance of the unvented attic due to the occupants' desired comfort temperatures.

The outdoor climates, applied to the simulation model, represent U.S. climate zones 1 to 7. The corresponding cities for each climate zone presented in given order are: Miami, FL; Austin, TX; Atlanta, GA; Baltimore, MD; Chicago, IL; Minneapolis, MN; and Fargo, ND.

The thermal resistance of the roof consists of either open- or closed-cell SPF with no assumed coating. The water vapor permeance (diffusion resistance factor) for the open and closed SPFs are 55.1 U.S. perm·in. (μ-value = 2.38) and 1.47 U.S. perm·in. (μ-value = 88.93), respectively. The required thickness, or actually the R-value, of the SPF depends on which U.S. climate zone the roof is to be constructed in and varies between R-30 and R-49 (5.28 and 8.63 $m^2 \cdot K/W$). The R-values are in compliance with regulation (ICC 2011) and implemented into the simulation model.

The unintended air leakages from the air distribution duct system into the attic space can be up to 20% or more (Lstiburek 2006; Polly et al. 2011). Though there are no requirements for duct airtightness if the air-handling unit and ducts are located inside the thermal envelope (ICC 2011), low and high leakage rates are still applied in the simulations to study the effect of different duct leakage rates. In this study, a 4% air leakage of the actual supply and return airflow rates is considered a low leakage rate; a high rate is considered 20%.

A leaky ceiling floor will enable air to exchange between the attic and the living space. The potential air movements are induced by air pressure differences, and the amount of air and the direction of the flow depends on the reigning conditions of temperatures, the ventilation system, or wind forces. Naturally, there are no regulations for the airtightness of the ceiling floor when the attic is located inside the thermal envelope. However, the impact of this phenomenon on the hygrothermal performance of the unvented attic is chosen to be investigated. The airtightness of the ceiling can vary greatly depending on the presence of recessed canister lights and other ceiling penetrations (Rudd 2005). In this study, two assumed conditions are chosen for the ceiling airtightness along the ceiling floor area (CFA): a low value is set to 0.4 cfm50/CFA (2.0 $l/[s \cdot m^2]$) and a high value is considered 2.0 cfm50/CFA (10.0 $l/[s \cdot m^2]$). Unpublished measurements made of the ceiling airtightness in a two-story residential building with an unvented attic in Knoxville, TN, USA, resulted in a cfm50/CFA of 1.04 (5.3 $l/[s \cdot m^2]$). This measured value of cfm50 coincides with the ranges chosen for the simulation model.

The generation of moisture inside a residential building can vary significantly (Christian and Trechsel 1994; Kalamees et al. 2006) depending on the user behavior and household composition and the number and type of appliances and their frequency of use (Yik et al. 2004). Two different moisture production rates are chosen, an average daily rate of 15.8 lb/day (7.2 kg/day) and a high rate of 33.6 lb/day (15 kg/day). The lower value represents a likely daily average of moisture production, based on simulations (Pallin 2012). The upper value agrees with the Residential Design Moisture Generation Rate of ANSI/ASHRAE Standard 160 (2011), Table 4.3.2, for a four-bedroom living space. The moisture production rate varies during the day, depending on the presence and frequencies of moisture-generative activities inside the household. In this study, the daily variation of the moisture production rate is estimated to follow the activity pattern according to the Daily Internal Heat Gain Profile of ASHRAE Standard 90.2 (2007), Table 8.8.1.

Combining all the above varying parameters results in 224 (i.e., 2 × 7 × 2 × 2 × 2 × 2) different scenarios of the unvented attic, where each scenario requires an iterative simulation run with the numerical MATLAB model and the two WUFI1D models. Each scenario is simulated until convergence criteria are fulfilled, in comparison with previous iteration; see Equations A.1 to A.2 in the appendix for details.

RESULT

After completing the simulations of the 224 different scenarios of unvented attics, the following measures of hygrothermal performances were taken into consideration: the variation of the moisture content (MC) in the OSB roof sheathing and the energy demand of the air-handling unit—and also identification of the most reliable and risky compositions of the varying parameters, in terms of the previously described performances.

The MC is a common performance indicator when estimating the service life of a wood-based building material (Straube et al., 2010). Commonly, a MC of 20%–25% is taken as critical upper limit to prevent decay of wooden materials (DIN 2012).The Mold Growth Index (MGI) is also a useful indicator when analyzing the condition of a wood-based material (Ojanen et al. 2011) and estimates the development of mold on a surface depending on the temperature and time of critical RH. In terms of an OSB sheathing with a closely attached upper shingle covering and with an underlying SPF, the surface exposure is very limited; hence, the issue of mold on the OSB surface is questionable. Instead, the MC is used as the performance indicator for moisture safety in this study.

There is a large disparity in the annual development of the MCs for the 224 different simulated OSB sheathings, as shown in Figure 2. Most of the OSB sheathings vary between

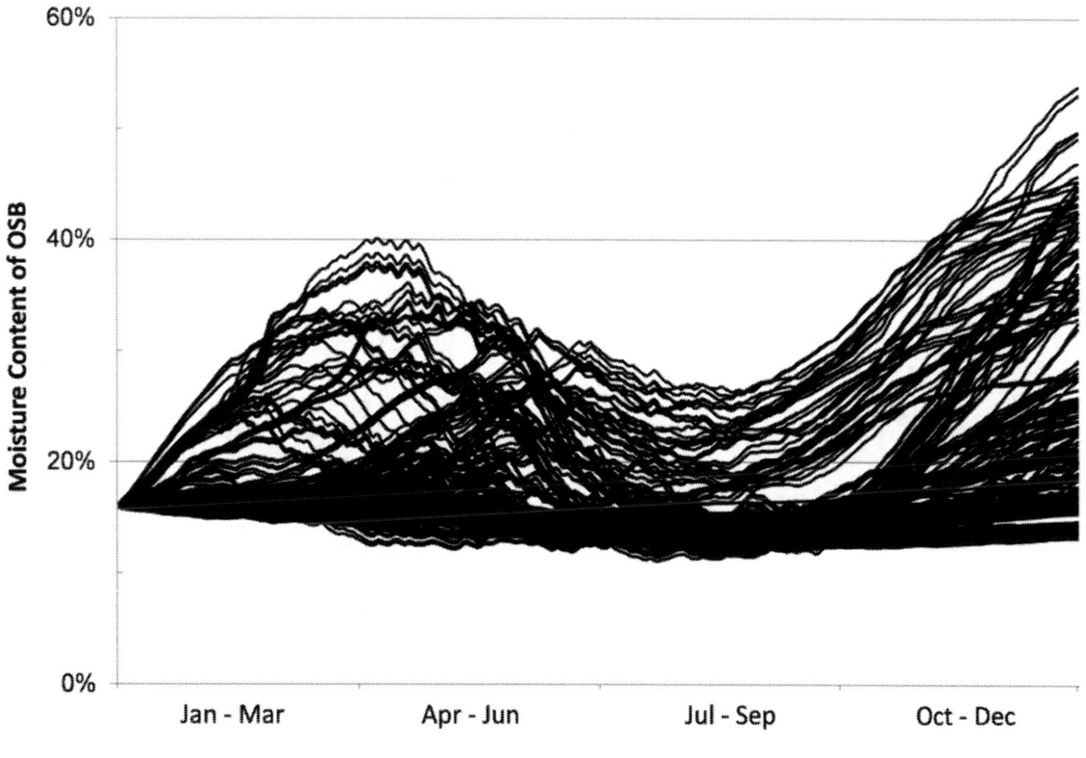

Figure 2 *Annual variation of the MCs in the OSB roof sheathing. The different curves depict the 224 different simulated scenarios of the OSB, including all varying parameters. According to the results, the MCs for most of the OSB sheathings vary between 13% and 17%, though several simulated scenarios result in critical levels of MC in terms of deterioration/rot and must be considered roof constructions with unacceptable hygrothermal performances.*

MCs of 13% and 17% annually; still, there are cases where the MC reaches high above accepted levels. The reason behind the assumed failure of these simulated OSBs is the composition of the varying parameters.

The composition of the parameters for the most reliable and the most risky roof constructions due to critical levels of MC are presented in Table 2. The different scenarios are presented for U.S. climate zones 1 to 7. There are some clear conclusions to make out of Table 2; e.g., every best-performing unvented attic roof is constructed with a closed SPF and with a low indoor moisture supply. The opposite is true for the most risky roof. In all cases, except for climate zone 4, a high duct leakage has a positive effect on the MC of the OSB, most likely due to the dehumidifying effect of the HVAC cooling coils, which, by a higher rate of air leakage, will have a higher influence on the vapor content of the attic air during the operating cooling mode. There are no clear patterns for the varying parameters of the thermostat setpoint temperatures and the ceiling airtightness. A detailed illustration of the annual variation of the MCs is presented in Figure 3 for climate zones 1 and 3, for the simulated best and worst scenarios according to Table 2.

The importance of the different varying parameters on the maximum MC of the OSB sheathing can be estimated with a sensitivity analysis. Since the values of the varying parameters are chosen based on their importance and not their likeliness of occurrence, the sensitivity analysis is not probabilistic and should therefore not be considered such. A common analysis method to compare the impact of varying input parameters on the result is to apply an Importance Index (Hamby 1994), which takes into account the variances of the parameters on the chosen indicator. This study uses a similar approach, in which the importance of the different varying parameters is estimated by measuring the average deviation of a varying parameter when remaining all parameters but one are fixed. Repeatedly, the disparity in the annual maximum of MC is compared when changing only the value of the investigated parameter. The average disparity of the parameter (in this study the MC) is referred to as D_p and expressed as follows:

$$D_p = \frac{1}{n/2} \sum_{i=1}^{n/2} \left| X_{i,p}^{A} - X_{i,p}^{B} \right| \qquad (1)$$

$$\text{for } A \in C, B \in C, A = -B$$

where

D_p = average disparity of one varying parameter, p ($p = 1{:}6$)

X = indicator or the analysis (in this case the maximum MC), %

n = total number of compositions, i.e., sets of different values of X ($n = 224$)

A is one of n compositions of the varying parameter, B is the opposite of A due to the chosen parameter p, and C is the set of all n compositions of varying parameters.

As an example, if a parameter $p = 3$ is varying in a binary sequence and A = 010110, then B = 011110 because B is considered the opposite of A in the position of the third parameter. In the case of this study, an example is the following: if we considered the importance of the moisture production, we'll look at the maximum MC of the first simulated scenario. Then we're comparing it with the MC in the simulated scenario where all the parameters are the same, except for the moisture production. We then summarize the disparities of the maximum MC when moving along all the simulated scenarios, where A is the scenario we are comparing with B. For an analysis with $n = 224$ simulated scenarios, D_p is based on the $n/2$ number of differences in MC; otherwise, a specific relative disparity is used twice, i.e., when the previous sequence of B becomes A.

Furthermore, D_p for each varying parameter is weighted with the parameter with the highest value, referred to as D_p^{max}; hence an relative disparity, D_p^{rel}, can be defined for each parameter as:

$$D_p^{rel} = \frac{D_p}{D_p^{max}} \qquad (2)$$

Consequently, the parameter with the highest D_p obtains a maximum relative value of 1.0. All other parameters receive a D_p^{rel} of either 1.0 or lower. This method enables a comparison of the varying parameters on their influence on a chosen performance indicator, which in this study is the maximum MC of the OSB sheathing for a simulated consecutive year, as presented in Figure 4. According to the analysis, the vapor permeance of the SPF is of greatest importance on the MC of the OSB—meaning, whether the SPF is closed or open, it will have the highest influence on the estimated conditions of the OSB sheathing in the roof construction. Other important parameters are the climate and the indoor moisture production. In this study, the chosen variation of the indoor moisture production was set to vary between a normal and high excess of moisture. If an expected lower value of the moisture production were been used instead (as for the duct leakage, the setpoint temperatures, and the ceiling airtightness), instead a larger value of D_p^{rel} would have been expected.

There is a low influence on the maximum MC of the OSB in terms of the thermostat setpoint temperatures and duct leakages, though the duct leakage seems to have some influence according to Table 2. However, these varying parameters must be assumed to possess a larger impact on the energy demand of the HVAC unit. Naturally, the thermostat setpoint temperatures govern the cycling of the HVAC and therefore the amount of energy required. Figure 5 illustrates the decrease in energy loss of the HVAC unit when changing the setpoint temperatures from 70°F/74°F (21.1°C/23.3°C) to 68°F/78°F

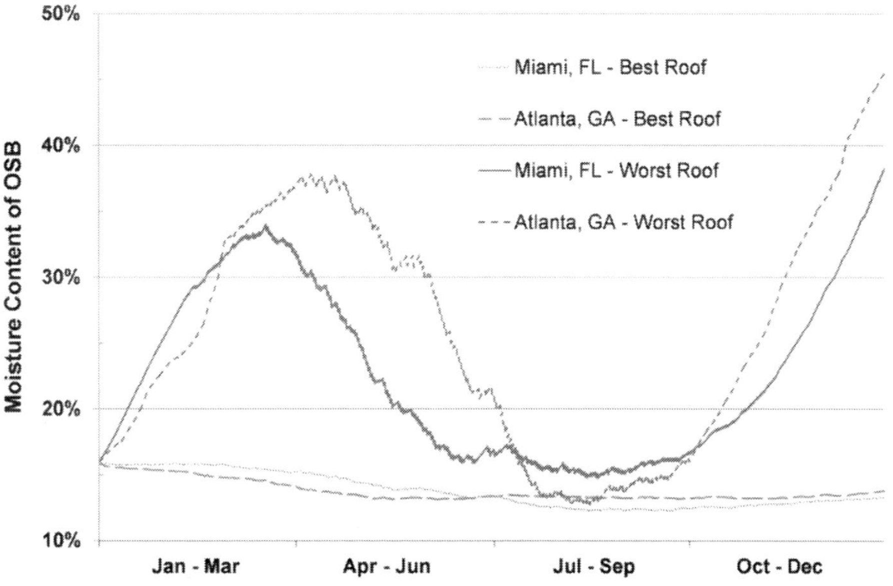

Figure 3 *The annual accumulation of moisture content (MC) in the OSB sheathing, for the best and worst simulated scenarios and for the chosen cities of U.S. climate zones 1 and 3. In the two OSB sheathings with the lowest accumulation of moisture, the end value is lower than the start value, which indicates that the MC must be assumed to decrease further if simulated for a longer period of time.*

Table 2. Compositions of the Varying Parameters with Lowest and Highest Risk of Deterioration of the OSB due to the End Value of MC, Presented for U.S. Climate Zones 1 to 7. In the Simulation, the Starting Value of the OSB MC is 16%.

Rank	Climate Zone	Direction	Setpoint	SPF	Duct Leakage	Ceiling Leakage	Moisture Production	MC End Value
Best	1	South	70/74	Closed	20%	2@50	Normal	14%
Worst		North	68/78	Open	4%	10@50	High	38%
Best	2	South	70/74	Closed	20%	2@50	Normal	13%
Worst		North	70/74	Open	4%	10@50	High	43%
Best	3	South	70/74	Closed	20%	2@50	Normal	13%
Worst		North	68/78	Open	4%	10@50	High	45%
Best	4	South	68/78	Closed	4%	10@50	Normal	14%
Worst		North	68/78	Open	4%	10@50	High	54%
Best	5	South	70/74	Closed	20%	2@50	Normal	14%
Worst		North	68/78	Open	4%	10@50	High	47%
Best	6	South	70/74	Closed	20%	2@50	Normal	14%
Worst		North	68/78	Open	4%	10@50	High	45%
Best	7	South	68/78	Closed	20%	2@50	Normal	13%
Worst		North	68/78	Open	4%	10@50	High	37%

(20°C/25.6°C). The analysis does not take the efficiency of the HVAC unit into consideration, nor the differences between heating and cooling; it solely considers the amount of energy needed, in a thermodynamic matter, to either cool or heat the air that is supplied into the living area. In this study, we refer

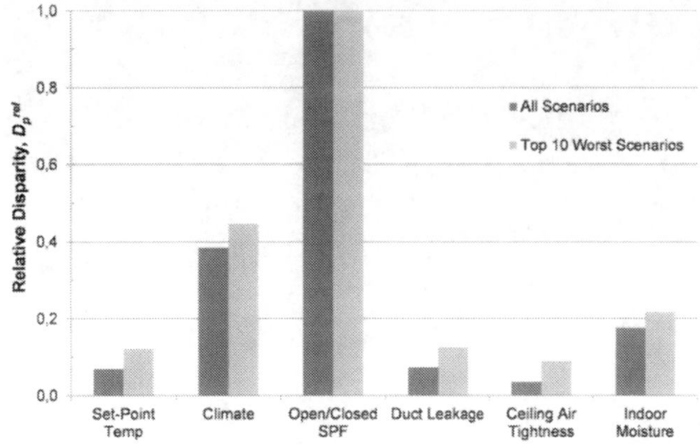

Figure 4 *The relative disparity,* D_p^{rel}*, for the varying parameters of this study. According to the analysis, the simulated maximum MC of the OSB will vary mostly depending on weather the SPF is vapor closed or open. Both the climate and the indoor moisture production will have rather high influence as well.*

to the reduced amount of energy as a factor of reduction (F_{red}) and express it as follows:

$$F_{red} = \frac{1}{n/2} \sum_{i=1}^{n/2} \frac{Q^B_{heating,n} + Q^B_{cooling,n}}{Q^A_{heating,n} + Q^A_{cooling,n}} \tag{3}$$

where

$Q_{heating}$ = energy required to heat the supply air to 105°F (40.6°C)

$Q_{cooling}$ = energy required to cool the supply air to 55°F (12.8°C), in which n, A, and B equal the definitions of Equation 1

Once again, simulation A represents the investigated scenario that is compared with the scenario where all parameters are fixed except for the setpoint temperatures, i.e., simulation B. Therefore A means the scenario with a setpoint of 70°F/74°F (21.1°C/23.3°C) and B means 68°F/78°F (20°C/25.6°C). A value of F_{red} close to zero means a low energy reduction if changing the range of indoor comfort temperatures.

The analyses of the F_{red} for the different U.S. climate zones are presented in Figure 5, which indicates that a warmer climate has a higher potential of energy savings compared to a colder climate. It is important to emphasize that the U.S. climate zones can only be compared with each other as relative reductions, e.g., a specific amount of energy in a 20% savings for one climate zone can actually be larger than a 30% savings for another.

The duct leakages from the air distribution system also prove to influence the cycling of the HVAC unit. Figure 6 illustrates the increase in energy demand of the HVAC unit when changing the assumed leakage rate from 4% to 20% in the supply and return duct system. As for the previous analysis with the setpoint temperatures, the efficiency of the HVAC unit is not taken into consideration, nor are the differences between heating and cooling. In comparison with the reduction of energy in Equation 3, the reduction becomes an increase since the relative difference in energy demand is instead referred to as F_{inc}.

The result of the analysis in Figure 6 shows that the leakage rate of the duct system does make a difference on the annual energy cost (AEC), despite the lack of regulations for

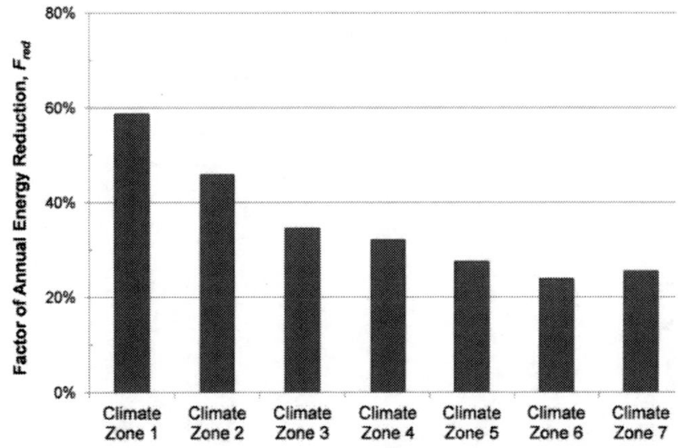

Figure 5 *Relative annual reduction of energy demand when changing the thermostat setpoint temperatures from 70°F/74°F (21.1°C/23.3°C) to 68°F/78°F (20°C/25.6°C). According to the results, warmer climate zones have a higher savings potential in HVAC energy demand if a wider range in setpoint temperatures is accepted.*

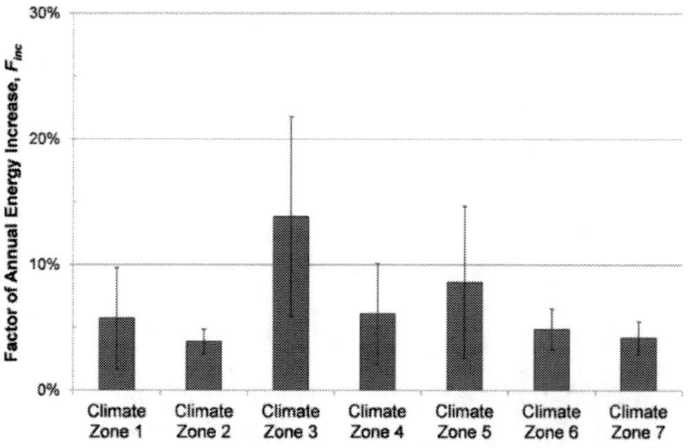

Figure 6 *Relative annual increase of energy demand when changing the assumed ventilation duct leakage from 4% to 20%. The results are presented for U.S. climate zones 1 to 7, where the increases will vary from 5% to 14%. A large uncertainty must be expected, as defined by the error bars, whose values represent the standard deviation.*

the airtightness of the duct system when situated inside the thermal envelope (ICC 2011, Section R.403.2.2) or accounted for (ASHRAE 2007, Section 8.8.4.5). The relative increase in energy required to heat or cool the air varies between 5% and 14% when comparing a 4% duct leakage with a 20% leakage rate, though with a large uncertainty, as seen in Figure 6.

Furthermore, the RH inside the attic was investigated due to the risk of mold growth. All of the 224 simulated unvented attics received nothing else than an index of zero on the MGI, which means that none of the simulated attic climates had any potential of developing mold on wood-based surfaces inside the attic. This result was expected since an unvented attic is embraced by the thermal envelope, but this result was nonetheless important for the verification of the simulation model. Therefore, the risk of decay in the OSB sheathing is considered most important in terms of moisture safety; thus, mold growth inside the attic space becomes less important.

CONCLUSION

In this study, 224 different compositions with 6 varying parameters were simulated for an unvented attic and the adjacent roof construction. The six varying parameters are the thermostat setpoint temperatures, the outdoor climate, the vapor permeance of the rigid spray foam insulation, the air leakage rate from the ventilation supply and return ducts, the airtightness of the ceiling floor, and the indoor moisture production. The impact due to the parameter variability on the hygrothermal performance of the roof and attic was investigated. Three different performance indicators were analyzed: the maximum moisture content (MC) of the OSB, the HVAC system energy demand, and the mold growth index (MGI) of the wood-based materials inside the attic space.

The MC in the OSB sheathing varies mostly due to whether the spray polyurethane foam (SPF) is vapor closed or open. Having an open SPF is actually a risk in all the investigated U.S. climate zones, 1 to 7, depending on the chosen values of the other varying parameters. Naturally, the outdoor climate will influence the MC of the OSB, but also the indoor moisture production rate has a significant impact.

A high air leakage rate from the air distribution duct system has a positive impact on the MC of the OSB sheathings due to the dehumidifying effect of the HVAC unit, though it has a negative influence in terms of the HVAC energy demand. On average, an increase in 5% to 14% in energy demand is predicted for the different climate zones when comparing a 4% with a 20% duct leakage rate, though both a higher and a lower increase can be expected. Consequently, this study shows that the rate of air leakage from the ventilation duct system can make a large impact on the annual energy demand of a residential building, despite the lack of any existing guidance or regulations for the airtightness of the duct system when constructed inside the thermal envelope. Furthermore, the thermostat setpoint temperatures have a very large impact on the annual energy demand. Changing from 70°F/74°F (21.1°C/23.3°C) to 68°F/78°F (20°C/25.6°C) has the highest impact in hotter climates, but colder climates also have a significant decrease in energy demand when accepting a wider span of setpoint temperatures.

The risk of developing mold on surfaces inside the attic space is negligible for all the simulated plausible cases of an unvented attic. However, it is important to emphasize that the outer roof construction is assumed to be airtight in the simulation model and that any deviation from that assumption will affect the risk of mold inside the attic space.

Finally, this study proves that a moisture-safe unvented roof can be constructed for each U.S. climate zone. A parameter study has shown which of the parameters has the largest impact on the hygrothermal performance, as illustrated in Figures 4 to 6, and which alternatives exist to improve the performances; see Table 2. Generally, a closed SPF is a good decision in terms of the risk of moisture damage in the OSB. The roof shingle system of the simulated roof construction is simulated watertight and with a rather high vapor permeance. A possible mechanism of solar-driven moisture between the laps of the shingles is therefore not taken into account. However, recent unpublished measurements at multiple Oak Ridge National Laboratory (ORNL) test facilities indicate that moisture transport through the roof shingle system is suspected. It is also important to point out the risk when enclosing an organic material, such as the OSB, between two rather vaportight materials such as the SPF and the roof shingles. If water or vapor reaches the OSB sheathing, the drying potential is very low and the moisture becomes trapped. This possible event has not been investigated in this study, though it should be considered as a potential risk that ought to be further studied. Minimizing the moisture-productive activities and appliances inside the living space may also be effective to increase the moisture safety of the OSB. With regards to energy, the setpoint temperatures accepted by the occupants and the reduced air leakages from the ventilation duct system make a significant difference on the HVAC system energy demand.

APPENDIX

In this appendix, a detailed description is given of the structure, mathematical expressions, iterative process, and convergence criteria behind the numerical MATLAB model of the studied roof construction. The description is divided into two parts: one for the heat transfer and one for the water vapor transfer, representing the hygrothermal performance if the complete unvented roof construction. The relations between temperatures/vapor contents and incident heat/moisture sources can be defined in a network scheme, where each node represents a temperature/vapor content, connected to each other with conductances. The thermal network, representing the heat transfer of the unvented attic, is illustrated in Figure 7, and the network representing the water vapor transfer is presented in Figure 9.

Nomenclature

Temperatures, K

T_e = outdoor air temperature

T_{in} = indoor air temperature

T_{attic} = attic air temperature

T_{supply} = ventilation supply air temperature

T_{return} = ventilation supply return air temperature

T_n = temperature at surface *n*

T_N^{fict} = fictitious radiation node temperature at node *N*

Conductance—Conductivity, W/K

K_{cond}^{duct} = heat conductance between the inner and outer ventilation duct surfaces

$$= \frac{2 \cdot \pi \cdot \lambda_{duct} \cdot L_{duct}}{\ln(r_2/r_1)}$$

where

- λ_{duct} = thermal conductivity of the duct insulation
- L_{duct} = total length of the ventilation ducts
- r_2 = outer radius
- r_1 = inner radius

$K_{cond}^{wall+slab}$ = mutual conductance of the thermal conductivity between the interior and exterior surfaces of the building envelope, except through the ceiling and the roof

K_{cond}^{ceil} = conductance, representing the heat conductivity between the upper and lower ceiling surfaces

Conductance—Convection, W/K

$K_{conv}^{sr,n}$ = conductance from natural convection at an interior surface *n*

$$= 2 \cdot \sqrt[3]{T_a - T_s} \cdot A_s$$

where

- T_a = ambient air temperature
- T_s = surface temperature
- A_s = area of surface *n*

$K_{conv}^{leak,ceil}$ = natural air convection-induced conductance between the indoor living and the attic space

$$= \rho_{air} \cdot c_{air} \cdot C \cdot \Delta P^{\exp}$$

where

- C = air leakage coefficient
- ΔP = convective induced pressure difference
- exp = air pressure exponent

and

$$\Delta P = z \cdot 3456\left(\frac{1}{T_{attic}} - \frac{1}{T_{in}}\right),$$

in which *z* is the vertical distance from the neutral pressure plane. ΔP can either be positive or negative, depending on the direction of the airflow.

$K_{conv}^{leak,c}$ = natural air convection-induced conductance, through building component c; see $K_{conv}^{leak,ceil}$ for computational methods

K_{airing} = forced-air convection due to airing, regulated by the occupants of the building

$K^{vent,supp}$, $K^{vent,ret}$ = heat conductances as a result of the mechanically induced air movements from and to the HVAC system

$K_{leak}^{vent,sup}$ = conductance due to air leakages in the supply duct system

$K_{inner}^{vent,ret}$ = exiting airflow conductance, from the inside into the return duct system

$K_{leak\ attic}^{vent,ret}$ = conductance due to air leakages from the attic and into the return duct system

$K^{vent,fresh}$ = conductance, representing the supplied outdoor air into the HVAC system

Conductance—Radiation, W/K

$K_{rad}^{sr\_n}$ = radiative surface conductance at surface *n*.

$$= \frac{\varepsilon_n \cdot A_n}{1 - \varepsilon_n} \cdot \sigma \cdot 4 \cdot T_n^3$$

where

- A_n = surface area of *n*
- ε = surface emissivity
- σ = Stefan-Boltzmann constant

K_{rad}^{nm} = conductance between fictitious radiative temperature nodes *m* and *n*

$$= A_n \cdot F_{nm} \cdot \sigma \cdot 4 \cdot \bar{T}_{nm}^3$$

where

- F_{nm} = view factor from surface *n* to *m*
- $\bar{T}_{nm}$ = average surface temperature of *n* and *m*

Heat Sources, W

I_{unit}^{HVAC} = excess of heat from the operating HVAC unit, acting as a heat source into the attic space

I_{pwr}^{HVAC} = power needed to heat or cool the air to a given supply temperature

I_{in} = heat load from indoor activities, appliances, and humans

The symbols for water vapor transfer are similar to those for heat transfer. In general, *T* is substituted with *v*, which is the volumetric vapor content, and *K* is replaced by *Kv*, which is the conductance for water vapor transfer. There are two additional symbols defined in the following subsection.

Moisture Sources and Sinks, lb (kg)

G_{in} = moisture load from indoor activities, appliances, and humans

S = reduction of moisture content in the supply air due to the dehumidifying effect of the cooling coils

Heat and moisture balances are calculated for each temperature and vapor node and for each given time step. The simulation model is programmed in a mathematical numerical tool, representing the complete building except for the roof construction. The numerical tool calls on WUFI1D, subsequently representing the roof construction. In order to make these two separate models work, an iterative process is required since the output parameters of one model serve as the input parameters of the other. The iterations continue until two convergence criteria are fulfilled. In this study the required convergences are 0.1% for the attic temperature and 1% for the vapor content of the attic air. These convergences are referred to as C_{temp} and C_{moist}, then:

$$C_{temp}^{max} \le \frac{T_{attic}(i+1,t) - T_{attic}(i,t)}{T_{attic}(i,t)} \le 0.1\% \qquad \text{(A.1)}$$

$$C_{moist}^{max} \le \frac{v_{attic}(i+1,t) - v_{attic}(i,t)}{v_{attic}(i,t)} \le 1.0\% \qquad \text{(A.2)}$$

for iteration number i and at the time step t. In order to fulfill the convergences, C_{temp} and C_{moist} shall not exceed the criteria for any of the simulated time steps.

Heat Transfer Balances

The temperature of each node is calculated by balancing the flux of energy into the nodes. The temperatures are dependent of each other and therefore determined explicitly, i.e., the computations are based on temperatures from previous simulated time steps. Once the first complete simulation iteration has been finished between the two separate models, the computations are based on the temperatures from the same time step but from the previous iteration. To clarify, the first iteration uses temperatures, when needed, from the previous time step; the subsequent iterations use the temperatures from the present time step but from the previous iteration. This approach is the reason behind the setup for convergences criteria in which the iterations continue until the results from the last two iterations converge.

Figure 7 illustrates the energy balances in the air temperature nodes of the roof and attic models, as shown in Figure 7.

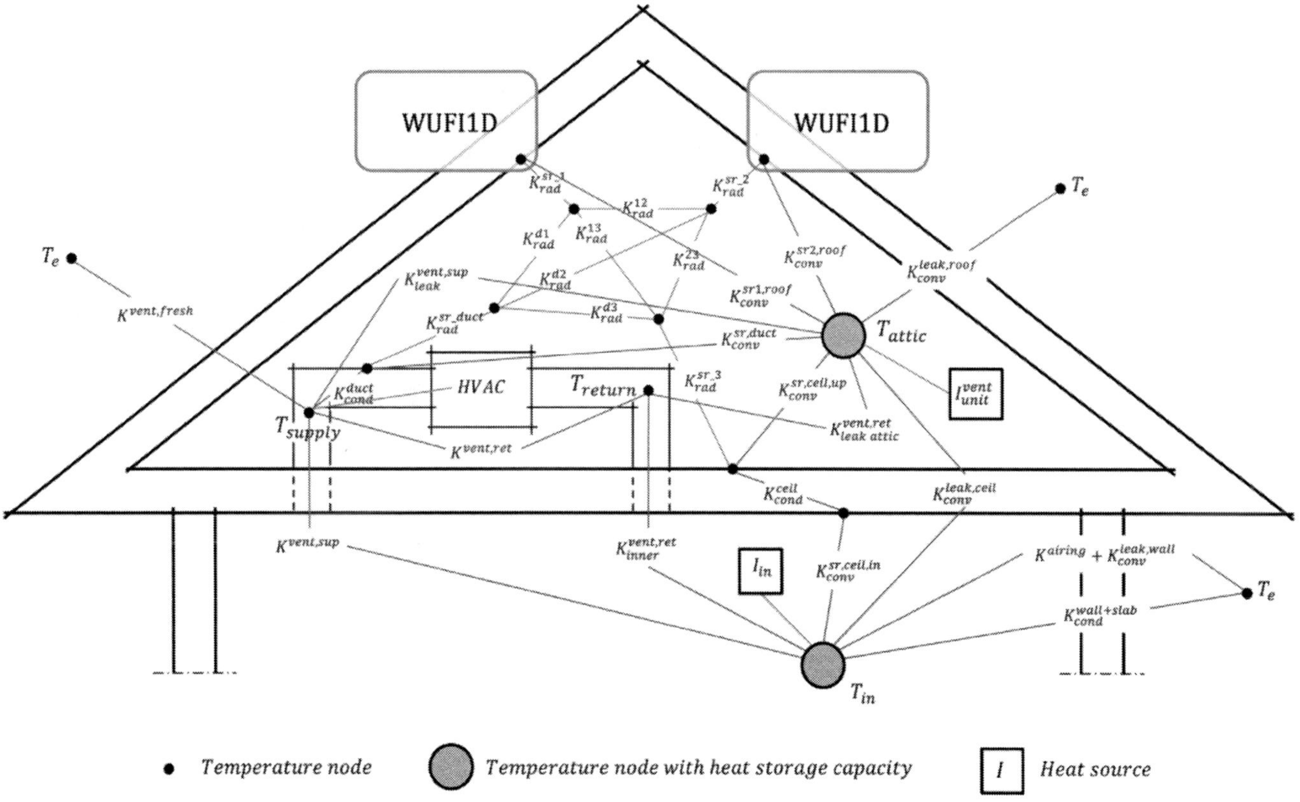

Figure 7 *Thermal network of the unvented attic, representing the overall heat transfer between the attic space, the air distribution system, and the indoor and outdoor environments. Each surface or air temperature is represented as a node, connected to each other with conductances.*

Ventilation Supply and Return Temperatures, T_{supply} and T_{return}

The thermostat setpoint temperatures determine whether the HVAC system is running or not. When the indoor temperature is exceeded by these setpoints, the HVAC system is started. The temperature of the ventilation supply air, T_{supply}, is either 55°F (12.8°C) or 105°F (40.6°C), depending on whether the HVAC unit is in cooling or heating mode. Still, there are unknown temperatures and a heat source involved in the energy balance of T_{supply}, hence a heat transfer network is required:

$$T_{supply} = \frac{\left(\begin{array}{c} T_{surf}^{duct} \cdot K_{cond}^{duct} + T_{return} \cdot K^{vent,\,ret} + T_e \\ \cdot K^{vent,\,fresh} + I_{pwr}^{HVAC} \end{array}\right)}{K_{supply}} \tag{A.3}$$

in which

$$K_{supply} = K_{cond}^{duct} + K^{vent,\,ret} + K^{vent,\,fresh} \tag{A.4}$$

Furthermore, the required power of the HVAC unit, I_{pwr}^{HVAC}, can be expressed as follows:

$$I_{pwr}^{HVAC} = \left(\begin{array}{c} T_{supply} \cdot K_{supply} - T_{surft}^{duct} \cdot K_{cond}^{duct} \\ - T_{return} \cdot K^{vent,\,ret} - T_e \cdot K^{vent,\,fresh} \end{array}\right) \tag{A.5}$$

where

$$T_{return} = \frac{T_{in} \cdot K_{inner}^{vent,\,ret} + T_{attic} \cdot K_{leak\ attic}^{vent,\,ret}}{K_{return}} \tag{A.6}$$

$$K_{return} = K_{inner}^{vent,\,ret} + K_{leak\ attic}^{vent,\,ret} \tag{A.7}$$

The Attic Temperature, T_{attic}

The attic temperature is determined by a coupled energy balance and a thermal capacity. In this study, the thermal capacity of the attic consists of the wooden ceiling joists and the air volume of the attic space. The relation between the heat capacity and the energy flux, Q (W), is defined as follows:

$$\frac{dT_{attic}}{dt} \cdot \Sigma(V \cdot \rho \cdot c)_{attic} = Q_{cond+conv} + Q_{Ra} + Q_I \tag{A.8}$$

$$Q_{cond+conv} = \left(\begin{array}{c} (T_{surf}^{left\ roof} - T_{attic}) \cdot K_{conv}^{sr1,\,roof} \\ + (T_{surf}^{right\ roof} - T_{attic}) \cdot K_{conv}^{sr2,\,roof} \\ + (T_{surf}^{ceil,\,up} - T_{attic}) \cdot K_{conv}^{sr,\,ceil,\,up} \\ + (T_{surf}^{duct} - T_{attic}) \cdot K_{conv}^{sr,\,duct} \end{array}\right) \tag{A.9}$$

$$Q_{Ra} = \left(\begin{array}{c} T_{supply} \cdot K_{leak}^{vent,\,sup} - T_{attic} \cdot K_{leak\ attic}^{vent,\,ret} \\ + (T_e - T_{attic}) \cdot K_{conv}^{leak,\,roof} \\ + (T_{in} - T_{attic}) \cdot K_{conv}^{leak,\,ceil} \end{array}\right) \tag{A.10}$$

$$Q_I = I_{unit}^{vent} \tag{A.11}$$

If existing, $K_{conv}^{leak,\,roof}$ and/or $K_{conv}^{leak,\,ceil}$ are assume to be in a balanced airflow, meaning there is an exchange of equal airflow in both directions.

For a specific time step $t = j, j-1, j+2 \ldots j+n$, the disparity in T_{attic} compared to the previous time step is defined as:

$$(T_{attic}^{j+1} - T_{attic}^{j}) = \frac{dt \left(\begin{array}{c} (T_{surf}^{left\ roof} - T_{attic}^{j}) \cdot K_{conv}^{sr1,\,roof} \\ + (T_{surf}^{right\ roof} - T_{attic}^{j}) \cdot K_{conv}^{sr2,\,roof} \\ + (T_{surf}^{ceil,\,up} - T_{attic}^{j}) \cdot K_{conv}^{sr,\,ceil,\,up} \\ + (T_{surf}^{duct} - T_{attic}^{j}) \cdot K_{conv}^{sr,\,duct} \\ + T_{supply} \cdot K_{leak}^{vent,\,sup} - T_{attic}^{j} \\ \cdot K_{leak\ attic}^{vent,\,ret} + (T_e - T_{attic}) \\ \cdot K_{conv}^{leak,\,roof} + (T_{in} - T_{attic}) \\ \cdot K_{conv}^{leak,\,ceil} + I_{unit}^{vent} \end{array}\right)}{\Sigma(V \cdot \rho \cdot c)_{attic}} \tag{A.12}$$

The Indoor Temperature, T_{in}

The temperature if the indoor environment is also determined by a coupled energy balance with a thermal capacity of the inner mass. The thermal inertia of the indoor environment consists of furniture and contents (ASHRAE 2007, Section 8.8.2) equivalent to the thermal properties of wood and gypsum. The indoor temperature and coupled energy balance is defined as:

$$\frac{dT_{in}}{dt} \cdot \Sigma(V \cdot \rho \cdot c)_{in} = Q_{cond+conv} + Q_{Ra} + Q_I \tag{A.13}$$

where

$$Q_{cond+conv} = (T_{ceil,\,low}^{sr} - T_{in}) \cdot K_{ceil,\,low}^{sr} + (T_e - T_{in}) \cdot K_{cond}^{wall+slab} \tag{A.14}$$

$$Q_{Ra} = \left(\begin{array}{c} T_{supply} \cdot K^{vent,\,sup} - T_{in} \cdot K_{inner}^{vent,\,ret} \\ + (T_{attic} - T_{in}) \cdot K_{conv}^{leak,\,ceil} + (T_e - T_{in}) \\ \cdot (K^{airing} + K_{conv}^{leak,\,ceil}) \end{array}\right) \tag{A.15}$$

$$Q_I = I_{in} \tag{A.16}$$

For a specific time step $t = j, j+1, j+2 \ldots j+n$, the disparity in T_{in} compared to the previous time step is defined as:

$$(T_{in}^{j+1} - T_{in}^{j}) = \frac{dt \begin{pmatrix} (T_{ceil,low}^{sr} - T_{in}) \cdot K_{ceil,low}^{sr} \\ + (T_e - T_{in}^{j}) \cdot K_{cond}^{wall+slab} + T_{supply} \\ \cdot K^{vent,sup} - T_{in}^{j} \cdot K_{inner}^{vent,ret} \\ + (T_{attic} - T_{in}) \cdot K_{conv}^{leak,ceil} \\ + (T_e - T_{in}) \cdot (K^{airing} + K_{conv}^{leak,ceil}) \end{pmatrix} + I_{in}}{\Sigma(V \cdot \rho \cdot c)_{in}} \tag{A.17}$$

Radiation Heat Exchange

The exchange of longwave radiation between the surfaces of the attic will affect not only each other but also all the heat transfer in the attic. The relation between the surfaces in terms of longwave radiation is illustrated in Figure 8.

The fictitious node temperatures, T_1^{fict} to T_4^{fict}, are defined as:

$$T_1^{fict} = T_1 \cdot K_{rad}^{sr_1} + T_2^{fict} \cdot K_{rad}^{12} + T_3^{fict} \cdot K_{rad}^{13} + T_4^{fict} \cdot K_{rad}^{14} \tag{A.18}$$

$$T_2^{fict} = T_2 \cdot K_{rad}^{sr_2} + T_1^{fict} \cdot K_{rad}^{12} + T_3^{fict} \cdot K_{rad}^{13} + T_4^{fict} \cdot K_{rad}^{14} \tag{A.19}$$

$$T_3^{fict} = \ldots$$

Furthermore, the surface temperatures, T_1 to T_4, are derived out of T_1^{fict} to T_4^{fict}. The surface temperature of the return duct system is assumed to be equivalent with the attic temperature; thus it is assumed to be negligible in this study.

Water Vapor Transfer Balances

As for the heat transfer, the vapor content of each node is calculated by balancing the flux into the nodes. The vapor contents are dependent of each other and therefore determined explicitly. Once the first complete simulation iteration between the two separate models has been finished, the computations are based on the vapor contents from the same time step but from the previous iteration, i.e., as the iterations continue, the vapor contents are determined more implicitly. The vapor contents of the supply and return air, v_{supply} and v_{return}, are both calculated based on first explicit then implicit values of the vapor contents of the attic and indoor air, v_{attic} and v_{in}.

Ventilation Supply and Return Air Vapor Content, v_{supply} and v_{return}

From Figure 9:

$$v_{return} = \frac{v_{in} \cdot Kv_{inner}^{vent,ret} + v_{attic} \cdot Kv_{leak\ attic}^{vent,ret}}{Kv_{return}} \tag{A.20}$$

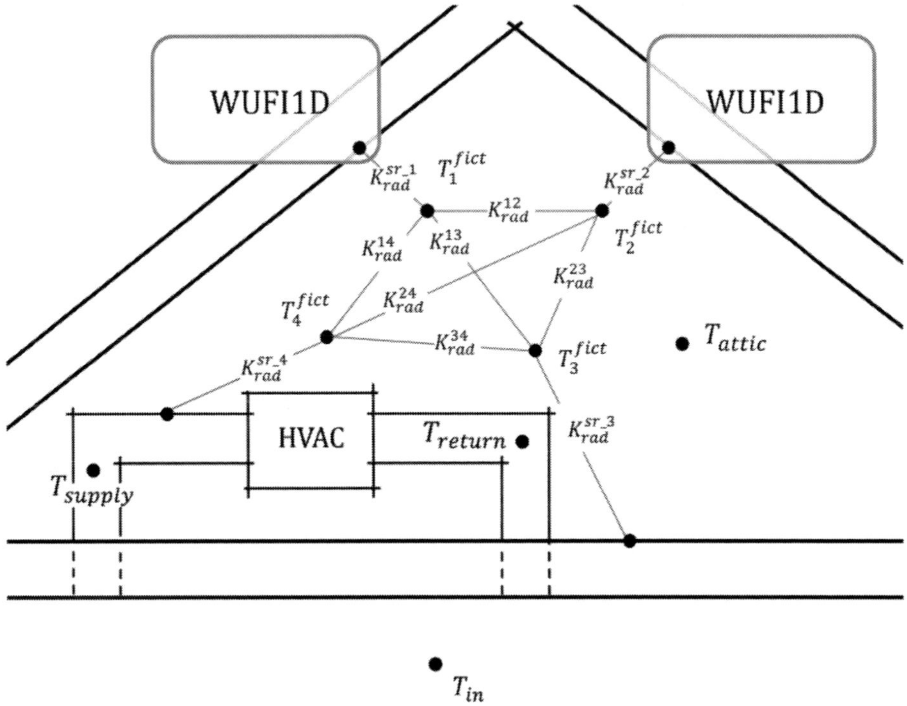

Figure 8 *Longwave radiation heat exchange between the surrounding surfaces of the attic space. The surface temperatures are determined with the help of fictitious node temperatures.*

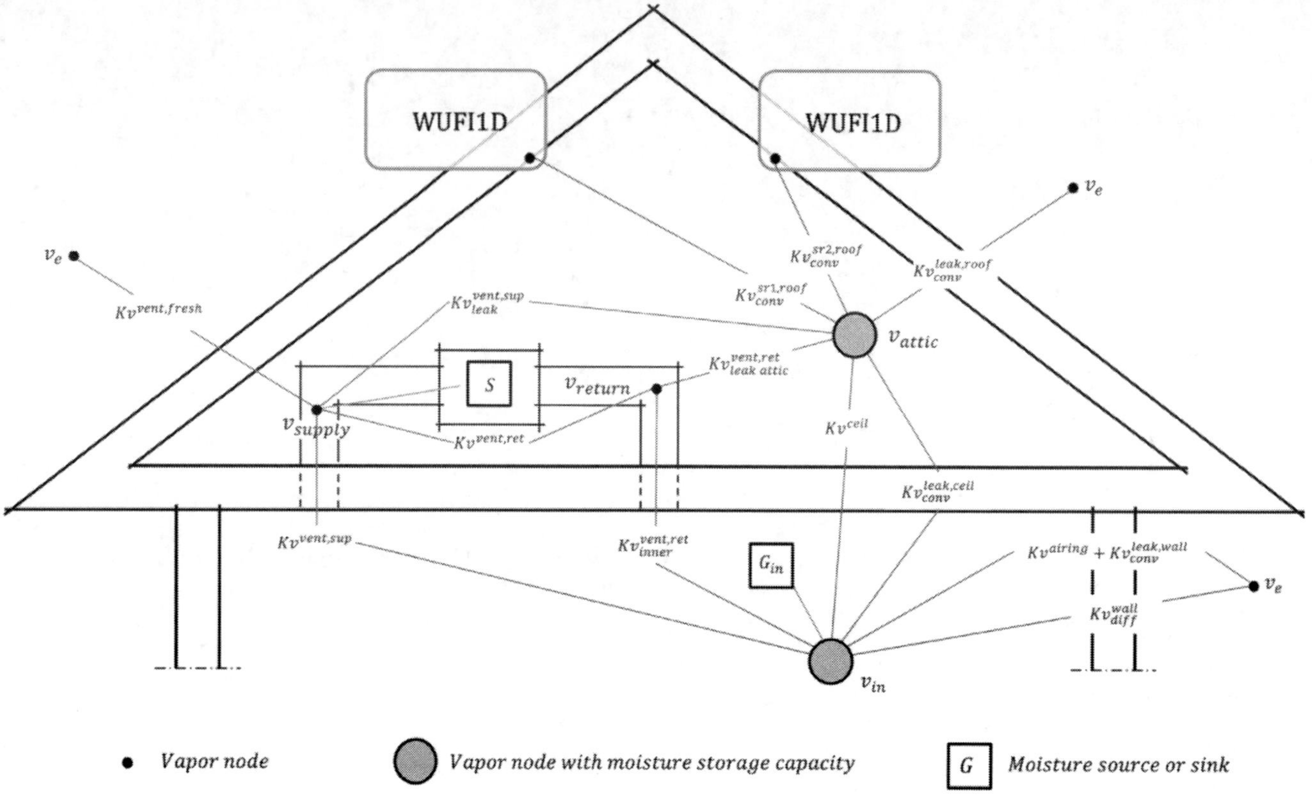

Figure 9 *Network of the unvented attic, representing the overall water vapor transfer between the attic space, the air distribution system, and the indoor and outdoor environments. Each vapor content of a surface or air volume is represented as a node, connected to each other with conductances.*

where

$$Kv_{return} = Kv_{inner}^{vent,\,ret} + Kv_{leak\ attic}^{vent,\,ret} \tag{A.21}$$

and

$$v_{supply} = \frac{v_e \cdot Kv^{vent,\,fresh} + v_{return} \cdot Kv^{vent,\,ret}}{Kv_{supply}} \tag{A.22}$$

where

$$Kv_{supply} = Kv^{vent,\,fresh} + Kv^{vent,\,ret} \tag{A.23}$$

In Figure 10, S represents the reduction of vapor content in the supply air due to the dehumidifying effect of the HVAC cooling coils; therefore, S is only operating when the HVAC system is in cooling mode.

If the HVAC system runs in cooling mode, the following is valid:

$$v_{res,\,sup} = \frac{v_e \cdot Kv^{vent,\,fresh} + v_{return} \cdot Kv^{vent,\,ret}}{Kv_{supply}} \tag{A.24}$$

$$v_{supply} = v_{res,\,sup} - S = \left(\frac{v_e \cdot Kv^{vent,\,fresh} + v_{return} \cdot Kv^{vent,\,ret}}{Kv_{supply}}\right) - S \tag{A.25}$$

The dehumidifying effect can be defined as the difference in vapor content between the residual vapor content, $v_{res,sup}$, and the dew point, v_{sat}, at the supplied cooling temperature, T_{supply}:

$$S = (v_{res,\,sup} - v_{sat}(T_{supply})) \tag{A.26}$$

Vapor Content of the Attic Air, v_{attic}

The wooden joists, supporting the ceiling floor, are assumed to buffer moisture in which d_e is the effective moisture penetration depth. This effective depth, together with the surface area, defines the moisture buffering volume and increases with increasing periodic time length, t, of varying air humidity:

$$d_e = \sqrt{a_v \cdot t} \tag{A.27}$$

with a moisture diffusivity a_v.

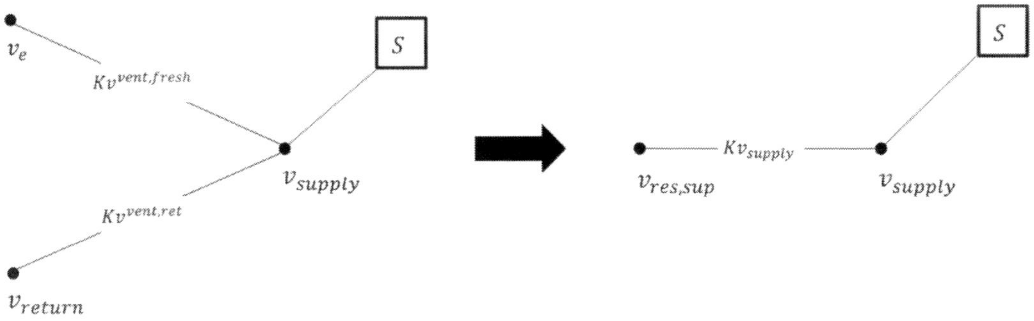

Figure 10 *A network reduction of the ventilation supply and return air vapor contents together with the dehumidifying effect of the HVAC cooling coils,* S.

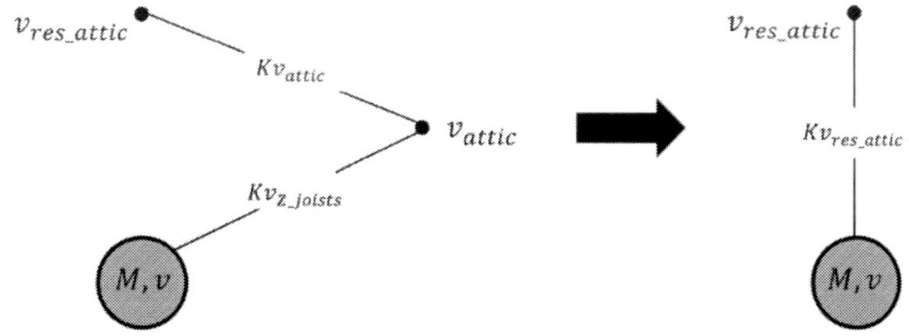

Figure 11 *A network reduction depicting the coupled water vapor transfer and buffering capacity inside the attic space.* M *is the total moisture weight of the buffering material and* v *is the vapor content.*

The coupled vapor balance and buffering capacity in the node can be expressed as:

$$\frac{dM}{dt} = \frac{dw}{d\varphi} \cdot \frac{d\varphi}{dt} \cdot A \cdot d_e \tag{A.28}$$

Basically, the left term is the total flux of moisture to and from the node and the right term is the combination of the change in total volumetric moisture content of a material, w, as a function of the relative humidity, φ, multiplied with the change of φ over the specific time, t, and finally multiplied with the volume of the moisture buffering material, $A \cdot d_e$.

According to Figure 9:

$$v_{res\_attic} = \frac{\begin{pmatrix} v_e \cdot Kv^{leak,\,roof}_{conv} + v_{in} \cdot (Kv^{leak,\,ceil}_{conv} + Kv^{ceil}) \\ + v_{supply} \cdot Kv^{vent,\,sup}_{leak} + v^{surf1} \\ \cdot Kv^{sr1,\,roof}_{conv} + v^{surf2} \cdot Kv^{sr2,\,roof}_{conv} \end{pmatrix}}{Kv_{attic}} \tag{A.29}$$

in which

$$Kv_{attic} = \begin{pmatrix} Kv^{leak,\,roof}_{conv} + Kv^{leak,\,ceil}_{conv} + Kv^{ceil} \\ + Kv^{vent,\,sup}_{leak} + Kv^{sr1,\,roof}_{conv} + Kv^{sr2,\,roof}_{conv} \end{pmatrix} \tag{A.30}$$

From the reduction of conductances in Figure 11, the summarized total conductance Kv_{res_attic} can be defined as:

$$\frac{1}{Kv_{res\_attic}} = \frac{1}{Kv_{attic}} + \frac{z}{A_{surf}} \tag{A.31}$$

$$\frac{1}{kv_{z\_joists}} = \frac{z}{A_{surf\_joists}} = \frac{z_{si} + \dfrac{d_{e\_joists}}{2 \cdot \delta_v}}{A_{surf\_joists}} \tag{A.32}$$

Furthermore, the change in the total volumetric moisture content of the buffering material is obtained:

$$\frac{dw_{joists}}{d\varphi} \cdot \frac{d\varphi}{dt} \cdot A_{surf\_joists} \cdot d_e = Kv_{res\_attic} \cdot (v_{res\_attic} - v_{joists}) \tag{A.33}$$

$$\frac{dw_{joists}}{dt} = \frac{Kv_{res\_attic} \cdot (v_{res\_attic} - v_{joists})}{A_{surf\_joists} \cdot d_e} = \frac{w_{joists,\,j+1} - w_{joists,\,j}}{t_{j+1} - t_j} \tag{A.34}$$

The total moisture content of the joists, w_{joists}, for the forthcoming time step, $j+1$, is then expressed as:

$$w_{joists,\,j+1} = \left(w_{joists,\,j} + \frac{Kv_{res\_attic}\cdot(v_{res\_attic}-v_{joists})\cdot(t_{j+1}-t_j)}{A_{surf\_joists}\cdot d_e}\right) \tag{A.35}$$

Hence, the vapor content of the joists, v_{joists}, for the forthcoming time step, $j+1$, is

$$v_{joists,\,j+1} = \varphi(w_{joists,\,j+1})\cdot v_{sat}(T_{attic,\,j+1}) \tag{A.36}$$

Finally, the vapor content of the attic air, v_{attic}, is obtained:

$$v_{attic} = \frac{v_{joists}\cdot Kv_{z\_joists} + v_{res\_attic}\cdot Kv_{attic}}{Kv_{z\_joists} + Kv_{attic}} \tag{A.37}$$

Vapor Content of the Indoor Air, v_{in}

Similar to the attic space, there is an assumed moisture buffering capacity of materials inside the living space. There are two assumed types of materials, wood and gypsum, in accordance with ASHRAE Standard 90.2 (2007), Section 8.8.2.

According to Figure 9, the indoor residual vapor content, v_{res_in}, is defined as:

$$V_{res\_in} = \frac{\begin{pmatrix} v_{attic}\cdot(Kv^{leak,\,ceil}_{conv} + Kv^{ceil}) + v_e \\ \cdot(Kv^{airing} + Kv^{wall}_{diff}) + v_{supply} \\ \cdot Kv^{vent,\,sup} + G_{in} \end{pmatrix}}{Kv_{in}} \tag{A.38}$$

in which

$$Kv_{in} = Kv^{leak,\,ceil}_{conv} + Kv^{ceil} + Kv^{airing} + Kv^{wall}_{diff} + Kv^{vent,\,sup} \tag{A.39}$$

As shown in Figure 12, the residual vapor contents of the two separated networks, $v_{res\_and\_wood}$ and $v_{res\_and\_gyps}$, are defined as:

$$v_{res\_and\_wood} = \frac{v_{res\_in}\cdot Kv_{in} + v_{gyps}\cdot Kv_{z\_in\_gyps}}{Kv_{in} + Kv_{z\_in\_gyps}} \tag{A.40}$$

and

$$v_{res\_and\_gyps} = \frac{v_{res\_in}\cdot Kv_{in} + v_{wood}\cdot Kv_{z\_in\_wood}}{Kv_{in} + Kv_{z\_in\_wood}} \tag{A.41}$$

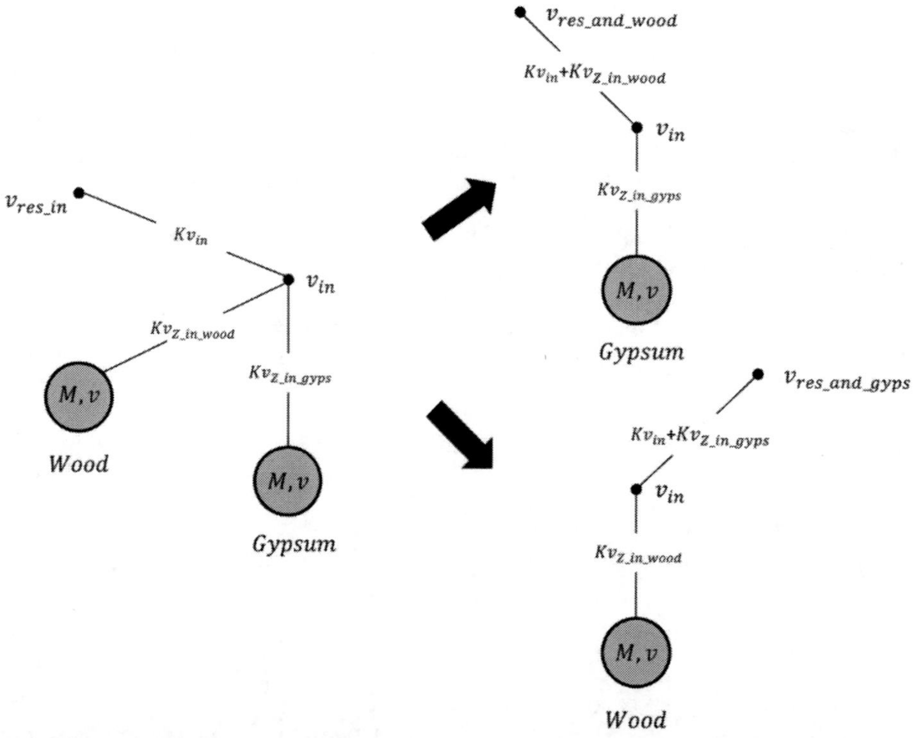

***Figure 12** A network separation of the water vapor transfer balance of the indoor space, where the two moisture buffering materials are extracted into two different networks.* M *is the total moisture weight of the buffering material and* v *is the vapor content.*

The total volumetric moisture content can be defined, for the wooden materials for the next coming simulation step, $j + 1$, as:

$$w_{wood,\, j+1} = w_{wood,\, j} + \frac{\left(\left(\frac{1}{Kv_{z\_wood}} + \frac{1}{Kv_{in} + Kv_{z\_in\_gyps}}\right)^{-1} \cdot (v_{res\_and\_wood} - v_{wood}) \cdot (t_{j+1} - t_j)\right)}{A_{surf\_wood} \cdot d_{e\_wood}} \quad \text{(A.42)}$$

The vapor content of the wood material is then:

$$v_{wood,\, j+1} = \varphi(w_{wood,\, j+1}) \cdot v_{sat}(T_{in,\, j+1}) \quad \text{(A.43)}$$

The same procedure is valid for the assumed moisture buffering materials of gypsum, where the volumetric moisture content for the next coming simulation step, $j + 1$, is:

$$w_{gyps,\, j+1} = w_{gyps,\, j} + \frac{\left(\left(\frac{1}{Kv_{z\_gyps}} + \frac{1}{Kv_{in} + Kv_{z\_in\_wood}}\right)^{-1} \cdot (v_{res\_and\_wood} - v_{gyps}) \cdot (t_{j+1} - t_j)\right)}{A_{surf\_gyps} \cdot d_{e\_gyps}} \quad \text{(A.44)}$$

and

$$v_{gyps,\, j+1} = \varphi(w_{gyps,\, j+1}) \cdot v_{sat}(T_{in,\, j+1}) \quad \text{(A.45)}$$

Finally, as shown in Figure 12, the vapor content of the indoor air is:

$$v_{in} = \frac{\left(v_{res\_in} \cdot Kv_{in} + v_{in\_wood} \cdot Kv_{z\_in\_wood} + v_{in\_gyps} \cdot Kv_{z\_in\_gyps}\right)}{Kv_{in} + Kv_{z\_in\_wood} + Kv_{z\_in\_gyps}} \quad \text{(A.46)}$$

REFERENCES

ASHRAE. 2005. *ASHRAE Handbook—Fundamentals*, Chapter 29, Residential Cooling and Heating Load Calculation. Atlanta: ASHRAE.

ASHRAE. 2007. ANSI/ASHRAE Standard 90.2-2007, *Energy-Efficient Design of Low-Rise Residential Buildings*. Atlanta: ASHRAE.

ASHRAE. 2011. Addendum a to ANSI/ASHRAE Standard 160-2009, *Criteria for Moisture-Control Design Analysis in Buildings*. Atlanta: ASHRAE.

ASHRAE. 2012. Addendum n to ANSI/ASHRAE Standard 62.2-2010, *Ventilation and Acceptable Indoor Air Quality in Low-Rise Residential Buildings*. Atlanta: ASHRAE.

Christian, J.E., and H. Trechsel. 1994. *Moisture control in buildings*. Philadelphia: ASTM.

DIN. 2012. DIN 68800-2 (2012-02), *Wood Preservation – Part 2: Preventive Constructional Measures in Buildings*. Berlin: Deutsches Institut für Normung.

Hagentoft, C.-E., A. Sasic Kalagasidis, M. Thorin, and S.F. Nilson. 2008. Mould growth control in cold attics through adaptive ventilation. Presented at the 8th Nordic Symposium on Building Physics, Copenhagen.

Hamby, D.M. 1994. A review of techniques for parameter sensitivity analysis of environmental models. *Environmental Monitoring and Assessment* 32(2):135–54. DOI: 10.1007/bf00547132

Hukka, A., and H.A. Viitanen. 1999. A mathematical model of mould growth on wooden material. *Wood Science and Technology* 33.

ICC. 2011. *2012 International Energy Conservation Code—Residential Energy Efficiency*. Country Club Hills, IL, USA: International Code Council, Inc.

Kalamees, T., J. Vinha, and J. Kurnitski. 2006. Indoor humidity loads and moisture production in lightweight timber-frame detached houses. *Journal of Building Physics* 29:219–46.

Künzel, H.M. 1995. Simultaneous Heat and Moisture Transport in Building Components. One- and two-dimensional calculation using simple parameters. Dissertation, University Stuttgart, IRB Verlag. Retrieved from www.WUFI.com

Lstiburek, J. 2006. Understanding attic ventilation. *ASHRAE Journal* 48(4):36–45.

MathWorks. 2013. MATLAB®. The MathWorks, Inc., Natick, MA.

Ojanen, T., R. Peuhkuri, H.A. Viitanen, K. Lähdesmäki, J. Vinha, and K. Salminen. 2011. Classification of material sensitivity –New approach for mould growth modeling. Presented at the 9th Nordic Symposium on Building Physics, Tampere, Finland.

Pallin, S. 2012. Probabilistic Risk Assessment of Energy Efficient Retrofitting Techniques – Focus on Multi-family Dwellings and the Effects of Changing Air Movements. Licentiate, Chalmers University, Gothenburg, Sweden.

Polly, M., M. Gestwick, M. Bianchi, R. Anderson, S. Horowitz, C. Christensen, and R. Judkoff. 2011. A method for determining optimal residential energy efficiency retrofit packages. For Building America—Building Technologies Program by National Renewable Energy Laboratory, Golden, CO, USA.

Rudd, A. 2005. Field performance of unvented cathedralized (UC) attics in the USA. *Journal of Building Physics* 29(2):145–69. DOI: 10.1177/1744259105057695

Russell, M., M. Sherman, and A. Rudd. 2007. Review of residential ventilation technologies. *HVAC&R Research* 13(2):325–48. DOI: 10.1080/10789669.2007.10390957

Straube, J., J. Smegel, and J. Smith. 2010. Moisture-safe unvented wood roof systems. Buildingscience.com.

Vose, D. 2008. *Risk Analysis: A Quantitative Guide*, 3rd ed. Hoboken, NJ: Wiley.

Yik, F.W.H., P.S.K. Sat, and J.L. Niu. 2004. Moisture generation through Chinese household activities. *Indoor and Built Environment* 13:115–31.

All-in-One Design Tool Solution for Passive Houses and Buildings—Monthly Energy Balance and Hygrothermal Simulation

Florian Antretter
Associate Member ASHRAE

Katrin Klingenberg

Matthias Pazold

ABSTRACT

The U.S. Department of Energy (DOE) outlined a path to advance high-performance and zero-energy homes in its Building America research program (DOE 2008). *That program—and other efforts in the Unites States over the last few years to tighten energy codes—follow closely the evolution of energy standard developments in Europe. That is, developments moved from conventional buildings to low-energy buildings, then to passive buildings, net zero, and recently even plus energy. The common denominator is low-load homes that require less energy from the start. Designing these homes requires specialized design tools to achieve targeted performance and quality assurance. The tools must enable the designer to meet the energy performance of the envelope and the minimized mechanical system and—critically—to achieve proper hygrothermal design of building skins before the detailed construction documents phase. The hygrothermal component reduces the risk of potentially catastrophic design flaws, thereby reducing risk for designers, builders, and homeowners.*

The passive design approach offers a solid baseline for all low-load homes. The simplified static balance-based method has provided early adopters a useful tool. But it lacks the capacity to assess certain dynamic factors of the energy balance, such as transient effects that occur under real conditions—for example, thermal lag time and related overheating in summer or the hygrothermal performance of the building envelope.

This paper outlines the incorporation of the balance-based method currently employed to design passive buildings into hygrothermal whole-building simulation software. A case study illustrates both methods as they are applied to an actual building. Additional results and insights from the dynamic building simulation are highlighted. The comparison of the balance-based method with the dynamic simulation, both performed within one tool based on the exact same building data and inputs, demonstrates benefits and drawbacks as well as improvement potential for tools and methods moving forward. While the static monthly balance method simplifies the required steps and the time required to design a passive house, it might not be fine-grained enough for all climate zones and building types. The results from the dynamic simulation are more granular and offer a more realistic depiction of the actual behavior of the building and its interaction with mechanical equipment in terms of comfort.

INTRODUCTION

Households account for a significant part of global energy consumption. In the USA, the domestic consumption in 2011 accounted for 22% of the total energy use (EIA 2012). Energy-efficient building standards can reduce this consumption. While in the USA on average all existing buildings (with the exception of strip malls) consume approximately 220 kWh/m^2/yr (69.7 kBtu/ft^2/yr) for heating, residential energy consumption for heating is on average significantly lower. Households in the state of New York, which has a cold climate, consume on average 95 kWh/m^2/yr (30.1 kBtu/ft^2/yr) (EIA 2005). For comparison, ENERGY STAR® (EPA 20123) and the *International Energy Conservation Code* (ICC 2012) have over the past three years put requirements in place that reduce this consumption further, by up to 30%. A house built to the most efficient choices to receive ENERGY STAR certification today

Florian Antretter *is the group manager of and* ***Matthias Pazold*** *is a scientist in the Hygrothermal Building Analysis group in the Hygrothermics department at the Fraunhofer Institute for Building Physics, Valley, Germany.* ***Katrin Klingenberg*** *is executive director at the Passive House Institute US (PHIUS), Urbana, IL.*

requires approximately 70 kWh/m^2/yr (22.2 kBtu/ft^2/yr) for heating (EPA 2012) in the state of New York. The certification criteria for a passive house built to meet the current European standards is much lower, 15 kWh/m^2/yr (4.75 kBtu/ft^2/yr). This energy reduction corresponds to 85% savings over the current average in the state of New York in terms of heating energy and illustrates how the use of energy-efficient technologies and design can significantly improve the current performance standards of the leading U.S. energy codes and programs.

When designing energy-efficient buildings, a designer often uses different approaches to tweak the building in terms of energy efficiency, comfort, and the hygrothermal performance of the components. Common approaches for the assessment—and also for the certification—of the energy demand rely on steady-state methods, using monthly balances, to compute the values used for the estimation of energy demand and for comparing different building designs under predefined boundary conditions. Those methods *exclude* all transient effects in buildings, such as the effects of thermal inertia. A comprehensive analysis of comfort conditions indoors is also not possible.

This causes the designer to switch to a whole-building model, using hourly data to evaluate the dynamic behavior of the building. Often, very new and innovative building component configurations are employed to meet the stringent energy criteria. Therefore, a third step is critical to performance and quality assurance: assessing the hygrothermal performance of envelope components. This third step evaluates the most critical components by employing a hygrothermal component simulation model.

This paper shows the development of a single tool, WUFI Passive (Fraunhofer IBP 2013), that integrates all three steps. A steady-state approach for assessing the energy performance is combined with the dynamic modeling features of a whole-building simulation model including a full-scale hygrothermal component analysis.

HISTORY OF THE PASSIVE HOUSE CONCEPT

Passive house is a quantitative, performance-based energy concept for buildings based on the understanding of the relationship between the influence of the thermal quality of the envelope and the resulting sizing of the mechanical system. The underlying principles were pioneered in the United States in the 1970s following the oil embargo crisis. The current *International Energy Conservation Code* (*IECC*) recognizes extremely-low-load homes, defining them as homes with a peak load smaller than 1 W/ft^2 for heating in section 101.5.2 (ICC 2012). According to the International Code Council (ICC), the *IECC* is the successor of the first 1975 *Model Energy Code* (*MEC*), from which this definition was originally adopted. William Shurcliff, a physicist at Harvard at the time, published many books on the passive solar and superinsulation concept in the late 1970s and early 1980s. Shurcliff first used the term *passive house* in his 1982 self-published book *The Saunders-Shrewsbury House*. In his later writings he defines the same set of physics principles and guiding energy metrics that characterize passive houses today. At the time the term *superinsulation* evolved as the most commonly used label for this set of principles in the sizable North American high-performance community. Shurcliff also noted in 1988 in his book *Superinsulated Houses and Air-To-Air Heat Exchangers* that this type of energy-efficient home construction was already a "mature technology" and that one might see further improvements in window technologies, vapor retarders, and compact minimized mechanical systems. In the early 1990s physicist Wolfgang Feist continued that research and development in Europe by further codifying the influence of highly improved envelope components (including triple-pane windows) on the minimization potential of the heat load in low-energy buildings for his research project Kranichstein (Feist 1992). His research facilitated those critical improvements predicted by Shurcliff to passive house components, components that soon became available on the European market at reasonable costs.

When Feist applied these principles and improved technologies in Germany in the 1990s, he also applied the same guiding energy metrics defined earlier in the USA. He found that designing a building in Central Europe to meet 1 W/ft^2 peak load (defined in the 1975 *MEC* and now in the *IECC*) resulted in an approximate annual heating demand of 15 kWh/m^2/yr. That figure became the defining energy metric for the standard. The metric quickly became successful in all of Europe and is now considered by many to be the world's leading standard in energy-efficient construction.

Shurcliff's original passive building definition outlined in a 1986 article mentioned one additional principle that was omitted in the later European definitions—that of the "prevention of moisture migration into cold regions within the wall, and other regions where much condensation could occur" (Shurcliff 1986). Shurcliff, working in the varied and often humidity-laden climates of the United States, was fully aware of the critical importance of good hygrothermal design of the envelope components when it comes to superinsulated and airtight walls and components.

In 2002 the first home using the European metrics and design tools for passive houses was built in the United States in Urbana, Illinois, by architect Katrin Klingenberg. Since then many projects have been completed in all U.S. climates except the southern Florida region. Collective experience has shown that the general principles hold true in all climates but that the standard would benefit from refinement. For example, in very cold climates (Fairbanks, Alaska, for example), home designs tend to require over-insulation with walls of 1 m (3.28 ft) thick, while in California comparatively very little insulation is required to meet the standard—leaving further, cost-effective efficiency untapped. This suggests it might be productive to propose relaxing the standard in the north and tightening it in the south for North America.

The most important lessons learned from the reawakening of passive house implementation in the United States over the past 11 years is that hygrothermal analysis is a critical necessity in North America's varied climate zones. Improper hygrothermal design can quickly lead to catastrophic failure in the structure. Designers and builders need a tool to manage this risk properly in conjunction with the energy optimization. With high solar radiation, the lack of comfort assessment is also a significant issue and has led to serious overheating problems in many projects. Nonetheless, efforts to rekindle the interest in the passive house concept in the U.S. are bearing fruit: the number of certified projects in the U.S. and Canada (PHIUS 2013) has been increasing steadily and is now reaching a point of exponential growth, as shown in Figure 1.

Passive House Basic Principles

The passive house concept is based on the following five main principles:

- Excellent continuous thermal insulation of the envelope components, including avoidance of thermal bridges
- Airtightness of the building envelope to minimize ventilation energy loads
- Avoidance of moisture migration into the envelope assembly, where condensation or other moisture problems could occur by means of airtightness, and accurate climate-specific placement of a vapor control layer
- High-performance windows and window areas specified and sized according to climate
- Constant fresh air supply, in most climates via a mechanical ventilation system with heat and/or moisture recovery

In addition, the space conditioning should be, but does not have to be, integrated into the balanced ventilation delivery system for optimal conditioning distribution. Eliminating a separate space-conditioning distribution system also optimizes the initial additional investment cost.

Passive building is a comprehensive approach to cost-effective, durable, comfortable, and sustainable construction. Under the current European protocol, buildings can use no

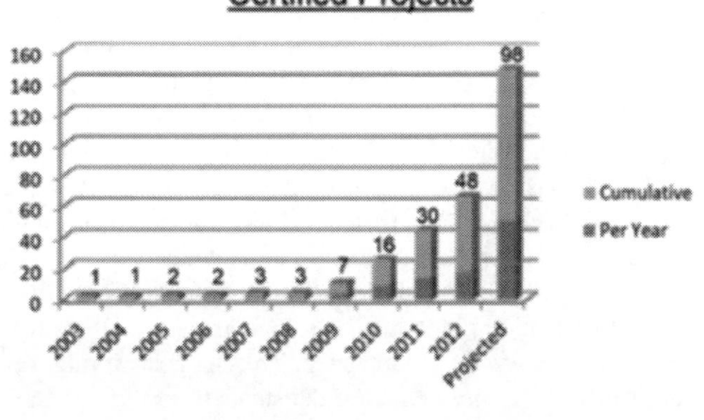

Figure 1 *PHIUS+ certified passive building projects over the past 10 years.*

more than 15kWh/m^2/yr (4.75 KBtu/ft^2/yr) for heating and cooling or have to have a peak load for heating and cooling not to exceed 10 W/m^2 (1 W/ft^2). These requirements for the envelope translate into about a 90% reduction in the energy needed to heat and cool a typical new U.S. house built to code while health and comfort in these homes are increased. In addition to the energy criteria, the European standard also defines a primary energy limit total for all space-conditioning and household energy consumed. This limit is 120 kWh/m^2/yr (38 kBtu/ft^{2}yr).

Passive house standards emphasize the importance of a high-performing building envelope that allows for the reduction or elimination of HVAC equipment first. Active photovoltaic production is only accounted for as savings after all energy criteria have been satisfied by envelope and efficient mechanical measures. The thus-optimized envelope and systems make for the most cost-effective baseline to reach net-zero energy or plus-energy home levels. Only a very small active system, typically a 2–3 kW (about 8.5 kBtu/h) size for a small 1000 ft^2 (92.9 m^2) residence, is required to attain zero site energy status.

In the first multifamily projects currently underway in the U.S., passive house construction appears to be almost cost-neutral (Cohen 2012). The single-family projects that have been built over the past five years in the U.S. are showing good initial results in terms of cost-effectiveness (Oram 2011) and should improve significantly with increased market availability of high-performance materials and equipment.

The Energy Balancing Method for Passive Houses

The underlying method of a passive house design is an accurate energy balancing of the thermal losses or gains through the envelope with the internal gains or losses of that building. In the heating case, in a cold climate the losses need to be almost equal to the internal gains, and only a small difference is allowed to be made up for by a very small auxiliary heating system. As per *IECC* section 101.5.2 (ICC 2012), if after the balancing process the peak load can be shown to be less than 1 W/ft^2 the home is exempt from needing a separate heating system as it is assumed that the internal heat gains in a typical residence will make up for the extremely low peak loss. The balancing is performed similarly for the cooling case, except that in the cooling case all internal gains are added to the thermal peak cooling load and are not helpful to offset it.

For the past decade, only one guiding simplified modeling tool has been on the market that was specifically created to facilitate this balancing process for structures following passive house design principles: the Passive House Planning Package (PHPP) from the Passivhaus Institute (PHI) in Darmstadt, Germany (Feist et al. 2007). The PHPP tool employes Excel© as its base and calculates the annual heating demand by monthly heat balances. For an estimation of the total annual heating demand, the monthly method is adequate. It is assumed that the monthly method has been validated by performing a transient calibration simulation (Feist et al. 2007).

There are limitations of the PHPP tool for North American users:

- While well validated for moderately cool and cold climates, the PHPP is clearly limited in assessing cooling loads and demands and related dehumidification requirements accurately. In those cases, dynamic simulation more accurately assessing thermal lag time should improve the model.
- Energy balancing in areas with very high solar radiation potential (hence with a greater percentage of internal gains as part of the overall energy balance) is also wanting. Early projects have frequently led to comfort issues such as overheating in the shoulder seasons. Comfort assessments concerning interior relative humidity by individual zone are also not possible.
- As of now the PHPP provides no path of risk assessment for the designer in regards to a very important factor: the appropriate envelope design for many different climate zones from very cold to hot and humid—hygrothermal assessments of superinsulated components are critical to avoid moisture concerns in walls. Widely varying climate combinations have distinctly different moisture-control layer requirements. Moisture-related issues in the U.S. market—mold problems, for example—have been widespread and well documented. Designers, builders, and homeowners need confidence that an airtight home will not present problems.
- Multizone modeling of more complex buildings such as, for example, a mixed-use larger building with commercial spaces on the first floor and apartments above, is not possible.

In addition, the user-friendliness of the PHPP tool is rather poor. The PHPP operator may make many "what-if?" design changes but is left to keep track of various optimization steps and must manually record the results of each option in the energy balance; there is no way to log and manage changes in one comprehensive file. The architect/designer in charge of the project needs to be closely involved with the PHPP to understand and remember the optimization process and assumed boundary conditions. Creating representations of this process to communicate results to clients also becomes very time consuming. For small single-family projects this is manageable but inconvenient. For larger project management, where the tool technician is not also the project architect, the tool becomes rather clumsy and time-consuming if not altogether unworkable.

SIMULATION METHODOLOGY

This section describes in short the basic equations for the calculation of the steady-state and dynamic models. Furthermore it describes how integration into one tool for energy, comfort, and hygrothermal component performance was implemented in the new WUFI Passive tool.

Monthly-Balance-Based Method

The monthly-balance-based method depends strongly on overall heat transfer coefficients, temperature difference, and considered time period. It is in accordance with DIN EN ISO 13790 (2008), in particular the simplified approach. The heat transfer coefficient, the reciprocal of the thermal resistance, is an important input value for opaque components. It is calculated from the thermal conductivity and layer thickness of the building envelope materials. For transparent components, the heat transfer coefficient is input data. The monthly heat losses across the building envelope are calculated by determining the heat transfer coefficients, areas of the components, and appropriate temperature differences according to component location. The external boundary condition of exterior walls can be the ambient air temperature; it can also be the ground temperature in case of a basement wall or the increased or decreased derivative of the ambient air temperature in cases of attached spaces exterior of the thermal envelope, such as garages.

With the temperature difference and a considered time period, the heating degree-hours are calculated (Equation 1) following Feist et al. (2007). Monthly heating degree-hours consider the hour count of a month. The period under consideration in the PHPP for the annual demand depends on the monthly difference between the heat losses and the heat gains. If this difference is greater than 0.1 kWh (about 0.341 kBtu) the month will be considered in the calculation of the total annual heating demand. This means that the period under observation could vary between the different cases. Ventilation heat losses are calculated considering the effective air change rate, building volume, effective heat recovery efficiency, and annual heating degree-hours as well.

$$G_t = (\vartheta_i - \vartheta_e) \cdot t / 1000 \tag{1}$$

where

G_t	=	heating degree-hours, kK·h
ϑ_i	=	interior temperature, °C
ϑ_e	=	exterior temperature, °C
t	=	period under review, h

Climate data contains information on the solar radiation for north, east, south, west, and horizontal. Each component is associated with a cardinal or horizontal direction. For transparent components, the solar heat gain is calculated, considering heat transmittance, shading reduction factors due to obstructions, overhangs, and reveals. The solar heat gain of opaque components is computed considering the exterior absorptivity and emissivity. The required heating demand, over a specified time period, is calculated by Equation 2 following Feist et.al. (2007) and in accordance with EN ISO 13790 (2008).

$$Q_H = (Q_{T,H} + Q_{V,H}) - (Q_{S,H} + Q_{I,H}) \cdot \eta_H \tag{2}$$

where

Q_H	=	heating demand
$Q_{T,H}$	=	transmission heat loss
$Q_{V,H}$	=	ventilation heat loss
$Q_{S,H}$	=	solar heat gain
$Q_{I,H}$	=	internal heat gain
η_H	=	heat gain utilization factor

The monthly utilization factors (Equation 3) indicate how much of the available heat gains can be used to counteract the heating demand during the heating period. It is calculated from the heat gain and loss ratio and a so-called time constant, depending on the internal heat capacity and the total heat loss coefficient of the building. For the time constant (Equation 4), a continuously heated building (heated more than 12 hours per day) is considered and the coefficients $a_0 = 1$ and $\tau_0 = 16$ are used.

$$\eta_H = \frac{1-\gamma_H^{a_H}}{1+\gamma_H^{a_H}} \tag{3}$$

where

γ_H	=	heat gain and loss ratio for heating demand $\left(\gamma_H = \frac{Q_T+Q_V}{Q_S+Q_I}\right)$, dimensionless
a_H	=	time constant for heating demand, h

$$a_H = a_0 + \frac{C/H_{L,H}}{\tau_0} \tag{4}$$

where

C	=	internal building heat capacity, Wh/K
$H_{L,H}$	=	total heat loss coefficient of the building for heating demand, W/K
a_0, τ_0	=	defined coefficients

The monthly losses are calculated using monthly heating degree-hours, the annual losses using annual heating degree-hours. The total heating demand, including the transmission heat losses (Equation 5) for all components and thermal bridges and the ventilation heat loss (Equation 6), is decreased by the total heat gain comprised of the solar (Equation 7) and internal heat gains, multiplied by an utilization factor. Monthly heating degree-hours are determined by multiplying the hour count of the month with a temperature difference. The difference between the interior setpoint and the ambient air temperature is used for components adjacent to ambient. The ground heating degree-hours are calculated using the interior setpoint difference to the ground temperature. If the difference between monthly heat loss and monthly heat gain is greater than 0.1 kWh (about 0.341 kBtu), then the heating degree-hours of that month are added to the annual heating degree-hours. To make projects comparable or verify specific limitations, all losses and gains and especially the total heat demand are specific to the treated floor area of the building.

$$Q_T = \Sigma_j(A_j \cdot U_j \cdot f_{T,j} \cdot G_{t,j}) + \Sigma_k(L_k \cdot \psi_k \cdot f_{T,k} \cdot G_{t,k}) \tag{5}$$

where

A_j	=	component area, m^2
U_j	=	thermal transmittance, $W/m^2 \cdot K$
$f_{T,j}$	=	reduction factor for decreased temperature difference, dimensionless
$G_{t,j}$	=	heating degree-hours, kK·h
L_k	=	thermal bridge length, m
ψ_k	=	linear thermal transmittance, W/m·K

$$Q_V = V_{RAX} \cdot [n_{V,Sys} \cdot (1-\eta_{SHX}) \cdot (1-\eta_{HR}) + n_{V,Res}] \cdot c \cdot G_{t,e} + V_{RAX} \cdot [n_{V,Sys} \cdot \eta_{SHX} \cdot (1-\eta_{HR})] \cdot c \cdot G_{t,g} \tag{6}$$

where

$G_{t,e}$	=	heating degree-hours against ambient air, kK·h
$G_{t,g}$	=	heating degree-hours against ground, kK·h
V_{RAX}	=	effective air volume, m^3
$n_{V,Sys}$	=	mean system air change rate, 1/h
$n_{V,Res}$	=	mean natural air change rate, 1/h
n_{HR}	=	effective ventilation heat recovery efficiency, dimensionless
n_{SHX}	=	effective recovery efficiency soil heat exchanger, dimensionless
c	=	specific heat capacity of air, J/kg·K

$$Q_s = \Sigma_\theta(r_\theta \cdot g_\theta \cdot A_{F,\theta} \cdot J_\theta) \tag{7}$$

where

θ	=	associated direction (north, east, south, west, horizontal), dimensionless
r_θ	=	radiation reduction factor (averaged per orientation), dimensionless
g_θ	=	g-value (used in Europe, averaged per orientation), dimensionless
$A_{F,\theta}$	=	window rough opening area (averaged per orientation), m^2
J_θ	=	global radiation, W/m^2

One of the passive house certification criteria is the total annual primary energy demand. To calculate the primary energy demand of a building, the electrical and nonelectrical demand of the mechanical system, including auxiliary energy, plug loads, appliances, and lighting, are summed up. The heating demand of the domestic hot-water production and distribution is taken into account, and if solar hot-water generation is used it is reduced by an estimated solar fraction. It is assumed that the energy use by any device or service is not necessarily continuous. The uses are either reduced by different usage or utilization factors stemming from predetermined utilization patterns or a certain frequency is assumed for each usage. If such energy use takes place within the thermal envelope it is added to the internal heat gains. Heat gains from people are already included in the heat gains. Then an annual specific internal heat gain is estimated. The total is then multiplied by the treated floor area and hours of the month, resulting in the total monthly internal heat gains.

In addition to the heating demand the cooling demand is calculated using a very similar algorithm. One difference is that the heat gains are not weighted by the utilization factor as the heat losses are. Here the factor indicates the usability of the heat losses as shown in Equation 8. Another difference is the calculation of the heating degree-hours. The interior temperature is set to the overheating limit temperature or summer interior temperature (slightly higher). Furthermore, the month July is divided into 1-, 4-, and 12-day peaks with increased ambient temperatures, and the rest of July results mostly in negative heating degree-hours (heat gains across the building envelope). The time constant for the cooling demand is, in contrast to the time constant for the heating demand, computed using the total heat loss coefficient for the cooling demand.

$$Q_C = (Q_{S,C} + Q_{I,C}) - (Q_{T,C} + Q_{V,C}) \cdot \eta_C \tag{8}$$

where

Q_C = cooling demand
$Q_{T,C}$ = transmission heat losses
$Q_{V,C}$ = ventilation heat losses
$Q_{S,C}$ = solar heat gains
$Q_{I,C}$ = internal heat gains
η_C = heat loss utilization factor

$$\eta_C = \frac{1 - \gamma_C^{a_C}}{1 + \gamma_C^{a_C + 1}} \tag{9}$$

where

γ_C = heat loss and gain ratio for cooling demand $\left(\gamma_C = \frac{Q_S + Q_I}{Q_T + Q_V}\right)$, dimensionless
a_C = time constant for cooling demand, h

A frequency of overheating is calculated for the entire year using the length of time period of when the maximum allowed summer temperature is exceeded. If this frequency is too high then additional shading measures are necessary to decrease summer overheating. Natural ventilation strategies can be employed to increase ventilation losses if the climate permits. If shading measures or natural ventilation strategies alone are not sufficient to decrease overheating then active cooling will become necessary.

For sizing of the mechanical system, peak heating and cooling loads are determined using the worst-case scenario climate conditions for summer and winter. The values used are different from ASHRAE's (2010) worst-case design temperatures. Here the worst case is determined using a 24-hour average to account for the delayed response time of well-insulated buildings to short-term extreme temperature swings. The conditions are provided as part of the climate data set. It contains two weather conditions to determine the peak heat load (one very cold day and one with very low solar radiation but with milder temperatures) and one weather condition to determine the peak cooling load. The transmission heat losses across the building envelope are calculated for this peak condition similarly to the annual demand calculations—so are the peak ventilation heat losses and the solar and internal peak heat gains. A utilization factor is not required.

Dynamic Hygrothermal Whole-Building Simulation

The dynamic hygrothermal simulation combines single-building components such as walls, floors, and roofs to be modeled as a whole building. Coupled heat and moisture transport is simulated for each opaque component composed of different layers of materials such as wood, insulation, membranes, or even air layers. This model was developed by Künzel (1994). It considers capillary action, diffusion, and vapor absorption and desorption. The conductive heat and enthalpy flow by vapor diffusion with phase changes strongly depends on the moisture field. The vapor flow is simultaneously governed by the temperature and moisture field due to the exponential changes of the saturation vapor pressure with temperature. Resulting differential equations are discretized by means of an implicit finite volume method. The component model was validated by comparing its simulation results with measured data of extensive field experiments (Künzel 1994). The temperature and moisture field within the component is simulated as a result of the model. The differential equation for heat and enthalpy flow is Equation 10 and the one for vapor flow in a component is Equation 11.

$$\frac{\partial H}{\partial \vartheta} \cdot \frac{\partial \vartheta}{\partial t} = \frac{\partial}{\partial x}\left(\lambda \frac{\partial \vartheta}{\partial x}\right) + h_V \frac{\partial}{\partial x}\left(\frac{\delta}{\mu} \cdot \frac{\partial p_{sat}}{\partial x}\right) \tag{10}$$

$$\rho_w \frac{\partial u}{\partial \varphi} \cdot \frac{\partial \varphi}{\partial t} = \frac{\partial}{\partial x}\left(\rho_w D_\varphi \cdot \frac{\partial u}{\partial \varphi} \cdot \frac{\partial \varphi}{\partial x}\right) + \frac{\partial}{\partial x}\left(\frac{\delta}{\mu} \cdot \frac{\partial p_{sat}}{\partial x}\right) \tag{11}$$

where

D_φ = liquid conduction coefficient, m^2/s
H = enthalpy of the moist building material, J/m^3
h_V = evaporation enthalpy of the water, J/kg
p_{sat} = water vapor saturation pressure, Pa
u = water content, m^3/m^3
ϑ = temperature, °C
λ = thermal conductivity of the moist building material, W/m·K
μ = water vapor diffusion resistance factor of a dry building material, dimensionless
ρ_w = density of water, kg/m^3

φ	=	relative humidity, dimensionless
t	=	time, s

Coupling all the envelope components leads to the multi-zone building model. A zone constitutes one or more rooms with the same indoor climate. The zone boundaries are the components. There is also an outdoor zone. If there are two zones attached to each other then the respective indoor climates of the neighboring zones become the exterior boundary conditions of the other. The outdoor climate is specified by location in the climate files, assuming that the building itself does not influence the climate. However, the indoor climate is influenced by the simulation results of the component and vice versa—the component simulation is influenced by the indoor climate. Considering this interaction, the indoor climate can be simulated. With every time step the zone temperature and humidity values are generated by solving heat and moisture balance equations (Equations 12 and 13) (Lengsfeld and Holm 2007).

Besides the heat and moisture flow across the building, the envelope internal heat and moisture sources and sinks are taken into account. They are caused by people, lighting, mechanical equipment, infiltration, and solar radiation. Such sources or sinks can occur not only in the zones of the building itself but can also occur in the building envelope component with a direct influence on the heat and moisture field of the component. Additionally, transparent components, such as windows, can be modeled more accurately. The solar transmission that passes through a transparent component is calculated taking into account the sun elevation and azimuth angle and the orientation and inclination of the component. The solar heat gain that results from global radiation is dependent on the sun radiation incidence angle. Therefore, the solar heat gain coefficient is a varying input for different incidence angles. The solar heat gain—due to diffuse radiation—is calculated using the hemispherical solar heat gain coefficient. Solar heat gain contributions that pass through the transparent components are apportioned out directly to the indoor air and to the inner surfaces of opaque components according to a defined percentage (user defined or estimated according to surface area). Besides the shortwave solar irradiation, the longwave balance is also considered for the opaque building components. Therefore, not only the solar heat gains but also the longwave irradiation losses can be calculated.

$$\rho \cdot c \cdot V \cdot \frac{d\vartheta_i}{dt} = \Sigma_j A_j \cdot \alpha_j(\vartheta_j - \vartheta_i) + \dot{Q}_{Sol} + \dot{Q}_{IHG} + n \cdot V \cdot \rho \cdot c \cdot (\vartheta_a - \vartheta_i) + \dot{Q}_{VAC} \qquad (12)$$

where

ρ	=	air density, kg/m^3
α_j	=	heat transfer coefficient, W/m^2·K
ϑ_a	=	ambient air temperature, °C
ϑ_j	=	component indoor surface temperatures, °C
ϑ_i	=	indoor air temperature, °C
t	=	time, s
A_j	=	component areas, m^2
c	=	specific heat capacity of air, J/kg·K
n	=	air change rate, h^{-1}
$\dot{Q}_{Sol}$	=	short wave solar irradiance, directly to indoor air, W
$\dot{Q}_{IHG}$	=	internal heat sources due to people, lighting, equipment, W
$\dot{Q}_{VAC}$	=	heat flow due to mechanical ventilation, W
V	=	zone volume, m^3

$$V \cdot \frac{dc_i}{dt} = \Sigma_j A_j \cdot \dot{g}_{wj} + n \cdot V(c_a - c_i) + \dot{W}_{IMS} + \dot{W}_{VAC} \qquad (13)$$

where

c_a	=	absolute humidity of ambient air, kg/m^3
c_i	=	absolute humidity of indoor air, kg/m^3
$\dot{g}_{wi}$	=	moisture flow from indoor surface to indoor air, kg/s·m^2
$\dot{W}_{IMS}$	=	internal moisture source, kg/h
$\dot{W}_{VAC}$	=	moisture flow due to mechanical ventilation, kg/h

The zone model was validated via cross-validation with other tools, experiments, and standards such as ANSI/ASHRAE Standard 140 (2007). The validation of both—the energetic and the hygric parts of the zone model—is described by Antretter et al. (2011). Currently the ideal mechanical system has the capacity to supply all minimized heating, cooling, humidification, dehumidification, and mechanical ventilation loads. As long as the system's capacity is sufficient, the indoor temperature and moisture can be maintained between defined design conditions and thus the hourly demand can be calculated. If the capacity is not sufficient then the temperature or moisture will rise above or fall below the specified design conditions. If there is no ideal mechanical equipment defined, a "free floating" indoor climate is simulated.

Every time step depends strongly on the previous steps because of the water content and thermal energy storage within the envelope and the air in the zones. A time step is characterized by these dynamic previous variables. New boundary conditions are created with each time step and varying input data such as the outdoor climate. Using these initialization values, the coupled heat and moisture transport is calculated and consequently the zone heat and moisture balance equations are created. Should these balances not be within an expected defined accuracy of the simulation then the indoor temperature and relative humidity are iteratively adapted.

Combining both Models into WUFI Passive

Both models, the monthly passive house calculation and the dynamic whole-building simulation, rely on user inputs and assumptions. Some inputs are predefined, such as specific building materials and their dimensions, location, and orientation. Some have to be estimated by measurements or experience.

The predefined input is fundamentally the same for both models though quite more detailed for the dynamic simulation—not only because of the additional consideration of moisture. For example, the moisture storage function provides the information of the water content depending on the relative humidity of a material. However, the building geometry, room and component dimensions, widths and heights, and roof inclination are the same. Fenestration parameters such as solar heat gain coefficients, frame geometries, shading reduction factors, and many other boundary conditions, such as the design indoor temperature, the overheating limit temperature, and the natural air change rate, are the same as well.

More different is the climate data. For the monthly method, only monthly mean values for temperatures and solar radiation are necessary. For the dynamic hourly simulation, hourly input data for the outdoor climate must be provided. For the more detailed radiation calculation, hourly diffuse and direct global solar radiation data are required. Additionally needed is information on wind velocity and the quantity of rainfall to calculate the driving rain on the external surfaces. Aside from predefined input data, some results of the steady-state method can be used for the dynamic simulation, such as, for example, the mechanical ventilation volume flow rates for summer and winter ventilation, the simplified effective heat recovery efficiency, and the space heating and cooling capacities of the mechanical equipment. Internal heat sources due to people, lighting, and household and mechanical equipment are the same but also have to be supplemented with moisture characteristics. For the monthly method it is possible to calculate these sources using a utilization factor depending on the average usage. For the dynamic hourly simulation this is a good first assumption, but it might be more realistic to create specific time schedules.

Basically, the main difference between the models is the level of detail. A much more detailed simulation needs more time to compute. The monthly method is fast. An ordinary personal computer can compute all results within less than a second. The dynamic results may need some minutes up to some hours, depending on the complexity of the building model. In the dynamic simulation not just the indoor air temperature is simulated but also the temperature of the surroundings of a room, e.g., to calculate the operative temperature, is computed and, once more, also the humidity. Assessments of comfort conditions become possible once those values have been generated. The predicted mean vote (PMV) or the predicted percentage of dissatisfied (PPD) is calculated hourly. Even if generally accepted boundary conditions are exceeded, one can assess how long they will be exceeded.

The combination of both models within WUFI Passive using many of the same initial inputs results in numerous positive synergy effects. On one hand it is possible to obtain very fast results using only the monthly method, including heating, cooling, electricity, and primary energy demand, and on the other, with some more calculation time, it is possible to get detailed information on risk of mould growth, rotting components, and of course interior comfort conditions.

To simulate a building, different inputs for the desired results are necessary, as shown in Figure 2. Typically a user starts with the passive house calculation. The first thing to do is to input the building geometry, including structure, material,

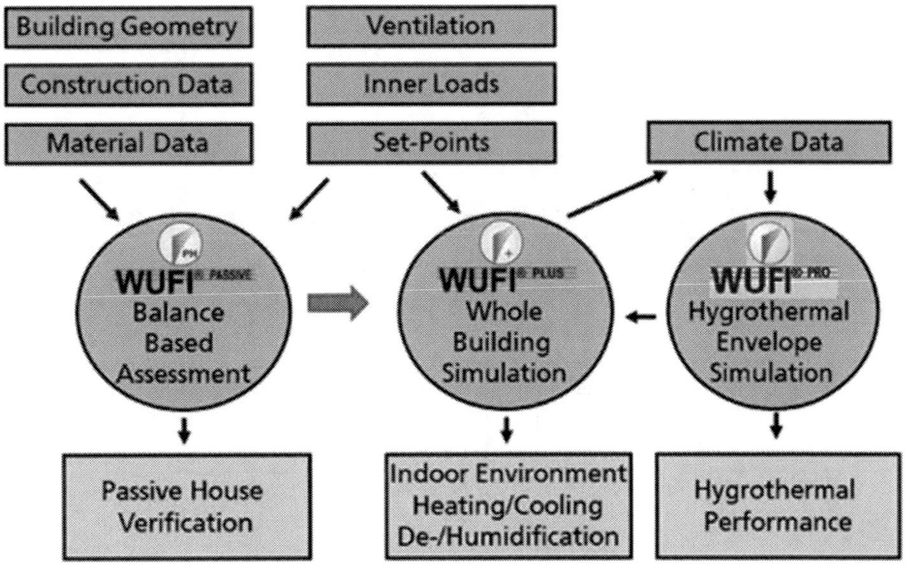

Figure 2 *Inputs, interaction of the different parts of the simulation engine, and outputs.*

location, and all essential passive house verification data. An import function allows importing a three-dimensional (3-D) building model from SketchUp, a free drawing program (Trimble 2013) (other commercial drawing programs can typically save files as a SketchUp file, which then can be imported as well). Another way is to use the integrated building wizard for more typical and simpler building geometries. It is important to designate inner and outer sides of each component correctly to ensure that boundary conditions get assigned accurately. The input using component lists and databases only is also possible; the model cannot be visualized in 3-D if this path is used. Once the geometry is set, thermal bridges and windows can be defined. The next step is to define the usage of the building (residential or nonresidential) as the input for inner loads is different for each type. Last but not least, the user has to define the mechanical equipment. Therefore, different systems for, e.g., heating, domestic hot-water production, and ventilation, can be defined as well as their distribution. The software gives feedback at all times during the entry process to inform the user about still-missing inputs. Once all inputs are complete, the heating demand and all other passive house verification results are calculated instantly.

For the dynamic hygrothermal simulation a user can switch to WUFI Plus mode. Some input screens change to the dynamic relevant input data. Some boundary conditions are not applied automatically because more detailed information may be required (like the indoor setpoint temperature, which can be defined via time schedule), but there is an option to choose them with one click. If a user has used building materials or assemblies provided within the database, no additional information is needed, but some additional parameters such as indoor moisture loads should be defined. The software will check all inputs for completeness and prompt the user for missing information before the simulation starts. During the simulation a user can monitor movies for the heat and moisture profiles of each component or the hourly heating demand. Once the simulation is finished, detailed reports and graphs illustrate the simulation results.

APPLICATION EXAMPLE

The above-described design and certification tool is used in this section to reassess an already certified passive house. The results for the steady-state method match the results of the previously used tool to certify passive houses. It is also shown that the combination of the steady-state methodology with a dynamic building simulation allows a more thorough analysis of the thermal and hygric performance of the building and its components.

Building Model and Boundary Conditions

A passive house designed for the Northern Central climate of the United States is used as a case study to demonstrate how to use the new tool and how to showcase the range of possible results. The building is located in Urbana, Illinois. Cold winters and relatively warm summers have to be taken into account while designing an energy-efficient building. The weather file for Chicago, Illinois, was used for the dynamic simulation. The hourly temperature and relative humidity conditions as well as mean, minimum, and maximum conditions are shown in Figure 3. Those conditions were the basis

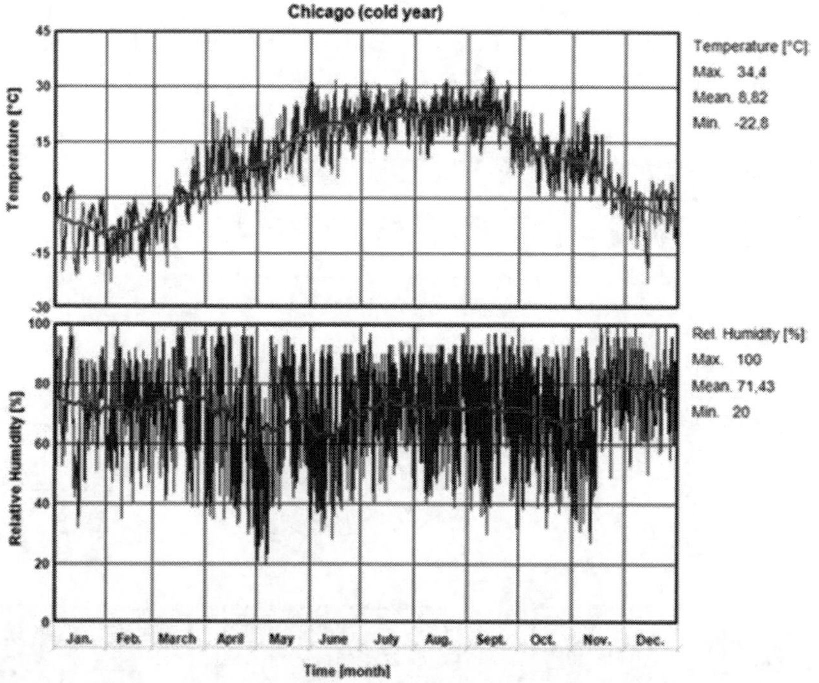

Figure 3 *Hourly weather data in Chicago, IL, representing the used weather file.*

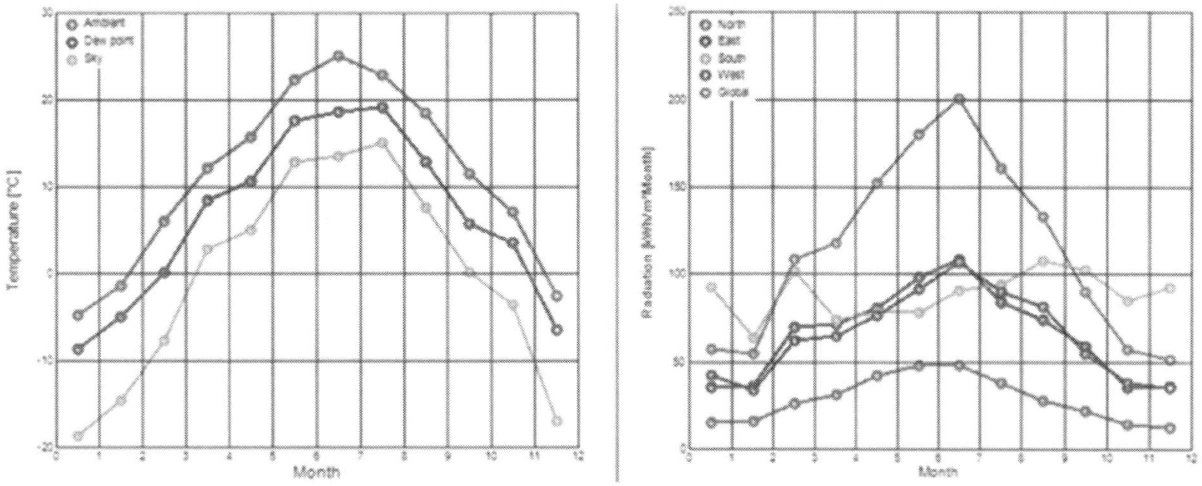

Figure 4 Screenshot of the monthly mean weather data used for the balance-based assessment.

for the dynamic simulation. The monthly method uses monthly conditions as described previously. Figure 4 shows the used monthly values.

The building is a two-story residential building with a relatively open floor plan. The first and second floors are connected by an open staircase. The home has three bedrooms and two bath and a discounted interior conditioned floor area (treated floor area, TFA) of approximately 1000 ft^2. The construction type is light frame insulated with high-density blown-in fiberglass and cellulose blown into the vented attic. The foundation is a fully insulated float foundation frost protected system—exposed finished concrete floors insulated to the ground with expanded polystyrene. The envelope is thermal-bridge free other than the installation thermal bridges from windows and doors and bridges due to mechanically necessary penetrations. Construction cost was \$128/ft^2 of TFA. A screenshot of the 3-D building representation in the software is shown in Figure 5.

The standard boundary conditions for passive house verification are used for the balance method.

The dynamic simulation uses the same assumptions for ventilation (0.05 ach for infiltration and a mechanical ventilation with 83% heat recovery of 0.32 ach), setpoints (temperature between 73°F and 77°F), and inner loads (4.2 kWh/d).

This is suitable for comparison purposes. To get more detailed information on the building performance under transient boundary conditions it is possible to apply schedules for day profiles on different days (e.g., weekday vs. weekend) or periods (e.g., holidays).

The building geometry was kept exactly the same, i.e., the same building model is used. Also, all component information is used for both models—the monthly balance and the dynamic assessment.

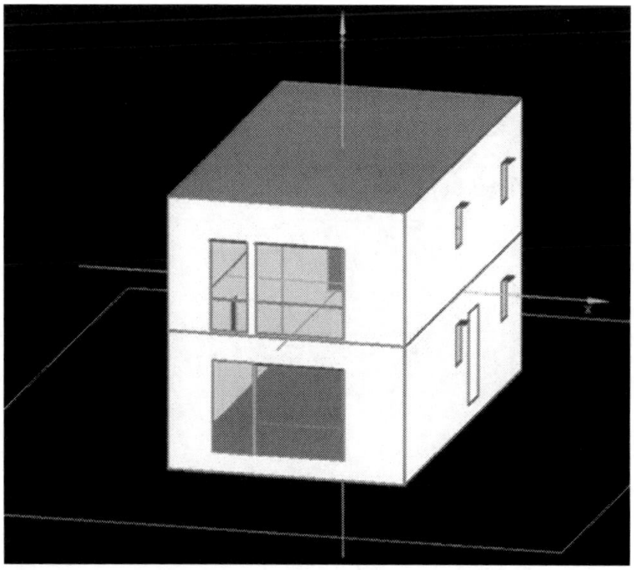

Figure 5 Screenshot of the 3-D building representation in the software.

Steady-State Results

In the steady-state mode, monthly balances are used to compute the heating and cooling energy demand. The heat flux for each component is simulated. This allows the identification of the highest contributors to heating and cooling loads and their improvement potential. Figure 6 shows the results for the contribution of different component types, ventilation, internal heat gains, solar gains, and heating or cooling to the overall energy balance.

All the information is available in graphical and numerical presentations. Figure 7 shows a screenshot of the actual results contributing to the heating and cooling energy balance.

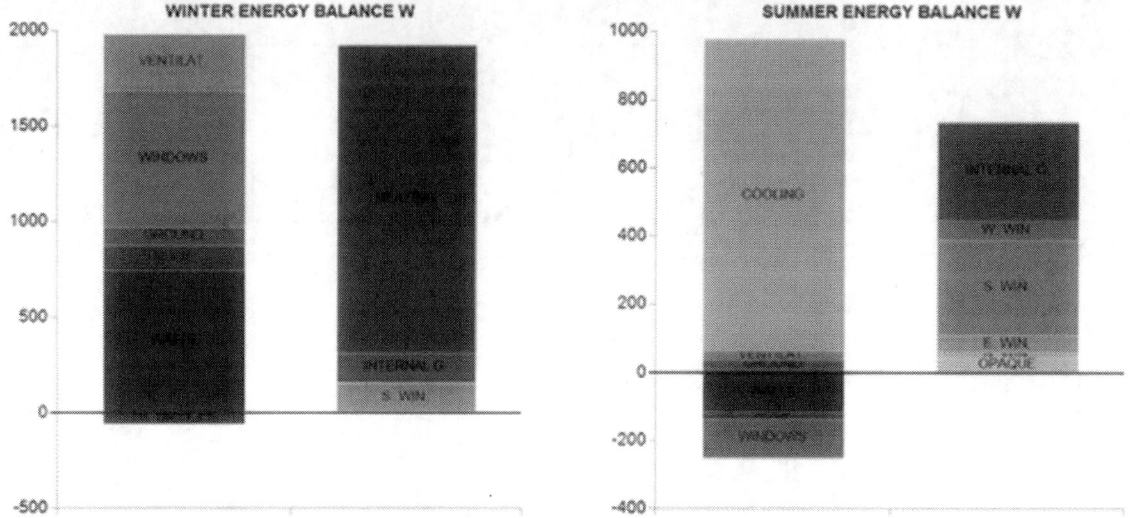

Figure 6 *Steady-state results for the winter and summer energy balance.*

ANNUAL HEAT DEMAND		
Transmission Losses :	3646,5	kWh/a
Ventilation Losses:	418,3	kWh/a
Total Heat Losses:	4064,8	kWh/a
Solar Heat Gains:	2156,9	kWh/a
Internal Heat Gains:	994,9	kWh/a
Total Heat Gains:	3151,8	kWh/a
Utilization Factor:	89,1	%
Useful Heat Gains:	2808,4	kWh/a
Annual Heat Demand:	1256,4	kWh/a
Specific Annual Heat Demand:	13,5	kWh/m²a

ANNUAL COOLING DEMAND		
Solar Heat Gains:	2355,7	kWh/a
Internal Heat Gains:	1290,5	kWh/a
Total Heat Gains:	3646,2	kWh/a
Transmission Losses :	2970,2	kWh/a
Ventilation Losses:	1986,9	kWh/a
Total Heat Losses:	4957	kWh/a
Utilization Factor:	62,3	%
Useful Heat Losses:	3088,7	kWh/a
Cooling Demand - sensible:	557,5	kWh/a
Cooling Demand - latent:	40,7	kWh/a
Annual Cooling Demand:	598,2	kWh/a
Specific Annual Cooling Demand:	6,4	kWh/m²a

Figure 7 *Screenshot of the annual heating and cooling demand contributions.*

Dynamic Results

A dynamic building simulation enables the user to assess the overall energy performance of a building and get a more detailed look at the transient interaction between building envelope and interior space. The implemented model also takes into account the coupled heat and moisture transport into the envelope components. Hence, the hygrothermal component performance can also be assessed.

Some example results from the dynamic simulation are shown in Figures 8–12. The energy performance can be assessed in more detail by looking at the hourly values. Thermal comfort values in the building can be computed according to common standards such as ANSI/ASHRAE Standard 55 (2010), and possible moisture-related problems can be identified by analyzing the time-dependent temperature and moisture distributions in each component layer.

Energy Demand

During the design phase of a building it is often necessary to divide the different contributions to the overall energy balance of a building. Figure 8 shows the monthly sum of all different heat fluxes. This example identifies that in the heating period the ventilation losses are small compared to the losses through opaque partitions and to the large losses through windows. The solar gains are high all year round. This leads to overheating/high cooling demand in the cooling period when gains through opaque wall assemblies and small inner gains are added to the balance. We find that a large first improvement for both heating and cooling demand can be achieved by using windows with a lower overall U-factor (reducing the losses in winter) and lower solar heat gain coefficient or/and better shading (reducing the gains in summer).

The dynamic model yields other important insights. A closer look at the hourly values in Figure 9 reveals some other shortcomings of the current design. Especially during the in-between seasons, the daily temperature swing reaches both the heating and the cooling setpoint temperatures. The graphs in Figure 9 illustrate clearly that on some sequential days heating and cooling is required on the same day. One design option would be to add some thermal inertia, to store the gained energy and release it later on. These effects can only be assessed with a dynamic simulation; a monthly balance gives no indication that this problem might occur.

Indoor Environment

Indoor environment, especially thermal comfort, can be assessed with different standards such as the North American standard ASHRAE Standard 55 (2010) or the international standard DIN EN ISO 15251 (2007). Both allow the assessment of the overall thermal comfort and localized thermal discomfort. The overall thermal comfort is in general assessed by the predicted mean vote (PMV) according to Fanger (1982), which describes the mean vote of people exposed to the climate conditions on a –3 to +3 scale with 0 being a neutral vote expressing the feeling that it is neither too warm nor too cold. The calculation of the PMV takes the air and the radiant temperature, the relative humidity of the air (all of which are a result of the simulation), the air velocity (which can be guessed in a regularly conditioned building), and some individual parameters such as clothing insulation and metabolic rate into account. Figure 10 shows the calculated PMV and the predicted percentage of dissatisfied (PPD) for the example building with the assumption of a 0.1 m/s (about 3.94 in./s) air speed, a clothing insulation of 0.9 clo, and a metabolic rate of 1 met. It shows that the conditions fluctuate between the two temperature setpoints, causing slightly cool conditions in winter with a PMV close to –1 for the assumed clothing level. During summer the mean vote is that it is slightly warm, but overall the conditions are in an acceptable range, coming in most of the time between –0.5 and +0.5.

Hygrothermal Component Assessment

The hygrothermal component assessment usually contains two parts. First, the total water content and the water content of different layers is assessed. Simply put, the total water content should not continuously increase over time and its annual fluctuations should remain within a certain limit. Figure 11 shows the total water content of the south-oriented exterior wall. Regular annual fluctuations are reached and the total water content does not increase over time.

Furthermore, the water content of different layers should remain below a certain limit. For example, one could analyze the mass percentage of the fiberboard sheathing to make sure that there is no long-term risk of wood rot. The drying capacity regarding built-in moisture can also be assessed.

Second, the hygrothermal conditions for critical locations can be assessed. This can be applied, for example, to assess the risk for mold growth on the interior surface of the exterior envelope using realistic interior conditions resulting from the zone simulation. A very helpful tool to assess the performance

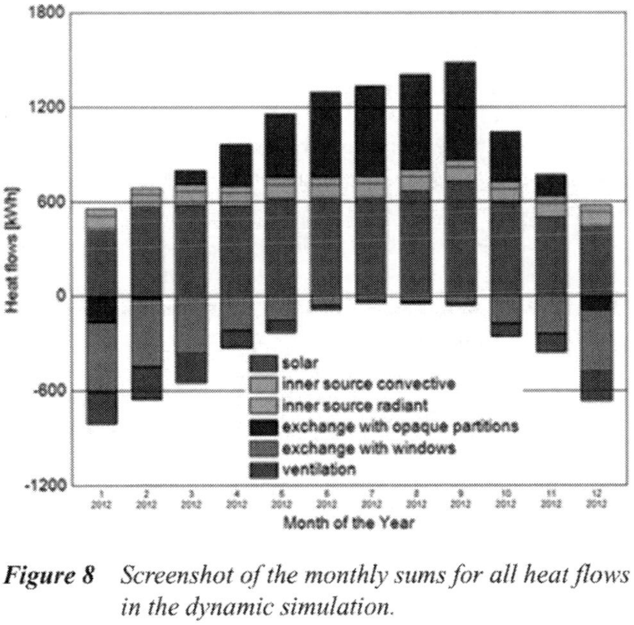

Figure 8 *Screenshot of the monthly sums for all heat flows in the dynamic simulation.*

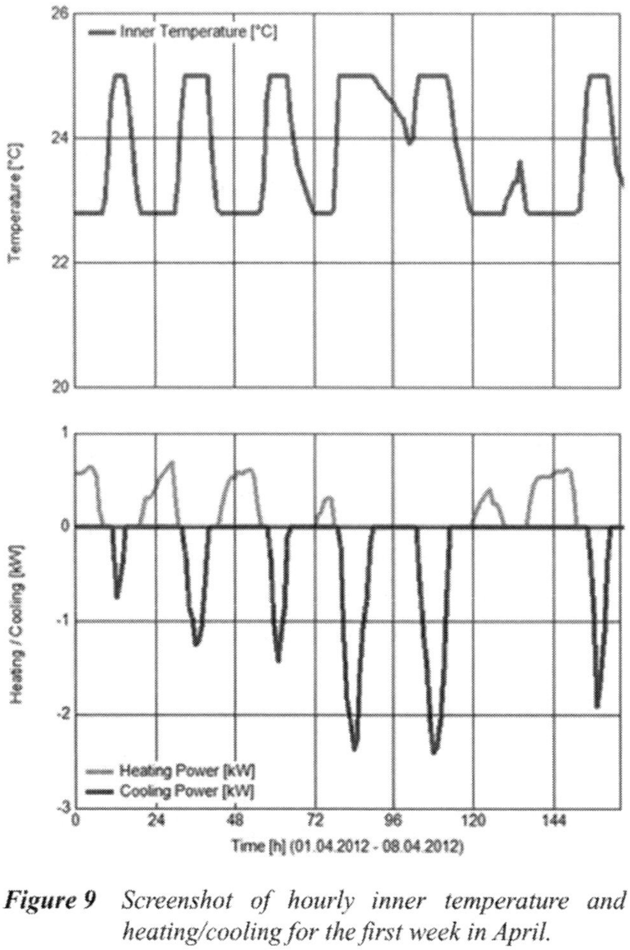

Figure 9 *Screenshot of hourly inner temperature and heating/cooling for the first week in April.*

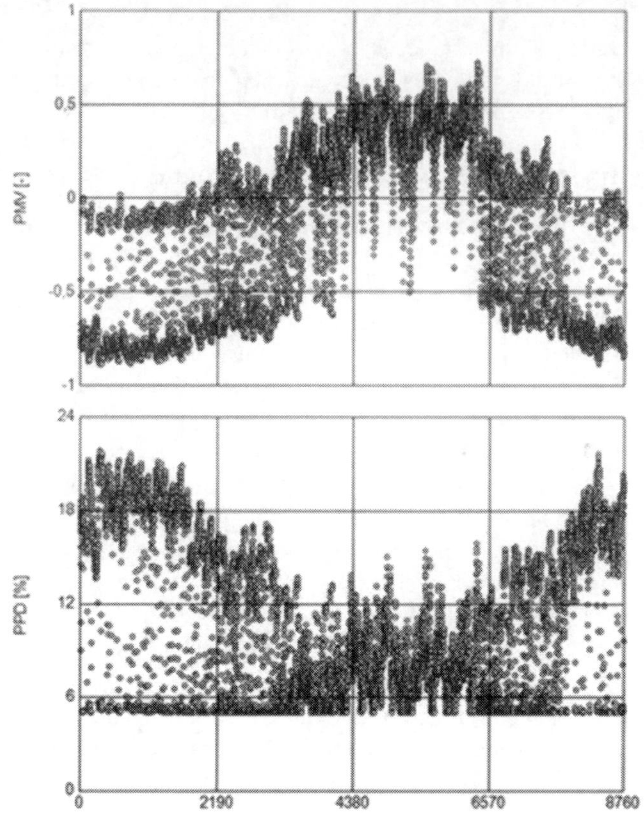

Figure 10 *Screenshot of the calculated PMV and PPD values for the example building.*

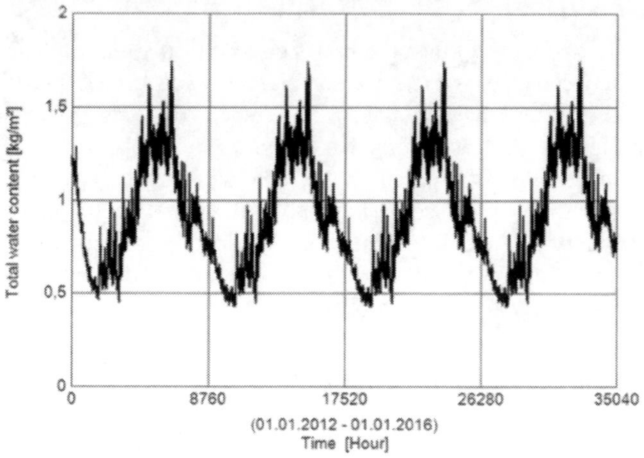

Figure 11 *Screenshot of the total water content of the south-oriented exterior wall.*

and the processes inside the components is the visualization of the temperature and humidity distribution across the building layers in the form of a movie. Figure 12 shows as an example a screenshot of the temperature and humidity in the different layers of the exterior wall assembly at a certain point in time.

DISCUSSION AND CONCLUSIONS

A concern about the new tool is that it will not allow access to the underlying formulas and calculation paths. But because it provides an organized self-guiding user interface, the kind of tracing and navigation that was necessary for a very large Excel calculator will likely no longer be necessary. Furthermore, the Excel output and export function of the new tool allows integration of all data into other spreadsheet calculators or custom optimizers.

Overall the benefits of a user-friendly interface and modeling synergies significantly outweigh any drawbacks. If passive buildings go mainstream, the majority of users will be architects who appreciate a simple interface and being guided through the tool. So far the percentage of architects taking up passive building design has been limited to those who already have a physics interest or strong mathematical inclination. Most architects like more guidance. Thus, the combination of both models into one tool has real potential to transform the passive building design process in North America by making very complex processes accessible to more design professionals.

Improvements start from a simple work-flow perspective; it organizes the input process along a clearly guided path of the familiar tree structure of the WUFI software family while providing constant feedback on missing data entries. It offers management help to optimize the design process in passive verification mode by allowing the modeler to store side by side an essentially infinite number of different cases. It organizes all required project and systems data for certification into one file and has the promise to simplify the certification process itself by eliminating opportunities for mistakes, essentially making critical quality assurance during the planning process much easier. The static calculation is fast and efficient, and outputs include both numerical and illustrative graphical representations of the results, which are very helpful in discussions with clients.

Even more significant are the improvements in regards to the design process. Previously the designer did not have enough information to dial in the correct window value combinations for greatly varying climates in the United States. The general recommendation of installed window U-factors of 0.85 $W/m^2/K$ (0.15 $Btu/ft^2 \cdot hr \cdot °F$) and a g-value/solar heat gain coefficient higher than 50% might lead to overheating and in some cases produces the need for heating and cooling during the same day during the in-between seasons. The designer had no feedback that would help to identify the most effective climate-specific combination of those values while ensuring thermal comfort as demonstrated in the previous application example (where the designer was forced to come up with a best guess).

Many designers in North America hence have been forced to master and use separate hygrothermal tools and secondary dynamic energy models to assess wall component appropri-

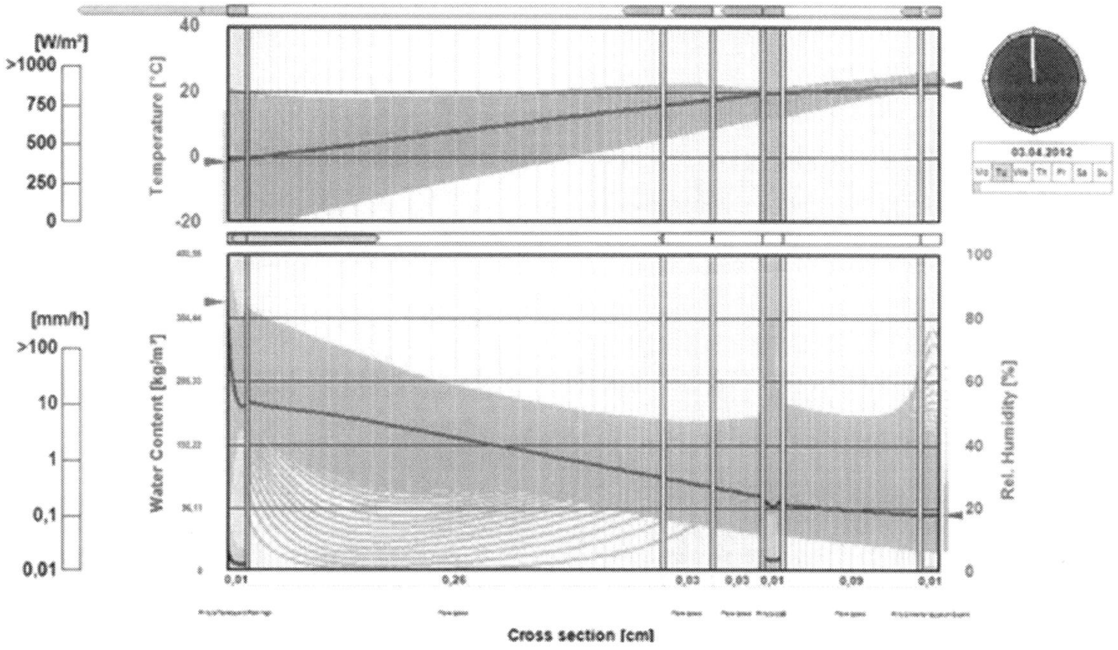

Figure 12 *Exemplary screenshot of the temperature and humidity visualization of a component.*

ateness by climate and thermal comfort by zone. All additional tools required some form of double entry of material properties, dimensions, and mechanical specifications. In the design process many of those pieces of information are still in flux and lead to the need of updating three models instead of one when a design change is made. This is not only labor intensive but also increases the likelihood of error.

In conclusion, the most significant improvement of WUFI Passive, aside from the more efficient and organized workflow, is the all-in-one risk management capability. The next-generation passive modeling for varying climates needs to include dynamic simulation to improve models in regards to cooling and dehumidification load accuracy as well as to improve the ability to predict comfort issues such as overheating and high relative indoor humidity. The few optimization conclusions highlighted over the course of this paper based on the dynamic model outputs of the built example in Urbana are excellent examples of how the dynamic results are informing the design process to ensure passive building designs in North America are the most efficient and comfortable.

And to conclude with a gaze into the future: as larger buildings with multiple zones are being modeled to meet the passive building energy metrics, it will become imperative to model and verify comfort in multiple zones in more complex buildings. WUFI Passive will meet those challenges on the horizon.

REFERENCES

Antretter, F., F. Sauer, T. Schöpfer, and A. Holm. 2011. Validation of a hygrothermal whole building simulation software. *Proceedings of Building Simulation 2011: 12th Conference of International Building Performance Simulation Association, Sydney, Australia.*

ASHRAE. 2007. ASHRAE Standard 140-2007, *Building Thermal Envelope and Fabric Load Tests*. Atlanta: ASHRAE.

ASHRAE. 2010. ASHRAE Standard 55-2010, *Thermal Environmental Conditions for Human Occupancy*. Atlanta: ASHRAE.

Cohen, A. 2012. Design of the Hickory Hall Dormitory. *Proceedings of the 7th Annual North American Passive House Conference.*

DIN. 2007. DIN EN ISO 15251:2007-08, *Indoor environmental input parameters for design and assessment of energy performance of buildings addressing indoor air quality, thermal environment, lighting and acoustics*. Berlin, Germany: Deutsches Institut for Normung.

DIN. 2008. DIN EN ISO 13790:2008, *Energy performance of buildings – Calculation of energy use for space heating and cooling*. (*Energieeffizienz von Gebäuden - Berechnung des Energiebedarfs für Heizung und Kühlung*). Berlin, Germany: Deutsches Institut for Normung.

DOE. 2008. Building America. Building Technologies Program, Research and Development. Washington, DC: U.S. Department of Energy. http://apps1.eere.energy.gov/buildings/publications/pdfs/corporate/myp08research_ch2.pdf

EIA. 2005. Residential Energy Consumption Survey—Detailed Tables. www.eia.gov/consumption/residential/data/2005/c&e/spaceheating/pdf/tablesh8.pdf

EIA. 2012. *Annual Energy Review 2011*. www.eia.gov/aer Washington, DC: U.S. Energy Information Administration.

EPA. 2012. *ENERGY STAR® Qualified Homes Version 3 (Rev 06), National Program Requirements*. Washington, DC: U.S. Environmental Protection Agency. www.energystar.gov/ia/partners/bldrs_lenders_raters/ES_Combined_Path_v_65_clean_508.pdf?f14e-e0e4

Fanger, P.O. 1982. *Thermal Comfort*. Malabar, Florida: Robert E. Krieger Publishing.

Feist, W. 1992. Passivhäuser in Mitteleuropa. Dissertation, University of Kassel.

Feist, W., et.al. 2007. *PHPP Handbook*. Darmstadt, Germany: Passivhaus Institute.

Fraunhofer IBP. 2013. *WUFI® Passive*, version 2.5.3. Fraunhofer Institute for Building Physics, Valley, Germany.

ICC. 2012. *International Energy Conservation Code*. Washington, DC: International Code Council.

Künzel, H.M. 1994. Simultaneous Heat and Moisture Transport in Building Components. Dissertation, University of Stuttgart. www.building-physics.com

Lengsfeld, K., and A. Holm. 2007. Entwicklung und Validierung einer hygrothermischen Raumklima-Simulationssoftware WUFI®-Plus. *Bauphysik* 29, Magazin 3.

Oram, S. 2011. 16th & Nebraska Passive House First Year Billing Analysis. Seattle, WA: Ecotope, Inc.

PHIUS. 2013. Certified PHIUS Projects. Passive House Institute US, Urbana, IL. www.passivehouse.us/projects.php

Shurcliff, W. 1982. The Saunders-Shrewsbury House. Self-published.

Shurcliff, W. 1986. Superinsulated houses. *Annual Reviews Energy* 11:1-24.

Shurcliff, W. 1988. *Superinsulated Houses and Air-To-Air Heat Exchangers*. Andover, MA: Brickhouse Publishing Company.

Trimble. 2013. SketchUp. Sunnyvale, CA: Trimble Navigation Limited.

Hygrothermal Material Properties for Soils in Building Science

Manfred Kehrer
Member ASHRAE

Simon Pallin

ABSTRACT

Hygrothermal performance of soils coupled to buildings is complicated because of the dearth of information on soil properties. However they are important when numerical simulation of coupled heat and moisture transport for below-grade building components are performed as their temperature and moisture content has an influence on the durability of the below-grade building component. Soils can be classified by soil texture. According to the Unified Soil Classification System (USCA), 12 different soils can be defined on the basis of three soil components: clay, sand, and silt.

This study shows how existing material properties for typical American soils can be transferred and used for the calculation of the coupled heat and moisture transport of building components in contact with soil. Furthermore a thermal validation with field measurements under known boundary conditions is part of this study, too. Field measurements for soil temperature and moisture content for two specified soils are carried out right now under known boundary conditions. As these field measurements are not finished yet, the full hygrothermal validation is still missing.

INTRODUCTION

Hygrothermal performance of soils coupled to buildings is complicated because of the dearth of information on soil properties. A computational approach for heat transfer through the ground has been well defined (EN-ISO-13370 2007), and simplified methods have been developed (Staszczuk et al. 2010). However, these approaches generally ignore the transfer of soil moisture, which is not negligible (Janssen et al. 2004).

The material properties of soils must be reliable for the performance of realistic hygrothermal calculations of soils coupled to buildings, but such material properties are not well defined in present hygrothermal calculation tools. Typical types of building construction that are greatly influenced by soils are basements, crawl spaces, and slabs on grade. Reliable hygrothermal performance indicators for these constructions are sorely needed; requests for this ranked within the top ten Building America Enclosure Research Ideas at the Enclosures STC—Residential Energy Efficiency Stakeholder Meeting held February 29, 2012, in Austin, Texas. The soil science literature contains an extensive amount of measurements on soil properties, but this information must be gathered as well as adapted for application to hygrothermal simulations in building science. Soil properties are important when analyzing and designing both new buildings and retrofits for which the outer boundary of the building enclosure will consist of soil materials. For basement retrofits aimed at energy efficiency, interior solutions to decrease the energy demand must be designed in consideration to existing soil properties. For exterior retrofits, the soil material can be replaced, if needed, with a more suitable filling material, though this approach applies only to basement walls. The soil material beneath the basement floor obviously cannot be replaced; hence the soil properties of this part of a building enclosure still must be taken into consideration.

The intention of this first part is to gather, comprehend, and adapt soil properties from soil science. The obtained information must be applicable to related tasks in building science and validated with hygrothermal calculation tools, where

***Manfred Kehrer** is a senior R&D staff member in the Building Technologies Research and Integration Center Engineering at Oak Ridge National Laboratory, Oak Ridge, TN. **Simon Pallin** is a member of the Division of Building Technology Department of Civil and Environmental, Chalmers University of Technology, Götenborg, Sweden.*

basic changes in the software code may be required in the future. Different types of basement construction will be created with a hygrothermal calculation tool, WUFI. Simulation results from WUFI are to be compared with existing thermal only measurements and will be accomplished with ongoing hygrothermal measurements.

The final outcome of the study will be the evaluation of several soil types in several climate zones for a number of basement assembly types. The study will define the type of soil, together with the type of building construction, considered most and least reliable with respect to energy consumption and moisture safety. Further, the study will determine (1) the influences that different soils have on total energy loss through the ground and (2) whether the performance of different soils can be measured solely by a comparison of soil properties.

HYGROTHERMAL SOIL PARAMETERS

Soils can be classified by soil texture. According to the Unified Soil Classification System (USCA), 12 different soils can be defined on the basis of three soil components: clay, sand, and silt (Noorallah 1999). The 12 soil textures in the USCA are presented in the soil texture triangle shown in Figure 1. This study will define the material properties of these 12 soil types.

The soil properties of interest for hygrothermal calculations are:

1. Dry bulk and particle density
2. Porosity
3. Thermal conductivity
4. Specific heat capacity
5. Moisture storage function
6. Liquid water transfer

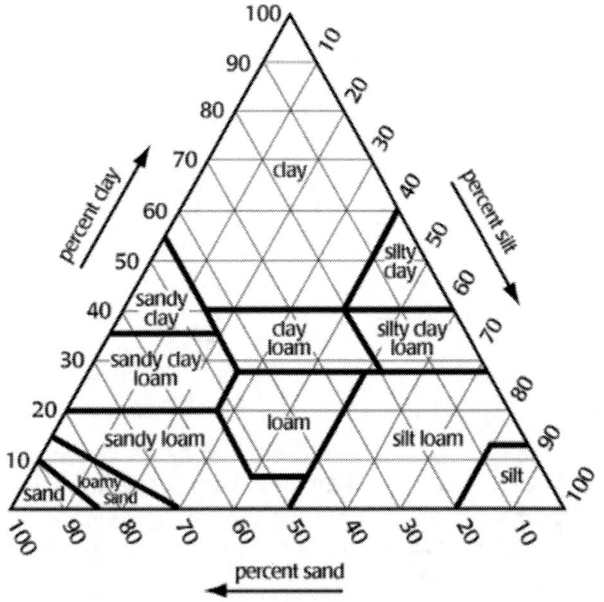

Figure 1 *Soil texture triangle representing the 12 classes of soil texture used by the Natural Resources Conservation Service of the US Department of Agriculture (Duble 2011; Noorallah 1999).*

These soil properties will be defined in this study for the 12 soil textures presented in Figure 1. These properties will be collected from multiple studies, measurements, and standards; and empirical formulas will be created and used to estimate their variations as a function of different parameters. The influence of gravity on water movement in soil is typically neglected because the water pressure head is usually much higher than the gravitational pressure (Gopalakrishnan and Manik 2007; Prunty 2009).

Moisture Storage Function

The van Genuchten equation for the estimation of soil hydraulic properties, Equation 1, has been widely used since it was first published in 1980 (Neto et al. 2011; van Genuchten 1980; Zhou and Yu 2005). In the equation, the soil water retention curve (WRC) defines the relationship between the volumetric water content of the soil as a function of the suction pressure, p_{suc}, expressed as an equivalent soil pressure head, h, (meters) in the above mentioned studies. The WRC can be obtained by use of the analytical expression of the effective saturation (also referred to as normalized water content), S_e:

$$S_e(h) = \frac{\theta - \theta_r}{\theta_s - \theta_r} = [1 + (\alpha|h|)^n]^{-m} \tag{1}$$

Hence the volumetric water content can be expressed as:

$$\theta(h) = \theta_r + \frac{\theta_s - \theta_r}{[1 + (\alpha|h|)^n]^m} \tag{2}$$

where

S_e = effective saturation, –
θ = volumetric water content, m^3/m^3
θ_r = residual water content, m^3/m^3
θ_s = volumetric water content at saturation, m^3/m^3
h = soil water pressure head, m
α and n are fitting parameters (m^{-1}) and (–)
m = $(1 - 1/n)$, –

The parameter n is related to the pore size distribution where a larger value corresponds to more uniform pore sizes (Zhou and Yu 2005). The value of α is related to the inverse of the bubbling point, representing the air pressure at the point of air entry into a pore (Schaap et al. 2001) and m is related to the asymmetry of the curve (Zhou and Yu 2005). The soil textures analyzed in this study (the type of soils for which essential properties for hygrothermal simulation will be defined) correspond to the 12 texture classes used by the Natural Resources Conservation Service of the US Department of Agriculture (Noorallah 1999). The soil texture classes are defined in Table 1 where the

Table 1. Average Hydraulic Parameters Required in Equation 1 to Calculate the Saturation Effect and in Equation 13 to Calculate the Liquid Diffusivity. The Values are Based on 560 Soil Samples (Schaap 2002)

Soil Texture	θ_r, m³/m³	θ_s, m³/m³	α, m⁻¹	n, –	m, –	K_s, m/s
Clay	0.10	0.46	1×10^{-4}	1.25	0.20	1.7×10^{-6}
Clay Loam	0.08	0.44	2×10^{-4}	1.42	0.29	9.3×10^{-7}
Loam	0.06	0.40	1×10^{-4}	1.47	0.32	1.4×10^{-6}
Loamy Sand	0.05	0.39	3×10^{-4}	1.75	0.43	1.2×10^{-5}
Sand	0.05	0.38	4×10^{-4}	3.18	0.69	7.4×10^{-5}
Sandy Clay	0.12	0.39	3×10^{-4}	1.21	0.17	1.3×10^{-6}
Sandy Clay Loam	0.06	0.38	2×10^{-4}	1.33	0.25	1.5×10^{-6}
Sandy Loam	0.04	0.39	3×10^{-4}	1.45	0.31	4.4×10^{-6}
Silt	0.05	0.49	1×10^{-4}	1.68	0.40	5.1×10^{-6}
Silty Clay	0.11	0.48	2×10^{-4}	1.32	0.24	1.2×10^{-6}
Silty Clay Loam	0.09	0.48	1×10^{-4}	1.52	0.34	1.3×10^{-6}
Silt Loam	0.07	0.44	1×10^{-4}	1.66	0.40	2.1×10^{-6}

input parameters of Equation 1 and Equation 2 are based on 560 soil samples obtained from a computer software for estimating soil hydraulic characteristics (Schaap 2002). The functions are be either measured directly or estimated indirectly through prediction from more easily measured data based using quasi-empirical models to estimate unsaturated hydraulic properties from surrogate soil data such as soil texture data and bulk density. Models of this type are called pedotransfer functions (PTFs) since they translate basic soil data into hydraulic properties and are state of the art in soil science.

The sorption isotherm describes the water content, w (kg/m³), of a material as a function of the relative humidity, $w(\phi)$ or w (RH), which is an essential parameter when making hygrothermal calculations with porous materials. Ongoing field measurements will show whether the approach in Equation 1 is valid in the capillary region only or also at relatively dry conditions. The WRC, thus the soil water pressure head, p_{suc}, can be expressed as the relative humidity with the Kelvin equation (Kiessl and Gertis 1980):

$$\varphi = e^{\left[\frac{-2\sigma}{r \cdot \rho_w \cdot R_D \cdot T}\right]} \tag{3}$$

and

$$p_{suc} = \frac{2 \cdot \sigma \cdot \cos\gamma}{r} \tag{4}$$

where

φ = relative air humidity, %

σ = surface tension coefficient, N/m

r = radius of capillary tube, m

ρ_w = density of pore water, kg/m³

R_D = water vapor gas constant, J/kg, K

= 461.5

T = temperature, K

p_{suc} = suction pressure, Pa

γ = contact angle, °

The contact angle, γ, between the water meniscus and the capillary walls can be assumed equal to zero (Hagentoft 2001) due to the hydrophilic properties of the pore wall in stony or wood based materials (Hens 2007), as for soils. Hence the radius of the capillary tube can be expressed as

$$r = \frac{2 \cdot \sigma}{p_{suc}} \tag{5}$$

The relative humidity in Equation 3 can be expressed as function of p_{suc}:

$$\varphi(p_{suc}) = e^{\left[\frac{-p_{suc}}{\rho_w \cdot R_D \cdot T}\right]} \tag{6}$$

The soil water pressure head, h, can be described as suction pressure, p_{suc}:

$$h \cdot g \cdot \rho_w = p_{suc} \tag{7}$$

Combining Equation 6 and Equation 7 results in:

$$h(\varphi) = \frac{-\ln\varphi \cdot R_D \cdot T}{g} \tag{8}$$

Further, the volumetric water content in Equation 2 are given as a function of φ.

$$\theta(\varphi) = \theta_r + \frac{\theta_s - \theta_r}{\left[1 + \left(\alpha\left(\frac{-\ln\varphi \cdot R_D \cdot T}{g}\right)\right)^n\right]^m} \qquad (9)$$

The relation between the water content, w, and the volumetric water content, θ, is:

$$w = \rho_w \cdot \theta \qquad (10)$$

Finally w can be expressed as a function of φ with the van Genuchten equation:

$$w(\varphi) = \rho_w\left[\theta_r + \frac{\theta_s - \theta_r}{\left[1 + \left(\alpha\left(\frac{-\ln\varphi \cdot R_D \cdot T}{g}\right)\right)^n\right]^m}\right] \qquad (11)$$

Obviously φ in Equation 6 cannot be described solely as a function of the p_{suc} when considering the moisture storage. Temperature also has an influence on the suction pressure, though proven almost insignificant in the temperature range between 0°C and 20°C (Figure 2). Therefore, in this study, the temperature is assumed to be 10°C. The resulting moisture storage functions of the 12 soil textures, based on Equation 11, are presented in Appendix A.

Hysteresis is defined as the difference in moisture storage capacity of a material depending on the state from which the material is moving from. A material that is in a drying process has a higher moisture storage capacity at a given ϕ than a material that is in a wetting process (Hagentoft 2001). The effect of hysteresis on hydraulic soil properties has been investigated (Šimunek et al. 1999) but will not be taken into account for the moisture storage capacity due to software limitations, though considered in the liquid diffusivity, defined later in this paper.

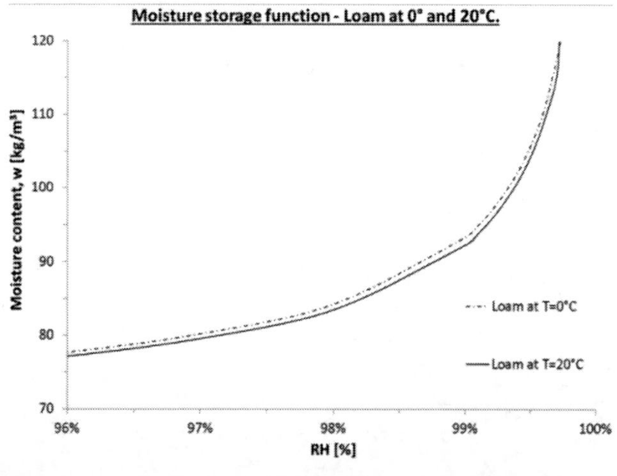

Figure 2 *Moisture storage function of loam at fixed temperatures of 0°C and 20°C. The moisture content of the soil is defined in the range 96%–100% rh.*

Liquid Water Transfer

The liquid diffusivity, D, can be defined for one-dimensional water flux as a function of $K(\theta)$ and WRC (Gopalakrishnan and Manik 2007; Hills et al. 1989; Prunty 2009):

$$D(\theta) = K(\theta)\frac{\partial h}{\partial \theta} \qquad (12)$$

in which K is hydraulic conductivity as a function of volumetric water content. Further, liquid diffusivity can be expressed as a function of the effective saturation, S_e, for unsaturated soils (van Genuchten 1980):

$$D(S_e) = \frac{(1-m)K_s}{\alpha m(\theta_s - \theta_r)} S_e^{\left(\frac{1}{2} - \frac{1}{m}\right)}\left[(1 - S_e^{\frac{1}{m}})^{-m} + (1 - S_e^{\frac{1}{m}})^m - 2\right] \qquad (13)$$

where

D = liquid diffusivity, (m²/s)

K_s = saturated hydraulic conductivity, (m/s)

All other variables are specified under Equation 2.

In addition, the value of liquid diffusivity varies according to whether the soil is being dried or wetted (i.e., an effect of hysteresis). The liquid diffusivity for a soil that is being dried is referred to as drying or drainage diffusivity, whereas a soil being wetted is referred to as wetting or absorption diffusivity. In this study, the terms drying diffusivity, D_{dry}, and wetting diffusivity, D_{wet}, are applied.

Equation 13 represents the drying liquid diffusivity (Mabirizi 2000); hence:

$$D(S_e) = D_{dry}(S_e) \qquad (14)$$

Consequently, the wetting diffusivity also needs to be established. There exist very few measurements of both drying and wetting diffusivities where the latter can be determined either by Mitchell's test method (Mabirizi and Bulut 2009) or by the relation between the drying and wetting diffusivities in which either of the unknown diffusivities can be estimated. In this study, a relation factor, β is applied:

$$D_{wet} = \frac{D_{dry}}{\beta} \qquad (15)$$

The effects of the disparity between the wetting and drying soil conditions and of the moisture storage function in Equation 11 have been investigated, using the input parameters of van Genuchten (Šim nek et al. 1999). According to the authors, the hysteresis can be taken into account for WRC of unsaturated soils by making the following assumption of the fitting parameter α:

$$\alpha_{wet} = 2 \cdot \alpha_{dry} = \frac{\alpha_{dry}}{\beta} \qquad (16)$$

This assumption is proven to work well for the moisture storage function in Equation 2 (Šim nek et al. 1999).

According to a study of 40 different samples comprising six different soil types, the relation factor, β, may vary between 0.6 and 3.9 (Mabirizi and Bulut 2009), where the overall average from the measurements is 1.75, but only parts of the 12 soil types were investigated in that study.

In this study, D_{dry} is determined with Equation 13 together with the input parameters defined for the 12 soil textures in Table 1. D_{wet} is determined with the relation factor of 1.75 in Equation 15. D_{dry} values for the defined soil texture classes are presented in Appendix B together with D_{wet} for clay and silt. The necessity of distinguishing the liquid diffusivity into D_{dry} and D_{wet} might be questionable due the small disparity of their distributions. Still, the two different diffusivities remain in this study.

The liquid diffusivities are presented together with the moisture content, w, which can be expressed as a function of effective saturation:

$$w(S_e) = (S_e(\theta_s - \theta_r) + \theta_r)\rho_w \quad (17)$$

from combining Equation 1 and Equation 10.

Vapor Diffusion Resistance Factor

The water vapor diffusion resistance factor, μ, that is, the rate of vapor diffusion through a material in comparison with stagnant air, is typically assumed to be 50 for soils (EN-12524 2000), which corresponds to a permeability of about 2.6 perm-in.

Dry Bulk Density, Particle Density and Porosity

The bulk density of soil is the combined weight of the soil solids, water, and air divided by the bulk volume (kg/m^3). The dry bulk density, ρ_b, which is an important material property, is the bulk density for a completely dry soil. The bulk density typically increases with soil depth on account of reduced organic materials and natural compaction (USDA 2008). Compaction is the phenomenon by which soils are compressed, which usually decreases the porosity (Richard et al. 2001) and thus the ability for water and vapor absorption. Other research has shown that the bulk density varies mainly down to 0.3 m (Kim et al.2010) where a 7% increase in bulk density can be expected from 0 to 1 m (Ray et al. 2004). In this study, the effect of compaction will not be taken into account when estimating soil material properties.

In addition, the bulk density depends on the soil type (Keller and Håkansson 2010). Table 2 presents ρ_b for the 12 specified soil textures, based on average values from 560 soil samples (Schaap 2002).

The particle density, ρ_s, in soils is defined as the mass of the solids divided by the volume of the solids. Consequently, without considering the volume of the pores or the mass of liquid and gas inside the pores, the particle density can be used to determine the porosity, n, in relation with the bulk density:

$$n = \left(1 - \frac{\rho_b}{\rho_s}\right) \geq \theta_s \quad (18)$$

Naturally, the volumetric water content at saturation, θ_s, cannot exceed the porosity if the swelling phenomenon is neglected.

The particle density of soils and earth materials is typically assumed to be 2650 kg/m^3 (Blanco-Canqui et al. 2006; Eshel et al. 2004) though variation exists between 2400 and 2900 kg/m^3 depending on the composition of minerals and organic components (Rühlmann et al. 2006). An existing study presents the particle densities from 176 experimental sites (Keller and Håkansson 2010) which include nine of the 12 defined soil textures. In this study these variations are applied. The particle density of the other three soil textures, which is not included, is assumed to equal 2650 kg/m^3. The particle densities and porosities for the 12 soil textures are presented in Table 2.

Thermal Conductivity and Heat Capacity

The thermal conductivity, k, of soils increases with the moisture content (Abu-Hamdeh 2003). Other factors that influence the conductivity, though slightly, are mineral composition, temperature, soil texture, and time (Becker and Fricke 1997). An empirical solution of k has been defined on

Table 2. Dry Bulk Density, Particle Density, and Porosity for the 12 Soil Textures Where the Bulk and Particle Density Are Obtained from Averages of 560 Soil Samples (Keller and Håkansson 2010; Schaap 2002)

Soil Texture	Dry Bulk Density, ρ_b, kg/m^3	Particle Density, ρ_s, kg/m^3	Porosity, n, %
Clay	1270	2620	52
Clay Loam	1360	2600	48
Loam	1290	2600	50
Loamy Sand	1510	2640	43
Sand	1580	2650	40
Sandy Clay	1400	2650	47
Sandy Clay Loam	1520	2620	42
Sandy Loam	1550	2580	40
Silt	1390	2650	48 (49)[a]
Silty Clay	1400	2610	47 (48)[a]
Silty Clay Loam	1280	2590	50
Silt Loam	1390	2600	45

a. The two values in parentheses are adjusted values used in this study because the volumetric water content at saturation, θ_s, cannot exceed the porosity, see Equation 18.

the basis of the available data collected from the literature as a function of effective saturation, S_e, of three different soil textures: sand, silt, and clay (Becker and Fricke 1997):

$$S_e(k) = \lambda_1[\sinh(\lambda_2 \cdot k + \lambda_3) - \sinh(\lambda_4)] \quad (19)$$

Hence, k can be defined as

$$k(S_e) = \frac{-\left(\lambda_3 - \alpha \sin h\left[\sin h(\lambda_4) + \frac{S_e}{\lambda_1}\right]\right)}{\lambda_2} \quad (20)$$

where

S_e = effective saturation, –
k = soil thermal conductivity, W/m·K
$\lambda_{1\text{-}4}$ = coefficients varying with soil type, –

The results of calculating the soil thermal conductivity by Equation 20 for clay, sand, and silt are presented in Appendix C, though as a function of water content, w. The water content is expressed as a function of effective saturation in Equation 17; hence k(w) becomes:

$$k(w) = \frac{-\left(\lambda_3 - \alpha \sin h\left(\sin h(\lambda_4) + \frac{(w \cdot \rho_w^{-1} - \theta_r)}{(\theta_s - \theta_r)\lambda_1}\right)\right)}{\lambda_2} \quad (21)$$

The soil thermal conductivities for clay, sand, and silt, both frozen and unfrozen, are presented in Appendix C in relation to soil water content.

Of the 12 soil textures defined by the USCA, sand, silt, and clay are the three major ones, with the other nine textures defined from them. No coefficients are presented for these nine textures (Becker and Fricke 1997) since, in this study, an interpolation has been made based on the distributions of the thermal conductivity of clay, sand, and silt. The interpolation is also based on the assumption that these three major textures are not pure clay, sand, and silt; instead, they are assumed to have their average composition given by Figure 1. The reason behind this approach is that the coefficients for the thermal conductivity presented in Table 3 are based on multiple samples within the range of clay, sand, or silt. Thus the coefficients of clay, sand, and silt represent average values. The assumed compositions of the three provided soil textures are presented in Table 4. The soil compositions are the basis for obtaining the thermal conductivities for intermediate soil textures with a system of linear equations. This method, together with results, is presented in Appendix D.

The resulting soil thermal conductivities for sand, loam, silt, and clay are presented in Figure 3. The distribution of loam rely on their defined portion of clay, sand, and silt, which then are implemented into the result of a system of linear equations described in Appendix C.

The specific heat capacity of dry soils, c_{dry}, can vary from 710 to 1550 J/kgK (Olchev et al. 2009), though most commonly c_{dry} is assumed to be about 850 J/kgK (Acs et al.1990; Kung and Steenhuis 1986) where the actual volumetric heat capacity of the soil, c, increases linearly with the soil moisture content:

$$c = \rho_b \cdot c_{dry} + w \cdot c_w \quad (22)$$

where ρ_b is the dry bulk density (kg/m^3), w is the water content, and c_w is the specific heat capacity of water (J/kgK).

Table 3. Coefficients for Clay, Sand, and Silt (Frozen and Unfrozen) Required to Estimate the Soil Thermal Conductivity by Equation 20 (Becker and Fricke 1997)

Soil Type	λ	λ_1	λ_2	λ_3	λ_4
Clay	Frozen	14.5	1.7	−2.5	−2.0
	Unfrozen	27.0	1.8	−1.5	−1.0
Sand	Frozen	10.0	1.7	−2.2	−1.6
	Unfrozen	6.8	2.8	−2.9	−1.5
Silt	Frozen	19.5	1.9	−1.8	−1.5
	Unfrozen	17.0	2.8	−2.6	−1.6

Table 4. Assumed Texture Compositions of Clay, Sand, and Silt. These Compositions are the Basis for Estimating the Unknown Conductivities for the Nine Soil Textures that Are Not Defined in Table 3

Average Composition	Clay	Sand	Silt
Portion of Clay	60%	5%	8%
Portion of Sand	20%	90%	8%
Portion of Silt	20%	5%	84%

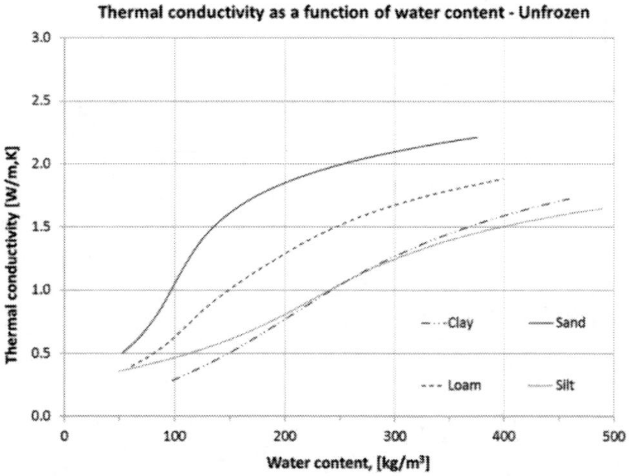

Figure 3 Thermal conductivity for sand, loam, silt, and clay as a function of the water content of the soils. The distribution of loam are based on determining their compositions and on the provided distributions of clay, sand, and silt (see Appendix C).

VALIDATION WITH THERMAL SOIL MEASUREMENT

Since 1987 the temperatures at different depths of a local soil have been measured at Fraunhofer Institute for Building Physics in Holzkirchen, Germany. Three different temperatures are measured hourly: at the surface and at the depths of 0.5 and 1.0 m. Simultaneously, at the same location, comprehensive weather data have been measured, which enables the thermal behavior of the soil to be validated by a computer simulation. The typical soil in the Holzkirchen area is loamy sand (LFU Bayern 2012), the properties of which are applied in the simulation model discussed previously.

In this study, the period from 2005 to 2009 has been simulated in WUFI 1D (Künzel 1995) for the comparison and thermal validation of the soil properties. The model has been designed to represent the semi-infinite conditions of one-dimensional heat transfer through the ground according to Figure 4.

Typically, the precipitation in a simulation model functions as a boundary condition in which the moisture load initially affects only the grid element closest to the border, e_1. If this element is saturated, it will not have further moisture storage capacity; hence, the surplus of moisture is neglected. This assumption is acceptable for a surface with drainage (e.g., a vertical wall) though not applicable in soils in which most of the precipitation will be absorbed eventually. Instead, in this model, the precipitation is distributed directly into the first four elements, e_1 to e_4, as an impregnated source (see Figure 4). This approach decreases the amount of precipitation neglected during the simulation run and serves as the best practice manageable with the current WUFI model.

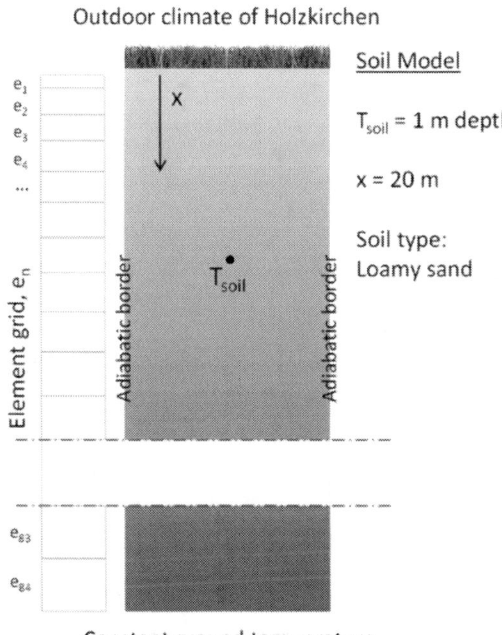

Figure 4 *Simulation model representing one-dimensional and semi-infinite hygrothermal transfer through the ground. The soil material in the simulation model is loamy sand from which the temperature variation at the depth of 1 m is compared with measurements. The temperature is assumed constant at a depth of 20 m, and the border moisture is assumed to be adiabatic.*

The temperature variations of the ground vary with the surface temperature, with the amplitude of the average temperature decreasing with ground depth. At the depth of 10 m, the annual temperature variation of a typical soil is about 4% of the temperature variation at the surface (Hagentoft 2001). At the depth of 20 m, the corresponding variation is 0.2%, which is the depth used in the simulation model of this study. The average ground temperature at Holzkirchen is estimated to 9.5°C (Heidreich 2006) which is the assumed constant temperature at the ground boundary (i.e., at the depth of 20 m). This approach simplifies the semi-infinite temperature variations of the ground for a model defined by boundaries.

The moisture transfer of the ground behaves accordingly. In this study, the starting moisture content of the soil is based on 10 years of pre-simulation where the moisture content of the lower boundary remains constant. This assumption does not enable moisture to pass the lower boundary; hence the moisture content of the simulation model, at different depths, varies solely on the basis of the variation of the moisture content at the upper surface of the ground. Future enhancements to the WUFI software will attend to this deficiency.

The results of the comparison between simulated and measured ground temperature in soil at Holzkirchen are presented in Figure 5. The slightly lower simulated ground temperature between July and October could be a result of a too high surface heat transfer coefficient. In the simulation model, the wind-driven heat transfer coefficient is assumed to represent the clean horizontal surface of a flat roof, though in reality there are other factors that should be taken into account. First, the surface of the ground is not clean and smooth. The texture of the grass will affect the surface wind velocity, thus the convection-induced heat transfer. Further, the measurement of the ground temperature is taken in the vicinity of other objects, such as buildings, which naturally also will have an effect. A brief analysis of the effects of these factors on the heat transfer coefficient was performed, revealing a better fit between the measured and the simulated temperature variations.

Additionally, neither snow coverage nor its effect on the surface thermal resistance and long- and short-wave radiation is taken into account in the simulation.

DISCUSSION AND CONCLUSION

The soil properties of interest in building science have been defined for 12 different soil textures. These properties can serve as input parameters when performing hygrothermal calculations of building constructions coupled to soil

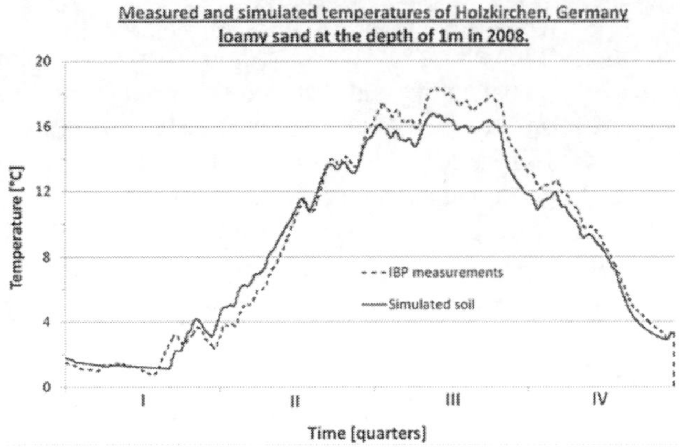

Figure 5 *Comparison between measured and simulated ground temperature at the depth of 1 m in Holzkirchen, Germany. The presented variations in temperature occurred during the year 2008, and the soil texture was loamy sand.*

materials. The reliability of the soil parameters will be further evaluated with measurements in ongoing hygrothermal measurements.

The moisture storage capacity, *w*, ranges from 49 kg/m^3 in loamy sand and 138 kg/m^3 in sandy clay to 375 kg/m^3 in sand and 489 kg/m^3 in silt. In sand, the moisture storage capacity increases insignificantly for relative humidity below 99.8%, which makes sand nonhygroscopic. Loamy sand shows similar behavior as sand with regard to the moisture storage function. Of the 12 defined soil textures, the moisture storage capacity increases most rapidly for clay.

The liquid diffusivity must be distinguished into two types, wetting and drying diffusivities, of which the latter is usually a little smaller. The drying diffusivity is commonly measured and expressed in existing literature; hence a relation factor for the wetting diffusivity has been established and applied in this study.

The thermal conductivities for clay, sand, and silt have been defined in existing studies. The other nine defined soil textures have been estimated by combining the thermal conductivities of these three soil components.

The comparison of measured and simulated ground temperature at the depth of 1 m shows rather good agreement, though improvements to the simulation model are required. Currently, WUFI does not allow full absorption of precipitation at the surface border if the outer element is moisture saturated. Further, the moisture transfer at the lower boundary (i.e., the border representing the infinite ground) should be improved to better simulate the existing moisture transfer mechanisms at this border. Consequently, changes in the existing software code WUFI must be implemented to consider the effects discussed regarding liquid uptake of precipitation and liquid flow through and out of the simulated system.

Field measurements for soil temperature and moisture content for two specified soils are carried out right now under know boundary conditions. These data will allow a full comparison between measurement and simulation. This will tell us whether the simplification in the material properties and in the calculation model is acceptable.

The heat transfer coefficient at the soil surface varies with a number of factors that are not yet included in the simulation model. Once these improvements are implemented, the simulations results are more likely to better fit the measurements. An example of a factor not yet included is the effect on the wind-driven heat transfer coefficient resulting from the texture of the grass and the influence of surrounding objects. In addition, snow coverage of the ground, along with its effect on the surface thermal resistance and long- and short-wave radiation, has not been taken into account yet.

ACKNOWLEDGMENTS

This study was sponsored by the U.S. Department of Energy Building Technologies Program under contract DE-AC05-00OR22725.

APPENDIX A—MOISTURE STORAGE FUNCTIONS BASED ON EQUATION 11

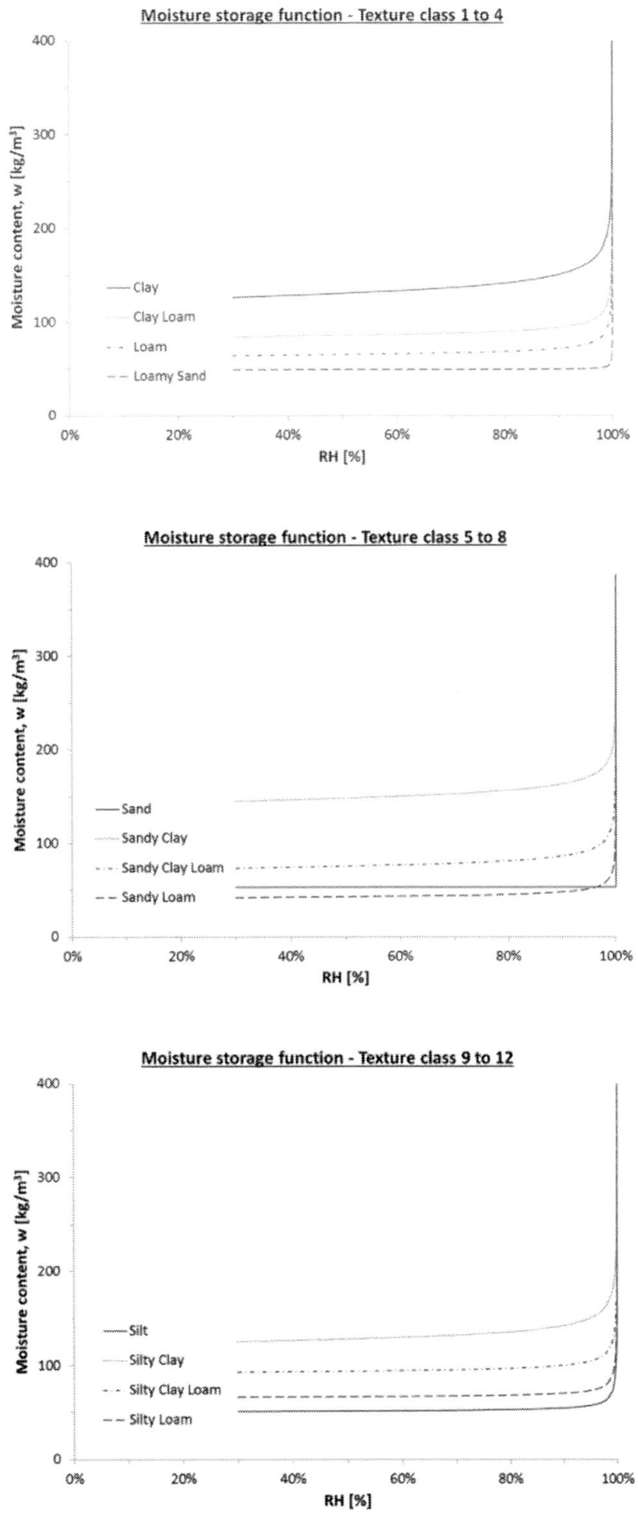

Figure A.1 *Moisture storage function of all texture classes.*

APPENDIX B—D_{dry} VALUES FOR ALL SOIL TEXTURE CLASSES AND D_{wet} VALUES FOR CLAY AND SILT

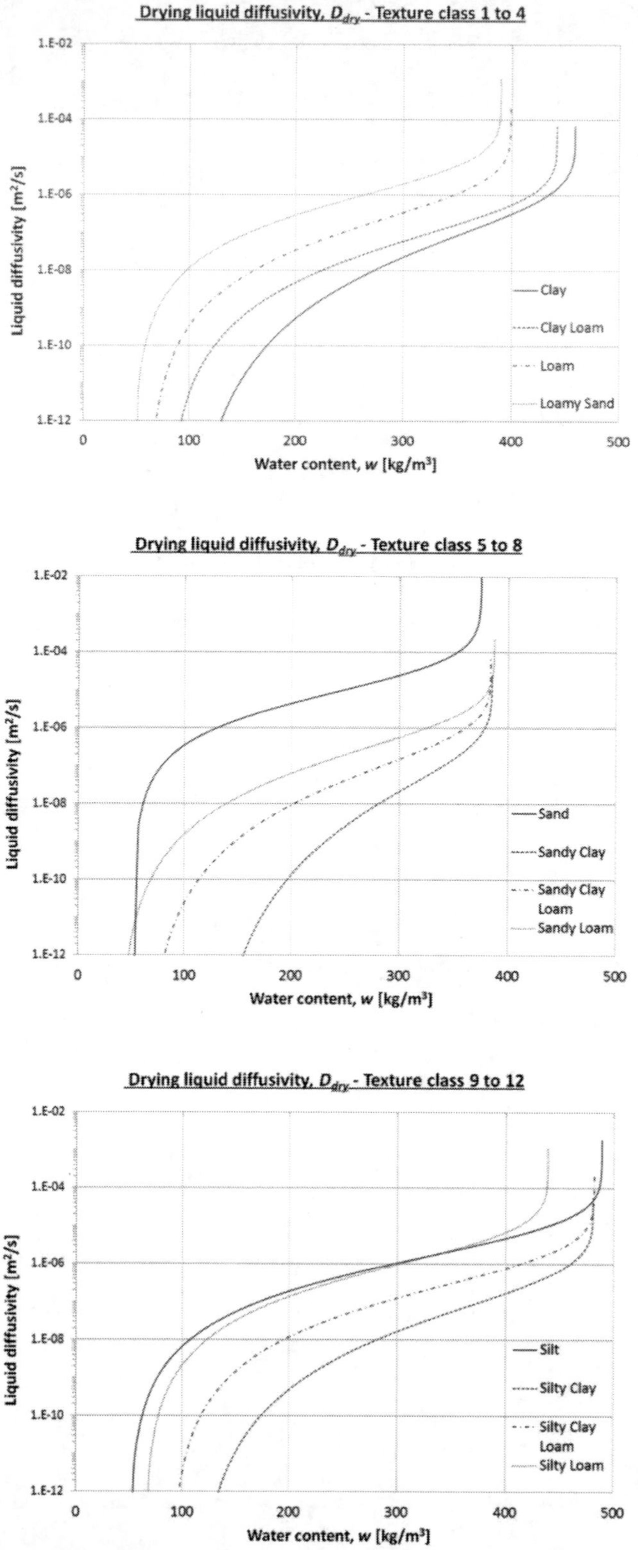

Figure B.1 *Dry values for all texture classes.*

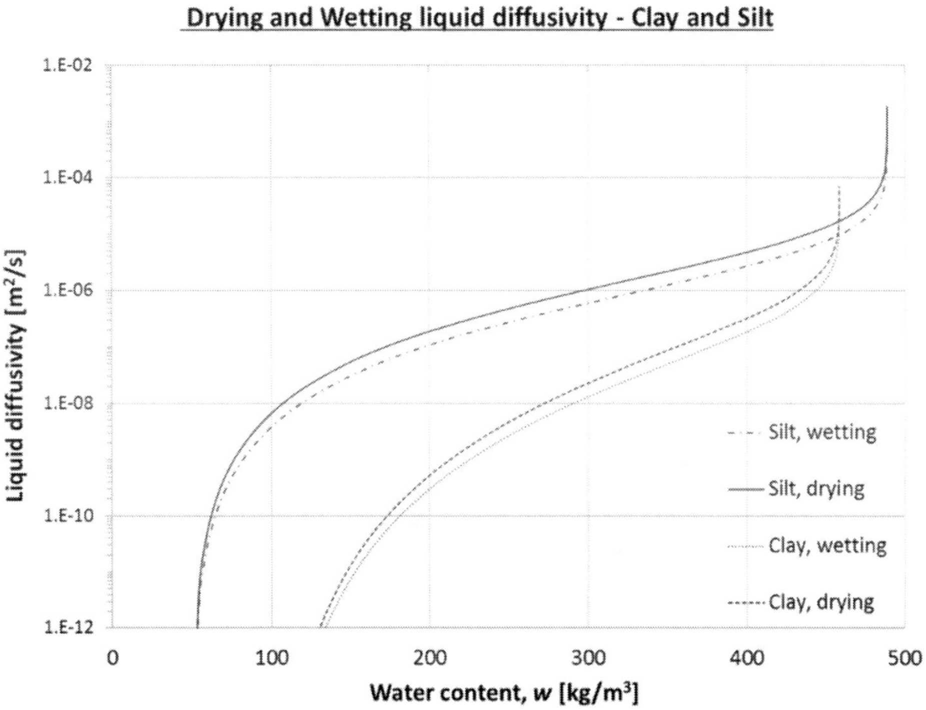

Figure B.2 *Comparison of* D_{dry} *and* D_{wet} *for silt and clay.*

APPENDIX C—SOIL THERMAL CONDUCTIVITIES AS A FUNCTION OF WATER CONTENT FROM EQUATION 20 FOR CLAY, SAND, AND SILT

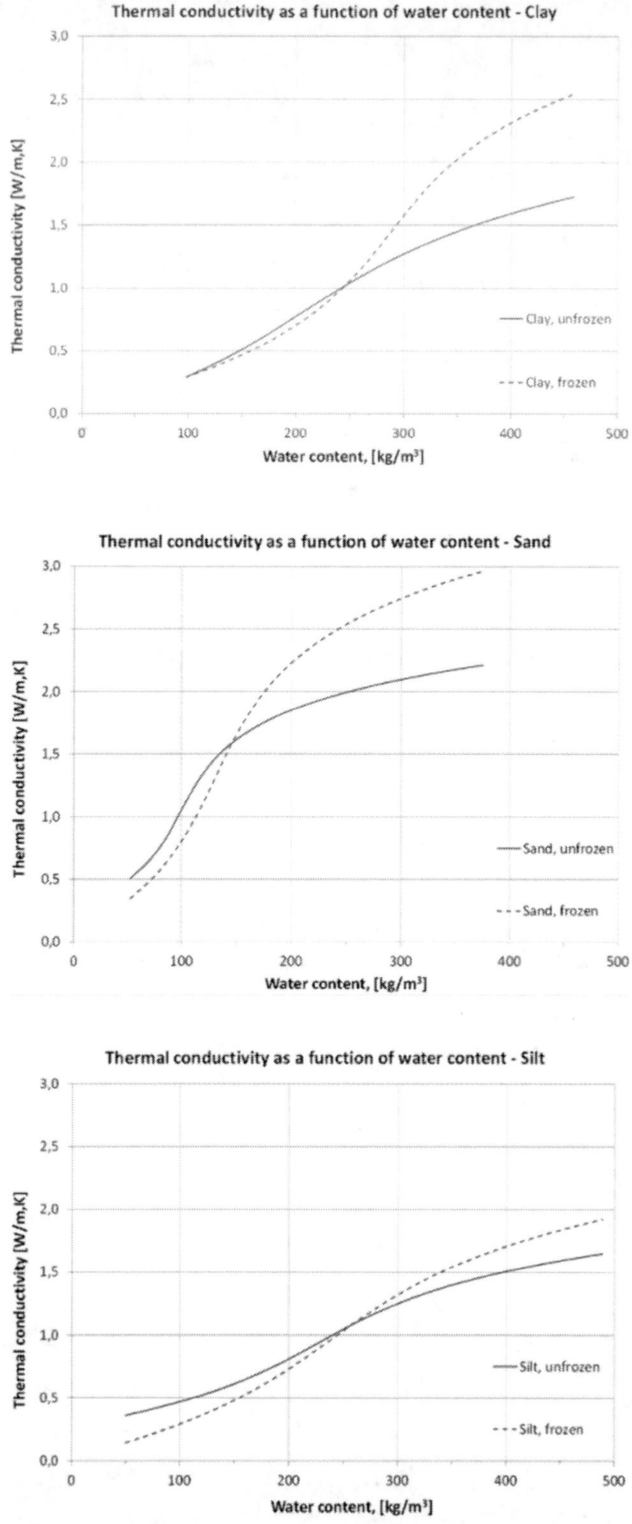

***Figure C.1** Thermal conductivity as a function of moisture content for all texture classes.*

APPENDIX D—COMPOSITIONS OF THE 12 USCS SOIL TEXTURES AND THEIR THERMAL CONDUCTIVITIES

Estimated compositions of the 12 soil textures defined by the Unified Soil Classification System (USDA 2008) are listed in Table 5.

The soil thermal conductivities as a function of water content, $k(w)$, have been estimated for clay, sand, and silt. The thermal conductivities of the other nine soil textures are estimated by use of a system of linear equations described here.

If clay, sand, and silt are considered to be the components of the 12 soil textures, then let

clay = X

sand = Y

silt = Z

The three principal soil textures clay, sand, and silt can described as proportions of the components X, Y, and Z; hence, the soil compositions (according to the upper part of Table 5) are as follows:

$$\text{Clay} = P_{clay}\cdot X + P_{sand}\cdot Y + P_{silt}\cdot Z = 60X + 20Y + 20Z$$

$$\text{Sand} = P_{clay}\cdot X + P_{sand}\cdot Y + P_{silt}\cdot Z = 5X + 90Y + 5Z$$

$$\text{Silt} = P_{clay}\cdot X + P_{sand}\cdot Y + P_{silt}\cdot Z = 8X + 8Y + 84Z$$

Assume that the three principal soil textures are clay = A, sand = B, and silt = C. Note that variable A is the soil of clay, whereas X is the clay component.

The solution becomes:

$$X = A/60 - Y/3 - Z/3$$

$$Y = B/90 - X/18 - Z/18$$

$$Z = C/84 - 2X/21 - 2Y/21$$

Further, the solution of X, Y, and Z as a function of A, B, and C is

$$X(A, B, C) = 94A/5375 - 19B/5375 - 17C/4300$$

$$Y(A, B, C) = -19A/21500 + 61B/5375 - C/2150$$

$$Z(A, B, C) = -17A/10750 - 4B/5375 + 53C/4300$$

The soil thermal conductivity for any arbitrary soil texture, k_{arb} (W/m·K), becomes

$$k_{arb} = P_{clay}\cdot X + P_{sand}\cdot Y + P_{silt}\cdot Z = P_{clay}\cdot X(A, B, C) + P_{sand}\cdot Y(A, B, C) + P_{silt}\cdot Z(A, B, C)$$

Example:

The thermal conductivity of clay loam, k_{cl}, can be expressed as

$$k_{cl} = P_{clay}\cdot X + P_{sand}\cdot Y + P_{silt}\cdot Z = 35\cdot X(A, B, C) + 30\cdot Y(A, B, C) + 35\cdot Z(A, B, C)$$

where A, B, and C must be calculated for each given step of moisture content, defined in Equation 21.

Table D.1 Assumed Texture Compositions of Clay, Sand, and Silt. These Compositions are the Basis for Estimating the Unknown Conductivities for the Nine Soil Textures that Are Not Defined in Table 3

Average Composition	Portion of Clay %	Portion of Sand %	Portion of Silt %
Clay	60	20	20
Sand	5	90	5
Silt	8	8	84
Estimated Composition	**Portion of Clay %**	**Portion of Sand %**	**Portion of Silt %**
Clay Loam	35	30	35
Loam	18	42	40
Loamy Sand	7	82	11
Sandy Clay	40	52	8
Sandy Clay Loam	27	58	15
Sandy Loam	10	65	25
Silty Clay	47	7	46
Silty Clay Loam	34	10	56
Silt Loam	15	20	65

REFERENCES

Abu-Hamdeh, N.H. (2003). Thermal Properties of Soils as affected by Density and Water Content. *Biosystems Engineering*, 86(1), 97–102. doi: 10.1016/s1537-5110(03)00112-0

Acs, F., Mihailovic, D., and Rajkovic, B. (1990). A coupled soil moisture and surface temperature prediction model. *Journal of Applied Meteorology*, 30(6).

Becker, R.B., and Fricke, B.A. (1997). Effects of Saturation and Dry Density on Soil Thermal Conductivity. Department of Mechanical and Aerospace Engineering, University of Missouri-Kansas City, MO, USA.

Blanco-Canqui, H., Lal, R., Post, W.M., Izaurralde, R.C., and Shipitalo, M.J. (2006). Organic carbon influences on soil particle density and rheological properties. *Soil Science Society of America Journal*, 70(4), 1407–1414.

Duble, R. (2011). Turfgrass Rootzones - Introduction - Soil Texture Triangle Retrieved 2012-06-11

EN-12524. (2000). Wärme- und feuchteschutztechnische Eigenschaften Tabellierte Bemessungswerte Baustoffe und -produkte: Deutsche Norm.

EN-ISO-13370:. (2007). EN ISO 13370:2007 Thermal performance of buildings - Heat transfer via the ground - *Calculation Methods TC 163 Thermal Performance and Energy Use in the Built Environment* - TC 163/SC 2 - Calculation methods: European standard.

Eshel, G., Levy, G.J., Mingelgrin, U., and Singer, M.J. (2004). Critical evaluation of the use of laser diffraction for particle-size distribution analysis. *Soil Science Society of America Journal*, 68(3), 736–743.

Gopalakrishnan, K., and Manik, A. (2007). A Mathematical Model for Predicting Isothermal Soil Moisture Profiles

Using Finite Difference Method. *International Journal of Computational and Mathematical Sciences* (1:2 2007).

Hagentoft, C.E. (2001). *Introduction to Building Physics*. Lund, Sweden: Studentlitteratur.

Heidreich, U. (2006). Nutzung oberflächennaher Geothermie zum Heizen und Kühlen eines Bürogebäudes. Paper presented at the Symposium Energetische Sanierung von Schul- und Verwaltungsgebäuden, FH Münster.

Hens, H. (2007). *Building Physics Heat, Air and Moisture - Fundamentals and Engineering Methods with Examples and Exercises*. Germany: Ernst and Sohn Verlag fur Architektur und technische Wissenschaften GmbH und Co. KG, Berlin.

Hills, R.G., Hudson, D.B., Porro, I., and Wierenga, P.J. (1989). Modeling One-Dimensional Infiltration into Very Dry Soils. 2. Estimation of the Soil-water Parameters and Model Predictions. *Water Resources Research*, 25(6), 1271–1282. doi: 10.1029/WR025i006p01271

Janssen, H., Carmeliet, J., and Hens, H. (2004). The influence of soil moisture transfer on building heat loss via the ground. *Building and Environment*, 39(7), 825–836. doi: 10.1016/j.buildenv.2004.01.004

Keller, T., and Håkansson, I. (2010). Estimation of reference bulk density from soil particle size distribution and soil organic matter content. *Geoderma*, 154(3–4), 398–406. doi: 10.1016/j.geoderma.2009.11.013

Kiessl, K., and Gertis, K. (1980). Feuchtetransport in Baustoffen: Eine Literaturauswertung zur rechnerischen Erfassung hygrischer Transportphänomene: Univerität Essen Gesamthochschule.

Kim, H., Anderson, S.H., Motavalli, P.P., and Gantzer, C.J. (2010). Compaction effects on soil macropore geometry and related parameters for an arable field. *Geoderma*, 160(2), 244–251. doi: 10.1016/j.geoderma.2010.09.030

Kung, S.K.J., and Steenhuis, T.S. (1986). Heat and moisture transfer in a partly frozen nonheaving soil. *Soil Science Society of America Journal*, 50(5), 1114–1122.

Künzel, H.M. (1995). Simultaneous Heat and Moisture Transport in Building Components. - One- and two dimensional calculation using simple parameters. Dissertation, University Stuttgart, IRB Verlag. Retrieved from www.WUFI.com

LFU_Bayern (2012). [Personal communication—Bayerisches Landesamt fuer Umweltschutz; Bodenkundliche Aufnahme].

Mabirizi, D. (2000). Wetting and Drying Unsaturated Soil Diffusivity Measurements in Laboratory. Master, Oklahoma State University.

Mabirizi, D., and Bulut, R. (2009). Unsaturated Soil Moisture Drying and Wetting Diffusion Coefficient Measurements in the Laboratory: School of Civil and Environmental Engineering, Oklahoma State University/Oklahoma Transportation Center.

Neto, D.D., van Lier, Q.D., van Genuchten, M.T., Reichardt, K., Metselaar, K., and Nielsen, D.R. (2011). Alternative Analytical Expressions for the General van Genuchten-Mualem and van Genuchten-Burdine Hydraulic Conductivity Models. [Article]. *Vadose Zone Journal*, 10(2), 618–623. doi: 10.2136/vzj2009.0191

Noorallah, G.J. (1999). Volume 01: *Introduction to Soil Science and Soil Resources*: Salman Productions Inc.

Olchev, A., Radler, K., Sogachev, A., Panferov, O., and Gravenhorst, G. (2009). Application of a three-dimensional model for assessing effects of small clear-cuttings on radiation and soil temperature. *Ecological Modelling*, 220(21), 3046–3056. doi: 10.1016/j.ecolmodel.2009.02.004

Prunty, L. (2009). Soil Water Thermal Liquid Diffusivity. *Soil Science Society of America Journal*, 73(3), 704–706. doi: 10.2136/sssaj2008.0097

Ray, C., Vogel, T., and Dusek, J. (2004). Modeling depth-variant and domain-specific sorption and biodegradation in dual-permeability media. *Journal of Contaminant Hydrology*, 70(1–2), 63–87. doi: 10.1016/j.jconhyd.2003.08.009

Richard, G., Cousin, I., Sillon, J.F., Bruand, A., and Guérif, J. (2001). Effect of compaction on the porosity of a silty soil: influence on unsaturated hydraulic properties. Effet du compactage sur la porosité d'un sol limoneux: conséquences sur les propriétés hydrauliques en non saturé. European Journal of Soil Science, 52(1), 49–58. doi: 10.1046/j.1365-2389.2001.00357.x

Rühlmann, J., Körschens, M., and Graefe, J. (2006). A new approach to calculate the particle density of soils considering properties of the soil organic matter and the mineral matrix. *Geoderma*, 130(3–4), 272–283. doi: 10.1016/j.geoderma.2005.01.024

Schaap, M.G. (2002). Rosetta: A computer program for estimating soil hydraulic parameters with hierarchical pedotransfer functions (Version 1.2). http://cals.arizona.edu/research/rosetta/rosetta.html.

Schaap, M.G., Leij, F.J., and van Genuchten, M.T. (2001). Rosetta: a computer program for estimating soil hydraulic parameters with hierarchical pedotransfer functions. *Journal of Hydrology*, 251(3–4), 163–176. doi: 10.1016/s0022-1694(01)00466-8

Šim nek, J., Kodešová, R., Gribb, M.M., and van Genuchten, M.T. (1999). Estimating hysteresis in the soil water retention function from cone permeameter experiments. *Water Resour. Res.*, 35(5), 1329–1345. doi: 10.1029/1998wr900110

Staszczuk, A., Radon, J., and Holm, A. (2010). Evaluation of simplified calculation method of heat exchange between building and ground.

USDA, N.R.C.S. (2008). Soil quality indicators—Bulk density (pp. 2). USDA, Natural Resources Conservation Service.

van Genuchten, M.T. (1980). A Closed-form Equation for Predicting the Hydraulic Conductivity of Unsaturated Soils. *Soil Science Society of America Journal*, 44(5), 7.

Zhou, J., and Yu, J. L. (2005). Influences affecting the soil-water characteristic curve. *Journal of Zhejiang University*, 6(8), 8.

Effect of Natural Convection on the Thermal Transmittance of Exterior Walls with Air-Permeable Insulation

Petteri Huttunen

Juha Vinha, PhD
Member ASHRAE

ABSTRACT

In the past few years, the Finnish building code's demands for energy efficiency has increased the thickness of insulation layers in building envelopes. Natural convection is considered in the building code by not allowing structures whose modified Rayleigh's numbers exceed the limit values. However, it has been unclear how considerably the natural convection can affect the heat loss through envelope in instantaneous or annual level. The aim of this study was to inspect the actual effect of natural convection in relatively thick, air-permeable vertical insulation layers by laboratory measurements and computational modeling.

Several U-factor measurements by calibrated hot box method were carried out to wood frame test walls with mineral wool or cellulose insulation. Attributes presumably affecting the convection process were varied. Poor workmanship's effect was studied with intentionally sloppy installation work and with intentional horizontal air gaps in the insulation layer. Careful installation with the convection barrier was used as a reference where the effect of convection was supposed to be negligible. Measured U-factors were compared to numerically calculated U-factors and to simplified U-factor calculations. Permeabilities and thermal conductivities of the used materials were measured for the calculations' initial data.

Measured and calculated results were in relatively good agreement. Results showed that during cold periods natural convection can slightly increase the actual U-factors of exterior walls. The convection barrier can be used effectively to restrict natural convection as demanded by the building code. However, the quality of workmanship should not be forgotten when pursuing a very low U-factor by increasing insulation thickness.

INTRODUCTION

Heat losses through the building envelope play a significant part in the overall energy consumption of small buildings in countries with cold climates. Such countries have typically developed long traditions in the use of thick insulation layers. In addition, during recent decades the growing demands in energy efficiency have led to efforts to save energy by using over 400 mm thick insulation layers in roofs and exterior walls. Also, Finnish building code's instructions for thermal insulation were revised in 2010, reducing the reference U-factor for exterior walls from 0.042 to 0.030 Btu/(h·ft^2·°F) (0.24 to 0.17 W/[m^2·K]).

A Finnish wood-frame house is typically insulated with mineral wool batts, but the cellulose insulation has also gained popularity due to its low price and also because it can be installed into vertical structures by wet-spraying. Manufacturers have managed to develop the thermal conductivities of both materials to competitive levels. However, both of these types of materials remain air-permeable, which has raised questions about the role of convection in the total heat transfer through envelope structures with even thicker insulation layers.

Currently pending building code revision will restrict the modified Rayleigh number (Ra_m) of prescribed exterior walls by not allowing wall structures with Ra_m exceeding the value of 2.5 when the temperature difference between indoors and outdoors is 90°F (50 K). Conformity of structures shall be

***Petteri Huttunen** is a research assistant and **Juha Vinda** is a professor of building physics in the Department of Civil Engineering, Tampere University of Technology, Tampere, Finland.*

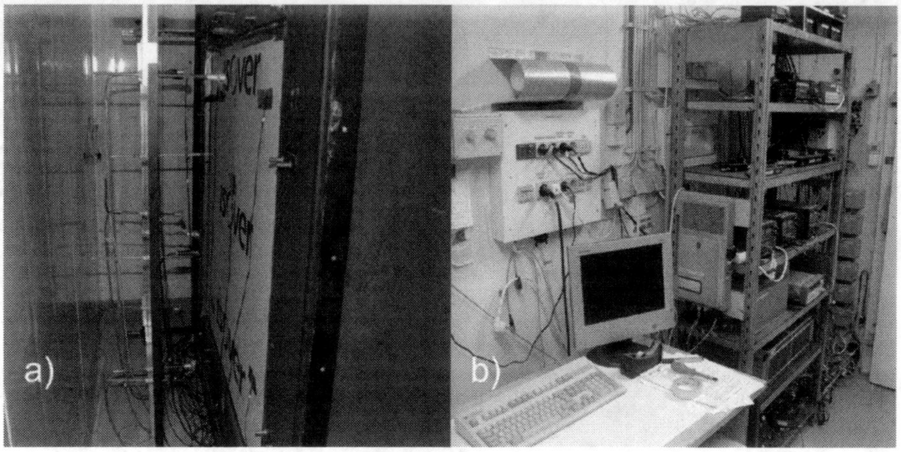

Figure 1 *(a) The hot box of the research equipment and (b) the monitoring room.*

proved by calculating Ra_m in accordance with international standard EN ISO 10456 (ISO 2007):

$$Ra_m = 3 \times 10^6 \frac{d \cdot k \cdot \Delta T}{\lambda} \quad (1)$$

In theory, for vertical structures heated from the side there is no critical value of Ra_m determining the onset of natural convection, but the convection may start as soon as the temperature difference is introduced. The purpose of this research was to investigate the effect of natural convection on the U-factor of typical wood frame wall structures in average winter outdoor temperatures and to evaluate the adequacy of the limit value of Ra_m. A goal was also to clarify the actual influence of poor workmanship, air gaps and convection barrier (CB) on the thermal resistance.

RESEARCH ARRANGEMENT

Evaluations of natural convection's magnitude were based on comparisons of U-factor measurements and software simulations. Convection's proportion of the total heat transfer through a structure can be described with Nusselt number (Nu) which is here defined as:

$$Nu = \frac{\phi_{cd} + \phi_{conv}}{\phi_{cd}} \quad (2)$$

Experimental Nusselt numbers were determined by assuming that the effect of natural convection is negligible when examining the U-factor of a test wall with convection barrier. Thus, the increment in total heat flux due to removal of convection barrier is assumed to be entirely due to natural convection and the Nusselt number can be calculated from two measurements:

$$Nu = \frac{U_{noCB}}{U_{CB}} \quad (3)$$

The laboratory equipment used in this research is called Building Physical Research Equipment and it is located in the building laboratory of Tampere University of Technology. The operational principle of the U-factor determination is calibrated hot box (CHB) method and the equipment can be used to carry out the standard U-factor test in accordance to international standard EN ISO 8990 (Vinha 1998). In addition, the equipment can be used to conduct building physically more comprehensive experiments where indoor and outdoor humidity conditions together with pressure difference are adjusted. However, only thermal adjustments and measurements necessary in U-factor determination were utilized in this research.

Test Walls

Due to limitations of dimensions of the equipment, the height of the test walls was 46.65 in. (1185 mm), which is less than half of the height of typical actual walls and the height-thickness ratio is known to be a major determinant of the natural convection process in finite vertical cavity. However, the test wall dimensions in this research represent critical circumstances which occur in actual building envelopes, for example, under or above a window. Test walls were insulated with either mineral wool batts or wet-sprayed cellulose insulation. Two insulation thicknesses were investigated: 7.67 in. and 12.60 in. (200 mm and 320 mm). Current reference U-factor 0.030 Btu/(h·ft^2·°F) [0.17 W/(m^2·K)] can be achieved with approximately 7.87 in. (200 mm) thick mineral wool or cellulose insulation layer when the structure has also a heat insulating wind shield. Modern low-energy houses may have over 15.75 in. (400 mm) thick insulation layers but the dimensions of the equipment ruled out test walls with overall thickness over 15 in. (380 mm). Therefore, the maximum insulation thickness of a test wall with wind shield was 12.60 in. (320 mm). Dimensioned drawing of the test walls is shown in Figure 2. Convection barriers were made of vapor permeable plastic sheets which were tightened and fixed with staples and duct tape as shown in Figure 4a. Temperature difference between indoors and outdoors was 63°F (35°C) (T_{in} = 68°F (20°C), T_{out} = 5°F [15°C]) in all measurements. To prevent the effect of small deviations in the actual surface resistances

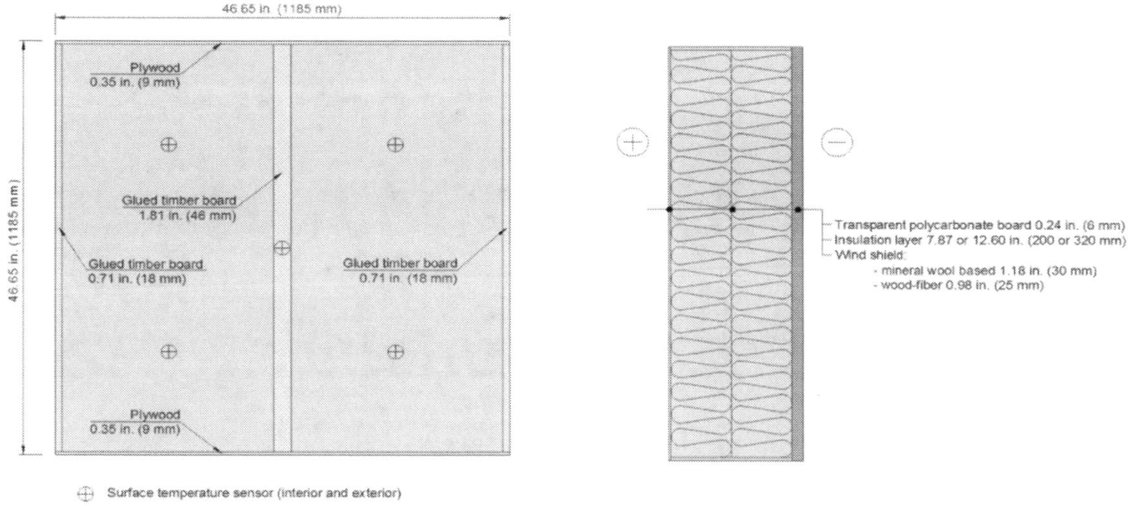

Figure 2 *Dimensioned drawing of the test walls. Surface temperature sensor positions are also illustrated.*

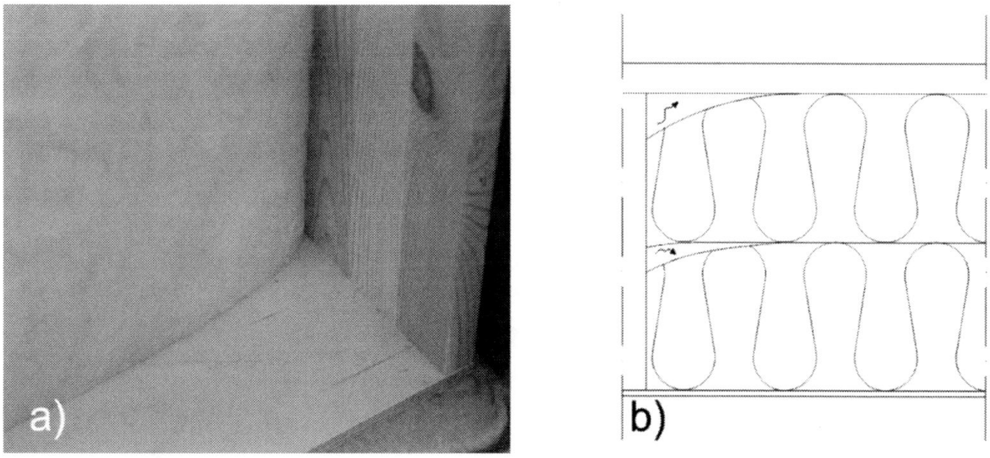

Figure 3 *(a) Rounded wool batt in a test wall and (b) an illustration of air gaps formed due to sloppy installation.*

between separate experiments, the U-factors used in comparative calculations were calculated from surface-to-surface thermal resistances measured in experiments and surface resistances according to standard EN ISO 6946 (2007).

Poor workmanship was introduced in part of the test walls by intentionally sloppy installation of the mineral wool batts. As a result of sloppy installation, the edges of batts may become rounded forming vertical air gaps as shown in Figure 3. Because these types of air gaps are normally hard to see after the installation, the inner surface of the test walls was made of transparent polycarbonate board. As reference for the sloppy installation, ideal test walls were built by extremely carefully installing the batts from the outside against the polycarbonate board and getting visual confirmation of the installations' quality and the absence of air gaps.

Negligent installation may also cause horizontal air gaps at upper or lower edges of wall insulation. In case of wet-sprayed cellulose insulation these may form due to creep of the material. Unexpectedly extensive creep turned up in the cellulose insulated test walls during the test series and its effect on test results is discussed later in this paper.

In total, 14 U-factor measurements were carried out and the attributes of test walls are listed in Table 1.

In Table 1 the insulation material is denoted in the first column (MW = mineral wool, CI = cellulose insulation) and the values of Ra_m according to the testing conditions in the third column.

Computational Studies

U-factors of part of the test walls were also determined with a commercial FEM-software COMSOL Multiphysics version 4.2a. The module used in the software uses Brinkman's equation, which can be used to model the fluid flow in porous media (Nield and Bejan 2006). With very low airflow velocities, the Forcheimer's drag can be neglected and in

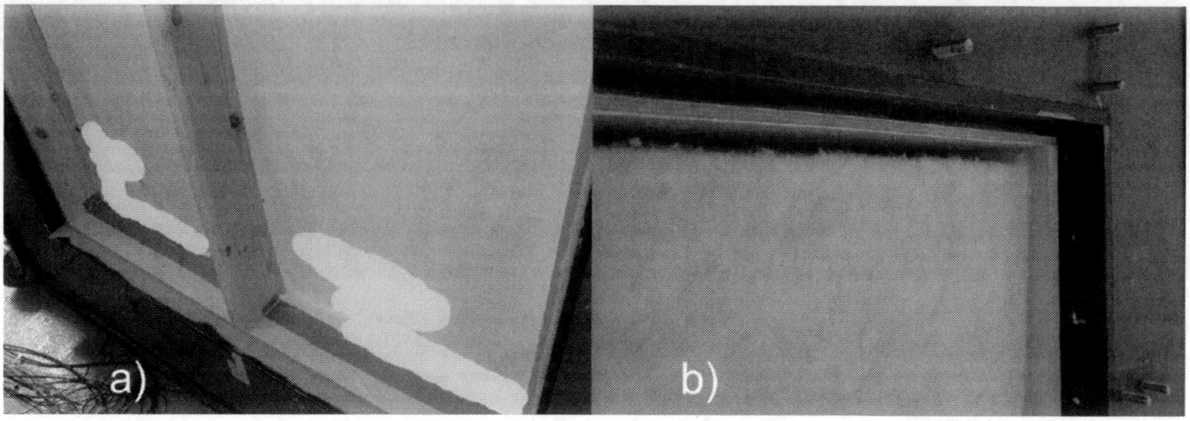

Figure 4 *(a) Convection barrier showing in a test wall from which the wind shield and half of the insulation is removed and (b) horizontal air-gap in a test wall from which the wind shield is removed.*

Table 1. Test Series of U-Factor Measurements

Test Wall	Insulation Thickness, in. (mm)	Ra_m	Installation	CB	Horizontal Air Gap
MW01	7.87 (200)	1.80	ideal	no	no
MW02	7.87 (200)	—	ideal	yes	no
MW03	7.87 (200)	1.80	sloppy	no	no
MW04	7.87 (200)	1.80	sloppy	no	yes
MW05	7.87 (200)	—	sloppy	yes	yes
MW06	12.60 (320)	3.08	ideal	no	no
MW07	12.60 (320)	—	ideal	yes	no
MW08	12.60 (320)	3.08	sloppy	no	no
MW09	12.60 (320)	3.08	sloppy	no	yes
MW10	12.60 (320)	—	sloppy	yes	yes
CI01	7.87 (200)	1.83	—	—	no
CI02	7.87 (200)	1.83	—	—	yes
CI03	12.60 (320)	3.13	—	—	no
CI04	12.60 (320)	3.13	—	—	yes

stationary state the governing equations of incompressible flow in porous media and heat transfer are reduced to:

Brinkman's equation:

$$\frac{\rho}{\varepsilon_p}\left[(\boldsymbol{u}\cdot\nabla)\frac{\boldsymbol{u}}{\varepsilon_p}\right] = \nabla\cdot\left[-p\boldsymbol{I}+\frac{\eta}{\varepsilon_p}\left(\nabla\boldsymbol{u}+(\nabla\boldsymbol{u})^T\right)-\frac{2\eta}{3\varepsilon_p}(\nabla\cdot\boldsymbol{u})\boldsymbol{I}\right] -\frac{\eta}{k}\boldsymbol{u}+\boldsymbol{F} \quad (4)$$

Continuity:

$$\nabla\cdot\boldsymbol{u} = 0 \quad (5)$$

Table 2. Measured Material Properties

Material	Thermal Conductivity (λ), Btu/(h·ft·°F) (W/[m·K])	Permeability (*k*), 10^{-9} ft^2 (10^{-9} m^2)
Mineral wool	0.020 (0.035)	39 (3.6)
Cellulose insulation	0.024 (0.042)	46 (4.3)
Glued-timber board	0.058 (0.10)	—
Plywood	0.14 (0.25)	—
Mineral-wool wind shield	0.029 (0.050)	—
Wood-fiber wind shield	0.017 (0.029)	—

Heat transfer by conduction and convection:

$$\rho C_p\boldsymbol{u}\cdot\nabla T = \nabla\cdot(\lambda\nabla T) \quad (6)$$

The density of air was expressed in the governing equations as a function of pressure and temperature. From the ideal gas law can be derived:

$$\rho = \frac{p}{R_{air}T} \quad (7)$$

Coupled simulations were carried out in *2D* and the effect of linear thermal transmittances due to vertical wood frame parts in the test walls were first determined with the same software and added afterwards to the U-factors based on *2D* simulations where the airflows were taken into account. Material properties used in the calculations were measured in accordance with the standards ISO 8301 (1991) (thermal conductivity) and ISO 9053 (1991) (air permeability). Measured values are shown in Table 2.

Thermal conductivity of polycarbonate used in the calculations was not measured. In calculations, a tabulated design

Table 3. U-Factor Results, Btu/(h·ft^2·°F) [W/(m^2·K)]

Test Wall	Measured	COMSOL	EN ISO 6946
MW01	0.0294 (0.167)	0.0294 (0.167)	0.0285 (0.162)
MW02	0.0291 (0.165)	0.0291 (0.165)	0.0285 (0.162)
MW03	0.0305 (0.173)	—	—
MW04	0.0308 (0.175)	—	—
MW05	0.0313 (0.178)	—	—
MW06	0.0229 (0.130)	0.0211 (0.120)	0.0195 (0.111)
MW07	0.0220 (0.125)	0.0199 (0.113)	0.0195 (0.111)
MW08	0.0224 (0.127)	—	—
MW09	0.0247 (0.140)	—	—
MW10	0.0247 (0.140)	—	—
CI01	0.0410 (0.233)	0.0370 (0.210)	0.0361 (0.205)
CI02	0.0454 (0.258)	—	—
CI03	0.0321 (0.182)	0.0259 (0.147)	0.0240 (0.136)
CI04	0.0326 (0.185)	—	—

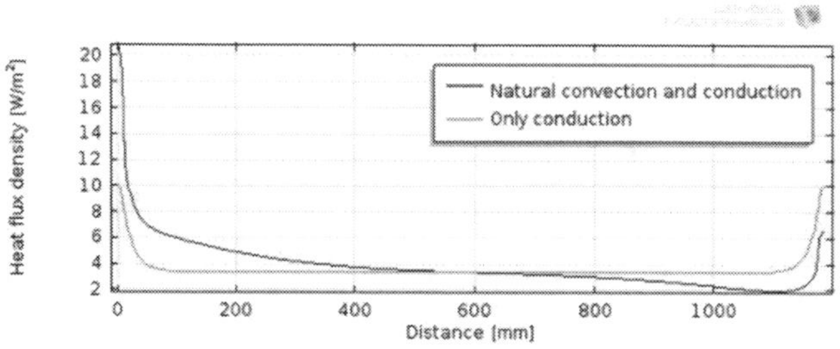

Figure 6 *Heat flux density as a function of distance from lower edge in test wall MW06 (interior surface).*

value (0.12 [Btu/(h·ft·°F)] 0.2 W/[m·K]) from the standard EN ISO 10456 (2007) was used. Permeability of mineral wool was measured in the direction parallel to the batts.

RESULTS

Measured, simulated, and standard calculated U-factors of the test walls are shown in Table 3.

Examples of visualized result data of simulations are shown in Figures 5 and 6.

Based on measured U-factors and software simulations Nusselt numbers were calculated for test walls with mineral wool (ideal installation, no air gaps). Because convection barrier was used only in mineral wool insulated test walls, only simulation based Nusselt numbers were calculated for cellulose insulated test walls. Results are shown in Table 4.

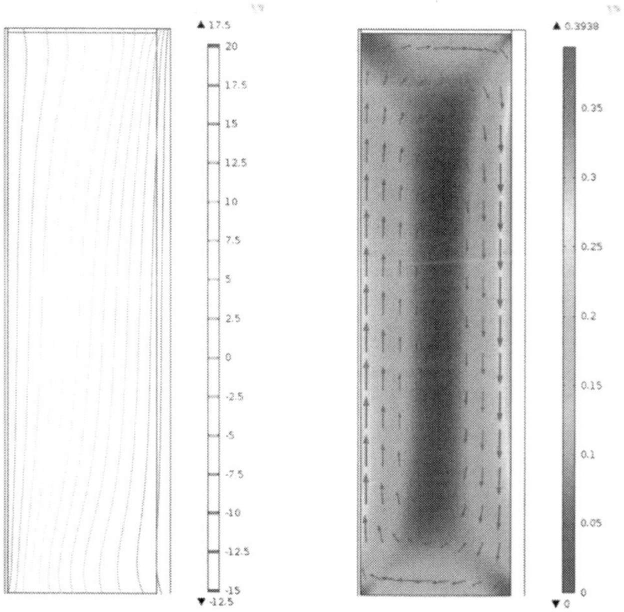

Figure 5 *Isotherms and airflow velocity field in test wall MW06.*

Table 4. Nusselt Numbers Based on Measurements and Simulations

Insulation Material and Thickness, in. (mm)	Measured	COMSOL
MW 7.87 (200)	1.01	1.01
MW 12.60 (320)	1.04	1.06
CI 7.87 (200)	—	1.02
CI 12.60 (320)	—	1.07

DISCUSSION

Based on U-factor measurements, it can be concluded that natural convection had relatively easily measurable effect on heat transfer in exterior walls when the temperature difference was only 63°F (35°C), yet the differences between all measured U-factors were fairly small. The total uncertainty of any U-factor measured with the equipment which was used in this research is ±4% with a safety factor of 2 (Vinha 1998). The majority of this uncertainty is caused by the inaccuracies involved in the calibration procedure of the equipment. Because the absolute value of error due to calibration inaccuracy is exactly the same in two sequentially determined U-factors, the differences between the U-factors of test walls with same thickness can be determined very accurately. In practice, the total uncertainty is almost impossible to determine accurately because of the problem of border effect involved in CHB-method. However, when studying the effect of natural convection by laboratory measurements, the border effect is hard to eliminate by using separate guarding box and measuring box because the natural convection alters

Figure 7 *The effect of creep in cellulose insulated test wall. Air gap is refilled in the left side.*

heat flux densities for the most part at the border regions of the test wall as shown in Figure 6.

Measurements and simulations of test walls with mineral wool insulation were in relatively good agreement. Differences between measured and numerically determined U-factors of the test walls with 12.60 in. (320 mm) thick insulation was assumed to be mainly caused by the border effect. From Table 3 can be seen that poor workmanship may in different cases increase U-factor 5% to 10% and use of convection barrier did not positively affect the test walls with horizontal air gaps.

Unexpectedly large U-factors were measured from all the cellulose insulated test walls. Although the cellulose insulation has somewhat greater permeability, the simulated effect of natural convection was of the same order as in the mineral wool test walls. Eventually, the large values could not be entirely explained, but it was clear that the unexpected creep of the insulation material had great effect. The test walls were stored in vertical position during the test series and in Figure 7 it is shown how air gaps were formed due to creep also in the test walls which were originally sprayed full. These air gaps were refilled with the insulation material but without the adhesive that was used in wet-spraying. Wet-spraying of the insulation material was carried out in other facilities and during the transports minor damages were caused to the test walls which may have also affected the results.

Computational studies were at first attempted in 3*D*, but the numerical solvers programmed in the software failed to reach sufficient level of convergence. Also, the ortotropy of mineral wool's permeability was neglected in the computational studies. These issues should be taken into account when planning future studies of modeling convection in porous insulation.

Repetitions of the U-factor measurements and more thorough statistical analysis were not made partly due to scheduling and financial reasons and because the test results did not show clear need for that.

CONCLUSION

A series of U-factor measurements and numerical determinations were carried out for typical Finnish wood frame exterior wall structures. Results showed that the effect of natural convection on the thermal transmittance is fairly small in exterior walls and can be eliminated by careful installation and convection barrier. Thus, convection barrier can be effectively used to restrict natural convection's effects as demanded by the building code. Air gaps formed due to poor workmanship do not dramatically increase the U-factor if they do not enable air circulation from hot side to cold side. When evaluating the results shown in this paper, it should be kept in mind that the height of the test walls was approximately only half of actual whole walls. On the other hand, in the Finnish climate the temperature difference can occasionally be much larger than in the measurements of this research.

The only certain conclusion that could be made from the cellulose insulated test walls was that the elastic behavior of wet-sprayed cellulose insulation materials should be investigated more.

ACKNOWLEDGMENTS

The studies shown in this paper are part of a larger Finnish three-year research project called "Future Envelope Assemblies and HVAC Solutions – FRAME". The project was mainly funded by Finland's Ministry of the Environment and Confederation of Finnish Construction Industries RT. Both organizations are gratefully acknowledged.

NOMENCLATURE

ρ	=	density of air, kg/m^3
Δ	=	difference, dimensionless
ε_p	=	porosity, dimensionless
d	=	thickness, m
λ	=	thermal conductivity, W/(m·K)
η	=	dynamic viscosity of air, Pa·s
k	=	permeability, m^2
$\boldsymbol{u}$	=	velocity field, m/s
$\boldsymbol{I}$	=	identity matrix, dimensionless
$\boldsymbol{F}$	=	volume force, N/m^3
C_p	=	specific heat capacity, J/(kg·K)
T	=	temperature, K
p	=	pressure, Pa
R_{air}	=	specific gas constant of air, J/(kg·K)
ϕ	=	heat flux, W

Subscripts

cd	=	conduction
conv	=	convection
CB	=	convection barrier

REFERENCES

ISO. 2007. EN ISO 6946, *Building Components and Building Elements – Thermal Resistance and Thermal Transmittance – Calculation Method*. European Committee for Standardization.

ISO. 1996. EN ISO 8990, *Thermal Insulation: Determination of Steady-State Thermal Transmission Properties, Calibrated and Guarded Hot Box.* European Committee for Standardization.

ISO. 2007. EN 10456, *Building Materials and Products, Hygrothermal Properties—Tabulated Design Values and Procedures for Determining Declared and Design Values.* European Committee for Standardization.

ISO. 1991. ISO 8301, *Thermal Insulation—Determination of Steady-state Thermal Resistance and Related Properties —Heat Flow Meter Apparatus*. International Organization for Standardization.

ISO. 1991. ISO 9053, Acoustics – Materials for Acoustical Applications—Determination of Airflow Resistance. International Organization for Standardization.

Nield D.A. and A. Bejan. 2006. *Natural Convection in Porous Media*, 3rd edition. Springer.

Vinha, J. 1998. Rakenteiden lämmöneristysominaisuuksien mittauslaitteisto (in Finnish). Licenciate thesis. Tampere University of Technology.

Moisture-Related Durability of Walls with Exterior Insulation in the Pacific Northwest

Jonathan Smegal

Joseph Lstiburek, PhD, PE
Fellow ASHRAE

John Straube, PhD, PEng
Associate Member ASHRAE

Aaron Grin, PEng

ABSTRACT

This paper compares the moisture-related performance of a wall with exterior insulation installed over oriented strand board (OSB) to the performance of a drained and ventilated wall with a 19 mm (0.75 in.) gap and a wall with direct-applied stucco over building paper. The data presented is a subset of experimental data from a multiphase, multiyear research project at the Vancouver Field Exposure Test Facility in Coquitlam, British Columbia, Canada. The analysis includes results from normal indoor operating conditions in a high stress exterior moisture environment (typical of the Pacific Northwest climate) and under intentional controlled liquid-water wettings to the interior and exterior of the sheathing. There were no measured or observed moisture-related durability concerns relating to the wood structural sheathing when 38 mm (1.5 in.) of low vapor permeance exterior insulation was installed.

INTRODUCTION

A multiphase, multiyear research project at the Vancouver Field Exposure Test Facility is being led by Building Science Corporation (BSC) and Gauvin 2000 Construction Limited.

A test hut was constructed in the fall of 2005, in Vancouver, British Columbia, Canada that permits the side-by-side construction and comparison of seven 1.0 × 2.4 m (38 in. × 96 in.) test wall panels on each cardinal orientation (for a total of 28 wall test panels) and three 3.6 × 7.2 m (12 × 24 ft) roof panels on each of the north and south-facing roof slopes (for a total of 6 roof test panels). All of the test panels are exposed to the same controlled indoor conditions. The objective of the test hut was to determine the performance of historical, current, and possible future wall assembly configurations under real world field conditions. Each wall or roof assembly is instrumented with different types of sensors to collect the desired assembly data (e.g., moisture content, temperature, relative humidity, etc.) and is stored on a central data logging system. On-site weather data is also collected with a small meteorological tower mounted on the roof. The climate at the test hut is very similar to Seattle WA, and typical of the Pacific Northwest.

The third phase of testing started December 17, 2009 and was completed on November 3, 2011. The primary focus of this phase was to simulate wetting events directly against the exterior surface of the wood sheathing instead of the interior of the sheathing as previously tested, while controlling the interior conditions to simulate representative wintertime relative humidity levels. These test conditions are described in more detail further in the paper.

This paper is divided into five main sections:

- Section 1 explains the background of the research, the objective of the paper, and how the analysis will be conducted.
- Section 2 describes the experimental plan including the test facility, monitoring instrumentation, and wetting systems.
- Section 3 discusses data analysis from the moisture monitoring instrumentation system. The measured sheathing moisture contents are compared following intentional wetting events, and under normal operating

***Jonathan Smegal** and **Aaron Grin** are senior associates at Building Science Consulting Inc., Waterloo, Ontario, Canada. **Joseph Lstiburek** is a principal at Building Science Corporation and an adjunct professor of Building Science at the University of Toronto, Toronto, Ontario. **John Straube** is a principal at Building Science Consulting Inc. and a professor of Building Science at the University of Waterloo, Waterloo, Ontario.*

conditions. Dew-point analysis is always conducted to determine the moisture durability risk to the walls from potential air-leakage condensation.

- Section 4 includes the visual observations from the test wall deconstruction conducted in November, 2011.
- Section 5 presents the conclusions based on the data analysis, and observations.

Objective

There are reported perceptions that installing low vapor permeance exterior insulation over wood sheathing will result in moisture-related durability issues of the wood based structural sheathing. This perception is important to address because installing exterior insulation will become more prevalent with the changing building codes, and requirements for continuous insulation. This paper analyzes the performance of a test wall on all four cardinal orientations constructed with 38 mm (1.5 in.) of exterior insulation installed over oriented strand board (OSB), in particular, when there is liquid water trapped directly against the sheathing between the sheathing and the exterior extruded polystyrene (XPS) insulation.

The performance results of the exterior insulated wall will be compared to the performance of two other common construction methods, side by side, tested over the same duration. The analysis will include the results from normal operating conditions (i.e., without intentionally adding water to the sheathing) in a high stress moisture environment (on the exterior that is typical of the Pacific Northwest United States climate), and under intentional controlled wettings to the interior and exterior of the sheathing.

Scope

This analysis focuses on the performance of an exterior insulated wood-framed residential wall on all four cardinal orientations. The performance results of the exterior insulated wall will be compared to the performance of two other wall systems. The comparison wall systems are as follows:

1. Ventilated stucco with an interior 6 mil polyethylene vapor control layer; and
2. Direct-applied stucco with an interior 6 mil polyethylene vapor control layer.

The direct-applied stucco is a wall construction that was more typical in the 1970s and 1980s. The prevalence of this wall design has decreased significantly due to a large number of moisture-related durability failures that have been attributed in part to the enclosure design, but it still constructed in some areas of North America. The ventilated stucco wall assembly has been constructed as a moisture durable alternative to direct-applied stucco. In the Vancouver area, ventilated stucco has become required by the building code for stucco assemblies and is a more typical choice in other regions as well.

Comparisons will be made between the test walls on each orientation, but comparisons of each wall with respect to all four orientations will not be conducted in this report.

Approach

The analysis of these walls will be done by comparing several criteria. These criteria include:

- Sheathing-wood moisture-content measurements under normal operating conditions;
- Sheathing-wood moisture-content measurements during the wetting and drying from a wetting event on the interior of the sheathing;
- Sheathing-wood moisture-content measurements during the wetting and drying from a wetting event on the exterior of the sheathing;
- Visual qualitative observations during deconstruction.

EXPERIMENTAL PLAN

The Test Facility

A 90 m^2 (9 × 10 m) (957 ft^2 [29 ft × 33 ft) field exposure test facility was designed and permits the side by side construction and comparison of seven 1.0 × 2.4 m (38 × 96 in.) test wall panels on each cardinal orientation for a total of 28 wall test panels. This building permits all of the test panels to be exposed to the same interior conditions. The test panels were also exposed to the same exterior conditions relative to their orientation, (i.e., all north walls can be expected to have the same exterior conditions).

Instrumentation

Each of the test walls was outfitted with a series of temperature, relative humidity (RH), and wood moisture content (MC) sensors. These sensors were continuously monitored and recorded throughout the testing period using a data acquisition system. Variations on a typical sensor package (illustrated in Figure 1) were used for each test panel.

MC pins were installed in the framing lumber and the wood sheathing (from the interior) in all wall systems. Wood MCs can be determined from electrical resistance of wood based on the Garrahan equation (Garrahan 1988; Onysko et al. 2008). These pins can be used to measure MC at any depth chosen because the pins are electrically insulated except for the tips. Measurements are most commonly taken at 6 mm (0.25 in.) tip depth. In this study, MCs were taken at two depths on the lower OSB sheathing near the wetting system. To correct the MC readings for temperature effects, a hole was drilled at each MC measurement to the same depth as the MC pins and a temperature sensor was installed inside.

Relative humidity sensors were installed in the middle of each stud cavity and in the drainage space of each wall. In wall 5 with the 0.75 in. (19 mm) gap, two RH sensors were installed between the OSB and 30 min building paper, and a third was installed on the exterior of the 30 min paper in the gap. In wall

2, two RH sensors were installed between the OSB and the 30 min building paper, and a third was installed on the exterior of the building paper in the stucco base coat. On wall 2, the RH sensors between the 30 min paper and the OSB formed very small isolated pockets between the paper and the OSB. On wall 7, two RH sensors were installed between the OSB and corrugated housewrap, and a third RH sensor was installed between the corrugated housewrap and XPS. On wall 7, the XPS had small grooves placed on the interior surface so the RH sensors did not form an unintended gap to the interior for the XPS. The relative humidity sensor was always installed in combination with a temperature sensor, both of which are protected by a vapor permeable, water resistant cover. Relative humidity and temperature sensors were installed at the midpoint of the stud space between the drywall and the sheathing, as well as drainage cavities behind the cladding.

One of the most significant advantages to using a test hut with removable/replaceable walls for analysis compared to instrumenting walls in existing occupied buildings is that the experimenter can deliberately and easily stress the walls with high moisture loads, either in terms of vapor (e.g., ≥50% interior RH) or liquid water (e.g., intentional wetting systems). In most cases, building owners are not interested in participating in research conducted on their enclosure walls by adding moisture. In addition, it is often difficult to determine the performance of a wall system without stressing the moisture tolerances of a wall to determine the comparative risk of certain construction techniques.

Enclosure-Wetting Systems

A wetting apparatus was installed on both the interior and exterior of each of the test walls to allow a known amount of water to be injected at a controlled time and location. The wetting apparatus consisted of a storage media installed directly against the interior and exterior surface of the wood sheathing (Figure 2), with a tube connecting a perforated section of the tube at the storage media to the interior of the test hut for access for water injections. This enables wetting to either the interior or exterior independently without opening and disturbing the wall system. The wetting system is designed to simulate a window leak, and is used to help determine the drying potential of a wall system. A photograph of an exterior-installed wetting apparatus is shown Figure 2. In the analysis graphs in Section 3, the exterior-wetting events are indicated by red vertical dashed lines, the interior-wetting events are indicated by blue vertical dashed lines. When an intentional wetting event occurred repeatedly over multiple days, only one line is shown indicating the first wetting. There were five intentional wetting events as shown in Table 1. For each wetting event, 45 mL (1.5 oz) was injected twice a day for five days. This is a total of ~450 mL (15 oz) into the wetting apparatus directly against the sheathing. Over the entire test period of 24 months ~1.8 L (60 oz) was injected against the exterior of the sheathing, and ~450 mL (15 oz) was injected against the interior surface of the sheathing. A review of previous research was conducted and there was no precedent for the volume of water to be added to a wetting apparatus in a wall system. The volume of water added during this study was determined based on previous wetting apparatus testing in a previous study. The 45 mL (1.5 oz) injected twice a day sufficiently wet the storage media, and produced a MC response in the sheathing.

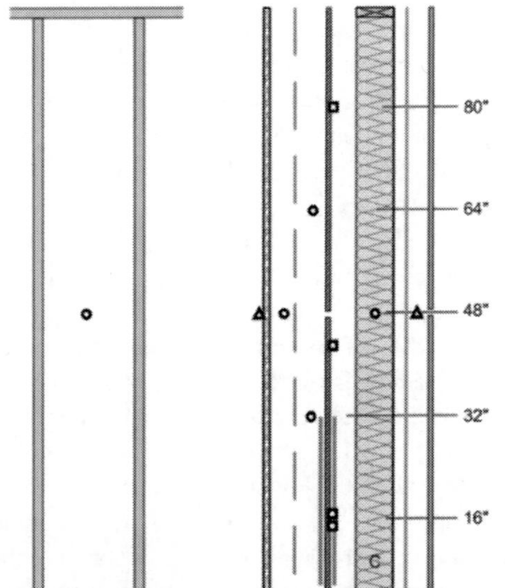

Assembly

Layer	Function	Description
A	Interior Finish	Painted ½" drywall
B	Vapor Control	Sealed 6 mil poly
C	Studspace	R20 fiberglass batt
D	Framing	2x6
E	Sheathing	½" OSB
F	Sheathing Membrane	1 layer 30 min paper
G	Drainage Gap	none
H	Cladding	¾" stucco

Sensor Package

Layer	Code	Location
A	taem	Interior finish, exterior face, 48"
C	r/tcnm	Studspace, interstitial, 48"
D	m/tdeb (siam)	Framing, exterior edge, bottom plate
D	m/tdib (siam)	Framing, interior edge, bottom plate
E	m/tenu	Sheathing, interstitial, 80"
E	m/tenm	Sheathing, interstitial, 48"
E	m/teel	Sheathing, exterior, 16"
E	m/teil	Sheathing, interior, 16"
G	r/tgiw	Drainage space, interior, 32"
G	r/tgie	Drainage space, interior, 64"
G	r/tgem	Drainage space, interstitial, 48"
H	them	Cladding, exterior surface, 48"

○ Relative Humidity and Temperature Sensor Package
□ Moisture Content Pins and Temperature Sensor Package
▲ Temperature Sensor
▪ Moisture Content Wafer and Temperature Sensor Package
= Moisture Content Pins

Figure 1 *Typical wall construction and sensor package configuration.*

Figure 2 *Exterior wetting apparatus.*

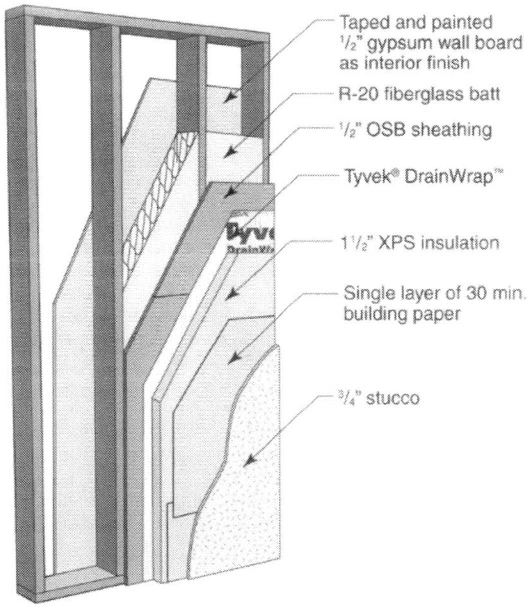

Figure 3 *Construction of wall 7 on all four orientations.*

Table 1. Intentional Wetting Event Schedule and Location for Phase III

	Location	Start Date	Amount
Wetting event 1	Exterior	July 5, 2010	450 mL (15 oz)
Wetting event 2	Exterior	August 30, 2010	450 mL (15 oz)
Wetting event 3	Interior	Jan 19, 2011	450 mL (15 oz)
Wetting event 4	Exterior	May 9, 2011	450 mL (15 oz)
Wetting event 5	Exterior	August 9, 2011	450 mL (15 oz)

Climate

To monitor the exterior weather conditions, a steel mast on the roof of the test facility supports an on-site weather station. The monitoring system continuously collects weather data including: temperature, relative humidity, wind speed and direction, rainfall, and solar energy on a horizontal surface. It is important to have on-site weather information for comparison and correlation purposes with the data.

The measured exterior temperature for the analysis period is shown in Figure 9, further in the paper in the dew point analysis section.

Vancouver is approximately 200 km from Seattle, WA, and subjected to a similar climate (Smegal et al. 2012). Vancouver and Seattle experience a similar annual distribution of relative humidity. The annual average rain amounts are also similar in both cities, with Vancouver's driving rain coming predominantly from the east, and Seattle's coming from the south.

OBSERVATIONS AND DATA ANALYSIS

This paper and analysis is limited to observations and data collected during the monitoring period (December 17, 2009 to November 3, 2011) for the exterior insulated wall 7 (Figure 3), traditional stucco wall 2 (directly applied to one layer of 30 min paper over wood sheathing) (Figure 4), as well as the ventilated stucco cladding wall 5 east and west (Figure 4).

As mentioned in the approach, there are several analysis criteria that will be used for comparing the performance of the test walls. These criteria include:

- Sheathing-wood moisture-content measurements under normal operating conditions;
- Sheathing-wood moisture-content measurements during the wetting and drying from an interior wetting event;
- Sheathing-wood moisture-content measurements during the wetting and drying from an exterior wetting event; and
- Qualitative visual observations during deconstruction.

For this paper, we will focus on the measured MC of the sheathing at the four measurement locations, the measured temperature at the interior of the sheathing, and the measured-interior dew point.

The four sheathing moisture measurements are:

- 16 in. (406 mm) from the bottom of the bottom plate at the interior edge of the sheathing;
- 16 in. (406 mm) from the bottom of the bottom plate at the exterior edge of the sheathing;
- 48 in. (1219 mm) from the bottom of the bottom plate at the center of the sheathing; and
- 80 in. (2032 mm) from the bottom of the bottom plate at the center of the sheathing.

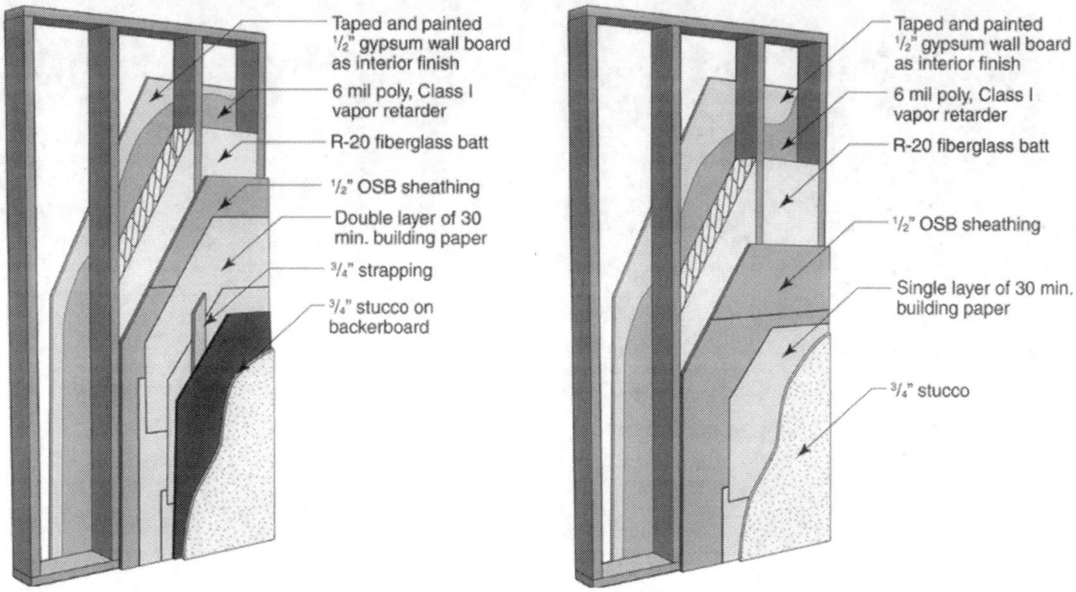

Figure 4 *Wall 5 (left) on the east and west orientation (ventilated stucco) and wall 2 (right) on all four orientations (direct-applied stucco).*

The two MC sensors at 16 in. will be used to evaluate the performance following wetting events as they are installed in the sheathing where it is in direct contact with the wetting system. The MC sensors at 48 in. and 80 in. from the bottom plate generally speaking are not affected by either the interior or exterior wetting events, and are a good indication of how the sheathing MC is affected under normal operating conditions.

Sheathing moisture is used as the performance criteria because this is the first location where vapor diffusion condensation would occur in a cold climate during the heating season, and also may experience moisture accumulation from the exterior in wet climates. In the case of these test walls, the most significant moisture risks were at the locations of the wetting systems, which are used to compare the sheathing moisture performance under significant moisture stresses. MCs of the sheathing will be used as a comparison rather than pass/fail criteria for the wall assembly. Generally, under normal conditions, the following criteria are used to assess the risk of various test wall assemblies:

1. Peak sheathing MC less than 20%, no mold growth—very little risk.
2. Peak sheathing MC between 20% to 28%—potential for mold growth eventually, depending on frequency and length of wetting, and temperatures during wetting. This design can be successful but conservative assessments usually require corrective action be taken.
3. Peak sheathing MC >28%—moisture-related problems are expected and this design is not recommended.

These MC thresholds are based on moisture physics and correlation of hygrothermal simulations with field investigations. These thresholds are not based on an existing standard, but on the extensive experience of the authors with respect to moisture-related issues, especially in cold climates.

The measured MC should be kept in context and good engineering judgment is required to determine the moisture risk to the sheathing. For example, elevated wood MCs in the cold winter months when the wood substrate is on the cold side of the assembly are much safer from a mold growth perspective than similar MCs in the summer, when the temperatures are in the correct range for optimal mold growth. Also, high MC for a short period followed by drying is not necessarily risky, as wood framed structures are able to manage high MCs for short periods without exceeding the safe storage capacity of the assembly.

The safe storage capacity is the amount of moisture an assembly is able to manage without suffering any moisture-related issues. The baseline wood MC is a factor in the safe storage capacity since the lower the wood MC is during normal operation (without wetting events), the more moisture the wood can handle before reaching durability risk. If the measured wood MC is consistently higher, even if there are no moisture durability risks, there is less moisture-buffering capacity in the wood before reaching moisture-related durability risk levels.

Because OSB is an engineered wood product that varies between manufacturers, and varies over time, it is difficult to assign a safe moisture storage capacity. It has also been observed during long-term testing that OSB may change its physical characteristics, such as vapor permeance and adsorption, after being subjected to high moisture environments such as elevated relative humidities, or moisture during construction.

Table 2. Walls for the Comparison of Exterior Insulation

	Wall 2 (All Orientations)	Wall 5 (East and West Only)	Wall 7 (All Orientations)
Interior finish	Latex paint on 0.5 in. (12.7 mm) drywall	Latex paint on 0.5 in. (12.7 mm) drywall	Latex paint on 0.5 in. (12.7 mm) drywall
Vapor control	6 mil polethylene	6 mil poly polethylene	Included as interior finish
Insulation	R-20 fiberglass batt	R-20 fiberglass batt	R-20 fiberglass batt
Framing	2 × 6 wood	2 × 6 wood	2 × 6 wood
Sheathing	7/16 in. (11.1 mm) OSB	7/16 in. (11.1 mm) OSB	7/16 in. (11.1 mm) OSB
Sheathing membrane	1 × 30 min bldg. paper	2 × 30 min building paper	Corrugated housewrap on OSB
Exterior insulation	None	None	1.5 in. (38 mm) XPS
Drainage/ventilation gap	None	0.75 in. (19 mm) strapping	Small gap formed by corrugated housewrap
Cladding	0.75 in. (19 mm) stucco on metal lathe	0.75 in. (19 mm) stucco on metal lathe on backerboard	0.75 in. (19 mm) stucco on paper-backed metal lathe

The wall system performance will also be evaluated qualitatively based on the photos and observations from the test wall deconstruction on November 2 and 3, 2011.

Wall Assemblies

Wall 7 was installed at all four orientations. The construction of wall 7 is shown in Figure 3. A corrugated housewrap was installed as the air and water barrier against the sheathing. A paper-backed metal lathe was installed on the exterior of the XPS as a bond break for the stucco. The performance of the exterior insulated wall is compared to two other walls for the purposes of this analysis. In this analysis, comparisons were only made between walls on the same orientation since they are typically subjected to the same exterior boundary conditions simultaneously. It is possible to conduct comparisons between walls on different orientations to see what effect the orientation has on the performance, but that is outside the scope of this analysis.

The first comparison wall (east/west 5) is representative of a well-ventilated and drained-stucco cladding typical of code approved current construction practices in Vancouver (Figure 4). This comparison was made to determine what performance differences exist between a moisture durable recommended wall assembly, and the more energy-efficient exterior-insulated wall assembly that is not as accepted by code officials for lack of a drainage gap.

The second wall for comparison (wall 2) is a more commonly constructed wall (though no longer code approved in Vancouver) with direct applied 0.75 in. stucco on one layer of 30 min building paper (Figure 4). This comparison was conducted to demonstrate any potential improvements in performance by including exterior insulation to the direct-applied stucco wall assemblies.

Wall 2 was implicated in many moisture-related building failures in the Vancouver area, and has been replaced by the ventilated stucco wall, wall 5 in the Vancouver/Lower Mainland area. The moisture-related failures were largely due to water management design deficiencies at penetrations and interfaces, but may have also resulted from enclosure design. The scope of this study does not include assessment of design deficiencies in the water management system or penetrations in the test walls. A wall comparison summary table is shown in Table 2.

Wall 7, with exterior insulation has a class III interior vapor control layer (latex paint on the drywall) as is allowed in the International Residential Code (IRC) Table R601.3.1 for climate zone 4C. The comparison walls use a polyethylene class I vapor control layer on the interior of the framing, as is required by code.

Comparison of Walls 5 and 7 on the East and West Orientation

The sheathing MC comparisons for walls 5 and 7 on west orientation are shown in Figures 5 and 6. Results on the east orientation were very similar and not included here. Figure 6 uses the same *y*-axis scale of 0%–40% as Figure 5 for comparison purposes between graphs. A reduction in scale for Figure 6 is not necessary to compare MCs more closely in the safe MC range. Wall 5 on the north and south orientation was constructed with fiber cement cladding, and is not compared to wall 7 because of the multiple construction variables between the walls.

Comparison on the west orientation (Figure 5) shows the measured MC at the interior and exterior edge of the lower OSB sheathing for walls 5 and 7. The results on the east orientation are similar and can be seen in Smegal et al. (2012). These comparisons demonstrate the moisture performance during both interior and exterior wettings, since both of these sensors are adjacent to the wetting system as indicated in Figure 1. Following each of the exterior wetting events on both orientations, wall 5 with a 0.75 in. ventilation cavity dries

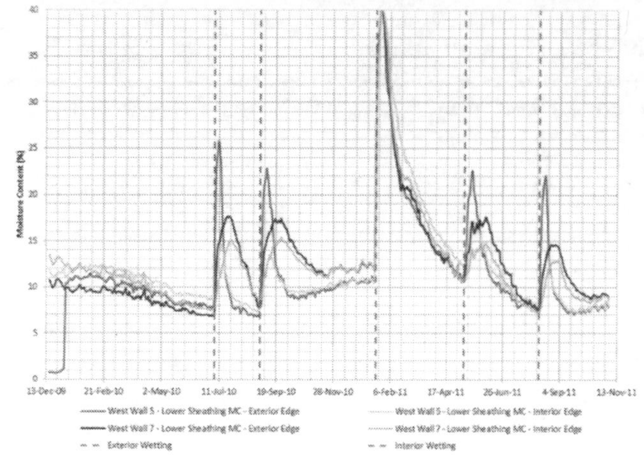

Figure 5 *Lower OSB sheathing measured MC comparison between west wall 5 (ventilated stucco) and west wall 7 (exterior insulation).*

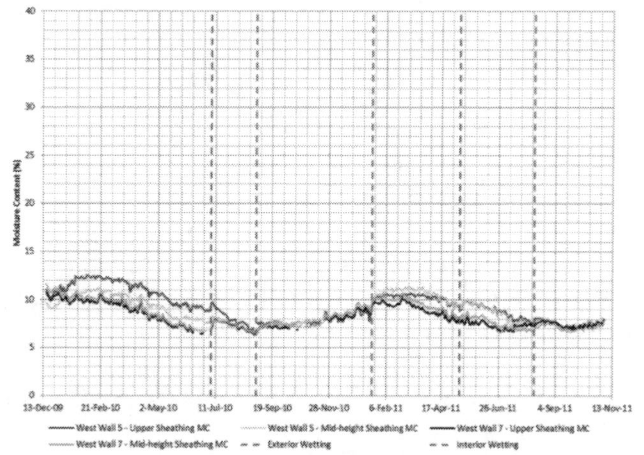

Figure 6 *Middle and upper OSB sheathing measured MC comparison between west wall 5 (ventilated stucco), and west wall 7 (exterior insulation).*

more quickly than wall 7, most likely because of wall 5's ventilation drying capability.

Following the interior wetting event (blue vertical line in Figures 5, 6, and 7, Jan 19, 2011) both walls perform very similarly, reaching the same MC and drying at similar rates. At the start of testing, under normal operating conditions without any wetting (prior to July 5, 2010), the walls have very similar MC measurements with no measured risk of moisture-related durability problems (see the criteria in the previous Climate subsection).

The middle and upper sheathing MC measurements were largely unaffected by the intentional wetting events because of the distance from the wetting system and are used as a comparison of sheathing moisture conditions without the influence of water leaks. Figure 6 shows the comparison between the midheight and upper sheathing MC sensors in walls 5 and 7 on the west orientation. The measured sheathing MC was dry and very

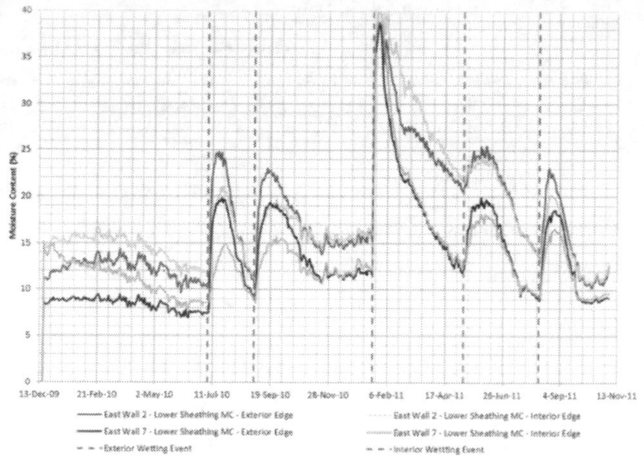

Figure 7 *East wall 2 (direct-applied stucco) and east wall 7 (insulating sheathing) lower OSB sheathing measured MC.*

similar in both wall 5 and wall 7 on both orientations at the midheight and upper sheathing MC sensors.

Comparison of Walls 2 and 7

Comparing the overall MC measurements at the lower measurement location (both interior and exterior) indicates the MC was higher on wall 2 relative to wall 7 for the same sensor (interior edge compared to interior edge, and exterior edge compared to exterior edge) for the entire monitoring period. An example of the lower MC location comparison is shown for the east orientation in Figure 7. All other orientations are similar and can be seen in Smegal et al. (2012).

Following the intentional interior wetting event (January 2011), wall 7 dried more quickly in all cases than wall 2. This is because there is very little drying potential to the exterior in both wall 2 and wall 7 as neither have an intentionally ventilated drainage gap. Wall 2 is constructed with class I polyethylene vapor control on the interior, so drying to the interior is not possible, whereas wall 7 has a class III interior vapor control layer (latex paint on the drywall) and is thus able to dry to the interior. Table 3 shows the amount of time (in days) following the interior wetting event until the measured MC was less than 20%. The next wetting event was May 5, 2011, so if the measured sheathing MC was still above 20% MC at that time, then a value of >106 days was used.

At the midheight and upper MC measurement locations on the east orientation (Figure 8), which is largely unaffected by the intentional wetting events, the MC measurements for wall 7 do not exceed 13%, and are always less than wall 2. Wall 2 measurements do not exceed 17%, which is also considered a safe level of moisture in the OSB sheathing. This is the same on the other three orientations. Figure 8 uses the same *y*-axis scale of 0%–40% as Figure 7 for comparison purposes between graphs. A reduction in scale is not necessary to compare MCs more closely in the safe MC range.

Comparison of Sheathing Temperatures and Interior Dew Point

The measured and calculated performance improvements to the building enclosure by using exterior insulation are not new ideas. In 1964, Neil Hutcheon demonstrated how the temperature gradients across a masonry wall changed when the insulation was moved from the interior of the structure to the exterior of the structure. He showed that condensation issues at 35% interior relative humidity were solved by moving the insulation to the exterior. These same principles have been more recently illustrated by Straube using more current construction practices (2011).

The analysis in this report discusses the potential for interior air-leakage condensation based on the measured-interior-air dew point and the measured-sheathing temperature. This analysis is a measure of moisture durability risk, and not actual moisture durability because it is assumed the vapor control of the enclosure is adequate and condensation will only occur if there is a path for air leakage from the interior to the sheathing, and the sheathing is below the dew point of the air.

Field experience with airtightness and blower door testing has demonstrated that the enclosure will always have some air leaks, and their location and size depends on the type and quality of construction. It is not uncommon to find air leakage pathways through electrical outlets, switches, and other interior finish penetrations. This means that with air-permeable cavity-filled insulation, which is the most commonly used, it is not uncommon to find evidence of moisture condensation on the interior surface of the exterior OSB behind penetrations.

The condensation potential is directly related to the interior conditions and the sheathing temperature. Generally speaking, dry air has less moisture available to condense so there will be less concern for lower humidity interior conditions. For this analysis, two time periods were used over the colder winter months, and these two periods are shown on the temperature graph in Figure 9. The interior temperature was approximately 20°C for the entirety of both comparison periods, with some small variations. The interior relative humidity was set to 40% for the winter months and maintained with a humidifier controlled by the data acquisition system. An interior relative humidity of 40% is not considered unrealistic in climate zone 4C but depends on the occupants, enclosure, airtightness, and HVAC system. It is difficult to distinguish between the three

Table 3. Comparison of Drying Rates for Wall 2 and Wall 7 Following Interior Wetting Event in January 2011

Orientation	Wall 2 (Direct-Applied Stucco)	Wall 7 (XPS Exterior Insulation)
North	>106 Days	68 Days
East	>106 Days	53 Days
South	68 Days	29 Days
West	95 Days	47 Days

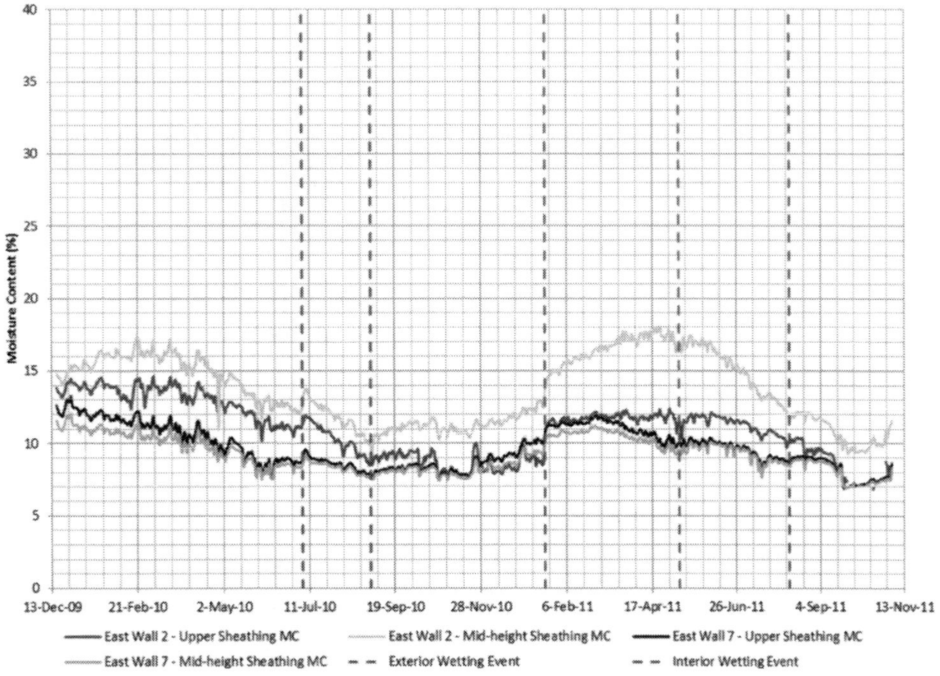

Figure 8 *East wall 2 (direct-applied stucco), and east wall 7 (XPS insulating sheathing) middle and upper OSB sheathing measured MC.*

interior temperature measurements in Figure 9 because the interior conditions of the test hut were uniform throughout.

The temperature and relative humidity are used to calculate the hourly dew point of the interior air, for comparison to the sheathing temperature. The temperature of the sheathing is taken at the midheight midthickness of the sheathing and not at the interior surface, but the difference in temperature is negligible over half the thickness of OSB.

The comparison results of this analysis are summarized numerically in Table 4 for both comparison time periods, and the results for the second time period are shown graphically in Figure 10. Wall 5 in this analysis was only constructed on the east and west orientations so wall 5 is not included for the north and south orientation in Table 4. The measured results show that there is a significant decrease in the number of measured potential hours of air-leakage condensation when 38 mm (1.5 in.) of XPS is installed as exterior insulation.

The degree of potential risk in terms of vapor diffusion and air leakage moisture condensation is proportional to the length of time that the sheathing temperature is continuously below the dew point without any drying potential, and the magnitude of the difference between the dew point and the sheathing temperature. This means that a sheathing temperature 10° below the dewpoint will condense more water than a sheathing temperature 2° below the dewpoint, all other factors being equal. If the temperature of the sheathing is below freezing, condensation will occur as frost or ice, and then melt when the sheathing temperature increases.

Figure 11 is an example of the measured-interior dew point and simultaneous measured-sheathing temperature for walls 5 and 7 on the east orientation. Only the east orientation was included in this report, because all other orientations resulted in similar graphs, and this was used as an example as an alternative method to numerical analysis in Table 4 and Figure 10. In Figure 11 the green line is the measured-sheathing

Table 4. Number of Hours of Potential Air-Leakage Condensation Based on the Interior Dew Point and Sheathing Temperature

Test Wall	Potential Hours of Condensation Dec. 18, 2009–May 30, 2010 (3936 Total Hours)	Potential Hours of Condensation Sept. 19, 2010–March 23, 2011 (4441 Total Hours)
North wall 2	1252	2417
North wall 7 (XPS)	91	551
East wall 2	1165	2293
East wall 5	519	1879
East wall 7 (XPS)	72	478
South wall 2	1050	1980
South wall 7 (XPS)	51	320
West wall 2	741	2011
West wall 5	716	1680
West wall 7 (XPS)	94	518

Note: Measure of moisture-related durability risk, not moisture-related durability.

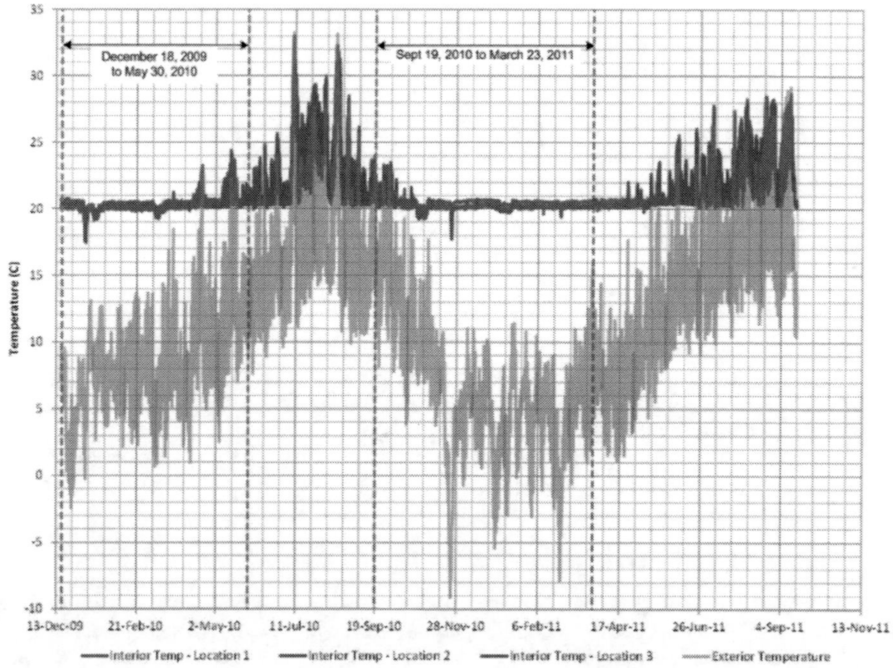

***Figure 9** Measured-interior and ambient temperatures.*

temperature of wall 7, the blue line is the sheathing temperature of wall 5. The black line is the measured-interior dew point. When the sheathing temperature falls below the dew-point temperature, condensation will occur if the interior air contacts the interior surface of the sheathing. East wall 2 was not included in Figure 11 because the graph became more difficult to interpret with the addition of east wall 2, and it was clear from Table 4 and Figure 10 that east wall 2 had the greatest number of potential condensation hours of the three comparison walls on the east orientation.

The graph shown in Figure 11 is a more visual method of displaying the results than Table 4 and illustrates the relative magnitude of the condensation more clearly than the table format. It is clear that even with 38 mm (1.5 in.) of exterior XPS insulation, there is still a potential for a small amount of condensation, but significantly less than most building code minimum approved wall systems. If condensation does occur in wall 7, it will be able to dry much more quickly to the interior because latex paint is the only form of vapor control, compared to walls 2, and 5 with a polyethylene vapor barrier.

TEST WALL DECONSTRUCTION

The deconstruction of phase III test walls was conducted on November 2 and 3, 2011. The deconstruction was conducted carefully and systematically so that each component of each wall system could be photographed, examined, and compared. Disassembly is critical because it was very informative to inspect all aspects of all layers of the test walls. It is important to see if the visual observations correlate to, and validate, the measured results (Lstiburek 2011).

It is important with this qualitative comparison analysis to keep in mind the boundary conditions that the walls were subjected to, both on the interior and exterior. Figure 9 shows the interior and exterior temperatures. The interior relative humidity was set to 40% but did get as high as 50% in the summer because there was no dehumidification, only ventilation.

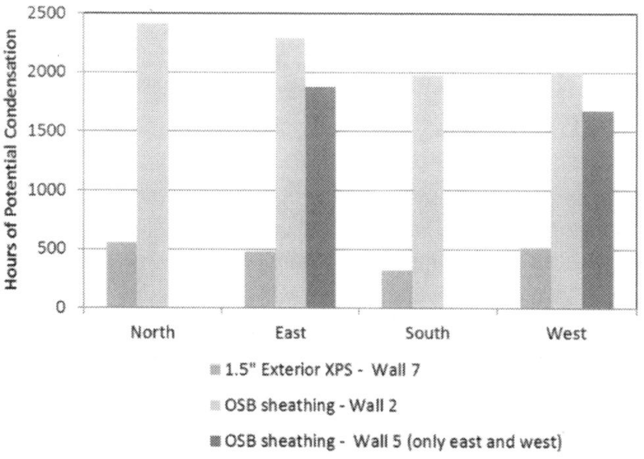

Figure 10 *Comparison of the hours of potential condensation between wall 2 and wall 7 from September 19, 2010, to March 23, 2011, using measured-interior dew point and sheathing temperature.*

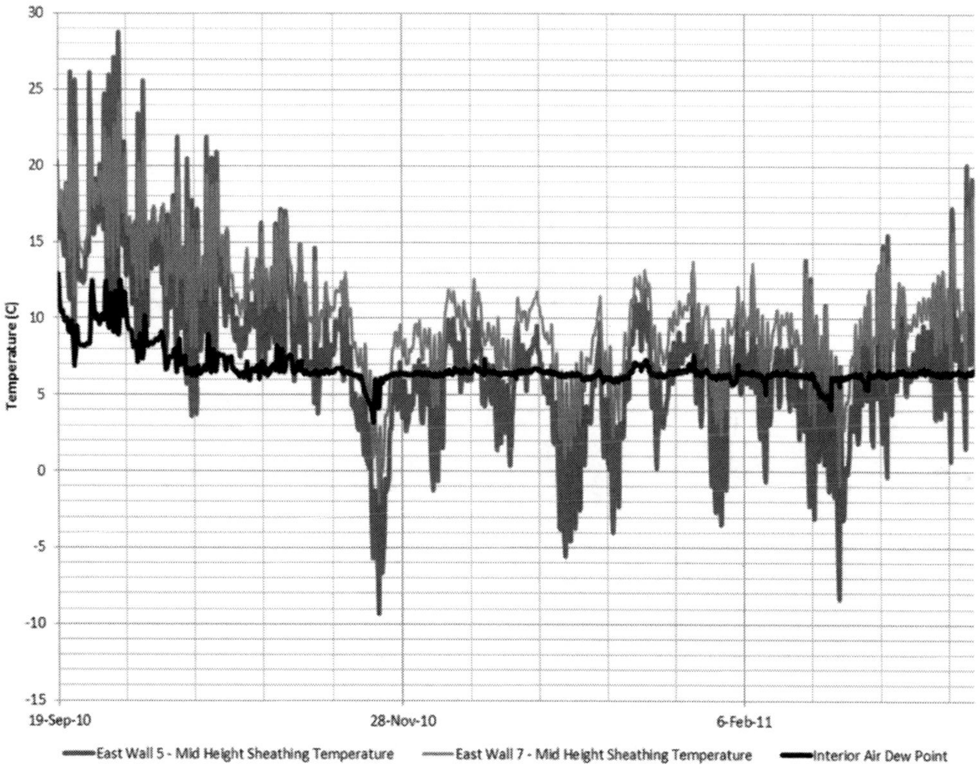

Figure 11 *Comparison of condensation potential between east walls 5 and 7, September to March 2010.*

The other important boundary condition was the intentional wetting events to the surface of the sheathing as indicated in Table 1.

The testing was conducted for approximately two years time, so the conditions will require extrapolation to determine what the OSB might look like after many years in service; although, it is difficult to extrapolate qualitative results conclusively. It is likely that walls experiencing the most staining would continue to experience staining and discoloration at a higher rate than other test walls during the life of the walls.

Following two years of testing, there was no observed physical deterioration of the OSB sheathing in any of the test walls, only staining of the surfaces.

North Orientation

On the north orientation, the exterior of the OSB on wall 2 was quite dark and stained, particularly in the vicinity of the wetting apparatus. Both walls 5 and 7 on the north orientation appeared to be in new/good condition. The area immediately behind the wetting apparatus was a little darker on wall 7 than wall 5, corresponding to the measured quicker drying rate of wall 5, due to ventilation drying. Overall, the rest of the north wall 7 was marginally better than north wall 5.

On the interior surface of the OSB, wall 2 had the most staining, again, surrounding the wetting apparatus. Walls 5 and 7 looked better than wall 2 and very similar to each other on the interior of the OSB.

East Orientation

Of the three comparison walls on the east orientation (Figure 12), wall 2, is the most stained in the area surrounding the wetting apparatus and on the upper portion of OSB. There are no visual differences in the photo between walls 5 and 7. Overall, wall 7 looked the best of all east walls, but both wall 5 and wall 7 looked like new over most of the surface of the OSB.

South Orientation

The exterior surface of the OSB on south wall 2 is the darkest and most stained of the three comparison walls. There was some staining in the vicinity of the wetting apparatus on both walls 5 and 7, but overall, the exterior surface of the sheathing showed no signs of moisture-related damage on wall 5, and looked like new on wall 7.

Figure 12 *Exterior OSB sheathing on the east orientation immediately following deconstruction.*

West Orientation

On the west orientation, wall 2 was not as dark and stained as the other orientations, but did have some staining of the OSB, particularly around the area of the wetting apparatus. The OSB on wall 5 was in good condition with staining only in the vicinity of the wetting apparatus. On wall 7 the OSB looked like new, except for the area immediately behind the wetting apparatus.

CONCLUSION

Wall 5 (East and West) and Wall 7

Following the interior wetting events, both wall 5 (east and west) and wall 7 dried very similarly. Neither wall outperformed the other during interior wetting events.

Following the exterior wetting events, wall 5 (east and west) dried more quickly than wall 7

MC monitoring of the midheight and upper sheathing demonstrated no significant performance differences between wall 5 and wall 7.

In dew-point analysis, wall 5 had significantly higher moisture durability risk with respect to interior air-leakage condensation than wall 7. This is not a durability problem unless interior air reaches the surface of the sheathing through air leaks

During the wall deconstruction, the OSB surfaces of walls 5 and 7 both appeared to be in good condition, with very little staining only around the wetting apparatus. Although, wall 7 did appear a little more pristine overall than wall 5.

Wall 2 and Wall 7

MC monitoring of the midheight and upper sheathing showed that the sheathing MC of wall 2 was always higher than wall 7, but did not exceed criteria for moisture-related durability concerns. The performance may have been better than expected to some based on the historical, large-scale moisture-related durability, but the scope of this test did not include penetrations in the assembly and design deficiencies in the water management details that were more prevalent historically.

Following the interior wetting event, wall 7 dried more quickly to safe levels than wall 2, because the vapor control layer on wall 7 was class III, and the vapor control layer on wall 2 was class I eliminating drying to the interior.

Following the exterior wetting events, the drying rates were relatively similar between walls 2 and 7. Generally, the MC was higher on wall 2, so even though the drying rates were similar, the MC of the sheathing on wall 2 remained elevated compared to wall 7.

In dew-point analysis, wall 2 had significantly higher moisture durability risks with respect to interior air-leakage condensation than wall 7. This is because the exterior insulation in wall 7 keeps the sheathing at a higher temperature, reducing the number of hours that the sheathing is below the interior-air dew point. This is not a durability problem unless interior air reaches the surface of the sheathing

Observations from the wall deconstruction showed that the exterior surface of the OSB for wall 2 was quite stained behind and around the wetting apparatus. There was also staining on the upper portion of the OSB on wall 2, as well as a general overall darker appearance to the entire OSB sheet compared to wall 7. The OSB on wall 7 looked like new except directly behind and under the wetting apparatus.

It was shown through measured data analysis and wall deconstruction that there were no moisture-related durability concerns of the wood structural sheathing in the Vancouver, BC, Canada climate (climate zone 4C [ICC 2012]) when 1.5 in. of exterior insulation was installed. Liquid water was intentionally placed in direct contact with the exterior surface of the sheathing behind the low vapor permeance XPS exterior insulation four times through the monitoring period as described in this paper and there were no moisture-related durability concerns measured or observed.

ACKNOWLEDGMENTS

BSCI would like to acknowledge Mark Gauvin of Gauvin 2000 Construction Ltd, for his time, resources, and patience to make the Vancouver test hut project a valuable, worthwhile research program.

REFERENCES

Garrahan, P. 1988. *Moisture Meter Correction Factors.* Forintek Ottawa, Canada: Canada Corp.

Hutcheon, N. 1964. Principles applied to an insulated masonry wall. CBD-50, National Research Council, Canada.

ICC. 2012. *International Energy Conservation Code*, Table C301.3(2). Washington, DC: International Code Council.

Lstiburek, J. 2012. Building sciences: Parthenon, eh! *ASHRAE Journal* 54(4):80–86.

Onysko, D., C. Schumacher and P. Garrahan. 2008. Field measurements of moisture in building materials and assemblies: pitfalls and error assessment. *Proceedings of Building Enclosure Science and Technology (BEST) Conference, June. Minneapolis, MN.*

Smegal, J., J. Lstiburek, J. Straube, and A. Grin. 2012. *Vancouver Field Exposure Facility: Phase III Exterior Insulation Analysis.*Somerville, MA: Building Science Corporation,.

Straube, J. 2011. *BSD-163: Controlling Cold-Weather Condensation Using Insulation.* Somerville, MA Building Science Corporation,.

Development of a New Hot Box Apparatus to Measure Building Enclosure Thermal Performance

C.J. Schumacher

J.F. Straube, PhD, PEng
Associate Member ASHRAE

D.G. Ober, PE

A.P. Grin, PEng

ABSTRACT

R-value is an effective, well established, and widely accepted metric for describing the thermal performance of building insulation materials but it is an incomplete metric for describing the thermal performance of building enclosure assemblies. In practice, many building designers have directly applied insulation R-value to the thermal performance of building enclosure assemblies; however, with the proliferation of new materials and enclosure systems, this simplification increasingly results in poor predictions of performance. More accurate predictions have become increasingly important as stricter energy codes and the use of compliance computer modeling have proliferated.

The thermal performance of any building enclosure assembly is highly dependent on the amount of insulation installed; however it can also be heavily influenced by a number of other factors such as air leakage, thermal bridging, operating conditions, moisture content, and installation defects. These factors are not properly or completely captured when insulation R-value is used as the sole metric for assembly thermal performance. A consortium of insulation manufacturers, representing all types of insulating materials and all regions of North America, undertook a series of research projects to examine these issues and their impact on the prediction of the thermal performance of building enclosures.

This paper documents the development of the primary apparatus used in the research: a new hot box designed to accurately measure the true thermal performance of modern wall assemblies. The approach is based on ASTM C1363. However, a number of improvements and performance extensions have been made to facilitate the research: a large, double-guarded meter box; systems to simulate heating and cooling climates without removing the test specimen; a closed-loop air transfer system that induces and measures air pressure differences across and movement through the assembly; a tracer gas system to measure unintentional and intentionally induced airflows; and the ability to measure with greater precision. All of these features allow quantification of wall assembly performance that is more representative of actual in-service conditions. The apparatus components, capacity, accuracy, calibration and validation are presented.

INTRODUCTION

R-value is an effective, well established, and widely accepted metric for describing the thermal performance of building insulation materials, but it is an incomplete metric for describing the thermal performance of building enclosure assemblies. In practice, building designers often directly apply R-value to the thermal performance of building enclosures. This practice has recently come into question as energy-cost and security issues have generated demand for building enclosures that exhibit higher levels of thermal performance. The market has responded with innovative insulation products and novel building enclosure systems such as various types of spray foam and spray-applied fibrous insulations, exterior insulated sheathing, structural insulated panel systems (SIPS), insulated concrete forms (ICF), radiant barrier systems (RBS), etc.

Because contemporary insulation materials and systems control heat flow in different, new, and non-traditional ways,

***C.J. Schumacher** is a founding principal and **A.P. Grin** is a research engineer at Building Science Consulting Inc., Waterloo, Ontario, Canada. **D.G. Ober** is founding principal of Ober Engineering & Consulting LLC, Charlotte, NC. **J.F. Straube** is a professor in the Department of Civil Engineering and the School of Architecture at the University of Waterloo, Waterloo, Ontario, Canada.*

they are more or less sensitive to thermal bridging, workmanship (i.e., quality of installation), internal convection, and through convection (i.e., infiltration, exfiltration, wind washing and re-entrant looping). The impact of such "anomalies and "defects" is not captured in the R-value metric. Furthermore, the discrepancy between the actual heat flow and that predicted by combining R-values increases as the absolute temperature, the temperature difference, and the net resistance to heat flow increase. These realizations have generated a growing interest in the development of a new metric for the thermal performance of building enclosures.

NEED FOR A NEW THERMAL METRIC

To improve upon the simple addition of material R-values, a new thermal metric must address the factors known to have a significant influence on heat flow. These factors can be grouped into five categories:

1. Thermal bridging
2. Changes in material properties (e.g., with temperature or time)
3. Airflow in and through building enclosures
4. Thermal mass (including phase changes)
5. Hygric storage (and the associated latent effects and conductivity changes)

This research paper addresses only the first three categories directly.

Thermal Bridging

Thermal bridging has been, and continues to be, extensively studied. Oak Ridge National Laboratory in particular has undertaken numerous measurements and simulations of full-scale walls and collated the results in a series of papers (Christian and Kosny 1995). They proposed several definitions for assembly R-value: center-of-cavity R-value, clear-wall R-value and whole-wall R-value. The most widely adopted of these, clear-wall R-value, accounts for thermal bridging through regular, necessary framing members (e.g., studs, girts, clips) in a clear section of wall that is free of penetrations (e.g., windows, doors, vents) and intersections (e.g., with roofs, foundations, other walls) Clear-wall R-value is used in some codes and standards (e.g., ASHRAE 90.1) to define the thermal performance of building enclosures. The development and adoption of clear-wall R-value can be viewed as evidence of the industry's interest in and willingness to accept an evolution of metrics for the thermal performance of building enclosures.

Bombino and Burnett (1999) used a steady-state 2D heat flow software package and the clear-wall R-value concept to emphasize the impact of thermal bridging through steel studs and demonstrate the benefit of continuous exterior insulation. Codes have evolved to acknowledge the impact of thermal bridging; many now provide different prescriptive requirements for installed R-values in walls that employ continuous exterior insulation.

Thermal bridging problems can be made more complex by air leakage and moisture issues. Many complex thermal bridging problems can now be addressed using sophisticated, commercially-available software for predicting steady-state 2D and 3D heat flow. These computer programs cannot predict air leakage effects, thermal expansion/contraction, or moisture effects. A physical test facility is required to simulate, study, and quantify these effects in order to fully assess the impact of thermal bridging on building performance.

Changes in Material Properties

It is well accepted that a material's thermal properties can change as a function of temperature, time, moisture content, etc.

The temperature dependency of material thermal properties can have significant practical implications. Figure 1 shows the measured conductivity versus temperature for a range of materials employed in this research project. Over the temperature range that is relevant to buildings, most insulation materials exhibit a near linear increase in apparent conductivity. This is expected and largely predictable because of the role of gas conductivity and radiation. However, some refrigerant-blown foam insulation materials have been shown to exhibit dramatic and non-linear relationships between temperature and apparent conductivity (Graham 2010, Schumacher 2013), as is the case for the polyisocyanurate insulation (shown in Figure 1) that was used for the thermal break in the wall specimen cartridge.

The US Federal Trade Commission produced a significant regulation in 1979 that has profoundly impacted the reporting of insulation thermal properties in the United States and Canada. The regulations were developed to promote the comparison of insulation materials under similar conditions. The FTC R-value rule (FTC 1979) requires testing to be conducted at a mean temperature of 75°F (24°C), although it does not specify a temperature difference or an orientation for the samples. However, ASTM C1058-03 *Standard Practice for Selecting Temperatures for Evaluating and Reporting Thermal Properties of Thermal Insulation*, which is referenced by the rule, provides only one set of test temperatures with a mean of 75°F (24°C): a hot side of 100°F ± 9°F (38°C ± 5°C) and a cold side of 50°F ± 9°F (10°C ± 5°C). This is not the only set of temperatures that are allowed, but it is by far the most commonly chosen by manufacturers for testing and reporting R-values. It seems appropriate to conduct material R-value tests over a wider range of temperatures that might reflect realistic building applications. For example, insulation for a wall assembly in a cold climate might be tested with –10°F (–23°C) on the cold side and 72°F (22°C) on the warm side. Alternatively, hot-climate roof insulation could be tested at a warm side of 140°F (60°C) and a cold side of 75°F (24°C).

Realistic temperatures should also be employed in testing the thermal performance of building enclosures. When build-

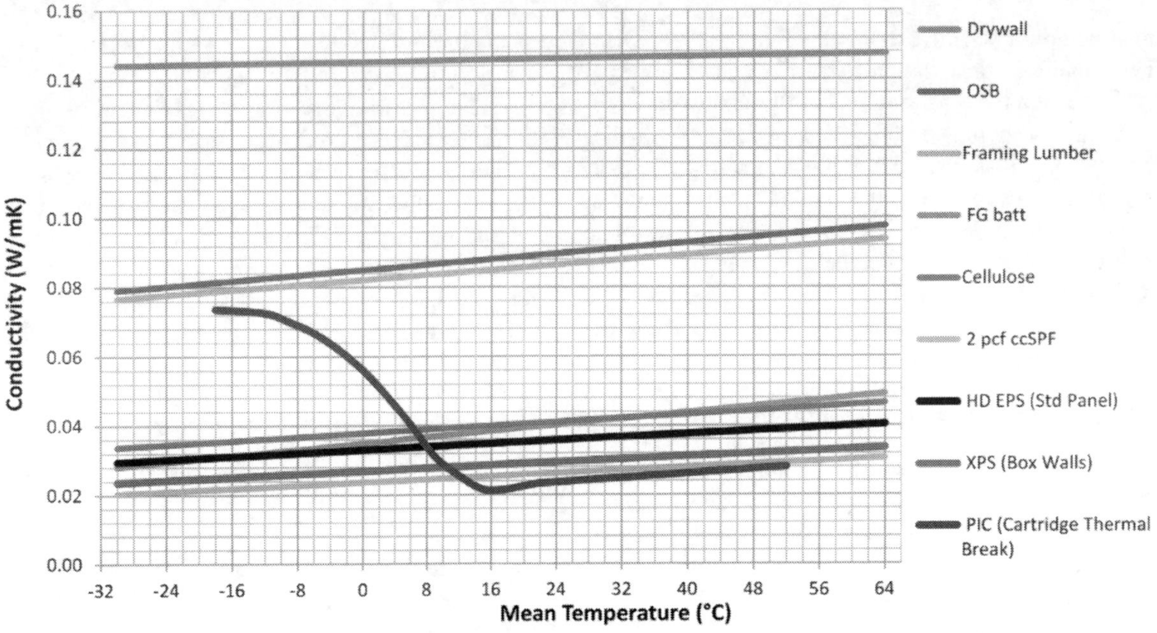

Figure 1 *Measured conductivity of selected materials used in the construction of the new hot box and in test wall specimens.*

ing enclosures are subjected to realistic temperature changes, materials expand and contract, gaps and paths open and close, and the thermal performance of the assembly changes. These effects are not easily predicted; hot box testing is often needed to assess their relevance to any comprehensive metric of assembly thermal performance.

The performances of some insulation materials also change with time. For example, some closed-cell foam insulation materials are of particular interest. ASTM C1303-07 *Standard Test Method for Predicting Long-Term Thermal Resistance of Closed-Cell Foam Insulation* (the LTTR method), was developed specifically to address gas replacement and provide an estimate of the long-term thermal performance of rigid foam insulations that use blowing agents.

Airflow in and through Building Enclosures

It has long been recognized that the control of airflow is a crucial and intrinsic part of heat and moisture control in modern building enclosures (Wilson 1963, Garden 1965). Air *can carry* heat with it and hence convection is one of the primary modes of heat transfer. Figure 2 shows four different airflow paths that influence the thermal performance of building enclosures.

Insulation material R-values do not account for any airflow through the material as a perfect air barrier is assumed somewhere in the assembly. For the purposes of building energy modeling, designers simply add the heat associated with predicted air leakage (i.e., flow path 1) to the predicted building enclosure heat flow (i.e., resulting from conduction). However, this additive approach may not properly account for the impact of air leakage. For example, Yarborough and

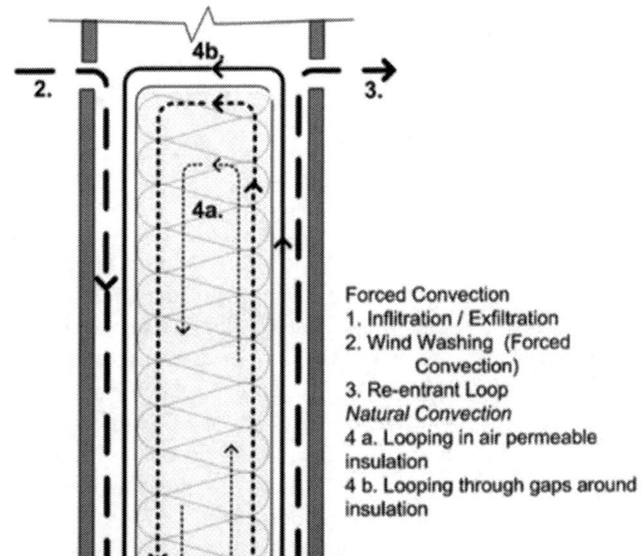

Figure 2 *Common convective airflow paths in and through enclosures.*

Graves (2006) conducted a bench top study of heat flow through air permeable insulation with imposed airflow using a modified ASTM C518 test approach. They found that the simple addition of conductive and air leakage heat flows is not sufficient to predict the measured thermal performance. This is expected because the temperature within air permeable insulation materials will vary as air flows vary through them, disturbing heat flow patterns. Airflow through diffuse and tortuous paths will result in long contact time, resulting in some heat recovery.

Lecompte (1991) conducted an extensive hot box study of the impact of convective loops (i.e., flow path 2 or 3) through small but realistic gaps between air impermeable board insulation and its substrate (i.e., continuous exterior insulation over sheathing). He measured a 30% increase in heat flow for a 1/4 in. (6 mm) gap and a doubling of heat flow for a 3/8 in. (9.5 mm) gap when considering an approximately R-10 assembly.

The impact of wind washing on thermal performance (i.e., flow path 2) has been studied in Scandinavia. Finnish research (Uvloskk 1996) demonstrated that heat loss due to wind washing in a test hut increased by 10% to 30% depending on wind speed.

Insulation products are not always installed as intended by their manufacturers or in a manner similar to how the R-value is tested. If the insulation is improperly installed or gaps form because of shrinkage or settling, flow paths can be formed. Brown et al. (1993) conducted a hot box study to assess the impact of gaps that can occur in the corners of stud spaces when batts are poorly fitted (i.e., flow paths 4a and 4b). For large gaps and large temperature differences, reductions in thermal performance of 25% to 33% were measured. Trethowen (1991) conducted a hot box study of natural convection looping around air impermeable insulation installed in a cavity (i.e., flow path 4b). In his experiment a 5/8 in. (15 mm) cavity was provided on both sides of an air impermeable e xpanded polystyrene (EPS) board insulation installed in a wood stud wall. A variable width of gap at the top and bottom of the stud space was used to investigate workmanship effects. With a gap of 1/8 in. at the top and bottom, heat flow was almost twice that of the sealed gap condition (a 50% reduction in thermal performance).

Previous full-scale physical (i.e., hot box) tests have focused on airflow paths 2, 3, and 4. Little physical testing has been conducted to assess the impact of flow paths that have air moving from one side of the wall to the other. For example, flow paths that combine 1, 2, and 3 are complex and increase the contact time and thus the potential impact on heat flow. Chebil et al. (2003) did investigate these impacts using a computer model and showed significant 8% to 15% changes in heat flow for reasonable ranges in air leakage, depending on flow path. The only comprehensive tests made of airflow through enclosure walls in a hot box were conducted by Jones, et al. (1995). They undertook the hot box measurements of 40 different assemblies with and without air leakage. The base wall was a traditional nominal R-12 2 × 4 wall with R-2.5 foam sheathing. The air leakage of the walls covered a wide range from as little as 0.3 l s/m^2 at 75 Pa to over 1.2 l s/m^2 at 75 Pa. Jones et al showed that: "Test results for the wall assemblies reveal that airflow rates as low as 0.2 L s/m^2 can produce a 46% increase in apparent thermal conductance". The influence of wind washing, although not studied in detail due to equipment limitations, was projected to reduce the thermal resistance by 18%. These percentages would be much higher for high-R walls.

NEED FOR A NEW HOT BOX

Physical measurement will be required for the development of metrics that more accurately capture the thermal performance of new building materials and systems. Laboratory testing is preferred because it permits the control necessary to isolate the influence of various parameters. However, few conventional hot boxes have ability to control important variables (e.g., temperature, air pressure, humidity) either simultaneously or within the range of interest. A new hot box apparatus is needed for this work.

DESIGN REQUIREMENTS FOR NEW HOT BOX

A new hot box apparatus was designed, constructed and commissioned to address the needs outlined above. Table 1 summarizes the design requirements that were established.

Table 1. Design Requirements for New Hot Box Apparatus

Consideration	Requirement	Motivation
Wall Area	100 square ft (9.3 m^2)	Sufficiently large to accept full-scale test wall specimens. Also to maximize signal to noise ratio.
Wall Thickness	2 to 16 in. (51 to 406 mm)	Accommodate new technologies from VIP to double stud walls.
Temperature Range	–4°F to +140°F (–20°C to +60°C)	Reproduce conditions that represent 10% of hours in cold climates through to 10% of hours on solar heated surfaces.
R-value Range	R15 to R60 (RSI-2.6 to 10.6)	Accommodate assemblies that are typical of those constructed now through to those that may be constructed in the foreseeable future.
Air Leakage Pressure Difference	±20 Pa	To reproduce conditions experienced by mid-rise buildings for 10% of the year.
Air Leakage Flow Rate	0.4 to 0.04 cfm/square ft. at 75 Pa	Accommodate assemblies that are typical of those constructed now through to those that may be constructed in the foreseeable future.

Several other capabilities were considered for future testing: the ability to assess moisture effects (both material conductivity and latent effects) and the ability to assess dynamic effects (including thermal mass and phase change materials).

FEATURES OF THE NEW HOT BOX

In general, the new hot box apparatus has been designed and constructed in accordance with ASTM C1363, *Standard Test Method for Thermal Performance of Building Materials and Envelope Assemblies by Means of a Hot Box Apparatus*. ASTM C1363 recognizes two configurations for hot boxes: guarded and calibrated. Figure 3 provides schematics of a conventional guarded hot box, a conventional calibrated hot box, and the new hot box. In all three apparatuses the test wall specimen is installed between a meter box and a climate box. The conventional guarded hot box and the new hot box include a third box, the guard box that seeks to eliminate the temperature difference across the walls, floor, and roof of the meter box.

The new hot box features a number of improvements beyond conventional ASTM C1363 hot boxes:

- A double guard (insulated guard box plus liquid guard loop) to improve control over the temperature differential across the meter box walls and minimize uncertainties,
- Equipment to control and measure both heating and cooling of the meter box,
- An air transfer system to induce infiltration/exfiltration,
- A tracer gas system to measure air exchange over all tests (not only those run with intentionally induced air pressure differences),
- A deeper meter box to permit the testing of wall-wall and wall-floor intersections at close to full scale,
- Draw-through fans to create more realistic airflow over the inside surface of the wall specimen, and
- A modified specimen frame or "cartridge" to control flow of heat and mass at the perimeter of the metered area of the test wall specimen.

METER BOX

The meter box has a depth of 1.5 m (5 ft) to permit testing of wall-wall and wall-floor intersections at full scale. This is significantly deeper than conventional hot boxes which are usually designed to minimize depth and meter box wall area in an effort to minimize heat loss across the meter box walls. The new hot box design uses a double guard and meter box wall insulation to offset the additional wall area associated with the increased depth of the box.

Construction

All three of the hot box boxes are constructed using custom assembled structural insulated panels comprising 11 mm (7/16 in.) good, one-side plywood adhered to either side of a solid layer of 100 mm (4 in) extruded polystyrene (XPS) insulation to create a stiff, strong, airtight wall with an unbridged, continuous thermal resistance of more than RSI-3.7 (R21). These SIPs are attached to the inside of a steel exoskeleton using fasteners that only penetrate the outer layer of plywood. The meter box walls are insulated with an additional RSI-1.76 (R10) of foil-faced polyisocyanurate insulation. The foil acts as an isothermal surface to which to fasten temperature sensors, and as a low-emissivity surface that ensures a uniform radiant exposure behind the insulated air baffles.

Air Circulation and Measurement

The insulated baffles are used to form consistent vertical airflow patterns over the interior faces of the test wall specimen to ensure predictable and uniform heat transfer coefficients. The baffles consist of RSI-0.88 (R5) insulation boards with a low-emissivity foil skin facing the inside of the meter box and a painted plastic skin facing the wall specimen. The low-emissivity foil skin and the insulation ensure that the baffle is at a constant temperature close to that of the air that is travelling across the face of the test wall specimen. The painted plastic skin ensures that the surface of the test wall specimen radiates to the baffle as a real wall would to its surrounding environment. Calibrated precision thermistors (±0.1°C) are used to measure temperatures at 24 points on the baffle surface, 24 corresponding points in the air stream, and 24–36 points on the interior surface of the wall test specimen.

Airflow in the baffle space is induced by a set of DC axial circulation fans at the top or the bottom of the baffle. The fan speed can be adjusted to draw the air through the baffle space at velocities representative of natural convection in real-world conditions, typically 0.3 m/s (1 fps). The lower fans are used to draw air in and down the wall during cold climate tests while the upper fans are used to draw air in and up the wall during hot climate tests. The use of draw-through fans (rather than blow-through) ensures that velocities over the test wall specimen are uniform and the flow is not turbulent. Uniformity of the flow velocity profile over the specimen was confirmed via hot wire anemometer traverses. The voltage and current to the circulation fans are measured across precision (±0.01%) resistors so that the power added to the meter box may be calculated.

Heating System and Energy Measurement

The temperature in the meter box is controlled by electric heat and hydronic cooling. Two heating arrays, each consisting of 48 heaters and 8 mixing fans, are installed in the upper and lower portions of the mixing part of the meter box as seen in Figure 4. The size, number, and distribution of the heaters and fans ensure that the temperature is relatively uniform throughout the meter box. The heaters and fans are divided into four separately measured banks. The voltage drop across each bank is measured, using the hot box monitoring and control system (MCS), across a voltage divider comprising eight 1 Mohm ±0.1% resistors installed in paral-

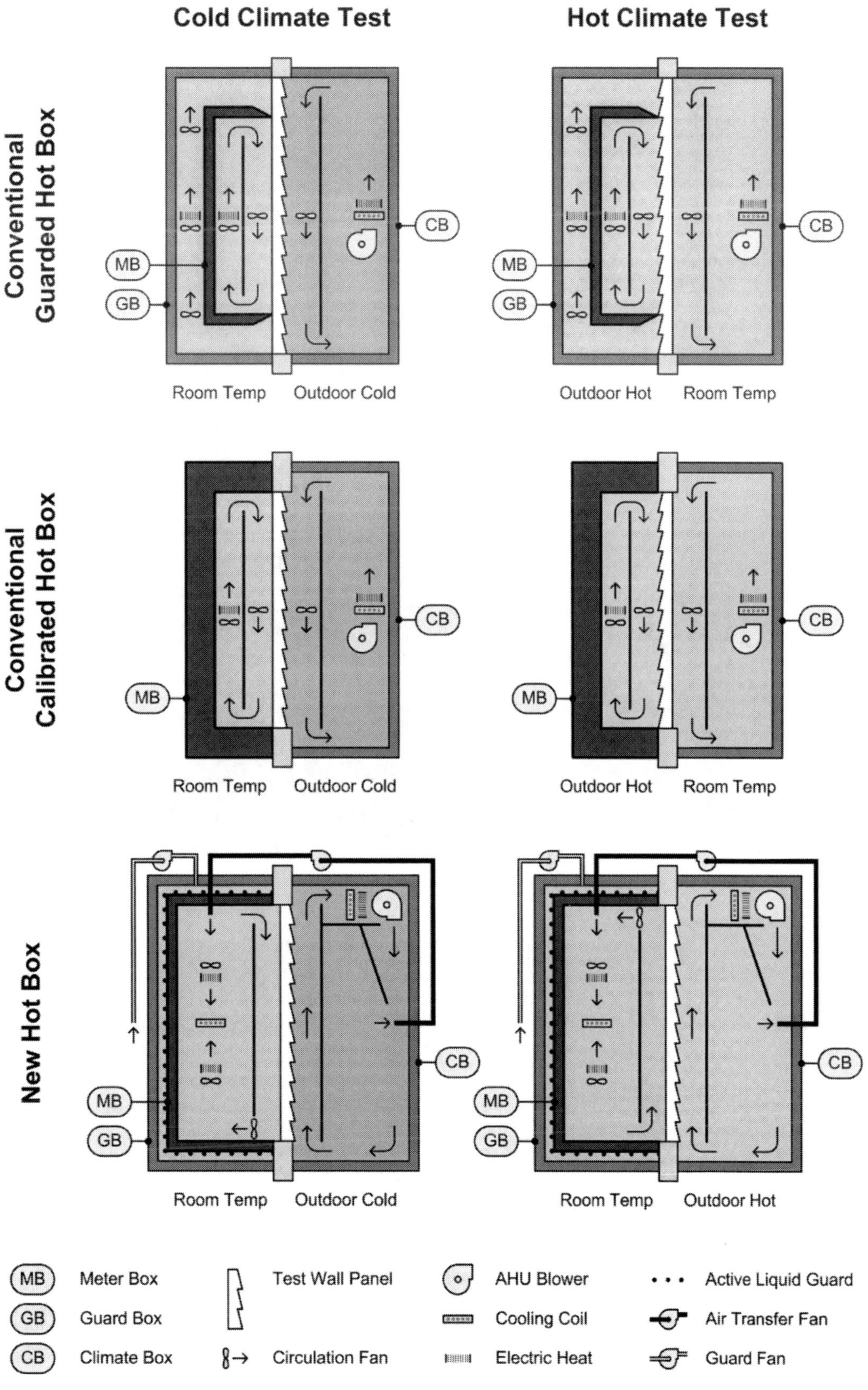

Figure 3 *Schematics of conventional guarded hot box, conventional calibrated hot box, and the new hot box apparatus developed as a result of this research.*

Figure 4 *Upper and lower heating arrays and cooling coil mounted on a movable rack in the meter box.*

lel with the resistance heaters. The monitoring and control system (MCS) also measures the current in each bank as the voltage drop across 1 ohm ±0.01% 7 W resistors with Kelvin connections. The uncertainty in power measurement is better than ±0.25% over the full range of heater output.

Cooling System and Energy Measurement

The new hot box apparatus achieves meter box cooling using a large, finned convection coil, mounted at mid-height in the air mixing part of the meter box. The large heat transfer area permits the removal of significant amounts of heat with only modest (e.g., 2°C or 3.6°F) temperature increases between the inlet and outlet of the coil. Distilled water is pumped from a chilled, constant temperature (±0.05°C) buffer tank, into the meter box, through the convection coil, and back out of the meter box. The flow rate (typically 0.5 to 1.5 lpm) is measured using a NIST traceable ±0.2% of reading flow meter and the supply and return temperatures are measured using a pair of precision thermistors (±0.1°C) and a pair of ultra-precision RTDs (±0.012ohm). These measurements can then be used to calculate the energy the cooling extracts from the meter box.

The cooling coil and the two heating arrays are mounted on a rack that can be moved forward or backward into the meter box as necessitated by the geometry of the test specimen. The overall accuracy of the cooling measurement system is estimated to be ±1.5%.

GUARD BOX

The new hot box employs a double guard: an insulated guard box surrounds the meter box and an active liquid guard loop is installed over the outside surface of the meter box as seen in the photograph of Figure 5. The guard box minimizes the influence of temperature changes in the lab and reduces spatial temperature gradients over the surface of the meter box. The liquid guard loop further reduces any spatial temperature gradients and all but eliminates any temperature difference between the inside and the outside of the meter box walls.

Figure 5 *View illustrating the double-guarded concept: a conventional guard box on the left and an active liquid guard installed on the meter box on the right.*

Construction

The guard box employs the same basic custom SIP in steel exoskeleton construction as the meter box: 11 mm (7/16 in.) good, one-side plywood skins adhered to either side of a solid layer of 100 mm (4 in) XPS insulation for a continuous thermal resistance of more than RSI-3.7 (R21).

Active Liquid Guard

The active liquid guard comprises six separate circuits of PEX piping that are attached to the outside of the meter box using aluminum heat transfer plates. The six circuits (one on each of the top, bottom, and sides, plus two on the back) are set up as secondary loops off of a primary loop that provides conditioning for the active liquid guard. Each circuit can be individually controlled with metering valves to allow the flows to be calibrated from time to time to ensure spatial uniformity of the temperature. The water flow of each loop has been designed to absorb or release the expected heat flow through the RSI-3.7 (R21) guard chamber walls (in the range of 2 to 4 W per loop) with a temperature rise of less than 0.005°C (0.009°F).

Delta T Measurement

The temperature difference or delta T across the meter box walls is measured by paired precision thermistor arrays that are applied to the inside and outside of each of the five faces of the meter box at a density of more than five sensors per square meter. In all, the temperature difference is measured at 176 locations. Using the aggregated differential temperature measurements, the hot box system controls the active liquid guard supply temperature, thereby reducing the average temperature difference across the meter box to less than 0.05°C (0.09°F). Typical average temperature difference during operation is better than 0.02°C (0.04°F).

Heating and Cooling

In the new hot box, the guard box is not conditioned using electric resistance heaters and circulation fans as is common in conventional guard boxes (which typically have no meter box cooling). Instead, the temperature in the guard box is controlled by the active liquid guard. The primary loop of the liquid guard is indirectly cooled using a plate heat exchanger. The primary loop flow rate is fixed so that the outlet temperature on the guard side of the exchanger is controlled by varying the temperature on the chiller side. This is accomplished through the use of a computer controlled three-way mixing valve. The fluid is slightly overcooled (typically 0.2°C or 0.36°F) then precisely heated to the target setpoint by a pair of pulse-controlled electric resistance band heaters.

CLIMATE BOX

The climate box has the same overall size as the guard box.

Construction

The climate box employs the same basic custom SIP-in-steel exoskeleton construction as the guard box: 11 mm (7/16 in.)

good, one-side plywood skins adhered to either side of a solid layer of 100 mm (4 in) XPS insulation for a continuous thermal resistance of more than RSI-3.7 (R21).

Air Circulation and Measurement

Insulated baffles are used to form consistent vertical airflow patterns over the exterior face of the test wall specimen and establish uniform heat transfer coefficients. The baffles use the same construction as the meter box.

Airflow in the baffle space is induced by a set of four air handling units, each with an ECM blower programmed to generate flow rates of 165, 212, 283, and 378 lps (350, 450, 600, and 800 cfm). The fan speed can be adjusted to push air through the baffle space at velocities representative of various wind speeds in real world conditions, typically 0.9 to 1.8 m/s (3 to 6 fps). The flow velocity profile can be tuned via a series of adjustable vanes located in the supply plenum at the back of the climate box. Uniformity of the flow velocity profile over the specimen was confirmed via hot wire anemometer traverses.

Heating and Cooling

The temperature in the climate box is controlled by a series of four fan coils connected to an 8 kW at –20°C (and 4.5 kW at –30°C) capacity air-cooled liquid chiller and a 3 kW hydronic heater. The oversized coils allow for a very small temperature drop across the coil during most test conditions. Reheat coils and individually-controlled tight-fitting dampers allow for individual defrost. This feature allows three fan coils to continue conditioning and circulating air while the fourth is defrosted. The reheat coils can be used for humidity control, and the low temperature drop cooling coils allow RH levels of 90% to 95% rh to be maintained over most of the temperature range. In general the chiller temperature is set to overcool and fine tuning of the temperature is accomplished using 1.5 kW of pulse-controlled electric resistance heat.

WALL SPECIMEN (SAMPLE) CARTRIDGE

Section 6.7.1 of ASTM C1363 requires the provision of a specimen frame to support the wall test specimen in position between the meter box and climate box and to insulate the perimeter of the specimen to reduce flanking heat transfer. In a conventional guarded hot box, the wall test specimen area extends beyond the perimeter of the meter box so the portion of the wall between the meter box and the climate box sees the same heat flow as the portion of the wall between the guard box and the climate box. This is an extremely effective method of minimizing flanking heat flows; however, when hollow (e.g., framed) walls are tested, it provides paths for air to flow not just between the climate box and the meter box, but also between these two boxes and the guard box.

The interaction between heat and airflow is of particular interest in this research program, hence a new specimen frame was designed to minimize flanking losses while eliminating airflow outside of the area of the wall test specimen.

Construction

The new hot box specimen frame or "cartridge" comprises alternating layers of 11 mm (7/16 in) plywood and 100 mm (4 in) XPS foam board glued up to create an exceptionally stiff sandwich panel as seen in Figure 6. Two 38 × 38 mm (nominal 2 × 2 in.) nailers are embedded in the cartridge to provide fastening support. A 100 mm (4 in) thick polyisocyanurate thermal break lines the entire rough opening of the cartridge. A 3 mm (1/8 in.) thick sacrificial MDF lining protects the polyisocyanurate insulation and facilitates air sealing between the cartridge and the test wall specimen. The finished opening of the cartridge and the size of the wall test

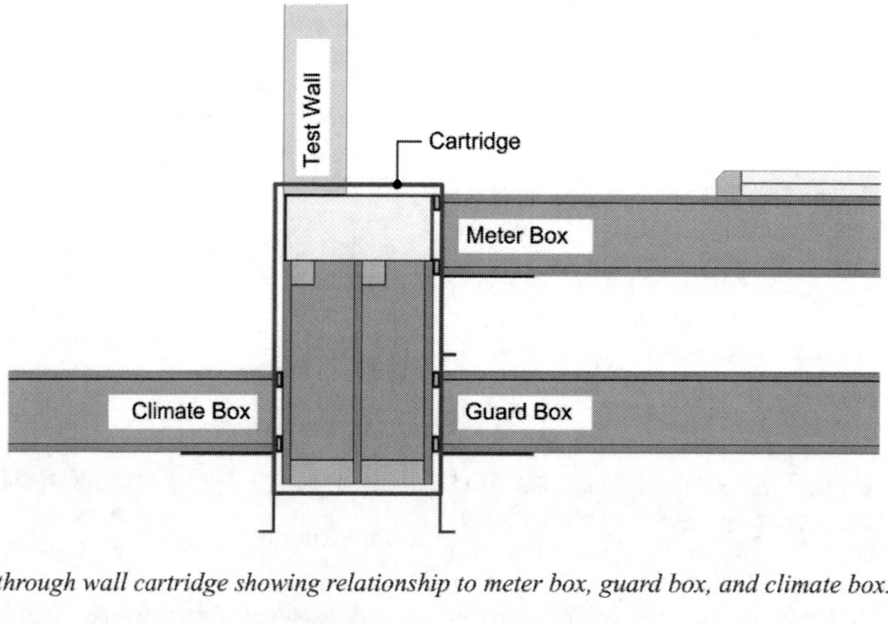

Figure 6 *Section through wall cartridge showing relationship to meter box, guard box, and climate box.*

specimen match the meter box opening: 3.66 m (12 ft) wide by 2.44 (8 ft) high.

Box-to-Box Air Sealing

Box-to-box air sealing is accomplished using two-stage joints comprising mated closed-cell foam gaskets. These air seals exist at three interfaces: between the meter box and the cartridge, between the guard box and the cartridge, and between the climate box and the cartridge. The gaskets are engaged, compressed and sealed using 20 tons of clamping force.

Sample Air Sealing

After the test specimen is installed in the cartridge, the perimeter of the test wall specimen is sealed (both inside and out) to the MDF cartridge liner using tapes and sealants as appropriate for the building materials in the test wall assembly.

PRESSURIZATION AND AIR TRANSFER SYSTEM

One of the most novel aspects of the new hot box is the air transfer system (ATS). The system, pictured in Figure 7, generates a pressure difference between the meter box and the climate box to drive airflow through available paths in the test wall specimen. The system comprises an inline fan, an inline heater, three high accuracy (±2% of reading) mass flow sensors (installed in parallel), and piping and valves to allow researchers to negatively pressurize (i.e., induce infiltration) or positively pressurize (i.e., induce exfiltration) the meter box. A guard fan is used to minimize the pressure difference between meter and guard boxes so that airflow only occurs between the meter and the climate boxes.

The system can accommodate flow rates up to 900 lpm (32 cfm) at pressures of up to 50 Pa. This imposes leakage rates of up to 100 lpm/m^2 (0.34 cfm/ft^2).

Heat transfer associated with the airflow is calculated using the measured flow rate, the heat capacity of air (at the measured pressure, temperature, and humidity) and the temperature difference between the delivered air temperature (measured using an ultra-precision RTD at ±0.012 ohm) and the air temperature in the meter box (measured using an array of precision thermistors at ±0.1°C).

TRACER GAS SYSTEM

The tracer gas system permits monitoring of the air change between the meter box and the climate box during all modes of testing. This system is vital to the research as the test wall specimens include realistic air leakage paths. Hence, airflow is induced not just by the air transfer system but also by pressure differences that naturally exist as a result of stack effect (temperature differences) and circulation fans.

The tracer gas system comprises a regulated and metered supply of 99% pure CO_2; four CO_2 sensors, one in each of the meter box, the guard box, the climate box, and the laboratory; and dilution system capable of supplying 2.5 lps (5 cfm) of dilution air at –40°C (–40°F) dew point. The meter box is

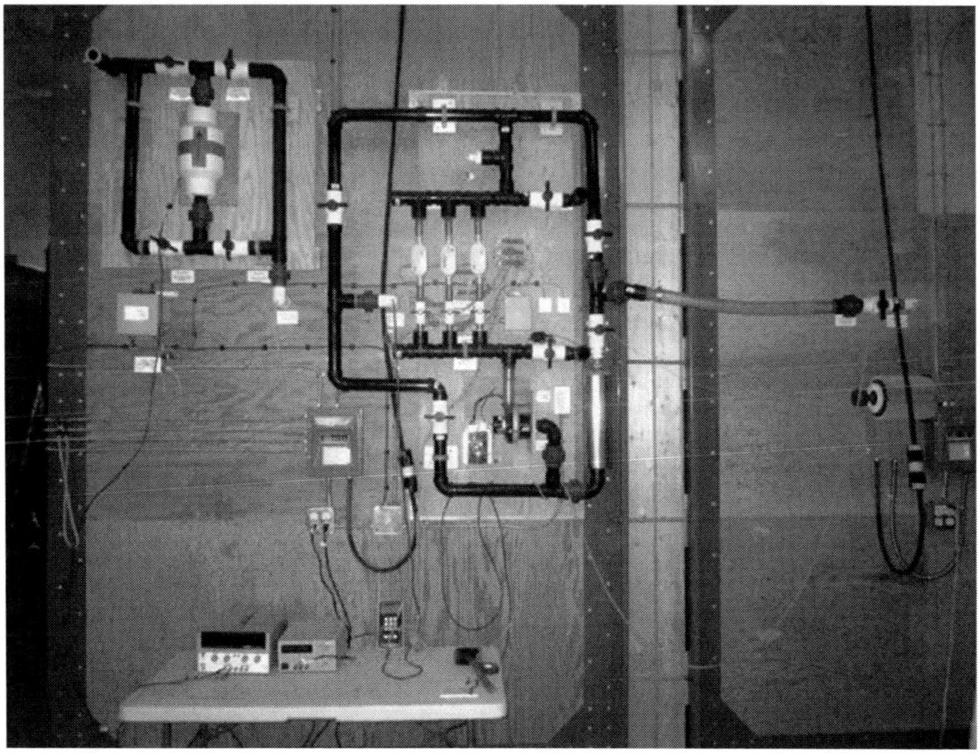

Figure 7 *Guard fan system on the left and air transfer system on the right.*

dosed to a concentration of 10,000 ppm; the air change rate between the meter box and the climate box is calculated using decay method described in *ASHRAE Handbook—Fundamentals* 2009, Chapter 16. The decay of CO_2 and correlated air transfer rate are shown in Figure 8.

TEMPERATURE MEASUREMENT (TM)

The sensors were fabricated using 10k ohm NTC precision thermistor components (Honeywell/Fenwall 192-103LET-A01) soldered to 28 AWG leads. The resulting temperature sensors are approximately the same size as the thermocouples that are typically used in hot box research.

From the manufacturer these sensors have a tolerance of ±0.2°C. The research team sought to reduce as many uncertainties as possible; hence sensors were individually calibrated over the range of temperatures in which they were to be used. The meter box guard sensors were calibrated at 16°C, 18°C, 20°C, 22°C, and 24°C while the baffle surface, air space and wall specimen sensors were calibrated at –30°C, –20°C, –10°C, 0,°C 10°C, 20°C, and 30°C.

Roughly 600 of the temperature sensors were fabricated to instrument the TM hot box and use on the test wall specimens for the first phase of research. Each sensor was assigned a unique serial number and calibrated in an aluminum calibration block set in a controlled temperature bath, a VWR 1157P, capable of maintaining the bath temperature within ±0.01°C of the setpoint. The aluminum calibration block further ensures the spatial and temporal stability of the temperature during calibration. A NIST traceable HH41 reference thermometer (±0.023°C or over the range of –20°C to 60°C) was inserted in a 100 mm (~4 in.) deep hole in the middle of the block. Sensors were calibrated in sets by inserting them in the 12 holes that circle the reference thermometer.

For each setpoint, the bath was brought to equilibrium and allowed to run for 15–20 minutes, after which five readings were taken at 1 minute intervals. For each reading, the time, the bath temperature, and the reference thermometer temperature were manually recorded. Meanwhile, the resistances of the thermistor-based temperature sensors were automatically measured and recorded using a Campbell Scientific CR1000 measurement and control system and a half-wheatstone bridge circuit with a precision (±0.01%) sense resistor. Sensor-specific calibration coefficients were determined for a third order polynomial equation. The resulting temperature sensor uncertainty is better than ±0.05°C (0.09°F).

ENERGY BALANCE

Figure 9 shows the energy balance on the meter box (MB). All heat (i.e., energy) flows are considered positive into the meter box except for the test sample heat flow which is considered positive when heat is flowing from the meter box to the climate box (CB).

Using the heat flows shown in Figure 9 results in the following energy balance:

$$Q_h + Q_c + Q_f + Q_{fl} + Q_{mw} + Q_{tas} + Q_L + Q_{ih} + Q_s = 0 \quad (1)$$

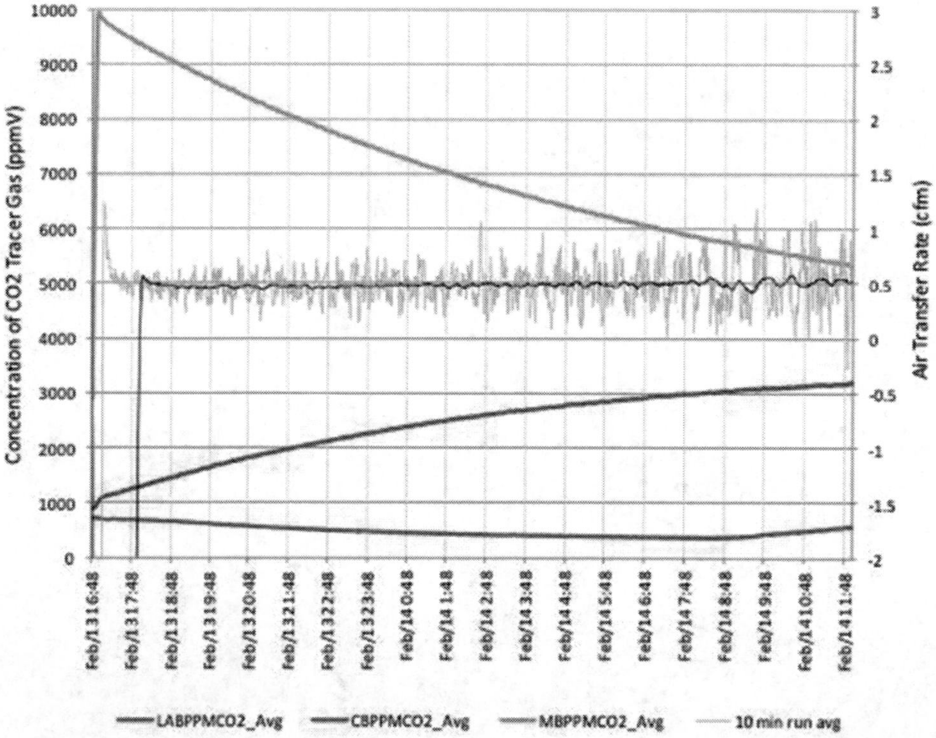

Figure 8 *CO_2 decay and air transfer rate measurement.*

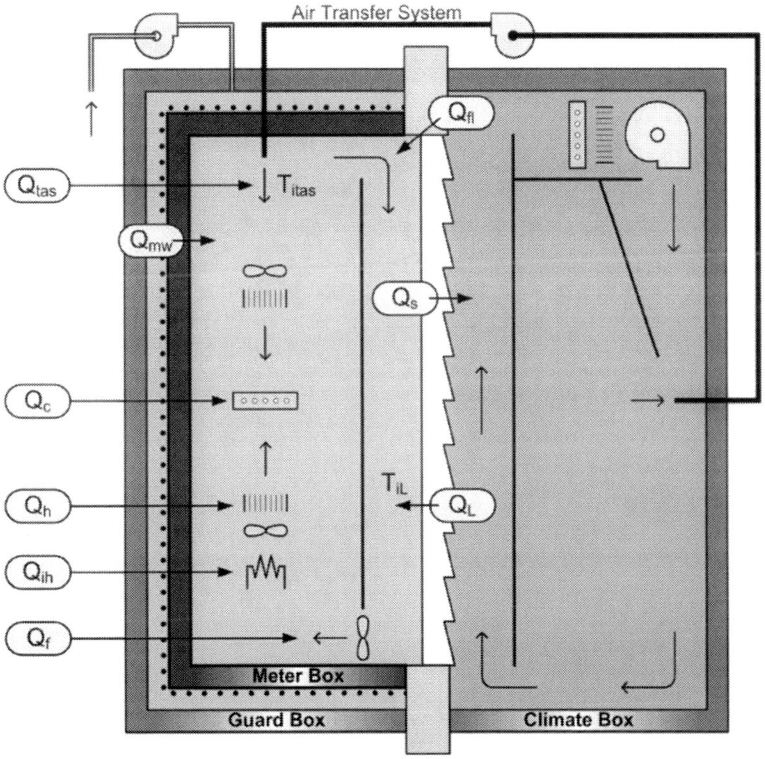

Figure 9 *Energy balance diagram for the new hot box apparatus.*

where

Q_s = heat flow through the test sample

Q_h = heat input from the electric heaters

Q_c = heat input or removal from the cooling system

Q_f = heat input from the circulation fans

Q_{fl} = flanking heat transfer through the sample cartridge and associated elements

Q_{mw} = heat exchange between the meter box and guard box through meter box walls, roof, and floor

Q_{tas} = heat flow due to air entering/exiting the meter box through the air transfer system

Q_L = heat flow due to air transfer through the sample in/out of the meter box

Q_{ih} = heat input into the independent verification heater

Temperatures and coefficients are defined as follows:

$T_{air\text{-}mb}$ = meter box air temperature

$T_{air\text{-}cb}$ = climate box air temperature

$T_{s\text{-}mb}$ = sample surface temperature on the meter box side of the sample

$T_{s\text{-}cb}$ = sample surface temperature on the climate box side of the sample

T_{itas} = air temperature at the MB/air transfer system entrance

T_{iL} = air temperature of the air at the MB/test sample interface

C_p = specific heat of air

m = mass flow rate

Temperature T_{itas} is the temperature of the air just as it enters/exits the air transfer system into the meter box. Air entering the air transfer system from the meter box represents air infiltration through the sample and is essentially at the meter box temperature. For air exfiltration air enters the meter box from the air transfer system. This needs to be included in the energy balance on the meter box whenever air flows from the air transfer system into the meter box.

Q_{tas} is calculated by

$$Q_{tas} = mc_p\,(T_{itas} - T_{air\text{-}mb}) \tag{2}$$

Q_L is calculated by

$$Q_L = mc_p\,(T_{iL} - T_{air\text{-}mb}) \tag{3}$$

For the standard hot box, $Q_{tas} = 0$, Q_L=0 and Q_{ih}=0.

For convenience Q_{tot} is defined as the heat flow that is measured during a test using the following formula

$$Q_{tot} = Q_h + Q_c + Q_f + Q_{mw} \tag{4}$$

Q_{net} is defined as needed in the sections below.

Flanking Heat Transfer Prediction

For most calibrated hot boxes Q_{fl} is determined through a series of calibration tests with samples of known thermal properties. While valid, this approach runs the potential risk of associating any other extraneous errors in measurement or design solely with Q_{fl}. Another approach that was used by one of the authors in a previous calibrated hot box design avoids this potential problem. Since in a well-designed, calibrated hot box the flanking heat flow is a conductive heat flow through materials whose material properties can be measured, it is possible to predict Q_{fl} using finite element programs and then incorporate the predicted Q_{fl} into the energy balance. Then one or more samples of known thermal properties are tested to verify the overall operation of the hot box. This approach eliminates any arbitrary calibration constants. An additional advantage of this approach is that any interactions between a test sample and the sample cartridge can be investigated prior to testing to assess the sample/cartridge interaction and that effect, if any, on Q_{fl}.

Q_{fl} is a function of the following major factors:

- Cartridge geometry (fixed)
- Material thermal properties (measured)
- Film coefficients
- Test sample thickness
- Temperature

A 2D finite element program was then used to predict Q_{fl}. Figure 10 shows the model used in the finite element program. A typical output is shown in Figure 11. The results from the numerical simulation were post processed to obtain the correct area weighting. Runs were made for several film coefficients and sample thicknesses. Table 2 and Table 3 show the variation of Q_{fl} for three levels of film coefficients and four sample thicknesses. Table 2 shows Q_{fl} in Watts per degree Kelvin, while Table 3 shows the data in total Watts. The interior film coefficient (h_{in}) was fixed at 7 W/m^2·K typical of the heat transfer coefficient on the indoor side of an assembly. The exterior film coefficient (h_{out}) was varied from 7 to 14 to 28 W/m^2·K. The panel thickness was modeled from 4 in. to 6.5 in. to cover the range of test wall specimens considered in the first phase of testing with the new hot box apparatus. There is a substantial effect on Q_{fl} for different test sample thicknesses but only a small effect due to the film coefficients.

MATERIAL PROPERTY DATA

After the hot box was assembled, the thermal properties of the cartridge materials were measured over a temperature range of –18°C to +42°C using the C518 machine.

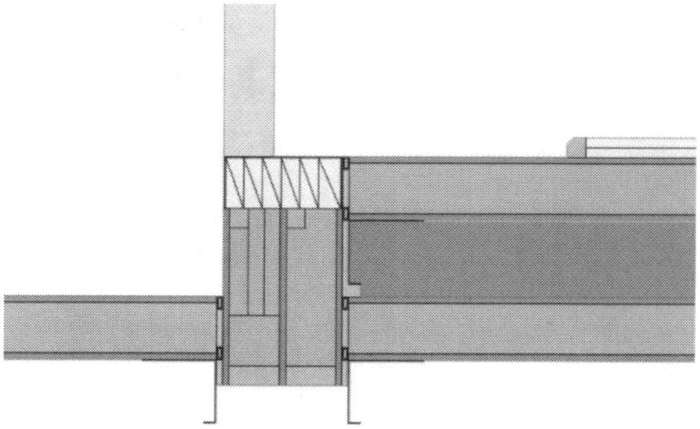

Figure 10 *Model of wall cartridge for predicting flanking using 2D finite element heat transfer program.*

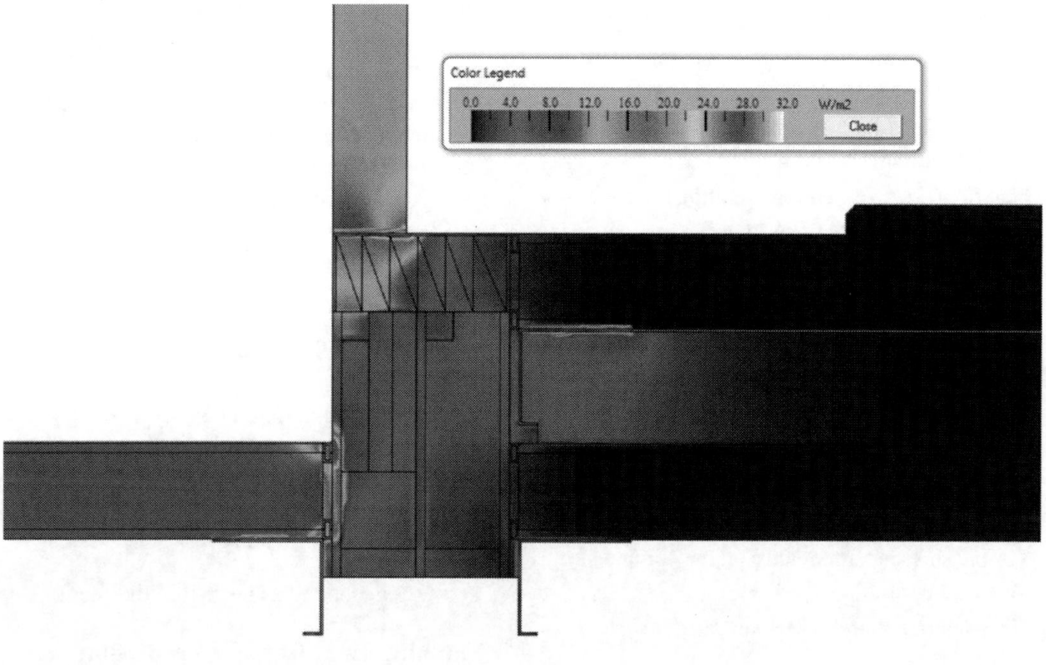

Figure 11 *Heat flux intensities predicted using 2D finite element heat transfer program.*

Table 2. Flanking Heat Flow (in W/K) Calculated from the Finite Element Program

Wall Thickness (in.)	(mm)	h_{in} ($W/m^2 \cdot K$)	h_{out} ($W/m^2 \cdot K$)	Climate Box Temperature 62°C	42°C	2°C	−18°C	−28°C
				Correction as Function of Air-Air delta T (W/K)				
			7	0.2816	0.2681	0.2835	0.4301	0.4756
4	101.6	7	14	0.2855	0.2717	0.2894	0.4393	0.4886
			28	0.2863	0.2740	0.2896	0.4444	0.4928
			7	0.2523	0.2425	0.2527	0.3737	0.4124
4.5	114.3	7	14	0.2558	0.2456	0.2578	0.3829	0.4213
			28	0.2573	0.2471	0.2596	0.3869	0.4261
			7	0.2303	0.2179	0.2444	0.3303	0.3556
5	127	7	14	0.2310	0.2204	0.2478	0.3364	0.3645
			28	0.2344	0.2238	0.2502	0.3416	0.3682
			7	0.1693	0.1603	0.1717	0.2454	0.2795
6.5	165.1	7	14	0.1711	0.1620	0.1737	0.2494	0.2841
			28	0.1718	0.1628	0.1747	0.2533	0.2869

Table 3. Flanking Heat Flow (in W) Calculated from the Finite Element Program

Wall Thickness (in.)	(mm)	h_{in} ($W/m^2 \cdot K$)	h_{out} ($W/m^2 \cdot K$)	Climate Box Temperature 62°C	42°C	2°C	−18°C	−28°C
				Correction as Function of Air-Air delta T (W/K)				
			7	11.27	5.36	−5.67	−17.21	−23.78
4	101.6	7	14	11.42	5.43	−5.79	−17.57	−24.43
			28	11.45	5.48	−5.79	−17.78	−24.64
			7	10.09	4.85	−5.05	−14.95	−20.62
4.5	114.3	7	14	10.23	4.91	−5.16	−15.32	−21.06
			28	10.29	4.94	−5.19	−15.47	−21.30
			7	9.21	4.36	−4.89	−13.21	−17.78
5	127	7	14	9.24	4.41	−4.96	−13.46	−18.22
			28	9.38	4.48	−5.00	−13.66	−18.41
			7	6.77	3.21	−3.43	−9.82	−13.97
6.5	165.1	7	14	6.84	3.24	−3.47	−9.97	−14.21
			28	6.87	3.26	−3.49	−10.13	−14.35

Thermal conductivity data was measured for the following materials and is shown in Figure 1.

- Expanded polystyrene (EPS) used for the standard samples
- Extruded polystyrene (XPS) used in the hot box walls
- Polyisocyanurate used in the cartridge thermal break
- Spruce/pine/fir (SPF) used in the cartridge thermal break
- Plywood used as an external surface for the hot box walls
- Hardboard (MBF) used as an air sealing surface for cartridge thermal break
- Steel used for the exoskeleton support structure

EPS Thermal Properties

The following equation was used to predict the thermal conductivity of the EPS for the standard sample versus mean temperature, T_m:

$$k = 0.032975 + 0.000116\ T_m\ \text{(W/m K)} \tag{5}$$

XPS Thermal Properties

XPS thermal properties used in the hot box are given by:

$$k = 0.026833 + 0.0001067\ T_m\ \text{(W/m K)} \tag{6}$$

Polyisocyanurate Thermal Properties

The polyisocyanurate used in the cartridge walls shows a substantial increase in thermal conductivity at temperatures below about 10°C. This is due to condensation of the blowing agent which increased the thermal conductivity. Considerable testing in using the C518 equipment was performed to characterize the thermal conductivity of this polyisocyanurate versus mean temperature. In particular, we were interested in the thermal conductivity at small temperature differences. This involved first verifying that the C518 calibration was valid for temperature differences as small as 3°C. Once this verification was complete, the polyisocyanurate was tested at each mean temperature at temperature differences of 12°C, 9°C, 6°C, and 3°C. Table 4 summarizes the thermal break conductivity measured at nineteen mean temperatures. This thermal conductivity data was then used in numerical heat transfer program to predict Q_{fl} as described above.

Other Thermal Properties

These properties are based on handbook values at 24°C since these components have a small effect on the overall thermal resistance.

$$\text{SPF} - k = 0.0862\ \text{(W/m K)}$$

$$\text{Plywood} - k = 0.010913\ \text{(W/m K)}$$

$$\text{Hardboard} - k = 0.1053\ \text{(W/m K)}$$

$$\text{Steel} - k = 50\ \text{(W/m K)}$$

Except for the polyisocyanurate all the materials show expected variation with mean temperature (i.e., lower thermal conductivities with lower mean temperatures).

STANDARD SAMPLE CONSTRUCTION FOR HOT BOX VERIFICATION

A nominal 12 × 8 ft standard sample was assembled, consisting of two layers of 2 in. thick 25 kg/m^3 EPS sheets. The EPS had been previously purchased and allowed to age for 6 months. The joints in each layer were staggered to minimize short circuiting through the joints and to provide structural integrity. The two layers were glued together. The exterior surfaces were then painted black to seal the sample against air leakage and to provide a high emissivity surface which simulates normal building construction emissivities. After assembly, thickness measurements were made around the perimeter of the sample. During assembly, samples of the EPS were retained for testing per C518. After assembly the finished sample was cut to the size of the opening in the sample cartridge to assure a tight fit.

The retained specimens were tested per C518 over a temperature range of –18°C to +52°C (0°F to 126°F) as discussed previously. A linear thermal conductivity versus mean temperature (T_m) curve was fit to the data.

VERIFICATION OF THE BASIC HOT BOX ENERGY BALANCE WITH NO AIR LEAKAGE

The verification sample was installed in the cartridge. Tests were run from –28°C to +62°C (–18°F to +144°F) with no air leakage. CO_2 tracer gas decay measurements confirmed that air leakage did not contribute to heat transfer.

Table 5 compares the hot box measurements on the verification sample with the assigned R-value from the C518 tests. In this table, $Q_{net} = Q_{tot} + Q_{fl}$ and Q_{std} is calculated as follows for the standard sample.

$$Q_{std} = kA(T_{out} - T_{in})/t \tag{7}$$

where

k = the standard sample thermal conductivity calculated using the C518 at the mean temperature of the hot box test

A = the sample area

t = the sample thickness

T_{out} = the surface temperature on the CB side of the sample

T_{in} = surface temperature on the MB side of the sample

Q_{std} is the expected heat flow and Q_{net} is the measured heat flow from the hot box tests. The agreement is within ±2% as shown in Table 5. Q_{fl} was calculated specifically for these tests and will vary slightly from the values shown in Table 1.

VERIFICATION OF COOLING SYSTEM ENERGY MEASUREMENTS

One of the unique features of this hot box is the capability to test under both winter and summer conditions where cooling is required for the meter box. The cooling energy measurements are generally not as accurate as the heating energy measurements. This is due to the requirement to measure cooling energy by measuring the mass flow rate of the cooling fluid and a relatively small temperature change in the fluid. To verify the accuracy of the cooling energy measurements, a test was first performed using the verification sample under cooling conditions and then adding an independent nominal 40 W electrical load in the meter box which required the same amount of cooling energy to offset the increased electrical load. Table 6 shows the energy measurement for the first test and the energy balance with the nominal 40W increased heat load. The 40W increased load, since it was an electrical load, could be measured very accurately. The cooling system has to offset this load.

As shown in the table, the Q_{net} for the baseline case is –72.0W. With added nominal 40W electrical heater the energy balance for this case is

$$Q_{net} = Q_{tot} + Q_{ih} = -109.5\text{ W} + 36.1\text{ W} = -73.4\text{ W} \quad (8)$$

where Q_{ih} is the measured energy input to the added independent nominal 40W heater.

The agreement on the total energy load is 1.9%.

VERIFICATION OF AIR LEAKAGE ENERGY LOADS

To verify that energy balances are accurate when air leakage is occurring, four tests were performed for infiltration and exfiltration using two different orifice sizes. First, the standard sample was tested and the energy use measured. This was done under summer conditions. Then a 1 in. diameter hole was carefully made in the sample and tests were repeated under the same temperature conditions for air infiltration (airflow from the CB to the MB) and for air exfiltration (airflow from the MB to the CB). For the 0.25 in. hole a plate with a 0.25 in. hole in it was placed over the 1in. hole on the CB side. The 1 in. diameter tests will be discussed in detail. Table 7 shows the measurements.

The energy balance for these cases is

$$Q_h + Q_c + Q_f + Q_{fl} + Q_{mw} + Q_{tas} + Q_L + Q_{ih} + Q_{ID} = 0 \quad (9)$$

$$Q_{net} = Q_{tot} + Q_{tas} + Q_L + Q_{ID}$$

For air infiltration Q_{iL} is calculated from

$$Q_{iL} = mc_p(T_{air\text{–}cb} - T_{air\text{–}mb}) \quad (10)$$

since $T_{iL} = T_{air\text{–}cb}$. For air exfiltration $Q_{iL} = 0$ since $T_{iL} = T_{air\text{–}mb}$.

For air exfiltration Q_{tas} is calculated from

$$Q_{tas} = mc_p(T_i - T_{air\text{–}mb}) \quad (11)$$

since T_i is measured. For air infiltration $Q_{tas} = 0$ since $T_i = T_{air\text{–}mb}$.

Table 4. Thermal Conductivity Measurements for Polyisocyanurate Insulation Used as Thermal Break in Sample Cartridge

Mean Temperature T_m (°C)	Thermal Conductivity (W/mK)
–18	0.0742
–12	0.0737
–9	0.0712
–6	0.0681
–3	0.0635
0	0.0576
3	0.0502
6	0.0420
9	0.0328
12	0.0254
15	0.0215
18	0.0216
21	0.0229
24	0.0236
30	0.0245
36	0.0255
42	0.0264
52	0.0280
58	0.0290

Solid (Baseline) test: the heat flow for the baseline case is –72.0 W.

Air Infiltration

For this test air was flowing from the CB to the MB through the hole. The flow rate was 61.5 lpm. Q_{net} is calculated as follows:

$$Q_{net} = Q_{tot} + Q_{tas}$$

$$= -97.5\text{W} + 61.5\text{ lpm} \times 1006\text{ J/kg·K} \times 1.183\text{ kg/m}^3 \times (42.04 - 21.90)\text{ C/ }(1000\text{ L/m}^3 \times 60\text{ min/s})$$

$$= 72.6\text{ W}$$

This is 2.2% different from the baseline case.

Table 5. Comparison of Hot Box Measurements and Expected Heat Flow through Standard Verification Panel

		Units	Test Conditions				
Nominal MB Temperature		°C	22	22	22	22	22
Nominal CB Temperature		°C	62	42	2	–18	–28
CB Air T		°C	61.5	41.9	2.0	–17.9	–27.9
Wall Surf T_{out}		°C	60.5	41.3	2.6	–16.6	–26.1
MB Air T		°C	22.1	22.0	21.5	21.2	21.1
Wall Surf T_{in}		°C	24.08	23.0	20.6	19.5	19.0
Measured Heat Flow	Q_{tot}	W	–141.7	–68.5	62.8	126.6	157.5
Flanking Heat Flow	Q_{fl}	W	11.78	5.63	–5.83	–17.79	–24.55
	Q_{net}	W	–129.94	–62.87	57.01	108.83	132.99
Standard Sample Heat Flow Based on Assigned Thermal Properties							
Standard Sample Heat Flow	Q_{std}	W	–127.76	–62.14	56.67	109.28	134.20
Difference	$Q_{net} - Q_{std}$	W	–2.18	–0.73	0.34	–0.45	–1.21
% diff			1.7%	1.2%	0.6%	–0.4%	–0.9%

Air Exfiltration

For this test air was flowing from the MB to the CB through the hole. The flow rate was 58.5 lpm. Q_{net} is calculated as follows:

$$Q_{net} = Q_{tot} + Q_{tas}$$

$$= -77.0\text{W} + 58.5 \text{ lpm} \times 1006 \text{ J/kg·K} \times 1.183 \text{ kg/m}^3 \times (27.5 - 21.83) \text{ C} / (1000 \text{ L/m}^3 \times 60 \text{ min/s})$$

$$= 70.6 \text{ W}$$

This is 2.0% different from the baseline case.

For the 0.25" orifice test the tests are within –3.2% and –2.2% respectively of the baseline case.

SUMMARY

The thermal performance of wall assemblies in buildings is complex. The thermal performance of building enclosures is heavily influenced by a number of factors such as insulation level, air leakage, thermal bridging, operating conditions, moisture content, and installation defects. Simple R-value metrics do not capture all of these factors. It has been recognized that there is a need to develop new thermal metrics that provide better ways to characterize these factors. It was recognized that more sophisticated testing would be required to understand these factors prior to developing the new metrics.

Table 6. Data for Verification of Cooling System Measurement

	Units	Test Conditions: Baseline	Test Conditions: Added Heater
Nominal MB Temperature	°C	22	22
Nominal CB Temperature	°C	42	42
Extra Heat Input	—	No	Yes
CB Baffle Inlet T	°C	42.04	42.04
	°F	107.7	107.7
MB Baffle Inlet T	°C	21.85	21.85
	°F	71.3	71.3
Wall Surf T_{out}	°C	41.55	41.55
	°F	106.8	106.8
Wall Surf T_{in}	°C	23.07	23.03
	°F	73.5	73.5
TASdmT	°C	NA	NA
	°F	NA	NA
Airflow	SLPM	—	—
	CFM	—	—
Q_{tot}	W	–72.0	–109.5
	btu/h	–245.6	–373.6
Q_{ih}	W	NA	36.1
	btu/h	NA	123.2
Q_{net}	W	–72.0	–73.44
	btu/h	–245.6	–250.4
Percent Difference		—	1.9%

The hot box described in this paper has been designed to meet the requirements of C1363. The hot box has the following features and capabilities.

1. Full wall sample size
2. Capability to test large thermal bridging configurations such as wall/wall, wall/floor, and wall/ceiling thermal bridges
3. Simultaneous measurement of heat flows that include air infiltration/exfiltration

Table 7. Data for Verification of Air Transfer System Measurement

		Solid Panel	1 in. Diameter Hole		0.25 in. Diameter Orifice	
	Units	**Test Conditions**				
Nominal MB Temperature	°C	22	22	22	22	22
Nominal CB Temperature	°C	42	42	42	42	42
Air Leakage Direction		none	infiltration	exfiltration	infiltration	exfiltration
CB baffle inlet T	°C	42.04	42.04	42.04	42.02	42.04
	°F	107.7	107.7	107.7	107.6	107.7
MB baffle inlet T	°C	21.85	21.90	21.83	21.93	21.85
	°F	71.3	71.4	71.3	71.5	71.3
Wall surf T_{out}	°C	41.55	41.56	41.51	41.50	41.52
	°F	106.8	106.8	106.7	106.7	106.7
Wall surf T_{in}	°C	23.07	23.10	23.05	23.15	23.05
	°F	73.5	73.6	73.5	73.7	73.5
TASdmT	°C	NA	21.60	27.50	21.20	25.00
	°F	NA	70.9	81.5	70.2	77.0
Airflow	SLPM	0.00	61.50	58.50	40.00	40.00
	CFM	0.00	2.17	2.07	1.41	1.41
Q_{tot}	W	–72.0	–97.5	–77.0	–85.2	–72.5
	btu/h	–245.6	–332.7	–262.7	–290.7	–247.4
Q_{tas}	W	NA	0	6.4	0	2.5
	btu/h	NA	0	21.9	0	8.5
Q_L	W	NA	23.9	0	15.5	0
	btu/h	NA	81.5	0	52.9	0
Q_{net}	W	–72.0	–73.6	–70.6	–69.7	–70.0
	btu/h	–245.6	–251.1	–240.8	–237.8	–238.8
Percent Difference	%	—	2.2%	–2.0%	–3.2%	–2.8%

4. Air leakage (infiltration/exfiltration) measurements as a function of outdoor air temperature that include contraction/expansion effects on specimen integrity
5. Measurements under both winter and summer outdoor temperatures without removing the sample

Future capabilities will include building assemblies that include moisture transfer and thermal mass effects.

A unique approach was used in formulating the energy balance on the meter box. For this facility the flanking heat flow was predicted using a finite element heat transfer program. The inputs for analysis included measured thermal property data for the materials used in the hot box construction. The other heat flows to the meter box were all measured. This approach eliminates the need to calibrate the box which can mask other errors in design or measurement. In order to verify the accuracy of this approach, a standard sample of known thermal properties was constructed to verify. Verification of three features of the hot box have been performed. Energy balances were verified for the following conditions:

1. Verification of heat flows under standard hot box conditions (no air leakage) that verified that the flanking heat flows were accurately modeled under both summer and winter conditions
2. Verification of the accuracy of the cooling measurements using an independent heat source in the meter box
3. Verification of the accuracy of the energy balances for both exfiltration/infiltration air leakages using two sizes of orifices.

The verification exercises suggest that the accuracy of the hot box is better than ±5% for measurements under both summer and winter conditions including infiltration/exfiltration air leakages.

REFERENCES

ASHRAE. 2009. *ASHRAE Handbook—Fundamentals*. Atlanta: American Society of Heating Refrigeration and Air Conditioning Engineers, Inc.

ASTM C518-04. 2004. *Standard Test Method for Steady-State Thermal Transmission Properties by Means of the Heat Flow Meter Apparatus*.

ASTM C1058-03. 2003. *Standard Practice for Selecting Temperatures for Evaluating and Reporting Thermal Properties of Thermal Insulation*.

ASTM C1303-07. 2007. ASTM C1303-07 *Standard Test Method for Predicting Long-Term Thermal Resistance of Closed-Cell Foam Insulation*.

ASTM C1363-05. 2005. *Standard Test Method for Thermal Performance of Building Materials and Envelope Assemblies by Means of a Hot Box Apparatus*.

Bombino, R. and Burnett, E.F.P. 1999. Design Issues with Steel-stud-framed Wall Systems, PHRC Research Series: Report No. 58: 44, The Pennsylvania Housing Research Center, University Park.

Brown, W.C., Bomberg, M.T., Ullet, J.M. and Rasmussen, J. Measured Thermal Resistance of Frame Walls with Defects in the Installation of Mineral Fibre Insulation, *J. of Thermal Insulation and Building Envelopes*, Vol 16, April 1993: 318-339.

Chebil, S., Galanis, N. and Zmeureanu, R. 2003. Computer Simulation of Thermal Impact of Air Infiltration Through Multi-layered Exterior Walls. Proc. of Eight International IBPSa Conference, Eindhoven, Netherlands, August 2003: 155-162.

Christian, J.E., and J. Kosny. 1995. Toward a national opaque wall rating label. *Thermal Performance of the Exterior Envelopes of Buildings* VI. Atlanta: American Society of Heating, Refrigerating and Air-Conditioning Engineers, Inc.

FTC. 1979. Home Insulation Rule, 16 CFR Part 460. Washington, DC: Federal Trade Commission.

Garden, G.K., 1965. Control of Air Leakage is Important. *Canadian Building Digest* 72, National Research Council of Canada, Ottawa, Canada.

Graham, M.S. 2010. R-value Concerns. Professional Roofing, May 2010: 24p. National Roofing Contractors Association, Rosemont, IL.

Jones, D.C., Ober, D.G. and Goodrow, J.T. 1995. Thermal Performance Characterization of Residential Wall Systems Using a Calibrated Hot Box with Airflow Induced by Differential Pressures, Airflow Performance of Building Envelopes, Components and Systems, ASTM STP 1255: 197-228, American Society for Testing and Materials, Philadelphia, PA.

Lecompte, J., 1990. Influence of Natural Convection in an Insulated Cavity on the Thermal Performance of a Wall, Insulation Materials, Testing and Applications, ASTM STP1030: 397-420. D.L. McElry and J.F. Kimpflen, Eds., Amercian Society for Tesitng and Materials, Philadelphia, PA.

Schumacher, C.J., 2013. Temperature Dependence of R-values in Polyisocyanurate Roof Insulation, Building Science Info Sheet 502, BuildingScience.com, Somerville, MA.

Trethowen, H.A., 1991. Sensitivity of Insulated Wall and Ceiling Cavities to Workmanship, *J. of Thermal Insul.*, Vol 15, October 1991.

Uvslokk, S., 1996. The Importance of Wind Barriers for Insulated Timber Frame Constructions, *J. Thermal Insul. & Building Envelopes*, Vol. 20, July 1996.

Wilson, A.G., 1963. Air Leakage in Buildings, *Canadian Building Digest* 23, National Research Council of Canada, Ottawa, Canada.

Yarbrough, D.W., Graves, R.S., 2006. The Effect of Airflow on Measured Heat Transport Through Wall Cavity Insulation, Proceedings at 2006 ASTM Symposium on Heat, Air, and Moisture Transport in Buildings, Toronto, Canada.

Coupling of Dynamic Thermal Bridge and Whole-Building Simulation

Florian Antretter
Associate Member ASHRAE

Jan Radon, DrIng

Matthias Pazold

ABSTRACT

The effect of thermal bridges on the overall energy performance of buildings is not fully taken into account in standard compliance methods. Steady-state methods are predominately deployed for thermal bridge analysis. One important issue that is often underestimated is the dynamic performance of thermal bridges under different exterior climate conditions.

This paper shows the coupling of a hygrothermal whole-building simulation software with a three-dimensional dynamic thermal bridge simulation. The basics of the hygrothermal whole-building model and the thermal bridge model are explained. The coupling of both models allows the validation of the software for standard conform computations.

The steady-state validation of the thermal bridge model was successful and allows the discussion of application areas in a dynamic simulation.

INTRODUCTION

With higher requirements on energetic envelope design and on airtightness, the ratio of thermal bridge losses compared to the overall losses under heating conditions increases. But the effect of thermal bridges on the overall energy performance of buildings is often not fully taken into account in standard compliance methods. In such methods, steady-state methods are predominately deployed for thermal bridge analysis. Dynamic interaction of the thermal bridge with the building zones is not adequately accounted for.

This was the basis for the coupling of a dynamic thermal bridge simulation with a hygrothermal whole-building simulation. In this combination, the dynamic effects on the thermal bridge can interact with the dynamics of overall building envelope and space. Besides the energy effects, the potential risk for harmful humidity conditions, e.g., on the coldest spots in a zone, can be assessed.

A critical review of existing literature is required to determine documented effects of thermal bridge simulation in combination with dynamic building simulation. In the next step, a model for three-dimensional objects is developed that needs to be implemented and validated in a whole-building simulation environment in order to produce reliable results. The implemented and validated model can then be used to demonstrate the effect of thermal bridges on the energy demand of buildings.

DYNAMIC SIMULATION OF THERMAL BRIDGES

In literature, the necessity of taking thermal bridges into account for both compliance methods and dynamic simulation is discussed controversially. Standard methods to compute the building energy demand are usually monthly-balance-based methods and take thermal bridges with their linear loss coefficient into account. This could be a simple approach to account for the thermal bridge effect in dynamic building simulation. But the dynamic behavior of the thermal bridge is not taken into account.

Martin et al. (2012a) describe two ways to account for thermal bridges in dynamic building simulation. The first is to solve the two-dimensional (2-D) and three-dimensional (3-D) heat transfer and the second is to use an equivalent building assembly that performs just as the dynamic thermal bridge.

***Florian Antretter** is the group manager of and **Matthias Pazold** is a scientist in the Hygrothermal Building Analysis group in the Hygrothermics department at the Fraunhofer Institute for Building Physics, Valley, Germany. **Jan Radon** is a professor at the University of Cracow, Poland.*

For the direct implementation the authors see further demand in solving time and simplification of the problem specification. By using equivalent assemblies the results are no longer that accurate (Martin et al. 2011, 2012b).

Kosny and Kossecka had already concluded in 2002 in a comparison of the equivalence method with detailed calculations and hot-box measurements that serious deviations in the demand calculations with dynamic simulations can be a result if only one-dimensional (1-D) components are taken into account.

Ascione et al. (2012) compare results of conventional approaches with detailed 2-D and 3-D simulation model results. It is described that different modeling approaches can lead to 20% divergence in the results for the energy demand of a typical office building in Italy. The authors recommend the development of dynamic building simulation software that provides modules to take into account thermal bridges according to the standard DIN EN ISO 10211 (DIN 2008).

Besides the assessment of the effect of a thermal bridge on the energy demand of a building, a detailed thermal bridge simulation also allows determination of surface temperatures and therefore the analysis of potential mold growth risk. To simulate realistic hygrothermal conditions inside a building, a hygrothermal whole-building model is required. With this background a new module to dynamically compute 2-D and 3-D thermal bridges was implemented into the existing hygrothermal whole-building simulation software WUFI® Plus (Fraunhofer IBP 2013). In the following the basics of both simulation methods and their coupling are explained.

HYGROTHERMAL WHOLE-BUILDING SIMULATION SOFTWARE WITH 3-D THERMAL BRIDGE SIMULATION

This section describes in short the used software model WUFI Plus. It furthermore shows how the 3-D elements are implemented in the whole-building simulation model.

The WUFI Plus software

WUFI Plus is a dynamic whole-building simulation model based on the hygrothermal envelope calculation model developed by Künzel (1994). The 1D coupled heat and moisture transfer in opaque building components is simulated. Moisture sources or sinks inside a component, capillary action, diffusion, and vapor absorption and desorption as a response to the exterior and interior climate boundary condition as well as thermal parameters are taken into account. The conductive heat and enthalpy flow by vapor diffusion with phase changes depends strongly on the moisture field. The vapor flow is simultaneously governed by the temperature and moisture field due to the exponential changes of the saturation vapor pressure with temperature. Resulting differential equations are discretized by means of an implicit finite volume method. The model was validated by comparing its simulation results with the measured data of extensive field experiments (Künzel 1994) and corresponds to the specifications of DIN EN 15026 (DIN 2007).

The hygrothermal behavior of the building envelope affects the overall performance of a building. Therefore, the components are coupled to a whole-building model (Holm et al. 2004). It can be discretized in different zones regarding one or more rooms with the same interior condition. Thus, the components define the zone boundaries and deliver the heat and moisture flow across the building envelope. The zones provide the boundary conditions for the components. Depending on all current heat and moisture flows across the zone boundaries and the previous states of the components and zones, the indoor climate is simulated iteratively. Heat and moisture balances within the zones are examined. As long as they are not satisfied, the indoor temperature and humidity are adapted for each iteration and time step (Lengsfeld and Holm 2007).

Additionally, there is one predefined zone, the outdoor zone, where the climate data is the input, e.g., obtained by measurements. This climate data mainly contains outdoor air temperature, relative humidity, wind speed and direction, normal rain, and solar radiation data, which influence the exterior building envelope. In the envelope, there can be transparent components, the fenestration. The solar radiation is not only absorbed by the exterior surfaces but a part of it passes through the transparent components and directly heats the indoor air and interior surfaces. Internal heat and moisture sources and sinks, due to people, lighting, and household and plant equipment, are considered too and are a part of the balance equations. Last but not least, the natural, mechanical, and interzonal ventilation influence the simulated indoor climate.

Ideal plant equipment systems provide space heating, cooling, humidification, dehumidification, and ventilation. Furthermore, minimum and maximum design conditions for the zones can be set. If the indoor climate exceeds those design conditions, the required demand to counteract the excess is calculated, as long as the plant equipment capability is sufficient.

Some results of the dynamic hygrothermal building simulation model are the mentioned indoor temperature and humidity for each calculated time step. Usually the time step size is one hour. Further results are the heating, cooling, humidification, dehumidification, and ventilation demands for each time step, summarized over the simulation period. Every heat and moisture profile across a component can be saved, including the surface conditions, which allows the assessment of moisture-related problems, e.g., wood rotting or mold growth. The predicted mean vote (PMV) and the predicted percentage of dissatisfied (PPD) are calculated. Even if some of those simulated values exceed custom limits, one can make a statement about how long or how often they will be exceeded. The zone model was validated via cross-validation with other tools, experiments, and standards such as ANSI/ASHRAE Standard 140 (2007). The validation for both the energetic and the hygric parts of the zone model is described by Antretter et al. (2011).

Coupling of Dynamic Building Simulation with 3-D Element

The WUFI® Plus software was developed with the ambition to be a reliable and easy-to-operate calculation tool for architects and engineers. It enables thermal, energy, and moisture simulations of buildings exposed to transient real climate conditions. The coupled heat and moisture 1D transfer calculation algorithm through multilayer assembly has been extensively verified by experimental measurements (Antretter et al. 2011). The thermal building envelope, however, includes places where processes can only be analyzed in 2-D or 3-D states. To close this gap, the software was supplemented by so-called 3-D objects. These objects are applied for 2-D and 3-D thermal bridges and some special cases, such as transient heat exchange of rooms with the ground. In fact, ground thermal coupling can also be regarded as a large-scale thermal bridge with some special assumptions.

Every real building includes some geometrical or structural (or both) thermal bridges. Transmission heat loss through these elements in steady-state calculations is accounted for by linear or point thermal transmittance coefficients. In the most simplified case, some lump value is added to the overall heat loss coefficient. In DIN EN 12831 (DIN 2003), for calculation of the design heat load only linear thermal bridges are included. The gross variety of thermal bridges and complexity of their calculations compared to 1-D calculations are the reason that these essential parts of the thermal envelope are frequently omitted in transient calculations. Hence, great development effort was invested in the addition of thermal bridges to the WUFI Plus software to make the process as easy as possible on one hand, preserving appropriate calculation accuracy on the other.

For the calculation of 3-D objects the finite volume technique is used (Eymard et al. 2000). This method, based on the thermodynamic law of energy conservation, with clear physical interpretation of heat flow and accumulation, is ideal for the calculation of thermal bridges. In this method, heat-transferring space is divided into small control volumes and the evaluation of integrals and fluxes is conducted via numerical methods.

The addition of a 3-D object starts with dividing the space into X, Y, and Z directions to prepare the geometry that reflects the location of materials and boundary conditions (Figure 1). Based on the WUFI database (Fraunhofer IBP 2013) or user input, relevant materials are stored in the upper list box. In the second list box all kinds of boundary conditions defined in the project—e.g., outer climate, inner zone air, file-based climate, optional climate—are already included. Materials and boundary conditions are assigned in a graphical way with the mouse after choosing the appropriate element from the list box. Any 2-D plane (X, Y, Z slice) can be accessed for assigning materials and boundary conditions. Simultaneously, the 3-D picture

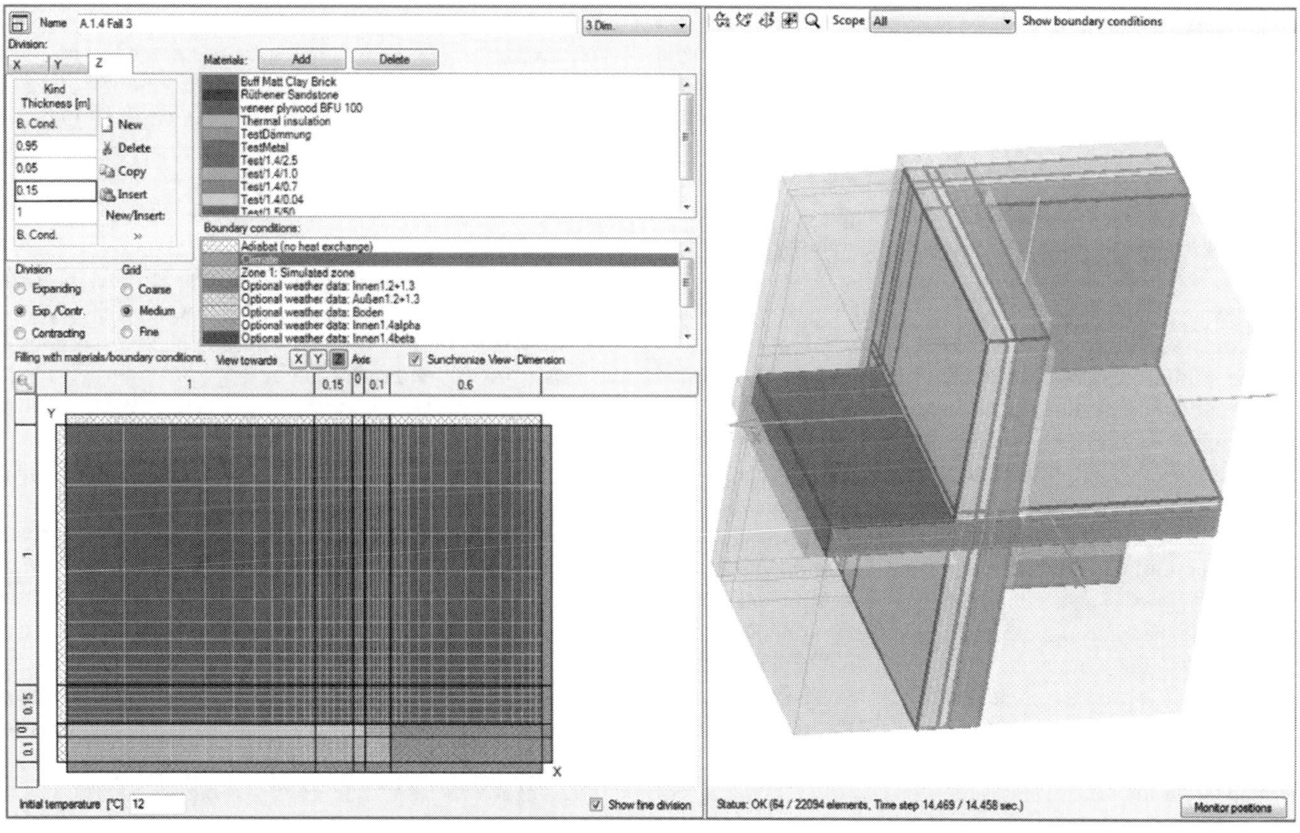

Figure 1 Addition of a 3-D object in WUFI Plus software—in this case a mesh generated for a concrete ceiling and balcony.

of the edited object is generated and updated in the window on the right. Boundary conditions are depicted transparent (such as air) but with different colors to allow identification.

The fine division into control volumes is done automatically by the program. The user can choose coarse, medium, or fine division to get more or less elements. The overall section is divided according to geometrical series with expanding, expanding/contracting, or contracting elements. This enables better adjustment of the mesh to the object and some kind of optimization. The initial series segment, scale factor, and number of terms are adapted depending on the division type. Figure 1 shows the adapted mesh by an example of a balcony.

Spatial division and number of control volumes have a crucial impact on both accuracy and calculation time. A more dense mesh should be applied in places where higher temperature gradients and rapid changes of boundary conditions are expected, and a less dense mesh should be applied towards adiabatic planes to reduce computation time.

High numbers of control volume elements and 3-D objects in one building lead to problems with random access memory (RAM) space when a personal computer is used for calculation. An implicit solving method would require the construction of a linear equation matrix with N × N numbers (N = number of control volumes). N grows exponentially with rank. If the space is divided into 50 elements in each direction, 2500 elements are used for a 2-D object and 125000 for a 3-D object. Using standard double numbers (8 bytes each), an appropriate matrix would take, accordingly, 47.7 MB of RAM for the 2-D object and 116.5 GB of RAM for the 3-D object. To avoid memory problems, the explicit method (forward Euler method, where the results of a current time step depend only on the results of the last time step and no iterations have been done) is used to solve for the temperature distribution. Numerical stability of this method requires a limitation of the time step width. This can be very "painful" when objects include materials with high thermal conductivity (e.g., metal). In such a case, at least for 2-D objects, the implicit method would be a better solution.

Once 3-D objects and remaining building data are ready the calculation can be started. Coupling with simulated zones is accomplished by boundary conditions. For every 3-D object any boundary condition can be applied (from a simulated and/or attached zone, outer air, or any optional climate). Boundary conditions vary from time step to time step. Temperature and relative humidity parameters of the air in the simulated zones are calculated iteratively. First, time step initial values, mostly design conditions, are assumed. Then the coupled heat and moisture flows in assemblies and 3-D objects are calculated. After that the heat and moisture balance is checked and, if necessary, the temperature and relative humidity are corrected. The process is repeated as long as calculated parameters comply with the defined accuracy for the given time step. The maximal viable time step for thermal bridges is usually much lower than the overall time step (mostly 1 hour). Thermal bridges are repeatedly calculated with small sub time steps starting with the initial temperature distribution obtained by the previous time step. After the iteration process is finished, the actual temperature distribution is used as the initial condition for the next time step. The heat exchange with 3-D objects is accounted for in the heat balance of simulated zones and has direct impact on the inner temperature, heating demand, cooling demand, etc. The temperature distribution in 3-D objects can be visualized for every cross-section during the calculation (see Figure 2).

VALIDATION OF THE 3-D THERMAL BRIDGE SIMULATION

The newly developed model for the calculation of 3-D elements is validated with the standard DIN EN ISO 10211 (DIN 2008), which provides information and validation examples for the detailed calculation of heat flows and surface temperatures of thermal bridges in building construction. Specified are geometrical 2-D and 3-D models of thermal bridges to compute the heat flows for assessing the energy losses of a building and the lowest surface temperatures to compute the risk for condensate. The standard also contains modeling rules.

Appendix A of the standard provides reference cases for the validation of simulation models—two 2-D cases and two 3-D cases. According to the standard, a model is an accurate model if the simulated results are within a defined accuracy of the given solutions.

The validation examples present steady-state results. Therefore, fixed boundary conditions as provided in the standard were used on the elements. The simulations were conducted until steady-state conditions in the elements were achieved.

Figure 3 shows case 1 of the standard, half of a quadratic stud with known surface temperatures, for which an analytical solution is possible, as well as its implementation in the software. Table 1 shows that all results are in the required accuracy range of less than 0.1 K deviations (listed in Table 2) from the given solution.

Figure 4 shows case 2 of the standard, a 2-D insulated aluminum profile covered with concrete and wood, as well as its implementation in the software. Table 3 shows that all results are in the required accuracy range of less than 0.1 K for temperature or 0.1 W/m for the heat flow deviations from the given solution (Table 4).

Figure 5 shows case 3 of the standard, a 3-D edge, and its implementation in the software. Table 5 shows that all results are in the required accuracy range of less than 0.1 K for temperature from the given solution. The heat flow calculation given in the standard has two flaws: first there are two wrong algebraic signs in Equations A.8 and A.9 (DIN 2008). The temperature boundary condition within room Alpha, below the ceiling (Figure 5), is higher than the temperature in the room Beta, above the ceiling. Following a heat flow from room Alpha to room Beta is expected and not, as given in the standard, a heat flow from room Beta to room Alpha. The second is a simple numerical mistake in Equation A.7, where 24.36 +

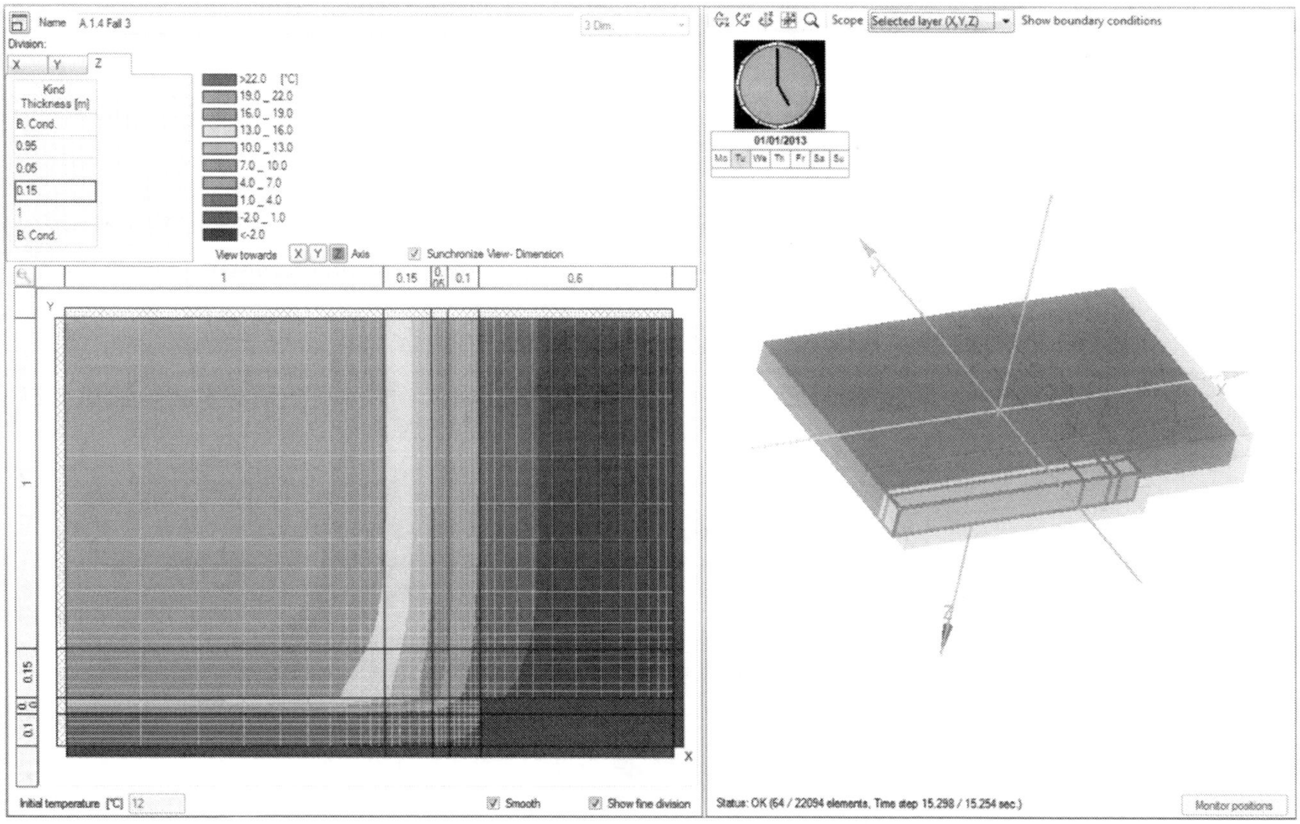

Figure 2 *Visualization of temperature distribution in 3-D objects during calculation in WUFI Plus software.*

35.62 should result in 59.98, not 58.98 as given in the standard (DIN 2008). The results and the corrected standard numbers are shown in Table 6. They are all within the 1% difference.

Figure 6 shows case 4 of the standard, a 3-D thermal bridge representing a steel bar penetrating an insulation layer, and its implementation in the software. Table 7 shows that the results are in the required accuracy range of less than 0.005 K for temperature, and Table 8 shows that the results are in the required accuracy range of less than 1% of the listed heat flow from the given solution.

It is shown that the implemented model can be validated against all four validation cases given in the standard. The validation was successful and the model can be regarded an accurate model conforming to the standard (DIN 2008).

DISCUSSION AND CONCLUSIONS

The use of energetic building simulation for the design of buildings increases. To achieve net-zero or plus-energy buildings, it is important to match the time-dependent energy production and demand. In this case, dynamic simulation software is essential for proper design. If the building simulation model uses in addition a component model that is able to compute the coupled heat and moisture transfer conclusions, not only energy demand but also comfort conditions and the hygrothermal performance of the components are possible.

Such a component model is usually 1-D. The literature, however, shows the necessity for a more detailed analysis that allows a detailed representation of 2-D and 3-D effects.

The coupling of a hygrothermal whole-building model with a transient 3-D thermal bridge calculation was described in this paper. The validation of the thermal bridge model under steady-state conditions according to DIN EN ISO 10211 (DIN 2008) was successful. One verification case in the standard needed adaption, as an application error was found in the standard and described in this paper.

The new combined software tool allows a broad range of applications. The effect of the dynamic thermal bridge performance on the overall energy demand of a building can be assessed. Furthermore, moisture-related problems on the thermal bridges can be assessed.

REFERENCES

Antretter, F., F. Sauer, T. Schöpfer, and A. Holm. 2011. Validation of a hygrothermal whole building simulation software. *Proceedings of Building Simulation 2011: 12th Conference of International Building Performance Simulation Association, Sydney, Australia.*

Ascione, F., N. Bianco, F. de' Rossi, G. Turni, and G.P. Vanoli. 2012. Different methods for the modelling of thermal bridges into energy simulation programs: Com-

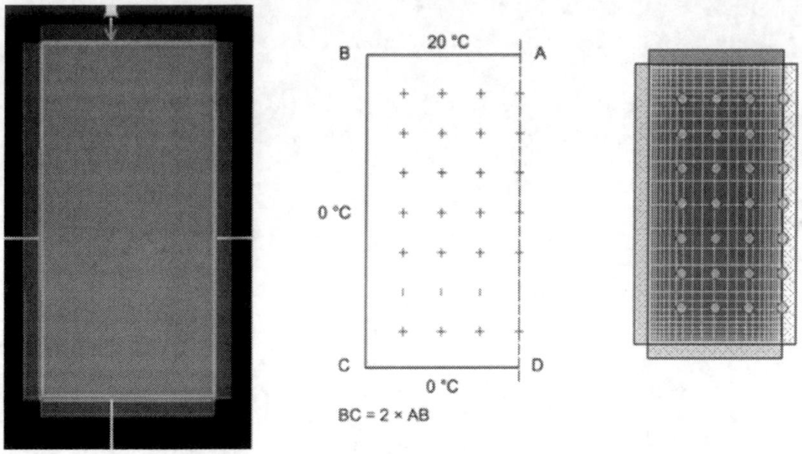

Figure 3 *Validation case 1 of DIN EN ISO 10211 (DIN 2008) and its implementation as a 3-D element in WUFI Plus.*

Table 1. Results of Case 1 of DIN EN ISO 10211 (DIN 2008)

WUFI | ISO 10211	WUFI | ISO 10211	WUFI | ISO 10211	WUFI | ISO 10211	Confirm
9.67 | 9.70	13.37 | 13.40	14.73 | 14.70	15.08 | 15.10	y | y | y | y
5.27 | 5.30	8.64 | 8.60	10.31 | 10.30	10.81 | 10.80	y | y | y | y
3.19 | 3.20	5.60 | 5.60	7.00 | 7.00	7.45 | 7.50	y | y | y | y
2.03 | 2.00	3.66 | 3.60	4.67 | 4.70	5.00 | 5.00	y | y | y | y
1.26 | 1.30	2.31 | 2.30	2.98 | 3.00	3.21 | 3.20	y | y | y | y
0.74 | 0.80	1.36 | 1.40	1.77 | 1.80	1.91 | 1.90	y | y | y | y
0.34 | 0.30	0.63 | 0.60	0.82 | 0.80	0.89 | 0.90	y | y | y | y

parisons of accuracy for flat heterogeneous roofs in Italian climates. *Applied Energy* 97:S.405–18.

ASHRAE. 2007. ASHRAE Standard 140-2007, *Building Thermal Envelope and Fabric Load Tests*. Atlanta: ASHRAE.

DIN. 2003. DIN EN 12831:2003, *Heizungsanlagen in Gebäuden – Verfahren zur Berechnung der Norm-Heizlast*. Berlin, Germany: Deutsches Institut for Normung.

DIN. 2007. DIN EN 15026:2007, *Wärme- und feuchtetechnisches Verhalten von Bauteilen und Bauelementen - Bewertung der Feuchteübertragung durch numerische Simulation*. Berlin, Germany: Deutsches Institut for Normung.

DIN. 2008. DIN EN ISO 10211:2008, *Wärmebrücken im Hochbau – Wärmeströme und Oberflächentemperaturen – Detailierte Berechnunge*. Berlin, Germany: Deutsches Institut for Normung.

Eymard, R., T.R. Gallouët, and R. Herbin. 2000. The finite volume method. *Handbook of Numerical Analysis*, Vol. VII, S. 713–1020. Amsterdam: North-Holland.

Fraunhofer IBP. 2013. WUFI® Plus, Ver. 2.5.3. Fraunhofer Institute for Building Physics, Valley, Germany.

Holm, A., J. Radon, H.M. Künzel, and K. Sedlbauer. 2004. Berechnung des hygrothermischen Verhaltens von Räumen. *WTA Schriftenreihe* H. 24, S. 81–94.

Künzel, H.M. 1994. Simultaneous Heat and Moisture Transport in Building Components. Dissertation, University of Stuttgart. www.building-physics.com

Kosny, J., and E. Kossecka. 2002. Multi-dimensional heat transfer through complex building envelope assemblies in hourly energy simulation programs. *Energy and Buildings* 34:S.445–54.

Lengsfeld, K., and A. Holm. 2007. Entwicklung und Validierung einer hygrothermischen Raumklima-Simulationssoftware WUFI®-Plus, *Bauphysik* 29, Magazin 3.

Martin, K., A. Erkoreka, I. Flores, M. Odriozola, and J.M. Sala. 2011. Problems in the calculation of thermal bridges in dynamic conditions. *Energy and Buildings* 43:S.529–35.

Martin, K., C. Escudero, A. Erkoreka, I. Flores, and J.M. Sala. 2012a. Equivalent wall method for dynamic characterisation of thermal bridges. *Energy and Buildings* 55:S.704–14.

Martin, K., A. Campos-Celador, C. Escudero, I. Gómez, and J.M. Sala. 2012b. Analysis of a thermal bridge in a guarded hot box testing facility. *Energy and Buildings* 50:S.139–49.

Table 2. Differences between WUFI Plus (Fraunhofer IBP 2013) and Case 1 of DIN EN ISO 10211 (DIN 2008)

Difference	Difference	Difference	Difference
0.03	0.03	0.03	0.02
0.03	0.04	0.01	0.01
0.01	0.00	0.00	0.05
0.03	0.06	0.03	0.00
0.04	0.01	0.02	0.01
0.04	0.04	0.03	0.01
0.04	0.03	0.02	0.01

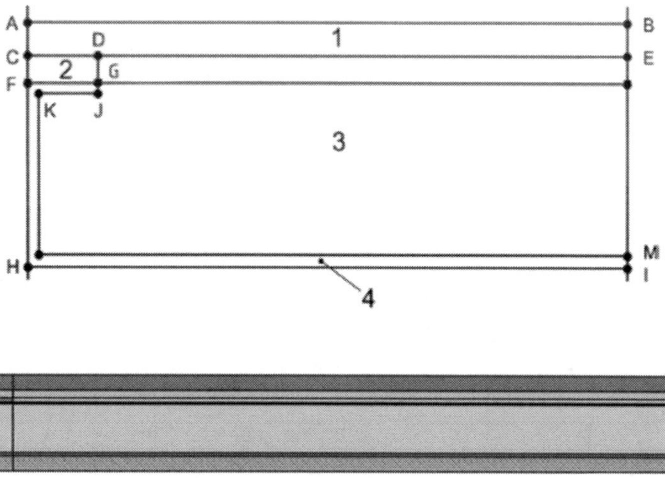

Figure 4 *Validation case 2 of DIN EN ISO 10211 (DIN 2008) and its implementation as a 3-D element in WUFI Plus.*

Table 3. Results of Case 2 of DIN EN ISO 10211 (DIN 2008)

WUFI Result	DIN EN ISO 10211 (DIN 2008)	Difference	Confirm
7.05	7.10	0.05	y
0.76	0.80	0.04	y
7.88	7.80	0.02	y
6.26	6.30	0.04	y
0.83	0.80	0.03	y
16.42	16.40	0.02	y
16.35	16.30	0.05	y
16.78	16.80	0.02	y
18.34	18.30	0.04	y

Table 4. Heat Exchange of Case 2 of DIN EN ISO 10211 (DIN 2008)

WUFI Result	DIN EN ISO 10211 (DIN 2008)	Difference
9.461	9.500	0.039

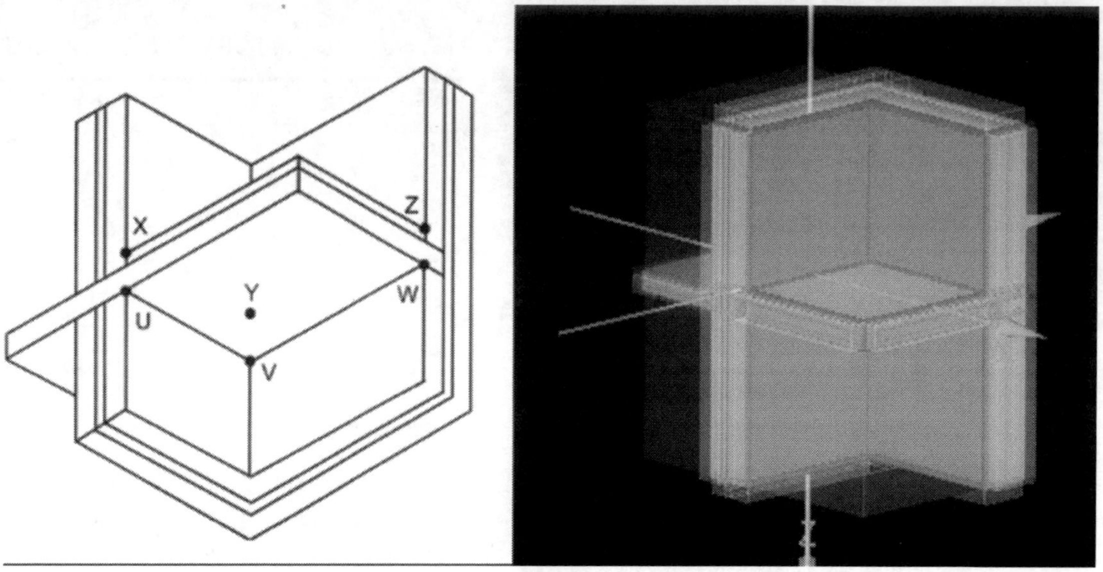

Figure 5 *Validation case 3 of DIN EN ISO 10211 (DIN 2008) and its implementation as a 3-D element in WUFI Plus.*

Table 5. Results of Case 3 of DIN EN ISO 10211 (DIN 2008)

WUFI Result	DIN EN ISO 10211 (DIN 2008)	Difference	Confirm
11.39	11.32	0.07	y
11.02	11.11	0.09	y

Table 6. Heat Exchange of Case 3 of DIN EN ISO 10211 (DIN 2008)

Watt	Outside	Room Alpha	Room Beta
WUFI	59.80713	45.97178	13.83534
DIN EN ISO	59.98000	46.09000	13.89000
1% of DIN	0.59980	0.46090	0.13890
Difference	0.17287	0.11822	0.05466

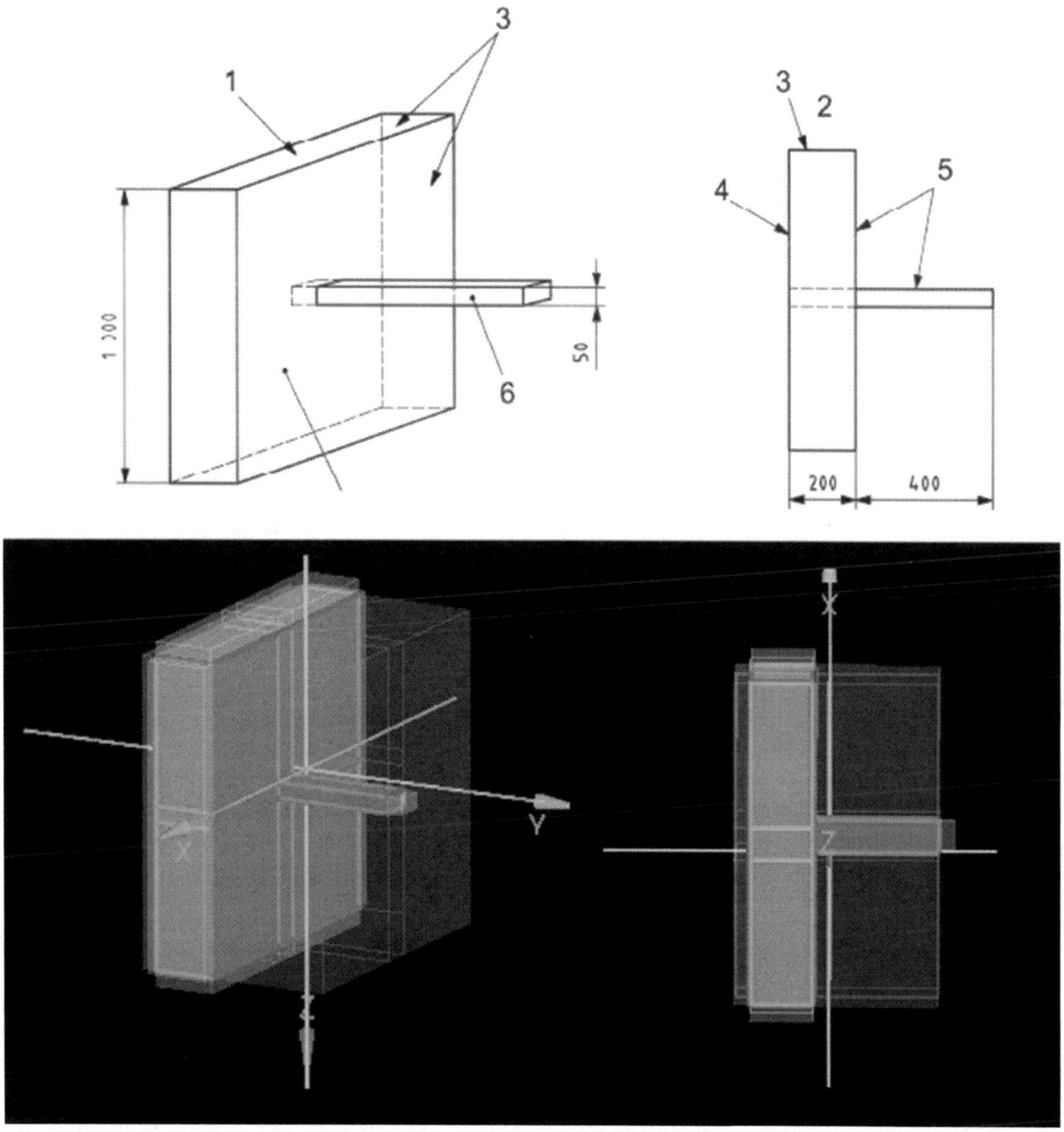

Figure 6 *Validation case 4 of DIN EN ISO 10211 (DIN 2008) and its implementation as a 3-D element in WUFI Plus.*

Table 7. Results of Case 4 of DIN EN ISO 10211 (DIN 2008)

WUFI Result	DIN EN ISO 10211 (DIN 2008)	Difference	Confirm
0.807169	0.805	0.002169	y

Table 8. Heat Exchange of Case 4 of DIN EN ISO 10211 (DIN 2008)

WUFI Result	DIN EN ISO 10211 (DIN 2008)	Difference	1% of ISO
0.5375	0.5400	0.0025	0.0054

Moisture Modeling: Effective Moisture Penetration Depth Versus Effective Capacitance

Jason Woods, PhD

Jon Winkler, PhD
Associate Member ASHRAE

Dane Christensen, PhD
Member ASHRAE

ABSTRACT

This paper compares two moisture models in the context of residential whole-building simulations—the effective moisture penetration depth (EMPD) model and the effective capacitance (EC) model. Previous research has shown that the EMPD model provides a more physically realistic response to humidity-load fluctuations than the EC model. But it is unclear how the choice of moisture model affects overall conclusions informing building design and equipment selection. In this study, the models are evaluated based on their ability to predict relative humidity, predict energy use, and to select HVAC equipment. Whole building simulations indicate that the EC and EMPD models are not equivalent, and that combining the capacitance of the building materials with the indoor air capacitance in the EC model is the root cause of this discrepancy. It means that the EC model cannot dampen humidity loads of different frequencies. This study also showed that a simulation using a certain value of EC in one city can agree with the EMPD results, but that the same EC value in a different city may not agree with the EMPD results. However, the results showed that some conclusions drawn from simulations using the EMPD model and the EC model may often be the same, even if the results differ.

INTRODUCTION

Humidity in residential buildings can cause problems, particularly in hot-humid climates. High relative humidity (RH) is associated with occupant discomfort (Fang et al. 1998), may lead to occupant health problems (Haverinen et al. 2003, Bornehag et al. 2004), and can damage building materials (Steeman, 2009). High efficiency constructions, such as high-insulation envelopes, can amplify these problems by reducing air-conditioner runtime (Rudd and Henderson 2007). Mechanical ventilation, as mandated by ASHRAE 62.2 (ASHRAE 2004), can also worsen the problem in hot-humid climates by introducing humid ventilation air (Walker and Sherman 2007). Several studies have since supported these findings, which show extensive periods of high indoor RH, which are possibly worse in high-performance, low energy homes (Henderson et al. 2008; Lstiburek 2009a; Lstiburek 2009b; Winkler et al. 2013).

Questions still remain on how to best control indoor RH. Modeling the transport and storage of moisture in buildings materials is important in answering these questions. Simulations tools such as WUFI (Künzel et al. 2001) can be used to predict and better understand moisture transport and storage inside wall assemblies. WUFI uses a finite-difference approach to solve for the profiles of temperature and moisture content inside the materials (Künzel 1995, Antretter et al. 2011). While useful when analyzing a wall assembly in detail, finite-difference methods take a long time to solve. In whole-building simulations, moisture modeling is often used to quickly analyze many design features for optimization purposes, as done by Winkler et al. (2013). Thus, this study is looking for alternatives to the finite-difference approach that achieve similar accuracy with a shorter runtime.

Current standard practice in whole-building energy simulations is to use an effective capacitance (EC) multiplier to account for the moisture buffering of partition walls, flooring, and furnishings (EPA 2001). While this EC multiplier is an

Jason Woods, Jon Winkler, *and* ***Dane Christensen*** *are research engineers in the Electricity, Resources, and Building Systems Integration Center at the National Renewable Energy Laboratory, Golden, CO.*

Published by ASHRAE 2013

improvement over neglecting the materials, it does not correctly model the transfer of moisture between the room air and the materials, since the materials and the room are combined into a single node. But it can be solved several orders of magnitude more quickly than the finite-difference approach.

Another approach is the effective moisture penetration depth (EMPD) model (Cunningham 1988; Kerestecioglu et al. 1990; Cunningham 1992), which has similar computation requirements as the effective capacitance approach, but is more physically realistic. A previous study compared the EMPD and EC models against an analytical solution for a simple test case and with a more advanced finite-difference model for two more complicated test cases. In each test case, Woods et al. (2013) found that a modified EMPD model (referred to there as $EMPD_2$) obtained closer results than the EC model.

Several researchers have looked at this moisture buffering effect with finite-difference modeling (Künzel et al. 2005; Rode and Grau 2008; Steeman et al. 2010; Tariku et al. 2011), EMPD modeling (Abadie and Mendonça 2009; Janssen and Roels 2009), or with tests on individual materials (Plagge et al. 2007; James et al. 2010). While some studies made simple comparisons between the EMPD and EC models (e.g., EPA 2001), it is unclear how their differences affect the overall conclusions from whole-building simulations. This is the focus of this study.

While there are other moisture-buffering related factors needed to ensure a simulation accurately represents a real building (e.g., moisture properties of materials), a physically accurate moisture model is a key factor. The purpose of this paper is to compare the EMPD and EC models for use in whole-building simulations. This paper describes the fundamental differences between these two models and compares their results to illustrate their ability to:

1. predict the indoor RH
2. estimate the energy use of temperature and humidity control
3. select appropriate air-conditioning and dehumidification equipment for a particular building

METHODS

The EC and EMPD models differ in the way they solve for moisture transfer into and out of walls and furnishings. This study explores the implications of these differences by using EnergyPlus, a whole-building energy simulation software (DOE 2012).

For this comparison, we simulate a 223 m^2 (2400 ft^2) house in two hot-humid cities: Houston and Orlando. These cities represent two major population centers in the Building America hot-humid climate zone (PNNL 2010), but provide some geographical variation. We use the same construction details as a recent study (Winkler et al. 2013), which is a home that represents 50% source energy savings over 2009 International Energy Conservation Code (IECC) requirements (ICC 2009). Key characteristics of this home are shown in Table 1. Internal gains from equipment, occupants, and domestic hot water use were simulated according to the Building America Housing Simulation Protocols (Hendron and Engebrecht 2010). The sensible gains varied based on the time of year and day of the week from 16.6–21.8 kWh/day, and latent gains varied from 3.4–3.7 kWh/day (4.95–5.3 kg of water per day or 10.9–11.7 lbs/day).

As mentioned in the Introduction, the simulations are used to compare how these models predict indoor RH, predict energy use, and select HVAC equipment. Specifically, this study investigates the following.

- **Indoor RH.** For predicting the indoor RH, we perform annual simulations without controlling humidity. We compare the RH profiles throughout the simulation period, and also look at the aggregate results that are often used to assess thermal comfort and the risk of mold growth (e.g., hours above 60% rh).
- **Energy Use.** We compare the energy use for cases without and with humidity control. The latter includes a 45 pints/day (21.3 L/day) dehumidifier to maintain the RH setpoint, with both 60% and 55% used for this setpoint. We also simulate a case with no moisture buffering for a baseline comparison.
- **HVAC Equipment Selection.** There are different strategies for controlling the RH in a building. Winkler et al. (2013) used the EMPD model to look at several equipment options for high-performance homes in hot-humid climates. We look at a few of the same equipment options to determine how those findings compare to the more commonly-used EC model. In addition to the baseline equipment in Table 1, we also simulate dehumidifiers of different sizes and a high efficiency, variable-speed air conditioner.

Table 1. Characteristics of High-Performance Home

Characteristic	Description
Foundation	Slab-on-grade, uninsulated
Walls	R-13, 2 × 4 16 in. on center + 1 in. of foam
Windows	U-Factor = 2.1 W/m^2·K (0.368 Btu/h·ft^2·°F); SHGC[a] = 0.3
Attic	Vented, R-30 blown-in fiberglass
Infiltration	0.00018 specific leakage area (3.7 air changes/h at 50 Pa)
Ventilation	Continuous, at ASHRAE Standard 62.2 rates 92 m^3/h (54 cfm)
HVAC	SEER 21 air conditioner, 78% AFUE[b] gas furnace, XX L/kWh dehumidifier (when used)
Ducts	Within conditioned space
Water heater	Gas tankless, (0.82 Energy Factor)
Appliances	ENERGY STAR®
PV	kW on unshaded rooftop

[a] SHGC = solar heat gain coefficient
[b] AFUE = annual fuel utilization efficiency

We run each of these simulations for both the EMPD and EC models. We perform two-year simulations and discard the results from the first year to eliminate any effects from the initial conditions. The EC model simulations were run in EnergyPlus v7.1, while the EMPD simulations were run using a custom version of EnergyPlus that includes a modified EMPD model (referred to as $EMPD_2$ in Woods et al. [2013]). The EC and EMPD models are described below.

Effective Capacitance (EC) Model

The fundamentals of the EC model are depicted in Figure 1a, which shows how a single node is used to represent the capacitance of the room air and its materials leading to the following equation:

$$\rho_{air}V(\mathrm{EC})\frac{d\omega_{room}}{dt} = \sum_{i=1}^{N}\dot{m}_{v,i} \tag{1}$$

On the left hand side, ρ_{air} is the density of air, V the volume of the room, ω_{room} the room air humidity ratio, and t time. The moisture capacitance of both the room air and the materials are lumped together with the effective capacitance, EC. For example, if EC = 20, the capacitance of the building materials is assumed to be 19 times larger than that of the air alone (1 unit from air + 19 units from materials = 20 units). It also assumed that the moisture content of the materials changes instantaneously with the room humidity, neglecting the finite mass transfer resistance between the room air and the materials. This last assumption is one of the key differences with the EMPD model described below.

The right hand side of Equation (1) includes N moisture flow rates ($\dot{m}_v$) from each source or sink i. Common sources and sinks of moisture (Figure 1a) are moisture gains or losses from ventilation (and infiltration), moisture losses from HVAC equipment, and moisture gains from internal sources (e.g., people).

Effective Moisture Penetration Depth (EMPD) Model

The EMPD model includes the same moisture sources and sinks as the EC model, but the capacitance of the room air is left unaltered and instead two material nodes are used to model the material moisture capacitance: a surface-layer node for short term fluctuations in humidity and a deep-layer node for longer term fluctuations (see Figure 1b). The node for the room air is separated from the node for the material with a mass-transfer resistance. A moisture balance on the room air gives:

$$\rho_{air}V\frac{d\omega_{room}}{dt} = \sum_{i=1}^{N}\dot{m}_{v,i} - \sum_{j=1}^{N}\frac{\omega_{room}-\omega_{surf,j}}{R_{1,j}} \tag{2}$$

Instead of using the EC multiplier on the room air, the last term on the right-hand-side accounts for the moisture transfer between the room air and each surface j. The surfaces are given a fictitious humidity ratio, ω_{surf}, which is the humidity ratio in equilibrium with the moisture content of the surface layer. The mass transfer resistance, R_1, includes both an air-side boundary layer resistance and a diffusive resistance inside the material. As opposed to the EC model, this resistance results in delayed buffering from each surface depending on the air velocity in the room and the moisture properties of the material.

As shown in Figure 1b, the EMPD model also includes a deep layer, which provides moisture buffering for humidity fluctuations on the order of weeks or months. For many building constructions and furnishings, this deep layer is generally needed to match the finite-difference model results (see Woods et al. [2013]). Note that the computation time needed for this EMPD model, which uses two nodes, is still much smaller than the computation time needed for finite-difference models, which use more than ten nodes (a finite-difference model with only two nodes would not be as accurate as this EMPD model). Finite-difference models also generally require more iterations to completely resolve the temperature and humidity profile in the wall.

Additional equations are needed to solve for the moisture content of the surface layer and deep layer nodes:

$$\rho_{matl}Ad_{EMPD}\left.\frac{du}{d\phi}\right)_j\frac{1}{\omega_{sat}}\frac{d\omega_{surf,j}}{dt} = \frac{\omega_{room}-\omega_{surf,j}}{R_{1,j}} - \frac{\omega_{surf,j}-\omega_{deep,j}}{R_{2,j}} \tag{3}$$

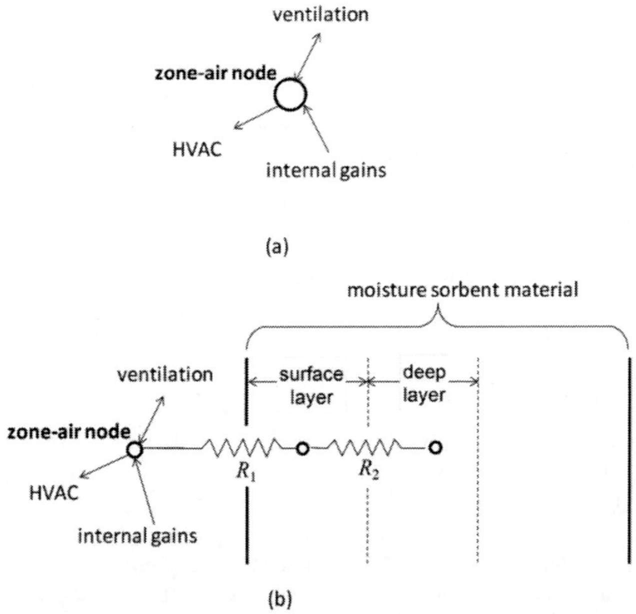

Figure 1 *Summary of transport resistances and capacity nodes for (a) effective capacitance model and (b) effective moisture penetration depth model.*

$$\left(\rho_{matl} A d_{EMPD-deep} \frac{du}{d\phi}\right)_j \frac{1}{\omega_{sat}} \frac{d\omega_{deep,j}}{dt} = \frac{\omega_{surf,j} - \omega_{deep,j}}{R_{2,j}} \quad (4)$$

where ρ_{matl} is the dry density of the material, u the moisture content of the material (kg/kg), $du/d\phi$ the slope of the moisture sorption curve, and $d_{EMPD\text{-}deep}$ and d_{EMPD} the effective penetration depths for the deep and surface layers, respectively. The transport resistances (R_1 and R_2) include the convective mass transfer resistance of the boundary layer and diffusive resistances of the material. These can be calculated with the equations in Woods et al. (2013).

Estimating the Models' Parameters

The EC and EMPD models use different parameters to represent the moisture buffering of materials. There is only one parameter for the EC model: the effective capacitance. In this study, we use values of 5, 15, 30, and 45. This extends beyond the range of 10 to 25 suggested previously (EPA 2001). We also simulate a home with no material moisture buffering (EC = 1).

In the EMPD model, there are six parameters for each buffering material: the surface area, moisture sorption curve, density, permeability, and two penetration depths (d_{EMPD} and $d_{EMPD\text{-}deep}$). While not explicitly required in the model, we also include a coating permeance for the paint on the gypsum walls and ceilings.

The simulated house includes the following moisture buffering materials:

- Painted gypsum on the interior surface of exterior walls and on partition walls
- Painted gypsum on ceilings
- Carpet flooring
- Soft (cotton) and hard (finished wood) furniture

The density and permeability of these materials are assumed constant and are listed in Table 2, while the moisture sorption slopes ($du/d\phi$) are functions of the RH, as shown in Figure 1. The penetration depths for the surface and deep layers are calculated based on these material properties, as discussed in Woods et al. (2013).

Calculating the penetration depths also requires an assumption for the short-term and long-term cycle times for the humidity fluctuations, which coincide with the latent loads on the building. These penetration depths are calculated with:

$$d_{EMPD} = \sqrt{\frac{\delta_{derm} p_{sat} \tau}{\rho_{mail} \frac{du}{d\phi} \pi}} \quad (5)$$

where δ_{perm} is the vapor permeability, p_{sat} the saturation vapor pressure, and τ the humidity cycle period. This study uses τ = 24 hours for d_{EMPD}, which buffers high-frequency loads such as internal gains, and τ = 10 days for $d_{EMPD\text{-}deep}$, which buffers low-frequency loads such as changes in the ventilation load due to the weather. Woods et al. (2013) used a thick block of concrete for their test case, which resulted in penetration depths that were always less than the material thickness. In this study, $d_{EMPD\text{-}deep}$ for the soft furnishings and the carpet is on the order of 100 mm and 35 mm, respectively, which would normally be thicker than the actual material thickness. It is unclear exactly how to address this issue. There are generally additional materials under the carpet, and possibly under or behind the soft furnishings. In this study, we adjusted the penetration depths so that they were on the same order of magnitude as the material thickness.

Unlike the EC model, the EMPD model requires inputs for the surface area of each moisture-buffering material. The wall and ceiling surface areas were based on the geometry of the house. We estimated material surface areas for the carpet and hard and soft furnishings using engineering judgment for typical furnishings in a three bedroom, 223 m^2 (2400 ft^2) house. These represented the medium buffering level shown in Table 3. The low and high buffering levels were scaled from the medium values to bound realistic furnishing levels.

While these two models are fundamentally different, we can compare a rough estimate of the overall moisture capacitance of the EMPD model with the EC values from above. Consider the surface areas for the medium buffering

Table 2. EMPD Moisture Properties

Material	Density, kg/m³ (lb/ft³)	Permeability, kg/m·s·Pa (gr/ft·h·in. Hg)	Coating Permeance, ng/s·m²·Pa (perms)	d_{EMPD}, mm (in.)	$d_{EMPD\text{-}deep}$, mm (in.)
Carpet	220[a] (14.2)	3E-11[a] (1.72)	n/a	11.1 (0.44)	16.8 (0.66)
Gypsum	800[a] (51.6)	2E-11[a] (1.14)	570[d] (10)	5.8 (0.23)	18.1 (0.71)
Wood furniture	512[a] (33)	2E-12[a] (0.114)	n/a	1.3 (0.05)	4.0 (0.157)
Soft furniture	250[b] (16.1)	5E-10[c] (28.6)	n/a	12.7 (0.5)	0.0 (0.0)

[a] Kumaran 1996
[b] Svennberg 2005
[c] Kniest et al. 1998
[d] Walls and ceilings are assumed to be painted with a 10-perm paint

level, the surface penetration depths, and the material densities. These allow us to calculate an effective mass of each material:

$$Effective\ Mass = \rho_{matl} A d_{EMPD} \tag{6}$$

Combining these with an average moisture sorption curve from Figure 2, we can calculate the mass of moisture that will be absorbed by each material for a given range in RH ($\Delta\phi$):

$$Moisture\ Absorbed = (Effective\ Mass)\Delta\phi\frac{du}{d\phi} \tag{7}$$

Assuming an air temperature of the space, we can convert the $\Delta\phi$ into a difference in humidity ratio ($\Delta\omega$), which leads to an effective air mass:

$$Effective\ Air\ Mass = \frac{Moisture\ Absorbed}{\Delta\omega} \tag{8}$$

If we use a range in rh of 40% to 60%, the effective air mass of the EMPD surface layers for the medium buffering case is around 9600 kg (21,160 lbs), which is around EC = 16. Including the deep layers increases this to EC = 38. Keep in mind, though, that the models are inherently different, and this conversion is just an approximation.

RESULTS AND DISCUSSION

The simulations described above are used to assess the appropriateness of the EC and EMPD models for use in whole-building simulations. The importance of differences between the two models depends on the goal of the simulation. The sections below consider the accuracy of RH predictions, energy use predictions, and the effects of moisture buffering on equipment selection.

Predicting Relative Humidity

High-frequency humidity loads are difficult to dampen because of the slow rate of transfer into sorptive materials. This is captured in the EMPD model with the transport resistance between the room air and the material surface layer. It uses actual permeabilities for the sorptive materials and their low-perm coatings, as well as the air-side resistance to moisture transfer. As noted previously, the EC model does not include a resistance between the room air capacitance and the capacitance of the materials, and this generally means the EC model will miscalculate the dynamics of high-frequency loads.

The two-layer EMPD model gives more realistic responses to lower frequency humidity loads. The EC model cannot dampen longer term humidity loads without also affecting the response to high frequency humidity loads. This is a key difference between the EMPD and EC models. They give different responses to humidity loads of different frequencies.

This is illustrated in Figure 3, which shows the annual RH traces for the high-buffering EMPD case and for effective capacitance values of 5 and 45. The figures show the results for both Houston (left) and Orlando (right). The key drawback of the EC model is that it cannot predict the response to all moisture loads correctly, regardless of the EC value chosen.

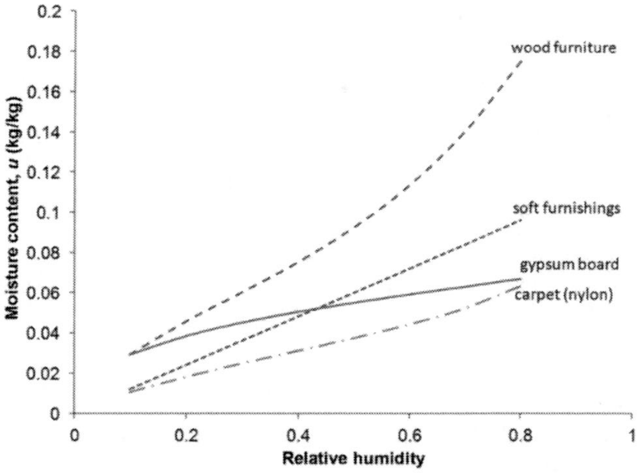

Figure 2 *Moisture sorption curves of materials. Wood furniture, gypsum board, and carpet are from Kerestecioglu et al. 1988 while soft furnishings is cotton sorption curve from Svennberg 2005.*

Table 3. Humidity Buffering Surface Areas (for EMPD Simulations)

	Moisture Buffering Level		
	Low	Medium	High
Carpet surface area, m^2 (ft^2)	67 (721)	145 (1,61)	212 (2282)
Soft furniture, m^2 (ft^2)	25 (269)	75 (807)	200 (2153)
Wood furniture, m^2 (ft^2)	25 (215)	60 (646)	150 (1615)
Painted gypsum (interior walls), m^2 (ft^2)	223 (2400)	223 (2,400)	223 (2,400)
Painted gypsum (exterior walls), m^2 (ft^2)	175 (1,80)	175 (1,880)	175 (1,880)
Painted gypsum (ceiling), m^2 (ft^2)	223 (2400)	223 (2400)	223 (2400)

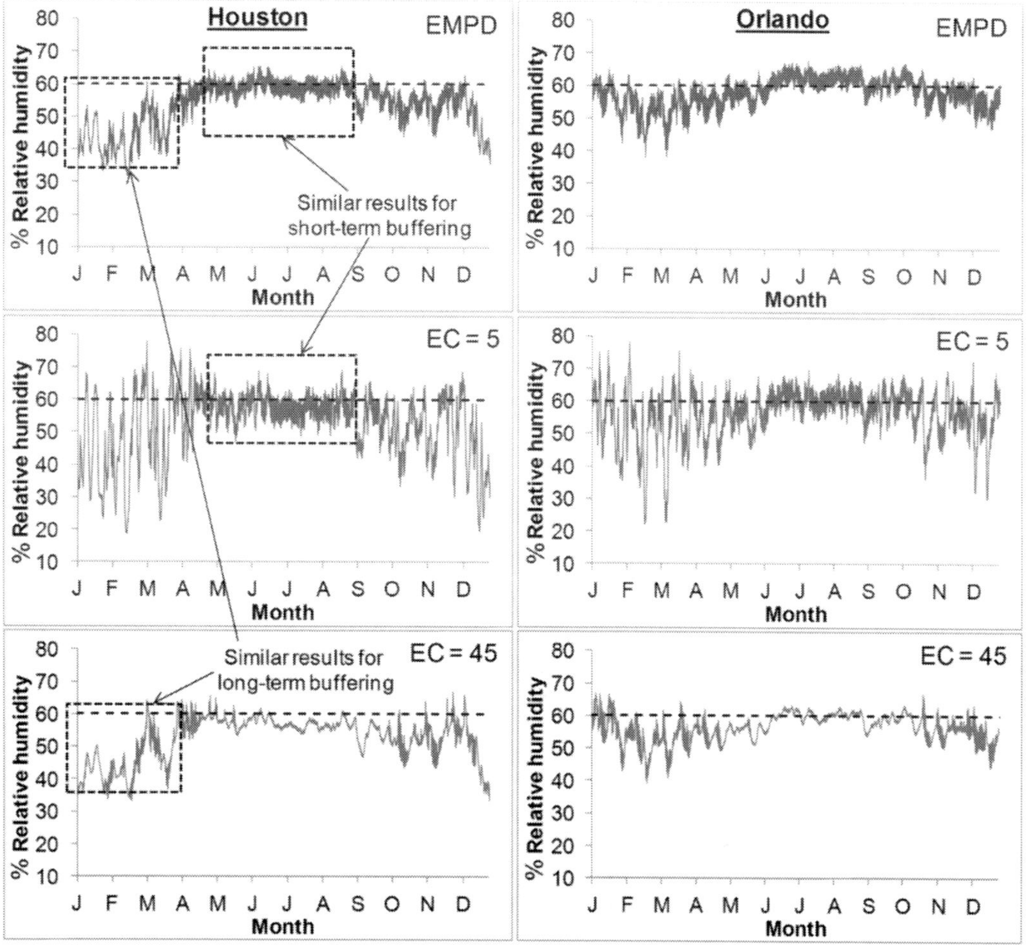

Figure 3 *Zone relative humidity for Houston (left) and Orlando (right). Top plot is the high-buffering EMPD case, while bottom two plots are the EC model at two different values for the effective capacitance.*

Figure 3 shows that the EC = 5 case generally predicts the correct response to the high-frequency humidity loads due to air conditioner moisture removal (summer), but it does not provide enough damping to the low-frequency humidity loads due to changes in the weather (winter and swing seasons). Alternatively, using EC = 45 generally matches the low frequency loads, but overdamps the high-frequency loads.

Figures 4 and 5 show more clearly the RH fluctuations for two months: February and August. The February results in Figure 4 illustrate a case with low-frequency humidity loads. The EC = 45 case resembles the EMPD response, which buffers these low frequency loads. Using EC = 5 provides little buffering for these low frequency fluctuations, resulting in large swings in the humidity not seen in the EMPD case. The August results (Figure 5), where the air conditioner is running intermittently to meet the sensible load, illustrate a case with high frequency humidity loads. The EC = 5 case resembles the EMPD response, providing buffering for high-frequency loads, while using EC = 45 provides too much buffering.

These results show that if the goal of the simulations is to predict the indoor RH throughout the year, the EC model will likely predict results much different from the EMPD model. If the goal is instead to predict high RH events to estimate the risk of moisture-related issues in the building, these differences may be less important. There will certainly still be differences, such as in Houston in August (Figure 5), where the EC = 45 case predicts much fewer hours above 60% rh than the EMPD or EC = 5 results. But what do these results tell us about the need for dehumidification or the risk of having moisture-related problems, such as mold growth?

While some guidelines (ASHRAE 2013) exist to indicate high probability of moisture problems, these guidelines refer to a maximum surface RH, as opposed to an average indoor RH. Since the EC model does not calculate surface conditions, we instead use the *indoor* RH metrics in Table 4 as surrogates, which have been used previously (Rudd and Henderson 2007; Fang et al. 2011; Winkler et al. 2013).

These data show some similarities. In both the EMPD and EC cases, the hours above 60% rh and the number of

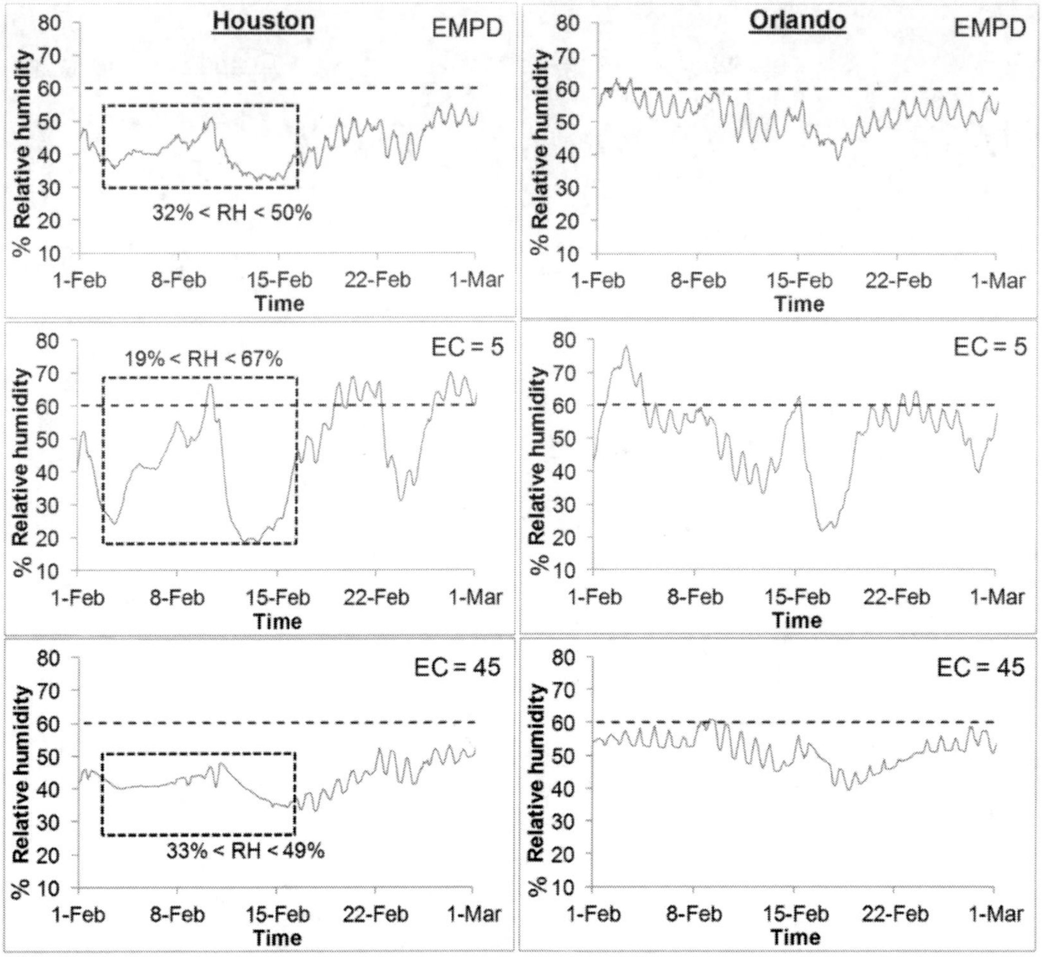

Figure 4 *Close-up view of February from Figure 3.*

excursions longer than 4 hours are high enough to warrant supplemental dehumidification. Both models also show similar trends: higher levels of buffering decrease the hours above 60% rh and decrease the number of excursions.

There are also some important differences. The hours above 60% rh and the number of excursions for the EMPD cases are higher than those for the EC cases. The EMPD model generally predicts many short excursions, compared to fewer, but longer, excursions predicted by the EC model. The resistance between the indoor air node and the material node (Figure 1b) slows the buffering in the EMPD model, which results in more short-term 'spikes' in humidity. On the other hand, an excursion may last several days or weeks in the EC model because of the small amplitude of the daily RH fluctuations (see, e.g., the EC = 45 case for Orlando in Figure 5). The lower amplitude of the RH profile means there are fewer times where the RH cycles back and forth between an RH greater than 60% and an RH less than 60%.

Since we have results for two cities, we can make a preliminary evaluation about consistency between the two models. If we interpolate between the different EC cases for Houston, EC = 10 and EC = 25 give about the same hours above 60% rh as EMPD low and EMPD high. In Orlando, EMPD low and EMPD high are about the same as EC < 5 and EC = 10. This indicates that converting between the two models is inconsistent across climates. This is due to the inherent differences in the models; the two models cannot be made equivalent.

However, if someone is interested in understanding the risks of moisture-related problems, the two models may lead to the same conclusions. Based on the results presented here, it appears that EC values of around 10 to 15 (which are commonly used) will lead to similar overall conclusions (e.g., there is or is not a risk of mold growth) as the EMPD model, particularly if hours above 60% is the relevant metric. This is true for the two cities and homes analyzed here, but may not hold universally. It depends on what parameters are selected for each model. As discussed in the Methods section, the EMPD model includes more realistic inputs, which can presumably give more reliable results if these material properties are known or can be measured. Future research will focus on these two areas: (1) generalizing this study to different buildings and different climates and (2) developing methods to measure the EMPD properties in-situ in a home.

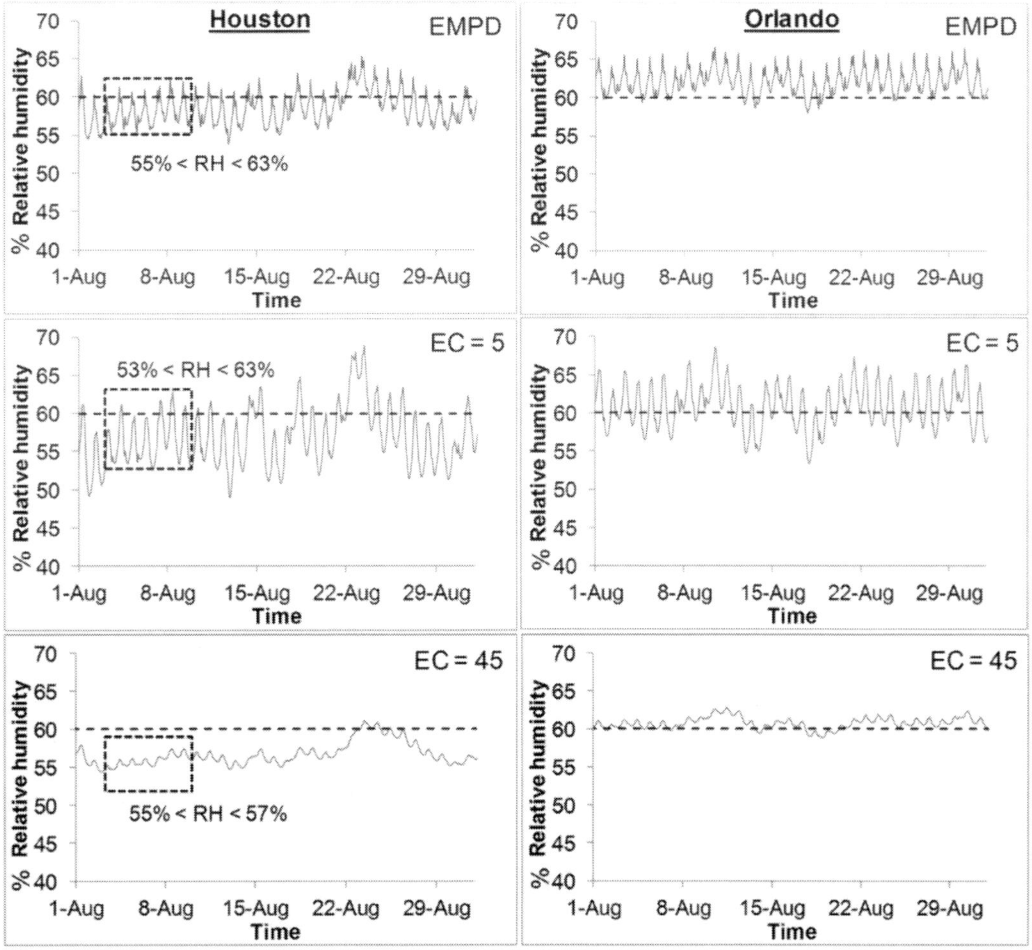

Figure 5 *Close-up view of August from Figure 3.*

Table 4. Indoor RH Metrics for the EMPD and EC Models for a Range of Moisture Capacitance Levels

		No Buffering	EMPD			Effective Capacitance			
			Low	Medium	High	EC = 5	EC = 15	EC = 30	EC = 45
Houston	Total Number of Hours with RH > 60%	2704	1598	1182	754	1,995	1084	670	494
	Number of Excursions [a]	265	206	196	166	166	78	54	49
	Average Excursion Length (h)	10.2	7.8	6	4.5	12.0	13.9	12.4	10.1
	Maximum Excursion Length (h)	83	43	43	41	85.5	83	64.3	61
	Number of Excursions > 4 h	216	155	106	45	124	44	33	29
	Number of Excursions > 8 h	140	68	43	22	71	29	26	19
Orlando	Total Number of Hours with RH > 60%	3266	2936	2755	2515	2,805	2356	2,055	1,993
	Number of Excursions	345	273	250	209	210	129	88	75
	Average Excursion Length (h)	9.5	10.8	11	12	13.4	18.3	23.4	26.6
	Maximum Excursion Length (h)	66	93	111	134	90.5	131	258	400
	Number of Excursions > 4 h	257	181	156	107	155	88	56	45
	Number of Excursions > 8 h	199	131	115	75	118	72	44	36

[a]An "excursion" is defined as a unique event when indoor RH exceeds 60%.

Predicting Energy Use

Differences between the two models can also impact the predicted energy use for air conditioners and dehumidifiers, which is important if a researcher wants to understand the energy use related to temperature and humidity control. Figures 6 and 7 show the required cooling energy for Houston and Orlando, respectively. Three cases are shown: uncontrolled humidity (no dehumidifier), humidity controlled to 60% rh, and humidity controlled to 55% rh.

For the case with no humidity control, the cases with moisture buffering predict slightly higher energy use than the case with no buffering. This is because the RH at the evaporator coil inlet for these cases is slightly higher. Figure 8 shows a histogram of the RH while the air conditioner is running. The air conditioner provides a certain level of sensible cooling to this air to meet the temperature setpoint, but since the humidity is higher for the EMPD and EC cases, the air conditioner simultaneously provides more latent cooling. Thus, it provides more total cooling, and uses slightly more energy, to maintain the same temperature setpoint.

In Figures 6 and 7, the second and third columns show the energy use when the RH is controlled to 60% and 55%, respectively. Since the temperature setpoints have not changed, the energy use is higher when controlling humidity. In general, the energy use predicted by the EC model is lower than that predicted by the EMPD model. If we compare the commonly used value of EC = 15 and the EMPD medium case, the EC model predicts 24% less dehumidification energy (4% less total cooling energy) in Houston and 35% less dehumidification energy (17% less total cooling energy) in Orlando. Note that the trend from EMPD low to EMPD high also differs from the trend from EC = 5 to EC = 45, particularly for the cases in Orlando.

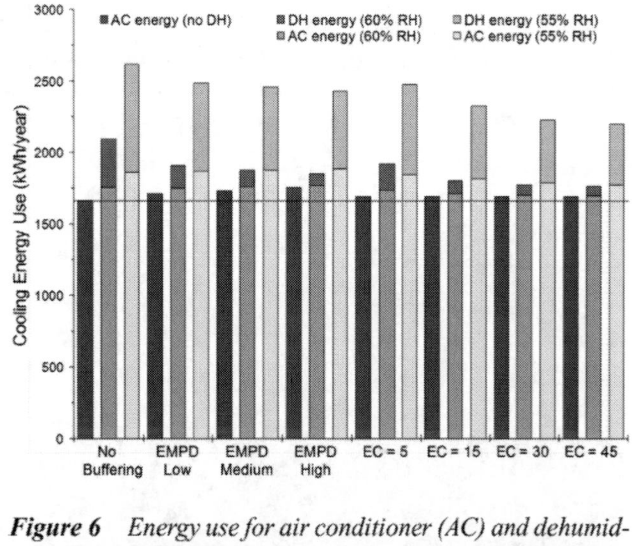

Figure 6 *Energy use for air conditioner (AC) and dehumidifier (DH) in Houston with different levels of moisture buffering. The cases with humidity control use a 45 pints/day (21.3 L/day) dehumidifier with an efficiency of 1.5 L/kWh. Humidistat setpoint for each simulation is indicated in parentheses.*

These differences can be seen more easily by using the no-buffering case as a benchmark for comparison. Table 5 shows relative differences in energy use compared to the no-buffering case. The trends with moisture buffering level correlate reasonably well with the hours above 60% rh when humidity is uncontrolled (Table 4). For example, adding moisture buffering to a home in Orlando has a smaller effect on energy use than it does in Houston, which is consistent with the effect of buffering level on hours above 60% rh.

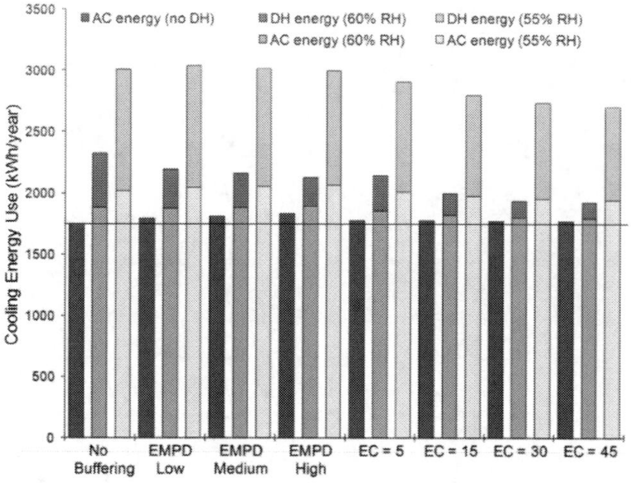

Figure 7 *Energy use for air conditioner (AC) and dehumidifier (DH) in Orlando with different levels of moisture buffering. The cases with humidity control use a 45 pints/day (21.3 L/day) dehumidifier with an efficiency of 1.5 L/kWh. Humidistat setpoint for each simulation is indicated in parentheses.*

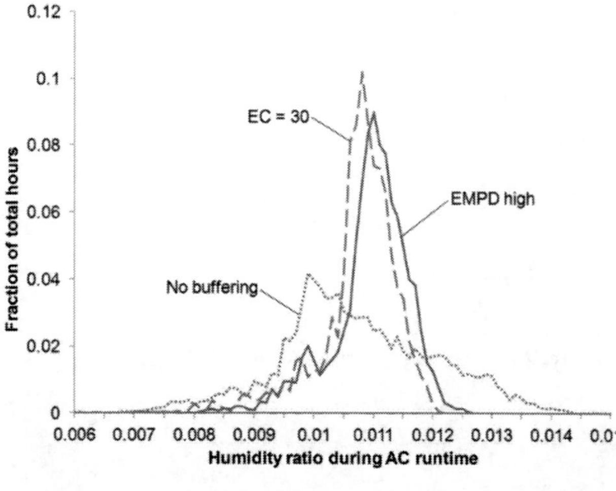

Figure 8 *Indoor humidity ratio frequency distribution for Houston while air conditioner is on. No dehumidification.*

Table 5. Energy Use Relative to No-Buffering Case

		60% rh Control		55% rh Control	
		Total Cooling Energy	DH Energy	Total Cooling Energy	DH Energy
Houston	EMPD Low	–9%	–54%	–5%	–18%
	EMPD Medium	–10%	–65%	–6%	–23%
	EMPD High	–11%	–75%	–7%	–28%
	EC = 5	–8%	–45%	–5%	–16%
	EC = 15	–14%	–73%	–11%	–33%
	EC = 30	–15%	–78%	–15%	–42%
	EC = 45	–16%	–80%	–16%	–44%
Orlando	EMPD Low	–6%	–28%	1%	0%
	EMPD Medium	–7%	–37%	0%	–3%
	EMPD High	–8%	–47%	0%	–6%
	EC = 5	–8%	–34%	–3%	–9%
	EC = 15	–14%	–59%	–7%	–17%
	EC = 30	–17%	–68%	–9%	–21%
	EC = 45	–17%	–70%	–10%	–23%

Similar to the results above, the difference between EMPD low and EMPD high do not coincide with EC = 5 and EC = 45, respectively, and the results are not consistent from one city to the other. The EMPD model results fall between EC = 5 and EC = 15 in Houston, while in Orlando they seem to agree for EC < 5 and EC = 10.

It is important to note that these simulations (both EMPD and EC) do not model the release or uptake of the heat of sorption when moisture is adsorbed by the materials. This effect can change the sensible heat ratio of the building and can affect the heating or cooling energy. Future research is planned to investigate this effect.

Selecting HVAC Equipment

This section discusses potential differences between the two models when selecting HVAC equipment. In particular, we look at dehumidifier sizing and the impacts of variable speed air conditioners.

Dehumidifier Sizing. The level of moisture buffering affects how large a dehumidifier will need to be to maintain the indoor RH. Figure 9 shows the dehumidifier size for different levels of buffering to limit the time spent above 60% rh to 100 hours and 400 hours. The solid lines are the dehumidifier sizes for EC levels from 1 to 45. The shaded bands correspond to the range of dehumidifier sizes for EMPD low to EMPD high. The dotted, horizontal line inside the band is the EMPD medium case, while the top and bottom dashed lines are for EMPD low and high, respectively. The figures indicate that dehumidifier size decreases with increased buffering for the EMPD model. This is true for the EC model initially, but not for higher levels of buffering.

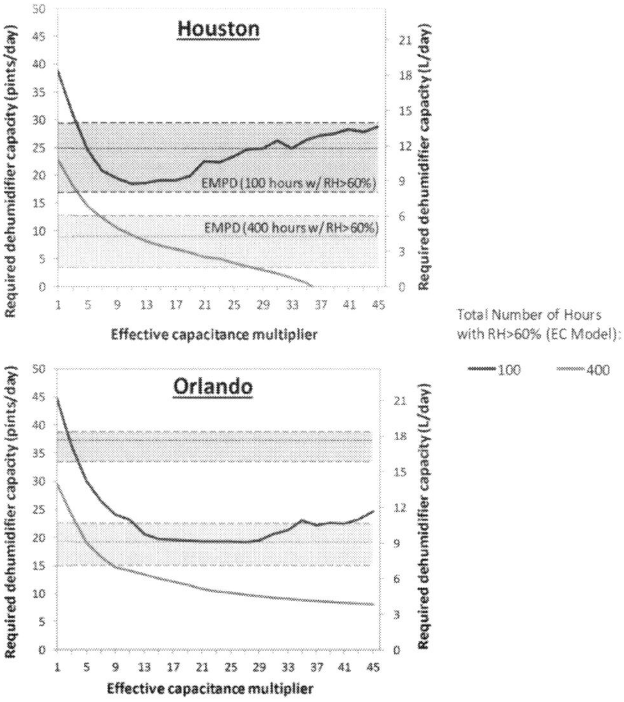

Figure 9 *Required dehumidifier size as a function of the level of moisture buffering. The EMPD band includes lines for the low, medium, and high buffering levels, with the low buffering line at the top.*

As an example, consider a case where a dehumidifier is sized to restrict the time above 60% rh to 100 hours per year. In both Houston and Orlando, using EC = 15 leads to a dehumidifier of approximately 9.5 L/day (20 pints/day). For EMPD medium, the required dehumidifier size is 11.8 L/day (25 pints/day) in Houston, but 17.5 L/day (37 pints/day) in Orlando. This is 300 to 500 W (1000 to 1700 Btu/h) of latent cooling capacity, which is on the order of 5% of the total cooling capacity of the air conditioner required for this low-load home.

Comparing the EMPD and EC results, we see that the estimated dehumidifier size may be different between the two models. The results also indicate inconsistencies between the two cities, similar to the RH statistics in Table 4. However, in the cases illustrated here, a modestly sized dehumidifier

(40 pints/day) will provide adequate dehumidification in both cities regardless of the level of moisture buffering.

Air Conditioners with Variable Speed Compressors. In a previous study, Winkler et al. (2013) used the EMPD model to look at variable speed air conditioners. They concluded that these air conditioners do not effectively manage latent loads during much of the year, with roughly 4500 hours above 60% rh in Houston and 5900 hours above 60% rh in Orlando. In this study, we ran the same analysis with the EC model, which confirmed that these conclusions will be the same regardless of the chosen moisture model. The RH for these simulations, whether performed with the EMPD or EC models, are higher than 60% throughout most of the year, and the fluctuations around the mean RH are less important when looking at hours above 60%.

CONCLUSIONS

This study used whole-building simulations of homes in hot-humid climates to compare the EC and EMPD moisture models. The variation of RH throughout the year was significantly different between these two models. This is due to the added transport resistances in the EMPD model, both between the zone air and the surface layer and between the surface layer and the deep layer. The second resistance enables the use of a deep layer to account for buffering of humidity loads of lower frequencies (on the order of weeks). The EC model is incapable of buffering these loads of different frequencies. The EC = 45 case matches the EMPD model for low-frequency loads, but overdamps high-frequency loads. The EC = 5 case matches the EMPD model for high-frequency loads, but underdamps low-frequency loads.

Regarding the simple moisture-risk metrics, the EMPD model predicts many short excursions above 60% rh, while the EC model predicts fewer excursions of longer duration. This is because the transport resistances in the EMPD model slow the buffering, resulting in more short-term spikes in humidity. However, if hours above 60% rh is used as a metric, the EMPD and EC models generally predict similar results, except for high EC values (EC > 30). Comparing the results from Houston and Orlando showed that a simulation using a certain value of EC in one city can agree with the EMPD results, but that the same EC value in the other city may not agree with the EMPD results.

The two moisture modeling strategies predict different cooling energy use. When humidity is uncontrolled, the buffering has a small effect, but when humidity is controlled, higher buffering generally results in lower energy use. The EMPD model predicted slightly larger air conditioner and dehumidifier energy use than most of the EC cases.

So which model should be used? Comparing each model's assumptions, the EMPD model is certainly more realistic. And this study showed that the simplifying assumptions in the EC model can lead to unrealistic results. But this question cannot be fully answered by this paper, and we hesitate to recommend one model over the other without further research and some experimental data.

The lack of experimental data is one limitation to this study. The moisture properties of the materials for the EMPD model were taken from material databases and the surface areas of the different materials were estimated based on engineering judgment. Future research will use an experimental method to measure, in-situ, the moisture properties of the materials in a home, and will also infer the surface areas of the materials.

A second limitation is that both the EMPD and EC models used in this study were isothermal. A next step in this research is to change the EMPD model to a non-isothermal model. This means the model will include the effects of the latent heat that is absorbed or released at the material surface due to water vapor sorption and desorption. It will also model the changing temperature at the wall surface due to this latent heat transfer. We plan to implement this updated model into the EnergyPlus software.

ACKNOWLEDGMENTS

This work was supported by the U.S. Department of Energy under Contract No. DE-AC36-08-GO28308 with the National Renewable Energy Laboratory.

NOMENCLATURE

A	=	surface area (m^2)
d_{EMPD}	=	effective moisture penetration depth (m)
EC	=	effective moisture capacitance (m^3/m^3)
m_v	=	water vapor flow rate (kg/s)
p_{sat}	=	saturation vapor pressure
R	=	mass transfer resistance (s/kg)
RH	=	relative humidity
t	=	time (s)
u	=	moisture content of material (kg/kg)
V	=	volume (m^3)
δ_{perm}	=	vapor permeability (kg/m·s·Pa)
ρ	=	density (kg/m^3)
ϕ	=	relative humidity
τ	=	cycle period used in penetration depth calculation (s)
ω	=	humidity ratio (kg/kg)

Subscripts

air	=	property of indoor air
deep	=	material deep layer
i	=	index of moisture flows
j	=	index of surfaces
matl	=	property of material
sat	=	saturated
surf	=	material surface layer

REFERENCES

Abadie, M.O. and K.C. Mendonca, 2009. Moisture performance of building materials: From material characterization to building simulation using the Moisture Buffer Value concept, *Build. Environ.* 44, 388-401.

Antretter, F., F. Sauer, T. Schoepfer, A. Holm. Validation of a hygrothermal whole building simulation software, in *Proceedings of Building Simulation 2011*: 12th Conference of International Building Performance Simulation Association. Sydney, 2011.

ASHRAE, 2004. ANSI/ASHRAE Standard 62.2-2004, *Ventilation and Acceptable Indoor Air Quality in Low-Rise Residential Buildings*. Atlanta: ASHARE.

ASHRAE, 2012. ASHRAE Position Document on Limiting Indoor Mold and Dampness in Buildings. Atlanta: ASHARE.

Bornehag, C.G., J. Sundell, T. Sigsgaard, 2004. Dampness in buildings and health (DBH): Report from an ongoing epidemiological investigation on the association between indoor environmental factors and health effects among children in Sweden, *Indoor Air*. 14, 59-66.

Cunningham, M.J., 1988. The moisture performance of framed structures—a mathematical model, *Build. Environ*. 23, 123-35.

Cunningham, M.J., 1992. Effective penetration depth and effective resistance in moisture transfer, *Build. Environ*. 27, 379-86.

DOE, 2012. EnergyPlus Engineering Reference, US Department of Energy.

EPA, 2001. Indoor Assessment Tool (IHAT) reference manual. US Environmental Protection Agency.

Fang, L., G. Clausen, P.O. Fanger, 1998. Impact of temperature and humidity on the perception of indoor air quality, *Indoor Air-International Journal of Indoor Air Quality and Climate*. 8, 80-90.

Fang, X., J. Winkler, D. Christensen, Using EnergyPlus to perform dehumidification analysis on Building America homes, *HVAC&R Res*. 17 (2011) 268-83.

Haverinen, U., M. Vahteristo, D. Moschandreas, A. Nevalainen, T. Husman, J. Pekkanen, 2003. Knowledge-based and statistically modeled relationships between residential moisture damage and occupant reported health symptoms, *Atmospheric Environment*. 37, 577-85.

Henderson, H., D. Shirey, C.K. Rice, 2008. Can conventional cooling equipment meet dehumidification needs for houses in humid climates?, in 2008 ACEEE Summer Study on Energy Efficiency in Buildings. Pacific Grove, CA.

Hendron, R. and C. Engebrecht, 2010. Building America House Simulation Protocols. National Renewable Energy Laboratory, TP-5500-49246.

ICC, 2009. International Code Council, *International Energy Conservation Code*.

James, C., C.J. Simonson, P. Talukdar, S. Roels, 2010. Numerical and experimental data set for benchmarking hygroscopic buffering models, *Int. J. Heat Mass Tran*. 53, 3638-54.

Janssen, H. and S. Roels, 2009. Qualitative and quantitative assessment of interior moisture buffering by enclosures, *Energ. Buildings*. 41, 382-94.

Kerestecioglu, A., M. Swami, R. Dabir, N. Razzaq, P. Fairey, 1988. Theoretical and Computational Investigation of Algorithms for Simultaneous Heat and Moisture Transport in Buildings. Florida Solar Energy Center, DOE/SF/16305-2.

Kerestecioglu, A., M. Swami, A. Kamel, 1990. Theoretical and computational investigation of simultaneous heat and moisture transfer in buildings: "Effective penetration depth" theory, *ASHRAE Transactions*. 96, 447-54.

Kniest, F.M., K.-U. Blank, N. Kalsdorf, 1998. Differences between air permeable and air-nonpermeable fabrics on the efficacy and comfort of mite and allergen-proof encasings, *Journal of Aerosol Science*. 29, Supplement 1, S1285-S86.

Kumaran, M.K., 1996. Final Report Task 3 Hygrothermal properties of building materials. Annex 24: Heat Air and Moisture Transfer in Insulated Envelope Parts, IEA.

Künzel, 1995. Simultaneous heat and moisture transport in building components. PhD thesis, Fraunhofer Institute of Building Physics Fraunhofer IRB.

Künzel, H.M., A. Karagiozis, and A. Holm. 2001. A hygrothermal design tool for architects and engineers, Chapter 9. In *ASTM Manual 40: Moisture Analysis and Condensation Control in Buildings Envelopes*. H.R. Treschsel, ed. West Conshohocken, PA: ASTM International.

Künzel, H.M., A. Holm, D. Zirkelbach, A.N. Karagiozis, 2005. Simulation of indoor temperature and humidity conditions including hygrothermal interactions with the building envelope, *Sol. Energy*. 78, 554-61.

Lstiburek, J., 2009a. Building Science Insights - Thermodynamics: It's Not Rocket Science. Building Science Corporation, Report BSI-02. Available at: http://www.buildingscience.com/documents/insights/bsi-021-thermodynamics-its-not-rocket-science/files/bsi-021_thermodynamics_rocket_science.pdf.

Lstiburek, J., 2009b. Building Science Digests – Moisture Control for New Residential Buildings. Report BSD-012. Available at: http://www.buildingscience.com/documents/digests/bsd-012-moisture-control-for-new-residential-buildings/files/bsd-012_moisture_control_new_bldgs.pdf.

Plagge, R., G. Scheffler, A. Nicolai, 2007. Experimental Methods to Derive Hygrothermal Material Functions for Numerical Simulation Tools, in Thermal Performance of the Exterior Envelopes of Whole Buildings X International Conference. Clearwater Beach, FL.

PNNL & ORNL, 2010. Guide to Determining Climate Regions by County, Pacific Northwest National Laboratory, PNNL-17211.

Rode, C. and K. Grau, 2008. Moisture buffering and its consequence in whole building hygrothermal modeling, *Journal of Building Physics*. 31, 333-60.

Rudd, A. and H. Henderson, 2007. Monitored indoor moisture and temperature conditions in humid-climate US residences, *ASHRAE Transactions*. 113, 435-49.

Steeman HJ, Van Belleghem M, Janssens A, De Paepe M. 2009. Coupled simulation of heat and moisture transport in air and porous materials for the assessment of moisture related damage. *Build. and Environ*. 44(10): 2176-84.

Steeman, M., A. Janssens, H.J. Steeman, M. Van Belleghem, M. De Paepe, 2010. On coupling 1D non-isothermal heat and mass transfer in porous materials with a multizone building energy simulation model, *Build. Environ*. 45, 865-77.

Svennberg, K., 2005. Sorption isotherms for textile fabrics and foam used in the indoor environment, Report TVBH-7227, Lund Institute of Technology. Presented at IEA Annex 41 Meeting.

Tariku, F., K. Kumaran, P. Fazio, 2011. Determination of indoor humidity profile using a whole-building hygrothermal model, *Building Simulation*. 4, 61-78.

Walker, I.S. and M.H. Sherman, 2007. Humidity Implications for Meeting Residential Ventilation Requirements. Lawrence Berkeley National Laboratory, LBNL-62182.

Winkler, J., J. Woods, D. Christensen, 2013. Equipment and operational settings for high performance homes in hot-humid climates. National Renewable Energy Laboratory, (in review) TP-5500-57440.

Woods, J., J. Winkler, D. Christensen, 2013. Evaluation of the effective moisture penetration depth (EMPD) model for estimating moisture buffering in buildings. National Renewable Energy Laboratory, TP-5500-57441.

Moisture Performance of Sealed Attics in Climate Zones 1 to 4

Mikael Salonvaara
Member ASHRAE

Achilles Karagiozis, PhD

Andre Desjarlais

ABSTRACT

Sealed (also called unvented) attics are insulated at the roof deck while the ceiling plane is uninsulated. Sealed attics are usually not directly conditioned, but instead the conditions float between indoor and outdoor temperature conditions. The moisture performance of sealed attics has been investigated using a whole house hygrothermal simulation model to understand the risks for high moisture content in the roof sheathing and for high humidity in the attic. The results have also been validated against field test data. The analyzed shingled roof decks have been insulated either with open-cell spray foam or air-permeable insulation (fibrous insulation). The concept of the air-permeable insulation to perform as the condensation-control layer is analyzed.

A parametric study was carried out varying attic air leakage both to the indoors and outdoors, water leakage to roof sheathing, vapor permeance of the insulation, as well as the indoor air-moisture loads and exterior climate. The results show that the sealed attic can experience elevated humidity at warm temperatures which can create favorable conditions for mold growth unless the attic is intentionally conditioned or via duct air leaks. The vapor permeance of the insulation layers (and if a vapor-retarding coating is present) was a key factor that controlled the moisture content of the roof sheathing together with the overall airtightness. The air impermeable but vapor permeable insulation was not sufficient alone to prevent condensation in the roof deck.

INTRODUCTION

Conventional attic construction involves providing insulation on the floor of the attic and ventilation of the attic space (vented attic) from the outside. When an air-distribution system is installed in the vented attic, any duct or air-handler leakage creates an energy penalty in addition to conduction and radiation losses. An alternative to conventional vented attics is an unvented attic (UVA) (also known as a sealed attic). Insulating the attic roof deck and blocking ventilation of the attic space to the outside moves the thermal boundary to the sheathing roof line. The air-distribution system is now located within a semiconditioned space, which one would expect increases its overall efficiency, durability, and maintainability. The sealed attic can be built as an UVA with insulation in direct contact with the roof deck or with a ventilated roof deck with an airtight interior layer in the attic side of the ventilation gap. The ventilated roof deck is not common in UVAs, but is an option for a cathedral ceiling. The cathedral ceiling is an overconditioned living space, whereas the sealed attic is only indirectly conditioned (no ventilation or intentionally heating/cooling supplied to the attic).

The conventional vented attics and the sealed attics both have clearly defined pros and cons.

A house with a vented attic typically has lower conduction heat loss and heat gain through the ceiling due to thicker insulation deployed on less heat transfer area than a house with an UVA approach, but the HVAC systems in the attic can cause an energy penalty. In a house with an UVA the HVAC system in the attic is inside the thermal boundary. The typical construction practice to build UVAs deploys spray foam sprayed directly under the roof sheathing. The UVA has its thermal boundary at the roof deck, whereas the vented attic has the thermal boundary at the ceiling level. The application to install the insulation in direct contact with the roof deck (UVA) results in a higher temperature gradient across the insulation

__Mikael Salonvaara__ is a research associate at Owens Corning, Jefferson, GA. __Achilles Karagiozis__ is Global Building Science Director at Owens Corning, Granville, OH. __Andre Desjarlais__ is group leader of the Building Envelope Group, Oak Ridge National Laboratory, Oak Ridge, TN.

Table 1. Common Features of Conventional Attics and UVAs

Feature	Conventional Attic	UVA
Conduction Loss	Small insulated area (ceiling). Easy to install thick insulation at low cost. Typical insulation: blown fiberglass or cellulose.	Larger insulated area (sloped roof deck). High R-value may be difficult to achieve. Typical insulation: spray foam (open or closed cell).
Airtightness	Airtightness plane is at the ceiling.	Airtightness plane is at the roof deck.
HVAC in the Attic	HVAC and air-distribution system is outside of thermal boundary. System efficiency is reduced with exposure to attic (≈outdoor) conditions.	HVAC is within thermal and air boundary. Duct/AHU leakage exchanges air between the attic and the living space.
Moisture Control	Attic ventilation removes attic moisture from water intrusion or from indoor air leaks	Roof deck may have difficulties in drying from water intrusion. Attic temperature and humidity are not moderated.

layer during the peak hours than when the insulation is installed on the floor of the attic. The lower temperature gradient in the vented attic (and thus the lower heat flow) through the insulation is the result of attic ventilation and the thermal radiation exchanges. Therefore, the insulation layers in the two systems perform under different temperature gradients. Additionally vented attics typically have more insulation and higher R-value than the roof deck in the sealed attics. Table 1 summarizes the common features of conventional and UVAs.

While providing energy benefits by bringing ductwork into conditioned space, the moisture performance of the UVAs has risks that deserve a closer look.

UVAs can be built with many different design options. The current building code requires the use of air-impermeable insulation only, or it allows for a combination of air impermeable and air-permeable insulation (ICC 2012). If air-permeable insulation is used in the UVA, the design has to include a condensation-control layer. The goal is to raise the temperature of the cold-side inner surface (i.e., underside of the roof deck, or inner foam surface) sufficiently that condensation will not occur, if interior air comes in contact with that surface. This is done by using what is referred to as air-impermeable insulation, such as rigid foam board or spray foam. The building code further sets vapor retarder requirements in climate zones (CZ) 5 and above. The air-impermeable insulation must be a vapor retarder (class II) or a class II vapor retarder coating has to be applied on the interior surface of the insulation in CZs 5 and above.

The building code asks for air-impermeable insulation such as rigid foam board or spray foam to be used as a condensation-control layer in all climates zones with the exception of the dry CZs 2B and 3B (ICC 2012). The moisture performance of the vented attic and the sealed attic insulated with spray foam are briefly presented here. In addition to investigating the moisture content of the roof deck, this paper focuses on the importance of airtightness of the attic to outdoors.

MOISTURE PERFORMANCE SIMULATIONS OF THE ROOF DECK IN VENTED AND UVAS

The analyses for the hygrothermal performance of the attics have to focus not only in the roof deck and its moisture performance but also on the conditions of the air in the attic itself. Previous studies such as Lstiburek and Schumacher have exclusively focused on the condensation control of the roof deck (2011). The temperature and humidity of the air in the UVA is not controlled but it is rather floating between the outdoor and indoor conditions. The impact of moisture intrusion events into the attic, for example, may create conditions that could promote mold growth. Moisture can enter the attic by diffusion through the roof deck, water leaks through the roof or by air leaks from outdoors (especially hot and humid climates) or from indoors.

UVAs are often allowed to have less insulation than what is generally required by the local building energy codes for vented attics. The lower R-value may in some areas be allowed prescriptively or the lower R-value has to be justified by trade-off calculations. When the heating and cooling systems and ducts are brought inside the thermal and the air barrier control boundary in the building enclosure, energy savings can be expected when the air handler and the ducts are not leaking to the outside.

UVAs are typically insulated with spray foam applied directly under the roof sheathing. There are different types of spray-foam insulation and they are typically divided into two main groups: open cell and closed-cell spray foam. Other qualities of spray foam products exist, but these two are considered most common. In terms of thermal and moisture performance, the two foam types have four major differences in material properties:

1. Open-cell foam is light density (0.5 lb/ft^3 [8 kg/m^3]) and soft, closed-cell foam is more rigid and has higher density (2 lb/ft^3 [32 kg/m^3])
2. Open-cell foam has a lower R-value per inch (R-3.5/in. [RSI-20/25.4 mm]) than closed-cell foam (R-6/in. [RSI-34/25.4 mm])
3. Open-cell foam is vapor permeable (20–50 perm-in. [29–73 ng/smPa]), closed cell is vapor tight (1–2 perm-in. [1.46–2.92 ng/smPa])
4. In order to be an air barrier, 3.5–5.5 in. (89–140 mm) of open-cell foam and 1 in. of closed cell is typically needed to satisfy the air barrier material requirement 0.02 L/s·m^2 at 75 Pa (0.004 cfm/ft^2 at 0.3 in. H_2O) to be considered air-impermeable material per code.

Most spray foams today need a fire-retardant coating when left exposed to the attic. The intumescent coating is applied directly on the foam. The coatings are typically low in permeance with permeance less than 1 perm and act as a vapor retarder. In the southern climates open-cell spray foam is more commonly used than closed-cell spray foam.

The effect of the water vapor-permeability of the spray-foam insulation on the roof deck moisture contents was investigated in the simulations by using two different permeances for the spray foam: spray foam classified as open-cell foam with permeability 23 perm-in. (33.58 ng/smPa) and one open-cell foam with higher permeability 54 perm-in. (78.84 ng/smPa). The open-cell foam was additionally simulated with an intumescent coating. Furthermore, the necessity of the code requirement for the condensation-control layer (air-impermeable insulation with a specified R-value) was evaluated by calculating the UVA without the air-impermeable insulation and with highly vapor-permeable fiberglass insulation installed in place of spray foam.

Simulations were carried out in selected cities in four different IECC CZs (Table 2).

Codes and Standards

International Residential Code (IRC) chapter R806.5 "Unvented attic and unvented enclosed rafter assemblies" requires air-impermeable insulation (ASTM E2178: Flow rate <0.02 L/sm^2 at 75 Pa [<0.004 cfm/ft^2 at 0.3 in. H_2O]) to be used in direct contact with the roof sheathing when insulating under the roof sheathing (ICC 2012). If air-permeable insulation is used in this scenario, then the air-permeable insulation has to be supplemented with a condensation-control layer (air-impermeable insulation) either on top of or under the roof sheathing. Only CZs 2B and 3B (dry climates) with tile roof are exempt. The required R-value for condensation control depends on the climate. The intent is to keep the surface where condensation would likely occur above 45°F (7.2°C) (monthly average) during the cold winter months. This would prevent or limit condensation in the roof in homes with typical moisture loads.

METHODS

The moisture performance of the attic was simulated in two ways: first the roof sheathing moisture content was analyzed with a building enclosure simulation model, and second the attic humidity was investigated using a whole house simulation model.

Building and Attic Selection

A typical residential wood frame home with an attic and shingles as roof cladding was selected for the study.

Simulation Model Description

The models used for the moisture analysis were WUFI-Pro Karagiozis and Künzel (2001) and WUFI-Plus which are members of the WUFI software family developed jointly by the Fraunhofer Institute for Building Physics and Oak Ridge National Laboratory. The North American versions of the software WUFI-Pro 5.1 is a one-dimensional hygrothermal (heat, air and moisture) simulation model which allows for realistic calculation of the transient coupled heat and moisture transport in multilayer building components exposed to natural weather. WUFI-Plus allows for modeling a whole building with the full interaction between the building enclosure parts and the indoor climate.

Table 2. CZs, Locations, and Applied R-Values of Roof Insulation

CZ	City, State	IECC 2012 R-Value	Modified R-Value for UVA Simulations
1	Miami, FL	R-30	R-21
2	New Orleans, LA Houston, TX	R-38	R-21
3	Atlanta, GA	R-38	R-21
4	Knoxville, TN	R-49	R-21
4	Baltimore, MD	R-49	R-21

Table 3. Minimum R-Values of Air-Permeable Insulation in UVAs (ICC 2012, Table R806.5)

CZ	Minimum Rigid Board or Air-Permeable Insulation R-value
2B and 3B tile roof only	None required
1, 2A, 2B, 3A, 3B, 3C	R-5
4C	R-10
4A, 4B	R-15
5	R-20
6	R-25
7	R-30
8	R-35

Climatic Boundary Conditions

Exterior. In hot and humid climates, there is less of a concern for condensation in the roof deck due to the mild or warm winter. In the summer the high outdoor humidity and air leakage can bring in moisture into the attic air.

Interior. During cold weather, the indoor moisture loads determine the indoor humidity together with the fresh air ventilation (or air leakage) rates. In the summer in hot and humid climates, the air moisture content is lower indoors than outdoors due to the dehumidification by the cooling system.

MODEL CALIBRATION

The focus of this study is to investigate the moisture performance of UVAs with special focus on air leakage and ventilation. The hygrothermal models were calibrated against measured attic data humidity and temperature data. Oak Ridge National Laboratory has been measuring the thermal and moisture performance of a test house in Knoxville, TN. Five months of temperature and relative humidity data were available from the beginning of August till the end of December for our model validation purposes. The house is a two-story unoccupied single family home with simulated interior sources for heat and moisture to represent realistic interior loads. The UVA has two main orientations North and South (with 16° deviation). The attic contains an HVAC system. The roof consists of the following layers starting from the outside:

1. Shingles
2. Felt underlayment
3. Oriented strand board (OSB) 0.5 in. (12.5 mm)
4. Open-cell spray foam 6 in. (152 mm) (installed in two installations)
5. Spray-on fiberglass 2 in. (51 mm)

Attic Airtightness Testing

The UVA is an indirectly conditioned space even though the whole space is inside the thermal and air barrier boundaries. It is important to understand the connections of the attic not only to the outside but also to the inside. In cooling climates, the stack effect works in the other direction than in the heating climate. In hot and humid (cooling) climates the stack effect and wind pressures will pull air from outside into the attic. On the other hand, the air leaks in the HVAC system (if located in the attic) effectively creating some conditioning in the attic. Additionally an imbalance in air leaks from the supply and return side may create a small pressure difference between the attic, the outdoors, and the living space. The airtightness of the ceiling plane plays a role in the air movements. A Building America project for UVAs in hot-dry climates noted that the whole building airtightness in a house increased by 50% when the attic hatch was opened (Building America 2003). Therefore, it was decided to do blower-door testing that would provide not only the whole building airtightness, but also individually the airtightness of the roof deck and the ceiling plane. Table 4 lists the volumes of the living space and the attic.

Table 4. Test House Areas and Volumes

Space	Area (ft^2 [m^2]), Volume (ft^3 [m^3])
Living space 1st & 2nd floors	2241 (209), 22293 (635)
Attic	1383 (129), 4564 (130)

Table 5. House and Attic Airtightness

Space	ACH50 (CFM50)
Whole house, attic hatch closed	4.1 (1525)
Whole house, attic hatch open	3.4 (1541)
Attic only, leakage to outdoors and indoors	18.9 (1440)
Attic only, leakage to outdoors only	4.4 (336)
Attic to indoors only (deducted from two previous measurements)	— (1104)

Note: The ACH50 values are for information only and are for the included volume of the space only (attic or whole house).

Several blower door tests were carried out for the house with two blower door systems. First, a whole house airtightness testing was carried out with the attic hatch closed. Second, a blower door system was installed in the attic hatch and the overall airtightness of the attic was performed with doors open to the outside in the living space to create outdoor pressure conditions both above the roof and below the ceiling. Third, the roof deck airtightness to the outside was performed by maintaining equal pressures in the attic and in the living space to eliminate the air leaks in the ceiling plane. These two tests allowed us to characterize the attic air leaks both towards the outdoors through the roof and towards the indoors through the ceiling plane. Results are presented in Table 5.

These results show that the attic can contribute to whole house air leakage by about 23%. Thus, we cannot ignore the air leakage between the attic and outdoors. It must be noted here that the ceiling plane is very leaky, which means that the guarded test as presented can have a big uncertainty and the results should be treated as such. Testing was carried out at multiple pressures and the authors feel that the results are representative. A set of whole building air leakage tests were carried out with the attic hatch closed or open. The attic pressure was monitored in the test with attic hatch closed. The attic zone pressure measured 7.7 Pa (0.03 in. H_2O) (attic-to-house) when the house interior was depressurized to −50 Pa (0.2 in. H_2O). The flow network analysis resulted in air leakage rate of 320 cfm50 (151 L/s at 50 Pa) for the roof deck, which is within 5% of the unguarded test result. An alternative test procedure is being developed to minimize the uncertainty caused by the leaky ceiling.

Pressure testing between the attic and the second floor showed that the air-conditioning system is providing a slightly negative pressure of 0.5 Pa (0.002 in. H_2O) in the attic, which shows an imbalance of airflow supply and return. This pressure difference causes air exchange between the attic and the living space. The ceiling plane was found to be very leaky and even small pressure differences cause air exchange. In effect, the air-conditioning system is providing some conditioning to the attic. Based on the ceiling airtightness characteristics, the airflow rate through the ceiling to the attic would be about 45 cfm (21 L/s) when the HVAC is on while the additional airflow rate from outdoors to the attic would be about 10 cfm (5 L/s) due to the pressure difference created by the air conditioning system.

Airflows in the Attic Due to Air Leakage in Hot Climates

What is the importance of understanding the air leakage of the attic itself? The house is fairly airtight with 3.4 ach50 overall. In hot climates, the stack effect works in the reverse fashion to the way it works in cold climates, i.e., the air tends to flow in from the top and out from bottom of the building when the outdoor air is hotter than the interior. Let's assume this house is in Houston, TX. Rough estimates for seasonal ventilation rates due to infiltration in Houston, TX give 0.29 ach for winter and 0.18 ach for summer (output from Tectite blower door software/Energy Conservatory with Houston weather) (Energy Conservatory 2007). Let's assume that we have this average ventilation rate in the summer and that the air infiltration occurs top-down as discussed above. In a top-down flow scenario, the whole house air leakage would flow from outdoors through the attic to the living space below. The amount of airflow is 18% of the whole house volume in an hour. For this house volume (including attic) the airflow rate is 0.18 1/h · 765 m^3 (27013 ft^3) = 138 m^3/h (38 L/s [82 cfm]). In terms of the attic volume this 138 m^3/h (38 L/s [82 cfm]) represents 138/130 ≈ 1 ach. It is clear that even if the whole-house air leakage is fairly small, the attic may still experience quite high air exchange with outdoor air. In hot and humid climates this air exchange would bring moisture into the attic which is one of the main research finding of this paper.

ORNL Attic Data

Figure 1 presents the measured and calculated temperature and relative humidity between the roof sheathing (OSB) and the spray-foam insulation. The figure shows simulation results with two different methods, which we will explain later in the paper. Locally-measured exterior weather data and interior conditions (attic temperature and humidity) were used in the simulations. During a period in October (80+ days from August 1st in Figure 1) a plastic sheet was placed on the roof for research purposes and this clearly increased the effect of solar radiation on the roof temperatures.

The test data reveals one behavior that is typically not consistent with simulation models. The wood handbook published by Forest Products Laboratory shows data for sorption isotherms for wood at different temperatures (Glass and Zelinka 2010). Figure 2 shows the modified sorption curves for OSB at different temperatures, which the author has created by simply scaling the level of the moisture content to match a measured sorption isotherm at room temperature. The higher the temperature, the lower the equilibrium moisture content of wood at different relative humidity. Figure 3 shows how the correct sorption isotherm for the high temperature will create as much as 25% higher vapor pressure in the material with the same moisture content. This means that even short times at a high temperature can create much higher vapor diffusion rates promoting drying of the roof deck. In airtight materials, total pressure inside the material can further increase vapor outflow. This effect is however not discussed here.

The water in the pore space of porous materials is in both liquid and vapor form. When the temperature of the material increases without the moisture content of the material changing, the liquid water evaporates creating a higher vapor pressure. A single sorption curve in modeling assumes that in this situation the relative humidity inside the material does not change because the moisture content does not change. However, the sorption curves have been measured for some materials such as wood at different temperatures /1/. The measurements show that a single sorption curve is not accurate at high temperatures above 60°C (140°F). Wood can absorb less water at high temperatures than at low temperatures at the same relative humidity. The effect of temperature-dependent sorption can be seen in the measured relative humidity data for this roof.

The results in Figure 1 have been calculated with two different methods: first with single sorption curve without temperature effects on the sorption, and second with multiple sorption curves that take into account the effect of temperature on the ability of the material to absorb and desorb moisture in the hygroscopic range. The hourly behavior of the relative humidity under the roof sheathing behaves in opposite ways depending on whether or not the temperature-dependent sorption is used or not. The relative humidity goes down when the temperature goes up if only a single sorption isotherm is used in the modeling. However, both the measurements as well as the simulations with the temperature-dependent sorption show that the relative humidity goes up when the temperature of the roof deck goes up. When looking at the daily average values, the hygrothermal model agrees quite well with the measured data for the north oriented roof whether the temperature dependent or a single sorption isotherm is used. In Figure 1 the quadratic mean of the difference in relative humidity between the simulated and the measured values was 7% and 9% for the north-oriented roof slope (15% and 21% for the south-oriented roof slope), for the simulations with and without the temperature dependent sorption, respectively. However, the south oriented roof shows much lower measured than simulated relative humidity. This could be due to several factors, one of them can be the temperature-dependent sorption isotherms that should be generated not only for the roof sheathing but also for other materials including even for the insulation layer. The results show the simulated results will match the measured data better when including temperature-dependent sorption isotherms for the roof sheathing (OSB) (Figure 4): the peaks and valleys of the fluctuations in the diurnal cycles match the measured ones. The actual difference (level) between the measured and simulated humidity on the interior surface of the roof sheathing is still quite high for the south-oriented roof (Figure 1B). The models used for building envelope analyses typically have material properties measured only at one temperature. This brief study shows that there is a need for more research to understand moisture performance at high temperatures.

This study focuses on predicting mold growth in the roof and in the attic and not on more severe damage such as wood rot. Moisture content in the roof deck is typically low due to high temperatures in CZs 1 to 4 unless the roof experiences water leaks. The moisture content measurements were not available for this study and they were not as much of interest because the conditions that describe conditions susceptible to mold growth are based on temperature, relative humidity, and time (Viitanen).

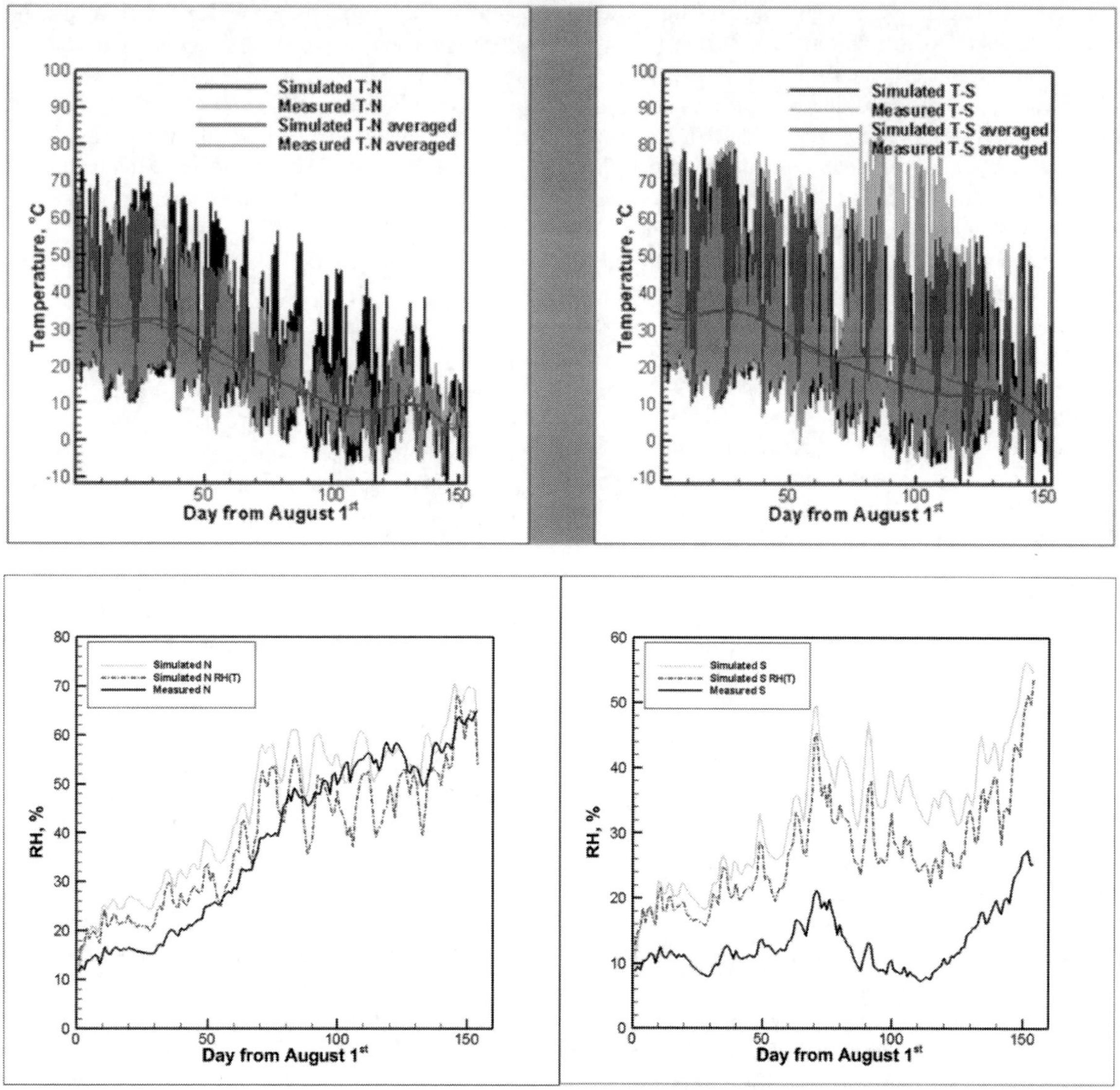

Figure 1 *Daily averages of the measured and the simulated temperature (top) and relative humidity (bottom) between the roof sheathing and the insulation in the test house:* A = *north orientation,* B = *south orientation. Results have been simulated with (Simulated N/S RH[*T*]) and without temperature dependent sorption.*

The effect of temperature-dependent sorption is less at low temperatures than at high temperatures. The condensation in the roof deck occurs at low temperatures, and therefore a single sorption isotherm as used in the hygrothermal models can be deemed sufficient for the purpose of studying condensation and moisture accumulation in the roof deck. The model calibration for the north-facing roof deck shows that the single curve captures the performance sufficiently in the long term. The south-facing roof deck has a low relative humidity and a high temperature which does not result in conditions that would be favorable for mold growth. The north-oriented roof deck would be the first to experience conditions susceptible for mold growth. The model that includes the temperature-dependent sorption is only for modeling building envelope parts and not whole building performance. The whole building model that was needed and used in the parametric analyses does not have the availability of temperature dependent sorption. Therefore, for those reasons the following condensation and attic humidity studies were made with a model using a single sorption curve only and without temperature effects on sorption isotherms.

SIMULATIONS IN CZS 1 TO 4

The focus of the simulations was to investigate two performance aspects of the UVA: the moisture content of the roof sheathing and the humidity of the attic space. These two are critical in terms of the durability and the safety of the building enclosure, and the indoor air quality. High moisture content of

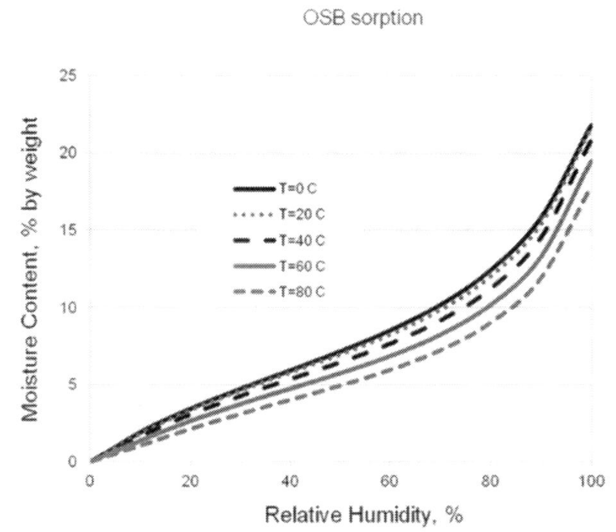

Figure 2 *OSB sorption curves as used in simulating the test house roof. The temperature-dependent sorption curves were estimated by scaling a curve measured at room temperature with the equation for wood in Wood Handbook/Forest Products Laboratory (Glass and Zelinka 2010).*

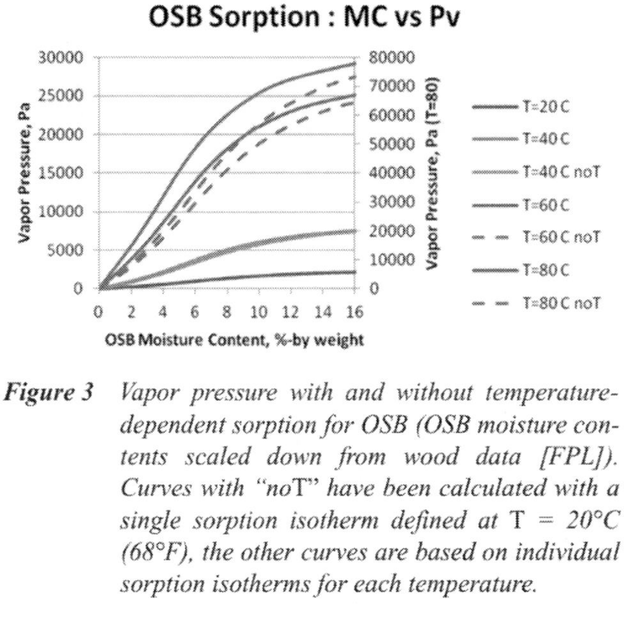

Figure 3 *Vapor pressure with and without temperature-dependent sorption for OSB (OSB moisture contents scaled down from wood data [FPL]). Curves with "noT" have been calculated with a single sorption isotherm defined at T = 20°C (68°F), the other curves are based on individual sorption isotherms for each temperature.*

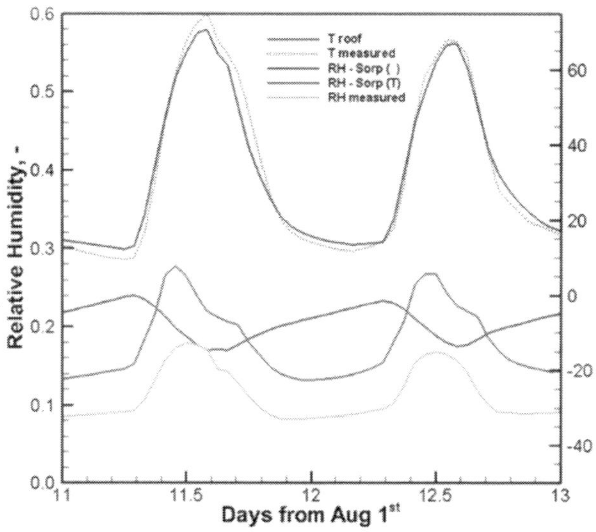

Figure 4 *The measured and the simulated relative humidity between the roof sheathing and the insulation for the south oriented roof. Temperature-dependent sorption is (RH – Sorp[T] or is not taken into account (RH-Sorp[ ]).*

the roof deck (over ~20% by weight) can reduce the structural strength of the roof sheathing, initiate mold growth on its surface and cause dimensional changes that in turn can cause visual degradation of the roof. High humidity (80% rh) in the attic air can cause mold growth on the surfaces exposed to the attic.

Description of the Attic and House

Two different types of roofs (UVA, vented attic) were simulated in four cities in the southern CZs 1–4. The cities were Miami, FL; New Orleans, LA; Atlanta, GA; and Baltimore, MD.

Simulation Set Description

The roofs are oriented to the north with a roof pitch 4/12 (18.4°).

The roof layers in the UVA are from outside to inside as in the roof for model calibration with the exception of the fiberglass insulation on top of the spray foam. The material layers from outside to inside are: shingles, roofing felt, 0.5 in. (12.5 mm) OSB and 6 in. (152 mm) open-cell spray-foam (or fiberglass).

The vented attic has the same materials except that instead of the spray-foam insulation under the roof sheathing there is blown-in fiberglass on top of the ceiling drywall. The attic is ventilated at rate 2 ach with an average attic height of 0.9 m (3 ft).

Indoor and Outdoor Weather. The indoor moisture loads were calculated by assuming a house with the volume 500 m^3 (17,623 ft^3) that has 12 L/day (26.7 lb/day) moisture production and ventilation rate 0.25 ach (including natural and mechanical ventilation). This results in 4 g/m^3 (0.00025 lb/ft^3) moisture load. The moisture loads were chosen to represent not the conditions in an average home but rather the design conditions that would cover most homes (ASHRAE 2009). The important number in the calculations is the moisture load (4 g/m^3 [0.00025 lb/ft^3]) which results from the moisture production and the ventilation rate. The same moisture load can be realized in a home with less moisture production and with less ventilation, i.e., a home with no mechanical ventilation would need less moisture production to end up with the same moisture load. In the summer, cooling system dehumidifies the indoor air to a maximum of 60% rh during cooling hours. Indoor temperature was allowed to float between 21°C

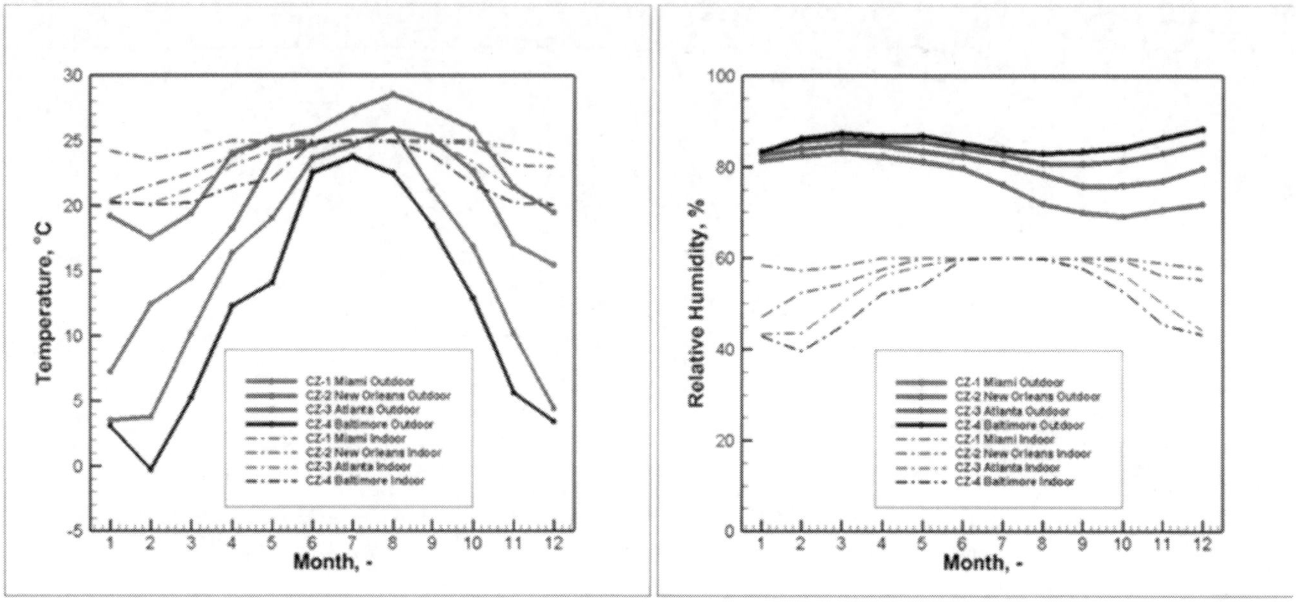

Figure 5 *Average monthly temperature (left) and relative humidity (right) for the weather and the indoor conditions used in the simulations.*

and 25°C (69.8°F and 77°F). The built-in cold year weather files in the WUFI software were used for the locations. The monthly average outdoor and indoor temperature and relative humidity are shown in Figure 5.

Results

Results are provided first for the moisture content of the roof sheathing both for the vented and the UVA configurations by varying the water vapor permeance of the insulation layers. The results are shown for the roof slope facing north, which is the most likely orientation to experience higher moisture contents due to lower temperatures than the south-facing roof slope. Then the air conditions inside an UVA in a hot and humid location are shown with different air leakage characteristics of the attic.

Moisture Contents of the Roof Sheathing. The moisture contents of the roof sheathing in the attics were monitored in the simulations for both the unvented and the vented attic. The moisture content of the OSB roof sheathing in the vented attic is shown in Figure 6, first without rain intrusion and second with 1% rain intrusion to the roof sheathing.

The vented attic performs well with low moisture contents in the roof sheathing with or without the rain intrusion. Even with 1% water intrusion, the moisture content of the OSB stays below 15% by weight at all times in all the four locations. The roof is capable of drying out water intrusion of 1% of the rain hitting the roof.

Figure 7 shows the moisture contents for the OSB in the UVA with open-cell spray foam (vapor permeance 23 perm-in. [33.58 ng/smPa]). The moisture contents are clearly higher than in the vented attic, except for the UVA with no water intrusion in Miami, which shows equal performance to the vented attic. Water intrusion into the roof sheathing at a rate of 1% from the rain hitting the roof elevates the moisture contents above 20% by weight in the OSB in all locations but Miami (CZ 1).

Spray foam is installed in the field and the quality can vary depending on many factors such as field conditions and installers' ability to control the spray equipment. As a result, the water vapor permeance of the installed insulation can vary too. Figure 8 shows the performance with a higher permeance 54 perm-in. (78.84 ng/smPa) for the spray-foam insulation. The moisture contents in the winter are higher with higher-permeance spray foam than with lower-permeance foam in all other climates but CZ 1 (Miami). Miami appears to have very little vapor drive from indoors towards the roof deck.

Spray foam must either have built-in fire-retarding chemicals or a fire-retardant spray will have to be added to allow for the insulation be left exposed in the attic. The intumescent coatings can be vapor retarders with low vapor permeance. In the simulations one of these products has been used together with the open-cell spray foam that has water vapor permeance 23 perm-in. (33.58 ng/smPa). Figure 9 shows the pros and cons of the coating. When the roof has no water intrusion, the additional vapor resistance (vapor permeance 1 perm-in. [1.46 ng/smPa] dry cup tests and 3 perm-in. [4.38 ng/smPa] wet cup) of the intumescent paint reduces the water vapor drive from the attic to the roof deck and the moisture contents of the OSB stay well below 10% after the drying from initial conditions. However, drying of the roof is greatly reduced as well, and if the roof has any water intrusion the moisture content of the OSB can quickly increase to risk levels (above 20%–26%).

An UVA can be insulated with an air-permeable insulation such as fiberglass if the local building code allows it in the

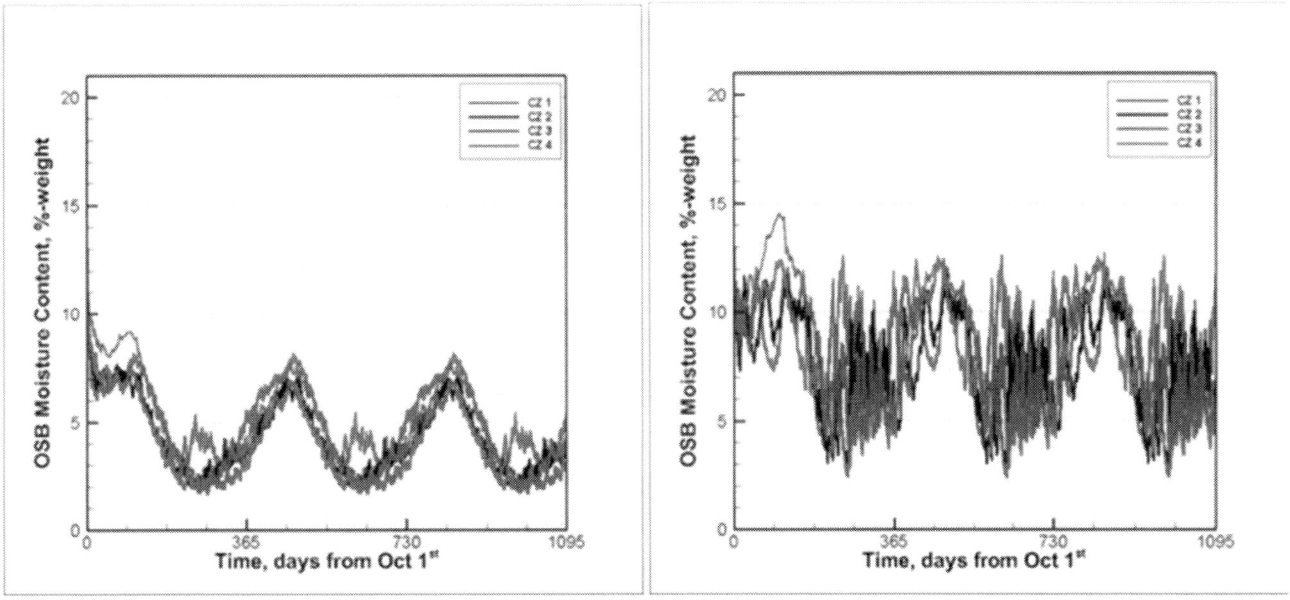

Figure 6 *Moisture content of the roof sheathing (OSB) in a vented attic with no rain intrusion (left) or with 1% rain intrusion to the roof sheathing (right).*

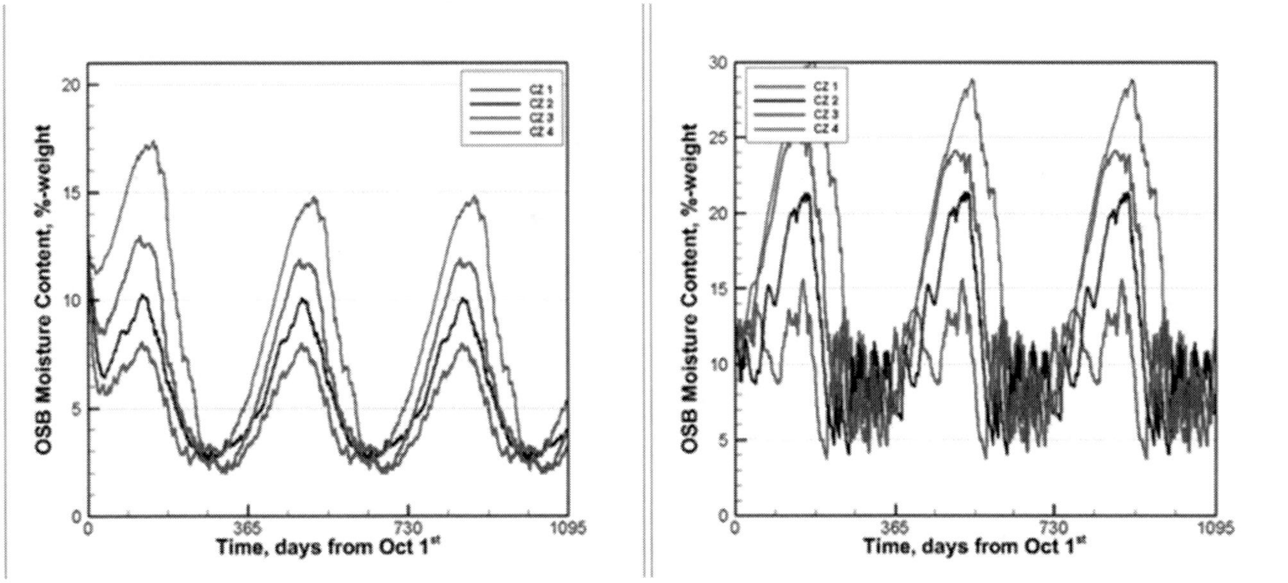

Figure 7 *Moisture content of the roof sheathing (OSB) in an UVA with no rain intrusion (left) or with 1% rain intrusion to the roof sheathing (right) when open-cell spray foam with vapor permeance of 23 perm-in. is used.*

location or CZ. The roof deck can and must be air sealed first to reduce air leakage between the attic and outdoors. Once the roof deck is air sealed blown-in fiberglass insulation is recommended to seamlessly and fully fill all gaps and edges below the roof deck. Dense-packed fiberglass insulation that is produced today has low air permeability that minimizes any air convection between the attic and the roof deck. Fiberglass insulation has high vapor permeability about 100 perm-in. (146 ng/smPa). The netting used for keeping the fiberglass in place can be tuned to have a desired water vapor permeance. Figure 10 shows the moisture content of the OSB roof sheathing with fiberglass insulation only without any additional vapor resistance in the netting. CZs 1 (Miami, FL) and 2 (New Orleans, LA) exhibit low moisture contents but CZ 3 (Atlanta, GA) with close to 20% moisture content and CZ 4 (Baltimore, MD) with above 30% moisture contents show signs of moisture stress. Figure 10 shows the results with netting that has vapor permeance of 10 perms. The additional vapor resistance has a beneficial effect of lowering moisture contents in the roof both without and with water intrusion.

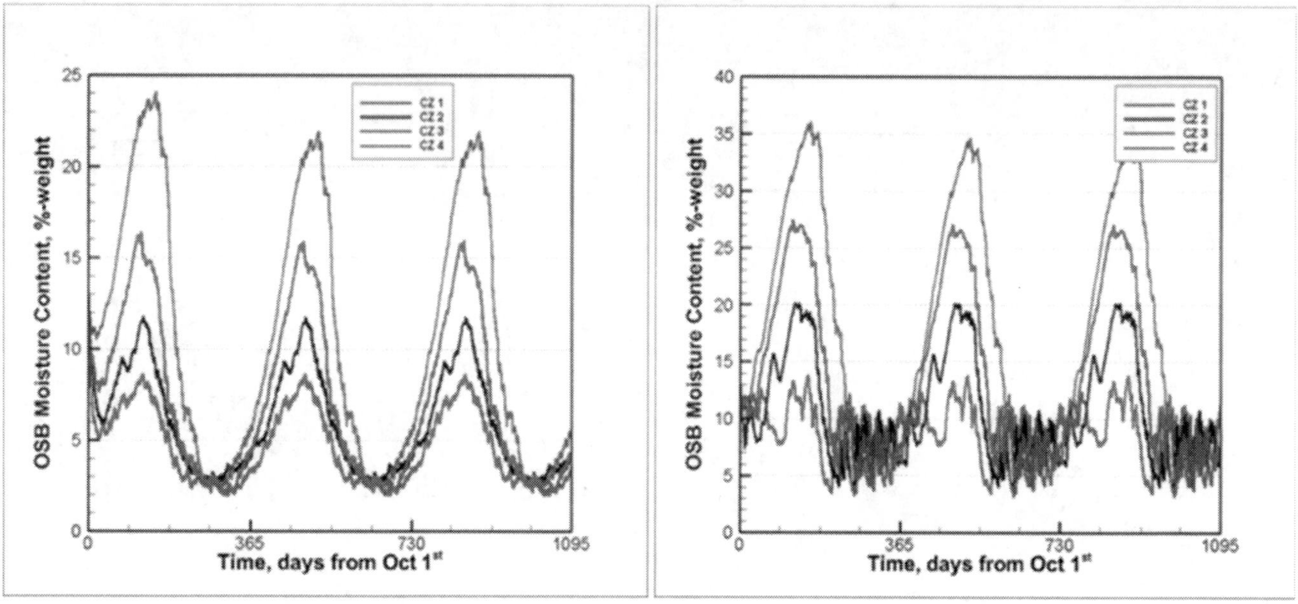

Figure 8 *Moisture content of the roof sheathing (OSB) in an UVA with no rain intrusion (left) or with 1% rain intrusion to the roof sheathing (right) when open-cell spray foam with vapor permeance of 54 perm-in is used.*

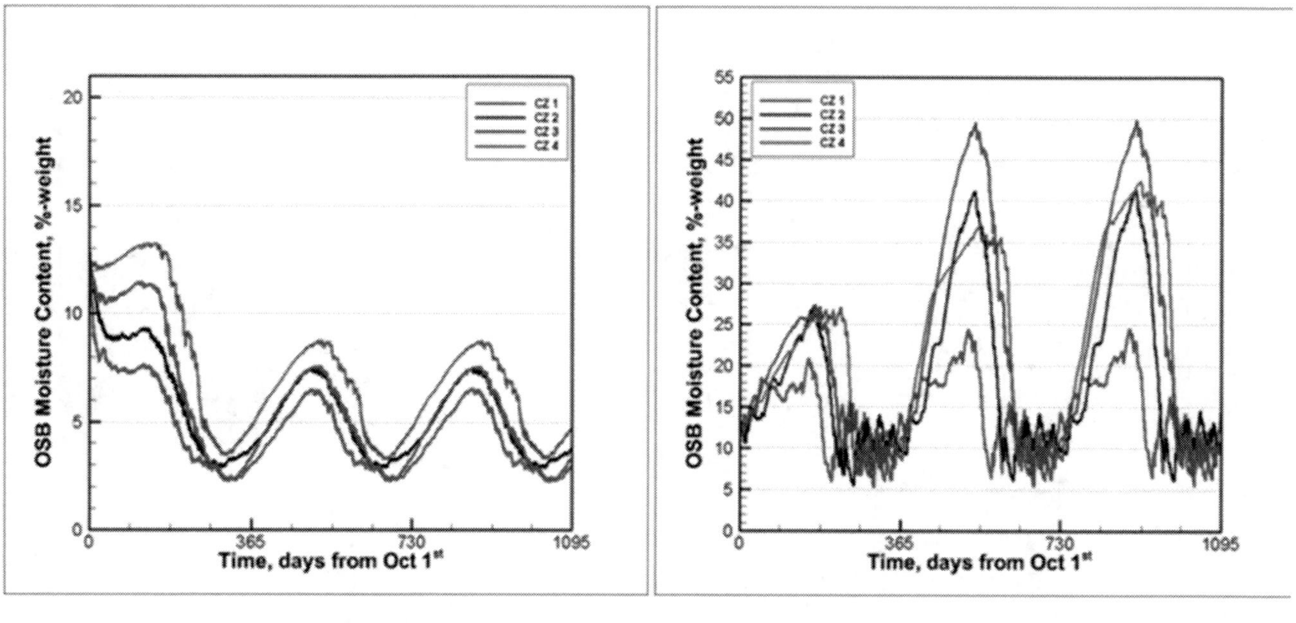

Figure 9 *Moisture content of the roof sheathing (OSB) in an UVA with no rain intrusion (left) or with 1% rain intrusion to the roof sheathing (right) when open-cell spray foam with vapor permeance of 23 perm-in is used together with an intumescent paint coating (permeance 1–3 perms from dry to wet).*

The building code requires the UVA systems built with air-permeable insulation to have R-5 air-impermeable insulation (in CZs 1–3 with the exception of tile roof in CZ 2B and 3B) as condensation-control layer (ICC 2012). In the light of the results shown above this requirement seems to be unnecessary.

Attic Humidity. Vented attics are ventilated by design whereas in the UVAs we attempt to achieve high airtightness to prevent any air exchange between the attic and outdoors. In cold climates some ventilation with outdoor air in the attic would remove moisture from the attic in the winter which would further reduce the possibility of moisture accumulation in the roof deck. In hot and humid climates the attic ventilation may not have the same effect and especially in summer time ventilating the attic with outdoor air could bring moisture into the attic elevating the humidity in the attic space. Therefore,

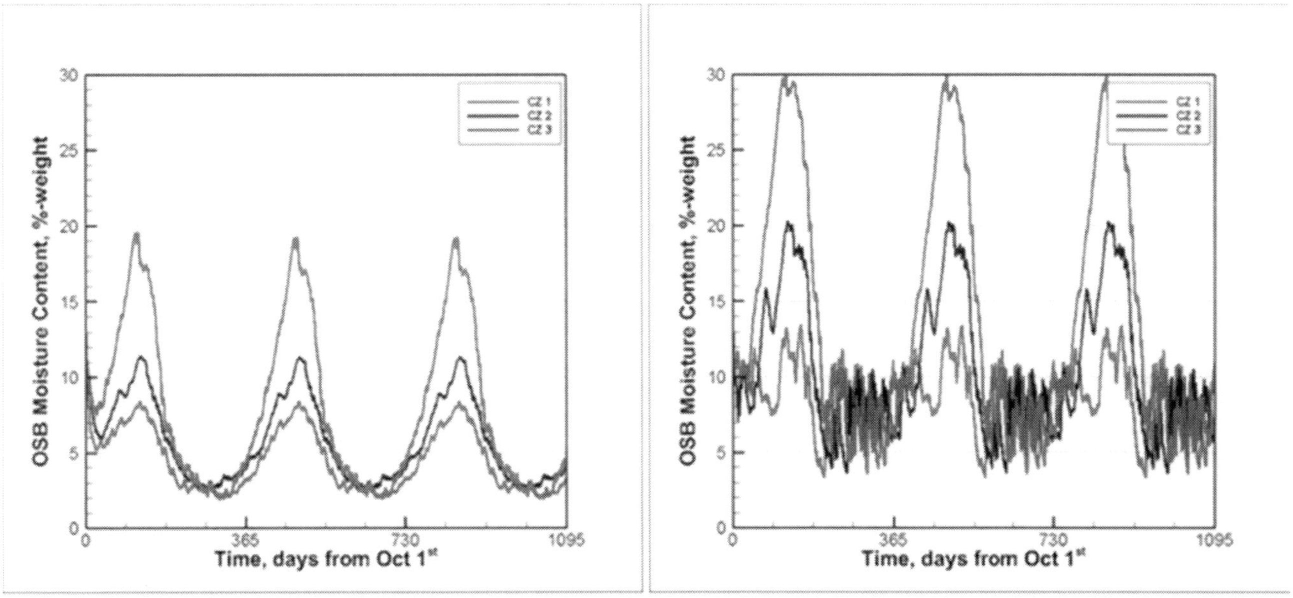

Figure 10 *Moisture content of the roof sheathing (OSB) in an UVA with no rain intrusion (left) or with 1% rain intrusion to the roof sheathing (right) when fiberglass insulation with vapor permeance of 100 perm-in is used.*

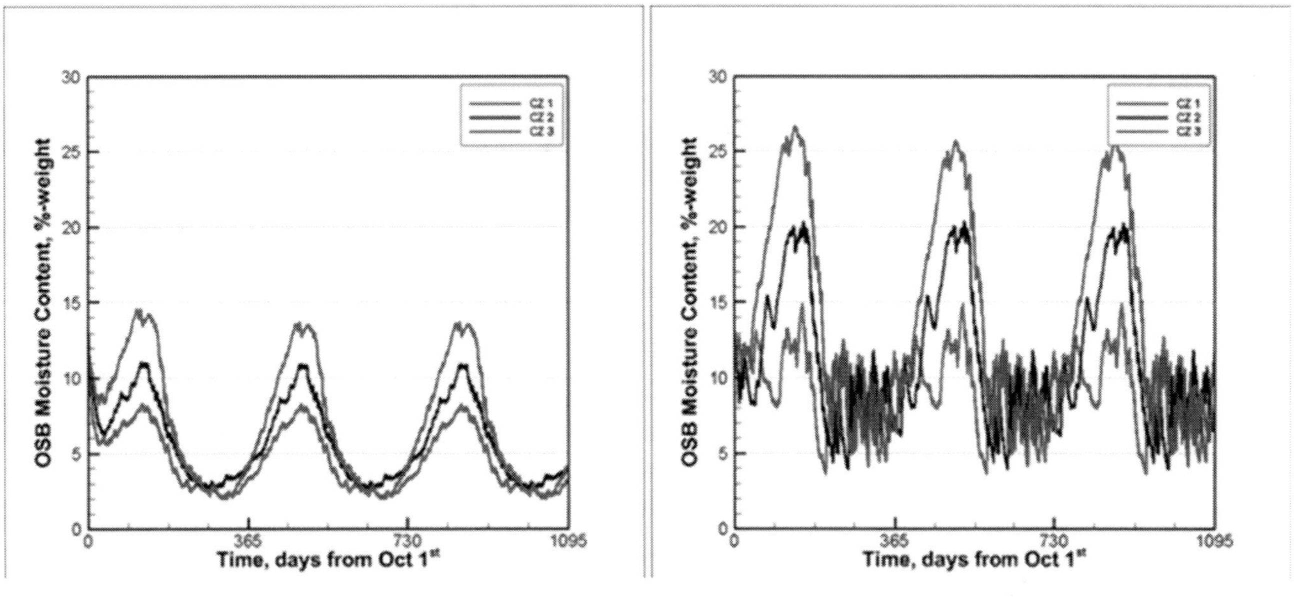

Figure 11 *Moisture content of the roof sheathing (OSB) in an UVA with no rain intrusion (left) or with 1% rain intrusion to the roof sheathing (right) when fiberglass insulation with vapor permeance of 100 perm-in. is used together with a netting on the attic side that has permeance of 10 perms.*

especially in the southern climates it is necessary to investigate the impact of the airtightness of the attic on the attic air temperature and humidity. Results are presented for an UVA in Houston, TX, which is in a hot and humid climate.

An UVA was simulated with the whole building simulation model WUFI-Plus with the focus on the air exchange between the indoors, the outdoors, and the attic. Figure 12 shows the attic air temperature and relative humidity when there is some air leakage between the attic and the indoors, and the attic and the outside. The air leaks from outdoors into the attic in the summer bring humid air into the unconditioned (or indirectly conditioned) attic space that has a temperature close to the indoor temperature. The additional moisture load into the attic can increase the relative humidity of the attic air to levels that are favorable to mold growth (over 80% rh). Note that this high humidity is inside the conditioned attic and the temperature in the attic is about the same as the indoor air temperature and thus the temperature conditions are always such that would allow for mold to grow. Earlier in the paper, the authors presented model calibration against measured attic data for the interior surface of the roof sheathing outside of the insulation adhered to the roof deck. The roof deck is exposed to a large temperature range, both low and high temperatures depending on the climate, and the relative humidity at these low and high temperatures has to be higher than at room temperature to allow for mold growth (Viitanen 1996).

Ideally, the attic would be conditioned to avoid uncontrolled high humidity. However, since the attic is not part of the living space, the interior surfaces of the attic are not finished (insulation is left exposed) and the attic space is not cleaned routinely it would be better for indoor air quality to avoid circulating air from the attic into the living space. Therefore, it is critical that in hot and humid climates the attic is air sealed to the outside carefully. Any holes created after installing the insulation have to be sealed with extreme care. Figure 13 shows how ventilating the attic with conditioned indoor air will reduce the attic humidity below the critical mold growth level 80% rh. The assumption in Figure 13 is that the airflows are balanced in the attic and that the HVAC does not pressurize or depressurize the attic. The more air leakage between the attic and the outdoors, the more conditioned air will be needed in the attic to maintain low humidity in the summer due to high outdoor humidity. If the HVAC system depressurizes the attic, the air leakage from outdoors to the attic will increase and the humidity level ends up higher.

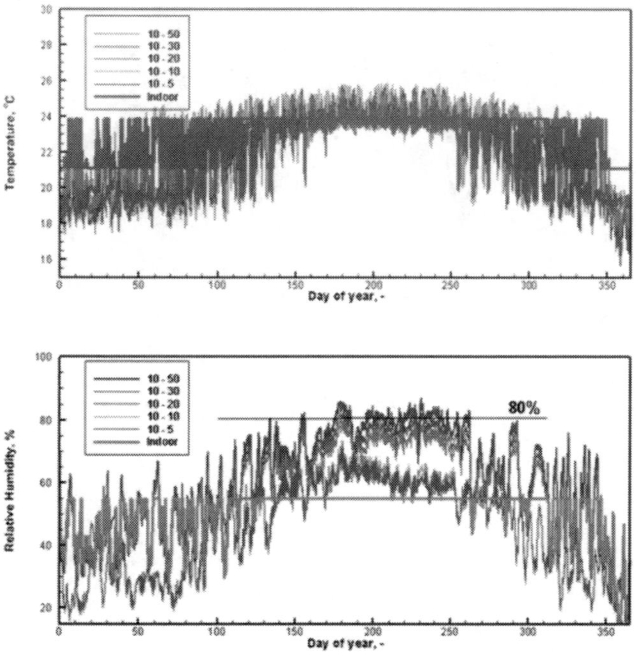

Figure 12 *Temperature (top) and relative humidity in the attic when the attic has different air leakage rates between indoors and outdoors. The first number in the line legends is the ventilation rate between the attic and the indoors (L/s) and the second number is the ventilation between the attic and the outdoors (air leakage in L/s). [1 L/s = 2.12 cfm]*

CONCLUSIONS

An UVA and a vented attic have been analyzed with hygrothermal models in terms of their moisture performance. The UVA has a lower tolerance to water penetration by allowing for less drying capability for the roof sheathing than the vented attic. The fire-retardant coatings, while they do reduce the condensation from indoor sources, further reduce the drying capability. The simulation results do not show the need for the condensation control insulation layer with R-value of R-5 in CZs 1 and 2 (ICC 2012).

The UVA is not directly conditioned but instead its temperature and humidity floats between that of indoors and outdoors. In hot and humid climates, or in any climate with hot and humid summer, the attic space can become humid even with seemingly low air leaks from outdoors to the attic. The UVA has no designed controls for temperature or humidity and the humidity in the attic can become high enough to allow for mold growth. Unless the house is very airtight overall (1–2 ach50) it is recommended to verify the airtightness of the roof deck by

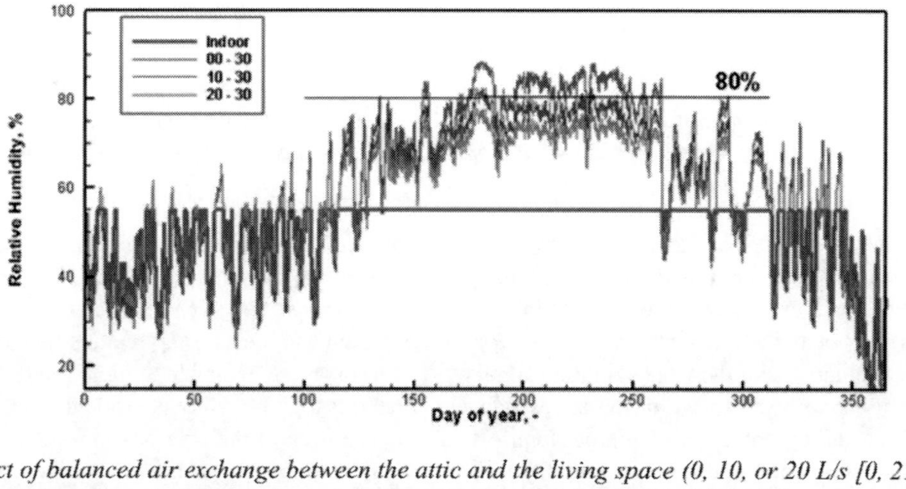

Figure 13 *The effect of balanced air exchange between the attic and the living space (0, 10, or 20 L/s [0, 21.2, or 42.4 cfm]) on the attic humidity when the outdoor air leakage is maintained at 30 L/s (63.6 cfm).*

carrying out either a guarded-blower door test or blower door tests with zone pressure measurements to find out the air leakage of the roof deck to the outside.

ACKNOWLEDGMENTS

Many thanks are due to Roderick Jackson, Philip Boudreaux, Tony Gehl and Diana, Hun (Oak Ridge National Laboratory) for their help and by allowing us to perform blower door testing of the test home and to analyze some of the measured attic temperature and humidity data.

REFERENCES

ASHRAE. 2009. ANSI/ASHRAE Standard 160-2009, *Criteria for Moisture-Control Design Analysis in Buildings*. Atlanta: ASHRAE.

Building America. 2003. *Thermal Performance of Unvented Attics in Hot-Dry Climates: Results from Building America*.

Energy Conservatory. 2007 Tectite Blower Door Software v3.2.

Glass, S.V. and S.L. Zelinka. 2010. Wood handbook—Wood as an engineering material. General Technical Report. FPL-GTR-190, Chapter 4, Madison, WI, U.S. Department of Agriculture, Forest Products Laboratory.

International Code Council. 2012. International residential code.

Karagiozis, A. and H.M. Künzel. 2001. WUFI-ORNL/IBP—A North American Hygrothermal Mode. *Proceedings for Performance of Exterior Envelopes of Whole Buildings VIII: Integration of Building Envelopes*, December 2–7, Clearwater Beach, FL.

Lstiburek, J. and C. Schumacher. 2011. Hygrothermal analysis of California attics. Research Report – 1110, Building Science Corporation.

Viitanen H. 1996. Factors affecting the development of mould and brown rot decay in wooden material and wooden structures. Effect of humidity, temperature and exposure time. Doctoral thesis, Department of Forest Products, The Swedish University of Agricultural Sciences, Uppsala.

Construction and Field Monitoring of Exterior Walls Using Vacuum Insulation Panels (VIPs) in a Net Zero Home

Anil Parekh, PEng
Member ASHRAE

Chris Mattock

Christopher J. McLellan

ABSTRACT

Vacuum insulation panels (VIPs) are an emerging technological breakthrough for Canadian residential construction. The high-insulating values of VIPs can contribute to reducing the thickness of highly-insulated building envelopes. The first known Canadian field demonstration of VIPs was located in Burnaby, British Columbia. The project team for the net-zero energy Harmony House Project has used exterior walls with VIPs. Using a number of mock-ups and field tests, the Harmony House team successfully installed 15 mm (9/16 in.) thick VIPs in the center of the stud cavity, covered by a 50 mm foil-faced isocyanurate foam board on the exterior and spray-foam on the interior. This provides an estimated effective insulation value, when averaged over the entire wall, of 6.8 $m^2 \cdot K/W$ (RSI) or 38.5 $ft^2 \cdot F/Btu$.

Design-stage thermal and hygrothermal analyses and the construction and monitoring of VIP wall assemblies showed that these systems do exhibit superior overall performance. Field construction of a home with VIPs demonstrated that builders in cold-climate countries can build thin-profile, low-heat loss wall systems.

Based on the construction and monitoring of VIP wall assemblies, it appears that VIPs are a viable option in the marine climate of Northwestern North America. Field performance of VIP walls show that the wall assembles must utilize a vented rain-screen cavity to protect from wind-driven rain penetration and to promote drying to the exterior; most importantly, north facing and other walls that do not receive exposure to solar radiation must use a preservative-treated exterior structural sheathing to minimize the possibility of fungal growth on the sheathing.

INTRODUCTION

In cold-climate housing markets, in order to achieve net zero energy (NZE) performance, the thermal characteristics of the building envelope have to be radically improved. For low-rise housing, a typical house built in Canada per the current energy efficiency regulations has an annual heat loss ranging from 80 to 130 GJ (~760 to 1230 therm) per year depending on the climate (Parekh & Kirney 2012). As shown in Figure 1, these heat losses must be further reduced by at least 50% or better to 40 to 65 GJ (~380 to 616 therm) per house to achieve NZE performance levels. With significantly reduced heat losses, the size of renewable energy systems can be minimized reducing overall costs.

By definition, NZE houses on an annual basis produce as much energy as they consume—all energy used in a home including that for space heating, hot water, ventilation, air conditioning, base electricity, and plug loads. This is typically achieved by significantly reducing heating, cooling, lighting, and appliance loads and supplying the energy needs of the building through on-site passive and active renewable energy systems. In order to reduce space heating and cooling loads to a level where on-site renewable energy systems can supply all the energy requirements, the building envelope is typically highly insulated and airtight. For low-rise residential buildings (three stories or less) in Canadian climates this requires insulation values in the range of RSI 4.2 (R-24)[1] for the basement or crawlspace floor slabs, RSI 5.3 to RSI 7.0 (R-30 to R-40) for

***Anil Parekh** is a senior research manager at CanmetENERGY and **Christopher McLellan** is a technical advisor at Office of Energy Efficiency, Natural Resources Canada, Ottawa, Ontario and **Chris Mattock** is president of Habitat Design + Consulting Ltd., Vancouver, British Columbia, Canada.*

below-grade and above-grade walls, and RSI 10.6 to RSI 14.1 (R-60 to R-80) for ceilings. Achieving these insulation levels in walls and ceilings using conventional insulations (such as mineral and glass fiber batts, rigid polystyrene boards) requires very thick assemblies in the range of 406 to 610 mm (16 to 24 in.). These thick assemblies can lead to:

- Reduced total usable interior floor area,
- Reduced-ground floor area for a given footprint area,
- Additional flashings and detailing around doors and windows if they are set to the inside face of exterior walls,
- Additional interior wall finishes and mouldings if the windows and doors are set to the outside of the exterior wall,
- Additional framing to accommodate thicker insulations, therefore high costs of construction, and
- Higher costs per unit liveable floor area.

One or more of the above conditions will impede the adoption and construction of NZE housing. Recently, there has been great emphasis on the development of high R-value insulating materials that can provide significant thermal resistance per unit thickness. High-density fiberglass insulation batts can raise the insulation level from RSI 0.024 to RSI 0.030 per mm thickness (R-3.5 to R-4.4 per in.), which is close to a 26% improvement. The re-engineered polymeric foam insulation boards can provide close to RSI 0.045 per mm (R-6.5 per in.), a 30% better insulation value than conventional rigid insulation sheathing. These developments are certainly encouraging; however, these are still incremental. In the last several years, vacuum insulation panels (VIPs) have been developed for use by the construction industry and hold the promise of significantly higher levels of thermal performance.

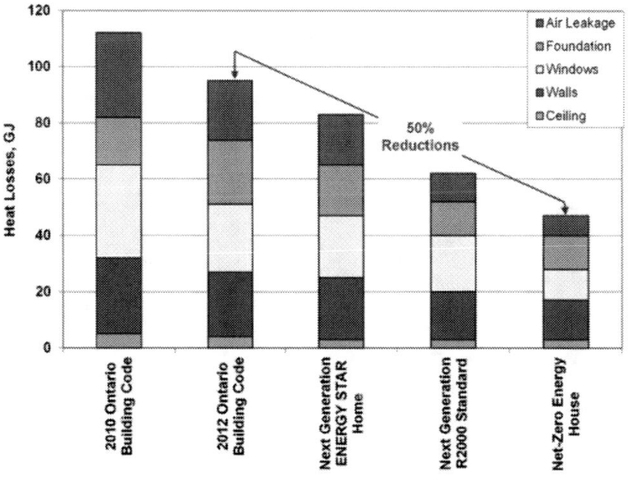

Figure 1 *Annual heat losses of a typical 186 m² (2000 ft²) 2 story home built in Ottawa, ON.*

[1] RSI–thermal resistance expressed in $m^2 \cdot K/W$ and R in $ft^2 \cdot F/Btu$

This paper documents the construction details of exterior above-grade walls with VIPs and provides assessments of hygrothermal performance of these walls over a period of one year. The annual energy performance of the net zero home using these VIPs is being evaluated over a period of two years and will be presented in a future paper.

VIPS

VIPs have been successfully integrated in wall assemblies and demonstrated in a number of European houses and buildings (Binz et al. 2005; Simmler et al. 2005; Erb et al. 2005) and Japanese houses (Super-E Japan 2008). In the Canadian context, VIPs are among the newest emerging technologies for residential construction. Used successfully for cryogenics and in the space industry, VIPs can achieve significantly higher insulation values with thinner wall construction than when using conventional insulation materials.

VIPs are either made with an open porous powder or a rigid porous core made from silica gel, compressed silica, or fiberglass board. The core is encased in a gas-impermeable skin of aluminum foil or aluminum/plastic composite laminate and placed under vacuum (less than 20 to 100 Pa [0.2 to 1 mbar]) to reduce convection and conduction heat transfer. The encased material is impregnated with desiccants and getters for absorbing any gases that may enter the VIP over time. As a result, the thermal insulating capacity of VIPs is up to ten times higher than current commercially available insulating materials. The VIPs are particularly useful in places where space is at a premium or where energy demand is high. The higher insulating values of VIPs can effectively reduce the thickness of building envelopes. The thermal resistance of VIP ranges from RSI 0.21 to RSI 0.31 per mm (R-30 to R-45 per in.) thickness (Mukhopadhyaya et al. 2008).

With technological improvements in the last several years, including improvements in core materials and manufacturing processes, the quality and durability of VIPs have improved significantly. Several manufacturers of VIPs are now claiming 40 to 60 year life spans (NRCC 2013). The recently published, ASTM Standard C1484-10 deals not only with product specifications and insulation performance tests but also puncture tests—which somewhat alleviate concerns regarding the handling of fragile VIPs, though the panels still should be handled carefully and protected during construction. When using VIPs in a building envelope assembly, it should be noted that:

- The wall area covered with VIPs form an absolute vapor barrier; however, the edges of VIPS should be properly lapped and sealed to prevent moisture transfer;
- All framing materials in the wall assembly must be dry before closing in; and
- The consequences of vacuum failure in VIPs could potentially lead to moisture accumulation in building envelope assemblies as well as radically reduced thermal performance.

VIPs should not be used as stand-alone insulation materials due to practical and construction-related considerations, and they work best when integrated and protected with polystyrene plastic insulation boards or other rigid protective layers.

The first known Canadian field demonstration of VIPs is in Burnaby, British Columbia near Vancouver, shown in Figure 2. This is a coastal marine climate with heating degree-days of 2925°C (5265°F). The Harmony House project team, successful winner of CMHC's EQuilibrium™ Sustainable Housing Demonstration Initiative, has built exterior walls using VIPs (CMHC 2010). The Harmony House is a two-story, 438 m^2 (4714 ft^2) home with a basement that includes a self-contained secondary suite and an attached garage.

HARMONY HOUSE WALL ASSEMBLIES WITH VIPS

The design goals for the Harmony House above-grade wall assemblies were as follows:

- To construct a wall no thicker than currently used by the housing industry in Canada. The most common exterior wall system used in Canada and the northern US is based on 38 × 140 mm (2 × 6 in.) framing. Builders and subtrades are used to working with this dimension of lumber and windows and doors are manufactured for walls of this thickness;
- To achieve an effective thermal insulation value for the entire assembly in the range of RSI 7.04 (R-40);
- To construct a durable wall assembly that will not be affected by moisture accumulation from rain penetration or condensation formation due to exterior or interior airborne moisture;
- To construct a wall assembly that will protect the VIP during and after construction; and
- To produce an airtight wall assembly resulting in a measured airtightness of 0.35 cm^2/m^2 (0.005 $in.^2/ft^2$) at 10 Pa normalized leakage area or lower.

These criteria were addressed by the wall design shown in Figure 3; the wall was constructed in the following manner starting from the exterior cladding and moving inwards:

- The exterior cladding is supported on vertical preservative-treated 12 × 50 mm (1/2 × 2 in.) plywood strapping that provides a vented rain-screen cavity allowing for both drainage and drying by natural convection.
- A layer of commercial grade spun-bonded polyolefin building wrap (water resistive barrier) stapled to the exterior face of the 12.7 mm (1/2 in.) thick plywood wall sheathing. This prevents wetting of the wall sheathing while also allowing for drying. The design-stage hygrothermal analysis predicted the potential for moisture accumulation in the north wall sheathing; therefore, all north wall sheathing was treated with a preservative.
- The 38 × 140 mm (2 × 6 in.) studs are spaced at 610 mm (24 in.) on center and advanced framing details were used to minimize lumber in the wall assembly, such as two stud corners, single top plates, and let-in lintels and rim joists. Advanced framing reduces the thermal bridging through framing enhancing the thermal characteristics of the wall. Conventional framing practices typically result in 23% of the wall surface area covered in framing, whereas in this case only 9% of the wall area was covered in framing.
- A 50 mm (2 in.) layer of foil-faced isocyanurate foam board was placed between the studs behind the exterior plywood wall sheathing; this provides an RSI 2.1 (R-12), which in Vancouver's climate ensures in the case of VIP failure that the interior face of the VIP will be kept above the indoor air dew point temperature.
- A 15 mm (9/16 in.) thick VIP with an insulation value of RSI 6.59 (R-37.4)[2] is then placed with double sided tape on the inside face of the foam board and sealed to the sides of the studs and to the wall top and bottom plates with urethane foam; this secures the VIP and helps contribute to the wall assembly airtightness. Placing the VIP near the center of the wall cavity depth provides the most protection possible from nails and screws during and after construction.

2. Based on test results according to ASTM C 518 (2010) produced by National Research Council of Canada on an 15.0 mm (9/16 in.) thick VIP.

Figure 2 *Harmony House Net Zero Energy House front; south elevation showing triple-glazed windows, photovoltaic array and solar domestic water heating system; and interior. (source: www.harmony-house.ca).*

- The inner layer of insulation consists of a castor bean oil-based low-density spray foam insulation which both insulates and air seals.
- The interior finish consists of 12.7 mm (1/2 in.) thick gypsum wall board which is air sealed to the wall framing using low-density closed-cell foam tape, screws and sealants using a technique called airtight drywall.

The design stage thermal and hygrothermal evaluations verified construction specifications, corner details, and also the long-term moisture and energy efficiency performance of wall assembly. The high-insulating values of this wall assembly, along with the placement of the vapor barrier formed by the VIP, led to concerns that the intent of the British Columbia Building Code would not always be met with in regard to ensuring no moisture accumulation in the assembly (BCBC 2006). A hygrothermal analysis was carried out using commercially available hygrothermal software to ensure code compliance when the VIPs were both new and if their performance degraded due to damage or aging; the findings are summarized as follows (Parekh and Mattock 2012):

- Heat transfer analysis was conducted using the two-dimensional heat transfer modeling program based on the finite-element method with THERM 5.2 (LBNL 2007). Temperature and heat-flux isotherms showed that the majority of heat flow occurs through the wood framing members. As the VIP has the highest thermal resistance, the largest thermal gradient occurs across the panel. The interior surface of the VIP could be expected to have a minimum surface temperature of 15.6°C (60°F) when the exterior temperature is –7°C (19.4°F), and well above normal indoor dew point temperatures of approximately 10°C (50°F) during the wintertime.
- Hygrothermal analysis of the VIP wall assembly was performed using WUFI 5.1 (ORNL 2010). Analyses showed that north facing walls and other walls that do not receive exposure to solar radiation are more prone to higher humidity for framing materials (sheathing and bottom headers). The VIP wall relies on cladding ventilation for drying of the exterior sheathing.

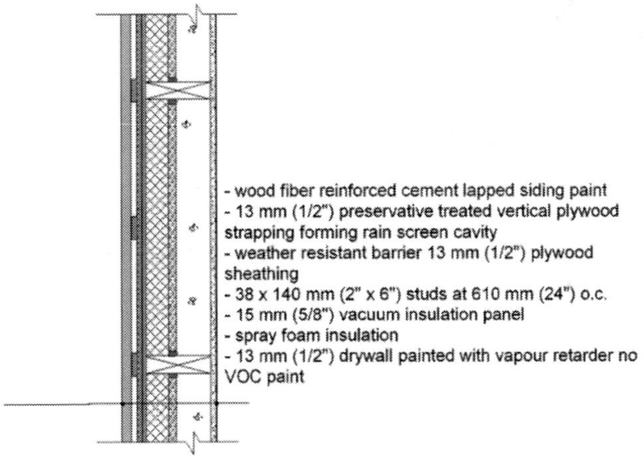

Figure 3 *Plan section through the Harmony House above-grade exterior wall assembly.*

CONSTRUCTION SEQUENCING

All exterior walls are framed using advanced framing techniques to minimize lumber and related thermal bridging. The 50 mm (2 in.) thick foil-faced isocyanurate foam board is installed between the studs behind the wall sheathing upon completion of framing and installation of the building wrap and a waterproof membrane on the roof. In order to minimize the possibility of damage, all electrical, plumbing, and heating and ventilation services were installed before the VIPs. All wiring in exterior walls is for electrical receptacles only; this allows the wiring to be kept low and makes for easy installation of the VIPs. All light switches, daylight sensors, and occupancy sensors in the home are wireless and therefore can be surface mounted later on the finished drywall, and therefore do not interfere with the installation of the VIPs. Figures 4 and 5 show photos taken during the construction.

Airtightness

The home was blower-door tested upon completion of the air barrier and was measured at 0.75 ach at 50 Pa and found to have a normalized leakage area of 0.35 cm^2/m^2 at 10 Pa exceeding the performance target of 1.0 ach.

EXTERIOR WALL PERFORMANCE MONITORING

Monitoring of the thermal and moisture conditions at various locations in the wall assembly was carried out over a 12 month period. The location of the sensors that measured temperature, wood moisture content, and relative humidity are shown in Figure 6. The sensors are wireless and communicate with an internet gateway located in the home. The sensors shown in Figure 7 have been calibrated prior to installation. Readings for all three parameters were taken every 5 min. Information measured and transmitted by the sensors passes through the gateway to a web site maintained by the sensor manufacturer. This allows the monitoring data to be accessed at any time and downloaded for analysis. If a longer monitoring period is desired, the battery powered wireless sensors will continue to operate for at least 10 more years and will allow for extended monitoring.

Wireless Sensors

The sensors are available in two variants both with identical electronics, but one with a larger profile and the other with a thinner profile.

Sensor Installation

During the wall construction process, sensors were installed in the main floor wall assemblies in the four cardinal orientations as shown in the Figures 8 and 9.

MONITORING RESULTS

Construction of the Harmony House was completed in late fall of 2011. The Burnaby, British Columbia climate is generally rainy during the fall; however, the construction crew took enough care to keep the wood-framing shielded from direct rain and water. The initial moisture content (MC) of sheathing and frame component ranged from as low as 7% to as high as 26%. Sensors were installed in the walls during the construction and the gathering of data started in January 2012.

The results of the monitoring of the wall assembly essentially confirmed the trends predicted by the hygrothermal modeling. The wall component that has proven the most critical and the best indicator of long-term durability in the Vancouver climate is the exterior structural sheathing.

The north wall sheathing experienced elevated MC levels ranging between 15.3% to 22%, as shown in Figure 10. This appears to be due to a combination of the fact that the wall is shaded over the entire year, no drying can occur to the interior, and the high thermal resistance of the wall assembly.

The south-facing wall-sheathing moisture contents ranged from 7.3% to 12% as shown in Figure 11. Solar exposure ensures drying of the sheathing.

East wall sheathing MC ranged from 9.9% to 20.1% as shown in Figure 12. Figure 13 shows the west wall sheathing MC ranged from 8.3% to 15.9%. In the cases of the east and west walls, there appears to be a slight drying trend in the wall sheathing year to year.

In the case of the north wall, while some drying occurred in the summer, the MC level then returned to the values seen

Figure 4 *Left photo showing north kitchen wall 38 × 140 (2 × 6) advanced framing and second floor rim joist. The let-in laminated veneer lumber (LVL) rim joist allows for the vacuum panels to run up past the edge of the upper floor. Right photo shows north side of north kitchen wall showing preservative treated plywood sheathing.*

Figure 5 *Left: Foil faced isocyanurate foam board placed between the 38 × 140 (2 × 6) studs behind the structural sheathing on north wall. Middle: wall after installation of the vacuum insulation panels. Right: interior spray foam application.*

the previous winter as the year progressed. The south wall sheathing started off being very dry, probably due to solar exposure during construction and then accumulates moisture over the winter, drying out again the following summer. All structural sheathings are below 26% MC throughout the year, which is the point at which fungal germination occurs.

Temperature, RH, and MC of Framing Wall Components

The MC, RH, and temperature at various locations in the four walls were recorded over the year. All framing in the east, west, and south walls was below, and in most cases well below, 20% MC for the entire year. Only the north wall bottom-plate

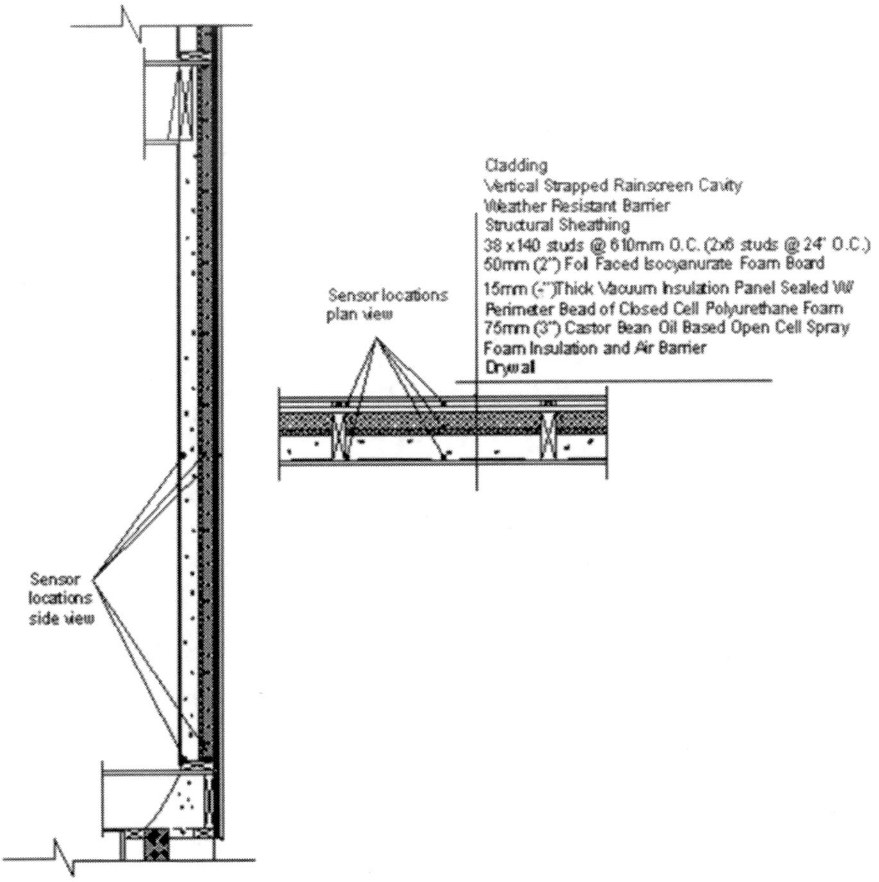

Figure 6 *Vertical and horizontal sectional views of the exterior wall assembly showing sensor locations.*

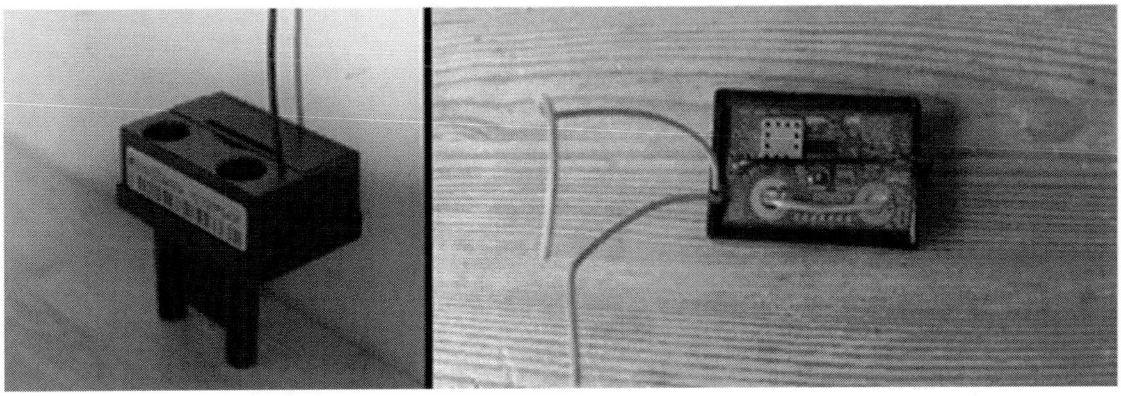

Figure 7 *On left wireless sensor: stainless steel screws pass through the sensor body and the posts on the bottom of the sensor to secure it to a wood substrate and to act as probes for moisture content readings; and on right thin profile sensor, the two wire leads are stripped at the ends and secured with stainless steel screws to measure moisture content.*

outer point exhibited MC levels consistently close to 20% MC, as shown in Figures 14 and 15. In all cases, the MC was below 26%, which is the beginning point for fungal germination.

Discussion

VIPs offer a breakthrough enhancement to the thermal performance of wall assemblies without increasing the thickness of those assemblies. The following lessons were learned from this project.

Thermal and hygrothermal analysis must be an integral part of the design of new high-performance wood-frame wall assemblies, particularly with high levels of insulation such as VIPs. This is to allow for optimization of the thermal aspects of the wall design and to ensure potential areas of moisture accumulation are identified and dealt with at the design stage.

The construction of wall mock-ups prior to construction is an extremely useful tool for refining the wall design and finalizing construction details and sequencing. Wall mock-ups usually include framing, insulations, a corner, a window, and plumbing and wiring. These mock-ups were built mainly to test and verify constructability.

Our construction mock-up also showed that VIPs should not be placed in direct contact with oriented strand board (OSB); this is due to the fact that OSB has small protruding shards of wood that can puncture some VIP skins.

Before specifying sealants, adhesives, and insulations to use in conjunction with VIPs, check with the manufactures to ensure long-term product compatibility or carry out field tests.

Due to the fact that the VIPs cannot be cut, and in the case of this project the most common VIP used in stud bays is 550 × 1500 mm (21.6 × 59 in.), a strict framing module of

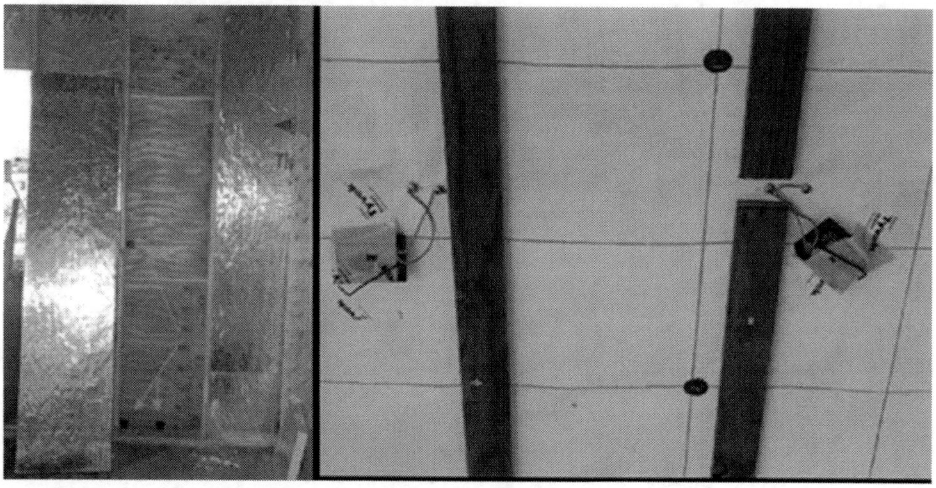

Figure 8 *Left: west wall showing foil faced isocyanurate foam board temporarily removed and three wireless sensors installed behind sheathing on the outside of the stud and bottom plate. Right: thin profile sensors installed on the exterior of the wall sheathing for measuring temperature and RH in the rain-screen cavity and moisture content.*

Figure 9 *Left: South wall showing installation of sensors on the inside of the studs and plates. Right: thin profile sensor installed on the inside face of the isocyanurate foam board at the center of the stud bay, to measure temperature and RH immediately outside the VIP.*

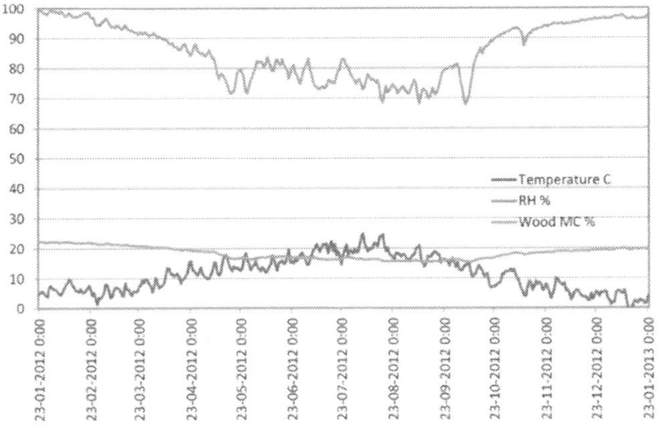

Figure 10 *North wall exterior sheathing MC, RH, and temperature for the rain-screen cavity for full year.*

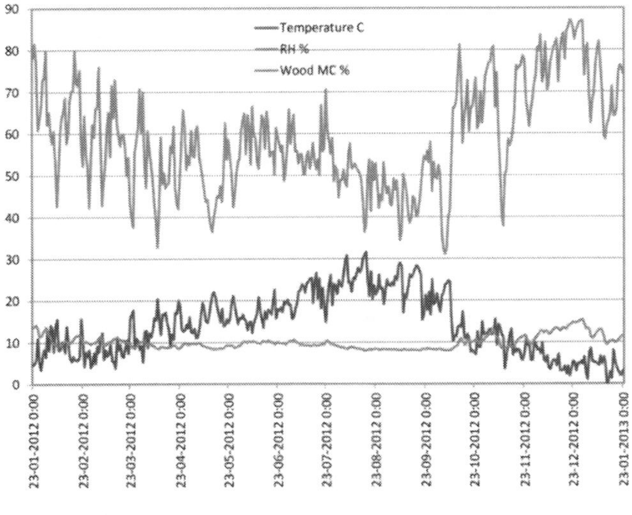

Figure 11 *South wall exterior sheathing MC, RH, and temperature for the rain-screen cavity.*

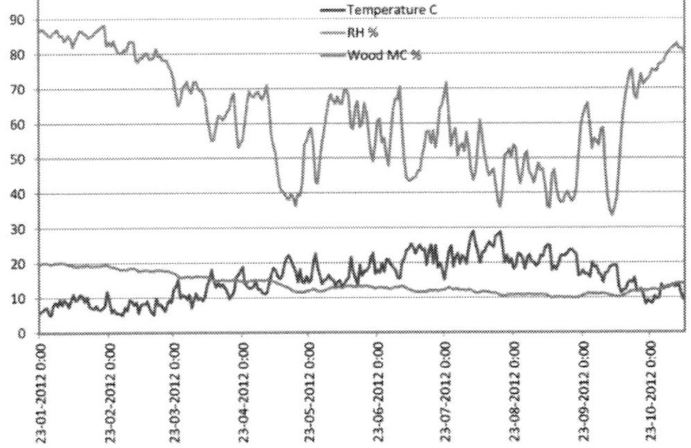

Figure 12 *East wall exterior sheathing MC, RH, and temperature for the rain-screen cavity.*

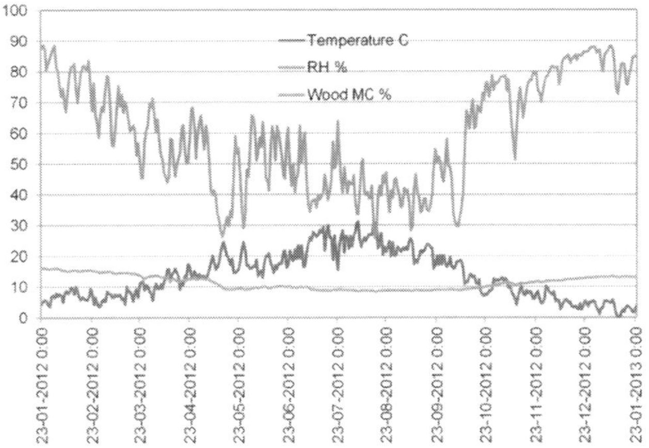

Figure 13 *West wall exterior sheathing MC, RH and temperature for the rain-screen cavity.*

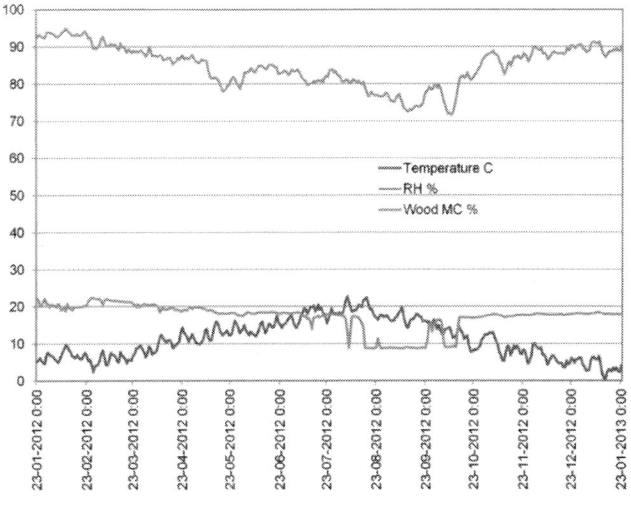

Figure 14 *North wall bottom plate outside face adjacent to stud.*

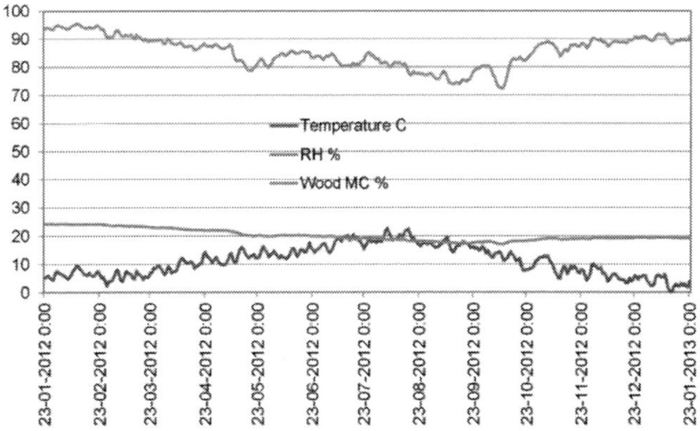

Figure 15 *North wall bottom plate outside face centre of stud bay.*

610 mm (24 in.) on center had to be maintained throughout. The designer must keep this in mind when developing the overall dimensions of the building and determining size and locations of windows, doors, and structural elements such as columns and beams.

Window sill and head heights were dictated by VIP dimensions. Again this will influence window dimensions and their vertical locations in wall assemblies.

The VIPs were overlapped vertically when the stud cavity was not tall enough to accommodate two panels for their full height.

The smaller VIPs used beneath windows were overlapped horizontally where the stud bay was not wide enough to accommodate two full-width panels.

In the case of sloped ceilings, triangular sections at the top of the stud cavities were only filled with the isocyanurate foam and a greater depth of spray foam.

At the design stage where possible, plumbing, wiring, and HVAC services should be confined to interior walls. Where services occur on exterior walls, care should be taken with location and installation of wiring and plumbing services to allow later installation of the VIPs.

The use of wireless switches and sensors greatly reduced the need for wiring in exterior walls and the potential for VIP damage.

CONCLUSIONS

VIPs can be integrated into 38 × 140 mm (2 × 6 in.) wall assemblies, with the thermal and hygrothermal performance being enhanced by the adoption of the advanced framing techniques, reducing the lumber content of the wall thereby reducing thermal bridging. For the application of VIPs in housing assemblies, it should be noted that:

- VIPs installed with lapped and sealed edges form an excellent vapor barrier assembly;
- All framing materials must be dry before the wall assembly is closed in; and
- The consequences of vacuum failure in the VIPs can be high in terms of thermal performance and potential moisture accumulation and must be accounted for in the wall assembly design.

Based on the design, modeling, construction, and monitoring of this wall assembly, it appears that a wall assembly of this type utilizing high-performance VIPs is viable option in the marine climate of northwestern North America for NZE housing with the following provisos:

- The wall assembly must utilize a vented rainscreen cavity to protect from wind-driven rain penetration and to promote drying to the exterior.
- The wall assembly must be very airtight to prevent leakage of interior air into the wall cavity that could cause interstitial moisture accumulation.
- North walls and other walls that do not receive exposure to solar radiation must use a preservative-treated exterior structural sheathing to minimize the possibility of fungal growth on the sheathing.

ACKNOWLEDGMENTS

The Harmony House Project is part of Canada Mortgage and Housing Corporation's (CMHC's) highly successful EQuilibrium™ Sustainable Housing Demonstration Initiative. Technical and financial contribution was provided by CMHC and Natural Resources Canada's Program on Energy Research and Development (PERD) as part of Built Environment Portfolio (BEP) subprogram on Building Envelopes.

REFERENCES

ASTM. 2010. ASTM C518-10, *Standard Test Method for Steady-State Thermal Transmission Properties by Means of the Heat Flow Meter Apparatus*. West Conshohocken, PA: ASTM International.

BCBC. 2006. *British Columbia Building Code 2006*, Table 9.25.1.2. Victoria, BC: Office of Housing and Construction Standards, BC Ministry of Energy and Mines.

Binz, A., A. Moosmann, et al. 2005. Vacuum insulation in the building sector: Systems and applications. http://www.ecbcs.org/docs/Annex_39_Report_Subtask-B.pdf.

CMHC. 2010. Harmony house. www.cmhc-schl.gc.ca/en/inpr/su/eqho/haho/index.cfm.

Erb, M. 2005. Vacuum insulation: Panel properties and building applications—Summary. http://www.ecbcs.org/docs/Annex_39_Report_Summary_Subtask-A-B.pdf.

LBNL. 2007. THERM—Two-dimensional building heat-transfer modeling. University of California, Berkeley: Lawrence Berkeley National Laboratory. http://windows.lbl.gov/software/therm/therm.html.

ORNL. 2010. WUFI—Moisture design tool. Oak Ridge, Tennessee: Oak Ridge National Laboratory. http://www.ornl.gov/sci/ees/etsd/btric/wufi_software.shtml.

Mukhopadhyaya P., K. Kumaran, N. Normandi, D. Reenan, J Lackey. 2008. High performance vacuum insulation panel: development of alternative core materials. *Journal of Cold Regions Engineering* 22:4.

NRCC. 2013. Integration of VIPS into Canadian wood frame walls—Performance assessment in the laboratory. Client Report, National Research Council Canada, Ottawa, ON.

Parekh, A. and C. Kirney. 2012. Thermal and mechanical systems descriptors for simplified energy use evaluation of Canadian houses, *Proceedings of SimBuild, Madison, WI.*

Parekh, A. and C. Mattock. 2012. Incorporation of vacuum insulation panels in a wood frame net zero energy home. *Proceedings of 7th International Cold Climate HVAC Conference, ASHRAE.*

Simmler H., H. Brunner, S. et al. 2005. Vacuum insulation panels—Study on vip components and panels for service life prediction of vip in building applications. IEA ECBSC Annex 39 Report.

Super-E Japan. 2008. Sapporo net-zero energy home Hokkaido, Japan http://www.super-e.co.uk/PDF/SE_SapporoCaseStudy_HokkaidoJapan.pdf.

Post-Occupancy Evaluation Framework for Multi-Unit Residential Buildings

Anik Teasdale-St-Hilaire, PhD, PEng

ABSTRACT

There is a growing awareness in the residential building industry about the significance of energy use in buildings and the desire to minimize energy consumption and costs. In the construction industry, the importance of building commissioning to verify the installation and performance of various building systems is rising.

Post-occupancy evaluation (POE) of buildings goes beyond commissioning. POE is the systematic assessment of building performance during service and typically includes analysis of the users' perceptions.

A literature survey revealed the lack of a comprehensive POE methodology for multi-unit residential buildings (MURBs) that addresses the major building systems, including the building envelope. To fill this gap, a POE framework has been developed for MURBs. As very limited information is available in the public domain on the performance of MURBs, the information obtained from POEs could be of use to identify avenues of improvements to the benefit of building owners and occupants and to develop benchmark data on performance indicators to support the development of building codes, regulations, and guidelines. Evaluating the performance of new, innovative building systems and practices applied to newly constructed or retrofitted MURBs can serve to show whether promised benefits are actually delivered. The POE protocol focuses on seven performance areas: energy efficiency, water use efficiency, indoor air quality, lighting and the visual environment, acoustics, thermal comfort, and building envelope performance. This paper presents the literature survey findings, the protocol's objectives, the qualitative and quantitative indicators used to gauge performance, and an overview of the methodology's various tasks, with emphasis on the building envelope performance assessment, and its interaction with other systems.

INTRODUCTION

There is a general tendency in the building industry to design and construct buildings to meet higher performance targets, due in part to more and more stringent building code requirements, and in part to better meet the needs and expectation of occupants. This is true for the energy and water consumption of buildings, as well as other performance areas such as lighting, thermal comfort, and the building envelope.

Very little information is available in the public domain about the actual, in-service performance of buildings in general, and multi-unit residential buildings (henceforth called *MURBs*) in particular. Such information could be useful to identify avenues of improvements to the benefit of building owners and occupants and to develop baseline data on performance indicators to support the development of building codes, regulations, and guidelines. Evaluating the performance of new, innovative building systems and practices applied to newly constructed or retrofitted MURBs can serve to show whether promised benefits are actually delivered, potentially speeding their adoption by the industry and driving regulatory changes.

Building commissioning is becoming a more and more common practice and has as its main objective to ensure that systems—primarily mechanical and electrical systems—are constructed and operated as designed. The bulk of commissioning activities are typically conducted prior to building

***Anik Teasdale-St-Hilaire** is a building science engineer at Morrison Hershfield Limited, Ottawa, Ontario.*

occupancy, with some commissioning projects also including follow-up reviews to address potential building performance issues prior to the expiration of warranty periods. As with post-occupancy evaluation (POE), commissioning has focused on non-residential buildings.

This paper starts with a definition of POE and presents key results of our literature survey on POE of MURBs. The development of a POE framework for MURBs—a work in progress—is outlined, including its objectives, performance areas, and specific performance indicators and metrics, where these exist. The paper discusses the building envelope performance assessment in greater details, including test methods. Finally, an overview of the framework's tasks is given, focusing on the performance assessment of the building envelope, and its interaction with different systems.

What is Post-Occupancy Evaluation?

Post-occupancy evaluation, or POE, is one of the best methods of obtaining performance data on existing buildings. Post-occupancy evaluation goes beyond building commissioning and is the systematic assessment of building performance during service, typically including but not limited to analysis of the users' perceptions, for various purposes. POE also commonly includes building monitoring to support the findings of the user surveys and assess the level of success of design strategies in meeting design performance targets or reference baselines. Post-occupancy evaluations have been undertaken for some 40 years (Leaman et al. 2010), although significant documentation on methodologies begins in the 1980s. POE fits in with building performance evaluation or verification, which includes a number of activities taking place during a building's life cycle to determine "whether facilities will work for the people that will use, occupy, or otherwise be impacted by them" (Mallory-Hill et al. 2012).

LITERATURE REVIEW

The POE tools available in the public domain which were reviewed during our literature survey are listed in Table 1, with their target building type. ASHRAE's "Performance Measurement Protocols for Commercial Buildings" (ASHRAE 2010c) was published just prior to the start of this work and was not included in the survey. The literature review was not intended to be exhaustive, but rather a sampling of available methodologies, focusing on their performance indicators, performance baseline or benchmark data, assessment techniques, monitoring instrumentation and methodologies, satisfaction surveys, and reporting techniques. Key findings relevant to the development of the protocol are highlighted here.

Two of the methodologies rely exclusively on the feedback obtained through surveys of occupants or various stakeholders (Baird 2010; Watson 1996, 2003; Watson and Fitzgerald 1998). The remainder of tools employs a combination of survey of stakeholders and qualitative and/or quantitative performance indicators to assess building performance.

Many of the surveys deal with assessment of non-technical areas, such as ease and effectiveness of building monitoring; occupant satisfaction with items such as washrooms and communal spaces (EcoSmart Foundation 2007); education on building features; office layout (Keen Engineering 2005); client satisfaction; and construction defects, construction cost, productivity, and profitability (HEFCE 2006).

The performance areas addressed in the tools also vary depending on the tools' objectives. For instance, the HEFCE (2006) tool is intended for higher education institutions to better understand both technical aspects (such as energy and water use during construction and during service) and non-technical aspects (like predictability of procurement costs and time and productivity) of the building life cycle from construction through building service. As another example, the EcoSmart tool seeks information on the implementation

Table 1. POE Protocol and Target Building Type

POE Protocol	Target Building Type
EcoSmart (EcoSmart Foundation 2007)	MURBs
Keen Engineering POE protocol (Keen Engineering 2005)	Office buildings
Watson (Watson 1996, 2003); Watson and Fitzgerald (1998)	Various
Higher Education Funding Council for England (HEFCE 2006)	Higher education buildings
Baird (Baird 2010)	Sustainable commercial and institutional buildings
Post-Occupancy Review of Buildings and their Engineering, or PROBE (Cohen et al. 1999; Leaman, Stevenson, and Boyce)	Offices, institutional industrial and government buildings
Soft Landings Framework (BSRIA 2009)	Non-specific
Building Research Establishment (BRE)'s Design Quality Method (Cook 2007)	Various, including schools, hospitals, and housing
Sanders (2010)	Housing
Birt and Newsham (2009)	Commercial and office buildings

of major technical design strategies to assess their success. As such, the tool focuses primarily on technical performance areas such as energy and water consumption. The tool also serves as a post-mortem on the design and construction process, with the goal of eliciting feedback to improve not only future buildings, but the design and construction process.

It is also of interest that a POE developed for a specific target building type cannot necessary be applied "as is" to another building type. For instance, the performance areas and associated indicators (such as interior design: office layout, visual privacy, ease of interaction, etc.) employed to assess the performance of office buildings are not necessarily applicable to MURBs. Moreover, many baselines and benchmarks applicable to specific building types cannot be used as a performance comparison for others, as shown in the variations in energy and water use benchmark data given for different commercial buildings (ASHRAE 2012).

Leaman et al. (2010) state that far more POE studies have been conducted on non-domestic buildings. Our findings with respect to POE tools were similar: of the ten tools reviewed, only two were developed specifically for residential buildings—EcoSmart (EcoSmart Foundation 2007), and Sanders' methodology (Sanders 2010). The EcoSmart protocol is an extensive tool comprised of 11 mandatory components addressing various aspects of residential building performance, including but not limited to energy consumption, water use, indoor air quality, thermal comfort, and lighting. Building envelope performance is not included. Also, while an occupant survey is included as one of the components, no survey has actually been developed for the protocol. The Sanders methodology is limited in that it has been developed strictly to evaluate the energy performance of dwellings. The literature also revealed that existing tools provide little information with respect to benchmark data for MURBs performance comparison.

OBJECTIVES OF THIS PROTOCOL

To fill the gaps in the existing POE methodologies for the performance assessment of MURBs, a new POE methodology was developed. There was a desire that the methodology be modular, and could be implemented for newly constructed buildings, newly retrofitted or rehabilitated buildings, or simply buildings where the performance is to be assessed. Also, the methodology would provide baseline or benchmark data, where such data are applicable and available.

The specific objectives of this methodology are the following:

1. To compare the overall performance of a building relative to baselines or benchmarks.
2. To determine the extent to which a building meets its occupants' and the building management team's needs and satisfaction.
3. To provide feedback on the building design, construction and commissioning process.
4. To evaluate the success of specific innovative measures and systems, and unique building features, compared to conventional technologies.

PERFORMANCE AREAS AND INDICATORS

The POE methodology for MURBs includes the performance assessment of seven areas referred to as performance areas:

1. Energy efficiency
2. Water use efficiency
3. Indoor air quality
4. Lighting and the visual environment
5. Acoustics
6. Thermal comfort, and
7. Building envelope performance

This section provides a general description of the performance areas, their relevance to building performance, and the qualitative and quantitative performance indicators or metrics used to assess performance.

Energy Efficiency

Energy efficiency is a key performance area in this methodology as it addresses the rising importance of energy efficiency, the reduction in energy consumption, associated costs, and environmental impact. Several performance indicators were established to evaluate the energy performance of MURBs:

- Annual energy consumption per square meter of conditioned space (gas, electricity, propane, diesel, fuel oil, etc.). Energy use is reported on an energy type basis and as aggregated total of different energy types.
- Energy consumption per square meter of conditioned space, normalized for weather conditions. Normalized energy use reported on an energy type basis and as aggregated total of different energy types.
- Disaggregated energy use by system or end use.

Normalization of energy data is conducted to account for varying weather conditions on energy consumption. Normalization can be conducted following the general procedures and guidance discussed in ASHRAE (2012, 2011). A good explanation of the process is given by Avina (2012). Note that this type of energy analysis should be conducted by an experienced professional or contractor.

While some large property owners and developers may have their own proprietary benchmark data, statistically valid data for MURBs is not readily available in the public domain at this time. The Canada Mortgage and Housing Corporation (CMHC) provides some information (HiSTAR database [CMHC 2001]), but information is limited and based on small sampling of buildings. Thus, finding comparable buildings may be difficult.

Water Use Efficiency

Research published by CMHC cites that one of the common inadequacies of multi-unit residential buildings is high domestic water consumption (Kernan et al. 2002). The importance of water use efficiency is reflected in the emergence of incentive programs such as Toronto Water's multi-unit toilet replacement program and commercial and residential washer rebate programs (Detta Colli). The efficient use of water impacts not only the withdrawal of this natural resource from rivers, aquifers and reservoirs, but also the infrastructure required to supply it and to treat it pre- and post-consumption.

The efficient use of water can be characterized using one of many quantitative performance indicators:

- Water consumption per occupied floor area
- Water consumption per unit
- Water consumption per occupant

Available benchmark data is typically given on a basis of volume of water per square meter and/or per unit. While some large property owners and developers may have their own proprietary data, statistically valid data for MURBs are not readily available in the public domain. CMHC does provide some information, but the information is limited and based on small samples of buildings (CMHC 2001).

Indoor Air Quality

Poor indoor air quality (IAQ) is a concern as it can adversely affect the health of occupants. Sources of contaminants include emissions from materials (particularly in new buildings), products used by occupants such as cleaners, and air leakage from garages and garbage chutes. Occupant activities such as bathing can cause high indoor humidity conditions if the moisture is permitted to accumulate. Mold growth stemming from water penetration or condensation on interior surfaces is another source of contaminants.

Quantitative indicators of indoor air quality performance include the following:

- Carbon dioxide (CO_2) elevation over exterior ambient conditions
  Carbon monoxide (CO)
- Total volatile organic compounds (TVOCs)
- Humidity ratio (or relative humidity [RH])
- Temperature
- Exhaust (ventilation) air flow
- Volatile Organic Compounds (VOCs)
- Particulates

Other than the applicable building code, there are various sources of baseline data for the performance indicators:

- ASHRAE Standard 62.1-2010 establishes maximum acceptable levels for a wide range of indoor air pollutants, and defines required ventilation rates (ASHRAE 2010a).
- Health Canada's "Exposure Guidelines for Residential Indoor Air Quality" (Health Canada 1989)
- Health Canada updates to 1987 report for CO, formaldehyde, ozone, toluene, and molds (Health Canada)

This POE methodology for MURBs recommends using the lowest published values for the maximum permissible TVOC concentrations.

Qualitative performance indicators include perceived odours and observations of water stains, visible water on surfaces, visible mold, and pests such as insects. Qualitative "measurements" rely on the judgment of a trained or knowledgeable person and generally there would be no numerical baseline or benchmark against which the "measurement" can be compared; rather, a subjective basis of comparison applies, such as "obvious tobacco, cooking or perfume smells on entering suite" (Morrison Hershfield 2012).

Lighting and the Visual Environment

Adequate levels of lighting, both natural and artificial, are important for the visual comfort and health of occupants. Lighting can also impact energy efficiency in that taking advantage of daylighting can reduce energy consumption, as can the use of energy efficient fixtures. However, the efficient use of resources is affected by the durability of fixtures.

There are several performance indicators for assessing the performance of the lighting system. Measurable indicators include:

- Illuminance
- Daylight factor (DF)
- Light trespass

Qualitative performance indicators include:

- Visibility
- Aesthetic
- Control flexibility
- Daylight
- Light pollution and light trespass

Sources of baseline information for lighting illuminance are found in the *Lighting Handbook* (Illuminating Engineering Society 2011) and the applicable building code. The Canada Green Building Council (CaGBC)'s *LEED Canada Reference Guide for Green Building* provides some guidance with respect to lighting pollution and trespass (CaGBC 2009). An average DF between 2% and 5% is generally considered the suitable range to balance good visibility and daylight appearance (NIBS 2012), although lower DF values may be appropriate for some spaces such as living rooms.

Acoustics

Noise from neighbouring units, common spaces, building mechanical and plumbing systems, building services, and the outdoors, can impact the comfort of occupants. Noises can be in the form of airborne sound transmission or vibration noises transmitted through the building structure.

Quantitative performance indicators to assess the acoustical performance of a building assembly (floor, party wall, etc.) and associated sources of benchmark data include the following:

- Noise from neighbouring units (and amenities) or other uses
  - Applicable building code requirements for sound insulation (Sound Transmission Class (STC) ratings)
  - Reducing Noise in Your Apartment, About Your Apartment Series (CMHC 2005).
- Noise from services (heating, ventilation, and air conditioning [HVAC], plumbing systems, and garbage chutes), including equipment that generates noise that radiates outdoors):
  - *ASHRAE Handbook—HVAC Applications*, Chapter 48 Noise and Vibration Control (ASHRAE 2011).
  - Legislation (municipal by-laws, provincial regulations).
- Noise from outdoors (e.g., rooftop equipment, traffic):
  - LU131 – Noise Assessment Criteria in Land Use Planning (Ontario Ministry of the Environment 1997).
  - Directive 038: Noise Control (Alberta Energy Resources Conservation Board 2007).
  - Applicable legislation and regional guidelines.
- Noise from services: elevators and garbage chutes
  - *ASHRAE Handbook—HVAC Applications*, Chapter 48 Noise and Vibration Control (ASHRAE 2011).
  - European noise codes for building services.

There are no codes in Canada for noise levels from building services and *ASHRAE Handbook—HVAC Applications* (ASHRAE 2011) does not address noise from elevators or from garbage chutes. A combination of European noise codes for building services and ASHRAE guideline levels (even though they are for HVAC noise and not elevators or garbage chutes) can give an idea of how loud the noise is relative to typical benchmarks for background sound.

Thermal Comfort

Poor thermal comfort due to building envelope air leakage, inadequate thermal performance of the envelope and thermal bridging, and an ineffective mechanical system is listed as one of several complaints cited by building occupants (Kernan et al. 2002). Poor thermal comfort can affect the health of building occupants, and can also impact the energy efficiency of the building, as building occupants may increase the thermostat setting or open the windows in the heating season to improve their comfort level.

Quantitative performance indicators for thermal comfort include the following:

- Air temperature
- Humidity ratio or RH
- Operative temperature

Operative temperature assessments include examination of air velocity, humidity levels, radiant loads, and asymmetries as described in ASHRAE Standard 55 (ASHRAE 2010b), the industry-wide most accepted standard for thermal comfort. These performance indicators characterize thermal comfort by quantifying temperature and humidity conditions that are outside the normal range of comfort. Sources of benchmark data include ASHRAE Standard 55, building codes, and/or project design specifications.

Qualitative indicators that rely on the judgment of a trained or knowledgeable person on the POE team, include:

- Drafts
- Hot/cold surfaces
- Uneven temperature within space (including stratification)
- Temperature swings

Building Envelope Performance

None of the existing POE tools surveyed in the literature covered the building envelope in a comprehensive manner. The performance of the building envelope is included within this methodology because it impacts several factors:

- Occupant health (poor thermal comfort due to air leakage and an inadequate thermal envelope; poor indoor air quality resulting from mold growth on building envelope surfaces, for example)
- Energy efficiency (e.g., heat loss due to low thermal resistance of envelope systems and localized thermal bridges; heat loss due to air leakage)
- Resource efficiency and environmental impact (filling of landfills and extraction of new resources associated with replacement of failed building envelope systems)
- Affordability and economic viability (costs associated with replacement of building envelopes caused by short material life cycles/low durability or premature failures)

The building envelope as a whole consists of many systems or assemblies, with each of these including several components. Together, the assemblies provide environmental separation. Given the complexity of the building envelope, several indicators can be used to assess the envelope's performance, some of which are generic, and some which depend on the specific materials, cladding materials in particular. The building envelope is expected to meet these performance criteria:

- Air leakage control (airtightness)
- Water tightness

- Control of heat transfer
- Vapour diffusion control
- Condensation resistance (of glazing systems)

Many of the above performance indicators can be assessed qualitatively by a trained and knowledgeable person. Typically, a visual review is conducted to identify the presence of building envelope issues such as water leakage and condensation on interior surfaces. Simple tools such as a psychrometer (to measure ambient temperature and humidity) and an infrared gun or digital thermometer with integrated thermocouple (to measure surface temperature where thermal bridging or condensation is suspected) may be used to supplement the visual review.

Smoke testing can be conducted as a diagnostic tool to help locate air barrier breaches using a simple smoke pencil or following a more sophisticated test method such as that described in ASTM E1186 (ASTM 2009). This ASTM standard also provides an infrared thermography methodology for detecting air leakage. Quantitative air leakage testing of windows and doors can be conducted using a test method such as ASTM E783 (ASTM 2010a). Envelope system and whole building air leakage testing are possible, but have several limitations and would be considered outside the scope of a standard POE methodology.

For both the quantitative and qualitative performance indicators, sources of baseline data include the project specifications, performance targets, and the applicable building code. In Canada, air leakage characteristics for air barrier materials must meet a minimum airflow rate of 0.02 L/s·m^2 at a 75 Pa pressure difference. Appendix A of the 2010 *National Building Code of Canada* (CCBFC 2010) provides recommendations for maximum air leakages rates of air barrier systems, varying with the interior ambient relative humidity, for opaque insulated building envelopes for most Canadian locations. However, air leakage testing of air barrier systems is not practical in completed buildings and as such is outside the scope of the POE. Air leakage of glazing systems such as windows and window walls are typically given in project specifications.

Water leakage testing of glazing systems can be conducted using ASTM E1105 (ASTM 2008) or a modification thereof, at the air pressure difference given in the project specifications, to confirm the performance or diagnose a water leakage problem for a higher level POE assessment. Performance criteria (i.e., test air pressure difference) would be given typically in project specifications.

Locating breaches in the waterproofing membrane for exposed membrane roofs can be conducted following the methodology outlined in ASTM C1153 (2010b); for exposed membrane and protected membrane roofs, flood testing is an option to confirm watertightness—refer to ASTM (2005).

While field determination of the U-factor for installed envelopes is not practical, infrared thermography is possible for qualitative assessment of thermal bridging and missing insulation using methods such as that given in ASTM C1060 (ASTM 2011).

Thermal performance and condensation resistance can be indirectly assessed using a glazing property detector (for glass thickness, cavity space width, presence and location of low-e and low-e product) for comparison with project requirements. Except for simple measurement of ambient temperature and humidity, the qualitative and quantitative tests are considered to be higher-level assessments, and as such they are to be conducted if warranted, to help determine a possible cause and resolution for a problem, or, alternatively, indicate the need for further, more specialized expertise and equipment. For premature and unexpected failure of a building envelope component (for example: brick cracking, metal corrosion, and sealant failure) a more in-depth investigation is suggested as this is would generally be beyond the scope of a POE.

OVERVIEW OF POE TASKS

The objectives of the POE can vary depending on the needs of the commissioning party, be it a MURB developer, apartment building owner, condominium board, or a researcher. Whether the MURB is a condominium or a rental apartment building, the owner(s) and/or their representative(s) need to be included at the start of the POE process, to make them aware of the purpose and process of a POE and the specific objectives. Their participation is also critical to maximize the participation of the individual suite occupants. Ideas should be sought from the owner(s) to encourage the participation of the occupants in the occupant satisfaction survey. The POE results' level of confidentiality, if any, should be discussed and agreed upon. The confidentiality of information obtained from individual suite occupants, such as occupant satisfaction survey results and energy bills, also needs to be addressed.

The POE assessment of a building requires technical expertise, depending somewhat on the level of assessment (see below). The POE should ideally include the design team so that the designers can provide important information on the goals that drove the design, specific performance targets, the strategies developed to meet these goals, and any challenges encountered.

A flowchart showing the overall outline of the methodology is provided in Figure 1. The methodology consists of four steps, each with one or more tasks. The four steps are the following:

1. Data collection from stakeholders
2. Surveys and interviews
3. Data collection from site
4. Information transfer

Each of the seven tasks included in the data collection from site applies to a different performance area (energy efficiency, indoor air quality, etc.). The POE methodology incorporates a modular format so that it can be customized to suit the objectives of the POE mandate, and to follow up on the findings of the occupant and/or building management surveys.

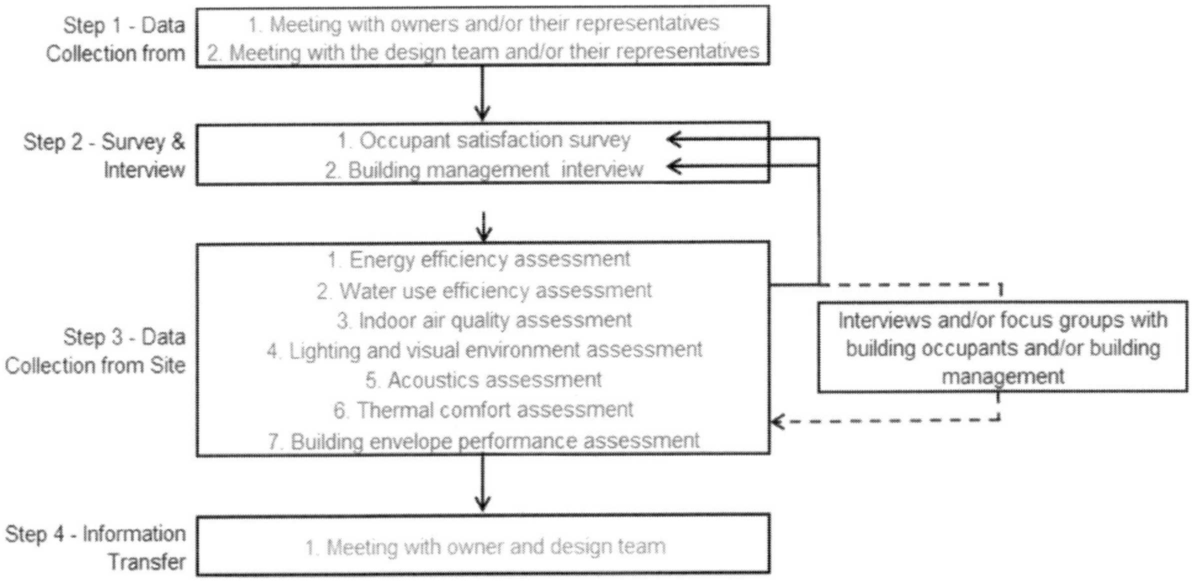

Figure 1 *Flowchart of POE methodology for MURBs.*

For each performance areas, there are two levels of assessments: Level 1 and Level 2. A performance area assessment begins with a Level 1 study. Generally, furthering the study with a Level 2 study is merited when the Level 1 assessment is inconclusive or cannot adequately explain a discrepancy with the design target, baseline, benchmark, or results of the occupant survey. A Level 2 study generally requires a higher level of expertise and more sophisticated and costly instrumentation.

The interviews with building management and building operators are intended to be conducted prior to the site data collection, so that the information gathered during these interviews can assist in selecting the number and location of suites to be monitored. However, in some cases, it may be useful to go back and revisit the building management and/or operators after the monitoring, in order to better understand the site data results.

Step 1—Data Collection from Stakeholders

Meeting with the Owners and/or Their Representatives. The purpose of this meeting is to establish a relationship between the owner and the POE team, to explain the purpose and process of the POE, confirm the performance areas to be assessed for the specific building, and to elicit necessary information to proceed with POE. Possible levels of release of the POE results are also to be discussed and formally agreed upon (confidential/publicly released without building name/publicly released). Findings from a POE may be released publicly to disseminate information to the construction industry or as a marketing strategy following a major retrofit in an apartment building, for example. General questions about the building performance and how it meets the owner's needs, as well as questions specific to the performance areas can be asked, such as:

- Detailed building occupancy (relevant to energy efficiency and water consumption performance areas)
- Building commissioning or any subsequent commissioning (relevant to energy efficiency performance area)

This is also an opportune time to obtain documentation from the owner, including building drawings and specifications, building commissioning reports, metered data for energy and water consumption, and/or any other relevant document pertinent to the POE. A customizable data collection form was formulated as part of the POE development to suit the needs of the POE team and the objectives of the POE.

Meeting with the Design Team and/or Their Representatives. Holding a meeting with the design team and/or their representatives is intended for a POE of a newly completed building or a building having undergone a major retrofit. Design team representatives present at the meeting should at least include the lead project architect and mechanical engineer.

During the meeting, the POE team will discuss the components that will be involved in the POE, and the possible levels of release of the POE results. The main objective of the meeting is to obtain information about the building's design goals and strategies, including but not limited to the following:

- Proposed use and occupancy
- Main design goals and strategies used to attain such goals (relating to performance areas to be evaluated)

- Any provisions made during design for the building to be ready for a post-occupancy evaluation (contractual provision, setting up files for utility bill analysis, instrumentation, etc.)
- Any innovative or high performance features
- Features of the building to be specifically evaluated in the occupant satisfaction survey

A data collection form was developed that can be customized and expanded to suit the needs of the POE team depending on the performance areas that will be evaluated for the MURB.

Step 2—Survey and Interview

Occupant Satisfaction Survey. While the need for an occupant satisfaction survey can be argued, in this POE tool for MURBs the occupant satisfaction survey is compulsory as it meets one of the objectives of the protocol—to bring information to the POE team about the building performance as perceived by the occupants. The occupant satisfaction survey is modular in form, enabling the user to extract questions to suit the building-specific POE objectives. A cover letter accompanying the survey describes the purpose of the survey, the party requesting the information, the person who should respond to the survey within each suite, and reassures the respondents that results will be kept confidential within the POE team. The format of the questionnaire (paper-based, computer-assisted, or multi-modal) and its advantages and limitations should be discussed with the owner(s) and/or their representatives, or the property manager during the start-up meeting. The POE protocol also makes recommendations with respect to strategies for maximizing the survey response rate, survey data processing, analysis and reporting, and provides ideas for further work, such as conducting one-on-on interviews and focus groups.

Building Management Interview. The interview with building management staff serves primarily to inform the staff about the POE, explain its purpose and process, and to obtain information about the history and operation of the building, including possible issues with the following, as is pertinent to the building-specific POE:

- Moisture
- Thermal comfort
- Lighting
- Mechanical and electrical system operation
- Acoustics
- Premature failures

The POE team should seek reports on the following:

- Previous repairs and replacements
- Outstanding deficiencies that may affect the building performance (e.g., water leakage through cracks in exposed concrete exterior walls)

In addition, information about the building is requested regarding operational goals, the metering plan (if relevant to the POE), and the most successful and least successful aspects of the building (e.g., an innovative domestic hot water heater results in quick delivery of hot water to suites; air leakage has been reported at the building's windows). A data collection form was developed to facilitate data entry for this task.

Step 3—Data Collection from Site

The protocol presents guidance on the expertise required to carry out each level of the POE for each performance area and level of assessment, as well as instrumentation, measurements and conditions, complementary data collection, interactions with other systems, the anticipated time and effort required, estimated minimum cost of typical equipment, and the need for further investigation.

For every POE performance area investigated (energy efficiency, water use efficiency, building envelope performance, etc.), a visual walk-through is to be conducted by the POE assessor, with the objective of confirming the construction of building systems in various parts of the building and reviewing their condition.

In-suite reviews should be done in the presence of the suite occupant(s) to permit feedback from them whenever possible. For some performance areas, there may be benefit in obtaining feedback from the occupant prior to setting up the physical monitoring. For instance, the building envelope technician or engineer should discuss condensation complaints with the occupants to better understand the nature of the complaint (time of occurrence, etc.) in order to select the most appropriate type of test and timing for the test.

General Guidance for Building Monitoring. As stated earlier, the occupant satisfaction survey is a compulsory component of the protocol. This is, in part, because the results of the occupant satisfaction survey can guide the POE team with respect to selecting the number and specific suites to be monitored in Step 3. For instance, one would expect that if a significant number of complaints are received with respect to a given performance area, the number of suites investigated on that issue would be larger than if the level of satisfaction was found to be good or neutral. The following considerations are given when selecting suites for monitoring:

- View all suites with low occupant satisfaction (a limit on the number may be placed if the suites are very numerous).
- View suites with similar layout/orientation with high or neutral satisfaction or no response.
- Also view a variety of suite layouts and locations (floor and orientation), as appropriate for that performance area.

Reviewing similar suites with varying levels of satisfaction allows the POE assessor to distinguish the presence of a genuine problem versus a highly sensitive occupant. Sampling the suites with good satisfaction or no response permits qualitative and/or quantitative verification of systems that appear to be functioning as designed and constructed.

Viewing a variety of suite layouts and locations provides a more representative "picture" of the building as a whole. The suite selection may depend on the performance area being investigated. Thermal comfort and building envelope performance are both strongly affected by the orientation of the suite's exterior walls and the percentage of glazing. Clustering the suites selected for thermal comfort and building envelope monitoring to those on the south side of the building, for example, will likely cause skewed results.

At first, the POE team is suggested to select a larger number of suites than what is strictly necessary, for each performance area. The suite selection should then be discussed with the building management staff to exclude suites that are unsuitable for the given assessment purpose. For instance, consider a suite with good occupant satisfaction with acoustics being initially selected for physical measurements. If it is then discovered that the neighbouring suites are vacant, the POE team should omit this suite from the monitoring protocol, as the occupants' perceptions are based on the vacancies rather than good acoustical performance.

The following section provides guidance specific to building envelope performance assessment. For this and other performance areas, two levels of assessment are provided: Level 1 is intended to be a simpler, less costly assessment method requiring a lower level of expertise and less sophisticated instrumentation and tools. Should the Level 1 assessment reveal performance issues that are not clearly understood based on the available information, the Level 2 assessment may be warranted.

Building Envelope Performance Assessment—Level 1

Expertise Required. The minimum level of expertise required is a building science technologist or professional with at least five years of experience.

Instrumentation Required. This assessment level relies mainly on visual observations of trained personnel. The instrumentation required to carry out the work includes a psychrometer for ambient air temperature and relative humidity spot measurements, an infrared surface temperature meter (or similar device), and a smoke pencil or smoke generator to qualitatively detect air leakage.

Measurements and Conditions. The assessment should be based on the issues identified in the occupant and building management satisfaction surveys. The assessor should review the issues raised in the surveys and select a representative number of suites, ensuring that a representative sample of suites from each type of unit and exposure are included. The sample size is at the discretion of the POE team, depending on the building configuration.

Complementary Data Collection. The following information could be of use to obtain an understanding of the building construction, its climatic exposure, and history:

- Architectural drawings and specifications
- Shop drawings and submittals of systems and components such as windows, doors, and roofing products
- Test reports of systems tested during construction, such as testing of glazing system for air and water leakage
- Building performance audit report, if available
- History with respect to outstanding construction deficiencies, failures since construction completion, repairs and replacements, etc.
- Building geographical location, features in the surrounding landscape and built environment (e.g., location on the top of a hill, protection from the weather by nearby buildings)
- Information on local climatic conditions (winter and summer design temperatures, annual rainfall, moisture index, etc.)

Interaction with Other Systems:

- There can be strong interactions between a building's envelope and a building's mechanical systems. For instance, the occurrence of condensation and/or mold formation on envelope interior surfaces can indicate inappropriate selection of glazing systems or thermal bridging with respect to the envelope, but it can also indicate a deficient heating system due to poor mechanical ventilation or air distribution. Alternatively, such issues can be caused by a low temperature setpoint.
- Building envelope design and construction will impact thermal comfort.
- Formation of mold growth on interior building envelope surfaces or within the building envelope, due to condensation or water penetration, can cause indoor air quality issues.

Time and Effort to Collect Data. Allow for two or more days on site, depending on the size of the building, plus reporting time.

Typical Minimum Equipment Cost. Purchase costs for the tools vary, but estimates are as follows: digital psychrometer for spot measurements, \$250; infrared surface temperature meter, \$100 to \$1,000; smoke pencil or smoke generator, negligible cost. As these are typical tools, building science firms are likely to own these tools.

Building Envelope Performance Assessment—Level 2

Expertise Required. The minimum level of expertise required is a mid- to senior-level building technologist, professional engineer, or architect.

Instrumentation Required. Simple, relatively low-cost, non-destructive testing measures are provided below and depend on what type of assessment is being carried out:

- Condensation of windows, walls, ceilings; mold growth on windows, walls, ceiling; ghost marking of wall framing:
  - Psychrometer for measurement of indoor air temperature and relative humidity and outdoor air temperature.
  - Infrared thermographic camera for measuring surface temperatures and for qualitative assessment of thermal bridging, missing insulation, and air leakage.
  - Temperature instrumentation (thermocouples) for quantitative measurement of interior surface temperatures on window frames, glass, walls, etc.
- Condensation of glazing assemblies: glazing property detectors for glass thickness, cavity space width, presence and location of low-e coating and low-e product.
- Water leakage testing for glazing systems: water spray rack, fan, and pressure manometer.
- Air leakage testing for diagnosing the location of air barrier breaches in the building envelope: smoke generator, a blower door, and pressure manometer. A flexible pipe may be useful to focus the smoke to localized area.
- Quantitative air leakage testing for determining the air leakage rate of installed glazing systems: fan, pressure manometer, and flow meter.
- Windows and doors:
  - Icicle formation and difficulty of operation: non-destructive assessment of fit of sash to frame and condition of weatherstripping, etc.
- Roofs:
  - Water leakage: for exposed membrane roofs, thermography and/or spray/flood testing of roof to locate leak; for protected membrane roofs, spray/flood testing.

If specific issues described below are observed with respect to claddings, glazing system and roofing systems, a more in-depth study would be required, which would generally be beyond the scope of a POE.

- Brick: cracks of bricks and/or mortar, efflorescence, bowing, water runoff and staining pattern, and ice accumulation.
- Pre-cast panels: cracks, bowing, water runoff and staining pattern, sealant failure, and staining below drainage openings.
- Metal panel: premature sealant failure, coating delamination, and corrosion staining.
- Exterior insulation and finish system (EIFS): premature sealant failure, failure of the EIFS finish coat (debond/re-emulsification), cracks, and debonding panels.
- Siding (wood, vinyl, cementitious cladding): bowing cracking, coating delamination, and staining.
- Glazing systems: fogging of insulated glazing units and cracking of insulated glazing units.
- Roofing systems: blistering of roof membrane, tenting of the membrane, displacement of ballast and insulation in a protected membrane roof (also called an inverted roof), missing shingles and other components, and displacement of components.

Measurements and Conditions. Conducting the measurements following these conditions could be of benefit to the assessor:

- When investigating issues such as formation of condensation, mold and ghost marking on building envelope interior surfaces, weather conditions (i.e., time of year) should replicate those identified in the occupant or building management survey as much as possible.
- Field tests should be conducted following standard test methods and practices where these exist. Refer to the "Building Envelope Performance" heading in the above "Performance Areas and Indicators" section.

Complementary Data Collection. In addition to the information reviewed during the Level 1 study, review of field review reports by the architect and/or the building envelope consultant may bring forward construction issues shedding light on matters raised during the POE.

Interaction with Other Systems. See the Level 1 procedure.

Time and Effort to Collect Data. For each test: one or more on site, depending on type of test, size of building, and complexity and number of tests.

Typical Minimum Equipment Cost. Purchase costs for the tools vary and depend on quality and number of features (costs shown for order-of-magnitude only): digital psychrometer for spot measurements, $250; infrared camera, $7,000; thermocouples and data logger, cost per data logger ranges from $400 to $1,700 and up, depending on the number of thermocouple inputs, cost of thermocouples is insignificant.

Costs of the above equipment, as well as for water and air leakage testing apparatus (water spray racks, fan, manometer, smoke generator, etc.) are commonly used equipment typically included in fee charged by consultant.

Further Work. A higher level of study is possible, and, if warranted, could include conducting exploratory openings and destructive testing. However, conducting further extensive investigation and the use of highly specialized expertise and equipment would generally be considered outside the mandate of a POE.

Step 4—Information Transfer

Meeting with the Owner and Design Team. The purpose of the final meeting with the owner and the design team is information transfer. The building owner, the entire

POE team, and the design professionals should be present at this meeting (if the latter are involved in the POE process).

The POE team presents the main results of the study and discusses how the actual building performance compares with predicted or design targets or industry benchmarks, focusing on building-specific design goals. Any successes, challenges or disparities related to these goals should be identified and discussed. Lessons learned pertaining to building design and operation should be identified and possible strategies for communicating these lessons to a wider design community should be discussed. Recommendations for further investigation should also be presented.

CONCLUSION

This paper presents a methodology for the post-occupancy evaluation of multi-unit residential buildings to meet specific objectives. Seven performance areas are included in the methodology: energy efficiency, water use efficiency, indoor air quality, lighting and the visual environment, acoustics, thermal comfort, and building envelope performance. The methodology is a modular, customizable tool that can be adapted to suit the objectives of the commissioning party. The overall process of the methodology is described step-by-step, from the initial data collection from stakeholders, to the surveys and interviews, data collection from site, and the information transfer.

The findings from POEs of MURBs can be fed back to the stakeholders to improve the operation of existing buildings, and the design, construction, and operation of future buildings. POE outcomes can also serve to provide evidence of the actual performance of innovative systems, equipment, and practices to hasten their adoption by the building industry. The performance data can lead to the determination of benchmarks, which could be used by the building industry and regulatory bodies in the development of building codes.

ACKNOWLEDGMENTS

This multidisciplinary project was not the work of one person. I wish to thank my project partners for their significant efforts. I also wish to acknowledge the contribution of Brian Chapnik of Howe Gastmeier Chapnik Limited on all aspects of this methodology pertaining to acoustics. I also wish to thank Canada Mortgage and Housing Corporation (CMHC) for providing the opportunity and framework for the development of this post-occupancy evaluation for MURBs and for presenting valuable feedback at various stages of the project. This project was funded by CMHC, but the views expressed are the views of the project partners and do not necessarily reflect the views of CMHC. CMHC's financial contribution to this report does not constitute an endorsement of its contents.

REFERENCES

Alberta Energy Resources Conservation Board. 2007. Directive 038; Noise Control. Calgary: Alberta Energy Resources Conservation Board, 56 p.

ASHRAE. 2010a. ANSI/ASHRAE Standard 62.1-2010. *Ventilation for Acceptable Indoor Air Quality*. Atlanta: ASHRAE, 54 p.

ASHRAE. 2010b. ANSI/ASHRAE Standard 55-2010. *Thermal Environmental Conditions for Human Occupancy*. Atlanta: ASHRAE, 40 p.

ASHRAE. 2010c. Performance Measurement Protocols for Commercial Buildings. ASHRAE, 298 p.

ASHRAE. 2011. *ASHRAE Handbook—HVAC Applications*. Atlanta: ASHRAE.

ASHRAE. 2012. Performance Measurement Protocols for Commercial Buildings: Best Practices Guide. Atlanta: ASHRAE, 258 p.

ASTM. 2005. ASTM D5957-98 (2005): *Standard Guide for Flood Testing Horizontal Waterproofing Installations*. West Conshohocken, PA: American Society for Testing and Materials, 5 p.

ASTM. 2008. ASTM E1105-00 (2008): *Standard Test Method for Field Determination of Water Penetration of Installed Exterior Windows, Skylights, Doors, and Curtain walls, by Uniform or Cyclic Static Air Pressure Difference*. West Conshohocken, PA: American Society for Testing and Materials, 7 p.

ASTM. 2009. ASTM E1186-03: *Standard Practices for Air Leakage Site Detection in Building Envelopes and Air Barrier Systems*. West Conshohocken, PA: American Society for Testing and Materials, 7 p.

ASTM. 2010a. ASTM E783-02: *Standard Test Method for Field Measurement of Air Leakage Through Installed Exterior Windows and Doors*. West Conshohocken, PA: American Society for Testing and Materials, 6 p.

ASTM. 2010b. ASTM C1153-10: *Standard Practice for Location of Wet Insulation in Roofing Systems Using Infrared Imaging*. West Conshohocken, PA: American Society for Testing and Materials, 6 p.

ASTM. 2011. ASTM C1060: *Standard Practice for Thermographic Inspection of Insulation Installations in Envelope Cavities of Frame Buildings*. West Conshohocken, PA: American Society for Testing and Materials, 7 p.

Avina, J. 2012. *An Introduction to Utility Bill Weather Normalization for Energy Contractors*. Retrieved May 29, 2013 from http://www.abraxasenergy.com/articles/intro-weather-normalization-contractors/

Baird, G. 2010. *Sustainable Buildings in Practice: What the Users Think*. Routledge, New York, 335 p.

Birt, B. and Newsham, G.R. 2009. Post-occupancy evaluation of energy and indoor environment quality in green buildings: a review. Report NRCC-51211. National Research Council, Ottawa, June, 8 p. Retrieved August 30, 2011 from http://www.nrc-cnrc.gc.ca/obj/irc/doc/pubs/nrcc51211.pdf.

BSRIA. 2009. The Soft Landings Framework for Better Briefing, Design, Handover and Building Performance In-Use. BSRIA, Bracknell, 30 p. Available at www.usablebuildings.co.uk.

CaGBC. 2009. *LEED Canada Reference Guide for Green Building*. Ottawa: Canada Green Building Council.

CCBFC. 2010. *National Building Code of Canada*. Ottawa: Canadian Commission on Building and Fire Codes, National Research Council of Canada.

CMHC. 2001. Analysis of the Annual Energy and Water Consumption of Apartment Buildings in the CMHC HiSTAR Database. Ottawa: Canada Mortgage and Housing Corporation, 4 p.

CMHC. 2005. Reducing Noise in Your Apartment, About Your House Series, 63904. Ottawa: Canada Mortgage and Housing Corporation, 10 p.

Cohen, R. et al. 1999. PROBE Strategic Review 1999: Report 1: Review of the PROBE Process. 41 p. Retrieved Aug. 18, 2011 from www.usablebuildings.co.uk.

Cook, M. 2007. *The Design Quality Manual: Improving Building Practice*. Blackwell Publishing, Oxford, 199 p.

Detta Colli, R. No date. TowerWise.ca: Reaching Higher for Energy Savings-High-Rise Inventive Programs, 12 p. Retrieved February 29, 2012 from http://www.towerwise.ca/incentivesbooklet.pdf.

EcoSmart Foundation. 2007. EcoSmart. Vancouver, B.C.: EcoSmart Foundation. Retrieved August 4, 2011 from http://www.ecosmart.ca/index.cfm?pk=3.

Health Canada. 1989. Exposure Guidelines for Residential Indoor Air Quality: A Report of the Federal-Provincial Advisory Committee on Environmental and Occupational Health. Ottawa: Health Canada, 28 p. Retrieved February 9, 2012 from http://www.hc-sc.gc.ca/ewh-semt/pubs/air/exposure-exposition/index-eng.php.

Health Canada. 2012. Residential Indoor Air Quality Guidelines. Retrieved February 9, 2012 from http://www.hc-sc.gc.ca/ewh-semt/air/in/res-in/index-eng.php.

Higher Education Funding Council for England. 2006. Guide to Post-Occupancy Evaluation. Bristol: Higher Education Funding Council for England, 33 p.

Illuminating Engineering Society of North America. 2011. *The IES Lighting Handbook*, 10th ed. New York: Illuminating Engineering Society of North America, 1087 p.

Keen Engineering. 2005. Post Occupancy Evaluation Project Report: Completion of Phase 1: POE Protocol Development. Keen Engineering, Vancouver, 46 p. Retrieved August 30, 2011. Available from http://www.sbtc.ca/Docs/POEProtocol1.pdf.

Kernan, P, Sheltair Group and Société Logique. 2002. Healthy High-Rise: A Guide to Innovation in the Design and Construction of High-Rise Residential Buildings. Ottawa: Canada Mortgage and Housing Corporation, 136 p.

Leaman, A., Stevenson, F. and Bordass, B. 2010. Building evaluation: practice and principles. *Building Research and Information*. Routledge, London, Vol. 38, no. 5, pp. 564-577.

Leaman, A., Stevenson, F. and Boyce, H. BUS Methodology occupant survey, Part of Technology Strategy Board Building Performance Evaluation, PowerPoint presentation, 54 slides. Retrieved August 18, 2011 from www.usablebuildings.co.uk.

Mallory-Hill, S., Preiser, W. F.E. and Watson, C., eds. 2012. Enhancing Building Performance. Sussex: Wiley-Blackwell, 360 p.

Merriam-Webster, 2012. http://www.merriam-webster.com/dictionary.

Morrison Hershfield. 2012. Development of a Post Occupancy Evaluation Performance Assessment Framework for Multi-Unit Residential Buildings. Unpublished report submitted to Canada Mortgage and Housing Corporation, Ottawa, ON, Canada.

NIBS. 2012. Daylighting: General Principles and Commitments. National Institute for Building Sciences. Retrieved August 22, 2012 from http://www.wbdg.org/references/mou_daylight.php

Ontario Ministry of the Environment. 1997. Noise Assessment Criteria in Land Use Planning: Publication LU-131. Toronto: Ministry of the Environment.

Sanders, C. 2010. Development of an Affordable Methodology for the Evaluation and Reporting of the Energy Performance of Occupied Dwellings – Final Report. Livingston: Scottish Government, Directorate for the Built Environment, Building Standards Division, 102 p. Retrieved September 2, 2011 from http://www.scotland.gov.uk/Topics/Built-Environment/Building/Building-standards.

Watson, C. 1996. Evolving Design for Changing Values and Ways of Life. IAPS14 International Association for People-Environment Studies, Stockholm, July, 10 p. Retrieved August 11 2011 from http://www.postoccupancyevaluation.com/publications/default.shtml 1996.

Watson, C. 2003. Review of Building Quality Using Post Occupancy Evaluation. Journal of the Programme on Educational Building, Organisation for Economic Co-operation and Development, Paris, February. Retrieved August 11, 2011 from http://www.postoccupancyevaluation.com/publications/default.shtml.

Watson, C. and Fitzgerald, C. 1998. Customer Focus in Design and Communication in a Major Project. Australasian Evaluation Society Conference, Melbourne, October, 9 p. Retrieved August 11, 2011 from http://www.postoccupancyevaluation.com/publications/default.shtml.

VIP Technology Review and Cost Study for Building Retrofits in Northern US Locations

Nitin Shukla, PhD

Ali Fallahi, PhD
Associate Member ASHRAE

Jan Kosny, PhD
Member ASHRAE

ABSTRACT

Vacuum insulation panel (VIP) is an emerging insulation technology that is slowly finding applications in buildings due to its energy saving potential. VIP consists of an evacuated nanoporous core material that is sealed off with an impermeable enclosure. Nanoporous structure of core in conjunction with reduced air pressure results in an ultra-low thermal conductivity of VIP that is almost 5–10 times smaller than the conventional insulations such as foams and fibers. In this paper, we first provide a brief discussion on heat transfer mechanisms that are responsible for reduced thermal conductivity of VIP. Next, we investigated the cost performance of VIP technology when implemented in the deep energy building retrofits in cold climate regions of the northern US. Deep energy retrofits in cold climate regions generally require higher target R-value envelopes. VIPs with significantly higher R-value per unit thickness than the conventional insulations save considerably on the cost associated with the loss of living space caused by the installation itself. In deep energy retrofits, thinner VIP installations may also eliminate the cost associated with alteration of window and door openings, which is otherwise necessary for thicker layers of conventional insulations. We compare the VIP retrofit method against a number of thermally equivalent conventional wall insulation techniques using foam sheathing, sprayed foam, fiber insulations, etc. The cost analysis indicates that for lower priced VIPs, the proposed method of VIP wall retrofit may become cost competitive with current deep retrofit strategies that employ foam insulations. We predict that in the near future, further advancements in VIP manufacturing technology, larger volume production, and higher reliability through quality control of cores and films will enhance VIP cost effectiveness as a retrofit option for providing local insulation and mitigating thermal bridging effects in building envelopes.

INTRODUCTION

Building share of primary energy consumption has increased from 33.7% (26.3 Quads) in year 1980 to 42.1% (40.4 Quads) in year 2010 (EERE 2010). The share is expected to rise in the future (EERE 2010), potentially causing strain on our energy resources and an increase in green house gas emissions. There is a need for more energy efficient buildings that exchange less thermal energy with the exterior environment. A simple way to realize this goal is to add thermal insulation to the building envelope, thereby retarding the energy flow through the envelope. Vacuum insulation panel (VIP) is an emerging insulation technology that holds huge promise in reducing building energy consumption due to its far superior insulation properties than the conventional foam and fiber based insulations (ECBCS 2005; Simmler et al. 2005a). In retrofit building applications, VIP technology becomes even more attractive because high thermal resistivity of VIP allows for application of thinner retrofit solutions, thus saving on usable space. Future success of VIP technology depends on its cost effectiveness in addition to thermal performance and durability as compared to existing conventional technologies.

VIP consists of a nanoporous core that is evacuated and enclosed in a container that is highly impervious to air and water vapor transmission. Typically, silica in pressed fumed, precipitated, or powder form is used as the core material, while a thin metal or a metalized polymer film is used to surround

Nitin Shukla *and* ***Ali Fallahi*** *are technical staff members and* ***Jan Kosny*** *is the Building Enclosure Program lead at Fraunhofer CSE, Cambridge, Massachusetts.*

and seal the core. Thermal conductivity, λ, of a typical VIP is ~0.004 W/m·K (0.0023 Btu/h·ft·°F), a factor of 5–10 lower than the conventional insulation such as extruded polystyrene (XPS, λ = 0.029 W/m·K or 0.0168 Btu/h·ft·°F), expanded polystyrene (EPS, λ = 0.036 W/m·K or 0.0208 Btu/h·ft·°F), polyisocyanurate (PIC, λ = 0.022 W/m·K or 0.0127 Btu/h·ft·°F), and polyurethane (PU, λ = 0.024 W/m·K or 0.0139 Btu/h·ft·°F).

In this paper, we will first discuss heat transport mechanisms behind the ultra-low thermal conductivity of VIP. Next, we will discuss cost competitiveness of VIP technology in a specific building retrofit application.

HEAT TRANSPORT IN VIP

Thermal conductivity, λ_T, of a porous material, such as conventional insulation and VIP, is a combination of following four heat transfer mechanisms (ECBCS 2005; Caps et al. 2001; Aegerter et al. 2011; Kwon et al. 2009): thermal conduction through the solid network, λ_S, thermal conduction through the entrapped gas, λ_G, convection of the gas component, λ_C, and radiation through the solid and pores, λ_R, i.e.,

$$\lambda_T = \lambda_S + \lambda_G + \lambda_C + \lambda_R \tag{1}$$

Typical pore size in these insulating materials is on the order of nanometer and micrometer. Brownian motion of the gas molecules is completely suppressed; convective heat transport, λ_C, can be safely assumed to be negligible (Aegerter et al. 2011). The porous structure of insulation materials offers only limited channels for heat conduction. This fact, combined with strong scattering of heat carriers (mainly phonons in case of silica core) in the amorphous structure of the core material, results in a significantly reduced solid thermal conductivity, λ_S (Fricke et al. 1989; Scheuerpflug et al. 1991). Lack of gas convection and minimization of solid conductivity are mainly responsible for low thermal conductivity of the conventional insulation, which typically ranges between 0.035 and 0.060 W/m·K (0.02 and 0.035 Btu/h·ft·°F).

Gas conduction, λ_G, in a porous media with an average pore size, d, is given by the following relation (Kennard 1938):

$$\lambda_G = \frac{\lambda_{G0}\Pi}{1 + 2\beta\frac{l}{d}} \tag{2}$$

where λ_{G0} is thermal conductivity of the free air, Π is the porosity, β is a constant that depends on the type of gas and temperature, and l is mean free path (MFP) of air molecules. MFP is inversely proportional to gas pressure, P_G, (Kennard 1938)

$$l \propto \frac{1}{P_G} \tag{3}$$

At atmospheric pressure and room temperature conditions, λ_{G0} = 0.026 W/m·K (0.015 Btu/h·ft·°F); a significant but unavoidable contributor to the heat transport in the conventional insulations. According to Equation 2, thermal conductivity will decrease with reduced gas pressure, offering a route for further minimization of thermal conductivity in porous materials. This may be achieved via evacuating and sealing off the porous material, and this concept is at the heart of VIP technology. According to Equation 2, a perfect vacuum condition will be required to eliminate the gas conduction; however, for highly porous and nanostructured VIP core, an order of magnitude reduction in gas conduction may be achieved even at a weak vacuum levels of 0.01–0.5 Pa (1.45e-6-72.5e-6 psi) (Simmler et al. 2005b).

For optically thick silica core, the thermal radiation propagates via diffusion (Aegerter et al. 2011), and radiative part of the thermal conductivity may be given by the following equation:

$$\lambda_R = \frac{16}{3}\frac{n^2\sigma T^3}{E} \tag{4}$$

where n is the index of refraction, T is the temperature, E is the extinction coefficient defined as the inverse of mean free path of photons or the ratio of optical thickness and geometrical thickness, and σ is the Stephen-Boltzman constant. In general, core materials such as fumed silica are transparent for a major part of the IR wavelength spectrum at room temperature (i.e., the extinction coefficient is low, leading to a large radiation contribution). To reduce the radiative component, optically contrasting materials known as opacifiers that absorb or scatter photons in IR spectrum are added to the core material. Carbon black, titanium oxide, iron oxide, and silicon carbide are some of the commonly used opacifiers.

To summarize, nanoporous structure of the core in conjunction with reduced air pressure is responsible for ultra-low thermal conductivity of VIP. Adding small amounts of opacifiers to the core material may further reduce the thermal conductivity to as low as 0.004 W/m·K (0.0023 Btu/h·ft·°F).

VIP COST ANALYSIS

Currently, higher cost of VIP has been a major barrier to its wide-scale adoption in the buildings. In retrofit projects where living space is valuable, VIP technology becomes a very cost attractive alternative because it provides significantly higher R-value per unit thickness, minimizing loss of the living space caused by the installation itself. For "deep" energy retrofit projects that aim to reduce building envelope-related energy consumption by at least 50%,[1] thinner VIP installations may avoid cost associated with alteration of window and door openings, which is otherwise necessary for thicker layers of conventional insulations. Installation of thinner VIP may be achieved by performing only minor adjustments to the window and door opening without causing any major structural changes to the envelope itself. On the other hand, conventional insulation due to its thickness may require

1. See National Grid (2013) for more information and Castle Square (2010) for an example.

special attachment methods (long nails or additional furring) and roof overhang adjustments. Another advantage of VIP is that it is nonflammable and may be packed between nonflammable protective layers to provide a nonflammable thermal insulation solution in buildings; this could be important if the United States adopts more restrictive building fire codes in the future, such as those that are already adopted in Europe, Japan, and China (ISO 2002a, 2002b, 2002c, 2002d, 2002e).

In this section, we investigate cost performance of VIP technology when implemented in the favorable scenario of deep energy building retrofits in cold climate regions. In deep energy retrofit projects in cold climate regions, target R-values for walls are usually higher (i.e., around 5.28 $m^2 \cdot K/W$ [R-30 $h \cdot ft^2 \cdot {}^\circ F/Btu$]). The higher target R-value results in increased living space or savings for VIP technology. Triple-deckers[2] and similar historic houses in New England region (Krim 1977) offer a large market potential for such deep retrofit applications. In fact, city of Boston alone has almost 9000 triple decker houses (City of Boston 2013). The primary goal of the study is to estimate and compare the cost of wall retrofit projects using current state-of-the-art VIP technology to that of the conventional foam applications. Foam insulations have been chosen for comparison because they are most commonly used in exterior wall retrofits and provide higher thermal resistance per unit thickness than fiber insulations.

Installed Cost Estimation

We estimate and compare the total cost[3] to deep energy wall retrofitting of a baseline house[4] using the conventional foam insulations (Waier et al. 2011).[5] The conventional foam insulations considered are rigid board (XPS, EPS, and PIC foams) and spray-applied (closed-cell PU foam) options. We assume that the rigid boards are installed in the form of an exterior sheathing, and the foam is sprayed either in between added studs or between exterior furrings. Although retrofit applications can apply combined cavity and exterior insulations to reach 6.16 $m^2 \cdot K/W$ (R-35 $h \cdot ft^2 \cdot {}^\circ F/Btu$) thermal resistance, the cost analysis considers only exterior installation as compared to VIPs. This is because for the cases considered in the study, the installed cost of combined cavity and exterior installation is higher than the installed cost of exterior insulation alone. Table 1 shows an example comparison.

2. The triple-decker (or three-decker) is a unique housing type characteristic of New England cities in early 20th century. Generally defined, the triple-decker is a free standing, wood frame structure on its own narrow lot, three stories high and one family unit at each floor.

3. The total cost is the sum of the bare material cost, the bare labor cost, and the bare equipment cost plus 10% for profit.

4. The baseline house is a two-story average class residential house with 2000 ft^2 floor surface area and 2700 ft^2 wall surface area (Waier et al. [2011], p. 29).

5. The main source of cost data to estimate total cost is the 2011 RSMeans Construction Cost Data and RSMeans Residential Cost Data.

We consider a new retrofit method of installing VIPs insulation that takes advantage of exterior cladding of a wall assembly to achieve continuity of thermal insulation in addition to the convenience of quick and easy installations. In this method, a 2.5 cm (1 in.) thick VIP is encapsulated between two layers of 2.5 cm (1 in.) XPS layers. Next, a 3 in. thick VIP-based EIFS is installed on the exterior sheathing using adhesives. For deep retrofit projects, VIP-based EIFS offers several advantages over the conventional foam based EIFS: higher R-value per unit thickness, improved thermal continuity, and avoidance of numerous added thickness-related costly tasks such as window alteration because it directly adheres to the existing wall. For example, in deep retrofits of walls using conventional foam insulations, adding thick layers of thermal insulation is necessary (often 15–25 cm and 6–10 in.). Thick layers of insulation usually require special attachment methods (long screws or additional furring). In addition, window and door openings need to be altered to allow window replacements closer to the exterior surface, reduce local thermal bridging, and improve water drainage, for example. Further work is very often necessary to adjust roof overhangs because the original wall surface moves notably toward the outside when the thick foam layers are installed. When using significantly thinner high performance VIPs, these tasks are not necessary.

We used the following approach to estimate the total cost of each retrofit strategy:

- A target R-value is estimated.
- An equivalent thickness is calculated to achieve the target R-value.
- A list of required retrofit tasks is determined. The retrofit work depending on insulation type and equivalent thickness includes a series of tasks from altering the existing walls and roof to installing the insulation and exterior cladding.
- The cost of each retrofit task and the installed cost are estimated.

It is important to recognize that the market for the VIP technology is immature, and the prices are expected to decrease as demand grows. Therefore, it is more appropriate to compare this technology with conventional insulation methods assuming thermal performance or target R-value as the criterion rather than finding an optimum balance between R-value and cost/savings.

Target R-Value

VIPs in the market have thermal resistance in the range of 4.4 to 8.8 $m^2 \cdot K/W$ (R-25 to R-50 $h \cdot ft^2 \cdot {}^\circ F/Btu$) for a 2.5 cm (1 in.) thick, panel depending on the core material type. For the purpose of this study, we assumed that the VIP has a nominal center-of-the-panel thermal resistance of R-40.[6]

6. Non-aged condition.

Table 1. Installed Cost Estimates for Conventional and Emerging Insulation Techniques to Retrofit the Wall of a Baseline House

VIP Retrofit Techniques Do Not Require All Retrofit Tasks because of Low Thickness. Costs of the Exterior Wall Finishes Are Not Included in Analysis for all Considered Cases of Deep Energy Retrofits; Only the Insulation Part of the Retrofit Strategy is Analyzed.

Insulation Type		Retrofit Tasks and Cost, ($/ft^2) *(Note numbers are included in brackets.)*																		Total Cost, ($/ft^2)
		Removal of Exterior Cladding (vinyl siding)	Furring	Framing (2x4; OC 16)	Framing (2x6; OC 16)	Window Relocation for deep window wells	Insulation Installation (material and labor)	Insulation protection layers (2 XPS layers)	EIFS with 1/2 in. cement board sheathing and 2.5#/S.Y metal lath substrate	Removal of shingles (two layers)	Removal of OSB Deck	Rafter extension (4 feet)	Fascia Boards Installation	Soffit Nailer	Replace-ment of OSB Deck	Replacement of shingles	Rake Overhang	Building Kraft Paper	Strapping Screws	Installation of exterior cladding (vinyl siding)	
Rigid Insulation		*[1]*	*[2]*			*[2]*	*[3]*			*[8]*	*[9]*	*[10]*	*[9]*	*[4]*	*[9]*	*[9]*	*[5]*	*[1]*	*[11]*	*[1]*	
	XPS (7 in.)	$0.59	$0.92	—	—	$0.52	$5.81	—	—	$0.04	$0.03	$0.129	$0.01	$0.01	$0.03	$0.09	$0.39	$0.22	$0.39	$7.79	$16.98
	EPS (9 in.)	$0.59	$0.92	—	—	$0.52	$5.04	—	—	$0.04	$0.03	$0.129	$0.01	$0.01	$0.03	$0.09	$0.39	$0.22	$0.39	$7.79	$16.21
	PIC (6.0 in.)	$0.59	$0.92	—	—	$0.52	$4.51	—	—	$0.04	$0.03	$0.129	$0.01	$0.01	$0.03	$0.09	$0.39	$0.22	$0.25	$7.79	$15.54
Sprayed Applied Foam Insulation		*[1]*	*[1]*	*[4]*	*[4]*	*[2]*	*[3]*			*[8]*	*[9]*	*[10]*	*[4]*	*[4]*	*[9]*	*[9]*	*[5]*	*[1]*	*[11]*	*[1]*	
	PU[1] (7.5 in.)	$0.59	—	—	$1.91	$0.52	$5.39	—	—	$0.04	$0.03	$0.129	$0.01	$0.01	$0.03	$0.09	$0.39	$0.22	$0.39	$7.79	$16.34
	PU[2] (7.0 in.)	$0.59	—	$1.50	—	$0.52	$5.78	—	—	$0.04	$0.03	$0.129	$0.01	$0.01	$0.03	$0.09	$0.39	$0.22	$0.39	$7.79	$16.32
	PU[3] (6.0 in.)	$0.59	$0.92	—	—	$0.52	$4.62	—	—	$0.04	$0.03	$0.129	$0.01	$0.01	$0.03	$0.09	$0.39	$0.22	$0.25	$7.79	$14.59
Combined Cavity and Exterior Insulations		*[1]*	*[1]*	*[4]*	*[4]*		*[12]*			*[8]*	*[9]*	*[10]*	*[4]*	*[4]*	*[9]*	*[9]*	*[5]*	*[1]*	*[11]*	*[1]*	
	Blown Cellulose (5 3/16 in.) *[1]*	$0.59		—	—	—	$1.88	—	—	—	—	—	—	—	—	—	—	—	—	—	
	XPS (3 in.)	—	$0.92			—	$2.30	—	—	—	—	—	—	—	—	—	—	—	—	$7.79	$13.48

Table 1. ***(Continued)*** **Installed Cost Estimates for Conventional and Emerging Insulation Techniques to Retrofit the Wall of a Baseline House**

VIP Retrofit Techniques Do Not Require All Retrofit Tasks because of Low Thickness. Costs of the Exterior Wall Finishes Are Not Included in Analysis for all Considered Cases of Deep Energy Retrofits; Only the Insulation Part of the Retrofit Strategy is Analyzed.

		Removal of Exterior Cladding (vinyl siding)	Furring	Framing (2x4; OC 16)	Framing (2x6; OC 16)	Window Relocation for deep window wells	VIP Insulation	Insulation protection layers for VIPs (2 XPS layers)	Installed cost of EIFS with 1/2 in. cement board sheathing and 2.5#/S.Y metal lath substrate (material and labor); cost of foam insulation not included	Removal of shingles (two layers)	Removal of OSB Deck	Rafter extension (4 feet)	Fascia Boards Installation	Soffit Nailer	Replacement of OSB Deck	Replacement of shingles	Rake Overhang	Building Kraft Paper	Strapping Screws	Installation of exterior cladding (vinyl siding)	
		[1]						*[6]*	*[7]*												
VIP Insulation	VIP1 (3.0 in.)	$0.59	—	—	—	—	$30.00	$2.42	$8.44	—	—	—	—	—	—	—	—	—	—	—	$41.45
	VIP2 (3.0 in.)	$0.59	—	—	—	—	$11.43	$2.42	$8.44	—	—	—	—	—	—	—	—	—	—	—	$22.88
	VIP3 (3.0 in.)	$0.59	—	—	—	—	$7.00	$2.42	$8.44	—	—	—	—	—	—	—	—	—	—	—	$18.45
	VIP4 (3.0 in.)	$0.59	—	—	—	—	$4.27	$2.42	$8.44	—	—	—	—	—	—	—	—	—	—	—	$15.72
	VIP5 (3.0 in.)	$0.59	—	—	—	—	$3.50	$2.42	$8.44	—	—	—	—	—	—	—	—	—	—	—	$14.95
	VIP6 (3.0 in.)	$0.59	—	—	—	—	$3.00	$2.42	$8.44	—	—	—	—	—	—	—	—	—	—	—	$14.45
	VIP7 (3.0 in.)	$0.59	—	—	—	—	$1.23	$2.42	$8.44	—	—	—	—	—	—	—	—	—	—	—	$12.68
	VIP8 (3.0 in.)	$0.59	—	—	—	—	$1.15	$2.42	$8.44	—	—	—	—	—	—	—	—	—	—	—	$12.60
	VIP9 (3.0 in.)	$0.59	—	—	—	—	$4.25	$2.42	$8.44	—	—	—	—	—	—	—	—	—	—	—	$15.70
	VIP10 (3.0 in.)	$0.59	—	—	—	—	$3.25	$2.42	$8.44	—	—	—	—	—	—	—	—	—	—	—	$14.70
	VIP11 (3.0 in.)	$0.59	—	—	—	—	$2.78	$2.42	$8.44	—	—	—	—	—	—	—	—	—	—	—	$14.23
	VIP12 (3.0 in.)	$0.59	—	—	—	—	$5.18	$2.42	$8.44	—	—	—	—	—	—	—	—	—	—	—	$16.63

Table 1. ***(Continued)*** **Installed Cost Estimates for Conventional and Emerging Insulation Techniques to Retrofit the Wall of a Baseline House**

VIP Retrofit Techniques Do Not Require All Retrofit Tasks because of Low Thickness. Costs of the Exterior Wall Finishes Are Not Included in Analysis for all Considered Cases of Deep Energy Retrofits; Only the Insulation Part of the Retrofit Strategy is Analyzed.

Notes:

PU^1: Closed cell polyurethane foam was applied inside of 2 × 4 wood studs.
PU^2: Closed cell polyurethane foam was applied inside of 2 × 6 wood studs.
Both cases: the added stud is installed from inside and standing off a few inches from existing stud.
PU^3: Closed cell polyurethane foam was applied in exterior furring.

VIP1: Glacier Bay (according to website in 2011: http://passivehousetoronto.blogspot.com/2011/03/vacuum-insulated-panels-and-prices.html).
VIP2: Dow Corning - Silica Core (based on price quotations obtained from Dow Corning's in 2011).
VIP3: Popular Mechanics (according to Popular Mechanics magazine, Oct 2009). (See www.popularmechanics.com/home/improvement/3455301 for more information.)
VIP4: Richard T. Bynum (according to Richard T. Bynum, Technology and Engineering obtained in 2011).
VIP5: Dow Corning - Mineral Wool Core (according to Dow Corning's quoting obtained in 2011).
VIP6: University of Chicago (personal communication with Dr. Alan Feinerman, Associate Professor, University of Illinois at Chicago in 2011).
VIP7: Thermal Vision (according to Thermal Vision company's quoting obtained in 2011).
VIP8: Dow Corning Future price (estimated by Dow Corning in 2011).
VIP9 – Based on the following paper: Alam, M., Singh, H., and Limbachiya, M. C. (2011). Vacuum Insulation Panels (VIPs) for building construction industry - A review of the contemporary developments and future directions. Applied Energy, 88(11), 3592–3602.
VIP10 - Qingdao Kerui New Environmental Materials Co. (according to company's quoting for R-30 VIP obtained in 2011).
VIP11 - Sinoarch Shanghai Co. (according to company's quoting obtained in 2011); http://www.alibaba.com/showroom/vip-insulation.html.
VIP12 - Hangzhou Sanyou Dior Insulation Materials MFG Co.,Ltd (according to company's price quotation obtained in 2011).

Aerogel1: According to Aspen Aerogel in 2011. The unit price is $2.75/$ft^2$ with 3/8 in. thickness from Aspen Aerogel.
Aerogel2: Assuming 80% cost drop in future.

[1] According to RSMeans Building Construction Cost Data 2011.
[2] According to Building Science Corporation (BSC). Estimated as $100/window × 14 windows of base case house / 2,700 (total façade area of the base case house).
[3] According to RSMeans Building Construction Cost Data 2011, which includes material, equipment, and labor costs; the cheapest combination of rigid board insulation thickness was assumed for the cost analysis (e.g., 9 in. was assumed as 3 × 3 in. layer instead of 4 × 2 + 1 in.). For PU, thermal bridging effect was taken into account while calculating required thickness to achieve target R-value.
[4] According to RSMeans Building Residential Cost Data 2011.
[5] According to William A. Zoeller from Steven Winter Associates, Inc., Norwalk, CT.
[6] VIPs are encapsulated between two layers of XPS protection layers to reduce lateral damages. The cost includes material and labor costs.
[7] According to RSMeans Building Construction Cost Data 2011. The cost estimated for EIFS with ½ in. cement board sheathing and 2.5#/S.Y metal lathe substrate (material and labor). The cost of 3 in. EPS foam insulation is not included because in VIP-based EIFS the foam insulation is replaced with 3 in. encapsulated VIPs.
[8] According to RSMeans Building Construction Cost Data 2011. Estimated as 2 × 40 ft (removed area) × 2 (shingle layers) × $0.35 (removal unit price /ft^2) × 2 (both eave sides) / 2,700 ft^2 (total façade area).
[9] According to RSMeans Building Construction Cost Data 2011. Estimated the same way as [8].
[10] According to RSMeans Building Construction Cost Data 2011. The rafter extension is 4 in.; 2 in. overlap and 2 in. extension from its existing length.
[11] According to http://www.bestmaterials.com/SearchResult.aspx?CategoryID=750.
[12] According to Steven Winter Associates, Inc. (https://www.google.com/search?q=Practical+Residential+Wall+Systems%3A+R-30+and+Beyond&ie=utf-8&oe=utf-8&aq=t&rls=org.mozilla:en-US:official&client=firefox-a). This is the cost of blown cellulose insulation including material and labor into 2 × 6 framing wall ($1,500 for 100 × 8 ft wall).

HEATING 7.3, a 3D heat transfer computer program, was used to model the effective R-value for the case where the VIP is sandwiched between two layers of XPS insulation layers (1 in. XPS + 1 in. VIP + 1 in. XPS) to protect it from impact and damage. An effective thermal resistance of 6.16 $m^2 \cdot K/W$ (R-35 $h \cdot ft^2 \cdot °F/Btu$) was determined by taking into account all thermal bridging effects of XPS and foil layers. This number also includes R-value reduction caused by possible inserts of XPS in the remaining areas where modular dimensions of VIPs do not exactly cover the whole wall. In the cost analysis, 6.16 $m^2 \cdot K/W$ (R-35 $h \cdot ft^2 \cdot °F/Btu$) is taken as the target R-value to find equivalent thickness of other insulations.

Equivalent Thickness

The VIP is sandwiched between two layers of XPS rigid board protection layers. The foam-VIP-foam panel considered has a total thickness of 7.5 cm (3 in.) For thermally equivalent conventional insulations, however, the thickness can vary between 15 and 22.5 cm (6 and 9 in.) For foam insulations, the effective R-value was calculated by taking into account the thermal bridging effect of exterior wood studs or furring. Figure 1 shows the equivalent thickness that is required to achieve 6.16 $m^2 \cdot K/W$ (R-35 $h \cdot ft^2 \cdot °F/Btu$) for different insulations.

Retrofit Tasks

A VIP-based EIFS can be quickly and easily installed. Because VIPs are sandwiched between foam, the overall installation and drainage systems for this technology are almost identical to those of conventional EIFS (exterior insulation and finish system) applications. In contrast, installing thermally equivalent foam insulation involves several additional costly and labor-intensive retrofit tasks. Moreover, because the equivalent thickness of foam insulation is greater than 15 cm (6 in.), the retrofit tasks are limited to not only the walls but also fenestration openings and roof overhangs that might need to be readjusted or extended. This adds to the cost as well. Table 1 shows the required wall retrofit tasks of a baseline house for VIPs and foam insulations.

Discussion

In the cost analysis, for foam insulation retrofits, the costs of material, installation, and labor were taken from the RSMeans Cost Data 2011 book. For VIP retrofits, the price quotations were obtained from several VIP manufacturers around the world and used as material cost. Data from manufacturers' websites and technical publications were also included. The labor cost for VIP-based EIFS installation was assumed to be the same as the cost of installing 7.5 cm (3 in.)

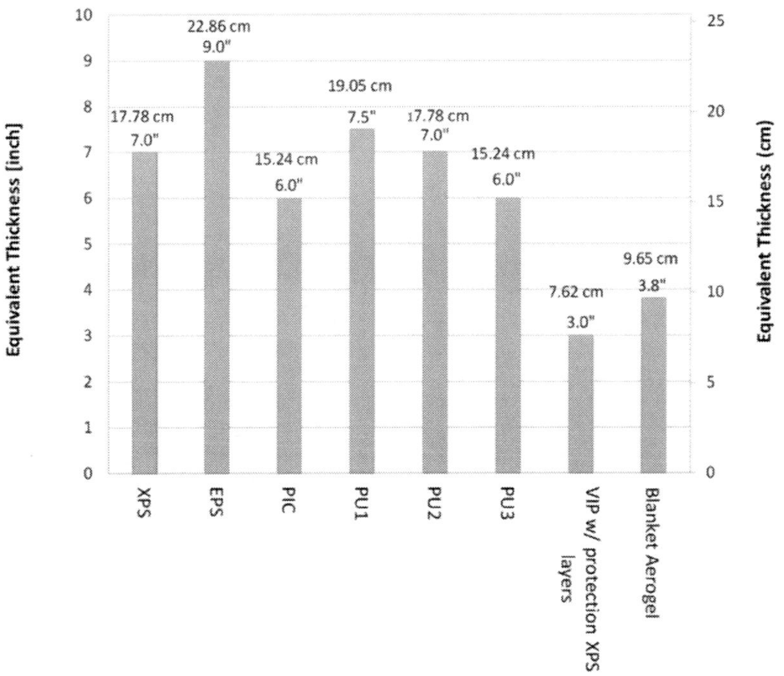

Figure 1 *Equivalent thicknesses of different exterior wall insulation systems to achieve a thermal resistance of 6.16 $m^2 \cdot K/W$ (R-35 $h \cdot ft^2 \cdot °F/Btu$). The thicknesses in this figure are rounded from the manufactured thickness (rigid board with at least 2.54 cm or 1 in. thickness and spray foam with at least 1.25 cm or 0.5 in. thickness). The equivalent thickness for PU takes into account the thermal bridging effect of wood studs. VIP encapsulated in XPS insulation has an effective thermal resistance of 6.16 $m^2 \cdot K/W$ (R-35 $h \cdot ft^2 \cdot °F/Btu$). The thermal resistivity of EPS, XPS, PU, and PIC insulations have been assumed to be 28, 35, 42, and 46 $m \cdot K/W$ (4, 5, 6, and 6.5 $h \cdot ft^2 \cdot °F/Btu$ in.), respectively.*

of foam insulation. There is a large variation in VIP prices among manufacturers and references.

Figure 2 shows cost estimates for wall retrofit methods with VIP and foam insulations.[7] VIP retrofit cost is the installed cost and not just the material cost. Wall retrofits with foam insulation costs between \$151 and \$183/m^2 (\$14 and \$17/ft^2), while VIP retrofits vary between \$129 and \$441/m^2 (\$12 and \$41/ft^2), depending on the manufacturer. This wide range in VIP-based EIFS retrofit cost reflects the large variability in the VIP prices across different manufacturers. Taking the lower range of VIP-based EIFS cost, a 6.16 m^2·K/W (R-35 h·ft^2·°F/Btu) VIP-based wall retrofit costs almost 15% less than that using foam insulation. In fact, it is interesting to note that majority of the VIP-based solutions covered in this study are very cost competitive with foam-based retrofits, and at the same time provide the convenience of easy and quick installation with minimum disruption to the occupants.

VIP is a fragile material and once punctured it may significantly lose its insulating capability due to loss of vacuum (ECBCS 2005). It is important that the VIP element is protected from external damage or impact by enclosing it within a damage-resistant material such as XPS, etc. as described in this paper. We predict that in the near future, further advancements in VIP manufacturing technology, larger volume production, and higher reliability through quality control of cores and films will enhance VIP cost effectiveness as a retrofit option for providing local insulation and mitigating thermal bridging effects in building envelopes.

7. Exterior cladding installation cost is not included in the cost estimation.

CONCLUSIONS

Vacuum insulation panel (VIP) is an emerging insulation technology that is slowly finding applications in buildings due to its energy saving potential. VIP technology involves evacuating a nanostructured porous material and enclosing it in an impermeable container. The nanoporous structure of core in conjunction with reduced air pressure results in an ultra-low thermal conductivity of VIP that is almost 5–10 times smaller than the conventional insulations such as foams and fibers. In this paper, we first provided a brief discussion on heat transfer mechanisms that are responsible for reduced thermal conductivity of VIP.

Next, we investigated the cost performance of VIP technology when implemented in the deep energy building retrofits in cold climate regions. Deep energy retrofits in cold climate regions generally require higher target R-value envelopes. VIPs with significantly higher R-value per unit thickness than the conventional insulations save considerably on the cost associated with the loss of living space caused by the installation itself. In addition, thinner VIP installations may also avoid cost associated with alteration of window and door openings, which is otherwise necessary for thicker layers of conventional insulations. VIP prices were collected from a variety of sources such as VIP manufacturer websites, techni-

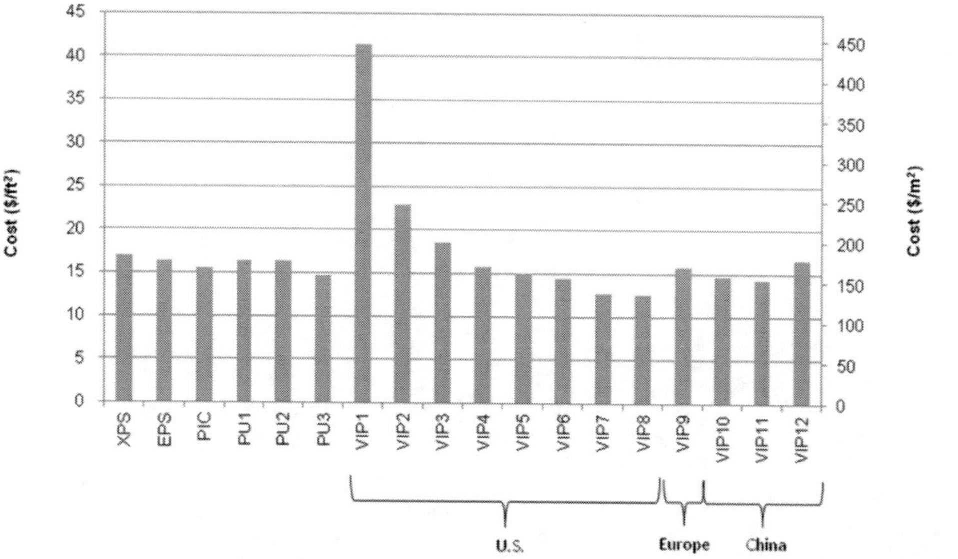

Figure 2 *Installed cost of different exterior wall insulation systems to achieve a thermal resistance of 6.16 m^2·K/W (R-35 h·ft^2·°F/Btu). The equivalent thickness for PU takes into account the thermal bridging effect of wood studs. VIP encapsulated in XPS insulation has an effective thermal resistance of 6.16 m^2·K/W (R-35 h·ft^2·°F/Btu). The thermal resistivity of EPS, XPS, PU, and PIC insulations have been assumed to be 28, 35, 42, and 46 m·K/W (4, 5, 6, and 6.5 h·ft^2·°F/Btu.in), respectively. The VIP cost was collected either directly from manufacturers or from published data on the internet for US, European, and Chinese markets.*

cal publications, and quotes from manufacturers in the United States, Europe, and China. A wide range in VIP prices was observed across these sources. The cost analysis indicated that for lower priced VIPs, the proposed method of VIP wall retrofit may become cost competitive with current deep retrofit strategies that employ foam insulations. In fact, we found that a 6.16 $m^2 \cdot K/W$ (R-35 $h \cdot ft^2 \cdot °F/Btu$) VIP-based wall retrofit costs almost 15% less than that using foam insulation considering the lower range of VIP-based EIFS cost.

Further advancements in VIP manufacturing technology, larger volume production, and higher reliability through quality control of cores and films may further improve VIP cost effectiveness as a retrofit option for local insulation and mitigation of thermal bridging effects in buildings envelopes.

REFERENCES

Aegerter, M.A., Leventis, N., and Koebel M.M. 2011. *Aerogels Handbook*. New York: Springer.

Caps, R., Heinemann, U., Ehrmanntraut, M., and Fricke, J. 2001. Evacuated insulation panels filled with pyrogenic silica powders: Properties and applications. *High Temperatures—High Pressures* 33:151–56.

Castle Square. 2010. Deep Energy Retrofit. Boston: Castle Square Tenants Organization. www.castledeepenergy.com/?page_id=185 for more information.

City of Boston. 2013. The 3D Program. www.cityofboston.gov/3d/

ECBCS. 2005. ECBCS/IEA Annex 39, Vacuum Insulation Panels, Study on VIP-components and Panels for Service Life Prediction of VIP in Building Applications (Subtask A).

EERE. 2010. *Buildings Energy Databook*. Washington, DC: U.S. Department of Energy, Energy Efficiency and Renewable Energy.

Fricke, J., Hümmer, E., Morper, H. J., and Scheuerpflug, P. 1989. Thermal properties of silica aerogels. *Revue de Physique Appliquée* 24(4):C487–C497.

ISO. 2002a. ISO 5660-1, *Reaction to fire tests—Heat release, smoke production and mass loss rate—Part 1: Heat release rate (cone calorimeter method)*. Geneva: International Organization for Standardization.

ISO. 2002b. ISO 11925-2, *Reaction to fire tests—Ignitability of building products subjected to direct impingement of flame—Part 2: Single-flame source test*. Geneva: International Organization for Standardization.

ISO. 2002c. ISO 1716, *Reaction to fire tests for building products—Determination of the heat of combustion*. Geneva: International Organization for Standardization.

ISO. 2002d. ISO 1182, *Reaction to fire tests for building products—Non-combustibility test*. Geneva: International Organization for Standardization.

ISO. 2002e. ISO 9239-1, *Reaction to fire tests for floorings—Part 1: Determination of the burning behaviour using a radiant heat source*. Geneva: International Organization for Standardization.

Kennard, E.H. 1938. *Kinetic Theory of Gases, with an Introduction to Statistical Mechanics*. New York: McGraw-Hill.

Krim, A.J. 1977. The three-deckers of Dorchester: An architectural historical survey. Paper submitted to Boston Redevelopment Authority. www.sidewalkmemories.org/archives/The%20Three-Deckers%20of%20Dorchester.pdf

Kwon, J. S., Hyo Jang, C., Jung, H., and Song, T. H. 2009. Effective thermal conductivity of various filling materials for vacuum insulation panels. *International Journal of Heat and Mass Transfer* 52:5525–32.

National Grid. 2013. Services and Rebates—Deep Energy Retrofit. National Grid USA Service Company, Inc. www1.nationalgridus.com/DeepEnergyRetrofit-MA-RES

Scheuerpflug, P., Morper, H.J., Neubert, G., and Fricke, J. 1991. Low-temperature thermal transport in silica areogels. *Journal of Physics D—Applied Physics* 24:1395–1403.

Simmler, H., Brunner, S., Heinemenn, U., et al. 2005a. Vacuum insulation panels: Study of VIP-components and panels for service life prediction of VIP in building applications. IEA/ECBCS Annex 39, Implementing agreement, Report on Subtask A.

Simmler, H., Brunner, S., Heinemann, U., et. al. 2005b. Vacuum Insulation Panels. Study on VIPcomponents and Panels for Service Life Prediction in Building Applications (Subtask A), Final Report for the IEA/ECBCS Annex 39 HiPTI-project (High Performance Thermal Insulation for Buildings and Building Systems).

Waier, P.R., Babbitt, C., Baker, T., Balboni, B., and Bastoni, R.A. 2011. *RSMeans Residential Cost Data 2011*, 30th Annual Addition.

AUTHOR INDEX

AUTHOR INDEX

Oak Ridge National Laboratory
P.O. Box 2008, Bldg. 3147
Oak Ridge, TN 37831-6070

ISBN 978-1-5108-2783-7

9 781510 827837

CURRAN ASSOCIATES INC.
proceedings.com